Minerals in Soil Environments

MEMORIAL

This book is offered as a memorial to Dr. Charles I. Rich who contributed much to the knowledge of minerals in soils and their influence on soil properties. His untimely death on 18 September 1975 came while he served on the organizing committee for the book.

Minerals in Soil Environments

Editorial Committee: J. B. DIXON, co-editor
S. B. WEED, co-editor
J. A. KITTRICK
M. H. MILFORD
J. L. WHITE

Organizing Committee: J. B. DIXON, chairman
M. E. HARWARD
M. L. JACKSON
M. M. MORTLAND
C. I. RICH (deceased)

Managing Editor: RICHARD C. DINAUER

Assistant Editors: JACQUELYN NAGLER
JUDITH H. NAUSEEF

Published by: Soil Science Society of America
Madison, Wisconsin USA
1977

A-n

Soil Science Society of America, Inc.
677 South Segoe Road, Madison, Wisconsin 53711 USA

Library of Congress Catalog Card Number: 77-080728

Standard Book Number: 0-89118-765-0

Printed in the United States of America

CONTENTS

Chapter 3   Carbonate, Halide, Sulfate, and Sulfide Minerals

H. E. DONER and WARREN C. LYNN

Chapter 4   Aluminum Hydroxides and Oxyhydroxides

PA HO HSU

Chapter 5   Iron Oxides

UDO SCHWERTMANN and REGINALD M. TAYLOR

Chapter 6   Manganese Oxides and Hydroxides

R. M. MC KENZIE

Chapter 7   Micas

DELVIN S. FANNING and V. Z. KERAMIDAS

Chapter 8   Vermiculites

LOWELL A. DOUGLAS

Chapter 9   Montmorillonite and Other Smectite Minerals

GLENN A. BORCHARDT

Chapter 10   Chlorites and Hydroxy Interlayered Vermiculite and Smectite

RICHARD I. BARNHISEL

Chapter 11   Kaolinite and Serpentine Group Minerals

J. B. DIXON

Chapter 12 Interstratification in Layer Silicates

BRIJ L. SAWNHEY

Chapter 13 Palygorskite (Attapulgite), Sepiolite, Talc, Pyrophyllite, and Zeolites

LUCIAN W. ZELAZNY and FRANK G. CALHOUN

Chapter 14   Silica in Soils: Quartz, Cristobalite, Tridymite, and Opal

LARRY P. WILDING, NEIL E. SMECK and LARRY R. DREES

Chapter 15   Feldspars, Olivines, Pyroxenes, and Amphiboles

P. M. HUANG

Chapter 16   Allophane and Imogolite

KOJI WADA

Chapter 17   Phosphate Minerals

WILLARD L. LINDSAY and PAUL L. G. VLEK

Chapter 18   Titanium and Zirconium Minerals

JOHN T. HUTTON

Chapter 19   Shrinking and Swelling of Clay, Clay Strength, and Other Properties of Clay Soils and Clays

KIRK W. BROWN

Chapter 20   Reactions of Minerals with Organic Compounds in the Soil

ROBERT D. HARTER

Chapter 21   Reactions of Minerals with Soil Humic Substances

M. SCHNITZER and H. KODAMA

CONTRIBUTORS

B. L. Allen

Professor of Soils, Plant and Soil Science Department, Texas Tech University, Lubbock, Texas

Richard I. Barnhisel

Associate Professor of Agronomy, Agronomy Department, University of Kentucky, Lexington, Kentucky

Glenn A. Borchardt

Associate Geochemist, California Division of Mines and Geology, San Francisco, California

John L. Brown

Principal Research Scientist and Head, Analytical Services Branch, Engineering Experiment Station, Georgia Institute of Technology, Atlanta, Georgia

Kirk Wye Brown

Associate Professor, Department of Soil and Crop Sciences, Texas A&M University, College Station, Texas

Frank G. Calhoun

Assistant Professor of Soil Taxonomy, Soil Science Department, University of Florida, Gainesville, Florida

Joe Boris Dixon

Professor of Soil Science, Department of Soil and Crop Sciences, Texas A&M University, College Station, Texas

Harvey E. Doner

Associate Professor of Soil Chemistry, Department of Soils and Plant Nutrition, University of California, Berkeley, California

Lowell A. Douglas

Professor of Soil Mineralogy, Department of Soils and Crops, Cook College, Rutgers University, New Brunswick, New Jersey

Larry R. Drees

Formerly Instructor, Department of Agronomy, The Ohio State University, Columbus, Ohio; now Assistant Professor, Department of Soil and Crop Sciences, Texas A&M University, College Station, Texas

Delvin S. Fanning

Associate Professor of Soil Science, Department of Agronomy, University of Maryland, College Park, Maryland

Robert G. Gast

Formerly Professor of Soil Chemistry, Department of Soil Science, University of Minnesota, St. Paul, Minnesota; now Chairman, Department of Agronomy, University of Nebraska, Lincoln, Nebraska

Ben F. Hajek

Associate Professor, Department of Agronomy and Soils, Auburn University, Auburn, Alabama

Robert D. Harter

Associate Professor of Soil Chemistry, University of New Hampshire, Durham, New Hampshire

Pa Ho Hsu

Professor of Soil Chemistry, Department of Soils and Crops, Rutgers-The State University of New Jersey, New Brunswick, New Jersey

P. M. Huang — Associate Professor, Department of Soil Science, University of Saskatchewan, Saskatoon, Saskatchewan, Canada

John T. Hutton — Principal Research Scientist, C.S.I.R.O. (Australia), Division of Soils, Adelaide, Glen Osmond, South Australia

Vissarion Z. Keramidas — Formerly Graduate Student, Department of Agronomy, University of Maryland, College Park, Maryland; now Soil Scientist, Laboratory of Soil Science, Aristotelian University, Thessaloniki, Greece

James A. Kittrick — Professor of Soils, Washington State University, Pullman, Washington

Hideomi Kodama — Senior Research Scientist, Soil Research Institute, Agriculture Canada, Central Experimental Farm, Ottawa, Ontario

Willard L. Lindsay — Professor of Agronomy (Soils), Department of Agronomy, Colorado State University, Fort Collins, Colorado

Warren C. Lynn — Soil Scientist, National Soil Survey Laboratory, Soil Conservation Service, U. S. Department of Agriculture, Lincoln, Nebraska

Thomas R. McKee — Formerly Research Assistant, Department of Oceanography, Texas A&M University, College Station, Texas; now Research Associate, Departments of Geology and Chemistry, Arizona State University, Tempe, Arizona

R. M. McKenzie — Research Scientist, C.S.I.R.O. (Australia), Division of Soils, Glen Osmond, South Australia

Charles I. Rich — Professor of Agronomy (Deceased), Department of Agronomy, Virginia Polytechnic Institute and State University, Blacksburg, Virginia

Brij L. Sawhney — Soil Chemist, Department of Soil and Water, The Connecticut Agricultural Experiment Station, New Haven, Connecticut

Morris Schnitzer — Principal Research Scientist, Soil Research Institute, Agriculture Canada, Central Experimental Farm, Ottawa, Ontario

Udo Schwertmann — Professor of Soil Science, Institut fur Bodenkunde, Technische Universitat Munchen, F. R. G., Freising-Weihenstephan, West Germany

Neil E. Smeck — Associate Professor, Department of Agronomy, The Ohio State University and The Ohio Agricultural Research and Development Center, Columbus, Ohio

K. H. Tan — Associate Professor of Soil Science, Agronomy Department, University of Georgia, Athens, Georgia

Reginald M. Taylor — Research Scientist, C.S.I.R.O. (Australia), Division of Soils, Glen Osmond, South Australia

Paul L. G. Vlek — Soil Scientist, International Fertilizer Development Center, Muscle Shoals, Alabama

Koji Wada — Professor of Soils, Faculty of Agriculture, Kyushu University, Fukuoka, Japan

Joe L. White — Professor of Soil Mineralogy, Department of Agronomy, Purdue University, West Lafayette, Indiana

Lawrence P. Wilding — Formerly Professor of Agronomy, The Ohio State University and The Ohio Agricultural Research and Development Center; now Professor of Soil Science, Department of Soil and Crop Sciences, Texas A&M University, College Station, Texas

Lucian W. Zelazny — Formerly Associate Soil Mineralogist, Soil Science Department, University of Florida, Gainesville, Florida; now Associate Professor of Agronomy, Department of Agronomy, Virginia Polytechnic Institute and State University, Blacksburg, Virginia

CONVERSION FACTORS FOR U. S. AND METRIC UNITS

To convert column 1 into column 2, multiply by	Column 1	Column 2	To convert column 2 into column 1, multiply by
	Length		
0.621	kilometer, km	mile, mi	1.609
1.094	meter, m	yard, yd	0.914
0.394	centimeter, cm	inch, in	2.54
	Area		
0.386	kilometer^2, km^2	mile^2, mi^2	2.590
247.1	kilometer^2, km^2	acre, acre	0.00405
2.471	hectare, ha	acre, acre	0.405
	Volume		
0.00973	meter^3, m^3	acre-inch	102.8
3.532	hectoliter, hl	cubic foot, ft^3	0.2832
2.838	hectoliter, hl	bushel, bu	0.352
0.0284	liter	bushel, bu	35.24
1.057	liter	quart (liquid), qt	0.946
	Mass		
1.102	ton (metric)	ton (U.S.)	0.9072
2.205	quintal, q	hundredweight, cwt (short)	0.454
2.205	kilogram, kg	pound, lb	0.454
0.035	gram, g	ounce (avdp), oz	28.35
	Pressure		
14.50	bar	lb/inch^2, psi	0.06895
0.9869	bar	atmosphere, atm	1.013
0.9678	kg(weight)/cm^2	atmosphere, atm	1.033
14.22	kg(weight)/cm^2	lb/inch^2, psi	0.07031
14.70	atmosphere, atm	lb/inch^2, psi	0.06805
	Yield or Rate		
0.446	ton (metric)/hectare	ton (U.S.)/acre	2.24
0.892	kg/ha	lb/acre	1.12
0.892	quintal/hectare	hundredweight/acre	1.12
	Temperature		
$\left(\dfrac{9}{5}\,°C\right) + 32$	Celsius $-17.8C$ $0C$ $100C$	Fahrenheit $0F$ $32F$ $212F$	$\dfrac{5}{9}\,(°F - 32)$
	Water Measurement		
8.108	hectare-meters, ha-m	acre-feet	0.1233
97.29	hectare-meters, ha-m	acre-inches	0.01028
0.08108	hectare-centimeters, ha-cm	acre-feet	12.33
0.973	hectare-centimeters, ha-cm	acre-inches	1.028
0.00973	meters^3, m^3	acre-inches	102.8
0.981	hectare-centimeters/hour, ha-cm/hour	feet^3/sec	1.0194
440.3	hectare-centimeters/hour, ha-cm/hour	U.S. gallons/min	0.00227
0.00981	meters^3/hour, m^3/hour	feet^3/sec	101.94
4.403	meters^3/hour, m^3/hour	U.S. gallons/min	0.227

Plant Nutrition Conversion—P and K

P (phosphorus) $\times 2.29 = P_2O_5$
K (potassium) $\times 1.20 = K_2O$

CONSTANTS, CONVERSION FACTORS, AND SI UNITS

Physical Constants

Quantity	Symbol	Formula
electronic charge	e	4.8029×10^{-10} statcoulomb (esu)
		1.6021×10^{-19} coulomb
Planck constant	h	6.6252×10^{-27} erg second molecule^{-1}
speed of light	c	2.997930×10^{10} cm s^{-1}
Avogadro number	L	6.0226×10^{23} mole^{-1}
Faraday constant	F	9.6491×10^{4} coulomb mole^{-1}
gas constant	R	8.31432 joule $^{\circ}$K^{-1} mole^{-1}
Boltzmann constant	$k=(R/L)$	1.38052×10^{-23} joule $^{\circ}$K^{-1}

Conversion Factors and Defined Constants

1 atm = 1.01325 bar = 1.0332×10^4 kg m^{-2} = 760 torr; 1 atm $\times 1.013250 \times 10^5$ = Pa
1 erg = 10^{-7} joule = 2.3901×10^{-8} defined calorie = 10^{-7} volt. coulomb
1 coulomb = 1/10 emu = (c/10) esu
1 weber m^{-2} = 10^4 gauss
1 gauss = c^{-2} esu; 1 gauss $\times 1.0000 \times 10^{-4}$ = tesla
standard gravitational constant, g = 980.665 cm s^{-2}
Kelvin temperature (K) = 273.15 + Celsius temperature.

SI Units and Symbols

Certain of the International System of Units are given below.† Only selected members are given. The system is abbreviated SI from the French name.

Base Units—SI is based on seven well-defined units which by convention are regarded as dimensionally independent:

Quantity	Unit	Symbol
length	metre	m
mass	kilogram	kg
time	second	s
electric current	ampere	A
thermodynamic temperature	kelvin	K
amount of substance	mole	mol
luminous intensity	candela	cd

Selected Derived Units

Quantity	Unit	Symbol	Formula
force	newton	N	kg·m/s^2
pressure, stress	pascal	Pa	N/m^2
energy, work, quantity of heat	joule	J	N·m
power, radiant flux	watt	W	J/s
quantity of electricity, electric charge	coulomb	C	A·s
electric potential, potential difference, electromotive force	volt	V	W/A
capacitance	farad	F	C/V
electric resistance	ohm	Ω	V/A
conductance	siemens	S	A/V
magnetic flux	weber	Wb	V·s
magnetic flux density	tesla	T	Wb/m^2

† From *Standards for Metric Practice*. ASTM designation: E 380–76. American Society for Testing and Materials, 1916 Race St., Philadelphia, PA 19103.

PERIODIC TABLE OF THE ELEMENTS

Group		IA	IIA	IIIB	IVB	VB	VIB	VIIB	VIII
		Representative elements		Transition elements-d					
Period	Valence shell	s^1	s^2	d^1s^2fx	d^2s^2	$(d^3s^2\S)$	$(d^5s^1\S)$	d^5s^2	$(d^6s^2\S)$
$n=1$	$1s$	1 H 1.008							
$n=2$	$2s2p$	3 Li 6.941	4 Be 9.012						
$n=3$	$3s3p$	11 Na 22.99	12 Mg 24.31						
$n=4$	$4s3d4p$	19 K 39.10	20 Ca 40.08	21 Sc 44.96	22 Ti 47.90	23 V 50.94	24 Cr 52.00	25 Mn 54.94	26 Fe 55.85
$n=5$	$5s4d5p$	37 Rb 85.47	38 Sr 87.62	39 Y 88.91	40 Zr 91.22	41 Nb 92.91	42 Mo 95.94	43 Tc 99.91	44 Ru 101.1
$n=6$	$6s4f5d6p$	55 Cs 133.0	56 Ba 137.3	57–71 †	72 Hf 178.5	73 Ta 180.9	74 W 183.9	75 Re 186.2	76 Os 190.2
$n=7$	$7s5f6d7p$	87 Fr 223	88 Ra 226.0	89–103 ‡					

† Lanthanide series		57 La 138.9	58 Ce 140.1	59 Pr 140.9	60 Nd 144.2	61 Pm 145	62 Sm 150.4
‡ Actinide series		89 Ac 227	90 Th 232	91 Pa 231	92 U 238	93 Np 237.0	94 Pu 242

§ Variable valence shells.

PERIODIC TABLE OF THE ELEMENTS

Transition elements-d			Representative elements						Noble gas elements
VIII		IB	IIB	IIIA	IVA	VA	VIA	VIIA	0
$(d^7s^2\S)$ $(d^8s^2\S)$		s^1d^{10}	s^2	s^2p^1	s^2p^2	s^2p^3	s^2p^4	s^2p^5	s^2p^6
									2 He 4.003
				5 B 10.81	6 C 12.01	7 N 14.01	8 O 16.00	9 F 19.00	10 Ne 20.18
				13 Al 26.98	14 Si 28.09	15 P 30.97	16 S 32.06	17 Cl 35.45	18 Ar 39.95
27 Co 58.93	28 Ni 58.71	29 Cu 63.54	30 Zn 65.37	31 Ga 69.72	32 Ge 72.59	33 As 74.92	34 Se 78.96	35 Br 79.91	36 Kr 83.80
45 Rh 102.9	46 Pd 106.4	47 Ag 107.9	48 Cd 112.4	49 In 114.9	50 Sn 118.7	51 Sb 121.8	52 Te 127.6	53 I 126.9	54 Xe 131.3
77 Ir 192.2	78 Pt 195.1	79 Au 197.0	80 Hg 200.6	81 Tl 204.4	82 Pb 207.2	83 Bi 209.0	84 Po 210	85 At 210	86 Rn 222

63 Eu 152	64 Gd 157.2	65 Tb 159	66 Dy 162.5	67 Ho 164.9	68 Er 167.3	69 Tm 168.9	70 Yb 173	71 Lu 175
95 Am 243	96 Cm 247	97 Bk 247	98 Cf 249	99 Es 254	100 Fm 253	101 Md 256	102 No 254	103 Lr 257

Mineral Equilibria and the Soil System[1]

JAMES A. KITTRICK, Washington State University, Pullman

I. INTRODUCTION

It is often helpful to simplify complicated processes in order to better understand their nature. A successful example of this compares the geochemical system of the earth's surface to a giant chemical engineering plant (Siever, 1974). That portion of the plant analogous to the soil system was designated as the liquid extraction unit. The liquid extraction analogy is appropriate to the process of dissolving unstable soil minerals, but the soil is also analogous to a reaction vessel (operating at room temperature and pressure) where new minerals may form.

As unstable primary minerals dissolve, they help to determine the composition of the soil solution (e.g., solution A in Fig. 1-1, *upper left*). As such minerals eventually become depleted (with the smaller particles disappearing first), the composition of the soil solution changes. These changes also take place more rapidly near the soil surface where rain waters are usually less saturated with respect to mineral constituents (e.g., solution A $\neq$ solution B in Fig. 1-1, *upper right*). With this solution composition change, a new secondary mineral may become stable and form in the reaction vessel. Any initial secondary minerals that may have formed may be unstable in this new solution and therefore gradually dissolve in the subsequent "liquid extraction" process.

In some places the process just described continues for long periods of time, but most soils are only short term features on a geologic time scale. Thus, some physical processes such as glaciation or erosion-deposition-uplift interrupt the cycle and redistribute the reactants. This is analogous to periodic stirring of the reaction vessel (Fig. 1-1, *lower left*) except that some of the reactants may temporarily leave the soil system reaction vessel, undergoing physical and chemical changes prior to returning to the vessel. Thus, a major part of the new feed from the pulverizer to the reaction vessel (Fig. 1-1, *lower right*) will be minerals in recycled rocks (see Garrels & Mackenzie, 1971, p. 330). These are the sedimentary, metamorphic, and igneous rocks formed from pre-existing sediments.

Typical consequences of the extraction-reaction process are summarized in Fig. 1-2. It is evident that the liquid extraction process has been highly selective, almost entirely dissolving the Ca feldspars, and the inosilicates and nesosilicates (silicates high in Fe and Mg). Very little primary quartz is dissolved and the supply is augmented by additions of quartz formed in recycled

[1]Contribution from the Department of Agronomy and Soils, Washington State University, Pullman, WA 99163. Published as Scientific Paper no. 4446, Project 1885. College of Agriculture Research Center.

1

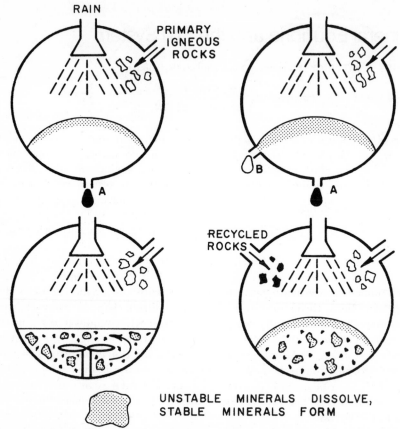

Fig. 1-1. Liquid extraction-reactor units analogous to the soil system. As unstable minerals dissolve (*upper left*), composition of the soil solution changes so that solution B ≠ solution A (*upper right*). Physical processes occasionally mix reactants (*lower left*) and recycled rocks (*lower right*) become a major part of the feed from the pulverizer (physical weathering).

rocks. Illite and probably most of the chlorite and carbonates are also added to the reactor through recycled rocks. A relatively modest amount of mineral formation actually takes place in the reaction vessel of the soil system. The amount of mineral formation that takes place in a single interval between stirring the contents of the reaction vessel is quite small for most soils. Thus, the dominant mineralogy of most soils is inherited.

Because soils are usually short term entities relative to the time required for dissolution and formation of most minerals, it might seem hopeless to observe mineral changes that may have taken place on the time scale of soil formation. This is true for the bulk mineralogy of most soils, but not necessarily true for the mineralogy of the clay-size fraction. For a given mineral, the rate of dissolution is dependent on exposed surface area, with smaller

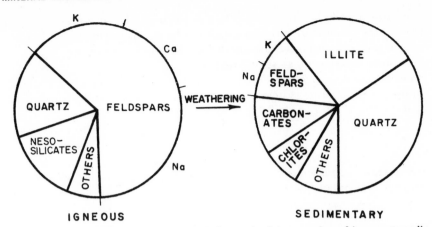

Fig. 1-2. Estimated average mineralogical changes in the conversion of igneous to sedimentary rock during weathering (wt %) (based upon Garrels and Mackenzie [1971], p. 245).

particles having the highest surface area per unit weight. Further, secondary minerals that precipitate obviously must appear as small particles at first, and some never grow very large. Thus, although the bulk mineralogy of most soils is inherited, there exist some good correlations between clay fraction mineralogy and the climate (mainly rainfall and temperature) in which the present soil resides (Jackson et al., 1948).

The relative dissolution rates of soil minerals have never been investigated systematically. This is partly because dissolution must take place in undersaturated solutions. What constitutes an undersaturated solution depends upon the composition of a saturated solution, which has been unknown for most soil minerals. It is also necessary to know the composition of a saturated solution in order to understand the precipitation of secondary minerals. Studies of soil mineral stability have been hindered by the low solubility, long equilibration times, and the generally complicated nature of many soil minerals. Consequently, there is at present only a modest amount of data available and even less agreement on the stabilities of various soil minerals. At this stage of our understanding it seems most worthwhile to discuss the general principles involved, and to illustrate them with some of the available soil mineral stability information.

II. FUNDAMENTALS OF MINERAL EQUILIBRIA

A. Phase Rule

Basset et al. (1966) define the number of components (C) as the minimum number of chemical species that must be specified in order that the composition of all phases in the system can also be specified. A phase (P) is

a macroscopically homogeneous region in a system. The degrees of freedom
(F) are the number of variables which can be changed independently for a
system without causing a change in the number of phases present.

Some systems are so simple that the relationship between phases, components, and degrees of freedom is obvious. For most systems it is necessary
to use the accounting system provided by the Gibbs phase rule, which states
that

$$F = C - P + 2$$

for systems *at equilibrium*. Under earth-surface conditions, temperature and
pressure can usually be considered to be constant, which removes two degrees of freedom from the system, so that

$$F = C - P$$

The challenge in the application of the phase rule is obviously not in the
arithmetic. It lies instead in determining the number of phases and particularly in determining the number of components present in a given system.

1. SOIL MINERAL EQUILIBRIA

Assume the estimated mineralogical analysis of the average sediment in
Fig. 1-2 to be an approximation of the composition of an average soil. Can
all of these minerals be in equilibrium with each other and also with the soil
solution under earth surface conditions? It is possible to answer the question with the phase rule without knowing the actual thermodynamic stability
of any of the individual minerals. The minerals of the average sediment of
Fig. 1-2 are listed in Table 1-1, along with their components. Although
there are arbitrary aspects to estimating mineralogical compositions of this

Table 1-1. Mineralogy and components of the average sedimentary rock
(based upon Garrels and Mackenzie [1971], p. 245).

Mineral	Components (not repeated in minerals above)
Quartz	SiO_2
Albite feldspar	Na_2O, Al_2O_3
K-feldspar	K_2O
Hematite	Fe_2O_3
Calcite	CaO, CO_2
Dolomite	MgO
Illite	H_2O
Chlorite	FeO
Montmorillonite	
Total 9	10

type (Garrels & Mackenzie, 1971, p. 244), the relative numbers of phases and components is appropriate.

From Table 1-1, we see that the minimum number of components that make up these minerals is 10. There are nine minerals plus the soil solution for a total of 10 phases, thus

$$F = C - P = 10 - 10 = 0.$$

According to the phase rule, if all of these minerals were in equilibrium with each other and with the soil solution, the composition of the soil solution would always remain constant (zero degrees of freedom). The system would be invariant because no solution composition variables could be independently changed without changing the number of phases. As many analyses document the inconstancy of the soil solution composition, it is likely that all soil minerals are seldom, if ever, simultaneously in equilibrium with the soil solution.

Even if most soil mineralogical systems are not at complete equilibrium, we can reasonably assume that they are altering in such a way as to more closely approach equilibrium. An understanding of this situation requires three main types of study: (i) the minerals involved; (ii) the stability of the minerals (so that mineral equilibria can be predicted), and (iii) kinetic studies (to determine how fast various mineral equilibria are approached). Obviously, much is known about the soil minerals involved (most of this book is devoted to that subject). A small start has been made on soil mineral stability and equilibria. Almost nothing is known at present about soil mineral kinetics. Before going further in our examination of soil mineral equilibria, it may be helpful to review a few thermodynamic tools.

B. Free Energy, Enthalpy, Entropy

Just as "money is the medium of exchange in commercial transactions, energy is the medium of exchange in chemical reactions" (Allen, 1966). *Thermodynamics* is the study of energy changes. It can be a complicated discipline, but fortunately key thermodynamic quantities do not change much over the temperature and pressure ranges of interest to soil mineral weathering. This simplifies the situation, so that it is usually necessary to venture into only the shallowest of thermodynamic waters.

A spontaneous process is one that occurs of its own accord, without external assistance. It occurs spontaneously because of an imbalance in two great natural tendencies. The first tendency is the spontaneous conversion of potential energy into work and heat. The second tendency is the spontaneous increase in randomness of the system. It is useful to consider each process separately and then in combination. As the major difficulty in elementary thermodynamics is not mathematical, but conceptual, some simplifying analogies will be presented.

1. ENTHALPY

Consider a brick on the top of a tall building. As the brick falls to earth it rapidly loses its potential energy of position. Similarly, the potential energy of attraction between isolated ions of opposite charge (Fig. 1–3) decreases rapidly as the ions come together from infinite distance at a temperature of absolute zero (a negative sign indicates that energy is given off in the process). When the two ions come close enough so that their electron fields interact, the potential energy of repulsion between the two ions increases rapidly. At equilibrium, the ions reside in a potential energy minimum that is the resultant of the curves for attraction and repulsion. The position of this potential energy well determines the equilibrium distance between ions, r_0. The energy given off as the ions come together is indicated by the depth, d, of the potential energy well, which is, of course, the energy that would be required to separate the ions to infinite distance again. For a crystal, this is the lattice (structural) energy or bond energy, or the enthalpy change upon bonding, ΔH. Thus, in a chemical reaction, the enthalpy change is the bond energy change (if no pressure-volume work is done in the process).

In the analogy with the brick on top of the building, the pavement served as a handy zero reference point (although more potential energy would have been released had the brick fallen into a hole in the pavement). Similarly, the zero reference point for enthalpy changes is the enthalpy of

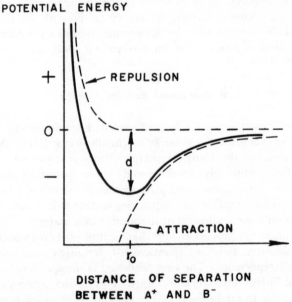

Fig. 1-3. Potential energy as a function of separation distance between A^+ and B^- at a temperature of absolute zero. Dotted lines show the attractive and repulsive contributions resulting in the potential energy well, d, at the equilibrium distance of separation, r_0 (after Swalin [1964]).

the elements in their most stable states under standard conditions (for ions, H^+ is chosen as zero). Absolute enthalpy values are unknown and unneeded.

The spontaneous approach of ions of opposite charge illustrates the first great natural tendency, which is the spontaneous conversion of potential energy into work and heat. One might suppose that all ions and atoms in the world should have found their way into their lowest respective potential energy wells. By this reasoning, all material in the world should be neatly arranged into crystalline substances. There are two reasons why this has not happened. The first is that the world has not yet come to equilibrium. Many substances that are not now crystalline, or in their most stable crystalline form, will be so at some point in the future. The second reason results from the second great natural tendency. It is linked to thermal motion, and is the reason why one of the conditions of Fig. 1-3 was that the temperature be absolute zero.

2. ENTROPY

Thermal motion is random motion, which spontaneously tends to increase the randomness of substances at temperatures above absolute zero. Consider two bulbs connected by a stopcock (Fig. 1-4). The bulbs are isolated from their surroundings with the left bulb in A containing an ideal gas (no attractive forces and negligible molecular volume) and the right bulb being evacuated. At absolute zero, there is no random motion and, hence, no pressure difference between the bulbs in A. At temperatures above absolute zero, molecular motion in the gas in the left bulb creates a pressure difference, so that when the stopcock is opened in B, gas inevitably flows from left to right. The temperature of the system remains constant and no work transfer or heat flow to the surroundings occurs. The process is always observed to occur in the same direction, never the reverse. What drives this spontaneous process? It is driven by the spontaneous increase in disorder or randomness of the system.

The measure of the randomness of a system is its entropy, S. The fact that randomness increases with temperature leads to an absolute value of entropy, which is zero for a perfect crystal at absolute zero. Although entropy is fundamentally different from energy, it is evident that entropy has an energy equivalent. In Fig. 1-4, it would require a certain amount of

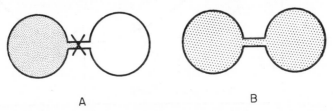

A   B

Fig. 1-4. Irreversible expansion of a perfect gas (*left*) into an evacuated bulb (*right*) in an isolated system. When the stopcock is open the process is always observed to go from A to B, never in the reverse direction (after Mahan [1963]).

energy to capture gas molecules in B and return them to their starting state in A.

Two factors have now been introduced that contribute to the direction of chemical change. There is a tendency for spontaneous change in a system to be favored by an enthalpy decrease ($\Delta H < 0$) and an entropy increase ($\Delta S > 0$). Clearly what is needed is some way to relate these two factors.

3. FREE ENERGY AND THE EQUILIBRIUM CONSTANT

The free energy (G) or net energy is defined so that, at constant temperature and pressure

$$\Delta G = \Delta H - T\Delta S. \qquad [1]$$

The absolute temperature T must be specified because the entropy change is temperature dependent. Fortunately, H and S do not vary much with temperature over a limited range, but G may vary considerably with temperature unless S happens to be very small. At relatively low temperatures, such as room temperature, ΔH is usually the most important factor in determining the directions of chemical reactions, whereas at high temperatures ΔS becomes more important. A spontaneous reaction is indicated by a decrease in G (negative ΔG) which requires a negative ΔH (energy given off) or a positive ΔS (randomness increased), or both.

Consider the chemical reaction

$$A^+ + B^- = AB. \qquad [2]$$

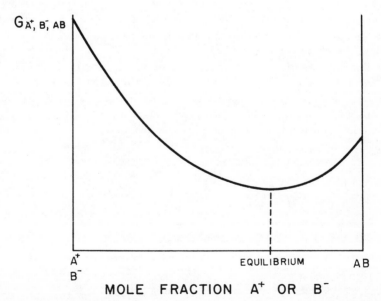

Fig. 1–5. The free energy of the constituents in reaction (1.2) (after Waser [1966]).

Reaction [2] will proceed until the free energy of the reaction mixture reaches a minimum (Fig. 1-5). At this point the reaction is at equilibrium. This is, of course, the same equilibrium point reached by the reverse of reaction [2]. At equilibrium, the rates of the forward and reverse reactions are equal, hence energy exchanges are also equal ($\Delta G = 0$).

Because absolute H values are unknown, absolute G values are also unknown. The standard zero reference state for G values is the most stable configuration of the element under standard conditions. The standard free energy of formation of a substance, ΔG_f°, is the ΔG in forming one mole of the substance from the stable elements, under standard conditions. Henceforth, for simplicity, the subscript and superscript on ΔG will be assumed unless otherwise specified. The standard free energy change of a reaction, ΔG_r, is the ΔG of the products minus the ΔG of the reactants, with all products and reactants at unit activity. Since few reactions occur under the very restricted conditions of unit activity, ΔG_r would be of little interest except for the Nernst relationship

$$\Delta G_r = -RT \ln K \qquad\qquad [3]$$

where R is the gas constant and K is the equilibrium constant.

It can be seen then that ΔG for the standard reaction (ΔG_r) can be related to the extent of reaction under general conditions through K. If ΔG_r is negative, K is larger than unity and the right hand side of the reaction is favored. If ΔG_r is positive, K is less than unity, and the left hand side of the equation is favored. Basically, ΔG_r is related to K because the extent to which any reaction will convert reactants into products depends upon the energy change for the reaction. This demonstrates that energy is the medium of exchange for chemical reactions, and we have come full circle.

III. GRAPHING THERMOCHEMICAL RELATIONS

At what silica level are gibbsite and kaolinite in equilibrium? This is a specific question concerning soil mineral stabilities that thermochemistry is well suited to answer. The first and usually the most important step in answering any thermochemical question is to formulate a suitable equation (the classic reference in this regard is Garrels and Christ, 1965). In this case it is evident that the equation will have to contain gibbsite, kaolinite, and H_4SiO_4, which are specifically mentioned in the question.

Thus

$$Al_2Si_2O_5(OH)_{4(kl)} + 5\ H_2O\ _{(1)} = 2\ Al(OH)_{3(gb)} + 2H_4SiO_{4(aq)},$$

$$K = (H_4SiO_4)^2,\ and$$

$$pK = 2pH_4SiO_4,$$

assuming the activities of water, gibbsite, and kaolinite to be unity. Then

$$\Delta G_r = 2\,\Delta G_{Al(OH)_3} + 2\Delta G_{H_4SiO_4} - 5\,\Delta G_{H_2O} - \Delta G_{Al_2Si_2O_5(OH)_4}$$

$$= 2\,(-275.2) + 2\,(-312.6) - 5\,(-56.7) - (-906.2)$$

$$= 14.1 \text{ kcal}$$

$$= 1.36 \text{ p}K \text{ (from Eq. [3] at 25C)}.$$

Therefore, $pK = 2pH_4SiO_4 = 10.3$ and

$$pH_4SiO_4 = 5.2. \qquad\qquad [4]$$

Thus the silica level at which gibbsite and kaolinite are in equilibrium is $1 \times 10^{-5.2}M$. [Sources of thermodynamic values will be discussed later in the chapter. Experience and confidence in such calculation can be acquired from related problems and answers found in Allen (1966), Bassett et al. (1966), and Garrels and Christ (1965)].

Applying this information to nature, one might check some soil solution analyses to see if their silica content is $1 \times 10^{-5.2}M$. But first, one must ask whether gibbsite and kaolinite are present in large amounts in these soils, whether gibbsite and kaolinite are both stable in these soils, and whether gibbsite and kaolinite react rapidly enough to control the soil solution equilibria that engendered the silica analyses. A negative answer to any of these queries probably means that gibbsite-kaolinite equilibria are inapplicable to the soils in question. Thermochemistry answers only specific questions. It is up to the investigator to critically determine whether the answers are applicable to that part of nature under consideration.

Another question that might be asked is: for what soil solution composition conditions are gibbsite and kaolinite stable? This is more general than the first question and also a good deal more complicated to answer. In this case it is necessary to consider all soil minerals that compete for the same elements as do gibbsite and kaolinite (essentially Al and Si). This might be done with several sets of simultaneous equations, but it would not be easy. A simpler way to answer the question would be with a graph displaying the stability relationships of the appropriate minerals. Graphs can provide the same information as equations and at the same time display unanticipated relationships that provide new insight. In some ways thermochemical data are analogous to letters of the alphabet; they have tremendous information potential, but must be arranged into some orderly sequence before they become useful. Equations might then be analogous to words; they can answer only simple specific questions. Graphs would be analogous to sentences; they are capable of answering more general, more complicated questions.

The obvious problem in graphing mineral stability relations is that ordinary graphs have two coordinates and, thus, can conveniently display only two variables. As most natural systems contain more than two variables, types of graphs that display more than two variables are often used, but with a corresponding decrease in convenience. It is important to be aware of the various types of graphs available because various graphs emphasize different aspects of existing information. Often only one type of graph is particularly well suited to a specific problem.

A. Variables and Coordinates Equal in Number

1. TWO VARIABLES—TWO COORDINATES

The relatively simple procedures used for constructing graphs of two variables on two coordinates are worth considering because they are fundamental to all other graph types. For example, consider the construction of a stability line for gibbsite.

$$Al(OH)_{3(gb)} + 3H^+_{(aq)} = Al^{3+}_{(aq)} + 3H_2O_{(l)}.$$

The two solution variables are Al^{3+} and H^+, although other Al ions and OH^- could have been selected as variables instead, had we chosen to construct the defining equation in a different manner. For this reaction

$$K = Al^{3+}/(H^+)^3 \text{ and } pK = pAl^{3+} - 3 \text{ pH}^+.$$

In terms of free energies

$$\Delta G_r = \Delta G_{Al^{3+}} + 3 \Delta G_{H_2O} - 3 \Delta G_{H^+} - \Delta G_{Al(OH)_3}$$

$$= -116.0 + 3(-56.7) - 3(0) - (-275.2)$$

$$= -10.9 \text{ kcal}$$

$$= 1.36 \text{ p}K.$$

Therefore, $pK = -8.0.$

If pH is chosen as the independent variable and Al^{3+} as the dependent variable, then

$$pAl^{3+} = 3pH^+ + pK.$$

The stability equation for gibbsite is now in the form of a linear equation with slope 3 and intercept pK (-8.0). A plot of this relationship is given in Fig. 1-6. The gibbsite stability line in this figure displays the equilibrium pH and pAl^{3+} relations of soil solutions for those cases where gibbsite controls the level of Al^{3+} in solution. This might occur where gibbsite is essentially the only Al mineral in the soil, where equilibria for gibbsite are much more rapid than for other Al minerals present, or where gibbsite is the least soluble

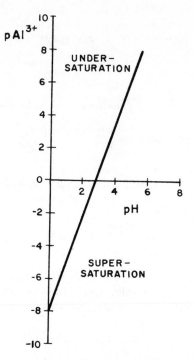

Fig. 1-6. A stability line for gibbsite.

of the Al minerals present. It is evident that gibbsite supports measurable quantities of Al^{3+} (pAl $<$ 5) only at low soil pH levels.

2. THREE VARIABLES–THREE COORDINATES

Extension of the two dimensional graph to three dimensions engenders a graph that is more difficult to construct and much more difficult to interpret. The technique is clearly described and throughly illustrated in Garrels and Christ (1965), so only a brief description of the steps will be given here. For the kaolinite stability plane on a three-dimensional graph having pH^+, pAl^{3+} and pH_4SiO_4 as the variables, the defining chemical equation would be

$$Al_2Si_2O_5(OH)_{4(kl)} + 6\ H^+_{(aq)} = 2Al^{3+}_{(aq)} + 2\ H_4SiO_{4(aq)} + H_2O_{(l)}. \qquad [5]$$

For this reaction, $pK = 2\ pAl^{3+} + 2\ pH_4SiO_4 - 6\ pH$ and

$$\Delta G_r = 2\ \Delta G_{Al^{3+}} + 2\ \Delta G_{H_4SiO_4} + \Delta G_{H_2O} - 6\ \Delta G_{H^+} - \Delta G_{Al_2Si_2O_5(OH)_4}$$

$$= 2(-116.0) + 2(-312.6) + (-56.7) - 6(0) - (-906.2)$$

$$= -7.7\ kcal$$

$$= 1.36\ pK.$$

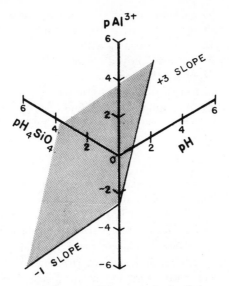

Fig. 1-7. A stability plane for kaolinite.

Therefore, $pK = -5.6$.

Constructing a three-dimensional graph is essentially a matter of joining two two-dimensional graphs. On the face where pAl^{3+} is plotted against pH_4SiO_4 (where pH = 0), we see from Eq. [5] et seq. that

$$pAl^{3+} = -pH_4SiO_4 -2.8.$$

This is the equation of a straight line of slope -1 and intercept -2.8 (Fig. 1-7). On the face where pAl^{3+} is plotted against pH^+ (where $pH_4SiO_4 = 0$), we see from Eq. [5] et seq. that

$$pAl^{3+} = 3pH^+ - 2.8.$$

This is the equation of a straight line of slope 3 and intercept -2.8 (Fig. 1-7). Stability planes for other minerals that involve one or more of these three variables could also be located on Fig. 1-7.

B. Variables Exceed Coordinates

1. SINGLE-VALUE VARIABLES

If it were desired to locate the montmorillonite stability plane in Fig. 1-7, this could be done for a fixed value of pMg^{2+} (and pFe^{3+} if this ion also occurred in the montmorillonite structure). The montmorillonite equation equivalent to Eq. [5] would contain pMg^{2+}, and a realistic level of this parameter could be inserted and combined with the intercept (as it is treated as a constant). Graphing would then proceed as for the case of three variables. Observing stability relationships with one or more variables fixed is the most common stratagem employed to deal with an excess of variables over coordinates. It is most effective when the fixed variables are also fixed or of only limited range in the natural system under investigation. If nature is not helpful, a series of graphs may be employed where the level of the fixed variable spans the range of interest in the natural system.

2. CONTOURS

Mapmakers have long placed three variables onto two coordinates through the use of contours. A similar approach for kaolinite might be to eliminate the pAl^{3+} vs. pH_4SiO_4 face in Fig. 1-7, and to draw instead a series of parallel kaolinite stability lines on the pAl^{3+} vs. pH^+ face, each corresponding to a particular pH_4SiO_4 level. Placement of these lines would proceed in the usual way, from equations similar to the [5] series, except that pH_4SiO_4 would be fixed for each corresponding contour level. A contour approach was used by Gardner (1970) to uncover some kaolinite-gibbsite-quartz relationships that might not have been evident with other types of graphs.

3. TWO VARIABLES ON ONE AXIS

Consider the equilibrium between microcline and kaolinite

$$2KAlSi_3O_{8(micro)} + 2H^+_{(aq)} + 9\,H_2O_{(l)} = Al_2Si_2O_5(OH)_{4(kl)}$$
$$+ 2\,K^+_{(aq)} + 4\,H_4SiO_{4(aq)} \qquad [6]$$

for which the equilibrium constant can be written as

$$pK = 4pH_4SiO_4 + 2pK^+ - 2\,pH^+. \qquad [7]$$

If it were desired to plot the equation of the microcline-kaolinite stability boundary on two coordinates, one alternative would be to place $pH^+ - pK^+$ on one axis and pH_4SiO_4 on the other axis. Equation [7] would then become

$$pH^+ - pK^+ = 2\,pH_4SiO_4 - 1/2\,pK$$

Solving for pK

$$\Delta G_r = \Delta G_{kl} + 2\,\Delta G_{K^+} + 4G_{H_4SiO_4} - 9\,\Delta G_{H_2O} - 2\,\Delta G_{micro}$$

$$= -906.2 + 2\,(-67.7) + 4\,(-312.6) - 9\,(-56.7) - 2\,(-892.8)$$

$$= 3.9\ kcal$$

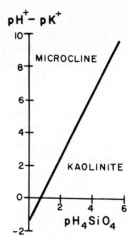

Fig. 1-8. The microcline-kaolinite stability boundary (the intersection of microcline and kaolinite stability planes).

$$= 1.36 \, pK.$$

Therefore $pK = 2.9$ and

$$pH^+ - pK^+ = 2 \, pH_4SiO_4 - 1.4.$$

This is the equation of a straight line of slope 2 and intercept -1.4 as shown in Fig. 1-8. Microcline and kaolinite are in equilibrium along this line, the defining Eq. [6] having been formulated to produce such a boundary. The high K^+ side of the boundary is the microcline stability area, since it is evident from Eq. [6] that an increase in K^+ will shift the equilibrium toward microcline. The parameter pH^+-pK^+ is more ambiguous to interpret than a single variable, but this arrangement permits the consideration of a variable that would otherwise have been placed on a third coordinate, or fixed and essentially eliminated as a variable. The 2-variable parameter was popularized by Garrels, and is frequently used in mineral stability diagrams.

4. VARIABLE DEFERRED

Notice that the microcline and kaolinite in Eq. [6] both contained Al, but that the equation was written in such a way that Al^{3+} did not appear as a variable. Eliminating a variable from the defining equations does not abolish the variable, it merely defers its consideration. Eliminating a variable in the defining equations amounts to considering only those relationships between phases and components that do not involve the eliminated variable. Suppose that in Fig. 1-8 Al^{3+} was a third parameter on an axis vertical to the plane of the paper. The intersection of the kaolinite and microcline planes has a vertical dimension, but because the viewer is looking directly down the Al^{3+} axis, it is not possible to view the Al^{3+} component of this intersection. It is similar to viewing a flat roof in a photograph (which compresses three dimensions into two) from directly overhead. Without extraneous visual clues (such as shadows), it is not possible to tell if the building under the roof is one story, two stories, or if in fact the roof is resting directly on the ground. Only relationships that do not involve the vertical variable are observed.

There are several situations where one might wish to eliminate a variable from defining equations. If one were interested in microcline-kaolinite relationships in natural waters, but had only analyses for pH, K^+, and H_4SiO_4, there would be no need to have an Al^{3+} axis on the subsequent stability diagram because none of the data points would have an Al^{3+} component.

IV. MINERAL EQUILIBRIA APPLIED TO SOILS

A. Subsystem Size

Most major soil minerals contain one or more elements in common. Rarely, if ever, can these minerals all be in equilibrium with each other. Of the group competing for common elements, individual minerals will usually be stable (least soluble) over a rather narrow range of solution compositions. Thus, it may be possible to understand the soil system over a small portion

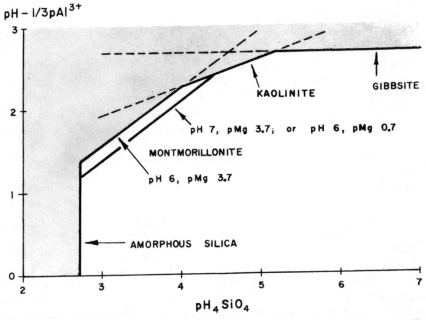

Fig. 1-9. Stability lines for a small subsystem. Contours for Belle Fourche montmorillonite are for the indicated conditions plus control of Fe^{3+} by hematite. The shaded portion represents supersaturation with respect to one or more of the minerals (after Kittrick [1971]).

of its solution composition range by considering the stabilities of only a few of its minerals.

1. SMALL SUBSYSTEMS

Consider Fig. 1-9, a two-dimensional graph with a two-ion parameter on the ordinate and a single illustrative contour line. Three ion variables are shown in this graph, with pH and pMg fixed at selected levels of interest (pMg 3.7 is the average content of natural waters) and pFe^{3+} assumed to be fixed by hematite equilibrium and pH. Only four minerals are considered explicitly, yet the graph appears to explain the formation of gibbsite, kaolinite, and montmorillonite, and their control of Al^{3+} as a function of H_4SiO_4. Briefly, as pH_4SiO_4 decreases (or as the H_4SiO_4 concentration increases) first gibbsite, then kaolinite, and finally montmorillonite becomes the most stable mineral of the group.

At this point it may be worthwhile to confirm that the most stable mineral is also the least soluble. At a pH_4SiO_4 of 3.0, the montmorillonite line[2] (pH 6, pMg 3.7) is below the metastable extension of the kaolinite stability line (dashed), which is below the metastable extension of the gibbsite stability line (dashed). Because the pH of the system is fixed, the lowest value of $pH -1/3\, pAl^{3+}$ means the highest value of pAl^{3+}, which is the lowest level of Al^{3+}. Gibbsite supports almost ten thousand times as much Al^{3+} as montmorillonite at pH_4SiO_4 of 3.0 and pH 7.0. At pH_4SiO_4 6.0, their relative solubilities are reversed.

---

[2]The dissolution equation for Belle Fourche montmorillonite (Kittrick, 1971) is as follows:

$$[(Si_{7.87}Al_{0.13})(Al_{3.03}Mg_{0.58}Fe^{3+}_{0.45})O_{20}(OH)_4]^{0.56-}$$
$$+ 7.48\,H_2O + 12.55H^+ = 3.16Al^{3+} + 0.58Mg^{2+} \qquad [A]$$
$$+ 0.45Fe^{3+} + 7.87H_4SiO_4.$$

Taking negative logarithms of activities, where K is the equilibrium constant and assuming the activity of montmorillonite and water to be unity:

$$pK = 3.16pAl^{3+} + 0.58pMg^{2+} + 0.45pFe^{3+} + 7.87pH_4SiO_4 - 12.55pH^+$$
$$= 1.09. \qquad [B]$$

Rearranging and dividing by 9.48 to isolate $pH-1/3pAl^{3+}$ and pH_4SiO_4 as graphing variables we obtain

$$pH-1/3pAl^{3+} = 0.83pH_4SiO_4-[0.12(pH-1/2pMg^{2+})$$
$$+ 0.14(pH-1/3pFe^{3+}) + 0.059pH^+ + 0.105pK]. \qquad [C]$$

According to [C], the $pH-1/3pAl^{3+}$ supported by Belle Fourche montmorillonite depends upon pH_4SiO_4, $pH-1/2pMg$, $pH-1/3pFe^{3+}$, pH and pK. The term $0.059pH^+$ of Eq. [C]

(footnote 2 continued on next page)

It is evident from the two montmorillonite stability lines (contours) that montmorillonite becomes more stable as the pH increases and as pMg decreases. This increased montmorillonite stability comes at the expense of kaolinite, whose stability range is correspondingly restricted. Where the montmorillonite and kaolinite stability lines intersect, these two minerals can be in equilibrium with each other. Montmorillonite-amorphous silica and kaolinite-gibbsite are also compatible pairs (notice that the intersection of the kaolinite-gibbsite stability lines occurs at a pH_4SiO_4 of 5.2 as predicted by Eq. 4). Incompatible pairs are amorphous silica-kaolinite, amorphous silica-gibbsite, and montmorillonite-gibbsite (unless solution conditions are such that montmorillonite becomes sufficiently stable to eliminate the kaolinite stability field entirely).

---

[2]Continued from p. 17.

arises from the excess charge of 0.56− on the montmorillonite unit cell in Eq. [A]. If [A] had been written with Na^+ as an exchangeable ion, for example, then Eq. [C] would contain the term $0.059(pH–pNa)$ instead of simply $0.059pH$. If [C] contained the term $pH–pNa$, it would imply that the solubility of montmorillonite depended upon the activity of the exchangeable Na^+ in solution. It seemed more reasonable to the author to write Eq. [A] in terms of montmorillonite structural ions, without regard to the type of exchangeable ion present.

Graphing Eq. [C] gives the solubility line for Belle Fourche montmorillonite in terms of $pH–1/3pAl^{3+}$ vs. pH_4SiO_4. The slope of this linear equation is 0.83, and the intercept is minus the quantity inside the brackets of [C]. With a known slope, it is only necessary to evaluate a point for conditions of interest, in order to locate the solubility line. For equilibrium with hematite ($pH–1/3pFe^{3+}$ = −0.32), pH 6.0 and pMg 3.7 ($pH–1/2pMg$ of 4.15) and pH_4SiO_4 3.0, Eq. [C] gives

$$pH–1/3pAl^{3+} = 0.83(3.0) – [0.12(4.15) + 0.14(-0.32) + 0.059(6.0)$$

$$+ 0.105(1.09)] \qquad\qquad [D]$$

$$= 1.6.$$

This point plus the slope of 0.83 establishes the upper stability line for Belle Fourche montmorillonite in Fig. 1-9. For a similar calculation at pH 7.0

$$pH–1/3pAl^{3+} = 0.83(3.0) – [0.12(5.15) + 0.14(-0.32) + 0.059(7.0)$$

$$+ 0.105(1.09)] \qquad\qquad [E]$$

$$= 1.4.$$

The Belle Fourche montmorillonite is more stable at pH 7.0 than at pH 6.0 and Eq. [E] provides the point necessary to locate the lower of the two Belle Fourche solubility lines in Fig. 1-9. To determine the pMg that would correspond to this lower line at pH 6.0, we can solve the following equation for pMg.

$$1.4 = 0.83(3.0) – [0.12(6.0–pMg/2) + 0.14(-0.32)$$

$$+0.059(6.0) + 0.105(1.09)]$$

yielding $pMg = 0.7$.

Perhaps the most important insight to be obtained from Fig. 1-9 is an appreciation that a variety of minerals can be formed in soils in response to a variety of soil solution environments. In any event, it should be clear that graphical display of even a few mineral stabilities often can evoke an understanding of relationships that might otherwise be obscured.

The main advantage in working with only a few minerals is the increased likelihood that necessary thermodynamic data are available. Further, a relatively small system can be more thoroughly investigated. The danger of small subsystems is that most soils have a complex mineralogy which may be impossible to explain by considering just a few minerals in isolation from the others. When considering a relatively simple soil mineral system, the best check of its utility is a thorough comparison with the salient features of the natural system.

2. LARGE SUBSYSTEMS

In theory, the optimum number of minerals to be considered is the number that actually compete for the elements of interest. At present the maximum number of minerals that can be considered is limited by the quality and even the existence of necessary thermodynamic data. Unfortunately, there are some investigations that cannot be attempted unless many minerals are considered, i.e., investigations of the mineralogical control of natural water composition (Kharaka & Barnes, 1973; Truesdell & Jones, 1973). If the entire spectrum of mineralogical controls were to be considered now, it would be necessary to include tenuous thermodynamic data for some minerals. The reliability of the programs for such calculations would of necessity be uneven, but should be improved as better data become available.

B. Thermochemical Data

Presently, the two best compilations of thermodynamic data are Robie and Waldbaum (1968), which went into a second printing in 1970, and a series of Technical Notes (the "270" series) from the National Bureau of Standards (1965-). The 270 series is entitled "Selected Values of Chemical Thermodynamic Properties," and is designed to replace Circular 500 by Rossini et al. (1952). An additional recent compilation of interest is by Naumov et al. (1971), as translated by G. J. Soleimani. Both Robie and Waldbaum (1968) and Naumov et al. (1971) contain considerable high temperature data, with the latter including an estimation of ΔG values of ions and molecules at elevated temperatures.

Accurate thermochemical data usually emerge from a confirmation process involving several investigators, independent methods, and an examination of compatibility with known relationships. The process often takes a long time. Any extensive compilation of thermodynamic values will contain

Table 1-2. Selected ΔG_f° values in kcal mole^{-1} listed by various authors.

Formula and name	Garrels and Christ (1965)	Robie and Waldbaum (1968)	Wagman et al. (1968)	Naumov et al. (1971)
SiO_2 (quartz)	-192.4	-204.6	-204.8	-204.7
H_4SiO_4(aq)	-300.3	-314.7‡	-314.7	-313.1
$Al(OH)_4^-$(aq)	-314.1†	--	-310.2	-312.0
Al^{3+}(aq)	-115.0	-116.0	-116.0	-117.6
$Al(OH)_3$(gibbsite)	-277.3	-273.5	-273.3	-276.5
$Al_2Si_2O_5(OH)_4$ (kaolinite)	-884.5	-902.9	-903.0	-903.0

† $\Delta G_{H_2AlO_3^-} + \Delta G_{H_2O}$.
‡ $\Delta G_{H_2SiO_3^-} + \Delta G_{H_2O}$.

data that are at different stages of the confirmation process and can thus vary widely in accuracy. Important variations can be expected between compilations of thermodynamic data selected by different authors at different times (Table 1-2). All too infrequently an informed reviewer [such as Parks (1972) for example] will critically evaluate the available thermodynamic data, selecting the most reliable from a mass of conflicting data.

The neat rows of data in thermochemical compilations tend to lend authenticity to what, in some cases, are low quality or even erroneous data. Those who select thermodynamic data for these compilations do the best they can, but this does not absolve the user of thermochemical data from the responsibility of checking the source of the data. Thermochemical data needed in calculations involving soil minerals are often nonexistent, and what is worse, sometimes highly inaccurate.

1. MINERALS

Free energy determinations by solubility methods require accurate measurement of all common ions under equilibrium conditions where the mineral is stable (or at least metastable). Some minerals are not stable at room temperature and pressure, or are stable only under solution composition conditions that contain submeasurable amounts of some ion, or involve unknown coordination constants. Equilibrium is usually reached slowly, so that more solubility determinations are flawed by lack of equilibrium than for any other single cause. The main weakness in calorimetric determinations of mineral enthalpies is the required correction for impurities, the nature and extent of which are often unknown.

Considering the difficulties in obtaining reliable thermochemical data, it is not surprising that in general the simpler the mineral the better the available data. For example, the thermochemical data for simple oxides and hydroxides such as quartz, hematite, and gibbsite are good. The stability of kaolinite is reasonably well known. A moderate amount of data are available for smectites, but there is very little agreement. There have been no experimental stability determinations to date for chlorites. Both smectites and chlorites share the problem of a wide range in chemical compositions and a variety of structural types. A stability determination on every possible series member would be a staggering task. It may be possible to calculate stability data on a series members from information on end members (Tardy & Garrels, 1974). Chlorites and some smectites share several other characteristics which make solubility determinations difficult. Their stability is often restricted to the neutral to slightly alkaline pH range. This engenders extremely small amounts of Al in solution, plus the likelihood of unknown hydroxy aluminum complexes. Further, Fe^{3+} and Fe^{2+} will generally be below detection levels under these conditions and the couple produced by such small amounts cannot reliably be determined by Eh measurements. Many of the smectite and all of the chlorite thermochemical values used in geochemical calculations to date are not based upon experimental determinations, but are theoretical values based upon assumed stabilities in certain geochemical environments. Unfortunately, these assumptions are not always stated in the original work and often appear to be unsuspected when quoted.

Exchangeable ions are also an important problem in stability determinations for smectites. There is no question that replacing one exchangeable ion with another involves an energy change. The question is whether such exchangeable ions act as common (or structural) ions in controlling mineral equilibria. For example, if a smectite contains exchangeable Ca^{2+}, does the activity of Ca^{2+} in solution affect the common ion equilibrium in the mass action expression (other than through ionic strength, ion pair formation, and carbonate phase equilibria)? If so, then smectite stability will change as the exchangeable ion composition changes. I have assumed that exchangeable ions do not act as structural ions whereas almost all other authors have assumed that they do. The situation urgently needs more facts and fewer assumptions.

2. IONS

It is not too surprising to find that chemists have rarely made stability determinations on minerals that are primarily of interest to earth scientists. If earth scientists want such information, they probably will have to determine much of it for themselves. What comes as a shock is the discovery that basic thermochemical information on some rather ordinary ions is lacking or in the early stages of confirmation and controversy.

The aluminum situation is a good example. Soil scientists are familiar with the uncertain nature of polynuclear hydroxy aluminum species in the near neutral pH range. They also must cope with organic aluminum complexes that are responsible for most of the dissolved aluminum in natural waters. In the face of these difficulties, it is discouraging to find that even the ΔG for Al^{3+} is uncertain. As indicated in Table 1-2, $\Delta G_{Al^{3+}}$ is variously quoted as -115, -116, and -117.6 kcal mole^{-1}. The commonly accepted value once was -115 kcal mole^{-1}, but recently this consensus value appears to have been replaced by -116 kcal mole^{-1}, for reasons that are as yet unapparent. The uncertainty in ΔG for Al^{3+} means that the ΔG for all minerals containing Al is similarly uncertain.

The uncertainty in Al^{3+} need not be a problem in geochemical calculations if the ΔG used for substances containing Al is based upon the same $\Delta G_{Al^{3+}}$. High quality ΔG compilations strive for such internal consistency in instances where absolute accuracy is not yet possible. Unfortunately, not all authors make such self-consistent comparisons. Evidentally it is too temptingly easy to use ΔG values from several sources without first checking the calculations of each author to determine if all used the same values for Al^{3+} and for other common ions and minerals.

C. Nonequilibrium Considerations

As it is obvious that not all soil minerals are in equilibrium with the soil solution, the equilibrium processes previously described often function primarily as a frame of reference for understanding the kinetics of alteration and formation of soil minerals. The activities of ions in the soil solution which are common to soil minerals arise from the relative rates of the dissolution and precipitation processes, coupled with the rates of solution addition or removal during water flow processes. The dissolution and precipitation processes appear to be the summation of many reactions whose rates are rather specifically dependent upon local conditions, as illustrated by the transformation of lepidocrocite to goethite (Schwertmann & Taylor, 1972a, 1972b) and of trioctahedral micas to vermiculite (Kittrick, 1973). The critical mass of data required to permit broad kinetic generalizations does not yet exist. However, there appear to be at least three tentative generalizations that can be mentioned.

First, where solutions have had appreciable contact with precipitating soil minerals, they appear to be in equilibrium with them. At least there are such indications for the soil solution (Weaver et al., 1971), spring waters (Garrels & Mackenzie, 1967), lake waters (Kramer, 1967; Sutherland, 1970), and the ocean (Mackenzie & Garrels, 1966; Helgeson & Mackenzie, 1970).

Second, it appears that the H^+ in percolating rain water reacts with soil minerals very quickly (Bricker et al., 1968; Johnson et al., 1968).

Finally, the conversion from an unstable mineral form to a stable one can be quite rapid where only slight bond shifts are required, such as in the α-β quartz transition. Even the dissolution of montmorillonite at room temperature with the subsequent precipitation of kaolinite can be observed in only a few years (Kittrick, 1970). The third tentative generalization is that the slowest processes at room temperatures seem to be those where some bonding changes are required, but where the overall structural change is not great. The conversion of halloysite to kaolinite, of amorphous silica to quartz, or of small particles to larger particles are all energetically favorable processes that usually display negligible rates at room temperature. The gradual conversion of gibbsite to boehmite to diaspore over hundreds of millions of years may be one example of such slow processes, where an approximate timescale is available (Kittrick, 1969). Unfortunately, the stability sequence indicated by thermodynamic data (Wagman et al., 1968) is contradicted by some hydrothermal data (see Chesworth, 1972).

For a soil mineral to be in equilibrium with the soil solution, it is only necessary that it control the activity of one of its common ions. Other common ions can be controlled by other minerals or by kinetic processes. For example, in Fig. 1–9, the pH of the system may be determined by a wide variety of kinetic and equilibrium factors, the pH_4SiO_4 by kinetics (Wollast, 1967; Kittrick, 1969), and the pAl^{3+} by the indicated mineral equilibria. Kinetics (and in some cases equilibria) are likely to be controlled by those rapidly reacting phases, including surface coatings, that might be called the *reactive fraction*. Combinations of kinetics and equilibria for various model systems can sometimes be treated on the same graphs by considering a series of reactions, each reversible with respect to the next, but irreversibly related to the initial state of the system (Helgeson et al., 1969). At this stage in our knowledge, calculation methods appear to be more advanced than the data available to use in them.

V. PAST, PRESENT, AND FUTURE

Prior to the middle 1960's, it was thought by many that soil minerals were different from ordinary substances, that they were essentially insoluble, that they could not be described in terms of elementary thermodynamics, and that they especially were not amenable to stability determinations by solubility methods. That these fears have been shown to be unfounded does not mean that the job ahead is easy. The soil mineral system is extensive and complicated, often requiring many stability determinations in order to understand even a modest portion of the system. Ion exchange and a wide range in both kinetics (reactivity) and crystallinity complicate the situation. The advantageous characteristic of the thermodynamic approach to understanding the soil mineral system is that each tested thermodynamic value can serve as a stepping stone to others. Understanding will come at an increasing pace as thermodynamic information is accumulated. Studies involving soil mineral weathering, chemical reactions in the soil, and plant nutrition will all benefit. Only ill-defined and unidentified phases are likely to escape thermodynamic characterization. Discouragements most likely to retard the thermodynamic approach to understanding the soil mineral system will most likely come from erroneous predictions, based upon unconfirmed thermodynamic data or upon neglect of the kinetic aspect of soil mineral reactions. At times the thermodynamic approach, with appropriate kinetic considerations, may be arduous but there do not appear to be any fundamental limitations to a marked improvement in our understanding of the soil mineral system through its use.

LITERATURE CITED

Allen, J. A. 1966. Energy changes in chemistry. Allyn & Bacon, Boston, Mass.

Bassett, L. G., S. C. Bunce, A. E. Carter, H. M. Clark, and H. B. Hollinger. 1966. Principles of chemistry. Prentice-Hall, Englewood Cliffs, New Jersey.

Bricker, O. P., A. E. Godfrey, and E. T. Cleaves. 1968. Mineral-water interaction during the chemical weathering of silicates. p. 128-142. In R. F. Gould (ed.) Trace inorganics in water. Adv. Chem. Ser. 73. American Chemical Society, Washington, D. C.

Chesworth, W. 1972. The stability of gibbsite and boehmite at the surface of the earth. Clays Clay Miner. 20:369-374.

Gardner, L. R. 1970. A chemical model for the origin of gibbsite from kaolinite. Am. Mineral. 55:1380-1389.

Garrels, R. M., and C. L. Christ. 1965. Solutions, minerals and equilibria. Harper & Row, New York.

Garrels, R. M., and F. T. Mackenzie. 1967. Origin of the chemical compositions of some springs and lakes. p. 222-242. In R. F. Gould (ed.) Equilibrium concepts in natural water systems. Adv. Chem. Ser. 67. American Chemical Society, Washington, D. C.

Garrels, R. M., and F. T. Mackenzie. 1971. Evolution of sedimentary rocks. W. W. Norton & Company, New York.

Helgeson, H. C., R. M. Garrels, and F. T. Mackenzie. 1969. Evaluation of irreversible reactions in geochemical processes involving minerals and aqueous solution. II. Applications. Geochim. Cosmochim. Acta. 33:455-481.

Helgeson, H. C., and F. T. Mackenzie. 1970. Silicate-seawater equilibrium in the ocean system. Deep Sea Res. 17:877-892.

Jackson, M. L., S. A. Tyler, A. L. Willis, G. A. Bourbeau, and R. P. Pennington. 1948. Weathering sequence of clay-size minerals in soils and sediments. I. Fundamental generalizations. J. Phys. Colloid Chem. 52:1237-1260.

Johnson, N. M., G. E. Likens, F. H. Bormann, and R. S. Pierce. 1968. Rate of chemical weathering of silicate minerals in New Hampshire. Geochim. Cosmochim. Acta. 32:531-545.

Kharaka, Y. K., and I. Barnes. 1973. Solution-mineral equilibrium computations. National Technical Information Service, Springfield, VA.

Kittrick, J. A. 1969. Soil minerals in the Al_2O_3-SiO_2-H_2O system and a theory of their formation. Clays Clay Miner. 17:157-167.

Kittrick, J. A. 1970. Synthesis of kaolinite at 25°C and 1 atm. Clays Clay Miner. 18: 261-267.

Kittrick, J. A. 1971. Montmorillonite equilibria and the weathering environment. Soil Sci. Soc. Am. Proc. 35:815-820.

Kittrick, J. A. 1973. Mica-derived vermiculites as unstable intermediates. Clays Clay Miner. 21:479-488.

Kramer, J. R. 1967. Equilibrium models and composition of the Great Lakes. p. 243-254. In R. F. Gould (ed.) Equilibrium concepts in natural water systems. Adv. Chem. Ser. 67. American Chemical Society, Washington, D. C.

Mackenzie, F. T., and R. M. Garrels. 1966. Chemical mass balance between rivers and oceans. Am. J. Sci. 264:507-525.

Mahan, B. H. 1963. Elementary chemical thermodynamics. W. A. Benjamin, New York.

National Bureau of Standards. 1965-. Selected values of chemical thermodynamic properties. Tech. Note 270. U. S. Department of Commerce. (270 series available from Superintendent of Documents).

Naumov, G. B., B. N. Ryzhenko, and I. L. Khodakovsky. 1971. Handbook of thermodynamic data (translated from the Russian by G. J. Soleimani for the U. S. Geological Survey, Water Resources Division, 1974). Available from NTIS, PB226722.

Parks, G. A. 1972. Free energies of formation and aqueous solubilities of aluminum hydroxides and oxide hydroxides at 25°C. Am. Mineral. 57:1163-1189.

Robie, R. A., and D. R. Waldbaum. 1968. Thermodynamic properties of minerals and related substances at 298.15°K (25.0°C) and one atmosphere (1.013 Bars) pressure and at higher temperatures. Geological Survey Bulletin 1259.

Rossini, F. D., D. D. Wagman, W. H. Evans, S. Levine, and I. Jaffe. 1952. Selected values of chemical thermodynamic properties. Nat. Bur. of Stand. Circ. 500, U. S. Dep. of Commerce.

Schwertmann, U., and R. M. Taylor. 1972a. The transformation of lepidocrocite to goethite. Clays Clay Miner. 20:151–158.

Schwertmann, U., and R. M. Taylor. 1972b. The influence of silicate on the transformation of lepidocrocite to goethite. Clays Clay Miner. 20:159–164.

Siever, R. 1974. The steady state of the earth's crust, atmosphere and oceans. Sci. Am. 230:72–79.

Sutherland, J. C. 1970. Silicate mineral stability and mineral equilibria in the Great Lakes. Environ. Sci. Technol. 4:826–833.

Swalin, R. A. 1964. Thermodynamics of solids. John Wiley & Sons, New York. 343 p.

Tardy, Y., and R. M. Garrels. 1974. A method of estimating the Gibbs energies of formation of layer silicates. Geochim. Cosmochim. Acta. 38:1101–1116.

Truesdell, A. H., and B. F. Jones. 1973. WATEQ, A computer program for calculating chemical equilibria of natural waters. Available from NTIS, Springfield, Va.

Wagman, D. D., W. H. Evans, V. B. Parker, I. Halow, S. M. Bailey, and R. H. Schumm. 1968. Selected values of chemical thermodynamic properties. Tables for the first 34 elements. Nat. Bur. of Stand. Tech. Note 270–3. U. S. Government Printing Office, Washington, D. C.

Waser, J. 1966. Basic chemical thermodynamics. W. A. Benjamin, New York.

Weaver, R. M., M. L. Jackson, and J. K. Syers. 1971. Magnesium and silicon activities in matrix solutions of montmorillonite-containing soils in relation to clay mineral stability. Soil Sci. Soc. Am. Proc. 35:823–830, 36:854.

Wollast, R. 1967. Kinetics of the alteration of K-feldspar in buffered solutions at low temperature. Geochim. Cosmochim. Acta. 31:635–648.

Surface and Colloid Chemistry

ROBERT G. GAST, University of Minnesota, St. Paul

I. INTRODUCTION

The surface and colloidal chemistry of soils is largely determined by two basic properties of soil minerals, their large surface area, and the presence of a surface electrical charge. Emphasis in this chapter will be on the origin and nature of the surface charge and its effect on several physical or chemical properties important in soils. These include ion exchange, particle interactions, single ion activity measurements, and anion adsorption.

There are many other aspects of the surface chemistry of minerals which could be included. It is emphasized that their omission is the combined result of space limitations and topic emphasis and should not be interpreted as any judgment as to their relative validity or merit. Also, while efforts have been made to include literature citations adequate to document the material presented, it is not necessarily intended that the chapter serve as a complete literature review. Hopefully, the citations given will provide an adequate entry into the literature dealing with surface phenomena.

II. CHARACTERISTICS OF CONSTANT CHARGE, VARIABLE POTENTIAL SURFACES

A. Origin and Extent of Charge

The charge on soil colloids may result from either structural imperfections in the interior of the crystal structure or preferential adsorption of certain ions on particle surfaces (van Olphen, 1963). Structural imperfections due to ion substitutions or site vacancies frequently result in a permanent charge on soil colloidal particles. This type of colloid, which has a completely polarizable interface, is considered to have a constant charge- variable potential-surface. Such colloids are typified by, but may not be limited to the so-called 2:1 type of clay minerals such as micas or smectites where the charge results from isomorphous substitution of octahedrally or tetrahedrally coordinated cations. The charge resulting from a given ion substitution may theoretically be either positive or negative (Ross & Hendricks, 1945). However, in the case of 2:1 type clay minerals which consist largely of Si^{4+} or Al^{3+} in tetrahedral coordination, and Fe^{3+}, Fe^{2+}, Al^{3+}, and Mg^{2+} in octahedral coordination, ion size limitations generally result in a substitution of cations of lower valence for those of higher valence resulting in a net negative charge on the clay structure (Pauling, 1930; Marshall, 1949).

chapter 2

Table 2-1. Properties of representative 2:1 type clay minerals (from Grim, 1953). Chemical composition is based on unit cell containing $O_{20}(OH)_4$ or 44 negative charges

Mineral name	Chemical composition		Charge deficit per unit cell			Molecular weight per unit cell	Structural charge, meq/100 g	Calculated surface area, m²/g†	Surface charge density	
	Tetrahedral layer	Octahedral layer	Tetrahedral layer	Octahedral layer	Total				esu/cm² (10⁴)	μcoul/ cm²
Pyrophyllite	(Si_8)	(Al_4)	0	0	0	720.4	0	[765.4]	0	0
Talc	(Si_8)	(Mg_6)	0	0	0	709.8	0	[801.3]	0	0
Montmorillonite	(Si_8)	$(Al_{3.34} \cdot Mg_{0.66})$	0	−0.66	−0.66	718.6	91.8	772.0	3.44	11.5
Beidellite	$(Si_6 \cdot Al_2)$	$(Al_{4.44})$	−2.0	+1.32	−0.68	730.1	93.1	775.8	3.47	11.6
Nontronite	$(Si_{7.34} \cdot Al_{0.66})$	(Fe_4^{3+})	−0.66	0	−0.66	835.2	79.0	694.4	3.29	11.0
Hectorite	(Si_8)	$(Mg_{5.34} \cdot Li_{0.66})$	0	−0.66	−0.66	747.0	88.5	769.2	3.32	11.1
Saponite	$(Si_{7.73} \cdot Al_{0.66})$	(Mg_6)	−0.66	0	−0.66	757.8	87.1	767.6	3.28	10.9
Muscovite	$(Si_6 \cdot Al_2)$	(Al_4)	−2.0	0	−2.0	718.2	278.5	[788.7]	10.2	34.0
Biotite	$(Si_6 \cdot Al_2)$	$(Mg + Fe^{2+})_6$	−2.0	0	−2.0	818.4	244.4	[730.4]	9.67	32.2
Phlogopite	$(Si_6 \cdot Al_2)$	(Mg_6)	−2.0	0	−2.0	756.3	264.4	[783.2]	9.76	32.5
Glauconite	$(Si_{7.30} \cdot Al_{0.70})$	$(Al_{0.94} \cdot Fe_{1.94}^{3+} \cdot Fe_{0.38}^{2+} \cdot Mg_{0.80})$	−0.7	−1.0	−1.70	786.3	216.2	[730.8]	8.55	28.5

† Surface areas in brackets are generally "not available" due to non-expanding structure.

‡ This table contains a set of traditional and somewhat idealized formulae. A smaller set of more contemporary layer charge values is given in Table 9-1.

On the basis of known ion size limitations, coordination numbers, and a knowledge of the chemical composition for a given 2:1 type clay mineral, it is possible to calculate the chemical formula, assign atoms to the structural planes and calculate unit cell dimensions, surface area, charge deficit per unit cell, surface charge density, and a theoretical cation exchange capacity. Example calculations of this type, similar to those outlined by Marshall (1949), Ross and Hendricks (1945), and van Olphen (1963) are given in the Appendix.

Since such calculations are applicable only to homogenous samples of a given clay mineral, they cannot be generally applied to the complex mixture of clay minerals found in most soils. However, as shown in Table 2-1, there are significant differences in the origin and magnitude of the permanent negative charge on the various clay minerals. These known differences can be combined with a knowledge of the mineralogical composition of soils to help explain and predict many of the chemical and physical properties of soils.

B. Determining Extent of Permanent Charge

Since the calculations outlined in the Appendix cannot be generally applied to heterogenous soil clays, it is necessary to measure the extent of the surface charge experimentally by determining the quantity of counter ions (cations) compensating the negative charge (Ensminger, 1944; Mehlich, 1948). The charge determined in this manner is commonly referred to as the cation exchange capacity (CEC) and, if appropriate precautions are taken, is a measure of the negative charge balanced by cations available for exchange by the saturating cations.

Cation exchange capacity measurements generally involve washing the soil sample with a given salt solution until the negative charge is compensated by only the added cation species. The quantity of compensating cations and hence the net negative charge or CEC can then be determined by either extracting with an excess of a second salt solution and determining the quantity of counter ions present (Jackson, 1969; Rich, 1961) or by labeling the saturating cation with an appropriate radioisotope and counting if directly without extraction from the soil (Beetem et al., 1962; Francis & Grigal, 1971). In either case, it is common to wash out the excess salt before extraction or counting. Other than the possible retention of excess salt and loss of sample by dispersion (Jackson, 1969; Rich, 1962), hydrolysis of the saturating cation during the washing procedure constitutes the most likely discrepancy between the net negative charge and the measured CEC.

As indicated previously, the charge or CEC measured in this manner is the charge neutralized by cations that are available for exchange with those in the saturating solution. Therefore, it is important to use a cation which is preferred or held more tightly by the surface than the cations being replaced. For this reason, Cs^+, Ca^{2+}, Mg^{2+}, or Sr^{2+} are commonly used (Sawhney et al., 1959). Even when an excess of a highly preferred cation is used, part of the net negative charge on the surface may be blocked in such a manner that the compensating cations are not replaced by those in the saturating solution. This frequently is the case in micas and illites where the K^+ is "fixed" against normal exchange processes. In these cases, the measured CEC is significantly less than the net charge deficit on the mineral structure.

C. The Electric Double Layer

In a clay-water system, the counter ions tend to diffuse into the bulk aqueous phase (due to their kinetic energy) until the counter potential set up by their departure restricts this tendency. The equilibrium distribution of counter ions assumes a diffuse layer which can be described by the Boltzmann function (Bolt, 1955; Overbeek, 1952; Parks, 1967). It is this diffuse layer of counter ions along with the negative charge on the particle surface which is referred to as the *electric double layer*.

The electrochemical potential, $\bar{\mu}$, of each ion is constant throughout such a system at equilibrium including the diffuse layer of counter ions near the clay surface (de Bruyn & Agar, 1962). For cations in such a system we then find that

$$\bar{\mu}_{+(x)} = \bar{\mu}_{+(x=\infty)} \tag{1}$$

where x is the distance between the cation in the diffuse layer and the solid-liquid interface, and $x=\infty$ refers to a distance far away from the solid surface where the concentrations of cations and anions are equal. The electrochemical potential is defined by:

$$\bar{\mu}_+ = \mu_+ + z_+ e \psi \tag{2}$$

where μ_+ is the chemical potential, z_+ is the valence, ψ is the electrical potential at the distance x from the surface, and e is the electronic charge. The chemical potential, μ_+, is related to the activity of the cation, a_+, by:

$$\mu_+ = \mu_+^0 + RT \ln a_+ \tag{3}$$

where μ_+^0 is the standard chemical potential, R is the gas constant, and T is the absolute temperature. Combining Eq. [1], [2], and [3] and remembering the $F = N_A e$, where F is the Faraday and N_A is Avogadro's number, gives the result

$$\psi_{(x)} - \psi_{(x=\infty)} = \frac{RT}{F} \ln \frac{a_{+(x=\infty)}}{a_{+(x)}}. \tag{4}$$

If $\psi_{(x=\infty)}$ is assumed to be zero, i.e. if all electrical potentials are referred to the potential of the bulk solution, and if the activity of the cation is assumed to be equal to its molar concentration it follows from Eq. [4] that

$$n_{+(x)} = n_+^0 \exp\left(-z_+e \ \psi_{(x)}/kT\right) \tag{5}$$

where n_+^0 is the concentration of the cation in bulk solution and k is the Boltzmann constant. Equation [5] is the Boltzmann distribution law which can be applied equally to the distribution of both cations and anions; i.e. there is a negative adsorption or repulsion of anions away from negatively charged surfaces which can also be theoretically described by electric double-layer theory (Bower & Goertzen, 1955; Bolt & Warkentin, 1958; Edwards & Quirk, 1962).

The net charge density, ρ, at any point in the diffuse layer of ions near the charged particle surface is given by (van Olphen, 1963)

$$\rho_i = \Sigma Z_i e n_i \tag{6}$$

which can be written in the form of Eq. [7] for symmetrical electrolytes

$$\rho = (z_+en_+ - z_-en_-) \tag{7}$$

where n_+ and n_- are the cation and anion concentrations, respectively. The integral of ρ from the surface to the bulk solution gives the total excess charge in the solution, per unit surface area, and is equal in magnitude but opposite in sign to the surface charge density, σ; i.e.;

$$\sigma = -\int_0^\infty \rho \ dx. \tag{8}$$

Poisson's equation, simplified for an infinite flat plate (Eq. [9]),

$$\frac{d^2\psi}{dx^2} = -\left(\frac{4\pi}{D}\right)\rho \tag{9}$$

relates the divergence of the electrical potential gradient to the charge density at a given point and can be combined with Eq. [8] to give Eq. [10] (Adamson, 1967; Verwey & Overbeek, 1948)

$$\sigma = \frac{D}{4\pi}\int_0^\infty \frac{d^2\psi}{dx^2} \ dx = -\frac{D}{4\pi}\left[\frac{d\psi}{dx}\right]_{(x=0)} \tag{10}$$

Here, D is the diabattivity (rather than the dielectric constant) (Grahame, 1947).

A quantitative relationship between surface charge density, electrolyte concentration, and electrical potential can be derived by first combining the Boltzmann equation (Eq. [5]) and the Poisson equation (Eq. [9]) along with Eq. [7] giving (Babcock, 1963, p. 472)

$$\frac{d^2\psi}{dx^2} = -\frac{4\pi}{D} \Sigma z_i en_i^0 \exp\left(-z_i\psi e/kT\right) \tag{11}$$

which may be written as

$$\frac{d^2\psi}{dx^2} = -\frac{4\pi|z_i|en_i^0}{D} \left\{ \left[\exp\left(-\frac{|z_i|e\psi}{kT}\right)\right] - \exp\left[\left(\frac{|z_i|e\psi}{kT}\right)\right] \right\} \qquad [12]$$

Writing

$$y = |z_i|e\psi/kT \qquad [13]$$

Eq. [11] becomes

$$\frac{d^2y}{dx^2} = -\frac{4\pi|z_i|en_i^0}{DkT} \left[\exp(-y) - \exp(y)\right]. \qquad [14]$$

Recalling that by definition

$$sinh(y) = [\exp(y) - \exp(-y)]/2 \qquad [15]$$

Eq. [14] becomes

$$\frac{d^2y}{dx^2} = \frac{8\pi z_i^2 e^2 n_i^0}{DkT} sinh(y). \qquad [16]$$

Babcock (1963) has shown that Eq. [16] can be reduced to

$$dy/dx = -2\kappa \, sinh(y/2) \qquad [17]$$

where

$$\kappa^2 = 8\pi z_i^2 e^2 n_i^0/DkT. \qquad [18]$$

Equation [17] can be integrated under appropriate limits to give the electrical potential, ψ, as a function of distance, x, from the surface. Also, from Eqs. [10] and [17] at $[d\psi/dx]_{x=0}$ the following relationship in the Gouy-Chapman theory of the electric double layer for a symmetrical electrolyte is obtained (Adamson, 1967; Babcock, 1963; van Olphen, 1963; Verwey & Overbeek, 1948; Overbeek, 1952)

$$\sigma = (2n^0Dkt/\pi)^{1/2} \, sinh(ze\psi_0/2kT) \qquad [19]$$

where ψ_0 is the electrical potential at the surface. This fundamental charge-electrolyte concentration-electrical potential relationship can be related to many chemical and physical properties of soils. In the case of colloidal particles with a permanent charge such as the 2:1 clay minerals, Eq. [19] shows that the surface potential decreases with increasing electrolyte concentration (Fig. 2-1). Also, if the electrolyte concentration is increased, the dif-

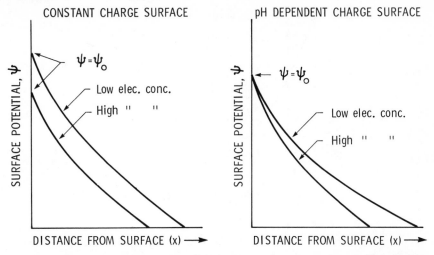

Fig. 2-1. Electric potential distribution with distance from constant- and pH dependent-
charge surfaces at two electrolyte concentrations according to Gouy-Chapman model.

fuse part of the electrical double layer is compressed and the electrical potential curves decay more rapidly with distance from the surface (van Olphen,
1963; Babcock, 1963). This fact has particular significance in flocculation-
dispersion phenomena.

The Gouy-Chapman equation (Eq. [19]) has limited quantitative application due to the assumption that ions in solution behave as point charges
and can approach the surface without limit (Bolt, 1955b). This limitation
particularly applies to variable charge surfaces where, as will be discussed
later, predicted concentrations of ions near the surface are unreasonably high
at large values of ψ_0 and n_0 (Eq. [5]). Application of Eq. [19] is also limited
for fixed charge surfaces even though the maximum number of counter ions
at the surface is determined by the surface charge density.

Stern (1924) introduced a modification of the Gouy-Chapman theory
such that the first layer of ions (Stern layer) is not immediately at the surface, but a distance, δ, away from it (Overbeek, 1952). As a result, the concentration and potential in the diffuse part of the double layer drops to
values low enough to warrant the approximation of ions as point charges.
Stern further considered the possibility of "specific adsorption" of ions and
assumed that these ions were located in the plane, δ. The stern theory of the
electrical double layer then assumes that the surface charge is balanced by the
charge in solution which is distributed between the Stern layer at a distance
δ from the surface and a diffuse layer which has a Boltzmann distribution
(Fig. 2-2). The total surface charge, σ, is balanced by the sum of the charge
in the two layers

$$\sigma = -(\sigma_1 + \sigma_2) \qquad [20]$$

where σ_1 is the Stern layer charge and σ_2 is the diffuse layer charge.

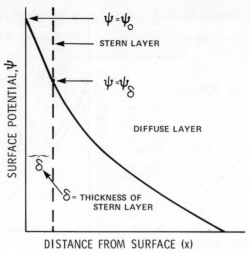

Fig. 2-2. Electric potential distribution with distance from charged surface according to Stern's model.

As outlined by Stern (1924) and van Raij and Peech (1972), the charge in the Stern layer is given by

$$\sigma_1 = \frac{N_i z e}{1 + (N_A \omega / Mn) \exp \left[-(z e \psi_\delta + \phi)/kT \right]} \qquad [21]$$

where N_1 is the number of available spots per cm^2 for adsorption of ions, N_A is Avogadro's number, M is the molecular weight of the solvent, ω is the solvent density, n is the electrolyte concentration, ψ_δ is the Stern potential or electrical potential at the boundary between the Stern layer and the diffuse layer, and ϕ is the specific adsorption potential.

The charge in the diffuse layer is given by the Gouy-Chapman theory (Eq. [19]) except that the reference is now the Stern potential instead of the surface potential,

$$\sigma_2 = (2n^0 DkT/\pi)^{1/2} \sinh(z e \psi_\delta / 2kT). \qquad [22]$$

Since a linear drop in potential across the Stern layer is assumed, the surface charge is also given by the Gauss equation for a molecular condensor,

$$\sigma = (D'/4\pi\delta)(\psi_0 - \psi_\delta) \qquad [23]$$

where D' is the diabattivity in the Stern layer. As shown in the sample calculations, Eq. [19]-[23] can be used to calculate surface charge-surface potential relationships for both constant charge and pH dependent charge surfaces.

III. CHARACTERISTICS OF CONSTANT POTENTIAL, VARIABLE
CHARGE SURFACES

A. Origin and Nature of pH-Dependent Charge

Exposure of oxide or hydrous oxide particles to water vapor generally results in physical or chemical adsorption of water on the surfaces; i.e., chemical adsorption differs from physical adsorption in that H_2O is split into H^+ and OH^- during adsorption to form an hydroxylated surface (Parks, 1967; Blyholder & Richardson, 1962). Further adsorption of water vapor results in patches and finally layers of hydrogen-bonded water (Parks, 1967; McCafferty & Zettlemoyer, 1970; Gast et al., 1974). As the thickness of the adsorbed water increases, its properties approach those of bulk liquid and eventually the solid may be considered immersed.

Hydroxylated surfaces can be expected on all oxide materials at equilibrium with an aqueous environment. Charge can develop on these hydroxylated surfaces either through amphoteric dissociation of the surface hydroxyl groups or by adsorption of H^+ or OH^- ions. Following Parks (1967), these reactions can be written as follows, where under-scored symbols refer to species forming part of the surface:

$$\underline{M}\,OH \rightleftharpoons \underline{M}\,O^- + H^+_{(aq)} \qquad\qquad [24]$$

$$\underline{M}\,OH \rightleftharpoons \underline{M}^+ + OH^-_{(aq)} \qquad\qquad [25]$$

$$\underline{M}^+ H_2O \rightleftharpoons \underline{M}\,OH_2 \qquad\qquad\qquad [26]$$

Since the probability of bare $\underline{M}^+$ existing at the surface is small, the basic dissociation or formation of a positively charged site probably occurs through a combination of the reactions shown in Eq. [25] and [26] (Mott, 1970),

$$\underline{M}\,OH + H_2O \rightleftharpoons \underline{M}\,OH_2^+ + (OH^-)_{(aq)} \qquad\qquad [27]$$

In both of the above mechanisms, the H^+ and OH^- ions which establish the surface charge are also those which would establish the potential of a reversible oxide or hydroxide electrode and hence are referred to as *potential determining ions* (PDI). Concentrations of the PDI and the net surface charge are obviously pH dependent and there will be a pH value at which the net surface charge is zero; i.e. the densities of positive and negative charge are equal. This pH is referred to as the *isoelectric point* (IEP) or *zero point(s) of charge* (ZPC) of the solid. Various criteria have been used to distinguish between ZPC and IEP (Parks, 1967; Mott, 1970; Lyklema, 1971; Wright & Hunter, 1973b). However, it will suffice for purposes here to consistently use ZPC when referring to the point of zero net surface charge.

The above picture is somewhat oversimplified, for as Mott (1970) has pointed out, the cation $\underline{M}$, having a charge of four or less, is generally octa-hedrally or tetrahedrally coordinated with O in the oxide structure. Con-sequently, O ions in the oxide crystal are sharing charge from several cations. Since the surface OH^- is coordinated to only one cation it will have a net negative charge the extent of which will depend on the structure. Therefore, the magnitude of the charge at any particular adsorption site will vary, but the charging mechanism is basically that given in Eq. [23]-[27] with the sum of all positive and negative charges being zero at the ZPC.

Since the net charge on such surfaces is dependent on the H^+ or OH^- activity, it can be made positive or negative by raising or lowering the pH. However, since the H^+ and OH^- ions which establish the surface charge are also potential determining ions, the surface potential, ψ_0, will also vary as the net surface charge varies. This can be illustrated by considering an oxide in equilibrium with its saturated solution. Let us assume that the net surface charge is zero in this system; the observed pH or H^+ activity, $(a_{H^+})_{ZPC}$, is then defined as the *zero point of charge*. Suppose now that the H^+ activity in solution is increased 10-fold by the addition of HCl. As a result, some H^+ ions are adsorbed on the oxide surface creating a net positive charge. If these are few compared to the other H^+ ions on the oxide surface, the chemi-cal potential of H^+ in the oxide is virtually unchanged. However, the electro-chemical potential, $\bar{\mu}_+$, of H^+ must be the same in both phases and as shown in Eq. [4], this can be true only if the surface potential has increased by 59 mV. From Eq. [4] it can also be seen that at the ZPC where (a_{H^+}) = $(a_{H^+})_{ZPC}$, the surface potential, ψ_0, is zero and will be positive for (a_{H^+}) > $(a_{H^+})_{ZPC}$ and negative for (a_{H^+}) < $(a_{H^+})_{ZPC}$. Although ψ_0 is zero at the ZPC, we cannot conclude that the total potential difference between the two phases is zero. If this were true, absolute potentials and single ion activities could be determined (de Bruyn & Agar, 1962) which of course is not the case. The existence of a zero point of charge only allows us to conclude that part of the total potential between the two phases due to the presence of free charges at the interface is zero. However, polarization effects may also con-tribute to the total potential between the two phases without involving any free charge or transfer of charge (de Bruyn & Agar, 1962).

The Nernst relationship (Eq. [28]-[29]), which can be derived from Eq. [4], can be used to calculate ψ_0 at a given pH if the ZPC is known (Van Raij & Peech, 1972).

$$\psi_0 = \frac{RT}{F} \ln [(a_{H^+})/(a_{H^+})_{ZPC}] \qquad [28]$$

$$= 59 \text{ (ZPC-pH) in mV at 25C (298K)} \qquad [29]$$

Use of this relationship is based on the assumption that the chemical po-tential of the potential determining ions, H^+ and OH^-, on the oxide surface is constant over the pH or surface potential range of interest (Levine & Smith, 1971; Wright & Hunter, 1973a). While this assumption is generally valid for an ionic solid, its application to oxides whose surface groups are weakly dissociated at the ZPC is limited to relatively narrow ranges of surface charge and potential. Levine and Smith (1971) and Wright and Hunter (1973a) have recently developed more elaborate models allowing for variations in the surface chemical potential of the PDI, but the agreement between the calculated and observed surface charge densities is still limited.

The charge developed by association or dissociation of H^+ or OH^- ions on oxide surfaces in the absence of a supporting electrolyte is balanced by the dissociating H^+ or OH^- ions themselves. However, excess negative or positive charge can be developed on oxide surfaces by the addition of HCl or NaOH and subsequent adsorption of H^+ or OH^-. The resulting surface charge is balanced by the added Cl^- or Na^+ ions. These counter ions will diffuse into the surrounding medium forming a diffuse double layer in the aqueous phase near the oxide surface. As in the case of the fixed charge surfaces, the electrochemical potential, $\bar{\mu}_+$, of each ion in the diffuse double layer is constant and consequently the considerations leading to Eq. [19] apply to these surfaces as well. However, we now see that the surface charge, σ, increases with both the electrolyte concentration, n_0, and the surface potential, ψ_0. Since ψ_0 is constant for a given pH we have the so-called constant potential, variable charge surfaces. Even though the surface potential at a given pH does not vary with electrolyte concentration, the electrical potential curves do decay more rapidly with distance from the surface at high concentrations (Fig. 2-1).

The pH of the ZPC for a given oxide will depend on the relative basic and acidic properties of the solid (de Bruyn & Agar, 1962). These acidic and basic properties in turn are a function of such variables as cation size and valence, the hydration state of the solid, and the geometrical arrangement of the ions (Parks, 1965). Strongly amphoteric oxides such as Fe_2O_3 and Al_2O_3 should have ZPC's near neutral pH while an acid oxide such as SiO_2 should have its ZPC at low pH values since it is a proton donor (de Bruyn & Agar, 1962; Parks, 1965, 1967; Parks & de Bryun, 1962). The ZPC of complex mixtures of oxides and hydrous oxides, as might be found in soils, will be a complex function of all the various components involved. In many cases organic matter, which has a low ZPC, will also be involved tending to lower the observed ZPC of the soil (van Raij & Peech, 1972).

B. Determining Surface Potential and Surface Charge on Oxides

If the pH at the ZPC for a given oxide is known, the surface potential can be calculated according to the Nernst equation (Eq. [28]). The surface charge density, σ, can in turn be calculated as a function of surface potential and electrolyte concentration using either the Gouy-Chapman equation (Eq. [19]) or the Stern theory equation (Eq. [20]-[23]). As indicated previously, the validity of such calculations depends on the validity of the assumption that the Nernst relationship applies to oxide surfaces.

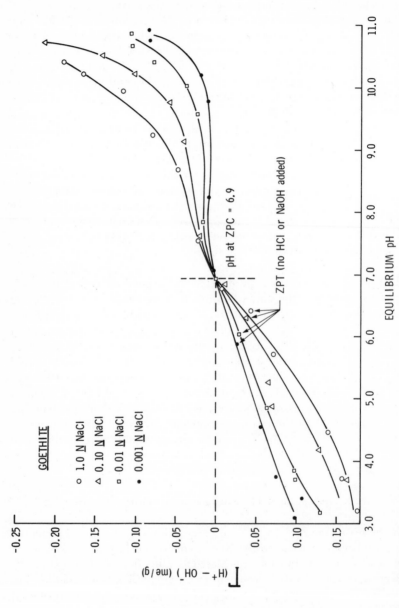

Fig. 2-3. pH titration curve for synthetic goethite and amorphous hydrated iron oxide showing zero points of titration (ZPT) and zero point of charge (ZPC). (continued on next page)

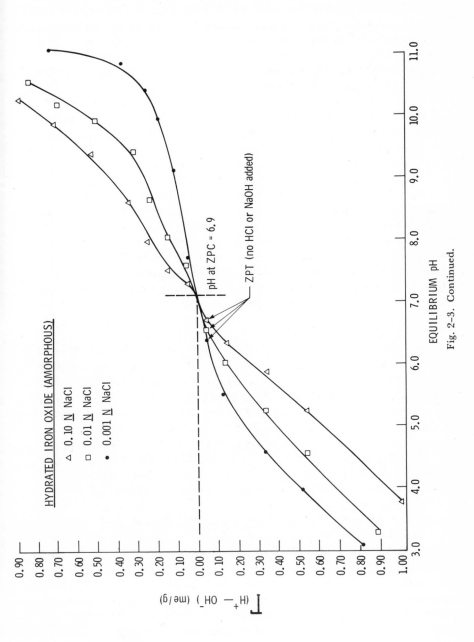

Fig. 2-3. Continued.

Alternatively, the surface charge on oxide and surfaces may be determined experimentally as a function of pH and electrolyte concentration by either conducting a pH titration curve or by determining the extent of counter ion retention (anion or cation). The pH titration curve may be conducted by initially placing a given amount of washed, electrolyte-free solid in solutions of "indifferent electrolyte" at varying concentrations. An indifferent electrolyte is defined as one in which the anion and cation retention by the oxide surface is the result of electrostatic forces; i.e. the anion or cation are not specifically adsorbed or incorporated in the surface, significantly changing the surface properties. Indifferent electrolytes include, but are not necessarily limited to the alkali metal cations in combination with such anions as Cl^- and NO_3^-. The oxide-electrolyte mixture is then titrated with an acid or base having ions common with the electrolyte used; i.e. if the solid were suspended in NaCl, then HCl and NaOH would be used.

The titration procedure provides a means of determining the net surface charge relative to that at the *zero point(s) of titration* (ZPT); i.e., the ZPT are the initial points plotted before the addition of any acid or base. In order to establish the absolute charge for a given pH and electrolyte concentration, it is necessary to determine the (ZPC). This is by definition the pH at which the titration curves for the different electrolyte concentrations intersect, i.e. the pH at which the net surface charge is zero at all electrolyte concentrations. To determine this intersection point of the titration curves, and hence the ZPC, the relative differences in surface charge of the oxide at the ZPT must be accurately plotted. This relative charge at the ZPT can be calculated from the change in pH of the electrolyte solutions of different concentrations on initially adding the oxide. The relative charge vs. pH at the ZPT is then the starting point for plotting the titration curves; the pH at which the curves intersect being the ZPC. The calculated charge at each pH can be adjusted for the difference in ZPC and ZPT to give the absolute value of the net surface charge. Examples of such titration curves are shown in Fig. 2-3 for goethite and amorphous hydrated iron oxide (Prasad, 1975).[1]

The ZPC can also be determined by the "solid titration" method of Bérubé and de Bruyn (1968). This involves immersing small amounts of the dry oxide in indifferent electrolyte solutions of known pH and ionic strength and observing the direction of the induced drift in pH. If the solid is added to a solution at the pH of the ZPC, no drift in pH will occur.

Alternately, the extent of positive or negative charge on oxide surfaces can be determined by measuring the excess counter ions retained by the solid at a given pH and electrolyte concentration. Experimentally, this involves repeatedly washing the solid with a solution of indifferent electrolyte at a given concentration and pH until there is no detectable change in either pH or electrolyte concentration. The excess counter ions retained by the solid can then be determined as in the regular CEC determination.

---

[1]Basawan Prasad. 1975. Charge characteristics and phosphorus adsorption on ferruginous soils and synthetic iron oxides. Unpublished Ph.D. Thesis. University of Minnesota.

IV. CHARGE CHARACTERISTICS OF MIXTURES OF CONSTANT- AND
pH-DEPENDENT CHARGE SURFACES

Soils in the temperate regions tend to be dominated by the crystalline or constant-charge clay minerals while the high oxide content of many extensively weathered tropical soils results in a dominance of oxides or hydrous oxides which have pH-dependent charge surfaces. All soils will, however, contain a mixture of both types of surface even though one type might tend to dominate over the other. These mixtures may result from any one or all of the following:

1) edge effects on crystalline clay minerals,
2) isomorphour substitution or site vacancies in Si–Al co-gels, or
3) oxide coatings or interlayers on crystalline clay minerals.

Edges of crystalline clay minerals contain ions not fully coordinated, much the same as oxide surfaces. These edges also adsorb H^+ and OH^- to form hydroxylated surfaces which in turn develop an electrical charge through amphoteric association or dissociation. Since all clay minerals have such edges, they all must have a mixture of constant and pH-dependent charge surfaces. However, in the case of the 2:1 type clay minerals the edges constitute only about 1% of the total available surface area (Dyal & Hendricks, 1950). Consequently, the pH-dependent charge on these minerals has limited effect on their overall behavior.

However, the 1:1 or kaolinitic type clay minerals tend to be aggregated into much larger particles through stacking in the c-axis direction. Since the internal surfaces do not tend to separate except through rather specific intercalation reactions, they are not available for surface reactions. Consequently, the edges of these minerals constitute a much greater fraction of the total surface area. These minerals have relatively low cation exchange capacities (Grim, 1968), and while much of the surface charge can be attributed to the amphoteric dissociation on the particle edges, it is difficult to rule out the possibility of some isomorphous substitution (Schofield & Samson, 1953).

Although not commonly recognized, ion substitutions and site vacancies may also exist in amorphous silica-alumina-iron co-gels resulting in an intrinsic or permanent negative charge (Parks, 1967). Such a permanent negative charge has been shown to exist in X-ray amorphous volcanic ash soils (Espinoza et al., 1975), but probably tends to become less evident with increased weathering.

Oxide and hydrous oxide coatings and interlayering of the crystalline clay minerals probably constitute the greatest source of complex mixture of constant- and pH-dependent charge surfaces in soils. Since the nature of such coatings and interlayers are dealt with in other chapters of this book, they will not be discussed extensively here except from the standpoint of their general effect on the charge characteristics of soils.

The presence of a permanent negative charge generally results in a titration curve where the ZPC is at a lower pH than the ZPT. This is a result of some of the H⁺ ions added during the titration exchanging with the counter ions associated with permanent negative charge. The shape of the titration curve and the extent that the ZPC is shifted below the ZPT is dependent on the relative amounts of permanent and pH-dependent charge. Maximum displacement of the ZPC from the ZPT will occur when the permanent negative charge is balanced entirely by counter ions other than H⁺. Since titration curves are usually conducted at fairly high concentrations of indifferent electrolyte such as NaCl, the permanent negative charge will be balanced predominantly by counter ions other than H⁺, tending to maximize the difference between the ZPT and ZPC. An example of a titration of volcanic ash soils showing a large displacement between ZPT and ZPC is shown in Fig. 2–4 (Espinoza et al., 1975).

Obviously the ZPC observed from intersection of the titration curves in complex mixtures of permanent and pH-dependent charge surfaces is not the

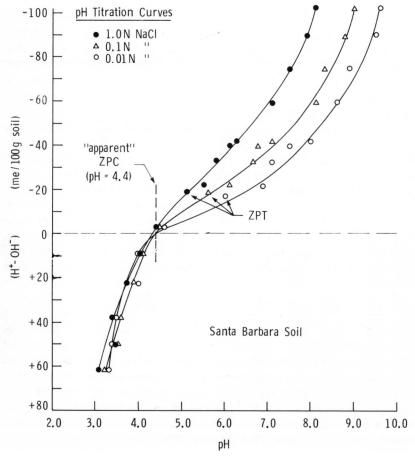

Fig. 2–4. pH titration curves for Santa Barbara soil derived from volcanic ash (from Espinoza et al., 1975).

ZPC of the pH-dependent surfaces alone and hence should probably be referred to as an "apparent" ZPC. Since there is relatively little permanent positive charge on soil colloids, the best measure of the extent of pH-dependent charge can be obtained from measuring anion retention as a function of pH and electrolyte concentration. Also, a better way of establishing that any observed displacement of the ZPC below the ZPT is due to a permanent negative charge is to determine cation retention at relatively low pH's where little pH-dependent negative charge would be expected.

V. ION EXCHANGE

The ion exchange process between the counter ions balancing the charge on soil colloids and the ions in the soil solution has the following general characteristics (Helfferich, 1962):

1) it is reversible,
2) it is diffusion controlled; i.e. the rate-limiting step is the diffusion of one counter ion against another,
3) it is stoichiometric, and
4) in most cases there is some selectivity or preference for one ion over the other by the surface.

All these characteristics can be readily demonstrated in the case of a pure montmorillonite suspension where (i) the quantity of counter ions is equal to the charge deficit on the crystal structure, (ii) one counter ion species can be stoichiometrically exchanged for another, and (iii) the exchange of one ion for another occurs within the time limits of detection if the suspension is rapidly stirred to eliminate concentration gradients.

While these general processes also apply to soil systems, they are often more complex and less obvious as in the case of "fixed" K^+ on mica surfaces. This K^+ can be considered as a counter ion since it does balance the fixed charge on the mica structure resulting from isomorphous substitution, and can be removed without altering the mineral structure itself. The exchange of another cation species for the interlayer K^+ is stoichiometric and the rate of exchange is controlled by the rate of diffusion in the interlayer region. Therefore, this exchange process has the characteristics indicated above in that it is reversible, stoichoimetric, and diffusion controlled. However, since the interparticle diffusion rates are very slow (Jacobs, 1963), and since the K^+ is so highly preferred over other cations most commonly found in soils, ion exchange on mica-like surfaces is generally considered separately from ion exchange on the "more available" colloidal surfaces.

The total diffusion process, which involves movement of both solution and counter ions against a concentration gradient, has been demonstrated to play a significant role in movement of nutrient ions to plant root surfaces (Barber et al., 1963). However, the extent that "ion exchange" diffusion, or stoichiometric exchange of one counter ion for another during movement along the colloid surface, contributes to the total movement or flux of ions through the soil depends on the relative quantities of ions in the soil solution vs. those in the electric double layer balancing the surface charge. Generally speaking, solution diffusion is a much faster process than "ion exchange" diffusion.

Ion exchange selectivity or the preference of one ion over another has been the subject of a great deal of study. As indicated previously, the counter ions are held primarily by electrostatic forces. If these ions could be treated as point charges, as assumed in applying the Poisson equation during derivation of the Gouy-Chapman equation (Eq. [19]), then there would be no preference between ions of equal valence. However, ions do have significantly different crystalline and hydrated sizes as shown in Table 2-2. Since electrostatic forces are involved in the retention of counter ions, it can be predicted from Coulomb's law (Pauley, 1953) that the ion having the smallest "effective" radius will be preferred. It has been frequently demonstrated that there is a definite relationship between ion size and ion selectivity (Pauley, 1953; Cloos et al., 1965; Jenny, 1932; Gast, 1969). Since cations present in soils are hydrated, ions with the smaller hydrated size or the larger crystalline size are preferred. This is reflected in the so-called Lytropic series which shows the following order of monovalent cation preference by soil colloids:

$$Cs > Rb > K > Na > Li.$$

Close relationships have been demonstrated between selectivity and various ion size parameters including polarizability, Debye-Huckle parameters of closest approach, and hydrated radii.

Since the basic equations for the electric double layer do not allow for variations in ion size, they do not predict any difference in ion distribution or selectivity between ions of equal valence; i.e. the fraction of the surface

Table 2-2. Ion sizes, polarizability, Debye-Huckle parameters of closest approach, and coordination numbers

Ion	Crystalline radii (Å) †	Crystalline radii (Å) ‡	Hydrated radii (Å)	Polarizability (Å) §	Debye-Huckle parameter (Å)	Coordination number¶
Li^+	0.78	0.60	3.82	0.079	4.32	6
Na^+	0.98	0.95	3.58	0.196	3.97	6, 8
K^+	1.33	1.33	3.31	0.876	3.63	8–12
Rb^+	1.49	1.48	3.29	1.407	3.49	8–12
Cs^+	1.65	1.69	3.29	2.452	--	12
Mg^{2+}	0.78	0.65	4.28	0.110	5.02	6
Ca^{2+}	1.06	0.99	4.12	0.523	4.73	6, 8
Sr^{2+}	1.27	1.13	4.12	0.880	4.61	8
Ba^{2+}	1.43	1.35	4.04	1.682	4.45	8–12
Al^{3+}		0.50				4, 6
Si^{4+}		0.41				4

† Baver, 1956.   ‡ Glasstone, 1946.   § Cloos et al., 1965.
¶ Berry and Mason, 1959.

charge neutralized by a given ion species is equal to its fraction in the bulk electrolyte phase away from the surface (Babcock, 1963). As indicated previously, ion size effects can theoretically be introduced in the Stern theory approach by allowing for variations in the specific adsorption potential, ϕ, in Eq. [21] (van Olphen, 1963). Selectivity would then be related to the different values of ϕ. The usefulness of such an approach for predicting selectivity is limited by the lack of an independent means of arriving at the values of ϕ. A more recent approach limits the use of the concept of a specific adsorption potential, ϕ, and the Stern equation (Eq. [21]) to cases where true specific adsorption occurs, i.e. cases where other than electrostatic forces are involved (Bowden et al., 1973). Ion size effects can still be introduced by assuming a distance of nearest approach or thickness of the Stern layer, δ, and the Stern layer potential, $\psi\delta$, for different ion species in Eq. [22] and [23].

Electric double-layer theory does predict differences in ion distribution and hence selectivity when ions of different valence are involved (Erickson, 1952; Bolt, 1955b; Lagerwerff & Bolt, 1959; Krishnamoorthy & Overstreet, 1949). Babcock (1963) has outlined the derivation of the following relationship, originally given by Erickson (1952), for calculating the fraction of the surface charge neutralized by monovalent cations in an electrolyte system of two symmetrical mono- and divalent salts:

$$\frac{\Gamma_1}{\Gamma} = \frac{r}{\Gamma \cdot \sqrt{\beta}} \sinh^{-1} \frac{\Gamma \cdot \sqrt{\beta}}{r + 4v_d\sqrt{m_2^0}} \qquad [30]$$

where
   Γ_1 = charge neutralized by monovalent cations,
   Γ = total surface charge density (meq/cm^2),
   r = the "reduced ratio" = $m_1^0/\sqrt{m_2^0}$, where m_1^0 and m_2^0 are the molar concentrations of the mono- and divalent cations in the bulk electrolyte solution, respectively,
   $\beta = 8000\pi F^2/DRT$, where $F = 2.982 \times 10^{-11}$ esu/meq and D is the dielectric constant. At 25C, $\beta = 1.08 \times 10^{15}$ cm mole/meq^2, and
   v_d = cosh y at the midplane between clay particles where

$$y = z_i e \psi /kT. \qquad [31]$$

Sample calculations using Eq. [30] are given in the Appendix. This relationship shows that for a given reduced ratio or "r" value in solution, the fraction of the surface charge neutralized by monovalent cations is a function of both the surface charge density, Γ, and m_2^0, or the total electrolyte concentration. As will be discussed later, the ratio of mono- to divalent cation saturation of soil colloids is particularly important in determining the dispersion or flocculation of soil particles under saline conditions, i.e. the greater the Na^+ saturation vs. $(Ca^{+2} + Mg^{+2})$ saturation, the more dispersed the soil will be, potentially creating undesirable physical properties.

The quantities Γ_1/Γ and "r" are related to quantities in other equations used at the U. S. Salinity Laboratory (Richards, 1954). The r value is similar to the sodium adsorption ratio (SAR) given by

$$SAR = M_{Na^+}/(M_{Mg^{2+}} + M_{Ca^{2+}})^{\frac{1}{2}}$$ [32]

except that the solution concentrations are in millimoles per liter in SAR rather than (moles per liter)$^{\frac{1}{2}}$ as in the case of r (Babcock, 1953). The quantity, Γ_1/Γ, is similar to the exchangeable sodium percentage (ESP) which is given by

$$ESP = (ES/CEC) \times 100$$ [33]

where ES and CEC are the exchangeable Na and cation exchange capacity, respectively.

VI. PARTICLE INTERACTIONS

Interactions between charged colloidal particles are the result of a balance between van der Waals attractive forces and several repulsive forces including electric double-layer repulsion and other short range repulsive forces (van Olphen, 1963). The van der Waals attractive forces between atoms, which are the result of dipole interactions, are small and decay rapidly with distance for a given pair of atoms. However, these attractive forces are additive for atoms in a surface and as a result that attraction between particles containing a very large number of atoms is equal to the sum of the attractive forces between all the atoms on the two particle surfaces. While this attractive force is inversely proportional to the seventh power of the distance for two atoms, it is inversely proportional to the third power of distance for two spherical particles. The attractive energy for such particles is then inversely proportional to the square of the distance of separation (van Olphen, 1963).

The short range repulsive forces are a combination of Born repulsive forces and specific adsorption forces. Born repulsive forces are due to the resistance on interpenetration of crystal structures. They become effective when protruding structural points or regions come into contact and consequently the forces are very short ranged. Specific adsorption forces are the result of molecular adsorption by the particle surface. In soils this largely involves water molecules either adsorbed on the surface or in the hydration shells of the counter ions. These adsorbed water molecules must be removed before the distance between the particles can become less than the thickness of the hydration shell. The work required for this desorption is reflected by a short-range repulsion between the particles (van Olphen, 1963; Ash et al., 1973).

As Norrish (1972) has pointed out, there is some question concerning the forces involved in the relatively long range swelling of smectite type clays. Low and co-workers (Davidtz & Low, 1970; Low & White, 1970) have proposed a mechanism of clay swelling that involves an increased structure of water near clay surfaces which is controlled by the mineral surfaces as well as cation hydration. More generally, however, these forces have been attributed to what are referred to as electric double-layer repulsive forces and this approach will be used here since, qualitatively at least, most of the swelling properties to be considered can be explained on this basis. Electric double-layer repulsive forces are generally considered to be the result of the work required to overcome the increase in free energy associated with the overlapping of the diffuse layers of counter ions as two charged particles approach each other. The ionic distribution and hence the repulsive force in the diffuse layer near a given surface is a function of the counter ion valence and electrolyte concentration in the solution phase, as described by the electrical double-layer equations, i.e. the thickness of the diffuse layer decreases with increasing valence and electrolyte concentration.

Norrish (1954) considered three stages of swelling of montmorillonite. However, for purposes here, particle interactions can be thought of as falling into two ranges or categories: (i) those involving the initial interactions as two charged particles approach each other in rather dilute aqueous suspension or gels, and (ii) those involving the short-range forces between particles. In the first case, the net balance of forces determines whether particles in suspension are dispersed or flocculated. This is largely a balance between the electrical double-layer repulsive forces and van der Waals attractive forces since the particles are flocculated before they approach the minimal distance involving the short range forces.

Since van der Waals attractive forces are essentially the same between different 2:1 type clays, the magnitude of the electric double-layer repulsive forces largely determines whether particles are dispersed or flocculated. As the electric double-layer equations indicate, and as is indicated in Fig. 2-5, these repulsive forces are a function of counter ion valence and the ionic strength of the bulk solution. Such flocculation-dispersion phenomena have major significance in many commercial and industrial applications. However, the counter ions in most soil systems are dominated by Ca^{2+}, Mg^{2+}, H^+, or Al^{3+}, all of which tend to reduce the electrical double-layer repulsive forces to the point that they are less than the van der Waals attractive forces. As a result, soil particles generally interact to form aggregates. The major exception is in the case of soils with high Na^+ saturation. Under such conditions, the repulsive forces may predominate over the attractive forces resulting in dispersion of the soil particles and poor soil physical properties. Such dispersion tends to occur at certain critical exchangeable sodium percentage or ESP values which explains much of the interest in ESP–SAR relationships. While there may be some variation between soils, particle dispersion, and poor soil physical properties can generally be expected at ESP values in the range of 15 or above.

As indicated above, the electric double-layer repulsive forces in most soil clays are less than the van der Waals forces, resulting in a net particle at-

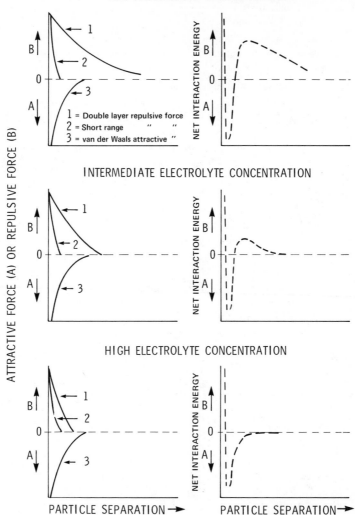

Fig. 2-5. Net interaction energy between particles as a function of particle separation at three electrolyte concentrations (after van Olphen, 1963).

traction. As a result, the clay particles approach each other until these net attractive forces are balanced by the short-range repulsive forces. As also indicated previously, the net attractive force is determined by the magnitude of the electric double-layer repulsive forces which will vary with the extent and location of the charge on the clay and with the size and valence of the saturating cation. Since the Born repulsive forces are very short ranged and essentially constant for different 2:1 type clays, the short-range repulsive forces are determined largely by the specific adsorption of water. As Greenland (1970) has pointed out, there is considerable controversy regarding the state of adsorbed water on 2:1 type clays. Most evidence indicates that water sorption is closely related to the hydration of the exchangeable cations (van Olphen, 1963; Norrish, 1954). However, as indicated above, interactions between water and the silicate surface may also play a role (Low, 1961) as suggested by recent evidence obtained using electron spin resonance (Clementz et al., 1973) which indicates that the silicate layers play a role in stabilizing some hydration complexes.

The observed c-axis spacing may vary between different 2:1 type clays for a given cation, due to a variation in the electrostatic forces between the charged surface and the counter ions; i.e. for a given cation, the electric double-layer repulsive forces and in turn the net attractive forces will vary between clay minerals having different surface charge densities. In turn, the spacing for a given 2:1 type clay mineral may vary between cations due to differences in the short-range repulsive forces associated with different hydration energies of the counter ions. This net balance of forces and the resultant c-axis spacing is used as a diagnostic tool for X-ray mineral identification of clay minerals with the distinctions being based largely on surface charge density although in some cases tetrahedral vs. octahedral charge location may also be a factor. More specifically, clay mineral identification based on these criteria is used to distinguish between the relatively low charge density smectites, and the higher charge density mica like minerals, including vermiculite and illite.

As a result of the relatively high charge density of micas, illites, and vermiculites, (Table 2–1), the electric double-layer repulsive forces are significantly reduced compared to those for smectites. Consequently, the van der Waals attractive forces exceed the repulsive forces for those cations having lower hydration energies, such as K^+, Rb^+, Cs^+, and NH_4^+, resulting in a dehydration of the counter ions. The c-axis spacing is then the 9.3Å of the alumino-silicate structure plus any additional particle separation due to the presence of the dehydrated counter ions. In the case of K^+, the observed spacing is 10Å which is somewhat less than the combined 9.3Å c-dimension of the mineral plus the 2.66Å diameter of the K^+ ion. The K^+ ions must then be partially positioned in the hexagonal oxygen rings of the clay surface.

Micas, and to a lesser extent illite, are by definition largely K^+ saturated. Consequently they have a 10Å c-axis spacing and since K^+ is not readily replaced by Mg^{2+}, do not expand on "saturation" with Mg^{2+} or other divalent cations. In contrast, cations in the interlayer regions of smectites are readily replaced by Mg^{2+} and the net balance of repulsive and attractive forces results in a c-axis spacing of about 18Å when glycerol solvated. Smectites differ from the higher charge density vermiculites in that the latter show a 14Å c-axis spacing when Mg-saturated in water which persists on glycerol solvation. In addition, vermiculite irreversibly collapses to 10Å when K-saturated.

The net balance for repulsive and attractive forces and hence the c-axis spacings can be controlled to a large extent in the laboratory by controlling the partial pressure of water and hence the degree of hydration of the counter ions. For example, smectites saturated with counter ions having hydration energies equal to or less than Na^+ can be dehydrated by evacuation to 10^{-5}-10^{-6} mm Hg resulting in c-axis spacings ranging from about 10Å for K^+, Rb^+, Cs^+, and NH_4^+ down to about 9.8Å for Na. Those cations having hydration energies above that for Na^+ cannot be dehydrated by evacuation alone. However, since the partial pressure of water in soil air is seldom $<$ 0.98, clay surfaces and counter ions are hydrated and "structure collapse" resulting in a trapping or fixation of counter ions only occurs in the case of the high charge density minerals saturated with cations having low hydration energies.

VII. DONNAN EQUILIBRIUM AND SINGLE ION ACTIVITY MEASUREMENTS

A Donnan system (Donnan, 1911) is one in which an electrolyte solution containing two or more diffusible ions is separated from a second phase containing the same diffusible ions plus nondiffusible, charged colloidal particles. The two phases are separated by a membrane or constraint which is permeable to the diffusible ions and water, but not to the charged colloidal particles. The membrane or constraint may be an actual permeable membrane such as cellulose acetate dialysis tubing or gravitational forces keeping the solution and colloidal phases separate. Such a system has the following three important properties:

1) an unequal distribution of ions between phases,
2) an osmotic pressure difference between phases, and
3) an electrical potential difference between phases.

As Davis (1942) pointed out, such a system at equilibrium is really at steady state and since the colloid cannot diffuse freely between phases, it is never at complete equilibrium with respect to the nondiffusible species.

Charged soil colloids along with their counter ions and the soil solution may be considered as Donnan systems and have the three properties listed above. The constraint separating the soil colloids from the equilibrium solution is either the gravitational force causing particles to settle from suspension or particle interactions which result in aggregate formation and hence separation of the solid and solution phases. Donnan equilibrium characteristics are present in all soils and have particular significance in at least two major areas, soil-plant root relationships and potentiometric measurements of single ion activities involving a reversible electrode in conjunction with a reference electrode.

The properties of Donnan systems can be illustrated by considering a relatively simple system involving a 10% by volume Na^+-saturated montmorillonite in equilibrium with a 0.01M NaCl solution. If the clay has an exchange capacity of 1 meq/g and, assuming no negative adsorption of Cl^- in the colloidal phase, the ionic distributions would be as indicated in Fig. 2–6.

The partial molar free energy $\bar{G}$ of all diffusible species (i.e. not the colloid) must be the same in all phases of such a system at equilibrium. For the above system then

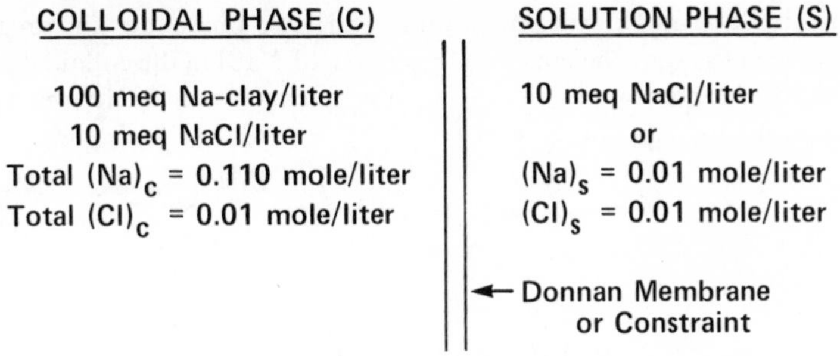

COLLOIDAL PHASE (C)   SOLUTION PHASE (S)

100 meq Na-clay/liter   10 meq NaCl/liter
10 meq NaCl/liter   or
Total $(Na)_c$ = 0.110 mole/liter $(Na)_s$ = 0.01 mole/liter
Total $(Cl)_c$ = 0.01 mole/liter $(Cl)_s$ = 0.01 mole/liter

← Donnan Membrane
or Constraint

Fig. 2-6. Ion distribution in a Donnan system consisting of 10% Na-clay suspension in equilibrium with $0.01N$ NaCl.

$$(\bar{G}_{NaCl})_c = (\bar{G}_{NaCl})_s \qquad [34]$$

where the subscripts, c and s, refer to the colloidal and solution phases, respectively. Following the definition given by Lewis and Randall (1923), the activity of a molecular species is given by

$$(\bar{G}_{NaCl} - \bar{G}^{\circ}_{NaCl}) = RT \ln a_{NaCl} \qquad [35]$$

where $\bar{G}^{\circ}$ is the partial molar free energy at a suitably chosen standard state. From Eq. [34] and [35] we get Eq. [36] if we select the same standard state in each phase,

$$(a_{NaCl})_c = (a_{NaCl})_s \qquad [36]$$

Introducing the concept of the mean ionic activity coefficient for NaCl, $\gamma\pm$, for relating concentrations and activities we get

$$a_{NaCl} = (\gamma\pm)^2 (M\pm)^2$$
$$= (\gamma\pm)^2 (M_{Na})(M_{Cl}) \qquad [37]$$

where M_{Na} and M_{Cl} are the molarities of Na^+ and Cl^- and $M\pm$ is the mean ionic molarity of NaCl.

Combining Eq. [36] and [37] and recognizing that $M_{Na} = M_{Cl}$ in a NaCl solution we get

$$(\gamma\pm)^2_c (M_{Na})_c (M_{Cl})_c = (\gamma\pm)_s (M_{Na})_s (M_{Cl})_s$$
$$(\gamma\pm)^2_c (M_{Na})_c (M_{Cl})_c = (\gamma\pm)^2_s (M\pm)^2_s \qquad [38]$$

Equation [38] can be arranged to give

$$(\gamma\pm)_c = \frac{(\gamma\pm)_s (M\pm)_s}{[(M_{Na})_c (M_{Cl})_c]^{1/2}}. \qquad [39]$$

Thus, if Na^+ and Cl^- concentrations in the two phases are known along with $(\gamma\pm)_s$, then $(\gamma\pm)_c$ or the mean ionic activity of NaCl in the colloidal phase can be calculated, but *not* the single ion activity coefficient of Na^+ or Cl^-.

Equation [36] can also be written as

$$(a_{Na})_c \ (a_{Cl})_c = (a_{Na})_s \ (a_{Cl})_s \qquad [40]$$

which leads to

$$(a_{Na})_c/(a_{Na})_s = (a_{Cl})_s/(a_{Cl})_c. \qquad [41]$$

Relationships for Donnan systems given to this point are thermo-dynamically rigorous. They do not, however, provide any means for evaluating the activity coefficient of the adsorbed Na^+ ions in the colloidal phase $(\gamma_{Na})_c$, which is the quantity of special interest in such systems. The reason for this particular interest can be illustrated by considering the Na-clay, NaCl system shown in Fig. 2–6.

If we assume, as a first approximation, that

$$(a_{Na})_s = (a_{Cl})_s = (a_{Cl})_c$$

and that the activity coefficients for these quantities are approximately unity, it follows that $(a_{Na})_c$ is

$$(a_{Na})_c = \frac{(a_{Na})_s \ (a_{Cl})_s}{(a_{Cl})_c} = \frac{(0.01) \ (0.01)}{0.01} = 0.01. \qquad [42]$$

However, the "average" concentration of (Na^+) is 0.110 moles/liter. Therefore, the value of the activity coefficient for Na^+ in the colloidal phase $(\gamma_{Na})_c$, is

$$(\gamma_{Na})_c = \frac{(a_{Na})_c}{(C_{Na})_c} = \frac{0.01}{0.110} = 0.091 \qquad [43]$$

i.e. the activity coefficient for Na^+ in the colloidal phase is much lower than in the solution phase.

While it may be generally accepted that the assumptions concerning the activities of single ions in solution phase are approximately correct, we know that the assumption concerning $(a_{Cl})_c$ cannot be absolutely correct for there is always negative adsorption in suspensions containing negatively charged colloids. Unfortunately, there is no way of conclusively establishing the true value of the activity of Cl^- in the colloidal phase allowing calculation of $(\gamma_{Na})_c$.

Potentiometric measurements have been used as an alternative approach for determining single-ion activities in such systems. The theoretical basis for this approach is based on the separation of the partial molar free energy, $(\bar{G})$, for the species, i, into a chemical and an electrical term (Guggenheim, 1929),

$$(\bar{G}) = \mu_i + z_i F \ \psi_i, \qquad [44]$$

where μ_i and ψ_i are the chemical potential and electrical potential, respectively. Since the partial molar free energy for a given species must be the same in the two phases, it follows that

$$(\mu_i + z_i F \ \psi)_c = (\mu_i + z_i F \ \psi)_s. \qquad [45]$$

The Donnan potential, E_m, is the electric potential between the two phases given by

$$(E_m = \psi_s - \psi_c). \qquad [46]$$

Combining Eq. [45] and [46] gives

$$E_m = (\psi_s - \psi_c) = [(\mu_i)_c - (\mu_i)_s]/z_i F. \qquad [47]$$

Remembering that

$$\mu_i - \mu_i^\circ = RT \ln a_i \qquad [48]$$

it follows from Eq. [47] and [48] that

$$E_m = \frac{RT}{zF} \ln \frac{(a_i)_c}{(a_i)_s}. \qquad [49]$$

If then the Donnan potential could be measured and if a reasonable assumption can be made concerning the value of $(a_{Na})_s$, the value of $(a_{Na})_c$ and $(\gamma_{Na})_c$ could be calculated.

The Donnan potential, E_m, is the electrical potential difference between the solution and colloidal phases of a system such as that shown in Fig. 2-6. This potential, however, can only be measured by introducing reference electrodes with salt bridges into the two phases and this adds irreversible processes to the equilibrium system (Overbeek, 1956). The reference electrode most commonly used is the calomel electrode which employs a saturated KCl salt bridge for transference of electrical current. Any time such an electrode is used there is a possibility of a liquid junction potential being developed at the salt bridge itself and contributing to the observed emf. So while the Donnan potential, E_m, between the phases can be rigorously defined in a theoretical sense, any practical definition must include possible contributions of liquid junction potentials at the two reference electrodes; i.e. the Donnan potential, E_m, can only be defined as the emf observed between the reference electrodes placed in the two phases as shown in Fig. 2-7.

The potential significance of being able to calculate the single ion activity and in turn activity coefficients of counter ions near charged surfaces using the measured value of E_m and Eq. [49] has generated a great deal of

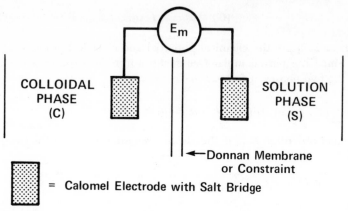

Fig. 2-7. Schematic illustration of the Donnan potential, E_m, between colloidal and solution phases.

interest in this subject (Marshall, 1948, 1956; Peech et al., 1953; Jenny et al., 1950; Coleman et al., 1951; Bower, 1961). At the same time, there has been a great deal of debate as to the relative contribution of the liquid junction potentials to the measured values of E_m.

Liquid junction potentials occur in cells with transference where the ionic species have different velocities; i.e., if the anion species move faster than the cation species, or vice-versa, an electrical potential quickly develops between the ions (MacInes, 1939). Saturated KCl is used in most liquid junctions since the velocities of K^+ and Cl^- are similar, giving a very small liquid junction potential. It is generally agreed that even though the liquid junction potential of a calomel electrode in a solution of strong electrolyte may not be zero, it is relatively small compared to values of E_m that can be observed. However, there is no general agreement concerning the possible contribution of the liquid junction of the reference electrode positioned in the colloidal phase. There is extensive evidence from both electrical conductance and diffusion measurements showing that mobilities of counter ions in the electrical double layer do have reduced mobilities which could result in unequal transference numbers and hence liquid junction potentials. Overbeek (1953, 1956), in an exhaustive treatment of the subject derived an expression (his Eq. [58], 1956) relating the Donnan potential to reduced mobilities of the counter ions and concluded that "the first and biggest effect on the Donnan emf is given by the nonideal behavior of the mobilities rather than of the activities."

It can be shown that there is good agreement between the measured Donnan potentials and those calculated using Overbeek's equation. This might be interpreted as meaning that most, if not all, of the observed Donnan potential is a liquid junction potential resulting from reduced mobilities of the counter ions in the colloidal phase. However, it must also be remembered that in deriving Overbeek's equation, it was necessary to make a basic assumption concerning concentrations vs. activities of the counter ions (Overbeek, 1956, p. 79) and that without this assumption, a different conclusion might have been reached.

Alternatively, Low (1954) concluded that while the liquid junction potential is present, it is not of sufficient magnitude to nullify the membrane potential measurement. Marshall (1956) has argued that the high concentrations of K^+ and Cl^- ions in the salt bridge eliminate any significant role of the counter ions and hence the liquid junction potential is negligible. It has been further argued that the good correlation between the low activity coefficients of the counter ions and their reduced mobilities is evidence that the observed emf is the true potential difference between phases and that the counter ion activities calculated using Eq. [49] are indeed an accurate reflection of the escaping tendency of the ions (Marshall, 1956).

At present there does not appear to be any rigorous way of resolving which of these interpretations is correct and the true significance of the Donnan potential remains in question. Unfortunately, the consequences of this uncertainty are more than academic for it enters into all measurements using a reversible, "single ion electrode" in conjunction with a reference electrode employing a salt bridge such as the calomel electrode. The most common such measurement is, of course, the pH or hydrogen ion activity measurement using the hydrogen glass-calomel electrode pair. However, many other single-ion electrodes have recently been developed which would be extremely valuable for studying soil systems if this uncertainty did not exist.

The existence of the Donnan potential and the uncertainty it introduces are observed in the so-called suspension effect in pH measurements; i.e. the difference in the measured pH between a sedimented soil and its equilibrium supernatant solution. This is illustrated in Fig. 2–8. As indicated, a different pH is observed if the electrode pair is placed in the supernatant solution rather than the soil sediment. However, an equivalent potential difference can be observed by placing the calomel electrodes alone in the two phases which by definition gives the Donnan potential. It immediately follows that for a given position of the calomel electrode, the same potential will be observed regardless of the position of the glass electrode. This, of course, must be the case since to be otherwise would require obtaining work from a system at equilibrium.

The uncertainties in single-ion activity measurements outlined above enter into all pH measurements in soil or charged colloidal systems. For example, when conducting a pH titration curve to determine the net charge and ZPC on oxide surfaces some decision must be made concerning positioning of the electrodes.

Since the uncertainties in interpretation outlined above do exist and apparently cannot be rigorously resolved, it is important that the method of measurement (position of the electrodes) be reported, especially if the results are to be used in theoretical calculations. In such cases it would appear advisable to report at least the solution phase pH since it is more reproducible. It may also be helpful to report the suspension pH as well in some instances. While the suspension effect is present in routine soil pH measurements, its

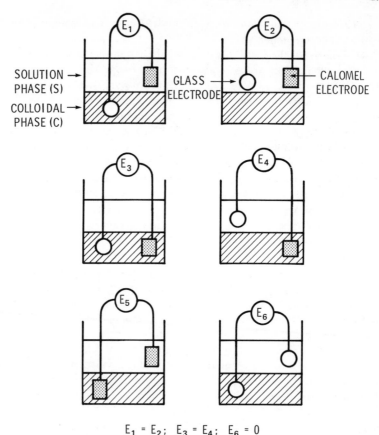

$$E_1 = E_2; \quad E_3 = E_4; \quad E_6 = 0$$

$(E_1 \text{ or } E_2) - (E_3 \text{ or } E_4) = E_5 = \text{Donnan Potential}, \; E_m = \text{suspension effect}$

Fig. 2–8. Schematic illustration of the suspension effect (after van Olphen, 1963).

implications have been reduced by a general standardization of procedures allowing at least a relative comparison of results. Also, measurement of pH in the presence of moderate to high salt concentrations (salt pH) greatly reduces the suspension effect by minimizing the role of the adsorbed ions in the system.

VIII. SPECIFIC ADSORPTION OF ANIONS

As indicated in the discussion of ion exchange, any permanent or pH-dependent charge on mineral surfaces must be balanced by an equivalent quantity of oppositely charged counter ions. These counter ions, held by electrostatic or coulombic forces, are located in either the Stern or Gouy-Chapman diffuse part of the electrical double layer. Adsorption of such ions is generally termed nonspecific (Hingston et al., 1967; Mott, 1970) and in the case of anions is limited to pH-dependent charge surfaces at pH's below the ZPC where an excess of positive charged sites exists. The term non-specific refers to adsorption which is essentially electrostatic attraction. Some selectivity does exist, but this can be explained in terms of factors affecting electrostatic attraction alone without involving other forces.

In contrast to the nonspecifically held counter ions, some anions are specifically adsorbed on mineral surfaces. Such specific adsorption of anions occurs irrespective of the sign of the net surface charge and in quantities out of proportion to their concentration or activity (Hingston et al., 1972). Adsorption is relatively insensitive to changes in ionic strength indicating that it is not directly affected by the properties of the diffuse double layer (Hingston et al., 1967). In addition, there is little if any detectable influence of the ZPC of the original oxide on specific adsorption although specific adsorption does shift the ZPC to more acid values (Hingston et al., 1970).

Plots of the amount of anion specifically adsorbed vs. equilibrium solution concentrations (at a given pH) commonly show adsorption isotherms having a shape similar to that for Langmuir isotherms as shown in Fig. 2-9 for the adsorption of pyrophosphate and fluoride by goethite (taken from Hingston et al., 1968a). However, the shape of these isotherms is attributed to other factors than those used in derivation of the Langmuir theory (Bowden et al., 1973).

Anions which most typically undergo specific adsorption on mineral surfaces are those resulting from the dissociation of weak acids such as phosphate, selenite, arsenate, silicate, and molybdate although specific adsorption of some anions of completely dissociated acids such as SO_4^- and F^- has been

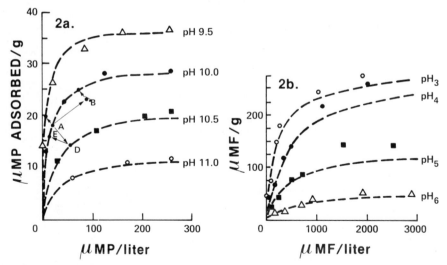

Fig. 2-9. Isotherms for adsorption of pyrophosphate and fluoride on goethite. Broken lines indicate Langmuir curves calculated from the maximum amount of adsorption at the given pH and the K_L calculated from the concentration of the anion at half the maximum (from Hingston et al., 1968a).

reported (Hingston et al., 1972). The mechanism involved in such specific adsorption is frequently referred to as ligand exchange whereby the anion displaces OH^- (or H_2O) from the surface and forms partly covalent bonds with the structural cations (Mott, 1970; Hingston et al., 1970; Breeuwsma & Lyklema, 1973). In soils, this type of adsorption is generally limited to mineral surfaces where oxygen atoms are only partly coordinated such as the surfaces of oxides and broken edges of the layer silicates; i.e. it does not occur on the planar surfaces of layer silicates. Specific adsorption can theoretically occur for any anion capable of coordination with the surface metal anions. However, since oxygen is the ligand commonly coordinated to the metal ions in soil minerals, the oxyanions are particularly involved in such reactions with phosphate being the anion of greatest significance from an agricultural standpoint.

Maximum anion adsorption at a given pH can be calculated from the adsorption data (Fig. 2-9) using the Langmuir equation (de Boer, 1968). These adsorption maxima have been plotted vs. pH to obtain curves referred to as adsorption envelopes (Hingston et al., 1967, 1968a, 1968b). Examples of adsorption envelopes for silicate and orthophosphate are shown in Fig. 2-10 (Hingston et al., 1970). In the case of silicate, which behaves as a monobasic acid (Cotton & Wilkinson, 1962) undergoing the following dissociation reaction,

$$H_4SiO_4 \rightleftarrows H_3SiO_4 + H^+,$$ [50]

the adsorption maximum occurs at approximately $pH = pK_A$ where

$$K_A = (H_3SiO_4^-)(H^+)/(H_4SiO_4)$$ [51]

with adsorption falling off rapidly at lower and higher pH's.

The factors responsible for variations in the adsorption maximum with pH and the shape of the adsorption envelopes have been extensively studied by Hingston and coworkers (Hingston et al., 1967; Hingston et al., 1968a; Hingston et al., 1970; Hingston et al., 1972). Based on earlier concepts, Hingston et al. (1972) derived the following expression relating the amount of adsorption, X_A, to the degree of dissociation of the acid, α,

$$X_A = 4 V_m \alpha(1 - \alpha)$$ [52]

where V_m is the maximum adsorption on the envelope. Since α is a function of H^+ concentration, this expression was then used to calculate the amount of adsorption as a function of pH. While these earlier concepts did provide an explanation as to why maximum adsorption occurs at approximately where $pH = pK_A$, the actual amount of adsorption on either side of the maximum is usually greater than that predicted by Eq. [52] (see Fig. 2-10).

More recently Bowden et al. (1973) have proposed an alternative ex-

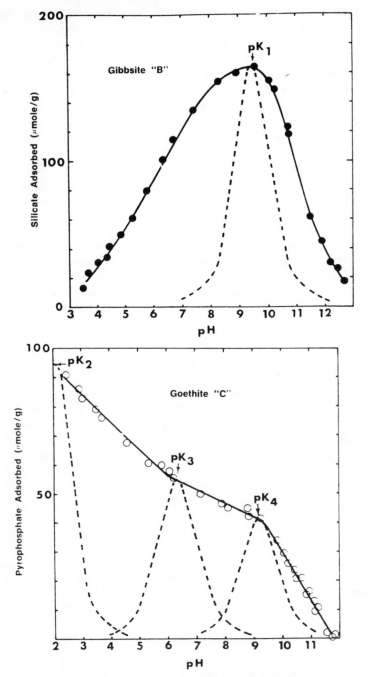

Fig. 2-10. Adsorption of silicate on gibbsite and pyrophosphate on goethite. Broken lines indicate the amount of adsorption predicted from Eq. [51] using pK for silicic acid = pH 9.5 and pK_2, pK_3, and pK_4 for phosphoric acid = pH 2, 6, and 9, respectively (from Hingston et al., 1970).

planation for the observed adsorption maximum near the point where pH = pK_A, based on the relative effect of pH change on the concentration of ionic species being adsorbed and the surface charge of the oxide surface. As the pH increases toward pK_A, there is a more rapid increase in the number of adsorbing species (i.e. for example $H_3SiO_4^-$) than in the rate of decrease of positive charge on the surface. As a result, adsorption increases with increasing pH until the pK_A is reached. Beyond this point, the surface tends to become increasingly negative at a rate greater than the rate of dissociation of uncharged molecules to form the adsorbing species. Consequently, maximum adsorption tends to occur at approximately the point where pH = pK_A. In the case of polybasic acids with large differences in pK such as phosphoric acid, a break occurs in the range of the pK values rather than a maximum. (See Chapter 5 on Iron Oxides by Schwertmann and Taylor in this book for additional discussion of this subject.)

The desorption of specifically adsorbed anions varies between complete reversibility and almost complete irreversibility. Such desorption is relevant to the uptake of plant nutrients such as phosphate. The reasons for this varying degree of reversibility are complex, but one suggestion is that anions which can be readily desorbed have one coordinated bond to the surface while those more strongly retained have two coordinated bonds (Hingston et al., 1974).

IX. APPENDIX—SAMPLE CALCULATIONS

Sample calculations are outlined below using the chemical analysis for Upton Wyoming montmorillonite reported by Grim (1968), p. 578. These include calculation of several chemical and physical properties of the clay including the chemical formula, unit cell dimensions, surface area, charge deficit per unit cell, theoretical cation exchange capacity, surface charge density, and surface potential. While many of these calculations cannot be applied directly to soil clays, they do show the principles involved and illustrate the origin of many of the structural and surface properties of clays.

Table 2-3. Outline of procedures used in calculating structural formula for clay from its chemical composition

Chemical component	Weight %	Molecular weight	Gram equiv. of cation	Proportion- ality factor	Gram equiv. cation per unit cell	Cation valence	Atoms per unit cell
SiO_2	55.44 ÷	60.06	× 4 = 3.6923	× 8.3543	30.8466 ÷	4 =	7.7116
Al_2O_3	20.14 ÷	101.94	× 6 = 1.1854	×	9.9032 ÷	3 =	3.3011
Fe_2O_3	3.67 ÷	159.70	× 6 = 0.1379	×	1.1521 ÷	3 =	0.3840
FeO	0.30 ÷	71.85	× 2 = 0.0084	×	0.0718 ÷	2 =	0.0351
MgO	2.49 ÷	40.32	× 2 = 0.1235	×	1.0318 ÷	2 =	0.5159
CaO	0.50 ÷	56.08	× 2 = 0.0178	×	0.1487 ÷	2 =	0.0743
K_2O	0.60 ÷	94.192	× 2 = 0.0127	×	0.1061 ÷	1 =	0.1061
Na_2O	2.75 ÷	61.994	× 2 = 0.0887	×	0.7410 ÷	1 =	0.7410
TiO_2†	0.10		5.2667				

† Ti is not included as part of the clay structure since it may be due to the presence of anatase (Grim, 1953).

The electric double-layer calculations presented were chosen to illustrate the basic surface charge–surface potential relationships. They admittedly have limited applications in themselves, and to a certain extent duplicate those given by van Olphen (1963). However, it is hoped that by outlining the calculations in detail, including the complex units involved, the student will be in a better position to understand and perhaps apply the more complex and rigorous treatments of the electrical double layer (Wright & Hunter, 1973a, 1973b) and its role in such phenomena as swelling pressures, anion exclusion (Kemper & Quirk, 1970) and ion adsorption (Bowden et al., 1973).

A. Calculation of Chemical Formula, Surface Area, Cation Exchange Capacity, and Surface Charge Density

Calculations are based on a unit cell consisting of 44 negative charges [i.e. $O_{20}(OH)_4$], which has eight tetrahedral and six octahedral sites. The procedure involves calculating the gram equivalents of each cation species per unit cell and assignment of those cations having coordination numbers of four and six (Table 2-2) to the tetrahedral and octahedral sites, respectively (see Table 2-3). Those cations having coordination numbers greater than seven, such as Ca, K, and Na, are considered to be counter ions balancing the charge deficit on the clay structure.

1) $44/5.2667 = 8.3543$ is the proportionality factor adjusting the gram equivalent of cations to the 44 required to balance the negative charge on the unit cell.

2) Assignment of cations in tetrahedral layer.
   All Si^{4+} atoms, which have a coordination number of 4, are assigned to the tetrahedral sites. The remainder of the 8 sites are then filled with Al^{3+} atoms which have coordination numbers of 4 or 6.
   a) Al^{3+} in tetrahedral layer $= (8.0 - 7.71) = 0.29$.
   b) Cations in tetrahedral layer $= (Si^{4+}_{7.71} Al^{3+}_{0.29})$.

3) Assignment of cations to octahedral layer.
   The remainder of the Al^{3+} atoms and all Mg^{2+}, Fe^{3+}, and Fe^{2+} atoms are assigned to the octahedral sites. Since any Mg or Fe that might be present in the sample either as counter ions or impurities would be incorrectly assigned to the octahedral sites it is important that these be removed from the sample prior to analysis.
   a) $Al^{3+} = (3.30 - 0.29) = 3.01$.
   b) Cations in octahedral layer $= (Al^{3+}_{3.01}, Fe^{3+}_{0.38}, Fe^{2+}_{0.04}, Mg^{2+}_{0.52})$.

4) The structural formula for the clay is then
   $(Si^{4+}_{7.71}, Al^{3+}_{0.29})(Al^{3+}_{3.01}, Fe^{3+}_{0.38}, Fe^{2+}_{0.04}, Mg^{2+}_{0.52}) O_{20}(OH)_4$.

5) Calculating charge deficit per unit cell.
   In the "ideal" structures of pyrophyllite and talc, 32 of the 44 negative charges per unit cell are balanced by cations in the tetra-

hedral sites and 12 by cations in the octahedral sites. The charge deficit per unit cell can then be calculated as the difference between the positive charge on cations in each layer and the negative charge of 32 or 12.

a) Tetrahedral layer

$$32 - [(7.71 \times 4) + (0.29 \times 3)] = 0.29$$

b) Octahedral layer

$$12 - [(3.01 \times 3) + (0.38 \times 3) + (0.04 \times 2) + (0.52 \times 2)] = 0.71$$

c) Total charge deficit per unit cell = 1.00

6) Milliequivalents of exchangeable cations per unit cell.
The sum of the counter ions is equal to the calculated net negative charge on the clay structure, i.e.,

$$(0.07Ca \times 2) + (0.11K) + (0.74Na) = 0.99.$$

In some instances this may not be true, usually due to the omission of H^+ as a counter ion. When the calculated net charge on the clay and the experimentally determined counter ions do not balance, some error is introduced into the calculations since the gram equivalent of cations is not correct. This can be corrected by repeated calculations, adjusting the H^+ until the two quantities balance.

7) The sum of the cations in the octahedral sites is 3.95, or close to 4 of the 6 sites are filled. Therefore, the clay is dioctahedral.

8) Molecular weight per unit cell = 729.5, not including the counter ions.

9) Calculated exchange capacity, (meq/g of clay), not including the weight of the counter ions:

$$CEC = (1/0.7295) \times 1.00 \text{ charge deficit per unit cell}$$

$$= 1.37 \text{ meq per g of clay,}$$

excluding the weight of the counter ions. This is significantly greater than the other smectite-type minerals listed in Table 2-1, due largely to a significant charge deficit in both the octahedral and tetrahedral layers.

10) Calculating surface area.
a) Calculating a and b dimensions of unit cell (Grim, 1953, p. 60),

$$b = 8.91 + 0.06r + 0.034s + 0.048t \text{ (in Angstroms)}$$

where

r = number of Al^{3+} atoms in tetrahedral coordination,
s = number of Mg^{2+} atoms in octahedral coordination,

$t =$ number of Fe^{3+} and Fe^{2+} atoms in octahedral coordination,
$b = 8.91 + (0.06)(0.29) + (0.034)(0.52) + (0.048)(0.42) =$
8.96Å, and
$a = b/\sqrt{3} = 8.96/1.73 = 5.18$Å.

b) Surface area per g of clay =
(1g/729.5) $\times$ 6.02 $\times$ 10^{23} $\times$ 2 $\times$ 8.96Å $\times$ 5.18Å $\times$ 10^{-20} m²/
Å² = 766 m²/g clay
c) The reader is referred to Grim (1968, p. 88) for additional
comments concerning unit cell dimensions.

11) Calculating surface charge density in esu/cm² and μcoul/cm² from
CEC in meq/g.

$$esu/cm^2 = (1.37 \text{ meq/g}) (2.89 \times 10^{11} \text{ esu/meq}) \frac{1}{10^4 \text{ cm}^2/\text{m}^2}$$

$$\frac{1}{766 \text{ m}^2/\text{g}} = 5.17 \times 10^4 \text{ esu/cm}^2$$

$$\mu coul/cm^2 = \frac{5.17 \times 10^4 \text{ esu/cm}^2}{3.0 \times 10^9 \text{ esu/coul}} = 1.72 \times 10^{-5} \text{ coul/cm}^2 = 17.2$$
μcoul/cm²

B. Calculating Surface Potential Using Gouy-Chapman Theory of the Electric Double Layer

Calculating surface potential, ψ_0, using Gouy-Chapman theory of the
electric double layer where $\sigma = 5.17 \times 10^4$ esu/cm² and the electrolyte concentration is 0.1 moles/liter of 1:1 electrolyte,

$$\sigma = (2nDkT/\pi)^{1/2} \sinh(ze\psi_0/2kT) \qquad \text{(Eq. 19)}$$

where
$\sigma =$ surface charge density in esu/cm²,
$n =$ counter ion concentration in ions per cm³ = normality $\times$ 10^{-3} $\times$ 6.02
$\times$ 10^{23},
$D =$ diabattivity (rather than dielectric constant which is dimensionless,
Grahame (1947), p. 472),
$= 80 \dfrac{\text{esu}^2}{\text{dyne cm}^2}$ or $\dfrac{\text{esu}^2}{\text{erg cm}}$ since 1 dyne = 1 erg cm,
$z =$ ion valence (dimensionless),
$k =$ Boltzmann constant = gas constant per ion or molecule = 1.38 $\times$ 10^{16}
erg/ion deg.
$T =$ temperature = 298°K at 25°C,
$\psi_0 =$ surface potential in erg/esu,
$e =$ electronic charge = 4.80 $\times$ 10^{-10} esu/ion, and
$\pi = 3.1415$ (dimensionless).

Let

$$A = \left(\frac{2nDkT}{\pi}\right)^{\frac{1}{2}}$$

$$= \frac{(2)(6.02 \times 10^{19}\,\text{ions/cm}^3)(80\,\text{esu}^2/\text{erg cm})(1.38 \times 10^{-16}\,\text{erg/ion deg})(298K)}{3.1415}$$

$$= 1.261 \times 10^8\,\text{esu}^2/\text{cm}^4)^{\frac{1}{2}} = 1.123 \times 10^4\,\text{esu/cm}^2$$

$$B = ze\psi_0/2kT$$

$$= \frac{(4.80 \times 10^{-10}\,\text{esu/ion})(\psi_0)}{(2)(1.38 \times 10^{-16}\,\text{erg/ion degree})(298K)} = (5.84 \times 10^3\,\text{erg/esu})(\psi_0)$$

Then

$$\text{Sinh } B = \frac{\sigma}{A} = \frac{5.17 \times 10^4\,\text{esu/cm}^2}{1.123 \times 10^4\,\text{esu/cm}^2} = 4.604$$

or

$$B = 2.23$$

and $\psi_0 = \dfrac{2.23}{5.84 \times 10^3\,\text{esu/erg}} = 3.82 \times 10^{-4}\,\text{erg/esu}.$

Converting the surface potential from erg/esu to mV where

$$\text{mV} = (\text{erg/esu})(10^{-7}\,\text{joules/erg})(3 \times 10^9\,\text{esu/coul})(1\,\text{volt coul/joul})$$
$$(10^3\,\text{mV/volt})$$
$$= (\text{erg/esu})(3 \times 10^5)\ \text{or erg/esu} = \text{mV}/(3 \times 10^5).$$

so

$$\psi_0 = (3.82 \times 10^{-4}\,\text{erg/esu})(3 \times 10^5) = 114.6\,\text{mV}.$$

Using the above formula and calculations, the surface potential at 1.0, 0.1, and 0.01N electrolyte concentrations is 60.0, 114.6, and 173.1 mV, respectively, reflecting the constant charge, variable surface potential nature of the 2:1 type clay minerals.

C. Calculating the Stern and Gouy Layer Charge and the Stern Potential for Upton Wyoming Montmorillonite

Calculating the Stern and Gouy layer charge, σ_1 and σ_2, and the Stern potential, $\psi\delta$, for Upton Wyoming montmorillonite having a surface charge density of $5.17 \times 10^4\,\text{esu/cm}^2$ and at 0.001 moles/liter 1:1 electrolyte concentration using the following equations and assigned values for the various parameters:

1) Units and assigned values:

a) $\sigma_1 = \dfrac{N_1 ze}{1 + \left(\dfrac{N_A \omega}{Mn}\right)\exp\left[-\left(\dfrac{ze\psi\delta + \Phi}{kT}\right)\right]}$ (Eq. 21) where

σ_1 = Stern layer charge in esu/cm^2,

N_1 = number of adsorption sites = 10^{15} sites per cm^2 (assumed),

z = ion valence (dimensionless),

e = electronic charge = 4.80×10^{-10} esu/ion,

N_A = Avogadro's no. = 6.02×10^{23} ions/mole,

ω = solvent density = 1 g/cm^3 for water,

M = molecular wt. of solvent = 18 g/mole for water,

n = counter ion concentration = 6.02×10^{17} ions/cm^3,

ψ_δ = Stern potential in erg/esu.

k = Boltzmann constant = 1.38×10^{-16} erg ion^{-1} deg^{-1},

T = 298K at 25C, and

Φ = specific adsorption potential = 0.1 electron volt = (0.1 eV)(1.6 $\times$ 10^{-12} erg/eV) = 1.6 $\times$ 10^{-13} erg (assumed).

b) $\sigma_2 = (2nDkT/\pi)^{\frac{1}{2}} \sinh(ze\psi\delta/2kT)$ (Eq. 22)

where σ_2 = Gouy or diffuse layer charge in esu/cm^2

The rest of symbols have same meaning and value as indicated above.

c) $\sigma = (D'/4\pi\delta)(\psi_0 - \psi\delta)$ (Eq. 23)

where

σ = total surface charge density = 5.17×10^4 esu/cm^2.

D' = diabattivity in Stern layer = 6 (esu^2/dyne cm^2)(assumed),

δ = thickness of Stern layer = 5Å = 5×10^{-8} cm or approximately the thickness of one water layer,

ψ_0 = surface potential in erg/esu, and

$\psi\delta$ = Stern potential in erg/esu.

The quantity $D'/4\pi\delta$ is the capacitance per cm^2 of the electric double layer and has units of esu^2/erg cm^2 = esu^2/dyne cm^3 in the following calculations. Capacitance can be expressed in farads/cm^2 since farads =

$$\frac{coul^2}{newton\ meter} = \frac{esu^2}{dyne\ cm}\ (1.11 \times 10^{-12})$$

d) $\sigma = \sigma_1 + \sigma_2$ (Eq. 20)

2) General procedure for calculations:

a) Calculate $(\psi_0 - \psi\delta)$ using Eq. 23.

b) Knowing that

$$\sigma = \sigma_1 + \sigma_2 = 5.17 \times 10^4\ esu/cm^2$$

substitute for σ_1 and σ_2 from Eq's. [21] and [22] and solve for $\psi\delta$.

   c) Knowing $\psi\delta$, solve Eq. [21] and [22] for σ_1 and σ_2.
   d) Knowing $\psi\delta$ and $(\psi_0 - \psi\delta)$ solve for ψ_0.

3) Calculations

   a) Calculating $(\psi_0 - \psi\delta)$ using Eq. [23].

$$5.17 \times 10^4 \text{ esu/cm}^2 = \frac{6 \text{ esu}^2/\text{erg cm}}{(4)(3.14)(5 \times 10^{-8} \text{ cm})} \ (\psi_0 - \psi\delta)$$

$$= 9.55 \times 10^6 \ \frac{\text{esu}^2}{\text{erg cm}^2} \ (\psi_0 - \psi\delta)$$

$$(\psi_0 - \psi\delta) = \frac{5.17 \times 10^4 \text{ esu/cm}^2}{9.55 \times 10^6 \text{ esu}^2/\text{erg cm}^2}$$

$$= 5.41 \times 10^{-3} \text{ erg/esu} \cdot 3 \times 10^5 = 1623 \text{ mV}$$

   b) Calculating $\psi\delta$ by solving

$$\sigma_1 + \sigma_2 = \sigma = 5.17 \times 10^4 \text{ esu/cm}^2$$

or

$$(2nDkT/\pi)^{\frac{1}{2}} \sinh(ze\psi\delta/2kT) + \frac{N_1 ze}{1 + \dfrac{N_A \omega}{Mn} \exp[-(ze\psi\delta + \Phi)/kT]}$$

$$= \sigma = 5.17 \times 10^4 \text{ esu/cm}^2$$

let
$A = (2nDkT/\pi)^{\frac{1}{2}} = (1.262 \times 10^6 \text{ esu}^2/\text{cm}^4)^{\frac{1}{2}} = 1.123 \times 10^3$
   esu/cm^2 for $n = 0.001$ moles/liter
$y = ze\psi\delta/2kT = 5.84 \times 10^3 \text{ esu/erg} \cdot \psi\delta$
$C = N_1 zeMn/N_A \omega =$

$$\frac{(10^{15} \text{ site/cm}^2)(1)(4.80 \times 10^{-10} \text{ esu/ion})(18 \text{ g/mole})(6.02 \times 10^{17} \text{ ions/cm}^3)}{(6.02 \times 10^{23} \text{ ions/mole})(1 \text{ g/cm}^3)}$$

$$= 8.64 \text{ esu/cm}^2$$

then

$$\frac{ze\psi\delta + \Phi}{kT} = \frac{ze\psi\delta}{kT} + \frac{\Phi}{kT} = 2y + \frac{\Phi}{kT} = 2y$$

$$+ \frac{1.6 \times 10^{-13} \text{ erg}}{0.4 \times 10^{-13} \text{ erg}} = 2y + 4.$$

Assuming that the value 1 in the denominator of Eq. [21] can be neglected then:

$$\sigma_1 = C \exp(2y + 4) = 8.64 \text{ esu/cm}^2 \exp(2y + 4)$$

$$\sigma_2 = A \sinh(y) = 1.123 \times 10^3 \text{ esu/cm}^2 \cdot \sinh(y)$$

Since

$$\sigma_1 + \sigma_2 = \sigma,$$

then

$$5.17 \times 10^4 \text{ esu/cm}^2 = 8.64 \text{ esu/cm}^2 \exp(2y + 4)$$

$$+ 1.123 \times 10^3 \text{ esu/cm}^2 \sinh(y)$$

Since there is no unique solution to the above equation for y, the correct value must be arrived at by the process of successive approximation. For the given values of A and C, the correct value is $y = 2.30$, so

$$y = ze\psi\delta/2kT = 5.84 \times 10^3 \text{ esu/erg} \cdot \psi\delta = 2.30$$

or

$$\psi\delta = \frac{(2.30)}{5.84 \times 10^3 \text{ esu/erg}} = (3.94 \times 10^{-4} \text{ erg/esu})(3 \times 10^5)$$

$$= 118.2 \text{ mV}$$

and
$$\sigma_1 = 8.64 \text{ esu/cm}^2 \exp(2y + 4) = 8.64 \text{ esu/cm}^2 \exp(8.60) =$$
$$4.69 \times 10^4 \text{ esu/cm}^2,$$
$$\sigma_2 = 1.123 \times 10^3 \sinh(y) = 1.123 \times 10^3 \text{ esu/cm}^2 \sinh(2.30),$$
$$= (1.123 \times 10^3 \text{ esu/cm}^2)(4.94) = 0.555 \times 10^4 \text{ esu/cm}^2, \text{ and}$$
$$\sigma_1 + \sigma_2 = (4.69 \times 10^4) + (0.555 \times 10^4) = 5.25 \times 10^4 \text{ esu/cm}^2$$
compared to the initial value of 5.17×10^4 esu/cm^2 used reflecting approximations made during calculations. The ratio of the change in the two layers is then

$$\sigma_1/\sigma_2 = 4.69/0.555 = 8.45$$

Since
$$\psi\delta = 118.8 \text{ mV and } (\psi_0 - \psi\delta) = 1623 \text{ mV},$$
then
$$\psi_0 = 1623 + 118.8 = 1741.8 \text{ mV}.$$

D. Calculating the Surface Potential and the Surface Charge Density for pH-Dependent Charge Surfaces

Calculating the surface potential, ψ_0, and the surface charge density, σ, for pH-dependent charge surfaces.

1) Calculating surface potential.

If the pH at the ZPC is known, and it can be assumed that the Nernst eq. applies, the surface potential can be calulated for a given pH using the relationship:

$$\psi_0 = 59 \ (ZPC - pH) \ \text{in mV at 25C (298K)} \quad (\text{Eq. 29})$$

i.e., the surface potential increases by 59 mV for each change in pH away from the ZPC.

2) Calculating the surface charge density for a given surface potential.

a) The surface charge density, σ, at a given pH or surface potential and electrolyte concentration can be calculated using the Gouy-Chapman equation (Eq. [19]). The calculations involved are essentially the same as those outlined previously. Results calculated in this manner can then be compared with those obtained from the pH titration curve (van Raij & Peech, 1972).

b) Alternatively, the Stern theory equations (Eq. [20-23]) have been used to calculate the surface charge density (van Raij & Peech, 1972; Blok & de Bruyn, 1970). Again, the calculations involved are similar to those outlined above and require assumptions concerning the value of D' and δ. Values of σ calculated for given values of D' and δ can in turn be compared with values obtained from pH titration curves.

c) As indicated earlier in this chapter, there is some question as to whether the concept of a specific adsorption potential and hence Eq. [21] should be applied to counter ion adsorption; i.e., adsorption where only electrostatic forces are involved (Bowden et al., 1973). Allowances for ion size effects can still be introduced, however, by assuming that the first layer of ions are a distance, δ, away from the surface and that they are at an electric potential, ψ_δ, compared to the surface potential, ψ_0. By using assumed values of D' and δ, fixing the electrolyte concentration, n, and assigning values of ψ_δ, Eq. [22] and [23] can be used to calculate values of the surface charge vs. pH or surface potential, ψ_0. This procedure introduces the Stern concept of finite ion size by assuming that the ions have a Boltzmann or diffuse distribution starting at a distance, δ, from the surface.

E. Calculating Γ_1/Γ or Fraction of Exchange Capacity Saturated with Sodium

Calculating Γ_1/Γ or fraction of exchange capacity saturated with Na for Upton Wyoming montmorillonite with surface charge density of 5.17×10^4 esu/cm^2 in equilibrium with solution containing $0.004\ M$ Ca^{2+} and an $r =$ Na/$\sqrt{\text{Ca}}$ value of 0.20.

1) $$\frac{\Gamma_1}{\Gamma} = \frac{r}{\Gamma\sqrt{\beta}}\ \sinh^{-1}\frac{\Gamma\sqrt{\beta}}{r + 4\,v_d\sqrt{\text{Ca}}}$$

where

Γ = surface charge density in meq/cm^2,

$$= \frac{5.17 \times 10^4\ \text{esu/cm}^2}{2.89 \times 10^{11}\ \text{esu/meq}} = 1.79 \times 10^{-7}\ \text{meq/cm}^2,$$

$r = 0.20$ (moles/liter)$^{1/2}$, and

$\beta = 8000\ \pi\ F^2/DRT$

with

$F = 2.892 \times 10^{11}$ esu/meq,

$D = 80$ esu^2/erg cm,

$R = 8.314 \times 10^7$ ergs/deg. mole, and

$T = 25$C (298K)

giving

$$\beta = \frac{(8000)(3.14)(2.892 \times 10^{11})^2\ \text{esu}^2/\text{meq}^2}{(80\ \text{esu}^2/\text{erg cm})(8.314 \times 10^7\ \text{erg/deg mole})(298K)}$$

$$= 1.06 \times 10^{15}\ \text{cm mole/meq}^2$$

v_d = where coshy at the midplane between clay particles, assumed to have a value of 1 for these calculations, and

$\sqrt{\text{Ca}} = (0.004\ \text{mole/liter})^{1/2} = 0.0632$ (mole/liter)$^{1/2}$.

2) $$\frac{r}{\Gamma\sqrt{\beta}} = \frac{0.20\ (\text{moles/liter})^{1/2}}{(1.79 \times 10^{-7}\ \text{meq/cm}^2)(1.06 \times 10^{15}\ \text{cm mole/meq}^2)^{1/2}}$$

$$= 0.0343\ (\text{no units}).$$

(In order for units in above calculations to cancel, liters must be considered as cm^3.)

3) $$\frac{\Gamma\sqrt{\beta}}{r + 4\,v_d\sqrt{\text{Ca}}}$$

$$= \frac{(1.79 \times 10^{-7}\ \text{meq/cm}^2)\,(1.06 \times 10^{15}\ \text{cm mole/meq}^2)^{1/2}}{0.2\ (\text{moles/liter})^{1/2} + (4)(1)(0.0632)\ (\text{moles/liter})^{1/2}}$$

$$= 12.86\ (\text{no units}).$$

(In order for units in above calculations to cancel, liters must be considered as cm^3.)

4) $\Gamma_1/\Gamma = 0.0343 \sinh^{-1} 12.86$

$= (0.0343)(3.25) = 0.111$, or under the conditions stated, the CEC of the clay would be 11.1% Na saturated.

LITERATURE CITED

Adamson, A. W. 1967. Physical chemistry of surfaces. 2nd ed. Interscience Publishers, New York.

Ash, Stuart, G., Douglas H. Everett, and Clay Radke. 1973. Thermodynamics of the effects of adsorption on interparticle forces. J. Chem. Soc., Faraday Trans. 2. 69: 1256-1277.

Babcock, K. L. 1963. Theory of the chemical properties of soil colloidal systems at equilibrium. Hilgardia 34(11):417-542.

Barber, S. A., J. M. Walker, and E. H. Vasey. 1963. Mechanisms for the movement of plant nutrients from the soil and fertilizer to the plant root. J. Agric. Food Chem. 11:204-207.

Baver, L. D. 1956. Soil physics. 3rd ed. John Wiley & Sons, Inc., New York.

Beetem, W. A., V. J. Janzer, and J. S. Wahlberg. 1962. Use of cesium-137 in the determination of cation exchange capacity. Geol. Surv. Bull. 1140-B.

Berry, L. G., and Brian Mason. 1959. Mineralogy. W. H. Freeman & Co., San Francisco.

Bérubé, Y. G., and P. L. de Bruyn. 1968. Adsorption at the rutile-solution interface. I. Thermodynamic and experimental study. J. Colloid Interface Sci. 27:305-318.

Blok, L., and P. L. de Bruyn. 1970. The ionic double layer at the ZnO/solution interface. III. Comparison of calculated and experimental differentiated capacities. J. Colloid Sci. 32:533-538.

Blyholder, G., and E. A. Richardson. 1962. Infrared and volumetric data on the adsorption of ammonia, water and other gases on activated ion (III) oxide. J. Phys. Chem. 66:2597-2602.

Bolt, G. H. 1955a. Ion adsorption by clays. Soil Sci. 79:267-276.

Bolt, G. H. 1955b. Analysis of the validity of the Gouy-Chapman theory of the electric double layer. J. Colloid Sci. 10:206-218.

Bolt, G. H., and B. P. Warkentin. 1958. The negative adsorption of anions by clay suspensions. Kolloid Z. 156:41-46.

Bowden, J. W., M. D. A. Bolland, A. M. Posner, and J. P. Quirk. 1973. General model for anion and cation adsorption at oxide surfaces. Nature Phys. Sci. 245:81-83.

Bower, C. A. 1961. Studies on the suspension effect with a sodium electrode. Soil Sci. Soc. Am. Proc. 25:18-21.

Bower, C. A., and J. O. Goertzen. 1955. Negative adsorption of salt by soils. Soil Sci. Soc. Am. Proc. 19:147-151.

Breeuwsma, A., and J. Lyklema. 1973. Physical and chemical adsorption of ions in the electrical double layer on hematite (α-Fe$_2$O$_3$). J. Colloid Interface Sci. 43:437-448.

Clementz, D. M., Thomas J. Pinnavaia, and M. M. Mortland. 1973. Stereochemistry of hydrated copper(II) ions in the interlamellar surfaces of layer silicates. An electron spin resonance study. J. Phys. Chem. 77:196-200.

Cloos, Paul, Jose J. Fripiat, George Poncelet, and Anny Poncelet. 1965. Comparison entre les proprietes d'echange de la montmorillonite et d'une resine vis-à-vis des cations alcalins et alcalino-terraux. II. Phenomene de selectivite. Bull. Soc. Chim. Fr. 42:215-219.

Coleman, N. T., D. E. Williams, T. R. Nielsen, and H. Jenny. 1951. On the validity of interpretations of potentiometrically measured soil pH. Soil Sci. Soc. Am. Proc. 15:106-110.

Cotton, F. A., and G. Wilkinson. 1962. Advanced inorganic chemistry. Interscience Pub. John Wiley & Sons, New York. 185 p.

Davis, L. E. 1942. Significance of Donnan equilibria for soil colloidal systems. Soil Sci. 54:199-219.

Davidtz, J. C., and P. F. Low. 1970. Relation between crystal-lattice configuration and swelling of montmorillonites. Clays Clay Miner. 18:325-332.

de Boer, J. H. 1968. The dynamical character of adsorption. Oxford Press, London.

de Bruyn, P. L., and G. E. Agar. 1962. Surface chemistry of flotation. p. 91-138. In D. W. Fuerstenau (ed.) Froth flotation, 50th Anniversary Volume: Am. Inst. of Mining, Metall., & Pet. Engineers, Inc., New York.

Donnan, F. G. 1911. Theorie der membrangleichgewichte and membran potentiale be; vorhandensein von nicht dialysierenden. Elektrolyten. A. Elecktrochem. 17: 572-581.

Dyal, R. S., and S. B. Hendricks. 1950. Total surface of clays in polar liquids as a characteristic index. Soil Sci. 69:421-432.

Edwards, D. G., and J. P. Quirk. 1962. Repulsion of chloride by montmorillonite. J. Colloid Sci. 17:872-882.

Ensminger, L. E. 1944. A modified method for determining base-exchange capacity of soils. Soil Sci. 58:425-432.

Erikson, E. 1952. Cation-exchange equilibria on clay minerals. Soil Sci. 74:103-113.

Espinoza, W., R. G. Gast, and R. S. Adams, Jr. 1975. Charge characteristics and nitrate retention by two andepts from south-central Chile. Soil Sci. Soc. Am. Proc. 39: 842-846.

Francis, C. W., and D. F. Grigal. 1971. A rapid and simple procedure using Sr-85 for determining cation exchange capacities of soils and clays. Soil Sci. 112:17-21.

Gast, R. G. 1969. Standard free energies of exchange for alkali metal cations on Wyoming bentonite. Soil Sci. Soc. Am. Proc. 33:37-41.

Gast, R. G., Edward R. Landa, and Gordon W. Meyer. 1974. The interaction of water with goethite (α-FeOOH) and amorphous hydrated ferric oxide surfaces. Clays Clay Miner. 22:31-39.

Glasstone, Samuel. 1946. Textbook of physical chemistry. D. Van Nostrand Co., Princeton, N. J.

Grahame, D. C. 1947. The electrical double layer and the theory of electrocapillarity. Chem. Rev. 47:441-501.

Greenland, D. J. 1970. Sorption of organic compounds by clays and soils. In Sorption and transport processes in soils. Soc. Chem. Ind. Monograph no. 37:79-91.

Grim, R. E. 1953. Clay mineralogy. 1st ed. McGraw-Hill, New York.

Grim, R. E. 1968. Clay mineralogy. 2nd ed. McGraw-Hill, New York.

Guggenheim, E. A. 1929. The conceptions of electrical potential difference between two phases and the individual activities of ions. J. Phys. Chem. 33:842-849.

Helfferich, F. 1962. Ion exchange. McGraw-Hill, New York.

Hingston, F. J., R. J. Atkinson, and J. P. Quirk. 1967. Specific adsorption of anions. Nature 215:1459-1461.

Hingston, F.J., R. J. Atkinson, A. M. Posner, and J. P. Quirk. 1968a. Specific adsorption of anions on goethite. Int. Congr. Soil Sci., Trans. 9th (Adelaide, Aust.) 1:669-678.

Hingston, F.J., A. M. Posner, and J. P. Quirk. 1968b. Adsorption of selenite by goethite. In R. F. Gould (ed.) Adsorption from aqueous solution. Adv. Chem. Series 79: 82-90.

Hingston, F. J., A. M. Posner, and J. P. Quirk. 1970. Anion binding at oxide surfaces—the adsorption envelope. Search. 1:324-327.

Hingston, F. J., A. M. Posner, and J. P. Quirk. 1972. Anion adsorption by goethite and gibbsite. I. The role of the proton in determining adsorption envelopes. J. Soil Sci. 23:177-192.

Hingston, F. J., A. M. Posner, and J. P. Quirk. 1974. Anion adsorption by goethite and gibbsite. II. Desorption of anions from hydrous oxide surfaces. J. Soil Sci. 25: 16-26.

Jackson, M. L. 1969. Soil chemical analysis—Advanced course. Publ. by author, Dep. of Soils, Univ. of Wisconsin, Madison 6, Wis.

Jacobs, D.J. 1963. The effect of collapse—inducing cations on the Cs sorption properties of hydrobiotite. p. 239-248. In Proc. Int. Clay Conf. (Stockholm, Sweden). Pergamon Press, New York.

Jenny, Hans. 1932. Studies on the mechanism of ionic exchange in colloidal aluminum silicates. J. Phys. Chem. 36:2217-2258.

Jenny, H., T. R. Nielson, N. T. Coleman, and D. E. Williams. 1950. Concerning the measurement of pH, ion activities, and membrane potentials in colloidal systems. Science 112:164-167.

Kemper, W. D., and J. P. Quirk. 1970. Graphic presentation of a mathematical solution for interacting diffuse layers. Soil Sci. Soc. Am. Proc. 34:347-350.

Krishnamoorthy, C., and R. Overstreet. 1949. Theory of ion-exchange relationships. Soil Sci. 68:307-315.

Lagerwerff, J. V., and G. H. Bolt. 1959. Theoretical and experimental analysis of Gapon's equation for ion exchange. Soil Sci. 87:217-222.

Levine, S., and A. L. Smith. 1971. Theory of the differential capacity of the oxide/ aqueous electrolyte interface. Discuss. Faraday Soc. 52:290-301.

Lewis, C. N., and M. Randall. 1923. Thermodynamics and the free energy of chemical substances. McGraw-Hill, New York.

Low, P. F. 1954. Ionic activity measurements in heterogeneous systems. Soil Sci. 77: 29-41.

Low, P. F. 1961. Physical chemistry of clay-water interaction. Advan. Agron. 13:269-327.

Low, P. F., and J. L. White. 1970. Hydrogen bonding and polywater in clay-water systems. Clays Clay Miner. 18:63-66.

Lyklema, J. 1971. General discussion. p. 318. *In* Surface chemistry of oxides. Discuss. Faraday Soc. 52.

MacInnes, D. A. 1939. The principles of electrochemistry. Reinhold, New York.

McCafferty, E., and A. C. Zettlemoyer. 1970. Entropy of adsorption and the mobility of water vapor on α-Fe_2O_3. J. Colloid Sci. 34:452-460.

Marshall, C. E. 1948. The electrochemical properties of mineral membranes. VII. The theory of selective membrane behavior. J. Phys. Colloid Chem. 52:1284-1295.

Marshall, C. E. 1949. The colloid chemistry of the silicate minerals. Academic Press, Inc., New York.

Marshall, C. E. 1956. Thermodynamic-quasithermodynamic, and nonthermodynamic methods as applied to the electrochemistry of clays. Clays Clay Miner. 4:288-300.

Mehlich, A. 1948. Determination of cation—and anion exchange properties of soils. Soil Sci. 6:429-445.

Mott, C. J. B. 1970. Sorption of anions by soils. *In* Sorption and transport processes in soils. Soc. Chem. Ind. Monograph no. 37:40-53.

Norrish, K. 1954. The swelling of montmorillonite. Discuss. Faraday Soc. 18:120-134.

Norrish, K. 1972. Forces between clay particles. Proc. 1972 Int. Clay Conf., Madrid. p. 375-383.

Overbeek, J. Th. G. 1952. Electrochemistry of the double layer. *In* H. R. Krugt (ed.) Colloid Sci. I:115-193. Elsevier Publ. Co., Amsterdam.

Overbeek, J. Th. G. 1953. Donnan-emf and suspension effect. J. Colloid Sci. 8:593-605.

Overbeek, J. Th. G. 1956. The Donnan equilibrium. Progr. Biophys. 6:58-94.

Parks, G. A. 1965. The isoelectric points of solid oxides, solid hydroxides, and aqueous hydroxo complex systems. Chem. Rev. 65:177-198.

Parks, G. A. 1967. Aqueous surface chemistry of oxides and complex oxide minerals. Isoelectric point and zero point of charge. *In* R. F. Gould (ed.) Adv. in Chem. Series 67:121-160.

Parks, G. A., and P. L. de Bruyn. 1962. The zero point of charge of oxides. J. Phys. Chem. 66:967-973.

Pauley, J. L. 1953. Prediction of cation-exchange equilibria. J. Am. Chem. Soc. 76: 1422-1425.

Pauling, L. 1930. The structure of micas and related minerals. Proc. U. S. Nat. Acad. Sci. 16:123-129.

Peech, M., R. A. Olson, and G. H. Bolt. 1953. The significance of potentiometric measurements involving liquid junction in clay and soil suspensions. Soil Sci. Soc. Am. Proc. 17:214-218.

Rich, C. I. 1961. Calcium determination for cation-exchange capacity determinations. Soil Sci. 92:226-231.

Rich, C. I. 1962. Removal of excess salt in cation-exchange capacity determinations. Soil Sci. 93:87-94.

Richards, L. A. ed. 1954. Diagnosis and improvement of saline and alkaline soils. USDA Handbook 60.

Ross, Clarence E., and Sterling B. Hendricks. 1945. Minerals of the montmorillonite group. Their origin and relation to soils and clays. Prof. Paper 205-B. U. S. Geological Survey, U. S. Dep. of the Interior.

Sawhney, B. L., M. L. Jackson, and R. B. Corey. 1959. Cation-exchange capacity determinations of soils as influenced by the cation species. Soil Sci. 87:243-248.

Schofield, R. K., and H. R. Samson. 1953. The deflocculation of kaolinite suspensions and the accompanying change-over from positive to negative chloride adsorption. Clay Min. Bull. 2:45–51.

Stern, O. 1924. Zur theorie der elecktrolytischen doppelschict. Z. Electrochem. 30: 508–516.

van Olphen, H. 1963. Introduction to clay colloid chemistry. Interscience Publishers, New York.

van Raij, Bernardo, and Michael Peech. 1972. Electrochemical properties of some oxisols and alfisols of the tropics. Soil Sci. 36:587–593.

Verwey, E. J. W., and J. Th. G. Overbeek. 1948. Theory of the stability of lyophobic colloids. Elsevier, New York.

Wright, H. J. L., and R. J. Hunter. 1973a. Adsorption at solid-liquid interfaces. I. Thermodynamics and the adsorption potential. Aust. J. Chem. 26:1183–1189.

Wright, H. J. L., and R. J. Hunter. 1973b. Adsorption at solid-liquid interfaces. II. Models of the electrical double layer at the oxide-solution interface. Aust. J. Chem. 26:1191–1206.

Carbonate, Halide, Sulfate, and Sulfide Minerals[1]

H. E. DONER, University of California, Berkeley

WARREN C. LYNN, USDA-SCS, Lincoln, Nebraska

Carbonate, halide, sulfate, and sulfide minerals are found in soil environments with a vast range of characteristics. Part I of this chapter is concerned with those minerals more commonly associated with soils of alkaline reaction and Part II with those minerals common to acid sulfate soils. This scheme of dividing the chapter is strictly for the convenience of discussion.

In Part I emphasis is placed on calcium, magnesium, and sodium carbonates and sulfates. Their genesis and equilibrium environments are discussed. In Part II a thorough discussion of the genesis and morphological features of sulfide and sulfate minerals is included. In both Parts I and II chemical and physical properties of the minerals are discussed along with methods of identification.

I. ALKALINE SOILS

Carbonate, halide, and sulfate minerals are commonly characterized by their relatively high solubilities and rates of dissolution compared to silica minerals in soils. Soils in arid and semiarid regions often contain evaporites because of little or no leaching. Obviously, minerals within this group (carbonates, halides, and sulfates) have large differences in solubilities. Calcite ($CaCO_3$), for example, is only slightly soluble compared to halite ($NaCl$).

Chemical examination of solutions extracted from soils of semiarid and arid regions reveals Na^+, Ca^{2+}, Mg^{2+}, K^+, Cl^-, SO_4^{2-}, HCO_3^-, and CO_3^{2-} as major ionic components. A survey of the literature indicates 30 to 40 possible minerals which could be formed from soil solutions containing these ions upon evaporation. Many of these minerals have been identified from marine or geologic deposits, but relatively few have been identified in soils.

Calcium carbonate, $MgCO_3$, and $CaSO_4$ have been the most extensively studied soil minerals considered in Part I. In the case of the more soluble minerals, studies have focused on their effects on soil solution and physical characteristics, rather than on actual identification of specific minerals. Extensive studies have been made on effects of the more soluble minerals as related to soil solution chemistry and soil physical properties. Practical applications of these investigations are found in such areas as soil reclamation, irrigation water quality and land use planning.

In this portion of the chapter, discussion will be focused on carbonate, halide, and sulfate minerals identified in soils. The only important halide

---

[1] Part I, Alkaline Soils, was prepared by H. E. Doner; Part II, Acid Sulfate Soils, was prepared by W. C. Lynn.

chapter 3

mineral in soils is halite (NaCl). Several sulfate minerals have been identified, including: gypsum ($CaSO_4 \cdot 2H_2O$), hemihydrate ($CaSO_4 \cdot \frac{1}{2}H_2O$), mirabilite ($Na_2SO_4 \cdot 10H_2O$), thenardite ($Na_2SO_4$), epsomite ($MgSO_4 \cdot 7H_2O$), hexahydrite ($MgSO_4 \cdot 6H_2O$), and bloedite [$Na_2Mg(SO_4)_2 \cdot 4H_2O$]. Calcite ($CaCO_3$), Mg-calcite ($Ca_{1-x}Mg_xCO_3$, where $0 < x > 0.5$), dolomite [$CaMg(CO_3)_2$], nahcolite ($NaHCO_3$), trona ($Na_3CO_3HCO_3 \cdot 2H_2O$), and soda ($Na_2CO_3 \cdot 10H_2O$) constitute most of the soil carbonate minerals. Occasionally the minerals aragonite ($CaCO_3$), magnesite ($MgCO_3$), hydromagnesite [$Mg_5(CO_3)_4(OH)_2 \cdot 4H_2O$], sodium iodate ($NaIO_3$), and thermonatrite ($Na_2CO_3 \cdot H_2O$) are reported. Future research will probably establish the importance of other minerals as well.

A. Natural Occurrence

1. SOILS

Very soluble minerals, such as halite and epsomite, are quickly removed from soils where precipitation is high enough to cause deep leaching. For these very soluble minerals to be present in soils requires the rate of accumulation for the minerals to be equal to or greater than their removal rate. Many of the arid and semiarid areas of the world have this climatic characteristic, ranging from the equator to the poles (Tedrow & Ugolini, 1966). A closer examination of the conditions for accumulation of these minerals and example areas may be more useful than an exhaustive list of geographic locales.

The closed basins east of the Sierra Nevada Mountains provide an example of an arid to semiarid climate. Rainfall is low enough that many localities have no effective external drainage to the ocean. Winter rains leach the most soluble minerals to lower elevations. Restricted infiltration may result in accumulation of the minerals with evaporation. Alternatively, these minerals may accumulate at the soil surface upon capillary rise of ground water and subsequent evaporation. In extreme cases, efflorescence of these minerals can be seen on soil surfaces.

Lacustrine deposits of Lake Bonneville, one of the great Pleistocene lakes in what is now northwest Utah, have a vast assemblage of evaporites and carbonates, such as, halite, gypsum, calcite, dolomite, aragonite, and magnesite (Graf et al., 1961). The latter mineral is less commonly found in soils. Lake Bonneville offers an extreme example of the effect of a large internal drainage basin and a long period for evaporation on evaporite and carbonate formation.

The dry valleys of the Antarctic also have soils high in evaporites and carbonates. The source of salts in the Victoria Valley region is thought to be from weathering of local rocks over a long period of time (Gibson, 1962; Tedrow & Ugolini, 1966). Besides the presence of minerals such as gypsum, mirabilite, thenardite, and calcite, there are also found salt deposits of sodium iodate ($NaIO_3$) and sodium nitrate ($NaNO_3$) (Gibson, 1962; Johannesson & Gibson, 1962). It is thought that iodine released from rocks during weathering was oxidized to form the iodate, more or less, in situ. The formation of nitrate minerals is beyond the scope of this chapter.

The less soluble minerals (e.g., calcite and dolomite) can be found from arid to humid areas. Because of their solubility characteristic, soil parent material becomes a more important factor in determining geographic distribution. These minerals may occur either from the original soil parent materials or as the result of pedogenic processes.

2. GEOLOGIC MATERIALS

Sedimentary deposits of marine evaporites have received extensive study. The Stassfurt series in northern Germany is a well-known deposit of which much theoretical study has been applied. Some of the minerals included in this deposit are halite, sylvite (KCl), kieserite ($MgSO_4 \cdot H_2O$), carnallite ($KMgCl_3 \cdot 6H_2O$), anhydrite ($CaSO_4$), polyhalite [$K_2MgCa_2(SO_4)_4 \cdot 2H_2O$], loeweite [$Na_2Mg(SO_4)_2 \cdot 5/2H_2O$], picrometrite [$K_2Mg(SO_4)_2 \cdot 6H_2O$], vanthoffite [$Na_6Mg(SO_4)_2 \cdot 5/2H_2O$], langbeinite [$K_2Mg_2(SO_4)_3$], and kainite [$MgSO_4 \cdot KCl \cdot 3H_2O$] (Stewart, 1963). High temperatures and pressures in these sedimentary deposits afford conditions for formation of a wide variety of minerals. An extensive list of additional evaporite minerals from other deposits can be found in Stewart's (1963) report.

Of the carbonate minerals in sedimentary rocks, the most abundant are calcite and dolomite, followed by siderite ($FeCO_3$) and ankerites (iron-bearing dolomites (Lippmann, 1973). Reducing conditions are required for formation of $FeCO_3$ minerals in which Fe occurs in the Fe^{2+} state. Aragonite may be found in more recent sediments and animal shells, but is not common to sediments of Tertiary age or older. Magnesite is also found in sedimentary deposits. Magnesium carbonate has a strong tendency to form hydrous minerals such as hydromagnesite [$Mg_5(CO_3)_4(OH)_2 \cdot 4H_2O$ or $Mg_4(CO_3)_3(OH)_2 \cdot 3H_2O$], artinite ($Mg_2(CO_3)(OH)_2 \cdot 3H_2O$), nesquehonite ($MgCO_3 \cdot 3H_2O$), and lansfordite ($MgCO_3 \cdot 5H_2O$). In contrast to magnesium carbonates, hydrated $CaCO_3$ minerals are only rarely found (Lippmann, 1973, p. 88–90). Two hydrated forms have been identified, $CaCO_3 \cdot 6H_2O$ and $CaCO_3 \cdot H_2O$. Both forms were identified from laboratory products; the latter, $CaCO_3 \cdot H_2O$, tentatively called monohydrocalcite, has also been found under natural conditions. Alkali carbonates, mainly of Na, are commonly found in lacustrine sediments. Lippman (1973) has prepared an extensive list of such minerals as found in various geologic deposits.

B. Equilibrium Environment and Conditions for Synthesis

1. CARBONATES

The formation of calcite in soil follows the reaction:

$$Ca^{2+}_{(aq)} + 2HCO^-_{3(aq)} = CaCO_{3(calcite)} + H_2O + CO_2. \qquad [1]$$

The concentration of Ca dissolved, according to Eq. [1], depends on the partial pressure of CO_2 and the temperature. The reaction of CO_2 and water can be shown by:

$$CO_{2(g)} + H_2O = CO_{2(aq)} + H_2O \qquad [2]$$

$$CO_{2(aq)} + H_2O = H_2CO_3 \qquad [3]$$

$$H_2CO_3 = H^+ + HCO_3^- \qquad [4]$$

$$HCO_3^- = H^+ + CO_3^{2-} \qquad [5]$$

The term $CO_{2(g)}$ represents the partial pressure of CO_2 gas, which is 3×10^{-4} bar in the normal atmosphere. The solubility of the gas in H_2O is highly temperature dependent (0.08, 0.03, and 0.02 moles CO_2/liter H_2O at 0, 25, and 40C (273, 298, and 313K), respectively, at 1.01 bar CO_2 pressure, as calculated from Hodgman, 1960, p. 556). The equilibrium constant for Eq. [3] is about 1.7×10^{-3} at 25C (298K) (Kern, 1960) showing that very little of the dissolved CO_2 is actually hydrated to H_2CO_3. In practice, Eq. [2], [3], and [4] are combined to give the apparent first dissociation constant for carbonic acid (H_2CO_2):

$$K_1 = \frac{(H^+)(HCO_3^-)}{CP_{CO_2}} = 4.28 \times 10^{-7} \text{ at 25C (298K)} \qquad [6]$$

(Wagman et al., 1968) where (H^+) and (HCO_3^-) are activities of the respective ions, C equals 0.0344 from Henry's Law (Harned & Davis, 1943), and P_{CO_2} is the partial pressure of CO_2 in bars. The second dissociation constant for carbonic acid is

$$K_2 = \frac{(H^+)(CO_3^{2-})}{(HCO_3^-)} = 4.68 \times 10^{-11} \text{ at 25C (298K)} \qquad [7]$$

(Wagman et al., 1968). Here, again, the parentheses denotes ion activity. The solubility product constant for calcite as reported by Langmuir (1968) is (Table 3-1):

$$K_{\text{sp calcite}} = (Ca^{2+})(CO_3^{2-}) = 3.98 \times 10^{-9} \qquad [8]$$

at 25C (298K) and 1.01 bar total pressure. Combining Eq. [6], [7], and [8] and taking the log gives

$$2 \text{ pH} + \log P_{CO_2} = 9.76 + \log [1/(Ca^{2+})]. \qquad [9]$$

The alignment chart in Fig. 3-1 shows the relationship between pH, CO_2 partial pressure, and Ca^{2+} activity (concentration) at 25C (298K) and 1.01 bar total pressure for solutions saturated with respect to calcite. As an ex-

Table 3-1. Solubility products of selected carbonates at 25C (298K) and 1.01 bar total pressure

Mineral	Chemical formula	k_{sp}	Reference
Calcite	$CaCO_3$	3.98×10^{-9}	Langmuir, 1968
Aragomite	$CaCO_3$	5.62×10^{-9}	Langmuir, 1964
Dolomite	$CaMg(CO_3)_2$	10^{-17}	Hsu, 1967
Nesquehonite	$MgCO_3 \cdot 3H_2O$	10^{-5}	Langmuir, 1965
Magnesite	$MgCO_3$	7.9×10^{-9}	Langmuir, 1965
Hydromagnesite	$Mg_5(CO_3)_4(OH)_2 \cdot 4H_2O$	6.3×10^{-31}	Langmuir, 1965
Lansfordite	$MgCO_3 \cdot 5H_2O$	3.47×10^{-6}	Langmuir, 1965
Nahcolite	$NaHCO_3$	0.82 moles/liter at $0°C$†	
Soda	$Na_2CO_3 \cdot 10H_2O$	0.752 moles/liter at $0°C$†	
Trona	$Na_2CO_3 \cdot NaHCO_3 \cdot 2H_2O$	0.58 moles/liter at $0°C$†	

† Solubility from *Handbook of chemistry and physics.* 41st ed. (Hodgman, 1960).

ample, for a solution pH 8.0 and CO_2 partial pressure of 0.001 bar, the Ca^{2+} activity (concentration) would be 6×10^{-4} molar.

Thus far, the discussion has centered around ideal aqueous systems containing clacite. This helps to understand the equilibrium and synthesis of carbonate minerals in soils. However, the solubility of $CaCO_3$ in several calcareous soils has been found greater than predictions based on calcite (Olsen & Watanabe, 1959). Soil components may affect the solubility of $CaCO_3$ by forming minerals other than calcite. For example, Mg^{2+} and SO_4^{2-} at low concentrations promote the formation of aragonite and vaterite, both of which are crystalline polymorphs of $CaCO_3$ (Doner & Pratt, 1968, 1969). Vaterite has not been identified in sediments (Lippmann, 1973, p. 67) and aragonite is the stable form of $CaCO_3$ only above 50C (323K) (Wray & Farrington, 1957). Formation of neither aragonite nor vaterite has been reported in soils. It is possible that aragonite and vaterite, if indeed formed in soils, are soon transformed to the more insoluble calcite form.

Calcium and magnesium carbonate systems present some particularly interesting problems in soil mineralogy. When $CaCO_3$ precipitates, the non-hydrated mineral, calcite, is commonly formed; however, Mg^{2+}, having a higher hydration energy than Ca^{2+}, commonly forms hydrated $MgCO_3$ minerals such as hydromagnesite $[Mg_5(CO_3)_4(OH)_2 \cdot 4H_2O]$, nesquehonite ($MgCO_3 \cdot 3H_2O$), lansfordite ($MgCO_3 \cdot 5H_2O$), and artinite $[Mg_2(CO_3)(OH)_2 \cdot 3H_2O]$ (for review see Lippmann, 1973). The main forms of $MgCO_3$ reported in soils are mixed with $CaCO_3$, rather than existing as separate $MgCO_3$ minerals. Sherman et al. (1962) reported pedogenic Mg-calcite and dolomite in Mg solonetz soils. More recently, St. Arnaud and Herbillon (1973) identified Mg-calcite which was pedogenically formed in the Cca and Ck horizons from Calciboroll, Haploboroll, and Natriboroll soils of central Saskatchewan, Canada. Formation of these Mg-calcites was thought to occur by dissolution of existing calcite and dolomite in the parent material, translocation of Ca^{2+}, Mg^{2+}, and HCO_3^- ions through the soil profile, and subsequent precipitation. Such transport of $CaCO_3$ in response to soil water movement has been well documented (Jenny, 1941, p. 122; Arkley, 1963).

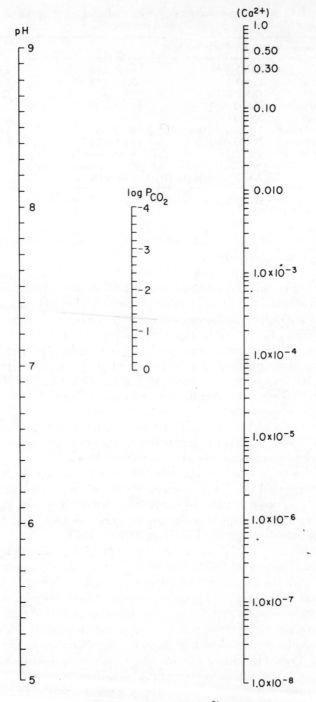

Fig. 3-1. An alignment chart relating pH, P_{CO_2}, and Ca^{2+} activity (concentration) for calcite from the equation: $2 \text{ pH} + \log P_{CO_2} = 9.76 + \log [1/(Ca^{2+})]$.

The precipitation of $CaCO_3$ and $MgCO_3$ is not simply the reverse process of dissolution of calcite and dolomite mixtures. There is a preferential precipitation of $CaCO_3$ (partitioning effect) which follows closely the equation:

$$Mg_c/Ca_c = k_{Mg} (Mg_l/Ca_l), \qquad [10]$$

where Mg_c and Ca_c are the molar ratios of Mg and Ca in calcite, Mg_l and Ca_l are the molar ratios of Mg and Ca in the equilibrium solution, and k_{Mg} is the distribution coefficient for partitioning of Mg between calcite and solution (St. Arnaud & Herbillon, 1973 after Winland, 1969). The value for k_{Mg}, according to best information, is about 0.02 in the range 0–20% $MgCO_3$ in calcite (Winland, 1969; Glover & Sippel, 1967). This indicates that only a small fraction of the total Mg is incorporated into calcite.

Consider a model for Mg-calcite formation similar to that described by St. Arnaud and Herbillon. A soil solution is equilibrated with dolomite and calcite in soil at pH 7 and P_{CO_2} of 0.03 bar. Dolomite and calcite would dissolve according to the reactions:

$$CaMg(CO_3)_2 \rightarrow Ca^{2+} + Mg^{2+} + 2CO_3^{2-} \qquad [11]$$

(dolomite)

$$CaCO_3 \rightarrow Ca^{2+} + CO_3^{2-} \qquad [12]$$

(calcite)

At equilibrium the ratio between Mg^{2+} and Ca^{2+} in solution would be (reference-Table 3–1 for solubility product constants):

$$\frac{(Ca^{2+})(Mg^{2+})(CO_3^{2-})^2}{(Ca^{2+})^2 (CO_3^{2-})^2} = \frac{K_{sp \text{ dolomite}}}{K_{sp \text{ calcite}}^2} = \frac{10^{-17}}{(3.98 \times 10^{-9})^2} = 0.6 \qquad [13]$$

At pH 7 and P_{CO_2} = 0.03 bar, the concentration of Ca^{2+} would be 1.9×10^{-3} moles/liter (reference, Fig. 3–1) and of Mg^{2+} would be 1.1×10^{-3} moles/liter, according to Eq. [13]. Assume this solution moved to a lower part of the soil profile and P_{CO_2} decreased to 0.001 bar (by drying of the soil) and pH increased to 7.8. Now calcite would precipitate. At equilibrium the Ca^{2+} concentration would be 1.4×10^{-3} moles/liter and the Mg^{2+} concentration would be only slightly $< 1.1 \times 10^{-3}$ moles/liter, as equilibrium no longer exists with dolomite and only a small amount precipitates with calcite (Eq. [10]). Using Eq. [10] and assuming the Ca^{2+} and Mg^{2+} concentrations to be 1.4×10^{-3} and 1.1×10^{-3} moles/liter, respectively, we can calculate the mole fraction of $MgCO_3$ in calcite as

$$Mg_c/Ca_c = 0.02 (1.1 \times 10^{-3}/1.4 \times 10^{-3}) = 0.02. \qquad [14]$$

This amounts to approximately 2 moles % $MgCO_3$ in calcite. Further drying of the soil would cause a further increase in the Mg^{2+} concentration in solu-

tion and ostensibly in the calcite crystal. Subsequent dissolution, translocation, and precipitation would increase the percent of $MgCO_3$ in calcite.

Sherman et al. (1962) visualized the reaction for dolomite formation as
a substitution of Mg for Ca:

$$2CaCO_3 + MgSO_4 \rightarrow CaCO_3 \cdot MgCO_3 \downarrow + CaSO_4$$

(Hardinger reaction)

Theirs is the only postulated occurrence of pedogenic dolomite.

Sodium carbonate formation in soils has been reviewed by Kelley (1951).
Although often not mentioned by mineral name in the literature, three forms
of Na_2CO_3 are associated with alkali soils, namely: nahcolite ($NaHCO_3$),
soda ($Na_2CO_3 \cdot 10H_2O$), and trona ($Na_2CO_3 \cdot NaHCO_3 \cdot 2H_2O$). The phase diagram, Fig. 3-2, shows the effects of CO_2 partial pressure and temperature on
these mineral forms. In interpreting the phase diagram, one must remember
that the boundary lines between stability fields are placed where the solution
is in equilibrium with two mineral phases. Changes in CO_2 partial pressure
and temperature can result in a transformation of one mineral to another.
Such equilibria are discussed by Szabolcs (1971). Besides the above Na_2CO_3
minerals, Kovda (1965) reports thermonatrite ($Na_2CO_3 \cdot H_2O$). This mineral
might be found in very dry environments.

The synthesis of Na_2CO_3 minerals can be separated into two mechanisms. One process is physicochemical, whereas the other is biological (Kelley,
1951; and Kozhevnikov, 1974). A soil leached with highly saline water
(e.g., a high concentration of NaCl), followed by water of lower salinity, will
result in hydrolysis of exchangeable Na and Na_2CO_3 formation as follows:

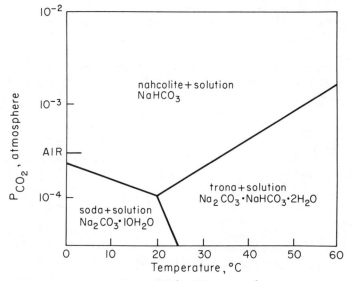

Fig. 3-2. Mineral stability fields for the $Na^+ - HCO_3^- - CO_3^{2-}$ system at various temperatures and CO_2 pressures (after Milton & Eugster, 1959).

$$CaX + 2NaCl \rightarrow Na_2X + CaCl_2$$

$$Na_2X + 2H_2O \rightarrow H_2X + 2NaOH$$

$$2NaOH + CO_2 \rightarrow Na_2CO_3 + H_2O$$

where X denotes soil cation exchange sites. Irrigation of saline-sodic soils with waters of low salt content can thus promote Na_2CO_3 formation. In calcareous soils, Ca^{2+} instead of H^+ will be the exchangeable cation.

Another Na_2CO_3 formation process is the weathering of igneous rocks to produce bicarbonates of Na, Ca, and Mg. Upon evapotranspiration of the resultant solution, $CaCO_3$ and $MgCO_3$ minerals will be precipitated, and the remaining solution will contain Na^+ and HCO_3^-. With further concentration of the solution by drying of the soil, Na_2CO_3 will form. Sodium carbonate will be formed only when the HCO_3^- concentration is greater than the Ca^{2+} plus Mg^{2+} concentration on an equivalent basis.

The biological formation of Na_2CO_3 in soils by the reduction of sulfate to sulfide was described by Whittig and Janitzky (1963). Figure 3-3, taken from their paper, indicates the biochemical and chemical reactions as they occur in the soil environment. The xeric climate at their study site in the Sacramento Valley of California promotes evapotranspiration in the summer. Salts, including HCO_3^-, migrate out of zones of biological reduction in drainageways into the surrounding soil at slightly higher elevations. Sodium carbonate accumulates in a band around the drainageway where the soil pH may range above 10. Field observations were substantiated by laboratory ex-

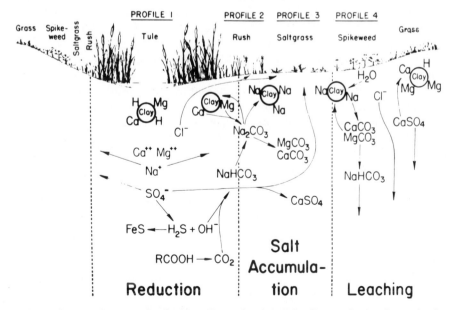

Fig. 3-3. Reactions associated with sulfate reduction and sodium carbonate formation in soils. Reproduced with permission, from L. D. Whittig & P. Janitzky, 1963—Plate II.

periments (Janitzky & Whittig, 1964). In coastal settings the HCO_3^- are flushed by the tides and the potential for Na_2CO_3 formation or for the neutralization of subsequent acid sulfate soils is lost. A more detailed discussion of acid sulfate soils is found later in this Chapter.

2. SULFATES (Ca)

Gypsum is approximately 100 times less soluble than other sulfate minerals normally found in soils. It is also the most common sulfate mineral in soils. In geologic deposits, several $CaSO_4$ minerals are found, but under conditions normal to the earth's surface, gypsum is the principal form. Hemihydrate ($CaSO_4 \cdot \frac{1}{2} H_2O$) is sometimes found at soil surfaces under extremely dry conditions (Kovda, 1946, p. 243). As with the deposition of $CaCO_3$ minerals, $CaSO_4$ is transported by the soil solution and gypsum precipitates when the solubility of the mineral is exceeded.

3. SULFATES (Mg and Na) AND HALIDES (NaCl)

The high solubility of $MgSO_4$, Na_2SO_4, and NaCl minerals causes their formation of the furthest extent of soil-water movement. Because of their high solubility, the crystalline forms of these minerals are commonly found at the soil surface, under conditions of extreme desiccation. They are not found at any appreciable depth in moist soil profiles (Kovda, 1946).

In a recent study of salt efflorescences in the Great Konya Basin, Turkey (Driessen and Schoorl, 1973) chemical, X-ray diffraction, and electron microscopy methods were used to identify the minerals halite (NaCl), mirabilite ($Na_2SO_4 \cdot 10H_2O$), thenardite (Na_2SO_4), bloedite [$Na_2Mg(SO_4)_3 \cdot 4H_2O$], epsomite ($MgSO_4 \cdot 7H_2O$), and hexahydrite ($MgSO_4 \cdot 6H_2O$). A phase diagram for the $MgSO_4$ system (Fig. 3-4) shows that thenardite is more stable at higher temperatures than mirabilite. Field observations confirm this conclusion. Mirabilite was always found a few centimeters below the soil surface, whereas thenardite was always found at the soil surface in summer. Thenardite transformed to mirabilite in the winter when it was cooler. Not only seasonal weather conditions changed the amount and kinds of such soluble minerals, but also daily fluctuations in temperature and humidity. Minerals present in the dry heat of day may dissolve or transform to other minerals during the high humidities at night (Kovda, 1946, p. 246).

C. Chemical Properties

1. CARBONATES

Phosphate fixation is a well-known phenomenon associated with calcareous soils. Monolayers of phosphate are thought to be formed at low phosphorus concentrations on calcite surfaces (Cole et al., 1953). Recently, Griffin and Jurinak (1973) proposed another mechanism for phosphate inter-

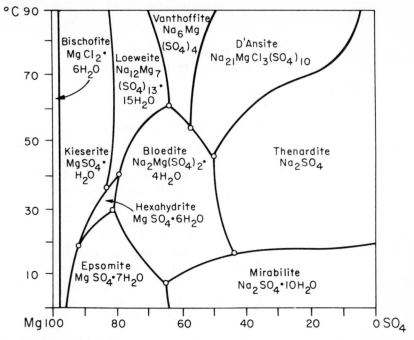

Fig. 3-4. Mineral stability fields for $MgSO_4$ – Na_2SO_4 in saturated NaCl solutions (after Braitsch, 1962).

action with calcite. According to their mechanism, phosphate is first adsorbed at specific sites on calcite. With aging, crystalline calcium phosphate is formed which results in further phosphate adsorption.

Trace metals are also known to be adsorbed by soil carbonate minerals. Jurinak and Bauer (1956) reported relative Zn adsorption levels of magnesite > dolomite > calcite. It was thought that Zn (0.83Å) replaced Mg (0.78Å) in the crystal structure. This assumes some Mg to be present in the calcite structure. Adsorption probably is associated with formation of some insoluble compound, since carbonate minerals have no exchange capacity in the same sense as found for layer silicate minerals.

Jurinak and Griffin (1972) reported that pure calcite adsorbed NO_3^-. This was thought to be due to physical adsorption, since their experiments were conducted at a pH slightly below the isoelectric point for calcite. Steric compatability of NO_3^- and CO_3^{2-} (both are planar ions) may promote adsorption, as Cl^- was not adsorbed under similar conditions. They found that soil $CaCO_3$ showed no adsorption of NO_3^-. It was suggested that either competing anions in solution or coatings on the carbonate mineral surfaces interfered with NO_3^- adsorption, for soil carbonate minerals are known to have surface coatings (Lahav & Bolt, 1963).

Soils containing Na_2CO_3 commonly have high pH values (pH 8.5 and greater). If the salt concentration is low, soil clays and organic matter tend to be dispersed as a result of the high pH and the presence of Na^+.

2. SULFATES AND HALIDES

Dissolution of gypsum results in both charged and uncharged Ca^{2+} and SO_4^{2-} species in solution. The reaction can be written as follows:

$$CaSO_4 \cdot 2H_2O = XCa^{2+} + XSO_4^{2-} + (1-X)CaSO_4^0 + 2H_2O$$

where X is $<$ "1" and the notation "$CaSO_4^0$" represents an ion-pair which is a soluble, but uncharged species. In studies concerned with the solubility of sulfate minerals, these ion-pairs are important in accounting for dissolved species (Nakayama, 1971). Magnesium and Na^+ are also known to form these ion-pairs or complexes. With Na^+, the negatively charged complexes $NaSO_4^-$ are of some importance.

It should be mentioned in connection with the above discussion that CO_3^{2-} and HCO_3^- also form ion-pairs and complexes with many cations common to soil solution. Chloride ion-pairs and complexes with Na^+, Ca^{2+}, and Mg^{2+} form extremely weak bonds, and probably are not an important consideration in most cases.

D. Physical Properties

1. CARBONATES

Calcium carbonate minerals may be present in soils as particles or coatings over or between other particles. Carbonate coatings in a soil matrix may form in layers from impeded water movement (Stuart & Dixon, 1973). The formation of these calcareous crusts may close soil pores and result in horizontal water movement.

2. SULFATES AND HALIDES

Sodium and magnesium sulfate and halide minerals may form crusts on soil surfaces. Figures 3-5 and 3-6 show scanning electron micrographs of bloedite and thenardite crystals, respectively, from Driessen and Schoorl's (1973) study. The overlapping crystals of bloedite gave the surface of the salt crust a glazed or "glassy-like" appearance. Growth of bloedite crystals toward the center of soil pores was reported to be effective in sealing the soil, thereby restricting evaporation and aeration (Fig. 3-7) (Dr. P. M. Driessen, personal communication). The more open network of thenardite crystals did not exhibit this sealing phenomenon.

E. Mineral Identification

The chemical compositions of some important soil carbonate, sulfate, and halide minerals were reported above. Crystalline structures of these

Fig. 3-5. A crust of overlapping bloedite crystals (×3000). (Driessen & Schoorl, 1973).

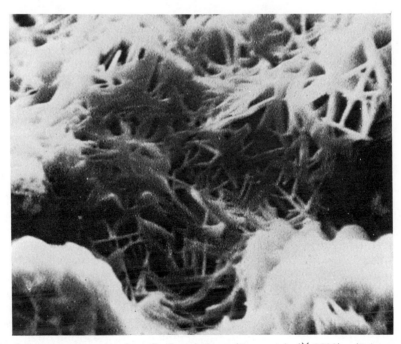

Fig. 3-6. A loose crust of needle-shaped thenardite crystals (× 3200). (Driessen & Schoorl, 1973).

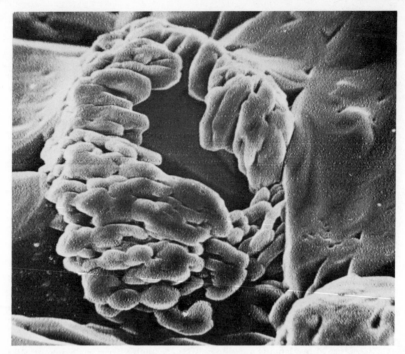

Fig. 3-7. Filling of a cavity by growth of bloedite crystals ($\times$ 720). (Dr. P. M. Driessen, personal communication).

minerals can be found in many reference books, e.g., Bragg et al., 1965; Lippmann, 1973.

Hilgard (1912, p. 76) writes,

> The presence of common salt may, as a rule, be detected by the taste, well known to every one; when this taste is very intense or somewhat bitterish, it indicates the presence of bittern. The presence of salt, however, is easily verified without the use of chemical reagents, by slowly evaporating some of the clear water leached from the soil in a clean silver spoon. If the last few drops are allowed to evaporate spontaneously, it will be easy to distinguish, even with the unaided eye, the square, cubical crystals, sometimes combined into cross-shape, which are characteristic of common salt.

Many minerals can be identified by their crystalline form and optical characteristics. The use of a microscope is sometimes necessary. Since Hilgard's time, many advances in identification of minerals have been made, and some are described below.

Calcium and magnesium carbonate minerals are often identified by X-ray diffraction (XRD) techniques from lists of d-spacings similar to those given in Table 3-2. When Mg substitutes for Ca in calcite, a shift to a lower $d(112)$ spacing is observed (Goldsmith et al., 1955). St. Arnaud and Herbillon (1973) observed this shift in d-spacing along with broad and often asymmetrical peaks for Mg-calcite in soil.

Table 3-2. Major X-ray diffraction spacings for some common carbonate, sulfate, and halide minerals

Calcite (5-0586)[†]		Aragonite (5-0453)		Dolomite (11-78)		Nahcolite (21-119)		Soda (natron) (15-800)	
$d(\text{Å})$	I/I_o	$d(\text{Å})$	I/I_o	$d(\text{Å})$	I/I_o	$d(\text{Å})$	I/I_o	$d(\text{Å})$	I/I_o
3.04	100	3.40	100	2.89	100	2.97	100	3.04	100
2.29	18	1.98	65	2.19	30	2.60	90	3.02	70
2.10	18	3.27	52	1.79	30	3.08	25	2.89	60
3.86	12	4.212	2	4.03	3	5.92	6	7.09	2

Trona (11-643)		Gypsum (21-816)		Hemihydrate (14-453)		Epsomite (8-467)		Hexahydrite (1-354)	
$d(\text{Å})$	I/I_o	$d(\text{Å})$	I/I_o	$d(\text{Å})$	I/I_o	$d(\text{Å})$	I/I_o	$d(\text{Å})$	I/I_o
2.66	100	2.87	100	3.00	100	4.21	100	4.40	100
3.08	80	4.28	90	6.01	95	5.35	26	2.92	60
		2.68	50	2.80	50	2.68	24	4.04	30
9.88	60	7.61	55	6.01	95	5.99	22		

Mirabilite (11-647)		Thenardite (5-631)		Bloedite (19-1215)		Halite (5-628)	
$d(\text{Å})$	I/I_o	$d(\text{Å})$	I/I_o	$d(\text{Å})$	I/I_o	$d(\text{Å})$	I/I_o
5.49	100	2.78	100	3.25	100	2.82	100
3.21	75	4.66	73	4.56	95	1.99	55
3.26	60	3.18	51	3.29	95	1.63	15
7.55	1	4.66	73	5.46	2	3.258	13

† Powder Diffraction File (1973) Joint Committee on Powder Diffraction Standards.

Infrared spectra of calcite and dolomite are characterized by three major and two minor adsorption bands, and of aragonite by three to five major and three minor absorption bands attributed to carbonate (Huang & Kerr, 1960). Identification of individual minerals can be made from band position. Staining techniques have also been successfully applied for identification of calcite, aragonite, dolomite, and Mg-calcite (Friedman, 1959).

Sodium carbonate minerals can be identified by XRD (Table 3-2) and from infrared spectra (Huang & Kerr, 1960). Gypsum often appears as fibrous crystals in soils. X-ray diffraction provides a strong tool for its identification as well as for the other, more soluble sulfate and halide minerals (Table 3-2).

F. Quantitative Determination

1. CARBONATES

Quantitative determination of soil carbonate minerals commonly is done by acidifying the soil and measuring the amount of CO_2 evolved (Holmgren, 1973; Bundy & Bremner, 1972). Various schemes for measuring the CO_2 evolved include weighing the amount of CO_2 absorbed by an alkaline media, manometric (volumetric) measurements, and measurement by titration of the amount of base neutralized by CO_2. These methods normally make no distinction between the various carbonate minerals, but instead give only a gross estimate of total inorganic carbonates.

Various methods have been devised to quantitatively separate calcite from dolomite. Turner and Skinner (1960) suggested the use of a method based on the difference in rates of dissolution for determining the amounts calcite and dolomite in mixtures of the two. Peterson et al. (1966) used a method of selective dissolution to dissolve first calcite and then dolomite.

St. Arnaud and Herbillon (1973) suggest a selective dissolution method to determine the amount of Mg from Mg-calcite. They used a H_2CO_3-$KHCO_3$ buffered (pH 6.55) system, which was standardized against known dolomite, calcite, and Mg-calcite mixtures. The method is strictly empirical.

2. SULFATES AND HALIDES

A quantitative estimate of soil gypsum can be made with a dissolution method outlined by Deb (see Hesse, 1972, p. 86–87). By making a saturation soil extract and determining the Ca^{2+} and SO_4^{2-} concentrations, followed by SO_4^{2-} determination on a 1:20 extract, the problems of cation exchange and of the presence of other soluble sulfate minerals are reduced. Khan and Webster (1968) suggest an XRD technique using KCl as an internal standard to analyze for gypsum in the presence of Na_2SO_4.

An estimate of the amount of the more soluble minerals of SO_4^{2-} and halide can be obtained by analyzing soil extracts for Ca^{2+}, Mg^{2+}, Na^+, Cl^-, and SO_4^{2-}. Obviously, this procedure does not identify mineral species, which would have to be done by independent methods, such as X-ray diffraction, infrared spectroscopy, and microscopy.

II. ACID SULFATE SOILS

This section of the chapter deals with the soil mineralogy involved in the formation of acid sulfate soils or "cat clays." The discussion is divided into three parts, namely, the soil genesis and the soil environment, the identification and characteristics of the minerals directly involved, and the effect of the particular soil environments on some other common soil minerals.

A symposium on acid sulfate soils was held in August 1972 at Wageningen, the Netherlands. The reader who wants a more detailed treatment of the subject should refer to the proceedings of the symposium (Dost, 1973).

A. Soil-Genesis and Soil Environments

The development of acid sulfate soils logically separates into two phases, the reduction phase in which sulfides (e.g., pyrite) accumulate and the oxidation phase in which acidity develops and jarosite is formed (Fig. 3–8). Several

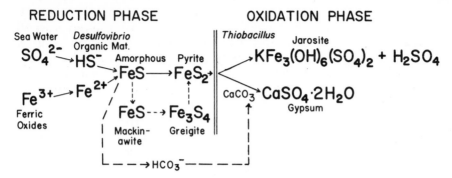

Fig. 3-8. Schematic diagram of mineral transformations in the development of acid sulfate soils.

ingredients are necessary for the reduction phase to proceed, i.e., sources of iron and sulfate, a redox potential generally around −400 mV, sulfate-reducing bacteria (*Desulfovibrio* spp. Simon-Sylvestre, 1960), an energy source for the microorganisms, and a mechanism to remove the bicarbonate formed in the process. Because of the sulfate requirement, sulfide accumulation is limited primarily to coastal margins (Kawalec, 1973), although some cases of sulfide accumulation are reported in inland settings (Poelman, 1973). Sulfidic sediments are commonly silty clays or silty clay loams high in organic matter that occur on tidal flats or in interdistributary basins of river deltas. Peats are often associated with sulfidic sediments, but the sulfide accumulation usually occurs in the mineral soil just beneath the contact with the peat (Benzler, 1973). If the supply of sulfate is adequate, the limiting factor is generally the supply of organic matter as an energy source for the sulfate-

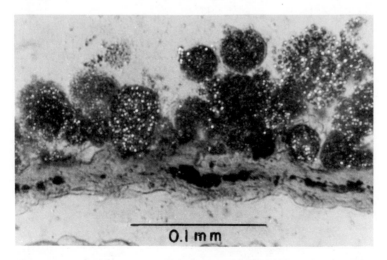

Fig. 3-9. Framboidal soil fabric of sulfidic materials. Reproduced, with permission from van Dam & Pons, 1973, p. 69. Pyrite crystallaria (framboids) are in various stages of decomposition. Some are intact and show the original spherical shape, surrounded by a pseudoskin. Others are decomposed and the individual microcrystals start to scatter.

reducing bacteria (Ricard, 1973). Not only are the organic matter and the microorganisms necessary for the process, but they strongly influence the distribution of sulfides in the soil fabric. The bacteria concentrate on the organic matter, and the H_2S (or HS^-) released by the organisms is precipitated in the immediate vicinity to form spherical aggregates containing pyrite microcrystallites (Fig. 3-9). This characteristic fabric is termed *framboidal* (van Dam & Pons, 1973). The first iron sulfide precipitate is usually amorphous (FeS) and undergoes diagenetic conversion to pyrite (FeS_2). Mackinawite (FeS) and greigite (Fe_3S_4) are possible intermediates. It should also be recalled that biological reduction of sulfate to sulfide was part of the system discussed previously in connection with Na_2CO_3 formation in soils (see Fig. 3-3).

If sulfidic sediments are drained and aerated, the sulfide minerals become unstable and the second or oxidation phase of acid sulfate soil formation commences. The soil generally shrinks irreversibly to form polygons about 2 m across separated by large cracks. Drainage is rather rapid. Acid soils with pH of 3.0-3.5 can develop within 1 to 2 years. The common oxidation product, jarosite, can form within 6 years and probably less. The jarosite appears as a yellow efflorescence around root channels and on natural surfaces. van Breeman and Harmsen (1975) describe the morphology of acid sulfate soils in Thailand that have typical features. Crystals are small (Fig. 3-10). Acid sulfate soils apparently can develop without the formation of jarosite. In such cases, the presence of 0.05% water soluble sulfate (plus the

Fig. 3-10. Jarosite crystals formed on soil ped surface. The crystal size is small and fairly uniform. Scanning electron micrograph by Dr. Kit Lee, School of Life Sciences, Univ. of Nebraska.

low pH) has been suggested as an indicator of an acid sulfate soil. If carbonates are present in the sediments, the acidity is neutralized and gypsum forms. Gypsum and jarosite can occur together. If the bicarbonate formed during the reduction phase is not removed from the system (commonly by tidal action), it would help neutralize the acidity developed during the oxidation phase.

Pyrites occur in spoil from mining operations. When pyritic spoil is exposed to the atmosphere, acid soils can develop through oxidation of the pyrites by the same mechanisms that operate in drained tidelands soils. The author is not aware, however, of any reports of jarosite formed in acid mine spoils. The acid soil environment can make revegetation and use of mine spoil areas a problem just as there are problems utilizing acid sulfate soils formed in coastal areas.

B. Soil Minerals Directly Involved

In the reduction phase, the soil minerals directly involved are amorphous FeS, mackinawite (tetragonal FeS), greigite (cubic Fe_3S_4), and pyrite (cubic FeS_2). The only distinctions likely possible in field observations are between pyrite and the other forms. Amorphous FeS (and the intermediates) are finely disseminated and impart a distinct black color to the sediment, particularly around organic materials and just beneath the surface of bottom sediments. Amorphous FeS reacts with cold HCl to give off the distinct odor of H_2S. Mackinawite and greigite reportedly react similarly with hot HCl. Pyrite commonly imparts a grayer color to the sediment and does not react with HCl.

Sulfidic fabrics as viewed in thin section are characteristically frambiodal (aggregates containing pyrite microcrystallites). Sulfides within the framboids are often of submicron dimensions (Fig. 3–9).

The crystalline phases of the pertinent sulfides can be identified by X-ray diffraction. The most intense characteristic lines for mackinawite, greigite, and pyrite are listed in Table 3–3.

Table 3-3. Major X-ray diffraction spacings for common minerals directly involved in the formation of acid sulfate soils. Card numbers refer to the ASTM X-ray diffraction card file.

Mackinawite (Card 15-37)		Greigite (Card 16-713)		Pyrite (Card 6-0710)		Jarosite (Card 10-443)		Natrojarosite (Card 11-302)	
$d(Å)$	I/I_o	$d(Å)$	I/I_o	$d(Å)$	I/I_o	$d(Å)$	I/I_o	$d(Å)$	I/I_o
5.03	100	5.72	8	3.128	35	5.94	30	5.94	40
2.97	80	3.50	30	2.709	85	5.09	40	5.57	50
2.31	80	2.98	100	2.423	65	3.11	60	5.06	100
1.838	40	2.470	55	2.2118	50	3.08	100	3.12	70
1.808	80	1.901	30	1.9155	40	2.547	30	3.06	80
1.725	40	1.746	75	1.6332	100	2.292	50	2.24	60
1.240	40	1.105	16	1.4448	25	1.978	50	1.98	60
1.055	60	1.1001	30	1.0427	25	1.823	50	1.83	50

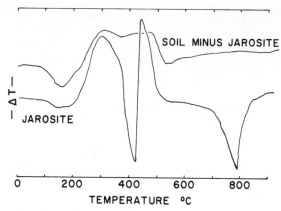

Fig. 3-11. Differential thermal analysis patterns of jarosite and of the nearby soil.

In the oxidation phase, jarosite $[KFe_3(OH)_6(SO_4)_2]$ and natrojarosite $[NaFe_3(OH)_6(SO_4)_2]$ are the most common S minerals (van Breemen, 1973). Gypsum is common if carbonates are present in the sediments. Jarosite or natrojarosite occurs as a characteristic yellow efflorescence around root channels and on structural surfaces and is readily identified in the field. Individual crystals, (possibly coatings on clay particles), generally 1 to 2 μm across, can be seen by scanning electron microscope (Fig. 3-10). Identification can be confirmed by X-ray diffraction. The principal spacings are given in Table 3-3. A typical DTA pattern is shown in Fig. 3-11. Gypsum in this setting commonly occurs as macroscopic euhedral crystals. Its identification can be confirmed by optical microscope or by X-ray diffraction. A red efflorescence that looks like hematite is often associated with the jarosite in acid sulfate soils. The material seems to be amorphous in most cases.

C. Effect on Other Minerals

The formation of acid sulfate soils is independent of the silicate mineralogy. In some acid sulfate soils, kaolinite predominates while in others it is mica (illite) and smectite (van Breemen, 1973). The relative abundance of silicate minerals is probably not altered much during the formation of acid sulfate soils. However, some alterations of the layer silicates do occur and should be noted.

In the reduction phase, iron is reduced and mobilized, which affects not only iron oxides, but also the chlorites and other silicate minerals that contain iron. Some experimental evidence (Lynn & Whittig, 1966) indicates that sufficient interlayer material can form in smectites to give a mineral that has a 14Å spacing after it is heated to 550C (823K). A swelling chlorite is present in dikeland soils of Nova Scotia (Brydon & Heystek, 1958). Chlorite, some of which may be authigenic, is fairly common in coastal marine sediments.

In the oxidation phase, the acidity influences the stability of layer silicates in general and of interlayers in particular. Amorphous material often represents a significant proportion of the clay fraction (Horn & Chapman, 1968). Chlorites and smectites apparently are degraded (Lynn & Whittig, 1966). Aluminum is solubilized in the acid environment and satisfies a large proportion of the exchange capacity which seems to lend physical stability to the soil (Frink, 1973).

III. SUMMARY

In Part I the kinds of carbonate, sulfate, and halide minerals commonly found in arid and semiarid soils are discussed. Although the possible number of different minerals is large, only a relatively small number have been identified in soils. Special attention is given to the formation of $CaCO_3$ and $MgCO_3$ minerals. Calcite is the most abundant $CaCO_3$ mineral. Reports of $MgCO_3$ minerals identified in soil are very limited. Apparently Mg, to a limited extent, substitutes for Ca during calcite formation. Dolomite is formed only under special conditions. These carbonate minerals may restrict water movement in soils and adsorb certain heavy metals and phosphate. Sodium carbonate formation may result from physico-chemical and biological mechanisms. Understanding these processes is particularly important in studies of sodic soil reclamation.

Gypsum is one of the most commonly identified sulfate minerals in soils. This is due to its relatively low solubility compared to Na_2SO_4 or $MgSO_4$. However, in areas of high salinity, efflorescence on the soil surface of Na_2SO_4, $MgSO_4$, and $NaCl$ minerals can be observed. Even though these crusts of salts are the result of evaporation, once formed they may restrict further losses of soil water.

Identification of many of the carbonate, sulfate, and halide minerals can be accomplished by visual observation of morphology and crystalline form. More refined instrumental techniques for their identification are discussed as well.

In Part II the minerals associated with acid sulfate soils are discussed. Sulfide minerals, such as pyrite, are formed under reducing conditions in the presence of organic matter. The activity of sulfate-reducing bacteria is extremely important in the distribution of sulfide minerals in the soil fabric. As a result of drainage, oxidation of sulfide minerals causes the formation of acid sulfate soils. Jarosite is an iron-bearing sulfate commonly associated with these soils.

Other iron minerals besides pyrite are associated with sulfidic sediments. Some distinctions can be made among these minerals by visual and simple chemical tests. As with other minerals, X-ray diffraction provides the most definitive tool for their identification. Jarosite and natrojarosite may be found as yellow efflorescence around root channels, and on structural surfaces.

The environments which produce acid sulfate soils may also affect the mineralogical characteristics of other minerals. A chlorite may be formed during the reduction phase. The strong acid conditions in soils from the oxidation of sulfide have a solubilization effect on many minerals.

IV. PROBLEMS

1) Derive an equation relating the activity of Ca to pH and CO_2 partial pressure in a solution in equilibrium with aragonite at 25C. Assume the solubility product constant for aragonite to be 5.62×10^{-9}.

2) Assume a water containing 2×10^{-5} moles/liter Mg^{2+} in equilibrium with calcite in a soil at pH 7.6 and $P_{CO_2} = 0.01$ bar. This solution is subsequently transported to a lower soil horizon where the $P_{CO_2} = 0.001$ bar and the pH is 8.4. What is the average mole percent $MgCO_3$ in the calcite after equilibration at the new P_{CO_2} and pH? Assume activity and concentration are equivalent.

[answer 0.4% $MgCO_3$]

LITERATURE CITED

Arkley, R. J. 1963. Calculation of carbonate and water movement in soil from climatic data. Soil Sci. 96:239-248.

Benzler, J. H. 1973. Probleme bei der kartierung von "maibolt" und "pulbererde"—besonderen formen der sulfatsauren boden—in den Marschgebieten Niedersachsen (BRD). p. 211-214. In H. Dost (ed.) Acid sulfate soils. ILRI Publ. 18, Vol. II. Int. Inst. for Land Reclam. and Improv., Wageningen.

Bragg, W. L., G. F. Claringbull, and W. H. Taylor. 1965. Crystal structure of minerals. The crystalline state, Vol. IV. Cornell University Press, Ithaca, New York. 409 p.

Braitsch, O. 1962. Entstechung und staffbestand der salzlagerstatten. Mineralogie und petrographie in einzeldurstellungten. III. Springer, Berlin.

Brydon, J. E., and H. Heystek. 1958. A mineralogical and chemical study of the dikeland soils of Nova Scotia. Can. J. Soil Sci. 38:171-186.

Bundy, L. G., and J. M. Bremner. 1972. A simple titrimetric method for determination of inorganic carbon in soils. Soil Sci. Soc. Am. Proc. 36:273-275.

Cole, C. V., S. R. Olsen, and C. O. Scott. 1953. The nature of phosphorus sorption by calcium carbonate. Soil Sci. Soc. Am. Proc. 17:352-356.

Doner, H. E., and P. F. Pratt. 1968. Solubility of calcium carbonate precipitated in montmorillonite suspension. Soil Sci. Am. Proc. 32:743-744.

Doner, H. E., and P. F. Pratt. 1969. Solubility of calcium carbonate precipitated in aqueous solutions of magnesium and sulfate salts. Soil Sci. Soc. Am. Proc. 33:690-693.

Dost, H. (ed.). 1973. Acid sulfate soils. Proceedings of the Int. Symp., 13-20 Aug. 1972, Wageningen, The Netherlands. Vol. I and II. Publ. 18, Int. Inst. for Land Reclam. and Improv., Box 45, Wageningen, The Netherlands.

Driessen, P. M., and R. Schoorl. 1973. Mineralogy and morphology of salt efflorescences on saline soils in the Great Konya Basin, Turkey. J. Soil Sci. 24:436-442.

Friedman, A. M. 1959. Identification of carbonate minerals by staining methods. J. Sediment. Petrol. 29:87-97.

Frink, C. R. 1973. Aluminum chemistry in acid sulfate soils. p. 131-168. In H. Dost (ed.) Acid sulphate soils. ILRI Publ. 18, Vol. I. Int. Inst. for Land Reclam. and Improv., Wageningen.

Gibson, G. W. 1962. Geological investigations in southenr Victoria Land, Antarctic. 8. Evaporite salts in the Victoria Valley region. New Z. J. Geol. Geophys. 5:361-374.

Glover, E. D., and R. L. Sippel. 1967. Synthesis of magnesium calcites. Geochim. Cosmochim. Acta 31:603-613.

Goldsmith, J. R., D. L. Graf, and O. I. Joensuu. 1955. The occurrence of magnesian calcite in nature. Geochim. Cosmochim. Acta 7:212-230.

Graf, D. L., A. J. Eardley, and N. F. Shimp. 1961. A preliminary report on magnesium carbonate formation in glacial Lake Bonneville. J. Geol. 59:219-223.

Griffin, R. A., and J. J. Jurinak. 1973. The interaction of phosphate with calcite. Soil Sci. Soc. Am. Proc. 37:847-850.

Harned, H. S., and R. Davis, Jr. 1943. The ionization constant of carbonic acid in water and the solubility of carbon dioxide in water and aqueous salt solutions from 0 to 50°. J. Am. Chem. Soc. 65:2030-2037.

Hesse, P. R. 1972. A textbook of soil chemical analysis. Chemical Publishing Co., Inc., New York. 520 p.

Hilgard, E. W. 1912. Soils. The MacMillan Co., London.

Hodgman, Charles D. (ed.-in-chief). 1960. Handbook of chemistry and physics. 41st ed. Chemical Rubber Publishing Co., Cleveland, Ohio.

Holmgren, G. S. 1973. Quantitative calcium carbonate equivalent in the field. Soil Sci. Soc. Am. Proc. 37:304-307.

Horn, M. E., and S. L. Chapman. 1968. Clay mineralogy of some acid sulphate soils on the Guinea coast. Int. Congr. Soil Sci., Trans. 9th (Adelaide, Aust.) III:31-40.

Hsu, K. Jinghwa. 1967. Chemistry of dolomite formation. p. 169-191. In George V. Chilingar, Harold J. Bissell, and Rhodes W. Fairbridge (eds.) Carbonate rocks: Physical and chemical aspects. Elsevier Publ. Co., New York.

Huang, C. K., and P. F. Kerr. 1960. Infrared study of carbonate minerals. Am. Mineral. 45:311-324.

Janitzky, P., and L. D. Whittig. 1964. Mechanisms of formation of Na_2CO_3 in soils. II. Laboratory study of biogenesis. J. Soil Sci. 15:145-157.

Jenny, H. 1941. Factors of soil formation: A system of quantitative pedology. McGraw-Hill Book Co., Inc. 281 p.

Johanneson, J. K., and G. W. Gibson. 1962. Nitrate and iodate in antarctic salt deposits. Nature 194:567-568.

Jurinak, J. J., and Norman Bauer. 1956. Thermodynamics of zinc adsorption on calcite, dolomite and magnesite-type minerals. Soil Sci. Soc. Am. Proc. 20:466-471.

Jurinak, J. J., and R. A. Griffin. 1972. Nitrate ion adsorption by calcium carbonate. Soil Sci. 113:130-135.

Kawalec, A. 1973. World distribution of acid sulphate soils. References and Map. p. 292-295. In H. Dost (ed.) Acid sulfate soils. ILRI Publ. 18, Vol. I. Int. Inst. for Land Reclam. and Improv., Wageningen.

Kelley, W. P. 1951. Alkali soils. Reinhold Publ. Co., New York.

Kern, D. M. 1960. The hydration of carbon dioxide. J. Chem. Ed. 37:14-23.

Khan, S. U., and G. R. Webster. 1968. Determination of gypsum in solonetz soils by an X-ray technique. Analyst (London) 93:400-402.

Kovda, V. A. 1946. Origin of saline soils and their regime. Vol. 1. Academy of Sciences of the U.S.S.R. Dokuchaev Soil Sci. Inst. L. I. Prasolov (ed.). Translated from Russian by Israel Program for Scientific Translations, Jerusalem, 1971. 509 p.

Kovda, V. A. 1965. Alkaline soda-saline soils. p. 15-48. In I. Szabolcs (ed.) Proceedings of the Symposium on Sodic Soils. Budapest, 1964.

Kozhevnikov, K. Ya. 1974. Factors responsible for sodium-carbonate formation in soils and parent materials. Sov. Soil Sci. 6:180-190.

Lahav, N., and G. H. Bolt. 1963. Interaction between calcium carbonate and bentonite suspensions. Nature (London) 200:1343-1344.

Langmuir, D. 1964. Thermodynamic properties of phases in the system $CaO-MgO-CO_2-H_2O$. Geol. Soc. Am. Spec. Papers. 82:120.

Langmuir, D. 1965. Stability of carbonates in the system $MgO-CO_2-H_2O$. J. Geol. 73: 730-754.

Langmuir, D. 1968. Stability of calcite based on aqueous solubility measurements. Geochim. Cosmochim. Acta 32:835-851.

Lippman, F. 1973. Sedimentary carbonate minerals. Springer-Verlag, New York. 229 p.

Lynn, W. C., and L. D. Whittig. 1966. Alteration and formation of clay minerals during cat clay development. Clays Clay Miner. 14:241–248.

Milton, C., and H. P. Engster. 1959. Mineral assemblages of the Green River formation. p. 118–150. In P. H. Abelson (ed.) Researches in geochemistry. John Wiley & Sons, New York.

Nakayama, F. S. 1971. Calcium complexing and the enhanced solubility of gypsum in concentrated sodium-salt solutions. Soil Sci. Am. Proc. 35:881–883.

Olsen, S. R., and F. S. Watanabe. 1959. Solubility of calcium carbonate in calcareous soils. Soil Sci. 88:123–129.

Petersen, G. W., G. Chesters, and G. B. Lee. 1966. Quantitative determination of calcite and dolomite in soils. J. Soil Sci. 17:328–338.

Poelman, J. N. B. 1973. Soil material rich in pyrite in non-coastal areas. p. 197–207. In H. Dost (ed.) Acid sulphate soils. ILRI Publ. 18, Vol. II. Int. Inst. for Land Reclam. and Improv., Wageningen.

Ricard, D. T. 1973. Sedimentary iron sulfide formation. p. 28–65. In H. Dost (ed.) Acid sulfate soils. ILRI Publ. 18, Vol. I. Int. Inst. for Land Reclam. and Improv., Wageningen.

St. Arnaud, R. J., and A. J. Herbillon. 1973. Occurrence and genesis of secondary magnesium-bearing calcites in soils. Geoderma 9:279–298.

Sherman, G. D., F. Schultz, and F. J. Alway. 1962. Dolomite in soils of the Red River Valley, Minnesota. Soil Sci. 94:304–313.

Simon-Sylvestre, G. 1960. Compounds of sulfur in soil, their evolution, relationship with microflora, and utilization by plants. Ann. Agron. (Paris) 11:309–330.

Stewart, F. H. 1963. Marine evaporites, Chapter Y in data of geochemistry. 6th ed., U. S. Geol. Survey Brief Paper 440-Y.

Stuart, D. M., and R. M. Dixon. 1973. Water movement and caliche formation in layered arid and semiarid soils. Soil Sci. Soc. Am. Proc. 37:323–324.

Szabolcs, I. 1971. Solonetz soils in Europe: Their formation and properties with particular regard to utilization. p. 9–34. In I. Szabolcs (ed.) European solonetz soils and their reclamation. Akademiai Kiado Budapest.

Tedrow, J. C. F., and F. C. Ugolini. 1966. Antarctic soils. p. 161–177. In J. C. F. Tedrow (ed.) Antarctic soils and soil forming processes. Am. Geophy. Union NAS-NRC Publ. 1418.

Turner, R. C., and S. I. M. Skinner. 1960. An investigation of the intercept method for determining the proportion of dolomite and calcite in mixtures of the two. Can. J. Soil Sci. 40:232–241.

van Breeman, N. 1973. Soil forming processes in acid sulphate soils. p. 66–130. In H. Dost (ed.) Acid sulphate soils. ILRI Publ. 18, Vol. I. Int. Inst. for Land Reclam. and Improv., Wageningen.

van Breeman, N., and K. Harmsen. 1975. Translocation of iron in acid sulfate soils: I. Soil morphology, and the chemistry and mineralogy of iron in a chronosequence of acid sulfate soils. Soil Sci. Soc. Am. Proc. 39:1140–1148.

van Dam, D., and L. J. Pons. 1973. Micromorphological observations on pyrite and its pedological reaction products. p. 169–196. In H. Dost (ed.) Acid sulphate soils. ILRI Publ. 18, Vol. II. Int. Inst. for Land Reclam. and Improv., Wageningen.

Wagman, D. D., W. H. Evans, V. B. Parker, I. Halow, S. M. Bailey, and R. H. Schurn. 1968. Selected values of chemical thermodynamic properties. Nat. Bur. Stand. Tech. Notes. p. 270–273.

Whittig, L. D., and P. Janitzky. 1963. Mechanisms of formation of sodium carbonate in soils: I. Manifestation of biological conversions. J. Soil Sci. 14:322–333.

Winland, H. D. 1969. Stability of calcium carbonate polymorphs in warm, shallow sea water. J. Sediment Petrol. 39:1579–1587.

Wray, J. L., and D. Farrington. 1957. Precipitation of calcite and aragonite. J. Am. Chem. Soc. 79:2031–2034.

Aluminum Hydroxides and Oxyhydroxides

PA HO HSU, Rutgers University, New Brunswick, New Jersey

I. NOMENCLATURE AND STRUCTURAL PROPERTIES

A number of crystalline aluminum hydroxides, oxyhydroxides, and oxides are found in nature or prepared in the laboratory. Gibbsite, one of the aluminum hydroxide polymorphs, is common in soils and bauxitic deposits. Aluminum oxyhydroxides are common in bauxitic deposits, particularly in Europe, but their existence in soils is less common. Anhydrous aluminum oxides are high temperature products and are present only as minor components in some igneous and metamorphic rocks. Their presence in soils, if any, must be very limited and can only be inherited from parent rocks. The present chapter will deal mainly with aluminum hydroxides, and to a lesser extent, with oxyhydroxides.

Compounds of $Al(OH)_3$ and $AlOOH$ compositions have frequently been referred to as hydrous oxides or oxide hydrates in the literature, particularly in earlier ones. This system of nomenclature implies that free water molecules are present in the structure. Evidence from X-ray diffraction analysis (Megaw, 1934), proton magnetic resonance (Glemser, 1959) and infrared absorption (Elderfield & Hem, 1973), however, has shown that "water" is present as OH^- radicals, but not as free water. Therefore, it has been recommended by the International Committee on Aluminum Hydroxides Nomenclature that $Al(OH)_3$ and $AlOOH$ be referred to as aluminum trihydroxides and aluminum oxide-hydroxides, respectively (Newsome et al., 1960). In recent literature, they are more commonly referred to as aluminum hydroxides and aluminum oxyhydroxides, respectively, which will be adopted in this chapter. It should be noted that alumina trihydrate and monohydrate, when they are used for aluminum hydroxide and oxyhydroxide, respectively, have frequently been denoted with Greek prefixes, but two different systems have been shown in the literature (Newsome et al., 1960):

Mineral	Alcoa	Weiser and Milligan
Gibbsite	α – alumina trihydrate	γ – alumina trihydrate
Bayerite	β – alumina trihydrate	α – alumina trihydrate
Boehmite	α – alumina monohydrate	γ – alumina monohydrate
Diaspore	β – alumina monohydrate	α – alumina monohydrate

In addition, gibbsite has also been referred to as hydrargillite in Europe.

Crystalline aluminum hydroxides exist in three polymorphs: gibbsite, bayerite, and nordstrandite. They are composed of the same fundamental units: two planes of close-packed OH^- ions with Al^{3+} sandwiched between them (Fig. 4-1). The Al^{3+} ions reside in 2/3 of the octahedral holes and

99

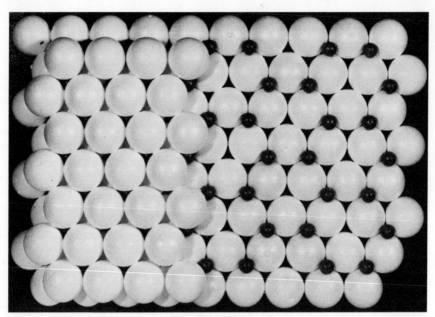

Fig. 4-1. Model illustrating the unit layer of aluminum hydroxide structure. OH^- ions are closely packed, with Al^{3+} ions sandwiched between them, occupying two thirds of the octahedral positions. Large white ball: OH^-ions; small black ball: Al^{3+} ion.

they are distributed in hexagonal rings. In the interior, each Al^{3+} shares six OH^- with three other Al^{3+}, and each OH^- is bridged between two Al^{3+} (Fig. 4-2). At the edge, however, each Al^{3+} shares only four OH^- with two other Al^{3+}, and the other two coordination sites are occupied by one OH^- ion and one H_2O molecule; neither is bridged between Al^{3+} ions. The entire unit may be conceived to be a giant molecule, but not discrete $Al(OH)_3$ molecules. The three polymorphs differ in their stacking of such units. In gibbsite (Megaw, 1934) the OH^- ions in one unit reside directly on the top of the OH^- ions of another such unit (Fig. 4-3a) because of the H-bond (Bernal & Megaw, 1935). In bayerite (Montoro, 1942), the OH^- ions in one unit reside in the depression of another such unit, i.e., they are closely packed (Fig. 4-3b), similar to the structure of brucite except that in the latter all octahedral holes are occupied by Mg^{2+} ions. In nordstrandite (van Nordstrand et al., 1956), gibbsite and bayerite arrangements alternate (Fig. 4-3c).

These three polymorphs are also different in their growth habits (Schoen & Roberson, 1970). Gibbsite crystallites of large size and high crystallinity have frequently been found in nature or prepared in the laboratory. They are usually well developed in the X and Y directions, but limited in the Z direction, and therefore appear as hexagonal plates (Fig. 4-4a and 4-5a). In contrast, bayerite crystallites usually take the form of a pyramid, with their long direction perpendicular to the basal plane. They often appear as prisms or rods under electron-microscopic observation (Fig. 4-4b and 4-5b). Nordstrandite crystallites often take the form of long rectangular prisms (Fig.

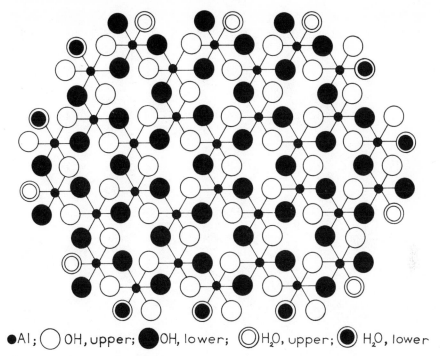

$\bullet$ Al; $\bigcirc$ OH, upper; $\bullet$ OH, lower; $\circledcirc$ H$_2$O, upper; $\circledbullet$ H$_2$O, lower

Fig. 4-2. Schematic representation of the unit layer of aluminum hydroxide. Edge, un-shared hydroxyls may protonate as a function of pH to form H$_2$O.

4-4c and 4-5c). The cross section of nordstrandite is hexagonal, and, as in bayerite, the Z axis parallels the long direction of the crystallite. The crystal thickness in the X direction is found to be between one-half and one-sixth of the thickness in the Y axis direction.

Crystalline aluminum oxyhydroxides are present in two polymorphs: diaspore and boehmite. The structure of diaspore has been determined using natural single crystals. The structure of boehmite is mainly derived from the corresponding Fe^{3+} compound lepidocrocite since no crystal of boehmite of sufficient size for single crystal analysis has been available. Van Oosterhout (1960) proposed a simple method to describe the structures of lepidocrocite and goethite. Lippens and Steggerda (1970) used the same model to describe the structures of boehmite and diaspore. In the direction of the X axis, there are the HO–Al–O chains (Figs. 4-6a and 4-6b). Two OH–Al–O chains can be placed in an antiparallel position (parallel to each other, but point in opposite directions) in such a way that the O atoms of the second chain are at the same level as the Al atoms of the first chain (Fig. 4-6c). Such an arrange-ment is presented schematically in Fig. 4-6d). Diaspore and boehmite are different in their arrangements of the double chains, as shown in Fig. 4-7. In both cases, the X axis is perpendicular to the plane of the drawing.

In addition, there is a poorly crystallized compound which has frequent-ly been referred to as pseudoboehmite or gelatinous boehmite. This com-

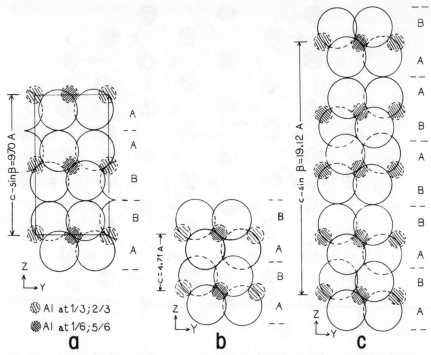

Fig. 4-3. Diagramatic representation showing the different stacking of three aluminum hydroxide polymorphs (modified from Schoen & Roberson, 1970): *(a)* gibbsite, *(b)* bayerite, and *(c)* nordstrandite.

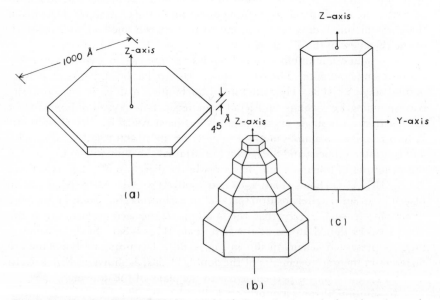

Fig. 4-4. Sketch of hypothetical crystals of three aluminum hydroxide polymorphs (modified from Schoen & Roberson, 1970): *(a)* gibbsite, *(b)* bayerite, and *(c)* nordstrandite.

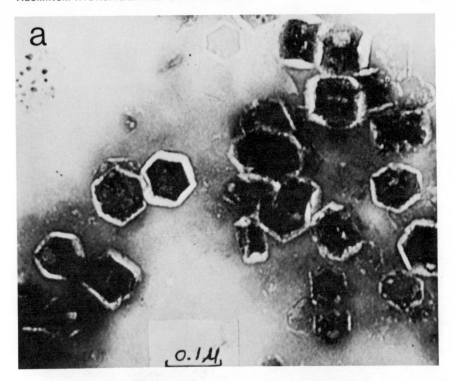

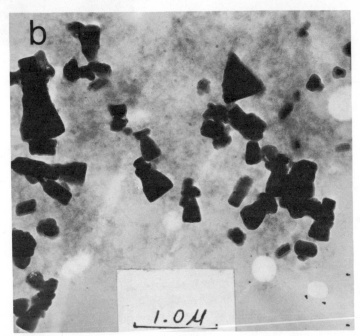

Fig. 4-5. Electron micrographs of aluminum hydroxide polymorphs (reproduced from Schoen & Roberson, 1970): *(a)* gibbsite, *(b)* bayerite, and *(c)* nordstrandite.

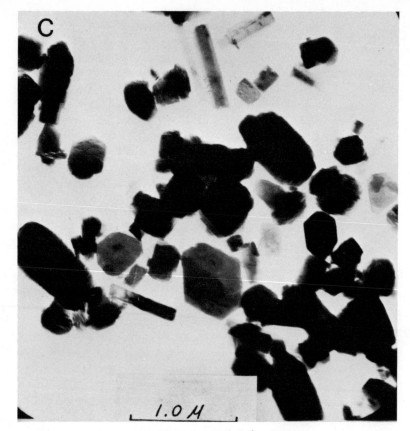

Fig. 4-5. Continued.

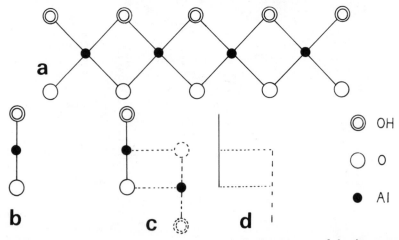

Fig. 4-6. Schematic representation of OH–Al–O chains in the structure of aluminum oxyhydroxide (after Lippens & Steggerda, 1970): *(a)* OH–Al–O chain parallel to the plane
of drawing; *(b)* OH–Al–O chain perpendicular to the plant of the drawing; *(c)* Two antiparallel OH–Al–O chains perpendicular to the plane of the drawing; and *(d)* Same as *c,*
but circles representing atoms omitted.

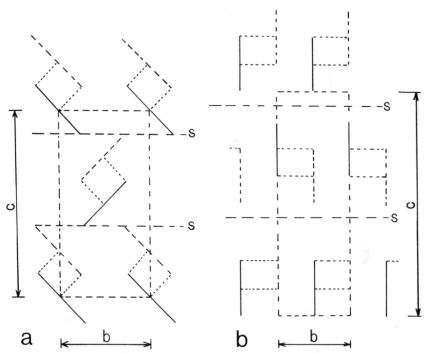

Fig. 4–7. Schematic representation of the crystal structure of aluminum oxyhydroxide polymorphs: *(a)* Diaspore; and *(b)* boehmite (after Lippens & Steggerda, 1970). *S* represents cleavage plane, and *b* and *c* are repeated distances along *Y* and *Z* axes, respectively.

pound yields an X-ray diffraction pattern very similar to boehmite, but its peaks are very diffuse and its strongest peak represents a spacing ranging from 6.4-6.9Å. This compound contains more water than boehmite and appears as gel-like material when observed under the electron microscope. It was suggested by Papée et al. (1958) that pseudoboehmite has the same atomic arrangement as boehmite at short distances, but a large excess of water was situated between the elemental units.

Although previous studies are largely confined to crystalline components, large amounts of noncrystalline aluminum hydroxides and oxyhydroxides are also present in nature. Such noncrystalline materials do not have definite composition or structure. They are only poorly defined, not clearly understood, and very loosely referred to in the literature. As will be discussed later in this chapter, noncrystalline and crystalline aluminum hydroxides are of essentially the same structure and chemical characteristics, differing only in particle size. Therefore, principles derived from the study of crystalline compounds should also be applicable to noncrystalline ones. The latter probably dominate the chemical reactions in soils because of their extremely small particle size. Although noncrystalline aluminum hydroxide in pure form is not stable for an extended period of time, many soil components can effectively retard or inhibit its crystallization.

Table 4-1. X-ray diffraction data for aluminum hydroxides and oxyhydroxides (compiled from Berry, 1974)

Gibbsite (Card no. 7-324)			Bayerite (Card no. 20-11)			Nordstrandite (Card no. 18-31)			Boehmite (Card no. 21-1307)			Diaspore (Card no. 5-355)		
dÅ	I/I_1	hkl	dÅ	I/I_1	hkl	dÅ	I/I_1	hkl	dÅ	I/I_1	hkl	dÅ	I/I_1	hkl
4.85	320	002	4.71	90	001	4.79	100	00$\bar2$	6.11	100	020	4.71	13	020
4.37	50	110	4.35	70	110, 020	4.33	25	110	3.164	65	120	3.99	100	110
4.32	25	200	320	30	111, 021	4.22	25	110	2.346	55	140, 031	3.214	10	120
3.306	16	112	2.699	4	121	4.16	20	200	1.980	6	131	2.558	30	130
3.187	12	11$\bar2$	2.464	2	031	3.896	15	20$\bar2$	1.860	30	051	2.434	3	021
3.112	8	10$\bar3$	2.356	4	002	3.604	10	20$\bar2$	1.850	25	200	2.386	5	101
2.454	25	021	2.222	100	201, 131	3.446	10	11$\bar2$	1.770	6	220	2.356	8	040
2.420	20	004	2.156	2	211	3.022	15	$\bar1\bar1$2	1.662	14	151	2.317	56	111
2.388	25	311	2.073	2	112, 022	2.867	5	112	1.527	6	080	2.131	52	121
2.285	6	31$\bar2$	1.983	4	221	2.481	15	$\bar3$12	1.453	16	231	2.077	49	140
2.244	10	022, 21$\bar3$	1.969	2	041	2.454	10	3$\bar1$0	1.434	10	002	1.901	3	131
2.168	8	312	1.917	2	122	2.393	35	004, 310	1.412	2	180	1.815	8	041
2.085	2	114	1.835	2	141	2.265	35	02$\bar2$	1.396	2	022	1.733	3	150
2.043	18	31$\bar3$	1.826	2	032	2.015	30	314	1.383	6	171	1.712	15	211
1.993	12	023	1.725	40	202, 132	1.904	20	312	1.369	2	260	1.678	3	141
1.960	2	12$\bar3$	1.695	2	2$\bar1$2	1.781	15	02$\bar4$	1.312	16	251	1.633	43	221
1.921	12	411	1.688	2	212	1.595	10	31$\bar6$	1.303	4	122	1.608	12	240
1.799	14	31$\bar4$	1.656	2	310	1.513	10	31$\bar4$	1.224	2	320	1.570	4	060
1.750	16	024	1.646	2	246	1.478	10	60$\bar2$	1.209	2	320	1.522	6	231
1.689	14	31$\bar4$	1.641	2	150	1.440	20	33$\bar0$, 33$\bar2$	1.178	4	280	1.480	20	160, 151
1.654	4	22$\bar4$	1.600	10	222, 042	1.431	5	224				1.430	7	250
1.638	2	42$\bar1$	1.572	8	320, 003	1.403	10	330				1.423	12	002
1.593	4	224, 551	1.554	8	311, 241	1.388	10	600				1.400	6	320
1.584	2	422, 51$\bar2$	1.492	2	321							1.376	16	061
1.573	4	230, 503												
1.555	2	404												
1.551	2	231												
1.486	2													
1.477	2													
1.457	10													
1.441	6													
1.409	6													

II. METHODS OF IDENTIFICATION AND DETERMINATION

Monomineralic aluminum hydroxide or oxyhydroxide of high crystallinity can be identified and determined with X-ray diffraction, thermal analysis, infrared absorption, or selective dissolution without much difficulty. The identification and determination of these components in soils, however, are more complicated because of the presence of many uncertain variables. A combination of several methods is frequently necessary.

A. X-ray Diffraction Analysis

General principles and techniques regarding the application of X-ray diffraction for the identification and estimation of minerals have been summarized by Zussman (1967). The specific discussion on their application to aluminum hydroxides and oxyhydroxides is referred to by Rooksby (1961).

Almost all recent X-ray diffraction examinations of aluminum hydroxides and oxyhydroxides have been made with a diffractometer, using Ni filtered Cu $K\alpha$ radiation. Random powder specimens are best prepared by placing the powder in a recess in a glass or plastic plate, compacting it under just sufficient pressure to cause cohesion without use of a binder, and smoothing off the surface. For routine identification in which the relative intensities of diffraction peaks are not of major concern, the specimen can be conveniently prepared by dispersing the sample in water and drying on a glass slide. The specimen can also be prepared by collecting the particles with a Millipore filter. The filter containing the particles is pasted onto a glass slide. The latter is particularly useful for suspensions of tiny, charged particles which cannot be collected for X-ray diffraction otherwise.

The X-ray diffraction data for aluminum hydroxides and oxyhydroxides are compiled in Table 4-1. The data for nordstrandite in different reports are not consistent in relative peak intensities. The data (Card no. 18–31) of Saalfeld and Mehrotra (1966) appear to be in the middle of all data available and thus are arbitrarily chosen in compilation. Other data for nordstrandite are reported by Newsome et al. (1960); Davis & Hill (1974). The strongest peaks for gibbsite and nordstrandite are 4.85Å and 4.79Å, respectively. Bayerite yields a very strong peak at 4.71Å, but its strongest peak is at 2.22Å. For specimens of high crystallinity, these polymorphs can be distinguished without difficulty with the present X-ray diffractometer. Gibbsite yields a series of peaks at 3.31, 3.11, 2.42, 2.04, 1.80, and 1.75Å, although weak in intensities, which are not shown by other polymorphs. The very strong diffraction peak at 2.22Å is unique for bayerite. Nordstrandite is also characterized by a number of peaks at 4.22, 4.16, 3.90, 3,60, and 3.45Å. With specimens of poor crystallinity, however, these diffraction peaks may not be completely revealed nor their relative intensities reproduced, and therefore their identification must be examined with caution. For example, the 4.35Å peak is frequently missing for poorly crystallized gibbsite developed from aging OH–Al solutions (Hsu, 1966) whereas the 4.71Å peak for bayerite is sometimes missing (Schoen & Roberson, 1970). It was interpreted in the latter report that bayerite is usually poorly developed in its 001 plane, forming pyramid or cone-shaped crystals (Fig. 4–4b). In contrast, gibbsite is usually present as a hexagonal plate (Fig. 4–4a), and therefore frequently yields a strong 4.85Å 002 diffraction peak, but weak 4.35Å 110 diffraction peak.

The strongest peaks for boehmite and diaspore are 6.11Å and 3.99Å, respectively. They are well separated from each other and from the three polymorphs of aluminum hydroxide. The poorly crystallized and defined pseudoboehmite yields a broad peak with its diffraction maximum representing a spacing varying from 6.4Å to 6.9Å with a preferentially oriented specimen, but additional broad peaks corresponding to boehmite were also observed in a randomly packed specimen. The broad 6.4–6.9Å peak is sometimes of a disproportionately large area, but accounts for only a small portion of the aluminum present (Hsu & Bates, 1964a).

The diffraction intensity falls off rapidly away from the Bragg angle for a given d-value and the spread of the diffraction peak bears an inverse relationship to the crystal size. The broadening of a diffraction peak is appreciable if the particle crystallite size is of the order of 1,000Å or less. Schoen and Roberson (1970) estimated the size of $Al(OH)_3$ crystallites based on this principle. Nevertheless, other sources for line broadening should also be considered. The existence of a range of d-values for a given set of crystal planes also will result in line broadening. This may occur through the presence of mechanical strain or through the lack of chemical homogeneity in the specimen. Mineral crystals having mosaic structure also lead to a broadening of the diffraction peak (Zussman, 1967).

The amount of gibbsite in soils can be approximately estimated with X-ray diffraction. The X-ray diffraction intensity is governed by a number of factors such as crystal size, crystallinity, impurities, chemical composition, and the presence of noncrystalline components, other than concentration. The estimation of aluminum hydroxide or oxyhydroxides in soils, however, is much less difficult than with other clay minerals because most of the interfering factors listed above can be eliminated by using an internal standard.

The major drawback of the X-ray diffraction technique is its low sensitivity. Gibbsite will not be detected if its content is 5% or less (Jackson, 1969).

B. Thermal Analysis

There are two common types of thermal analysis: differential thermal (DTA) and thermogravimetric (TGA or DTGA). The former records the enthalpy change during heating and has been widely employed for the identification and estimation of aluminum hydroxide in soils (Bryant & Dixon, 1964; Fieldes, 1956; Glenn & Nash, 1964; Dixon, 1966).

The literature on the differential thermal analysis of aluminum hydroxides and oxyhydroxides, as well as the general principles and techniques, has been reviewed by MacKenzie (1970; 1972). The major advantage of this technique is its sensitivity. The presence of 1% of gibbsite in soils can be unmistakeably detected. The amount of gibbsite in soils can also be estimated quantitatively by comparing the 300–330C (573–603K) endothermic curve for unknowns with those for standards. The major drawback of this technique is its requirement of critical control of operation conditions, particularly the rate of heating and the atmospheric composition. When heated in static air at a rate of $10C \, min^{-1}$, both gibbsite and bayerite (Fig. 4–8) yield an endothermic peak in the range from 300–330C (573–603K) (Mackenzie, 1972). The peak temperature for bayerite is in general slightly lower than that for gibbsite. Such a small difference can be distinguishable if experimental conditions are critically controlled (Barnhisel & Rich, 1965). These endothermic peaks are due to the transformation of $Al(OH)_3$ to Al_2O_3. With large crystals, two other

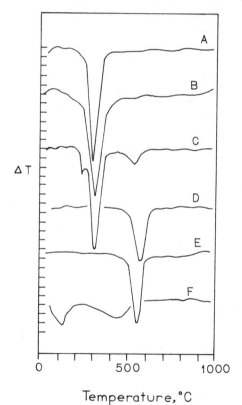

Temperature, °C

Fig. 4–8. DTA curves for aluminum hydroxides and oxyhydroxides [modified from MacKenzie, 1970; static air, heating rate 10C per min. Each division on the ΔT axis is equivalent to 2C; (A) bayerite, synthetic, 0.2 g; (B) gibbsite from Puente Arce, Santander, Spain, 0.35 g; (C) gibbsite, synthetic, coarse crystals, 0.26 g; (D) diaspore from Sverlovsk, USSR, 0.3 g; (E) boehmite from Swiss, Missouri, USA; and (F) boehmite, synthetic, fine particulate and possibly poorly crystalline, 0.2 g.

endothermic peaks, at 250 and 550C (523 and 823K), may also be observed (Fig. 4-8, curve C). The former is due to the transformation of $Al(OH)_3$ to AlOOH, and the latter to the decomposition of the AlOOH developed. For AlOOH originally present in the specimen the first endothermic peak at 250C (523K) is not observed. The DTA pattern for nordstrandite has not been reported, but is presumed to be similar to those of gibbsite and bayerite because of their similar structures. The 300–330C (573–603K) peak does not overlap with any major components in soils except FeOOH. The endothermic peak temperature for FeOOH varies from 190–490C (363–763K), depending on crystallinity and particle size. The presence of these two minerals in soils, however, can be distinguishable by carrying out three DTA curves (MacKenzie & Robertson, 1961; Mackenzie, 1972): (i) untreated specimen; (ii) specimen pretreated with sodium dithionite to remove FeOOH, leaving $Al(OH)_3$ intact; and (iii) specimen pretreated with NaOH to remove $Al(OH)_3$, leaving FeOOH intact. DTA curves for a lateritic soil from Guma, Sierra Leone, are shown in Fig. 4-9. The endothermic peak at 320C (593K) in curve A suggests the presence of $Al(OH)_3$ or FeOOH or both. The reduction of this peak to a small hump after the NaOH treatment (curve C) suggests the presence of $Al(OH)_3$. The small hump was attributed to the presence of FeOOH. The 320C (593K) peak in curve B appears not much different from that in A. Nevertheless, a considerable amount of material was dissolved during dithionite treatment. When the weight loss was allowed in calculation, the peak was found to be noticeably smaller in B than in A. Therefore, the authors concluded the presence of FeOOH. In curve C, a large amount of material was dissolved in the NaOH treatment. An appropriate amount of reference material was added to the sample to compensate for the weight loss, and therefore curves A and C can be compared directly.

The usefulness of DTA for identifying AlOOH in soils is limited. Dia-

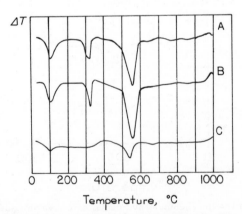

Fig. 4-9. DTA curves for a soil from Guma, Sierra Leone after different treatments (0.2 g specimen size; static air; heating rate 10C/min; after MacKenzie and Robertson, 1961): *(A)* untreated; *(B)* after treatment with sodium dithionite; and *(C)* after treatment with sodium hydroxide, thereafter diluted with reference material to compensate for the amount dissolved.

spore and boehmite yield endothermic peaks at 540C (813K) and 450-580C (723-853K), respectively. The great variation in the peak temperature for boehmite is due to the diversity of crystallinity and particle size (Fig. 4-8, curves E and F). The 550C (823K) peak overlaps with that of kaolinite and therefore creates the difficulty for definite identification. Although kaolinite can be distinguished by an exothermic peak at 1,000C (1,273K), the peak intensity is too weak for a clear-cut identification.

Differential thermal analysis also provides a sensitive technique for monitoring the crystallization of $Al(OH)_3$ in pure systems, but the identification of noncrystalline $Al(OH)_3$ in soils is difficult because: (i) The noncrystalline aluminum hydroxide has not been clearly defined. It is difficult to assess the reliability of the information in the literature and to standardize the procedure for identification; and (ii) so-called noncrystalline aluminum hydroxide yields endothermic peaks in the range from 100C (373K) to slightly above 200C (473K). This range of temperature overlaps with the endothermic peaks for smectite, vermiculite, halloysite, allophane, etc.

Thermogravimetric (TGA) or differential thermogravimetric analysis (DTGA) measure the weight change as a specimen is heated. The general principles and techniques have been summarized by Garn (1965). Aluminum hydroxides and oxyhydroxides lose water during heating and their thermal reactions are known. Therefore, TGA and DTGA are suitable for the identification and estimation of these components. An agreement between the results from TGA or DTGA and those from DTA can greatly assist interpretation. In their study of the formation of Al–Fe mixed hydroxide, Tullock and Roth (1975) reported that under their experimental conditions, gibbsite lost 82.3% of water in a span from 30C below to 30C above the peak maximum at 265C (538K). In contrast, noncrystalline aluminum hydroxide lost its water below 200C (473K). They then suggested that the amount of noncrystalline aluminum hydroxides can be estimated by subtracting gibbsite from the total amount of aluminum that is retained by the clay.

C. Infrared Absorption Analysis

Each of the polymorphs of aluminum hydroxides and oxyhydroxides shows a series of characteristic OH-stretching and bending bands in the infrared region. The OH-stretching absorption spectra have commonly been used for identification (Fig. 4-10 and 4-11).

Gibbsite shows three absorption bands and another doublet. With instruments of poor resolution, the last doublet may appear as a single band. The exact frequencies reported by different workers are slightly different from one another (Table 4-2). The observation of a complete set of characteristic peaks can be considered as evidence for the presence of well-crystallized gibbsite. Fieldes et al. (1972) suggested that the amount of gibbsite can be estimated by comparing the height of the 3490 cm^{-1} peak in an unknown

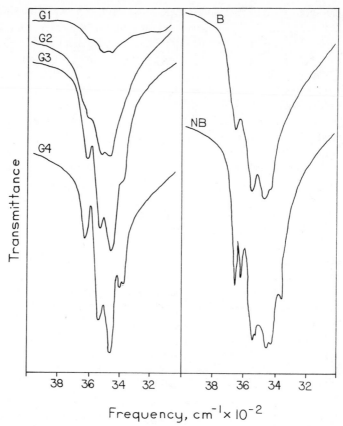

Fig. 4-10. Infrared absorption spectra of the OH-stretching region of aluminum hydroxide polymorphs (modified from Elderfield & Hem, 1973): *G*1 to *G*3: gibbsite of increasing crystallinity obtained from aging of OH–Al solutions (*G*1, 11 days; *G*2, 24 months; *G*3, 45 months) *G*4: natural well-crystallized gibbsite from Guyana; *B*. bayerite, synthetic; and *NB*: mixture of nordstrandite and bayerite, synthetic.

with a series of standards containing known gibbsite contents. Their results are in good agreement with those obtained with X-ray diffraction and differential thermal analyses. The following remarks, however, should be considered for identification and particularly quantitative estimation:

1) Elderfield and Hem (1973) showed that, with increasing crystallinity, each of the absorption frequencies shifted slightly (Table 4-3) and its resolution increased (Fig. 4-10). The characteristic absorption bands for the poorly crystallized specimen are not clearly distinguishable. The interpretation of absorption spectra therefore must be made with caution. On the other hand, such displacement of frequency with crystallinity may be considered to be a useful criterion for studying the process of aluminum hydroxide crystallization.

2) The 3622 cm^{-1} absorption band superimposes with kaolinite group

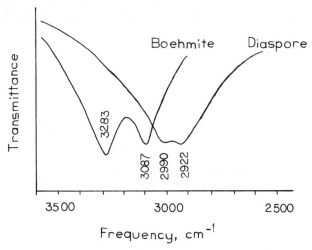

Fig. 4-11. Infrared absorption spectra of the OH–Al region of aluminum oxyhydroxide polymorphs (modified from Ryskin, 1974; recorded at 300K).

and micaceous clays. Farmer and Russell (1967) showed that the water in the interlayer space coordinated to cations on clay surfaces may yield a broad absorption band in the range from 3000 to 3630 cm^{-1}. If such interlayer water is present in sufficient amounts, the absorption band characteristic for gibbsite may not be observed even though gibbsite of high crystallinity is present.

3) The identification of bayerite and nordstrandite in soils or bauxitic deposits has not been reported, presumably because these two polymorphs has been found in nature only relatively recently and are usually present in a small amount. An exceptionally large amount of nordstrandite was found in Jamaica bauxitic deposits only recently (Davis & Hill, 1974). The absorption spectrum for nordstrandite is less than certain because pure specimens are difficult to obtain. The product of laboratory preparation of nordstrandite is usually a mixture with bayerite or gibbsite.

Boehmite shows two OH-stretching bands at 3087 and 3283 cm^{-1}; diaspore shows two OH-stretching bands at 2922 and 2990 cm^{-1} (Ryskin, 1974).

Table 4-2. Hydroxyl stretching frequencies of gibbsite reported by different investigators

Reference	OH-stretching frequencies, cm^{-1}			
Frederickson, 1955	3618	3518	3428	3361/3378
Kolesova & Ryskin, 1959	3617	3520	3428	3380
Fieldes et al., 1972	3625	3540	3490	N.R.†
Elderfield & Hem, 1973‡	3622	3527	3460	3384/3396

† An absorption band is shown in the original graph, but the exact frequency is not reported.
‡ The data for the well-crystallized gibbsite are compiled here.

Table 4-3. Hydroxyl stretching frequencies in cm^{-1} of gibbsite of varying crystallinity
(after Elderfield & Hem, 1973)†

$G1$	$G2$	$G3$	$G4$	$G5$	$G6$
			cm^{-1}		
3590	3600	3610	3622	3623	3623
3500	3510	3515	3529	3527	3527
3470	3560	3450	3460	3460	3460
--	--	3390	3396	3396	3396
--	--	--	3384	3385	3384

† Samples $G1$, $G2$, and $G3$ are synthetic precipitates; $G4$ is well crystallized natural gibbsite from Guyana; $G5$ is also well crystallized gibbsite from Minas Gerais, Brazil. $G6$ is synthetic crystals produced by the Bayer process.

D. Selective Dissolution

A number of procedures have been reported for selective dissolution of certain soil components (Mitchell et al., 1974; Jackson, 1969). This technique is based on the differences in the rate of dissolution or solubility. Soil components are complex and many of them are only slightly different from one another with respect to their solubility or rate of dissolution in a given extractant. Furthermore, within each mineral group, different species may differ in their rates of dissolution, depending on particle size, crystallinity, and impurities. Therefore, it is difficult, if not impossible, to design a procedure for specifically dissolving any particular component without a destructive effect on others. Nevertheless, this method can yield valuable information if all experimental conditions are carefully controlled and all limitations carefully considered, particularly for studying the chemical reaction in simple laboratory systems.

Hashimoto and Jackson (1960) reported that, by boiling a clay specimen in 0.5N NaOH for 2.5 min, gibbsite and all noncrystalline aluminum silicates, free alumina, and free silica can be dissolved. In samples which are known not to contain such noncrystalline material, the data from such dissolution can be considered to be the maximum level of gibbsite present. On the other hand, if the amount of gibbsite can be determined with other techniques, such as DTA, a substraction of gibbsite from the total amount of aluminum dissolved may facilitate the interpretation of the noncrystalline components dissolved.

In his studies of aluminum hydroxide crystallization, this writer found that all noncrystalline aluminum hydroxides and basic aluminum salts in his studies, such as sulfate, phosphate, and silicate, can be rapidly dissolved in 1N HCl in less than 20 min at room temperature. Free silica, however, is not dissolved. A combination of this 1N HCl-20 min extraction with the boiling 0.5N NaOH-2.5 min treatment may provide better information regarding the nature and amount of noncrystalline components in soils. It should be noted that all the noncrystalline aluminum hydroxides studied by this writer were aged in the presence of mother liquid or water. A repeated cycling of wet and dry conditions that is likely to occur in nature may produce some noncrystalline components that are more resistant to acid. This possibility remains to be tested.

III. MECHANISM OF ALUMINUM HYDROXIDE FORMATION

A number of processes for preparing crystalline $Al(OH)_3$ have been reported (Lippens & Steggerda, 1970). Gibbsite is commonly produced on an industrial scale by the Bayer process, in which a concentrated sodium aluminate solution is cooled slowly in the presence of crystal seeds. Large crystals of high crystallinity can be produced with this process. Aluminum hydroxides can also be prepared in the laboratory by bubbling CO_2 through a concentrated sodium aluminate solution. The reaction product is gibbsite if the system is kept above pH 12, and bayerite if between pH 9 and 12. Bayerite and nordstrandite have frequently been prepared by hydrolysis of aluminum alcoholates or amalgamated aluminum foil, or by neutralization of an aluminum salt solution with a base. Most of these preparations were carried out in alkaline medium under conditions which show little resemblance to the ordinary weathering process in soils. In contrast, gibbsite, the most common polymorph of aluminum hydroxides in nature, is known to occur only in highly weathered, acidic soils. Several questions then arise: Why do aluminum hydroxides crystallize most easily in alkaline medium in laboratory preparations? How does gibbsite develop in acidic soils? Why does gibbsite not form in alkaline or neutral soils?

The initial step of weathering probably is the hydrolytic breaking of the Si-O-Al linkage in the primary minerals, releasing Al^{3+} to the solution. Among those reported processes of $Al(OH)_3$ crystallization, the neutralization of Al^{3+} with a base is probably most similar to that encountered in nature. Because of this, earth scientists have endeavored to investigate various aspects of the Al-OH interactions. It is now known that by adding NaOH to an aluminum chloride or nitrate solution, crystalline aluminum hydroxides can be developed at NaOH/Al mole ratio = 3 in a matter of hours, but the initial reaction products are clear solution at NaOH/Al mole ratio = 2.4 and below. Clear solutions can be obtained up to NaOH/Al = 2.7 if the NaOH is added very slowly[1] (Hsu & Bates, 1964a). Most investigators have now recognized that the hydrolyzed species in these partially neutralized solutions are positively charged polynuclear OH-Al complexes, but the proposed composition have been controversial. Recent studies revealed that these OH-Al polymers are metasable and eventually convert to crystalline aluminum hydroxide after prolonged aging, accompanied by a decrease in solution pH

---

[1]If the rate of NaOH addition is not slow enough, the product may appear turbid initially but turn clear after standing. The initial turbidity may be attributed to the substitution of Cl^- for OH^- in the polymeric structure.

(Hsu, 1966; Hem & Roberson, 1967; Schoen & Roberson, 1970; Ross & Turner, 1971; Smith, 1971; Smith & Hem, 1972). The results available suffice to describe an approximate outline of the process of aluminum hydroxide formation in soils although much detailed chemistry remains obscure.

A. Structure of OH–Al Polymers

Although the existence of OH–Al polymers has long been recognized, surprisingly little attention has been paid to their structures, their mechanism of formation, and their relation with the crystallization of aluminum hydroxide. It was first proposed by Brosset et al. (1954) and later followed by this writer (Hsu & Rich, 1960; Hsu & Bates, 1964a) and others (Hem & Roberson, 1967; Schoen & Roberson, 1970; Smith & Hem, 1972) that the OH–Al polymers are in fact fragments of crystalline aluminum hydroxide. Crystalline aluminum hydroxide is a stable and slightly soluble compound in which Al^{3+} ions are distributed in hexagonal rings, connected with OH-bridges (Fig. 4-2). A necessary requirement for a stable structure is that the repulsion between Al^{3+} ions should be approximately balanced by the attraction of Al-OH-Al linkages. This principle should also hold for the OH–Al polymers in solution. That is, if polymerization takes place, the arrangement of Al^{3+} and OH^- ions in the product should follow a pattern similar to $Al(OH)_3$ solids. In the $Al(OH)_3$ structure, however, each OH^- at the edge is linked to only one Al^{3+}, and is therefore in rapid equilibrium with the H^+ in solution. The isoelectric point of aluminum hydroxide is approximately pH 9.2 (Parks, 1965). Therefore, aluminum hydroxide particles large or small, are always positively charged in acidic medium, but the average net positive charge per aluminum atom decreases with increases in particle size. When the particles are extremely small with a high net positive charge per aluminum atom, they behave as true solutes and are defined as soluble polynuclear complexes. Nevertheless, the changes in particle size and the average net positive charge per Al atom are gradual. Some particles are small enough to maintain a stable suspension that cannot be separated from solution by centrifugation at 1,000 g for 30 min, but yielded well-defined X-ray diffraction peaks characteristic for aluminum hydroxide when collected with a Millipore filter (Hsu, 1973).

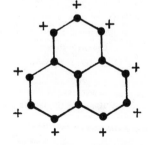

Fig. 4-12. Schematic representation of a positively charged OH–Al polymer of composition $[Al_{13}(OH)_{30}]^{9+} \cdot 18H_2O$ (modified from Hsu & Bates, 1964a). Two OH are shared between adjacent Al (black dots). Each edge Al is coordinated by 4 OH^- and 2 H_2O.

Well-crystallized gibbsite particles are also positively charged in an acidic medium. The fundamental difficulty is the lack of clear-cut criteria, in principle as well as in practice, to distinguish true solute from colloidal particles, or colloidal particles from precipitates. The schematic illustration of a positively charged OH–Al polymer of composition $[Al_{13}(OH)_{30}]^{9+}$ is shown in Fig. 4-12. For the sake of simplicity, the coordination of H_2O molecules are not shown in the composition.

If we accept this reasoning, many of the proposed compositions such as

$$[Al_2(OH)_2]^{4+},\ [Al_3(OH)_7]^{2+},\ [Al_4(OH)_{10}]^{2+},\ \text{and}\ [Al_8(OH)_{20}]^{4+}$$

do not seem to be likely. In a structure such as

$$
\begin{array}{ccc}
 & OH & \\
\diagup & & \diagdown \\
Al & & Al \\
\diagdown & & \diagup \\
 & OH & \\
\end{array}
$$

the repulsion between Al^{3+} ions may be too strong to be stable. Being aware of this possibility, Grunwald and Fong (1969) suggested the following structure:

$$
\begin{array}{ccc}
 & OH-H_2O & \\
\diagup & & \diagdown \\
Al & & Al \\
\diagdown & & \diagup \\
 & H_2O-OH & \\
\end{array}
$$

Nevertheless, the Al–OH–H_2O–Al linkages should not be very strong and therefore the proposed dimeric $[Al_2(OH)_2]^{4+}$ should not be the dominant species in the system, if it does exist.

Polynuclear OH–Al complexes containing 13 Al^{3+} ions have been suggested to be $[Al_{13}(OH)_{32}]^{7+}$ by Aveston (1965) using ultracentrifugation and potentiometric titration, and to be $[Al_{13}O_4(OH)_{24}(H_2O)_{12}]^{7+}$ by Johansson (1960; 1962) based on the structure of crystalline basic aluminum sulfate prepared thereof. It should be noted that Johansson prepared his OH–Al solutions at 80–90C (353–363K) and aged them in the presence of a large excess of sulfate. This writer found that the basic aluminum sulfate prepared from the reaction of sulfate with the room temperature OH–Al solutions can be of different compositions, depending on experimental conditions. Most of them were amorphous to X-ray diffraction. Those that yielded X-ray diffraction peaks were too small for single crystal analysis. He believes that the 13-Al species should be more logically depicted as a triple ring of composition $[Al_{13}(OH)_{30}]^{9+}$ or $[Al_{13}(OH)_{32}]^{7+}$ as suggested by Aveston (1965), if two nonbridged OH^- ions are present in each polynuclear unit.

B. Hydrolysis and Polymerization

Although many investigators have recognized the existence of polynuclear OH-Al complexes in partially neutralized solutions, some others reported that their results could be satisfactorily interpreted by assuming only monomeric species to be the hydrolysis products (Frink & Peech, 1963; Schofield & Taylor, 1954; Holmes et al., 1968). These two different interpretations are not really conflicting, but rather complementary to each other (Frink, 1972; Patterson & Tyree, 1973). A scrutiny of the literature indicates that those who suggested the monomeric hydrolysis species invariably prepared their solutions by simply dissolving an aluminum salt in water. Those advocates of polynuclear complexes, except Grunwald and Fong (1969), prepared their solutions involving the addition of base for pH adjustment.

When an aluminum salt is dissolved in water, the initial hydrolysis reaction must yield a series of monomeric species—$Al(OH)^{2+}$, $Al(OH)_2^+$ and $Al(OH)_3$—and there must be a rapid equilibrium among them:

$$Al^{3+} + H_2O = Al(OH)^{2+} + H^+ \qquad [1]$$

$$Al(OH)^{2+} + H_2O = Al(OH)_2^+ + H^+ \qquad [2]$$

$$Al(OH)_2^+ + H_2O = Al(OH)_3 + H^+ \qquad [3]$$

The hydrolysis constant for reaction [1] has been calculated by a number of workers with excellent agreement, all in the neighborhood of 1×10^{-5}, even though the existence of $Al(OH)_2^+$ and $Al(OH)_3$ were not considered. Ignoring $Al(OH)_2^+$ and $Al(OH)_3$ is practically acceptable when their activities are too small to affect the constancy of K_1, but this assumption is not always valid. Frink and Peech (1963) showed that the degree of hydrolysis increased with increased dilution. In solutions $0.0005M$ in Al or lower, the neglect of $Al(OH)_2^+$ resulted in a deviation in K_1. More important, their data also showed that the $(Al^{3+})(OH^-)^3$ ionic products of their solutions varied from $10^{-32.22}$ to $10^{-33.57}$. Therefore, all of the solutions they studied were supersaturated with respect to gibbsite which has a solubility product of $10^{-34.05}$ (Kittrick, 1966; Singh, 1974). In other words, the activity of the monomeric $Al(OH)_3$ molecule, although small, already exceeds that permitted by the solubility of gibbsite. Nevertheless, the solubility of crystals is always lower than that of corresponding nuclei (Nielsen, 1964; Kolthoff et al., 1969). An aluminum solution which is only slightly supersaturated with respect to gibbsite may be still undersaturated with respect to nuclei. Spontaneous precipitation of gibbsite from such a solution may not take place at all or may be preceded by a long induction period. The addition of NaOH not only increases solution pH but the localized high alkalinity could help pull monomeric species together; then polymerization of monomeric $Al(OH)_3$ molecules begins,

$$mAl(OH)_3 = Al_m(OH)_{3m} \qquad [4]$$

Under constant stirring, the localized high alkalinity disappears as solution pH homogenizes. If the resultant solution is supersaturated with respect to $Al(OH)_3$ nuclei, the nuclei thus formed will not be redissolved but will immediately become positively charged because of the dissociation of the edge OH^-,

$$Al_m(OH)_{3m} + nH^+ = [Al_m(OH)_{3m-n}]^{n+} \qquad [5]$$

Because Al^{3+} has a higher net positive charge per Al atom than the OH–Al polymer, additional NaOH will react with the former and transform it to the latter. Therefore, the OH–Al polymers are limited in size and charge in the early part of neutralization, up to at least NaOH/Al = 2.1. The compositions proposed in the literature, however, have a net positive charge per Al atom varying from 0.5 to 1. Such variation probably is due to the differences in sample preparation, methods of identification, or the assumption upon which the interpretation is based, or all of them. Even the same set of data can be interpreted in different ways. For example, Brosset (1952) first interpreted his potentiometric-titration data by assuming an infinite series of polynuclear complexes with the formula $Al[Al(OH)_3]_n^{3+}$, but later (Brosset et al., 1954) revised his interpretation to favor the single $[Al_6(OH)_{15}]^{3+}$ species, although the existence of an infinite series with the formula $Al[Al_2(OH)_5]_n^{3+n}$ was not ruled out. Later the model was further revised to be $[Al_{13}(OH)_{32}]^{7+}$ (Sillen, 1959) or a mixture of $[Al_{13}(OH)_{34}]^{5+}$ and $[Al_7(OH)_{17}]^{4+}$. Likewise, on the evidence of colloid coagulation, Matijevic (Matijevic & Težak, 1953) at first suggested a dimer, but later suggested an octomeric $[Al_8(OH)_{20}]^{4+}$ (Matijevic et al., 1961).

In his studies of Al-resin (Hsu & Rich, 1960) and Al-vermiculite systems (Hsu & Bates, 1964b), this writer found 1 + charge per Al atom, corresponding to the composition $[Al_6(OH)_{12}]^{6+}$. In his study of the neutralization of $AlCl_3$ or $Al_2(SO_4)_3$ in the absence of resin or vermiculite, however, he observed 0.8+ charge per Al atom, corresponding to the composition $[Al_{10}(OH)_{22}]^{8+}$ (Hsu & Bates, 1964a). Because only limited samples were analyzed and the precision of analysis was poor, he was not certain about the composition at the time of writing. Therefore, a mechanism was proposed by assuming $[Al_6(OH)_{12}]^{6+}$ to be the initial hydrolysis product at NaOH/Al = 2.1 and below. With additional results obtained in later studies, he now believes that in the absence of Dowex 50 resin or vermiculite, the hydrolyzed products are likely $[Al_{10}(OH)_{22}]^{8+}$ (double ring) or $[Al_{13}(OH)_{30}]^{9+}$ (triple ring) at NaOH/Al = 2.1 and below. The revised mechanism is shown in Fig. 4-13.

C. Crystallization of Aluminum Hydroxide

At about NaOH/Al = 2.2 to 2.4, the inclusion of Al in the basic ring units is complete. Additional OH^- then transforms these ring units into larger polymers which gradually increase in size and decrease in net positive charge

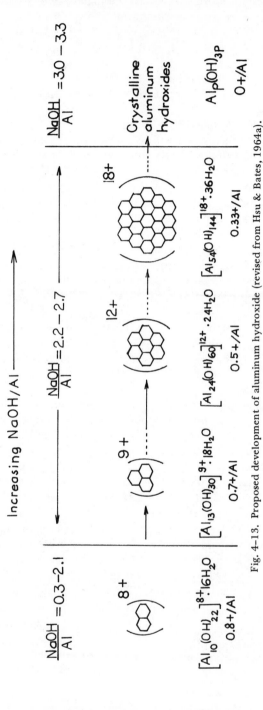

Fig. 4-13. Proposed development of aluminum hydroxide (revised from Hsu & Bates, 1964a).

per aluminum atom with increased NaOH/Al ratio. In addition to the growth in X and Y directions, the polymers may also grow in their Z direction by stacking one octahedral sheet on top of another. At NaOH/Al ratio of 3, aluminum is completely neutralized, and then crystalline $Al(OH)_3$ forms rapidly in a matter of hours.

On the basis of this interpretation, $Al(OH)_3$ does not exist as a simple molecule as shown by its formula, but as a compound occupying an end-member position in a series of polymers and characterized by zero positive charge per aluminum atom. Before the ratio of 3 is reached, the residual positive charge of the polymers must be balanced by counter-anions, giving rise to basic salts. The nature and concentration of the counter-anions play a critical role in governing the stability of the basic salts formed. With anions such as Cl^-, ClO_4^-, and NO_3^- (Hsu, 1966; Ross & Turner, 1971), which do not have strong affinity for aluminum, the basic salts are highly soluble and yield clear solutions. Also, the OH–Al polymers continue to hydrolyze and polymerize into larger units during prolonged aging unless a high concentration of the counter-anion is present. Repeated hydrolysis and polymerization also result in a gradual increase in size and decrease in the net positive charge per aluminum atom, following the same process as the rapid precipitation of $Al(OH)_3$ at NaOH/Al = 3 (Hsu, 1966). These two cases, however, differ in the following three aspects:

1) Although the OH^- added as NaOH is almost completely and immediately taken up by all aluminum (Al^{3+} as well as hydroxy-aluminum polymers), the spontaneous hydrolysis of the hydroxy-aluminum polymers is very slow. This slow hydrolysis is not noticeable in ordinary laboratory preparation when sufficient NaOH is added to convert all the aluminum into $Al(OH)_3$, but becomes important in cases where there is not sufficient NaOH added for the formation of $Al(OH)_3$ and the resultant hydroxy-aluminum solutions are allowed to stand for a long period of time.

2) In cases where the slow hydrolysis of hydroxy-aluminum polymers is barely noticeable, the formation of aluminum hydroxide requires a complete neutralization of Al^{3+} by OH^- with the resultant medium being slightly alkaline. In cases where the slow hydrolysis of OH–Al polymers does take place during aging, the formation of $Al(OH)_3$ is associated with a release of H^+ ion to the solution, resulting in an acidic medium.

3) The concentration of Cl^- in the solution has little effect on the uptake of OH^- (added as NaOH) by hydroxy-aluminum polymers; however, it does show a very significant effect on the rate of spontaneous hydrolysis of polymers. The spontaneous hydrolysis of OH–Al polymers can be noticeably accelerated by removing Cl^- from the system and noticeably retarded or even prohibited by adding extra Cl^- to the solution.

In the presence of anions that have strong affinity for Al, such as sulfate (Hsu, 1973), phosphate (Hsu, 1975), and silicate (Luciuk & Huang, 1974), precipitation of basic salts occurs rapidly after mixing with the counter-anions linked in the first coordination sphere. Because of steric reasons, the tetrahedral, polyvalent anions tend to link OH-Al polymers together, but in distorted arrangement. Therefore, most such basic salts are amorphous. Because of their affinity for Al, these anions prevent or at least retard the OH-Al polymers from further hydrolysis and polymerization into crystalline $Al(OH)_3$.

Many expandable clay minerals can also hold OH-Al polymers in their interlayer space (Rich, 1968), analogous to the formation of basic aluminum salts, but the consequence is complex. At one extreme, Libby vermiculite holds OH-Al polymers very tightly and prevents their crystallization into gibbsite (Hsu & Bates, 1964b). At the other extreme, Upton, Wyoming smectite holds OH-Al polymers only very weakly. Instead, it promotes the crystallization of gibbsite because the surface of this mineral may serve as templets to help the ordering of OH-Al polymers into position or to help push the counter-anions away from the polymers (Jackson, 1963; Barnhisel & Rich, 1963, 1965). The templet hypothesis is in accord with the results that the basicity of the interlayer OH-Al polymers is not always identical to that present in the original solution, but is related to the characteristic of the clay minerals. When Libby, Montana vermiculite was treated with OH-Al solutions of basicity = 2.25 or above, the basicity of the initial interlayer OH-Al was high, but later, during aging, it gradually decreased, indicating the gradual breakdown of the polymer (Hsu & Bates, 1964b). In contrast, the initial basicity of the interlayer OH-Al polymers in smectite and illite always fell in a range between 2.5 and 2.7 regardless of the basicity of the OH-Al polymers present in the original solutions (Turner, 1965; Turner & Brydon, 1967; Hsu, 1968). It should also be noted that both vermiculite and smectite are group names and each of them consists of a large number of members which may not hold the OH-Al polymers with the same strength. Even with the same specimen, the mineralogical surface is not necessarily homogeneous with respect to its negative charge distribution (Hsu, 1968). This explains why when Arizona smectite was treated with a small amount of OH-Al solution, the interlayer Al was stable. When the same smectite was treated with a large amount of OH-Al solution, however, gibbsite rapidly developed during aging (Turner, 1965; Turner & Brydon, 1967).

Many organic acids in soils can complex aluminum and probably are more effective in governing aluminum hydroxide crystallization than any inorganic components studied but they have received little attention. A recent report (Kwong & Huang, 1975) showed that the crystallization of aluminum hydroxide is greatly inhibited in the presence of citric acid.

D. Polymorphs of Aluminum Hydroxides

Aluminum hydroxide is known to exist in three polymorphs, but the factors governing their development have not been clear. Hsu (1966) suggested that rapid precipitation favored the bayerite structure while slow crystal growth favored the gibbsite structure. Barnhisel and Rich (1965) reported that the alkaline medium favored bayerite structure whereas the acidic medium favored gibbsite. A mixture of nordstrandite with bayerite or gibbsite or both was obtained in the intermediate pH range. Schoen and Roberson (1970) proposed an interpretation for the role of pH in governing the development of the three polymorphs. According to Bernal and Megaw (1935), each OH^- ion in gibbsite is split to give a tetrahedral charge distribution because of the strong hydrolyzing power of the Al^{3+} ion, with one $\frac{1}{2}+$ and three $\frac{1}{2}-$ charges distributed at the corners. Such distribution of electrostatic charge permits an ionic bond to form with the tetrahedral hydroxyl ions in the superposed aluminum hydroxide layers. The attraction of these ionic hydroxyl bonds requires that the hydroxyl ions arranged themselves vertically above one another perpendicular to the layers (Fig. 4–3a). In contrast, because of the mild hydrolyzing power of Mg^{2+}, the OH^- ions in brucite remain cylindrical and unable to form the hydroxyl ionic bond as in gibbsite. Instead, the OH^- ions of two adjacent magnesium hydroxide unit layers are linked with van der Waal's force and are closely packed.

Schoen and Roberson (1970) further advanced this concept by assuming that the hydrolyzing power of Al^{3+} in solution is related to pH. In an acidic medium, aluminum is present mainly as monomeric Al^{3+} or $Al(OH)^{2+}$. These cationic species have strong hydrolyzing power and consequently favor the formation of gibbsite. In an alkaline medium, aluminum is present as $Al(OH)_4^-$. The hydrolyzing power of Al^{3+} is much nullified by the excess negative charge. Therefore, aluminum is precipitated in the form of bayerite which is structurally similar to brucite. This interpretation appears to be interesting, but it cannot explain the fact that gibbsite can be prepared above pH 12 on an industrial scale whereas bayerite forms between pH 9 and 12 when a supersaturated sodium aluminate solution is allowed to cool.

The nature and concentration of anions may play an important role in governing the formation of nordstrandite. Hsu and Bates (1964a) found that a mixture of bayerite and nordstrandite was obtained when $Al_2(SO_4)_3$ was neutralized with NaOH, but pure bayerite was obtained when $AlCl_3$ was neutralized. Davis and Hill (1974) reported that in Jamaica, nordstrandite is predominant in high silica bauxite deposits, whereas gibbsite occurred in low silica bauxite. Barnhisel and Rich (1965) also pointed out the possible role of anions.

IV. OCCURRENCE OF GIBBSITE IN SOILS

Among the three polymorphs of aluminum hydroxides, only gibbsite is common in soils and bauxite deposits. Bayerite (Bentor et al., 1963; Gross & Heller, 1963) and nordstrandite (Wall et al., 1962; Hathaway & Schlanger, 1962, 1965; Goldbery & Loughnan, 1970; Davis & Hill, 1974) were found in soils or bauxites only recently. The results of laboratory studies summarized in the preceding section suggest that slow aging of OH–Al polymers in acidic mediums usually yields gibbsite whereas rapid precipitation of aluminum in neutral or alkaline mediums yields bayerite or nordstrandite or both, although the exact mechanism governing their distribution is not clear. The development of aluminum hydroxide in soils is a slow process and takes place mostly in acidic mediums. These conditions favor the development of gibbsite.

Gibbsite is one of the major minerals in many Oxisols (Laterites, Lateritic soils, and Latosols) which usually occur in humid tropical or subtropical uplands. The major diagnostic characteristic of Oxisols is the existence of an oxic horizon in which iron or aluminum or both are greatly enriched. The most Al-rich great soil groups under the Oxisol order are Gibbsiaquox, Gibbsihumox, and Gibbsiorthox which have sheets or gravel-size aggregates containing 30% or more gibbsite. Gibbsite and the various polymorphs of iron oxyhydroxides and oxides commonly exist together as the end products of advanced weathering. In general, iron oxides and oxyhydroxides appear earlier and more often than gibbsite in the process of soil genesis. Many oxic horizons contain goethite or hematite without gibbsite, but pure gibbsitic oxic horizons are rare. Even Al-rich members usually contain considerable amounts of goethite or hematite or both.

Gibbsite is also a common, but minor component in many Ultisols (Red-Yellow Podzolic soils) which is widely spread in humid tropical, subtropical, and temperate regions (Rich et al., 1959; Mohr et al., 1972).

Andosols are developed from volcanic ash. They are generally considered to be relatively young soils and contain 50% or more amorphous components. They are classified under the order of Inceptisol, according to the U. S. Seventh Approximation of soil classification. Nevertheless, gibbsite has been found to be a major mineral in some Andosols in Japan (Wada & Aomine, 1966).

Bauxite deposits, dominated mostly by gibbsite and to a lesser degree by boehmite or diaspore, are the products of extremely advanced weathering. The extensive study of the origin of bauxite has shed much light on the genesis of gibbsite in soils. The primary aluminum silicates in igneous rocks are high-temperature products and are not stable under the present environment. The general consensus is that under favorable conditions, those primary aluminum silicates will eventually weather to aluminum hydroxide but the detailed process has not been clear (Gordon et al., 1958; Keller, 1964).

Some investigators (Harrison, 1934; Abbott, 1958; Young & Stephen, 1965; Sherman et al., 1967) suggested that the primary aluminum silicates weathered directly to gibbsite. This hypothesis was based on the observation that in many bauxite deposits, gibbsite and unaltered rocks were separated by a boundary less than the thickness of a knife edge. Also observed was that microscopic gibbsite replaced feldspar while preserving the outline and original cleavage of the parent crystal (Sherman et al., 1967; Young & Stephen, 1965). Many other investigators, however, suggested that the formation of gibbsite from primary aluminum silicates passed through clay minerals as intermediates (Mead, 1915; Harrison, 1934; Allen, 1952; Bates, 1962). It has been reported that the abundance of kaolinite and gibbsite is frequently inversely related, with gibbsite the dominant mineral in highly weathered soils and bauxites, and kaolinite the dominant mineral in less weathered ones. In some cases, gibbsite and unaltered rocks were separated by a zone of kaolinite varying from a few centimeters to many meters thick. Bates (1962) presented evidence showing the feldspar-halloysite-gibbsite transition in the weathering of andesites in Hawaii. In the least weathered specimens, halloysite started to appear along with unaltered feldspar. In the more weathered specimens, halloysite became dominant, accompanied by the destruction of feldspar and the appearance of gibbsite. In highly weathered specimens, gibbsite and iron oxides were abundant.

These two mechanisms are not mutually exclusive nor totally independent. Keller (1958) postulated that the formation of gibbsite and kaolinite is a precipitation reaction and is controlled by pH and the concentration of Al and Si in solution. This concept was further advanced by construction of solubility diagrams for gibbsite and kaolinite (Garrels & Christ, 1965). The initial step of chemical weathering involves the displacement of the balancing bases with H^+, which is followed by the breakdown of the Al-O–Si linkage, releasing Al and Si to solution. The formation of gibbsite can be very rapid once Si is separated from Al. Nevertheless, if the precolating water moves slowly, its Si concentration may gradually accumulate and recombine with Al to form kaolinite, following the solubility product principle. Therefore, whether the primary aluminum silicates weather directly to gibbsite or through a kaolinite intermediate is governed by the intensity of leaching, which is affected by a number of factors such as rainfall, temperature, parent rocks, topography, ground-water table, vegetation and time. A search of literature indicates that all studies (Harrison, 1934; Abbott, 1958; Young & Stephen, 1965; Sherman et al., 1967) that showed evidences supporting the direct gibbsite formation concept were made under conditions of high rainfall, tropical or subtropical temperature, basic or intermediate rocks, and upland. All of these conditions provided an environment of excessively good drainage. In contrast, the presence of a kaolinite intermediate was observed under conditions where the drainage condition is only moderately good. In his studies under humid tropical climate, Harrison (1934) found that basic or intermediate rocks weathered directly to gibbsite, but acidic rocks weathered through kaolinite to gibbsite. Basic rocks contain Al-rich fedlspathoids and plagioclase that are highly susceptible to chemical weathering. Basic rocks also contain ferromagnesium minerals such as olivine and pyroxene that are usually the first components to decompose in chemical weathering. The decomposition of these minerals yields a porous texture—a condition for good drainage. In contrast, highly siliceous acidic rocks such as granite, gneiss, and schist are normally compact and resistant to weathering. Consequently, the percolating water in general moves slowly in the weathering of these rocks. On the other hand, Wolfenden (1961) found that a thick layer of kaolinite was present between bauxite and the basic parent rocks in Sarawak under humid tropic climate. In this case, the formation of kaolinite was due to the presence of a high ground-water table which impeded drainage. Gordon et al. (1958) also found that kaolinite formed within the ground-water table underneath bauxite.

For the sake of convenience, the above discussion is purposely over-simplified by assuming kaolinite to be the sole product of Al–Si interaction. The situation in soils is more complicated. The following reactions should be considered:

1) The initial reaction products of Si–Al interaction are likely amorphous aluminum silicates of varying composition. These amorphous materials are able to age not only to kaolinite, but also other minerals such as halloysite although the exact conditions for their crystallization are not known. Also, the rate of crystallization is extremely slow. The initial amorphous products may not be significant in old bauxite, but could be significant in relatively younger soils.

2) Aluminum may react with the relics of the partially altered parent minerals, sulfate, phosphate, or organic matter to form a variety of transition products. The formation of these transition products requires that sufficient concentrations of their constituents are present in solution. They may account for the slow crystallization of gibbsite in some young soils, particularly when their drainage conditions are poor.

3) Although kaolinite will desilicate to gibbsite under continuing leaching, there is the possibility that the clay minerals may be transferred and redeposited elsewhere, constituting today's sediments.

4) Direct formation of gibbsite from mica-type minerals has not been reported. It has been suggested that mica weathers first to vermiculite, then to smectite, to pedogenic 2:1 to chlorite intergrade, to kaolinite and eventually to gibbsite (Jackson, 1964). This sequence of weathering represents a continual process of desilication. In this sequence, the intermediate products vermiculite and smectite are expandable and can retain Al^{3+} and OH–Al polymers tightly. Therefore, gibbsite will not form until all the expandable 2:1 clay minerals are decomposed. Many neutral and alkaline soils are dominated by vermiculite and smectite.

Many geologists (Gordon et al., 1958) emphasized that the development of gibbsite was most favorable under alternating wet and dry climate. In contrast, Sherman (1949) showed that Low Humic Latosol was developed in a climate having a definite dry season while Humic Latosol and Hydrohumic Latosol were developed under a continuously wet climate. He suggested that the duration of the wet season and the total amount of rainfall really governed the development of gibbsite. One should note that, for the same amount of rainfall, the weathering and leaching may proceed to a greater depth when the rainfall is concentrated in a short period than when it is evenly distributed throughout the year. Sherman (1949) also showed that ferruginous Hydrohumic Latosol was developed under climates having a definite dry season. If the rainfall occurs every day and the solum is wet throughout the year, iron tends to be present in its reduced form which can be lost rapidly due to leaching, resulting in low-iron bauxitic soils. Under alternating wet and dry conditions, the fluctuating ground-water table tends to concentrate iron along with Al resulting in laterites or lateritic soils.

The genesis of gibbsite in soils is a slow process. In correlating time with the abundance of gibbsite and other reaction products, however, the following possibilities should be considered:

1) Climate is not constant with time. Although meteorological measurements do not reach into history for any great length of time, there is much geological and botanical evidence to show that climate has changed through time. Evidence is steadily accumulating to show that climatic change may be important within the history of many of the soils we presently observe on the earth's surface (Buol et al., 1973). An excellent example is the occurrence of Oxisols in Queensland, Australia, where the present environment does not favor the development of this soil. Whitehouse (1940) indicated that the Oxisols there were the products of two humid periods in the Pliocene.

2) Many geological episodes have taken place on the earth's crust in the past. Some of these episodes may have involved local variations of temperature or pressure which affected the nature of chemical weathering and the genesis of gibbsite.

3) Feldspar, muscovite, and biotites are all high temperature products. The formation of gibbsite from these minerals is irreversible.

Therefore, there may not be any simple relationship between age and the occurrence of gibbsite. We should be cautious about applying thermodynamic principles to the study of the genesis of gibbsite in soils.

V. FORMATION OF ALUMINUM OXYHYDROXIDES IN SOILS

The conditions for the development of AlOOH in soils are not clearly understood. It was first observed by Brown et al. (1953) and later corroborated by others (deBoer et al., 1954a, 1954b; Brindley & Choe, 1961; and Sato, 1962, 1964) that the thermal decomposition of aluminum hydroxide proceeds along two paths. With coarse aluminum hydroxide crystals, boehmite is the initial dehydration product which later changes to anhydrous alumina at higher temperatures. Fine aluminum hydroxide particles, however, convert directly to anhydrous alumina without the intermediate product boehmite. Also, the endothermic peak for the $Al(OH)_3$-AlOOH transformation was observed only with coarse aluminum hydroxide crystals (Fig. 4-8). Because boehmite can be developed rapidly under hydrothermal conditions (Ervin & Osborn, 1951; Kennedy, 1959), it was then postulated (de Boer et al., 1954a) that with coarse particles, an impervious layer of Al_2O_3 forms at the surface which prevents the escape of H_2O from the interior, creating a condition resembling hydrothermal preparations. According to this interpretation, boehmite should not form under our present soil environment. Its presence in soils or bauxites must have originated from hydrothermal reactions in the distant past. On the other hand, Keller (1964) reported that gibbsite, boehmite, and diaspore are present together in some bauxites. This suggests that the conditions for their formation should not be very different. Some investigators (Garrels & Christ, 1965; Kittrick, 1969; Chesworth, 1972; Parks, 1972) made attempts to predict the direction of reaction by calculating the change in free energy. The results have been inconclusive, however. The free energies of formation reported for gibbsite, boehmite, and diaspore are slightly different from each other. The calculated free energy of reaction was found to be either positive or negative, depending on the free energies of formation chosen for calculation. In all cases, the calculated free energy of change was nearly zero, whereas the total free energy of formation involved was quite large (Table 4-4).

An ideal experiment for clarifying the understanding would be to expose $Al(OH)_3$ and $AlOOH$ in our present atmosphere and examine whether or not $Al(OH)_3$ changes to $AlOOH$ or vice versa. This experiment is not practical because of the slowness of the reaction. Some indirect evidence, however, suggests that $Al(OH)_3$ is more stable than $AlOOH$ at least in aqueous medium. It has been repeatedly reported that by neutralizing an aluminum salt solution with a base, the initial precipitate was pseudoboehmite which gradually changed to aluminum hydroxide later. Hsu and Bates (1964a) found that when a dilute aluminum chloride solution was neutralized with dilute NaOH, aluminum hydroxide was formed at the very beginning. Pseudoboehmite was developed in the presence of sulfate (Hsu & Bates, 1964a) or a high concentration of indifferent electrolytes such as NaCl (Hsu, 1967; Chesworth, 1972). It is likely that most of the earlier reports in which pseudoboehmite was observed initially were carried out in high salt concentrations. Hsu also provided evidence illustrating the gradual transformation of pseudoboehmite to aluminum hydroxide during aging. Although pseudoboehmite is not identical with the typical boehmite, both of them consist of Al-O-Al linkages. The above results clearly indicate that the Al-O-Al linkage is less stable than Al-OH-Al linkages in aqueous alkaline solutions. The difference between pseudoboehmite and typical boehmite should be a matter of rate, but not the basic nature of the reaction.

Despite the belief that boehmite is less stable than gibbsite in soils, the former might be developed under the following two conditions:

1) The results of Hsu (1967) and Chesworth (1972) suggest that pseudo-boehmite may form in a saline environment. The pseudoboehmite thus formed may persist in nature for a very long period of time if it is dried before its conversion to $Al(OH)_3$. It has been reported that so-called boehmite in most of the European bauxites actually consists of submicroscopic particles which are closer to pseudoboehmite rather than typical boehmite (Lippens & Steggerda, 1970).

Table 4–4. Free energy of reaction for the dehydration of gibbsite to boehmite or diaspore at 25C (298K), 1 bar. The free energy of formation of H_2O is 56.7 kcalories/mole in all calculations

G_f°			ΔG_r°		
Gibbsite	Boehmite	Diaspore	Gibbsite-boehmite	Gibbsite-diaspore	Reference
——————————— kcalories/mole ———————————					
-277.3	-217.5	--	3.1	--	Garrels & Christ, 1965
-274.2	-218.2	-220.0	-0.7	-2.5	Kittrick, 1969
-274.49	-217.67 ± 3.51	-219.45	-3.38 to (+3.62)	-1.66	Apps, 1970[†]
-276.5	-218.26	-220.2	1.54	-0.4	Naumov et al., 1971
-273.49 ± 0.31	-217.67 ± 3.51	--	-4.70 to (+2.94)	--	Chesworth, 1972
-275.3 + 0.2	-218.7 ± 0.2	-219.5 ± 0.5	-0.5 to (+0.3)	-0.2 to -1.6	Parks, 1972

† J. A. Apps. 1970. The stability field of analcime. Ph.D. Thesis, Harvard University.

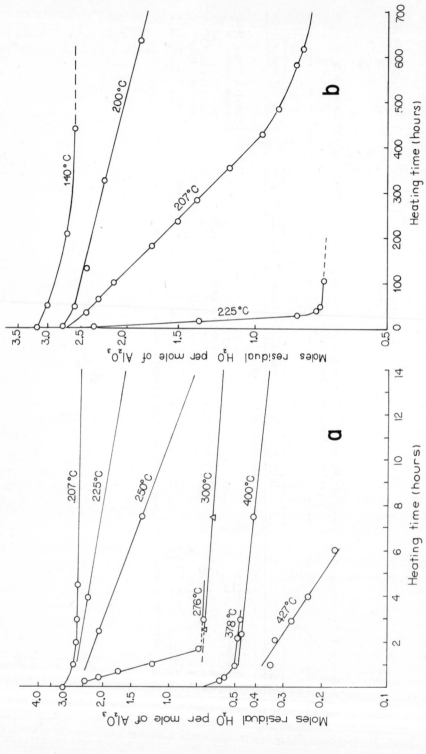

Fig. 4-14. Thermal dehydration of gibbsite in relation to temperature and time (modified from Newsome et al., 1960). (*A*) up to 14 hours; (*B*) up to 640 hours.

2) Even well-crystallized $Al(OH)_3$ can be dehydrated to AlOOH at 150C (423K). This temperature is not high in the sense of metamorphism. It is possible to dehydrate $Al(OH)_3$ to AlOOH even though metamorphism of other minerals is not noticeable. Furthermore, amorphous aluminum hydroxide can be dehydrated to AlOOH at an even lower temperature.

The failure to observe boehmite in the thermal decomposition of fine $Al(OH)_3$ particles probably is related to the rate of heating. Newsome et al. (1960) presented detailed results showing that the rate of $Al(OH)_3$ dehydration was greatly dependent on temperature (Fig. 4-14). At 375C (648K), the water content was reduced to 0.6 mole per mole of Al_2O_3 in a few minutes. The rate of dehydration decreased rapidly with progressively decreasing temperature. At 207C (480K), the water content was reduced to 0.8 mole per mole of Al_2O_3 after 640 hours, but the end of dehydration was still not in sight. It is not impossible to obtain boehmite from fine $Al(OH)_3$ crystals if the specimen is heated at a temperature somewhat below 207C (480K), but the rate of conversion must be very slow. In many differential thermal analyses the rate of heating was maintained at 10C per min. In the thermal decomposition studies of Brown et al. (1953) and others, the specimens were maintained at each temperature level for several hours and weighed. With such rates of heating, the $Al(OH)_3$-AlOOH conversion will not have a chance to take place before the specimen temperature is high enough to form anhydrous alumina. The validity of this interpretation remains to be tested.

VI. REACTIONS OF ALUMINUM HYDROXIDES IN SOILS

A. Reaction with Anions

Many anions can be retained by aluminum hydroxides. The strength of their retention is an important factor governing their mobility in soils. Among all anions, phosphate has been most extensively studied and the interpretation most controversial. The reaction has been suggested to be either adsorption or precipitation. It has long been assumed that the phosphate retained by aluminum hydroxide is of low availability to plant growth, but this assumption probably has been oversimplified and overemphasized. The reactions with other anions have been interpreted mostly to be surface adsorption, but the possibility of precipitation should not be ruled out for some anions. It has been suggested that adsorption and precipitation originate from essentially the same reaction force (Hsu, 1965). Most evidence for either adsorption or precipitation was obtained under typical conditions, but these two processes are not always distinguishable.

The adsorption of sulfate by aluminum hydroxide greatly increases with decreasing pH, but is always less tightly held than phosphate over the entire range of pH studied (Chao et al., 1964; Harward & Reisenauer, 1966; Geb-hardt & Coleman, 1974). Therefore, the adsorbed sulfate is completely re-placeable by phosphate, but not vice versa. This may also account for the increasing sulfate movement in soils after phosphate fertilization (Chao et al., 1962). Similar results were observed in the precipitation of phosphate and sulfate using OH–Al polymers (Hsu, 1975; Yuan & Hsu, 1971). The adsorp-tion of silicic acid (or silicate) shows a maximum of pH 9 to 9.5 and rapidly decreases on either side of the maximum (Jones & Handreck, 1967; Mc-Keague & Cline, 1963; Obihara & Russell, 1972). In acidic mediums the ad-sorption of silicic acid is very limited. Therefore, the application of silicate frequently increases the phosphate availability in neutral or alkaline soils. The reaction with molybdate (Jones, 1957), borate (Hatcher et al., 1967; Sims & Bingham, 1968a, 1968b), selenite (Hingston et al., 1972), arsenate (Huang, 1975; Jacob et al., 1970), and flouride (Huang & Jackson, 1966; Samson, 1952) have also been reported.

In aluminum hydroxide, each outermost Al^{3+} shares four OH^- ions with two other Al^{3+} ions and the remaining two coordination sites are occupied by either H_2O or nonbridged OH^-, depending on the pH (Fig. 4-2). Chemical adsorption of any anion must involve a ligand exchange with either H_2O or OH^-. Although this concept has been suggested repeatedly in earlier reports on the adsorption of phosphate, sulfate or silicate, it did not gain general recognition until the detailed investigations of Hingston et al. (1967, 1972, and 1974). These workers suggested three types of adsorptions:

1) Nonspecific adsorption. Some anions such as NO_3^-, ClO_4^-, and Cl^- can only be adsorbed by positively charged surfaces, being loosely held in the diffuse layer.

2) Specific adsorption of anions of completely dissociated acids, such as SO_4^{2-} and F^-. These anions are chemically adsorbed, involving a ligand exchange with the surface H_2O. Their adsorption can take place only on the acidic side of the ZPC (zero point of charge).

3) Specific adsorption of anions of incompletely dissociated acids, such as phosphate and silicate (or silicic acid). They suggested that the adsorption of such anions can take place at the ZPC or on its nega-tive side, involving an exchange with ligand OH^-. The requirement for such adsorption is the presence of both a proton donor (acid) and a proton acceptor (salt). With monobasic acid, the adsorption shows a maximum near its pK value. With polybasic acids, the ad-sorption shows breaks at individual pK values.

In order to explain the relative mobility of different anions in soils, this writer proposes another scheme by grouping them into four categories on the basis of their affinity for Al^{3+}:

1) Anions with weak affinity, such as NO_3^-, ClO_4^-, and Cl^-. These anions can be present only in the diffuse layer except in completely dehydrated systems.

2) Anions with moderate affinity, such as sulfate. This anion is able to be chemically adsorbed on the surface of aluminum hydroxide involving an exchange with the ligand H_2O. Its ability to exchange with the ligand OH^- is very limited. The adsorption can take place only on the acidic side of the ZPC.

3) Anions with a strong affinity, such as phosphate. The affinity of this ion for Al^{3+} is strong enough to remove OH^- from the surface or to liberate H^+ from H_3PO_4, $H_2PO_4^-$, or HPO_4^{2-}. Therefore, phosphate can be adsorbed at the ZPC or on its alkaline side. The species adsorbed on the surface is not necessarily the same as the species in solution.

4) Anions of very strong affinity, such as F^-. The anions in this category not only can remove the nonbridged OH^- at the edge, but can also break the interior Al–OH–Al linkages, resulting in the destruction of the aluminum hydroxide structure. Hingston et al. (1972) showed that the maximum amount of F^- adsorbed can be three times as much as Cl^- or SO_4^{2-} adsorbed, on an equivalent basis. It is likely that part of the F^- removed from solution is present as some aluminum fluoride precipitates. The rate of aluminum hydroxide breakdown should vary with pH, time of reaction, and the concentration of F^-. The Al–OH–Al linkage is much weaker and easier to break at low pH than at high. Therefore, more F^- is removed from solution at low pH than at high pH. Even at pH 9, the amount of F^- removed was only slightly less than the maximum amount of Cl^- or SO_4^{2-} removed on an equivalent basis.

It should be noted that phosphate is also known to induce the decomposition of $Al(OH)_3$, giving rise to the formation of aluminum phosphate precipitates, but all published evidence was obtained from reaction of concentrated phosphate solutions. Presumably the reaction is very slow with dilute phosphate solutions.

The affinities of different anions for Al^{3+} are not known for certain. A simple titration experiment, however, provides a qualitative, but reliable measure for this purpose. In this experiment, $0.02M$ $AlCl_3$ was titrated with NaOH in the presence of NaCl, Na_2SO_4, NaH_2PO_4, and KF. It is shown in Fig. 4-15 that the titration curves are displaced from right to left following the order: chloride, sulfate, phosphate, and fluoride. This displacement of titration curves may reflect the increasing competition of these anions with OH^- for Al^{3+}. Furthermore, in the presence of chloride and sulfate, crystalline $Al(OH)_3$ was observed at NaOH/Al mole ratio = 3 within a few hours after preparation (Hsu & Bates, 1964a), but was not observed in the presence of either phosphate or fluoride even after many months of aging. It should also be noted that a slight change in the OH^- activity in solution could yield a significant displacement in the titration curve. A calculation of the OH^- in solution versus that bound by Al^{3+} shows that even F^- is only a weak competitor relative to OH^- for Al^{3+}.

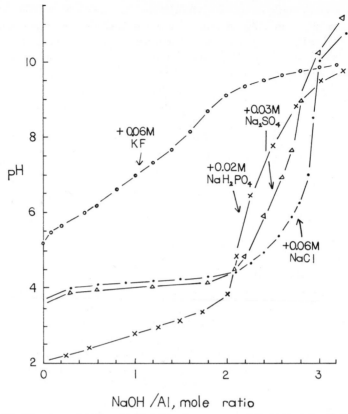

Fig. 4-15. Titration of $AlCl_3$ with NaOH in the presence of NaCl, Na_2SO_4, NaH_2PO_4, or KF($0.02M$ in Al).

B. Phosphate Retention Versus Fixation

The reaction with phosphate deserves additional discussion because of its voluminous literature and its importance to fertility management. The term "fixation" was originated from the observation that some soils indicated favorable response to phosphate fertilizer despite a large amount of phosphate present. Most laboratory research in the past, however, dealt with the removal of phosphate from solution (i.e., retention), but the results were assumed to be fixation. It should be noted that the phosphate retained by soils is not necessarily fixed. If the phosphate retained by soils could be released to soil solution later at a reasonable rate for plant uptake, then retention is advantageous rather than disadvantageous to fertility management. The failure to distinguish these two different consequences has contributed much of the confusion in the literature. Because of this, many investigators have discontinued the use of the term fixation. The confusion, however, remains.

That phosphate can be adsorbed by $Al(OH)_3$ has been proven beyond any doubt, but there is little information on the availability of the adsorbed phosphate. We can only speculate by reasoning that phosphate might be fixed through adsorption under the following two conditions:

1) The availability of the adsorbed phosphate should be related to the degree of surface saturation. When the surface is nearly saturated, the adsorbed phosphate might be highly available to plants. When the surface is nearly bare, the adsorbed phosphate may be very difficult for plants to obtain (Hsu, 1965).

2) During the processes of weathering and soil genesis, fresh aluminum hydroxide (or iron oxide) is continuously added to the system. The adsorbed phosphate may be rendered unavailable to plants if it is coated with another layer of aluminum hydroxides or iron oxides. This corresponds to the formation of occluded phosphate (Bauwin & Tyner, 1957).

According to the first possibility, only a small fraction of the adsorbed phosphate may be considered to be fixed. Much of it is temporarily reserved in soils and would be released to soil solution later for plant growth. Fixation may not be a serious problem except in soils with unusually large amounts of surface reactive aluminum hydroxides or iron oxides. So-called adsorption capacity is meaningless as far as fixation is concerned. According to the second possibility, the fixation is a slow process except in soils of frequent changing oxidation-reduction status in which the formation of fresh iron oxide coating could be rapid (Hsu, 1964). In most soils, however, much of the adsorbed phosphate may be taken up by plants before it can be coated by another layer of aluminum hydroxide or iron oxide. Russell (1973) summarized in his review that in countries using large amounts of phosphate fertilizer it was becoming increasingly difficult to find soils showing response to phosphate fertilizer, other than to starter doses of water-soluble phosphate to some phosphate-demanding crops. R. L. Flannery (personal communication) has found that Freehold sandy loam soil in New Jersey which had a history of high phosphate fertilizer, contained adequate available phosphate for the production of good yields of corn, soybeans, and alfalfa for 10 or more years without additional phosphate fertilization. Yields of crops grown on these soils did not differ significantly where 0 to 225 kg of phosphate (P_2O_5) per ha had been applied annually over a 10-year period.

In a solution of high phosphate concentration, the decomposition of aluminum hydroxide in association with the formation of aluminum phosphate as separate phases were observed with light or electron microscope or chemical analysis. Nevertheless, all the precipitates formed at room temperature, whether crystalline or noncrystalline, are fairly available to plant growth. On the other hand, variscite, a slightly soluble compound, can be prepared easily at about 100C (373K). It has often been assumed that the room temperature precipitates would eventually change to insoluble compounds after prolonged aging. This assumption is the basis of the precipitation concept of fixation, but has never been closely examined. This writer recently prepared nearly 100 samples by mixing $AlCl_3$, NaH_2PO_4, and NaOH in varying combinations. The concentration of Al ranged from 0.001M to 0.05M. The P/Al mole ratio ranged from 0.2 to 12. The NaOH/Al ratio ranged from 0 to 3.5, with the final acidity ranging from pH 2 to 8. These samples were aged at room temperature and examined periodically up to 1 year. Thus far no precipitate which can be considered to be difficultly soluble has been observed in any of these preparations. Taranakite and crystalline Ca-Al mixed phosphate were observed by reaction of aluminum hydroxide with concentrated K^+, NH_4^+ or acid calcium phosphate solutions (Taylor et al., 1964, 1965; Taylor & Gurney, 1965), but these compounds hydrolyzed to noncrystalline aluminum phosphates upon dilution (Taylor & Gurney, 1964) and were good sources of phosphate for plant growth (Taylor et al., 1963). Although a definite conclusion cannot be made until the investigation is extended to cover all possible conditions of precipitation and longer time of aging, the evidence obtained thus far suggests that those room-temperature aluminum phosphate precipitates are more logically considered to be slow-releasing phosphate sources rather than the products of fixation.

C. Formation of Mixed Hydroxides

Magnesium-aluminum hydroxide can be prepared by neutralizing a system containing active forms of Al and Mg to pH 8.2–8.5 (Turner & Brydon, 1962; Gastuche et al., 1967; Hunsaker & Pratt, 1970). The structure of Al-Mg mixed hydroxide is similar to that of brucite, with part of the Mg replaced by Al (Gastuche et al., 1967; Brown & Gastuche, 1967). At maximum, one of three Mg^{2+} can be replaced by Al^{3+}. The excess positive charge created from substitution is balanced by a sheet of anions and H_2O molecules in their interlayer space. This yields a basal spacing of 7.55–8.35Å. Gastuche et al. (1967) suggested that the basal spacing is related to the extent of substitution. With increased substitution of Al^{3+} for Mg^{2+}, attractive force between layers is increased and basal spacing is reduced. Clark and Ross (1969) reported that such positively charged Al-Mg mixed hydroxide can be developed in the interlayer space of montmorillonite. This product is resistant to heating up to 600C (873K) and can swell to 17.1Å when saturated with ethylene glycol vapor. They therefore described this product as swelling chlorite.

The formation of Al-Fe mixed hydroxide has also been reported (Gastuche et al., 1964; Tullock & Roth, 1975). For two cations to form solid solution, their ionic sizes should differ by $< 15\%$. Also, the individual hydroxides should have the same structure. Nevertheless, the compound $Fe(OH)_3$ is not stable, but rapidly dehydrated to FeOOH at room temperature. Therefore, the Al-Fe mixed hydroxides reported probably are only short lived. In contrast, the substitution of Al for Fe in goethites has been observed in Australian soils (Norrish & Taylor, 1961). AlOOH and FeOOH are known to have identical structures.

D. Stabilization of Soil Aggregates

Weldon and Hide (1942) found that the amount of sesquioxides extracted from a well-aggregated fraction of several prairie soils was considerably greater than that extracted from the poorly aggregated fraction. Kroth and Page (1947) showed that the stable aggregates were heavily coated with iron and aluminum oxides and the removal of these oxides destroyed the aggregates.   It was stressed in these two reports that sesquioxides acted as cementing agents. Saini et al. (1966) analyzed 24 soil samples from the upper B horizons of podzols in New Brunswick, Canada, and found that the partial regression coefficient relating aluminum "oxides" with aggregation was 1.84 times as large as that for iron oxides. The precipitation of aluminum hydroxides in situ was found to reduce the swelling of soils (Deshpande et al., 1964), Na-montmorillonite (El Rayah & Rowell, 1973) and Na-illite or kaolinite (El Swaify & Emerson, 1975); or to reduce the permeability of a Na-soil (El Rayah & Rowell, 1973); or to reduce the resistance of clay to dispersion in dilute NaCl solution (El Swaify & Emerson, 1975). These synthesis studies also showed that aluminum hydroxides are more effective than iron oxides in maintaining the stability of soil aggregates. It should be noted that the procedure of preparation is very critical to the nature and properties of the hydroxides and oxides prepared and the magnitude of their influence on aggregation.

The mechanism of aggregation in the presence of aluminum hydroxides has not been clearly presented in the literature but two types of effects are conceivable from reasoning. Small OH–Al polymers, which can be considered to be fragments of aluminum hydroxide solid, can be held more tightly than ordinary exchangeable cations in the interlayer space of expandable clay minerals (Chapter 10). Their presence may lead to a reduction in the swelling and the expansion of clay particles by bonding adjacent silica sheets together and by displacing interlayer cations of high hydration power, such as Na^+ and Li^+, thereby promoting aggregation.

Aluminum hydroxides and OH–Al polymers may also react with clay particles on their external surfaces and cement them together. The positive charges of aluminum hydroxides are distributed at the edge. Because of the spatial restriction, it is not possible to have all the positive charges of one aluminum hydroxide unit entirely satisfied by one clay particle. Instead, when one aluminum hydroxide unit is attached to one clay particle, it tends to link another. Using a special electron microscope technique, Jones and Uehara (1973) were able to show the presence of amorphous gel hull linkage between clay particles. Such amorphous gel hull must include noncrystalline aluminum hydroxide. The presence of aluminum hydroxide as a thin layer around illite particles has also been reported by El Swaify and Emerson (1975).

Because aluminum hydroxide decreases its positive charge with increasing pH or particle size, its cementing effectiveness must follow the same trend. El Swaify and Emerson (1974) showed that aluminum hydroxide exerted little effect on aggregation at its point of zero charge. In principle, well-crystallized aluminum hydroxide may also be able to act as a cementing agent in acidic mediums, but its magnitude may be negligible as compared with noncrystalline material. Iron oxides have a stronger tendency than aluminum hydroxide to crystallize. Therefore, the former is generally much less effective than the latter in its cementing effectiveness except in frequently alternating oxidized and reduced soils. Because the aluminum hydroxides and iron oxides in soils greatly vary in crystallinity and particle size, a poor correlation between their contents and aggregation does not necessarily indicate that they do not exert any effect on aggregation.

Organic matter, particularly polysaccharide, has also been known to be closely associated with soil aggregation. Nevertheless, soil organic matter is negatively charged and is thus not likely to react directly with clay particles. It has been suggested that organic matter may promote soil aggregation through the following linkage: clay-(Al, Fe)-organic matter-(Al, Fe)-clay (Edwards & Bremner, 1967). According to this hypothesis, the organic matter in soil samples should not be decomposed in assessing the effectiveness of aluminum hydroxides or iron oxides in aggregation.

LITERATURE CITED

Abbott, A. T. 1958. Occurrence of gibbsite on the island of Kauai, Hawaiian Islands. Econ. Geol. 57:842-853.
Allen, V. T. 1952. Petrographic relations in some typical bauxite and diaspore deposits. Bull. Geol. Soc. Am. 63:649-688.
Aveston, J. 1965. Hydrolysis of the aluminum ion: ultracentrifugation and acidity measurements. J. Chem. Soc. 4438-4443.
Barnhisel, R. I., and C. I. Rich. 1963. Gibbsite formation from aluminum-interlayers in montmorillonite. Soil Sci. Soc. Am. Proc. 27:632-635.
Barnhisel, R. I., and C. I. Rich. 1965. Gibbsite, bayerite, and nordstrandite formation as affected by anions, pH and mineral surfaces. Soil Sci. Soc. Am. Proc. 29:531-534.
Bates, T. F. 1962. Halloysite and gibbsite formation in Hawaii. Clays Clay Miner. 9: 315-328.
Bauwin, G. R., and E. H. Tyner. 1957. The nature of reductant-soluble phosphorus in soils and soil concretions. Soil Sci. Soc. Am. Proc. 21:245-257.
Bentor, Y. K., S. Gross, and L. Heller. 1963. Some unusual minerals from the "mottled zone" complex, Israel. Am. Mineral. 48:924-930.
Bernal, J. D., and H. D. Megaw. 1935. The function of hydrogen in intermolecular forces. Proc. Roy. Soc. London, A. 151:384-420.
Berry, L. G. (ed.). 1974. Selected powder diffraction data for minerals. Joint Committee on Powder Diffraction Standards, Swarthmore, Pa.
Brindley, G. W., and J. O. Choe. 1961. The reaction series, gibbsite $\rightarrow$ chi alumina $\rightarrow$ kappa alumina $\rightarrow$ corundum. Am. Mineral. 46:771-785.
Brosset, C. 1952. On the reactions of the aluminum ion with water. Acta Chem. Scand. 6:910-940.
Brosset, C., G. Biederman, and L. G. Sillen. 1954. Studies on the hydrolysis of metal ions XI. The aluminum ion, Al^{3+}, Acta Chem. Scand. 8:1917-1926.
Brown, J. F., D. Clark, and W. W. Elliott. 1953. The thermal decomposition of the alumina trihydrate, gibbsite. J. Chem. Soc. (London) 84-88.
Brown, G., and M. C. Gastuche. 1967. Mixed magnesium-aluminum hydroxides. II. Structure and structural chemistry of synthetic hydroxycarbonates and related minerals and compounds. Clay Miner. 7:198-201.

Bryant, J. P., and J. B. Dixon. 1964. Clay mineralogy and weathering of a red yellow podzolic soil from quartz mica schist in the Alabama Piedmont. Clays Clay Miner. 12:509-521.

Buol, S. W., F. D. Hole, and R. J. McCracken. 1973. Soil genesis and classification. The Iowa State Univ. Press, Ames.

Chao, T. T., M. E. Harward, and S. C. Fang. 1962. Movement of S^{35} tagged sulfate through soil columns. Soil Sci. Soc. Am. Proc. 26:27-32.

Chao, T. T., M. E. Harward, and S. C. Fang. 1964. Iron or aluminum coatings in retention to sulfate adsorption characteristics of soils. Soil Sci. Soc. Am. Proc. 28:632-635.

Chesworth, W. 1972. The stability of gibbsite and boehmite at the surface of the earth. Clays Clay Miner. 20:369-374.

Clark, J. S., and G. J. Ross. 1969. Reaction of mixed magnesium-aluminum and clacium-aluminum hydroxides with Wyoming bentonite. Can. J. Earth Sci. 6:47-53.

Davis, C. E., and V. G. Hill. 1974. Occurrence of nordstrandite and its possible significance in Jamaica bauxites. Travaux 11:61-70. (In English).

de Boer, J. H., J. M. H. Fortuin, and J. J. Steggerda. 1954a. The dehydration of alumina hydrates. Proc. K. Ned. Akad. Wet., B 57:170-180. (In English).

de Boer, J. H., J. M. H. Fortuin, and J. J. Steggerda. 1954b. The dehydration of alumina hydrates. II. Proc. K. Ned. Akad. Wet., B 57:434-444. (In English).

Deshpande, T. L., D. J. Greenland, and J. P. Quirk. 1964. Role of iron oxides in the bonding of soil particles. Nature (London) 201:107-108.

Dixon, J. B. 1966. Quantitative analysis of kaolinite and gibbsite in soils by differential thermal and selective dissolution methods. Clays Clay Miner. 14:83-89.

Edwards, A. P., and J. M. Bremner. 1967. Microaggregates in soils. J. Soil Sci. 18:64-73.

Elderfield, H., and J. D. Hem. 1973. The development of crystalline structure in aluminum hydroxide polymorphs on ageing. Mineral. Mag. 39:89-96.

El Rayah, H. M. E., and D. L. Rowell. 1973. The influence of iron and aluminum hydroxides on the swelling of Na-montmorillonite and the permeability of Na-soil. J. Soil Sci. 24:137-144.

El Swaify, S. A., and W. W. Emerson. 1975. Changes in the physical properties of soil clays due to precipitated aluminum and iron hydroxides: I. Swelling and aggregate stability after drying. Soil Sci. Soc. Am. Proc. 39:1056-1063.

Ervin, G., and E. F. Osborn. 1951. The system Al_2O_3-H_2O. J. Geol. 59:381-394.

Farmer, V. A., and J. D. Russell. 1967. Infrared absorption spectrometry in clay studies. Clays Clay Miner. 15:121-141.

Fieldes, M. 1956. Clay mineralogy of New Zealand soils. Part 4. Differential thermal analysis. N. Z. J. Sci. Tech. Sec. B 38:533-570.

Fieldes, M., R. J. Furkert, and N. Wells. 1972. Rapid determination of constituents of whole soils using infrared absorption. J. Soil Sci. 15:615-627.

Frederickson, L. D. 1955. Characterization of hydrated aluminas by infrared spectroscopy. Anal. Chem. 26:1883-1885.

Frink, C. R. 1972. Aluminum chemistry in acid sulfate soils. p. 131-168. In H. Dost (ed.) Acid sulfate soils. Proc. Int. Symposium, Wageningen, 1972. Int. Inst. for Land Reclam. Improv. Publ. 18.

Frink, C. R., and M. Peech. 1963. Hydrolysis of the aluminum ion in dilute aqueous solutions. Inorg. Chem. 3:473-478.

Garn, P. D. 1965. Thermoanalytical methods of investigation. Chap. 9, 10, and 11. Academic Press, New York.

Garrels, R. M., and C. L. Christ. 1965. Solutions, minerals and equilibria. Harper & Row, New York. 450 p.

Gastuche, M. C., T. Bruggenwert, and M. M. Mortland. 1964. Crystallization of mixed iron and aluminum gels. Soil Sci. 98:281-289.

Gastuche, M. C., G. Brown, and M. M. Mortland. 1967. Mixed magnesium-aluminum hydroxides. I. Preparation and characterization of compounds formed in dialyzed systems. Clay Miner. 7:177-192.

Gebhardt, H., and N. T. Coleman. 1974. Anion adsorption by allophanic tropical soils: II. Sulfate adsorption. Soil Sci. Soc. Am. Proc. 38:259-262.

Glemser, O. 1959. Studies of some hydroxides and hydrous oxides. Binding of water in some hydroxides and hydrous oxides. Nature (London) 183:943-944.

Glenn, R. C., and V. Nash. 1964. Weathering relationship between gibbsite, kaolinite, chlorite and expansible layer silicates in selected soils from lower Mississippi. Clays Clay Miner. 12:529-548.

Goldbery, R., and F. C. Loughnan. 1970. Dawsonite and nordstrandite in the Permian Berry Formation of the Sydney Basin, New South Wales. Am. Mineral. 55:477-490.

Gordon, M., J. I. Tracey, and M. W. Ellis. 1958. Geology of the Arkansas bauxite region. U. S. Geol. Surv. Prof. Paper 299.

Gross, S., and L. Heller. 1963. A natural occurrence of bayerite. Mineral. Mag. 33:723-724.

Grunwald, E., and D. W. Fong. 1969. Acidity and association of Al ions in dilute aqueous acid. J. Phys. Chem. 73:650-653.

Harrison, J. B. 1934. The katamorphism of igneous rocks under humid tropical conditions. Imp. Bur. Soil Sci. Rothamsted Exp. Stn., Harpenden, England.

Harward, M. E., and H. M. Reisenauer. 1966. Reactions and movements of inorganic sulfur. Soil Sci. 101:326-335.

Hashimoto, I., and M. L. Jackson. 1960. Rapid dissolution of allophane and kaolinite-halloysite after dehydration. Clays Clay Miner. 7:102-113.

Hatcher, J. T., C. A. Bower, and M. Clark. 1967. Adsorption of boron by soils as influenced by hydroxy-aluminum and surface area. Soil Sci. 104:422-426.

Hathaway, J. C., and S. O. Schlanger. 1962. Nordstrandite from Guam. Nature (London) 196:265-266.

Hathaway, J. C., and S. O. Schlanger. 1965. Nordstrandite (Al_2O_3-$3H_2O$) from Guam. Am. Mineral. 50:1029-1037.

Hem, J. D., and C. E. Roberson. 1967. Form and stability of aluminum hydroxide complexes in dilute solution. U. S. Geol. Surv. Water-Supply Paper 1827 A.

Hingston, F. J., R. J. Atkinson, A. M. Posner, and J. P. Quirk. 1967. Specific adsorption of anions. Nature (London) 215:1459-1461.

Hingston, F. J., A. M. Posner, and J. P. Quirk. 1972. Anion adsorption by geothite and gibbsite. I. The role of the proton in determining adsorption envelopes. J. Soil Sci. 23:177-192.

Hingston, F. J., A. M. Posner, and J. P. Quirk. 1974. Anion adsorption by goethite and gibbsite. II. Desorption of anions from hydrous oxide surfaces. J. Soil Sci. 25:16-26.

Holmes, L. P., D. L. Cole, and E. M. Eyring. 1968. Kinetics of aluminum ion hydrolysis in dilute solutions. J. Phy. Chem. 72:301-304.

Hsu, Pa Ho. 1964. Adsorption of phosphate by aluminum and iron in soils. Soil Sci. Soc. Am. Proc. 28:474-478.

Hsu, Pa Ho. 1965. Fixation of phosphate by aluminum and iron in acidic soils. Soil Sci. 99:398-401.

Hsu, Pa Ho. 1966. Formation of gibbsite from aging hydroxy-aluminum solutions. Soil Sci. Soc. Am. Proc. 30:173-176.

Hsu, Pa Ho. 1967. Effect of salts on the formation of bayerite versus pseudoboehmite. Soil Sci. 103:101-110.

Hsu, Pa Ho. 1968. Heterogeneity of montmorillonite surface and its effect on the nature of hydroxy-aluminum interlayers. Clays Clay Miner. 16:303-311.

Hsu, Pa Ho. 1973. Effect of sulfate on the crystallization of aluminum hydroxide from aging hydroxy-aluminum solutions. p. 613-620. In J. Nicolas (ed.) Proc. 3rd Int. Congr. on Studies of Bauxite and Aluminum Oxides-Hydroxides (Nice, France).

Hsu, Pa Ho. 1975. Precipitation of phosphate from solution using aluminum salts. Water Res. 9:1155-1161.

Hsu, Pa Ho, and T. F. Bates. 1964a. Formation of X-ray amorphous and crystalline aluminum hydroxides. Mineral. Mag. 33:749-768.

Hsu, Pa Ho, and T. F. Bates. 1964b. Fixation of hydroxy-aluminum polymers by vermiculite. Soil Sci. Soc. Am. Proc. 28:763-769.

Hsu, Pa Ho, and C. I. Rich. 1960. Aluminum fixation in a synthetic cation exchanger. Soil Sci. Soc. Am. Proc. 24:21-25.

Huang, P. M. 1975. Retention of arsenic by hydroxy-aluminum on surfaces of micaceous mineral colloids. Soil Sci. Soc. Am. Proc. 39:271-274.

Huang, P. M., and M. L. Jackson. 1966. Fluoride interaction with clays in relation to third buffer range. Nature (London) 211:779-780.

Hunsaker, V. E., and P. F. Pratt. 1970. The formation of mixed magnesium-aluminum hydroxides in soil materials. Soil Sci. Soc. Am. Proc. 34:813-816.

Jackson, M. L. 1963. Aluminum bonding in soils: A unifying principle in soil science. Soil Sci. Soc. Am. Proc. 27:1-10.

Jackson, M. L. 1964. Chemical composition of soils. p. 71-141. In F. E. Bear (ed.) Chemistry of the soil. Reinhold, New York.

Jackson, M. L. 1969. Soil chemical analysis-advanced course. 2nd ed. Published by the author. Madison, Wis.

Jacob, L. W., J. K. Syers, and D. R. Keeney. 1970. Arsenic sorption in soils. Soil Sci. Soc. Am. Proc. 34:750-754.

Johansson, G. 1960. On the crystal structures of some basic aluminum salts. Acta Chem. Scand. 14:771-773.

Johansson, G. 1962. On the crystal structure of the basic aluminum sulfate. 13 Al_2O_3-$6SO_3 \cdot XH_2O$. Ark. Kemi. 20:321-342.

Jones, L. H. P. 1957. The solubility of molybdenum in simplified systems and aqueous soil suspensions. J. Soil Sci. 8:313-327.

Jones, L. H. P., and K. A. Handreck. 1967. Silica in soils, plants and animals. Adv. Agronomy 19:107-149.

Jones, R. C., and G. Uehara. 1973. Amorphous coatings on mineral surfaces. Soil Sci. Soc. Am. Proc. 37:792-798.

Keller, W. D. 1958. Argillation and direct bauxitization in terms of concentrations of hydrogen and metal cations at surfaces of hydrolyzing silicates. Bull. Am. Assoc. Petrol. Geol. 42:233-245.

Keller, W. D. 1964. The origin of high alumina clay minerals. A review. Clays Clay Miner. 12:129-151.

Kennedy, G. C. 1959. Phase relations in the system Al_2O_3-H_2O at high temperature and pressures. Am. J. Sci. 257:563-593.

Kittrick, J. A. 1966. The free energy of formation of gibbsite and $Al(OH)_4^-$ from solubility measurements. Soil Sci. Soc. Am. Proc. 30:595-598.

Kittrick, J. A. 1969. Soil minerals in the Al_2O_3-SiO_2-H_2O system and a theory of their formation. Clays Clay Miner. 17:157-167.

Kolesova, V. A., and I. I. Ryskin. 1959. Infrared absorption spectrum of hydrargillite $Al(OH)_3$. Opt. Spectrosc. 7:165-167 (English translation).

Kolthoff, I. M., E. B. Sandell, E. J. Meehan, and S. Bruckenstein. 1969. Quantitative chemical analysis. Chap. 10. MacMillan, London.

Kroth, E. M., and J. B. Page. 1947. Aggregation formation in soils with special reference to cementing substances. Soil Sci. Soc. Am. Proc. 11:27-34.

Kwong, Ng Kee, and P. M. Huang. 1975. Influence of citric acid on the crystallization of aluminum hydroxides. Clays Clay Miner. 23:164-165.

Lippens, B. C., and J. J. Steggerda. 1970. Active alumina. p. 171-211. In B. G. Linsens (ed.) Physical and chemical aspects of adsorbents and catalysts. Academic Press, New York.

Luciuk, G. M., and P. M. Huang. 1974. Effect of monosilicic acid on hydrolytic reactions of aluminum. Soil Sci. Soc. Am. Proc. 38:235-244.

MacKenzie, R. C. 1970. Oxides and hydroxides of higher valency elements. p. 271-302. In R. C. MacKenzie (ed.) Differential thermal analysis, Vol. 1. Academic Press, New York.

MacKenzie, R. C. 1972. Soils. p. 267-297. In R. C. MacKenzie (ed.) Differential thermal analysis. Vol. 2. Academic Press, New York.

MacKenzie, R. C., and R. H. S. Robertson. 1961. The quantitative determination of halloysite, goethite and gibbsite. Acta Univ. Carol., Geol. Suppl. 1:139-149.

Matijevic, E., and B. Težak. 1953. Detection of polynuclear complex aluminum ions by means of coagulation measurement. J. Phy. Chem. 57:951-954.

Matijevic, E., K. G. Mathai, R. H. Ottewill, and M. Kerker. 1961. Detection of metal ion hydrolysis by coagulation. III. Aluminum. J. Phy. Chem. 65:826-830.

McKeague, J. A., and M. G. Cline. 1963. Silica in soils. Adv. Agron. 15:339-396.

Mead, W. J. 1951. Occurrence and origin of the bauxite deposits of Arkansas. Econ. Geol. 10:28-54.

Megaw, H. D. 1934. The crystal structure of hydrargillite $Al(OH)_3$. Z. Krist. 87:185-204. (In English).

Mitchell, B. D., V. C. Farmer, and W. J. McHardy. 1964. Amorphous inorganic materials in soils. Adv. Agron. 16:327-383.

Mohr, E. C. J., F. A. Van Baren, and J. Van Schuylenborgh. 1972. Tropical soils. Mouton-Ichtiar Baru-Van Hoeve. The Hague. 481 p.

Montoro, V. 1942. Structura cristillina della bayerite. Ric. Sci. Prog. Tec. 13:565-571. (Cited from Lippens & Steggerda, 1970).

Naumov, G. B., B. N. Ryzhenko, and R. T. L. Khodakousky. 1971. Handbook of thermodynamic data. Atomizdat, Moscow. (English transl., 1974. Nat. Tech. Infor. Serv., Dep. of Commerce, Springfield, Va. 328 p.)

Newsome, J. W., H. W. Heiser, A. S. Russell, and H. C. Stumpf. 1960. Alumina Properties. Aluminum Co., of Am. Tech. Rep. no. 10, 2nd ed.

Nielsen, A. E. 1964. Kinetics of precipitation. MacMillan, New York.

Norrish, K., and R. M. Taylor. 1961. The isomorphous replacement of iron by aluminum in soil goethites. J. Soil Sci. 12:294-306.

Obihara, C. H., and E. W. Russell. 1972. Specific adsorption of silicate and phosphate. J. Soil Sci. 23:105-117.

Papée, D., R. Tertian, and R. Biais. 1958. Constitution of gels and crystalline hydrates of alumina. Bull. Soc. Chim. France 1301-1310.

Parks, G. A. 1965. The isoelectric points of solid oxides, solid hydroxides, and aqueous hydroxy-complex systems. Chem. Rev. 65:177-198.

Parks, G. A. 1972. Free energies of formation and aqueous solubilities of aluminum hydroxides and oxide hydroxides at 25°C. Am. Mineral. 57:1163-1189.

Patterson, J. H., and S. Y. Tyree, Jr. 1973. Light scattering study of the hydrolytic polymerization of aluminum. J. Coll. Interface Sci. 43:389-398.

Rich, C. I., L. F. Seatz, and G. W. Kunze. 1959. Certain properties of selected southeastern United States soils and mineralogical procedures for their study. Southern Regional Bull. 61 for Cooperative Regional Research Project S-14. Va. Agric. Exp. Stn., Blacksburg, Va.

Rich, C. I. 1968. Hydroxy interlayers in expansible layer silicates. Clays Clay Miner. 16: 15-30.

Rooksby, H. P. 1961. Oxides and hydroxides of aluminum and iron. p. 354-392. In G. Brown (ed.) The X-ray identification and crystal structure of clay minerals. Minerological Society, London.

Ross, G. J., and R. C. Turner. 1971. Effect of different anions on the crystallization of aluminum hydroxide in partially neutralized aqueous aluminum salt systems. Soil Sci. Soc. Am. Proc. 35:389-392.

Russell, E. W. 1973. Soil conditions and plant growth. Chap. 23, Longman, London.

Ryskin, Y. I. 1974. The vibration of protons in minerals; hydroxyl, water and ammonium. p. 137-181. In V. C. Farmer (ed.) Infrared, spectra of minerals. Minerological Society, London.

Saalfeld, H., and B. B. Mehrotra. 1966. Zur struktur von nordstrandit $Al(OH)_3$. Naturwiss. 53:128-129. (Cited from Berry, 1974).

Saini, G. R., A. A. MacLean, and J. J. Doyle. 1966. The influence of some physical and chemical properties on soil aggregation and response to VAMA. Can. J. Soil Sci. 46:155-160.

Samson, H. R. 1952. Fluoride adsorption by clay minerals and hydrated alumina. Clay Miner. Bull. 1:266-270.

Sato, T. 1962. Thermal transformation of alumina trihydrate, bayerite. J. Appl. Chem. 12:553-556.

Sato, T. 1964. Thermal transformation of alumina trihydrate, hydrargillite. J. Appl. Chem. London, 14:303-308.

Schoen, R., and E. C. Roberson. 1970. Structures of aluminum hydroxide and geochemical implication. Am. Mineral. 55:43-77.

Schofield, R. K., and A. W. Taylor. 1954. Hydrolysis of aluminum salt solutions. J. Chem. Soc. (London) 4445-4448.

Sherman, G. D. 1949. Factors influencing the development of laterite and lateritic soils in the Hawaiian Islands. Pacif. Sci. 3:307-314.

Sherman, G. D., J. G. Cady, H. Ikawa, and N. E. Blumsberg. 1967. Genesis of the Bauxitic Hailu Soils. Hawaii Agric. Exp. Stn. Tech. Bull. no. 56.

Sillen, L. G. 1959. Quantitative studies of hydrolytic equilibria. Quart. Rev. 13:146-168.

Sims, J. R., and F. T. Bingham. 1968a. Sorption of boron by layer silicates, sesquioxides and soil materials. II. Sesquioxides. Soil Sci. Soc. Am. Proc. 32:364–369.

Sims, J. R., and F. T. Bingham. 1968b. Sorption of boron by layer silicates, sesquioxides and soil minerals. III. Iron and aluminum coated layer silicates and soil minerals. Soil Sci. Soc. Am. Proc. 32:369–373.

Singh, S. S. 1974. The solubility product of gibbsite at $15°$, $25°$, and $35°C$. Soil Sci. Soc. Am. Proc. 38:415–417.

Smith, R. W. 1971. Relations among equilibrium and non-equilibrium aqueous species of aluminum hydroxy complexes. Adv. Chem. 106:250–279.

Smith, R. W., and J. D. Hem. 1972. Effect of aging on aluminum hydroxide complexes in dilute aqueous solutions. U. S. Geol. Surv. Water Supply Paper 1827-D.

Taylor, A. W., W. L. Lindsay, E. O. Huffman, and E. L. Gurney. 1963. Potassium, and ammonium taranakites, amorphous aluminum phosphate, and variscite as sources of phosphate for plants. Soil Sci. Soc. Am. Proc. 27:145–151.

Taylor, A. W., E. L. Gurney, E. C. Moreno. 1964. Precipitation of phosphate from calcium phosphate solutions by iron oxide and aluminum hydroxide. Soil Sci. Soc. Am. Proc. 28:49–52.

Taylor, A. W., and E. L. Gurney. 1964. The dissolution of calcium aluminum phosphate, $CaAlH(PO_4)_2 \cdot 6H_2O$. Soil Sci. Soc. Am. Proc. 28:63–64.

Taylor, A. W., and E. L. Gurney. 1965. Precipitation of phosphate by iron oxide and aluminum hydroxide from solutions containing calcium and potassium. Soil Sci. Soc. Am. Proc. 29:18–22.

Taylor, A. W., E. L. Gurney, and A. W. Frazier. 1965. Precipitation of phosphate from ammonium phosphate solutions by iron oxide and aluminum hydroxide. Soil Sci. Soc. Am. Proc. 29:317–320.

Turner, R. C. 1965. Some properties of aluminum hydroxide precipitated in the presence of clay. Can. J. Soil Sci. 54:331–336.

Turner, R. C., and J. E. Brydon. 1962. Formation of double hydroxides and the titration of clays. Science 136:1052–1054.

Turner, R. C., and J. E. Brydon. 1967. Effect of length of time of reaction on some properties of suspensions of Arizona bentonite, illite and kaolinite in which aluminum hydroxide is precipitated. Soil Sci. 103:111–117.

Tullock, R. J., and C. B. Roth. 1975. Stability of mixed iron and aluminum hydrous oxides on montmorillonite. Clays Clay Miner. 23:27–32.

van Nordstrand, R. A., W. P. Hettinger, and C. D. Keith. 1956. A new alumina trihydrate. Nature (London) 177:713–714.

van Oosterhout, G. W. 1960. Morphology of synthetic submicroscopic crystals of α- and γ-FeOOH and of γ-Fe$_2$O$_3$ prepared from FeOOH. Acta Crystallogr. 13:932–935.

Wada, K., and S. Aomine. 1966. Occurrence of gibbsite in weathering of volcanic materials at Kuroishibaru, Kumamoto. Soil Sci. Plant Nutr. 12:151–157.

Wall, J. R. D., E. B. Wolfenden, E. H. Beard, and T. Deans. 1962. Nordstrandite in soil from West Sarawak, Borneo. Nature (London) 196:264–265.

Weldon, T. A., and J. C. Hide. 1942. Some chemical properties of soil organic matter and of sesquioxides associated with aggregation in soils. Soil Sci. 54:343–352.

Whitehouse, F. W. 1940. The lateritic soils of Western Queensland. Queensl. Univ. Dep. Paper 2, no. 1 (Cited from Mohr, et al., 1972).

Wolfenden, E. B. 1961. Bauxite in Sarawak. Econ. Geol. 56:972–981.

Young, A., and I. Stephen. 1965. Rock weathering and soil formation on high-altitude plateous of Malawi. J. Soil Sci. 16:322–333.

Yuan, W. L., and Pa Ho Hsu. 1971. Effects of foreign components on the precipitation of phosphate by aluminum. In S. H. Jenkins (ed.) Adv. in Water Pollution Res., Proc. 5th Int. Conf. 1–16.

Zussman, J. 1967. X-ray diffraction. p. 261–334. In J. Zussman (ed.) Physical methods in determinative mineralogy. Academic Press, New York.

Iron Oxides

UDO SCHWERTMANN, Technische Universität München, Freising-Weihenstephan, West Germany

REGINALD M. TAYLOR, CSIRO, Glen Osmond, South Australia

I. INTRODUCTION

During the weathering of a primary (magmatic) rock to a soil the iron, predominantly bound in silicates in the reduced (bivalent) state, will be released through a combined hydrolytic and oxidative reaction of the following type:[1]

$$-Fe^{2+}-O-Si- + H_2O = -Fe^{3+}-OH + HO-Si- + e^-$$

in which atmospheric oxygen takes up the electron

$$(O_2 + 4e^- = 2O^{2-}).$$

Due to the extremely low solubility of Fe^{3+} oxides[2] in the normal pH range of soils, the iron released will be precipitated as oxide or hydroxide. Only a small part of the oxidized iron will generally be incorporated into secondary-layer silicate clay minerals and/or complexed by organic matter.

The reversibility of the oxidation-reduction reaction of iron plays an important role in its behavior in soils. If soil oxygen becomes deficient due to excess water, microorganisms may utilize Fe^{3+} oxides as final electron acceptors to accomplish their oxidative decomposition of organic matter. The Fe^{3+} is thereby reduced to Fe^{2+}, which, being generally more soluble, accelerates the dissolution of the oxide. Migration of this solubilized Fe to zones of oxidation induces re-oxidation and subsequent reprecipitation.

Reduction- or chelation-solution and oxidation-precipitation reactions make it difficult to position iron oxides correctly in a mineral weathering-stability sequence. They participate in nearly all stages of weathering and indicate an advanced stage only if strongly accumulated under aerobic conditions (stage 12 in Jackson's (1967) scale of 13).

Even at low concentrations in a soil, iron oxides have a high pigmenting power and determine the color of many soils. Thus, soil color, as determined by the type (Schwertmann & Lentze, 1966) and distribution of iron oxides within a profile, is helpful in explaining soil genesis and is also an important criterion for naming and classifying soils (e.g. Red-Yellow Podzolic, Braunerde, sols rouges tropicaux, terra rossa, krasnozem, etc.).

---

[1] Superscript signs and Arabic numbers are employed uniformly in this book to indicate charge on an atom (for example Fe^{2+} or Fe^{3+}) instead of the Roman numeral designation employed in some publications and preferred by the authors of this chapter.

[2] For brevity this term will be used throughout the chapter as a group name for oxides, oxyhydroxides, and hydroxides of Fe.

The chemical nature and generally high specific surface area of iron oxides in particles and as coatings on other particles make them efficient sinks for anions such as phosphate, molybdate, and silicate as well as trace elements like Cu, Pb, V, Zn, Co, Cr, and Ni, some of which are essential for plant growth. The mechanism of this association is uncertain as the ions can be adsorbed on the surface or incorporated into the oxide structure (Kühnel et al., 1974; Norrish, 1975).

Iron oxides also affect soil structure and fabric, often being responsible for the formation of aggregates and cementation of other major soil components, giving rise to granules, nodules, pipe-stems, plinthites, ortsteins, or bog iron ores.

II. FORMS IN SOILS

The various crystalline iron oxides and their properties are given in Table 5-1. With the exception of magnetite, all have been identified as products of pedogenesis. When formed in this way, however, the properties of these crystalline oxides can differ markedly from the data given. This is attributed mainly to adsorption or incorporation into the iron oxide structure of foreign ions present in the soil environment, which can effect crystal development (possibly through structural strains). For example, the density of soil iron oxides is usually lower and crystal morphology often less well expressed than in synthetic oxides. Particle size and degree of isomorphous substitution, for example, can affect line width and line position, respectively, in X-ray diffraction analysis, as well as the temperatures of reaction during thermal analysis.

1. GOETHITE

Goethite is the most frequently occurring form of iron oxide in soils. Thermodynamically, it has the greatest stability under most soil conditions. Goethite occurs in almost every soil type and climate region and is responsible for the yellowish-brown color of many soils. It may be dispersed evenly throughout the soil, or as with all the oxides be concentrated in certain horizons or structural forms, nodules, pipe-stems etc. This concentration may occur in the initial development of the soil, or may proceed from a uniformly dispersed system due to a change in soil-forming factors. The high concentrations of goethite may assume dark brown and even black colors, but they generally give the yellow-brown streak typical of goethite.

Synthetic goethite is usually acicular (Fig. 5-1a) with the Z-axis crystal direction lying along the needle axis (van Oosterhout, 1960). This acicular morphology is also found in soils (Fig. 5-2a), although the needles are usually poorly developed (Fig. 5-2b). However, scanning electron micrographs show that the needle-like crystals are more prevalent where they have been able to develop in a void (Fig. 5-3 and Eswaran & Raghu Mohan, 1973), possibly due to reduced interference from the immediate environment.

Table 5-1. General characteristics of iron oxide minerals

Mineral properties	Mineral name					
	Hematite	Maghemite	Magnetite	Goethite	Lepidocrocite	Ferrihydrite
Formulae	α-Fe$_2$O$_3$	γ-Fe$_2$O$_3$	Fe$_3$O$_4$	α-FeOOH	γ-FeOOH	Fe$_5$HO$_8\cdot$4 H$_2$O† Fe$_5$(O$_4$H$_3$)$_3$‡
Crystal system	Rhombohedral	Isometric or tetragonal	Isometric	Orthorhombic	Orthorhombic	Rhombohedral
Cell dimensions (Å)	a_O = 5.04 c_O = 13.77	a_O = 8.34	a_O = 8.39	a_O = 4.65 b_O = 10.02 c_O = 3.04	a_O = 3.88 b_O = 12.54 c_O = 3.07	a_O = 5.08 c_O = 9.4
Density (g cm^{-3})	5.26	4.87	5.18	4.37	4.09	3.96
Standard free energy of formation $\Delta G°$ (kcal/mole)	-177.7#	-163.6§§	-243.1§§	-117.0††	-114.0††	-166.5§§
Solubility product‡‡ (pFe + 3 pOH)	42.2-43.3	40		43.3-44.0	40.6-42.5	37.0-39.4

Diagnostic characteristics

Mineral properties	Hematite	Maghemite	Magnetite	Goethite	Lepidocrocite	Ferrihydrite
X-ray spacings (Å)	2.70, 3.68, 2.52	2.52, 2.95	2.53, 2.97	4.18, 2.69, 2.45	6.26, 3.29, 2.47	2.5, 2.2, 1.97, 1.71, 1.5
DTA peaks (°C)	Nil	Exotherm 600-800	See footnote¶	Endotherm 280-400	Endotherm 300-350 exotherm 370-500	Endotherm 150, loss of adsorbed H$_2$O
Infrared spectroscopic peaks (cm^{-1})	345, 470, 540	400, 450, 570 590, 630	400, 590	890, 797	1026, 1161, 753	Nil
Color (Munsell)	5R-2.5YR bright red	Reddish brown	Black	7.5YR-10YR yellowish-brown	5YR-7.5YR orange	5YR-7.5YR reddish-brown
Usual crystal morphology	Hexagonal plates	Cubes	Cubes	Acicular	Laths, serrated elongated plates	Spherical

† After Towe and Bradley (1967).
‡ After Chukrov et al. (1973).
‡‡ After Garrels & Christ (1965).
¶ Magnetite converts via maghemite or directly to hematite, depending on particle size.

\# After Robie & Waldbaum (1968).
†† After Mohr et al. (1972).
‡‡ Depends on particle size.
§§ At 25C and 1 bar total pressure.

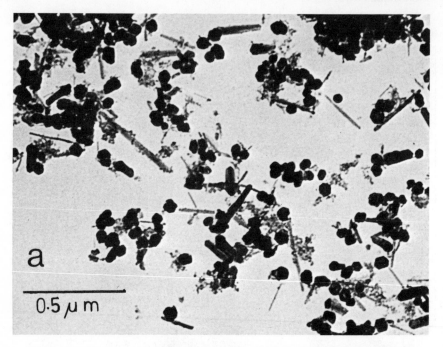

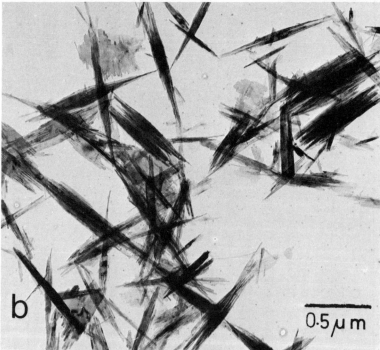

Fig. 5-1. Electron micrograph of synthetic iron oxides *(a)* goethite (needles), hematite (hexagonal plates), and ferrihydrite (small spheres), *(b)* lepidocrocite, *(c)* maghemite.

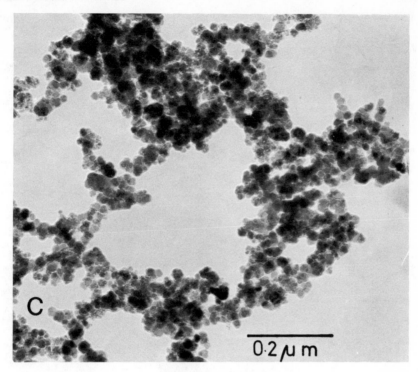

Fig. 5-1. Continued.

Fig. 5-2. Electron micrograph of soil goethites *(a)* From a laterite on serpentinite, New Caledonia (obtained by courtesy of W. Schellmann) *(b)* From a bog iron ore in a gley, Federal Republic of Germany. In contrast to Fig. 5-2a this goethite is very poorly crystalline (strong X-ray diffraction line broadening) and possibly associated with ferrihydrite (small spherical particles of 50–100Å diameter) acting as an iron source for further goethite formation (U. Schwertmann, unpublished) (Photo H. Ch. Bartscherer, Physik. Inst. München-Weihenstephan, Federal Republic of Germany).

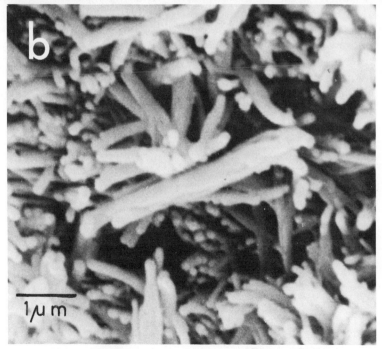

Fig. 5-3. Scanning micrographs of soil goethites. *(a)* From a pore coating in a laterite on acid igneous rock, Cameroun (obtained by courtesy of G. Boquier) *(b)* From a laterite on serpentinite, Brazil (obtained by courtesy of R. A. Kühnel, Photo Techn. Phys. Dienst. v. Landb. Wageningen, The Netherlands).

2. HEMATITE

In many reddish soils, goethite is associated with hematite which is the second most frequent soil iron oxide. Goethite is apparently not restricted to given climatic regions, but hematite appears to be absent in soils recently formed under a humid temperate climate such as in northern and mid-Europe and the northern part of the American continent.

Hematite colors the soil red and has a greater pigmenting effect than goethite so that even low concentrations of it in goethitic soils change the hue to redder than 5YR. The pigmenting effect of hematite is particularly high when it occurs in a finely dispersed form. In denser accumulations, it generally appears much darker. As with goethite, typical crystal shapes of hematite are only weakly expressed, and isodimensional particles much < 0.1 μm in diameter prevail (Fig. 5-4). These particles sometimes show a hexagonal outline whereas hexagonal plates at various stages of development are common for synthetic hematites (Fig. 5-1a).

3. LEPIDOCROCITE

Lepidocrocite occurs in soils less frequently than goethite or hematite (Van der Marel, 1951; Brown, 1953; Schwertmann, 1959b). It forms from the oxidation of precipitated Fe^{2+} hydroxy compounds and appears to be restricted to hydromorphic soils where the presence of Fe^{2+} is generated due to oxygen deficiency. In fact, the occurrence of lepidocrocite is indicative of hydromorphic conditions. It commonly is found in gleys and pseudogleys

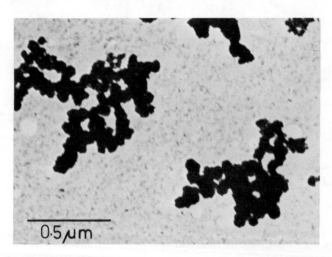

Fig. 5-4. Electron micrograph of a soil hematite from a Terra rossa, Algeria (Beutelspacher & van der Marel, 1968).

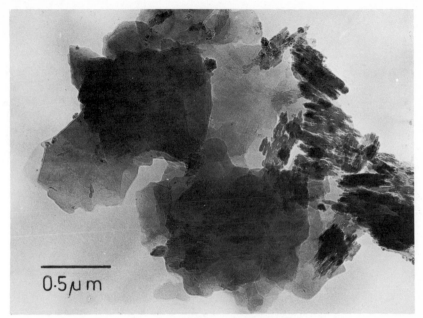

Fig. 5-5. Electron micrograph of a soil lepidocrocite (with layer silicates) from a pseudogley on schist, Federal Republic of Germany (Schwertmann, 1973).

(surface water gleys), particularly those high in clay, but it has not been reported in calcareous hydromorphic soils where goethite forms instead.
Goethite and lepidocrocite are often associated.

Macroscopically, lepidocrocite occurs as bright orange mottles or bands.
Under electron microscopy the crystals resemble those formed synthetically
by oxidation of Fe^{2+} salt solutions at pH 5-7 and ambient temperature. They
appear as highly serrated elongated plates, 0.1-0.7 μm in length and of variable thickness (Fig. 5-1b, 5-5), or as laths or starlike twins (Schwertmann,
1973a). The lath axis is the crystallographic Z-axis direction (van Oosterhout, 1960).

4. MAGHEMITE

Maghemite is especially common in highly weathered soils of tropical
and subtropical climates of Hawaii, Australia, India, Africa, and Lebanon,
and also occurs in soils of temperate regions such as Holland, Germany,
Japan, Russia, and Canada (for literature, see Taylor & Schwertmann, 1974b).
Occurrences in soils formed from basic ignous rocks seem to prevail in all
these regions.

The mineral is reddish-brown and ferromagnetic. It may be finely dispersed or in concretions frequently in association with hematite. Highly
opaque isodimensional-particles similar to synthetic maghemite can be observed under the electron microscope (Fig. 5-1c and 5-6).

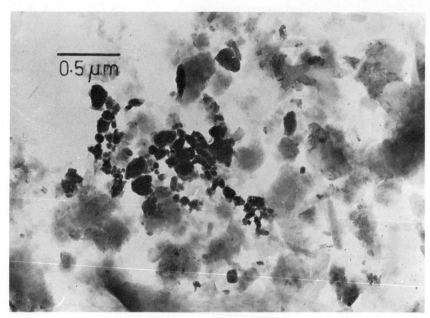

Fig. 5-6. Electron micrograph of a soil maghemite (with smectite) from a Red Mediterranean soil on basalt, Lebanon (U. Schwertmann, unpublished, Photo: H. Ch. Bartscherer).

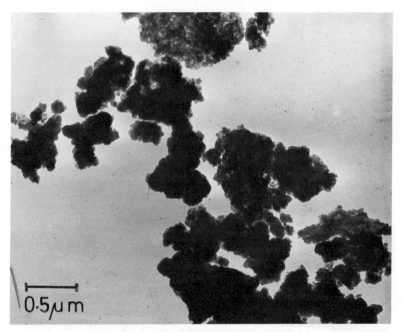

Fig. 5-7. Electron micrograph of a ferrihydrite from a ferruginous precipitate in a drainage ditch transecting a gley, Federal Republic of Germany (U. Schwertmann & Fischer, 1973).

5. FERRIHYDRITE

Ferrihydrite (accepted 1971 as a mineral by the Nomenclature commission of International Mineralogical Association, previously but incorrectly called "amorphous ferric hydroxide") has been found in particular environments associated with soils such as drainage ditches and small, slow running water courses (Chukhrov et al., 1973; Schwertmann & Fischer, 1973) in various countries (USSR, Hawaii, Germany, Australia, Austria). It also occurs in lichen weathering crusts on basalt (Jackson & Keller, 1970). It is often associated with goethite in bog iron ores and most probably constitutes the Fe^{3+} oxide of B_{Fe} horizons of true podzols. Ferrihydrite appears as a rusty, voluminous precipitate rich in adsorbed water and often rich in adsorbed inorganic ions and organic matter. It forms very small (50–100Å diameter) spherical particles (Fig. 5-1a, 5-7) with a high surface area (200–350 m^2/g). These particles are generally highly aggregated and are 90–100% soluble in ammonium oxalate (Schwertmann & Fischer, 1973).

6. OTHER MINERALS

In addition to the above minerals there are some which, although not listed in Table 5-1, should receive a brief mention.

a. **Green Rust**—Green Rust, defined from corrosion studies (Dasgupta & Mackay, 1959), forms from aerial oxidation of aqueous Fe^{2+} solutions (Feitknecht & Keller, 1950). This $Fe^{2+}Fe^{3+}$ hydroxy compound is a precursor of lepidocrocite, and may be responsible for the greenish-blue colors in anaerobic soils. Ponnamperuma (1967) has suggested on the basis of E_h, pH, and Fe determinations in submerged soils, that black "hydromagnetite," $Fe_3(OH)_8$, governs the redox potential of soils under anaerobic conditions. Although this compound, in contrast to green rust, has not yet been identified it is normally used in stability diagrams.

b. **Akaganeite and δ-FeOOH**—Akaganeite has the same composition as goethite and lepidocrocite, but is structurally different and is designated as β-FeOOH. It has been identified in mineral deposits (Mackay, 1962), but not as yet in soils. A fourth crystallographic FeOOH modification, δ-FeOOH, is ferromagnetic and formed on vigorous oxidation of $Fe(OH)_2$, e.g. by H_2O_2, under alkaline conditions. A similar poorly ordered nonmagnetic compound was called δ'-FeOOH by Chukhrov et al. (1976). Both forms were found by these authors in soil and marine concretions.

c. **Limonite**—Limonite is no longer accepted as a mineral name, but has previously been used in soil literature to describe brown, rusty iron oxide accumulations. It generally consisted of fine-grained goethite, but may have contained other forms as well. Equally incorrect is the term "ferric oxide hydrate" for fine-grained material with a large amount of adsorbed water.

III. CRYSTAL STRUCTURE AND CHEMISTRY

The basic structural unit of iron oxides is an octahedron in which the Fe atom is surrounded by 6 O+OH ions in either hexagonal (α forms) or cubic (γ forms) close packing. Tetrahedrally coordinated Fe sheets are also present in magnetite and maghemite. The various forms differ mainly in their arrangement of the octahedra (Wells, 1975). In goethite (α form, isostructural with diapore) the Fe-O-OH octahedra form infinite double bands linked by O-H-O bonds (Fig. 5-8) 2.65Å in length with the H atoms displaced slightly from the O-O axis. In contrast, the Fe-O-OH octahedra of lepidocrocite (γ form, isostructural with boehmite) are arranged in complex layers, linked by O-H-O bonds producing a more open structure than that of goethite. The Fe-O octahedra in hematite form an infinite, three-dimensional network (corundum structure).

Magnetite and maghemite have an inverse spinel structure containing 32 O and 24 Fe sites per unit cell of which 8 are coordinated tetrahedrally and 16 octahedrally (designated below with superscripts IV and VI, respectively). In magnetite all 24 sites are occupied, the divalent Fe generally being confined to the octahedral sites ($^{VI}Fe_8^{3+} + {}^{VI}Fe_8^{2+} + {}^{IV}Fe_8^{3+} + O_{32}$). In fully oxidized maghemite (γ Fe$_2$O$_3$) produced from magnetite, the number of Fe sites per 32 O is reduced to 21 1/3. If the oxidation is carefully carried out, the resultant Fe site vacancies may be ordered, giving rise to a tetragonal modification of the original cubic structure of magnetite which causes superlattice lines in X-ray diffraction patterns.

However, in pedogenic maghemites these additional lines are often absent, suggesting a random arrangement of the vacant sites (Taylor & Schwertmann, 1974b). In maghemite derived from oxidation of precipitated magnetite, Fe^{2+} may be randomly distributed between the two sheets (Haag, 1935).

Only a tentative structure for ferrihydrite has been suggested. The oxygen arrangement resembles that in hematite, however, some O is replaced by OH_2 and less Fe occurs in the octahedral sites, leading to a higher O/Fe ratio (see formula in Table 5-1) and the absence of the strong diffraction lines of hematite (Towe & Bradley, 1967; Chukhrov et al., 1973).

Since soil iron oxides generally form from solution, foreign ions also present in the soil solution may be incorporated into the oxide structures to varying extents. The intimate association of the different oxides in soils and the inability to separate them from the other soil materials make study of foreign ions in iron oxides extremely difficult.

Norrish and Taylor (1961) separated soil goethites from highly weathered Australian soils by boiling in 5N NaOH, and showed that they contained up to 30 mole % Al substituting for Fe. This caused a reduction in the size of the unit cell, lowering the (111) X-ray diffraction spacing from 2.452 to 2.426Å. The Al content increased with decreasing particle size and reduced solubility during dithionite reduction. All soil goethites investigated in that study contained some Al. The same was found for goethite in ooids (Correns

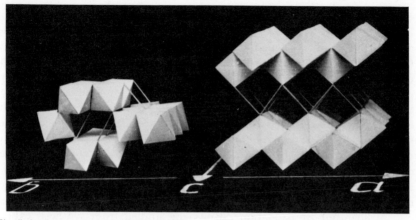

Fig. 5–8. Structural arrangement of FeO_6-octahedrons in goethite (*left*) and lepidocrocite (*right*). Oxygens occupy the corners of each octahedron and Fe atoms its centre. The lines between the octahedrons indicate O–H–O bonds. The letters refer to the crystallographic axes. (From Hiller, 1966. Courtesy of Verlag Chemie, Weinheim, W. Ger.).

& von Engelhardt, 1941; Schneiderhöhn, 1964; Schellmann, 1964). In accordance with these natural occurrences synthetic goethites were prepared containing up to 30 mole % Al (Thiel, 1963).

Aluminum also replaces Fe in natural hematites and enters the hematite structure if synthesized in the presence of Al thereby reducing the cell parameters (Caillère et al., 1960; Wefers, 1967).

The change in cell size with foreign ion incorporation is not related to the extent of the substitution alone since ions of varying atomic radii may be involved. As a result, composition cannot always be inferred from unit cell parameters. Goethites contain Si, Ti, V, Zn, Cu, Mo, and P (Norrish, 1975) in addition to Al, and relatively large amounts of these elements are sometimes removed during selective iron oxide extractions.

Goethite synthesized in the presence of Si contained Si in its structure (Schwertmann & Taylor, 1972a) to a maximum of about 0.2% SiO_2 (U. Schwertmann, unpublished). Soil maghemites, as opposed to γ-Fe_2O_3, contained between 5 and 13% of the Fe as Fe^{2+} in a series of samples examined (Taylor & Schwertmann, 1974b), giving a composition intermediate between magnetite and pure maghemite. However, the unit cell size was not related to Fe^{2+} content as expected, possibly due to Al and/or Ti (Fitzpatrick & Le Roux, 1976) again substituting for Fe in the structure.

IV. FORMATION

A. General

Elucidation of the conditions and modes leading to formation of the various iron oxides may indicate present and/or past conditions prevailing in a particular soil environment. Thermodynamic considerations and synthesis experiments can be used. Under equilibrium conditions, compounds will form which are thermodynamically most stable. However, conclusions deduced from stability diagrams constructed from known thermodynamic data (Table 5-1), often disagree with what actually occurs in a soil. This may be due either to differences in free energy data arising from particle size or impurity effects in soil phases, or to the nonattainment of equilibrium between these phases and the environmental solution due to slow reaction rates. The latter effect often occurs when soil minerals are inherited from an earlier weathering cycle where conditions differed from the present, for example, the subjection of a highly weathered tropical soil to drier conditions as has occurred in many parts of Australia (Veen, 1973). Furthermore, transformation of initially kinetically favored metastable phases to more stable ones may be extremely slow.

Mechanisms of formation can also be studied from synthesis experiments. The usefulness of such results in explaining particular mineral associations in soils increases as the conditions during synthesis approach those in soils.

B. Stability Relationships

Almost all thermodynamic data indicate that goethite is the only stable oxide under most soil conditions. Lepidocrocite, ferrihydrite, and maghemite are apparently less stable. However, some disagreement exists concerning hematite-goethite stability relationships. Berner (1969) arrived at a slightly higher stability for hematite. The standard free energy (ΔG_r°) of the reaction

$$\text{hematite} + H_2O = 2 \text{ goethite}$$

is slightly positive (+0.40 kcal/mole). However, Mohr et al. (1972) proposed a ΔG_r° value of –0.2 to –0.1 kcal/mole and Wefers (1966) a ΔH_{298} value of –1 kcal/mole. This difference is probably due to particle size and/or crystallinity differences between the two phases. Langmuir and Whittemore (1971) calculated the dependence of the Gibbs free energy of formation of goethite from hematite plus water as a function of particle size from surface enthalpies of the two forms (Ferrier, 1966). They found that for equal particle sizes, goethite is more stable than hematite for particles $> 760\text{Å}$ (taken as cube edge length), whereas hematite is more stable for smaller particles. Goethite of this particle size (760Å) has a surface area of about 20 m^2/g, assuming a density of 4.28 g cm^{-3}. However, surface areas in the range 23-177 m^2/g have been calculated for various soil goethites (Taylor & Schwertmann, 1974), indicating a smaller size than 760Å. These sizes agree with earlier published results where particle sizes down to 200Å were observed by electron microscopy. Also surface areas of 20-90 m^2/g (Norrish & Taylor, 1961), and particle sizes 150-200Å have been obtained from Mössbauer spectroscopy (Karpachevski et al., 1972). These goethites should theoretically convert to hematite. However, this does not appear to happen.

Activity measurements in soil solutions and natural waters have shown the solid phase in equilibrium with the solution to be a less stable (more soluble) iron oxide than goethite or hematite. Bohn (1967) found a $pK_{sp}(pFe + 3 pOH)$ of 39 for several soils, a value which is in the lower range of ferrihydrites (Feitknecht & Michaelis, 1962). Values for pK_{sp} in natural waters varied between 37.1 and 43.5, the latter being in the range of goethite (Langmuir & Whittemore, 1971). This variation is due to the degree of aging of the ferrihydrite, the extent of its conversion to goethite and/or hematite, and the particle sizes of these phases. Thus, the existence of metastable phases shows that, due to its dynamic nature, the Fe system in most soils is not in equilibrium. The usefulness of thermodynamic data to explain Fe oxide occurrences in soils is therefore limited.

C. Modes of Formation: Synthesis Experiments

In contrast to layer silicate clay minerals, the various forms of iron oxides can easily be synthesized in the laboratory at ambient temperatures and normal pH values. Results of syntheses are summarized in Fig. 5-9 excluding those results obtained under conditions far from those expected in soils, e.g., temperatures $> 100°C$, strongly alkaline solutions etc.

Reactions include dissolution and precipitation, hydrolysis, oxidation and reduction, complex formation, and dehydration. In addition to Fe concentration, pH, E_h and temperature are important factors to be considered. Among the types of transformations between solid species, reconstructive transformations by dissolution and reprecipitation are more frequent, but topotactic (one or two phase) reactions also occur (Mackay, 1960). The relative rates of such reactions also appear to be quite important and are therefore included in these discussions where possible. However, these transformation paths and reaction rates are prone to external environmental influences such as the retardation or acceleration of nucleation and crystal growth induced by the presence of various anions and cations.

1. THE Fe^{3+} SYSTEM

Fast hydrolysis of Fe^{3+} salt solutions at room temperature yields ferrihydrite if its reasonably high solubility product (Table 5-1) is exceeded. Goethite (and sometimes lepidocrocite) is formed through nucleation from solution if hydrolysis proceeds *slowly* under conditions where the comparatively lower solubility product of goethite (Table 5-1), but not that of ferrihydrite, is exceeded; e.g., if Fe^{3+} salt solutions are only partly neutralized $(OH/Fe<2)$ (Feitknecht & Michaelis, 1962; Atkinson et al., 1972a; Fordham, 1970; Hsu, 1972; Hsu & Ragone, 1972). In these solutions, goethite crystals

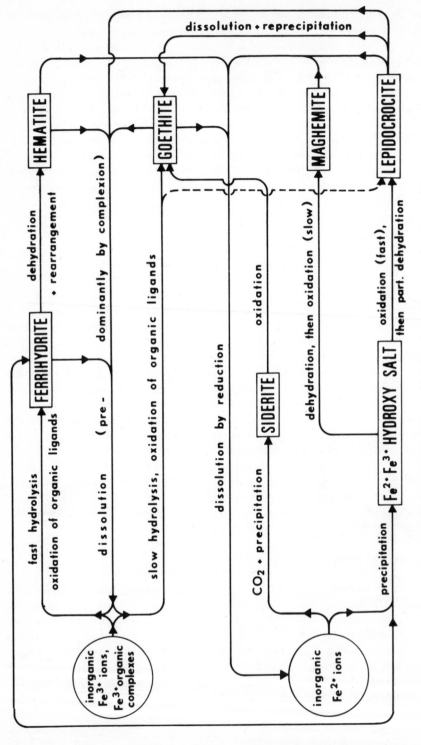

Fig. 5–9. Possible pathways of iron oxide formation under near pedogenic conditions.

are formed from small soluble entities such as $[Fe(OH)_2]^+$ which feed the growing FeOOH crystal (Knight & Sylva, 1974). Goethite can also form in the same way by dissolution of ferrihydrite and subsequent nucleation from solution. In contrast, Murphy (1973)[3] believes that FeOOH crystals grow in solution from tiny (15–30Å) spherical amorphous Fe polycations which first form linear arrays and later coalesce to needle- or lath-shaped crystals.

In the transformation of ferrihydrite to hematite as a competitive reaction to the via solution formation of goethite, the small ferrihydrite particles (70–100Å) first form larger aggregates in which hematite nucleation is facilitated. The hematite crystals then grow at the expense of the ferrihydrite particles and finally form crystal aggregates or coalesce into larger single crystals (Fischer & Schwertmann, 1975). Ferrihydrite, therefore, appears to be a necessary precursor for hematite, but not for goethite.

Since goethite formation from ferrihydrite is via solution, and since solubility of ferrihydrite increases with decreasing pH, goethite formation is faster and the proportion of goethite in a goethite–hematite mixture is higher as the pH decreases from 7 to 4. Goethite is also favored over hematite if the ferrihydrite suspension is seeded with goethite crystals (U. Schwertmann, unpublished). Hematite formation, on the other hand, is favored by those factors which support aggregation and dehydration of the ferrihydrite, such as high concentrations of ferrihydrite and higher temperatures (Schwertmann & Fischer, 1966). As expected this process is very slow at room temperature, but aging experiments conducted over 15 years yielded complete crystallization to hematite (and goethite) at pH 3–9, even in the presence of excess water (Schwertmann, 1965). This shows that neither drying nor high temperature is required for hematite formation from ferrihydrite.

Contrary to published contentions (Gallagher et al., 1974), goethite and hematite cannot be interconverted by simple solid state dehydration-rehydration reactions at ambient temperatures. This can only occur by a reconstructive dissolution-reprecipitation reaction. Since these minerals are extremely stable, their dissolution, which is essential for transformation, is greatly facilitated by reduction and/or complexion solution of the Fe.

2. THE Fe^{2+}-Fe^{3+} SYSTEM

In contrast to the above reactions involving only Fe^{3+} we must consider systems in which Fe is additionally present as Fe^{2+} ions or precipitates. Due to the much lower first hydrolysis constant ($10^{-8.3}$ vs. $10^{-2.2}$), Fe^{2+} ions are much more likely to occur in soils than Fe^{3+} ions if the E_h is sufficiently low. If not oxidized the Fe^{2+} ions may be precipitated as an Fe^{2+} compound or, more likely, as a mixed $Fe^{2+}Fe^{3+}$ compound. However, at pH 7 and with the Fe^{2+} ion concentrations occurring in soils, $Fe(OH)_2$ will not precipitate. Instead, either mixed $Fe^{2+}Fe^{3+}$ hydroxy compounds (e.g., Green Rust), siderite,

---

[3]P. J. Murphy. 1973. Formation and characterization of hydrolysed ferric ion solutions. Ph.D. Thesis. University of Western Australia, Nedlands, W. A. 6009.

$FeCO_3$, or Fe^{2+} sulphides will occur, depending on the pH, the partial pressure of CO_2, and the sulphide concentration (Mohr et al., 1972).

The $Fe^{2+}Fe^{3+}$ hydroxy compound is formed by a reaction of Fe^{2+} ions with ferrihydrite. At zero or low partial pressures of CO_2, its formation is more favored by high pH, high total Fe concentrations in solution, low oxidation rates, and, within certain limits, by high Fe^{3+}/Fe^{2+} ratios. The Fe^{3+} oxides are formed from these $Fe^{2+}Fe^{3+}$ compounds by a one or two phase topotactic oxidative transformation (Bernal et al., 1959), sometimes including dehydration. The order and rate of these processes are important factors in determining the type of Fe^{3+}-oxide formed. They lead either to lepidocrocite or maghemite (Taylor & Schwertmann, 1974b). *Slow* oxidation rates, higher temperatures [20-40C (293-313K)], higher pH values (within pH 6-8), higher total Fe concentrations and the presence of Fe^{3+}, favor the formation of maghemite rather than lepidocrocite. The probable unifying principle explaining these influences is the formation and subsequent dehydration of crystalline $Fe^{2+}Fe^{3+}$ hydroxy compounds before further oxidation proceeds. Both processes appear to be essential for maghemite formation. Magnetite is possibly an intermediate product during maghemite formation.

On *fast* oxidation, the $Fe^{2+}Fe^{3+}$ hydroxy compound will be fully oxidized before partial dehydration and lepidocrocite will be formed (Feitknecht, 1959; Misawa et al., 1973). The presence of high concentrations of CO_2 in the air during oxidation of an Fe^{2+} chloride solution leads to the formation of goethite rather than lepidocrocite (Schwertmann, 1959), possibly explaining the formation of goethite from siderite as found in nature.

The transformation of lepidocrocite into its stable polymorph, goethite, proceeds via solution as shown by electron microscopy and nucleation experiments, provided that the dissolution rate of lepidocrocite and the nucleation rate of goethite are not too low (Schwertmann & Taylor, 1972b).

3. IRON SYSTEMS WITH FOREIGN COMPOUNDS

Pedogenic mineral formation normally takes place in systems in which large numbers of interfering foreign compounds can be expected. Their presence can affect the solubility and rates of dissolution, precipitation, and crystallization of iron oxides. These effects on the crystallization may show up as an acceleration, retardation, or even inhibition of nucleation, and may cause a different crystal morphology or even a change in the mineral phases present.

The most pronounced influence in a soil system comes from organic compounds. For example, higher concentrations of hydroxycarbonic acids, such as citrate, are particularly effective in inhibiting the transformation of ferrihydrite to well-crystallized oxides. This was seen in the H_2O_2 and microbial oxidation of Fe^{3+} citrate solutions at pH 3-8 (Fischer & Ottow, 1972). However, low concentrations of citrate lead to the formation of lath-shaped rather than hexagonal hematite crystals (Schwertmann et al., 1968). The effect of oxalate was again different, and due to nucleation effects (Schwertmann, 1969/1970; Fischer & Schwertmann, 1975), favored the transformation of ferrihydrite to hematite rather than goethite.

Similar inhibitory effects are induced by inorganic anions, especially those with hydroxyl groups such as phosphate (Scheffer et al., 1957) and silicate (Schellmann, 1959). Similarly, goethite formation by oxidation of $FeSO_4$ solution in the presence of phosphate (R. M. Taylor & U. Schwertmann, unpublished data) and lepidocrocite formation from $FeCl_2$ solution in the presence of silicate (Schwertmann & Thalmann, 1976), did not occur until almost all the phosphate or silicate was removed from solution by ferrihydrite formation. Furthermore, silicate retarded the transformation of lepidocrocite to goethite by restricting goethite nucleation (Schwertmann & Taylor, 1972a). Removal of electrolytes from the systems can therefore accelerate crystallization (Gastuche et al., 1964).

In contrast, Fe^{2+} ions markedly increased the dissolution rate of metastable Fe^{3+} oxides, accelerating their transformation into more stable forms. This has been demonstrated for ferrihydrite → goethite transformations (Fischer, 1973), and for lepidocrocite → goethite transformations (Schwertmann & Taylor, 1972b).

D. Mode of Formation: Soils

Although directed towards soil conditions, the results of synthesis experiments as described above are significant only if they predict or explain what is observed in soils.

1. HEMATITE-GOETHITE

In humid temperate zones, hematite is usually not being formed in soils. In warmer climates, however, hematitic soils are widespread and the higher temperature is surely an important factor. In many regions, however, hematitic and nonhematitic soils are closely associated, with the latter generally located in wetter (depressions) and/or cooler (higher altitudes) areas. These environmental conditions most probably inhibit hematite formation through higher concentrations of organic compounds which complex the Fe released and thereby prevent the formation of ferrihydrite, a necessary precursor of hematite. Conversely, in soils in which the Fe released is not readily complexed, ferrihydrite, and consequently hematite, may be formed. This is associated with rapid decomposition of organic matter due to higher temperatures, neutral pH (calcareous soils) and good aeration (see also Fauck, 1974).

Accordingly, hematite in sediments and palaeosols exposed to a subsequent cooler and wetter climate, can frequently transform to goethite as evidenced by a yellow color penetration from the top into the redder subsoil (Schwertmann, 1971). Here the hematite is dissolved by reduction and/or complex formation, and subsequent oxidation and/or precipitation under the new environmental climate causes the neo-formation of goethite (or lepido-crocite), but never hematite. The active participation of organic matter in this process is demonstrated by the occurrence of a yellow zone adjacent to recent roots in a red soil, where under the influence of organic matter sup-plied from the roots, hematite is transformed to goethite.

2. FERRIHYDRITE

Since ferrihydrite rich in organic carbon can be easily synthesized by microbial oxidation of Fe^{3+} citrate solutions (Fischer & Ottow, 1972), a similar mode of formation was suggested for the natural carbon rich samples (Schwertmann & Fischer, 1973). Waters percolating through acidic surface horizons carry organic compounds which cause the dissolution of Fe in the soil. On exposure to oxidation conditions, such as in drainage ditches or springs the soluble Fe compounds are attacked by microorganisms and the oxide is rapidly precipitated. Further crystallization is often prevented by the extensive adsorption of environmental impurities, particularly organics, due to the high surface area of the precipitate (Schwertmann, 1966). Eventually, however, ferrihydrite will transform to a more stable crystalline form of goethite or hematite, depending again on the subsequent environ-mental conditions.

In soils, ferrihydrite could be expected where Fe is oxidized and pre-cipitated in the presence of much organic matter, such as in B horizons of Spodosols. Direct identification by X-ray diffraction techniques is not yet possible, however, due to its low crystallinity and the general presence of other soil material. The high proportion of oxalate soluble Fe in such hori-zons (Blume & Schwertmann, 1969) indicates ferrihydrite.

3. LEPIDOCROCITE

This orange oxyhydroxide has been found in hydromorphic soils of temperate humid climate. The morphology of soil lepidocrocite crystals (Fig. 5-5), the common presence of Fe^{2+} ions in such soils, and their pH range suggest that this mineral possibly formed in soils in a similar manner as in laboratory synthesis. Moreover, the absence of lepidocrocite in calcareous soils, although hydromorphic, agrees with laboratory results which indicate that the introduction of CO_2 into the system favors formation of goethite rather than lepidocrocite. A green crystalline $Fe^{2+}Fe^{3+}$ hydroxy compound, (Green Rust), which in synthesis oxidizes to and appears to be the precursor of lepidocrocite, has not been identified in soils. However, hydromorphic soils under strong anaerobic conditions quite often exhibit this green colora-tion which readily becomes rust colored on exposure to air.

The transformation of soil lepidocrocite to goethite appears to be a very slow process due to its low solubility and dissolution rate. In accordance with synthesis experiments, this transformation is retarded by silicate and ac-celerated by Fe^{2+} ions in solution (Schwertmann & Taylor, 1973).

4. MAGHEMITE

The formation of maghemite in soils has been explained by (i) aerial oxidation of magnetite, indicated by pseudomorphs of maghemite after mag-netite (Bonifas, 1959), (ii) the dehydration of lepidocrocite, and (iii) the thermal transformation of various oxides in the presence of organic matter (Van der Marel, 1951; Schwertmann & Heinemann, 1959). Since maghemite also occurs abundantly in soils whose parent material contains only traces, if any magnetite, and since firing can only be of very localized significance, an-other mode of formation must be considered (Oades & Townsend, 1963). On the basis of synthesis studies, an $Fe^{2+}Fe^{3+}$ hydroxy compound is probably a necessary precursor from which maghemite is formed by slow oxidation and simultaneous dehydration. The relative rates of these two processes de-termine whether lepidocrocite or maghemite is formed (Taylor & Schwert-mann, 1974b). The effect of dehydration might explain the widespread oc-currence of maghemite in soils of warmer climates. An ample supply of Fe^{2+} from the weathering of primary silicates appears to favor its formation, in line with synthesis experiments, and explains its preferential occurrence in soils of basic igneous rocks (W. Schellmann, Bundesanstalt für Geowissen-schaften und Rohstoffe, Hannover-Buchholz 3, West Germany, private com-munication.).

V. PROPERTIES RELEVANT TO SOILS

A. Surface Structure

Reactions between iron oxides and components of the soil solution take place at the solid-solution interface. Therefore, the structure and reactivity of this surface is an important property of iron oxides.

If not structurally hydroxylated, as on the surface of goethite and lepidocrocite, the surface Fe atoms in an aqueous medium complete their coordination shell of nearest neighbors through a reaction with water as fol-lows (Breeuwsma, 1973):

$$
\begin{array}{ccccc}
& \text{O} & & \text{OH} & \text{OH} \\
& \diagup\ \diagdown & & | & | \\
\text{Fe} & & \text{Fe} + \text{H}_2\text{O} = \text{Fe} & & \text{Fe} \\
& \diagdown\ \diagup & & \diagdown\ \diagup & \\
& \text{O} & & \text{O} &
\end{array}
$$

This "chemisorbed" water leads to a complete hydroxylation of the sur-face. For hematite, Breeuwsma (1973) suggests a surface hydroxylation density between 4.5 and 9 OH groups per 100Å^2. Hydroxyl groups having varying properties exist not only between different minerals, but also on sur-

faces of the same mineral, e.g., singly (Fe–OH) or doubly bound OH(Fe–O–Fe)
                                                                                   H

(Onoda & de Bruyn, 1966; Boehm, 1971). However, for practical purposes it can be generally assumed that the hydroxylated surfaces of iron oxides in soils have similar properties.

B. Water Adsorption

The hydroxylated oxide surface physically adsorbs additional water molecules in amounts depending on the type of oxide and, to a greater extent, on the surface area. In fact, water adsorption is widely used to determine surface area from a monolayer coverage (BET method). The first layer appears to be tightly bound by goethite (Gast et al., 1974) and by hematite (Breeuwsma, 1973) and adsorbed anions do not appear to influence the adsorption, at least in the case of hematite (Jurinak, 1966). Water adsorption is probably due to hydrogen bonding by structural (or surface) protons as suggested below (Breeuwsma, 1973):

$$
\begin{array}{cccc}
 & & \overset{\displaystyle H \qquad H}{\underset{\displaystyle \diagdown \quad \diagup}{}} & \\
 & & O & \\
 & & \diagup \; \diagdown & \\
\overset{\displaystyle H}{\underset{\displaystyle O}{}} & \overset{\displaystyle H}{\underset{\displaystyle O}{}} & \overset{\displaystyle H}{\underset{\displaystyle O}{}} & \overset{\displaystyle H}{\underset{\displaystyle O}{}} \\
| & | & | & | \\
Fe & Fe + H_2O = Fe & & Fe \\
 \diagdown & \diagup & \diagdown & \diagup \\
 & O & & O
\end{array}
$$

In contrast to goethite and hematite, ferrihydrite looses its physisorbed water completely on outgassing at 25C (Gast et al., 1974). Both chemically and physically adsorbed water essentially determine the reactivity of oxide surfaces (surface charge, adsorption, catalytic activity, electrical conductivity etc.).

C. Surface Charge

As a result of their hydroxylated surface, the surface charge and potential of iron oxides are determined by the concentration of H^+ and OH^- ions in solution. The surface charge is created by an adsorption or desorption of H^+ (or a desorption or adsorption of OH^-, respectively) in the potential determining layer consisting of surface O, OH, and OH_2 groups. H^+

and OH^- are therefore called *potential determining ions*. The following model has been proposed by Parks and de Bruyn (1962):

$$
\begin{array}{c}
\underset{OH_2}{\overset{OH_2}{\diagup}} \\
Fe \\
\underset{OH_2}{\diagdown}
\end{array}
\Bigg|^+
A^-
\underset{+\ H^+}{\overset{+OH^-}{\rightleftharpoons}}
\begin{array}{c}
\overset{OH_2}{\diagup} \\
Fe \\
\underset{OH}{\diagdown}
\end{array}
\Bigg|^0
\underset{+\ H^+}{\overset{+OH^-}{\rightleftharpoons}}
\begin{array}{c}
\overset{OH}{\diagup} \\
Fe \\
\underset{OH}{\diagdown}
\end{array}
\Bigg|^-
C^+
$$

The pH at which equal amounts of H^+ and OH^- ions are adsorbed is called the *zero point of charge* (ZPC). From this model it is obvious that the Fe atoms do not participate directly in charge development. An excess of surface charge of the potential determining layer is balanced by an equivalent amount of anions (A^-) or cations (C^+), located in the outer part of an electric double layer. In addition to pH, the charge also depends on the concentration and valence of the electrolyte in the equilibrium solution. (A detailed description is given in Chapter 2.)

Values recently reported for the ZPC of synthetic hematite (Atkinson et al., 1967; Breeuwsma, 1973) and goethite (Atkinson, 1969)[4] lie in the range of pH 7.5-9.3 without a significant difference between the two minerals. This indicates a similar surface structure and activity with regards to the Fe–OH groups. Hematite in an aqueous environment appears to have a "goethite" surface (Onoda & de Bruyn, 1966; Boehm, 1971).

For 10 different goethite preparations (specific surface area 14-77 $m^2/$g) with ZPC values of pH 7.6-8.9, the surface charge density at pH 4 in $1M$ NaCl is between 1.8 and 4.9 $\mu eq/m^2$ (Atkinson, 1969).[4] Values of 1.6-4.9 reported for hematites (specific surface area 34-45 m^2/g) are in the same range (Atkinson et al., 1967; Breeuwsma, 1973). Parks (1965) gave a general ZPC range for different oxides of the type M_2O_3 (which includes all Fe^{3+} oxides) as pH 6.5-10.4, and observed that the ZPC could vary somewhat even for the same oxide. A few ZPC data exist for lepidocrocite (pH 5.4-7.3), maghemite (pH 6.7), and ferrihydrite (pH 8.1, Kinniburgh et al., 1975). The ZPC values are generally lower for natural samples than those found for the synthetic ones (Parks, 1965).

D. Ion Adsorption

As mentioned in the previous section, the pH-dependent charge of iron oxides (in contrast to those layer silicates that have permanent charge from isomorphous substitution) is balanced by an equivalent amount of anions or cations (counter ions), electrostatically held in the outer diffuse electric double layer. This type of adsorption which treats the counter ion as a point

---

[4] R. J. Atkinson. 1969. Crystal morphology and surface reactivity of goethite. Ph.D. Thesis. University of Western Australia, Nedlands, W. A. 6009.

Table 5-2. Specific adsorption of anions and cations by iron oxides

| Oxide | Ion | Conditions of adsorption | | | | Amount adsorbed | | Authors |
		pH	Electrolyte	Temperature, °C	Equilibrium concentration, mmole/liter	μmole/g	μmole/cm² ×10⁴	
Hematite	Phosphate	5	No	20	0.25	34	1.6	Breeuwsma, 1973
Hematite	Molybdate	4	No	22	0.1	50	5.0	Reyes & Jurinak, 1967
Hematite	Silicate	5	No	25	0.3	38	‡‡	Beckwith & Reeve, 1963
Hematite	Lead	4.8	No	‡	18	385	‡‡	Hildebrand & Blum, 1974
Goethite	Phosphate	5	0.01M NaCl	‡	0.5	80	2.5	Hingston et al., 1968
Goethite	Molybdate	5	0.01M NaCl	20-23	2.0†	300	3.7	Hingston et al., 1974
Goethite	Silicate	5	0.01M NaCl	20-23	1.2†	160	2.0	Hingston et al., 1974
Goethite	Sulphate	5	0.01M NaCl	20-23	1.5-3.0†	60	0.73	Hingston et al., 1973
Goethite	2,4 D	5	0.01M NaCl	20	0.5	100	2.4	Watson et al., 1973
Goethite	Zinc	7	0.1M NaCl	25	0.1†	102	1.4	Bowden et al., 1974
Goethite	Lead	4.8	No	‡	18	65	‡‡	Hildebrand & Blum, 1974
Ferrihydrite	Molybdate	5	No	‡	10⁻³	200	‡‡	Reisenauer et al., 1962
Ferrihydrite	Borate	7.3	No	‡	72†	24	‡‡	Sims & Bingham, 1967
Ferrihydrite	Copper	5.5	0.05M CaCl₂	30	0.16	126§	‡‡	McLaren & Crawford, 1974
Ferrihydrite	Lead	4.8	No	‡	18	385	‡‡	Hildebrand & Blum, 1974
Ferrihydrite	Zinc	5.5	1M NaNO₃	20-27	1.0	95¶	‡‡	Kinniburgh & Jackson, 1974#
Ferrihydrite	Calcium	8.0	1M NaNO₃	20-27	1.0	96¶	‡‡	Kinniburgh et al., 1975

† Initial concentration.
‡ Not given.
§ Calculated Langmuir maximum.
¶ No adsorption maximum.
# D. G. Kinniburgh and M. L. Jackson, 1974. Zinc adsorption by iron hydrous oxide gels. Agron. Abstr. p. 122.

charge is termed *nonspecific*. In a multicomponent system of ions of equal valency these ions are adsorbed in simple proportion to their equilibrium activity in solution. Examples of such ions would be the anions Cl^-, NO_3^-, ClO_4^- and most of the alkali and alkaline earth cations. In this particular situation, replacement of one ion by another at constant ionic strength would not change the electrophoretic mobility of the particles. The definition of nonspecific adsorption can be extended to include the effect of ion size because the size (distance of closest approach to the surface) of the ion will determine the potential distribution in the diffuse layer and hence the distribution of counter ions.

In contrast, other anions and cations are held much more strongly by the oxide surface because these ions penetrate the coordination shell of the Fe atom, exchange their OH and OH_2 ligands and are bound by covalent bonds directly via their O and OH groups to the structural cation. This type of adsorption in the potential determining layer has been termed *chemisorption* or *specific adsorption*. It is further characterized by the ion adsorbing on neutral surfaces and surfaces with a charge of the same sign as the ion. These ions will reverse the sign of the surface charge which nonspecifically adsorbed ions do not (for further details see Chapter 2).

Specific adsorption of phosphate by iron oxides is most widely accepted and is an important concept in soil and environmental studies. However, specific adsorption of other anions, such as silicate (McKeague & Cline, 1963), molybdate, arsenate, selenate, sulphate, and organic anions are equally important. Some metal cations, e.g., Zn, Cd, Cu, Pb and even the small Li and Mg ions (Breeuwsma, 1973), can be specifically adsorbed.

The amount of a particular ion adsorbed depends mainly on its concentration in, and the pH of, the equilibrium solution. At a given pH, adsorption increases with increasing concentration. In many cases this relation may be described by one of the well-known adsorption isotherms (Langmuir, Freundlich, and others). Polyvalent anions such as HPO_4^{2-} in fact may create an additional negative charge (increase in CEC) which is countered by cations in the diffuse double layer.

Maximum adsorption values for various ions (from isotherms) on iron oxides have been provided (Table 5-2). At pH 4-5 the values range between 30 to 300 μmole/g of oxide which compares favorably with cation adsorption on clay minerals. As expected, sulphate shows the lowest adsorption, being only partly specifically adsorbed. The apparent difference between values for the other ions and between the different iron oxides decreases significantly if the adsorption is expressed per unit surface area. This suggests that the variation between different oxide samples originates mainly from their surface area differences rather than from composition.

Metals are preferentially adsorbed by goethite in the order Cu > Zn > Co > Pb > Mn (Grimme, 1968; Hildebrand & Blum, 1974). Grimme (1968) showed that adsorption at any pH increased with a decreasing solubility product or hydrolysis constant of the corresponding metal hydroxide. He took this as a measure of the metal (Me) affinity to the hydroxylated surface of the oxide. Most probably the hydrolized form, $MeOH^+$, has a higher affinity to the oxide surface than Me^{++}, explaining the increased adsorption with increasing pH because $[MeOH^+]/[Me^{2+}]$ increases. Among the simple organic anions strongly adsorbed by iron oxides are citrate (Schwertmann et al., 1968) tartrate, oxalate, and malate (Nagarajah et al., 1970). Such anions can compete with adsorbed phosphate in the same way as silicate for the iron oxide adsorption sites in soils, and can therefore increase phosphate availability to plants (Hingston et al., 1968).

The structure of the ion-surface bond in specific adsorption is not completely known. Mono(a) and bidentate(b) complexes may exist explaining different bonding strengths (b > a), and the partial irreversibility of phosphate sorption (Hingston et al., 1974).

(a)   (b)

(a) A structure with charge $2-$: an Fe bearing OH_2 groups bonded through bridging O atoms to a second Fe bearing OH_2, with a monodentate phosphate group ($O=P$ with OH and O) linked via O.

(b) A structure with charge $1-$: an Fe bearing OH_2 groups and a second Fe bearing OH_2, bridged through O atoms to a bidentate phosphate group (P with two bridging O and one O).

Infrared absorption studies (Parfitt et al., 1975) favor the bidentate form in which mostly single coordinated rather than double or triple coordinated OH groups of the oxide surface are involved.

A theoretical description of the pH dependency of specific adsorption of conjugate acids or bases by oxide surfaces was recently given by Bowden et al. (1974). This explanation is based mainly on the combined effect of the change with pH in the concentration of the ionic species being adsorbed and the surface charge of the oxide. As an example, this may be explained for the H_4SiO_4-$H_3SiO_4^-$ pair at low concentration as follows. At several pH units below the pK_a, (9-10), the acid is present mainly as H_4SiO_4. With increasing pH, $[H_3SiO_4^-]$ increases along an S-shaped curve, whereas the surface charge decreases more or less linearly. Therefore, assuming that the neutral molecule is not adsorbed to any significant degree (which might not be the case) the amount of Si adsorbed depends mainly on $[H_3SiO_4^-]$ and thus increases strongly with pH. As long as the rate of increase of $[H_3SiO_4^-]$ with pH is higher than the rate of decrease of positive surface charge, silicate adsorption will continue to increase. It reaches a maximum where the two rates are equal. This is somewhere near

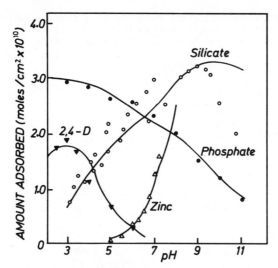

Fig. 5-10. Adsorption of orthophosphate, silicate, zinc, and 2,4-D in NaCl on goethite surface as a function of pH (Bowden et al., 1974).

the pK of the conjugate acid (Fig. 5-10). Beyond this point the silicate adsorption maximum decreases because the positive (negative) charge decreases (increases) faster with pH than the $[H_3SiO_4^-]$ increases.

For polyvalent anions such as phosphate a break instead of a maximum occurs in the range of their pK values as one charged species changes to the next. In contrast to Hingston et al. (1968), Breeuwsma (1973) did not find breaks at pK values for P adsorption on hematite. He suggests that, depending on pH, OH_2 molecules are exchanged by phosphate from positive surface groups and OH^- ions from uncharged or negative groups. This should result in a ratio of phosphate adsorbed to OH released between 0 at low pH and 1 at high pH which was confirmed experimentally. The increase in P adsorption with decreasing pH was explained by an increasing ratio of OH_2^+/OH surface groups.

Molybdate adsorption approaches zero at a lower pH than phosphate in accord with the lower pK values of the former (Barrow, 1970). Boron adsorption shows a flat maximum at around pH 9 (Sims & Bingham, 1967).

Extrapolation of these results to soils is difficult, although the amount of P incorporated per unit surface in goethitic iron oxide accumulations is in the range that has been found for P adsorption on pure synthetic goethites, (Table 5-3), indicating that the amounts of P in the accumulations is related to the goethite surface area. Indirect evidence concerning the importance of soil iron oxides in specific adsorption comes from the general observation that soils rich in iron oxides (e.g., Oxisols) fix large amounts of fertilizer P, or show deficiencies in available molybdenum. Furthermore, substantial proportions of the total P and other heavy metals are often extracted by free iron removal treatments, and after deferration the adsorption capacity of the soil for these elements often decreases.

Table 5-3. Comparison of P adsorbed and P incorporated per unit surface area between synthetic goethites (Atkinson et al., 1972b) and goethite rich soil formations, respectively (Taylor & Schwertmann, 1974a)

Sample	Content		Specific surface†	P
	Fe	P		
	— % —		m²/g	mg/m²
Goethites (synth.)	--	--	14–77	0.078‡
Weathering crust				
Hard	47.2	0.13	23	0.076
Soft	25.9	0.22	70	0.087
Bog iron ores				
(5 samples)	38.7‡	0.84‡	108–177	0.069–0.145

† For soil goethites determined from X-ray diffraction line broadening (Schwertmann, 1973b).
‡ Average figure.

E. Soil Aggregation

Cementation effects of iron oxides in such formations as bog iron ores, lateritic iron crusts, and concretions are well known for their detrimental effects on root penetration and development. Although the beneficial effect of iron oxide on soil aggregation is still an open question (Desphande et al., 1964), a significant correlation between water stability of aggregates and iron oxide content (Kemper, 1966) certainly points in this direction. Experiments with synthetic iron oxides also clearly demonstrate the aggregating ability of iron oxides. Recent results obtained by Blackmore (1973) suggest that freshly prepared ferrihydrite is specifically effective, whereas crystalline oxides (as measured from the ratio of oxalate soluble to total Fe) are much less effective (Table 5-4). Small additions of ferrihydrite (up to 3%) produce strong aggregation of silt. Adsorption of iron oxides on kaolinite surfaces has been demonstrated by various workers who showed that, particularly at low pH, positively charged iron oxide particles may be bound to a negatively

Table 5-4. The influence of synthetic iron oxide addition to a loess soil (86% silt) on the formation of 1–2 mm water stable aggregates after air drying (after Schahabi & Schwertmann, 1970).

Fe-oxide	Amount added, % Fe	Oxalate solubility†, % of total Fe	Aggregates 1–2 mm, %
No addition	0.0	--	18
Ferrihydrite	1.0	100.0	48
Ferrihydrite	2.3	100.0	83
Lepidocrocite	1.0	68.0	53
Goethite	1.0	0.1	17
Goethite	0.6	93.0	77
Hematite	1.0	0.1	20

† Taken as a measure of degree of crystallinity (Schwertmann, 1964).

charged clay silicate surface (Greenland & Oades, 1968; Follett, 1965; Fordham, 1970).

VI. DETERMINATION

Because of the diversity of available diagnostic criteria (Table 5-1), pure crystalline iron oxides can generally be detected. The most useful and widely used procedures are X-ray diffraction and differential thermal analysis (DTA), although other techniques, such as infrared and Mössbauer spectroscopy (Gangas et al., 1972) are increasingly being applied. However, in soils many of these diagnostic techniques cannot be used. Serious limitations to the efficiency of X-ray diffraction identification arises from low concentrations, the diffuseness of patterns caused by small particle sizes and/or poor crystallinity, and the nature of the other minerals present. Therefore a certain minimum concentration is quite often required for qualitative detection. Concentration procedures such as particle size separation (iron oxides are often concentrated in the fine fractions) or dissolution of kaolin minerals in $5N$ NaOH (Norrish & Taylor, 1961) are therefore recommended. In X-ray diffraction it is preferable to employ $CoK\alpha$ or $FeK\alpha$ radiation to avoid the high fluorescence background caused by the more energetic Cu radiation on a ferruginous sample. A comparison of the sample before and after removal of the Fe oxides (e.g., by dithionite treatment) is often helpful.

1. GOETHITE

Goethite can be recognized by its strong X-ray diffraction lines at 4.18, 2.69, and 2.44Å. The 4.18Å line, the strongest, cannot generally be resolved from the 4.26Å line of quartz, especially in low goethite-high quartz samples. Moreover, in soils the usual silicate clays have tailed ($0kl$) lines in this region making identification more difficult, especially with low resolution equipment. Presence of the 2.69Å line is taken as an indication that goethite and/ or hematite is present, since the strongest line of the latter at 2.70Å is coincident. However, if sufficient amounts of goethite are present the 2.44Å line is also visible.

If the sample is free of hematite, a semiquantitative estimation of goethite can be made from the integrated peak area of the 2.69Å line and the mass absorption coefficient of the sample by comparing the results with those for pure goethites (Norrish & Taylor, 1962). In the presence of hematite the increase in intensity of this line after heating the sample for 1 hour at 330C (603K) can be used (Taylor & Graley, 1967). The endothermic peak in the differential thermogram between 280 and 400C (553 and 673K) is a reasonably reliable indicator. In soil goethites, especially in poorly crystalline samples, the endotherm is generally located in the lower part of this range. The goethite content can be estimated by the area of this endothermic peak (Schwertmann & Fanning, 1976). Hematite does not interfere in this case.

2. HEMATITE

Hematite can often be recognized in the field by the bright red color imparted to soils even at low concentrations. In X-ray diffraction analysis, a line doublet at 2.70Å and 2.52Å is highly diagnostic. At higher concentrations, as in concretions, a weaker line at 3.68Å appears, which is sometimes difficult to resolve from the (002) kaolinite line. Recently Mössbauer spectroscopy has been successfully used to detect hematite crystals as small as 40Å in diameter in a red Mediterranean soil of Greece (Gangas et al., 1972).

3. LEPIDOCROCITE

An orange color in the field gives a good first indication of the presence of lepidocrocite. Due to its better crystallinity, it can be recognized by its X-ray diffraction line at 6.25Å at concentrations of only a few percent. Distinction from boehmite is made by dithionite treatment, which causes dissolution of lepidocrocite but not boehmite. Electron micrographs of lepidocrocite generally exhibit characteristic morphologies (Fig. 5-1b, 5-5). With DTA, an endothermic peak between 300–350C (573–623K), which cannot be differentiated from that due to goethite dehydroxylation, is immediately followed by a somewhat weaker exotherm arising from a transformation to maghemite. The presence of ferromagnetism after heating (Scheffer et al., 1959) is, however, not a reliable indication of initial lepidocrocite, as other Fe oxides heated in the presence of organic matter may also form maghemite (Schwertmann, 1959b).

4. MAGHEMITE

Maghemite can often be concentrated with a hand magnet and identified by X-ray diffraction peaks at 2.95Å and 2.52Å. Although these spacings differ slightly from those of magnetite in pure minerals, it is difficult to distinguish between them in soils due to the effects of isomorphous substitution and the Fe^{2+} content, which cause variations in the unit cell size (Taylor & Schwertmann, 1974b). The reddish-brown color and the low Fe^{2+} content are further supporting evidence for maghemite rather than magnetite. In the presence of hematite, the 2.52Å line is no longer diagnostic due to the interference from the coincident hematite line. In this case a change in the intensity ratio of the 2.70 and 2.52Å lines from that expected for pure hematite, accompanied by a line at 2.95Å, indicates that maghemite is present.

5. FERRIHYDRITE

Ferrihydrite can only be detected by X-ray diffraction in high concentrations because of its very weak and broad lines.

6. DIFFERENTIAL DISSOLUTION TREATMENTS

These are widely used to determine iron oxides in soils. The dithionite-bicarbonate-citrate treatment (Mehra & Jackson, 1960) results in almost complete solution of iron oxides without differentiation between various crystalline forms. Particle size and crystallinity effects may, however, influence the reducibility and dissolution rate. Coarse and highly crystalline samples therefore require several treatments. Maghemite in a mixture with other crystalline iron oxides can be preferentially dissolved by $0.2M$ oxalic acid (R. M. Taylor & U. Schwertmann, unpublished), whereas it often tends to be concentrated during initial dithionite extractions.

An acid ammonium oxalate treatment is used extensively for measuring the "amorphous" proportion of the iron oxides (Tamm, 1922; Schwertmann, 1964, 1973b; McKeague & Day, 1966). Goethite and hematite are only slightly attacked, if at all, as long as photochemical reduction is excluded by extracting in the dark and sources of Fe^{2+} such as siderite, are absent. Maghemite and lepidocrocite (Pawluk, 1972) appear to be partly soluble whereas ferrihydrite dissolves completely (Schwertmann & Fischer, 1973). Iron bound to organic matter will also be dissolved by oxalate. In spite of these limitations, the ratio of oxalate to dithionite iron oxides is widely used for soil genesis studies (McKeague & Day, 1966; Blume & Schwertmann, 1969; McKeague et al., 1971). Also the difference between dithionite- and oxalate-soluble iron was significantly correlated with the amounts of goethite determined by X-ray diffraction or DTA for goethite in soil samples (Schwertmann, 1959a). "Amorphous" iron oxide contents in typical soils have been determined by a technique employing eight consecutive alternating extractions with $8N$ HCl and $0.5N$ NaOH and extrapolation of the resultant extraction curves (Segalen, 1968).

VII. CONCLUSIONS

Iron oxides occur in almost all soils. They may (i) determine soil color; (ii) create adsorption sites for various anions and cations, particularly those of importance in plant nutrition and environmental pollution; (iii) promote aggregation and cementation; and (iv) reflect some aspects of soil genesis. Their mineralogy is reasonably simple, although the reasons for the existence of different forms in various soils can only come from a thorough knowledge of all the pathways leading to their formation. These are still not fully understood, and further synthesis experiments under near pedogenic conditions should be carried out, especially in the presence of soil compounds which might interfere. Previous results cited in this survey satisfactorily explain iron oxide occurrences in several pedogenic environments.

ACKNOWLEDGMENT

The authors gratefully acknowledge the valuable discussion with Dr. A. Posner, University of Western Australia. They also thank Mr. D. G. Kinniburgh, University of Bristol, for supplying unpublished data.

LITERATURE CITED

Atkinson, R. J., A. M. Posner, and J. P. Quirk. 1967. Adsorption of potential determining ions at the ferric oxide aqueous electrolyte interface. J. Phys. Chem. 71:550–557.

Atkinson, R. J., A. M. Posner, and J. P. Quirk. 1972a. Crystal nucleation in Fe(III) solutions and hydroxide gels. J. Inorg. Nucl. Chem. 30:2371–2381.

Atkinson, R. J., A. M. Posner, and J. P. Quirk. 1972b. Kinetics of isotopic exchange of phosphate at the α-FeOOH-aqueous solution interface. J. Inorg. Nucl. Chem. 34: 2201–2211.

Barrow, N. J. 1970. Comparison of the adsorption of molybdate, sulfate and phosphate by soils. Soil Sci. 109:282–288.

Beckwith, R. S., and R. Reeve. 1963. Soluble silica in soils I. Aust. J. Soil Res. 1:157–168.

Bernal, J. O., D. R. Dasgupta, and A. L. Mackay. 1959. The oxides and hydroxides of iron and their structural interrelationships. Clay Miner. Bull. 4:15–30.

Berner, R. A. 1969. Goethite stability and the origin of red beds. Geochim. Cosmochim. Acta 33:267–273.

Beutelspacher, H., and H. W. van der Marel. 1968. Atlas of electron microscopy of clay minerals and admixtures. Elsevier Publ. Co., New York.

Blackmore, A. V. 1973. Aggregation of clay by the products of iron (III) hydrolysis. Aust. J. Soil Sci. 11:75–82.

Blume, H. P., and U. Schwertmann. 1969. Genetic evaluation of the profile distribution of aluminum, iron and manganese oxides. Soil Sci. Soc. Am. Proc. 33:438–444.

Boehm, H. P. 1971. Acidic and basic properties of hydroxylated metal oxide surfaces. Discuss. Farad. Soc. 52:264–275.

Bohn, H. L. 1967. The $(Fe)(OH)^3$ ion product in suspensions of acid soils. Soil Sci. Soc. Am. Proc. 31:641–644.

Bonifas, M. 1959. Contribution a l'étude géochimique de l'altération latérique. Mem. Serv. Carte Géol. Alsace Lorraine 17. 159 p.

Bowden, J. W., M. D. A. Bolland, A. M. Posner, and J. P. Quirk. 1974. Generalized model for anion and cation adsorption at oxide surfaces. Nature (London) 245: 81–82.

Breeuwsma, A. 1973. Adsorption of ions on hematite (α-Fe$_2$O$_3$). Med. Landbouwhogeschool, Wageningen 73-1. 124 p.

Brown, G. 1953. The occurrence of lepidocrocite in British soils. J. Soil Sci. 4:220–228.

Caillère, S., L. Gatineau, and St. Hénin. 1960. Préparation à base témperature d'hématite alumineuse. Compt. Rend. Acad. Sci. Seance, Mai 1960:3677–3679.

Chukhrov, F. V., B. B. Zvyagin, L. P. Ermilova, and A. I. Gorshkov. 1973. New data on iron oxides in the weathering zone. Proc. Int. Clay Conf., 1972 (Madrid) 1:397–404.

Chukhrov, F. V., L. P. Ermilova, B. B. Zvyagin, and A. I. Gorshkov. 1976. Genetic system of hypergene iron oxides. p. 275–286. In S. W. Bailey (ed.) Proceedings of the International Clay Conference, 1973 (Mexico City). Applied Publishing, Ltd., Wilmette, Ill.

Correns, C. W., and W. von Engelhardt. 1941. Röntgenographische Untersuchungen über den Mineralbestand sedimentärer Eisenerze. Nachr. Akad. Wiss. Göttingen. Math.-Phys. Kl. 213:131–137

Dasgupta, D. R., and A. L. Mackay. 1959. β-Ferric oxyhydroxide and green rust. J. Phys. Soc. Japan 14:932–935.

Deshpande, T. L., D. J. Greenland, and J. P. Quirk. 1964. Role of iron oxides in the bonding of soil particles. Nature (London) 201:107–108.

Eswaran, H., and N. G. Raghu Mohan. 1973. The microfabric of petroplinthite. Soil Sci. Soc. Am. Proc. 37:79–82.

Fauck, R. 1974. Les facteurs et les mecanismes de la pédogenese dans les sols rouges et jaunes sur sable et grès en Afrique. Cah. O.R.S.T.O.M. Ser. Pedol. 12:69–72.

Feitknecht, W., and G. Keller. 1950. Über Hydroxide und basische Salze des 2 wertigen Eisens und deren dunkelgrüne Oxydationsprodukte. Z. Anorg. Chem. 262:61–68.

Feitknecht, W. 1959. Über die Oxydation von festen Hydroxyverbindungen des Eisens in wässrigen Lösungen. Z. Elektrochem. 63:34–43.

Feitknecht, W., and W. Michaelis. 1962. Über die Hydrolyse von Eisen (III)-perchlorat-Lösungen. Helv. Chim. Acta 45:212–224.

Ferrier, A. 1966. Influence de l'etat de division de la goethite et de l'oxyde ferrique sur le chaleurs de réaction. Rev. Chim. Mineral. 3:587–615.

Fischer, W. R. 1973. Die Wirkung von zweiwertigem Eisen auf Auflösung und Umwandlung von Eisen(III)-hydroxiden. p. 37–44. In E. Schlichting and U. Schwertmann (ed.) Pseudogley and Gley: Genesis and use of hydromorphic soils. Int. Soc. Soil Sci., Trans. (Stuttgart, W. Ger., 1971) Comm. V and VI. Verlag Chemie Weinheim/Bergstr, W. Ger.

Fischer, W. R., and J. C. G. Ottow. 1972. Abbau von Eisen(III)-citrat in durchlüfteter wässriger Lösung durch Bodenbakterien. Z. Pflanzenernaehr. Bodenkd. 131:243–253.

Fischer, W. R., and U. Schwertmann. 1975. The formation of hematite from amorphous iron(III)-hydroxide. Clays Clay Miner. 23:33–37.

Fitzpatrick, R. W., and J. Le Roux. 1976. Pedogenic and solid solution studies on iron-titanium minerals. p. 585–599. In S. W. Bailey (ed.) Proc. of the International Clay Conference, 1975 (Mexico City). Applied Publishing, Ltd., Wilmette, Ill.

Follett, E. A. C. 1965. The retention of amorphous colloidal "ferric hydroxide" by kaolinites. J. Soil Sci. 16:334–341.

Fordham, A. W. 1970. Sorption and precipitation of iron on kaolinite III. Aust. J. Soil Res. 8:107–122.

Gallaher, R. N., H. F. Perkins, and K. H. Tan. 1974. Chemical and mineralogical changes in glaebules and enclosing soil with depth in a plinthite soil. Soil Sci. 117:336–342.

Gangas, N. H., A. Simopoulos, A. Kostikas, N. J. Yassoglou, and S. Filippakis. 1972. Mössbauer studies of small particles of iron oxides in soil. Clays Clay Min. 21:151–160.

Garrels, R. M., and C. L. Christ. 1965. Solutions, minerals and equilibria. Harper & Row, New York.

Gast, R. G., E. R. Landa, and G. W. Meyer. 1974. The interaction of water with goethite (α-FeOOH) and amorphous hydrated ferric oxide surfaces. Clays Clay Miner. 22:31–39.

Gastuche, M. C., T. Bruggenwert, and M. M. Mortland. 1964. Crystallization of mixed iron and aluminum gels. Soil Sci. 98:281–289.

Greenland, D. J., and J. M. Oades. 1968. Iron hydroxides and clay surfaces. Int. Congr. Soil Sci., Trans. 9th (Adelaide, Aust.) I:657–668.

Grimme, H. 1968. Die Adsorption von Mn, Co, Cu und Zn durch Goethit aus verdünnten Lösungen. Z. Pflanzenernaehr. Bodenkd. 121:58–65.

Haag, G. 1935. Die Krystallstruktur des magnetischen Ferrioxyds, γ-Fe$_2$O$_3$. Z. Phys. Chem. B. 29, 95–103.

Hildebrand, E. S., and W. E. Blum. 1974. Lead fixation by iron oxides. Naturwissenschaften 61:169–170.

Hiller, J. E. 1966. Phasenumwandlungen im Rost. Werkst. Korrosion, Verlag Chemie, Weinheim/Bergstr, W. Ger. 17:943–951.

Hingston, F. J., R. J. Atkinson, A. M. Posner, and J. P. Quirk. 1968. Specific adsorption of anions on goethite. Int. Congr. Soil Sci., Trans. 9th (Adelaide, Aust.)I:669–678.

Hingston, F. J., A. M. Posner, and J. P. Quirk. 1974. Anion adsorption by goethite and gibbsite: II. J. Soil Sci. 25:16–26.

Hsu, P. H. 1972. Nucleation, polymerization and precipitation of FeOOH. J. Soil Sci. 23:409–419.

Hsu, P. H., and S. E. Ragone. 1972. Ageing of hydrolized iron(III) solutions. J. Soil Sci. 23:17–31.

Jackson, M. L. 1967. Chemical composition of soils. p. 71–141. *In* F. E. Bear (ed.) Chemistry of the soil. 2nd ed. Reinhold Publ. Corp., New York.

Jackson, T. A., and W. D. Keller. 1970. A comparative study of the role of lichens and "inorganic" processes in the chemical weathering of recent Hawaii lava flows. Am. J. Sci. 296:446–460.

Jurinak, J. J. 1966. Surface chemistry of hematite: Anion penetration effect on water adsorption. Soil Sci. Soc. Am. Proc. 30:559–562.

Karpachevski, L. G., V. F. Babanin, T. S. Gendler, A. A. Opallenko, and R. N. Kuznin. 1972. Identification of ferruginous soil minerals by Mössbauer-spectroscopy. Sov. Soil Sci. 4(10):110–120.

Kemper, W. D. 1966. Aggregate stability of soils from Western United States and Canada. USDA Tech. Bull. No. 1355.

Kinniburgh, D. G., J. K. Syers, and M. L. Jackson. 1975. Specific adsorption of trace amounts of calcium and strontium by hydrous oxides of iron and aluminum. Soil Sci. Soc. Am. Proc. 39:464–470.

Knight, R. J., and R. N. Sylva. 1974. Precipitation in hydrolysed iron(III) solutions. J. Inorg. Nucl. Chem. 36:591–597.

Kühnel, R. A., D. van Hilten, and H. J. Rooda. 1974. The crystallinity of minerals in alteration profiles: an example on geothite in laterite profiles. Delft Progr. Rep. Ser. E 1:1–8. *See also* Clays Clay Miner. 23:349–354.

Langmuir, D., and D. O. Whittemore. 1971. Variations in the stability of precipitated ferric oxyhydroxides. p. 209–234. *In* R. F. Gould (ed.) Nonequilibrium systems in natural water chemistry. Adv. Chem. Ser. No. 106.

Mackay, A. L. 1960. Some aspects of the topochemistry of the iron oxides and hydroxides. p. 571–583. *In* J. H. de Boer (ed.) Reactivity of solids. Proc. 4th Int. Symp. Reactivity of Solids (Amsterdam).

Mackay, A. L. 1962. β-Ferric oxyhydroxide-akaganeite. Mineral. Mag. 33:270–280.

McKeague, J. A., and J. H. Day. 1966. Dithionite- and oxalate extractable Fe and Al as aids in differentiating various classes of soils. Can. J. Soil Sci. 46:13–22.

McKeague, J. A., and M. G. Cline. 1963. Silica in soil solutions. Canad. J. Soil Sci. 43: 83–96.

McKeague, J. A., J. E. Brydon, and N. M. Miles. 1971. Differentiation of forms of extractable iron and aluminum in soils. Soil Sci. Soc. Am. Proc. 35:33–38.

McLaren, R. G., and D. V. Crawford. 1974. Studies on soil copper III. J. Soil Sci. 25: 111–119.

Mehra, O. P., and M. L. Jackson. 1960. Iron oxide removal from soils and clays by a dithionite-citrate system buffered with sodium bicarbonate. Clays Clay Miner. 7: 317–327.

Misawa, T., K. Hashimoto, and S. Shimodaira. 1973. Formation of $Fe(II)_1 Fe(III)_1$ intermediate green complex on oxidation of ferrous ion in natural and slightly alkaline sulphate solutions. J. Inorg. Nucl. Chem. 35:4167–4174.

Mohr, E. C. J., F. A. van Baren, and J. van Schuylenborgh. 1972. Tropical soils. 3rd ed. Monton-Ichtior Baru-Van Hoeve, The Hague-Paris-London.

Nagarajah, S., A. M. Posner, and J. P. Quirk. 1970. Competitive adsorption of phosphate with polygalacturonate and other organic anions on kaolinite and oxide surfaces. Nature (London) 228:84–85.

Norrish, K. 1975. Geochemistry and mineralogy of trace elements. *In* A. R. Egan and D. J. O. Nicholas (ed.) Proc. Waite Inst. Symposium on Trace elements in Soil-Plant-Animal Systems (Adelaide, 1974). Academic Press.

Norrish, K., and R. M. Taylor. 1961. The isomorphous replacement of iron by aluminum in soil goethites. J. Soil Sci. 12:294–306.

Norrish, K., and R. M. Taylor. 1962. Quantitative analysis by X-ray diffraction. Clay Miner. Bull. 5, 28:98–109.

Oades, J. M., and W. N. Townsend. 1963. The detection of ferromagnetic minerals in soils and clays. J. Soil Sci. 14:179–187.

Onoda, G. Y., and P. L. de Bruyn. 1966. Proton adsorption at the ferric oxide aqueous solution interface I. Surf. Sci. 4:48–63.

Parks, G. A. 1965. The isoelectric points of solid oxides, solid hydroxides and aqueous hydroxy complex systems. Chem. Rev. 65:177–198.

Parks, G. A., and P. L. de Bruyn. 1962. The zero point of charge of oxides. J. Phys. Chem. 66:967–973.

Parfitt, R. L., R. J. Atkinson, and R. St. C. Smart. 1975. The mechanism of phosphate fixation by iron oxides. Soil Sci. Soc. Am. Proc. 39:837–841.

Pawluk, S. 1972. Measurement of crystalline and amorphous iron oxides. Can. J. Soil Sci. 52:119–123.

Ponnamperuma, F. N., E. M. Tianco, and T. A. Loy. 1967. Redox equilibria in flooded soils. I. The iron hydroxide systems. Soil Sci. 103:374–382.

Reisenauer, H. M., A. A. Tabikh, and P. R. Stout. 1962. Molybdenum reaction with soils and the hydrous oxides of Fe, Al, and Ti. Soil Sci. Soc. Am. Proc. 26:23–27.

Reyes, E. D., and J. J. Jurinak. 1967. A mechanism of molybdate adsorption on α-Fe_2O_3. Soil Sci. Soc. Am. Proc. 31:637–641.

Robie, R. A., and D. R. Waldbaum. 1967. Thermodynamic properties of minerals and related substances at 298.15°K (25°C) and one atmosphere (1.013 bars) pressure and at higher temperatures. Geol. Surv. Bull. 1259:256.

Schahabi, S., and U. Schwertmann. 1970. Der Einfluß von synthetischen Eisenoxiden auf die Aggregation zweier Lößbodenhorizonte. Z. Pflanzenernaehr. Bodenkd. 125:193–204.

Scheffer, F., E. Welte, and F. Ludwieg. 1957. Zur Frage der Eisenoxidhydrate in Böden. Chem. Erde. 19:51–64.

Scheffer, F., B. Meyer, and U. Babel. 1959. Magnetische Messungen als Hilfe zur Bestimmung der Eisenoxide im Boden. Beitr. Miner. Petrogr. 6:371–387.

Schellmann, W. 1959. Experimentelle Untersuchungen über die sedimentäre Bildung von Goethit und Hämatit. Chem. Erde 20:104–135

Schellmann, W. 1964. Zur Rolle des Aluminiums in Nadeleisenerz-ooiden. Neues Jahrb. Mineral., Monatsh.:49–56.

Schneiderhöhn, P. 1964. Über das Vorkommen des Aluminiums in einer ooidischen Eisenerze enthaltenden marinen Schichtfolge. Beitr. Mineral. Petrogr. 10:141–151.

Schwertmann, U. 1959a. Über die Synthese definierter Eisenoxyde unter verschiedenen Bedingungen. Z. Anorg. Allg. Chem. 298:337–348.

Schwertmann, U. 1959b. Die fraktionierte Extraktion der freien Eisenoxide in Böden, ihre mineralogischen Formen und ihre Entstehungsweisen. Z. Pflanzenernaehr., Dueng., Bodenkd. 84:194–204.

Schwertmann, U. 1964. Differenzierung der Eisenoxide des Bodens durch photochemische Extraktion mit saurer Ammoniumoxalat-Lösung. Z. Pflanzenernaehr., Dueng., Bodenkd. 105:194–202.

Schwertmann, U. 1965. Zur Goethit- und Hämatitbildung aus amorphem Eisen(III)-Hydroxid. Z. Pflanzenernaehr., Dueng., Bodenkd. 108:37–45.

Schwertmann, U. 1966. Inhibitory effect of soil organic matter on the crystallization of amorphous ferric hydroxide. Nature (London) 212:645–646.

Schwertmann, U. 1969/70. Der Einfluß organischer Anionen auf die Bildung von Goethit und Hämatit aus amorphem Fe(III)-hydroxid. Geoderma 3:207–214.

Schwertmann, U. 1971. Transformation of hematite to goethite in soils. Nature (London) 232:624–625.

Schwertmann, U. 1973a. Electron micrographs of soil lepidocrocites. Clay Miner. 10:59–63.

Schwertmann, U. 1973b. Use of oxalate for Fe extraction from soils. Can. J. Soil Sci. 53:244–246.

Schwertmann, U., and D. S. Fanning. 1976. Iron-manganese concretions in a hydrosequence of soils in loess in Bavaria. Soil Sci. Soc. Am. J. 40:731–738.

Schwertmann, U., and W. R. Fischer. 1966. Zur Bildung von α-FeOOH und α-Fe_2O_3 aus amorphem Eisen(III)-hydroxid. Z. Anorg. Allg. Chem. 346:137–142.

Schwertmann, U., and W. R. Fischer. 1973. Natural "amorphous" ferric hydroxide. Geoderma 10:237–247.

Schwertmann, U., W. R. Fischer, and H. Papendorf. 1968. The influence of organic compounds on the formation of iron oxides. Int. Congr. Soil Sci., Trans. 9th (Adelaide, Aust.) I:645–655.

Schwertmann, U., and B. Heinemann. 1959. Über das Vorkommen und die Entstehung von Maghemit in nordwestdeutschen Böden. Neues Jahrb. Mineral. Monatsh. 174–181.

Schwertmann, U., and W. Lentze. 1966. Bodenfarbe und Eisenoxidform. Z. Pflanzenernaehr., Dueng., Bodenkd. 115:209–214.

Schwertmann, U., and R. M. Taylor. 1972a. The influence of silicate of the transformation of lepidocrocite to goethite. Clays Clay Miner. 20:159-164.

Schwertmann, U., and R. M. Taylor. 1972b. The transformation of lepidocrocite to goethite. Clays Clay Miner. 20:151-158.

Schwertmann, U., and R. M. Taylor. 1973. The in vitro transformation of soil lepidocrocite to goethite. p. 45-54. *In* E. Schlichting and U. Schwertmann (ed.) Pseudogley and Gley: Genesis and use of hydromorphic soils. Int. Soc. Soil Sci., Trans. Comm. V & VI, (Stuttgart, W. Ger., 1971). Verlag Chemie Weinheim/ Bergstr, W. Ger.

Schwertmann, U., and H. Thalmann. 1976. The influence of [Fe(II)], [Si] and pH on the formation of lepidocrocite and ferrihydrite during oxidation of aqueous $FeCl_2$ solution. Clay Miner. 14:189-200.

Segalen, P. 1968. Note sur une méthode de détermination des products minéraux amorphes dans certain sols a hydroxydes tropicaux. Cah. O.R.S.T.O.M. Ser. Pedol. 6:105-126.

Sims, J. R., and F. T. Bingham. 1967. Retention of boron by layer silicates, sesquioxides and soil materials. Soil Sci. Soc. Am. Proc. 32:364-369.

Tamm, O. 1922. Um bestämning ow de oorganiska komponenterna i markens gelkomplex. Medd. Statens Skogsförsökanst. 19:385-404.

Taylor, R. M., and A. M. Graley. 1967. The influence of ionic environment on the nature of iron oxides in soils. J. Soil Sci. 18:341-348.

Taylor, R. M., and U. Schwertmann. 1974a. The association of P with iron in ferruginous soil concretions. Aust. J. Soil Res. 12:133-145.

Taylor, R. M., and U. Schwertmann. 1974b. Maghemite in soils and its origin. Parts I & II. Clay Mineral. 10:289-298, 299-310.

Thiel, R. 1963. Zum System $\alpha FeOOH$–$\alpha AlOOH$. Z. Anorg. Allg. Chem. 326:70-78.

Towe, K. W., and W. F. Bradley. 1967. Mineralogical constitution of colloidal hydrous ferric oxides. J. Colloid. Interface Sci. 24:384-392.

van der Marel, H. W. 1951. Gamma ferric oxide in sediments. J. Sediment. Petrol. 21: 12-21.

van Oosterhout, G. W. 1960. Morphology of synthetic submicroscopic crystals of α and γ-FeOOH and of α-Fe_2O_3 prepared from FeOOH. Acta Crystallogr. Sec. B. 13: 932-935.

Veen, A. W. L. 1973. Evaluation of clay mineral equilibria in some clay soils (usterts) of the Brigalow lands. Aust. J. Soil Res. 11:167-184.

Watson, J. R., A. M. Posner, and J. P. Quirk. 1973. Adsorption of the herbicide 2,4-D on goethite. J. Soil Sci. 24:503-511.

Wefers, K. 1966. Zum System Fe_2O_3-H_2O. Ber. Dtsch. Keram. Ges. 43:677-684.

Wefers, K. 1967. Phasenbeziehungen im System Al_2O_3-Fe_2O_3-H_2O. Z. Erzbergbau Metallhuettenwes. 20:13-19, 71-75.

Wells, A. F. 1975. Structural inorganic chemistry. 4th ed. Clarendon Press, Oxford.

Manganese Oxides and Hydroxides

R. M. MC KENZIE, CSIRO, Glen Osmond, South Australia

I. INTRODUCTION

The manganese oxide and hydroxide minerals are important constituents in soils for two reasons. Manganese is an essential element for the nutrition of plants and animals, and to a large extent the oxidation and reduction of the manganese oxides controls the amount of the element available. Oxidizing conditions may reduce the availability of manganese to such an extent that deficiency occurs, while reducing conditions may lead to the accumulation of toxic levels of manganese in the soil.

The manganese oxides and hydroxides have a high sorption capacity for heavy metals, and they may control the availability of certain trace elements, some of which are essential to plants and animals.

The mineralogy of manganese is complicated by the large number of oxides and hydroxides formed, in which substitution of Mn^{2+} and Mn^{3+} for Mn^{4+} occurs extensively. The manganese ions may be oxidized or reduced without changing position. When the valence of a sufficient number of ions has been changed the structure becomes mechanically unstable and rearranges into a new phase. The substitutions result in changes in the average Mn–O bond lengths, with consequent changes in the unit cell size, and are accompanied by the substitution of some O^{2-} by OH^- to maintain electrical neutrality. There is therefore a continuous series of compositions from MnO to MnO_2 within which there are a number of stable and metastable arrangements of atoms to form the well known minerals, many of which may cover a wide range of compositions.

II. STRUCTURAL PROPERTIES AND MINERAL IDENTIFICATION

A. Structure and Composition

The minerals to be discussed here are summarized in Table 6-1. This list does not include all of the known oxides, and some of the minerals listed are not true oxides in that cations other than manganese are an essential part of the structure. The compositions shown are ideal formulae, and are subject to different degrees of variation.

Only pyrolusite and ramsdellite can be regarded as true modifications of manganese dioxide (Giovanoli, 1969; Giovanoli & Staehli, 1970).

a. **Pyrolusite**—Pyrolusite is the most stable form of MnO_2, and it crystallizes in the rutile structure. In this structure, single chains of MnO_6 octahedra are formed by sharing edges. These chains lie parallel to the Z-axis

Table 6-1. Some manganese oxide minerals (adapted from McKenzie, 1972a)

Mineral name	Other names	Composition
Pyrolusite	β-MnO$_2$, polianite	MnO$_2$
Ramsdellite	--	MnO$_2$
Nsutite	γ-MnO$_2$, ρ-MnO$_2$	variable
Birnessite	δ-MnO$_2$, manganous manganite	variable
Buserite	10Å manganite	variable
Todorokite	--	variable
Cryptomelane	α-MnO$_2$	K$_2$Mn$_8$O$_{16}$
Hollandite	α-MnO$_2$	Ba$_2$Mn$_8$O$_{16}$
Coronadite	α-MnO$_2$	Pb$_2$Mn$_8$O$_{16}$
Lithiophorite	--	(Al,Li)MnO$_2$(OH)$_2$
Groutite	α-MnOOH	MnOOH
Manganite	γ-MnOOH	MnOOH
Partridgeite	α-Mn$_2$O$_3$	Mn$_2$O$_3$
Hausmannite	--	Mn^{2+}Mn$_2^{3+}$O$_4$

and are linked to each other in the X and Y directions by sharing corners. The oxygen atoms are in hexagonal close packing, with every alternate octahedron occupied by a manganese atom in both the X and Y directions.

b. **Ramsdellite**—Ramsdellite is a rare modification of MnO$_2$, crystallizing in the disapore type. This structure also consists of chains of MnO$_6$ octahedra, but in this case the chains are doubled by sharing edges in the Y direction. The oxygen atoms are in hexagonal close packing, with pairs of occupied and unoccupied octahedra in the Y direction, while every alternate octahedron in the X direction is filled.

c. **Nsutite**—This mineral was named after a variety occurring in the Nsuta deposits in Ghana (Sorem & Cameron, 1960). It covers a group of oxides which includes several modifications of γ-MnO$_2$ and ρ-MnO$_2$. The structure is described by Giovanoli (1969) as structural intergrowths of pyrolusite and ramsdellite. The composition was given as Mn$_{(1-x)}^{4+}$Mn$_x^{2+}$O$_{(2-2x)}$(OH)$_{2x}$ by Zwicker et al. (1962).

d. **Cryptomelane, Hollandite, and Coronadite**—These minerals form a family of oxides sometimes grouped under the name α-MnO$_2$. They are characterized by the presence of a large foreign cation as an essential constituent of the structure. The foreign ion is potassium in cryptomelane, barium in hollandite, and lead in coronadite. Synthetic forms have been prepared by Maxwell and Thirsk (1954) with Ba, Ca, K, Na, and NH$_4$ as the foreign ion. The structure is described by Bÿstrom and Bÿstrom (1950, 1951) as consisting of double chains of MnO$_6$ octahedra running in the Z-direction. The double chains are combined in a three-dimensional framework by sharing corners, with tunnels running in the Z-direction, in which the foreign ions are situated. The structures of pyrolusite, ramsdellite, and cryptomelane are compared in Fig. 6-1.

Bÿstrom and Bÿstrom (1950) give the general formula as A$_{2-y}$B$_{8-z}$X$_{16}$, where A is Ba, K, or Pb; X is O or OH; and B chiefly Mn^{4+}. Although the

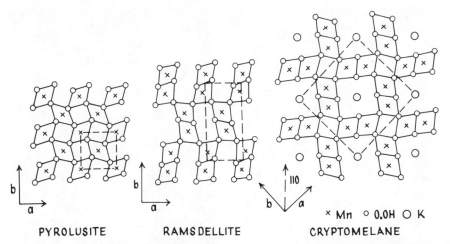

PYROLUSITE   RAMSDELLITE   × Mn ○ 0,OH ○ K

                                                   CRYPTOMELANE

Fig. 6-1.   The structures of pyrolusite, ramsdellite, and cryptomelane projected on the *XY* plane (after Gattow & Glemser, 1961).

presence of the large cation is necessary to prevent the structure from collapsing, the amount is variable, and is never more than enough to fill half of the A sites. The short A–A distance would make the structure unstable if all of the A sites were filled. Figure 6-2a shows an electron micrograph of a synthetic cryptomelane.

   e.   **Birnessite**—Birnessite refers to a group of oxides, some members of which have been known as δ-MnO_2, manganous-manganite, 7Å manganite, Mn(III) manganate(IV), manganous(II) manganate(IV), and NaMn(II,III) manganate(IV). The structure is described by Giovanoli et al. (1970b) as a double layer. The main layers consist of sheets of $Mn^{4+}O_6$ octahedra linked by sharing edges. The sheets are separated by a distance of about 7Å, and give rise to the basal reflections in the X-ray diffraction pattern. The structure of the intermediate layers is not known in detail. They consist of Mn^{2+} and Mn^{3+} co-ordinated to OH^- ions and water molecules, and may contain foreign ions such as Na, K, Ca, and others. These oxides occur in disperse form, and are nonstoichiometric. The composition of the type mineral was given by Jones and Milne (1956) as $(Na_{0.7}Ca_{0.3})Mn_7O_{14} \cdot 2.8H_2O$. Giovanoli et al. (1970a, b) gave the composition of three synthetic members as $Na_4Mn_{14}O_{27} \cdot 9H_2O$; $Mn_7O_{13} \cdot 5H_2O$ and $Mn_7O_{12} \cdot 6H_2O$. Figure 6-2b is an electron micrograph of platelets of synthetic birnessite.

   f.   **Buserite**—Buserite has been accepted as a mineral name for the 10Å manganite described by Buser (1959) in ocean floor nodules. The synthetic form has a platy habit, and may have a layer structure similar to birnessite, with a basal spacing of about 10Å (Giovanoli et al., 1973).

   g.   **Todorokite**—Giovanoli et al. (1971) proposed that todorokite is not a pure mineral species, but could be regarded as a partly decomposed buserite admixed with birnessite and manganite. However, some mineralogists do not accept this view, and they regard todorokite to be identical to the 10Å man-

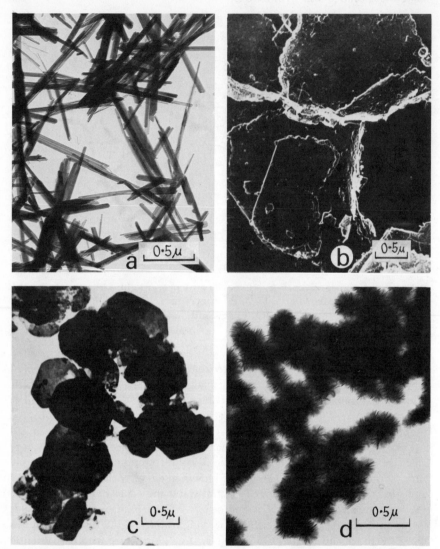

Fig. 6-2. Electron micrographs of manganese oxides: *(a)* Synthetic cryptomelane; *(b)* Platelets of synthetic birnessite, carbon replica, chromium shadowed (after Giovanoli et al., 1971); *(c)* Synthetic lithiophorite (after Giovanoli et al., 1973); and *(d)* Synthetic birnessite, balls of needles.

ganite described by Buser (1959), (Burns et al., 1975). Further work is required to resolve the todorokite-buserite problem. The structure of todorokite is not known.

    h. Lithiophorite—Lithiophorite has a layer structure with alternate sheets of $Mn^{4+}O_6$ and $(Al,Li)(OH_6)$ octahedra. Lithium occupies one-third of the Al sites (Wadsley, 1952). Although the separation of the MnO_6 sheets is 10Å, the mineral does not belong to the buserite group. Giovanoli et al.

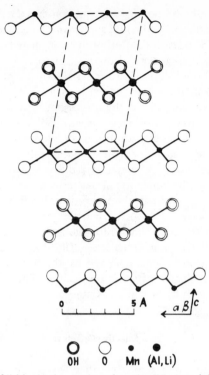

O O •   ●
OH   0   Mn   (Al, Li)

Fig. 6-3. The layers of lithiophorite, projected on the *XZ* plane (after Wadsley, 1952).

(1973) based this conclusion on the fact that only limited substitution can occur in lithiophorite and the mineral is essentially stoichiometric, whereas a much wider range of variability is possible in the buserite group. The structure of lithiophorite is shown in Fig. 6-3, while Fig. 6-2c shows the morphology as hexagonal platelets.

     i. **Manganite**—Manganite, γ-MnOOH, is a trivalent oxyhydroxide, and is the stable polymorph of MnOOH. The structure was determined by Dachs (1963), and resembles that of pyrolusite. Hydrogen bonding occurs between the OH group in edge-shared octahedra in one chain and the corner-shared oxygens in adjacent chains (Burns & Burns, 1975).

B. X-ray Diffraction

     Due to the small amounts of the manganese minerals in soils, X-ray diffraction methods can be used only where natural segregations of the minerals occur, such as in nodules or veins. Even then the diffuse nature of the X-ray diffraction patterns of some of the minerals, and the coincidence of diagnostic lines with those of other minerals necessitates some concentration pretreatment to permit positive identifications to be made. Taylor et al.

(1964) removed kaolin by boiling the sample in NaOH. When quartz was the main contaminant, the mineral could sometimes be concentrated by magnetic separation. They also used a flotation method in which the ground sample was treated with H_2O_2. Particles of the manganese oxide tended to be carried to the surface by bubbles of O_2. Comparison of the diffraction patterns before and after treatment led to positive identification in most cases.

The diffraction patterns of the birnessite, buserite, and nsutite groups show considerable variation. Changes in unit cell dimensions and line broadening brought about by substitution, nonstoichiometry, disorder, and small crystalline size make positive identification difficult in some cases.

The birnessite group is characterized by a strong basal reflection at about 7Å, and its higher orders. The patterns of the more crystalline varieties contain many lines, but in the more disordered forms usually found in soils, only four lines are present. When the disordered varieties occur in a very finely divided form, the basal reflections disappear altogether, and only two lines at 2.4 and 4.1Å remain (Bricker, 1965).

The buserite group is characterized by the basal reflection at 10Å, and again a considerable degree of variation occurs. The patterns of three varieties of nsutite are given by Sorem and Cameron (1960).

C. Selective Dissolution

The manganese oxides may be dissolved by reducing agents such as quinol, dithionite, hydroxylamine hydrochloride, or oxalate, and these reagents have been used in attempts to fractionate the manganese into "available" and "nonavailable" forms. Reagents of this type do not differentiate between the various mineral types, and some may dissolve iron oxides as well as manganese.

In acid solution the manganese oxides are reduced to manganous salts by H_2O_2. This treatment does not dissolve iron oxides, and the method was used by Taylor and McKenzie (1966) to study the chemical composition of manganese oxides in soils.

III. NATURAL OCCURRENCE

A. In Soils

The manganese oxides occur in soils as coatings on other soil particles, deposited in cracks and veins and as nodules up to 2 cm in diameter. The nodules often exhibit a concentric layering suggestive of seasonal growth, and they contain oxides of both iron and manganese as well as other soil constituents.

The manganese oxides in soil are often reported to be amorphous, e.g. McKeague et al. (1968), and very little data is available on mineral forms in soils. Tiller (1963) reported the occurrence of lithiophorite in krasnozems in Tasmania and Taylor et al. (1964) reported that, of 28 samples of nodules from soils in various parts of Australia, 10 were birnessite, 10 were lithiophorite, 3 were hollandite, 1 was pyrolusite, 1 was todorokite, and 1 contained both lithiophorite and hollandite. The minerals in the remaining two could not be identified. The sample of pyrolusite was found at a depth of 3 m in an alluvial clay underlying the soil of the Adelaide Plains, South Australia. The sample was unusual in that it was very soft and appeared to be essentially pure pyrolusite, uncontaminated by other soil material.

In a further seven samples from Bermuda, Europe, and the Middle East countries, Taylor (1968) identified four as birnessite, and one as lithiophorite. Eswaran (1973) identified lath-shaped crystals of manganite in a laterite in India. W. Sung (private communication) has identified lithiophorite in manganese nodules in soils in Hawaii.

B. In Sediments

Manganese and iron become separated in the geochemical cycle by the oxidation and hydrolysis of iron at values of Eh and pH at which manganese is still quite soluble. However, mixed deposits can form by coprecipitation of manganese with the iron oxides.

Bricker (1965) has observed the oxidation sequences rhodochrosite (manganous carbonate) $\rightarrow$ manganite $\rightarrow$ pyrolusite, and rhodochrosite $\rightarrow$ birnessite $\rightarrow$ oxidized birnessite $\rightarrow$ manganoan nsutite $\rightarrow$ nsutite in ore deposits, and Zwicker et al. (1962) have shown that a naturally occurring nsutite is readily transformed to pyrolusite. In the absence of foreign ions, the first formed minerals will recrystallize to nsutite and eventually to pyrolusite, which is the thermodynamically stable form. The presence of the appropriate cations leads to the formation of lithiophorite or one of the cryptomelane group of minerals.

IV. FORMATION IN SOILS

A. Oxidation and Reduction

Manganese is oxidized rapidly by air under alkaline conditions. At near neutral values of pH, oxidation is slow, and the first product is hausmannite and then manganite. Oxidation is catalyzed by fine particles in soils and also by MnO_2 (Hem, 1963; Morgan & Stumm, 1964). The higher oxides may also be formed by the disproportionation of hausmannite and manganite under acid conditions.

The biological oxidation of manganese has been demonstrated many times (Bromfield, 1958; Harriya & Kikuchi, 1964; Schweisfurth & Gattow, 1966; Zavarzin, 1968), and a number of the bacteria and fungi responsible for this oxidation have been identified.

In soils reduction occurs under anaerobic conditions, when bacterial oxidation of organic matter proceeds at such a rate that the dissolved O_2 supply is depleted. The bacteria then use the higher oxides of manganese as a source of O_2.

B. Formation of the Minerals in Soils

Taylor et al. (1964) observed that lithiophorite occurred mainly in neutral to acid subsurface soils, whereas birnessite, although found in both acid and alkaline soils, was more common in alkaline surface horizons.

A large number of studies have been made on the synthesis of these oxides, and the conversion of one form to another. These studies, together with the oxidation sequences observed in ore deposits can be extrapolated to formation of the minerals in soils.

a. **Pyrolusite and Nsutite**—Studies on the transformation of synthetic birnessite to nsutite and finally to pyrolusite, and the formation of nsutite or pyrolusite by hydrolysis of $MnCl_4$ have been made by Zwicker et al. (1962), Buser et al. (1954), Cole et al. (1947), McKenzie (1971), Gattow and Glemser (1961), and Glemser et al. (1961). These studies show that the presence of large amounts of foreign ions prevents the formation of nsutite, while even small amounts of foreign ions prevent the formation of pyrolusite. In soils, foreign ions derived from the weathering of other minerals will always be present, and would probably limit the formation of nsutite and pyrolusite.

b. **Birnessite and Buserite**—These minerals may be synthesized by the oxidation of $Mn(OH)_2$ in alkaline suspension. Oxidation with air or O_2 in the presence of NaOH leads to a sodium buserite, which on dehydration gives birnessite (Giovanoli et al., 1970a). When the alkali is KOH a potassium birnessite is formed directly (McKenzie, 1971). In both of these methods the product consists of thin platelets (Fig. 6-2b). Birnessite is also formed by the addition of HCl to boiling $KMnO_4$. In this case the product consists of balls of fine needles, as shown in Fig. 6-2d, but the X-ray diffraction pattern is identical to that of the platelet form.

The formation of birnessite by bacterial oxidation has been shown by Schweisfurth and Gattow (1966), and it is also formed by disproportionation of hausmannite (Bricker, 1965). It is probable that direct oxidation, biological oxidation, and disproportionation all contribute to its formation in soils.

c. **Lithiophorite**—Giovanoli et al. (1973) synthesized lithiophorite by hydrothermal treatment of $Mn_7O_{13}\cdot 5H_2O$ (a birnessite) with aluminum and lithium hydroxides, and Wadsley (1950a, b) prepared a similar material by substituting Al and Li into a sodium birnessite, and hydrothermal treatment of the product. The occurrence of lithiophorite in acid soils suggests that it is formed by alteration of birnessite.

d. **The Cryptomelane Group**—Birnessites containing large foreign cations are readily transformed to cryptomelane type minerals upon heating. If the concentration of foreign ions is < about 1% the conversion product is nsutite, but materials containing more than this level are converted to cryptomelane by boiling (Gattow & Glemser, 1961; Glemser et al., 1961; Maxwell & Thirsk, 1954). At high concentrations of K the conversion to cryptomelane is more difficult (McKenzie, 1971), and there appears to be an upper limit of about 7% of K in cryptomelane. The formation of this group of minerals in soils probably occurs by the transformation of birnessite containing an appropriate concentration of suitable foreign cations.

e. **Todorokite**—Todorokite has been synthesized by McKenzie (1971) from a Na buserite, and its formation in soils is probably by a similar process.

V. CHEMICAL PROPERTIES

A. Chemical Composition

The chemical composition of some manganese oxides from soils has been determined by Taylor et al. (1964). The minerals could not be separated from other soil constituents, and they were therefore extracted from soil nodules by treatment with H_2O_2 at pH 3. The extracts were evaporated to dryness and ignited before analysis. Table 6-2 shows mean analyses of a number of samples. Aluminum and lithium are higher in lithiophorite, and barium is higher in hollandite, as these elements are essential constituents of the respective minerals. Individual samples show a considerable range of concentrations, and chemical composition is not a useful means of differentiating the various minerals.

Table 6-2. Mean analyses of some manganese minerals in soils (data from Taylor & McKenzie, 1966)

Sample	Lithiophorite	Birnessite	Hollandite
		%	
Mn_3O_4	73.2	83.2	65.3
Al_2O_3	11.5	1.0	5.7
SiO_2	3.8	1.4	6.7
Fe_2O_3	4.7	0.94	6.6
CoO	1.5	0.57	1.4
NiO	0.13	0.43	0.21
CaO	0.20	3.2	5.7
MgO	0.25	1.1	1.0
BaO	3.6	4.5	6.0
Li_2O	0.16	0.06	0.12
Na_2O	0.20	0.09	0.14
K_2O	0.32	0.22	0.37

B. Surface Reactions

The manganese minerals in soils are finely divided, and have large surface areas. The zero points of charge of the types of oxides occurring in soils range from 1.5 for birnessites to 4.6 for the cryptomelane group (Healy et al., 1966). In soil environments they carry a high negative charge in all but extremely acid conditions, and have high sorption capacities, particularly for heavy metals. This leads to the accumulation of relatively high concentrations of heavy metals in the manganese oxides in soils. Taylor and McKenzie (1966) have compared analyses of whole soils and manganese nodules separated from them, and some of their results are shown in Table 6-3. Hydrogen peroxide extracts of these and other nodules show values of Co up to 1.4%, Ni 0.14%, V 0.3%, Pb 0.25%, Cr 0.24%, and Zn 0.12%.

Studies on Co in the manganese oxides by Taylor and McKenzie (1966) and McKenzie (1967, 1970, 1972a, b) show that Co is adsorbed to a greater extent than other heavy metals. Cobalt appears to replace manganese in the surface layers of the structure, and in many soils almost all of the Co in the soil is associated with the manganese minerals. This result has been confirmed recently by McKenzie (1975), using the electron probe microanalyzer. This accumulation is of considerable practical importance in agriculture, as Adams et al. (1969) have shown that the availability of Co to plants is controlled to a large extent by the amount of manganese in the soil. The presence of even moderate amounts of mineralized manganese can lead to Co deficiency in pastures, and to fixation of Co applied as fertilizer.

The high content of Pb in the manganese oxides in soils is also of considerable interest. Norrish (1975), using the electron probe microanalyzer, has observed Pb at levels up to 2% in accumulations of manganese oxides in soils. Lead is an essential constituent of the mineral coronadite, and it is interesting to speculate whether the uptake of Pb by plants would be influenced by the manganese oxides.

Table 6-3. Concentration of elements in whole soils, and in manganese nodules separated from them (after McKenzie, 1972a)

	Mn	Co	Ni	V	Pb	Ba	Cr	Zn
				ppm				
Whole soil	92	5	22	47	14	180	86	12
Nodules	3,600	82	67	97	100	140	120	30
Whole soil	2,500	20	8	49	10	220	26	12
Nodules	72,000	380	41	110	50	2,300	30	30
Whole soil	1,300	6	14	49	14	180	37	20
Nodules	40,000	110	39	88	34	1,600	31	33

VI. CONCLUSIONS

Birnessite, lithiophorite, and hollandite are the most common of the crystalline manganese oxide minerals in soils, and todorokite and pyrolusite are less common. These minerals occur as coatings on other soil particles, deposited in cracks and veins, and mixed with iron oxides and other soil constituents in nodules. The individual crystallites are small, and they have large surface areas.

These minerals have two important functions in soils. They act as a supply of manganese for plant nutrition, and oxidation and reduction reactions can reduce the supply of manganese to such an extent that deficiency occurs, or increase it to such high levels that it becomes toxic to plants.

The manganese oxide minerals in soils have a high sorption capacity, particularly for heavy metals, and they may accumulate large amounts of these elements. This accumulation is particularly pronounced in the case of Co, and can lead to Co deficiency in pastures, and to fixation of Co applied as fertilizer.

VII. FURTHER READING

General

McKenzie, R. M. 1972. The manganese oxides in soils—a review. Z. Pflanzenernaehr. Bodenkd. 131:221–42.

Crystal Structures

Povarennykh, A. S. 1972. Crystal chemical classification of minerals. Plenum Press, New York-London.

Mineral Forms in Soils

Taylor, R..M., R. M. McKenzie, and K. Norrish. 1964. The mineralogy and chemistry of manganese in some Australian soils. Aust. J. Soil Res. 2:235–48.

Accumulation of Heavy Metals

Jenne, E. A. 1968. Controls on Mn, Fe, Co, Ni, Cu, and Zn concentrations in soils and water: the dominant role of hydrous Mn and Fe oxides. Adv. Chem. Series, 73: 337–87.
McKenzie, R. M. 1970. The reaction of cobalt with manganese dioxide minerals. Aust. J. Soil Res. 8:97–106.
Taylor, R. M., and R. M. McKenzie. 1966. The association of trace elements with manganese minerals in Australian soils. Aust. J. Soil Res. 4:29–39.

LITERATURE CITED

Adams, S. N., J. L. Honeysett, K. G. Tiller, and K. Norrish. 1969. Factors controlling the increase of cobalt in plants following the addition of a cobalt fertilizer. Aust. J. Soil Res. 7:29–42.

Bricker, O. P. 1965. Some stability relations in the system Mn–O_2–H_2O and one atmosphere total pressure. Am. Mineral. 50:1296–1354.

Bromfield, S. M. 1958. The properties of a biologically formed manganoxide, its availability to oats and its solution by root washings. Plant Soil 9:325–37.

Burns, R. G., and Virginia M. Burns. 1975. Structural relationships between the manganese(IV) oxides. p. 306–327. In A. Kozawa and R. J. Brodd (ed.) Proc. Int. Symp. Manganese Dioxides. (Cleveland, October 1975) Electrochem. Soc.

Burns, Virginia M., R. G. Burns, and W. K. Zwicker. 1975. Classification of natural manganese dioxide minerals. p. 288–305. In A. Kozawa and R. J. Brodd (ed.) Proc. Int. Symp. Manganese Dioxides. (Cleveland, October 1975) Electrochem. Soc.

Buser, W. 1959. The nature of the iron and manganese compounds in manganese nodules (abs): 1st Int. Oceanographic Congr. (New York). Preprints p. 962–964. Am. Assoc. for the Adv. of Sci., Washington, D. C.

Buser, W., P. Graf, and W. Feitknecht. 1954. Beitrag zur Kenntnis der Mangan(II)-manganite und des δMnO_2. Helv. Chim. Acta 37:2322–33.

Býstrom, A., and A. M. Býstrom. 1950. The crystal structure of hollandite, the related manganese oxide minerals, and α-MnO_2. Acta Cryst. 3:146–54.

Býstrom, A., and A. M. Býstrom. 1951. The position of the barium atoms in hollandite. Acta Cryst. 4:469.

Cole, W. F., A. D. Wadsley, and A. Walkley. 1947. An X-ray diffraction study of manganese dioxide. Trans. Electrochem. Soc. 92:133–58.

Dachs, H. 1963. Neutron-und Röntgenuntersuchungen am Manganit. Z. Kristallogr. Mineral. 118:303–326.

Eswaran, H., and N. G. Raghu Mohan. 1973. The microfabric of petroplinthite. Soil Sci. Soc. Am. Proc. 37:79–82.

Gattow, G., and O. Glemser. 1961. Darstellung und Eigenschaften von Braunsteinen. II. Die γ-and η-gruppe der braunsteine. Z. Anorg. Allg. Chem. 309, 20–36.

Giovanoli, R. 1969. A simplified scheme for polymorphism in the manganese dioxides. Chimia 23:470–72.

Giovanoli, R., H. Buehler, and K. Sokolowska. 1973. Synthetic lithiophorite: electron microscopy and X-ray diffraction. J. Microsc. Paris 18:271–84.

Giovanoli, R., W. Feitknecht, and F. Fischer. 1971. Ueber oxidhydroxide des vierwertigen mangans mit schichtengitter 3. Reduktion von mangan(III)-manganat(IV) mit zimtalkohol. Helv. Chim. Acta 54:1112–1124.

Giovanoli, R., and E. Staehli. 1970. Oxide und oxyhydroxide des drei-und vierwertigen Mangans. Chimia 24:49–61.

Giovanoli, R., E. Staehli, and W. Feitknecht. 1970a. Ueber oxidhydroxide des vierwertigen mangans mit schichtengitter I. Natriummangan(II,III) manganat(IV). Helv. Chim. Acta. 53:209–20.

Giovanoli, R., E. Staehli, and W. Feitknecht. 1970b. Ueber oxidhydroxide des vierwertigen mangans mit schichtengitter 2. Mangan(III)-manganat(IV). Helv. Chim. Acta 53:453–64.

Glemser, O., G. Gattow, and H. Meisiek. 1961. Darstellung und Eigenschaften von Braunsteinen 1. Die δ-Gruppe der Braunsteine. Z. Anorg. Allg. Chem. 309:1–19.

Harriya, Y., and T. Kikuchi. 1964. Precipitation of manganese by bacteria in mineral springs. Nature 202:416–17.

Healy, T. W., A. P. Herring, and D. W. Fuerstenau. 1966. The effect of crystal structure on the surface properties of a series of manganese dioxides. J. Colloid Interface Sci. 21:435–444.

Hem, J. D. 1963. Chemical equilibria and rates of manganese oxidation. U. S. Geol. Surv. Water Supply Paper 1667-A.

Jones, L. H. P., and A. A. Milne. 1956. Birnessite, a new manganese oxide mineral from Aberdeenshire, Scotland. Mineral Mag. 31:283–88.

McKeague, J. A., A. W. H. Damman, and P. K. Heringa. 1968. Iron-manganese and other pans in some soils of Newfoundland. Can. J. Soil Sci. 48:243–253.

McKenzie, R. M. 1967. The sorption of cobalt by manganese minerals in soils. Aust. J. Soil Res. 5:235–46.

McKenzie, R. M. 1970. The reaction of cobalt with manganese dioxide minerals. Aust. J. Soil Res. 8:97–106.

McKenzie, R. M. 1971. The synthesis of cryptomelane and some other oxides and hydroxides of manganese. Mineral. Mag. 38:493–502.

McKenzie, R. M. 1972a. The manganese oxides in soils—a review. Z. Pflanzenernaehr. Bodenkd. 131:221–42.

McKenzie, R. M. 1972b. The sorption of some heavy metals by the lower oxides of manganese. Geoderma 8:29–35.

McKenzie, R. M. 1975. An electron microprobe study of the relationships between heavy metals and manganese and iron in soils and ocean floor nodules. Aust. J. Soil Res. 13:177–188.

Maxwell, K. H., and H. R. Thirsk. 1954. Electromotive force and strucutral relations in MnO$_2$ system. Int. Comm. Electrochem. Thermodyn. and Kinetics. Proc. 6th meeting—Poitiers 1954. Butterworth, New York.

Morgan, J. J., and W. Stumm. 1964. The role of multivalent metal oxides in limnological transformations as exemplified by iron and manganese. 2nd. Int. Conf. Water Pollution Research (Tokyo) p. 103–18.

Norrish, K. 1975. Geochemistry and mineralogy of trace elements. p. 55–81. In D. J. Nicholas and A. R. Egan (ed.) Trace elements in soil-plant-animal systems. Waite Agric. Res. Inst. Jubilee Symposium. Academic Press, New York.

Schweisfurth, R., and G. Gattow. 1966. Untersuchungen ueber Roentgenstruktur und zusammensetzung mikrobiell gebildeter braunsteine. Z. Allg. Mikrobiol. 6:303–8.

Sorem, R. K., and E. N. Cameron. 1960. Manganese oxide and associated minerals of the Nsuta deposits, Ghana, West Africa. Econ. Geol. 55:278–310.

Taylor, R. M. 1968. The association of manganese and cobalt in soils—further observations. J. Soil Sci. 19:77–80.

Taylor, R. M., and R. M. McKenzie. 1966. The association of trace elements with manganese minerals in Australian soils. Aust. J. Soil Res. 4:29–39.

Taylor, R. M., R. M. McKenzie, and K. Norrish. 1964. The mineralogy and chemistry of manganese in some Australian soils. Aust. J. Soil Res. 2:235–48.

Tiller, K. G. 1963. Weathering and soil formation on dolerite in Tasmania with particular reference to several trace elements. Aust. J. Soil Res. 1:79–90.

Wadsley, A. D. 1950a. A hydrous manganese oxide with exchange properties. J. Am. Chem. Soc. 72:1881–84.

Wadsley, A. D. 1950b. Synthesis of some hydrated manganese minerals. Am. Mineral. 35:485–99.

Wadsley, A. D. 1952. The structure of lithiophorite (Al,Li)MnO$_2$(OH)$_2$. Acta Cryst. 5:676–80.

Zavarzin, G. A. 1968. Bacteria in relation to manganese metabolism. In T. R. G. Gray and D. Parkinson (ed.) Int. Symp. Ecol. Soil Bact. (1966) 612–31. Liverpool University Press, Liverpool, UK.

Zwicker, W. K., W. O. J. G. Meijer, and H. W. Jaffe. 1962. Nsutite—a widespread manganese oxide mineral. Am. Mineral. 47:246–66.

Micas

DELVIN S. FANNING, University of Maryland, College Park

V. Z. KERAMIDAS, University of Thessaloniki, Greece

I. INTRODUCTION

Micas are 2:1 phyllo-(layer) silicate minerals with tightly held interlayer cations balancing a high layer charge. Technically, according to the Clay Minerals Society (CMS) (1971) and AIPEA (Association Internationale Pour L'Etude Des Argilles) (1972) Nomenclature Committees, micas are a group of minerals distinct from brittle micas. The status of the other terms, such as illite and glauconite, is in question. Materials that have been identified by all of these terms will be discussed. Primary attention is given to those micaceous minerals with potassium as the interlayer cation, since these are the abundant and important micas in most soils.

The micas in soils are largely, if not entirely, primary. This means that they are mainly inherited by soils from soil parent materials and are not known to form to any significant extent in soils. Micas are abundantly present in many rocks (e.g., shales, slates, phyllites, schists, gneisses, granites) and in sediments derived from these and other rocks. Soils containing macroscopic mica, as in the Piedmont region of the eastern United States, often sparkle in the sunshine as light reflects from the mica flakes. Perhaps this is the source of the word mica (*L. micare*, to shine) and of the German word for mica, which is *glimmer.*

Micas serve as precursors for other 2:1 layer silicate minerals, especially vermiculites, to which micas are transformed by replacement of the "non-exchangeable" interlayer cations (mainly potassium) by exchangeable cations with water of hydration. Much attention has focused on this subject in recent years and the factors controlling the rates and products of these transformations are being elucidated (Norrish, 1973).

Micas are often present in soils as components of particles that have been partially transformed to expansible 2:1 minerals—so that the mica is interstratified with the other minerals (B. L. Sawhney, Chapter 12, this book) or are present in particles as mica cores surrounded by expanded zones (e.g., Jackson, 1964). These mica components can significantly affect the properties of the particles, especially with respect to their cation exchange and selectivity behavior.

Through K release, micas are the most important natural source of K for growing plants in most soils. Soils vary in their natural K supplying power depending on the kinds, amounts, and the particle size of the micas present, and upon factors that affect K release and fixation such as wetting and drying of the soil. Dioctahedral micas, such as muscovite, are normally much more resistant to weathering than trioctahedral micas, such as biotite,

chapter 7

and release K at much slower rates. Aspects of these and other subjects about micas are discussed in this chapter.

II. STRUCTURAL PROPERTIES AND MINERAL IDENTIFICATION

A. Structures, Formulas, and Nomenclature

1. TETRAHEDRA, OCTAHEDRA, PLANES, SHEETS, LAYERS, INTERLAYERS, UNIT STRUCTURES

Micas are typical phyllosilicate minerals in that they contain continuous two-dimensional tetrahedral sheets of composition T_2O_5 (or T_4O_{10}), where T represents the tetrahedral cations (Table 7-1). In these minerals three of the four oxygens of each tetrahedron are shared with other tetrahedra. This sharing results in linked hexagonal or ditrigonal "rings" of tetrahedra (Fig. 7-1). In describing the structures of micas and many other phyllosilicate minerals the oxygens shared between tetrahedra are often referred to as the basal oxygens (Fig. 7-2). A fourth oxygen of each tetrahedron is not shared with other tetrahedra. In the structures of micas and most other phyllosilicate minerals, these unshared apical oxygens point from the basal oxygens into the layers.

The proper terminology (CMS Nomenclature Committee, 1971a, b; AIPEA Nomenclature Committee, 1972) is to refer to *planes* of atoms (or ions), to *sheets* as combinations of planes, and to *layers* as combinations of sheets. For example, the planes of anions and cations are pointed out in Fig. 7-2 for a general structure representing a $1M$ polytype muscovite as viewed along the Y axis. Here individual tetrahedral, octahedral, and interlayer cations are represented, but the individual anions are not. Figure 7-3 shows individual anions and also the orientation of tetrahedra.

Micas are 2:1 type layer silicate minerals. This means that each layer is composed of two tetrahedral sheets and one octahedral sheet. The octahedral sheets are composed of octahedral cations and the apical tetrahedral oxygens plus hydroxyls (Table 7-1, Fig. 7-2 and 7-3). Each octahedral cation is co-ordinated with six anions (normally four apical oxygens and two hydroxyls), three from each of the two planes of apical oxygens plus hydroxyls, to give the eight-sided polyhedra. Each plane of apical oxygens plus hydroxyls is part of both a tetrahedral and an octahedral sheet; thus neither sheet can be removed without disrupting the other.

The zones between the layers, where the basal oxygens of two layers face each other, are referred to as *interlayers*. In micas, the interlayers are normally occupied by unhydrated cations (Table 7-1), with the distinction between some of the different minerals (e.g., paragonite vs. muscovite, Table 7-1) being made on the basis of the element present in the interlayer.

The total assemblage of a layer plus interlayer material is referred to as a *unit structure*.

Table 7-1. Formulas and layer charge for some dioctahedral and trioctahedral micas (or closely related minerals, depending on definition of mica)

Mineral and references†	Layer charge (X)	Interlayer	Cations		Anions
			Tetrahedral	Octahedral	
Dioctahedral					
Ideal Muscovite	1	K	Si_3Al	Al_2	$O_{10}(OH)_2$
Ideal Paragonite	1	Na	Si_3Al	Al_2	$O_{10}(OH)_2$
Ideal Margarite	2	Ca	Si_2Al_2	Al_2	$O_{10}(OH)_2$
Glauconite (1)	0.71	$K_{0.65}Na_{0.06}$	$Si_{3.87}Al_{0.13}$	$Al_{0.63}Fe^{3+}_{0.82}Fe^{2+}_{0.19}Mg_{0.36}$	$O_{10}(OH)_2$
Glauconite (2)	0.89	$K_{0.76}(NaCa)_{0.13}$	$Si_{3.42}Al_{0.57}$	$Al_{0.4}Fe^{3+}_{0.87}Fe^{2+}_{0.48}Mg_{0.4}$	$O_{10}(OH)_2$
Illite (3)	0.64	$K_{0.61}Na_{0.02}Ca_{0.01}$	$Si_{3.34}Al_{0.66}$	$Al_{1.29}Fe^{3+}_{0.41}Fe^{2+}_{0.19}Mg_{0.18}Ti_{0.04}$	$O_{10}(OH)_2$
Illite (4)	0.82	$K_{0.64}Na_{0.02}Ca_{0.16}$	$Si_{3.38}Al_{0.62}$	$Al_{1.4}Fe^{3+}_{0.39}Mg_{0.22}$	$O_{10}(OH)_2$
Trioctahedral					
Ideal Biotite	1	K	Si_3Al	$(Mg, Fe^{2+})_3$	$O_{10}(OH)_2$
Ideal Phlogopite	1	K	Si_3Al	Mg_3	$O_{10}(OH)_2$
Ideal Clintonite (5)	2	Ca	$SiAl_3$	Mg_2Al	$O_{10}(OH)_2$
Lepidolite (6)	0.919	$K_{0.862}Ca_{0.006}Na_{0.051}$	$Si_{3.532}Al_{0.468}$	$Al_{1.285}Fe^{3+}_{0.006}Ti_{0.005}Mn^{2+}_{0.032}Mg_{0.003}Li_{1.584}$	$O_{10}(OH)_{0.12}F_{1.88}$

† Reference and source of material: (1) Cretaceous sample from Vidono formation, Anzoategui, Venezuela; Warshaw (1957)—ASTM Card 9-439. (2) Franconia formation, Wisconsin; Burst (1958b). (3) Beavers Bend, Illinois; Gaudette et al. (1966). (4) Silurian red shale in limestone, Montana; Weaver & Pollard (1973). (5) Radoslovich (1963)—called Xanthophyllite in original reference. (6) Lilac colored, Grosmont, Western Australia; Rausell-Colom et al. (1965).

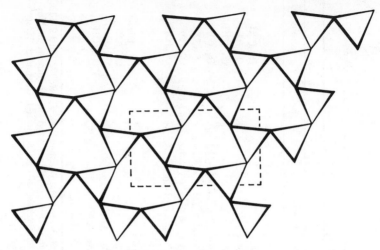

Fig. 7-1. Diagram showing the basal oxygen surface of margarite. Each triangle represents the base of a tetrahedron. The "rings" of tetrahedra are ditrigonal rather than hexagonal because of rotation of the tetrahedra (adjacent tetrahedra are rotated in opposite directions) relative to a true hexagonal configuration. The tetrahedra are also tilted so that the surface is corrugated, rather than planar. Heavy lines represent elevated tetrahedral edges (apical oxygens pointed down). After Bailey (1966).

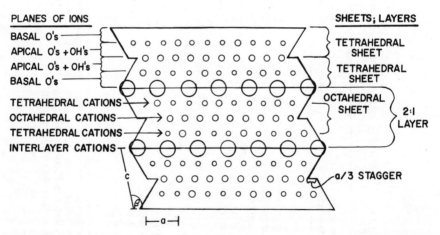

Fig. 7-2. Idealized view of a trioctahedral 1M polytype mica along the Y axis. Individual cations are shown, but individual anions are not. Within each 2:1 layer the upper tetrahedral sheet is shifted relative to the lower sheet by $a/3$. In other polytypes the shift is not in the same direction in successive layers.

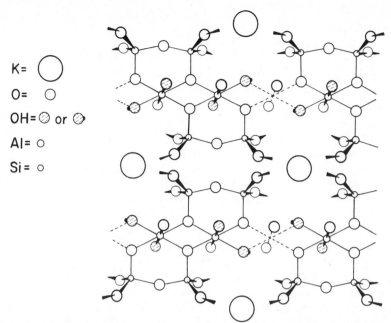

K =
O =
OH = or
Al =
Si =

Fig. 7–3a. Idealized slice through a muscovite (dioctahedral) structure. Only two of the six tetrahedra that surround a given potassium in each tetrahedral sheet are shown. The two tetrahedral sheets of a given layer are shown as superimposing for sake of simplicity of presentation—actually they are shifted relative to each other as shown in Figs. 2, 5, and 6.

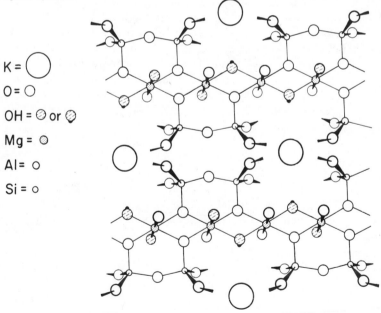

K =
O =
OH = or
Mg =
Al =
Si =

Fig. 7–3b. Idealized slice through a phlogopite (trioctahedral) structure. Protons of hydroxyls point directly toward interlayer cation in this structure, but not in the dioctahedral structure.

2. LAYER CHARGE

An important aspect of the structures and formulas is that the overall positive charge of the interlayer cations must balance the negative *layer charge* (designated as X). The layer charge of phyllosilicates is normally expressed on a formula unit basis (Table 7-1). *Micas* are defined as having $X \sim$ 1. In muscovite, $K(Si_3Al)Al_2O_{10}(OH)_2$, for example, the negative layer charge results from the isomorphous replacement of one out of each of four Si^{4+} in the pyrophyllite formula, $Si_4Al_2O_{10}(OH)_2$, with an Al^{3+},—pyrophyllite (like talc) is a mineral that has $X \sim 0$ and unoccupied interlayer. *Brittle micas* such as margarite (Table 7-1) have $X \sim 2$,—note that two out of four tetrahedral cations are Al in the margarite formula (Table 7-1) with the layer charge per formula unit balanced by a divalent interlayer cation.

Minerals of the vermiculite group have $X \sim 0.6$–0.9 and minerals of the smectite group have $X \sim 0.25$–0.6 (CMS Nomenclature Committee, 1971a). The higher layer charge of micas helps to explain why the interlayer cations are held in a "nonexchangeable" form, whereas most interlayer cations of vermiculites and smectites are hydrated and exchangeable under normal temperature and moisture conditions. Certain weakly hydrated ions such as K^+ may be "fixed" by vermiculite thus forming a structure analogous to mica. As has been pointed out by Kittrick (1966, 1969a, b), the hydration and exchangeability of the interlayer cations depends upon the balance between the electrostatic forces of attraction between the ion and the layer and the forces tending to hydrate the ions and the negative charge sites. In micas the attractive forces between the layer charge and the interlayer cation are normally larger than in vermiculites and smectites. However, minerals with vermiculite-like X-ray characteristics that appear to have the layer charge of the parent mica have been formed from muscovite by artificial weathering (Scott & Reed, 1965). In this restricted case, the layer charge distinction between micas and vermiculites appears to break down, since a mineral might be called vermiculite based on X-ray diffraction patterns, or mica based on layer charge.

3. DIOCTAHEDRAL VS. TRIOCTAHEDRAL STRUCTURES

Another important aspect of phyllosilicate structures is whether they are dioctahedral or trioctahedral (Fig. 7-3 and 7-4, and Table 7-1). In trioctahedral structures all of the octahedral cation sites are occupied—amounting to three octahedral cations per formula unit (Table 7-1). This is the common configuration when most of the octahedral cations are divalent. In dioctahedral structures, on the other hand, only two out of every three octahedral cation sites are occupied (Fig. 7-3a and 7-4). This leaves one out of three sites vacant. This is the common configuration when most of the octahedral cations are trivalent.

Bailey (1966) in reviewing the literature on detailed structure determinations of layer silicate minerals, including structures for several micas, noted

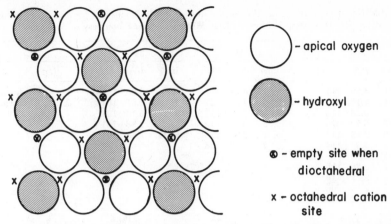

- apical oxygen

- hydroxyl

⊗ - empty site when dioctahedral

x - octahedral cation site

Fig. 7-4. Octahedral cations shown relative to the position of apical oxygens and hydroxyls of an underlying tetrahedral sheet with ideal hexagonal symmetry. Both di- and trioctahedral cases are illustrated.

that the vacant octahedral cation site of dioctahedral minerals was found to be ordered (not randomly distributed) in all of the numerous dioctahedral structures that had been reported.

Serratosa and Bradley (1958) found that the O-H bond is commonly oriented normal to the plane of the 2:1 layers in trioctahedral minerals, whereas in the dioctahedral structures it tends to be shifted (relative to the trioctahedral structures) toward the empty octahedral cation site (Fig. 7-3). Bassett (1960) hypothesized that since the H of the OH is closer to the K in the trioctahedral case, the K is not held as tightly as in dioctahedral K micas where the H is farther away from the K. This apparently is an important aspect of the greater weatherability of the trioctahedral micas (Norrish, 1973) and is to be discussed in more detail in later sections.

4. CRYSTALLOGRAPHIC AXES AND ANGLES, AND UNIT CELLS

The CMS Nomenclature Committee (1971a) has pointed out with reference to crystallographic axes and unit cell repeat distances that "strictly speaking, X, Y, and Z should be used to refer to crystallographic axes and a, b, c to the repeat distances along these axes (i.e. unit cell lengths)." The crystallographic angles are α (between Y and Z axes), β (between X and Z), and γ (between X and Y). The crystallographic parameters for various micas determine their crystal system. As discussed below in connection with polytypes, many micas are monoclinic ($\alpha = 90°$, $\gamma = 90°$, and $\beta \neq 90°$).

The *unit cell* is the true repeating structural unit. The a and b dimensions are about the same from mica to mica if the minerals are indexed on a monoclinic or orthorhombic cell. Small differences do exist, however, such as the generally smaller b dimension of the dioctahedral as opposed to the trioctahedral minerals. The c dimension varies, (Table 7-2) depending particu-

Table 7-2. Diagnostic X-ray diffraction lines and cell dimensions for muscovite mica polytypes. Data supplied by The Clay Minerals Society, courtesy of S. W. Bailey

2M$_1$			1M			3T			1M$_d$
Int.†	d(Å)	hkl	Int.†	d(Å)	hkl	Int.†	d(Å)	hkl	d(Å)
1. 1	4.29	111	1. 1½	4.35	11$\bar{1}$	1. 3	3.86	10$\bar{1}$4	1. ~3.66 absent‡
2. 1	4.10	022	2. 1	4.12	02$\bar{1}$	2. 2½	3.61	10$\bar{1}$5	2. ~3.07 absent‡
3. 3	3.88	11$\bar{3}$	3. 5	3.66	112	3. 3	3.10	10$\bar{1}$7	
4. 3	3.72	02$\bar{3}$	4. 5	3.07	112	4. 3½	2.88	10$\bar{1}$8	
5. 3	3.49	11$\bar{4}$	5. 1	2.93	11$\bar{3}$	5. ½	2.68	10$\bar{1}$9	
6. 3	3.20	114	6. 2	2.69	023				
7. 3½	2.98	025							
8. 3	2.86	115							
9. 2½	2.79	11$\bar{6}$							
10. 1	2.14	20$\bar{6}$							
3	2.12	135, 043							

$$a = 5.189\text{Å} \qquad a = 5.208\text{Å} \qquad a = 5.196\text{Å} \qquad a = 5.20\text{Å}$$
$$b = 8.988 \qquad b = 8.995 \qquad b = 5.196 \qquad b = 9.00$$
$$c = 20.069 \qquad c = 10.275 \qquad c = 29.971 \qquad c\sin\beta = 10.00$$
$$\beta = 95.4° \qquad \beta = 101.6° \qquad \beta = 90°$$
$$\gamma = 120°$$

† Relative intensity (larger number indicates more intensity).
‡ These lines are present for the ordered 1M polytype. With increasing disorder, these lines become broader until they are absent when the symbol 1M$_d$ is assigned.

larly on the number of layers between true repeats, as discussed below in connection with polytypes.

The intersections of repeating evenly spaced parallel planes (either populated with atoms or ions or parallel to populated ones) with the crystallographic axes may be designated by means of their Miller indices. Three whole numbers (h, k, l) refer to the distance between intersections with the X, Y, and Z axes, respectively, in terms of the inverse of fractions of the unit cell dimensions (a, b, and c). For example the Miller index, 221, means that the set of planes intersects the X axis at every ½ a, the Y axis at every ½ b, and the Z axis at every unit of c (1/1). 130 means that the set of planes intersects the X axis at every unit of a (1/1), the Y axis at every 1/3 b and does not intersect the Z axis at infinity, i.e., it is parallel to Z. Sometimes used are general terms such as the "00l planes." The 00l refers to all planes that intersect the Z axis, but do not intersect the X and Y axis. The 00l planes are the ones that are "seen" in X-ray diffraction patterns of layer silicates when parallel orientation specimens are employed.

Note that the Z axis of layer silicate minerals transects the layers, whereas the X and Y axes are parallel to the sheets and layers. The 001 d spacing for monoclinic micas is different from the c dimension of the unit cell of the mineral because $\beta \neq 90°$. For these minerals, $c = [d(001)/(\sin\beta)]$.

5. POLYTYPES AND OTHER STRUCTURAL DETAILS

"*Polymorphism* is the ability of the same chemical compound to exist in more than one structural form" (Bailey, 1967). Examples of polymorphs are diamond and graphite (for C); calcite, aragonite and vatterite (for $CaCO_3$);

and quartz, tridymite, and cristobalite (for SiO_2). Polytypism was considered by Bailey (1967) to be a special case of polymorphism, restricted to layer-like structures, in which the different structures result from different stacking sequences of identical layers. Thus, polytypism has sometimes been called one-dimensional polymorphism. In much of the earlier literature (e.g., Hendricks & Jefferson, 1939; Smith & Yoder, 1956), what are now being called polytypes of micas were called polymorphs. Bailey (1967) has pointed out that the different polytypes of a given group of layer silicates (e.g., micas) have essentially the same lateral (a and b) unit cell dimensions, differing only because of indexing on different types of unit cells. The polytypes often differ in their c dimension, where the true repeat distance is some integral multiple of the unit structure thickness (Table 7-2).

The different mica polytypes are determined primarily by differences in the octahedral sheet of different layers, since the tetrahedral sheets of adjacent layers essentially superimpose in the interlayer in order to fit around the interlayer cation. Bailey (1967) has shown (Fig. 7-5) that there are two different sets of octahedral cation sites that may be occupied and three different directions that one tetrahedral sheet may be shifted relative to the other for each set—after accounting for positions that are equivalent because of symmetry operations. Thus, there are possibilities for different sets of cation sites being occupied and for different directions of shift or stagger in successive layers.

Smith and Yoder (1956) have shown that there are only six mica polytypes with true periodicities from 1 to 6 layers if only one stacking angle is permitted in a given crystal (Fig. 7-6). The stacking angle is the angle between vectors (Fig. 7-6) that show the direction of intralayer shift (Fig. 7-5) in successive layers. For example, the 1M polytype has the same set of octahedral cation sites occupied (Set I) and the same shift direction in each layer. The stacking angle for this case is 0°. The 2Or polytype has Set I octahedral cation sites occupied in one layer and Set II sites occupied in the next; the intralayer shift direction in the second layer is opposite of that in the first and the stacking angle is 180°.

Smith and Yoder adopted a simple nomenclature system for their polymorphs that now is used to designate mica polytypes. The first symbol gives the number of layers in the true repeat (unit cell) unit and the second gives the (crystal system) symmetry of the structure. The 1M polytype has a true repeat every layer and is monoclinic. There are two two-layer monoclinic (2M) polytypes ($2M_1$ and $2M_2$). Also, Or represents orthorhombic; T, trigonal; and H, hexagonal.

Of the theoretical mica polytypes only certain ones seem to be common in nature (Smith & Yoder, 1956; Radoslovich, 1959; 1960b; Bailey, 1966, 1967). The $2M_1$ polytype appears to predominate for dioctahedral micas with 1M and 3T less abundant. Smith and Yoder (1956) found in muscovite laboratory synthesis experiments that the 1Md (d indicating disordered) polytype formed first at low temperatures. With longer runs or somewhat higher temperatures the 1M polytype formed and at still higher temperatures the $2M_1$ polytype formed. Thus, $2M_1$ muscovite is usually

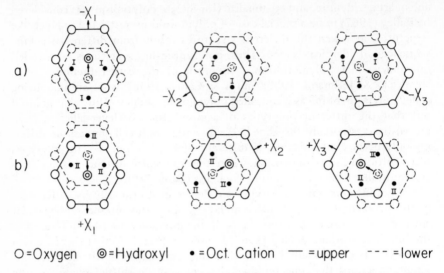

O = Oxygen   ◎ = Hydroxyl   • = Oct. Cation   — = upper   --- = lower

Fig. 7-5. Possible octahedral cation sites and intralayer (one tetrahedral sheet relative to other) shifts along the negative and positive directions, respectively, of three X axes. After Bailey (1967).

interpreted to have formed in environments of high thermal energy or pressure—such as in medium to high grade metamorphic rocks, igneous rocks, and most pegmatites (Bailey, 1966).

Bailey (1967) has pointed out that certain workers have found from muscovite synthesis experiments that small compositional deviations from the ideal formula may stabilize the 1M and 3T structures. Also, the trioctahedral micas are usually found to be 1M. Thus, Bailey (1967) states that it is evident that many of the natural mica structures are neither polytypic nor polymorphic (since to be so the minerals would have to have the same chemical composition). Thus, it may be more appropriate to say that 1M, 2M, 3T, etc. represent different structure types without calling them polytypes or polymorphs. Alternatively polytypes should be defined in such a way that they would not have to be polymorphs. A Joint Committee of the International Union of Crystallography and the International Mineralogical Association has proposed that the definition of polytypism be modified to permit minor deviations in overall composition and symmetry between individual layers (accepted by AIPEA Nomenclature Committee, 1972).

Other details about mica structures and the way they interact to control the polytype formed have been described by Radoslovich (1959, 1960a, b) and Takeuchi (1966). In many of the micas studied, the tetrahedra are rotated or rotated and tilted compared to an ideal structure in which the tetrahedra are linked up to give true hexagonal "rings." This rotation and tilting is illustrated in Fig. 7-1 for margarite and would apparently be similar, although not as marked, for many other dioctahedral minerals.

The main reason for the rotation and tilting of the tetrahedra apparently is to cause a better fit between the tetrahedral and octahedral sheets (since

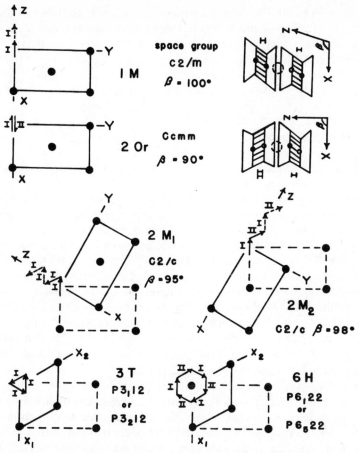

Fig. 7-6. Six simple mica polytypes are illustrated. Arrows represent vectors that show the direction of intralayer shift (Fig. 7-5) in successive layers. Roman Numerals indicate which set of octahedral cation sites are occupied in the respective layers. Insets (upper right) show side views of layers for 1M and 2Or polytypes. After Bailey (1967) as modified from Smith and Yoder (1956).

the apical oxygens are part of both). If unconstrained the octahedral sheet would be smaller than the tetrahedral sheet, particularly for dioctahedral minerals—based on comparison with other minerals such as gibbsite and considering the size of the ions involved (Bailey, 1966; Radoclovich, 1960a, 1963).

The rotation of the tetrahedra has the effect of giving ditrigonal, rather than hexagonal, "rings" around the interlayer cation (Fig. 7-1). This ditrigonal nature of many micas apparently causes the polytypes that have stacking angles of $0°, 120°,$ and $240°$ (1M, $2M_1$, 3T) to occur more frequently than those with stacking angles of $60°, 180°,$ and $300°$ ($2M_2$, 2Or, 6H) (Radoslovich, 1959).

The tilting in addition to the rotation has the effect of giving the plane of basal oxygens a corrugated nature (Fig. 7-1). Apparently the tilting takes place primarily with dioctahedral structures, since the fit between the octahedral and tetrahedral sheets requires more structural accommodation than in trioctahedral structures. In the corrugated structures the interlayer cation is also moved slightly in the $X-Y$ plane away from the center of the "ring" (Takeuchi, 1966). In trioctahedral structures the basal oxygens apparently are more nearly co-planar with the interlayer cations in the centers of the rings.

6. ILLITE AND GLAUCONITE

Illitic and glauconitic type micas have caused many nomenclature problems for clay mineralogists.

Illite was proposed (Grim et al., 1937) as a name for the "mica occurring in argillaceous sediments." This material was thought to be similar to materials that had been called sericite, hydromica, and Glimmerton by previous workers and to differ from muscovite primarily in having poorer crystallinity, lower K_2O content, and higher water content. The term has been widely used in geologic, mineralogic, and pedologic literature, although some workers have avoided the use of the term because of disagreement over exactly what illite is or should be. The term is still being used (e.g., see Weaver & Pollard, 1973). Also illitic mineralogy families are recognized in the U. S. soil classification system for soils in clayey particle-size families that have clay (< 2 μm) with "more than half illite (hydrous mica) by weight and commonly > 4 percent K_2O" (Soil Survey Staff, 1972).

The CMS Nomenclature Committee (1971a) has taken the position that the status of illite (or hydromica), sericite, etc. (glauconite by inference) must be left open because it is not clear whether or at what level such materials would enter the proposed classification scheme. It was felt that many materials to which these terms have been applied may be interstratified. Bailey (1966) rendered a somewhat similar opinion. His interpretation was that "the consensus of recent studies of sedimentary *illite* is that it is a heterogeneous mixture of detrital $2M_1$ muscovite, detrital mixed layer micaceous weathering products partly reconstituted by K-adsorption or by diagenetic growth of chloritic interlayers, plus true 1Md and 1M micas—some having mixed layering also."

Gaudette et al. (1966) defended the term illite, indicating that their reexamination of some "illites" showed them to have less potassium than well-crystallized micas and that some of them were not essentially mixed-layer structures. Their < 2 μm Beavers Bend illite (formula given in Table 7-1) was considered to have no detectable mixed layering although the X-ray diffraction pattern that they present seems to indicate the presence of a small amount of quartz and studies of particle size subfractions of this material would be desired to prove the absence of discrete and interstratified expanded and chlorite layers.

The term glauconite has been used (i) for small greenish earthy pellets (also called glauconite pellets) of heterogeneous mineralogy, and (ii) for a micaceous mineral (Burst, 1958a, b). It is the second usage that is examined here.

Mineralogical glauconite is much like illite in that many of the materials that have been designated by this term appear to have expanded 2:1 layers and possibly also chlorite layers interstratified with mica layers (Burst, 1958a, b; Hower, 1961; Manghnani & Hower, 1964; Bentor & Kastner, 1965; Tapper & Fanning, 1968). However, glauconites contain more iron and less aluminum in their octahedral sheets than illites; and glauconites are of the 1M or 1Md polytype (Tapper & Fanning, 1968, and many others), whereas the $2M_1$ polytype seems most prevalent in illites. Much evidence (discussed in a later section) indicates that mineralogical glauconite is an authigenic, usually marine, mineral. In some occurrences it can be quite well crystallized, to the point that it can be shown to be a 1M mica (Warshaw, 1957)[1], although crystals good enough for detailed structure analyses by single crystal X-ray methods apparently have not been found.

Normally the term glauconite has been attached to the mineral assemblage including the micaceous and the interstratified nonmicaceous layers. Thus, Burst (1958a, b) has used the degree of disorder and amount of interstratified expanded layers (which appear to go together) as criteria in classifying glauconites. Another possibility would be to use glauconite as a term for only the mica (10Å) layers of these materials as Tapper and Fanning (1968) did in a quantitative mineralogical analysis of a glauconitic material by the methods of Alexiades and Jackson (1966).

A main problem that arises, regardless of what the illitic and glauconitic materials are called, concerns their layer charge. If the layer charge is as low as the formulas that have been derived for them indicate (Table 7-1, also many more similar formulas can be found in the literature—see Weaver & Pollard, 1973), then these minerals would be vermiculites ($X \sim 0.6$-0.9) by the AIPEA (1972) and CMS (1971a) Nomenclature Committees' definition— probably an unsatisfactory classification for most earth scientists. Also, if the formulas (Table 7-1) are correct, a considerable part of the layer charge in the illites and glauconites originates in the octahedral sheet, whereas most of the layer charge originates in the tetrahedral sheet in the other mica minerals.

Some workers (e.g., Raman & Jackson, 1966) have presented data that indicate that the formulas mentioned above are wrong because of considerable interstratified or discrete chlorite, vermiculite, and montmorillonite, as well as other minerals in the materials. They would claim that these other minerals have not been taken into account in developing the formulas. Raman and Jackson (1966) and Alexiades and Jackson (1966) claim to have found and measured these other minerals by means of qualitative and quantitative analyses including the determination of chlorite by thermogravimetry.

---

[1]C. M. Warshaw. 1957. The mineralogy of glauconite. Ph.D. Thesis. The Pennsylvania State Univ., University Park.

A possible criticism of the work of Raman and Jackson is that their X-ray diffraction patterns indicate less chlorite than their thermogravimetric analyses.

For quantitative mineralogical analyses based on allocation of elements from elemental analyses, Alexiades and Jackson (1966) have chosen K_2O content values for soil micas (10% K_2O) and glauconite (9% K_2O) that are intermediate between ideal micas (11.8% K_2O in muscovite) and the values commonly reported for illites and glauconites. Inherent in this approach is an assumed intermediate layer charge.

The problems discussed above show that defining micas (and other layer silicates) in terms of their layer charge is somewhat unsatisfactory. Layer charge is very difficult to determine when the minerals in question occur as components of complex mixtures (the common case in soil clays). The definitions based on layer charge, even if theoretically sound, are not operational enough for application in most soil mineralogical analyses.

More information is needed before sound conclusions can be reached about the layer charge of sedimentary (illitic and glauconitic) micas. However, K-bearing 10Å 2:1 layer silicate minerals (minerals closely resembling the mica structure) may exist that have a layer charge too low for the present definition of micas. These materials may also have more of their charge originating in the octahedral sheet than the more true micas. If so, the terms illite and glauconite would seem to have precedence as names for these materials.

B. X-ray Diffraction

Identification of the mica group minerals can be achieved with X-ray diffraction patterns of their randomly or parallel oriented specimens. Muscovite, biotite, and illite X-ray diffraction patterns are characterized mainly by two intense peaks in the region of 10 and 3.3Å and a relatively weaker peak at 5Å. Generally the appearance of a peak at 10Å for Mg-saturated 25C (298K) specimens, which corresponds to the distance between the center of octahedral sheets in adjacent layers, indicates the presence of the mica group minerals. The three main peaks remain unaltered upon glycerol solvation and upon K saturation and heating up to 500C (773K) or higher.

The type of cations occupying the octahedral positions influences the relative intensities of the (00l) diffraction lines and particularly the 5Å one. In the case of muscovite (aluminous octahedral sheet) the 5Å line is relatively strong, whereas in iron-rich micas, such as biotite, glauconite, and lepidomelane (ferruginous octahedral sheet) the 5Å peak is weak or missing (compare Fig. 7-7a and 7-7b).

Very small peaks at about 11, 5.5, and 3.7Å in the patterns for muscovite and biotite (Fig. 7-7) indicate regularly interstratified expanded layers in a small part of the mica employed. Similar peaks from mica flakes have been presented by Alexiades (1970).

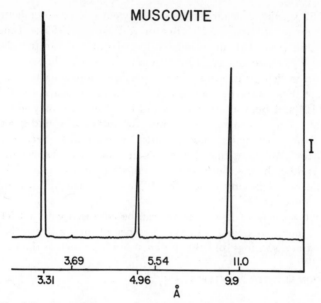

Fig. 7-7a. X-ray diffraction pattern of parallel oriented muscovite flakes.

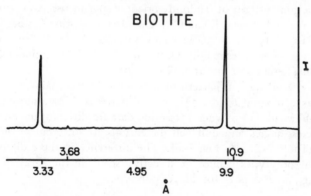

Fig. 7-7b. X-ray diffraction pattern of parallel oriented biotite flakes from Bancroft, Ontario, Canada.

1. DISTINCTION BETWEEN DI- AND TRIOCTAHEDRAL MICAS

Distinction between di- and trioctahedral micas is based on the fact that the b dimension in dioctahedral micas is smaller. The b dimension is usually found by looking at the 060 spacing in X-ray diffraction patterns of randomly oriented specimens. As Grim et al. (1951) pointed out, the 060 spacing for dioctahedral forms is close to 1.50Å. For the trioctahedral forms 060 lies between 1.525 and 1.54Å. The larger b dimension in trioctahedral micas can be accounted for by the larger size of Mg^{2+} and Fe^{2+} ions as opposed to Al^{3+} ions and by the larger number of octahedral cations.

The 060 spacing of glauconites (generally considered dioctahedral) falls within the range of 1.51–1.52Å (Bentor & Kastner, 1965). Tapper and Fanning (1968) reported an average value of 1.518Å for the glauconite studied and Tyler and Bailey (1961) found values lying between 1.511 to 1.515Å for iron-rich glauconites. Two possible reasons have been proposed for the large 060 values of glauconites. First, as suggested by Tyler and Bailey (1961) and Bentor and Kastner (1965) glauconite may not be truly dioctahedral. They calculated glauconite formulas and found greater than two out of three octahedral sites occupied with cations. Secondly, larger 060 values for glauconite than for muscovite should be expected, even if glauconite is dioctahedral, because the octahedral sheet of glauconite is ferruginous and Fe^{3+} (and any Mg^{2+} or Fe^{2+}) are larger than Al^{3+} which is more abundant in muscovite.

The 060 spacing of layer silicates can be confused with a 1.541Å diffraction maximum of quartz. In the presence of quartz it may be worthwhile to look at d spacings for other b dimension lines such as the 020.

2. DISTINCTION BETWEEN POLYTYPES

Differentiation of different mica polytypes is usually done by means of X-ray diffraction patterns of randomly oriented specimens. X-ray diffraction lines, in the region between 4.3 and 2.7Å are used to distinguish between 1M, $2M_1$, and 3T polytypes, as suggested by the Clay Minerals Society (Table 7–2). This region includes mainly ($02l$) and ($11l$) diffraction lines except for the one strong basal order line at 3.32Å.

The X-ray diffraction pattern of $2M_1$ muscovite contains more diffraction lines as compared to the 1M and 3T patterns. For a more complete data, see ASTM files of powder diffraction data for different micas. Crystallographic axes, angles and unit cell dimensions are different in the various polytypes (Table 7–2, also Fig. 7–6). The difference in the c dimension of the different polytypes cannot be directly "seen" in the X-ray diffraction patterns—e.g., no 20Å peak for 2-layer polytypes.

3. QUANTITATIVE INTERPRETATION OF X-RAY DIFFRACTION PATTERNS

Relative intensities of X-ray diffraction peaks can give an estimate of the amount of mineral present in the sample and X-ray diffraction patterns can serve as a semiquantitative method when their interpretation is exercised with caution.

Accurate quantitative determination of the components of complex mixtures, by means of their X-ray diffraction patterns, is nearly impossible, however. The reason is that imperfections of crystals greatly influence their diffraction characteristics. The term "imperfections" refers to the mixed layer structures (interstratification), the degree of crystallinity, and the degree of orientation in parallel oriented specimens. The X-ray diffraction patterns of different particle-size fractions of a Varna B2 horizon (Fig. 7–8)

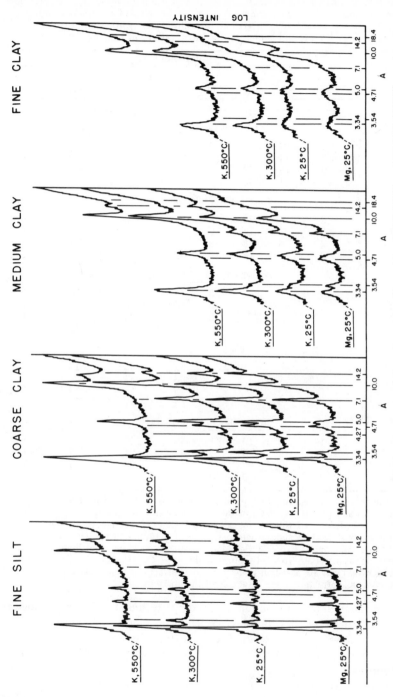

Fig. 7–8. X-ray diffraction patterns of Varna B2 horizon fine silt and clay fractions. Mg, 25C (298K) and K, 25C (298K) specimens were glycerol solvated. From D. S. Fanning, 1964. Mineralogy as related to the genesis of some Wisconsin soils developed in loess and in shale-derived till. Ph.D. Thesis, Univ. of Wisconsin, Madison.

Table 7-3. Estimated mineralogical composition of size fractions (carbonate, organic matter and free iron oxide removed) from four horizons of a Varna soil (now classified as Elliott-Aquic Argiudoll; fine, illitic, mesic) from southeastern Wisconsin. Data from Fanning and Jackson (1965) and Fanning (1964)[3],†

Horizon‡	NaFl	KFl	Qr	Chl	Mi	Exp. 2:1 (Vr+Mt)	Amor.	TiO$_2$
								%
Fine silt (2-5 μm)								
Ap	7	10	46	2	30	2	2	1
B2	8	14	39	5	28	3	2	1
C1	9	10	39	6	31	2	2	1
C3	11	11	38	5	32	1	1	1
Coarse clay (0.2-2 μm)								
Ap	2	4	20	8	45	16	4	1
B2t	2	3	17	10	50	12	5	1
C1	2	6	18	13	49	9	2	1
C3	3	3	16	16	51	7	3	1
Medium clay (0.08-0.2 μm)								
Ap				3	44	38	14	1
B2t				5	46	38	10	1
C1				8	53	30	8	1
C3				6	58	25	10	1
Fine clay (<0.08 μm)								
Ap					32	53	15	
B2t					38	51	11	
C1					43	43	14	
C3					46	41	13	
Total clay (<2 μm)¶								
Ap	1	2	10	5	42	30	9	1
B2t	1	2	10	7	48	24	7	1
C1	1	3	8	9	47	23	7	1
C3	2	2	9	11	53	16	6	1

† Mica analyses are based on 10% K$_2$O in mica after accounting for K$_2$O in K feldspars.
‡ The depth and pH (1:1 in H$_2$O) of the horizons were: Ap (0-20 cm, pH 6.7), B2t (20-41 cm, pH 7.5), C1 (58-94 cm, calcareous), C3 (117-140 cm, calcareous).
§ Abbreviations: NaFl = sodium feldspar, KFl = potassium feldspar, Qr = quartz, Chl = chlorite, Mi = mica, Exp. 2:1 (Vr+Mt) = vermiculite and montmorillonite, Amor. = amorphous materials, TiO$_2$ = titanium oxides.
¶ Data for total clay are based on values for clay subfractions weighted for the proportion of each subfraction in the total clay of the horizon.

when considered in conjunction with complete quantitative mineralogical analyses of these fractions (Table 7-3) based on allocation of elements from elemental analysis etc. (Jackson, 1969) illustrate the problems of trying to estimate mica content from X-ray diffraction patterns. A stronger 10Å peak with Mg saturation and glycerol solvation (indicating mica presence) was obtained for the fine silt fraction with 28% mica than for the medium clay with 46% mica because of interstratification of the mica with expanded (vermiculite and montmorillonite) layers in the finer fraction.

C. Differential Thermal Analysis

In examining the DTA curves of the mica minerals, emphasis will be given (i) to use of DTA curves as a diagnostic aid, (ii) to reactions responsible for the appearance of each peak, (iii) to phases formed on heating, and (iv) to differences in DTA curves as related to chemical composition and structural properties of each mineral.

1. GENERAL REMARKS

Peaks in DTA curves of most of the micas, when they are heated up to 1,200C (1,473K) originate mainly from expulsion of structural hydroxyls and accompanying structural rearrangements; recrystallization; oxidation of Fe^{2+}; and, in the case of lepidolites, loss of fluorine.

Dehydroxylation involves the conversion of M-OH (where M = an octahedral cation) to M-O and M $(2OH^- \rightarrow O^{-2} + H_2O)$ and in most of the micas occurs gradually and without any significant structural change, so that the endothermic dehydroxylation peaks are not sharp and generally are smaller in size than those given by other clay minerals. The temperature at which the structural water is driven off depends, to a large extent, on the type of the octahedral cations. In magnesian minerals dehydroxylations occur at higher temperatures than in aluminian species.

Particle size markedly affects the shape of the DTA curve. It was postulated by Arens (1951) that when the particle size is above 20 μm , the surface area is too small for dehydroxylation processes to occur rapidly enough to give a distinct DTA peak. As reported by MacKenzie and Milne (1953), grinding of muscovite greatly affects its thermal behavior and care must be taken in sample preparation. Illites and glauconites give dehydroxylation peaks at lower temperatures than other micas (Fig. 7-9), perhaps mainly because of smaller particle size.

Quantitative estimations of the different micas from their DTA patterns is not reliable in view of the variation in size of the dehydroxylation peak because of F for OH substitution, which is appreciable (MacKenzie, 1970). Also, as Roy (1949) reported, the weight loss on decomposition by heating, of mica type minerals, is not just a simple rapid chemical change, but depends upon temperature and is a function of time. Since the rate of dehydration, because of the extensive internal surfaces of layer type minerals, depends on the sample weight (Holt et al., 1958) one must be careful in making quantitative interpretations of DTA patterns.

2. DTA OF DIOCTAHEDRAL MICAS

Muscovite DTA curves are characterized by two endothermic peaks at 800-900C (1,073-1,173K) and about 1,100C (1,373K) and one small, hardly visible, exothermic peak at about 350C (623K) (curves A and B, Fig. 7-9). The exothermic peak can be related to a release of strain in the structure, but

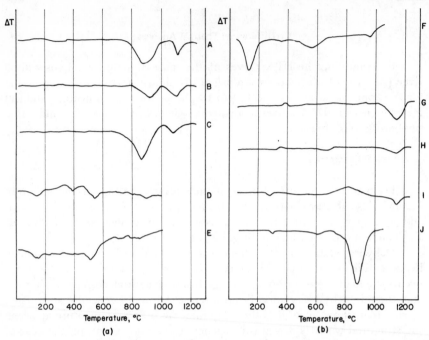

Fig. 7-9. Differential thermal analysis curves for micas. *(A)* muscovite, Mama, Eastern
Siberia, USSR; *(B)* muscovite, Madras, India; *(C)* paragonite; Monte Campoine, Switzer-
land; *(D)* Fithian illite, and *(E)* Beavers Bend illite, Illinois; *(F)* glauconite, New Jersey;
(G) phlogopite, Aldan, Eastern Siberia, USSR; *(H)* biotite, Kransnoyark, Siberia,
USSR; *(I)* biotite, Kolsk peninsula, USSR; *(J)* lepidolite, Kalbinsk Mount, Kazakhstan,
USSR. Curves redrawn from various sources. Curves *A, B, C, D, G, H, I, J*, from Mac-
kenzie (1970), Curve *E* from Gaudette et al. (1966), and Curve *F* from Tedrow (1966).

its effect is too small to be of any use as a diagnostic criterion (MacKenzie,
1970).

The first endothermic peak represents dehydroxylation (Grim et al.,
1951) or dehydroxylation accompanied with some structural rearrangement
(Tsvetkov & Valyashikhina, 1956). Above 980C (1,253K) the muscovite
structure starts disintegrating and the second endothermic peak represents recrystallization into γ-Al_2O_3 and/or spinel (Roy, 1949). According to Zwetsch
(1939) muscovite at 1,050C (1,323K) yields, γ-Al_2O_3, α-Al_2O_3 and leucite
and Tsvetkov and Valyashikhina (1956) reported the recrystallization into
leucite, corundum, and cristobalite.

According to Roy (1949) no significant change occurs in the crystal
structure of muscovite upon heating to 940C (1,213K) except for a slight expansion in the *c* dimension, as indicated by X-ray diffraction. This is true
even though dehydroxylation is complete. Holt et al. (1958) to explain the
stability of the muscovite structure up to 940C (1,213K), postulated that as
the two hydroxyls adjacent to the aluminum ion are driven off, leaving an
oxygen ion, the aluminum ion is shifted only a short distance to be in a stable
four-coordinated position with oxygen.

The aforementioned researchers in studying the rate of thermal dehydration of muscovite concluded that an increasing energy of activation of the hydroxyl ions to form water, is required with heating time. This means that the energy barrier for the hydroxyls to form water becomes higher as the reaction proceeds and an increasing amount of energy is needed for activation. According to a model that they proposed, the normal muscovite structure is in a strained condition. As dehydroxylation proceeds the strain is relieved, rendering the remaining hydroxyls more reluctant to leave the structure. The increasing tendency for the hydroxyls to remain in the structure must be surmounted by a corresponding increase of the activation energy.

The strain apparently can be removed by grinding muscovite up to 0.3 mm, which may explain the differences found in DTA and TG (thermogravimetry) curves of muscovite ground to different degrees of fineness (Lodding & Vaughan, 1965).

The paragonite DTA curve is similar to that of muscovite (Tsvetkov & Valyashikhina, 1956). According to the same researchers, finer particle size of the paragonite sample studied (curve C, Fig. 7-9) accounts for the sharper and larger first endothermic peak with respect to the second as compared to the curves (A and B, Fig. 7-9) shown for muscovite.

DTA curves for illites show a low temperature endothermic peak at about 100 to 200C (373-473K), corresponding to loss of interlayer water, a second medium size dehydroxylation peak at about 550C (823K), and a small endothermic peak at about 900C (1,173K) combined with a slight exothermic reaction (curves D and E, Fig. 7-9). The temperature range and the size of the second and third endothermic peaks show considerable variation in different samples (Grim et al., 1951). Gaudette et al. (1966) reported that the second endothermic reaction for Marblehead illite lies between 600-700C (873-973K) and MacKenzie et al. (1949) presented DTA patterns for Ballater Aberdeenshire illites which showed a sharp well-defined peak at 713C (986K). The small exothermic peak at about 900C (1,173K) indicates the development of a high temperature phase (Gaudette et al., 1966), probably the formation of a spinel.

The crystalline structure of illite persists up to about 900C (1,173K), followed by destruction of the structure at higher temperatures and formation of a spinel and mullite (Grim & Bradley, 1940; Hill, 1953; Grim & Kulbicki, 1957).

Glauconite DTA curves are similar to those of illites. Tedrow (1966) presented a clay-size glauconite DTA curve (curve F, Fig. 7-9) having endothermic reactions at 150 (423K), 560 (833K), and 925-975C (1,198-1,248K). The small peak at 360C (633K) is probably due to traces of goethite. The large peak at 150C (423K) represents dehydration of interlayer water and indicates the presence of interstratified expanded layers.

3. DTA OF TRIOCTAHEDRAL MICAS

Thermal behavior of trioctahedral micas is controlled primarily by the type of octahedral cations and the amount of F for OH substitution and secondarily by the type of interlayer cations. In general, DTA curves for trioctahedral micas are reminiscent of those of dioctahedral in that they show a gradual loss of structural hydroxyls. Peaks are relatively small in size and the structure of the mineral persists up to about 1,200C (1,473K).

Differential thermal analysis curves for phlogopite (curve G, Fig. 7-9) show a very small exothermic peak at about 350-400C (623-673K), generally more pronounced than for muscovite, and a relatively small endothermic peak at about 1,050-1,200C (1,323-1,473K) (Tsevtkov & Valyashikhina, 1956). The endothermic peak represents loss of structural hydroxyls and, as is common for magnesian minerals, it occurs at much higher temperatures than the equivalent peak on the curve for muscovite (aluminous octahedral sheet). Variation in the degree of F for OH substitution influences the peak temperature and its size as well. Upon heating of phlogopite, spinel formation follows dehydration and the spinel phase persists as the only crystalline phase up to 1,550C (1,823K) (Roy, 1949).

Lepidolite gives a single, sharp, well-defined endothermic peak at about 850-900C (1,123-1,173K), which represents the loss of volatiles (F and OH) (Roy, 1949; Mielenz et al., 1954), and as is common with the micas, a very slight exothermic peak at about 350C (623K) (curve J, Fig. 7-9).

Biotite, the most common of the trioctahedral micas, gives a DTA pattern similar to that of phlogopite, having an endothermic peak in the region of 1,000-1,200C (1,273-1,473K), which represents loss of hydroxyls (curves H and I, Fig. 7-9). Peak temperature is slightly lower in samples with higher Fe^{2+} content (MacKenzie, 1970). As would be expected for a mica, two other peaks appear, one at 350C (623K) and a larger exothermic peak at 700-800C (973-1,073K) which, according to MacKenzie (1970), is related to oxidation of structural Fe^{2+}. Phases identified on heating biotite up to 1,200C (1,473K), are leucite, γ-Fe_2O_3 and spinel (Grim & Bradley, 1940). Roy (1949) reported that, on heating biotite to about 1,100C (1,373K), the phases developing are an iron rich spinel, leucite, and mullite.

4. DTA CURVES OF MICAS AS A DIAGNOSTIC TOOL

Generally, in the thermal investigation of soil clays, the appearance of a high temperature endothermic peak in the range of 800-1,000C (1,073-1,273K) is indicative of the presence of the more crystalline micas (muscovite, paragonite, biotite, lepidolite), whereas the appearance of two endothermic peaks, one at about 150-240C (423-513K) and the other at 500-700C (773-973K) indicates the presence of the illitic type micas if discrete expansible 2:1 minerals or kaolinite are not present. The low temperature endotherm at 150-240C (423-513K), representing loss of water held on surfaces and on hydrated interlayer cations, in samples that also contain mica, is indicative of the presence of interstratified vermiculite and montmorillonite layers.

In conclusion, DTA curves, while not perhaps enabling positive identification of the different micas, can serve a diagnostic purpose in conjunction with other techniques (X-ray diffraction, chemical analysis, etc.).

D. Thermogravimetry

Quantitative information can be obtained from the DTA curves for there is a linear relationship between the peak area on the differential curve and the amount of the reacting material. Frequent deviations from linearity with DTA peak area dictate, however, that a thermogravimetric approach be used for absolutely quantitative measurements. In addition, TG curves can give insight into the chemical composition of the mica minerals because the mode of dehydration depends to a large extent upon the structural and chemical composition of the mineral under investigation.

In the case of illites and glauconites weight loss results from the vaporization of water held by surfaces and hydrated interlayer cations, as well as from the expulsion of structural hydroxyls. In the case of muscovite, biotite, and lepidolites, weight loss is mainly attributed to loss of structural hydroxyls and volatilization of fluorine. Muscovite and biotite TG curves show a gradual loss of OH water between 400 and 800C (673 and 1,073K) and the total OH water loss in this temperature range amounts to 4.5 and 4.0%, respectively (Jackson, 1969).

In Fe-bearing micas, oxidation of FeO to Fe_2O_3 may lead to a weight gain that must be taken into account when quantitative determinations based on water loss are made.

The water held by surfaces and hydrated cations is driven off at temperatures up to 300C (573K) and the OH water at temperatures from 300 to 950C (573 to 1,223K). The temperature at which dehydroxylation occurs is controlled by the type of octahedral cations and follows the order: $Mg >$ $Al > Fe$ (Jackson, 1969).

Thermogravimetry as a method for the quantitative determination of soil micas does not seem very promising because of the variation in the water content of the different species (illites, glauconites). Also, variation of the dehydration properties of each mineral with particle size, type of interlayer cations, type of octahedral cations and degree of F for OH substitution imposes limitations on the dehydration techniques of analysis. For more information about thermogravimetry the reader is referred to Jackson (1969) and to K. H. Tan and B. F. Hajek (Chapter 26, this book).

E. Infrared Spectroscopy

X-ray diffraction techniques are concerned with the long-range atomic periodicity in crystals whereas the short-range atomic arrangement of neighboring atoms is often not discernible by the X-ray diffraction techniques. Infrared spectroscopy "sharpens the eye" of the researcher enabling him to "see" into the atomic kingdom, i.e. interatomic distances, bond angles, bond forces, and chemical substitutions. In the case of micas, useful information can be obtained about the degree of isomorphous substitution in the tetrahedral and octahedral sheets, direction of the bonds, and orientation of the OH bonds.

Infrared spectra of most micas show strong absorption peaks in four regions: (i) 3600 to 3700 cm^{-1} (hydroxyl absorption), (ii) 1100 to 900 cm^{-1}, (iii) 800 to 650 cm^{-1}, and (iv) 550 to 470 cm^{-1}. According to Stubican and Roy (1961) the spectral assignments for trioctahedral phyllosilicates (octahedral cations Mg^{2+} and Fe^{2+}) are as follows:

(Si–O) stretching	1100–900 cm^{-1}
(Si–O) "unassigned"	668 cm^{-1}
(Si–O–Si) bending	460–430 cm^{-1}

Infrared spectra of different micas are shown in Fig. 7–10 and the difference between spectra of trioctahedral biotite and dioctahedral paragonite and muscovite are easily discernible (compare curve A with curves B and C). Curves D and E, of two lepidolites, show the change in infrared spectra with different degree of Al for Si substitution in the tetrahedral sheet. Changes in the 1084 to 1130 cm^{-1} peak and the 992 to 962 cm^{-1} one should be noted.

The OH stretching vibrations are generally at higher (3660–3710 cm^{-1}) wave numbers for trioctahedral composition particularly with Mg as an important part of the composition (White, 1971). The (AlAl) and (AlFe^{3+}) dioctahedral OH stretching vibrations occur at 3620 to 3630 cm^{-1}.

Trioctahedral layers are sensitive to angle of incidence (0–60° from the normal) of the infrared beam in the 3600 to 3700 cm^{-1} region while dioctahedral layers are not. Variation in the intensity of the absorption peak with changes in orientation of the specimen was first observed and investigated by Tsuboi (1950) who concluded that the hydroxyl ions in muscovite are oriented obliquely to the cleavage plane. Most studies of hydroxyl orientation are predicated on the availability of single crystals for observation, and calculation of the angle of orientation normally requires the use of a polarized infrared beam (Tsuboi, 1950; Vedder & McDonald, 1963). Nevertheless, the use of a nonpolarized beam can give valuable information about the hydroxyl ion orientation because when the dipole is oriented parallel to the direction of propagation of the beam, no absorption takes place. If the hydroxyl ion is oriented so that the dipole has a component lying perpendicular to the direction of propagation or parallel to the direction of vibration of the infrared beam, then absorption takes place, provided, of course, that the frequency of vibration matches the frequency of the bond.

Serratosa and Bradley (1958) using a pure trioctahedral mica, phlogopite, showed that the OH stretching band at 3700 cm^{-1} is highly sensitive to the orientation of a single crystal with respect to the infrared beam, and concluded that the OH ions are oriented perpendicular to the cleavage plane. On the other hand, the angle of incidence had no effect on the intensity of the OH absorption bond of muscovite, indicating that the hydroxyl ions are inclined to the cleavage plane and apparently different OH are oriented in different directions. The same results were obtained by Basset (1960) who

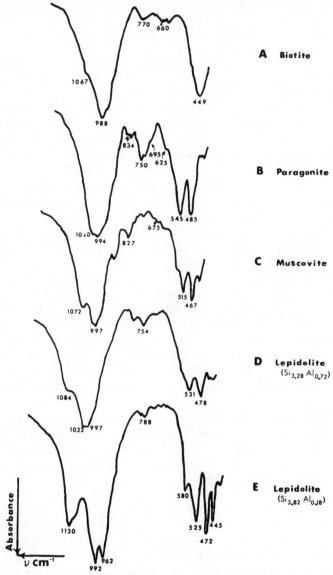

Fig. 7-10. Infrared absorption spectra for different micas (redrawn from Lyon, 1967). Hydroxyl region not shown. Wave number scale is not the same for all specimens.

proposed that the key factor responsible for the variable susceptibility to weathering (at least to K release as discussed in the section on simple weathering transformations) of the different micas is the orientation of the hydroxyl ion.

F. Electron Microscopy

Various techniques that use an electron beam, instead of a light beam, for illuminating the object have been used to obtain morphological, structural, and chemical data about soil micas. Morphology reflects the structural order of a crystalline substance and electron microscopy is the best tool to examine external surface morphology, particularly when dealing with clay size particles. The crystallographic characteristics, habit, and crystal size of micas depend upon two opposite forces. These are: (i) the strain developed from the close packing of smaller structural units within each layer, and (ii) the efficiency of interlayer bonds to counteract this strain in order for a large three-dimensional crystal to be produced. In micas, strong interlayer bonds tend to overcome the intralayer strain, allowing crystals to grow to large size with distinct crystal faces and crystallographic characteristics (Bates, 1964).

Micas in electron micrographs appear thin and platy and in the case of K- or Na-bearing micas the crystals show well-defined boundaries. Illites appear also as thin flakes, but, because of weaker interlayer bonds, the particles are not as morphologically distinct and in some cases resolution of the boundaries of each flake is not attained (Bates, 1964). In some cases illites show a crude hexagonal habit and the thinnest flake is approximately 30Å (Grim, 1968). An illite, composed of lath-shaped particles, was described by Weaver (1953). It occurred in the fine fraction of quartzites and had the properties of other illites. Also, Rex (1966) reported authigenic K mica crystals from sandstones which were lath-shaped (also called needle-shaped by Rex).

Electron microscopy has been used, in weathering studies of micas, to observe changes in the edges of the particles and relics produced by weathering (Jackson et al., 1946). Surface morphology of micaceous vermiculites has been examined by various investigators by means of the replica technique (e.g., Roth et al., 1968, 1969).

Chemical changes during artificial weathering of micas and micaceous vermiculites have been studied by means of the electron microprobe technique (Sawhney & Voight, 1969). Scanning electron microscopy can also be used in the examination of the external morphology of weathered micas and also to study the distribution of elements near the weathering front.

High resolution electron microscopy has been used in the examination of the internal periodicity of a natural and $BaCl_2$ treated muscovite and also for the observation of the frayed edges of a naturally weathered muscovite (Brown & Rich, 1968). High resolution electron micrographs showing mica layers in complex interstratification and intergradation with other layer silicates have been presented by Lee et al. (1975).

III. WEATHERING AND SYNTHESIS RELATIONSHIPS

A. Physical Weathering

Although chemical weathering of micas receives more attention, physical weathering which results in reduction of particle size is important with micas because many mica properties are particle size dependent. Fölster et al. (1963) noted that the formation of illitic clay during the genesis of some Alfisols developed in loess corresponded almost quantitatively to mechanical disintegration of mica minerals originally present in the silt.

Physical weathering of a micaceous material is also illustrated by the breakdown of sand-size glauconite pellets to clay during the genesis of soils developed in materials where glauconite occurs primarily as pellets. Tapper (1968)[2] found by thin section studies that the percentage of glauconite pellets in the sand fraction of a Collington soil (Typic Hapludult) decreased with approach to the soil surface. The C horizon sand contained 27% pellets (the remainder was mainly quartz), whereas the B horizon contained 20% pellets, and the Ap horizon had 5%. Near the surface the pellets had broken down into clay-size glauconite, which accumulated in the B horizon by eluviation-illuviation. This gave a sandy clay loam B horizon. The parent material of these soils has loamy sand textures that change to a sandy clay loam if the glauconite of the pellets is broken down by rubbing (field) or dispersion (lab). The natural pellet breakdown near the soil surface may have resulted partly from tillage manipulation, however, it probably was also enhanced by other types of soil mixing phenomena and by chemical weathering.

B. Simple Transformation to Expansible 2:1 Minerals

The transformation of K-bearing micas to expansible 2:1 minerals by exchanging the K with hydrated cations has received much attention in recent years (see review by Norrish, 1973), particularly since Barshad (1948) showed that vermiculite could be formed from biotite in this manner. Earlier evidence of this transformation had been reported by Gruner (1934) and Mehmel (1938), although Gruner considered the K to be replaced by water. This is termed a simple transformation because a considerable portion of the mica structure (the 2:1 layer) is retained intact as a transformation product.

1. SIMPLE TRANSFORMATION MODEL

The transformation of micas to expansible 2:1 minerals may be represented by the following equation:

---

[2]M. Tapper. 1968. A mineralogical characterization of soils developed in glauconitic parent materials in Southern Maryland. M.S. Thesis. Univ. of Maryland, College Park.

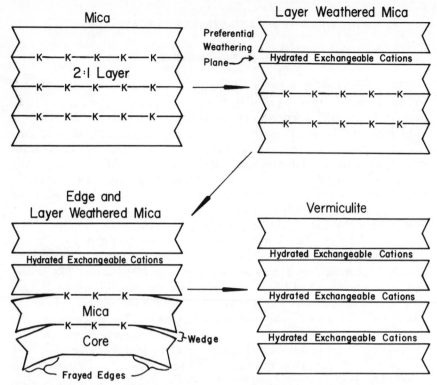

Fig. 7-11. Diagrams illustrating layer weathering and edge weathering of micas by exchanging interlayer potassium with hydrated exchange cations. Mica particles would normally be much wider in relation to layer thickness than the zone represented in the diagram.

$$\text{Micas} \xrightarrow{\text{$-$K + hydrated exchangeable cations}} \text{Expansible 2:1 minerals}$$
$$\text{(vermiculites and smectites)}$$

Structural changes during such transformation are represented in Fig. 7-11.

The transformation has been viewed to take place by (i) layer weathering (preferential weathering planes of Jackson et al., 1952) and (ii) edge weathering (Mortland, 1958; Rausell-Colom et al., 1965; Scott & Smith, 1967). Both models apparently apply.

In the layer weathering model, some interlayers are opened up all the way through a mica particle while others remain essentially or entirely closed (Fig. 7-11). Layer weathering leads to interstratified mica–vermiculite or mica–smectite (Jackson et al., 1952; Huff, 1972). The common presence of these materials in soils and sediments (B. L. Sawhney, Chapter 12, this book) lends credence to applicability of this model, although similar materials could also be produced by the preferential uptake of K by certain interlayers of expansible 2:1 minerals.

In some cases every other interlayer appears to open giving regular inter-stratification. A model for this, involving the shifting of protons of structural hydroxyls toward the opened interlayer, causing the K in adjacent interlayers to be more tightly held was presented by Norrish (1973) and is described in detail by B. L. Sawhney (Chapter 12, this book). It has also been suggested, at least as early as Jackson et al. (1952), that the regular interstratification may be related to differential release of K by alternate interlayers of two-layer polytypes.

The edge weathering model (or frayed edge model mentioned by Jackson et al., 1952) also applies in many instances. Here many interlayers are opened simultaneously along edges and fractures of mica particles, although the possibility of certain interlayers opening preferentially still exists. The opened interlayers, at least at first, are opened only part way across the mica particles, however. This gives rise to what has been termed mica cores (the part not opened) and frayed edges (the opened edges or partially opened edges—since some interlayers may still remain entirely closed). Also, the term *wedge* has been given to the place where an opened interlayer of the frayed edge joins the mica core (Fig. 7-11). Considerable significance has been attached to the wedge cation exchange sites as pertains to cation fixation and selectivity as will be discussed later in the section on chemical properties. Beautiful pictures showing the frayed edges on artificially weathered macro mica particles have been presented by Rausell-Colom et al. (1965) and Scott and Smith (1967) (e.g. Fig. 7-12). Also it has been shown by electron microprobe techniques (Rausell-Colom et al., 1965; Sawhney & Voigt, 1969) that the K was preferentially replaced at the edges of artificially weathered mica particles with the core zones remaining essentially unaltered.

Layer weathering appears to be most common with small mica particles, whereas edge weathering is most common in larger particles (Scott, 1968; Huff, 1972; Norrish, 1973).

One of the changes that usually appears to take place, at least under natural conditions, with the mica to expansible 2:1 mineral transformation is a reduction in layer charge (Norrish, 1973). The reason for and the mechanism of the layer charge reduction are not well understood, however. Based on a correlation of decreasing layer charge with the increasing structural Fe^{3+} for natural vermiculites, Norrish (1973) thought that, in spite of some data to the contrary, the layer charge reduction accompanying the transformation of micas to vermiculites was best related to oxidation of structural Fe^{2+} to Fe^{3+} during or following the opening of the interlayer. This mechanism has been suggested for many years, even for enhancing the release of K from micas, however, recent studies (e.g., Barshad & Fawzy, 1968; Gilkes et al., 1972a, 1972b) have shown that the oxidation actually retards K release from biotite as will be discussed in a later section. Also, some studies indicate that the oxidation of Fe^{2+} in layer silicates is accompanied by loss of protons (H^+) from hydroxyls (e.g., Roth et al., 1968, 1969; Veith & Jackson, 1974). If such a model, protons and electrons leaving the structure simultaneously during oxidation, pertained then no change in layer

Fig. 7-12. Mica flakes (25 mm on each side) after 18 months in sodium tetraphenylboron
    solutions showing edge weathering. *(A)* biotite, and *(B)* muscovite. Note greater
    weathering of biotite. Indentations from handling with tweezers, along the left hand
    side of biotite flake, show how the weathered (vermiculitized) part of the mineral has
    become softer. After Plate 6 of Scott and Smith (1967).

charge would take place upon oxidation as long as the protons lost equaled
the electrons lost.

Studies by Roth et al. (1968, 1969) supported such a model and indicated for micaceous vermiculites that such reactions were reversible. Both
protons and electrons were thought to leave the structure during oxidation
(by H_2O_2) and return upon reduction (with dithionite) with essentially no
change in layer charge in either case. However, a considerable increase in
CEC was observed with an initial reduction (dithionite) treatment that the
authors attributed to the opening of exchange sites previously blocked by
the free iron oxides.

More recently it has been noted (e.g. Farmer et al., 1971; Gilkes et al., 1972a, 1972b; Veith & Jackson, 1974) that octahedral cations may be ejected from the 2:1 layers during oxidation; thus this must also be taken into account in studying layer charge changes during oxidation of structural Fe (Veith & Jackson, 1974). Roth and Tullock (1973), who worked with nontronite (an Fe-rich dioctahedral smectite), have shown that the model for proton and electron migration, during the oxidation and reduction of Fe-bearing layer silicate minerals, may be more complicated than other papers have indicated.

Another mechanism for layer charge reduction is proton incorporation into the structure (e.g., by combining with apical oxygens converting them to OH as suggested by Raman & Jackson, 1966), without corresponding electron changes. Also to be considered as a mechanism for reducing layer charge is the possible exchange of Si for Al in the tetrahedral sheet during mica weathering. Formulas for vermiculite and montmorillonite nearly always show these minerals to have less Al relative to Si in tetrahedral coordination than in micas (e.g., Jackson, 1964; also see formula for mica to vermiculite transformation given in the section on the mica-vermiculite-kaolinite stability diagram). Sridhar and Jackson (1974) have proposed layer charge reduction by this means (Si replacement of Al and Fe in tetrahedral sheets) during a natural transformation of phlogopite to saponite. However, the saponite may have formed by precipitation of most or all of its constituents from solution.

All of these layer charge changes based on formulas derived from chemical analyses remain open to some question due to the difficulty of getting pure samples of the minerals for analysis, assumptions involved, etc. The determination of layer charge is also fraught with other problems such as the blocking of exchange sites by hydroxy-interlayers (see Chapter 10 by R. I. Barnhisel, this book), variation in layer charge between different layers in the same material, etc. Thus many problems still remain with regard to understanding layer charge changes as micas are transformed to expansible 2:1 minerals. However, it seems to be generally agreed by most workers that the expansible 2:1 minerals that result from the opening up of micas have, or develop, lower layer charge than their parent micas (Leonard & Weed, 1970a; Norrish, 1973).

Another change that may accompany the release of K from K-mica interlayers under acid conditions is the break-up of the 2:1 layer itself (Boyle et al., 1967; Sawhney & Voigt, 1969). This break-up appears to take place, at least at first, along the octahedral sheet—leaving the tetrahedral sheets dangling, so to speak, as amorphous "silica relics." Sawhney and Voigt (1969) have presented electron microprobe evidence documenting the production of these "dangling tetrahedral sheets" at the edge of biotite artificially weathered under acid conditions.

Still another change that appears to accompany the transformation is a change in the *b* crystallographic dimension—which tends to decrease after the removal of K with some dioctahedral micas and to increase with some trioctahedral micas (Leonard & Weed, 1970b).

2. EVIDENCE FOR SIMPLE TRANSFORMATIONS

The evidence for the simple transformation is manifold. Evidence comes from (i) artificial weathering studies in which micas have been transformed to expansible 2:1 minerals through the extraction of interlayer K with various chemical and biological agents, and (ii) studies of the distribution of clay minerals in soils developed under natural conditions that show the transformation to have taken place as a function of depth, time, rainfall, etc.

The transformation has been accomplished artificially by a variety of agents (Norrish, 1973). One of the first modern studies was by Barshad (1948), who exchanged the K from biotite by leaching with $MgCl_2$ solution. Many inorganic and also some organic cations have been used to replace the K (Norrish, 1973; Rausell-Colom et al., 1965), with Ba apparently the most effective of the inorganic cations used. Many workers today employ sodium tetraphenylboron (apparently first employed by Hanway, 1956) since the tetraphenylboron precipitates K. This maintains a low level of K in the extracting solution and speeds up the reaction. As reviewed by Norrish (1973), the transformation has also been made by molten salts (White, 1954; Tomita & Sudo, 1971), organic cations—apparently very effective (Weiss et al., 1956; Mackintosh & Lewis, 1968; Mackintosh et al., 1971), organic acids—which may extract octahedral as well as interlayer cations (Boyle et al., 1967; Sawhney & Voigt, 1969), and fungi—which served as a K sink during the exchange of Na for K in experiments by Weed et al. (1969).

Usually the expanded minerals formed by the artificial weathering have given the X-ray and other (e.g., DTA—Barshad, 1954b; Scott & Reed, 1962a) characteristics of vermiculite. However, in a few instances smectitic minerals have apparently been formed from illites, based on X-ray diffraction data (White, 1951; Huff, 1972). A more recent case of the transformation of glauconites and illites to smectites has been presented by Robert (1973). These data support other lines of evidence, such as the formulas mentioned previously, that glauconites and illites may have different layer charge characteristics than other micas.

The evidence of the transformation of micas to expansible 2:1 minerals from mineralogical analyses of field soil profiles is also convincing. However, there are often complications, since the expansible 2:1 minerals formed can undergo other transformations, particularly under acid conditions, that can either destroy them or develop hydroxy-interlayers in them—tending to convert them to pedogenic chlorite. Thus, depth functions that illustrate the transformation of micas to expansible 2:1 minerals are best developed in soils where the pH of the soil is rather high. The data of Table 7–3 illustrate such a situation for a Mollisol developed in fine-textured glacial till derived from Paleozoic shales and dolomite in southeastern Wisconsin (Fanning, 1964;[3] Fanning & Jackson, 1965). The mica in this soil was dioctahedral (Fanning, 1964,[3] as checked on the medium clay) and probably typical of

---

[3]D. S. Fanning. 1964. Mineralogy as related to the genesis of some Wisconsin soils developed in loess and in shale-derived till. Ph.D. Thesis, Univ. of Wisconsin, Madison.

that in illitic Paleozoic sedimentary rocks of Eastern United States. Mica increases with depth, particularly in the fine and medium clay, whereas the expansible 2:1 minerals increase toward the soil surface. The implication is that the mica has been transformed to the expansible 2:1 minerals, although part of the increase apparently also relates to transformation of trioctahedral chlorite (also present in the parent material) to expansible 2:1 minerals.

Other workers have also reported soil mineralogical composition depth functions that imply the transformation of micas to expansible 2:1 minerals (Cady, 1960; Johnson et al., 1963; Douglas, 1965; Post & White, 1967). Often, however, other transformations, such as the formation of hydroxy aluminum interlayers in the vermiculite under acid conditions (e.g. Douglas, 1965) have also taken place so that the overall weathering transformations were more complicated.

The greater K removal near the surface, in some cases at least, probably comes about because once the percolating solutions achieve a certain K concentration they are incapable of removing K from micas deeper in the profile. This is based on reasoning from artificial weathering studies, which show that K release ceases after the K concentration of the solution reaches a critical level (e.g., Scott & Smith, 1966; also see discussion near the end of the subsection below on factors affecting the rate and extent of the transformation).

Wells and Riecken (1969) have found that the mica content of the B horizon clay of some prairie soils from across the midwest U. S. (developed in loess) decreased, while the montmorillonite content increased, with increasing rainfall. This implies more transformation with increased leaching in these soils.

3. FACTORS AFFECTING RATE AND EXTENT OF SIMPLE TRANSFORMATION

The many factors that have been found to affect the mica to expansible 2:1 mineral transformation may be grouped under (i) nature of mineral; (ii) mica particle size; and (iii) characteristics of the surrounding environment.

Perhaps of greatest interest here is the effect of the *nature of the minerals*. Perhaps it is best to start with lepidolites, since these F and Li containing micas have been found to be least susceptible to K release in artificial weathering studies (Rausell-Colom et al., 1965; Leonard & Weed, 1970a; Norrish, 1973) although they apparently are trioctahedral (Table 7-1). Trioctahedral micas, with the exception of fluorine-rich lepidolites, generally release their K more readily than dioctahedral micas. The F of lepidolites is found in structural formulas in place of the OH of other micas (Table 7-1) and the great tightness with which lepidolites hold their K is generally attributed to this substitution. The great electronegativity of F and the absence or near absence of protons in lepidolite structures causes very strong bonding between the 2:1 layers and the K since the F occurs in the layers just above and below the K. Note also that F in other mica minerals retards K release—see critical K levels for three phlogopites of varying F content studied by Rausell-Colom et al. (1965) (Table 7-5).

With mica minerals that contain OH it appears that the proximity of the

proton (H^+) of the OH to the K causes a weaker bonding (relative to lepidolites) because positively charged ions (K^+ and H^+) occur in proximity. Within
the OH-bearing micas, the dioctahedral micas hold their K much more tightly
than the trioctahedral micas (Rausell-Colom et al., 1965; Scott & Smith,
1966; and many others). This difference is about two orders of magnitude
for muscovite as opposed to biotite and phlogopite according to Leonard
and Weed (1970a) (also see critical K levels given in Table 7-4). Apparently

Table 7-4. Critical solution K concentrations (levels) under various conditions[†]

Mineral or soil	Particle size	K depleted, %	Replacement solution	Temp. °C	Solution critical K level, ppm	References[†]
			Trioctahedral minerals			
Phlogopite	10-20 μm	15	1N NaCl	25 (298K)	23	1
Biotite	10-20 μm	15	1N NaCl	25 (298K)	11	1
Biotite	<1 μm	50	0.1N BaCl$_2$	80 (353K)	0.9	2
Biotite	1-2 μm	50	0.1N BaCl$_2$	80 (353K)	2.4	2
Biotite	2-5 μm	50	0.1N BaCl$_2$	80 (353K)	8.3	2
Biotite	5-20 μm	50	0.1N BaCl$_2$	80 (353K)	19.5	2
Biotite	20-50 μm	50	0.1N BaCl$_2$	80 (353K)	23.0	2
Phlogopite	<1 μm	50	0.1N BaCl$_2$	80 (353K)	6.9	2
Phlogopite	1-2 μm	50	0.1N BaCl$_2$	80 (353K)	14.0	2
Phlogopite	2-5 μm	50	0.1N BaCl$_2$	80 (353K)	31.3	2
Phlogopite	5-20 μm	50	0.1N BaCl$_2$	80 (353K)	35.5	2
Biotite 4	<50 μm	‡	1M NaCl	25 (298K)	2.3	3
Biotite 3	<50 μm	‡	1M NaCl	25 (298K)	7.8	3
Biotite	2-50 μm	‡	$10^{-3}M$ Ca(HCO$_3$)$_2$	25 (298K)	0.35	4
			Dioctahedral minerals			
Muscovite	10-20 μm	15	1N NaCl + 0.01M EDTA	25 (298K)	0.1	1
Illite	<2 μm	15	1N NaCl + 0.01M EDTA	25 (298K)	1	1
Glauconite	<50 μm	‡	1M NaCl, pH4.8-5.0	25 (298K)	0.6	3
Muscovite	2-50 μm	‡	$10^{-3}M$ Ca(HCO$_3$)$_2$	25 (298K)	0.19	4
			Soil			
Harps B2	<2 mm	‡	$10^{-3}M$ Ca(HCO$_3$)$_2$	25 (298K)	0.47	4
Harps B3	<2 mm	‡	H$_2$O, pCO$_2$ = 0.001 bar	25 (298K)	0.86	4

† Reference: (1) Scott & Smith, 1966. (2) Reichenbach, 1973. (3) Newman, 1969. (4) Henderson et al., 1976.
‡ Critical K level read from graph of solution K content vs. extraction number on flat portion of
graph representing partial K depletion.

Table 7-5. Critical potassium content of solution (ppm) as affected by nature and
concentration of replacing ions as chloride solutions at 122C (395K) for chopped
flakes of three trioctahedral micas of varying F content. Data after
Rausell-Colom (1965)

Sample no.[†]	F content	Replacement solution								
		Li	Na	Ca	Sr	Mg	Ba	Na	Ca	Ba
	%	—————— 1N ——————						—— 0.01N ——		
1	0.24	36	58	70	55	117	258	4	17	38
4	1.7	24	37	59	63	79	180	3	10	18
8	4.15	6	15	16	9	29	94	2	3	8

† After Rausell-Colom (1965). No. 1 = brown biotite; no. 4 = brown phlogopite; no. 8
= clear, pale, brown phlogopite.

the greater tightness with which the K is held in the dioctahedral minerals is related to the greater distance between the H and K, due to the tendency of the proton to shift toward the empty octahedral cation site in dioctahedral structures. This was discussed previously in the section on dioctahedral vs. trioctahedral structures and in the section on infrared analysis. However, the unique way that the K is "locked" into the $2M_1$ muscovite structure has also been suggested as being responsible for its stability (Radoslovich, 1963; Leonard & Weed, 1970a). The fact that K was released more rapidly by muscovite after heating to dehydroxylation temperatures (Scott et al., 1973) also seems to indicate that something besides OH orientation may be involved in the stability of dioctahedral structures.

A difference in OH orientation has also been cited as the cause of K being held more tightly by biotites after the oxidation of structural Fe than before (Barshad & Fawzy, 1968; Gilkes et al., 1972a, 1972b, 1973). From the investigations by Farmer et al. (1971) and Gilkes et al. (1972a, 1972b, 1973) it appears that, upon oxidation, some of the octahedral cations of biotite are expelled. The expulsion gives empty octahedral cation sites toward which the protons of structural OH apparently shift (away from the K). Data showing the greater tightness with which the K is held after oxidation have been presented by Barshad and Fawzy (1968), by Gilkes et al. (1973), and Gilkes and Young (1973), and by Scott et al. (1973). Also, vermiculites with Fe^{3+} have greater affinity for K than do those with Fe^{2+} (Barshad & Kishk, 1970). Data showing progressively less structural Fe with greater degrees of Fe oxidation by Br_2 have been presented by Gilkes et al. (1972a, 1972b, 1973). Infrared data indicating the shift in OH orientation following oxidation have been presented by Juo and White (1969) and Gilkes et al. (1972a, 1972b, and 1973).

If protons from structural OH are lost from the biotite during oxidation, as has been suggested by Ross and Rich (1974) and Veith and Jackson (1974)—this could also account for increased bond strength following oxidation since the OH^- would be converted to O^{2-}. This mechanism was favored by Ross and Rich (1974), to explain the increased K selectivity that they found following oxidation of Fe in biotite, since they observed very little expulsion of octahedral cations upon oxidation.

Earlier proposals that oxidation of mica structural Fe enhances weathering (Arnold, 1960) seem to be disproven by the more recent studies cited above. An early study by Dennison et al. (1929) agreed with the more recent findings, indicating that biotite was more resistant to weathering following oxidation.

Layer charge, however, does appear to be an independent factor that can affect the tightness with which K and other ions are bonded to the 2:1 layers. The bonding strength increases with increasing layer charge (Barshad, 1954a; Scott & Smith, 1966). Also layer charge originating in the tetrahedral sheet may be more effective in holding K than charge originating in the octahedral sheet.

The effect of *particle size* upon the transformation of K micas to expansible 2:1 minerals has been examined in artificial weathering studies by many workers, (e.g., Barshad, 1954b; Reed & Scott, 1962; Scott, 1968; Reichenbach & Rich, 1969; Mackintosh et al., 1971; Reichenbach, 1973; Ross & Rich, 1973a). These studies have shown that smaller mica particles release their K more rapidly than larger particles, at least until a final, difficult to extract, part of the K is reached. For example, Scott (1968), employing sodium tetraphenylboron, found that 0.2–0.7 μm muscovite reached a point at which no further K was extracted in about 10 days, whereas about 1,000 days were required for 50–60 μm muscovite. The fine particles in the study by Scott (1968) released a large part (about 50% for 0.2–0.7 μm muscovite) of their total K nearly instantaneously whereas silt-sized particles showed almost no instantaneous K release. The instantaneous release was attributed to layer weathering based on interstratification of mica and expanded layers found upon X-ray diffraction examination of the fine particles following the K release. Following the initial very rapid release by layer weathering, further K release apparently is mainly edge weathering in both fine and coarser size fractions.

The K release by edge weathering is a diffusion controlled process and the rate of this release also increases with decreasing particle size. Mathematical expressions for this K release by edge weathering have been developed by Mortland (1958), Mortland and Ellis (1959), Reed and Scott (1962), Rausell-Colom et al. (1965), and Quirk and Chute (1968). The slower release by coarser as opposed to finer particles by edge weathering was explained (Reed & Scott, 1962) in terms of (i) the smaller peripheral surface with larger particles, per unit weight of material, across which K ions may diffuse, and (ii) the greater distance that K ions must diffuse, with larger particles, for a given fraction of K removed.

In spite of their more rapid rate of release by edge weathering, small K mica particles normally contain a small, but significant part of their K in a form that is nearly impossible to extract by normal extraction treatments (Scott, 1968; Norrish, 1973), whereas all of the K is normally extractable from larger particles. The lower fraction of total K extractable from finer particles has also been attributed to the layer weathering that appears to take place preferentially with the finer particles. When a material undergoes layer weathering the K remaining in interlayers adjacent to the interlayers from which the K has been released becomes more difficult to release—leading to interstratification of mica and expanded layers (Bassett, 1959; Scott, 1968). The model presented by Norrish (1973) involving the shifting of the protons of structural OH towards the opened interlayers and away from the K in adjacent closed interlayers (see B. L. Sawhney, Chapter 12, this book) again seems to be the best explanation developed to explain the stability against extraction of the K in the remaining nonexpanded layers. Data have not been presented, however, to confirm this model. Apparently the small particle size and development of interstratification in illitic micas explains the considerable portion of the K not extracted from these materials by normal tetraphenylboron methods (Scott & Reed, 1962b; Smith & Scott, 1966). However, the residual K could be extracted if ultrasonic treatment was combined with the tetraphenylboron-EDTA extraction (Smith & Scott, 1966).

Rich and co-workers have discussed the differences between K release and absorption characteristics of different K-mica particle sizes in terms of the coarser particles undergoing greater bending upon expansion (Ross & Rich, 1973a, 1973b). However, no adequate explanation has been proposed for the greater tendency of the smaller particles to expand by layer weathering. Perhaps the areal extent of the layers (Jonas & Roberson, 1960) somehow affects the initial expansion, so as to give layer weathering with fine particles. Ross and Rich (1973a) have suggested that splitting (layer weathering) occurs first at crystal growth steps (Brown & Rich, 1968). Perhaps this would happen preferentially with fine particles.

That micas in particle size fractions of soils behave like the micas used in artificial weathering studies is supported by the study of Smith et al. (1968). However, K feldspars occurring mainly in silt fractions of the soils examined did not release their K and thus maximum K release to tetraphenylboron occurred in the coarse clay fraction.

In addition to the nature of the mineral and its particle size, the *characteristics of the environment* in which a mica occurs also influence the rate and extent of its transformation to expansible 2:1 minerals. Among the environmental characteristics that are important are (i) the nature and activity of various ions in the soil solution, (ii) temperature, (iii) Eh, and (iv) wetting and drying.

The activity of the mica interlayer cation (only K is considered here) in solution around mica particles markedly influences the rate and extent of transformation. Tables 7-4 and 7-5 show the critical or equilibrium concentrations of K for various conditions. When the K concentration is less than the critical value, K is replaced from the interlayer by other cations from the solution, but when the K concentration is greater than the critical value the mica-expansible 2:1 mineral takes K from the solution. Note (Table 7-4) that the critical K level is highly mineral dependent—being much higher for the trioctahedral minerals. Levels for muscovite are so low that even the K impurities in laboratory chemicals or dissolved from glassware are often sufficient to prevent any K release (Scott & Smith, 1966).

The nature and concentration of the replacing cation, however, and other factors influence the critical K level. Of the cations tested in chloride solutions by Rausell-Colom et al. (1965) (Table 7-5), the critical K levels were highest for Ba and lowest for Li and decreased in the order $Ba > Mg > Ca = Sr > Na > Li$ for the same concentration of these ions and with a constant mica particle size. The activity of all of these replacing ions in the solution phase must be much greater than that of the K in order for significant K release to occur. This is indicative of the great selectivity of the mica-vermiculite interlayer for certain weakly hydrated cations such as K, Rb, and NH_4.

The pH also influences the transformation. Lower pH may enhance K release particularly with trioctahedral minerals (Scott & Smith, 1966), however, this may be because the 2:1 layer itself is attacked. Thus, under acid conditions the transformation may become complex rather than simple.

Under natural conditions the activity of various ions in the soil solution is governed by other minerals in the soil system, by eluviation-illuviation phenomena, the partial pressures of gases present (particularly CO_2, Henderson et al., 1976) and biological activity.

Several experiments in which the K release from micas has been promoted by biological activity have been reported (Mortland et al., 1956; Boyle et al., 1967; Weed et al., 1969; Sawhney & Voigt, 1969). The organisms presumably lower the K content of the soil solution and thus their action may be likened to that of tetraphenylboron in artificial weathering studies, although the overall action of organisms is more complex when organic acids are produced, etc. (Boyle et al., 1967; Spyridakis et al., 1967; Sawhney & Voigt, 1969).

Leaching promotes the transformation, if the chemistry of the leaching waters favors K release, by carrying away the reaction products.

Increasing extraction temperature has been shown to increase the rate of transformation. For example, after 9 days, Sr from $1N$ $SrCl_2$ had diffused about 0.07 mm into biotite flakes at 20C (293K) and about 0.7 mm at 122C (395K) (Rausell-Colom et al., 1965). Under conditions of leaching of biotite with $0.1N$ NaCl the rate of K release appeared directly proportional to temperature in the range from 20-50C (293-323K) (Mortland, 1958). Under similar leaching conditions, Mortland and Ellis (1959) found the log of the rate constant for K release for fixed K in vermiculite to be directly proportional to the inverse of the absolute temperature.

Preheating of micas to high temperatures [up to 1,000C (1,273K)] prior to tetraphenylboron extraction (Scott et al., 1973) was found to enhance the rate of K extraction from muscovite, to decrease the rate for biotite (presumably because of oxidation of Fe—see earlier explanation under effect of nature of mica minerals) and to have little effect on phlogopite except at very high temperatures. The more rapid rate with muscovite following heating was unexplained and apparently contrary to existing theory—see discussion following the paper by Scott et al. (1973).

Wetting and drying has been known for many years to affect exchangeable K levels. The change after drying, however, has not always been in a consistent direction. A model was proposed by Scott and Smith (1968) that permits either an increase or a decrease in readily available (exchangeable) K following drying of a soil containing mica minerals. Scott and Smith envisioned that drying may cause K fixation in wedge exchange sites at the same time that K is being released by layer weathering (exfoliation). The appearance of macroscopic mica grains upon drying supported this model. According to the model, exchangeable K levels may increase or decrease, after a soil has been dried, depending upon the process that dominates.

From earlier discussions, that pointed out the greater tenacity with which K is held by biotite after oxidation of its structural Fe, it appears that biotite should be more stable in environments that oxidize Fe than in environments that reduce it.

C. Mica-Vermiculite-Kaolinite Stability Diagram

Conditions leading to the complete decomposition of the mica structure followed by the formation of precipitation products (complex transformation) have not received as much attention as those leading to the simple transformation reaction. However, under acid conditions the 2:1 layer may decompose as rapidly or more rapidly than the release of K from the interlayer —as indicated by the release of octahedral cations from Fithian illite to dilute salt solutions at pH 3 (Feigenbaum & Shainberg, 1975).

A stability diagram for dioctahedral mica-vermiculite-kaolinite has been developed by Henderson et al. (1976) (Fig. 7–13) by applying the principles described by J. A. Kittrick (Chapter 1, this book). This diagram is open to question, particularly regarding the reaction of muscovite to vermiculite that forms the basis for much of the diagram. However, the stability relationships shown are generally consistent with the field occurrence of the minerals. Thus, the diagram provides a useful framework for discussion of the conditions that favor complex as opposed to simple transformations.

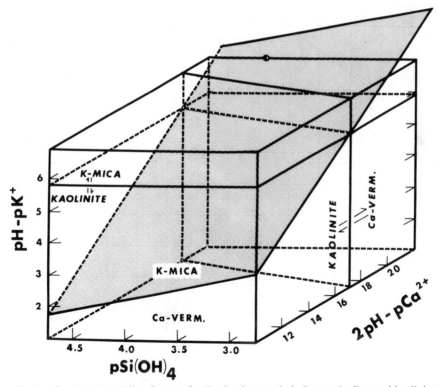

Fig. 7–13. Mineral stability diagram for K-mica (muscovite), Ca vermiculite, and kaolinite at 1 bar and 25C (298K). The shaded plane represents the mica-vermiculite system in the absence of kaolinite (kaolinite formation is not considered in this case except to indicate that the mica and vermiculite are metastable where the shaded plane falls within the kaolinite field). After Henderson et al. (1976).

The join between muscovitic mica and vermiculite in the diagram was developed based upon the reaction:

$$2.7K (Si_3Al)Al_2O_{10}(OH)_2 + 1.8Si(OH)_4 + 1.05Ca^{2+} + 0.6H^+$$
$$\text{(muscovite)}$$
$$\rightarrow 3Ca_{0.35}(Si_{3.3}Al_{0.7})Al_2O_{10}(OH)_2 + 2.7K^+ + 3.6H_2O$$
$$\text{(Ca vermiculite)}$$

This reaction was considered to occur in the presence of free carbonates such that Al would remain in the layers of the neo-formed phases. Under such conditions it was proposed that the transformation of muscovite to vermiculite would be controlled by the solute activities of H^+, K^+, Ca^{2+}, and $Si(OH)_4$.

The equilibrium constant was calculated from the expression:

$$K_{eq} = \frac{(K^+)^{2.7}}{(H^+)^{0.6}(Si(OH)_4)^{1.8}(Ca^{2+})^{1.05}}.$$

This equation was transformed to pH, pCa^{2+} etc. forms to develop the stability diagram. The equilibrium constant was determined from the concentration of the ions in a Blount soil B horizon, the clay fraction of which was considered to be dominated by dioctahedral mica and vermiculite. The Blount soil clay mineralogy is much like that of the Varna soil (Table 7-3 and Fig. 7-8), since the Blount is essentially the forested (Alfisol) counterpart of the Varna (Mollisol).

The diagram shows the mica-vermiculite system in two different situations, (i) where kaolinite formation is not considered (shaded plane in the diagram—in this case both mica and vermiculite are metastable over large parts of the diagram), and (ii) where kaolinite formation is considered (in this case the shaded plane should be ignored where it falls within that part of the diagram where kaolinite is the stable phase).

Kaolinite was found to be the stable phase relative to both muscovite and dioctahedral vermiculite over the range of ion activities that would be expected in most soils most of the time. Mica would be expected to transform to kaolinite rather than to vermiculite except at very high pH and very low pCa^{2+} (high Ca activity) conditions. Mica is stable at high pH and low pK^+ (high K^+ activity). Also, in some soils montmorillonite, rather than kaolinite may be the stable phase relative to dioctahedral micas and vermiculites (Weaver et al., 1971; Henderson et al., 1976). Both transformations (to kaolinite and to montmorillonite) would seem to involve complex transformations—breakdown into ionic constituents and reprecipitation. However, the possibility of transformation to montmorillonite by K exchange and layer charge reduction also exists.

The stability diagram predictions of Henderson et al. (1976) essentially agree with those of Kittrick (1973) who found vermiculites to be fast forming, but unstable, mica weathering intermediates. Kittrick mainly considered trioctahedral micas and vermiculites, but the conclusions of the two studies seem to coincide. The existence of vermiculite as a weathering product produced by simple transformations in many soils, even though it is unstable relative to kaolinite and/or montmorillonite appears to be a rate of reaction phenomenon. Apparently the rate of the K exchange reaction is more rapid than the rate at which the 2:1 layer is broken down in many soils (particularly in soils of high base status—Mollisols, Alfisols, Eutro great group soils of other orders, etc., Soil Survey Staff, 1975). Thus, in these soils expansible 2:1 minerals formed by simple transformation are common mica weathering intermediates and these minerals formed from mica (but also from trioctahedral chlorite, Coffman & Fanning, 1975) may account for a large part of the soil CEC.

In other more weathered soils the vermiculite stage is either by-passed during weathering, meaning that the 2:1 layer breaks down as rapidly as K is released, or vermiculite has formed from mica at an earlier stage of soil development and has subsequently been destroyed. This is not meant to imply that no mica-formed vermiculite exists in more highly weathered soils, but that is it not very abundant in these soils (e.g., most Ultisols, Oxisols, and Dystro great groups of Inceptisols).

Not considered in the above discussion is the possibility of the formation of hydroxy-Al interlayers in the expansible 2:1 minerals (R. I. Barnhisel, Chapter 10, this book) that may stabilize the expanded 2:1 mineral structure. Jackson (1964) has placed pedogenic dioctahedral vermiculite-chlorites and chlorites close to kaolinites in terms of their weathering susceptibility in leaching environments. Stability diagrams including these phases have not been developed.

D. Pedogenic Mica

Although most of the mica minerals in soils appear to be inherited from soil parent materials, the possibility of mica synthesis in soils must be considered.

Mica synthesis in surface horizons of certain high rainfall soils in Hawaii was proposed by Swindale and Uehara (1966) and Juang and Uehara (1968). Their hypothesis was based on observations that mica was absent in the basaltic parent material of the soils and that the content of mica in the soil in the < 2 μm clay increased with approach to the soil surface. It was thought the K was concentrated in the surface horizons by biocycling, giving rise to an ionic environment in which mica could form.

However, when ionic concentrations of soil solutions were plotted on a stability diagram for muscovite, kaolinite, and gibbsite (developed by Garrels and Christ, 1965) kaolinite was shown to be the stable phase—except for one sample where gibbsite was stable (Swindale & Uehara, 1966). This disagreement with the field occurrence, which seemed to indicate mica formation, was considered possibly due to the solutions not representing true equilibrium solutions or to the "pedogenic" mica having a different free energy of formation, thus a different equilibrium constant, than the mica employed by Garrels and Christ (Swindale & Uehara, 1966). Juang and Uehara (1968) mentioned that the mica might possibly be of eolian origin and that further studies of this possibility were needed.

Subsequently, studies by Dymond et al. (1974) have shown that the mica in the Hawaiian soils studied by Swindale and Uehara (1966) and Juang and Uehara (1968) is of eolian origin, apparently traveling long distances to Hawaii (probably mainly from the continents) as tropospheric dust. Thus, the mica and the quartz in these soils apparently originates as tropospheric dust (Rex et al., 1969; Syers et al., 1969; Dymond et al., 1974). Dymond et al. found the K–Ar (potassium–argon) age of the mica to be about 200 million years whereas the age of the volcanic rocks in which the soils were developed was < 3.5 million years. The age of the oldest rocks on Oahu Island, where the soils occur, was 8.5 million years. Thus, the mica, at least the bulk of it, could not have formed in the soils. Other data presented by Dymond et al. supported an eolian origin too.

Still the possibility of mica formation in soils cannot be ruled out, even if it does not apply to the Hawaiian case. A case of mica increasing toward the soil surface, with vermiculite increasing with depth, in clays of arid soils of Iran was reported by Majhoory (1975). A similar case was reported by Nettleton et al. (1973) for dryland soils of southwestern USA. Nettleton et al. considered the mica to be forming from vermiculite or vermiculite-biotite by uptake of K and NH_4 released by plant decay. In arid region soils, accumulation of K at the soil surface by biocycling would seem to lead to ionic conditions suitable for mica formation (high pH and low pK^+, Fig. 7-13), particularly when the soil dries out tending to concentrate the ions, as little K should be lost by leaching. However, an eolian or other parent material explanation would have to be checked before mica formation could be verified in any soil. Movement of particles through the atmosphere is known to be intense in arid regions.

A bona fide case of authigenic mica formation at temperatures that might prevail in soils was reported by Rex (1966). Needle-like mica crystals, apparently with an illite-like composition, that precipitated with kaolinite from solutions in arkosic sandstones were reported. Apparently K feldspars dissolved and first kaolinite and then kaolinite and mica precipitated. Electron-micrographs showed that some of the mica crystals had grown from kaolinite surfaces. The mica in this case apparently formed in solutions that were not marine.

Another possibility for mica formation in soils is by K fixation by expansible 2:1 minerals. Niederbudde et al. (1969a) reported an instance where illite apparently had formed in Bt horizon fine clay from beidellitic smectite (thought to be at least partially smectite that originally formed from illite) by long continued K fertilization by man. The K was considered preferentially adsorbed by the smectite occurring in the < 0.2 μm clay rather than being adsorbed by vermiculite which also occurred in these soils.

Niederbudde (1972, 1975, 1976) has collected further evidence (primarily K/Ca exchange data for soil clay fractions as a function of depth, but also smectite vs. mica contents in clay fractions as a function of soil depth) that has convinced him that illite forms in some soils by K uptake by beidellitic smectite. The model that he suggests has K being released by more "true" micas in silt fractions with illitic micas forming by K uptake by beidellitic smectites, occurring in the finer clay fractions.

E. Glauconite Formation and Weathering

Although glauconites form primarily under marine conditions, they are of interest to soil mineralogists because the temperatures at which glauconites form seem similar to soil temperatures. Thus an understanding of the conditions and mechanisms of glauconite formation, and the properties of glauconites, may give some insight into possible formation and properties of pedogenic micas. Also, glauconites are of interest because of soils developed in glauconitic sediments.

Most clay mineralogists who have studied (mineralogical) glauconite consider it to form under marine conditions. This origin is supported by the 1M or 1Md polytype of glauconite (Warshaw, 1957; Tyler & Bailey, 1961; Tapper & Fanning, 1968) which is the mica polytype expected to form at low temperatures (Bailey, 1966). Also, K-Ar and Rb-Sr ages of glauconites from various sedimentary geologic formations, ranging in geologic age from Cambrian through Pliocene, have been found to be similar or younger than the expected geologic age of the formations (Hurley et al., 1960). This indicates that the glauconite formed during or following sedimentation. Although glauconites in some sediments apparently have been eroded from others (Owens & Minard, 1960) there is no evidence that glauconites are other than low temperature minerals.

The chemistry of sea water favors the formation of mica (glauconite). Sea water contains about 380 ppm dissolved K, 400 ppm Ca, and 3 ppm Si (Mason, 1966). Assuming sea water pH of 8.2, $pH - pK^+ = 6.2$, $2pH - pCa^{2+} = 14.4$, and $pSi(OH)_4 = 4.0$. These values are in the mica stability field of Henderson et al. (1976) (Fig. 7-13). Although the exact $pH - pK^+$ values for glauconite stability are not known, it is seen that the formation of micas in sea water seems possible because of the high pH and low pK^+ (high dissolved K).

Rather than precipitating completely from dissolved constituents, glauconite has usually been pictured as forming from sediments in contact with sea water. Several workers have proposed the formation of glauconite from montmorillonites—by K absorption (Burst, 1958a; Hower, 1961; Porrenga, 1966). However, such an origin does not explain the high iron content of glauconites. As suggested by Takahashi (1939) and supported by Tapper and Fanning (1968), glauconite formation from a variety of starting materials seems likely, given a marine environment and a slow sedimentation rate—so that the materials have time to reach or approach equilibrium with the sea water.

For Maryland glauconitic deposits (of Cretaceous and Tertiary age) it seems likely that kaolinite and iron oxides eroded from the Piedmont region were important components of the sediments from which the glauconite formed. The only significant detrital mineral remaining in the deposits is quartz, although some K feldspar does appear to be present too. An approximate reaction for the glauconite formation from these materials is presented below:

$$\left.\begin{array}{l}\text{Kaolinite +}\\\text{free iron oxides +}\\\text{K (from sea water)}\end{array}\right\} \xrightarrow{\substack{\text{Fe mobilized by reduction}\\\text{with organic materials?}}} \left\{\begin{array}{l}\text{Glauconite +}\\\text{excess Al (to sea}\\\text{water as aluminate?)}\end{array}\right.$$

If the Fe is reduced (during glauconite formation) apparently it is oxidized again by the time the glauconites are sampled on land areas because most of the structural Fe in glauconite is found as Fe^{3+} (Table 7-1). Perhaps reactions leading to glauconite formation are catalyzed by passage of muds containing the reacting minerals through the digestive tracts of marine organisms, since glauconite often occurs in "fecal" pellets. On the other hand, vermiform pellets (Tapper & Fanning, 1968) may represent vermiculite, or mica, or possibly kaolinite "worms" that have somehow been transformed to glauconite. It is admitted that the above reaction and statements are speculative, but perhaps they may serve as models to be tested by future students.

Glauconite has a low amount of its layer charge originating in its tetrahedral sheet according to formulas that have been derived (Table 7-1, also Robert, 1973). In this regard, glauconite is similar to smectite—a 2:1 phyllosilicate that forms in some soils. With this low tetrahedral charge it appears that glauconites are more likely to alter to iron rich smectite (nontronite) upon K release, especially if the structural Fe is oxidized, whereas other micas more commonly alter to vermiculite (Robert, 1973). Under acid conditions glauconites appear to weather to kaolinite and goethite (Wolff, 1967).

Glauconite was considered by Jackson et al. (1948) and Jackson (1964) in the weathering sequence of clay-size minerals to resemble biotite, belonging to the most easily weathered class of phyllosilicate minerals. It now appears that glauconite is more resistant to weathering than biotite in a leaching environment—probably because glauconite is dioctahedral. High contents of glauconite occur in some Typic Hapludults that apparently provide a favorable chemical weathering environment (Tapper, 1968[2]).

IV. NATURAL OCCURRENCE

An understanding of the conditions and modes of mica formation and weathering in soils, covered in the previous section, provides insight into the distribution of various kinds of micas in soils. Knowledge of geology and of the geologic and pedologic processes that tend to accumulate mica minerals in certain size fractions and soil horizons is also helpful in understanding mica occurrence.

A. Occurrence in Soil Parent Materials

Micas are the third most extensive group of minerals (after feldspars and quartz) in granite, and in sialic (acid) rocks in general, but are less extensive in most mafic rocks (Holmes, 1965). The sialic rocks are more extensive than the mafic on land areas of the world (King, 1962).

The mica occurring in igneous and metamorphic rocks is mainly macroscopic, occurring in tabular sheets or books with perfect cleavage parallel with the 2:1 layers. Much of this mica probably has formulas close to the ideal ones given for muscovite, biotite, etc. in Table 7-1. Muscovite and biotite are the most extensive micas in igneous and metamorphic rocks. Phlogopite occurs as a product of metamorphism of magnesian limestones or dolomitic limestones and also in some serpentinitic rocks (Hurlbut, 1959). Lepidolites, although of considerable academic interest in artificial weathering studies, are of very restricted extent—occurring in pegmatites where they are associated with other Li-bearing minerals (Hurlbut, 1959; Pough, 1960).

Macroscopic micas are particularly abundant in certain metamorphic rocks (schists, phyllites, gneisses) and they dominate in mica schists where sheets several centimeters across may occur. Mica schist is by far the most extensive kind of schist because it develops by metamorphosis of argillaceous rocks (shales, slates) that are the most abundant of sediments (Holmes, 1965). Mica-bearing metamorphic rocks are very extensive in parts of the Piedmont region of the eastern USA.

Close to igneous and metamorphic rock source areas, macroscopic mica, occurring in flakes, is also common in coarse-textured sediments and sedimentary rocks (e.g., sands, sandstones). Since micas are softer than most of the other minerals with which they are commonly associated (see section on Physical Properties) they tend to be readily broken down physically during transport and sedimentation processes. Thus, micas are generally more extensive in fine-grained sediments and sedimentary rocks (clays, shales) than in coarser textured sedimentary materials (e.g., sandstones).

Shales, and slates derived from them, are usually rich in the fine grained illitic-type micas (e.g., Van Houten, 1953; Grim et al., 1957). Illitic micas are also important clay minerals in limestones and thus in some soils developed in residuum from limestone. For example, in studies of the clay mineralogy of a large number of Pennsylvania soils, Johnson (1970) found illitic mica as the predominant subsoil clay mineral where soils were derived from shales and limestones. Although much illite in sedimentary rocks appears to be detrital, some illites in these rocks have also been developed or affected by authigenic or diagenetic processes. Increased pressures and temperatures from deep burial in rock columns appears to contribute to the illite genesis in these rocks (Weaver & Pollard, 1973). Glauconites occur mainly in sediments deposited under marine conditions—see earlier section on formation of glauconites.

The amount and kind of mica occurring in more recent unconsolidated sediments (glacial materials, loess, alluvium) depend upon the origin of the sediments (Frye et al., 1962; Willman et al., 1963). For example, clays of fine-textured glacial tills derived largely from Paleozoic shales and limestones in areas around the Great Lakes are rich in illitic mica (Fanning & Jackson, 1965; illustrated by data of Fig. 7-8 and Table 7-3).

B. Occurrence in Soils

1. OCCURRENCE AS A FUNCTION OF DEGREE OF WEATHERING

Since micas in most soils originate mainly from soil parent materials and tend to weather to other minerals with time, they are generally more prevalent in the clay mineralogy of younger, less weathered soils (Entisols, Inceptisols, Mollisols, Aridisols, Alfisols) and are less prevalent in more weathered soils (Ultisols, Oxisols) (Jackson et al., 1948, 1952; Hseung & Jackson, 1952; Jackson, 1959, 1964). Also micas tend to occur more as discrete mica particles in the less weathered soils, if such particles are present in the soil parent material, whereas in more weathered materials the mica is more commonly interstratified with expansible 2:1 minerals that may also be partially chloritized (Jackson et al., 1952; Jackson, 1964).

Trioctahedral micas are uncommon in soils that have undergone much weathering. Thus the mica in clays of most soils is predominantly dioctahedral. Some dioctahedral micas appear to persist in the clay fraction even in highly weathered soils—at least as interstratified mica layers (Hseung & Jackson, 1952). This probably accounts for small amounts of K found in clays of these soils. Muscovite is resistant enough to weathering so that up to 6% of this mineral may occur in the 20-200 μm fraction of oxic horizons, although only 3% of other weatherable minerals are allowed in this fraction (Soil Survey Staff, 1975). The oxic horizon is the most weathered diagnostic horizon recognized in *Soil Taxonomy* (Soil Survey Staff, 1975).

2. OCCURRENCE AS A FUNCTION OF SOIL DEPTH

As discussed in the subsection (under weathering and synthesis) on evidence of the transformation of micas to expansible 2:1 minerals, the mica content of the clay fraction of soils which developed in nearly uniform parent material often increases with depth, reflecting greater mica weathering at the soil surface. There are many exceptions to this pattern, however. Niederbudde (1976) has reported instances where the mica content of the soil clay fraction was higher in the solum than in underlying C horizons apparently because of K uptake by montmorillonite (mica formation) during soil formation. Also, mica may show a maximum at the soil surface because of recent additions of windborne mica to soils naturally low in mica.

When the mica occurs as clay, or is broken down physically into clay (subsection on physical weathering) it accumulates by lessivation (eluviation plus illuviation of clay) in the Bt horizons. Thus, on a whole soil basis, the Bt horizon of such soils may contain more mica than either the A or C horizons. Also the leaching out of more soluble minerals, such as carbonates, increases the mica content of the leached horizons relative to the unleached. Thus, in areas with calcareous parent materials and Bt horizons, a common situation in some glaciated areas, the Bt horizon usually contains more mica than other horizons because of carbonate leaching and lessivation.

In soils developed from materials that are rich in coarse mica from noncalcareous materials (e.g., granite) the maximum mica content on a whole soil basis may occur in the C horizon because of greater mica weathering in the A and B horizons. An example of this was reported by Alexiades et al. (1973).

Other differences in the distribution of mica with soil depth are related to lithologic discontinuities in soil profiles and thus are controlled by soil parent material. For example, in the Tustin soil series (Arenic Hapludalfs, clayey, mixed, mesic) which occur in clayey lacustrine sediments overlain by sands in Wisconsin, the clays are expected to be richer in mica than the sands.

3. OCCURRENCE AS A FUNCTION OF PARTICLE SIZE

In most soils, micas tend to occur preferentially in certain size fractions. Often, with soils developed in sediments, micas are most abundant in the coarse clay (0.2-2 μm) fraction (e.g., Table 7-3). This appears to develop because (i) micas in coarser fractions are broken down physically during transport or soil-forming processes, whereas other minerals such as quartz are more resistant to physical disintegration and accumulate in the coarser fractions; and (ii) because the micas in the fine (< 0.2 μm) clay have undergone more transformation to expansible 2:1 layers and because neo-formed minerals in soils (some montmorillonite and kaolinite) tend to occur preferentially in the finer fractions.

In soils developed in materials containing predominantly macroscopic mica, the mica tends to occur mainly in coarse particles in the rock and saprolite. Weathering breaks down the large mica particles such that a larger proportion of mica occurs in silt and clay-size particles in the A and B horizons than in C and R horizons (Alexiades et al., 1973). In such soils, sand and silt-size vermiculite formed from mica may also occur—particularly in the deeper horizons where the particle size of the micaceous material has not been reduced as much by weathering (Alexiades et al., 1973).

In soils with argillic horizons the particle size spectrum of mica occurrence may shift toward finer sizes in the argillic horizon (relative to overlying and underlying horizons) because of accumulation of mica of finer particle sizes.

V. CHEMICAL PROPERTIES

A. Properties Related to Cation Exchange

1. CATION EXCHANGE CAPACITY (CEC)

The CEC of micas is small relative to that of expansible 2:1 minerals such as montmorillonites and vermiculites because the interlayer cations of these other minerals are usually exchangeable, whereas those of micas are not. Micas have a high layer charge balanced by "fixed" interlayer cations that are not exchanged in standard CEC determination methods. By such a view, micas exhibit cation exchange only on external faces of particles and even the planar external surfaces could be considered to be half-layers of vermiculite. At any rate, the contribution to CEC from external surfaces of the mica particles should normally be small, but would increase with decreasing particle size.

The CEC of illites, often given as about 40 meq/100 g, presumably results partly from interstratified vermiculite and smectite layers, which should be measured and reported as vermiculite and smectite by the quantitative mineralogical analysis methods of Alexiades and Jackson (1966).

2. EXCHANGE OF MICA INTERLAYER CATIONS

In the section on simple mica weathering transformations, it was explained that under certain conditions the unhydrated mica interlayer cations (usually K) can be exchanged by hydrated cations and that after this exchange the mica becomes an expanded 2:1 mineral—that would have a high CEC. This expanded material is usually identified as vermiculite or smectite, although technically the expanded material might still qualify as mica based on its layer charge characteristics. Thus, the interlayer cations of micas may be exchanged, given the proper conditions, but once these cations have been exchanged the mineral usually is no longer called mica. The conditions under which mica interlayer cations (chiefly K) may be exchanged have been considered earlier in the section on weathering and synthesis relationships and are mentioned again in regard to the Q/I relationship.

3. SELECTIVITY OF THE WEDGE AND EXTERNAL PLANAR CATION EXCHANGE SITES FOR CERTAIN CATIONS

Selectivity occurs when exchange materials adsorb ions by cation exchange in a proportion (ratio) different from the ratio of the activities of the ions in the medium (usually a solution) from which they are adsorbed. Mathematical equations have been developed to relate the ratio at which the ions are adsorbed to the ratio of the ions in solution. In these equations, demonstrated below, the (Gapon) selectivity coefficient (k or k_G) expresses the degree of selectivity or preference of the solid for one of the ions relative to the other. Usually two kinds of ions (e.g., K vs. Na) are compared.

The example equation given below for K and Na is slightly modified (different symbols) from Bolt et al. (1963):

$$K^+_{ads}/Na^+_{ads} = k(a_{K_s}^+/a_{Na_s}^+).$$

This equation says that the ratio at which the two ions are adsorbed, K^+_{ads} / Na^+_{ads}, is equal to the ratio of the activity of the ions in solution multiplied by the selectivity coefficient. The larger the value of k the more selective the adsorbing material is for the ion in the numerator. Similar equations may be used to compare other ions, however, divalent ions enter the right side of the equation according to the square root of their activity and trivalent ions are entered according to the cube root, etc. An example for K and Ca follows:

$$K^+_{ads}/Ca^{2+}_{ads} = k\left[a_{K_s}^+/(a_{Ca_s}^{2+})^{\frac{1}{2}}\right].$$

Selectivity is of particular interest with mica-expansible 2:1 minerals (edge-weathered mica) because such materials have such a high affinity (thus selectivity) for large monovalent cations such as K (Sawhney, 1972). This selectivity is usually attributed to the wedge cation exchange sites (Fig. 7–11) and the fact that the large monovalent cations such as K, Rb, and Cs are weakly hydrated. These ions lose their shell of hydration more easily than other cations, thus they enter the wedges preferentially. Once within the wedge these ions may enter the hexagonal or ditrigonal holes in the base of the tetrahedral sheets and become "fixed." When this happens the wedges collapse and the structure essentially becomes mica again (although extraction of the K again may be easier than initially) and the mica core (Fig. 7–11) grows. Thus whether K is released or adsorbed depends upon the ratio of ions in the (soil) solution and upon the selectivity of the exchange sites.

Since there are many other kinds of cation exchange sites in most soils, besides those in the wedges of mica-expansible 2:1 minerals, it is difficult to measure the selectivity of just the wedge sites. Bolt et al. (1963) developed a method by which they measured the selectivity of the wedge (called edge-interlayer by Bolt et al.) sites of an illitic soil. Their data indicated that the wedge sites had a selectivity coefficient of about 500 for K compared to Na for the soil studied, although the wedge exchange sites were only a small part (about 4%) of all the exchange sites. The selectivity coefficient for planar (external surface) sites of the same illite was estimated as 2.

The wedge exchange sites of edge-weathered micas also have a high selectivity for Rb, Cs, and NH_4 according to the studies on selectivity that have been reviewed by Sawhney (1972).

Dolcater et al. (1972) have shown that layer charge characteristics are also important in determining the magnitude of the selectivity coefficient.

4. THE Q/I RELATIONSHIP AND ITS CONNECTION TO POTASSIUM AND MICA CHEMISTRY

The Q/I (quantity-intensity approach) has been applied by Beckett (1964a, 1964b) and others (Niederbudde et al., 1969a, 1969b; Murthy et al., 1975) to assess the ability of soils to supply K to plants and to describe the exchange of K from soils by other ions, particularly Ca.

To develop the Q/I graph (an ideal graph is presented in Fig. 7-14) fixed quantities of solutions of the same or similar Ca activity, but of varying K activity, are shaken with fixed quantities of a soil for some period (e.g., 1 hour) and then allowed to stand in contact with the soil for some hours by some workers. Then the solution is separated from the soil by filtering or centrifuging and is analyzed for the two kinds of ions. From the content of K in the solution before and after contact with the soil the quantity of K adsorbed or released by the soil (Δ K) (expressed in meq/100 g) is determined. This is plotted as the ordinate (Q) of the Q/I graph (Fig. 7-14). From the concentration of the ions in the final solution, and making corrections to give activities, the activity ratio AR $= a_{K^+}/(a_{Ca^{2+}})^{1/2}$ or $a_{K^+}/(a_{Ca^{2+}} + a_{Mg^{2+}})^{1/2}$ is calculated. This allows positioning of points along the abscissa (I) of the graph. In some ways it would seem more appropriate to use the initial activity ratio rather than the final ratio for plotting the graph, since whether

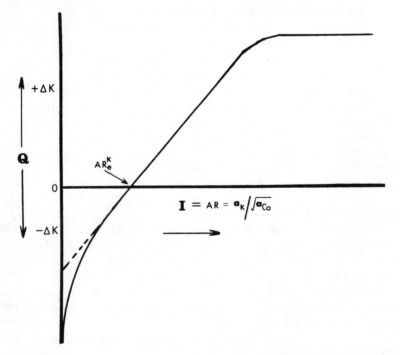

Fig. 7-14. Idealized Q/I diagram for soil containing exchangeable K and some mica. Based on relationships described by Beckett 1964b and others.

K^+ is adsorbed or released would seem to depend more upon the initial ratio, however, those who have used the Q/I approach have apparently used the final ratio, following Beckett (1964b).

The form of the graph for those soils studied, which usually have been fairly high base status soils in which K and Ca would be important exchangeable cations, has been of the form illustrated in Fig. 7-14. The graphs usually have a nearly linear sloping part which may intercept the AR axis (where $\Delta K = 0$). For the Lower Greensand soil from England, studied by Beckett (1964b), this intercept occurred at AR = 0.0135 (moles/liter)$^{1/2}$. This is considered the equilibrium ratio labeled AR_e^K or AR_o. The slope of the graph is thought to be a measure of the ability of the soil to buffer its soil solution K level and has been called the Potential Buffering Capacity (PBC^K). The intercept of a line drawn tangent to the curve at $\Delta K = 0$ and extended to the $Q(\Delta K)$ axis has been taken by Murthy et al. (1975) as a measure of the labile K (readily available K reserve) in the Texas soils that they studied and this intercept has been called $-\Delta K°$.

Interestingly, Murthy et al. (1975) found a high correlation between the $-\Delta K°$ value and whole soil mica content (the sum of the amount of mica in all size fractions where the mica in each fraction was measured by the method of Alexiades et al., 1966) for the soils that they studied. This seems to imply that the solution K level in the linear part of the graph may have been buffered by mica K. However, the $-\Delta K°$ value found by these workers was less than the amount of "exchangeable" K measured by standard methods, which may indicate that the buffering was not by mica interlayer K. Earlier a correlation between the K-supplying ability of soils and the amount of clay-size mica was reported (Milford & Jackson, 1966).

Beckett (1964a) considered the K that does the buffering in the "linear sloping" part of the graph to be exchangeable K, not mica core K. The work of Niederbudde et al. (1969a, 1969b) indicated that the K doing the buffering in this range was associated with montmorillonite (or montmorillonite converted to illite by K adsorption) in the soils that they studied.[4]

At more extreme AR values (low and high), the exchange model is different. At high AR (and also high, for the system being studied, $+\Delta K$ values) the graph may become parallel to the AR axis (Fig. 7-14). The explanation for this behavior is that all the exchange sites are occupied with K.

Of more interest with regard to micas, however, is what happens at very low AR values. Here it is seen that the graph becomes asymptotic to the $-\Delta K$ axis. This means that if the activity ratio is maintained low enough then all of the mica interlayer K should be released. This is in agreement with what has been shown in artificial weathering studies. It also appears that the AR value at which the graph would tend to become asymptotic to the $-\Delta K$ axis

---

[4]It has been pointed out to the authors by E. A. Niederbudde that in some cases the linear sloping part of the graph would not intersect the I axis, at least for some soils. Thus no $-\Delta K$ value would be possible unless the tangent was taken to the curve at ΔK_o rather than to the linear sloping part of the graph. If the curve is not linear where the graph cuts the I axis this would also seem to indicate that mica K could be buffering the K level at equilibrium in this range rather than the "exchangeable" K.

would be higher for trioctahedral micas since these micas support higher critical K levels than dioctahedral micas (Table 7-4).

Not mentioned in the above discussion is the rate at which the K is adsorbed or released or whether the points plotted on the Q/I graph represent equilibrium values. Equilibrium seems doubtful if the soil-solution contact time is only one hour, as in the studies by Murthy et al. (1975). However, if the points on the sloping part of the graph that intercepts the AR axis result from "normal" cation exchange they probably result from systems close to equilibrium. Thus K representing the sloping part below the AR axis would probably be supplied to growing plants quite readily. On the other hand, mica interlayer K should be released more slowly. Since it is also released only at very low AR values, it is not readily available to plants—although it does represent a reserve supply that slowly becomes available as K is removed from the soil. Thus micas have sometimes been called "soil potassium buffers" (Newman, 1969).

B. Chemical Composition

For the macro-chemical composition of micas the reader is referred to formulas of the various mica minerals (Table 7-1) and to accompanying discussion in the section on nomenclature, structures and formulas.

Micas usually contain small amounts of various minor elements (e.g., Zn, Ni, etc.) that may become available to plants on weathering. Quite complete chemical analyses for several micas are given by Rausell-Colom et al. (1965).

VI. PHYSICAL PROPERTIES

Micas exhibit outstanding cleavage along planes parallel to the basal $(00l)$ crystallographic planes. Thus micas usually occur as flakes. These flakes orient themselves parallel to surfaces of materials upon which micas sediment—this applies to the settling out of macroscopic mica flakes in an alluvial sediment as well as to the settling of clay-size mica particles, from a laboratory prepared suspension, on a glass slide for X-ray diffraction analysis. The common orientation of mica particles in a soil material enhances soil slippage in directions parallel to the cleavage. Thus it would seem that rock and soil slides parallel to bedding planes in sediments and sedimentary rocks may owe something to the orientation of micas and other clay minerals in the materials.

It was pointed out in earlier sections that greater physical weathering of mica minerals than of some other silicate minerals is related to the relative hardness of the minerals. Hardness values by the Mohs scale for some micas and some other silicate minerals with which they are often associated (after Hurlbut, 1959) are given below.

Mineral	Hardness
Quartz	7
Microcline	6
Albite	6
Biotite (mica)	2.5-3
Muscovite (mica)	2-2.5

Micas are known to be nonconductors of heat and electricity. Thus micas, particularly muscovite, are used commercially as insulator materials.

Sheets of most micas are quite flexible, however the brittle micas derive their name from the fact that they are more brittle.

For the optical characteristics of micas the reader is referred to texts on optical mineralogy (e.g., Winchell & Winchell, 1961).

VII. QUANTITATIVE DETERMINATION

The method for quantitative determination of micas most employed by soil mineralogists is that of Alexiades and Jackson (1966). (The method with latest revisions etc. is also given by Jackson, 1969.) This method only measures the K micas present, since it is based on the allocation of K from elemental analysis for this element (Jackson, 1969, Chapter 11). In order to make the proper allocation of K to micas (normally done on the basis of 10% K_2O in soil micas) the K attributable to feldspars must be determined (Kiely & Jackson, 1964).

The method must be applied to specific particle sizes because the determination of feldspars involves selective dissolution of the micas through pyrosulfate fusion. During the fusion a certain amount of feldspars also dissolve and a correction is made for the amount dissolved. The amount of feldspars that dissolve increases with decreasing particle size (Kiely & Jackson, 1969).

A method for direct determination of micas on a whole soil basis would be desirable—particularly if it is found that soil properties are correlated with whole soil mica content (Murthy et al., 1975). To date no method has been developed for direct mica determination on a whole soil basis.

VIII. CONCLUSIONS

Micas have been defined by Nomenclature Committees of the United States and International Clay Minerals Societies as a group of 2:1 layer silicate minerals having a layer charge of about one per 10 oxygen formula unit, e.g., muscovite—$K(Si_3Al)Al_2O_{10}(OH)_2$. Each layer is composed of four planes of anions and three planes of cations, grouped into two tetrahedral sheets and one octahedral sheet. Interlayer cations, K in most of the important and extensive micas in soils, balance a layer charge that originates in the tetrahedral sheet in ideal micas. The term "brittle micas" is used for a separate group of 2:1 minerals that has a layer charge of about 2 per formula unit, e.g., margarite—$Ca(Si_2Al_2)Al_2O_{10}(OH)_2$.

The common mica polytypes are 1M and $2M_1$. Both of these polytypes are monoclinic. The 1M polytype has a true repeat every layer, the $2M_1$ every other layer. Both of these polytypes, and the 3T polytype—also sometimes found in nature, have intralayer stacking angles that are multiples of 120°. Other polytypes that are based on intralayer stacking angles that are multiples of 60° (2Or, $2M_2$, and 6H) are almost never found in nature. This is considered to be because the six-fold rings of tetrahedra in the tetrahedral sheets are ditrigonal rather than true hexagonal. The ditrigonal nature comes about because the tetrahedra are rotated and tilted—relative to a true hexagonal configuration to allow a better fit between the tetrahedral and octahedral sheets. Most "high temperature" dioctahedral micas are of the $2M_1$ polytype. Most trioctahedral micas are of the 1M polytype, as are most "low-temperature" dioctahedral micas such as glauconite—although low temperature forms often tend to be disordered ($1M_d$), largely because of interstratified expanded layers.

The terms illite and glauconite, although not presently recognized by the Nomenclature Committees as terms for mica minerals, have been much used by some earth scientists for mica-like clay minerals occurring in sedimentary materials. Published formulas for these materials, although possibly erroneous, show them to have a layer charge that is too low for the present definition of micas. These formulas also show a considerable part of the layer charge originating in the octahedral sheet.

Practically, micas are identified by 10Å spacings (distance between the centers of successive interlayers or between the centers of the planes of octahedral cations in successive layers) as measured by XRD (X-ray diffraction) or HREM (high-resolution electron microscopy), usually employing Mg-saturated specimens at room temperature—sometimes also with glycerol or ethylene glycol solvation. There are conflicts with the (theoretical) layer charge definition in the case of high charge expanded layers, such as some micas "opened-up" by K extraction (i.e. micas by layer charge, but termed vermiculite by XRD or HREM). Also, there probably are conflicts with some of the illitic and glauconitic "micas" (i.e. largely mica by XRD or HREM, but partly or wholly not mica by layer charge). Interstratified vermiculite and/or smectite, and/or chlorite, and/or "expanded mica," and other problems make accurate quantitative measurement of mica in soil materials by XRD nearly impossible. Quantitatively, micas are measured by allocation of K from elemental analyses, usually on the basis of 10% K_2O in micas after accounting for K in feldspars.

The layer character of micas controls the properties of these minerals. Micas usually occur as flakes or sheets that tend to orient themselves with their faces parallel to each other or to other surfaces. Also weathering proceeds along the interlayers and along the octahedral sheets with the tetrahedral sheets being more resistant to decomposition in most environments.

Simple transformation to expansible 2:1 minerals takes place by replacement of the K or other "fixed" interlayer cations with "exchangeable" cations. This may take place by either "layer weathering,"—selective replacement of K from individual layers, or by "edge weathering,"—K-release from many layers simultaneously so as to give a particle with "frayed-edges" and a "mica-core." Such simple transformations have been extensively studied. The factors controlling the rate and extent of the simple transformation may be grouped under (i) mineral, (ii) particle size, and (iii) environmental characteristics.

Dioctahedral K micas release their K much less readily than trioctahedral K micas. To explain this a model based upon the orientation of the OH in the mica structures has received much attention and support. In the trioctahedral minerals the OH are oriented with the H pointing toward the K in the structure. This proximity of the positively charged H to the K causes the K to be less strongly held than in the dioctahedral structures, where the H occurs farther away from the K because of an inclination toward the empty octahedral cation site. Biotite, a trioctahedral mica, releases its K less readily after oxidation of structural Fe. This apparently results from reorientation of OH after ejection of octahedral cations from the structure during oxidation—although conversion of the OH to O is an alternative explanation. Lepidolites, micas in which F takes the place of the OH in other micas, hold their interlayer K more tightly than all other micas.

Small mica particles (e.g., clay) release K more rapidly than coarse particles. The small particles, however, retain a significant portion of their K in an essentially nonextractable condition—whereas all of the K may be extracted from coarser particles. The difference appears to be caused by a large amount of layer weathering with small particles, whereas edge weathering is predominant with coarse particles. A theory suggests that, with layer weathered particles, the OH bonds shift toward the opened interlayer such that the K is held more tightly in adjacent interlayers.

Of the environmental conditions affecting the simple transformation of K micas, the concentration of K, and of ions that might replace K from the interlayer, in solution around the mica are very important. A critical K level exists for each condition (kind of mica, concentration of other ions, temperature, etc.). When the solution K concentration is above the critical level, no K is released and K may be taken from the solution by the particle. When the K concentration is below the critical level K is released from the particles. "Wedge" cation exchange sites, occurring in the interlayer where expanded 2:1 layers close down into the "mica core"—common in edge weathered particles, have a very high selectivity for K and other weakly hydrated cations such as NH_4, Rb, and Cs. Thus, for other more strong hydrated cations to replace K from a mica interlayer, the other cations must occur at a concentration many times greater than the concentration of K.

Under acid conditions and/or in the presence of organic acids and chelators for octahedral cations, the 2:1 layer may break down as rapidly as mica interlayer cations are exchanged. Under these conditions complex transformations of micas take place. The mica decomposition products may precipitate as kaolinite, montmorillonite, and as oxide and hydroxide minerals depending upon kind of mica mineral and the environmental conditions.

Stability diagrams, and the conditions under which glauconite is expected to form, show that for micas to form the environmental pH should be high and the activity of K in solution should be high. Most mica in soils appears to be primary—inherited from soil parent materials. Mica formation in most soils seems unlikely, although gradual transformation of beidellitic smectite to "mica" by K uptake (e.g., from that added as fertilizer) has been suggested. Micas in soils are considered important resources because they serve as "soil potassium buffers."

IX. ACKNOWLEDGMENTS

Thanks are given to S. W. Bailey for reviewing the section on crystal structures and nomenclature and for assistance with Fig. 7-1, 7-5, and 7-6, and Table 7-2; to A. D. Scott for reviewing an outline of the manuscript and for assistance with Fig. 7-12; to E. A. Niederbudde for reviewing the section on the Q/I relationship; to M. L. Jackson and R. M. Weaver for assistance with Fig. 7-13; and to three anonymous reviewers for many useful suggestions.

X. PROBLEMS AND EXERCISES

1) Draw a cross-section of a mica structure and point out (i) a plane or planes of basal oxygens, (ii) a plane or planes of apical oxygens and hydroxyls, (iii) a plane or planes of tetrahedral and octahedral cations, (iv) a tetrahedral sheet, (v) an octahedral sheet, (vi) a 2:1 layer, and (vii) an interlayer. Even better, construct a three-dimensional model of a mica structure of a particular polytype. This may be done using styrofoam balls held together with sections of pipe cleaners. Very small plastic balls to represent H ions may be embedded into the surface of large styrofoam balls to show OH orientation.

2) Make a simple drawing that shows the difference in OH orientation and in the proximity of the H to the K in di- as opposed to trioctahedral OH bearing micas. Explain, with reference to the diagram, current theory relating to OH orientation as to why the dioctahedral micas hold their K (with reference to replacement with hydrated cations) more tightly than the trioctahedral.

3) How do micas or "partially opened" micas serve as "soil potassium buffers"? Why may potassium fertilization be required for high crop yields even though a soil may be rich in mica?

4) How can micas be differentiated from other layer silicate minerals from X-ray diffraction patterns of parallel oriented specimens? Mention any necessary cation saturation, solvation, and heat treatments.

5) How can dioctahedral micas be differentiated from trioctahedral micas? What problems might a soil scientist encounter in attempting to make this distinction for mica in a soil clay fraction? Dioctahedral micas are much more common than trioctahedral micas in the clay fractions of most soils. Why is this true?

6) Develop arguments for and against recognizing illite and glauconite as terms for mica, or for some other minerals.

7) What crystal structure factors or relationships give rise to mica polytypes? What are the apparent reasons for the 1M, $2M_1$, and 3T polytype micas being much more common in nature than the 2Or, $2M_2$, and 6H?

8) Small K-mica particles release K more rapidly than large particles (of the same mineral), but usually contain a significant fraction of K that is essentially nonextractable—whereas all of the K may be extracted from the large particles. What theoretical explanations are given for these relationships?

9) "Partially opened" mica particles are said to have a high selectivity for K. (i) What does this mean, and (ii) why is this behavior exhibited?

10) What are the sources of the peaks seen in DTA patterns of micas? Include explanation for the endotherm often seen at about 150C (423K).

11) Describe characteristics of the soil environment that would control the rate and extent of K release from mica minerals.

12) Describe conditions that would favor mica (i) simple transformations, (ii) complex transformations, and (iii) formation in soils.

13) Describe mechanisms that affect (increase, decrease, or cause to remain the same) the layer charge of micas as they undergo simple transformations.

XI. SUPPLEMENTAL READING

Radoslovich, E. W. 1975. Micas in macroscopic forms. p. 27-57. In J. E. Gieseking (ed.) Soil components. Vol. 2. Inorganic components. Springer Verlag. New York.

Rich, C. I. 1968. Mineralogy of soil potassium. p. 79-108. In V. J. Kilmer, S. E. Younts, and N. C. Brady (ed.) The Role of potassium in agriculture. Am. Soc. Agron., Madison, Wis.

Rich, C. I. 1972. Potassium in soil minerals. p. 3-19. In Proc. 9th Colloquium (Landshut, W. Ger.). Int. Pot. Inst., Berne, Switzerland.

Rich, C. I., and H. G. V. Reichenbach. 1975. Fine-grained micas in soils. p. 59-95. In J. E. Gieseking (ed.) Soil components. Vol. 2. Inorganic components. Springer Verlag, New York.

LITERATURE CITED

AIPEA (Association Internationale Pour L'Etude Des Argiles), Nomenclature Committee. 1972. Report of Committee. AIPEA (Int. Clay Miner. Soc.) Newsletter no. 7:8–13.

Alexiades, C. A. 1970. The soil clay minerals. Univ. of Thessaloniki Press, Thessaloniki, Greece. (In Greek).

Alexiades, C. A., and M. L. Jackson. 1966. Quantitative clay mineralogical analysis of soils and sediments. Clays Clay Miner. 14:35–52.

Alexiades, C. A., N. A. Polyzopoulos, N. A. Koroxenides, and G. S. Axaris. 1973. High trioctahedral vermiculite content in the sand, silt and clay fractions of a Gray Brown Podzolic soil in Greece. Soil Sci. 116:363–375.

Arens, P. L. 1951. A study of the DTA of clays and clay minerals. Excelsior, foto-offset, Wageningen, Netherlands.

Arnold, P. W. 1960. Nature and mode of weathering of soil-potassium reserves. J. Sci. Food Agric. 11:285–292.

Bailey, S. W. 1966. The status of clay mineral structures. Clays Clay Miner. 14:1–23.

Bailey, S. W. 1967. Polytypism of layer silicates. p. SB1A–SB28A. In Short course lecture notes, layer silicates. Am. Geol. Inst., Washington, D. C.

Barshad, I. 1948. Vermiculite and its relation to biotite as revealed by base exchange reactions, X-ray analyses, differential thermal curves and water content. Am. Mineral. 33:655–678.

Barshad, I. 1954a. Cation exchange in micaceous minerals. I. Soil Sci. 77:463–472.

Barshad, I. 1954b. Cation exchange in micaceous minerals. II. Replaceability of ammonium and potassium from vermiculite, biotite, and montmorillonite. Soil Sci. 78:57–76.

Barshad, I., and F. M. Fawzy. 1968. Oxidation of ferrous iron in vermiculite and biotite alters fixation and replaceability of potassium. Science 162:1401–1402.

Barshad, I., and F. M. Kishk. 1970. Factors affecting potassium fixation and cation exchange capacities of soil vermiculite clays. Clays Clay Miner. 18:127–137.

Bassett, W. A. 1959. The origin of the vermiculite deposit at Libby, Montana. Am. Mineral. 44:282–299.

Bassett, W. A. 1960. Role of hydroxyl orientation in mica alteration. Bull. Geol. Soc. Am. 71:449–456.

Bates, T. F. 1964. The application of electron microscopy in soil clay mineralogy. p. 125–147. In C. I. Rich and G. W. Kunze (ed.) Soil clay mineralogy. The Univ. of North Carolina Press, Chapel Hill.

Beckett, P. H. T. 1964a. Studies on soil potassium. I. Confirmation of the ratio law: Measurement of the potassium potential. J. Soil Sci. 15:1–8.

Beckett, P. H. T. 1964b. Studies on soil potassium. II. The "immediate" Q/I relations of labile potassium in the soil. J. Soil Sci. 15:9–23.

Bentor, Y. .K., and Miriam Kastner. 1965. Notes on the mineralogy and origin of glauconite. J. Sediment Petrol. 35:155–166.

Bolt, G. H., M. E. Summer, and A. Kamphorst. 1963. A study of the equilibria between three categories of potassium in an illitic soil. Soil Sci. Soc. Am. Proc. 27:294–299.

Boyle, J. R., G. K. Voigt, and B. L. Sawhney. 1967. Biotite flakes: alteration by chemical and biological treatment. Science 155:193–195.

Brown, J. L., and C. I. Rich. 1968. High-resolution electron microscopy of muscovite. Science 161:1135–1137.

Burst, J. F. 1958a. "Glauconite" pellets: Their mineral nature and applications to stratigraphic interpretations. Bull. Am. Assoc. Petrol. Geol. 42:310–327.

Burst, J. F. 1958b. Mineral heterogeneity in "glauconite" pellets. Am. Mineral. 43:481–497.

Cady, J. G. 1960. Mineral occurrence in relation to soil profile differentiation. Int. Cong. Soil Sci., Trans. 7th (Madison, Wis.). IV:418–424.

Clay Minerals Society (CMS), Nomenclature Committee. 1971a. Summary of national and international recommendations on clay mineral nomenclature. Clays Clay Miner. 19:129–132.

Clay Minerals Society (CMS), Nomenclature Committee. 1971b. Report of nomenclature committee. Clays Clay Miner. 19:132–133.

Coffman, C. B., and D. S. Fanning. 1975. Maryland soils developed in residuum from chloritic metabasalt having high amounts of vermiculite in sand and silt fractions. Soil Sci. Soc. Am. Proc. 39:723-732.

Dennison, I. A., W. H. Fry, and P. L. Gile. 1929. Tech. Bull. USDA no. 128.

Dolcater, D. L., M. L. Jackson, and J. K. Syers. 1972. Cation exchange selectivity in mica and vermiculite. Am. Mineral. 57:1823-1831.

Douglas, L. A. 1965. Clay mineralogy of a Sassafras soil in New Jersey. Soil Sci. Soc. Am. Proc. 29:163-167.

Dymond, J., P. E. Biscaye, and R. W. Rex. 1974. Eolian origin of mica in Hawaiian soils. Geol. Soc. Am. Bull. 85:37-40.

Fanning, D. S., and M. L. Jackson. 1965. Clay mineral weathering in southern Wisconsin soils developed in loess and in shale-derived till. Clays Clay Miner. 13:175-191.

Farmer, V. C., J. D. Russell, W. J. McHardy, A. C. D. Newman, J. L. Ahlrichs, and J. Y. H. Rimsaite. 1971. Evidence for loss of protons and octahedral iron from oxidized biotites and vermiculites. Mineral. Mag. 38:121-127.

Feigenbaum, S., and I. Shainberg. 1975. Dissolution of illite—a possible mechanism of potassium release. Soil Sci. Soc. Am. Proc. 39:985-990.

Fölster, H., B. Meyer, and E. Kalk. 1963. Parabraunerden aus primär carbonathaltigem Würm-Löss in Niedersachsen II. Profilbilanz der zweiten Folge bodengenetischer Teilprozess: Tonbildung, Tonverlagerung, Gefügeverdichtung, Tonumwandlung. Z. Pflanzenernaehr, Dueng., Bodenkd 100:1-12.

Frye, John C., H. D. Glass, and H. B. Willman. 1962. Stratigraphy and mineralogy of the Wisconsinan loesses of Illinois. Ill. State Geol. Surv. Circ. 334.

Garrels, R. M., and C. L. Christ. 1965. Solutions, minerals, and equilibria. Harper and Row, New York.

Gaudette, H. E., J. L. Eades, and R. E. Grim. 1966. The nature of illite. Clays Clay Miner. 13:33-48.

Gilkes, R. J., R. C. Young, and J. P. Quirk. 1972a. The oxidation of octahedral iron in biotite. Clays Clay Miner. 20:303-315.

Gilkes, R. J., R. C. Young, and J. P. Quirk. 1972b. Oxidation of ferrous iron in biotite. Nature 236:89-91.

Gilkes, R. J., R. C. Young, and J. P. Quirk. 1973. Artificial weathering of oxidized biotite: I. Potassium removal by sodium chloride and sodium tetraphenylboron solutions. Soil Sci. Soc. Am. Proc. 37:25-28.

Gilkes, R. J., and R. C. Young. 1974. Artificial weathering of oxidized biotite. III. Potassium uptake by subterranean clover. Soil Sci. Soc. Am. Proc. 38:41-43.

Grim, R. E. 1968. Clay mineralogy. McGraw-Hill, New York.

Grim, R. E., and W. F. Bradley. 1940. Investigation of the effect of heat on the clay minerals illite and montmorillonite. J. Am. Ceram. Soc. 23:242-248.

Grim, R. E., W. F. Bradley, and G. Brown. 1951. The mica clay minerals. p. 138-172. In G. W. Brindley (ed.) X-ray identification and structures of clay minerals. Mineral. Soc. G. Br., monograph.

Grim, R. E., W. F. Bradley, and W. White. 1957. Petrology of the Paleozoic shales of Illinois. Ill. State Geol. Surv. Rep. of Investigations 203.

Grim, R. E., R. H. Bray, and W. F. Bradley. 1937. The mica in argillaceous sediments. Am. Mineral. 22:813-829,

Grim, R. E., and G. Kulbicki. 1957. Etude des reactions de hautes temperatures dans les mineraux argileaux an moyen des rayons. I. Bull. Soc. Fr. Ceram. 36:21-28.

Gruner, J. W. 1934. The structure of vermiculites and their collapse on dehydration. Am. Mineral. 19:557-575.

Hanway, J. J. 1956. Fixation and release of ammonium in soils and certain minerals. Iowa State Coll. J. Sci. 30:374-375.

Henderson, J. H., H. E. Doner, R. M. Weaver, J. K. Syers, and M. L. Jackson. 1976. Cation and silica relationships of mica weathering to vermiculite in calcareous Harps soil. Clays Clay Miner. 24:93-100.

Hendricks, S. B., and M. E. Jefferson. 1939. Polymorphism of the micas with optical measurements. Am. Mineral. 24:729-771.

Hill, R. D. 1953. The rehydration of fired clay and associated minerals. Trans. Brit. Ceram. Soc. 52:589-613.

Holmes, A. 1965. Principles of physical geology. The Ronald Press Co., New York.

Holt, J. B., I. B. Cutler, and E. M. Wadsworth. 1958. Rate of thermal dehydration of muscovite. J. Am. Ceram. Soc. 41:242-246.

Hower, John. 1961. Some factors concerning the nature and origin of glauconite. Am. Mineral 46:313-334.

Hseung, Y., and M. L. Jackson. 1952. Mineral composition of the clay fraction: III. Of some main soil groups of China. Soil Sci. Soc. Am. Proc. 16:294-297.

Huff, W. D. 1972. Morphological effects on illite as a result of potassium depletion. Clays Clay Miner. 20:295-301.

Hurlbut, C. S. 1959. Dana's manual of mineralogy. 17th ed. John Wiley & Sons, New York.

Hurley, P. M., R. F. Cormier, J. Hower, H. W. Fairbairn, and W. H. Pinson, Jr. 1960. Reliability of glauconite for age measurements by K-Ar and Rb-Sr methods. Bull. Am. Assoc. Petrol. Geol. 44:1793-1808.

Jackson, M. L. 1959. Frequency distribution of clay minerals in major great soil groups as related to the factors of soil formation. Clays Clay Miner. 6:133-143.

Jackson, M. L. 1964. Chemical composition of soils. p. 71-141. In F. E. Bear (ed.) Chemistry of the soil. Reinhold Publishing Corp., New York.

Jackson, M. L. 1969. Soil chemical analysis—Advanced course. 2nd ed. Published by author. Dep. Soil Science, Univ. of Wisconsin, Madison.

Jackson, M. L., Y. Hseung, R. B. Corey, E. J. Evans, and R. C. Vanden Heuvel. 1952. Weathering of clay-size minerals in soils and sediments. II. Chemical weathering of layer silicates. Soil Sci. Soc. Am. Proc. 16:3-6.

Jackson, M. L., W. Z. Mackie, and R. P. Pennington. 1946. Electron microscope applications in soils research. Soil Sci. Soc. Am. Proc. 11:57-63.

Jackson, M. L., S. A. Tyler, A. L. Bourbeau, and R. P. Pennington. 1948. Weathering sequence of clay-size minerals in soils and sediments. I. Fundamental generalizations. J. Phys. Colloid Chem. 52:1237-1260.

Johnson, L. J. 1970. Clay minerals in Pennsylvania soils. Relation to lithology of the parent rock and other factors. Clays Clay Miner. 18:247-260.

Johnson, L. J., R. P. Matelski, and C. F. Engle. 1963. Clay mineral characterization of modal soil profiles in several Pennsylvania counties. Soil Sci. Soc. Am. Proc. 27:568-572.

Jonas, E. C., and H. E. Roberson. 1960. Particle size as a factor influencing expansion of the three-layer clay minerals. Am. Mineral. 45:828-838.

Juang, T. C., and G. Uehara. 1968. Mica genesis in Hawaiian soils. Soil Sci. Soc. Am. Proc. 32:31-35.

Juo, A. S. R., and J. L. White. 1969. Orientation of the dipole moments of hydroxyl groups in oxidized and unoxidized biotite. Science 165:804-805.

Kiely, P. V., and M. L. Jackson. 1964. Selective dissolution of micas from potassium feldspars by sodium pyrosulfate fusion of soils and sediments. Am. Mineral. 49:1648-1659.

King, L. 1962. Morphology of the earth. Hafner Publishing Co., New York.

Kittrick, J. A. 1966. Forces involved in ion fixation by vermiculite. Soil Sci. Soc. Am. Proc. 30:801-803.

Kittrick, J. A. 1969a. Quantitative evaluation of the strong-force model for expansion and contraction of vermiculite. Soil Sci. Soc. Am. Proc. 33:222-225.

Kittrick, J. A. 1969b. Interlayer forces in montmorillonite and vermiculite. Soil Sci. Soc. Am. Proc. 33:217-221.

Kittrick, J. A. 1973. Mica-derived vermiculites as unstable intermediates. Clays Clay Miner. 21:479-488.

Lee, S. Y., M. L. Jackson, and J. L. Brown. 1975. Micaceous vermiculite, and mixed-layered kaolinite-montmorillonite examination by ultramicrotomy and high resolution electron microscopy. Soil Sci. Soc. Am. Proc. 39:793-800.

Leonard, R. A., and S. B. Weed. 1970a. Mica weathering rates as related to mica type and composition. Clays Clay Miner. 18:187-195.

Leonard, R. A., and S. B. Weed. 1970b. Effects of potassium removal on the b-dimension of phlogopite. Clays Clay Miner. 18:197-202.

Lodding, W., and H. P. Vaughan. 1965. Determination of strain energy in muscovite by simultaneous differential thermal analysis-thermogravimetric analysis. p. 191. In J. P. Radfern (ed.) Proc. 1st Int. Conf. on Thermal Analysis (Aberdeen, Scotland). MacMillan & Co., London.

Lyon, R. J. P. 1967. Infrared absorption spectroscopy. p. 371–403. In J. Zussman (ed.) Physical methods of determinative mineralogy. Academic Press, New York.

MacKenzie, R. C. (ed.). 1970. Differential thermal analysis. Vol. 1, Fundamental aspects. Academic Press, New York.

MacKenzie, R. C. 1970. Simple phyllosilicates based on gibbsite and brucite-like structures. p. 497–537. In R. C. MacKenzie (ed.) Differential thermal analysis. Academic Press, New York.

MacKenzie, R. C., and A. A. Milne. 1953. The effect of grinding on micas. I. Muscovite. Mineral. Mag. 30:178–185.

MacKenzie, R. C., G. F. Walker, and R. Hart. 1949. Illite in decomposed granite at Ballater Aberdeenshire. Mineral. Mag. 28:704–713.

Mackintosh, E. E., and D. G. Lewis. 1968. Displacement of potassium from micas by dodecylammonium chloride. Int. Congr. Soil Sci., Trans. 9th (Adelaide, Aust.) Vol. II:695–703.

Mackintosh, E. E., D. G. Lewis, and D. J. Greenland. 1971. Dodecylammonium–mica complexes–I. Factors affecting the exchange reaction. Clays Clay Miner.19:209–218.

Mahjoory, R. A. 1975. Clay mineralogy, physical, and chemical properties of some soils in arid regions of Iran. Soil Sci. Soc. Am. Proc. 39:1157–1164.

Manghanani, M. H., and J. Hower. 1964. Glauconites: cation exchange capacities and infrared spectra. Am. Mineral. 49:586–598.

Mason, B. 1966. Principles of geochemistry. 3rd ed. John Wiley & Sons, New York.

Mehmel, M. 1938. Ab und Umbau am Biotit. Chem. Erde II:307–332.

Mielenz, R. C., N. C. Schieltz, and M. E. King. 1954. Thermal gravimetric analysis of clays and clay-like minerals. Clays Clay Miner. 2:285–314.

Milford, M. H., and M. L. Jackson. 1966. Exchangeable potassium as affected by mica specific surface in some soils of North Central United States. Soil Sci. Soc. Am. Proc. 30:735–739.

Mortland, M. M. 1958. Kinetics of potassium release from biotite. Soil Sci. Soc. Am. Proc. 22:503–508.

Mortland, M. M., and B. Ellis. 1959. Release of fixed potassium as a diffusion controlled process. Soil Sci. Soc. Am. Proc. 23:363–364.

Mortland, M. M., K. Lawton, and G. Uehara. 1956. Alteration of biotite to vermiculite by plant growth. Soil Sci. 82:477–481.

Murthy, A. S. P., J. B. Dixon, and G. W. Kunze. 1975. Potassium-calcium exchange equilibria in sandy soils containing interstratified micaceous clay. Soil Sci. Soc. Am. Proc. 39:552–555.

Nettleton, W. D., R. E. Nelson, and K. W. Flach. 1973. Formation of mica in surface horizons of dryland soils. Soil Sci. Soc. Am. Proc. 37:473–478.

Newman, A. C. D. 1969. Cation exchange properties of micas. I. The relation between mica composition and potassium exchange in solutions of different pH. J. Soil Sci. 20:357–373.

Niederbudde, E. A., A. Schwarzmann, and U. Schwertmann. 1969a. Tonmineralbedingter K-Haushalt einer gedüngten Parabraunerden aus Würm-Geschiebemergel. Z. Pflanzenernaehr. Dueng., Bodenkd 124:212–224.

Niederbudde, E. A., B. Todorcic, and E. Welte. 1969b. Veranderungen von K-Formen und K-Ca Aktivataten-verhaltnissen des Bodens durch K-Düngung und K-Entzug der Pflanzen. Z. Pflanzenernaehr. Dueng., Bodenkd 123:85–100.

Niederbudde, E. A. 1972. Changes in K/Ca exchange properties of clay in loess-derived soils in soil formation. p. 103–107. In Proc. 9th Colloquium (Landshut, W. Ger.). Int. Pot. Inst., Berne, Switzerland.

Niederbudde, E. A. 1975. Veranderungen von Dreischicht-Tonmineralen durch natives K in holozänen Lössböden Mitteldeutschlands und Niederbayerns. Z. Pflanzenernaehr. Bodenkd 2:217–234.

Niederbudde, E. A. 1976. Umwandlungen von Dreischichtsilikaten unter K-Abgabe und K-Aufnahme. Z. Pflanzenernaehr. Bodenkd. 139:57–71.

Norrish, K. 1973. Factors in the weathering of mica to vermiculite. p. 417–432. In J. M. Serratosa (ed.) 1972 Proc. Int. Clay Conf., Div. de Ciencias, Madrid.

Owens, J. D., and J. P. Minard. 1960. Some characteristics of glauconite from the Coastal Plain formations of New Jersey. U. S. Geol. Surv. Prof. Paper 400, Part B. p. 430–432.

Porrenga, D. H. 1966. Clay minerals in recent sediments of the Niger Delta. Clays Clay Miner. 14:221-233.

Post, D. F., and J. L. White. 1967. Clay mineralogy and mica-vermiculite layer charge density distribution in the Switzerland soils of Indiana. Soil Sci. Soc. Am. Proc. 31:419-424.

Pough, F. H. 1960. A field guide to rocks and minerals. 3rd ed. The Riverside Press, Cambridge, Mass.

Quirk, J. P., and J. H. Chute. 1968. Potassium release from mica-like clay minerals. Int. Congr. Soil Sci., Trans. 9th (Adelaide, Aust.) II:671-681.

Radoslovich, E. W. 1959. Structural control of polymorphism in micas. Nature 183:253.

Radoslovich, E. W. 1960a. The structure of muscovite $KAl_2(Si_3Al)O_{10}(OH)_2$. Acta. Crystallogr. 13:919-932.

Radoslovich, E. W. 1960b. Hydromuscovite with the $2M_2$ structure—a criticism. Am. Mineral. 45:894-898.

Radoslovich, E. W. 1963. The cell dimensions and symmetry of layer-lattice silicates IV. Interatomic forces. Am. Mineral. 48:76-99.

Raman, K. V., and M. L. Jackson. 1966. Layer charge relations in clay minerals of micaceous soils and sediments. Clays Clay Miner. 14:53-68.

Rausell-Colom, J. A., T. R. Sweatman, C. B. Wells, and K. Norrish. 1965. Studies in the artificial weathering of mica. p. 40-72. In E. G. Hallsworth and D. V. Crawford (ed.) Experimental pedology. Butterworths, London.

Reed, M. G., and A. D. Scott. 1962. Kinetics of potassium release from biotite and muscovite in sodium tetraphenylboron solutions. Soil Sci. Soc. Am. Proc. 26:437-440.

Reichenbach, H. G. V. 1973. Exchange equilibria of interlayer cations in different particle size fractions of biotite and phlogopite. In J. M. Serratosa (ed.) 1972 Proc. Intern. Clay Conf., Div. de Ciencias, Madrid.

Reichenbach, H. G. V., and C. I. Rich. 1969. Potassium release from muscovite as influenced by particle size. Clays Clay Miner. 17:23-29.

Rex, R. W. 1966. Authigenic kaolinite and mica as evidence for phase equilibria at low temperatures. Clays Clay Miner. 13:95-104.

Rex, R. W., J. K. Syers, M. L. Jackson, and R. N. Clayton. 1969. Eolian origin of quartz in soils of Hawaiian Islands and in Pacific pelagic sediments. Science 163:277-279.

Robert, M. 1973. The experimental transformation of mica toward smectite; relative importance of total charge and tetrahedral charge. Clays Clay Miner. 21:167-174.

Ross, G. J., and C. I. Rich. 1973a. Effect of particle thickness on potassium exchange from phlogopite. Clays Clay Miner. 21:77-81.

Ross, G. J., and C. I. Rich. 1973b. Effect of particle size on potassium sorption by potassium-depleted phlogopite. Clays Clay Miner. 21:83-87.

Ross, G. J., and C. I. Rich. 1974. Effect of oxidation and reduction on potassium exchange of biotite. Clays Clay Miner. 22:355-360.

Roth, C. B., M. L. Jackson, E. G. Lotse, and J. K. Syers. 1968. Ferrous-ferric ratio and CEC changes on deferration of weathered micaceous vermiculite. Isr. J. Chem. 6: 261-273.

Roth, C. B., M. L. Jackson, and J. K. Syers. 1969. Deferation effect on structural ferrous-ferric ratio and CEC of vermiculites and soils. Clays Clay Miner. 17:253-264.

Roth, C. B., and R. J. Tullock. 1973. Deprotonation of nontronite resulting from chemical reduction of structural ferric iron. p. 107-114. In J. M. Serratosa (ed.) 1972 Proc. Int. Clay Conf., Div. de Ciencias, Madrid.

Roy, R. 1949. Decomposition and resynthesis of micas. J. Am. Ceram. Soc. 32:202-209.

Sawhney, B. L. 1972. Selective sorption and fixation of cations by clay minerals: A review. Clays Clay Miner. 20:93-100.

Sawhney, B. L., and G. K. Voigt. 1969. Chemical and biological weathering in vermiculite from Transvaal. Soil Sci. Soc. Am. Proc. 33:625-629.

Scott, A. D. 1968. Effect of particle size on interlayer potassium exchange in mica. Int. Congr. Soil Sci., Trans. 9th (Adelaide, Aust.) II:649-660.

Scott, A. D., F. T. Ismail, and R. R. Locatis. 1973. Changes in interlayer potassium exchangeability induced by heating micas. p. 467-479. In J. M. Serratosa (ed.) 1972 Proc. Int. Clay Conf., Div. de Ciencias, Madrid.

Scott, A. D., and M. G. Reed. 1962a. Chemical extraction of potassium from soils and micaceous minerals with solutions containing sodium tetraphenyl boron: II. Biotite. Soil Sci. Soc. Am. Proc. 26:41–45.

Scott, A. D., and M. G. Reed. 1962b. Chemical extraction of potassium from soils and micaceous minerals with solutions containing sodium tetraphenyl boron: III. Illite. Soil Sci. Soc. Am. Proc. 26:45–58.

Scott, A. D., and M. G. Reed. 1965. Expansion of potassium depleted muscovite. Clays Clay Miner. 13:247–261.

Scott, A. D., and S. J. Smith. 1966. Susceptibility of interlayer potassium in micas to exchange with sodium. Clays Clay Miner. 14:69–81.

Scott, A. D., and S. J. Smith. 1967. Visible changes in macro mica particles that occur with potassium depletion. Clays Clay Miner. 15:357–373.

Scott, A. D., and S. J. Smith. 1968. Mechanism for soil potassium release by drying. Soil Sci. Soc. Am. Proc. 32:443–444.

Serratosa, J. M., and W. F. Bradley. 1958. Determination of the orientation of OH bond axes in layer silicates by infrared adsorption. J. Phys. Chem. 62:1164–1167.

Smith, J. V., and H. S. Yoder. 1956. Experimental and theoretical studies of mica polymorphs. Mineral. Mag. 31:209–235.

Smith, S. J., L. J. Clark, and A. D. Scott. 1968. Exchangeability of potassium in soils. Int. Congr. Soil Sci., Trans. 9th (Adelaide, Aust.) II:661–669.

Smith, S. J., and A. D. Scott. 1966. Extractable potassium in grundite illite I. Method of extraction. Soil Sci. 102:115–122.

Soil Survey Staff. 1972. Soil series of the United States, Puerto Rico, and the Virgin Islands. Their taxonomic classification. USDA-SCS, Washington, D. C.

Soil Survey Staff. 1975. Soil taxonomy: A basic system of soil classification for making and interpreting soil surveys. USDA Handb. 436. U. S. Government Printing Office, Washington, D. C.

Spyridakis, D. E., G. Chesters, and S. A. Wilde. 1967. Kaolinization of biotite as a result of coniferous and deciduous seedling growth. Soil Sci. Soc. Am. Proc. 31:203–210.

Sridhar, K., and M. L. Jackson. 1974. Layer charge decrease by tetrahedral cation removal and silicon incorporation during natural weathering of phlogopite to saponite. Soil Sci. Soc. Am. Proc. 38:847–851.

Stubican, V., and R. Roy. 1961. Isomorphous substitution and infrared spectra of layer lattice silicates. Am. Mineral. 46:32.

Swindale, L. D., and G. Uehara. 1966. Ionic relationships in the pedogenesis of Hawaiian soils. Soil Sci. Soc. Am. Proc. 30:726–730.

Syers, J. K., M. L. Jackson, V. E. Berkheiser, R. N. Clayton, and R. W. Rex. 1969. Eolian sediment influence on pedogenesis during the Quaternary. Soil Sci. 107:421–427.

Takahashi, Jun-Ichi. 1939. Synopsis of Glauconite. p. 503–512. In Parker Davies Trask (ed.) Recent marine sediments. Am. Assoc. Petrol. Geol.

Takeuchi, Y. 1966. Structures of brittle micas. Clays Clay Miner. 13:1–25.

Tapper, M., and D. S. Fanning. 1968. Glauconite pellets: similar X-ray patterns from individual pellets of lobate and vermiform morphology. Clays Clay Miner. 16:275–283.

Tedrow, J. C. F. 1966. Properties of sand and silt fractions in New Jersey soils. Soil Sci. 101:24–30.

Tomita, K., and T. Sudo. 1971. Transformation of sericite into an interstratified mineral. Clays Clay Miner. 19:263–270.

Tsuboi, M. 1950. On the positions of the hydrogen atoms in the crystal structure of muscovite as revealed by the infrared absorption study. Chem. Soc. Japan Bull. 23:83–88.

Tsvetkov, A. I., and E. P. Valyashikhina. 1956. Trudy Inst. Geol. rudn. Mestorozh., no. 4 (As cited in Mackenzie, R. C., 1970).

Tyler, S. A., and S. W. Bailey. 1961. Secondary glauconite in the Biwabic iron-formation of Minnesota. Econ. Geol. 56:1033–1044.

Van Houten, F. B. 1953. Clay minerals in sedimentary rocks and derived soils. Am. J. Sci. 251:61–82.

Vedder, W., and R. S. McDonald. 1963. Vibrations of the OH ions in muscovite. J. Chem. Phys. 38:1583–1590.

Veith, J. A., and M. L. Jackson. 1974. Iron oxidations and reduction effects on structural hydroxyl and layer charge in aqueous suspensions of micaceous vermiculites. Clays Clay Miner. 22:345-353.

Weaver, C. E. 1953. A lath shaped non-expanded dioctahedral 2:1 mineral. Am. Mineral. 38:279-287.

Weaver, C. E., and L. D. Pollard. 1973. The chemistry of clay minerals. Elsevier Scientific Publishing Co., Amsterdam.

Weaver, R. M., M. L. Jackson, and J. K. Syers. 1971. Magnesium and silicon activities in matrix solutions of montmorillonite-containing soils in relation to clay mineral stability. Soil Sci. Soc. Am. Proc. 35:823-830.

Weed, S. B., C. B. Davey, and M. G. Cook. 1969. Weathering of mica by fungi. Soil Sci. Soc. Am. Proc. 33:702-706.

Weiss, A., A. Mehler, and U. Hoffman. 1956. Kationenaustausch und interkrisstalline Quellungsvermogen bei den Mineralen der Glimmergruppe. Z. Naturforsch. 11: 435-438.

Wells, K. L., and F. F. Riecken. 1969. Regional distribution of potassium in the B horizon clay of some Prairie loess soils of the Midwest. Soil Sci. Soc. Am. Proc. 33: 582-587.

White, J. L. 1951. The transformation of illite into montmorillonite. Soil Sci. Soc. Am. Proc. 15:129-133.

White, J. L. 1954. Reactions of molten salts with layer lattice silicates. Nature (London) 174:799-800.

White, J. L. 1971. Interpretation of infrared spectra of soil minerals. Soil Sci. 112:22-31.

Willman, H. B., H. D. Glass, and J. C. Frye. 1963. Mineralogy of glacial tills and their weathering profiles in Illinois. Part I. Glacial tills. Ill. State Geol. Surv. Circ. 347.

Winchell, A. N., and H. Winchell. 1961. Elements of optical mineralogy. Part II. Descriptions of minerals. John Wiley & Sons, New York.

Wolff, R. G. 1967. X-ray and chemical study of weathering glauconite. Am. Mineral. 52:1129-1138.

Zwetsch, A. 1934. Röntgenuntersuchungen in der Keramik. Ber. Dtsch. Keram. Ges. 14: 2-14. (As cited in R. E. Grim, 1968. Clay Mineral., p. 336.)

Vermiculites

LOWELL A. DOUGLAS, Rutgers University, New Brunswick, New Jersey

I. INTRODUCTION

The recognition of vermiculite as an important component of soils is a relatively recent observation. Kelley et al. (1939) and Alexander et al. (1939) in separate reviews of the minerals in soil clays did not mention vermiculite. MacEwan reported the occurrence of vermiculite in soils in 1948 (MacEwan, 1948). Today, vermiculite is known to be very widely distributed (Table 8-1), and has the widest particle size distribution of the secondary phyllosilicates, having been found in all fractions from fine clay through coarse sand. Vermiculite has the ability to fix K and must be considered when plant-K soil reactions are investigated.

Concepts of vermiculite have been confused because two different vermiculites are common. Both vermiculites are based on ideal-mica structures, trioctahedral vermiculite being an alteration product of biotite and dioctahedral being an alteration product of muscovite. Dioctahedral vermiculite is sometimes called soil vermiculite because it was first found in soils (Brown, 1953) and although sometimes found in marine sediments, it occurs most frequently in soil environments.

In this chapter an attempt will be made to distinguish between dioctahedral and trioctahedral vermiculite. The term *vermiculite* will be used to include both trioctahedral and dioctahedral vermiculite. *Vermiculites* will usually be used to specifically include both the trioctahedral and dioctahedral varieties. However, it should be recognized that when only clay-size minerals are studied, it is sometimes impossible to distinguish between the di- and trioctahedral forms (Grim, 1968, p. 31).

II. DISTRIBUTION IN SOILS

The diverse soil environments in which vermiculites have been found are illustrated in Table 8-1. Vermiculites are found under polar and tropical temperatures, in deserts and in areas of high rainfall. They have been reported in all of the major soil great groups, although found more often in soils of temperate and subtropical climates than in soils of tropical areas. Soil vermiculites are nearly always reported to accompany; or as an alteration product of: muscovite, biotite, or chlorite. Most authors state that *all* vermiculites are alteration products of mica or chlorite. The two outstanding exceptions are: Smith (1962) who found vermiculite replacing feldspar, and Barshad and Kishk (1969) who reported that some dioctahedral vermiculites may be synthesis products. In soils, dioctahedral vermiculite is more com-

chapter 8

Table 8-1. Vermiculite distribution among soils†

Soil classification	Geographical location	Type of octahedral sheet	Associated minerals	Citation
Duric Haploxeralf	California	tri–D‡	P, K, Q, F	Rhoades (1967)
Typic Xerachrept	California	tri–D‡	P, K, Q, F	Rhoades (1967)
Highfield series	Pennsylvania	tri–D	C, N.S.	Johnson (1964)
Non-Calcic Brown	California	tri–D	HB, B	le Roux, Cady, & Coleman (1963)
Ultic Hapludalfs	Maryland	tri–A	N.S.	Coffman & Fanning (1975)
Dystric Eutrochrepts	Maryland	tri–A	N.S.	Coffman & Fanning (1975)
Typic Dystrandept	Columbia, South America	tri–D	S, C, H, K, A	Cortes & Franzmeier (1972)
Placic Cryandept	Columbia, South America	tri–D	S, C, H, K, A	Cortes & Franzmeier (1972)
Red-Yellow Podzolic	Japan	tri–D	Q, C, I, K	Kato (1965)
Andosol	Columbia, South America	tri–D	H, A, Cr	Calhoun et al. (1972)
Red-Yellow Podzolic	Alabama	di–D	K, G, S, C	Bryant & Dixon (1964)
Red-Yellow Podzolic	Virginia	di–D	K	Rich & Obenshain (1955)
Paleudults	Florida	di–A	K, G, Q	Carlisle & Zelazny (1973)
Gray–Brown Podzolic	Indiana	di–D	I, C, S	Klages & White (1957)
Gray–Brown Podzolic	New Jersey	di–D	I, K, Q	Hathaway (1955)
Red Loam (Latosolic)	Victoria, Australia	di–D	K, Go, I	Jones et al. (1964)
Solonchak	Utah	di–D	I, S, Ca, D	Güven & Kerr (1966)
Podzol	Yam-Alín Mountain Range, Russia	di–D	I, K	Sokolova & Shostak (1969)
Podzol	New Brunswick, Canada	di–D	I, S, C	Kodama & Brydon (1968)
Oxbow Orthic Black Chernozemic	Saskatchewan, Canada	N.S.	K, S, I, F, Q, C, A	Huang & Lee (1969)
Oxbow Low-Humic Eluviated Gleysol	Saskatchewan, Canada	N.S.	K, S, I, F, Q, C, A	Huang & Lee (1969)
Arctic Brown	Banks Island, Canada	tri–A	S, K, Q	Tedrow & Douglas (1964)
Lateritic Red Earth	Taiwan	di–D	I, K	Barshad & Kishk (1969)
Gray–Brown Podzolic	California	di–D	I, H, G, Q	Barshad & Kishk (1969)
Brown Podzolic	California	di–D	I, H, G, Q	Barshad & Kishk (1969)
Podzolic	California	di–D	I, H, G, Q, F	Barshad & Kishk (1969)
Podzolic	California	di–D	I, C, K, F	Barshad & Kishk (1969)
Red Podzolic	California	di–D	I, H, F	Barshad & Kishk (1969)
Prairie	California	di–D	S, F	Barshad & Kishk (1969)
Non-Calcic Brown	California	di–D	I, S, H, Q, F	Barshad & Kishk (1969)
Brown Forest	California	di–D	S, H, F	Barshad & Kishk (1969)

† N.S. = not stated
　A = assumed
　D = actually determined
　di = dioctahedral
　tri = trioctahedral

K = kaolinite
G = gibbsite
Q = quartz
Cr = cristobalite
I = illite

T = talc
P = pyrophyllite
F = feldspar
Go = goethite

HB = hydrobiotite
B = biotite
C = chlorite
S = smectite

D = dolomite
H = halloysite
A = amorphous
Ca = calcite

‡ Information is the result of a personal communication with the author (Rhoades, 1967).

mon than trioctahedral vermiculite (Jackson, 1959), probably indicating the relative stability of the muscovite and biotite structures, or a stability promoted by the presence of hydroxy-Al interlayers.

Trioctahedral vermiculite may comprise a significant amount of the sand and silt fractions of soils (Alexiades et al., 1973; Coffman & Fanning, 1974). Dioctahedral vermiculite is seldom observed as discrete crystallites > 5 μm. Medium-size silt fractions may contain muscovite-dioctahedral vermiculite-interstratified minerals. In a muscovite $\rightarrow$ dioctahedral vermiculite weathering sequence, as the particle size of the sample decreases, the ratio of dioctahedral vermiculite to muscovite increases (Rich, 1958). In many soils the sum of mica plus vermiculites (Douglas, 1965) or chlorite plus vermiculites (Kapoor, 1972) in each soil horizon is a constant, and the ratio of vermiculites to mica decreases from the soil surface downward. The reaction mica $\rightarrow$ vermiculite depends on a strong leaching environment to remove solution potassium so that the reaction can proceed. Consequently, the intensity of the reaction is greatest in the A horizon, and the intensity decreases deeper in the profile. It would be anticipated that this reaction rate would decrease drastically in glei soils, while the tendency is towards a mica $\rightarrow$ smectite alteration in glei soils. Mixed-layer minerals (biotite/vermiculite, muscovite/vermiculite, vermiculite/smectite, etc.) are common (B. L. Sawhney, Chapter 12, this book) in these weathering sequences.

Under intensive weathering environments, as evidenced by acid soil conditions, vermiculites invariably have a hydroxy-Al interlayer (Rich, 1958). These hydroxy-Al interlayered vermiculites may be either trioctahedral (Kato, 1965) or dioctahedral (Rich & Obenshain, 1955). In soils where kaolinite has long been thought to be the most abundant clay mineral (Red-yellow Podzolics, Udults, etc.), it is now recognized that hydroxy-Al-interlayered dioctahedral vermiculite is often the most abundant clay mineral in surface horizons, provided that mica was in the parent material. The amount of vermiculite decreases with depth in the profile. The amount of kaolinite increases with depth in the profile. The amount of kaolinite in these soils is relatively low in A horizons, but increases with depth so that kaolinite is the most abundant clay mineral at depth (Carlisle & Zelazny, 1973; Douglas, 1965; Rich & Obenshain, 1955). Two explanations have been offered for this kaolinite-hydroxy-Al-interlayered vermiculite relationship. One proposal states that both kaolinite- and hydroxy-Al-interlayered vermiculite might be formed. However, in the environments of near-surface horizons, vermiculite with extensive hydroxy-Al interlayers appears to be more resistant to further weathering than kaolinite (Carlisle & Zelazny, 1973; Rich & Obenshain, 1966; Zelazny et al., 1975[1]). The second possible explanation of this kaolinite-vermiculite relationship is that the soil conditions are more favorable for the formation of kaolinite in B2 and C1 horizons (Krebs & Tedrow, 1958; Rich & Obenshain, 1955) than in A horizons.

The alteration sequence: muscovite $\rightarrow$ vermiculite $\rightarrow$ smectite has been

---

[1]L. W. Zelazny, V. W. Carlisle, and F. G. Calhoun. 1975. Clay mineralogy of selected paleusdults from the Lower Coastal Plain. Agron. Abstr. p. 180.

observed in Podzolic soils (Brown & Jackson, 1958; Douglas & Lee, 1971;[2] Gjems, 1963; Ross & Mortland, 1966). In these soils the smectite is usually limited to A horizons and the vermiculite is concentrated in the B horizon and usually has a hydroxy-Al interlayer.

Vermiculite can form in many different soil environments, from a variety of parent minerals. For vermiculite to persist under intense weathering conditions or for long periods of time, the normal weathering sequence is: mica → vermiculite → hydroxy-aluminum-interlayered vermiculite.

III. STRUCTURE AND COMPOSITION

The structure of trioctahedral vermiculite is well known (Gruner, 1934; Hendricks & Jefferson, 1938; Mathieson, 1958; Mathieson & Walker, 1954; Shirozu & Bailey, 1966), with excellent information from single-crystal investigations (Table 8-2). The assumption is made that because the powder X-ray diffraction pattern and chemical composition of clay-size trioctahedral vermiculite is similar to that of larger crystals, the structures are the same.

Most of our knowledge of the structure of trioctahedral vermiculite has been obtained from macrocrystalline materials. The structure of dioctahedral vermiculite is not as well understood because this mineral has been found only as small particles, too small for single crystal X-ray examination. Since powder X-ray diffraction patterns and the chemical composition are similar to potassium-depleted muscovite, and the mineral is often found associated with muscovite, it has been generally assumed that dioctahedral vermiculite has a similar structure and composition to muscovite, excluding K. However, Barshad and Kishk (1969) and Kirkland and Hajek (1972) have accumulated some data describing the structure and composition of soil vermiculites. Macrocrystalline dioctahedral vermiculite apparently is not found in nature.

The composition of some trioctahedral vermiculites is shown in Table 8-3. The general formula for this mineral is:

$$(Mg \cdot Fe)_3(Al_x \cdot Si_{4-x})O_{10}(OH)_2 \cdot 4H_2O\ Mg_x,$$

---

[2]Lowell A. Douglas and Ya-Yuey Hwang Lee. 1971. Mica weathering in podzols. Agron. Abstr. p. 109.

Table 8-2. Cell constants for trioctahedral vermiculite

	Gruner (1934)	Hendricks and Jefferson (1938)	Mathieson and Walker (1954)	Shirozu and Bailey (1966)
a	5.3	5.33	5.33	5.349Å
b	9.2	9.18	9.18	9.255Å
c	28.57–28.77	28.85	28.90	28.89Å
β	97°09′	93°15′	97	97°07′
Space group	Cc or C2/c	Cn or C2/n	Cc	C2/c

Table 8-3. Chemical composition of trioctahedral vermiculites

Composition		Source	Citation
Octahedral	Tetrahedral		
$(Mg_{2.36}Fe^{3+}_{0.48}Al_{0.16})$	$(Al_{1.28}Si_{2.72})\,O_{10}(OH)_2$ $\cdot\,4.32\,H_2O\,Mg_{0.32}$	Kenya	Mathieson and Walker (1954)
$(Mg_{2.83}Fe^{3+}_{0.01}Al_{0.15})$	$(Al_{1.14}Si_{2.86})\,O_{10}(OH)_2$ $\cdot\,4.72\,H_2O\,(Mg_{0.48}K_{0.01})$	Liano, Texas	Shirozu and Bailey (1966)
$(Mg_{2.69}Fe^{3+}_{0.23}Fe^{2+}_{0.08})$	$(Al_{0.85}Si_{3.15})\,O_{10}(OH)_2^{\ -0.62}$	South Africa	Kittrick (1973)
$(Mg_{2.61}Fe^{3+}_{0.29}Fe^{2+}_{0.01}Al_{0.08})$	$(Al_{0.81}Si_{3.19})\,O_{10}(OH)_2^{\ -0.44}$	Libby, Montana	Kittrick (1973)

with x = 0.5 to 1.5. Most of the Fe in trioctahedral vermiculite is in the ferric state, occupying octahedral sites although ferrous Fe has been found in tetrahedral sites (Taylor et al., 1968). The octahedral sheet carries a positive charge of 0.1 to 0.9 and the tetrahedral sheet has a negative charge of 0.9 to 1.4, resulting in a total negative charge of 0.5 to 1.0 (Foster, 1963). This charge deficiency is balanced by exchangeable cations; usually Mg, sometimes accompanied by small amounts of Ca; located between the 2:1 layers (Fig. 8-1). A double layer of water molecules is also located in the interlayer space, giving vermiculite its 14Å basal spacing.

Although no definitive study has been made, there are indications of ordering of some of the octahedral cations (Veith & Radoslovich, 1963; Radoslovich, 1963) in trioctahedral vermiculite. Taylor et al. (1968) used Mossbauer spectroscopy to show that two trioctahedral vermiculites showed no tendency towards ordering. Shirozu and Bailey (1966) showed that Llano vermiculite has tetrahedral ordering of silicon and aluminum.

The dimensions of both octahedral and tetrahedral sheets vary with the number of octahedral sites actually occupied and with the size and charge of the cations occupying these sites (Bailey, 1966). In the unconstrained state the tetrahedral sheet of vermiculite would have a b-parameter larger than the corresponding parameter of the octahedral sheet. The tetrahedral sheet can be reduced to fit a smaller octahedral sheet by rotation of tetrahedra, forming a ditrigonal structure (Fig. 8-2 and Table 8-4) instead of the "ideal" hexagonal symmetry. In dioctahedral structures the difference between b-parameters of unconstrained tetrahedral and octahedral sheets is larger than in trioctahedral structures. In dioctahedral structures, fit between tetrahedral and octahedral sheets is accomplished by a simultaneous rotation and tilt of tetrahedra. This tilt of tetrahedra tends to shorten the distance of surface tetrahedral oxygens, to interlayer cations.

Radoslovich (1962) has calculated the effect of composition on b-dimension, for montmorillonite and trioctahedral vermiculite:

$$b = (8.944 + 0.096\,Mg + 0.096\,Fe^{3+} + 0.037\,Al_{tetrahedral}) \pm 0.012A.$$

The water layers and cations in interlayer positions in trioctahedral vermiculite have been subjected to extensive study. This exchangeable Mg

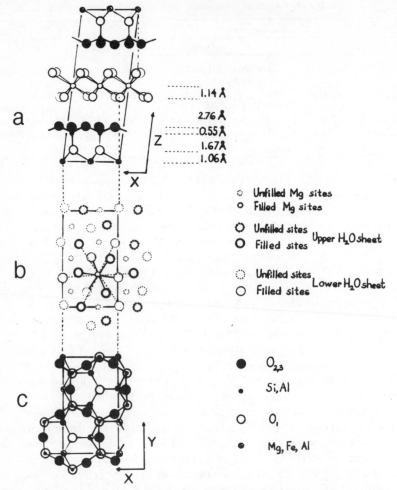

Fig. 8-1. The crystal structure of Mg-vermiculite. *(a)* projection on the *X Z* plane. *(b)* projection of the interlayer region showing interlayer water, normal to the *X Y* plane. *(c)* projection of silicate layer normal to the *X Y* plane (after Mathieson, 1958).

ions are located exactly equidistant from the adjacent tetrahedral sheets. Only 1/3 of the possible interlayer cation sites are occupied (Fig. 8-1b). The water molecules form two layers arranged in a distorted hexagonal pattern around the Mg ions. In Kenya vermiculite there is H-bonding between tetrahedral oxygen and water, and weak H-bonding within the water net, controlling the water layer's symmetry (Mathieson & Walker, 1954; Mathieson, 1958). In Llano vermiculite the position of interlayer Mg ions is controlled by the tetrahedral ordering of aluminum. Interlayer Mg ions are positioned vertically between the ordered aluminum tetrahedra (Shirozu & Bailey, 1966). The water molecules then form a hydration hexagonal pattern around the Mg ions.

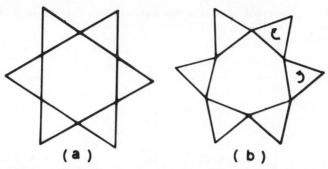

(a)   **(b)**

Fig. 8-2. The ditrigonal arrangement of silica tetrahedral in vermiculite. *(a)* ideal arrangement, *(b)* actual arrangement, arrows show direction of rotation of tetrahedra (after Rich, 1968).

The actual nature of "soil" vermiculite is less well known. In some cases the investigators have identified the platy mineral that is the parent material for the vermiculite, as muscovite, biotite, chlorite, etc. The assumption is then made that the soil vermiculite has inherited most of its structure from the parent mineral. Most of our knowledge of soil dioctahedral vermiculite has been gained in this manner. Early investigators, such as the classical study of Rich and Obenshain (1955), found dioctahedral vermiculite as clay, and muscovite in larger size fractions. The assumption that the dioctahedral vermiculite was formed by removal of K from muscovite seems justified. However, there is virtually no literature describing the actual structure and composition of this mineral. Such studies have been made most difficult because dioctahedral vermiculite is reported only as clay-size particles accompanied by other mineral species.

Vermiculites from soils developed from acidic rocks (Table 8-5) contained considerable Al and had Al substituted for Si in tetrahedral positions. These vermiculites, midway between ideal dioctahedral and trioctahedral populations, had properties usually attributed to trioctahedral micas. Vermiculites found in soils developed from mica-free basic rocks (Table 8-6) were Al deficient, with all of the charge located in octahedral sites. This type of vermiculite represents a new mineral species. Barshad and Kishk (1969) also described some vermiculites whose composition was intermediate between the high-Al and low-Al types, and interpreted these data as mixtures of the two types of soil vermiculites.

Table 8-4. Composition, cell dimensions and tetrahedral rotations for some trioctahedral vermiculites (Radoslovich, 1962)

Octahedral					Tetrahedral			b		Tetrahedral rotation, degree
Al	Fe^{2+}	Fe^{3+}	Mg	Other	Si	Al	Fe^{3+}	Observed	Calculated	
								Å		
0.22	0.08	0.46	1.92	0.11	2.72	1.28		9.222	9.376	$10°24'$
0.170	0.023	0.232	2.239		2.837	1.163		9.244	9.348	$8°35'$
		0.037	0.365	2.238	0.056	2.853	1.024	0.123	9.253	9.312	$6°27'$

Table 8-5. Structural formulas for Na-saturated vermiculites associated with mica
(after Barshad & Kishk, 1969)

	Sample number				
Composition	1	2	3	4	5
	Octahedral				
Al	1.44	1.47	1.24	0.93	1.12
Fe^{3+}	0.16	0.21	0.20	0.44	0.29
Ti	0.14	0.13	0.18	0.24	0.16
Mg	0.27	0.25	0.35	0.46	0.75
Mn	0.05	--	0.02	0.07	0.03
H	0.30	0.34	0.25	0.26	0.19
Σ	2.36	2.40	2.24	2.40	2.65
	Tetrahedral				
Si	2.90	2.87	3.24	2.89	2.69
Al	1.10	1.13	0.76	1.11	1.31
Σ	4.0	4.0	4.0	4.0	4.0
	Interlayer				
Na	0.61	0.64	0.64	0.67	0.60
	Layer charge				
Octahedral	+0.49	+0.49	+0.12	+0.44	+0.73
Tetrahedral	−1.10	−1.13	−0.76	−1.11	−1.31
Total	−0.61	−0.64	−0.64	−0.67	−0.58
	Cation exchange capacity, meq/100 g				
	166	173	166	182	144

Kirkland and Hajek (1972) described the vermiculite found in several Paleudults (Table 8-7). In all cases the vermiculite was diocthaedral, with Al substitution for Si, varying from 0.30 to 1.18 in tetrahedral sites.

IV. FORMATION OF VERMICULITE

Vermiculites will not form from the solidification of a magma, and it is almost universally assumed that the mineral can only be formed by alteration of a mica! Two investigators have presented evidence contradictory to this generally accepted hypothesis. Smith (1962) observed with the petrographic microscope trioctahedral vermiculite as a replacement product of feldspars. Barshad and Kishk (1969) found dioctahedral vermiculite in several soil clays which did not contain mica in their parent material and said that the vermiculite in these soils was a "synthesis" product. The "state-of-the-art" evidence is that vermiculites, trioctahedral and dioctahedral, are formed by the alteration of micas. Trioctahedral vermiculite is often reported in close association with pegmatites, and was once thought to be of hydrothermal origin. However, the predominant evidence indicates that trioctahedral vermiculite in rocks is formed by supergene alteration from phlogopite or biotite (Bassett,

Table 8-6. Structural formulas for Na-saturated vermiculites associated with montmorillonite (after Barshad & Kishk, 1969)

Composition	Sample number			
	8	9	10	11
Octahedral				
Al	0.33	--	0.88	--
Fe^{3+}	0.49	0.46	0.41	0.89
Mg	1.26	1.88	0.67	1.23
Mn	0.08	0.02	0.08	0.03
H	0.19	0.08	0.04	0.04
Σ	2.22	2.36	2.04	2.15
Tetrahedral				
Si	4.0	4.0	4.0	4.0
Σ	4.0	4.0	4.0	4.0
Interlayer				
Na	0.68	0.82	0.63	0.81
Layer charge				
Octahedral	−0.68	−0.82	−0.63	−0.81
Tetrahedral	0.0	0.0	0.0	0.0
Total	−0.68	−0.82	−0.63	−0.81
Cation exchange capacity, meq/100 g				
	168	207	161	204

Table 8-7. Interlayered dioctahedral vermiculite formulas (from Kirland & Hajek, 1972)

Soil series	Horizon	Tetrahedral		Octahedral			Interlayer		Exchange-able X^+	OH/Al ratio
		Si	Al	Mg	Fe	Al	Al	OH		
Malbis	Ap	3.70	0.30	0.14	0.23	1.63	1.66	4.91	0.38	2.96
	B21t	3.58	0.42	0.22	0.25	1.53	1.82	5.17	0.34	2.84
	B22t	3.38	0.62	0.20	0.21	1.59	1.84	4.99	0.29	2.71
Benndale	Ap	2.92	1.08	0.30	0.29	1.40	1.70	4.27	0.55	2.51
	B22t	2.83	1.17	0.22	0.32	1.46	1.84	4.71	0.58	2.56
	B23t	3.39	0.61	0.15	0.21	1.64	1.39	3.78	0.36	2.72
Lucedale (M)	Ap	3.35	0.65	0.12	0.20	1.68	1.34	3.67	0.45	2.74
	B1	3.27	0.73	0.16	0.23	1.62	1.18	3.02	0.38	2.57
	B22t	3.23	0.77	0.24	0.23	1.54	1.00	2.35	0.35	2.35
Lucedale (P)	Ap	3.13	0.87	0.20	0.20	1.60	1.70	4.45	0.42	2.62
	B1	2.82	1.18	0.26	0.27	1.47	1.30	2.99	0.54	2.30
	B22t	3.35	0.65	0.18	0.19	1.63	0.90	2.24	0.37	2.49
Average		3.24	0.76	0.20	0.24	1.56	1.45	3.79	0.41	2.61

1963; Roy & Romo, 1957). Vermiculite will not form at temperatures above 200-300C (473-573K). At these temperatures a "chlorite-like" mineral, rather than vermiculite, is formed. In soils chlorite may alter to vermiculite (Smith, 1962; Johnson, 1964).

Borchardt et al. (1966) showed that trioctahedral vermiculite pseudo-morphs after mica are common in many soils. Trioctahedral vermiculite in soils is formed by alteration of biotite, phlogopite, or chlorite either in the soil or in the parent material (Barshad, 1948; Johnson, 1964).

The alteration of micas to vermiculites may be subdivided into separate steps, or factors, affecting alteration: (i) Release of K, (ii) Oxidation of Fe, and (iii) Hydroxyl orientation.

A. Release of Potassium

The rate of release of K from interlayer positions in mica and replacement by other ions may be considered to be a diffusion process, with diffusion of K out and diffusion to the vacated spot by the counter ion (Chute & Quirk, 1967; Reed & Scott, 1962). A biotite disk when placed in an appropriate salt solution will alter so that in a short time a vermiculite halo will appear to encircle a biotite core. Alteration proceeds parallel to the X and Y directions with little apparent alteration parallel to the Z direction in the crystal. The release of K from micas is discussed in Chapter 7, this book (D. S. Fanning and V. Z. Keramidas). Only those factors which influence the properties of vermiculites will be considered in this chapter.

Potassium may diffuse from all interlayers, with mica going directly to vermiculite. In other cases K may diffuse along a specific 001 plane, but often not from adjacent 001 planes. The latter causes a wedge-shaped, K-free zone in the mica (Rich, 1964). This process produces the widely reported (Jackson, 1963; Bassett, 1959; Le Roux et al., 1963) mica-vermiculite-interstratified mineral mixtures. Small amounts of K in solution will prevent the replacement of K by Mg. The reaction biotite → vermiculite will proceed only when leaching is efficient and removes K (Bassett, 1963).

B. Oxidation of Iron

The Fe in biotite is usually in the reduced state and most of the Fe in vermiculite is in the oxidized state (Farmer et al., 1971; Weaver & Pollard, 1973, p. 100; Taylor et al., 1968). Some of the ions in the octahedral positions are lost during the alteration process,[3] resulting in an increase in the number of dioctahedral sites in the crystal (Farmer et al., 1971) increasing the rotation of tetrahedra, and thus decreasing the length of the b axis. Ferrous Fe may be dissolved and lost via the soil solution or precipitation. The loss of Fe is greater from micas which have high Fe^{2+}/Mg^{2+} ratios (Ismail, 1969). Ismail presents a strong argument that contrary to theory, under neutral and alkaline conditions the decrease in surface charge upon oxidation

---

[3]Ya-Huey Hwang. 1970. The release of K from biotite. M.S. Thesis, Rutgers University, New Brunswick, New Jersey.

of Fe in biotite is large enough to allow the formation of montmorillonite. Under acidic conditions the oxidation of Fe is balanced by a decrease in total charge of the octahedral sheet through loss of octahedral Fe and Mg. Under acidic conditions the surface charge remains large and vermiculite is formed. The simultaneous oxidation and expulsion of Fe from a mica may result in either amorphous interlayer oxides or a crystalline external phase of β-FeOOH on the vermiculite.

The oxidation of ferrous Fe accompanying the biotite $\rightarrow$ vermiculite transformation must be an electron-proton transfer. The structure could not maintain its integrity and accommodate atmospheric oxygen. The initial stages of oxidation of ferrous ions is compensated by a reversible loss of protons from octahedral hydroxyl groups. Subsequently, there is an irreversible loss of octahedral Fe. Veith and Jackson (1974) expressed these reactions as:

$$[(Fe^{2+})_2\,Mg_4O_4(OH)_4]^{\pm 0} \rightleftharpoons [(Fe^{3+})_2Mg_4O_4(OH)_2O_2]^{\pm 0} + 2e^- + 2H^+ \quad [1]$$

and

$$[(Fe^{2+})_5\,MgO_4(OH)_4]^{\pm 0} \rightarrow [(Fe^{3+})_4O_4(OH)_4]^{\pm 0} + 5e^- + Fe^{3+} + Mg^{2+} \quad [2]$$

As reaction [2] progresses, the reversibility of reaction [1] decreases. There is considerable debate regarding the nature of the oxidation reaction. One school believes that mica must expand (loss of K) prior to oxidation or expulsion of Fe. A second school believes that Fe may be oxidized either prior to or simultaneous with the loss of K.

C. Hydroxyl Orientation

Bassett (1960), in his classical study of the orientation of hydroxyls in phlogopite, biotite, and muscovite, did much to explain the differences in alteration rates between these three micas. In trioctahedral micas hydroxyl ions sit in a hole above three divalent ions (Fe^{2+} or Mg^{2+}). This causes a uniform charge distribution below the hydroxyl, and repels the hydrogen portion of the hydroxyl. In trioctahedral micas the hydrogen end of the hydroxyl is pointed directly towards the K. In dioctahedral micas the cations in the octahedral sheet are trivalent (Al) with only two-thirds of the available octahedral positions occupied; each hydroxyl ion sites above two cations and a hole. The octahedral cations orient the hydroxyl so that the hydrogen ion points in the general direction of the interlayer K ion, but tilted towards the vacant octahedral site. Thus, in trioctahedral micas the K sits in a volume strongly affected by the repulsive forces (polarity) of the hydrogen portion of the hydroxyl. In dioctahedral micas the interlayer K is more distant from the H portion of the hydroxyl, and is acted on by a higher retention field.

Dioctahedral micas alter to dioctahedral vermiculite slowly, and trioctahedral micas alter to trioctahedral vermiculite rather rapidly.

D. Total Charge

There is a marked difference between the charge characteristics of dioctahedral and trioctahedral vermiculites. Both muscovite and biotite ideally have a total charge of 1 per formula unit. It is usually assumed that both micas lose some charge as they alter to the appropriate vermiculite which has an apparent total charge of 0.5 to 0.9 per formular unit. Although the alteration of dioctahedral mica has not been extensively studied, the evidence is that when muscovite alters to dioctahedral vermiculite there is no actual decrease in total charge. The apparent loss of total charge is due to interlayer hydroxy-Al groups balancing some of the total charge (Cook & Rich, 1963).

Total charge may change during the formation of trioctahedral vermiculite by several different reactions, the oxidation of octahedral ions increasing the net positive charge, the lowering of interlayer charge, by the expulsion of octahedral cations, or by a change in hydroxyl content. Cation exchange capacity of trioctahedral vermiculite may be increased by treating vermiculite with sodium citrate-sodium dithionite. The Fe may then be oxidized to the ferric state with H_2O_2, however, the cation exchange capacity is not affected by this latter treatment because of the deprotonation $OH^- \rightarrow O^{2-} + H^+$ (Eq. [1]) occurring simultaneously with the oxidation step (Roth et al., 1968).

In addition to total charge, the distribution of charge between octahedra and tetrahedra decrees whether the alteration will proceed as mica $\rightarrow$ triocta-

Table 8–8. Transformations of different micas by extraction of K and Fe^{2+} oxidation (from Robert, 1973)

Total charge	Mica type, source	Type of behavior	
		After K extraction	After oxidation
Trioctahedral			
>0.9	Phlogopite	Vermiculite (high charge 0.9–0.8)	Vermiculite
	Biotite	Vermiculite (0.7)	Vermiculite (low charge 0.5)
Dioctahedral			
<0.9	Glauconite (Cormes)	Smectite	Smectite
	Glauconite (Villers)	Vermiculite	Smectite
	Illite (Grundite)	Vermiculite Smectite	Smectite
	Illite (Fithian)	Vermiculite	Smectite
	Illite (Puy)	Vermiculite	Vermiculite
>0.9	Phengite	Vermiculite high charge	
	Sericite	Vermiculite high charge	
	Muscovite	Vermiculite high charge	

hedral vermiculite or as mica → smectite. Robert (1971) described a highly oxidized biotite alteration product with a total charge of 0.5 per formula unit, typical of a smectite; but upon glycerol expansion this mineral sorbed only one glycerol layer, typical of a vermiculite. In this case some of the charge was located in tetrahedral sites. In a series of experiments where micas were first treated to remove K and then oxidized with H_2O_2. Robert (1973) showed (Table 8-8) that vermiculites may have either a high total charge, or a lower total charge with much of the charge originating in tetrahedral sites.

Several investigators have suggested, and Walker (1957,1958) presented, data indicating that a continuum might exist between trioctahedral vermiculites and smectites. However, Harward and Brindley (1965) and Harward et al. (1969) have shown that although there is some variation in total charge within trioctahedral vermiculites, vermiculites and smectites comprise two separate populations with no continuum between these minerals.

E. Stability of Vermiculites

Markos (1975)[4] has calculated the variation in trioctahedral vermiculites one might expect to find in equilibrium with biotite (Fig. 8-3). The vermiculites, V1, V2, V3, and V4 and the changes from the parent biotite are shown in Table 8-9. Stability of biotite compared with the four vermiculites is indicated by the shaded area of Fig. 8-3. The breaks in the lines V2, V3, and V4 are due to the influence of dominating Al species in the different pH ranges. It is interesting to note that the vermiculite more common under normal soil conditions (V4) involves the most complicated reactions (Table 8-9).

The revelation of Barshad and Kishk (1969) that some soil clay dioctahedral vermiculites may be precipitation products raises serious questions regarding the possible extrapolation of data obtained from trioctahedral vermiculites to explain the reactions of soil clay dioctahedral vermiculites. As Kittrick (1973) has pointed out,

> The major clay minerals of soils and sediments compete for a relatively small group of elements during their formation. In theory at least, each clay mineral forms only under solution conditions where it is the least soluble of the clay minerals competing for that group of elements. Thus, because of the stability of other minerals, the stability field of a particular mineral represents a much more restricted chemical environment than does a solution saturated with respect to that mineral alone.

Kittrick (1973) found a trioctahedral vermiculite to be unstable under acid conditions, where stabilities of montmorillonite, kaolinite and gibbsite had been determined. He then tried to locate a montmorillonite-vermiculite-amorphous silica triple point. The montmorillonite-vermiculite-amorphous silica mixtures went to the montmorillonite-magnesite-amorphous silica triple

---

[4]G. Markos. 1975. Representation of some soil mineral reactions in Eh-pH and ΔG_R-pH fields. Agron. Abstr. p. 175.

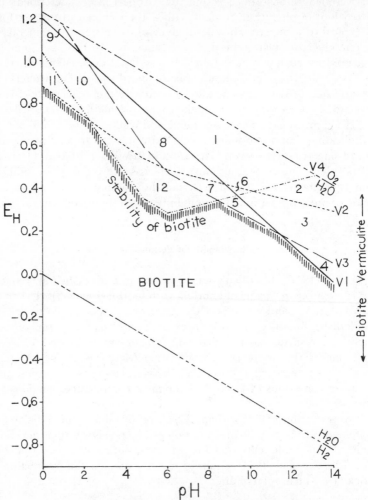

Fig. 8-3. Diagram showing the relationship of biotite to vermiculite. For the composition of the several phases see Table 8 (after Markos, 1975[4]).

point (Fig. 8-4). There was no stability area for vermiculite. Trioctahedral
vermiculite is unstable under the conditions encountered in soils.

Kittrick (1973) proposed that the presence of large amounts of trioctahedral vermiculite might be explained by a series of rate-controlled reactions.
The critical rates are slow for mica dissolution, rapid for K removal and Fe
oxidation, and slow for vermiculite dissolution. In chemical terms, mica-
derived trioctahedral vermiculites may be considered to be fast-forming unstable intermediates. This mode of formation tends to explain why the mica-
vermiculite alteration often proceeds with little, if any, reduction in particle
size (Borchardt et al., 1966). If Barshad and Kishk's (1969) assumptions are
correct, dioctahedral vermiculite must have a stability field in the soil en-

Table 8-9. Composition of the vermiculites and changes from biotite shown in the fields of Markos' phase diagram (Fig. 3) (from Markos, 1975)[4]

| | | Change from the biotite | | | |
| | | | Octahedral site | | |
No.	Composition†	Interlayer site	Mg	Fe	Tetrahedral site
V1	$Mg_{0.75}4H_2O(Mg_2Fe^{3+}_{0.5})(AlSi_3)O_{10}(OH)_2$	(all K is exchanged by Mg and coordinated water)	No change	Some loss	No change
V1	$Mg_{0.34}4H_2O(Mg_2Fe^{3+}_{0.6}Al_{0.2})(AlSi_3)O_{10}(OH)_2$		No change	Some replaced by Al	No change
V3	$Mg_{0.65}4H_2O(Mg_2Fe^{3+}_{0.7})(Al_{1.4}Si_{2.6})O_{10}(OH)_2$		No change	Some replaced by Al	Some Si replaced by Al
V4	$Mg_{0.35}4H_2O(Mg_{1.8}Fe^{3+}_{0.6}Al_{0.4})(Al_{1.3}Si_{2.7})O_{10}(OH)_2$		Some of both replaced by Al		Some Si replaced by Al

† All the mineral compositions here are fictive, but all are within the limits of compositional variations of their respective group. The numbered fields represent the possible combinations of the four vermiculites:

1—mainly V1 but in the higher pH ranges also V2 and V4 and below pH 2 V3 are present.
2—V2
3—V1 and V3
4—V1
5—V3
6—V1 and V4
7—V3 and V4
8—V2 at higher and V3 at the lower pH range
9—V1
10—V4 and V2
11—V2
12—V4

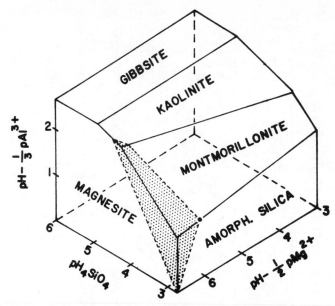

Fig. 8-4. Diagram indicating a possible vermiculite stability plane (shaded) (after Kittrick, 1973).

vironment, contrary to the lack of such a field for trioctahedral vermiculite. Many investigators believe that mica derived dioctahedral vermiculite may be an unstable intermediate, analogous to the trioctahedral vermiculite discussed by Kittrick (above). There is considerable evidence that hydroxy-Al inter- layered dioctahedral vermiculite is more stable than kaolinite in soils (Carlisle & Zelazny, 1973, Zelazny et al., 1975[1]). However, this dioctahedral vermicu- lite may be unstable with very slow reaction rates.

V. IDENTIFICATION OF VERMICULITES

Data that will be useful in describing vermiculites may be obtained by a large variety of methods. X-ray diffraction, thermal analysis, infrared anal- ysis, and electron microscopy will be discussed in this section. However, in the study of soil clay vermiculites other methods may be extremely useful such as; total chemical analysis, Mössbauer spectroscopy, neutron diffrac- tion, etc.

Most of the data used to characterize vermiculites has been obtained from macrocrystalline trioctahedral vermiculite. By studying macrocrystal- line trioctahedral vermiculite, investigators have been able to use a material of known crystallinity and purity. The present "state of the art" requires that the soil scientist intelligently and carefully extrapolate from macro- crystalline materials to soil clays, recognizing that much (possibly most) of the soil-clay vermiculite is dioctahedral.

A. X-ray Diffraction

X-ray diffraction is the technique most often used to characterize or identify vermiculite. Excellent data (Table 8-10) are available from triocta-hedral vermiculite; comparable data are not available for dioctahedral vermic-ulite.

The differentiation between dioctahedral and trioctahedral vermiculite may be made based on the 060 diffraction spacing, near 1.50Å for dioctahe-dral structures and 1.52 to 1.54Å for trioctahedral structures. Both kaolinite and quartz have diffraction lines in this region. The interfering diffraction line from kaolinite may be eliminated by heating the sample to 600C (873K) prior to X-ray studies. This heat treatment causes only a slight shift in the 060 diffraction spacing of vermiculites. Since most vermiculites are the al-teration product of micas, micas in the same sample with a vermiculite may be used to indicate whether the vermiculite is dioctahedral or trioctahedral. In a 10Å mica the ratio of 001/002 diffraction lines can sometimes be used to indicate the nature of the octahedral sheet and the chemical composition of the octahedral sheet (Grim et al., 1951).

Several minerals common in soil clays have basal diffraction lines in the 14-15Å region, such as vermiculites, chlorites and smectites. The interlayer water layers in vermiculite (Fig. 8-2b) are influenced by the cations in inter-layer position (Table 8-11), resulting in different basal spacings with different cations. This property is useful when one wishes to find out if a 14-15Å dif-fraction line is caused by vermiculite or some other clay mineral. When Mg-saturated, vermiculite's 002 diffraction line will be near 14.3Å, and upon K saturation this spacing collapses to 10.2Å. This process is complicated by hydroxy-Al interlayers in vermiculites. Hydroxy interlayers tend to increase the temperature at which a K-saturated vermiculite will collapse (Hsu & Bates, 1964, Carstea, 1967[5]). Aluminum-interlayered vermiculites have been reported which require a temperature of 300C to cause collapse. Potassium-saturated, hydroxy-Al-interlayered vermiculite does not collapse to 10.2Å, but rather to 11 to 11.5Å. The diffraction pattern from basal spacings of hydroxy-Al interlayered smectites is essentially the same as that from a vermiculite with similar interlayers (Pawluk, 1963; Tamura, 1958; Weed & Nelson, 1962). Hydroxy-Al interlayers may be removed by a 3- to 6-hour extraction with sodium citrate at pH 7.3 at 100C (373K) (Tamura, 1958), or by extraction with NH_4F (Rich & Obenshain, 1955). Appropriate studies can then be made to determine if the clay mineral is a vermiculite or a smectite. Special care must be used in suites that contain biotite because some of the pretreatments often used to prepare soil clays for X-ray diffraction may alter some of the biotite to vermiculite (Douglas, 1967).

Organic molecules may replace the two water layers of vermiculite (Walker, 1957). Diffractograms of clays that have been treated with polar

---

[5]D. D. Carstea. 1967. Formation and stability of Al, Fe, and Mg interlayers in mont-morillonite and vermiculite. Ph.D. Thesis. Oregon State Univ., Corvallis.

Table 8-10. X-ray diffraction data from trioctahedral vermiculite†

Joint committee (1974)			Gruner (1934)			
dÅ	Observed I/I_I	hkl	dÅ	Observed I/I_I	Theoretical intensity	hkl
14.2	100	002	14.15	10	451	002
7.14	15	004	7.07	0.3	21	004
4.76	10	006	4.72	0.5	13	006
4.57	60	020	4.61	1	55	020
4.41	10	$\bar{1}12$				
4.35 4.25	10b	022,112				
3.56	25	008	3.537	3.5	137	008
2.85	30	0.0.10	2.830	4	106	0010
			2.655		25	130
			2.655		13	$20\bar{2}$
			2.640⎫		39	$13\bar{2}$
2.615	50	$13\bar{2}$,200+	2.639⎬	1	18	200
2.570	50	132	2.581⎭		102	132
2.525	45	$20\bar{2},20\bar{4}$	2.580		53	$20\bar{4}$
			2.539⎱	1.5	302	134
			2.538⎰		151	202
2.430	5	134	2.437		28	134
2.380	35b	0.0.12				
2.365		$20\bar{6}$+	2.436		13	$20\bar{6}$
			2.379⎱	3.5	454	136
			2.378⎰		219	204
			2.358		13	0012
2.265	5	220,136+	2.254		27	$13\bar{6}$
2.20	5b	$20\bar{6},20\bar{8}$+	2.253		14	$20\bar{8}$
2.170			2.190⎱	1	214	$13\bar{8}$
			2.190⎰		103	206
2.080	5b	138	2.062⎱	1	92	138
			2.062⎰		48	$20\bar{1}0$
2.04	10b	0.0.14	2.021	1	104	0014
2.01		208				
			2.000⎱	1	57	$13\bar{1}0$
			1.999⎰		29	208
1.975	5	$1.3.\bar{1}0$	1.879		19	1310
1.82	5b	2.0.12	1.879		11	$20\bar{1}2$
1.79		$1.3.\bar{1}\bar{2}$+	1.821⎱	1	103	$13\bar{1}2$
			1.821⎰		54	2010
			1.768		6	0016
1.725	10b	1.3.12	1.710		0	1312
1.715		$31\bar{2}$+				
1.695	5	314				
1.665	15	$2.0.\bar{1}4$	1.710		0	$20\bar{1}4$
			1.659⎱	2.5	302	1314
			1.659⎰		151	2012
			1.572⎫		34	0018
			1.562⎬	0.5	69	$13\bar{1}4$
			1.561⎭		34	2016
1.543	10	2.0.14				
			1.536⎫		205	060
			1.536⎪		412	332
1.528	70	$2.0.\bar{1}6$,060	1.527⎬	4	250	330
			1.527⎪		334	$16\bar{4}$
			1.516⎪		214	$13\bar{1}6$
			1.516⎭		108	2014

(continued on next page)

Table 8-10. Continued.

Joint committee (1974)			Gruner (1934)			
dÅ	Observed I/I_I	hkl	dÅ	Observed I/I_I	Theoretical intensity	hkl
1.514	25	33$\bar{2}$,330				
1.502	15	1.3.16,33$\bar{4}$				
			1.501⎱ 1.501⎰	0.5	35 36	332 33$\bar{6}$
			1.431⎱ 1.431⎰	1.5	405 203	131$\bar{6}$ 201$\bar{8}$
			1.415	1	62	0025
			1.392		0	1318
			1.392		0	2016
			1.350⎱ 1.350⎰	0.5	67 61	338 33$\overline{12}$
			1.317⎱ 1.317⎰	2.5	405 204	131$\bar{8}$ 20$\bar{2}$0
			1.286	1	58	0022

† Random particle orientation.

Table 8-11. Basal spacings (Å) of air dried trioctahedral vermiculite saturated with various cations (after Barshad, 1948)

Saturating cation	Mg	Ca	Ba	Li	Na	K	NH$_4$	Rb	Cs
South Africa vermiculite	14.33	15.07			12.56	10.42	11.14		
North Carolina vermiculite	14.33	15.07	12.56	12.56	12.56	10.42	11.24	11.24	11.97

organics are used to differentiate between vermiculites and smectites. Considerable controversy existed regarding the relationship of vermiculites and smectites. Several investigators (Weiss et al., 1955; Hoffmann et al., 1956; Walker, 1957, 1958), suggested that trioctahedral vermiculites and smectites comprise a continuous series based on chemical composition and charge characteristics. Harward et al. (1969) showed that vermiculites would accommodate only one interlayer of either ethylene glycol or glycerol. The basal spacing of the organic-vermiculite complex was sensitive to both the way the organic was applied and the interlayer exchange cation (Table 8-12). Harward et al. (1969) also observed a series of organic-vermiculite complexes which produced much weaker diffraction lines. However, they found no consistent evidence for more than one layer of the organic in interlayer space in trioctahedral vermiculite. Heat treatments are not needed for differentiation of vermiculites and smectites. The identification of vermiculites and smectites is usually based on the 14Å diffraction line of vermiculite, and the 17-18Å diffraction line of smectites, when ethylene glycol or glycerol is added to a Mg-saturated clay.

Table 8-12. Basal spacings of trioctahedral vermiculites upon solvation with glycerol or ethylene glycol (after Harward et al., 1969)

Treatment	Ca saturated dÅ	Mg saturated dÅ
H_2O at 54% relative humidity	14.82	14.29
Glycerol vapor	13.62	14.11
Glycerol liquid	14.33	14.17
Ethylene glycol vapor	15.01	14.22
Ethylene glycol liquid	15.29	14.25
Glycerol vapor, 200C (473K)	14.04	--
Glycerol liquid, 200C (473K), 4 hours	14.07	14.08
Ethylene glycol, liquid, 150C (423K), 4 hours	13.5	

Vermiculite can be distinguished from chlorite by vermiculite's decrease in basal spacing when potassium saturated, sometimes with 300C (573K) heat treatment. The basal spacing of chlorite is stable to above 500C (773K).

B. Infrared Analysis

A definitive study of the infrared spectra of vermiculite has not been attempted. Much work has been completed relating absorption bands to structural features (Russell et al., 1970; Vedder & Wilkins, 1969) that would facilitate such an investigation.

Infrared spectra have enabled the unravelling and understanding of much of the vermiculite structure and its reactions. Bassett (1960) utilized infrared spectra to explain the differences in weathering rates of biotite and muscovite, and the structural reason why K is held more tightly by muscovite than by biotite, or more tightly by dioctahedral vermiculite than by trioctahedral vermiculite. Farmer et al. (1971) have used infrared absorption spectra to differentiate between hydroxyls coordinated with $R^{2+}R^{2+}$, $R^{2+}R^{3+}$, and $R^{3+}R^{3+}$ octahedral pairs in vermiculites.

C. Thermal Analysis

Differential thermograms show the effects of saturating cations on interlayer water in trioctahedral vermiculite (Fig. 8-5). Natural Mg-saturated and Ca-saturated trioctahedral vermiculites have two low-temperature endothermic peaks, one near 150C (423K) and another near 240C (513K), representing loss of water. The 150C (423K) endothermic peak accounts for the loss of 10 moles H_2O per absorbed cation. This water is interlayer water not directly coordinated to an exchange cation. Two moles of water per cation are lost between 150C (423K) and 250C (523K) and one mole of water per cation is lost between 250C (523K) and 500C (773K).

Barium-, Li-, and Na-saturated trioctahedral vermiculites have a single

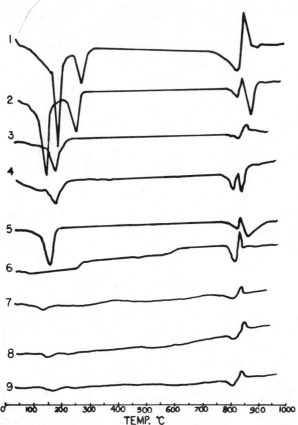

Fig. 8–5. Differential thermal curves of trioctahedral vermiculite. *(1)* Natural, *(2)* Calcium saturated, *(3)* Barium saturated, *(4)* Magnesium saturated, *(5)* Lithium saturated, *(6)* Ammonium saturated, *(7)* Potassium saturated, *(8)* Rubidium saturated, *(9)* Cesium saturated (after Barshad, 1968).

low-temperature endothermic peak, near 140–160C (413–433K) (Barshad, 1948). Barium-saturated vermiculite loses 6 moles of water per Ba cation in this endothermic range, and an additional two moles when heated to 600C (873K). Lithium- and Na-saturated vermiculites lose 3 moles of water per interlayer cation in the low endothermic range, and one additional mole between 150C (423K) and 500C (773K). The total water loss for K-, Rb-, and Cs-saturated vermiculites, below 600C (873K) was one mole per adsorbed ion. The 800C (1073K) endothermic peak corresponds to the loss of structural hydroxyls, and the 840C (1113K) exothermic peak represents the formation of a new phase probably enstatite (Gaudette, 1964), but there may be much variation in the products formed by this high temperature phase change.

Walker (1955) has described the changes of state of the interlayer water layers of a Mg-saturated trioctahedral vermiculite during dehydration. When Mg-vermiculite is placed in water there is a gradual increase in the basal spacing to 14.81Å. Essentially all interlayer water sites are occupied in the 14.81Å phase. In this phase the water molecules adopt a regular hexagonal pattern in which each water molecule site is located directly over an octahedral site of an adjacent silicate layer. Upon dehydration there is a gradual loss of water and shift to the 14.36Å phase (Fig. 8-1). Further loss of water causes an abrupt contraction to 13.82Å. This 13.82Å phase still has two layers of water. In this phase the interlayer cations have migrated to sites near the silicate layer surfaces. The further removal of water results in an abrupt contraction to 11.59Å. This phase has a single complete sheet of water between silicate layers. Removal of water molecules from the 11.59Å phase eventually causes a contraction to a 9.02Å phase in which there is no interlayer water.

It should be noted that dehydration of a K-saturated vermiculite is thought to be irreversible, while a dehydrated Mg-saturated vermiculite will absorb water and reexpand to the 14Å phase.

D. Electron Microscopy

Electron microscopy has not proven to be a particularly useful method of studying soil vermiculites. Kishk and Barshad (1969) showed that the mode of formation of vermiculite affected its morphology (Fig. 8-6). Clay vermiculites formed by the alteration of micas inherit their form from the parent minerals, and appear as large transluscent irregular polygonal sheets with some of the edges frayed; however, a large proportion of these particles appears to be thin pseudohexagonal flakes. Vermiculite formed by synthesis from gels are mostly small hexagonally shaped sheets resembling kaolinite. Since soil-clay dioctahedral vermiculite is often associated with kaolinite, electron-microscopic analysis may not differentiate between soil kaolinite and soil vermiculite.

Transmission electron microscopy, utilizing replica techniques, has been used to study the surface morphology of trioctahedral vermiculites (Raman & Jackson, 1964). Whereas micas usually have very smooth surfaces, Raman and Jackson (1964) found small humps, prominent crystallographic steps on the basal cleavage planes, marginal rolling of the layers and layer buckling. The surface morphology of these trioctahedral vermiculites was related to interlayer cation species, and the hydration of interlyear cations. Scanning electron microscopy has been used (Jackson & Sridhar, 1974; Sridhar & Jackson, 1973) to study the expulsion of saponite from trioctahedral vermiculite in the weathering sequence: biotite → vermiculite → saponite.

Modern electron microscopes utilizing nondispersive X-ray fluorescence analysis attachments should offer an excellent tool for the study of the micromorphology of vermiculite associated with mineral alteration. Sridhar and Jackson (1973) utilized comparable equipment on a microprobe to study the distribution of Na, Cs, Rb, NH_4, and K in vermiculites.

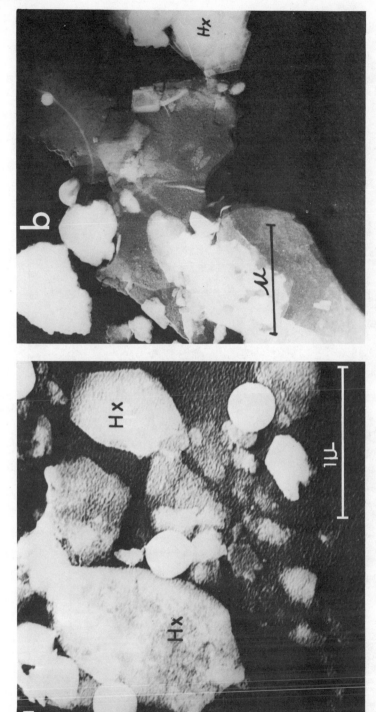

Fig. 8–6. Electron micrographs of vermiculite. *(Page 281)* Vermiculite, illite and kaolinite formed by alteration of muscovite and biotite. *(Page 282)* Vermiculite and montmorillonite formed by synthesis. *(Page 283)* Vermiculite formed by synthesis. Note the similarity to kaolinite (Micrographs, pages 281 and 282, from Kishk & Barshad, 1969. Micrographs, page 283, courtesy of Dr. Barshad).

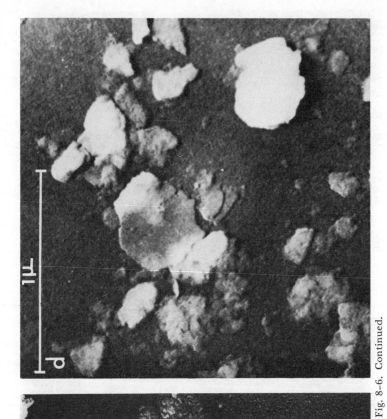

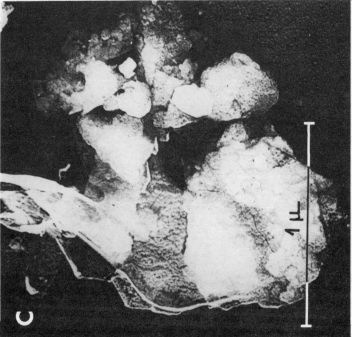

Fig. 8-6. Continued.

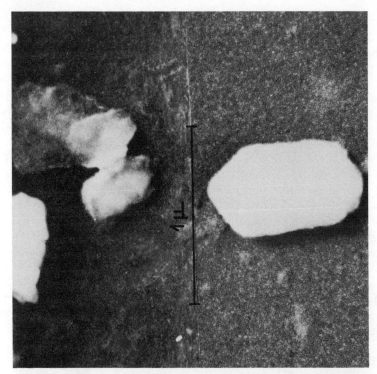

Fig. 8–6. Continued.

E. Quantitative Determination of Vermiculite

Quantitative analysis of soil-clay minerals is often considered more of an art than a science. In many soil genesis studies only the relative amount of a mineral in each of the several horizons of a soil profile needs to be determined. Peak heights or peak areas from X-ray diffractograms may provide adequate information (Johnson et al., 1963; Norrish & Taylor, 1962). X-ray diffraction is not suited for some soil-clay quantitative studies because of the variability of structure factors in these materials (Douglas, 1973).

Vermiculite in soils is determined by measuring the loss of cation exchange (CEC) capacity when a soil is K-saturated (Alexiades & Jackson, 1965). Cation exchange capacity is determined by washing with $CaCl_2$ solutions and replacement of the Ca with $MgCl_2$ solutions, after which the Ca CEC is determined. The sample is then washed with KCl solutions and dried overnight at 110C (283K). Exchangeable K is replaced with NH_4Cl solutions and is then determined. The amount of vermiculite in the sample is calculated from the difference in the two CEC values, assuming the vermiculite CEC assigned to interlayer positions is 154 meq/100 g. Coffman and Fanning (1974) have used this method to determine the total vermiculite in soils, not limiting the determination to clay-size vermiculite. This method for the quantitative determination of vermiculite is dependent on the CEC associated with innerlayer surfaces. Hydroxy-Al interlayers reduce the charge associated with these surfaces (Barnhisel, 1965).[6] In some cases it may be advisable to remove hydroxy-Al interlayers (Rich, 1966; Rich & Obenshain, 1955; Tamura, 1958) before determining the amount of vermiculite in the sample.

Rich (1966) has developed a technique that may be useful in some semiquantitative studies. A fluoride solution ($0.4N$ NH_4F, $0.1N$ HCL, $1N$ NH_4Cl) is used to dissolve clay-size allophane, halloysite, kaolinite, Mg-rich montmorillonite, biotite, and trioctahedral vermiculite. This treatment concentrates dioctahedral mica and dioctahedral vermiculite; it also removes hydroxy-Al interlayers from the vermiculite. A small amount of the dioctahedral mica and vermiculite is also dissolved by this process.

VI. ION EXCHANGE

A. Cation-Exchange Capacity

The cation-exchange capacities of several soil vermiculites are shown in Tables 8-5 and 8-6. Alexiades and Jackson (1965) arrived at the often quoted average value of 159 meq/100 g for the CEC of trioctahedral vermiculite. They prorated this CEC to 154 meq/100 g for interlyaer sites and 5 meq/100 g on external surfaces. The variation in the cation-exchange capaci-

---

[6]R. I. Barnhisel. 1965. The formation and stability of aluminum interlayers in clays. Ph.D. Thesis. Virginia Polytechnic Instit., Blacksburg.

ties of trioctahedral vermiculites is directly attributable to variations in total charge between different vermiculite specimens.

The CEC of dioctahedral vermiculite has not been extensively investigated. Barshad and Kishk (1969) calculated that in their dioctahedral vermiculites the CEC varied between 144 and 207 meq/100 g (on an ignited weight basis). If one assumes that the densities of vermiculites are proportional to the densities of the micas from which the vermiculite is derived, it is possible to estimate the CEC of dioctahedral vermiculite from the specific gravity of biotite (3.02) and muscovite (2.885). The CEC of dioctahedral vermiculite should approximate 3.02/2.885, or 1.05 times the CEC of trioctahedral vermiculite. However, Cook and Rich (1963) concluded that "dioctahedral micas apparently lose no negative charge on expansion to vermiculite-like minerals." This total charge should result in a CEC of dioctahedral vermiculite near 250 meq/100 g.

Vermiculites include the minerals with the largest CEC of the mineral fraction of soils; however, the presence of hydroxy-Al interlayers may considerably reduce the effective CEC. In mixtures of clay minerals in soils small amounts of minerals with large CEC can dominate the exchange properties of the soil and the accompanying uptake of nutrients by plants (Brown, 1955).

B. Cation Exchange

Interlayer water layers facilitate both the easy migration of cations in interlayer space and the ease of ion exchange. In ion exchange reactions involving vermiculites and ion pairs from Ca, Mg, Sr, Ba, Na, or Li, complete exchange of the sorbed ion by the counter ion is easily accomplished. In sand size vermiculite, when K, NH_4, Rb, or Cs is the displacing ion in such an exchange, some of the sorbed ions are not replaced (Frysinger, 1960). As this latter type of exchange reaction starts, the displacing ion replaces ions near the rim of the vermiculite particle, causing contraction parallel to the Z axis, displacing the interlayer water layers in the zone where exchange has taken place. This contraction isolates small expanded regions from which cations cannot migrate because of the surrounding contracted zones. The mineral then has isolated expanded and isolated contracted zones. As particle size decreases cation exchange becomes more complete, especially in acid soils where hydroxy-Al islands in interlayer space props the interlayer open.

C. Ion Fixation

The selective sorption or fixation of K, NH_4, Rb, or Cs has been observed by many investigators (Krishnomoorthy & Overstreet, 1950; Wilklander, 1950; Page et al., 1963, 1967). Early investigators thought this fixation was caused by the close fit of the cation within the hexagonal cavities in tetrahedral sheets (Barshad, 1948, 1950; Wear & White, 1951) of silicate layers. It is now recognized that the low hydration energy of the cation is the major factor in cation selectivity and fixation (Kittrick, 1966; Norrish, 1954; Rich, 1964; Shainberg & Kemper, 1966). Cations with low hydration energies, such as K, NH_4, Rb, or Cs, cause interlayer dehydration and are therefore, fixed in interlayer positions. Conversely, cations with high hydration energy, such as Ca, Mg, and Sr, produce hydrated, expanded interlayers and are readily available to enter into exchange reactions.

Because of interest in radioactive fallout and in the disposal of radioactive effluents, much of the work developing models of selective sorption has utilized cesium. The model that has been developed describes a general process involving the interaction of cations with low hydration energy and vermiculite. Jacobs and Tamura (1960) found that prior treatment affected the amount of Cs that would be sorbed by a vermiculite, and that at low Cs concentrations more Cs was sorbed by a vermiculite that had collapsed than by a Na-saturated vermiculite. That observation lead to a line of work that eventually found that expanded trioctahedral vermiculite has two types of selective sorption: (i) one with a high order selectivity, and (ii) one with a low order selectivity. One type of selectivity was found to be associated with microscopic zones of "mica" within the vermiculite, local zones within the vermiculite that have collapsed, leading to the "frayed edge" (Sawhney, 1972) or "wedge-shaped interlayers" (Rich, 1964) theories. Rich (1964) found that selective sorption and desorption in vermiculite is a function of cation radius and thickness of wedge-shaped zones adjacent to the mica-cores in vermiculite (Fig. 8-7). Potassium ions located near the apex of the wedge were displaced in increasing amounts, with Mg less than Ba, and Ba less than Ag. In Fig. 8-7, the position of these ions in the wedge is shown. The smaller ions, which could migrate into the apex, were the most efficient as displacing cations. Conversely, the apex of the wedge is the position where bonds may be satisfied over the shortest distance. These positions are the positions where cations with low hydration energies are selectively absorbed.

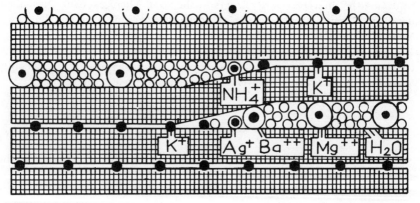

Fig. 8-7. A wedge-shaped vermiculite zone in mica showing the "fit" of several cations in the wedge (after Rich, 1964).

In these exchange reactions the selectivity coefficient probably varies with pH, K concentration, and counter-ion concentration (Rich, 1964).

Hydrogen ions are very effective as counter ions migrating into wedge-zones and replacing K. Much of the exchange in vermiculites observed when $MgCl_2$ solutions have been used as counter-ion sources has actually been a pH, hydrogen ion, effect caused by hydrolysis (Rich, 1964; Rich & Black, 1964). The second type of selectivity is the result of the interaction of cations in interlayer spaces with the tetrahedral sheet of the vermiculite structure. Adjacent tetrahedral sheets include pseudohexagonal holes into which unhydrated ions can migrate. This enables the charges that are satisfied by innerlayer cations, to be satisfied over relatively short distances. A second interaction in this fixing phenomenon is the relative proportion of dioctahedral and trioctahedral sites in trioctahedral vermiculite. Gilkes et al. (1973) found that as Fe is oxidized in the biotite → vermiculite alteration, octahedral cation vacancies are generated causing trioctahedral sites to become more dioctahedral, with appropriate tilting of structural hydroxyls. The availability of structural K to plants from biotite and trioctahedral vermiculite will be determined by the extent of oxidation of octahedral Fe.

Sawhney (1970) found that ion selectivity in many clay minerals is concentration dependent. At low concentrations of K, wedge sites are more selective for K from a K/Ca solution than plainer sites, however, at higher concentrations plainer (interlayer) sites were more effective in fixing K.

Cation fixation in dioctahedral vermiculites is not as well understood as fixation in trioctahedral structures. Fixation by dioctahedral materials has been studied (Dolcater et al., 1972; Rich, 1964) enough to give us confidence in carefully extrapolating much that has been learned from trioctahedral structures to dioctahedral structures.

It should be noted that while wedge-zones are zones of high selectivity, they comprise a relatively small volume in most vermiculites. In vermiculitic soils these wedge-zones influence the fixation of K. Fixation is primarily due to plainer surfaces.

Fixation must be guarded against in the determination of cation-exchange capacity. An ion (K^+, NH_4^+, Cs^+, etc.) which tends to be fixed by vermiculites should not be used in procedures where it will eventually have to be displaced from the mineral. Exchange determinations requiring a high degree of accuracy are difficult because very small ($<$ ppm levels) amounts of K^+ may confound the reaction. In these cases reagents must be carefully screened to be sure they are K^+ free.

VII. THE SOIL-CLAY-VERMICULITE PROBLEM

Our knowledge of the reactions of clay-size vermiculites has been obtained by studying coarser fractions, often specimen-type minerals, and assuming that soil clays are similar and react the same. However, there is a small amount of evidence that indicates that these assumptions are not true.

The CEC of clay-size Libby vermiculite is less than the exchange capacity of coarser separates (Baweja, 1974). A vermiculite regolith sample from Llano County, Texas, was fractionated into several size fractions. The vermiculite in the coarser fractions was trioctahedral while that in the finest fractions was dioctahedral (Kerns & Mankin, 1967). Chemical composition and CEC of these crystallites were shown to be directly correlated with particle size. Kerns and Mankin (1967) offered two possible explanations for these observations: (i) The heterogeneity may represent two phases inherited from the parent rock; and (ii) the dioctahedral material was an alteration product of the trioctahedral vermiculite.

The status of soil-clay vermiculite is further complicated by the almost universal agreement that all vermiculite is a mica alteration product, in spite of reports of vermiculites that are not mica alteration products and new mineral species in soils.

Most soil-clay vermiculite is probably dioctahedral, with properties assigned to it by extrapolation from coarser trioctahedral vermiculite or from K-depleted muscovite. Such assumptions must be used with great care.

VIII. SUPPLEMENTARY READING

Barshad, I., and F. M. Kish. 1969. Chemical composition of soil vermiculite clays as related to their genesis. Contrib. Mineral. Petrol. 24:136–155.

Rich, C. I. 1968. Mineralogy of soil potassium. p. 79–108. In V. J. Kilmer et al. (eds.) The role of potassium in agriculture. Am. Soc. of Agron.

Robert, M. 1971. Étude éxperimentale de l'évolution des micas (biotites)—I. Les aspects du process de vermiculitisation. Ann. Agron. 22:43–93.

Robert, M. 1971. Étude éxperimentale de l'évolution des micas (biotites). II. Les autres possibilités d'evolution des micas et leur place par rapport a la vermiculitisation. Ann. Agron. 22:155–181.

Walker, G. F. 1975. Vermiculite. p. 155–190. In John E. Gieseking (ed.) Soil components. Vol. 2, Inorganic components. Springer-Verlag, New York.

LITERATURE CITED

Alexander, L. T., S. B. Hendricks, and R. A. Nelson. 1939. Minerals present in soil colloids: II. Estimation in some representative soils. Soil Sci. 48:273–279.

Alexiades, C. A., and M. L. Jackson. 1965. Quantitative determination of vermiculite in soils. Soil Sci. Soc. Am. Proc. 29:522–527.

Alexiades, C. A., N. A. Polyzopoulos, N. A. Koroxenides, and G. S. Axaris. 1973. High trioctahedral vermiculite content in the sand, silt, and clay fractions of a Gray Brown Podzolic soil in Greece. Soil Sci. 116:363–375.

Bailey, S. W. 1966. The status of clay mineral structures. Clays Clay Miner. 14:1–23.

Barshad, I. 1948. Vermiculite and its relation to biotite as revealed by base exchange reactions, X-ray analysis, differential thermal curves and water content. Am. Mineral. 33:655–678.

Barshad, I. 1950. The effect of the interlayer cations on the expansion of the mica type of crystal lattice. Am. Mineral. 35:225–238.

Barshad, I., and F. M. Kishk. 1969. Chemical composition of soil vermiculite clays as related to their genesis. Contrib. Mineral. Petrol. 24:136–155.

Bassett, W. A. 1959. The origin of vermiculite deposits at Libby, Montana. Am. Mineral. 44:282–299.

Bassett, W. A. 1960. Role of hydroxyl orientation in mica alteration. Bull. Geol. Soc. Am. 71:449–456.

Bassett, W. A. 1963. The geology of vermiculite occurrences. Clays Clay Miner. 10:61–69.

Baweja, A. S., L. P. Wilding, and E. O. McLean. 1974. Mineralogy and cation exchange properties of Libby vermiculite separates as affected by particle-size reduction. Clays Clay Miner. 22:253–262.

Borchardt, G. A., M. L. Jackson, and F. D. Hole. 1966. Expansible layer silicate genesis in soils deplicted in mica pseudomorphs. p. 175–185. In L. Heller and A. Weiss (eds.) Proc. Int. Clay Conf., 1966, Vol. 1, Israel Program for Scientific Translations, Jerusalem.

Brown, B. E., and M. L. Jackson. 1958. Clay mineral distribution in the Hiawatha sandy soils of northern Wisconsin. Clays Clay Miner. 5:213–226.

Brown, D. A. 1955. Ion exchange in soil-plant root environments: II. The effect of type of clay mineral upon nutrient uptake by plants. Soil Sci. Soc. Am. Proc. 19:296–300.

Brown, G. 1953. The dioctahedral analogue of vermiculite. Clay Miner. Bull. 2:64–70.

Bryant, J. P., and J. B. Dixon. 1964. Clay mineralogy and weathering of a Red-Yellow Podzolic soil from quartz mica schist in the Alabama Piedmont. Clays Clay Miner. 12:509–521.

Calhoun, F. G., V. W. Carlisle, and C. Luna Z. 1972. Properties and genesis of selected Columbian Andosols. Soil Sci. Am. Proc. 36:480–485.

Carlisle, V. W., and L. W. Zelazny. 1973. Mineralogy of selected Florida Paleudults. Soil Sci. Soc. Fla. Proc. 33:136–139.

Chute, J. H., and J. P. Quirk. 1967. Diffusion of potassium from mica-like clay minerals. Nature (London) 213:1156–1157.

Coffman, C. B., and D. S. Fanning. 1974. "Vermiculite" determination on whole soils by cation exchange capacity methods. Clays Clay Miner. 22:271–283.

Coffman, C. B., and D. S. Fanning. 1975. Maryland soils developed in residium from chloritic metabasalt having high amounts of vermiculite in sand and silt fractions. Soil Sci. Soc. Am. Proc. 39:723–732.

Cook, M. G., and C. I. Rich. 1963. Negative charge of dioctahedral micas as related to weathering. Clays Clay Miner. 11:47–65.

Cortes, A., and D. P. Franzmeier. 1972. Weathering of primary minerals in volcanic ash-derived soils of the central cordillera of Columbia. Geoderma 8:165–176.

Dolcater, D. L., M. L. Jackson, and J. K. Syers. 1972. Cation exchange selectivity in mica and vermiculite. Am. Miner. 57:1828–1831.

Douglas, Lowell A. 1965. Clay mineralogy of a Sassafras soil in New Jersey. Soil Sci. Soc. Am. Proc. 29:163–167.

Douglas, Lowell A. 1967. Sodium-citrate-dithionite-induced alteration of biotite. Soil Sci. 103:191–195.

Douglas, Lowell A. 1973. Factors limiting the use of standard minerals in the X-ray diffraction analysis of clays. p. 88–96. In C. L. Grant et al. (eds.) Advances in X-ray analysis, Vol. 17, Plenum Press, New York.

Farmer, V. C., J. D. Russell, W. J. McHardy, A. C. D. Newman, J. L. Ahlrichs, and J. Y. H. Rimsaite. 1971. Evidence of loss of protons and octahedral iron from oxidized biotites and vermiculites. Mineral. Mag. 38:121–137.

Foster, M. D. 1963. Interpretation of the composition of vermiculites and hydrobiotites. Clays Clay Miner. 10:70–89.

Frysinger, G. R. 1960. Cation exchange behavior of vermiculite-biotite mixtures. Clays Clay Miner. 8:116–121.

Gaudette, H. E. 1964. Magnesium vermiculite from the Twin Sisters Mountains, Washington. Am. Mineral. 49:1754–1763.

Gilkes, R. J., R. C. Young, and J. P. Quirk. 1973. Artificial weathering of oxidized biotite: I. Potassium removal by sodium chloride and sodium tetraphenylboron solutions. Soil Sci. Soc. Am. Proc. 37:25–28.

Gjems, O. 1963. A swelling dioctahedral clay mineral of a vermiculite-smectite type in the weathering horizons of Podzols. Clay Min. Bull. 5:183–193.

Grim, R. E., W. F. Bradley, and G. Brown. 1951. The mica clay minerals. p. 138–172. In G. Brindley (ed.) X-ray identification and structures of clay minerals. Mineralogical Society of Great Britain. London.

Grim, R. E. 1968. Clay mineralogy. 2nd ed. McGraw-Hill Book Co., New York. 596 p.

Gruner, J. W. 1934. Vermiculite and hydrobiotite structures. Am. Mineral. 19:557-575.

Guven, Necip, and Paul F. Kerr. 1966. Selected Great Basin playa clays. Am. Mineral. 51:1056-1067.

Harward, M. E., and G. W. Brindley. 1965. Swelling properties of synthetic smectites in relation to lattice substitutions. Clays Clay Miner. 12:209-222.

Harward, M. E., D. D. Carstea, and A. H. Sayegh. 1969. Properties of vermiculites and smectites: Expansion and collapse. Clays Clay Miner. 16:437-477.

Hathaway, J. C. 1955. Studies of some vermiculite-type clay minerals. Clays Clay Miner. 3:74-86.

Hendricks, S. B., and M. E. Jefferson. 1938. Crystal structure of vermiculites and mixed vermiculites-chlorites. Am. Mineral. 23:851-863.

Hoffmann, U., A. Weiss, G. Koch, A. Mehler, and A. Scholz. 1956. Intracrystalline swelling, cation exchange, and anion exchange of minerals of the montmorillonite group and of kaolinite. Clays Clay Miner. 4:273-287.

Hsu, P. H., and T. F. Bates. 1964. Fixation of hydroxy-aluminum polymers by vermiculite. Soil Sci. Soc. Am. Proc. 28:763-769.

Huang, P. M., and S. Y. Lee. 1969. Effect of drainage on weathering transformations of mineral colloids of some Canadian prairie soils. Proc. Int. Clay Conf. (Tokyo). Vol. 1:541-551.

Ismail, F. T. 1969. Role of ferrous iron oxidation in the alteration of biotite and its effect on the type of clay minerals formed in soils of arid and humid regions. Am. Mineral. 54:1460-1466.

Jackson, M. L. 1959. Frequency distribution of clay minerals in major great soil groups as related to the factors of soil formation. Clays Clay Miner. 6:133-142.

Jackson, M. L. 1963. Interlayering of expansible layer silicates in soils by chemical weathering. Clays Clay Miner. 11:29-46.

Jackson, M. L., and K. Sridhar. 1974. Scanning electron microscopy and x-ray diffraction study of natural weathering of phlogopite through vermiculite to saponite. Soil Sci. Soc. Am. Proc. 38:843-851.

Jacobs, D. C., and T. Tamura. 1960. The mechanism of ion fixation using radio-isotope techniques. Int. Congr. Soil Sci., Trans. 7th (Madison, Wis.) II:206-214.

Johnson, L. J. 1964. Occurrence of regularly interstratified chlorite-vermiculite as a weathering product of chlorite in a soil. Am. Miner. 49:556-572.

Johnson, L. J., R. P. Matelski, and C. F. Engle. 1963. Clay mineral characterization of modal soil profiles in several Pennsylvania counties. Soil Sci. Soc. Am. Proc. 27:568-572.

Joint Committee on Powder Diffraction Standards. 1974. Selected powder diffraction data for minerals. Data book. Swarthmore, Pennsylvania.

Jones, L. H. P., A. A. Milne, and P. M. Attiwel. 1964. Dioctahedral vermiculite and chlorite in highly weathered Red Loams in Victoria, Australia. Soil Sci. Soc. Am. Proc. 28:108-113.

Kapoor, B. S. 1972. Weathering of micaceous clays in some Norwegion Podzols. Clay Miner. 9:383-394.

Kato, Yoshiro. 1965. Mineralogical study of weathering products of granodiorite at Shinshiro City (IV) Mineralogical compositions of silt and clay fractions. Soil Sci. Plant Nutr. Tokyo 11(2):16-27.

Kelley, W. P., W. H. Dove, A. O. Woodford, and S. M. Brown. 1939. The colloidal constituents of California soils. Soil Sci. 48:201-255.

Kerns, R. L., Jr., and C. J. Mankin. 1967. Compositional variation of a vermiculite as related to particle size. Clays Clay Miner. 15:163-177.

Kirkland, D. L., and B. F. Hajek. 1972. Formula derivation of Al-interlayered vermiculite in selected soil clays. Soil Sci. 114:317-322.

Kishk, F. M., and I. Barshad. 1969. The morphology of vermiculite clay particles as affected by their genesis. Am. Mineral. 54:849-857.

Kittrick, J. A. 1966. Forces involved in ion fixation by vermiculite. Soil Sci. Soc. Am. Proc. 30:801-803.

Kittrick, J. A. 1973. Mica-derived vermiculites as unstable intermediates. Clays Clay Miner. 21:479-488.

Klages, M. G., and J. L. White. 1957. A chlorite-like mineral in Indiana soils. Soil Sci. Soc. Am. Proc. 21:16-20.

Kodama, H., and J. E. Brydon. 1968. A study of clay minerals in podzol soils in New Brunswick, Eastern Canada. Clay Miner. 7:295-309.

Krebs, R. D., and J. C. F. Tedrow. 1958. Genesis of Red-Yellow Podzolic and related soils in New Jersey. Soil Sci. 85:28-37.

Krishnamoorthy, C., and R. Overstreet. 1950. An experimental evaluation of ion exchange relationships. Soil Sci. 69:41-53.

MacEwan, D. M. C. 1948. Les mineraux argileux de quelques sols ecossais. Verres Silic. Ind. 13:41-46.

Mathieson, A. McL. 1958. Mg-vermiculite: a refinement and re-examination of the crystal structure of the 14.36Å phase. Am. Mineral. 43:216-227.

Mathieson, A. McL., and G. F. Walker. 1954. Crystal structure of magnesium-vermiculite. Am. Mineral. 39:231-255.

Norrish, K. 1954. The swelling of montmorillonite. Discuss. Faraday Soc. 18:120-134.

Norrish, K., and R. M. Taylor. 1962. Quantitative analysis by X-ray diffraction. Clay Miner. Bull. 5:98-109.

Page, A. L., F. T. Bingham, T. J. Ganje, and M. J. Garber. 1963. Availability and fixation of added potassium in two California soils when cropped to cotton. Soil Sci. Soc. Am. Proc. 27:323-326.

Page, A. L., W. D. Burge, T. J. Ganje, and M. J. Garber. 1967. Potassium and ammonium fixation in vermiculite soils. Soil Sci. Soc. Am. Proc. 31:337-341.

Pawluk, S. 1963. Characteristics of 14Å clay minerals in the B horizons of podzolized soils of Alberta. Clays Clay Miner. 13:74-82.

Radoslovich, E. W. 1962. The cell dimensions and symmetry of layer-lattice silicates: II. Regression relations. Am. Mineral. 46:617-636.

Raman, K. V. Venkata, and M. L. Jackson. 1964. Vermiculite surface morphology. Clays Clay Miner. 12:423-429.

Reed, M. G., and R. D. Scott. 1962. Kinetics of potassium release from biotite and muscovite in sodium tetraphenylborn solutions. Soil Sci. Soc. Am. Proc. 26:437-440.

Rhoades, J. D. 1967. Cation exchange reactions of soil and specimen vermiculites. Soil Sci. Soc. Am. Proc. 31:361-365.

Rich, C. I. 1958. Muscovite weathering in a soil developed in the Virginia Piedmont. Clays Clay Miner. 5:203-212.

Rich, C. I. 1964. Effect of cation size and pH on potassium exchange in Nason soil. Soil Sci. 98:100-105.

Rich, C. I. 1966. Concentration of dioctahedral mica and vermiculite using a fluoride solution. Clays Clay Miner. 14:91-98.

Rich, C. I., and W. R. Black. 1964. Potassium exchange as affected by cation size, pH, and mineral structure. Soil Sci. 97:384-390.

Rich, C. I., and S. S. Obenshain. 1955. Chemical and clay mineral properties of a Red-Yellow Podzolic soil derived from muscovite schist. Soil Sci. Soc. Am. Proc. 19: 334-339.

Robert, M. 1971. Étude expérimentale de l'évolution des micas (biotites)—I. Les aspects du processus de vermiculitisation: Ann. Agron. 22:43-93.

Robert, M. 1973. The experimental transformation of mica toward smectite; relative importance of total charge and tetrahedral substitution. Clays Clay Miner. 21:167-174.

Ross, G. J., and M. M. Mortland. 1966. A soil beidellite. Soil Sci. Soc. Am. Proc. 30: 337-343.

Roth, C. B., M. L. Jackson, E. G. Lotse, and J. K. Syers. 1968. Ferrous-ferric ratio and CEC changes on deferration of weathered micaceous vermiculite. Israel Chem. 6: 261-273.

Roux, F. H. le, J. G. Cady, and N. T. Coleman. 1963. Mineralogy of soil separates and alkali-ion exchange-sorption. Soil Sci. Soc. Am. Proc. 27:534-538.

Roy, Rustum, and L. A. Romo. 1957. Weathering studies. 1. New data on vermiculite. J. Geol. 65:603-610.

Russell, J. D., V. C. Farmer, and B. Velde. 1970. Replacement of OH by OD in layer silicates, and identification of the vibrations of these groups in infrared spectra. Mineral. Mag. 37:869-879.

Sawhney, B. L. 1969. Cesium uptake by layer silicates: effect on interlayer collapse and cation-exchange capacity. Int. Clay Conf. Proc. (Tokyo) Vol. 1:605-611.

Sawhney, S. L. 1970. Potassium and cesium ion selectivity in relation to clay mineral structure. Clays Clay Miner. 18:47–52.

Sawhney, S. L. 1972. Selective sorption and fixation of cations by clay minerals: a review. Clays Clay Miner. 20:93–100.

Scott, A. D., and M. G. Reed. 1962. Chemical extraction of potassium from soils and micaceous minerals with solutions containing sodium tetraphenylboron. II. Biotite. Soil Sci. Soc. Am. Proc. 26:41–45.

Shainberg, I., and W. D. Kemper. 1966. Hydration status of adsorbed cations: Soil Sci. Soc. Am. Proc. 30:707–713.

Shirozu, H., and S. W. Bailey. 1966. Crystal structure of a two-layer Mg-vermiculite. Am. Mineral. 51:1124–1143.

Smith, W. W. 1962. Weathering of some Scottish basic igneous rocks with reference to soil formation. J. Soil Sci. 13:202–215.

Sokolova, T. A., and R. V. Shostak. 1969. Weathering of dioctahedral mica in a podzolic soil. Soviet Soil Sci. no. 6, p. 719–728. (Translated from Pochvovedeniye, 1969, no. 11, p. 106–115).

Sridhar, K., and M. L. Jackson. 1973. Fixing cation interaction with blister-like osmotic swelling on vermiculite cleavages. Clays Clay Miner. 21:369–377.

Tamura, T. 1958. Identification of clay minerals from acid soils. J. Soil Sci. 9:141–147.

Taylor, G. L., A. P. Ruotsala, and R. O. Keating, Jr. 1968. Analysis of iron in layer silicates by Mossbauer spectroscopy. Clays Clay Miner. 16:381–391.

Tedrow, J. C. F., and L. A. Douglas. 1964. Soil investigations on Banks Island. Soil Sci. 98:53–65.

Vedder, W. 1964. Correlations between infrared spectrum and chemical composition of mica. Am. Mineral. 49:736–768.

Vedder, W., and R. W. T. Wilkins. 1969. Dehydroxylation and rehydroxylation, oxidation and reduction of micas. Am. Mineral. 54:482–509.

Veith, J. A., and M. L. Jackson. 1974. Iron oxidation and reduction effects on structural hydroxyl and layer charge in a aqueous suspensions of micaceous vermiculites. Clays Clay Miner. 22:345–353.

Veith, L. G., and E. W. Radoslovich. 1963. The cell dimensions and symmetry of layer-lattice silicates. III. Octahedral ordering. Am. Mineral. 48:62–75.

Walker, G. F. 1955. The mechanism of dehydration of Mg-vermiculite. Clays Clay Miner. 4:101–115.

Walker, G. F. 1957. On the differentiation of vermiculite and smectites in clays. Clay Miner. Bull. 3:154–163.

Walker, G. F. 1958. Reactions of expanding-lattice clay minerals with glycerol and ethylene glycol. Clay Miner. Bull. 3:302–313.

Wear, J. I., and J. L. White. 1951. Potassium fixation in clay minerals as related to crystal structure. Soil Sci. 71:1–14.

Weaver, C. E., and L. D. Pollard. 1973. The chemistry of clay minerals. Elsevier Scientific Publishing Co., Amsterdam. 213 p.

Weed, S. B., and L. A. Nelson. 1962. Occurrence of chlorite-like intergrade clay minerals in Coastal Plain, Piedmont, and mountain soils of North Carolina. Soil Sci. Soc. Am. Proc. 26:393–398.

Weiss, A. A., G. Koch, and U. Hoffman. 1955. Saponite. Ber. Dtsch. Keram. Ges. 32: 12–17.

Wilklander, L. 1950. Fixation of potassium by clays saturated with different cations. Soil Sci. 69:261–268.

Montmorillonite and Other Smectite Minerals

GLENN A. BORCHARDT, California Division of Mines and Geology, San Francisco

I. INTRODUCTION

The smectite group includes some of the most important clay minerals in soils and sediments. The smectites include all minerals formerly classified in the montmorillonite group of expansible layer silicates. In soil environments, three smectites—montmorillonite, beidellite, and nontronite—are important. Smectite, along with vermiculite, is responsible for a large portion of the cation exchange capacities (CEC) in soils of temperate climates. Smectite is the model most often used to explain the application of the diffuse double-layer theory to soil properties. Smectite is responsible for most of the shrinking and swelling that occurs in soils. In seasonally wet and dry climates, this causes some highly smectitic soils (Vertisols) to become naturally "self-plowing". Smectites are responsible for the adhesive property that helps to prevent soil erosion. On the other hand, they adsorb large quantities of water that decrease the strength of soils causing very destructive landslides as well as soil creep. Other moderately weathered soils contain significant quantities of smectite, particularly those in poorly drained environments with above-neutral pH values. Smectites adsorb natural organic compounds as well as herbicides and pesticides. Their expansive nature and the negative charge of smectites cause them to be extremely reactive in soil environments.

II. STRUCTURAL PROPERTIES AND MINERAL IDENTIFICATION

The structure of smectite was first determined by Hofmann et al. (1933) with further refinements by Marshall (1935) and Hendricks (1942) (Fig. 9-1). It resembles the mica and vermiculite structures already discussed in Chapters 7 and 8, in that all three have an octahedral sheet in coordination with two tetrahedral sheets in which oxygen atoms are shared. Cationic substitution occurs in octahedral or tetrahedral sheets, and the corresponding differences in properties and chemical composition are used to classify the smectites (Table 9-1). The dioctahedral smectites may form as a result of weathering, whereas the trioctahedral smectites, such as hectorite, saponite, and sauconite, appear to be inherited from the parent material, and are only rarely found in soils. Most soil smectites have been called "montmorillonite" though they contain significant amounts of tetrahedral Al and octahedral Fe. It is useful to visualize them through the use of simplified ideal formulas that represent the end members: montmorillonite, beidellite, and nontronite (Table 9-1). Almost all ranges of composition appear in the literature for at least two reasons: (i) smectites generally occur with other minerals whose

Chapter 9

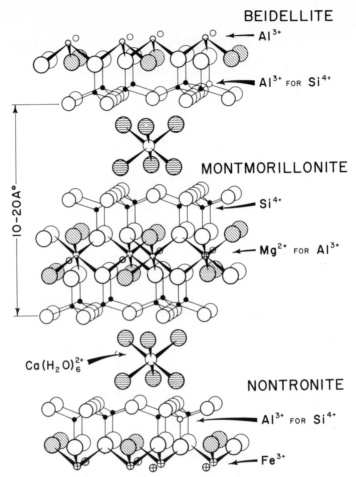

Fig. 9-1. The crystal structure of smectite illustrating three common types of substitution (modified from Brindley & MacEwan, 1953).

removal is difficult to accomplish, and (ii) most smectites have compositions varying from the idealized formulas.

A. X-ray Diffraction

The expansive characteristics of smectites are affected by the nature of adsorbed ions and molecules. As a result, X-ray conditions must be carefully controlled for proper identification. The factors requiring control include: slide preparation (Gibbs, 1965), orientation (Theisen & Harward, 1962), cation saturation and solvation (Harward & Brindley, 1964), and relative humidity (Harward et al., 1968). Without these special procedures, basal spacings of smectite may vary from 10Å to 20Å, making interpretation difficult or impossible.

Table 9–1. Idealized formulas of smectites found in soils (MacEwan, 1961) and implied data for unhydrated clays†

	Tetrahedral cations	Octahedral cations	Coordinated anions	Exchangeable cation	H_2O	CEC due to substitution	Formula weight
					%	meq/100 g	g/mole
Montmorillonite	Si_4	$Al_{1.5}Mg_{0.5}$	$O_{10}(OH)_2$	$Ca_{0.25}$	4.88	135.5	369
Beidellite	$Si_{3.5}Al_{0.5}$	Al_2	$O_{10}(OH)_2$	$Ca_{0.25}$	4.87	135.2	370
Nontronite	$Si_{3.5}Al_{0.5}$	Fe_2	$O_{10}(OH)_2$	$Ca_{0.25}$	4.21	117.0	428

† A substitution of 0.5 Mg or Al is probably the maximum occurring in smectites (Harward & Brindley, 1964). Values lower than this result from Fe^{3+} substitution for Al^{3+} in the octahedral sheet as well as from the impurities found in most samples. Soil smectites are considered to have an average CEC of 105 meq/100 g (plus 5 meq/100 g due to edge charge) (Alexiades & Jackson, 1965). This gives a negative charge per half cell of about 0.39 for montmorillonite. Many earlier authors (e.g., p. 28, Table 2–1) gave a charge of 0.33 due to substitution (CEC of 91 meq/100 g for montmorillonite) (MacEwan, 1961) but this now appears to be a minimum value.

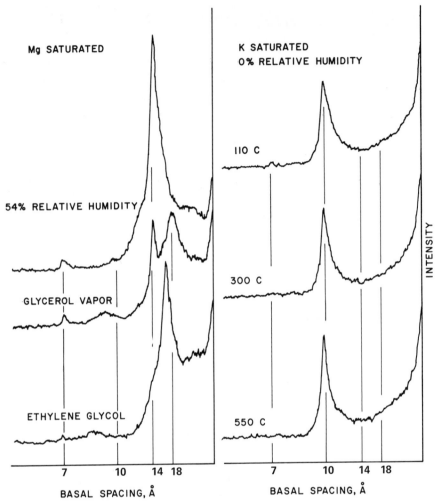

Fig. 9-2. X-ray diffraction patterns of an alluvial soil containing both montmorillonite and beidellite (DMG no. 547/72, Stevens Creek, CA).

The X-ray behavior of a soil clay containing both montmorillonite and beidellite, is illustrated in Fig. 9-2. Magnesium-saturated smectite gives a 15Å peak at 54% relative humidity. Beidellite gives a 14Å peak after glycerol solvation from the vapor, whereas montmorillonite expands to 18Å. Both expand to 17Å after ethylene glycol solvation. Unfortunately, the presence of chloritic intergrades (Chapters 10 and 12) may sometimes impede the expansion of montmorillonite in glycerol but not in ethylene glycol. Potassium-saturated smectites at 0% relative humidity give a 10Å peak that collapses further on heating if chloritic intergrades are absent.

Greene-Kelly (1955) found that montmorillonite could be distinguished from beidellite and nontronite by heating Li$^+$-saturated clay at 220C. Subsequent liquid glycerol solvation results in a 9.5Å X-ray peak for montmoril-

Fig. 9-3. Transmission electron micrograph of a soil smectite, Houston Black Clay <0.08 μm fraction (From Carson et al., 1976.).

lonite while nontronite and beidellite expand to 17.7Å. Apparently, Li^+ enters vacant octahedral positions only if tetrahedral substitution is absent, as in montmorillonite.

B. Electron Optical Properties

Smectite is notoriously nonphotogenic, but special preparation procedures such as Pt shadowing may reveal characteristically thin flakes with irregular outlines (Fig. 9-3). This contrasts with the hexagonal shape of the kaolinite particles often present as a contaminant with soil montmorillonites. Occasionally, 60° oriented striations can be seen, indicating three-dimensional order. More often, such indications of crystallinity can be traced to mica or other minerals common in soil clays. A peculiar lath shape sometimes has been associated with the Fe-rich smectite, nontronite.

Scanning electron microscopy of smectite (Fig. 9-4) shows a polygonal pattern resembling dessication cracks also characteristic of dried smectitic soils. Sample preparation obviously has a great deal to do with the results of electron microscopic studies of smectite.

C. Thermal Analysis

Like other hydrous minerals, smectites lose water when heated. The water originates from two different sources: (i) adsorbed water lost below 200C, and (ii) hydroxyl water lost at higher temperatures. At 110C, adsorbed water is largely associated with the exchangeable cations, and its release during heating is dependent upon hydration energy. Hydroxyl water

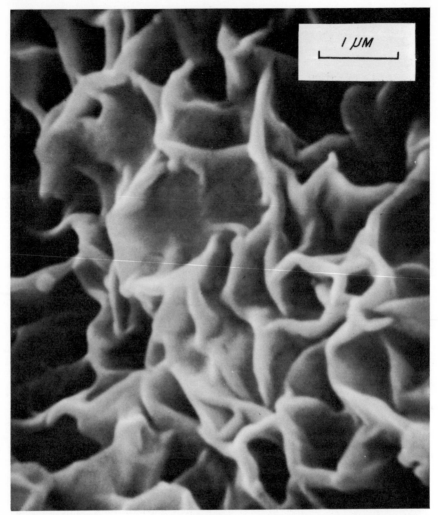

Fig. 9-4. Scanning electron micrograph of Wyoming bentonite (montmorillonite) (Reprinted with permission from Pergamon Press. Bohor & Hughes; 1971, Fig. 6b, p. 49-54.).

(Table 9-1) is lost from the structure when two hydroxyls combine to form one water molecule, leaving behind an oxygen atom.

1. THERMOGRAVIMETRIC ANALYSIS (TGA)

Heating of smectites produces both gradual and rapid changes in weight at temperatures below 1000C (Fig. 9-5). The greatest loss of hydroxyl water occurs at lower temperatures for nontronite ($<$ 550C) than it does for montmorillonite and beidellite. Amounts lost are also less for nontronite (4.21% vs. 4.88% for montmorillonite and beidellite, Table 9-1). X-ray diffraction of nontronite samples heated at 550C may show effects of dehydroxylation, whereas other smectites may not.

2. DIFFERENTIAL THERMAL ANALYSIS (DTA)

The DTA of smectites forces rapid dehydroxylation which occurs at temperatures generally between 500C and 700C (Fig. 9–5). Montmorillonite has a tendency to have endotherms above those of beidellite, but this is not diagnostic. Most other 2:1 layer silicates have endotherms in the same region. Exotherms above 850C result when new anhydrous minerals form. These have been studied in some detail, and attempts have been made to use them as diagnostic criteria (Grim & Kulbicki, 1961).

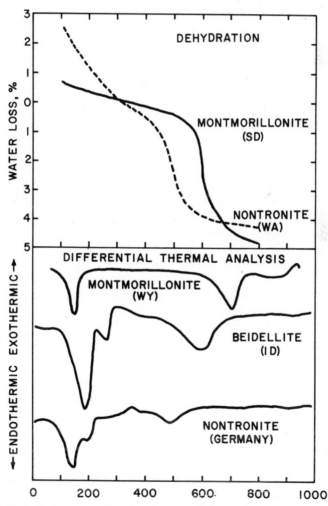

Fig. 9–5. Dehydration curves (TGA) and differential thermal analysis (DTA) of smectites found in soils (From Ross & Hendricks, 1945; Mackenzie, 1957).

D. Infrared Analysis (IR)

Infrared analysis of smectites has provided considerable information on their structural properties and the mechanisms of interactions with inorganic and organic chemicals. The principal absorption bands (Table 9-2 and Fig. 9-6) may aid in characterizing soil smectites. Silicate minerals have many strong Si-O stretching bands near 600 and 1000 cm^{-1} complicated by considerable variation due to Al substitution for Si (White, 1971). In smectites, hydroxyl bending vibrations cause absorption characteristic of the environment of the octahedral sheet (Table 9-2). When only Al is present in the octahedral sheet, the absorption is near 920 cm^{-1}. When only Fe is present, absorption is near 820 cm^{-1}. Various combinations of Al, Fe, and Mg result in intermediate values.

Hydroxyl stretching vibrations absorb in the 3000 to 3800 cm^{-1} region. These bands appear to broaden as a result of Al substitution for Si in the tetrahedral sheet (Farmer & Russell, 1967). In addition, these absorption peaks apparently reflect the type of cation saturation and the hydration conditions of the interlayer space. For nontronite, the presence of octahedral Fe rather than Al, produces a noticeable shift of the OH stretching band to a much lower frequency (Fig. 9-6).

The surface oxygens of montmorillonite are weak electron donors. They form weak H bonds whose strength is partially dependent upon the amount of tetrahedral charge. Due to this situation and their close association with the exchangeable cations, water molecules are more acidic in the interlayer space. That is, they readily contribute protons to other molecules. For example, in reactions with ammonia, NH_3 gas is readily converted to NH_4^+ when adsorbed on smectite surfaces (Mortland & Raman, 1968).

Selected IR studies have detected evidence for H^+ migration from exchangeable positions into the tetrahedral sheet. Heating of Al- and H-saturated montmorillonite causes the 1700 cm^{-1} band to disappear as a result of this migration (Yariv & Heller-Kallai, 1973). Deuterium can be used to replace H as an additional aid in IR studies (Russell & Fraser, 1971). Many smectite-organic complexes have been studied by using IR (Mortland, 1970). Infrared analysis will no doubt continue to be invaluable for the study of smectites and their many interactions in soils.

Table 9-2. Infrared bands for smectites (OH bending).

Mineral	Bond	White, 1971	Farmer & Russell, 1967
		cm^{-1}	
Montmorillonite	AlAl-OH	950	920
	AlMg-OH	844	–
	AlFe^{3+}-OH	887	845–890
Beidellite	AlAl-OH	935	–
Nontronite	FeFe-OH	–	820

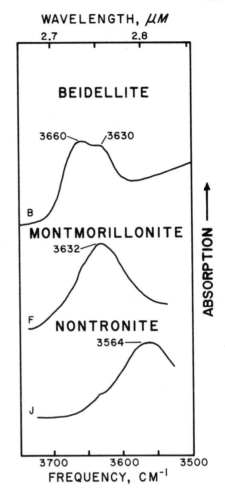

Fig. 9-6. Infrared hydroxyl absorption bands (OH stretching) of randomly oriented samples of beidellite, montmorillonite, and nontronite (after Farmer & Russell, 1967).

E. Elemental Analysis

Chemical analysis of rocks and soils provides information whose value is in direct proportion to the accuracy of the mineralogical analysis. The possibility of a continuous variation in the montmorillonite-beidellite-nontronite series (Fig. 9-7) compounds the problem of chemical allocation in soil clays. Procedures for calculations of smectite formulas in soils have been described (Sawhney & Jackson, 1958): (i) the elemental contents of all other minerals must be subtracted from the total analysis, (ii) all silica is allocated to the tetrahedral sheet, (iii) 22 negative charges (10 oxygen atoms and 2 hydroxyls) are assumed per half unit cell, (iv) up to 1 Al^{3+} atom is included with Si to get 4 tetrahedral atoms per half unit cell, (v) the remaining cations, usually Al^{3+}, Fe^{3+}, and Mg^{2+}, are allocated to the octahedral sheet, and (vi) exchangeable Ca^{2+}, Mg^{2+}, Na^+, K^+, and H^+ are used to satisfy the negative

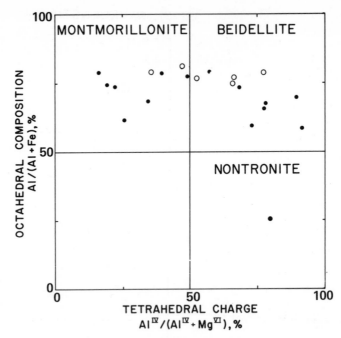

Fig. 9-7. Octahedral composition and relative tetrahedral charge of some soil smectites (●, Sawhney & Jackson, 1958; ○, Carson & Dixon, 1972). Both beidellite and nontronite have more charge due to tetrahedral (IV) substitution than octahedral (VI) substitution (Greene-Kelly, 1955). Nontronite has more Fe than Al in the octahedral sheet, while the occurrence of an Fe analog of montmorillonite has not been established. Nontronite is rarely found in soils except as an early weathering product of basic rocks (Sherman et al., 1962).

charge produced by isomorphous substitution. Difficulties with the method generally result from the presence of impurities such as silica relics, iron oxides, hydroxy interlayers, unweathered mica cores, etc. It is best used in conjunction with quantitative mineralogical analysis.

Some of the Ti commonly included in chemical formulas of clay minerals is simply anatase or rutile impurity (Raman & Jackson, 1965). Other elements have been ignored because of their status as trace elements. Gallium and germanium, for example, always occur with Al and Si, respectively. They have no significant impact on the chemical formulas of smectites, but their existence may be valuable for solving problems concerning smectite genesis.

F. Selective Dissolution Analysis (SDA)

Selective dissolution of smectites with acids results in (i) replacement of exchangeable cations with H_3O^+, followed by (ii) removal of octahedral Al, Mg, and Fe, followed by (iii) removal of tetrahedral Si and Al. Alkali attacks the Si sheet first. Nontronite is particularly susceptible to boiling alkali treatments used to remove noncrystalline material (Dudas & Harward, 1971). The destruction of nontronite is further accelerated by treatments to remove iron oxides. Heating to 550C, as is done in the SDA of kaolinite, may cause nontronite to dehydroxylate and dissolve in subsequent boiling alkali treatment.

Often, the presence of soil smectites is masked by large amounts of noncrystalline material. In such cases, a rigorous chemical treatment may be the only way to reveal the layer silicate component. Beidellite and low Fe montmorillonite will survive such treatment, while nontronite may be dissolved.

G. Other Methods of Smectite Identification

Optical absorption occurs in the ultraviolet and visible regions due to Fe and Ti (Karickhoff & Bailey, 1973). Only nontronite and other Fe-rich smectites show absorption in the 180–800 nm region studied. Information about the site of substitution and valence of Fe can be determined.

Mossbauer spectroscopy has been effective for studying the structure of Fe-rich smectites (Taylor et al., 1968). Nontronite had no tetrahedral Fe and the octahedral Fe was all in the 3+ state in the particular sample tested. Two different octahedral sites can be seen in three sheet structures by using Mossbauer spectroscopy (i) O_a, in which the two hydroxyls of the octahedron are on opposed apexes and (ii) O_b, in which the two hydroxyls are on adjacent tips of the octahedron.

III. NATURAL OCCURRENCE

A. Geographic Extent

Smectitic soils include $> 350,000$ km^2 of Vertisols distributed over the world (Buol et al., 1973). Smectite is especially common in level to gently sloping soils of alluvial plains. A tentative map of smectite occurrence in surface horizons of soils has been published by Gradusov (1974) (Fig. 9-8). Examples of the geographic extent of smectitic soils are given below:

1. HOUSTON BLACK CLAY OF SOUTHERN U.S.

Perhaps the most often cited example of smectitic soil is Houston Black clay (Kunze & Templin, 1956). Houston Black clay and similar soils cover an estimated area of 2.5 million acres (10,000 km^2) in parts of Texas, Mississippi, and Alabama occurring in outcrops of limey, argillaceous rocks associated with Upper Cretaceous marine clays, marl, and chalk. Houston Black clay has about 60% of < 2 μm material along with 5–70% CaCO$_3$ which indicates that weathering has been very limited.

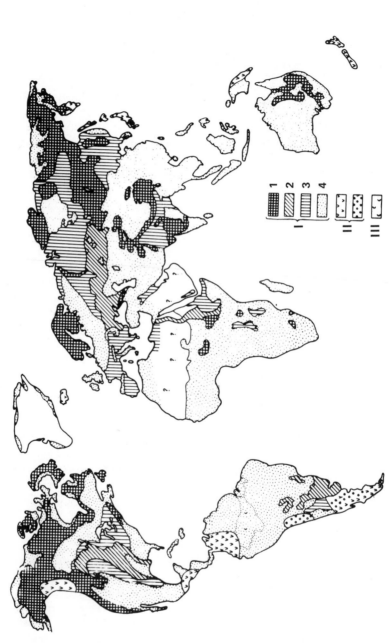

Fig. 9–8. Sketch map of the world showing areas of soils with different proportions of smectite in clay fractions of their A1 and/or A2 horizons and the general character of the smectite, plus major areas of some other soils: (I) Soils with smectite—1:70% or more; 2:50–70%; 3:30–50%; 4: 40% or less.   (II) Soils formed in ashfalls—1:humid areas; 2:arid areas.   (III) Soils of the Amazon basin.   (Reproduced from Geoderma, vol. 12, Gradusov, 1974.).

2. VERTISOLS OF SUDAN

The Sudan and the Nile River in Africa have $>$ 10,000 km^2 of Vertisols containing large amounts of smectite. The 40–60% montmorillonite found in clay fractions of these soils is attributed to weathering of basic volcanic rocks in the headwaters of the Nile and to deposition in a Pleistocene lake (El Abedine et al., 1971; El-Attar & Jackson, 1973).

3. RED AND BLACK COMPLEX OF AUSTRALIA AND INDIA

In Australia, smectite (40%) occurs in the black, low-lying, poorly drained soils, while the reddish, well-drained uplands contain primarily kaolinite (25%) and chlorite (30%) in clay fractions (Hosking, 1940; Briner & Jackson, 1970). This so-called "red and black complex" developed entirely from basalt. In India, the black smectitic soils formed from basalt, but the reddish soils formed from granite (Agrawal & Ramamoorthy, 1970).

4. SMECTITE IN LOESS OF THE MIDWEST, U.S.

The principal clay mineral in loess of Illinois, Kansas, Nebraska, Iowa, and Missouri is smectite (Beavers, 1957). This reflects the relative abundance of volcanic products found in the sedimentary rocks that were glaciated northwest of the loess deposition area. Four different zones of varying smectite content have recently been recognized within Woodfordian loesses of the midwest (Kleiss & Fehrenbacher, 1973). This explains some of the differences between soil horizons formerly attributed to soil formation.

5. SMECTITE IN CALIFORNIA

Cretaceous and Cenozoic shales occurring along the coast of California form soils containing large amounts of smectite (Alfors et al., 1973). Smectite controls many of the physical properties of these soils. Consequently, winter rains cause extensive areas of landsliding and soil expansion wherever smectite is found in abundance. In some cases, whole formations of bentonite have been involved in very deep landslides in California (Kerr & Drew, 1969).

6. SMECTITE IN VOLCANIC ASH SOILS

Volcanic ash, deposited in lakes, commonly alters to smectite. Smectites have been reported in soils formed in volcanic ash (Uchiyama et al., 1962; Chichester et al., 1969). However, other clay minerals such as imogolite, halloysite, and kaolinite commonly form in volcanic ash soils, as well. The smectites in volcanic ash soils often are associated with mica or vermiculite present in the initial material. Thus far, concrete evidence for neoformation (precipitation from solution) of smectite in volcanic ash soils has not been obtained.

7. SMECTITE IN OTHER SOILS

Smectite often weathers from mica. As a result, the ubiquity of mica almost guarantees that at least small amounts of smectite are present in most soils.

B. Soil Environment

Smectite in soils may occur in any one of the 4 B's: (i) Basins, (ii) B horizons, (iii) Super B horizons, as in Vertisols, and (iv) Beneath basic rocks (Jackson, 1965). Each of these environments has the high Si and basic cation potentials necessary for smectite formation or preservation (Jackson, 1968). Leaching is at a minimum in poorly drained basins most typical of smectite occurrences. The occurrence of smectite in soils has been reviewed extensively (Jackson & Sherman, 1953).

1. SMECTITE AND POORLY DRAINED ENVIRONMENTS

The red and black complex of Australia and India mentioned above is a good example of the poor drainage required for smectite preservation. Even areas of very high rainfall such as Hawaii have smectite in the lowlands with kaolinite occurring in the well-drained uplands.

The type of smectites that might be present in a particular soil is partially determined by the parent rock. Basic rocks containing high Fe commonly produce nontronite or Fe-rich smectite. An excellent example is the weathering of serpentinite (Wildman et al., 1968). The large amount of Mg in the rock might be expected to form a high Mg smectite such as saponite. Iron-rich smectite forms instead, thus illustrating the general rule that basic rocks tend to weather to nontronite.

Beidellite would be expected to occur as a weathering product of rocks containing micas and chlorites, because these already have the tetrahedral substitution required for the beidellite structure. Montmorillonite, on the other hand, might be expected to form pedogenically from solutions high in Si, Al, and Mg. Montmorillonite can be found with each of the other soil smectites as well as in environments devoid of micas and chlorites.

It is important to recognize that the occurrence of smectite in a basin environment does not necessarily mean that it has formed in situ. Reduced leaching may simply lead to the preservation of smectite analogous to the preservation of a maple log in a peat bog.

2. SMECTITE AS AN INTERMEDIATE IN WELL-DRAINED ENVIRONMENTS

The B horizons of soils may contain smectites though they are well drained. The pH, Si, and Al or Mg potentials may reflect the presence of smectite rather than the conditions necessary for its formation. In humid climates, it is doubtful that smectite can be more than a transitory weathering product under well-drained conditions. It may be inherited directly from the parent material, such as in certain loess deposits of the midwest (Kleiss & Fehrenbacher, 1973).

In other situations, to be discussed more fully later, micaceous materials may weather to vermiculite and then to smectite through K loss and layer charge reduction. Smectites formed in this way are likely to contain mixed layers and to form chlorite or kaolinite as the Si necessary for smectite preservation is leached away. Biotite, in particular, forms vermiculite and kaolinite under humid conditions while montmorillonite and vermiculite form under arid conditions (Ismail, 1970).

Smectites in arid regions may be transitory in nature and considerably dependent upon the parent material. Buol (1965, p. 47) states: "Contrary to popular belief, soils from arid and semiarid regions have not been shown, by the limited data available, to contain a predominance of expanding lattice minerals in their clay-size separates." Buol doubts that montmorillonite could have formed in certain arid regions, because calcite and gypsum are still present. On the other hand, the low amounts of rainfall in deserts appears to assure an environment likely to preserve smectite and may actually aid in its formation (Gal et al., 1974).

C. Sedimentary Occurrence

1. SMECTITE AND MARINE DIAGENESIS

The primary reaction of marine diagenesis involves the conversion of smectite to mica or chlorite under pressure of burial. Smectite does not occur in sediments that have undergone burial at depths greater than about 4 km (Weaver, 1959). Mica is therefore the most abundant mineral in shales, followed by lesser amounts of smectite, chlorite, and kaolinite. Smectite is scarce in sediments older than Mesozoic (230 million years old) largely because the chances for metamorphism or diagenesis increase with age.

Potassium, Mg, Fe, and Al are removed from sea water by smectites during diagenesis and during laboratory experiments (Carroll & Starkey, 1960). Potassium and Mg are always adsorbed preferentially over Na.

2. SMECTITE IN SEDIMENTARY ROCKS

Smectite is found in sedimentary rocks formed after periods of above average volcanism, such as the Permian, Triassic, and Jurassic (Weaver, 1959). Cretaceous sedimentary rocks of California are notable for their high smectite contents. The Texas gulf coast has Tertiary and Cretaceous rocks containing smectite.

The adsorptive properties of smectite may account for the observed relationship with petroleum deposits (Weaver, 1960). For example, Tertiary rocks are high in montmorillonite and in petroleum. Weaver (1960) estimates the relative amounts of smectite in sedimentary rocks as follows: Pliocene

= Miocene > Oligocene > Permian > Pennsylvanian = Mississippian = Ordovician > Eocene = Upper Cretaceous > Jurassic > Triassic > Lower Cretaceous > Silurian > Cambrian.

Very little smectite is expected in regions having geological formations affected by deep burial or twisted and contorted by metamorphism. Rock formations whose regional dip is very gentle might be expected to have smectite. Houston Black clay, for example, appears to be derived from outcrops of nearly level sedimentary rock.

3. BENTONITE: SMECTITE FROM VOLCANIC ASH

Bentonite (sometimes used incorrectly as synonomous with the term *montmorillonite*) is a rock term referring to an altered deposit of volcanic ash usually in prehistoric lakes or estuaries. The ash contains glassy material of very high Si content necessary for smectite formation. Bentonite beds are most common in Cenozoic and Mesozoic rocks but also have been found in all Paleozoic systems except Silurian (Weaver, 1963). Much is still unknown about the mechanism of bentonite formation. Marine formation is indicated for Wyoming bentonite, which is largely Na saturated whereas others, presumedly fresh water derived, are Ca saturated. Other minerals such as apatite, zircon, and biotite found in bentonite beds point to their volcanic origin. The fact that apatite has survived the alteration, supports the hypothesis of formation under conditions of poor drainage and high base content. Bentonites or their derivatives may contribute to the parent materials of many soils, particularly in the western U.S.

4. SMECTITE IN FRESH WATER SEDIMENTS

The sediments of rivers, reservoirs, lakes, and deltas reflect the smectite contents of the soils in their respective drainage areas. In discussing the high smectite content of the Mississippi River, Johns and Grim (1958, p. 197) stated:

> Soils of the Missouri drainage basin in the High Plains region have developed in large part from Mesozoic and Cenozoic sediments which contain montmorillonite and are in places bentonitic. The vast loess deposits of this region contribute in large measure to stream sediments and these have been shown over wide areas to contain montmorillonite as the chief clay mineral.

IV. EQUILIBRIUM ENVIRONMENT AND CONDITIONS FOR SYNTHESIS

A. Conditions for Laboratory Synthesis

Montmorillonite was first synthesized by Noll (1930) at 300C and 87 atm pressure under alkaline conditions. Sedletski (1937) was the first to synthesize montmorillonite at room temperature and pressure. Sodium silicate and sodium aluminate were mixed, and washed with $MgCl_2$ solution and distilled water to remove excess alkali. The gel was then kept in a moist condition for 4 years, after which an X-ray pattern for montmorillonite was obtained. Under identical conditions, about 500 times more smectite forms at 100C than at 0C in the same period of time. Most laboratory synthesis procedures rely on elevated temperatures and pressures to obtain sufficient quantity of material.

Harward and Brindley (1964) prepared smectites at 300–360C and 1055 kg/cm^2 from gels with varying amounts of tetrahedral and octahedral charge. An attempt was made to synthesize smectites with cation exchange capacities (CEC) between 45 and 270 meq/100 g but only CEC's between 95 and 135 meq/100 g could be obtained. Natural smectites appear to have an equally narrow range of charge substitution.

B. Transformation and Formation in Soils

1. TRANSFORMATION FROM MICA

Similarities in their sheet structures have long led to the conclusion that smectites could be derived from micas by depotassication (Bray, 1937). Smectites weathered from micas are likely to be tetrahedrally substituted, approaching the beidellite end member in chemical composition. The ubiquity of micas thus assures a wide distribution of soil smectites. The term *transformation smectite* has been proposed for smectites derived from micaceous minerals (Robert, 1973).

Vermiculite may transform from mica initially, but it is considered a fast-forming, unstable intermediate (Kittrick, 1973). Vermiculite disappears from soils by forming smectite or other minerals at rates dependent upon environmental conditions.

Laboratory "weathering" of micaceous vermiculite illustrates blister-like areas of low charge density (Fig. 9-9). These areas swell like smectite when Li$^+$ saturated and washed with water. Later, upon removal from the vermiculite surface, the blister-like material gives an 18Å X-ray peak characteristic of smectite (Sridhar et al., 1972). No more conclusive proof of smectite transformation from micas seems necessary.

The following equation describes the weathering of phlogopite mica to saponite, a trioctahedral smectite:

$$(Si_3Al)(Mg_3)O_{10}(OH)_2K + 0.5Si^{4+} + 0.25Ca^{2+} \longrightarrow$$
Phlogopite mica

$$(Si_{3.5}Al_{0.5})(Mg_3)O_{10}(OH)_2Ca_{0.25} + 0.5Al^{3+} + K^+$$
Saponite

The equation illustrates the essential changes necessary for smectite transformation from a mica. They are: (i) depotassication as in vermiculite formation already discussed, (ii) dealumination of the tetrahedral sheet,

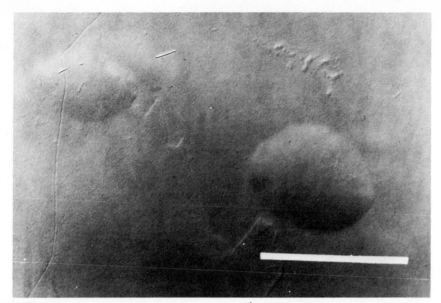

Fig. 9-9. Electron micrograph of Pt-C replica of Li$^+$-treated Montana vermiculite, showing smectite "blisters" in low charge areas (white bar = 1 μm) (Reprinted with permission from Mineralogical Society of America. Sridhar et al., 1972; Fig. 3b, p. 1838.).

followed by (iii) silication of the tetrahedral sheet. Without these extensive changes, micas are unable to possess the low-charge characteristics necessary for them to exhibit the properties of smectites. Varying amounts of ferrous Fe are usually present in the octahedral sheets of soil micas. Oxidation of this Fe and incorporation of protons within the structure may also decrease the layer charge of micas. Loss of Al from tetrahedral sheets appears essential for transformation of smectite from mica.

What type of environment would promote the transformation of phlogopite to saponite? First of all, the concentration of K$^+$ must be low, such as in rainwater and in soil solutions of heavily cropped soils. Second, Si(OH)$_4$ concentrations must be high as provided by mafic minerals with Si potentials higher than that of smectite (Huang, 1966). Third, Al concentrations must be low, as they are in soils with pH's above 6 or 7. In addition, tetrahedral Al is generally much less stable at room temperature and pressure than octahedral Al (Jackson, 1963). Thus, soils with low K$^+$ and Al^{3+} and high Si(OH)$_4$ and Ca^{2+} or Mg^{2+} activities and pH above 6 are likely to contain transformation smectites. At a soil pH < 6, vermiculite may weather directly to kaolinite (Ismail, 1970) or chlorite.

2. TRANSFORMATION FROM CHLORITE

Mafic chlorite is highly unstable in soils (weathering index = 4) (Jackson, 1964). It would be expected to lose its hydroxide interlayer at pH levels below 6 and under severe leaching conditions. These are not conditions that would be expected to preserve smectite indefinitely. Indeed, laboratory weathering of chlorite only resulted in the formation of non-crystalline material (Adams et al., 1971). Obviously, smectite transforms from chlorite under leaching and weathering conditions more restrictive than those for the transformation of mica to smectite.

3. FORMATION OF SMECTITE FROM SOLUTION

Smectites that precipitate directly from soil or matrix solutions could be called *neogenetic smectites*. These provide a particularly challenging problem for soil science. Unfortunately, micaceous parent materials are ubiquitous, and the polygenetic nature of most soils makes it difficult to eliminate transformation from mica as the mode of formation. The debate on the formation of smectites in volcanic ash is typical.

Chichester et al. (1969) found smectite in < 2 μm clay from coarse pumiceous soil in Oregon. The neogenesis of smectite in an otherwise well-drained soil was attributed to physical entrapment and concentration of Si and bases in the fine pores of pumice. Later, Borchardt et al. (1971) found traces of biotite in the silt fraction and hypothesized that this provided the 2:1 type layer structure. Dudas and Harward (1975) then demonstrated that biotite could be found in rock fragments torn from volcano walls and subsequently deposited along with the otherwise biotite-free volcanic ash. The presence of transformation smectite does not preclude formation of smectite by neogenesis, but it does make interpretation difficult.

To answer questions concerning smectite neogenesis, the first requirement is to account for all other possibilities for transformation to smectite. The parent material must be fully accounted for, and aeolian, alluvial, colluvial, or other forms of contamination must be assessed. This can sometimes be done simply by measuring the amounts of "unweatherable" sands and the ratios of coarse to fine particles for various minerals (Borchardt et al., 1968), as well as trace element ratios (Borchardt & Harward, 1971) or contents. For example, mica-derived smectites are likely to contain high amounts of boron (Kantor & Schwertmann, 1974).

After eliminating all other possible contributions to its presence, consideration can be given to the likely formation of neogenetic smectite from soil solutions. Most of what is known about smectite stability comes from evaluations of the environments in which smectites are found and from thermodynamic data determined on nearly pure specimens.

The stability of Wyoming montmorillonite, like that of other smectites, is largely dependent upon the Si activity and the pH (Fig. 1-9). Many of the soils analyzed have Si potentials very near the kaolinite-montmorillonite join at about 25 ppm Si (Weaver et al., 1971). Presumedly, kaolinite could form during wet months when the soil is being actively leached, and montmorillonite could form during dry months when Si (and Mg) concentrations would be relatively high. If drainage was restricted, only montmorillonite would be formed.

C. Smectite Weathering

The most common mechanism of smectite weathering involves pedogenic chlorite formation (see Chapters 10 and 12). Aluminum and iron form hydroxy interlayers in smectites, generally at pH values slightly below those necessary for smectite stability.

The well-known equation for mica weathering (mica $\rightleftharpoons$ vermiculite $\rightleftharpoons$ montmorillonite) implies a reversible reaction in which vermiculite forms from montmorillonite. Although addition of K to smectite may collapse the structure temporarily, the increase in layer charge is not sufficient to form a true mica (Arifin et al., 1973). It would appear unlikely that enough tetrahedral Al could be returned to the smectite structure to form mica under the temperature and pressure conditions in soils.

There is some evidence for the direct transformation of montmorillonite to kaolinite during weathering (Altschuler et al., 1963). In fact, mixed-layer kaolinite-montmorillonite has been reported (Sakharov & Drits, 1973). Also, kaolinite has been formed from montmorillonite in the laboratory by adding Al–OH polymers and heating at 220C in $1N$ HCl for 7 days (Poncelet & Brindley, 1967). The reaction involves inversion of Si tetrahedra and removal of Mg^{2+} and Fe^{3+} from octahedral positions. The transformation of smectite to kaolinite and iron oxides appears likely in the more permeable and highly leached soils of the red and black complex (Weaver et al., 1971).

V. CHEMICAL PROPERTIES

A. Cation Exchange

The presence of exchangeable cations in smectites was first suggested in 1935 by a soil scientist (Marshall, 1935). A significant portion of the fertility of moderately weathered soils can be traced to the presence of smectites. This is due to their ability to hold fertilizer cations such as K^+ and NH_4^+ against the effects of leaching by rainfall. Except for organic matter, only vermiculite has a higher CEC, about 160 meq/100 g as opposed to 110 meq/100 g for smectite.

Most monovalent and divalent cations are completely exchangeable from smectite by other cations. The large CEC arises from substitution of Al^{3+} for Si^{4+} in tetrahedral sheets and substitution of divalent cations like Mg^{2+} and Fe^{2+} for trivalent cations like Al^{3+} and Fe^{3+} in octahedral sheets. However, the negative charge is low enough to allow separation of the smectite layers.

The natural saturation of soil smectites is usually Ca^{2+} and Mg^{2+}. In acid soils, H_3O^+ causes smectite to lose Al^{3+} from the structure (Coleman & Harward, 1953). This Al^{3+} replaces the H_3O^+ on the exchange exhibiting many of the properties of both ions. Subsequent hydroxylation of the Al^{3+} produces nonexchangeable hydroxy Al polymers that may produce pedogenic chlorite, as discussed in Chapters 10 and 12. Iron may also become non-

Table 9-3. Reactions of various elements and chemical compounds with smectites
in soil environments

Element or compound	Type of reaction	Reference
Ca^{2+}, K^+, Li^+, NH_4^+	Cation exchange	Chapter 2, this book
Ni^{2+}, Be^{2+}	Cation exchange	McBride and Mortland (1974)
H_3O^+	Cation exchange, replaces octahedral Al	Coleman and Harward (1953)
Na^+	Cation exchange, replaced from exchange by octahedral Mg and Al	Bar-On and Shainberg (1970)
Fe^{3+}	Cation exchange, hydrolysis, and dissolution of smectite	Thomas and Coleman (1964)
Al^{3+}, Mg^{2+}, Fe^{3+}	Cation exchange, hydroxy interlayer formation	Chapters 2, 10, and 12, this book
$Hg(OH)_2$	Interlayer formation	Blatter (1973)
$Ca(OH)_2$	Dissolution, cementation, and adsorption	Ingles and Metcalf (1972)
NH_3	Proton acceptance	Mortland and Raman (1968)
CO_2, N_2	External adsorption with inter-lamellar adsorption under special circumstances	Fripiat et al. (1974), Knudson and McAtee (1974)
B	Anion exchange at pH 7 as $B(OH)_4^-$ or adsorption as $B(OH)_3$	Schalscha et al. (1973)
F^-	Formation of fluoraluminate, fluorferrate, and hydroxyl ion	Huang and Jackson (1965)
$H_2PO_4^-$	Anion exchange	Stout (1940)
$Diquat^{2+}$ (a herbicide)	Cation exchange	Dixon et al. (1970)
1,3-dichloropropene (a pesticide)	Adsorption	Bailey and White (1964)
Surfactant	Physico-chemical adsorption	Hower (1970)
Humic and fulvic acid	Interlamellar adsorption	Chapters 20 and 21, this book Mortland (1970)
Cellulose and Glucose	Stabilizing effect	Sorensen (1972)
Polysaccharides	Adsorption	Clapp and Emerson (1972)
Protein	Intercalation	Harter and Stotzky (1973)

exchangeable in this way (Table 9-3). Sodium-saturated smectite may also lose octahedral Al^{3+} and Mg^{2+} when Na electrolyte concentrations are low (Bar-On & Shainberg, 1970).

1. EDGE CHARGE AND pH-DEPENDENT CHARGE

The edges of the smectite structure (Fig. 9-1) have broken bonds which may contribute to the CEC. At pH 7, the CEC due to this "edge charge" may be 5 meq/100 g or more depending upon fineness of the smectite particle. Actual CEC measurements therefore include this negative charge as well as the amount due solely to isomorphous substitution (Table 9-1).

The charge due to substitution has been called "permanent" charge while that due to the broken O bonds and edge hydroxyls has been called "pH-dependent charge." This is because many of the broken bonds attract H^+ rather than other cations when the pH is below neutrality. At pH's > 7, the dearth of H^+ in the soil solution allows other cations to take part in the exchange reaction. As a result, the measured value for CEC increases as the pH increases. Other clay minerals, hydroxy Al polymers, organic matter, and noncrystalline minerals generally exhibit greater proportions of pH-dependent CEC than smectites.

2. FIXATION AND SELECTIVITY

Potassium fixation has been reported for soils containing smectites. The fixation, however, is due to other minerals such as vermiculite, weathered micas, and certain noncrystalline minerals that are invariably present in these soils as well. Actually, a commonly accepted definition of smectite specifically rules out minerals that fix K (Alexiades & Jackson, 1966). Wyoming montmorillonite showed no preference for K even when the amount of K saturation was very low (Carson & Dixon, 1972). "Transformation smectites" and beidellites are likely to have residual mica cores surrounded by "wedge sites" that may be primarily responsible for cation fixation. For this reason, K fixation has often been related to the amount of tetrahedral charge (Wear & White, 1951), but the significance of this has been disputed recently (Kodama et al., 1974).

Aluminum, Mg, and Fe interlayer formation (Chapters 10 and 12) is a form of cation fixation. The Al system, in particular, is greatly affected by drying and wetting (Westfall et al., 1973).

Size considerations limit the preference of expansible layer silicates for various cations. When two cations are added to a soil solution in equal amounts, soil smectites may adsorb one to a greater extent than the other. In general, the higher the charge on the cation and the smaller the atomic radius, the greater will be the preference that smectites will show for a particular cation.

Selectivity coefficients are usually determined by using the Gapon equation. For the K–Ca system this would be written

$$k_G = (K_{ads})(Ca_o)^{1/2}/(Ca_{ads})(K_o)$$

where

K_{ads} = K adsorbed on clay, meq/100 g,
Ca_{ads} = Ca adsorbed on clay, meq/100 g,
K_O = K concentration in equilibrated solution, moles/liter,
Ca_O = Ca concentration in equilibrated solution, moles/liter, and
k_G = selectivity coefficient for this particular clay, (liters/mole)$^{1/2}$.

Potassium selectivity increases upon oxidation of the ferrous Fe in the octahedral sheet (Kishk & El-Sheemy, 1974). Oxidation causes the OH$^-$ ion nearest the Fe^{3+} to change from perpendicular to the basal plane to a more inclined position, thus placing the K$^+$ in a more negative environment. The hydrated cations, such as Ca^{2+}, would not be significantly affected, because size considerations prevent them from approaching as close as K$^+$ to the seat of the negative charge.

Ion selectivity is reflected in the concept of the symmetry value, which is defined as the percentage of cation released when the replacing ion is added in amounts equal to the adsorbed ion (i.e. the CEC, for monoionic systems). The high affinity of montmorillonite for divalent cations is illustrated in Table 9–4. If this smectite had a CEC of 100 meq/100 g, the addition of 100 meq of NaCl to 100 g of Ca-saturated clay would result in the displacement of only 11.4 meq of Ca by Na. Potassium chloride would displace 25.2 meq and $MgCl_2$ would displace 40.4 meq. Trivalent salts would displace even greater amounts of the original cation. These considerations become very important when small amounts of various salts are added to soils in the field. Of course, laboratory saturation of clays is achieved simply by applying highly concentrated solutions and relying upon the law of mass action.

Symmetry values also have implications for natural environments. For example, some bentonites are naturally saturated primarily with sodium. Obviously, very high salt concentrations would be required to achieve predominant Na saturation under natural conditions. Even sea water has < 1 meq of NaCl/ml.

Two or more exchangeable cations adsorbed on the surface of smectites may not necessarily be randomly distributed. Demixing may occur in which different cations on the exchange complex separate into distinct regions (Fink et al., 1971). Research is continuing in an effort to determine if de-

Table 9–4. Symmetries for montmorillonite saturated with various cations at pH 7.55 for 10 days (Carroll, 1959).

Original cation saturation	Replacing salt					
	NaCl	NH$_4$Cl	KCl	MgCl$_2$	CaCl$_2$	BaCl$_2$
	symmetry†					
Na	--	54.5	63.6	82.4	87.8	88.0
K	34.0	48.5	--	71.6	--	60.0
NH$_4$	32.7	--	49.1	73.8	78.8	78.6
Ca	11.4	20.2	25.2	40.4	--	47.7
Ba	11.5	21.8	23.5	44.3	46.4	--

† % of adsorbed cation displaced by an amount of salt equivalent to the CEC.

mixing occurs by domain or rather by alternating layers. Perhaps alternation may be involved in the rapid decrease in hydraulic conductivity after Na saturation reaches 15% (Shainberg & Caiserman, 1971).

B. Anion Exchange

Anion exchange in smectites generally is < 5 meq/100 g. Anions of the appropriate size may replace structural hydroxyls only at the edges of smectite crystals (Bingham et al., 1965). This is why the anion exchange capacity (AEC) of smectite is so low. Most soils will have hydroxy-Al polymers on mineral surfaces in addition to other minerals that contribute more to the AEC.

C. Reaction Kinetics

"Kinetics is as important as equilibria in weathering" (Kittrick, 1971). This is especially true of reactions involving smectites. Montmorillonite stability is considered independent of the solution activity of exchangeable ions (Kittrick, 1971). The rate of the interchange reaction between adsorbed H_3O^+ and octahedral cations is affected by the amount of tetrahedral Al^{3+} (Barshad & Foscolos, 1970). The kinetics of the conversion of Na montmorillonite to Mg^{2+} and Al^{3+} saturation have been studied in detail (Shainberg, 1973).

D. Molecular Sorption

Molecular sorption on smectites may involve chemical as well as physical effects. Even neutral molecules, such as hydrocarbons, exhibit polarity interactions with the net negative charge of smectites and the positive charge of associated exchangeable cations. The large internal surface area of smectite (up to 800 m^2/g) provides most of the adsorption surface in many soils.

In addition, the external surface area of smectite is generally larger than that of other soil minerals due to its very small particle size. The external surface area of smectite is usually measured by physical adsorption of N_2. but a polar molecule such as ethylene glycol, is used to measure the internal surface area as well. However, recent work shows some interlamellar adsorption of N_2 and CO_2 when smectite is Na or Cs saturated (Knudson & McAtee, 1974; Fripiat et al., 1974).

Molecular adsorption often involves interactions with exchangeable cations and their associated waters of hydration. For example, NH_3 readily accepts a proton from water molecules surrounding cations in the interlamellar space, thus forming exchangeable NH_4^+ (Mortland & Raman, 1968). Organic compounds containing N are likely to display interactions of the same type.

E. Other Chemical Properties

Soil organic matter, pesticides, and herbicides form complexes with smectites as discussed in Chapters 20 and 21. Surfactants also interact with smectites (Hower, 1970). Proteins intercalate smectite (Harter & Stotzky, 1973). Polysaccharides with positive, neutral, and negative charges interact with smectite to form aggregates important in soil stability (Clapp & Emerson, 1972).

The CEC of smectite like that of other clays, decreases when heated at temperatures over 100C (Nishita & Haug, 1972). Benzidine was once used for the identification of smectites through the development of a colored, clay-organic complex, but such complexes also form with other soil constituents (Theng, 1971).

F. Smectites and Soil Fertility

The smectite structure itself contributes very little to fertility, although cropping of soils containing smectite causes Mg and Fe to be released from the octahedral sheet (Christenson & Doll, 1973). Nevertheless these are usually available from other phases in the soil. Transformation smectites are likely to be associated with mica cores containing K that could become available for plant growth. The high CEC of smectite is available to hold fertilizer cations such as K and NH_4, macronutrients such as Ca and Mg, and micronutrients such as Cu and Zn. Unfortunately, smectite in large amounts may produce unfavorable hydraulic conditions, and root growth may be restricted. Moreover, the environmental conditions necessary for smectite preservation sometimes produce high alkali and alkaline conditions that interfere with plant growth.

VI. PHYSICAL PROPERTIES

A. Shrink-Swell

Although other clays shrink and swell with changes in moisture content, these changes are minor compared to those of smectite which is able to adsorb several times its weight in water.

1. THEORETICAL ASPECTS OF HYDRATION AND DEHYDRATION

The diffuse double layer theory explains the influence of solution environment upon swelling in smectite (Sposito, 1973). Calcium-saturated smectite swells from 10Å to a maximum of 20Å (Norrish, 1954), but Na and Li smectite theoretically swell to infinity. At the liquid limit, Na montmorillonite may have water films between 100 and 200Å thick (Mielenz & King, 1955). Smectites in soils may not expand nearly as much as this for several reasons: (i) they are generally interstratified with other nonexpansive clay minerals such as micas, chlorites, and chloritic interlayers; and (ii) they are seldom monoionic and are usually saturated with Ca, Mg, and K rather than Na or Li. Nevertheless, considerable soil expansion and contraction does occur. The liquid limit and plasticity index of soils generally are related to the amount of smectite present (White, 1949).

2. EVALUATION OF EXPANSION IN SMECTITIC SOILS

The plasticity index, that is, the liquid limit minus the plastic limit, is the most common indicator of soil expansive properties. Another measure of expansion is the coefficient of linear extensibility, which is the ratio of the volume of a soil after wetting to the volume of that soil before wetting (Anderson et al., 1973). Swelling or uplift pressures generated by smectites may approach values as high as 10 kg/cm^2 (tons/feet^2) (Mielenz & King, 1955). Indeed, the annual amplitude of vertical movement was found to be > 5 cm in soils high in smectite (Yaalon & Kalmar, 1972). El Abedine and Robinson (1971) found that seasonal drying causes 10% shrinkage at the surface decreasing to zero at 120 cm in Vertisols of the Sudan.

3. SOIL STABILIZATION

The expansive properties of smectitic soils may be ameliorated by treatment with lime or other chemicals (Ingles & Metcalf, 1973). Lime, $Ca(OH)_2$, produces a decrease in the plasticity index, first, by increasing the plastic limit and second, by producing hydrated Ca–Al silicate cementing agents. Large organic cations produce similar decreases in expansive properties (Davidson, 1949), but their use for soil stabilization has been limited because of their high expense (Brandt, 1972).

4. SMECTITES AND THE HOME OWNER

After the most suitable land has been developed, homes are built in areas in which smectitic soils produce problems (Alfors et al., 1973). Modern soil and geologic surveys furnish maps delineating areas of expansive soils and landslides. These surveys are particularly useful to home builders and home owners. Special construction methods may sometimes be used in soils affected by expansion or creep (Fig. 9–10), but landslide prone soils should be avoided unless the soil is shallow and can be removed entirely.

Soil mineralogists are often asked questions concerning garden soils containing high amounts of smectite. Agricultural extension people are able to aid in this area (e.g., Aljibury, 1974). Briefly, the physical properties of smectitic garden soils can be improved as follows: (i) handle when moist but not wet; (ii) add about 10 cm of organic matter, such as compost, to the sur-

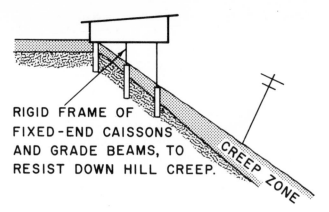

Fig. 9-10. House foundation designed to resist creep often associated with smectitic soils (after Scott & Schoustra, 1968).

face and mix into the top 20 cm of soil; (iii) use mulch to decrease cultivation and evaporation; and (iv) water such soils very slowly. In addition, apply sufficient N fertilizer to replace the N used by the decomposing organic matter.

B. Water Retention

The macro effects of shrinkage and swelling have been studied on a micro scale as well. Barshad (1949) noticed that X-ray spacings of smectite are related to the number of adsorbed water layers between the smectite layers. The water of hydration of the exchangeable cation (Fig. 9-1) forms the first layer with additional layers held with less energy (Barshad, 1960). Hendricks et al. (1940) proposed that water was retained on the surface of smectite in the form of a hexagonal net. However, this theory does not fully account for the effects of the hydrated cations. Various other theories have been presented, but none are completely satisfactory. Any theory on the structure of water adsorbed on clay surfaces must be consistent with these facts: (i) the water molecules are oriented parallel to the clay surface, (ii) adsorbed water is birefringent, (iii) has a lower temperature of maximum density, (iv) greater proton dissociation, (v) greater viscosity, (vi) lower vapor pressure, (vii) lower freezing point, and (viii) a lower density than normal water (Low & White, 1970).

Water movement in smectitic soils proceeds mainly through large cracks and pores. For example, the measured value for hydraulic conductivity in a swelling clay soil increased from 0.3 cm/day to 2.5 cm/day when the size of the soil sample was increased from 70 cm to about 10 m in diameter (Ritchie et al., 1972). This was because the smaller sample did not contain as great a proportion of large cracks and pores as the large one. Soil permeability increases as the salt concentration increases (Quirk & Schofield, 1955). This is particularly important in irrigated soils high in smectite, especially when Na salts are present.

C. Cohesion and Adhesion

1. LANDSLIDES

Smectite minerals are involved in a large proportion of landslides, even though the shear strength of smectite soils is much higher at a given water content than for soils containing other clay minerals. This is largely because of the high water-adsorption capacity of smectites as compared to other clays. Landslide control methods commonly use various types of drainage systems to avoid introducing water into a slide and attempt to remove water already introduced (Zaruba & Mencl, 1969). Retaining walls and other structures erected to stop a landslide are usually unsuccessful because of the high pressures developed by sliding soil.

Recent work involves the application of chemicals to landslides through headwall cracks and boreholes in an effort to increase the shear strength (Arora & Scott, 1974). Lime has been successful in landslide stabilization, but insolubility limits its usefulness. Other chemicals also are useful for landslide stabilization. These include salts of divalent cations such as $CaCl_2$, salts of trivalent metals such as $AlCl_3$, soluble phosphates such as $(NH_4)_2HPO_4$, acids such as H_3PO_4, and large organic cations such as Armac T, a fatty acid amine. Possibly, intercalation could occur in soil smectites under appropriate conditions as it does in laboratory experiments. For example, "chlorite" can be formed from smectite treated with $AlCl_3$ that is partially neutralized with NaOH (see Chapters 10 and 12).

2. SOIL CREEP

Soil creep differs from landslides in that movement increases with nearness to the soil surface. Smectite is nearly always involved in soils with noticeable soil creep. This is because creep results from expansion and contraction due to moisture changes accentuated by climates with alternating wet and dry seasons. Those who build on soils subject to creep can sometimes avoid destruction by anchoring foundations below the zone of annual moisture change and allowing the soil to flow around them (Fig. 9-10).

3. EROSION

Erodability of smectitic soils by wind and water may be inversely related to the tensile strength. The tensile strength of montmorillonite can be measured in a unique way because it can form thin films. Tensile strength depends upon the saturating cation with strengths decreasing in the following order: $Fe > K \geqslant Na > Al > Ca$ (Dowdy & Larson, 1971). The addition of polyvinyl alcohol to these films increases tensile strength. Interest has recently been renewed in the use of such chemicals for soil conditioning.

D. Particle Size Distribution

Smectites usually occur only in the clay fraction (< 2 μm diameter) of a soil. Quantitative analyses of nine soils of north-central U.S. showed that there was an average of 3% montmorillonite in the 2–0.2 μm fraction, 17% in the 0.2–0.08 μm fraction, and 25% in the < 0.08 μm fraction (Borchardt et al., 1966). This is a consideration when analyzing soil or clay suspensions in which minerals other than smectite occur. For example, methods of X-ray slide preparation that rely on gravity sedimentation to achieve preferred orientation are invalid (Gibbs, 1965). On the other hand, analyses of the smallest size fractions may allow one to observe smectite minerals not otherwise detected.

VIII. QUANTITATIVE DETERMINATION

A. X-ray Diffraction

The prospects for quantitative determination of smectite with the use of X-ray diffraction appear discouraging. The best that can be hoped for is a rough estimate of relative variations between samples of nearly identical genesis. Brindley (1961) has given the difficulties encountered and the considerations necessary for such valiant attempts.

The basic requirements for good semiquantitative estimates of smectite include:

1) Saturation with a known cation.
2) Proper slide preparation such as the smear-on-paste technique (Theisen & Harward, 1962). This is much favored over sedimentation methods that deposit the finer grained smectites on slide surfaces with coarser grained micas and kaolinites beneath them. Sedimentation methods cause up to 250% error (Gibbs, 1965).
3) Solvation with water (Sayegh et al., 1965) or with an organic such as ethylene glycol, glycerol, (Harward & Brindley, 1964) amine reagent (Rex & Bauer, 1964), or polyvinylpyrrolidone (Levy & Francis, 1975).
4) Consideration of X-ray scattering factors. An X-ray peak at 17Å will have four times the intensity of a peak at 10Å when a 2Θ compensating device is not used. No doubt, in many cases, this results in an exaggeration of smectite contents in soils.

B. Glycerol and Ethylene Glycol Sorption

Glycerol produces a duo-interlayer in montmorillonite and a monointerlayer in vermiculite and beidellite (Fig. 9–2) under certain conditions (Harward & Brindley, 1964; Harward et al., 1968). It has been used with moderate success for specific surface and smectite determinations (Mehra & Jackson, 1959; Moore & Dixon, 1970). Other organic chemicals produce inter-layer coverage which also happens to be influenced by the cation saturation (Barshad, 1952).

Ethylene glycol produces duo-interlayer coverage independent of ex-changeable cations if a special technique is used (Eltantawy & Arnold, 1974). This is a great advantage over other sorption methods that do not give values independent of the CEC. Unfortunately, other clay minerals also sorb ethyl-ene glycol. Nevertheless, the method would appear to be most promising as a check on other methods of smectite determination.

C. Cation Exchange Capacity Methods

In the clay mineral fraction of most soils, only vermiculite has a higher CEC than smectite. Recent advances have been made in using this property for a somewhat indirect determination of smectite (Alexiades & Jackson, 1965; Alexiades & Jackson, 1966; Coffman & Fanning, 1974).

With soil clays, certain advantages of CEC methods are apparent. Weathering of mica in soils produces layer silicate complexes that may con-tain mica, vermiculite, smectite, and chlorite all interleaved within a single particle (Jackson, 1964, p. 93; Borchardt et al., 1966). Physical separation of these four components cannot be achieved in such cases, and therefore chemical methods are necessary.

An allocation of CEC to smectite is contingent upon analyses of the other components of a soil. First, a regular CEC determination is made with Ca being replaced by Mg (CaEC). Next, the clay is K saturated, oven dried at 110C, and the amount of K replaced by ammonium is determined (K/EC). The difference between these two values is usually attributed to vermiculite and the remainder to smectite, noncrystalline material, and edge charge.

In a set of example soils tested, only variations in smectite CEC would produce significant variations in mineral percentages (McNeal, 1968). The CEC methods use 110 meq/100 g for the CEC of smectite. Harward & Brind-ley (1964) attempted to synthesize smectites with CEC's intermediate be-tween smectite and vermiculite. Their lack of success supports a discontinui-ty useful for quantitative determinations.

Studies of the CEC of noncrystalline material will further improve the determination of smectite. Presently, the Si-to-Al ratio of the noncrystalline material is used in estimating the amount of CEC to subtract from K/EC in the calculation of smectite. When the ratio is similar to that of the smectite, it is assumed that the noncrystalline material is a precursor or weathering product of smectite with the same CEC. Analysis of K/EC after selective dissolution of noncrystalline material (Rengasamy et al., 1975) would avoid these assumptions if one could be assured of no other changes in the ex-change characteristics after so drastic a treatment (Dudas & Harward, 1971). Perhaps, a more direct determination of smectite awaits development of a method specific for interlayer CEC.

D. Other Methods of Quantitative Determination

Density separation of clays continues to be of interest (Francis et al., 1970) in spite of the problem of layer silicate complexes in soils. These would have to be virtually destroyed prior to any physical separation. Density separations may still be useful for certain soils in which each clay mineral occurs in discrete particles.

IX. CONCLUSIONS

Smectite minerals play a major role in establishing the chemical and physical properties in many soils of temperate regions throughout the world. Their most important chemical properties include: high CEC, ion selectivity, and molecular sorption. Their most important physical properties include: expansion and collapse, retention of large quantities of water, high cohesion and adhesion, small particle size, and extremely large surface area. The chemical properties of smectite aid in the maintenance of soil fertility, although their physical properties make cultivation and management of soils difficult. Recent studies have emphasized interactions between smectites and trace elements, proteins, polysaccharides, herbicides, and pesticides. The huge economic loss produced by soil smectites has stimulated research in ways of ameliorating detrimental physical properties that cause soil expansion, soil creep, and landslides.

Most smectites are stable in neutral, poorly drained environments and in soils where the leaching of Si and bases is slow. Two basic types of smectites sometimes can be distinguished: (i) transformation smectites formed by loss of ions from existing silicate structures such as mica, and (ii) neogenetic smectites formed from ions precipitated directly from solution. A change in the soil environment to an acid pH or to a condition of rapid leaching causes smectites to become unstable. Weathering proceeds by dissolution, by conversion to kaolinite, or, more commonly, by polymerization of hydroxy Al and other cations between expanding layers of smectites to form pedogenic chlorite. Future research on soil smectites will continue to improve our understanding of soil environments and the beneficial changes we hope to produce in them.

X. PROBLEMS AND EXERCISES

1) What is the theoretical cation exchange capacity of a Na-saturated smectite that contains half beidellite and half nontronite?
2) Is there a relationship between soil smectites and latitude? Which state probably has the largest acreage of smectitic soils?
3) Compare the likely physical properties and fertility of smectitic soils developed from a basalt and from a muscovite granite.

4) Describe the interactions between a transformation smectite and the nutrients required for plant growth.

5) Discuss the mode of action for various chemicals that might be used to ameliorate the shrink-swell behavior of smectitic soil?

6) The clay fraction of an expansive soil had a CEC of 126.2 meq/100 g when determined with Ca^{2+} replaced by Mg^{2+}. This decreased to 75.4 meq/100 g after K^+ saturation, oven drying, and replacement with NH_4^+. Only insignificant amounts of the clay dissolved when it was boiled for 2.5 min in $0.5N$ KOH. What is the mineralogical composition of this sample according to the assumptions of Alexiades and Jackson (1966)?

XI. SUPPLEMENTARY READING LISTED BY SUBJECT

General References

Grim, R. E. 1968. Clay mineralogy. 2nd ed. McGraw-Hill, New York.

MacEwan, D. C. M. 1961. Montmorillonite minerals. p. 143–207. In G. Brown (ed.) The X-ray identification and crystal structures of clay minerals. Mineralogical Society, London.

McKeague, J. A., and M. G. Cline. 1963. Silica in soils. Adv. Agron. 15:339–396.

Marshall, C. E. 1964. The physical chemistry and mineralogy of soils. Volume 1: Soil materials. John Wiley & Sons, New York.

Origin of Smectite

Harder, Hermann. 1972. The role of magnesium in the formation of smectite minerals. Chem. Geol. 10:31–39.

Millot, Georges. 1970. Geology of clays. Springer-Verlag, New York.

Ross, G. J., and M. M. Mortland. 1966. A soil beidellite. Soil Sci. Soc. Am. Proc. 30: 337–343.

Sherman, G. D., and G. Uehara. 1956. The weathering of olivine basalt in Hawaii and its pedogenic significance. Soil Sci. Soc. Am. Proc. 20:337–340.

Tourtelot, H. A. 1974. Geologic origin and distribution of swelling clays. Bull. Assoc. Eng. Geol. 11:259–275.

Physico-chemical Properties

Baver, L. D. 1930. Relation of the amount and nature of exchangeable cations to the structure of a colloidal clay. Soil Sci. 29:291–309.

El Swaify, S. A., S. Ahmed, and L. D. Swindale. 1970. Effects of adsorbed cations on physical properties of Tropical Red and Tropical Black Earths. II. Liquid limit, degree of dispersion, and moisture retention. J. Soil Sci. 21:188–198.

Gieseking, J. E. 1939. Mechanism of cation exchange in the montmorillonite-beidellite-nontronite type of clay minerals. Soil Sci. 47:1–13.

Gillott, J. E. 1968. Clay in engineering geology. Elsevier/North-Holland, Amsterdam.

Grim, R. E. 1962. Applied clay mineralogy. McGraw-Hill, New York.

LITERATURE CITED

Adams, W. A., L. J. Evans, and H. H. Abdulla. 1971. Quantitative pedological studies on soils derived from Silurian mudstones. III. Laboratory and in situ weathering of chlorite. J. Soil Sci. 22:158–165.

Agrawal, R. P., and B. Ramamoorthy. 1970. Morphological and chemical characteristics of alkali and normal soils from black and red soils of India. Geoderma 4:403–415.

Alexiades, C. A., and M. L. Jackson. 1965. Quantitative determination of vermiculite in soils. Soil Sci. Soc. Am. Proc. 29:522–527.

Alexiades, C. A., and M. L. Jackson. 1966. Quantitative clay mineralogical analysis of soils and sediments. Clays Clay Miner. 14:35–52.

Alfors, J. T., J. L. Burnett, and T. E. Gay, Jr. 1973. Urban geology master plan for California. Calif. Div. Mines Geol. Bull. 198.

Aljibury, F. K. 1974. Managing clay soils in the home garden. OSA no. 267, Agric. Extension, Univ. of California, Davis.

Altschuler, Z. S., E. J. Dwornik, and H. Kramer. 1963. Transformation of montmorillonite to kaolinite during weathering. Science 141:148–152.

Anderson, J. U., K. E. Fadul, and G. A. O'Connor. 1973. Factors affecting the coefficient of linear extensibility in Vertisols. Soil Sci. Soc. Am. Proc. 37:296–299.

Arifin, H. F. Perkins, and K. H. Tan. 1973. Potassium fixation and reconstitution of micaceous structures in soils. Soil Sci. 116:31–35.

Arora, H. S., and J. B. Scott. 1974. Chemical stabilization of landslides by ion exchange. Calif. Geol. 27:99–107.

Bailey, G. W., and J. L. White. 1964. Soil-pesticide relationships: Review of adsorption and desorption of organic pesticides by soil colloids, with implications concerning pesticide bioactivity. J. Agric. Food Chem. 12:324–332.

Bar-On, P., and I. Shainberg. 1970. Hydrolysis and decomposition of Na-montmorillonite in distilled water. Soil Sci. 109:241–246.

Barshad, I. 1949. The nature of lattice expansion and its relation to hydration in montmorillonite and vermiculite. Am. Mineral. 34:675–684.

Barshad, I. 1952. Factors affecting the interlayer expansion of vermiculite and montmorillonite with organic substances. Soil Sci. Soc. Am. Proc. 16:176–182.

Barshad, I. 1960. Thermodynamics of water adsorption and desorption on montmorillonite. Clays Clay Miner. 8:84–101.

Barshad, I., and A. E. Foscolos. 1970. Factors affecting the rate of the interchange reaction of adsorbed H^+ on the 2:1 clay minerals. Soil Sci. 110:52–60.

Beavers, A. H. 1957. Source and deposition of clay minerals in Peorian loess. Science 126:1285.

Bingham, F. T., J. R. Sims, and A. L. Page. 1965. Retention of acetate by montmorillonite. Soil Sci. Soc. Am. Proc. 29:670–672.

Blatter, C. L. 1973. Hg-complex intergrades in smectite. Clays Clay Miner. 21:261–263.

Bohor, B. F., and R. E. Hughes. 1971. Scanning electron microscopy of clays and clay minerals. Clays Clay Miner. 19:49–54.

Borchardt, G. A., and M. E. Harward. 1971. Trace element correlation of volcanic ash soils. Soil Sci. Soc. Am. Proc. 35:626–631.

Borchardt, G. A., M. E. Harward, and E. G. Knox. 1971. Trace element concentration in amorphous clays of volcanic ash soils in Oregon. Clays Clay Miner. 19:375–382.

Borchardt, G. A., F. D. Hole, and M. L. Jackson. 1968. Genesis of layer silicates in representative soils in a glacial landscape of southeastern Wisconsin. Soil Sci. Soc. Am. Proc. 32:399–403.

Borchardt, G. A., M. L. Jackson, and F. D. Hole. 1966. Expansible layer silicate genesis in soils depicted in mica pseudomorphs. Int. Clay Conf., Jerusalem. Vol. 1:175–185.

Brandt, G. H. 1972. Soil physical property modifiers. p. 691–729. In C. A. I. Goring, and J. W. Hamaker (ed.) Organic chemicals in the soil environment. Marcel Dekker, New York.

Bray, R. H. 1937. Chemical and physical changes in soil colloids with advancing development in Illinois soils. Soil Sci. 43:1–14.

Brindley, G. W. 1961. Quantitative analysis of clay mixtures. p. 489–516. *In* G. Brown (ed.) The X-ray identification and crystal structures of clay minerals. Mineralogical Society, London.

Brindley, G. W., and D. M. C. MacEwan. 1953. Structural aspects of the mineralogy of clays and related silicates. p. 15–59. *In* A. T. Green and G. H. Stewart (ed.) Ceramics—A symposium. The British Ceramic Soc., Stoke-on-Trent.

Briner, G. P., and M. L. Jackson. 1970. Mineralogical analysis of clays in soils developed from basalts in Australia. Isr. J. Chem. 8:487–500.

Buol, S. W. 1965. Present soil-forming factors and processes in arid and semiarid regions. Soil Sci. 99:45–49.

Buol, S. W., F. D. Hole, and R. J. McCracken. 1973. Soil genesis and classification. Iowa State Univ. Press, Ames.

Carroll, Dorothy. 1959. Ion exchange in clays and other minerals. Geol. Soc. Am. Bull. 70:749–780.

Carroll, Dorothy, and H. C. Starkey. 1960. Effect of sea-water on clay minerals. Clays Clay Miner. 7:80–101.

Carson, C. D., and J. B. Dixon. 1972. Potassium selectivity in certain montmorillonitic soil clays. Soil Sci. Soc. Am. Proc. 36:838–843.

Carson, C. D., J. A. Kittrick, J. B. Dixon, and T. R. McKee. 1976. Stability of soil smectite from a Houston Black Clay. Clays Clay Miner. 24:151–155.

Chichester, F. W., C. T. Youngberg, and M. E. Harward. 1969. Clay mineralogy of soils formed on Mazama pumice. Soil Sci. Soc. Am. Proc. 33:115–120.

Christenson, D. R., and E. C. Doll. 1973. Release of magnesium from soil clay and silt fractions during cropping. Soil Sci. 116:59–63.

Clapp, C. E., and W. W. Emerson. 1972. Reactions between Ca-montmorillonite and polysaccharides. Soil Sci. 114:210–216.

Coffman, C. B., and D. S. Fanning. 1974. 'Vermiculite' determination on whole soils by cation exchange capacity methods. Clays Clay Miner. 22:271–283.

Coleman, N. T., and M. E. Harward. 1953. The heats of neutralization of acid clays and cation exchange resins. J. Am. Chem. Soc. 75:6045–6046.

Davidson, D. T. 1949. Large organic cations as soil stabilizing agents. Bull. 168, Iowa Eng. Exp. Stn.

Dixon, J. B., D. E. Moore, N. P. Agnihotri, and D. E. Lewis, Jr. 1970. Exchange of diquat^{2+} in soil clays, vermiculite, and smectite. Soil Sci. Soc. Am. Proc. 34:805–808.

Dowdy, R. H., and W. E. Larson. 1971. Tensile strength of montmorillonite as a function of saturating cation and water content. Soil Sci. Soc. Am. Proc. 35:1010–1014.

Dudas, M. J., and M. E. Harward. 1971. Effect of dissolution treatment on standard and soil clays. Soil Sci. Soc. Am. Proc. 35:134–140.

Dudas, M. J., and M. E. Harward. 1975. Inherited and detrital 2:1 type phyllosilicates in soils developed from Mazama ash. Soil Sci. Soc. Am. Proc. 39:571–577.

El Abedine, A. Z., and G. H. Robinson. 1971. A study on cracking in some Vertisols of the Sudan. Geoderma 5:229–241.

El Abedine, A. Z., G. H. Robinson, and A. Commissaris. 1971. Approximate age of the Vertisols of Gezira, Central Clay Plain, Sudan. Soil Sci. 111:200–207.

El-Attar, H. A., and M. L. Jackson. 1973. Montmorillonitic soils developed in Nile River sediments. Soil Sci. 116:191–201.

Eltantawy, I. M., and P. W. Arnold. 1974. Ethylene glycol sorption by homoionic montmorillonites. J. Soil Sci. 25:99–110.

Farmer, V. C., and J. D. Russell. 1967. Infrared absorption spectrometry in clay studies. Clays Clay Miner. 15:121–142.

Fink, D. H., F. S. Nakayama, and B. L. McNeal. 1971. Demixing of exchangeable cations in free-swelling bentonite clay. Soil Sci. Soc. Am. Proc. 35:552–555.

Francis, C. W., Tsuneo Tamura, W. P. Bonner, and J. W. Amburgey, Jr. 1970. Separation of clay minerals and soil clays using isopycnic zonal centrifugation. Soil Sci. Soc. Am. Proc. 34:351–353.

Fripiat, J. J., M. I. Cruz, B. F. Bohor, and Josephus Thomas, Jr. 1974. Interlamellar adsorption of carbon dioxide by smectites. Clays Clay Miner. 22:23–30.

Gal, M., A. J. Amiel, and S. Ravikovitch. 1974. Clay mineral distribution and origin in the soil types of Israel. J. Soil Sci. 25:79–89.

Gibbs, R. J. 1965. Error due to segregation in quantitative clay mineral X-ray diffraction mounting techniques. Am. Mineral. 50:741–751.

Gradusov, B. P. 1974. A tentative study of clay mineral distribution in soils of the world. Geoderma 12:49–55.

Greene-Kelly, R. 1955. Dehydration of the montmorillonite minerals. Mineral. Mag. 30: 604–615.

Grim, R. E., and G. Kulbicki. 1961. Montmorillonite: High temperature reactions and classification. Am. Mineral. 46:1329–1369.

Harter, R. D., and G. Stotzky. 1973. X-ray diffraction, electron microscopy, electrophoretic mobility, and pH of some stable smectite-protein complexes. Soil Sci. Soc. Am. Proc. 37:116–123.

Harward, M. E., and G. W. Brindley. 1964. Swelling properties of synthetic smectites in relation to lattice substitutions. Clays Clay Miner. 13:209–222.

Harward, M. E., D. D. Carstea, and A. H. Sayegh. 1968. Properties of vermiculites and smectites: Expansion and collapse. Clays Clay Miner. 16:437–447.

Hendricks, S. B. 1942. Lattice structure of clay minerals and some properties of clays. J. Geol. 50:276–290.

Hendricks, S. B., R. A. Nelson, and L. T. Alexander. 1940. Hydration mechanism of the clay mineral montmorillonite, saturated with various cations. J. Am. Chem. Soc. 62:1457–1464.

Hofmann, U., K. Endell, and D. Wilm. 1933. Kristallstruktur und Quellung von Montmorillonit. Z. Krystallogr. 86:340–348.

Hosking, J. S. 1940. The soil clay mineralogy of some Australian soils developed on granitic and basaltic parent materials. J. Counc. Sci. Ind. Res. (Aust.) 13:206–216.

Hower, W. F. 1970. Adsorption of surfactants on montmorillonite. Clays Clay Miner. 18:97–105.

Huang, P. M., and M. L. Jackson. 1965. Mechanism of neutral fluoride solution with layer silicates and oxides of soils. Soil Sci. Soc. Am. Proc. 29:661–665.

Huang, P. M. 1966. Mechanism of neutral fluoride interaction with soil clay minerals and silica solubility scale for silicates common in soils. Ph.D. Thesis, University Wisconsin-Madison (Diss. Abstr. 26:6269–6270).

Ingles, O. G., and J. B. Metcalf. 1973. Soil stabilization: Principles and practice. Halsted Press, New York.

Ismail, F. T. 1970. Biotite weathering and clay formation in arid and humid regions, California. Soil Sci. 109:257–261.

Jackson, M. L. 1963. Aluminum bonding in soils: A unifying principle in soil science. Soil Sci. Soc. Am. Proc. 27:1–10.

Jackson, M. L. 1964. Chemical composition of soils. p. 71–141. In F. E. Bear (ed.) Chemistry of the soil. Reinhold Pub. Co., New York.

Jackson, M. L. 1965. Clay transformations in soil genesis during the Quaternary. Soil Sci. 99:15–22.

Jackson, M. L. 1968. Weathering of primary and secondary minerals in soils. Int. Congr. Soil Sci., Trans. 9th (Adelaide, Australia) 4:281–292.

Jackson, M. L., and G. D. Sherman. 1953. Chemical weathering of minerals in soils. Advan. Agron. 5:219–318.

Johns, W. D., and R. E. Grim. 1958. Clay mineral composition of Recent sediments from the Mississippi River delta. J: Sediment. Petrol. 28:186–200.

Kantor, W., and U. Schwertmann. 1974. Mineralogy and genesis of clays in red-black soil toposequences on basic igneous rocks. J. Soil Sci. 25:67–78.

Karickhoff, S. W., and G. W. Bailey. 1973. Optical absorption spectra of clay minerals. Clays Clay Miner. 21:59–70.

Kerr, P. F., and I. M. Drew. 1969. Clay mobility, Portuguese Bend, California. Calif. Div. Mines Geol., Short Contributions to California Geology, Special Report 100: 3–16.

Kishk, F. M., and H. M. El-Sheemy. 1974. Potassium selectivity of clays as affected by the state of oxidation of their crystal structure iron. Clays Clay Miner. 22:41–47.

Kittrick, J. A. 1971. Montmorillonite equilibria and the weathering environment. Soil Sci. Soc. Am. Proc. 35:815–820.

Kittrick, J. A. 1973. Mica-derived vermiculites as unstable intermediates. Clays Clay Miner. 21:479–488.

Kleiss, H. J., and J. B. Fehrenbacher. 1973. Loess distribution as revealed by mineral variations. Soil Sci. Soc. Am. Proc. 37:291–295.

Knudson, M. I., Jr., and J. L. McAtee, Jr. 1974. Interlamellar and multilayer nitrogen sorption by homoionic montmorillonites. Clays Clay Miner. 22:59–65.

Kodama, Hideomi, G. J. Ross, J. T. Iiyama, and Jean-Louis Robert. 1974. Effect of layer charge location on potassium exchange and hydration of micas. Am. Mineral. 59: 491–495.

Kunze, G. W., and E. H. Templin. 1956. Houston Black clay, the type Grumusol: II. Mineralogical and chemical characterization. Soil Sci. Soc. Am. Proc. 20:91–96.

Levy, Rachel, and C. W. Francis. 1975. A quantitative method for the determination of montmorillonite in soils. Clays Clay Miner. 23:85–89.

Low, P. F., and J. L. White. 1970. Hydrogen bonding and polywater in clay-water systems. Clays Clay Miner. 18:63–66.

McBride, M. B., and M. M. Mortland. 1974. Copper (II) interactions with montmorillonite: Evidence from physical methods. Soil Sci. Soc. Am. Proc. 38:408–414.

MacEwan, D. C. M. 1961. Montmorillonite minerals. p. 143–207. In G. Brown (ed.) The X-ray identification and crystal structures of clay minerals. Mineralogical Society, London.

Mackenzie, R. C. (ed.). 1957. The differential thermal investigation of clays. Mineralogical Society, London.

McNeal, B. L. 1968. Limitations of quantitative soil clay mineralogy. Soil Sci. Soc. Am. Proc. 32:119–121.

Marshall, C. E. 1935. Layer lattices and the base exchange clays. Z. Krystallogr. 91: 433–449.

Mehra, O. P., and M. L. Jackson. 1959. Specific surface determination by duo-interlayer and mono-interlayer glycerol sorption for vermiculite and montmorillonite analysis: Soil Sci. Soc. Am. Proc. 23:351–354.

Mielenz, R. C., and M. E. King. 1955. Physical-chemical properties and engineering performance of clays. Calif. Div. Mines Geol. Bull. 169. Clays Clay Miner. 1:196–254.

Moore, D. E., and J. B. Dixon. 1970. Glycerol vapor adsorption on clay minerals and montmorillonitic soil clays. Soil Sci. Soc. Am. Proc. 34:816–822.

Mortland, M. M. 1970. Clay organic complexes and interactions. Advan. Agron. 22:75–117.

Mortland, M. M., and K. V. Raman. 1968. Surface acidity of smectites in relation to hydration, exchangeable cation, and structure. Clays Clay Miner. 16:393–398.

Nishita, H., and R. M. Haug. 1972. Some physical and chemical characteristics of heated soils. Soil Sci. 113:422–430.

Noll, W. 1930. Synthese von Montmorilloniten. Chem. Erde. 10:129–154.

Norrish, K. 1954. The swelling of montmorillonite. Discuss. Faraday Soc. 18:120–134.

Poncelet, G. M., and G. W. Brindley. 1967. Experimental formation of kaolinite from montmorillonite at low temperatures. Am. Mineral. 52:1161–1173.

Quirk, J. P., and R. K. Schofield. 1955. The effect of electrolyte concentration on soil permeability. J. Soil Sci. 6:163–178.

Raman, K. V., and M. L. Jackson. 1965. Rutile and anatase determination in soils and sediments. Am. Mineral. 50:1086–1092.

Rengasamy, P., V. A. K. Sarma, and G. S. R. Krishna Murti. 1975. Quantitative mineralogical analysis of soil clays containing amorphous materials: A modification of the Alexiades and Jackson procedure. Clays Clay Miner. 23:78–80.

Rex, R. W., and W. R. Bauer. 1964. New amine reagents for X-ray determination of expandable clays in dry samples. Clays Clay Miner. 13:411–418.

Ritchie, J. T., D. E. Kissel, and Earl Burnett. 1972. Water movement in undisturbed swelling clay soil. Soil Sci. Soc. Am. Proc. 36:874–879.

Robert, M. 1973. The experimental transformation of mica toward smectite: Relative importance of total charge and tetrahedral substitution. Clays Clay Miner. 21: 167–174.

Ross, C. S., and S. B. Hendricks. 1945. Minerals of the montmorillonite group, their origin and relation to soils and clays. U. S. Geol. Surv. Prof. Paper 205B:23–79.

Russell, J. D., and A. R. Fraser. 1971. Infrared spectroscopic evidence for interaction between hydronium ions and lattice OH groups in montmorillonite. Clays Clay Miner. 19:55–59.

Sakharov, B. A., and V. A. Drits. 1973. Mixed-layer kaolinite-montmorillonite: A comparison of observed and calculated diffraction patterns. Clays Clay Miner.21:15–17.

Sawhney, B. L., and M. L. Jackson. 1958. Soil montmorillonite formulas. Soil Sci. Soc. Am. Proc. 22:115–118.

Sayegh, A. H., M. E. Harward, and E. G. Knox. 1965. Humidity and temperature interaction with respect to K-saturated expanding clay minerals. Am. Mineral. 50:490–495.

Schalscha, E. B., F. T. Bingham, G. G. Galindo, and H. P. Galvan. 1973. Boron adsorption by volcanic ash soils in southern Chile. Soil Sci. 116:70–76.

Scott, R. F., and J. J. Schoustra. 1968. Soil mechanics and engineering. McGraw-Hill, New York.

Sedletski, I. D. 1937. Genesis of minerals from soil colloids of the montmorillonite group. C. R. Acad. Sci., U.S.S.R. 17:375–377.

Shainberg, I. 1973. Rate and mechanism of Na-montmorillonite hydrolysis in suspensions. Soil Sci. Soc. Am. Proc. 37:689–694.

Shainberg, I., and A. Caiserman. 1971. Studies on Na/Ca montmorillonite systems: 2. The hydraulic conductivity. Soil Sci. 11:276–281.

Sherman, G. D., H. Ikawa, G. Uehara, and E. Okazaki. 1962. Types of occurrence of nontronite and nontronite-like minerals in soils. Pac. Sci. 16:57–62.

Sorensen, L. H. 1972. Stabilization of newly formed amino acid metabolites in soil by clay minerals. Soil Sci. 114:5–11.

Sposito, Garrison. 1973. Volume changes in swelling clays. Soil Sci. 115:315–320.

Sridhar, K., M. L. Jackson, and J. K. Syers. 1972. Cation and layer charge effects on blister-like osmotic swelling of micaceous vermiculite. Am. Mineral. 57:1832–1848.

Stout, P. R. 1940. Alterations in the crystal structure of clay minerals as a result of phosphate fixation. Soil Sci. Soc. Am. Proc. 4:177–182.

Taylor, G. L., A. P. Ruotsala, and R. O. Keeling, Jr. 1968. Analysis of iron in layer silicates by Mossbauer spectroscopy. Clays Clay Miner. 16:381–391.

Theisen, A. A., and M. E. Harward. 1962. A paste method for preparation of slides for clay mineral identification by X-ray diffraction. Soil Sci. Soc. Am. Proc. 26:90–91.

Theng, B. K. G. 1971. Mechanisms of formation of colored clay-organic complexes. A review. Clays Clay Miner. 19:383–390.

Thomas, G. W., and N. T. Coleman. 1964. The fate of exchangeable iron in acid clay systems. Soil Sci. 97:229–232.

Uchiyama, N., J. Masui, and Y. Onikura. 1962. Montmorillonite in a volcanic ash soil. Soil Sci. Plant Nutr. 8:13–19.

Wear, J. I., and J. L. White. 1951. Potassium fixation in clay minerals as related to crystal structure. Soil Sci. 71:1–14.

Weaver, C. E. 1959. The clay petrology of sediments. Clays Clay Miner. 6:154–187.

Weaver, C. E. 1960. Possible uses of clay minerals in the search for oil. Clays Clay Miner. 8:214–227.

Weaver, C. E. 1963. Interpretive value of heavy minerals from bentonites. J. Sediment. Petrol. 33:343–349.

Weaver, R. M., M. L. Jackson, and J. K. Syers. 1971. Magnesium and silicon activities in matrix solutions of montmorillonite-containing soils in relation to clay mineral stability. Soil Sci. Soc. Am. Proc. 35:823–830.

Westfall, D. G., C. D. Moodie, and H. H. Cheng. 1973. Effects of drying and wetting on extractable aluminum in strongly acid soils and in aluminum-saturated soils. Geoderma 9:5–13.

White, J. L. 1971. Interpretation of infrared spectra of soil minerals. Soil Sci. 112:22–31.

White, W. A. 1949. Atterberg plastic limits of clay minerals. Am. Mineral. 34:508–512.

Wildman, W. E., M. L. Jackson, and L. D. Whittig. 1968. Iron-rich montmorillonite formation in soils derived from serpentinite. Soil Sci. Soc. Am. Proc. 32:787–794.

Yaalon, D. H., and D. Kalmar. 1972. Vertical movement in an undisturbed soil: Continuous measurement of swelling and shrinkage with a sensitive apparatus. Geoderma 8:231–240.

Yariv, S., and L. Heller-Kallai. 1973. I. R. evidence for migration of protons in H- and organo-montmorillonites. Clays Clay Miner. 21:199–200.

Zaruba, Q., and V. Mencl. 1969. Landslides and their control. Elsevier, New York.

Chlorites and Hydroxy Interlayered Vermiculite and Smectite

RICHARD I. BARNHISEL, University of Kentucky, Lexington

I. INTRODUCTION

Hydroxy forms of Al and Fe play significant roles in modifying the chemical and physical properties of minerals, in particular vermiculite and smectite. Hydroxy interlayered minerals may be considered to be a form of a solid-solution series. In this system the "pure end-members" consist of smectite or vermiculite at one end and chlorite at the other. Hydroxy interlayered minerals exist in between these end-members. Similar to a solid-solution series, the chemical composition of hydroxy interlayered minerals is variable and is a function of the environment in which they form. However, unlike most solid-solutions, the chemical composition may be a dynamic parameter reflecting changes in the environment.

Numerous articles have been published in which the role of Al or hydroxy-Al in modifications of surface and mineralogical properties and cation-exchange capacity of clays have been discussed. Modifications of these characteristics depend upon the degree to which filling of the interlayer space with hydroxy-Al occurs. Equilibrium between Al in soil solution and hydroxy-Al interlayered clays may influence a wide range of chemical reactions in soils, including adsorption of phosphate, formation of Al-phosphate compounds, and the formation of separate stable mineral phases such as gibbsite. These reactions are subjects of other chapters and hence will be only briefly discussed here. Major emphasis in this chapter will be placed on the identification, natural occurrences, laboratory synthesis, and properties of hydroxy-Al interlayers occurring in smectite and vermiculite. Hydroxy-iron and other hydroxy metal oxides which may form interlayers in vermiculite and smectite will be only briefly discussed.

II. ORIGIN AND SOURCES OF CHLORITE AND HYDROXY INTERLAYERED MINERALS

The minerals that comprise the chlorite group are commonly green-colored specimens which occur as fine-grained earthy masses. These minerals are associated with metamorphic rocks with low- to medium-grade regional metamorphism. Chlorites in soil are largely inherited as primary minerals found in metamorphic or igneous rocks or occur as alternation products from minerals such as hornblende, biotite, and other ferro-magnesian minerals. The abundance and frequency of occurrence of chlorites in soils are rela-

Chapter 10

tively low, and the geographic distribution is related to the parent material. The low frequency may be due to: the low stability of chlorite, or to the difficulty of distinguishing small amounts of chlorite in the presence of kaolinite, vermiculite, and smectite, especially if these latter minerals contain hydroxy-Al (or -Fe) interlayers.

Minerals of the chlorite group and their classification have been reviewed by Bailey (1975). Chlorites have diverse chemical compositions and some chlorites exhibit X-ray diffraction patterns that may be mistaken for minerals having 00l reflections of 7Å rather than 14Å. The trioctahedral minerals with a 7Å layer have been described in the literature as "7Å" chlorites or serpentines and include antigorite, cronstedtite, berthierine (also called chamosite), and amesite. The serpentines are discussed by J. B. Dixon in Chapter 11 of this book.

Hydroxy interlayered vermiculite and smectite occur in soils as a product of weathering. They may result from degradation of chlorite or deposition of hydroxy materials within the interlayer spaces of expansible layer silicates. Geographic distribution of hydroxy interlayered minerals is wide and may occur in the soils of several orders of the soil classification system, (Soil Survey Staff, 1975). However, their frequency of occurrence is greatest in the Ultisols and the Alfisols. Distribution of hydroxy interlayered clays within the solum may be uniform or restricted to one horizon. More frequently interlayering is greatest in the surface horizon and decreases with depth.

Reports of the occurrence of Al-interlayered minerals in soils throughout the world have been reviewed by Rich (1968). For example, Pearson and Ensminger (1949), MacEwan (1950), Brown (1953), and Rich and Obenshain (1955) recognized a 14Å soil mineral which exhibited properties similar to chlorite at room temperatures. However, the 14Å peak shifted toward a diffuse 10Å spacing when the mineral was heated and therefore did not fit the properties of either chlorite or vermiculite and smectite. Similar minerals have been reported in sediments (Grim & Johns, 1954).

III. STRUCTURAL PROPERTIES AND MINERAL IDENTIFICATION OF CHLORITE AND HYDROXY INTERLAYERED VERMICULITE AND SMECTITE

A. Idealized Structures

1. CHLORITE

The crystal structure of minerals of the chlorite group has been the subject of numerous articles. In general these minerals have been simply referred to as chlorite. Publications by Pauling (1930) and McMurcy (1934) served as the bases for subsequent crystal structure refinements, e.g.: Robinson and Brindley (1948), Brindley et al. (1950), and Bailey and Brown

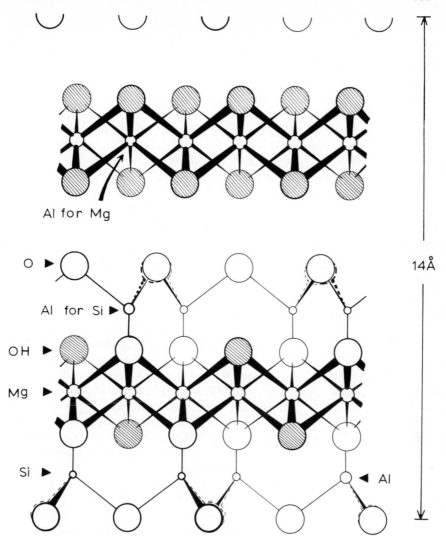

Fig. 10-1. The idealized crystal structure of trioctahedral chlorite (Figure modified from Brindley et al., 1950, and Jackson, 1964).

(1962). A comprehensive discussion of these structures may be found else-where, e.g., Brown (1961) and Bailey (1975). The layer structure of triocta-hedral chlorite, is presented in Fig. 10-1. This structure is composed of four sheets of polyhedra. Three of the sheets are chemically bonded together to form a 2:1 layer that is structurally similar to mica, consisting of two tetra-hderal sheets, one on each side of an octahedral sheet. The thickness of the 2:1 portion of the chlorite structure is about 10Å. The remaining portion of the chlorite structure has been described in the literature in several ways: an interlayer hydroxide sheet, an octahedral layer, a "brucite," or "gibbsite" layer.[1] The term interlayer hydroxide sheet, is the present accepted expression for this part of the chlorite structure (Bailey et al., 1971). This interlayer hydroxide sheet differs from the octahedral sheet in layer silicates in that it does *NOT* have a plane or planes of atoms shared with the adjacent tetrahedral sheet(s). The summation of the 2:1 layer and the interlayer hydroxide sheet results in a chlorite layer having a thickness of about 14Å.

The chemical composition of the octahedrally-coordinated cations that make up the two types of octahedral sheets for the minerals in the chlorite group is not the same even for one specific chlorite structure. More frequently either $Al(OH)_3$ or $Mg(OH)_2$ dominates the chemical composition of the interlayer hydroxide sheet in most chlorites, but cations of Fe, both Fe^{2+} and Fe^{3+}, Mn, Cr, Cu, V, Li, and Ni have been reported to occur as a part of chlorite structures. The "hydroxide" portion of the chlorite structure, is positively charged and the 2:1 part is negatively charged. The structural chemical composition of trioctahedral chlorite may be written as:

$$(R^{2+}, R^{3+})_3 (Si_{4-x}Al_x) O_{10} (OH)_2 \cdot (R^{2+}, R^{3+})_3 (OH)_6,$$

where
   $R^{2+} = Mg^{2+}, Fe^{2+}, Mn^{2+}, Ni^{2+}$, and
   $R^{3+} = Al^{3+}, Fe^{3+}, Cr^{3+}$.

The chlorite group may be divided into three subgroups as the result of differences in the chemical composition of the interlayer hydroxide and the octahedral sheets. These include: dioctahedral chlorite; di,trioctahedral chlorite; and trioctahedral chlorite (Bailey et al., 1971). Dioctahedral chlorite is dioctahedral in both the 2:1 layer and the interlayer hydroxide sheet. Di,trioctahedral chlorite is dioctahedral in the 2:1 layer, but trioctahedral in the interlayer hydroxide sheet, and trioctahedral chlorite is trioctahedral in both the 2:1 layer and the interlayer hydroxide sheet. Structural formulas for di,trioctahedral and dioctahedral chlorite may be written by changing the appropriate subscripts of the ions designated with letter R. In the above formula, if R^{2+} is written first followed by R^{3+}, then the subscript should be three, and this "octahedral" portion would be trioctahedral. For designation of a dioctahedral sheet, R^{3+} should precede R^{2+} in the formula and the subscript would be two.

2. HYDROXY INTERLAYERED VERMICULITE AND SMECTITE

The layer structures of hydroxy interlayered minerals are largely dependent upon three factors: (i) the structure of the basic or 2:1 portion of the mineral, e.g., vermiculite, smectite, chlorite, etc.; (ii) the degree of interlayer filling; and (iii) the chemical composition of the hydroxy materials that occur within the interlayer portion of these clay minerals. The structures of vermiculite, smectite, etc. are presented in other chapters and therefore will

---

[1]The names brucite and gibbsite, although referred to in the literature as being a part of chlorite and other silicate clay mineral structures, are separate minerals with chemical compositions $Mg(OH)_2$ and $Al(OH)_3$, respectively.

not be discussed here. However, differences in these structures may affect interlayer formation, and these factors will be presented later in this chapter. Hydroxy interlayered minerals may be described as chlorite-like minerals but the interlayer hydroxide sheet of the chlorite structure is incomplete.

A six-membered ring structure, $Al_6(OH)_{15}^{3+}$, has been proposed for the interlayer material in hydroxy interlayered clays (Hsu & Rich, 1960; Jackson, 1960) after a model given by Brosset et al. (1954) for a hydrolysis product of Al. The six-membered ring structure (see Fig. 4–2 in Chapter 4 by P. H. Hsu, this book) is perhaps the smallest polymer that may exist in clays that is not subject to exchange by other cations. In otherwords, it is "fixed" by the clay and referred to as nonexchangeable Al. Larger, more complex polymers of hydroxy-Al (see Fig. 4–13 in Chapter 4 by P. H. Hsu, this book) have been proposed by Jackson (1962, 1963) and Hsu and Bates (1964a, 1964b). Progressive filling of the interlayer space by such hydroxy-Al polymers may take place either by adding more polymers composed of a six-membered ring structure or by progressive building of larger, more polymerized, structures which eventually may form a complete layer and the mineral becomes chlorite and in this case an "aluminous" chlorite. Because of the variable nature of hydroxy interlayered clay minerals, a structural chemical composition would have little meaning. It would have a form similar to that given for chlorite.

As the result of the difficulty in identifying the presence and layer structure of hydroxy interlayered forms of clay minerals in soils, numerous names have been used in the literature to describe the interlayered material. The diagnostic methods used in identification of clay mineral species may also result in placement of hydroxy interlayered clays with other minerals, e.g., chlorite, interstratified mica-chlorite, etc. Those listed below are more or less synonymous.

1) A 14Å mineral, Pearson and Ensminger (1949)
2) A dioctahedral analogue of vermiculite, Brown (1953)
3) A chlorite-like mineral, Klages and White (1957)
4) A dioctahedral vermiculite mineral, Rich and Obenshain (1955)
5) Interstratified chlorite-vermiculite, Jackson et al. (1954), Dixon and Seay (1957)
6) Intergradient, Dixon and Jackson (1959)
7) Intergradient chlorite-expansible layer silicate, Dixon and Jackson (1962)
8) Intergradient chlorite-vermiculite, Bryant and Dixon (1964)
9) Intergrade, Dixon and Jackson (1962)
10) 2:1–2:2 intergrade, Malcolm et al. (1968)
11) Intergradient chlorite-vermiculite-montmorillonite, Dixon and Jackson (1960)
12) Hydroxy-Al or aluminum interlayers, Rich and Obenshain (1955), Sawhney (1958), Rich (1960a)
13) Chloritized expansible layer silicate, Glenn and Nash (1964)
14) Chlorite-vermiculite-intergrade, McCracken and Weed (1963)
15) 14Å montmorillonite-vermiculite-chlorite intergrade, Nash (1963).

B. Identification by X-ray Diffraction

1. CHLORITE

Chlorite has a $d(001)$-spacing of 14.0 to 14.4Å, depending on the species. Substitutions of Al for Si in tetrahedral sites and Fe^{2+} for Mg or Al in octahedral sites of various chlorite species are largely responsible for the range in $d(00l)$-spacings. This d-spacing is not significantly affected by ionic saturation, solvation with glycerol or ethylene glycol, and by heat treatment; however, peak intensities may be altered by heat treatments. Upon heating to temperatures as high as 550C (823K), the peak intensity of the 001 reflection (14Å) increases by as much as two to five times and at the same time, the peak intensities of higher ordered $00l$ reflections decrease in intensity.

Basal reflections from chlorite crystals may coincide with reflections from vermiculite, montmorillonite or other smectites, and from hydroxy interlayered forms of these clay minerals under certain conditions of ionic saturation and heat treatment. The 001 reflection for chlorites is most generally used for identification. Identification may become more complicated if the chlorite is Fe-rich because the 001, 003, and other "odd" ordered reflections have weak intensities, whereas the "even" reflections are strong. In such a case, especially if there were a small amount of the Fe-rich chlorite present in the sample, it could be mistaken for kaolinite. Therefore, a series of X-ray diffraction patterns from specimens subjected to chemical and heat treatments is normally needed to verify the presence and to determine the amount of chlorite in a sample.

Identification of chlorite in a mixture with kaolinite is difficult, especially for soil samples and recent sediments. Chlorite-kaolinite differentiation may be obtained for well-crystallized clays by the separation of the 004 reflection of chlorite and the 002 reflection of kaolinite (Bradley, 1954). However, as pointed out by Brindley (1961), these peaks from soil clays will tend to be broad and little or no resolution of the 004 chlorite and the 002 kaolinite reflections should be expected. Differentiation may be obtained at higher ordered reflections; however, in routine analysis, X-ray scanning is normally terminated at about a d-spacing equivalent to 3Å. If chlorite is present in a low concentration, peak intensities of chlorite may not be sufficient for identification. Intercolation of kaolinite with ammonium nitrate may also be used to differentiate between kaolinite and chlorite (Andrew et al., 1960). Such treatment expands kaolinite and the 001 peak shifts from 7.16Å to 11.6Å.

If kaolinite and chlorite are well-crystallized, thermal treatments may be used to differentiate between these minerals. Chlorite maintains a 14Å d-spacing for heat treatments of several hours at 550C (823K) whereas kaolinite decomposes. As discussed earlier, the intensities of the chlorite reflections will change with this treatment. However, this technique may not be valid for poorly crystallized soil clays, as chlorites have been reported to decompose at temperatures as low as 450C (723K) whereas kaolinite was not appreciably affected (Brindley, 1961). However, the chlorite, in this case, could have been a degraded variety and similar to hydroxy interlayered forms of vermiculite or smectite.

X-ray identification of chlorite in presence of smectite is accomplished by saturating the specimen with Ca^{2+} or Mg^{2+} ions and the addition of glycerol or ethylene glycol. This latter treatment expands smectite to 17–18Å whereas chlorite remains unaffected. A K-saturation treatment will result in at least partial collapse of smectite to about 12Å when X-rayed at room temperature [25C (298K)] in the absence of glycerol or ethylene glycol and if the smectite does not contain appreciable amounts of hydroxy interlayers. Upon heating to 110C (383K) the K-saturated smectite will collapse to 10Å if hydroxy interlayers are not present.

Differentiation between chlorite and vermiculite can be accomplished with K-saturation. Vermiculite containing little or no hydroxy interlayering will collapse to 10Å upon K-saturation at room temperature. Therefore, the persistence of a 14Å peak after K-saturation and heat treatment of 300C (573K) and subsequently at 550C (823K) confirms the presence of chlorite. Collapse to 10Å on heating confirms smectite or vermiculite.

Chlorite may also exist as a part of regularly or randomly interstratified minerals. These include mica-chlorite, vermiculite-chlorite, smectite-chlorite, and chlorite-swelling chlorite, the X-ray properties of which have been elaborated upon by B. L. Sawhney in Chapter 12 of this book. The presence of these interstratified minerals presents an additional problem of mineral identification which may be further complicated if hydroxy interlayered forms of vermiculite or smectite are present. When the concentration of chlorite is low ($< 25\%$) one must rely on additional methods for mineralogical identification, such as thermal and chemical analysis, including data from closely associated samples (such as adjacent soil horizons) where these specimens may produce distinct X-ray characteristics for one particular mineral species.

2. HYDROXY INTERLAYERED SMECTITE AND VERMICULITE

The X-ray diffraction properties of hydroxy interlayered clay mineral forms are intermediate between those of the "end-members," vermiculite or smectite, and chlorite. Similarities between the X-ray patterns of hydroxy interlayered forms and the end-members are related to the degree of filling that occurs within the interlayer space. X-ray patterns from Mg-saturated glycerol or ethylene glycol treated samples of hydroxy interlayered minerals are similar to those for vermiculite, and chlorite; hence, this treatment is not useful except to perhaps differentiate chlorite from smectite and swelling chlorite. The response to heat treatment of K-saturated samples is the most useful X-ray diffraction criterion in recognizing the presence of hydroxy interlayed forms. However, as the degree of filling of the interlayer space with hydroxy polymers approaches completeness it becomes increasingly more difficult to differentiate this mineral from chlorite. Similar problems

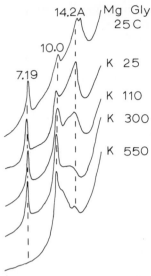

Fig. 10-2. Smoothed X-ray diffraction tracings of clay from the Ap horizon of a Maury silt loam soil. (Unpublished data from sample S72 Ky 105-6 obtained near site reported by Hutcheson, 1963).

exist when only a few polymers occupy the interlayer space, as X-ray diffraction patterns then approach those of vermiculite or smectite.

In most cases, X-ray patterns obtained at room temperature for hydroxy interlayered clays would be similar to those of chlorite when K-saturated. However, incomplete filling results in the lack of stability of the 14Å peak when these K-saturated specimens are heated from 25C (298K) to 110, 300, and 550C (383, 573, and 823K). The temperature required to obtain collapse (or partial collapse) of the 14Å peak toward 10Å may be used to estimate relative degree of filling; the higher the temperature to obtain a shift of the 14Å peak, the larger the degree of filling. Even with low levels of hydroxy interlayering, collapse to exactly 10Å upon heating to 550C (823K) is not usually obtained and the d-spacing is in the range of 10.2-10.5Å. Therefore, the magnitude of the shift in d-spacing of the 001 reflection upon K-saturation and heating to 550C (823K) is also a measure of the degree of filling; small changes from 14Å indicate large amounts of filling.

An example of X-ray diffraction patterns from a soil clay exhibiting properties of hydroxy interlayered vermiculite is illustrated in Fig. 10-2. The upper tracing of a Mg-saturated, glycerol-solvated specimen was obtained from the 2-0.2 μm fraction of the Ap horizon of the Maury silt loam soil.[2] This specimen was mounted on ceramic tiles by the method described by C. I. Rich and R. I. Barnhisel in Chapter 23 of this book. The other tracings were obtained from a sample after K-saturation and progressive heat treatments. A small 10Å peak in the Mg-saturated pattern indicates that the

---

[2]Unpublished data by the author for sample (S72Ky 105-6) collected near the site described by Hutcheson (1963). The Maury soil has been classified as a Typic Paleudults, clayey, mixed mesic.

soil sample contains some mica. Presence of the 14.2Å peak in the tracing from the K-saturated sample indicates a hydroxy interlayered form of vermiculite or smectite, or chlorite. The 7Å peak also indicates that kaolinite may be present, although this peak also corresponds with the 002 reflections from interlayered smectite or vermiculite and chlorite. As the specimen was heated to 110C (383 K) and then to 300C (573K), the 14Å peak shifted toward 10Å, indicating a hydroxy interlayered form of vermiculite or smectite. The increase in intensity of the 10Å peak between the 25C (298K) and 110C (383K) treatment may indicate the presence of some vermiculite. The pattern from the 550C (823K) heated sample indicates that the sample contains a significant amount of kaolinite since the 7Å peak is no longer present. This treatment also indicates the presence of some chlorite since a 14Å peak remained. The hump or shoulder on the low angle side of the 10Å reflection may be from the hydroxy interlayered form of vermiculite. This type of a pattern has been described in the literature as an increased central scattering (MacEwan, 1950) and a broadening of the 10Å peak (Jackson, 1962). X-ray diffraction patterns from samples with partial or nonuniformly filled interlayer space may exhibit a sequence of 00l reflections similar to a type of interstratification (Cotton, 1965[3]; Barnhisel & Rich, 1966).

C. Identification by Thermal Analysis

1. CHLORITE

Dehydration curves presented by Nutting (1943) for chlorite indicate that very little weight loss occurred until chlorite samples were heated beyond 500C (773K) (Fig. 10-3). Between 500-550C (773-823K) weight loss

---

[3]S. B. Cotton. 1965. Hydrolysis of aluminum in synthetic cation resins and dioctahderal vermiculite. Ph.D. Thesis, Virginia Polytechnic Inst., Blacksburg, Va.

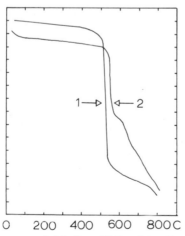

Fig. 10-3. Dehydration curves of chlorites, (1) Penninite, Paradise Range, Nevada; (2) chlorite, Danville, Virginia (after Nutting, 1943).

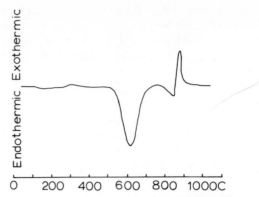

Fig. 10-4. Differential thermal analysis tracing of a < 2 μm sample of chlorite from Colorado (Obtained from Wards Natural Science Est.) Heating rate, 10C (283K)/min.

occurred as the result of the formation of water from the dehydroxylation of the 2:1 octahedral sheet and the interlayer hydroxide sheet. Gradual loss of weight was observed as the sample was heated further to 880C (1073K). The curves of Nutting were obtained by heating the sample at a given temperature until no further weight loss occurred. When chlorites are heated in a DTA apparatus, one should expect the large weight loss [observed in the temperature range of 500-550C (773-823K)] to produce an endothermic peak, but displaced to a higher temperature. A DTA tracing in Fig. 10-4 illustrates this dehydroxylation reaction, although the source of the chlorite specimens used in these two experiments was not the same. Structural breakdown of the chlorite crystals is associated with the weight loss upon heating.

The temperature at which loss of water from the chlorite structure takes place has been shown to be related to the cation dominating both the interlayer hydroxide sheet and the octahedral sheet of the 2:1 layer. (Table 10-1). The dehydration studies illustrated here are from well-crystallized chlorites. Soils would not be expected to contain this quality of chlorite, and chlorite should not be expected to dominate the clay mineralogy of most soils. The role of thermal analysis in identification of chlorite may be limited to verification of other mineralogical data. Chlorite determination from water loss in thermal gravimetric analysis has been proposed by Alexiades and Jackson (1966).

Table 10-1. Effect of chemical composition of chlorite on the dehydration temperature (after Caillère and Hénin, 1960).

Dominate cation	Hydroxide sheet	Octahedral sheet of 2:1 layer
	°C	
Mg^{2+}	640 (913K)	820 (1093K)
Al^{3+}	500 (773K)	750 (1023K)
$Fe(II)^{2+}$	430 (803K)	530 (803K)
$Fe(III)^{3+}$	250 (523K)	--

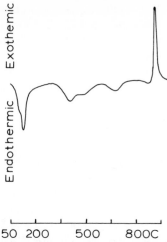

Fig. 10-5. Differential thermal analysis tracing of synthetic hydroxy-Al interlayered montmorillonite. OH/Al molar ratio = 1.50, heating rate 12C (285K)/min (after Barnhisel & Rich, 1963).

2. HYDROXY INTERLAYERED FORMS OF SMECTITE AND VERMICULITE

For soil clays, DTA techniques are often not diagnostic for hydroxy interlayered minerals, particularly for situations where these minerals are present in low concentrations. Such techniques have been used to confirm the presence of hydroxy materials if these minerals dominate the mineralogy (Rich & Obenshain, 1955; Glenn & Nash, 1964) or if prepared synthetically (Shen & Rich, 1962; Barnhisel & Rich, 1963; Brydon & Kodama, 1966). Two endothermic peaks at about 360 (633K) and 470C (743K) have been attributed to hydroxy interlayers (Fig. 10-5). Weight losses found in TGA tracings were associated with hydroxy interlayer formation and stability Barnhisel & Rich, 1963).

D. Identification by Infrared Analysis

1. CHLORITE

Characterization of chlorites by infrared absorption resulted in the assignment of numerous absorption bands that were related to the chemical composition (Hayashi & Oinuma, 1965). The most useful region was that between 450 (823K) and 900 (1173K) cm^{-1} and as the amount of Al increased in the interlayer hydroxide sheet with the associated decrease of Mg and Fe, the wave number of the absorption peak in the region of 540 and 560 cm^{-1} increased.

The presence of absorption bands in the OH-stretching region (3000-4000 cm^{-1}) has been related to the two types of hydroxyl sheets (Farmer,

1974). The use of D_2O in synthetic systems to produce chlorite-like minerals may provide further evidence that the OH ions of the interlayer hydroxide sheet and the octahedral sheet can be differentiated in infrared absorption spectra (Ahlrichs, 1968).

Infrared absorption may be a useful means of determining the amount of substitution of Al for Si as well as a technique to measure the Fe content of chlorite. The Si–O stretching band in the region of 110 to 950 cm^{-1} appears to be most diagnostic (Tuddenham & Lyon, 1959; Hayashi & Oinuma, 1967).

2. HYDROXY INTERLAYER FORMS OF SMECTITE AND VERMICULITE

The OH-stretching region has been used to substantiate the presence of hydroxy interlayers in smectite (Brydon & Kodama, 1966; Weismiller et al., 1967). The absorption bands near 3600 cm^{-1}, and 3700 cm^{-1} were observed in hydroxy-Al interlayered montmorillonite. The research of Ahlrichs (1968) expanded the use of infrared as evidence for interlayering of metal complexes to vermiculite and hectorite systems. The OH-stretching frequencies associated with the metal use have values of 3660, 3690, and 3710 cm^{-1} for Ni-, Al-, and Mg-interlayers, respectively, in montmorillonite. Infrared absorption results of Ahlrichs (1968) corroborate the proposed six-membered polymer ring structure of hydroxy interlayers.

E. Cation-exchange Properties of Chlorite and Hydroxy Forms of Vermiculite and Smectite

The cation-exchange capacity (CEC) of chlorite is relatively small with a range in values between 10–40 meq/100 g. The magnitude of the charge of the 2:1 portion of chlorite is similar to that of mica. This negative charge is largely due to substitution of Al^{3+} and occasionally of Fe^{3+}, or Cr^{3+} for Si^{4+} of the tetrahedral portion of the structure. The hydroxide interlayer sheet of chlorite has a positive charge as the result of substitution of trivalent cations such as Al^{3+} or Fe^{3+} for divalent cations such as Mg^{2+} or Fe^{2+}. This substitution partially neutralizes the negative charge of the 2:1 portion. The degree to which substitution occurs is a partial explanation for a range of CEC values. If the hydroxide sheet is partially removed upon weathering, the CEC of the residual material would increase.

The CEC of hydroxy interlayered forms of vermiculite and smectite has a wide range. It may approach low values as filling of the interlayer space becomes complete or it may approach the CEC of vermiculite or smectite.

The CEC of soil clays has been increased by extraction of interlayer-Al (Rich & Obenshain, 1955; Klages & White, 1957; Sawhney, 1960b, 1960c; Dixon & Jackson, 1962; Rich & Cook, 1963; Clark, 1964a, 1964b; Frink, 1965; Reneau & Fiskell, 1970; Lietzke & Mortland, 1973). The hydroxy interlayers may affect the CEC by (i) occupancy of exchange sites, (ii) physical blocking of the exchange reaction by not allowing the saturating cation to come in contact with the exchange sites, and (iii) influence of hydroxy material on the pH and vice versa. The last factor may interact with the methods used to measure the CEC and thereby be a partial explanation for the pH-dependent CEC of soils.

Coleman and Thomas (1967) concluded that the CEC of acid soils is largely dependent upon the method of determination, and in particular the pH of the extracting solution. In general, the higher the pH of this solution, the higher the CEC, and the degree of pH-dependent charge is related to the amount of hydroxy-Al naturally present in the soil or that formed in the process of the CEC determination. As the pH of the soil containing hydroxy interlayered minerals is increased, the amount of hydrolysis of the Al polymers is increased, or neutralization of the positively charged polymers forms neutral Al $(OH)_3$. If the pH is increased to high values (pH $>$ 7.5) during the measurement of the CEC, $Al(OH)_3$ may be further hydrolyzed to form an aluminate ion $Al(OH)_4^{1-}$. The aluminate ion may also contribute to the CEC of the system.

F. Characterization of Hydroxy Interlayer Materials by Use of Conductometric and Potentiometric Titration

The conductometric and potentiometric methods have largely been used to determine the presence and amount of Al^{3+} (as well as H_3O^+). Several researchers have used these methods to characterize the form of Al, or its degree of hydrolysis, in association with hydroxy interlayers. Schwertmann and Jackson (1964) described titration curves of hydronium-aluminum clays, and attributed the third buffer region to hydroxy-Al interlayers. Sawhney and Frink (1966) proposed that the third buffer region was produced by an alumino-silicate complex liberated by acid decomposition of the clay during the preparation of the acid clays and hydroxy-Al formation. When a base of a divalent rather than a monovalent cation is used for the titer the forms of Al are more easily differentiated (Coleman & Thomas, 1967; Kissel & Thomas, 1969; Rich, 1970; Dewan & Rich, 1970).

G. Removal of Hydroxy Interlayers

Various reagents have been used to remove cementing agents from soils to improve X-ray identification of the crystalline minerals present. Many of these extracting agents have also been used to remove hydroxy interlayers. The reagents used are assumed to extract the hydroxy material without materially altering the clay mineral structure. Methods that have been reported to at least partially remove hydroxy interlayers are listed below:

KOH + KCl—Brown, 1953

NH_4F—Rich and Obenshain, 1955

NH$_4$-citrate—Klages and White, 1957
F-resin—Huang and Jackson, 1966
Citric Acid—Klages and White, 1957
Na-citrate—Tamura, 1957, 1958; Sawhney, 1960b, 1960c
Na-dithionate-citrate-bicarbonate—Tamura, 1955; Arshad et al., 1972
400C (673K) dehydroxylation-NaOH—Dixon and Jackson, 1959, 1962
NH$_4$F + HCl + NH$_4$Cl—Rich and Cook, 1963; Rich, 1966

Analysis of the extract should be made in order to determine if the re-agent used extracted a portion of crystalline minerals. When interlayers are removed, the remaining clays respond to ionic saturation and heat treatments that result in X-ray patterns similar to standard smectite or vermiculite.

IV. LABORATORY SYNTHESIS OF HYDROXY INTERLAYERED VERMICULITE AND SMECTITE

Several methods have been used to prepare synthetic hydroxy interlayers in clays. Ojbectives of these experiments were: (i) to produce a mineral similar to chlorite-like minerals found in soils; (ii) to determine the physical nature of hydroxy material, i.e., size and/or shape of polymerized hydroxy complexes; (iii) to determine the chemical composition of this material, i.e., the OH/Al molar ratio of interlayer material; and (iv) to evaluate the effect of the amount of these hydroxy materials on the chemical, physical, and mineralogical properties of the clay minerals.

The early work of Caillère and Hénin (1949, 1950), Longeut-Escard (1950), and Youell (1960), as well as that of Slaughter and Milne (1960), was conducted in an attempt to prepare chlorite from smectite by the addition of hydroxy complexes of Al, Mg, Fe, and Ni. Although these chlorite-like minerals were not as heat stable, and did not exhibit the same X-ray diffraction intensities for the various 00l reflections as natural chlorite, they were similar in many ways.

Hydroxy interlayers in vermiculite and smectite also have been pre-pared to produce minerals similar to naturally-occurring interlayered clay minerals.

In early work (Slaughter & Milne, 1960), hydroxy interlayered clays were prepared in which the cations were neutralized by the addition of equivalent amounts of an appropriate base such as NaOH. In later experiments, a wide range of the OH/cation ratios have been included (Sawhney, 1960a; Shen & Rich, 1962; Barnhisel & Rich, 1963; Coleman & Thomas, 1964; Coleman et al., 1964; Hsu & Bates, 1964a; Schwertmann & Jackson, 1964; Turner, 1965; Turner & Brydon, 1962, 1965, 1967a, 1967b; Singh, 1967; Hsu, 1968a; Sawhney, 1968; Barnhisel, 1969; Gupta & Malik, 1969a, 1969b; Carstea et al., 1970a, 1970b, 1970c; Herrera & Peech, 1970; Hunsaker & Pratt, 1970a, 1970b; Richburg & Adams, 1970; Colombera et al., 1971; Blatter, 1973; Brown & Newman, 1973). The above methods may be grouped into two general categories, (i) precipitation of the hydroxy-complex in the presence of clay, and (ii) precipitation of the complex and then adding it to the clay. As long as the reagents are added slowly to dilute clay suspensions, interlayers can be obtained by either method.

The stability upon aging of synthetic hydroxy interlayers has been shown to be a function of the OH/cation molar ratio, time of aging, method of preparation of hydroxy interlayers, and the specific clay mineral into which the interlayers are introduced.

The stability of synthetic interlayers is also influenced by other factors: (i) amount of hydroxy material (Turner & Brydon, 1965, 1967a; Brydon & Kodama, 1966); (ii) anions or salt present during formation or preparation of interlayers (Barnhisel & Rich, 1965; Singh & Brydon, 1967); (iii) addition, or formation of gibbsite in the system (Turner & Brydon, 1967b; Barnhisel, 1970[4]); (iv) temperature at which aging is conducted (Carstea, 1968; Singh, 1972); (v) particle size of clay (Carstea, 1968); (vi) the nature of clay surfaces treated with hydrous oxide (Jackson et al., 1973). On aging, synthetic interlayers decrease the CEC of the mineral and may result in the formation of crystalline hydrous oxide minerals such as gibbsite. Gibbsite formed within a few weeks in systems with OH/Al molar ratios of 3.0 and in 3 months for systems with a ratio of 2.25. Interlayers with OH/Al molar ratios of 1.50 and less were stable even when aged for 6 months (Barnhisel & Rich, 1963). However, Carstea et al. (1970a) reported interlayers to increase in stability with aging up to 1 year in montmorillonite systems prepared by the same method, but with an OH/Al molar ratio of 2.43. No gibbsite was formed in this system.

Interlayers have been formed in the laboratory in which hydroxy-Fe, -Ni, and -Mg were added to smectite (Slaughter & Milne, 1960; Coleman et al., 1964; Carstea, 1968). These hydroxy metal-smectite complexes produced X-ray diffraction patterns similar to chlorite as well as to Al-interlayered smectites. Iron interlayers have also been found in soil clays (Singleton & Harward, 1971).

A. Nature of Hydroxy Interlayers Fixed by the Clay

Brosset et al. (1954) suggested a six-membered ring structure for the hydrolysis product of Al, having a formula of $Al_6(OH)_{15}^{3+}$ and an OH/Al molar ratio of 2.5. As discussed earlier, a similar structure was proposed for hydroxy-Al polymers fixed by exchange resins (Hsu & Rich, 1960; Jackson, 1960). It has since been suggested by many researchers that hydroxy interlayers in clays may consist of these or similar polymers (Jackson, 1962, 1963; Hsu & Bates, 1964a, 1964b).

Although a wide range of OH/Al molar ratios for the entire system has been used in preparing synthetic interlayered clays, evidence has accumu-

---

[4]R. I. Barnhisel. 1970. Stability of gibbsite in the presence of Al-interlayered soil clays. Agron. Abstr. p. 169.

lated which suggests that the OH/Al molar ratio of the "fixed" or nonexchangeable Al occurs in a rather narrow range of about 2.5 to 2.7. In systems in which the overall OH/Al molar ratio is < 2.5, Al may exist in two
forms, Al^{3+} and a polymer with an OH/Al molar ratio between 2.5 and 2.7.
Likewise, for systems with an "overall" OH/Al molar ratio > 2.7, Al exists as
the polymer described above, and/or as $Al(OH)_3$. If the OH/Al ratio is sufficiently high $Al(OH)_4^{-1}$ is produced, however, interlayer aluminate would
be remote. Turner (1965) reported that if the OH/Al molar ratio of the
"whole" system exceeded 2.7, then the polymers were metastable and the
solubility of gibbsite would control the OH/Al ratio (Turner & Brydon,
1965, 1967b). It has been suggested that Al^{3+} is the only major exchangeable Al species from partially neutralized Al-treated clay systems (Thomas,
1960; Frink & Peech, 1963; Barnhisel & Rich, 1963; Hsu, 1968a; Dewan &
Rich, 1970).

Several researchers have reported data for the reduction of the CEC of
clays as the result of adsorption of hydroxy-Al (Hsu, 1968a, 1968b; Brydon
& Kodama, 1966; Turner, 1965, 1967; Kozak & Huang, 1971; Huang & Lee,
1969; Brown & Newman, 1973; Carstea et al., 1970a). Reduction in CEC as
a function of hydroxy-Al adsorbed or fixed can also be used to determine
the composition of interlayers. Data from systems in which gibbsite or other
crystalline compounds were reported cannot be used since differentiation of
the amount of Al "fixed" by the clay and that existing as a separate crystalline phase is uncertain. When the amount of Al fixed was plotted against the
reduction in CEC, a linear relationship was found (Fig. 10-6). The slope of
this line gave a value of 2.7 for the OH/Al molar ratio of the adsorbed
hydroxy-Al interlayers. The regression line does not pass through zero sug-

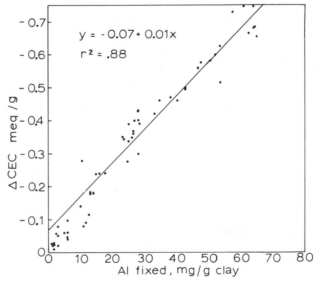

Fig. 10-6. Relationship of Al fixed by clays and the resultant decrease in cation-exchange capacity.

gesting that at low levels of Al adsorption, the OH/Al molar ratio is slightly lower. If adsorption values of < 20 mg Al fixed per g of clay were used, the calculated OH/Al molar ratio would approach 2.5.

Data used in Fig. 10-6 include a few values for kaolinite and illite systems, however the correlation coefficient and the overall OH/Al molar ratio for the adsorbed hydroxy-Al polymers were unchanged as a result of using these data. Unfortunately, it was not possible to use data from other published articles because of insufficient information. However, the conclusion as to the narrow range of OH/Al molar ratios for fixed Al is supported by several other observations in both natural and synthetic systems (Schwertmann & Jackson, 1964; Singh & Brydon, 1967; de Villiers & Jackson, 1967a, 1967b; Richburg & Adams, 1970; Kirkland & Hajek, 1972).

B. Comparison of Synthetic Vs. Naturally Interlayered Clays

In most cases, smectite with synthetic interlayers resembled natural hydroxy interlayered smectite in mineralogical and chemical properties. This has not been the case however, for vermiculite, particularly with respect to mineralogical properties. Synthetically prepared hydroxy interlayered vermiculite reported in the literature most frequently did not remain at a 14Å d-spacing when K-saturated or when subjected to heat treatments. Naturally interlayered vermiculites can be heated to at least 110 or 300C (383–573K) with little change in the d-spacing, whereas synthetically prepared clays tend to collapse toward a 10Å d-spacing at a lower heat treatment (Barnhisel, 1965).[5] Interlayer formation in vermiculite may need more controlled conditions than for smectite (Barnhisel & Rich, 1966). Using infrared techniques, Ahlrichs (1968) demonstrated that hydroxy-Al interlayers did not form in vermiculite where the same methods were employed as those commonly used to produce interlayers in smectite.

Several factors may be responsible for the lower stability of synthetic interlayers in vermiculite. The formation of an "atoll" structure (Frink, 1965) in contrast to a uniform distribution of hydroxy-Al polymers is illustrated in Fig. 10-7. The concept of two types of structures is supported by the relationship of surface area and CEC (Barnhisel, 1969) and by the work of Sawhney (1968), Novak et al. (1971), and Kozak and Huang (1971). The "atoll" structure results in a steric blocking of exchange sites occurring within the center of the particles and insufficient numbers of props. Thus, hydroxy-Al cannot support or maintain the crystal structure at a 14Å d-spacing when heated and K-saturated and the structure collapses toward 10Å. This latter mineralogical property is significantly different from synthetically interlayered smectite systems.

The interlayer space between sheets of vermiculite particles is less than for smectite and the particles of vermiculite are generally larger and most

---

[5] R. I. Barnhisel. 1965. The formation and stability of aluminum interlayers in clays. Ph.D. Thesis. Virginia Polytechnic Inst., Blacksburg, Va.

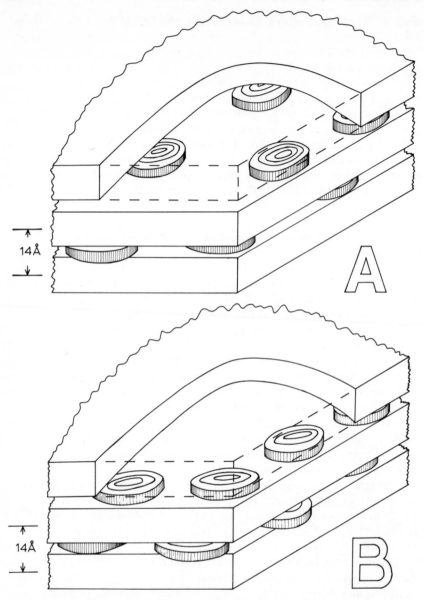

Fig. 10–7. Illustration of the distribution of hydroxy-Al polymers in the interlayer space of 2:1 clay minerals: *(A)* a uniform distribution; *(B)* an "atoll" arrangement (Figure modified from Dixon & Jackson, 1962).

likely more rigid. The charge density of vermiculite is higher than smectite and this may result in more frequent deposition of polymers along the crystal edges; the OH/Al ratio of polymers in vermiculite may be smaller than in smectite. Each of these factors may contribute to an interlayer material that undergoes dehydration at a lower temperature. In contrast, in natural systems interlayers probably formed over long time periods so that the above factors are diminished in importance.

Although optimum conditions for hydroxy interlayer formation are not clearly known, Rich (1968) summarized the favorable conditions as follows: "Moderately active weathering must be in progress or have occurred to furnish Al ions. Furthermore, the pH should be moderately acid, about pH 5.0, organic matter content should be low, and there should be frequent wetting and drying of the soil."

V. CHANGES IN PROPERTIES OF SMECTITE AND VERMICULITE AS A RESULT OF HYDROXY INTERLAYERING

A. Physical Properties

The swelling of Na-montmorillonite in NaCl solutions was reduced by the formation of hydroxy-Fe and -Al interlayers (El Rayah & Rowell, 1973). Swelling was reduced to a greater degree with hydroxy-Al than hydroxy-Fe. In addition, the tensile strength, liquid limit, shear-stress, and shrink-swell properties of montmorillonite clay systems may be significantly altered by hydroxy interlayering and the accompanying changes in cation saturation (Dowdy & Larson, 1971; Gray & Schlocker, 1969; Davey & Low, 1971; Kidder & Reed, 1972).

The removal of hydroxy interlayer material has resulted in increased swelling of clays. This was demonstrated by Tamura (1957) in which a soil clay exhibited the 18Å X-ray diffraction properties of smectite after interlayer removal and saturation with Ca and glycerol.

B. Anion Retention

The presence of hydroxy interlayers in vermiculite and smectites offers potential sites for adsorption of anions. This adsorption may be of a form in which the anions can be readily replaced, i.e. anion exchange, or the formation of a meta-stable compound in which the anions are difficult to replace, i.e. compounds with low solubilities.

Phosphorus readily reacts with interlayer hydroxides and hydroxy-Al polymers to form an insoluble aluminum phosphate compound (Hsu, 1964; 1965). Also, reactions have been observed for sulfate (Singh, 1967; Singh & Brydon, 1970; Kodama & Singh, 1972). Retention of sulfate in an exchangeable ionic form on hydroxy-Al has been used to explain the relatively slow downward movement of sulfate in acid red soils (Liu & Thomas, 1961; Chang & Thomas, 1963; Chao et al., 1963, 1965). The retention of other anions such as borate by hydroxy-Al compounds in soils also has been reported (Sims & Bingham, 1967; Hatcher et al., 1967). The relative movement of anions through soils also may be affected by modification of the electrical properties of the clay upon hydroxy-Al interlayer formation.

The result of such a modification may change diffusion rates of both cations and anions (Brown et al., 1968).

C. Changes in Potassium Selectivity and Fixation

Potassium selectivity and fixation by clays has been discussed in a recent review by Thomas and Hipp (1968). Since hydroxy interlayers result in the reduction of the effective CEC of the clay minerals, the interlayers should decrease their K fixation capacity. Fixation may also be reduced by the presence of exchangeable Al (Rich & Obenshain, 1955; Scott et al., 1957; Cook & Hutcheson, 1960; Carter et al., 1963; Rich & Lutz, 1965; Somasiri & Huang, 1974; Nagasawa et al., 1974). Prevention of interlayer collapse by Al-interlayer props inhibits K fixation (Rich, 1960a, 1960b). Effects of hydroxy interlayers on the selectivity of some clay minerals for K have also been studied (Rich, 1964; Rich & Black, 1964; Dolcater et al., 1968; Kozak & Huang, 1971; Murdock & Rich, 1972; Lietzke and Mortland, 1973).

VI. SUPPLEMENTAL READING LIST

Bailey, S. W. 1975. Chlorites. p. 191–262. *In* J. E. Gieseking (ed.) Soil components Vol. 2. Springer-Verlag, New York.

Brindley, G. W. 1961. The chlorite minerals. *In* G. Brown (ed.) X-ray identification and structure of clay minerals. Mineralogical Soc. London. Chapt. VI.

Jackson, M. L. 1962. Interlayering of expansible layer silicates in soils chemical weathering. Clays Clay Miner. 11:29–46.

Jackson, M. L. 1964. Chemical composition of soils. Chapter 2, p. 71–141. *In* F. E. Bear (ed.) Chemistry of the soil. Reinhold Publishing Corp., New York.

Rich, C. I. 1968. Hydroxy interlayers in expansible layer silicates. Clays Clay Minerals. 16:15–30.

LITERATURE CITED

Ahlrichs, J. L. 1968. Hydroxyl stretching frequencies of synthetic Ni-, Al-, and Mg-hydroxy interlayers in expanding clays. Clays Clay Miner. 16:63–71.

Alexiades, C. A., and M. L. Jackson. 1966. Quantitative clay mineralogical analysis of soils and sediments. Clays Clay Miner. 14:35–52.

Andrew, R. W., M. L. Jackson, and K. Wada. 1960. Intersalation as a technique for differentiation of kaolinite from chloritic minerals by X-ray diffraction. Soil Sci. Soc. Am. Proc. 24:422–424.

Arshad, M. A., R. J. St. Arnaud, and P. M. Huang. 1972. Dissolution of trioctahedral layer silicates by ammonium oxalate, sodium dithionate-citrate-bicarbonate and potassium pyrophosphate. Can. J. Soil Sci. 52:19–26.

Bailey, S. W. 1975. Chlorites. p. 191–262. *In* J. E. Gieseking (ed.) Soil components Vol. 2. Springer-Verlag, New York.

Bailey, S. W. (chairman), G. W. Brindley, W. D. Johns, R. T. Martin, and M. Ross. 1971. Summary of national and international recommendations on clay mineral nomenclature by 1969-70 Clay Min. Soc. Nomenclature Comm. Clays Clay Miner. 19:129–132.

Bailey, S. W., and B. E. Brown. 1962. Chlorite polytypism. I. Regular and semi-random one-layer structures. Am. Mineral. 47:819–850.

Barnhisel, R. I. 1969. Changes in specific surface areas of clays treated with hydroxy-aluminum. Soil Sci. 107:126–130.

Barnhisel, R. I., and C. I. Rich. 1963. Gibbsite formation from aluminum-interlayers in Montmorillonite. Soil Sci. Soc. Am. Proc. 27:632–635.

Barnhisel, R. I., and C. I. Rich. 1965. Gibbsite, bayerite and nordstrandite formation as affected by anions, pH and mineral surfaces. Soil Sci. Soc. Am. Proc. 29:531–534.

Barnhisel, R. I., and C. I. Rich. 1966. Preferential hydroxyaluminum interlayering in montmorillonite and vermiculite. Soil Sci. Soc. Am. Proc. 30:35–39.

Blatter, C. L. 1973. Hg-complex intergrades in smectite. Clays Clay Miner. 21:261–263.

Bradley, W. F. 1954. X-ray diffraction criteria for the characterization of chloritic material in sediments. Clays Clay Miner. 2:324–335.

Brindley, G. W., B. M. Oughton, and K. Robinson. 1950. Polymorphism of the chlorites. I. Ordered structures. Acta. Crystallogr. 3:408–416.

Brindley, G. W. 1961. The chlorite minerals. Chapter VI. p. 242–291. *In* G. Brown (ed.) X-ray identification and structure of clay minerals. Mineralogical Soc. London.

Brosset, C., G. Biedermann, and L. G. Sillen. 1954. Studies on the hydrolysis of metal ions. XI. The aluminum ion, Al^{3+}. Acta Chem. Scand. 8:1917–1926.

Brown, G. 1953. The dioctahedral analogue of vermiculite. Clay Miner. Bull. 2:64–70.

Brown, G. (ed.). 1961. The X-ray identification and crystal structures of clay minerals. Mineralogical Society, London.

Brown, D. A., R. E. Phillips, L. O.Ashlock, and B. D. Fuque. 1968. Effect of Al^{3+} and H^+ upon the simultaneous diffusion of ^{85}Sr and ^{86}Rb in kaolinite clay. Clays Clay Miner. 16:137–146.

Brown, G., and A. C. D. Newman. 1973. The reactions of soluble aluminum with montmorillonite. J. Soil Sci. 24:339–354.

Bryant, J. P., and J. B. Dixon. 1964. Clay mineralogy and weathering of a Red-Yellow Podzolic soil from quartz mica schist in the Alabama Piedmont. Clays Clay Miner. 12:509–521.

Brydon, J. E., and H. Kodama. 1966. The nature of aluminum hydroxide-montmorillonite complexes. Am. Mineral. 51:875–889.

Caillère, S., and S. Hénin. 1949. Formation of chlorite from montmorillonite. Mineral. Mag. 28:612–620.

Caillère, S., and S. Hénin. 1950. Mecanisme d'evolutions des mineraux phylliteux. Int. Congr. Soil Sci., Trans. 4th (Amsterdam, Neth.) I:96–98.

Caillère, S., and S. Hénin. 1960. Relationship between the crystallochemical constitution of phyllites and their dehydration temperature application in the case of chlorites. Bull. Soc. Fr. Ceram. 48:63–67.

Carstea, D. D. 1968. Formation of hydroxy-Al and -Fe interlayers in montmorillonite and vermiculite: Influence of particle size and temperature. Clays Clay Miner. 16:231–238.

Carstea, D. D., M. E. Harward, and E. G. Knox. 1970a. Comparison of iron and aluminum hydroxy interlayers in montmorillonite and vermiculite: I. Formation. Soil Sci. Soc. Am. Proc. 34:517–521.

Carstea, D. D., M. E. Harward, and E. G. Knox. 1970b. Comparison of iron and aluminum hydroxy interlayers in montmorillonite and vermiculite: II. Dissolution. Soil Sci. Soc. Am. Proc. 34:522–526.

Carstea, D. D., M. E. Harward, and E. G. Knox. 1970c. Formation and stability of hydroxy-Mg interlayers in phyllosilicates. Clays Clay Miner. 18:213–222.

Carter, D. L., M. E. Harward, and J. L. Young. 1963. Variation in exchangeable K and relation to intergrade layer silicate minerals. Soil Sci. Soc. Am. Proc. 27:283–287.

Chang, M. L., and G. W. Thomas. 1963. A suggested mechanism for sulfate adsorption by soils. Soil Sci. Soc. Am. Proc. 27:281–283.

Chao, T. T., M. E. Harward, and S. C. Fang. 1963. Cationic effects on sulfate adsorption by soils. Soil Sci. Soc. Am. Proc. 27:35–38.

Chao, T. T., M. E. Harward, and S. C. Fang. 1965. Exchange reactions between hydroxyl and sulfate ions in soils. Soil Sci. 99:104–108.

Clark, J. S. 1964a. Some cation-exchange properties of soils containing free oxides. Can. J. Soil Sci. 44:203–211.

Clark, J. S. 1964b. Aluminum and iron fixation in relation to exchangeable hydrogen in soils. Soil Sci. 98:302–306.

Coleman, N. T., and G. W. Thomas. 1964. Buffer curves of acid clays as affected by the presence of ferric iron and aluminum. Soil Sci. Soc. Am. Proc. 28:187–190.

Coleman, N. T., and G. W. Thomas. 1967. The basic chemistry of soil acidity. In R. W. Pearson and F. Adams (ed.) Soil acidity and liming. Agronomy 12:1–41. Am. Soc. Agron., Madison, Wis.

Coleman, N. T., G. W. Thomas, F. H. LeRoux, and G. Bredell. 1964. Salt-exchangeable and titratable acidity in Bentonite-sesquioxide mixtures. Soil Sci. Soc. Am. Proc. 28:35–37.

Colombera, P. M., A. M. Posner, and J. P. Quirk. 1971. The adsorption of aluminum from hydroxy-aluminum solutions on to Fithian illite. J. Soil Sci. 22:118–128.

Cook, M. G., and T. B. Hutcheson, Jr. 1960. Soil potassium reactions as related to clay mineralogy of selected Kentucky soils. Soil Sci. Soc. Am. Proc. 24:252–256.

Davey, B. G., and P. F. Low. 1971. Physico-chemical properties of sols and gels of Na-montmorillonite with and without adsorbed hydrous aluminum oxide. Soil Sci. Soc. Am. Proc. 35:230–236.

de Villiers, J. M., and M. L. Jackson. 1967a. Cation exchange capacity variations with pH in soil clays. Soil Sci. Soc. Am. Proc. 31:473–476.

de Villiers, J. M., and M. L. Jackson. 1967b. Aluminous chlorite origin of pH-dependent cation exchange capacity variations. Soil Sci. Soc. Am. Proc. 31:614–619.

Dewan, H. C., and C. I. Rich. 1970. Titration of acid soils. Soil Sci. Soc. Am. Proc. 34: 38–44.

Dixon, J. B., and M. L. Jackson. 1959. Dissolution of interlayers from intergradient soil clays after preheating at 400°C. Science 129:1616–1617.

Dixon, J. B., and M. L. Jackson. 1960. Mineralogical analysis of soil clays involving vermiculite-chlorite-kaolinite differentiation. Clays Clay Miner. 8:274–284.

Dixon, J. B., and M. L. Jackson. 1962. Properties of intergradient chlorite-expansible layer silicates of soils. Soil Sci. Soc. Am. Proc. 26:358–362.

Dixon, J. B., and W. A. Seay. 1957. Identification of clay minerals in the surface horizons of four Kentucky soils. Soil Sci. Soc. Am. Proc. 21:603–607.

Dolcater, D. L., E. G. Lotse, J. K. Syers, and M. L. Jackson. 1968. Cation exchange selectivity of some clay-sized minerals and soil materials. Soil Sci. Soc. Am. Proc. 32:795–798.

Dowdy, R. H., and W. E. Larson. 1971. Tensile strength of montmorillonite as a function of saturating cation and water content. Soil Sci. Soc. Am. Proc. 35:1010–1014.

El Rayah, H. M. E., and D. L. Rowell. 1973. The influence of iron and aluminum hydroxides on the swelling of Na-montmorillonite and the permeability of a Na-soil. J. Soil Sci. 24:137–144.

Farmer, V. C. 1974. The layer silicates. Chapt. 15. p. 331–363. In V. C. Farmer (ed.) The Infrared spectra of minerals. Mineralogical Society, Monograph 4. London.

Frink, C. R. 1965. Characteristics of aluminum interlayers in soil clays. Soil Sci. Soc. Am. Proc. 29:379–382.

Frink, C. R., and M. Peech. 1963. Hydrolysis and exchange reactions of the aluminum ion in hectorite and montmorillonite suspensions. Soil Sci. Soc. Am. Proc. 27:527–530.

Glenn, R. C., and V. E. Nash. 1964. Weathering relationships between gibbsite, kaolinite, chlorite, and expansible layer silicates in selected soils from the lower Mississippi coastal plain. Clays Clay Miner. 12:529–548.

Gray, D. H., and J. Schlocker. 1969. Electrochemical alteration of clay soils. Clays Clay Miner. 17:309–322.

Grim, R. E., and W. D. Johns. 1954. Clay mineral investigation of sediments in the northern Gulf of Mexico. Clays Clay Miner. 2:81–103.

Gupta, G. C., and W. U. Malik. 1969a. Transformation of montmorillonite to nickel-chlorite. Clays Clay Miner. 17:233–239.

Gupta, G. C., and W. U. Malik. 1969b. Chloritization of montmorillonite by its copre-cipitation with magnesium hydroxide. Clays Clay Miner. 17:331-338.

Hatcher, J. T., C. A. Bower, and M. Clark. 1967. Adsorption of boron by soils as in-fluenced by hydroxy aluminum and surface area. Soil Sci. 104:422-426.

Hayashi, H., and K. Oinuma. 1965. Relationship between infrared absorption spectra in the region of 450-900 cm^{-1} and chemical composition of chlorite. Am. Mineral. 50:476-483.

Hayashi, H., and K. Oinuma. 1967. Si-O absorption bands near 1000 cm^{-1} and OH ab-sorption bands of chlorite. Am. Mineral. 52:1206-1210.

Herrera, R., and M. Peech. 1970. Reaction of montmorillonite with iron (III). Soil Sci. Soc. Am. Proc. 34:740-742.

Hsu, Pa Ho. 1964. Adsorption of phosphate by aluminum and iron in soil. Soil Sci. Am. Proc. 28:474-478.

Hsu, Pa Ho. 1965. Fixation of phosphate by aluminum and iron in acidic soils. Soil Sci. 99:398-402.

Hsu, Pa Ho. 1968a. Heterogenity of montmorillonite surface and its effect on the na-ture of hydroxy-aluminum interlayers. Clays Clay Miner. 16:303-311.

Hsu, Pa Ho. 1968b. Heterogenity of montmorillonite surface aluminum polymers by vermiculite. Soil Sci. Soc. Am. Proc. 28:763-769.

Hsu, Pa Ho, and T. F. Bates. 1964a. Fixation of hydroxy-aluminum polymers by ver-miculite. Soil Sci. Soc. Am. Proc. 28:763-769.

Hsu, Pa Ho, and T. F. Bates. 1964b. Formation of X-ray amorphous and crystalline aluminum hydroxides. Mineral Mag. 33:749-768.

Hsu, Pa Ho, and C. I. Rich. 1960. Aluminum fixation in a synthetic cation exchanger. Soil Sci. Soc. Am. Proc. 24:21-25.

Huang, P. M., and M. L. Jackson. 1966. Fluorite interaction with clays in relation to third buffer range. Nature (London) 211:779-780.

Huang, P. M., and S. Y. Lee. 1969. Effect of drainage on weathering transformations of mineral colloids of some Canadian prairie soils. Proc. Int. Clay Conf. (Tokyo) I: 541-551.

Hunsaker, V. E., and P. F. Pratt. 1970a. Formation of mixed magnesium and aluminum hydroxides in soil materials. Soil Sci. Soc. Am. Proc. 34:813-816.

Hunsaker, V. E., and P. F. Pratt. 1970b. The solubility of mixed magnesium-aluminum hydroxide in various aqueous solutions. Soil Sci. Soc. Am. Proc. 34:823-825.

Hutcheson, T. B., Jr. 1963. Chemical and mineralogical characterization and comparison of Hagerstown and Maury soil series. Soil Sci. Soc. Am. Proc. 27:74-78.

Jackson, M. L. 1960. Structural role of hydronium in layer silicates during soil genesis. Int. Congr. Soil Sci., Trans. 7th (Madison, Wis.) II:445-455.

Jackson, M. L. 1962. Interlayering of expansible layer silicates in soils chemical weather-ing. Clays Clay Miner. 11:29-46.

Jackson, M. L. 1963. Aluminum bonding in soils: A unifying principle in soil science. Soil Sci. Soc. Am. Proc. 27:1-10.

Jackson, M. L. 1964. Chemical composition of soils. Chapter 2, p. 71-141. In F. E. Bear (ed.) Chemistry of the soil. Reinhold Publishing Corp., New York.

Jackson, M. L., S. Y. Lee, J. L. Brown, I. B. Sachs, and J. K. Syers. 1973. Scanning elec-tron microscopy of hydrous oxide crusts intercalated in naturally weathered mica-ceous vermiculite. Soil Sci. Soc. Am. Proc. 37:127-131.

Jackson, M. L., L. D. Whittig, R. C. Vanden Heuvel, A. Kaufman, and B. E. Brown. 1954. Some analysis of soil montmorin, vermiculite, mica, chlorite, and interstratified layer silicates. Clays Clay Miner. 2:218-240.

Kidder, G., and L. W. Reed. 1972. Swelling characteristics of hydroxyaluminum inter-layered clays. Clays Clay Miner. 20:13-20.

Kirkland, D. L., and B. F. Hajek. 1972. Formula deriviation of Al-interlayered vermicu-lite in selected soil clays. Soil Sci. 114:317-322.

Kissel, D. E., and G. W. Thomas. 1969. Conductimetric titrations with Ca(OH)$_2$ to esti-mate the neutral salt replaceable and total soil acidity. Soil Sci. 108:117-179.

Klages, M. G., and J. L. White. 1957. A chlorite-like mineral in Indiana soils. Soil Sci. Soc. Am. Proc. 21:16-20.

Kodama, H., and S. S. Singh. 1972. Hydroxy aluminum sulfate-montmorillonite com-plex. Can. J. Soil Sci. 52:209-218.

Kozak, L. M., and P. M. Huang. 1971. Adsorption of hydroxy-Al by certain phylosilicates and its relation to K/Ca cation exchange selectivity. Clays Clay Miner. 19: 95–102.

Lietzke, D. A., and M. M. Mortland. 1973. The dynamic character of a chloritized vermiculitic soil clay. Soil Sci. Soc. Am. Proc. 37:651–656.

Liu, M., and G. W. Thomas. 1961. Nature of sulfate retention by acid soils. Nature (London) 192:384.

Longeut-Escard, J. 1950. Fixation des hydroxydes par la montmorillonite. Int. Congr. Soil Sci., Trans. 4th (Amsterdam, Neth.) III:40–44.

MacEwan, D. M. C. 1950. Some notes on the recording and interpretation of X-ray diagrams of soil clays. J. Soil Sci. 1:90–103.

McCracken, R. J., and S. B. Weed. 1963. Pan Horizons in southeastern soils: Micromorphology and associated chemical, mineralogical and physical properties. Soil Sci. Soc. Am. Proc. 27:330–334.

McMurchy, R. C. 1934. The crystal structure of the chlorite minerals. Z. Kristallogr. Mineral. 88:420–432.

Malcolm, R. L., W. D. Nettleton, and R. J. McCracken. 1968. Pedogenic formation of montmorillonite from a 2:1–2:2 intergrade clay mineral. Clays Clay Miner. 16: 405–414.

Murdock, L. M., and C. I. Rich. 1972. Ion selectivity in three soil profiles as influenced by mineralogical characteristics. Soil Sci. Soc. Am. Proc. 36:167–171.

Nagasawa, K., G. Brown, and A. C. D. Newman. 1974. Artificial alternation of biotite into a 14Å layer silicate with hydroxy-aluminum interlayers. Clays Clay Miner. 22: 241–252.

Nash, V. E. 1963. Chemical and mineralogical properties of an Orangeburg profile. Soil Sci. Soc. Am. Proc. 27:688–693.

Novak, R. J., H. L. Motto, and L. A. Douglas. 1971. The effect of time and particle size on mineral alteration in several Quaternary soils in New Jersey and Pennsylvania, U.S.A. p. 211–224. In D. H. Yaalon (ed.) Paleopedology—origin, nature, and dating of Paleosols. Int. Soc. Soil Sci. and Israel Univ. Press, Jerusalem.

Nutting, P. G. 1943. Some standard thermal dehydration curves of minerals. U. S. Geol. Surv. Prof. Paper. 197E. p. 197–216.

Pauling, L. 1930. The structure of the chlorites. Proc. Nat. Acad. Sci. 16:578–582.

Pearson, R. W., and L. E. Ensminger. 1949. Types of clay minerals in Alabama soils. Soil Sci. Soc. Am. Proc. 13:153–156.

Reneau, R. B., and J. G. A. Fiskell. 1970. Selective dissolution effects on cation-exchange capacity and specific surface of some tropical soil clays. Soil Sci. Soc. Am. Proc. 34:809–812.

Rich, C. I. 1960a. Aluminum in interlayers of vermiculite. Soil Sci. Soc. Am. Proc. 24: 26–32.

Rich, C. I. 1960b. Ammonium fixation by two Red-Yellow Podzolic soils as influenced by interlayer-Al. Int. Congr. Soil Sci. Trans. 7th (Madison, Wis.) IV:468–475.

Rich, C. I. 1964. Effect of cation size and pH on potassium exchange in Nason soil. Soil Sci. 98:100–106.

Rich, C. I. 1966. Concentration of dioctahedral mica and vermiculite using a fluoride solution. Clays Clay Miner. 14:91–98.

Rich, C. I. 1968. Hydroxy interlayers in expansible layer silicates. Clays Clay Minerals. 16:15–30.

Rich, C. I. 1970. Conductometric and potentiometric titration of exchangeable aluminum. Soil Sci. Soc. Am. Proc. 34:31–38.

Rich, C. I., and W. R. Black. 1964. Potassium exchange as affected by cation size, pH and mineral structure. Soil Sci. 97:384–390.

Rich, C. I., and M. G. Cook. 1963. Formation of dioctahedral vermiculite in Virginia soils. Clays Clay Miner. 11:96–106.

Rich, C. I., and J. A. Lutz, Jr. 1965. Mineralogical changes associated with ammonium and potassium fixation in soil clays. Soil Sci. Soc. Am. Proc. 29:167–170.

Rich, C. I., and S. S. Obenshain. 1955. Chemical and clay mineral properties of a Red-Yellow Podzolic soil derived from muscovite schist. Soil Sci. Soc. Am. Proc. 19: 334–339.

Richburg, J. S., and F. Adams. 1970. Solubility and hydrolysis of aluminum in soil solution and saturated-paste extracts. Soil Sci. Soc. Am. Proc. 34:728–734.

Robinson, K., and G. W. Brindley. 1948. Crystal structure of chlorite minerals. Proc. Leeds Philos. Lit. Soc., Sci. Sect. 5:102–108.

Sawhney, B. L. 1958. Aluminum interlayers in soil clay minerals, montmorillonite and vermiculite. Nature (London) 182:1595–1596.

Sawhney, B. L. 1960a. Aluminum interlayers in clay minerals, montmorillonite and vermiculite: Laboratory synthesis. Nature (London) 187:261–262.

Sawhney, B. L. 1960b. Aluminum interlayers in clay minerals. Int. Congr. Soil Sci. Trans. 7th (Madison, Wis.) IV:476–481.

Sawhney, B. L. 1960c. Weathering and aluminum interlayers in a soil Catena: Hollis-Charlton-Sutton-Leicester. Soil Sci. Soc. Am. Proc. 24:221–226.

Sawhney, B. L. 1968. Aluminum interlayers in layer silicates. Effect of OH/Al ratio of Al solution, time of reaction, and type of structure. Clays Clay Miner. 16:157–163.

Sawhney, B. L., and C. R. Frink. 1966. Potentiometric titration of acid montmorillonite. Soil Sci. Soc. Am. Proc. 30:181–184.

Schwertmann, U., and M. L. Jackson. 1964. Influence of hydroxy aluminum ions on pH titration curves of hydronium-aluminum clays. Soil Sci. Soc. Am. Proc. 28:179–183.

Scott, A. D., J. L. Ahlrichs, and G. Stanford. 1957. Aluminum effect on potassium fixation by Wyoming bentonite. Soil Sci. 84:377–387.

Shen, M. J., and C. I. Rich. 1962. Aluminum fixation in montmorillonite. Soil Sci. Soc. Am. Proc. 26:33–36.

Sims, J. R., and F. T. Bingham. 1967. Retention of Boron by layer silicates, sesquioxides and soil materials. I. Layer silicates. Soil Sci. Soc. Am. Proc. 31:728–732.

Singh, S. S. 1967. Sulfate ions and ion activity product $(Al)(OH)^3$ in Wyoming bentonite suspensions. Soil Sci. 104:433–438.

Singh, S. S. 1972. The effect of temperature on the ion activity product $(Al)(OH)^3$ and its relation to lime potential and degree of base saturation. Soil Sci. Soc. Am. Proc. 36:47–50.

Singh, S. S., and J. E. Brydon. 1967. Precipitation of aluminum by calcium hydroxide in the presence of Wyoming bentonite and sulfate ions. Soil Sci. 103:162–167.

Singh, S. S., and J. E. Brydon. 1970. Activity of aluminum hydroxy sulfate and the stability of hydroxy aluminum interlayers in montmorillonite. Can. J. Soil Sci. 50:219–225.

Singleton, P. C., and M. E. Harward. 1971. Iron hydroxy interlayers in soil clay. Soil Sci. Soc. Am. Proc. 35:838–842.

Slaughter, M., and I. H. Milne. 1960. The formation of chlorite-like structures from montmorillonite. Clays Clay Miner. 8:114–124.

Soil Survey Staff. 1975. Soil taxonomy: A basic system of soil classification for making and interpreting soil surveys. Agric. Handb. no. 436. U. S. Government Printing Office, Washington, D. C.

Somasiri, S., and P. M. Huang. 1974. Effect of hydrolysis of aluminum on competitive adsorption of potassium and aluminum by expansible phyllosilicates. Soil Sci. 117:110–116.

Tamura, T. 1955. Weathering of mixed-layer clays in soils. Clays Clay Miner. 4:413–422.

Tamura, T. 1957. Identification of the 14Å clay mineral component. Am. Mineral. 42:107–110.

Tamura, T. 1958. Identification of clay minerals from acid soils. J. Soil Sci. 9:141–147.

Thomas, G. W. 1960. Forms of aluminum in cation exchanges. Int. Congr. Soil Sci., Trans. 7th (Madison, Wis.) II:364–369.

Thomas, G. W., and B. W. Hipp. 1968. Soil factors affecting potassium availability. p. 269–281. In V. J. Kilmer (ed.) The role of potassium in agriculture. ASA, CSSA, SSSA Publishers, Madison, Wis.

Tuddenham, W. M., and R. J. P. Lyon. 1959. Relation of infrared spectra and chemical analysis of some chlorites and related minerals. Anal. Chem. 31:377–380.

Turner, R. C. 1965. Some properties of aluminum hydroxide precipitated in the presence of clays. Can. J. Soil Sci. 45:331–336.

Turner, R. C. 1967. Aluminum removed from solution by montmorillonite. Can. J. Soil Sci. 47:217–222.

Turner, R. C., and J. E. Brydon. 1962. Formation of double hydroxides and the titration of clays. Science 136:1052–1053.

Turner, R. C., and J. E. Brydon. 1965. Factors affecting the solubility of $Al(OH)_3$ precipitated in the presence of montmorillonite. Soil Sci. 100:176-181.

Turner, R. C., and J. E. Brydon. 1967a. Effect of length of time of reaction on some properties of suspensions of Arizona bentonite, illite, and kaolinite in which aluminum hydroxide is precipitated. Soil Sci. 103:111-117.

Turner, R. C., and J. E. Brydon. 1967b. Removal of interlayer aluminum hydroxide from montmorillonite by seeding the suspension with gibbsite. Soil Sci. 104:332-335.

Weismiller, R. A., J. L. Ahlrichs, and J. L. White. 1967. Infrared studies of hydroxy-aluminum interlayer material. Soil Sci. Soc. Am. Proc. 31:459-463.

Youell, R. F. 1960. An electrolytic method for producing chlorite-like substances from montmorillonite. Clay Miner. Bull. 4:191-195.

Kaolinite and Serpentine Group Minerals

J. B. DIXON, Texas A&M University, College Station, Texas

I. INTRODUCTION

Kaolinite is one of the most widespread clay minerals in soils. It is most abundant in soils of warm moist climates and is a prominent constituent of oceanic sediments in the equatorial belt. Halloysite is formed rapidly in soils of volcanic origin. Kaolinite and halloysite are products of acid weathering. Kaolinite, halloysite, and the less common dickite and nacrite are 1:1 layer structured alumino-silicates of the same ideal composition except that in hydrated halloysite there is water between the layers. Kaolinite and halloysite have low surface area and low cation and anion exchange capacities.

Serpentines are 1:1 layer structured silicate minerals too. Magnesium, iron, aluminum, and other ions may occupy their octahedral positions making mineralogical properties much more complex than for kaolinite and halloysite. Serpentines are typically coarser than clay size and are relatively rare minerals in soils. Serpentines occur in altered ultrabasic rocks and commonly they weather to smectite. The structural similarities of kaolinite and serpentine subgroups bring together in one taxonomic group two classes of minerals with contrasting chemical properties.

The coverage of this chapter is largely devoted to kaolinite and halloysite because of their greater overall importance than other members of this mineral group. Also, the literature on serpentines is too sparse for thorough coverage and much of it is in conflict.

Type	Group	Subgroup	Species
1:1	Kaolinite–serpentine	Kaolinites (Dioctahedral)	Kaolinite Halloysite[‡] Dickite Nacrite
		Serpentines (Trioctahedral)	Chrysotile Lizardite Antigorite
	Other minerals having a trioctahedral structure similar to serpentine are:		Amesite[§] Cronstedtite[§] Berthierine[§] (chamosite) Greenalite[§] Garnierite[§]

[†] Adapted from Brindley (1961) and Bailey et al. (1971).
[‡] Degree of hydration or expansion should be stated where possible.
[§] Bailey, S. W., personal communication.

chapter 11

Taxonomy of the kaolinite-serpentine group minerals is described in the preceding tabular information adapted from Brindley (1961) and from Bailey et al. (1971). The classification given represents that approved by the Nomenclature Committee of the Clay Minerals Society and inclusion of additional minerals with structures similar to the serpentine subgroup (S. W. Bailey, personal communication).

II. STRUCTURAL PROPERTIES OF KAOLINITE

Kaolinite is composed of a tetrahedral and an octahedral sheet which constitute a single 7Å layer in a triclinic unit cell (Brindley & Robinson, 1946). Two-thirds of the octahedral positions are occupied by Al ions and the tetrahedral positions are occupied by Si ions. Aluminum ions are arranged in an orderly distribution occupying two rows of sites parallel to the X-axis (Fig. 11-1). Every third row of octahedral sites is vacant (Bailey, 1966). Hydroxyl ions make up the surface plane of octahedral anions and one third of the inner (shared) plane of anions in each 7Å layer (Fig. 11-1). The surface hydroxyls bond through their hydrogens to the oxygen sheet of the adjacent layer. Halloysite, dickite, and nacrite contain the same basic 7Å layer as kaolinite, but the stacking sequence of layers is different in each mineral. Serpentine minerals have a 7Å layer and they have Fe, Mg, and other ions in addition to Al as octahedral cations. Because only two of every three octahedral positions are occupied by Al in the kaolinite subgroup members, they are termed *dioctahedral*. All of the octahedral positions are filled in the serpentine minerals and they are termed *trioctahedral*.

Repulsive forces between adjacent Al cations in a given dioctahedral

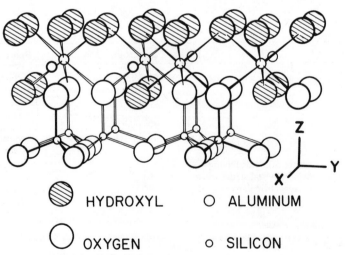

⊘ HYDROXYL   ○ ALUMINUM

○ OXYGEN   ○ SILICON

Fig. 11-1.   Idealized structural diagram of kaolinite layer viewed along the *a*-axis. Modified from Brindley and MacEwan (1953).

OCTAHEDRA IN NACRITE

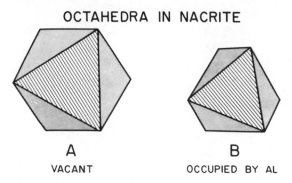

A
VACANT

B
OCCUPIED BY AL

Fig. 11-2. Relative size and distortion of octahedra in nacrite where vacant and occupied by Al ions (redrawn from Bailey, 1966).

sheet contribute to distortion of octahedra, thinning of the octahedral sheet, and expansion along X and Y axes. Resultant octahedra of anions around Al cations are smaller than those around vacant sites (Fig. 11-2). Repulsive forces are weaker among lower charged cations in trioctahedral clays than in dioctahedral clays containing trivalent Al. Where all octahedral sites are occupied as by Mg^{2+} or Fe^{2+}, anions are more uniformly arranged about cations than where vacancies occur in dioctahedral minerals. Anions are pulled inward normal to the trioctahedral sheet making it similar in thickness to the dioctahedral sheet in spite of the presence of larger cations such as Mg^{2+} and Fe^{2+} in the former (radius of Al^{3+} = 0.50Å; radius of Mg^{2+} = 0.65Å; radius of Fe^{2+} = 0.80Å; Evans, 1964).

Octahedral sheet thickness			
Dioctahedral		Trioctahedral	
Kaolinite	Dickite	Amesite	Cronstedtite
2.10Å	2.04Å	2.02Å	2.19Å

The octahedral sheet apparently is dominant in determining the dimensions of a silicate layer structure. Alternate rotation of the tetrahedra reduces the dimensions of the tetrahedral sheet (Fig. 11-3) and improves the fit between sheets of the two types of polyhedra (Bailey, 1966).

The early work of Brindley and Robinson (1946) showed broadening of the basal reflections (001, 002, etc.), disappearance of certain prismatic reflections, and the formation of a wedge-shaped diffraction band in the more disordered kaolinite minerals. The disorder was identified as random stacking displacements in the a-direction. Halloysite has disordered stacking of layers in both a- and b-directions. Broadening of X-ray diffraction peaks is produced by small crystals.

The stacking of kaolinite layers is controlled (Newnham, 1961) by three structural features: (i) repulsion between highly charged Si and Al cations

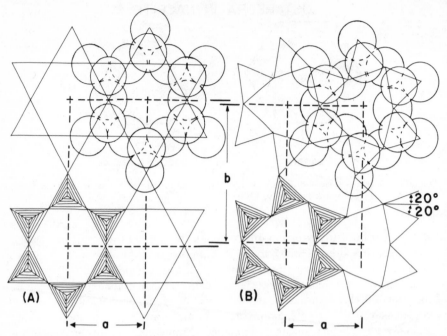

Fig. 11-3. Arrangement of Si–O tetrahedra in (A) an idealized hexagonal sheet structure and (B) a ditrigonal arrangement as found in kaolinite. The rotations of tetrahedra in (B) are alternately +20° and –20° with respect to the ideal positions resulting in reduction of the b-dimension. The upper parts of (A) and (B) show the oxygen atoms and the lower parts show tetrahedral faces connecting the centers of oxygen atoms. Silicon atoms are not shown. From Brindley and Gibbon (1968).

which tend to avoid superposition, (ii) rotation of basal oxygens toward hydroxyls in the adjacent layer to strengthen interlayer bonds and, (iii) fitting of the corrugation in the basal oxygen surface with that in the adjacent hydroxyl surface.   Dickite and nacrite are the only two-layer forms of the 108 possible ones that satisfy all three criteria and are considered by Newnham to be the most stable configurations.   Kaolinite is the only one-layer form to satisfy the first two criteria, but it does not satisfy the corrugation criterion.

Three-fourths of the hydroxyls within crystals of kaolinite occur in an inner surface position (Fig. 11-4, [C]) contiguous to the basal plane of oxygens of the next layer.   A second structural hydroxyl position [D], termed the inner hydroxyl, is within the kaolinite layer in the plane of atoms shared by the tetrahedral and octahedral sheets.   It is centered over the vacant six-member oxygen ring.   This latter hydroxyl position [D] is occupied by the other fourth of the hydroxyls in kaolinite crystals.   The vibration direction of the protons in the inner-surface hydroxyls is almost 90° to the plane of the kaolinite layer (Wada, 1967).   These hydroxyls have infrared absorption maxima at 3695, 3670, and 3650 cm^{-1} (Ledoux & White, 1966).   A change from OH to OD can be accomplished by D_2O washing when the kaolinite has

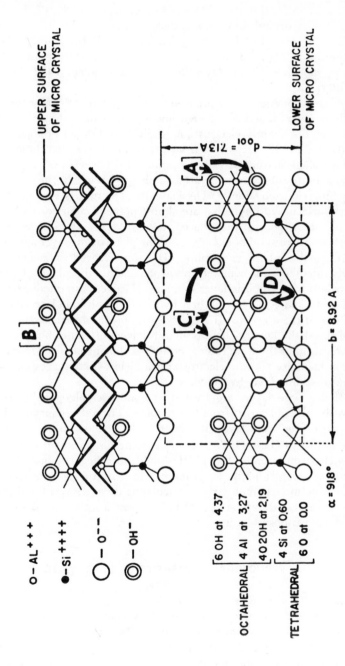

Fig. 11-4. Projection of the structure of kaolinite on the (100) plane showing the stacking of successive layers in a micro crystal (After Brindley). [A] and [B] indicate "outer hydroxyls;" [C] designates "inner-surface hydroxyls;" [D] indicates "inner-hydroxyls." From Ledoux and White (1966).

been previously expanded by intercalation with potassium acetate or certain other organic molecules. Since substituting D for H changes the frequency of the infrared absorption, the reactive OH group can be identified.

The inner hydroxyls of kaolinite, dickite, nacrite and halloysite minerals are largely unreactive to OD-exchange (Wada, 1967). In kaolinite, the inner OH is pointed at a vacant octahedral site at an angle of about 15° (0.26 radians) to the plane of the layer (Serratosa & Bradley, 1958; Ledoux & White, 1964).

Hydrogen bonding has formerly been considered the major force binding kaolinite layers together. Recent investigations have shown that OH–O bonding is predominantly electrostatic where both O's are bonded to other cations (Giese, 1973). Departure from a traditional view that these H-bonds involved covalent sharing of H has stimulated a reassessment of the structures of kaolinite subgroup minerals assuming the bonds to be electrostatic involving fully ionized atoms. The kaolinite structure is viewed as having two inner surface hydroxyls that are almost perpendicular to the layer and bonded to the oxygens of the adjacent layer. According to Giese, the third inner surface hydroxyl dipole vibrates in the plane of the layer and does not contribute to the interlayer bonding. Dickite is viewed as having all three hydroxyls oriented approximately perpendicular to the plane of the layer and contributing to the interlayer bonding. Nacrite is intermediate between kaolinite and dickite and has all three hydroxyls involved in interlayer bonding, but one of these hydroxyls is only slightly involved because it is inclined to the plane of the layer. There is an unresolved difference between the infrared indication of the third interlayer hydroxyl being involved in interlayer bonding in kaolinite and the electrostatic model of Giese suggesting a planar vibration for an interlayer hydroxyl.

Kaolinite type layers are treated as condenser plates by Cruz et al. (1972) with surface charges from OH dipoles. They also conclude that the cohesion energy between layers is primarily electrostatic in good agreement with the results of Giese (1973). There is some disagreement over whether or not such strong bonds through H should be called H-bonds, but Giese maintains that long H-bonds are primarily electrostatic and consequently fit the Coulombic model employed for calculating the interlayer bond energies. Cruz et al. suggest that interlayer H in 1:1 minerals plays a role somewhat like interlayer cations in mica.

III. STRUCTURAL PROPERTIES OF HALLOYSITE

Halloysite is a member of the kaolinite subgroup with two interlayer water molecules per formula unit and a 10Å spacing when fully hydrated. Tubular morphology has frequently been considered characteristic of halloysite. The reporting of a platy-hydrated halloysite by de Souza Santos et al. (1966) indicates that expansion must also be determined when tubular particles are absent, in case platy halloysite is present.

Chukhrov and Zvyagin (1966) investigated several dehydrated halloy-sites from the Soviet Union. Based on electron and X-ray diffraction data, they concluded that halloysite has a monoclinic unit cell composed of two 7Å layers differentiating it from kaolinite, which has a single 7Å layer in a triclinic unit cell. X-ray diffraction data do not always permit determination of the number of layers per unit cell, therefore, the proposed criterion of Chukhrov and Zvyagin is not always practical. The real structure of halloy-site is often disordered permitting determination of only the layer thickness. Honjo et al. (1954) reported a triclinic "tubular kaolinite" suggesting that all halloysites may not be monoclinic. The data on the structure of halloysites indicate that the presence of interlayer water is still the best criterion for halloysite.

IV. MORPHOLOGY OF KAOLINITE SUBGROUP MINERALS

Kaolinite has the characteristic platy morphology and fine particle size which typify clay minerals. Hexagonal plates in clays are characteristic of kaolinite, but their presence does not constitute proof of its presence. Other minerals such as vermiculite may exhibit similar morphology. Kaolinite plates retain their external morphology even after heating to 1000C (1273K) where crystals of mullite begin to form (McConnell & Fleet, 1970). Kaolinite crystals are altered structurally by dry grinding even for as little as 1 min (Hayes, 1963).

In 1950, Bates et al. showed that halloysite is often tubular in contrast to the platy morphology of kaolinite. During early studies by Bates (1958), lath-shaped and polyhedral particles of halloysite were discovered. Spheroidal particles of halloysite were shown in clays of Japan (Sudo & Takahashi, 1956). Also, spheroidal halloysite has been identified in soils of Guatemala (Askenasy et al., 1973) and of Mexico (Dixon & McKee, 1974b), in weathered rock of Nevada (K. C. McBride et al., 1976[1]), and in a Cretaceous sedimentary clay of Minnesota (Parham, 1970).

Although the curved forms of halloysite have been found in many loca-tions, platy forms of halloysite have been rare. A platy halloysite was re-ported in veins in gibbsitic bauxite of Brazil (de Souza Santos et al., 1966). The sample from Brazil was hydrated (10.0Å natural state, 10.9Å glycolated) and the plates were commonly about 0.5 μm wide. The hydrated sample when stored in water for several weeks began to curl and dispersed poorly. After 8 months of aging, the Brazilian sample curled markedly into poorly formed rolls resembling tubular halloysite. Some globular particles appeared to be present also. The rolling of platy halloysite particles in solution indi-cates that this mineral is very labile when hydrated.

Hope and Kittrick (1964) tested the hypothesis that surface tension pro-

---

[1]K. C. McBride, J. B. Dixon, and T. R. McKee. 1976. Spheroidal and tubular halloy-site in a volcanic deposit of Washoe County, Nevada. 25th Annual Clay Miner. Conf. Abstr., Corvallis, Oregon.

duces tubular rolls of 1:1 layers. Hydrated halloysite from Eureka, Utah was composed of irregularly shaped platy particles resembling smectite after freeze drying and was tubular after air drying from water. Thin kaolinite plates produced by prolonged boiling of kaolinite in nitrobenzene remained platy on freeze drying and they rolled into tubes on air drying. The results support the view that halloysite tubes can form from 1:1 structural layers by air drying from water. These results indicate that platy halloysite morphology is very labile and is not likely to survive in soils except under continuously wet conditions. The results further imply that some tubular halloysites in nature probably began as platy halloysites.

A platy mineral termed *tabular halloysite* by Kunze and Bradley (1964) was found in a Katy soil of Texas. The sample was composed of small plates, mostly $< 0.2 \mu m$ e.s.d., with diffraction peaks at 7.56Å unsolvated and 7.76Å solvated with glycol. The platy morphology was stable on drying and in the electron microscope. High resolution electron microscopy examination of Katy $< 0.08 \mu m$ clay reveals lattice fringes of orderly crystals mostly 50 to 100Å thick (J. B. Dixon & J. L. Brown, 1975[2]). The presence of frayed crystal edges with higher than 7.2Å spacings and interstratification of 2:1 layers (9.8Å) accounts for some of the X-ray diffraction effects at 7.7Å. The relative contribution of interstratification, frayed crystal edges, and crystal size on the broad 7.7Å X-ray diffraction peak requires further investigation. The platy particle morphology of the Katy $< 0.08 \mu m$ clay, its resistance to appreciable expansion and collapse, and the observed interstratification imply some similarity to the kaolinite-montmorillonite interstratification reported by Schultz et al. (1971).

The halloysite tubes (Fig. 11-5a, f open-shafted arrows) probably originated from the rolling of thin plates to form tube-like morphology as suggested by the examples given earlier. Growth of tubes may occur at edges such as those shown in Fig. 11-5a, but little growth subsequent to curling probably occurred in this sample because rolled particles are mostly small in cross-section. In addition to tubular particles, flattened particles composed of a folded thick packet of layers also were observed in Colorado halloysite in sectioned particles (see F in Fig. 11-5f) and in a replica (Fig. 11-5e). The folded particles had less frequent layer separations than the tubular ones; thus being suggestive of folded kaolinite. In view of the unlikely prospect of thick particle folding, it appears that growth may have proceeded on the exposed basal face subsequent to folding of a thin particle. Other particles in the tubular halloysite sample that have the largest (ca 0.2 μm) cross-sections (Dixon & McKee, 1974a) had flat prism faces around the tube axis implying that particle growth proceeded faster on flat crystal faces. The plate (P in Fig. 11-5b) appears to be hexagonal, possibly kaolinite, growing tangential to the halloysite sphere. These observations support the hypothesis that growth of halloysite is faster on basal crystal faces than on edges and that

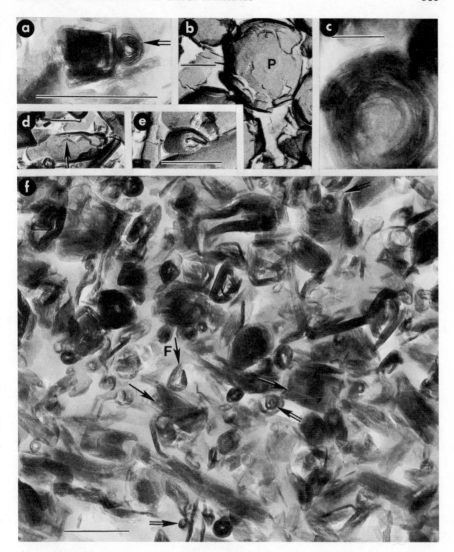

Fig. 11-5. Tubular halloysite from Wagon Wheel Gap, Colorado, and spheroidal halloysite from Redwood County, Minnesota external and internal morphology shown by
transmission electron micrographs. Micrographs (a) and (f) are of sectioned specimens.
Micrographs (b), (d) and (e) are of replicas. Micrograph (c) is of a specimen mounted
on carbon support film. Arrow (F) designates a folded particle and the other solid
arrows in (f) identify layer separations. Open shafted arrows designate rolled particles.
All micrographs were taken at 100 kV except (c) which was taken at 800 kV. The
reference lines are 0.2 μm (Dixon & McKee, 1974a).

such growth is more orderly than inside small tubes which were probably
produced by curling of thin plates.

Halloysite has long been thought to trap interlayer water to 350C
(623K) (Brindley & Robinson, 1946) suggesting interlayer porosity. Thin

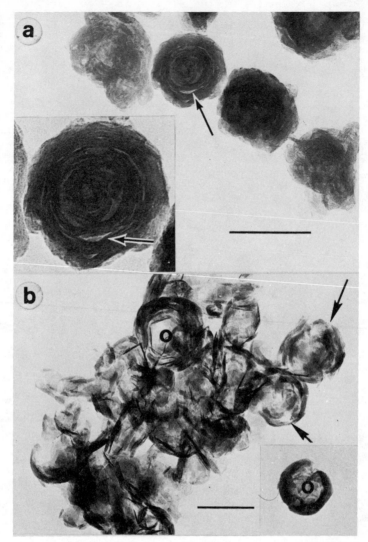

Fig. 11-6. Spheroidal halloysite particles before and after 0.5N NaOH boiling treatment showing (a) layer separation (arrows) and (b) open centers, "O's," and cavities (arrows) indicating halloysite dissolution by 0.5N NaOH treatment. Particles viewed are from 2-0.2 μm clay from Alotenango soil of Guatemala from 50-80 and 80-150 cm depths in (a) and (b), respectively. Reference lines represent 0.2 μm. Micrographs were taken at 100 kV (Askenasy et al., 1973).

sections of halloysite tubes and transmission electron micrographs have shown internal layer separations in tubes. A high voltage transmission micrograph (Fig. 11-5c) shows open interlayer spaces in dehydrated spheroidal halloysite from Minnesota. Similar interlayer separations are shown in spheroidal halloysite from a Guatemalan soil before and after NaOH treatment to remove amorphous material (Fig. 11-6).

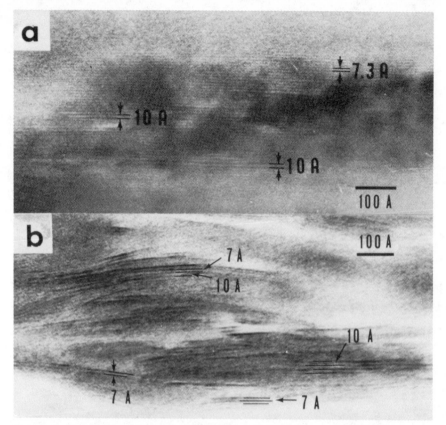

Fig. 11-7. High resolution electron micrographs of thin sections showing electron optical fringes (a) indicative of inclusion of mica in kaolinite and (b) interstratification of kaolinite and other layer silicates. Electron optical fringes shown indicate the spacings of basal planes viewed from the edge. From Lee et al. (1975a); Lee et al. (1975b).

Zones of mica occur in kaolinite from Georgia (Lee et al., 1975a), and zones of mica-vermiculite occur in fire clay (poorly crystalline) type kaolinite from France (Fig. 11-7a). These observations were based on electron optical fringes produced by clays sectioned perpendicular to the basal plane. The same technique has been employed to show kaolinite-2:1 layer silicate interstratification (Fig. 11-7b) in more disordered crystals from Mexico.

V. FORMATION OF KAOLINITE SUBGROUP MINERALS

A. Equilibrium Environment and Conditions for Synthesis

Formation of kaolinite in the presence of smectite at 25C (298K) provides an example of mineral precipitation under conditions much like those in field soils. Several samples of three reference smectites were acidified with

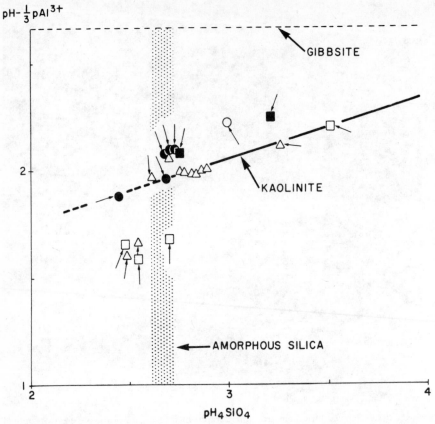

Fig. 11-8. Composition of solutions equilibrated with montmorillonites from Belle Fourche, South Dakota (□), Otay, California (○) and Aberdeen, Mississippi (△) after 3 to 4 years of equilibration. Solid symbols indicate kaolinite formation, and open symbols indicate no detectable kaolinite. Size of the symbols indicates analysis precision. Arrows indicate the direction of sample equilibration as shown by previous analyses. Dashed solubility lines indicate metastable area. From Kittrick (1970).

HCl and adjusted with solutions of Si and Al to provide a range of conditions prior to incubation (Kittrick, 1970). Analyses were made periodically to determine solution compositions. After 3 to 4 years, the formation of kaolinite was shown in some samples. A stability line for kaolinite derived from data obtained at equilibrium and an equation for the precipitation of kaolinite are shown in Fig. 11-8.

A stability line for kaolinite is drawn from Eq. [1] for the precipitation of kaolinite from Al^{3+} and H_4SiO_4.

$$2Al^{3+} + 2H_4SiO_4 + H_2O = 6H^+ + Si_2Al_2O_5(OH)_{4(kaolinite)} \qquad [1]$$

The equilibrium constant, K, for the reaction is:

$$K = (H^+)^6/(Al^{3+})^2 (H_4SiO_4)^2 \qquad [2]$$

and parentheses denote activity. An activity of unity is assumed for kaolinite and water.

In the negative log form

$$pK = 6pH - 2pAl^{3+} - 2pH_4SiO_4. \tag{3}$$

Dividing by 6 and rearranging gives

$$pH - 1/3pAl^{3+} = 1/3pH_4SiO_4 + 1/6 \, pK. \tag{4}$$

The slope of the kaolinite stability line in Fig. 11-8 is 1/3 and the intercept is 1/6 pK from Eq. [4]. The intercept is based on the sample with the highest pH_4SiO_4, the point (□) at the far right in Fig. 11-8. The last analysis of this sample had a pH of 3.47, pH_4SiO_4 of 3.53 (assuming all dissolved Si as H_4SiO_4) and pAl^{3+} of 3.78 (corrected from a molar concentration of pAl = 3.59 for Al equilibria and influence of ionic strength). Thus, employing Eq. [3] gives:

$$pK = 6(3.47) - 2(3.78) - 2(3.53)$$

$$pK = 6.20.$$

From Eq. [4] the intercept of the kaolinite stability line in Fig. 11-8 is 1/6 pK or 1.03.

The stability line in Fig. 11-8 represents the boundary between the areas of supersaturation (above) and undersaturation (below) for kaolinite. Kaolinite was detected by X-ray diffraction analysis only in solutions saturated or supersaturated with respect to kaolinite (solid data points). None of the data points indicated supersaturation with respect to gibbsite. All of the undersaturated solutions and all of the supersaturated solutions, except one unexplained example, appear to be progressing toward the kaolinite stability line. One solution in which kaolinite has formed has a composition on the metastable extension (dashed) of the kaolinite stability line. Solutions at saturation with respect to kaolinite tend to alter along the line and this is particularly evident for the six closely spaced Aberdeen analyses shown.

By calculating the free energy of formation, ΔG, equivalent to the stability line for the kaolinite derived above, comparison can be made with the stability of other kaolinite subgroup minerals in the literature. From Eq. [1]

$$\Delta G_r = \Delta G_{(kaolinite)} - 2\Delta G_{Al^{3+}} - 2\Delta G_{H_4SiO_4} - \Delta G_{H_2O} \tag{5}$$

where ΔG_r is the standard free energy of reaction. And $\Delta G_r = -RT \ln K = 1.364 \, pK$ or 8.46 kcal/mole (35.4 kJ/mole) at 25°C (298K), hence

$$\Delta G_{(kaolinite)} = 8.46 + 2(-115.0) + 2(-313.0) + (-56.7)$$
$$= -904.2 \text{ kcalories per mole } (-3.783 \text{ MJ/mole}).$$

The ΔG of -904.2 kcalories suggests a slightly greater stability for the synthetic kaolinite than does the ΔG of -903.8 kcalories (-3.781 MJ/mole) for a kaolinite of reportedly near maximum crystallinity. Kittrick (1969, 1970) suggests that the ΔG values for kaolinite precipitated in soils and sediments at 25C (298K) probably will range from -898.6 kcalories (3.760 MJ/mole) for halloysite to the initial precipitation level of -904.2 kcalories. From the above calculation, it is evident that ΔG values for the constituents of kaolinite such as Al^{3+} influence the ΔG for kaolinite that a given author may calculate. Since these primary ΔG values are not uniformly agreed upon, one must be cautious about using calculated ΔG values until the primary ΔG values employed have been compared (see Table 1-2 and accompanying discussion of this book).

The synthetic kaolinite (formed in the presence of montmorillonite) decomposed on heating at different temperatures near 500C (773K) at the same rate as a reference kaolinite. Crystallinity of the precipitated kaolinite was examined by X-ray diffraction analysis of a random powder sample and the diffuse band obtained near 4.4Å indicates relatively poor crystallinity. The first-order basal reflection was sharp and was enhanced by preparing a mount with preferred orientation which suggests platy morphology. No spacings above 7.15Å indicative of halloysite were obtained, but the weak X-ray reflections characteristic of halloysite prevent a positive statement on its absence.

B. Synthesis of Kaolinite and Halloysite at Low Temperature

Kaolinite crystallization at room temperature from Al and Si solutions or gels has been obtained to a limited extent. Kaolinite was formed from gels of Al and Si at low pH where Al was in sixfold coordination (De Kimpe et al., 1961). An Al-oxalate complex (Siffert, 1962) and the presence of fulvic acid (Linares & Huertas, 1971) have led to the formation of kaolinite from solutions that contained Si and Al. Kaolinite was synthesized at pH values from 2 to 9 when the ratio of SiO_2 to Al_2O_3 in solution was from 1 to 10 in the presence of an Al-fulvic acid complex. Kaolinite and halloysite were synthesized in very limited amounts from solutions containing Al and Si in the absence of organic molecules (Hem & Lind, 1974). The identity of the kaolinite was based on its hexagonal morphology. The rolled morphology and solubility were used to identify halloysite. No X-ray diffraction pattern could be obtained. Quercetin was added to solutions of Al and Si and the yield of hexagonal crystals was increased to about 5% in 6 months incubation (Hem & Lind, 1974). Quercetin is a yellow plant pigment available in pure form and it is similar in structure to organic substances in some natural waters. Quercetin accelerates the formation of kaolinite possibly because it prevents the formation of gibbsite or other hydroxy-Al crystal phases at pH

6.5 to 8.5. Also, quercetin may facilitate formation of the Al–O–Si bonds of kaolinite because the quercetin-Al complex contains Al–O bonds.

C. Kaolinite Formation Induced by Plant Growth

Kaolinite was formed from biotite where tree seedlings were grown for 7 and 13 months in sand cultures containing small amounts of 2 to 100 μm biotite flakes (Spyridakis et al., 1967). The rapid weathering of the biotite apparently occurred because of the pH range of 4.2 to 5.4, reduced K, Mg, and Fe concentration, fluctuating wet and dry conditions, and the presence of organic substances provided by the plants. The effectiveness of the seedlings in forming kaolinite is as follows: white cedar (*Thuja occidentalis*) > hemlock (*Tsuga canadensis*) > white pine (*Pinus strobus* L.) > white spruce (*Picea glauca*) > red oak (*Quercus rubra* L. *Q. borealis* Mich. x. f.) > hard maple (*Acer saccharum*).

D. Kaolinite Formation from Si–Al Gels at Elevated Temperatures

An unusual thin crumpled form of kaolinite was obtained after 7 days of aging silica–alumina gels at 200C (473K) and under neutral conditions. These hydrothermal experiments of Rodrique et al. (1972) indicate that precipitation of boehmite, which contains sixfold coordinated Al, retards kaolinite formation. The presence of fourfold coordinated Al did not deter kaolinite formation. Kaolinite crystallized most effectively where the gel contained more Si than does kaolinite, thus suggesting to the authors that kaolinite formation involved Si subtraction.

E. Kaolinite Formation From Hydroxy–Al Interlayered Montmorillonite

Kaolinite formation was induced by hydrothermal treatment of hydroxy-Al interlayered montmorillonite at 220C (493K). The kaolinite was expanded to approximately 8.5Å during the first few days of reaction. After 14 days of hydrothermal treatment or heating at 300C (573K) following hydrothermal treatment, the kaolinite collapsed to 7.2Å (Poncelet & Brindley, 1967). The early product had a spacing suggestive of halloysite or an halloysite-kaolinite interstratified mixture even though only platy material was reported.

F. Summary

From the various laboratory syntheses of kaolinite, it is concluded that kaolinite can be precipitated from gels or solutions of silica and alumina over a wide pH range. Precipitation of boehmite retards kaolinite formation in simple experiments suggesting that hydroxy precipitates of Al may be a factor in retarding kaolinite formation in natural environments. Time of kaolinite formation from solid or solution phases can be short (days to months) depending in part on the energy input into the system. The formation of kaolinite by plants from biotite which is trioctahedral and hydrothermally from gels containing Al in fourfold and sixfold coordination suggests that coordination of the Al in the initial material may not be important in kaolinite formation in soils. Rapid kaolinite formation apparently depends on maintaining Al in solution with Si or intimate dispersion of the two ions in an unstable silicated phase. The presence of organically complexed Al facilitates precipitation of kaolinite from solution.

VI. OCCURRENCE OF KAOLINITE SUBGROUP MINERALS IN SOILS AND IN WEATHERED RESIDUA

A. Kaolinite Increase with Depth in Ultisols

Kaolinite tends to become depleted in surface soil horizons in the coarse clay and silt fractions when weathering is advanced. This property is illustrated by the Madison soil (Ultisol order) formed from mica schist in the Piedmont Region of Alabama (Bryant & Dixon, 1963). Kaolinite is 17% of the 2-0.2 μm clay of A1 and A3 horizons (0-13 cm), 26% of the B1, B21, and B22 horizons (13-74 cm) and 36% in the B3 horizon (74-102 cm). Kaolinite was relatively uniform in the 0.2-0.08 μm clay with depth ranging from 27 to 33%. The percentage of gibbsite followed a trend with depth similar to that of kaolinite. The complementary minerals mica, vermiculite-chlorite intergrade (partially interlayered vermiculite), and quartz had an inverse trend in the 2-0.2 μm clay and were relatively uniform with depth in the 0.2-0.08 μm clay. The greater abundance of kaolinite in subsoil than surface horizon soil clays is further illustrated in three of four Ultisols formed from limestone residuum, cherty limestone residuum, and loamy Coastal Plain deposits of Alabama (Dixon, 1966). It appears that age difference of the two Coastal Plain deposits may account in part for the failure of one profile to show greater kaolinite concentration in the subsoil clay, because one profile occurred nearer the coast and is assumed to be younger. A survey of properties of 45 Ultisol pedons from the Coastal Plains of the Southeastern USA further indicates that kaolinite is more abundant in clay fractions of B than A horizons (Fiskell & Perkins, 1970).

B. Kaolinite Distribution in Soils

Since kaolinite occurs in most soils, a few diverse examples will be cited to illustrate the varied environments where it occurs. In their weathering sequence, Jackson et al. (1948) reported kaolinite in the following soils:

Miami B horizon from Wisconsin, Hagerstown of Missouri, Alamance from North Carolina, Cecil and Susquehanna B horizons from Alabama, Lufkin from Mississippi, and Nipe clay from Puerto Rico. This group of soils represents a weathering range that traverses youthful soils with about 15% kaolinite (in < 0.2 μm clay fraction except for Nipe < 2 μm) on the northern extremity to about 85% in the Cecil in an intermediate weathering example and about 10% of the Nipe clay where weathering intensity is maximal for this group of soils. As weathering intensity (also time) increases, kaolinite eventually gives way to the more stable oxide minerals.

Kaolinite was found in small amounts in C horizons of Alluvial and Brown Wooded soils and in B and C horizons of a Grey Wooded soil of the Northwest Territories of Canada (Wright et al., 1967). The occurrence of kaolinite in C horizon material implies formation of much of the kaolinite prior to present soil profile development. These soils occur in a region with a mean annual temperature of $-4C$ (24F, 269K) and the subsoils may not thaw out until July even though they do not have a permafrost. The mean annual precipitation is 300 mm (11.8 inches) and the soils are always cool and often dry. Although kaolinite was a small component of these soils, its presence in such a mild weathering environment illustrates its widespread occurrence in soils and their parent materials.

C. Soils with Abundant Kaolinite Content

Kaolinite is one of the most abundant clay minerals in Ultisols of the southeastern United States (Fig. 11-9). Fiskell and Perkins (1970) studied the mineralogy of 45 pedons of Ultisols from the Coastal Plain of Alabama, Florida, South Carolina, North Carolina, and Virginia and reported kaolinite as a major component of coarse (2-0.2 μm) and fine (< 0.2 μm) clay fractions. Most of this kaolinite is considered by the writer to be inherited from the acid sediments from which the soils formed. This view is supported by the observation that even the Vertisols and associated smectitic Alfisols and Inceptisols of the Black Belt region of Alabama and Mississippi contain an appreciable amount of well-crystallized kaolinite inherited from the parent Cretaceous chalk or associated clayey deposits (Dixon & Nash, 1968).

Kaolinite is abundant in four Ultisols and Oxisols of Brazil studied by Lepsch and Buol (1974) and Alfisols and Ultisols of southern Nigeria (Gallez et al., 1975). Soil profiles studied from several areas of Nigeria are high in kaolinite (Jungerius & Levelt, 1964; J. A. Ogunwale, 1973[3]). The abundance of kaolinite throughout the soil profiles and in the parent sedimentary rocks points to the importance of this mineral being inherited in soils of Nigeria (Jungerius & Levelt, 1964).

---

[3] J. A. Ogunwale. 1973. The genesis and classification of soils from sandstones of various lithological origin in Nigeria. Ph.D. Thesis. Univ. of Ibadan, Ibadan, Nigeria. 275 p.

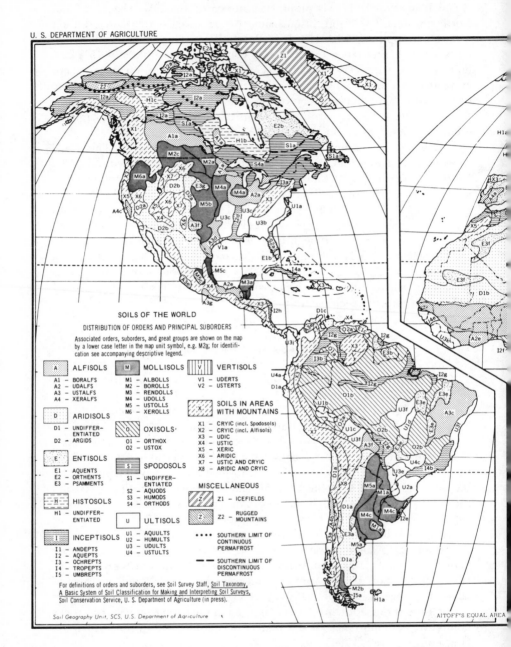

Fig. 11-9. A small-scale map of the world based on the current system of soil taxonomy used in the United States of America (by Soil Survey, SCS-USDA. From Kellogg, 1975).

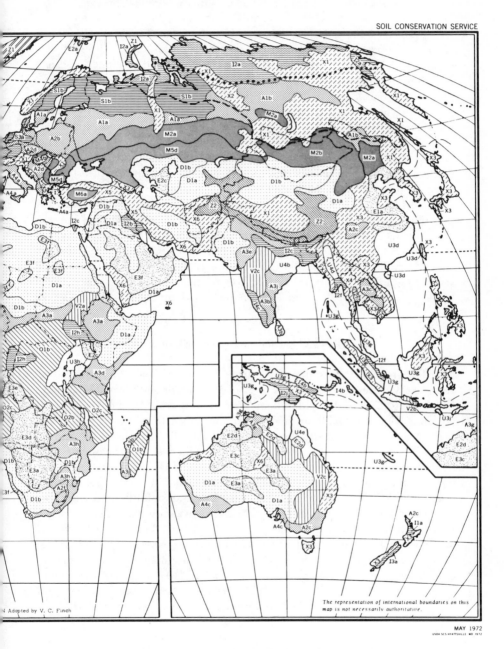

SOIL CONSERVATION SERVICE

N Adapted by V. C. Finch

The representation of international boundaries on this
map is not necessarily authoritative.

MAY 1972
USDA SCS HYATTSVILLE MD 1972

1 000   0   1 000   2 000   3 000 Miles

1 000   0   1 000   2 000   3 000 Kilometers

Approximate Scale (along Equator)

D. Kaolinite Determination in Young Soils: An Analytical Problem

Studies of loessial soils, Tama of Wisconsin and Loring of Mississippi, afford an opportunity to compare kaolinite formation under different temperatures, 7.8C (46F, 281K) vs. 17.8C (64F, 291K) and precipitation levels, 812 mm (32 inches) vs. 1,270 mm (50 inches) (Glenn et al., 1960; Glenn, 1960). It is surprising perhaps that the amounts of kaolinite reported are so similar in view of the contrasting weathering environments. For example, the values reported for A horizons of both soils are all between 8 and 13% kaolinite for 2-0.2 μm and 0.2-0.08 μm samples and similar kaolinite values are reported for B horizons. Some dissolution of kaolinite by the NaOH method employed to remove amorphous material was suspected in the Mississippi study because silica/alumina ratios of the dissolved material were similar to kaolinite. Adjustments for dissolved kaolinite would increase the percentage of kaolinite reported for both Tama and Loring soils, but the relative amount cannot be determined from the reports. These results indicate a need to refine methods of kaolinite determination in the presence of smectite. A similar study of Elliott, a Mollisol (formerly classified Varna soil series, see Table 7-3 of this book) soil formed from shale-derived till in Wisconsin reported little or no kaolinite present (Fanning & Jackson, 1964).

E. Stability of Kaolinite in Weathering Environment

The frequency of kaolinite occurrence is due partially to its formation from many different minerals. For example, Garrels and Christ (1965) have shown that the stability field of kaolinite is bounded by the stability fields of gibbsite, K-mica, K-feldspar, and amorphous silica (Fig. 11-10) as derived from suggested mineral equilibria. This stability diagram shows the association of kaolinite with the other four common soil constituents based on their relative solubility with respect to H_4SiO_4, K^+, and H^+ concentration in the equilibrium solution. Analyses of waters from many arkosic (feldspathic) sediments plotted almost all together in the kaolinite stability field suggesting that the direction of the weathering reactions was toward kaolinite formation in those environments. Figure 11-10 is a qualitative illustration of mineral sequences one might expect if equilibrium were attained (Garrels & Christ, 1965). It is of interest also that kaolinite may form from mica in equilibrium with sea water. Curtis and Spears (1971) plotted the ground water data from many different rock types on a H_4SiO_4 vs. pH diagram and found that they also fell in the kaolinite stability field as discussed later.

F. Kaolinite Influence on Hardening of Laterite (Plinthite)

The term *laterite* as used here refers to the ferruginous weathered materials that have or are capable of hardening more or less irreversibly on dry-

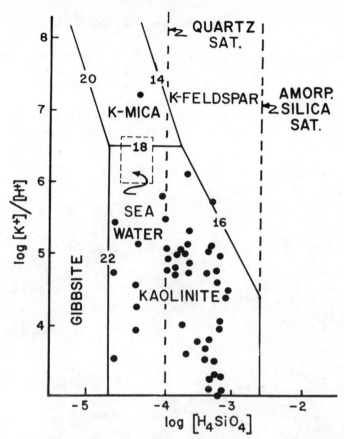

Fig. 11-10. Stability relations of some phases in the system K_2O-Al_2O_3-SiO_2-H_2O at 25C (298K) and 1 bar, as functions of $[K^+]/[H^+]$ and $[H_4SiO_4]$. Solid circles represent analyses of waters from arkosic sediments. Figure from Garrels and Christ (1965).

ing. In the United States, the term plinthite is applied to much of the material called laterite in other countries, but plinthite is applied only to the unhardened material.

Kaolinite is a major mineral in soils and weathered residua of tropical areas such as West Africa where laterite forms. But, kaolinite is usually less abundant in hardened laterite. The hardening process of laterite appears to involve the removal or segregation of kaolinite and crystallization of iron oxide usually as goethite (Sivarajashingham et al., 1962). Fripiat and Gastuche (according to Sivarajashingham et al., 1962) have found that kaolinite has a substantial capacity to adsorb and immobilize iron. Therefore, kaolinite would presumably have to be saturated before goethite could be crystallized. Saturation of kaolinite by iron is thought to occur by increases in iron from outside sources (such as from ground water) or by reduction of kaolinite by weathering to gibbsite. Appreciable gibbsite and quartz persist in soft and hard laterite.

G. Formation and Distribution of Halloysite in Soils

Halloysite is a fast-forming member of the kaolinite subgroup primarily in soils from volcanic deposits. Youthful soils of Japan contain halloysite mostly formed from volcanic glass and feldspar (Aomine & Wada, 1962). Halloysite spheres developed in porous volcanic glass of a young soil of Mexico (Dixon & McKee, 1974b). Halloysite formed in the early stages of weathering of soils in Mazama tephra of Oregon (Dudas & Harward, 1975).

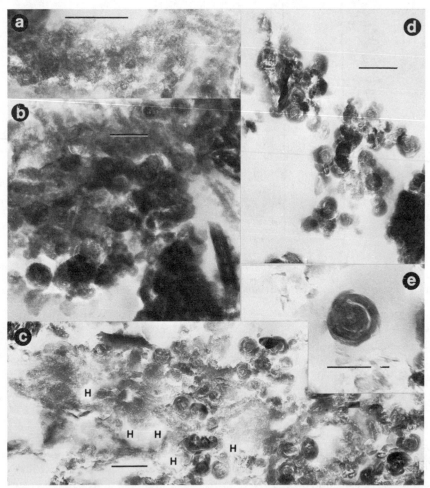

Fig. 11-11. Transmission electron micrographs of thin sections of 2–0.2 μm Mexican soil clay illustrating stages of spheroidal halloysite formation: (a) Small poorly organized particles in porous weathered glass, (b) Poor to well-developed larger spheroids in porous glass matrix, (c) Well-developed halloysite spheroids and similar size holes (H) in a matrix of glass, (d) Loose aggregates of halloysite spheroids, and (e) Single halloysite spheroid in the mounting resin. Scale is 0.2 μm. (From Dixon and McKee, 1974b).

Thin sections of 2–0.2 μm clay particles separated from a young volcanic ash soil near the Paricutin volcano of Mexico reveal stages of halloysite formation (Fig. 11-11). The porous granular matrix of glass contains circular particles ca 0.02 μm in diameter with layer structure (Fig. 11-11a). In the more advanced stages (b, c) the circular particles are almost 0.2 μm in diameter, have thicker walls, and are less transparent to the electron beam. Several large well-formed spheroids and holes of similar size are visible in a single particle of glass (c) thus showing the halloysite particles in the proposed environment of formation. Gradual weathering away of the glass reveals spheroids held loosely together (d) and, finally, an isolated sphere (e) surrounded by the mounting resin.

In addition to the above examples, halloysite is reported in soils from volcanic material in Guatemala (Askenasy et al., 1973), New Zealand (Birrell et al., 1955), and Bali, Indonesia (Dixon, unpublished). Halloysite tubes were found in the cold semiarid climate of antarctica (Claridge, 1965). The source of the tubes is not known, but the volcanic activity of that region and the suggested amorphous material imply an origin similar to that in volcanic deposits of other localities. Halloysite has been identified in the Ekiti soil of Nigeria formed from Pre-Cambrian rock composed of quartz, biotite, plagioclase, and hornblende (Gallez et al., 1975). Tubular particles suggestive of halloysite are reported sometimes in small amounts in soil clays of the Southeastern USA, Portugal (Bramao et al., 1952), and Nigeria (Ogunwale, 1973[3]). Halloysite, owing to its curved disorderly particles, gives weak X-ray diffraction patterns sometimes precluding identification by this method. Parham (1969) lists 27 recent studies from many parts of the world where halloysite has been identified mostly in weathered volcanic and other igneous deposits.

Halloysite is more soluble than kaolinite and has a less negative ΔG value (Kittrick, 1969) which supports the widely accepted view that halloysite gives way to kaolinite in long term weathering processes.

VII. KAOLINITE OCCURRENCE IN SEDIMENTARY ROCKS AND IN SEDIMENTS

A. Soil Minerals—A Major Source of Ocean Sediments

There is a close relationship between mineral composition of clay from soils and clay of pelagic (deep sea) surface (geologically Recent) sediments located at the same latitude and this relationship is particularly evident for kaolinite and gibbsite. These two weathering products occur in the highest concentrations in soils and in oceanic sediments at low latitudes. A major kaolinitic belt of sediments extends from the Alfisols, Ultisols, and Oxisols of north and eastern South America to the coast of Africa (compare Fig. 11-9 and 11-12). Also, Biscaye (1965) reported terrestrial material is the major source of Recent oceanic chlorite, illite (clay mica), and smectite from

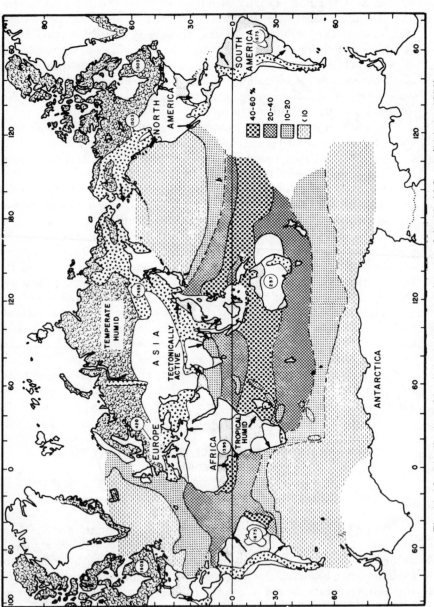

Fig. 11-12. Distribution of kaolinite in bottom sediments of the World Ocean (Lisitzin, 1972).

his study of 500 sediment samples from the Atlantic Ocean and adjacent seas and oceans. Exceptions occur where smectite is authigenic from volcanic glass deposits in the southwestern Indian Ocean and chlorite has formed in sediments near Hawaii according to Swindale and Fan (1967). Also, currents along the ocean bottom move detrital material in certain areas such as the southern South Atlantic Ocean. Most of the terrigenious material is transported by rivers to the ocean. Dusts such as those of the Harmattan haze of Africa also contribute to the westward distribution of kaolinite and other minerals from that continent (Biscaye, 1965). Higher concentrations of kaolinite sometimes occur in sediments from smaller rivers such as the São Francisco of South America whereas the world's largest river, the Amazon, has a more diverse mineralogy attributed to the influence of a wider population of weathering environments in the source area.

The eastern part of the Pacific Ocean is low in kaolinite probably because of lack of source areas in the mountainous west coast of the Americas. The western Pacific Ocean has a higher concentration of kaolinite than its eastern counterpart but data are scanty on the Asiatic borderland (Griffin et al., 1968).

Kaolinite is abundant in sediments of the Indian Ocean near Madagascar (Griffin et al., 1968) and according to Lisitzin (1972), kaolinite is abundant in a belt approximately from $0°$ to $30°S$ from Africa to Australia and extending eastward from northern Australia and New Guinea into the Pacific Ocean (Fig. 11-12).

Parham (1966) found that kaolinite usually decreases in amount from shoreward or fresh-water to the marine environment. He also noted that kaolinite, illite and chlorite are not restricted throughout geologic time in their occurrence in unmetamorphosed rocks; their stability does not appear to be time related. There is some evidence to the contrary in that kaolinite is found less commonly in older rocks, but other reasons may account for the paucity of kaolinite.

B. Kaolinite Formation in Rocks and Sediments

The content of silica in ground waters from several sources (Table 11-1 and Fig. 11-13) plotted on the solubility diagram of gibbsite and kaolinite clearly falls in the kaolinite stability field indicating that gibbsite could exist in these natural environments only as a metastable phase. Curtis and Spears (1971) reviewed the properties of several kaolinitic rocks including the Ayrshire Bauxitic Clay of Scotland which is mostly kaolinite with boehmite and diaspore. This Carboniferous Age deposit in many parts appears to be ancient in situ lateritic soil that has subsequently been silicified almost completely to kaolinite. Curits and Spears propose that the flint clays which break with subconchoidal fracture and have unusual hardness for argillaceous rocks are formed by silication of alumina minerals originally present in the sediment. The consequent increase in volume and crystal intergrowth are proposed to

Table 11-1. Groundwater analyses. U. S. Geol. Surv. Prof. Pap. 440-F. Water of low mineral content associated with common rock-types. These dilute waters were selected from environments in which the waters were most likely to be atmospheric precipitation that was then influenced primarily by reactions with the rocks in which they are found (including associated soil zones). From Curtis and Spears (1971)

Table no.	Group description	No. of samples	pH $\bar{x}$	S	ppm SiO$_2$ $\bar{x}$	S	μM H$_4$SiO$_4$ $\bar{x}$	S
1	Granite, rhyolite, and similar rock types	15	7.15	±0.48	37	±19	616	±316
2	Gabbro, basalt, and ultramafic rock types	16	7.52	±0.79	41	±13	683	±216
3	Andesite, diorite, and syenite	4	7.55	±0.24	20	± 9	333	±150
4	Sandstone, arkose, and graywacke	17	7.48	±0.82	23	±18	383	±300
5	Siltstone, clay, and shale	18	7.27	±1.25	28	±24	466	±400
6	Limestone	14	7.50	±0.37	13	± 6	216	±100
7	Dolomite	6	7.72	±0.33	14	± 7	233	±117
8	Miscellaneous sedimentary rocks (ironstone, chert, phosphorite, lignite, gypsum)	5	7.20	±0.55	18	± 6	300	±100
9	Quartzite and marble	7	7.27	±0.60	11	± 5	183	± 83
10	Slate, schist, gneiss, and miscellaneous metamorphic rocks	15	7.09	±0.84	23	±19	383	±316
11	Unconsolidated sand and gravel	20	7.65	±0.69	31	±17	516	±283

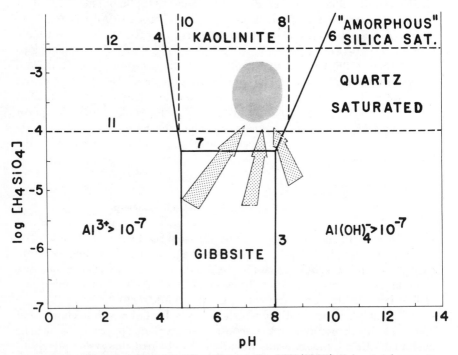

Fig. 11-13. Kaolinite/gibbsite stability diagram for 25C (298K), 1 bar total pressure (system H$_2$O–SiO$_2$–Al$_2$O$_3$). Heavy shading shows the composition of natural groundwaters. Light-shaded arrows represent soil/precipitate water interaction. (Contribution of H$_3$SiO$_4^-$ to silica solubility at high pH is not documented). From Curtis and Spears (1971).

contribute to the hardness and breakage properties of flint clays. Where the silication has gone to completion producing monomineralic kaolinite rock, greater hardness is the logical result in contrast to the softer Aryshire deposits that contain residual boehmite and diaspore.

The kaolinite in the Upper Cretaceous Age deposits of the Tuscaloosa formation in the Southeastern USA appears to have formed from sediments containing feldspars and muscovite. Alteration to kaolinite has, in part at least, occurred since deposition (Jonas, 1964). Bates (1964) suggested that much of the kaolinite may have been deposited during sedimentation and that the hardness of the kaolinite may be related to the salinity of the depositional environment.

Authigenic kaolinite and delicate mica needles have developed epitaxially (mutually symmetrically oriented) in clays from several sandstones taken in quarries in the interior of the USA (Rex, 1966). The growth of kaolinite and mica simultaneously supports the proposal of Garrels and Christ (1965) that the stability planes of the minerals join. These sandstones may have been exposed to water less saline than sea water for a long time and hence the results might not apply to a marine environment. These findings further support the hypothesis that kaolinite forms in rocks and may be inherited in soils formed from them.

VIII. INTERSTRATIFICATION OF KAOLINITE

Interstratification of montmorillonite in kaolinite (or halloysite) has been reported in "acid" clays of Japan (Sudo & Hayashi, 1956), in clays of Florida (Altschuler et al., 1963), in clays of Mexico (Schultz et al., 1971), in a clay from Poland (Wiewiora, 1971), and in clays of three Scottish soils (Wilson & Cradwick, 1972). A weathered clay and a hydrothermally altered one containing kaolinite–montmorillonite mixed layering were reported by Sakharov and Drits (1973).

Cradwick and Wilson (1972) have calculated theoretical diffraction patterns for assumed kaolinite–montmorillonite interstratified mixtures and the theoretical patterns are in substantial agreement with compositions of clays from Japan (Sudo & Hayashi, 1956) and from Mexico (Schultz et al., 1971). The kaolinite–montmorillonite components are segregated in the interstratified mixture in the Becal clay, but the clays from Ticul and Tepakan, Mexico and from Japan are randomly interstratified (Cradwick & Wilson, 1972). Altschuler et al. (1963) reported kaolinite–montmorillonite regular interstratification in clay of Florida and proposed a mechanism of acid weathering whereby silica was stripped from montmorillonite and the exposed octahedral oxygens were hydroxylated forming a kaolinite-like layer. Silica was presumed to be leached in the ground water. Montmorillonite in clayey sands has weathered to kaolinite extensively in Florida making it a reaction of regional importance.

Kaolinite–montmorillonite interstratification is of analytical importance

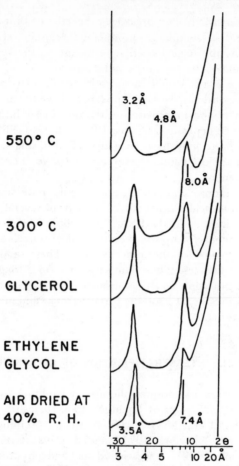

Fig. 11-14. Smoothed X-ray diffractometer traces from oriented aggregates of <0.25 μm
Ticul clay showing interstratification in kaolinite. Redrawn from Schultz et al. (1971).

because it produces subtle changes in X-ray diffraction patterns that may be
overlooked or attributed to halloysite. Schultz et al. (1971) demonstrated
gradual shifts of 7Å and 3.5Å peaks of randomly interstratified kaolinite and
montmorillonite. The most diagnostic feature is the peak of 8.0 to 8.4Å produced by a sample heated to 300C (573K), thus giving an intermediate reflection between the 7.15Å peak of kaolinite and the 9.6Å spacing of fully
collapsed montmorillonite (Fig. 11-14).

IX. ISOMORPHOUS SUBSTITUTION IN KAOLINITE

The presence of Fe in kaolinites of eight tropical ferruginous soils of
India was reported by Rengasamy et al. (1975). Their major criterion was
the consistent SiO_2/R_2O_3 molar ratios of 2.00 to 2.18 compared to the ideal
of 2.00 for kaolinite. The calculated unit cell composition included sub-

stitution of 0.05 to 0.41 Fe^{3+} atoms for Al per two octahedral sites. These Indian soil kaolinites were as small as 0.01 μm across the flakes with varying iron content from different horizons of the same soil profile which led the authors to conclude that the kaolinite precipitated from weathering solutions. Isomorphous substitution reported in kaolinite from Pugu, Tanganyika had a cation composition of $(Si_{1.982}Al_{0.018})$ $(Al_{1.958}Fe^{3+}_{0.039}Mg_{0.003})$. The Pugu sample was a relatively pure kaolinite fraction of thin hexagonal plates (Robertson et al., 1954). The observation of "extremely thin" clouds in the electron micrographs, considered to be hydrated iron by the authors, raises the question of smectite contamination and merits further study.

Isomorphous substitution of Fe^{3+} and possibly Fe^{2+} in the octahedra of kaolinite has been reported by Jefferson et al. (1975) and Maldon and Meads (1967) based on Mössbauer spectroscopic data and incomplete Fe dissolution when the clay is treated with HCl. The iron-stained kaolinites of Jefferson et al. (1975) contained a maximum of 2.46% Fe extractable with concentrated HCl (total Fe was not reported). Most of the iron was ferric. The occurrence of octahedral Fe^{3+} in kaolinite was detected in several samples. Since all of the Jefferson et al. samples except two from South Carolina contained a mica impurity, an argument for the presence of octahedral iron in kaolinite is most reliably based on the mica-free samples. The kaolinite clay samples from South Carolina had Mössbauer spectra with doublets that had parameters compatible with high-spin Fe^{3+} in a six-coordinated site both before and after a 21-day cleaning treatment with concentrated HCl. Electron spin resonance studies have also led to the conclusion that Fe^{3+} occurs in the octahedra of kaolinite (Angel & Hall, 1972). Titanium in the structure of kaolinite has been reported by Weiss and Range (1966).

In other investigations, an average of 86% of the original titanium present in ten kaolinite samples from several locations was recovered in the residue (mostly as anatase) after kaolinite dissolution by H_2TiF_6 (Dolcater et al., 1970). Some of the 14% titanium dissolved may have been in the kaolinite structure or it may have been in a soluble amorphous form. No titanium was present in three of the kaolinite samples investigated.

Some Fe_2O_3 may occur in halloysite because summation of Al_2O_3 and Fe_2O_3 gave an R_2O_3 value near the theoretical 34.6% for halloysite with $2H_2O$ (Askenasy et al., 1973). The soil halloysite sample was impure precluding a conclusion on the actual structural content of the halloysite.

X. CATION EXCHANGE PROPERTIES OF KAOLINITE

Permanent negative charge on kaolinite from isomorphous substitution has been postulated because exchangeable Na is retained under acid conditions (Schofield & Samson, 1954). Weiss and Russow (1963) report that exchangeable cations are bound only on the tetrahedral basal plane of kaolinite and that Al^{3+} for Si^{4+} substitution is confined to the outer tetrahedral sheet. The distribution of cationic colloidal AgI sorbed on crystal faces is shown by electron microscopy examination of thin kaolinite plates. Follett (1965) observed positively charged colloidal ferric hydroxide (i.e. hydrated ferric oxide) adsorbed on kaolinite flakes analogous to the colloidal AgI of Weiss and Russow (1963). Where untreated kaolinite was mixed with ferric hydroxide treated kaolinite desorption of the Fe-colloids occurred on some plates revealing the profile of a hexagonal kaolinite flake. This clean hexagonal desorption pattern indicates that no Fe-colloids are present on either side of the flake because colloids on both sides are shown in the transmission electron micrographs thus confirming the findings of Weiss and Russow that positive colloids are adsorbed on only one face of kaolinite. Further investigation of adsorption of Fe-colloid on quartz revealed that the silica surface sorbed the Fe-colloids, but less than kaolinite. Gibbsite sorbed very few Fe-colloid particles further suggesting the primary involvement of the silica surface in positive colloid adsorption.

Cashen (1966) proposes that Al released from the kaolinite structure is adsorbed on permanent charge sites as a large trivalent ion $Al_6(OH)_{15}^{3+}$. Further, the large trivalent cation bonded to permanent cation exchange sites on planar sites of kaolinite is credited with causing a maximum in viscosity commonly reported when the acid kaolinite is 80–90% neutralized. Cashen argues the proposed role of the $Al_6(OH)_{15}^{3+}$ ion is not in conflict with the recognized positive edges of kaolinite flakes (e.g., as shown by adsorption of negative gold colloids), but they are small in number compared to ions entering into cation exchange reactions on planar surfaces.

Other findings may contribute to understanding isomorphous substitution and cation exchange capacity in kaolinite. Smectite interstratification with kaolinite and mica zones in kaolinite particles have been reported (Altschuler et al., 1963; Lee et al., 1975a, 1975b). Hence, the permanent negative charge attributed to kaolinite could be a property of inclusions in impure kaolinite. Montmorillonite and mica in kaolinite particles and the recognized alteration of these minerals to kaolinite offer a logical mechanism for isomorphous substitution in kaolinite due to residual charge in octahedra and tetrahedra incorporated into the kaolinite structure from the parent mineral if such occurs. Internal pores have been reported in individual kaolinite crystals (Lee et al., 1975a) and well-crystallized halloysite particles (T. R. McKee et al., 1973[4], Dixon & McKee, 1974a). These pores influence rates of ion movement and surface area which in turn influence cation exchange properties of kaolinite subgroup minerals.

XI. POSITIVE CHARGE AND ANION RETENTION BY KAOLINITE

The positive charge on kaolinite has generally been attributed to sites on the edges of the plates that become positive by acceptance of protons in the acid pH range (Schofield & Samson, 1953). Evidence for positive edge charge

---

[4]T. R. McKee, J. B. Dixon, and D. F. Harling. 1973. High resolution electron microscopy investigations and lattice image comparisons of contrasting halloysites. 22nd Clay Miner. Conf., Banff, Alberta, Canada. Abstr. p. 25.

Table 11-2. Calculated edge-areas and edge exchange-capacities for various kaolinites before and after free iron removal. From Sumner and Reeve (1966)

Sample	Untreated			Deferrified	
	Free-iron oxides as Fe_2O_3	Edge† exchange-capacity	Edge-area	Edge† exchange-capacity	Edge-area
	%	meq/100 g	m^2/g	meq/100 g	m^2/g
Zetterlite	0.3	0.7	1.4	0.3	0.6
Peerless	0.2	1.3	2.6	0.4	0.8
Peerless (Cashen, 1959)	--	--	3.3	--	--
Cornwall	0.3	0.5	1.0	0.2	0.4
Dixie	0.4	1.4	2.8	0.8	1.6
Riedel de Haen	0.3	0.5	1.0	0.3	0.6
Devon	0.2	0.6	1.2	0.4	0.8
Kent sand	1.8	3.0	6.0	2.0	4.0
Rocky Gully (Quirk, 1960)	<2.0‡	3.0	6.0	--	--
Natal soil	15.0	7.8	15.6	2.0	4.0

† Based on Cl⁻ repulsion.
‡ <2% goethite.

was vividly shown by electron micrographs of negative gold colloids adsorbed to the edges of kaolinite flakes (Marshall, 1964). These positive edge sites have been described as

$$-Al\begin{matrix} \diagup OH \\ \diagdown OH \end{matrix}$$

groups at the edge of octahedral sheets (Muljadi et al., 1966).

Edge areas for kaolinite plates were calculated based on one positive charge per $33Å^2$ (Schofield & Samson, 1953) assuming each octahedral oxygen and OH at the edge has a +½ charge under acid conditions (Sumner & Reeve, 1966). The edge area values (Table 11-2) range from 0.4 to 4.0 m^2/g for deferrated clay and soil samples and for values calculated for hexagonal plates for crystal dimensions taken from Grim (1953). Agreement between the calculated range of values and determined values is good. A set of "edge" exchange capacity and "edge" area values also is reported for clay and soil sample before deferration illustrating the markedly higher values where iron oxides were present. The isoelectric pH is about 4 for 5 of the 8 deferrated kaolinite samples as indicated by sorption isotherms. The possible influence on Cl⁻ adsorption of contaminant minerals identified other than free iron oxides was not discussed. A recent report by Rand and Melton (1975) placed the isoelectric point of kaolinite edge surface at pH 7.3 based on viscosity measurements.

Kaolinite has historically been recognized for its reactivity with phosphate ions. The abundance of kaolinite in extensive areas of Ultisols, Alfisols, and Oxisols makes its reactivity with phosphate important. Muljadi et al. (1966) assessed this property in a series of reports on phosphate adsorption isotherms. They propose that phosphate is adsorbed by kaolinite on three types of sites that differ energetically and correlate with different regions of an adsorption isotherm.

Region I: At low phosphate concentration ($< 10^{-4}M$ P) when isotherm rises steeply and stays close to Y-axis. Very high affinity sites for phosphate.

Region II: Begins where isotherm becomes convex to Y-axis at approximately $10^{-4}M$ P.

Region III: The linear part of the isotherm at medium to high concentrations (10^{-3} to $10^{-1}M$ P).

Regions I and II are the first to sorb phosphates and they are proposed to represent two hydroxyl sites associated with Al at the edge of a kaolinite crystal (also gibbsite and pseudoboehmite) octahedral sheet and the Al associated with exchange sites of kaolinite. Region III may represent a type of site large in number compared to the amount sorbed or possibly a poorly crystalline region of the clay surface. The phosphate adsorption on kaolinite was largely reversible with respect to concentration for Regions II and III, but for Region I, the steep part of the curve, was not reversible even after 7-days equilibration in distilled water. Much of the increase in phosphate sorption with temperature was not reversible in the 20 to 2C (293–275K) and 40 to 2C (313–275K) ranges; thus implying the activation of new permanent sorption sites with increased temperature.

The observations (see earlier) of positively charged AgI, ferric hydroxide and hydroxy-Al groups on kaolinite basal surfaces illustrates sources of positive charge that alter the behavior of the negative oxygen surface. Cation exchange capacity will be reduced and the kaolinite crystal will have more positive properties.

The possible occurrence of positive charge on kaolinite from isomorphous substitution was discussed briefly by Sumner and Reeve (1966). Conceptually, at least, Ti^{4+} substitution for Al^{3+} or excess Al^{3+} in the octahedral sheet might give positive charge but no proof of these substitutions in kaolinite has been reported. However, a positive octahedral sheet in chlorite has been recognized.

XII. FLOCCULATION AND DISPERSION OF KAOLINITE

The flocculation of kaolinite in the absence of salt is attributed to the bonding of the positive crystal edges to the negative plate surfaces forming "edge-to-face" flocculation (Schofield & Samson, 1954). A second type of flocculation involving face-to-face bonding is induced by salts, particularly those with di- or trivalent cations (also Al ions, previous section).

Dispersion of clay particles occurs because of their repelling negative charges produced by the dissociation of the exchangeable cations in water.

The extent of dissociation of the exchange ions is influenced by their valence and degree of hydratability. Sodium is usually selected as the exchange ion for clays because it is monovalent and highly hydrated.

The low negative charge of kaolinite makes it difficult to disperse. The presence of positive crystal edge sites and exchangeable cations held firmly in a Stern layer (Hunter & Alexander, 1963) may also contribute to the tendency of kaolinite to flocculate. By raising the pH with NaOH, kaolinite increases greatly in surface charge density with the greater increase occurring between about pH 8.2 and 10.9 (Street & Buchanan, 1956). Also, the apparent viscosity (Fig. 11-15) of kaolinite suspensions reaches a minimum in NaOH at about pH 10.5 (Street, 1956; Street & Buchanan, 1956). These conditions of pH and Na saturation are essentially those employed by Hunter and Alexander (1963) to obtain a disperse kaolinite system except that they employed Na_2CO_3 buffer rather than NaOH. Also, pH values of 9.5 to 10.5 are commonly employed for kaolinitic soil dispersion and it appears from the viscosity and layer charge data that the higher pH is the better dispersing environment although dissolution of aluminosilicates at high pH must be considered as a possible deleterious effect.

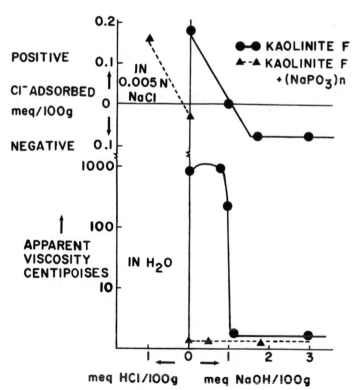

Fig. 11-15. Chloride sorption by kaolinite treated and untreated with metaphosphate measured in 0.005N NaCl and viscosity measured in distilled water. From Schofield and Samson (1954).

Kaolinite also can be dispersed even at neutral pH by treating with polyphosphate which alters the viscosity (Fig. 11-15) and the positive crystal edges as was vividly shown in early work by Thiessen (Marshall, 1964) where colloidal gold particles dotted the edges of kaolinite plates shown in electron micrographs and were not attracted following treatment with polyphosphate. The charge on kaolinite also is increased by phosphate treatment, e.g. from -1.36×10^{-2} coulomb/m^2 in $0.0172M$ NaCl treatment to -2.08×10^{-2} coulomb/m^2 in $0.0172M$ NaH$_2$PO$_4$ (Hunter & Alexander, 1963). When tetrasodium pyrophosphate solutions were used to disperse kaolinite, Al, Si, and Fe were released into solution (Bidwell et al., 1970) thus illustrating the need to minimize the exposure of kaolinite to phosphate dispersants.

Organic clay dispersants are in common use in the ceramic industry although they have been exploited in few soils investigations. Because inorganic dispersants adversely affect firing quality of clay castings and other fired products, organic dispersants are employed in these concentrated clay systems. These organic dispersants are mostly macromolecular substances with ionizing groups distributed throughout the chain. Sodium polyacrylate is an example that is in common use (Mitra et al., 1973). Bidwell et al. (1970) suggest that the polyacrylate is adsorbed on the positively charged edges of kaolinite.

XIII. PHYSICAL PROPERTIES OF KAOLINITE AND HALLOYSITE

The surface area of kaolinite consists largely of external planar surfaces (00l) and edge surfaces. Where large books of kaolinite occur, there are indentations that expose a few interlayer regions, but most of the surface area of kaolinite clays is external to the plates. Internal pores and imperfections have been shown in sectioned samples of kaolinite and of halloysite.

Ormsby et al. (1962) reported the surface area of kaolinites to range from 5.0 to 14.5 m^2/g of different particle size fractions as determined by a glycerol adsorption method, and the values correlated directly with CEC values (Table 11-3). Crystallinity was better among the finer fractions of a given kaolinite. Surface area values for five kaolinite samples analyzed by Schofield and Samson (1954) were 6 to 39 m^2/g by an N$_2$ BET method and 8 to 25 m^2/g by a negative adsorption method (Table 11-4).

Nitrogen-BET surface areas of halloysites studied by J. B. Dixon et al. (1974[5]) ranged from 21 to 43 m^2/g. The range of values for halloysites is above those reported by Ormsby et al. (1962) and is in the range of those reported by Ormsby and Shartsis (1960) and by Schofield and Samson (1954) for kaolinites. These results imply internal porosity of kaolinite or impurities that were overlooked in light of the considerable internal porosity reported for halloysites (Dixon & McKee, 1974a).

---

[5]J. B. Dixon, T. R. McKee, and L. W. Zelazny. 1974. Morphology and surface area of halloysites. 23rd Annual Clay Miner. Conf. (Cleveland, Ohio) Abstr. p. 26.

Table 11-3. Summary of cation-exchange capacity, surface-area, and crystallinity data. from Ormsby et al. (1962)[†]

Kaolinite	Property	Particle-size fraction, μm[‡]					
		44-10	10-5	5-2	2-1	1-0.5	0.5-0.25
A	Cation-exchange capacity	1.32	1.28	1.47	2.03	2.33	3.72
	Surface area	5.02	5.86	6.70	8.60	8.80	8.82
	Crystallinity ratio	0.27	0.23	0.35	0.57	0.64	0.68
H	Cation-exchange capacity	1.48	1.48	1.60	2.20	2.44	4.40
	Surface area	6.09	6.59	7.66	8.86	10.06	11.75
	Crystallinity ratio	0.19	0.20	0.23	0.38	0.40	0.48
B	Cation-exchange capacity	1.64	1.74	1.92	2.34	2.69	4.19
	Surface area	6.28	6.80	8.49	9.82	10.62	14.04
	Crystallinity ratio	0.16	(§)	0.22	0.28	0.31	0.44
E	Cation-exchange capacity	1.58	1.62	1.61	2.60	2.92	3.94
	Surface area	6.32	6.19	7.91	9.66	11.00	12.45
	Crystallinity ratio	0.25	0.22	0.23	0.22	0.26	0.25
I	Cation-exchange capacity	2.69	3.34	2.71	3.22	3.76	4.62
	Surface area	17.48	14.46	13.85	14.32	16.27	17.47

[†] Cation-exchange capacity, meq Mn^{2+}/100 g (average of three determinations; standard deviation, 0.15). Surface area by glycerol retention, m^2/g (average of two determinations; standard deviation, 0.41). Crystallinity ratio = $I_{(02\bar{1})}/I_{(060)}$; each value is the result of three determinations of each of the two peaks used.
[‡] Equivalent spherical diameter, μm.
§ Not determined; X-ray patterns for samples B(10-5 μm) and I were so poor that $I_{(02\bar{1})}$ could not be estimated.

Table 11-4. Surface area of five kaolinites by negative adsorption and Brunauer, Emmett, and Teller methods. After Schofield and Samson (1954)

Kaolinite	Area by negative adsorption	Area by B.E.T. method
	———————— m^2/g ————————	
F	15	--
C	8	6
M	9	10
G	18	16
S	25	39

The density of kaolinite is 2.630 g/cm^3 whether calculated from crystallographic data or measured in water (Deeds & van Olphen, 1963). A value of 2.605 was obtained where density was measured in hydrocarbons. The lower density is attributed to pores not penetrated by the hydrocarbons yet penetrated by water because of its higher adsorptive energy. Much higher percentages of inaccessible pores were reported for vermiculite and smectites than for kaolinite (Deeds & van Olphen, 1963).

XIV. IDENTIFICATION OF KAOLINITE AND HALLOYSITE

Kaolinite is identified by the 7.2 and 3.57Å diffraction maxima if neither chlorite nor expanded vermiculite is present. Chlorite may be indicated by separation of the chlorite peak at 3.54Å from the kaolinite peak at 3.57Å (Bradley, 1954). Well-crystallized chlorite is shown by the persistence, and commonly by reinforcement, of the 14Å diffraction peak after heating at 550 to 600C (823-873K) which destroys the 7.2Å reflection of kaolinite and reduces or destroys the chlorite 7Å and higher-order reflections (Brindley, 1961).

Heating soil clays at 300C (573K) or higher normally shifts the 14Å peak from intergradient chlorite–vermiculite (hydroxy interlayered vermiculite) to about 10 to 12Å thereby eliminating the slight contribution of its second order to the 7.2Å position. The 060 spacing of kaolinite at 1.49Å may permit separation from trioctahedral chlorite (or serpentines) that have higher 060 reflections of 1.53 to 1.55Å. Care must be taken to avoid the 060 reflection of muscovite mica that has a spacing of 1.50Å and of quartz which has a diffraction maximum at 1.54Å.

Additional evidence for kaolinite is the approximately 550C (823K) endothermic DTA peak and the sharp infrared absorption maxima. Unfortunately, other minerals may present OH endotherms and infrared absorption maxima that overlap those of kaolinite.

Halloysite tubes and spheroids give weaker basal reflections than platy phyllosilicates because of the common orientation of the latter type particles when dried or smeared on a planar surface. A diffraction peak near 4.45Å is from portions of layers perpendicular to the plane of the aggregate (assuming the clay has been prepared on a planar mount such as a glass slide) thus suggesting layer curvature which is a useful indicator of halloysite; when the intensity of the 4.45Å peak is half or more that of the basal peak at 7 to 11Å halloysite is indicated (Brindley, 1961). At 300C (573K) hydrated halloysite collapses to about 7.2Å kaolinite spacing, but reflections are still weak by comparison with kaolinite. Air-dried halloysite may give a broad peak at 7Å that reinforces on heating. Glyceration of hydrated halloysite is avoided for those specimens that are to be heated because residue persists in the interlayer giving a spacing between 7 and 10Å (Askenasy et al., 1973). Tubes or spheroids in electron micrographs are good indicators of halloysite even in trace amounts. Concentration of the curved forms may be required for identification by X-ray diffraction.

XV. QUANTITATIVE ANALYSIS OF KAOLINITE SUBGROUP MINERALS

Several approaches to quantitative analysis of kaolinite subgroup minerals are important because the success of the method chosen is dependent in part on the associated minerals and each method is subject to different interferences. A selective dissolution analysis (SDA) method (2.5 min boiling in 0.5N NaOH) used to dissolve amorphous material and (separately) to dissolve dehydroxylated kaolinite (Hashimoto & Jackson, 1960) gave satisfactory results on 2-0.2 μm and 0.2-0.08 μm soil clays from three Ultisols (Dixon, 1966). Apparently some smectite in a fourth Ultisol dissolved contributing to a higher kaolinite value than either DTA or X-ray diffraction peak intensity indicated. Dissolution of fine montmorillonite in the NaOH was recognized as a problem by Hashimoto and Jackson. Dissolution of halloysite prior to dehydroxylation by the 0.5N NaOH boiling has been suggested (Askenasy et al., 1973). Although the NaOH SDA method is subject to some errors, it sets an upper limit on kaolinite content where no correction for amorphous material is required and can be a useful method for estimating kaolinite content particularly in highly weathered clay samples composed of coarse well-crystallized components.

Differential thermal analysis is a useful method of quantitative kaolinite analysis where a satisfactory reference mineral sample is available and coincidental thermal reactions can be avoided. Gibbs (1965) has proposed separating minerals from the sample in question and this would appear to be the most certain way to have a reference mineral with the same thermal properties as the unknown. Purification of soil kaolinite for use as a standard may not be possible; a type of kaolinite, with similar peak temperature and shape to that of the unknown, may be an acceptable alternative (Dixon, 1966). Some correction for mica and other impurities in the reference kaolinite may be required even after appropriate fractionation. In soils containing smectite a correction for overlap of the smectite and kaolinite hydroxyl endotherms may be required (Carson & Dixon, 1972). Measuring a representative triangular area under the DTA peak may avoid error from interlayer hydroxyl loss from poorly crystalline chlorite.

One of the chief deterrents to quantitative X-ray diffraction analysis of kaolinite is the variability in particle orientation. A proposed method of controlling preferred particle orientation of kaolinite employs mixing the clay with an organic powder cement which is subsequently melted and ground to a fine powder (Brindley & Kurtossy, 1961). The authors emphasize the importance of selecting a reference kaolinite with a similar peak breadth to that of the unknown which would be particularly difficult for soil clays which differ in peak breadth from soil to soil and for different particle sizes. This method would more likely succeed in analysis of large uniform populations of samples than for most soil clays.

Quantitative analysis of kaolinite and halloysite based on NH_4Cl intercolated between structural layers has been suggested by Wada (1963) who observed that this salt is not sorbed appreciably by 2:1 and chloritic minerals. Other workers attempting to intercalate kaolinite and halloysite with salts have had different degrees of success depending on sample crystallinity, humidity control, and possibly other factors. The potential merit of this method is that the extent of the reaction can be monitored by X-ray diffraction.

An entirely satisfactory method for quantitative analysis of kaolinite subgroup minerals has not been developed. The analyst should provide evidence of the reliability of his results, for example, by determining all constituents by independent methods and obtaining a summation near 100% (Alexiades & Jackson, 1966).

XVI. SERPENTINE MINERALS

Serpentines rarely occur in soils as indicated by the small literature on the subject (e.g. Wildman et al., 1968). Serpentinite rocks are not extensive types (Beutelspacher & van der Marel, 1968) and serpentines apparently do not form in soils. Serpentines can have an important influence on weathering products in soils (Wildman et al., 1968). The principal occurrences of serpentine minerals are in altered ultrabasic rocks such as dunites, pyroxenites, and peridotites (Deer et al., 1966).

Chrysotile, lizardite, and antigorite are the most widely recognized serpentine minerals. Berthierine has priority over chamosite as the name for the 1:1 type layer silicate commonly found in ironstones (Bailey et al., 1971). These minerals have analogous structures to kaolinite minerals but contain other ions (Fe^{2+}, Fe^{3+}, Mg^{2+}, Ni^{2+}) substituted for octahedral Al and in a few cases other ions are substituted for Si (e.g. Al^{3+}, Fe^{3+}, and Ge^{4+}). Amesite has a higher substitution of Al than other serpentines with a formula approximating $(Mg_2Al)(SiAl)O_5(OH)_4$ (Brindley, 1961). It has layer-to-layer ionic attraction in addition to the usual OH–O bonding, thus reducing the layer thickness to about 7.0Å. Berthierine (sometimes called chamosites) and cronstedtite also exhibit contraction of the layer along the c-axis.

Antigorite is generally considered to be a serpentine, but it differs from the typical 1:1 layer silicate by inversion of the layer at each length of the a-axis producing a corrugated layer. The a-axis of antigorite is unusually long, often about 43Å, which gives characteristic electron diffraction patterns of single crystals.

Powder diffraction patterns rarely provide sufficient information for identification of individual serpentine minerals. Where exacting identification is not made, as by electron diffraction of single crystals, the general term *serpentine* is used with the morphology (platy, fibrous, etc.) as determined by electron microscopy.

XVII. SERPENTINE COMPOSITION

The composition of serpentines is a subject of controversy (Page, 1968; Whittaker & Wicks, 1970) because data on properly identified samples are inadequate. Chrysotiles have Mg and Si contents close to the theoretical $[Si_2Mg_3O_5(OH)_4]$, but the H_2O lost above 110C (383K) is higher. They tend to have less Fe and Al substitution than antigorite and lizardite. Substitution for tetrahedral Si appears to be greater for lizardite than for most antigorites and chrysotiles.

XVIII. WEATHERING OF SERPENTINITE ROCKS

Serpentine minerals are unstable in the weathering environment and tend to alter to other minerals during soil formation. Serpentinite rocks weather to iron-rich smectite in soils in the mountains of central California (Wildman et al., 1968). Under the accelerated weathering of a tropical rain forest, serpentinite weathers to a 7.5 m lateritic profile of goethite and gibbsite in the surface zone, to chlorite in the intermediate zone, and to a few centimeters of smectite adjacent to the rock (Schellman, 1964). Kanno et al. (1965) found varying amounts of expansible layer silicates including smectite in a Brown Forest soil formed from serpentinite rock in Japan.

The four serpentinite-derived soils of California reported by Wildman et al. (1968) contain < 0.2 μm clay that is almost pure smectite. Ferric iron is the most abundant octahedral ion in the smectite of the three most weathered soils. Very little serpentine is present in the < 0.2 μm clay of the B22t horizons except in the soil from New Idra where rainfall is only 380 mm (15 inches). Serpentinite weathering to smectite resulted in a 5- to 10-fold loss in Mg and a concentration of Si, Fe, and Al. The Fancher soil contained an olivine (forsterite) in the parent rock, and the Al content was similar in rock and < 0.2 μm clay. After allocation of the elements to a montmorillonite-vermiculite formula, there was 3 to 17% unallocated silica which apparently occurred as amorphous SiO_2 presumably formed by incongruous dissolution of the serpentine.

The formation of an iron-rich montmorillonite during the weathering of serpentine involves considerable structural alteration and change in ion composition from parent to product mineral. The montmorillonite apparently formed from ions in solution or in gels.

XIX. STABILITY OF SERPENTINE

The apparent standard free energy of three California serpentines was calculated from solubility data at 25C (298K) (Wildman et al., 1971). Based on the end-member formula, $Si_2Mg_3O_5(OH)_4$, the average value is −962.9 kcalories/mole (−4.029 MJ/mole). A stability diagram developed from the free energy value indicates that a serpentine mineral is not stable below a pH of about 7.0 in a $0.1M$ Mg^{2+} solution (Fig. 11–16). Serpentine is not stable below pH 8 in a $0.001M$ Mg^{2+} solution assuming log $[Si(OH)_4]$ = −2.6 as the saturation solubility of monomeric silicic acid at 25C (298K) and 1 bar (Garrels & Christ, 1965). If a lower $Si(OH)_4$ concentration is assumed the stability field for serpentine is reduced. From the stability diagram it appears that serpentine is unstable in the range of pH and Mg^{2+} and $Si(OH)_4$ activities encountered in most soils.

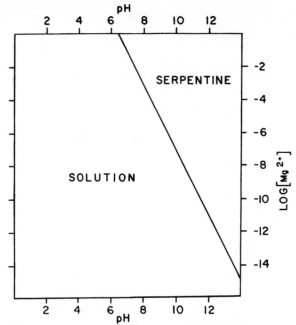

Fig. 11-16. Stability of serpentine expressed in terms of pH and activity of log Mg^{2+} at 25C (298K) and 1 bar. Activity of $Si(OH)_4$ is $10^{-2.6}$. From Wildman et al. (1971).

XX. MORPHOLOGICAL PROPERTIES OF SERPENTINE-SUBGROUP MINERALS

Serpentines occur in plates, laths, and tubes, sometimes all in one sample. A synthetic serpentine has a "cone-in-cone" structure and telescope-like sections of apparently concentric tubes (Bates, 1958). Stubby serpentine tubes, some visibly hollow, were reported from Ontario, Canada (Davis et al., 1950). Several other serpentines with the morphologies given above have been found in European locations and in England by Beutelspacher and van der Marel (1968).

Chrysotile is usually fibrous, although samples with largely lath-like morphology have been reported (Whittaker & Zussman, 1971). Chrysotile tubes may or may not be filled with solid material. Chrysotile tubes from Globe, Arizona have an inside diameter (ID) of about 50Å and an outside diameter (OD) of about 130Å of the thinnest tubes which gives a minimum wall thickness of 40Å. In investigations by Yada, sectioned chrysotile fibers viewed along the particle axes are shown to be made up of packets of layers rolled in a spiral fashion (Whittaker & Zussman, 1971). Lattice fringes reveal positions of the 7Å layers and radial (020) planes.

Lizardite is usually a very fine-grained serpentine with platy crystals. The only good lizardite cleavage is basal, thus making it the most ideal platy serpentine (Whittaker & Zussman, 1971).

Replicas of fracture surfaces and particles viewed in electron micrographs by Bates (1958) show that minerals of this subgroup are not typically clay size. An unusual deposit of almost pure antigorite and composed of particles about 2 to 0.5 μm in diameter occurs in a talc mine in Castor, Parana, Brazil (Brindley & de Souza Santos, 1971). This antigorite is white, platy, and of nearly ideal composition. This antigorite clay gives the first-order reflection of 7.2Å and the characteristic long a-spacing usually at 43.5Å, shown by electron diffraction.

XXI. THERMAL PROPERTIES OF SERPENTINES

Serpentines typically give two major DTA peaks, the large one being due to OH⁻ loss, usually occurs at about 600 to 650C (873-923K) for magnesian varieties. An exothermic peak occurs following the endotherm (Fig. 11-17). For some specimens the endothermic peak and the exothermic peak are unresolved and for others the two peaks are separated by over 100C. The low temperature peaks, produced by a loss of hygroscopic water, only appear on curves for altered samples which may be contaminated by colloidal material and, as for kaolinite, they are always small. These small adsorbed water peaks imply a low CEC and low surface area.

Nickel varieties of serpentine are termed *garnierites*. Garnierite is often

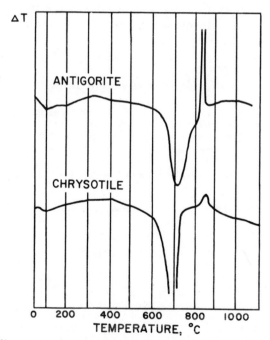

Fig. 11-17. Differential thermal analysis curves for serpentines: A-antigorite, Antigorio, Piedmont, Italy; B-chrysotile, Montville, New Jersey, USA. From Mackenzie (1970).

used as a general term, analagous to bauxite and limonite, because garnierites are rather poorly organized mixtures, commonly of serpentine and talc-like layers (Brindley & Hang, 1973). The nickel garnierties, composed mostly of the 7Å-type layers, lose hydroxyl water at 550 to 700C (823–973K).

XXII. SYNTHESIS OF SERPENTINE MINERALS

Synthesis of serpentine minerals has been investigated at temperatures from 150 to 900C (423–1,173K) and pressures from 500 to 30,000 psi (3.45 to 207 MPa) (Roy & Roy, 1968), and effective methods of synthesis were developed. Recently Mg serpentines were synthesized at 450 to 500C (723–773K) and 5,000 to 25,000 psi (34.5 to 172 MPa) (Iishi & Saito, 1973). More of the Mg-antigorite, in relation to chrysotile and lizardite, was formed at higher temperatures and pressures consistent with the association of antigorite with high grade metamorphic environments.

Gillery with Hill (1959) reported the hydrothermal synthesis of serpentine and chlorites from gels in terms of the composition $(Si_{4-x}Al_x)(Mg_{6-x}Al_x)O_{10}(OH)_8$. Serpentines are formed below 450C (723K) for $0 < x < 2.0$ and chlorites are formed above 500C (773K) from compositions with $x > 0.5$. A fibrous serpentine was produced at $x = 0$, but not at $x = 0.25$. Two other polytypes of serpentine with platy morphology were formed at low and high Al compositions. These results support the view that temperature has the greatest effect on whether serpentine or chlorite is produced and that composition has the greatest effect on which of three serpentine polytypes is produced.

XXIII. SUPPLEMENTAL READING

Jackson, M. L. 1969. Soil chemical analysis—advanced course. 2nd ed., 8th printing. 1973. Published by the author. Dep. of Soil Science, Univ. of Wisconsin, Madison.

Millot, G. 1970. Geology of clays. Springer-Verlag, New York. 429 p.

Rich, C. I., and G. W. Kunze (eds.). 1964. Soil clay mineralogy. The Univ. of North Carolina Press, Chapel Hill. 330 p.

Swindale, L. D. 1975. The crystallography of minerals of the kaolin group. p. 121–154. In J. E. Gieseking (ed.) Soil components, Vol. 2, Inorganic components. Springer-Verlag, New York.

Thorez, J. 1975. Phyllosilicates and clay minerals. G. Lelotte, Dison, Belgium. 578 p.

LITERATURE CITED

Alexiades, C. A., and M. L. Jackson. 1966. Quantitative clay mineralogical analysis of soils and sediments. Clays Clay Miner. 13:35–52.

Altschuler, Z. S., E. J. Dwornik, and H. Kramer. 1963. Transformation of montmorillonite to kaolinite during weathering. Science 141:148–152.

Angel, B. R., and P. L. Hall. 1972. Electron spin resonance studies of kaolins. In J. M. Serratosa and A. Sanchez (ed.) Int. Clay Conf. Proceedings (Madrid) 1:47–60.

Aomine, S., and K. Wada. 1962. Differential weathering of volcanic ash and pumice, resulting in formation of hydrated halloysite. Am. Mineral. 47:1024–1048.

Askenasy, P. E., J. B. Dixon, and T. R. McKee. 1973. Spheroidal halloysite in a Guatemalan soil. Soil Sci. Soc. Am. Proc. 37:799-803.

Bailey, S. W. 1966. The status of clay mineral structures. Clays Clay Miner. 14:1-24.

Bailey, S. W. (Chairman), G. W. Brindley, W. D. Johns, R. T. Martin, and M. Ross. 1971. Summary of national and international recommendations on clay mineral nomenclature by 1969-70 Clay Min. Soc. Nomenclature Committee. Clays Clay Miner. 19:129-132.

Bates, T. F. 1958. Selected electron micrographs of clays and other fine-grained minerals. Circ. no. 51. College of Mineral Industries, The Pennsylvania State Univ., University Park. 61 p.

Bates, T. F. 1964. Geology and mineralogy of the sedimentary kaolins of the southeastern United States—A review. Clays Clay Miner. 12:177-194.

Bates, T. F., F. A. Hildebrand, and A. Swineford. 1950. Morphology and structure of endellite and halloysite. Am. Mineral. 35:463-484.

Beutelspacher, H., and H. W. van der Marel. 1968. Atlas of electron microscopy of clay minerals and their admixtures. Elsevier Publishing Co., New York. 333 p.

Bidwell, J. I., W. B. Jepson, and G. L. Toms. 1970. The interaction of kaolinite with polyphosphate and polyacrylate in aqueous solutions—some preliminary results. Clay Miner. 8:445-459.

Birrell, K. S., M. Fieldes, and K. I. Williamson. 1955. Unusual forms of halloysite. Am. Mineral. 40:122-124.

Biscaye, P. E. 1965. Mineralogy and sedimentation of recent deep-sea clay in the Atlantic Ocean and adjacent seas and oceans. Geol. Soc. Am. Bull. 76:803-832.

Bradley, W. F. 1954. X-ray diffraction criteria for the characterization of chloritic material in sediments. Clays Clay Miner. 2:324-334.

Bramao, L., J. G. Cady, S. B. Hendricks, and M. Swerdlow. 1952. Criteria for the characterization of kaolinite, halloysite and related minerals in clays and soils. Soil Sci. 73:273-287.

Brindley, G. W. 1961. Kaolin, serpentine, and kindred minerals. p. 51-131. In G. Brown (ed.) The X-ray identification and crystal structures of clay minerals. Mineral. Soc., London.

Brindley, G. W., and D. L. Gibbon. 1968. Kaolinite layer silicate structure: relaxation by dehydroxlation. Science 162:1390-1391.

Brindley, G. W., and P. T. Hang. 1973. The nature of garnierites—I. Structures, chemical compositions and color characteristics. Clays Clay Miner. 21:27-40.

Brindley, G. W., and S. S. Kurtossy. 1961. Quantitative determination of kaolinite by x-ray diffraction. Am. Mineral. 46:1205-1215.

Brindley, G. W., and D. M. C. MacEwan. 1953. Structural aspects of the mineralogy of clays and related silicates. p. 15-59. In A. T. Green and G. H. Stewart (eds.) Ceramics—a symposium. The British Ceramic Soc., Stoke-on-Trent.

Brindley, G. W., and K. Robinson. 1946. Randomness in the structures of kaolinitic clay minerals. Trans. Faraday Soc. 42B:198-205.

Brindley, G. W., and P. de Souza Santos. 1971. Antigorite—its occurrence as a clay mineral. Clays Clay Miner. 19:187-192.

Bryant, J. P., and J. B. Dixon. 1963. Clay mineralogy and weathering of red-yellow podzolic soil from quartz mica schist in the Alabama Piedmont. Clays Clay Miner. 12:509-521.

Carson, C. D., and J. B. Dixon. 1972. Potassium selectivity in certain montmorillonitic soil clays. Soil Sci. Soc. Am. Proc. 36:838-843.

Cashen, G. H. 1966. Electric charge on clays. J. Soil Sci. 17:303-316.

Chukhrov, F. V., and B. B. Zvyagin. 1966. Halloysite, a crystallochemically and mineralogically distinct species. Int. Clay. Conf. Proc. (Jerusalem) 1:11-25.

Claridge, G. G. C. 1965. The mineralogy and chemistry of some soils from the Ross Dependency, Antarctica. New Zeal. J. Geol. Geophys. 8:186-220.

Cradwick, P. D., and M. J. Wilson. 1972. Calculated x-ray diffraction profiles for interstratified kaolinite-montmorillonite. Clay Miner. 9:393-406.

Cruz, M., H. Jacobs, and J. J. Fripiat. 1972. The nature of the interlayer bonding in kaolin minerals. In J. M. Serratosa and A. Sanchez (ed.) Int. Clay Conf. Proc. (Madrid) 1:35-46.

Curtis, C. D., and D. A. Spears. 1971. Diagenetic development of kaolinite. Clays Clay Miner. 19:219-228.

Davis, D. W., T. G. Rochaw, F. G. Rowe, M. L. Fuller, P. F. Kerr, and P. K. Hamilton. 1950. Electron micrographs of reference clay minerals. Am. Petrol. Instit. Proj. 49, Rep. no. 6. 17 p.

Deeds, C. T., and H. van Olphen. 1963. Density studies in clay-liquid systems, part 2: Application to core analysis. Clays Clay Miner. 10:318-328.

Deer, W. A., R. A. Howie, and J. Zussman. 1966. An introduction to the rock-forming minerals. John Wiley & Sons, Inc., New York. 528 p.

De Kimpe, C., M. C. Gastuche, and G. W. Brindley. 1961. Ionic coordination in alumino-silica gels in relation to clay mineral formation. Am. Mineral. 46:1370-1381.

Dixon, J. B. 1966. Quantitative analysis of kaolinite and gibbsite in soils by differential thermal and selective dissolution methods. Clays Clay Miner. 14:83-90.

Dixon, J. B., and T. R. McKee. 1974a. Internal and external morphology of tubular and spheroidal halloysite particles. Clays Clay Miner. 22:127-137.

Dixon, J. B., and T. R. McKee. 1974b. Spherical halloysite formation in a volcanic soil of Mexico. Int. Congr. Soil Sci., Trans. 10th (Moscow) VII:115-124.

Dixon, J. B., and V. E. Nash. 1968. Chemical, mineralogical and engineering properties of Alabama and Mississippi Black Belt soils. South. Coop. Ser. no. 130. Auburn Univ., Auburn, Alabama. 69 p.

Dolcater, D. L., J. K. Syers, and M. L. Jackson. 1970. Titanium as free oxide and sub-stituted forms in kaolinites and other soil minerals. Clays Clay Miner. 18:71-80.

Dudas, M. J., and M. E. Harward. 1975. Weathering and authigenic halloysite in soil de-veloped in Mazama ash. Soil Sci. Soc. Am. Proc. 39:561-566.

Evans, R. C. 1964. An introduction to crystal chemistry. Cambridge at the Univ. Press. 410 p.

Fanning, D. S., and M. L. Jackson. 1964. Clay mineral weathering in Southern Wisconsin soils developed in loess and in shale-derived till. Clays Clay Miner. 13:175-191.

Fiskell, J. G. A., and H. F. Perkins. 1970. Selected Coastal Plain soil properties. South. Coop. Ser. Bull. no. 148, Univ. Florida, Gainesville. 141 p.

Follett, E. A. C. 1965. The retention of amorphous, colloidal ferric hydroxide by kao-linites. J. Soil Sci. 16:334-341.

Gallez, A., A. S. R. Juo, A. J. Herbillon, and F. R. Moormann. 1975. Clay mineralogy of selected soils in Southern Nigeria. Soil Sci. Soc. Am. Proc. 39:577-585.

Garrels, R. M., and C. L. Christ. 1965. Solutions, minerals and equilibria. Harpers & Row, Inc., New York. 450 p.

Gibbs, R. J. 1965. Error due to segregation in quantitative clay mineral x-ray diffraction mounting techniques. Am. Mineral. 50:741-751.

Giese, R. F., Jr. 1973. Interlayer bonding in kaolinite, dickite and nacrite. Clays Clay Miner. 21:145-149.

Gillery, F. H. 1959. The x-ray study of synthetic Mg-Al serpentines and chlorites. Am. Mineral. 44:143-152.

Glenn, R. C. 1960. Chemical weathering of layer silicate minerals in loess-derived Loring silt loam of Mississippi. Int. Congr. Soil Sci., Trans. 7th (Madison, Wis.) IV:523-531.

Glenn, R. C., M. L. Jackson, F. D. Hole, and G. B. Lee. 1960. Chemical weathering of layer silicate clays in loess-derived Tama silt loam of Southwestern Wisconsin. Clays Clay Miner. 8:63-83.

Griffin, J. J., H. Windom, and E. D. Goldberg. 1968. The distribution of clay minerals in the world ocean. Deep-Sea Res. 15:433-459.

Grim, R. E. 1953. Clay mineralogy. McGraw-Hill Book Co., Inc., New York. 384 p.

Hashimoto, I., and M. L. Jackson. 1960. Rapid dissolution of allophane and kaolinite-halloysite after dehydration. Clays Clay Miner. 7:102-113.

Hayes, J. B. 1963. Kaolinite from Warsaw geodes, Keokuk Region, Iowa. Iowa Acad. of Sci. 70:261-272.

Hem, J. D., and C. D. Lind. 1974. Kaolinite synthesis at $25°C$. Soil Sci. 184:1171-1173.

Honjo, G., N. Kitamura, and K. Mihama. 1954. A study of clay minerals by means of single crystal electron diffraction diagrams—the structure of tubular kaolin. Clay Miner. Bull. 2:133-141.

Hope, E. W., and J. A. Kittrick. 1964. Surface tension and the morphology of halloysite. Am. Mineral. 49:859-866.

Hunter, R. J., and A. E. Alexander. 1963. Surface properties and flow behavior of kaolinite. Part I: Electrophoretic mobility and stability of kaolinite sols. J. Colloid Sci. 18:820-832.

Iishi, K., and M. Saito. 1973. Synthesis of antigorite. Am. Mineral. 58:915-919.

Jackson, M. L., S. A. Tyler, A. L. Willis, G. A. Bourbeau, and R. P. Pennington. 1948. Weathering sequence of clay-size minerals in soils and sediments. I. Fundamental generalizations. J. Phys. Colloid Chem. 52:1237-1260.

Jefferson, D. A., M. J. Tricker, and A. P. Winterbottom. 1975. Electron-microscopic and Mössbauer spectroscopic studies of iron-stained kaolinite minerals. Clays Clay Miner. 23:355-360.

Jonas, E. C. 1964. Petrology of the Dry Branch, Georgia kaolin deposits. Clays Clay Miner. 12:199-206.

Jungerius, P. D., and T. W. M. Levelt. 1964. Clay mineralogy of soils over sedimentary rocks in Eastern Nigeria. Soil Sci. 97:89-95.

Kanno, I., Y. Onikura, and S. Tokudode. 1965. Genesis and characteristics of Brown Forest soils derived from serpentine in Kyushu, Japan: Part 3. Clay Mineralogical characteristics. Soil Sci. Plant Nutr. 11:225-234.

Kellogg, C. E. 1975. Agricultural development: soil, food, people, and work. Soil Sci. Soc. Am., Inc., Madison, Wis. 233 p.

Kittrick, J. A. 1969. Soil minerals in the Al_2O_3-SiO_2-H_2O system and a theory of their formation. Clays Clay Miner. 17:157-167.

Kittrick, J. A. 1970. Precipitation of kaolinite at $25°C$ and 1 ATM. Clays Clay Miner. 18:261-268.

Kunze, G. W., and W. F. Bradley. 1964. Occurrence of a tabular halloysite in a Texas soil. Clays Clay Miner. 12:523-527.

Ledoux, R. L., and J. L. White. 1964. Infrared study of the OH groups in expanded kaolinite. Science 143:244-246.

Ledoux, R. L., and J. L. White. 1966. Infrared studies of the hydroxyl groups in intercalated kaolinite complexes. Clays Clay Miner. 13:289-316.

Lee, S. Y., M. L. Jackson, and J. L. Brown. 1975a. Micaceous occlusions in kaolinite observed by untramicrotomy and high resolution electron microscopy. Clays Clay Miner. 23:125-129.

Lee, S. Y., M. L. Jackson, and J. L. Brown. 1975b. Micaceous vermiculite, glauconite, and mixed-layered kaolinite-montmorillonite examination by ultramicrotomy and high resolution electron microscopy. Soil Sci. Soc. Am. Proc. 39:793-800.

Lepsch, I. F., and S. W. Buol. 1974. Investigations in an oxisol-ultisol toposequence in S. Paulo State, Brazil. Soil Sci. Soc. Am. Proc. 38:491-496.

Linares, J., and F. Huertas. 1971. Kaolinite: Synthesis at room temperature. Science 171:896-897.

Lisitzin, P. 1972. Sedimentation in the world ocean. Soc. Paleontol. Mineral. Spec. Publ. no. 17, Tulsa, Oklahoma. p. 13D.

McConnell, J. D. C., and S. G. Fleet. 1970. Electron optical study of the thermal decomposition of kaolinite. Clay Miner. 8:279-290.

Mackenzie, R. C. (ed.). 1970. Differential thermal analysis. Academic Press, New York. 775 p.

Maldon, P. J., and R. E. Meads. 1967. The solid state substitution by iron in kaolinite. Nature (London) 215:844-846.

Marshall, C. E. 1964. The physical chemistry and mineralogy of soils. John Wiley & Sons, Inc., New York. p. 327-328.

Mitra, N. K., M. Mukerjee, D. Biswas, and P. K. Bhaunik. 1973. Studies on the rheological characteristics of ciay-polyelectrolyte interactions. Indian J. Technol. 11:250-254.

Muljadi, D., A. M. Posner, and J. P. Quirk. 1966. The mechanism of phosphate adsorption by kaolinite, gibbsite and pseudoboehmite. J. Soil Sci. 17:212-228.

Newnham, R. E. 1961. A refinement of the dickite structure and some remarks on polymorphism in kaolin minerals. Mineral. Mag. 32:683-704.

Ormsby, W. C., and J. M. Shartsis. 1960. Surface area and exchange capacity relation in a Florida kaolinite. J. Am. Ceram. Soc. 43:44-47.

Ormsby, W. C., J. M. Shartsis, and K. H. Woodside. 1962. Exchange behavior of kaolins and varying degrees of crystallinity. J. Am. Ceram. Soc. 45:361-366.

Page, N. J.   1968.   Chemical differences among the serpentine "polymorphs."   Am. Mineral. 53:201–215.

Parham, W. E.   1966.   Lateral variations of clay mineral assemblages in modern and ancient sediments. Int. Clay Conf. (Jerusalem) 1:135–145.

Parham, W. E.   1969.   Formation of halloysite from feldspar: low temperature, artificial weathering versus natural weathering. Clays Clay Miner. 17:13–22.

Parham, W. E.   1970.   Clay mineralogy and geology of Minnesota's kaolin clays. Minn. Geol. Surv. Spec. Publ. 10. Univ. Minn., Minneapolis. 142 p.

Poncelet, G. M., and G. W. Brindley.   1967.   Experimental formation of kaolinite from montmorillonite at low temperatures. Am. Mineral. 52:1161–1173.

Rand, B., and I. E. Melton.   1975.   Isoelectric point of the edge surface of kaolinite. Nature (London) 257:214–216.

Rengasamy, P., G. S. R. Krishna Murti, and V. A. K. Sarma.   1975.   Isomorphous substitution of iron for aluminum in some soil kaolinites. Clays Clay Miner. 23:211–214.

Rex, R. W.   1966.   Authigenic kaolinite and mica as evidence for phase equilibria at low temperatures. Clays Clay Miner. 13:95–104.

Robertson, R. H. S., G. W. Brindley, and R. C. Mackenzie.   1954.   Kaolin clay from Pugu, Tanganyika. Am. Mineral. 39:118–139.

Rodrique, L., G. Poncelet, and A. Herbillon.   1972.   Importance of the silica subtraction process during the hydrothermal kaolinization of amorphous silico-aluminas. Int. Clay Conf. Proc. (Madrid) 1:187–198.

Roy, D. M., and R. Roy.   1958.   An experimental study of the formation and properties of synthetic serpentines and related layer silicate minerals. Am. Mineral. 39:957–975.

Sakharov, B. A., and V. A. Drits.   1973.   Mixed-layer kaolinite-montmorillonite: a comparison of observed and calculated diffraction patterns. Clays Clay Miner. 21:15–17.

Schellman, V. W.   1964.   Zur lateritischen Verwitterung von Serpentinit. Geol. Jahrb. 81: 645–678. (English Abstr.).

Schofield, R. K., and H. R. Samson.   1953.   The deflocculation of kaolinite suspensions and the accompanying change-over from positive to negative chloride adsorption. Clay Miner. Bull. 2:45–51.

Schofield, R. K., and H. R. Samson.   1954.   Flocculation of kaolinite due to the attraction of oppositely charged crystal faces. Disc. Faraday Soc. 18:135–145.

Schultz, L. G., A. O. Shepard, P. D. Blackmon, and H. C. Starkey.   1971.   Mixed-layer kaolinite-montmorillonite from the Yucatan Peninsula, Mexico. Clays Clay Miner. 19:137–150.

Serratosa, J. M., and W. F. Bradley.   1958.   Determination of the orientation of OH^- bond axes in layer silicates by infrared absorption. J. Phys. Chem. 62:1164–1167.

Siffert, B.   1962.   Some reactions of silica in solution: Formation of clay. Reports of the Geol. Map Serv. Alsace-Lorraine no. 21. (Translated from French by Israel Program for Scientific Translations, Jerusalem, 1967.) 100 p.

Sivarajasingham, S., L. T. Alexander, J. G. Cady, and M. G. Cline.   1962.   Laterite. Adv. Agron. 14:1–60.

Souza Santos, P. de, H. de Souza Santos, and G. W. Brindley.   1966.   Mineralogical studies of kaolinite-halloysite clays: Part IV. A platy mineral with structural swelling and shrinking characteristics. Am. Mineral. 51:1640–1648.

Spyridakis, D. E., G. Chesters, and S. A. Wilde.   1967.   Kaolinization of biotite as a result of coniferous and deciduous seedling growth. Soil Sci. Soc. Am. Proc. 31:203–210.

Street, N.   1956.   The rheology of kaolinite suspensions. Aust. J. Chem. 9:467–479.

Street, N., and A. S. Buchanan.   1956.   The Z-potential of kaolinite particles. Aust. J. Chem. 9:450–466.

Sudo, T., and H. Hayashi.   1956.   Types of mixed layer minerals from Japan. Clays Clay Miner. 4:389–412.

Sudo, T., and H. Takahashi.   1956.   Shapes of halloysite particles in Japanese clays. Clays Clay Miner. 4:67–79.

Sumner, M. E., and N. G. Reeve.   1966.   The effect of iron oxide impurities on the positive and negative adsorption of chloride by kaolinites. J. Soil Sci. 17:274–279.

Swindale, L. D., and P. F. Fan.   1967.   Transformation of gibbsite to chlorite in ocean bottom sediments. Science 157:799–800.

Wada, K. 1963. Quantitative determination of kaolinite and halloysite by NH$_4$Cl retention. Am. Mineral. 48:1286–1299.

Wada, K. 1967. A study of hydroxyl groups in kaolin minerals utilizing selective deuteration and infrared spectroscopy. Clays Clay Miner. 7:51–61.

Weiss, A., and K. J. Range. 1966. Übber titan im gitter von kaolin. Int. Clay Conf. Proc. (Jerusalem) 1:53–66.

Weiss, A., and J. Russow. 1963. Über die Lage der austauschbaren Kationen bei Kaolinit. Int. Clay Conf. Proc. (Stockholm) 1:203–213.

Whittaker, E. J. W., and F. J. Wicks. 1970. Chemical differences among the serpentine "polymorphs": A discussion. Am. Mineral. 55:1025–1047.

Whittaker, E. J. W., and J. Zussman. 1971. The serpentine minerals. p. 159–191. *In* J. A. Gard (ed.) The electron-optical investigation of clays. Mineral. Soc. Monogr. 3, London.

Wiewiora, A. 1971. A mixed-layer kaolinite-smectite from lower Silesia, Poland. Clays Clay Miner. 19:415–416.

Wildman, W. E., M. L. Jackson, and L. D. Whittig. 1968. Iron-rich montmorillonite formation in soils derived from serpentinite. Soil Sci. Soc. Am. Proc. 32:787–794.

Wildman, W. E., L. D. Whittig, and M. L. Jackson. 1971. Serpentine stability in relation to formation of iron-rich montmorillonite in some California soils. Am. Mineral. 56:587–602.

Wilson, M. J., and P. D. Cradwick. 1972. Occurrence of interstratified kaolinite-montmorillonite in some Scottish soils. Clay Miner. 9:435–437.

Wright, J. R., A. Leahey, and H. M. Rice. 1967. Chemical, morphological, and mineralogical characteristics of a chronosequence of soils on alluvial deposits in the Northwest Territories. p. 257–269. *In* J. V. Drew (ed.) Selected papers in soil formation and classification. SSSA Spec. Publ. no. 1. Soil Sci. Soc. of Am., Madison, Wis.

Interstratification in Layer Silicates

BRIJ L. SAWHNEY, The Connecticut Agricultural Experiment Station, New Haven, Connecticut

I. INTRODUCTION

The layer silicates mica, vermiculite, smectite, chlorite, and kaolinite contain tetrahedral and octahedral sheets as their basic structural units. Tetrahedral sheets consist of an array of Si and Al ions surrounded by four oxygens, and octahedral sheets consist of Mg, Fe, or Al ions surrounded by six oxygens and hydroxyls. Mica, vermiculite, and smectite are each composed of an octahedral sheet sandwiched between two tetrahedral sheets while chlorite consists of two pairs of alternating tetrahedral and octahedral sheets and kaolinite consists of one such pair. Because of structural similarities, these minerals often occur in a mixed order of stacking in which an individual crystal may consist of two or more layer silicates. These assemblages of layer silicates are commonly referred to as mixed-layer or interstratified minerals. Their formation has a pronounced effect on potassium and phosphate reactions in soils. For example, it has been shown that potassium is more tightly held by interstratified mica-vermiculite layers than by the individual components. Similarly, hydroxy interlayer formation in vermiculite and smectite increases the phosphate fixation capacity of these minerals. Mixed layers also reveal changes in environmental conditions since they are formed readily by only minor changes in the environment whereas most other mineral alterations occur very slowly. Mixed-layer minerals containing various combinations of layer silicates are abundant in the clay and silt fractions of soils and sediments; hence an understanding of their characteristics is essential in mineralogical studies of soils and sediments.

Mixing of layers commonly occurs in three different ways: (i) regular interstratification of layers, (ii) random interstratification of layers, and (iii) zonal segregation of layers where sufficient numbers of each kind occur together so that X-ray diffraction analysis reveals them to be separate phases. More than one type of interstratified layers may occur in a soil or sediment sample. Zonal segregation is similar to a mechanical mixture of crystallites of individual minerals and hence will not be discussed here.

II. REGULAR INTERSTRATIFICATION

Regular interstratification consists of alternate distribution of the component layers within the crystal; for example, two layers A and B corresponding to two separate layer silicates will constitute a regularly interstratified structure when they form layer sequences of the type ABABAB. . .within the crystallites. The repeat distance of the unit cell of the regularly interstratified structure is equal to A + B, the sum of the unit cell thicknesses of the individual components. Regular interstratification of A and B layers can also occur as AABAABAAB. . . . .

Some regularly interstratified layers have been given specific mineral names while others are described by their component layers. Minerals consisting of regularly interstratified mica-smectite layers, chlorite-swelling chlorite layers, or biotite-vermiculite layers are called rectorite, corrensite, and hydrobiotite, respectively. Other structures such as alternating chlorite and vermiculite layers or talc and saponite layers are described as regularly interstratified chlorite-vermiculite and talc-saponite, respectively.

A. Characterization of Regularly Interstratified Layers

Regularly interstratified minerals are readily identified by their basal 001 diffraction peak, corresponding to the sum of the spacings of the individual components, and subsequent peaks of higher integral orders on their X-ray diffraction patterns. Individual components in the mineral are identified from diffraction patterns of the specimen obtained after saturation with different cations, various heat treatments, or treatments with organic liquids. Most common treatments include saturation of separate samples of the specimen with Mg and K ions followed by treatment of the Mg-saturated sample with ethylene glycol or glycerol, and by successive heat treatments at 100, 300, and 500–600C (373, 573, and 773–873K) of the K-saturated sample. These procedures are discussed in other chapters for the identification of individual layer silicates and will be illustrated below in the characterization of interstratified minerals.

B. Examples of Regularly Interstratified Minerals and Interpretation of Their X-ray Diffraction Patterns

1. MICA-VERMICULITE

Regularly interstratified mica-vermiculite layers when saturated with Mg are characterized by a 24Å basal diffraction peak and its higher integral orders, corresponding to the sum of the spacings of mica (10Å) and vermiculite (14Å). Treatment of the Mg-saturated sample with glycerol does not change the diffraction peaks. Saturation of the sample with K and mild heat treatment (100C; 373K) collapses the vermiculite component and produces a diffraction pattern with a 10Å peak and higher integral orders.

Gruner (1934) described hydrobiotite as containing regularly interstratified layers of biotite and vermiculite. Chemical and X-ray diffraction analyses of specimens containing both hydrobiotite and biotite indicated a genetic relationship between the two minerals. Gruner concluded that hydrobiotite is an alteration product of biotite. Later, Bassett (1959) and

Boettcher (1966) in their studies of the vermiculite deposits at Libby, Montana observed that hydrobiotite was widely distributed in these deposits and concluded that the biotite had been altered to hydrobiotite and vermiculite by the action of hydrothermal solutions.

Regularly interstratified mica-vermiculite layers have also been observed as a weathering product of mica in soils. Coleman et al. (1963) found that in a number of California soils derived from granite-diorite, biotite had altered to hydrobiotite and vermiculite. X-ray diffraction analysis of biotite flakes weathered to different degrees showed that the flakes of specific gravity 2.85–2.95 contained mostly biotite whereas more weathered flakes of specific gravity 2.65–2.70 were almost pure hydrobiotite. Lighter fractions which were still more weathered contained increasing amounts of vermiculite. Similarly, Kapoor (1972) observed that in some Norwegian podzols, mica, which is dominant in the lower soil horizons, had weathered to hydrobiotite in the surface (A2) horizon.

Alteration of biotite to hydrobiotite involves replacement of K by Mg or Ca in alternate interlayer zones of biotite, producing regular interstratification of mica-vermiculite layers. Also, during the weathering of biotite, Fe in the octahedral sheet is oxidized. Although the exact manner in which K is removed from alternate interlayer zones of biotite is not known, mechanisms proposed for this regular interstratification are discussed in the last section of this chapter.

2. MICA-SMECTITE

Regularly interstratified mica-smectite layers are characterized by the same basal diffraction peaks as Mg-saturated mica-vermiculite layers. However, when Mg-saturated samples are treated with glycerol, the 24Å peak increases to 27Å.

Two well-known examples of regularly interstratified mica-smectite layers are allevardite described by Caillere et al. (1950), and rectorite described by Bradley (1950).

Brindley (1956) conducted a detailed X-ray diffraction analysis of an allevardite sample. The air-dry sample in its natural state of ion saturation exhibited a 24.62Å basal diffraction peak and its higher integral orders. On immersion in water the basal peak increased to 28.34Å still with higher integral orders. These spacings correspond to alternate layer sequences of collapsed (10Å) and expanded (18Å) layers and could be ascribed to regular interstratification of mica-smectite. Brindley confirmed the regular interstratification of the two types of layers using one-dimensional Fourier synthesis of the basal diffraction peaks. Using this method, he did not find evidence for any interlayer cations in the expanded layers. Therefore, he concluded that the mineral consisted of regularly interstratified mica-type (10Å) layers and uncharged pyrophyllite-like layers which expanded to 18Å on hydration.

Bradley (1950), on the other hand, suggested that a rectorite sample

Table 12-1. X-ray diffraction peaks (Å) from rectorite and allevardite

	Rectorite		Allevardite	
Treatment	Brown & Weir (1963)	Bradley (1950)	Brown & Weir (1963)	Brindley (1956)
Air dried	25.0	25.0	24.6	24.6
300C (573K)	19.3	--	19.2	--
560 or 600C (833 or 873K)	19.5	19.5	19.2	19.2
In water	28.5	25.0	28.5	28.4
Ethylene glycol	26.4	26.6	26.5	26.5
Glycerol	27.3	--	27.4	--
Mg-saturated	25.0	--	25.0	--
Mg-saturated + Glycerol	27.5	--	27.4	--

consisted of alternate sequences of pyrophyllite-like and vermiculite-like layers. He suggested that pyrophyllite-like layers did not expand while vermiculite-like layers expanded on hydration and glycol treatment.

Detailed chemical analysis, cation exchange measurements, X-ray examination, electron microscopy, and infrared absorption spectra by Brown and Weir (1963) showed that allevardite and rectorite are indeed the same mineral. They suggested that the name rectorite has priority over allevardite because rectorite had been proposed for the mineral as early as 1891. Their results as well as those of Bradley and of Brindley are given in Table 12-1. Based on chemical and exchange capacity measurements, Brown and Weir allocated the excess Na, which Brindley had assumed to be either an impurity or sorbed on mineral surfaces, to exchange sites in smectite. Thus, the expanded layers in the interstratified minerals are considered to be smectite and not pyrophyllite. The 19.5Å basal reflections from the heated rectorite samples result from alternate layers of smectite (9.5Å) and mica (10Å).

Heystek (1954) described a regularly interstratified mica-smectite produced by hydrothermal alteration of a shale by intrusion of dolerite sheets. The mineral gave intense 25.8Å and 12.4Å basal diffraction peaks which increased to 28.1Å and 13.6Å on treatment with ethylene glycol. Tomita and Dozono (1973) analyzed a sample from a hydrothermally altered andesite which showed all the characteristics of regularly interstratified mica-smectite layers except that the mineral had an unusual tendency of rehydrate. Normally, smectite is irreversibly dehydrated on heating to 640C (913K) while the smectite component of this interstratified structure rehydrated completely even after heating at 800C (1073K).

3. CHLORITE-VERMICULITE

Regularly interstratified chlorite-vermiculite is characterized by a 28Å basal diffraction peak and its higher integral orders in samples saturated with Mg and treated with glycerol. On K-saturation and heating at 500C (773K), the 28Å peak decreases to 24Å, since the vermiculite layers collapse to 10Å while the 14Å spacing due to chlorite remains unaltered.

A regularly interstratified mineral with a well-defined 29Å basal diffraction peak and 10 integral orders was found by Bradley and Weaver (1956) in the clay residue of the Upper Mississippian limestone of Colorado. The mineral collapsed to 24Å on heating to 550C (823K) and was described by the authors as a 1:1 regularly interstratified chlorite-vermiculite. However, since the basal (001) peak increased from 29Å to 31Å on ethylene glycol treatment, the interstratification should probably be described as chlorite-smectite. This interstratified mixture appears to have been chlorite originally. Removal of alternate brucitic layers by water and bases resulted in the formation of the interstratified mineral.

Occurrence of regularly interstratified chlorite-vermiculite layers in several shale samples was reported by Heystek (1956). Similarly, regularly interstratified chlorite-vermiculite layers formed by hydrothermal alteration of andesitic and dacitic rocks in the Goldfield district of Nevada have been described by Harvey and Beck (1962). The interstratified vermiculite mineral is thought to be formed in the zone of least alteration, while increased intensity of alteration results in the formation of smectite.

Regularly interstratified chlorite-vermiculite layers have also been found as weathering products in soils. Johnson (1964) observed the pronounced development of this mineral in fine silt separated from a soil developed from greenstone. Samples saturated with Mg with or without treatment with glycol gave 28.5Å basal diffraction peaks and higher integral orders. Saturation with K decreased the basal spacing to 24.3Å. When the sample was heated at 475C (748K), the intensity of 001 (23.9Å) and 002 (12.3Å) peaks increased while the intensity of 003, 004, 005, and 006 decreased (Fig. 12-1). These changes in intensities were attributed to dehydroxylation of the hydroxide sheet in the chlorite component. The sample provides an example of an highly pure interstratified chlorite-vermiculite mixture in soils. The calculated and observed intensities of the peaks are in good agreement (Table 12-2). Formation of chlorite-vermiculite interstratification in soils appears to result from the removal of the hydroxide sheet from alternate layers of chlorite during weathering (Jackson, 1963).

4. CHLORITE-SMECTITE

Regularly interstratified chlorite-smectite is similar to interstratified chlorite-vermiculite except that on saturation with Mg and glycerol, the chlorite-smectite complex gives a 32Å basal diffraction peak with integral higher orders.

Bailey and Tyler (1960) identified a regularly interstratified chlorite-smectite mineral in Lake Superior iron ores. The smectite in the mixed layer was of the dioctahedral type. Dioctahedral smectite interstratified with chlorite has also been reported as an alteration product of shale and siltstones caused by fluids accompanying intrusions (Blatter et al., 1973). Chlorite-smectite mixed layers have been reported from several Japanese clays formed by hydrothermal alteration of tertiary rocks (Sudo, 1954; Sudo & Hyashi,

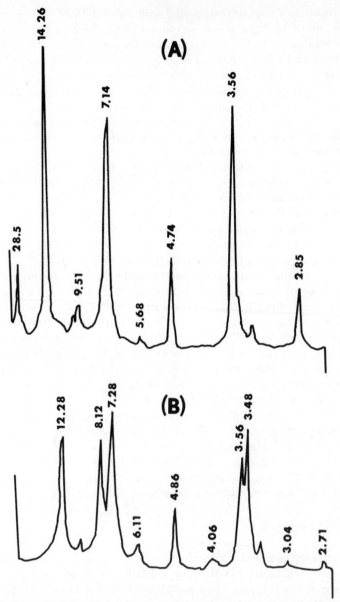

Fig. 12-1. X-ray diffraction patterns of parallel oriented silt samples [(A) Mg-saturated and (B) K-saturated] from the Highfield soil, showing regular interstratification of chlorite-vermiculite layers.   Units are in Å.   (Reprinted with permission from Mineralogical Society of America. Johnson, 1964; Fig. 1, p. 561.)

1956).   Earley et al. (1956) reported a detailed investigation of a sample of regularly interstratified chlorite-smectite which in its natural state gave a 30.5Å basal reflection that increased to 32.7Å on glycerol treatment and decreased to 24Å on heating to 500C (773K) (Table 12-3).

Table 12-2. X-ray diffraction peaks and their intensities from K-saturated samples, showing regular interstratification of chlorite-vermiculite layers†

Peak index	Relative intensities		
	Air dried		475C (748K)
	Observed	Calculated	Observed
001	--	4	15
002	81	54	100
003	76	33	4
004	13	12	2
005	43	21	--
006	6	11	2
007	100	100	15
008	4	23	4
009	4	15	4
0010	--	0	
0011	--	2	
0012	10	23	

† Reprinted with permission from Mineralogical Society of America (Johnson, 1964; Fig. 1, p. 560).

Table 12-3. X-ray diffraction peaks (Å) and their intensities from natural, glycerol- and heat-treated samples containing regularly interstratified chlorite-smectite layers (Earley et al., 1956)

Peak index	Natural		Glycerol-treated		Heated (500C, 773K)
	Å	I_{obs}	Å	I_{obs}	Å
001	30.5	668	32.7	516	21.6
002	14.5	1370	16.1	616	12.1
003	9.7	266	--	--	7.97
004	7.25	286	8.01	312	5.91
005	--	--	--	--	4.77
006	4.85	170	5.34	34	--
007	--	--	4.60	120	3.38
008	3.64	284	--	--	2.96
009	--	--	3.56	298	--
0010	2.93	104	--	--	--

5. CHLORITE-SWELLING CHLORITE

Regularly interstratified chlorite-swelling chlorite layers give a 28Å basal diffraction peak with integral higher orders similar to chlorite-vermiculite layers. However, on glycerol treatment, the peak increases to 31Å and its submultiples similar to a chlorite-smectite. Chlorite-swelling chlorite interstratification differs from the other two types in that the 28Å basal spacing persists after heating to 500C (773K) whereas the vermiculite and smectite components in the other two chlorite mixtures collapse to 10Å, producing a 24Å basal spacing and its submultiples. Regularly interstratified chlorite-swelling chlorite mineral has been named corrensite (Lippman, 1954).

Regularly interstratified chlorite-swelling chlorite layers were observed by Stephen and MacEwan (1951) in the weathered crust of the hornblende-rich rocks of the Malvern Hills. The untreated samples gave a 28Å basal diffraction peak and its submultiples which on glycol treatment increased to 32Å and its submultiples. On heating at 540C (813K), a sharp 14Å peak developed which is the second-order peak of a 28Å basal spacing. Honeyborne (1951) and Martin-Vivaldi and MacEwan (1957) have described several clays from Jura and Catalan Coastal ranges which have characteristics similar to regularly interstratified chlorite-swelling chlorite.

Other observations of regularly interstratified chlorite-swelling chlorites include their presence in the veins of talc formed by the action of hydrothermal fluids on serpentine masses (Alietti, 1958) and in clays of basic igneous rocks (Smith, 1960). Sutherland and MacEwan (1962) described a regularly interstratified sample in which the 14Å series of diffraction peaks remained unchanged after heating to 700C (973K).

Post and Janke (1974) have recently observed that the slaty strata of the western foothills of the Sierra Nevada contain chlorite which in the altered slate has completely changed to the regularly interstratified chlorite-swelling chlorite mineral.

The term swelling chlorite is used here to describe the naturally-occurring mineral that has a 17Å spacing, as smectite, after glycerol treatment and a 14Å spacing, as chlorite, after heating at 500–600C (773–873K). Swelling chlorite may form by incorporation of hydroxy-Al, Fe, or Mg interlayers in smectite or vermiculite or by partial removal of these interlayers from chlorite. Expansion and collapse of the interlayers in the resulting layer silicate would depend on its layer charge and the degree of hydroxy interlayer formation. Layer silicates with different expansion and collapse characteristics occur in nature and have been produced in the laboratory. These interlayered minerals are referred to as intergrades and are discussed elsewhere in the book.

6. OTHER REGULARLY INTERSTRATIFIED MINERALS

Alietti (1958) described that hydrothermal fluids had overrun serpentine rock altering it to talc which then weathered to three different mixed-layer assemblages. X-ray diffraction patterns of a dark green mineral which could be separated manually gave a basal spacing of 24.8Å with 10 integral higher orders. The basal spacing increased to 27.1Å on glycerol treatment, indicating regularly interstratified 17Å and 10Å layers. On heating at 400C (673K), the basal spacing from the mixed layers decreased to 9.6Å. The mixture was identified as regularly interstratified saponite and talc layers. A small layer of clayey material gave a basal peak at 29.4Å and 10 integral higher orders. Glycerol treatment increased the basal peak to 31.7Å. Heating at 550C (823K) reduced the peak to 23.5Å, indicating a regular interstratification of saponite and chlorite; saponite collapsed to 9.5Å on heating while chlorite remained at 14Å. The mineral which formed the main constituent of greyish black material associated with calcite and quartz showed chlorite-vermiculite interstratification.

Altschuler et al. (1963) in their investigation of the weathering of smectite to kaolinite in central Florida observed that partly transformed clay in these localities gave an integral series of basal diffraction peaks at 21.5Å. The peak increased to 25Å on glycol treatment and decreased to 17Å on 300C (573K) heating, indicating regularly interstratified kaolinite (7Å) layers and smectite layers that collapsed to 10Å on heating.

III. RANDOM INTERSTRATIFICATION

Random interstratification of layer silicates in soil clays and sediments is far more common than regular interstratification. It is also more complex because layer silicates can mix in different proportions in random fashion. Furthermore, more than two types of layer silicates may be randomly interstratified within the crystal. Their X-ray diffraction patterns are correspondingly more complex. Since different layers in randomly interstratified mixtures do not have any periodicity, they do not form definite unit cell repeat distances, but produce average spacings depending upon the proportions and spacings of the components.

A. Characterization of Randomly Interstratified Layers

In contrast to regularly interstratified structures, randomly interstratified layers give X-ray diffraction patterns with nonintegral diffraction peaks. Having recognized random interstratification from the presence of nonintegral diffraction peaks, further characterization involves: (i) identification of the component layers, (ii) determination of the proportion of each component in the mixed layer assemblage, and (iii) the manner in which the layers are distributed within the mixture, i.e. the nature of the succession of layers or their degree of randomness. The discussion here will be primarily concerned with the more common two-component systems with only a brief description of multiple-layer systems. Three methods for characterization of randomly interstratified mixed-layer minerals are discussed below.

1. MERING'S METHOD

This method involves visual examination of the positions of the X-ray diffraction peaks. It is the most convenient and probably the most commonly used method for characterizing randomly interstratified layer silicates in soil clays. Unfortunately, X-ray diffraction patterns of soil clays and sediments, which frequently contain several other minerals besides the mixed-layer assemblage, give only a few basal reflections and often only a single reflection from a mixed-layer mineral (MacEwan et al., 1961). In such cases characterizing the mixed layers is obviously difficult.

Mering's method (1949) is based upon the principle that the diffraction peaks from mixed-layer minerals are intermediate between diffraction peaks of their components. Although the positions of the diffraction peaks correspond to the reciprocal spacing of the components, the scale of X-ray diffraction diagrams and of the reciprocal lattice are nearly linearly related below $\theta = 20°$. The positions of the diffraction peaks from the mixed layers would, therefore, be related to the peaks from their components in the same way as the peaks from the reciprocal spacings. Thus, the diffraction peaks from a random mixture of two types of layers would be intermediate between the two nearest peaks of its components. The position of the peak would be determined by the relative amounts of the two components.

Identification of layer silicates in a random mixture is first made from the X-ray diffraction patterns of the specimen following diagnostic treatments commonly used for the individual layer silicate minerals. Then, the proportions of the two components in the mixture can be conveniently determined by comparing the position of the diffraction peaks from the mixture with the positions of the peaks from pure minerals. One can simply mark the positions of the peaks along a straight line and join the position of each observed peak from the mixture to the closest peaks from pure minerals. Then the distances of the positions of peaks of the mixture from the nearest positions of peaks from the pure minerals indicate the proportions of the components. The procedure is illustrated in Fig. 12-2 using an example of a naturally-occurring interstratified mixture discussed by Weaver (1956).

The X-ray diffraction pattern of a sample saturated with K gave diffraction peaks at 11.0Å, 5.25Å, and 3.24Å. The sample treated with ethylene

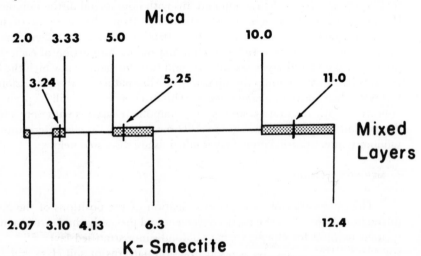

Fig. 12-2. Diagramatic illustration of Mering's method (1949) for interpreting X-ray diffraction patterns from randomly interstratified layer silicates. Numbers indicate X-ray diffraction peaks (Å) from mixed layers and from individual components.

glycol gave peaks at 13.3Å, 5.3Å, and 3.30Å. On heating the sample at 550C (823K) a series of diffraction peaks at 10Å and its submultiples were recorded. The apparent basal spacing of 11.0Å indicates interstratification of mica and a mineral with a higher spacing. The lack of collapse to 10Å on K saturation precluded vermiculite while complete collapse to 10Å on heating at 550C (823K) precluded the presence of chlorite. Clearly, the sample is an interstratified mixture of mica and smectite. To determine the proportion of mica and smectite in the mixture, the following procedure is used. In Fig. 12-2 the positions of the diffraction peaks from the mixed layers are marked on the straight line (indicated by arrows), while the positions of the peaks from mica and from K-saturated smectite are indicated above and below the line. The shaded areas indicate which adjacent peaks from the individual components combine to produce the observed peaks.

Since the 11.0Å peak is produced by the combination of 10Å and 12.5Å peaks from mica and smectite, their amounts in the mixture should have a ratio of 1.5:1 or about 60% mica and 40% smectite. Calculations using other combination peaks yield similar estimates of the mica fraction. Mering's method, thus, provides reasonable semiquantitative estimates of the two components in the interstratified mixture.

2. HENDRICKS AND TELLER TREATMENT

Brown and MacEwan (1950) calculated the diffraction patterns of a number of commonly occurring mixtures of mica with smectite, vermiculite, or kaolinite, assuming varying proportions of the two components in the mixture. The positions and intensities of the calculated diffraction peaks can be compared with the observed peaks from the mixture under investigation to identify the component layers and their proportions.

The diffraction patterns were calculated by Brown and MacEwan using the formula developed by Hendricks and Teller (1942) for the effect of mixing of different layers on the diffraction intensities. This is termed the mixing function Φ and is given by

$$\Phi = \frac{2f\,(1-f)\,\sin^2 \pi(d_2 - d_1/d')}{1 - 2f\,(1-f)\,\sin^2 \pi(d_2 - d_1/d') - f\cos 2\pi(d_1/d') - (1-f)\cos 2\pi(d_2/d')}$$

where d_1 = higher spacing, d_2 = lower spacing, d' = measured spacing, and f = fraction of layers with higher spacing.

To obtain the total intensity of the observed diffraction peak, the mixing function is multiplied by $|F|^2$, the squared modulus of the layer structure factor, and by Θ, the Lorentz Polarization factor which takes into account the polarization of the diffracted radiation and the form of the specimen. The factor Θ is given by

$$\frac{1 + \cos^2 2\theta}{\sin^2 \theta \, \cos \theta}$$

for a powder specimen and

$$\frac{1 + \cos^2 2\theta}{\sin 2\theta}$$

for a single crystal or an oriented specimen. Brown and MacEwan used Θ for oriented specimens.

Their calculations were based on the assumptions that (i) the compo-

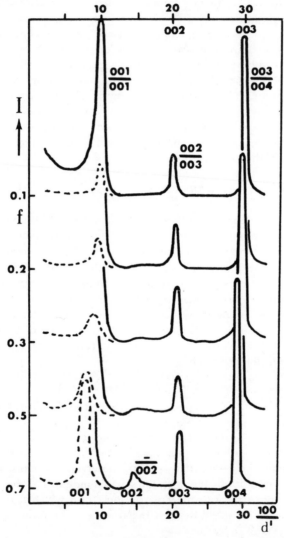

Fig. 12-3. Diffraction curves for random interstratification of 10Å and 14Å layers, showing positions of the combination peaks with varying proportions of the two components. Fractions of the 14Å layers are indicated. (Reprinted with permission from *Journal of Soil Science.* Brown & MacEwan, 1950; Fig. 4, p. 245.)

nent layers have the same structure factor, (ii) the mixing of layers is completely random, and (iii) the crystallites are infinitely large.

Although different layers do not have exactly the same structure factor nor are the crystallites infinitely large, the method offers a reasonable estimation of the individual components. Use of the same structure factor for different layers is based on the assumption that the major components involved in X-ray scattering in the various silicates are similar. Interlayer cations and water layers were considered to have low scattering power. MacEwan (1958) concluded that finite size of a crystallite would only broaden the diffraction peaks and not change their positions. Dense interlayer material as in chlorite, however, would contribute significantly to X-ray scattering, resulting in different structure factor estimates than used for silicate layers alone in the calculations of Brown and MacEwan. Despite these limitations, the curves are useful, and very convenient, for the analysis of most interstratified mixtures and have been used extensively.

An example of a calculated diffraction diagram (Brown & MacEwan, 1950) for random interstratification of two minerals with 10Å and 14Å spacings and with varying proportions of the two components is shown in Fig. 12-3. The figure shows that the combination peak from $001_{(10Å)}/001_{(14Å)}$ increases in spacing as the proportion of the 14Å component increases. However, the 002/003 combination peak moves to a lower spacing. Curves showing the migration of the combination peaks from the commonly occurring interstratified mixtures mica-vermiculite, mica-smectite, and mica-kaolinite are given in Fig. 12-4.

Generally, the movement of the peak positions follows an S-shape curve. When the two neighboring peaks from which the combination peak arises are close together, movement of the combination peak is linearly related to the composition of the mixture, as 001/001 in Fig. 12-4A. As the neighboring peaks become farther apart, the S-shape becomes more marked, as 001/001 in Fig. 12-4C. Peaks become more diffuse in the middle region of the S-shape and may disappear if the middle region is nearly vertical.

Some diffraction profiles considering finite crystallite size and different structure factors have been calculated by Mering (1950) and MacEwan (1958), but no detailed calculations of scattering over a wide range have yet been made. Reynolds (1967) using a fast digital computer for calculations of diffraction patterns from interstratified layers has shown that even interlayer cations and water or organic liquids used for characterization of different layers can cause significant errors in estimates from the Hendricks and Teller treatment. The errors are large especially at low diffraction angles. For accurate determinations of the components, therefore, both the structure factors for the individual layers as well as size of the crystallites need to be considered.

Reynolds (1967) and Reynolds and Hower (1970) have calculated diffraction patterns over a varying number of silicate layers per crystallite and for different structure factors of layers with different interlamellar material. The computer-simulated diffraction profiles revealed that if primary peaks of

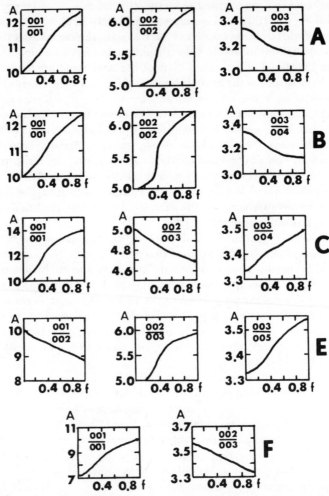

Fig. 12–4. Curves showing migration of combination peaks from randomly interstratified
layers: (*A*) 10/12.4Å (dioctahedral), (*B*) 10/12.4Å (trioctahedral), (*C*) 10/14Å (di-
and trioctahedral), (*E*) 10/17.7Å (di- and trioctahedral), (*F*) 7.14/10Å (dioctahedral).
(Reprinted with permission from *Journal of Soil Science*. Brown & MacEwan, 1950;
Fig. 10, p. 249.)

the two components are widely separated as the 001 (10Å) peak of mica and
the 001 (17Å) peak of glycol-treated smectite, then the combination peak
001/001 appears at 17Å rather than at an intermediate spacing. The peak is
merely diminished in intensity as the proportion of mica increases. They
concluded that earlier curves showing the 001/001 relationship for randomly
interstratified mica-smectite (10Å/17Å) cannot be satisfactorily used for the
determination of the two components. They recommended that when estimates of these two components are to be made from peak positions, higher
angle peaks should be used. Alternatively, the interstratified mixture could

be saturated with K to collapse the smectite layers to 12.4Å [at about 50% relative humidity (RH)] and then migration curves could be used to obtain the proportions of the two components. Their calculations also indicated that virtually all mica-smectite interstratifications with < 35% smectite are ordered structures while mixtures with > 35-40% smectite are almost always randomly interstratified.

3. DIRECT FOURIER TRANSFORM METHOD

In the previous treatments, the composition of the interstratified mixtures was determined indirectly by comparing the observed diffraction patterns with the calculated diffraction patterns from the assumed interstratified mixtures. In the direct Fourier transform method, the positions and intensities of the observed diffraction peaks are used to calculate the proportions of different layers in the interstratified mixture. The method requires some knowledge of the nature of layers present in the mixture, as do other methods. This can be readily obtained by employing common diagnostic treatments for identifying individual layer silicates. The problem then reduces to finding, say in a mixture of two-layer silicates, the probability of a layer at a distance R from any other layer. This can be obtained by using the following relationship calculated by MacEwan (1956) and MacEwan et al. (1961).

$$W_R = \frac{a}{\pi} \Sigma_s i(s) \cos (\mu_s R)$$

where
   W_R = the distribution function of interlayer distances,
   a = the mean thickness of a layer,
   $i(s) = I_s/\Theta |F_s|^2$,
   $\mu_s = 2/d_s'$
   I_s = the integrated intensity under the peak, and
   d_s' = the apparent spacing.
Θ and $|F_s|^2$ are the Lorentz polarization and the layer structure factor at the peak described earlier.

Cosine series calculated in this way represent the Fourier transform in which the heights of the fundamental peaks for individual components give their relative proportions. The heights of the combination peaks give the frequency of occurrence of the corresponding interlayer distances and hence the degree of randomness of layers. The method of calculation is illustrated in detail by MacEwan (1956). An example of the Fourier transform calculated from basal diffraction peaks given by a mixture of 10Å (layer A) and 17.8Å (layer B) spacings (MacEwan, 1956) is given in Fig. 12-5. The relative heights of the peaks A (0.48) and B (0.52) represent the proportions of the two components in the mixture. These estimates agree with predictions from the Hendricks and Teller treatment. The combination peaks that appear in the transform correspond to combinations of spacings of the individual com-

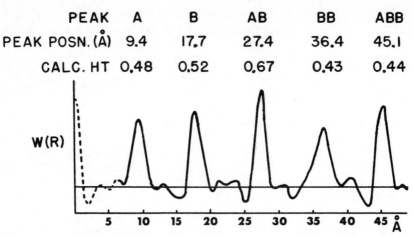

PEAK	A	B	AB	BB	ABB
PEAK POSN. (Å)	9.4	17.7	27.4	36.4	45.1
CALC. HT	0.48	0.52	0.67	0.43	0.44

Fig. 12-5. Fourier transform of basal diffraction peaks from randomly interstratified 10Å and 17Å layers. (Reprinted with permission from MacEwan, D. M. C. 1956. Fourier transform methods for studying X-ray scattering from lamellar systems. I. Kolloid Z. 149:96–108; Fig. 3, p. 103.)

ponents; for example, the 27.4Å peak arises from 9.3 + 17.7; the 36.4Å peak from 17.7 + 17.7; the 44.8Å peak from 9.4 + (2 × 17.7) and so on.

If the interstratification of the two layers were completely random, the height of the combination peak would be given by

$$h_{AB} = P_A P_{AB} + P_B P_{BA}$$

where P_A and P_B are the proportions of A and B in the mixture and P_{AB} is the probability of B following A and so on. For the randomly mixed layers with $P_A = 0.48$, $P_B = 0.52$, $P_{AB} = 0.52$ and $P_{BA} = 0.48$

$$h_{AB} = (0.48 \times 0.52) + (0.52 \times 0.48) = 0.50.$$

The higher value of the peak (0.67) obtained in the Fourier transform indicates a small tendency for layers to alterate. The height of the combination peak for complete alternation of layers (ABABAB. . . .) would be 1.0. Since the composition of the component layers is often not known, the choice of the structure factors to be used is difficult.

B. Examples of Randomly Interstratified Minerals and Interpretation of Their X-ray Diffraction Patterns

Although no agreement on the terminology of randomly interstratified minerals exists, they are generally described in terms of their component layers. The dominant component is listed first when describing the mixed layers. For example, an interstratified mixture of chlorite and mica with dominant chlorite layers is referred to as chlorite-mica interstratification. For comparable proportions of layers, 1-1 or 50–50 chlorite-mica or mica-chlorite interstratification is used.

Randomly interstratified layer silicates are frequently observed in soils (Jackson & coworkers; for example, see Whittig & Jackson, 1955; Schmehl & Jackson, 1956) and in sediments (Weaver, 1956). Weaver estimated that over 70% of the 6,000 sedimentary rock samples examined by him contained some interstratified layers. Only selected examples of random interstratification will be discussed here.

1. MICA-VERMICULITE

Mica-vermiculite mixed layers can be readily identified from a basal diffraction peak intermediate between 10Å and 14Å from Mg-saturated samples. Glycerol or glycol treatment does not alter the spacings. On K-saturation and heating to 100C (373K), the layers of the vermiculite component collapse to the mica spacing, resulting in a 10Å diffraction peak and a series of higher orders.

Analyses of most mica-vermiculite interstratifications have been confined to visual examination of the diffraction patterns. Barshad (1949) observed a broad diffraction peak at 11.77Å from a vermiculite sample from Libby, Montana and concluded that the peak resulted from a random mixture of 75% biotite and 25% vermiculite layers. Walker (1949) analyzed biotite flakes from a fresh outcrop and from a soil developed on this parent material. Biotite in the fresh outcrop gave a 10Å basal diffraction peak while moderately weathered biotite from the soil profile gave broad peaks at 11.8–13.6Å, indicating randomly interstratified biotite-vermiculite layers. Highly weathered flakes in the soil profile gave an integral series of a 14Å diffraction peak characteristic of vermiculite. Walker concluded that weathering of biotite initially produces mixed-layer biotite-vermiculite and, as weathering proceeds, vermiculite is formed.

Determinations of the relative proportions of mica and vermiculite in interstratified mixtures in soils of temperate regions, using the peak migration curves of MacEwan et al. (1961), are often uncertain because of the formation of hydroxy-interlayers in the vermiculite component of the mixture. These interlayers have strong X-ray scattering effects whereas the curves of MacEwan et al. were calculated assuming negligible scattering from interlayer positions. Tamura (1956) found that in the D horizon of a New England soil, where hydroxy interlayers were not present in vermiculite, use of different combination peaks, as 001/001 and 003/004, gave equal proportions of mica and vermiculite in an interstratified mixture. But as the hydroxy-interlayers increased with proximity to the soil surface, estimates of the two types of layers were different from different combination peaks.

MacEwan et al. (1961) have suggested the use of an "effective squared structure factor" instead of a single structure factor in such cases. The "effective squared structure factor",

$$|F|^2_{eff} = (|F_1|^{p_1} \cdot |F_2|^{p_2})^2$$

where F_1 and F_2 are structure factors of individual layers present in the mixture in proportions p_1 and p_2. Because of the uncertainties in determining the extent of hydroxy interlayers in soil clays and hence, in the value of the structure factor, the procedure appears to have limited application.

2. MICA-SMECTITE

Mica-smectite interstratification can be characterized in the same manner as mica-vermiculite except that on glycerol or glycol treatment the intermediate peak increases in spacing. Mica-smectite mixed layers are by far the most common of interstratified mixtures, especially in sediments and sedimentary rocks. Bystrom (1954) noted that the principal mineral in Ordovician bentonite beds was composed of mica-smectite mixed layers. Using the peak migration curves she found that the upper thin beds contained mica-smectite in a 3:2 ratio, while the lower beds contained the two components in a 1:4 ratio. The presence of the two components in only these two ratios over such a vast extent of the beds suggests that certain ratios are more stable. Cole (1966) and Gilkes and Hodson (1971) also described interstratified mica-smectite mixtures occurring in 3:2 proportions. However, there are a number of other examples where the mica-smectite ratio varies over a wide range: Bradley (1945) reported that bravasite is made up of random stacking of mica and smectite layers in almost equal proportions. Similarly, Hamilton (1967) found examples of partially ordered mica-smectite mixed layers in equal proportions in sediments from New South Wales. A range in composition of mica-smectite layers has been reported in sediments by Weaver (1956), Heller-Kallai and Kalman (1972), and in hydrothermally altered rocks of New Zealand by Steiner (1968). Perry and Hower (1970) and Foscolos and Kodama (1974) in their studies of buried sediments found a range of mica-smectite ratios in interstratified mixtures. In both these studies, it was found that the mica content of mixed layers increased while the smectite content decreased with depth of burial of the sediments. From a large number of analyses of shales, Weaver (1956) concluded that mica and smectite occur in all possible ratios in random mixtures.

Examples of mica-smectite interstratification as well as complex mixed layers in soils include studies by Whittig and Jackson (1955), Chichester et al. (1969), Mills and Zwarich (1972), and others.

3. MICA-CHLORITE

Random interstratification of mica-chlorite is characterized by basal spacings intermediate between 10Å and 14Å which do not change either on K-saturation or Mg-saturation and glycerol treatment. Lucas and Ataman (1968) observed that in the Jura basin, a progressive mineralogical variation from degraded mica to well-crystallized chlorite occurs through intermediate stages of mixed-layer mica-chlorite structures.

4. CHLORITE-VERMICULITE

Chlorite-vermiculite interstratification is characterized by a 14Å basal diffraction peak from a Mg-saturated sample and a peak intermediate between 10Å and 14Å from K-saturated samples since the vermiculite component collapses to 10Å on K-saturation. Weaver (1956) described a number of chlorite-vermiculite interstratifications. The presence of chlorite-vermiculite mixed layers was deduced from the relative intensities of the 14Å and 7Å peaks from the untreated material since their intensities were intermediate between those expected from chlorite and from vermiculite. Weaver found that on heating the sample to 400C (673K), approximately 50% of the layers collapsed to 10Å resulting in combination peaks at 12.6Å (001/001), 8.0Å (001/002), 4.9Å (002/003), and 3.49Å (003/004). After the sample was heated at 550C (823K) the temperature at which OH is removed from the chlorite and the 001 peak decreases to 13.8Å, the 001/001 combination peak decreased from 12.6Å to 11.6Å. Similarly, Heystek (1956) observed that the X-ray diffraction pattern of the glycol-treated clay fraction from a shale gave a 14.3Å diffraction peak and its submultiples. However, when the sample was heated to 500C (773K), the 001/001 combination peak decreased to 11.8Å, suggesting a 50:50 interstratified chlorite-vermiculite.

As soil clays often contain vermiculite with hydroxy interlayers which when saturated with K give peaks intermediate between 10Å and 14Å, the interlayered vermiculite may be confused with interstratified chlorite-vermiculite. However, collapse of hydroxy interlayers often occurs gradually with increasing temperatures, and the diffraction peak decreases accordingly. On the other hand, vermiculite in chlorite-vermiculite mixed layers should collapse readily to 10Å on K-saturation and the combination peak thus produced should remain unaltered on heating to 400C (673K) (Weaver, 1956). Stephen (1952) reported that clay of a biotite-rich soil contained a randomly interstratified chlorite-vermiculite mixture. Saturation of the sample with K gave a combination peak at 12Å, representing a randomly interstratified mixture of chlorite and vermiculite layers in approximately equal proportions.

5. CHLORITE-SMECTITE

The principal difference between chlorite-vermiculite and chlorite-smectite is that 001/001 combination peak for Mg-saturated chlorite-smectite shifts to higher spacings on glycerol treatment while that from chlorite-vermiculite remains at 14Å. Grim et al. (1960) observed a range of mixed-layer chlorite-smectite sequences in clay partings in ore beds in New Mexico evaporite. An example of a predominately dioctahedral chlorite-smectite was found in the C horizon of a soil of the Alberni soil series (Brydon et al.,

1961). The interstratified mixture gave a basal diffraction peak at 14.5Å
which increased to 15.5Å on glycol treatment and decreased to 11.5Å on
heating the sample. These treatments indicated a 40:60 ratio of chlorite to
smectite in the interstratified mixture. The aluminous chlorite in clays of the
surface soil was assumed to have been formed from the interstratified mixture by precipitation of hydroxy-Al in interlayers of the smectite component.
Mixed layers containing dioctahedral chlorite and smectite have also been reported by Sudo and Kodama (1957).

6. CHLORITE-SWELLING CHLORITE

Chlorite-swelling chlorite interstratification can be distinguished from
chlorite-smectite interstratification from its basal diffraction peak at about
13.8Å after heating at 500C (773K), while the chlorite-smectite diffraction
peak decreases below this spacing after similar heating. Honeyborne (1951)
and Martin-Vivaldi and MacEwan (1957) investigated mixed clays from
coastal ranges. In their natural form the two samples gave 14.5Å and 13.6Å
basal diffraction peaks that increased to 16.4Å and 14.8Å, respectively, on
treatment with glycerol. After heating at 500C, the peaks decreased to 13.7–
13.8Å, indicating a chlorite-swelling chlorite interstratification. The sample
studied by Honeyborne appeared to be a 1:1 random interstratification of
chlorite-swelling chlorite, while that studied by Martin-Vivaldi and MacEwan
apparently contained a smaller proportion of swelling chlorite.

7. KAOLINITE-SMECTITE

A sparsity of reports of interstratified kaolinite-smectite has been due
to the fact that definitive diagnostic criteria for their identification have only
recently been developed. Schultz et al. (1971) used the position of the basal
diffraction peak resulting from the combination of 7.15Å (001 of kaolinite)
and 7.7Å (002 of air-dried smectite) peaks. The combination peak at 7.45Å
in some Mexican clays was interpreted as a 1:1 random interstratification of
kaolinite-smectite. The most diagnostic feature of these mixed layers is the
diffraction peak near 8Å after heating at 300C (573K), resulting from a combination of 7.15Å (001 kaolinite) and 9.6Å (001 of heated smectite). The
7.45Å peak from untreated mixed layers could be confused with halloysite.
However, the layer thickness of halloysite decreases to 7.2Å after similar heat
treatment.

Shimoyama et al. (1969), Wiewiora (1971), and Wilson and Cradwick
(1972) have reported random interstratifications of kaolinite-smectite in clay
deposits and soil clays from Japan, Poland, and Scotland, respectively. Cradwick and Wilson (1972) calculated diffraction profiles to determine the composition of the kaolinite-smectite interstratifications from Poland, Japan, and
Mexico. The layer proportions predicted by the above method were in close
agreement with the proportions reported for clays from Japan and Poland,
but kaolinite content of Mexican clays was found to be higher than reported

earlier. Sakharov and Drits (1973) described two examples of hydrothermally altered clays containing kaolinite-smectite mixed layers in 3:1 ratio.

8. THREE COMPONENT MIXED-LAYER SYSTEMS

Weaver (1956) reported that three-component mixed layers containing illite, chlorite, and smectite were not uncommon in shales. These interstratifications, however, can be reduced to two component systems and then can be treated in the usual manner. An example given below from Weaver (1956) illustrates the procedure. X-ray diffraction patterns of a naturally occurring three-component system gave a 12Å basal diffraction peak after saturation with Mg. This spacing was considered to result from a combination of 60% 10Å (illite) and 40% 14Å (smectite or chlorite) components. The presence of vermiculite in the shale was not considered. After heating at 400C (673K), the basal spacing decreased to 11Å, indicating that a portion of the 14Å component collapsed to 10Å. Thus, a different two component system containing 10Å (illite + smectite) and 14Å (chlorite) developed on heating. The 11Å combination peak from this system indicated the presence of 25% chlorite layers. The presence of chlorite layers was verified from diffraction pattern obtained after the sample was heated at 550C (823K). Thus, the mixed layers were interpreted as 60% illite, 25% chlorite, and 15% smectite. Foscolos and Kodama (1974) studied a three-layer mixture containing mica, smectite, and vermiculite; Fourier transforms obtained by using a computer showed that the interstratified mixture in the sample from a depth of 1,798 m (5,900 feet) contained 47% mica, 34% smectite, and 19% vermiculite, whereas the sample from 3,078 m (10,100 feet) contained 82% mica and only 9% each of smectite and vermiculite. An increase in mica content with depth in buried sediments has also been observed by Perry and Hower (1970). Triangular diagrams to estimate ratios of the components from the first-order spacing of three component mica-smectite-chlorite interstratifications have been developed by Jonas and Brown (1959).

IV. FORMATION OF INTERSTRATIFIED MINERALS

Mixed-layer minerals occurring in nature are formed primarily by hydrothermal alteration or by weathering involving partial removal of interlayer K from micas or removal of hydroxide interlayers from chlorite accompanied by possible structural changes which decrease the layer charge on minerals. Conversely, uptake of K and similar ions and formation of brucitic or gibbsitic interlayers in expanding-layer silicates, vermiculite, and smectite form mixed-layer minerals.

For understanding the processes and reactions involved in the formation of interstratified minerals in nature and for the development of procedures for their identification, investigations of the formation of interstratified minerals in the laboratory are necessary. Most laboratory experiments have been concerned with either the removal of K from micas or the uptake of K and similar cations by vermiculite and smectite at room temperatures. These investigations have been reviewed recently by Sawhney (1972) and by Norrish (1973). Mixed-layer minerals have also been produced synthetically at elevated temperatures (Mumpton & Roy, 1956) and by heat treatments of vermiculite (Walker, 1956) and chlorite (Brindley & Chang, 1974). The discussion here will be restricted to the laboratory formation of mixed layer minerals by removal or uptake of K and similar ions. Formation of hydroxy interlayers in vermiculite and smectite is discussed in the chapter containing intergrade minerals.

A. Mica-Vermiculite

1. REMOVAL OF INTERLAYER K

Experiments on weathering of micas have produced either mica-vermiculite mixed layers or vermiculite, depending upon the severity of the method used for extraction of K. Extraction of K from biotite with $MgCl_2$ solutions produced primarily vermiculite (Barshad, 1948) whereas extraction of K from biotite with NaCl solution produced a 12Å spacing ascribed to randomly interstratified mica-vermiculite layers, although most layers expanded to 14Å in these experiments (Mortland, 1958). Progressive removal of K from micas with sodium tetraphenylboron (NaTPB) produced increasing amounts of randomly interstratified mica-vermiculite mixtures (DeMumbrum, 1959; Scott, 1968).

In several other investigations, removal of K from micas by dilute salt solutions produced interstratified mica-vermiculite layers with a tendency toward ordered structures. Biotite treated with a dilute solution of $CuCl_2$ (Bassett, 1959) produced a regularly interstratified mica-vermiculite structure. Similarly, extraction of K from mica with dilute solutions of $SrCl_2$ (Rausell-Colom et al., 1965), $MgSO_4$ (Hoda & Hood, 1972) or Na_2CO_3 (Baweja et al., 1974) resulted in basal diffraction peaks at 25Å and 12Å, indicative of ordered mica-vermiculite layers. Extraction of a soil mica with dilute $MgCl_2$ solution (Rich & Cook, 1963) also produced a 25Å basal spacing.

Regularly interstratified mica-vermiculite layers were produced by Tomita and Sudo by treatment of mica with molten $LiNO_3$ (1971) and by extracting K from preheated micas with acid or with NaTPB (Tomita and Dozono, 1972, 1973).

2. UPTAKE OF K AND SIMILAR IONS

Uptake of K by vermiculite and in some instances by K-depleted micas has resulted in the formation of regularly interstratified mica-vermiculite layer sequences. Experiments by Sawhney (1967, 1969a) show that increasing sorption of K or Cs ions by Ca-saturated vermiculite collapsed alternate layers, forming regularly interstratified mica-vermiculite layer sequences. As shown in Fig. 12–6, at about 45% saturation with K, a vermiculite sample

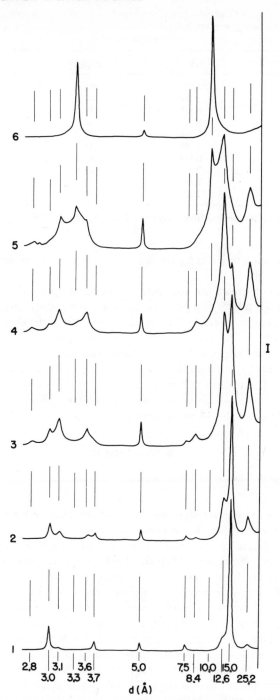

Fig. 12–6. X-ray diffraction patterns of Ca-saturated vermiculite (*1*), showing conversion of vermiculite (15Å) to mica-like (10Å) layers through collapse of alternate interlayers on K sorption (*2, 3, 4, 5, 6*). (Reprinted with permission from Pergamon Press. Sawhney, 1967; Fig. 2, p. 79.)

Table 12-4. Observed and calculated structure factors, $|F|$, of 00l diffraction peaks (Å)
from regularly interstratified mica-vermiculite layers formed by K sorption
by vermiculite[†]

l	Å	$|F|_{obs}$	$|F|_{cal}$
001	25.20	93.6	93.6
002	12.60	147.9	72.5
003	8.40	30.2	2.6
004	6.30	tr[‡]	10.7
005	5.04	59.2	57.1
006	4.20	tr	4.2
007	3.60	48.9	9.6
008	3.15	118.9	105.0
009	2.80	31.4	11.8
0010	2.50	46.9	25.2

[†] Reprinted with permission from Pergamon Press (Sawhney, 1967; Fig. 2, p. 79).
[‡] tr = trace.

was almost completely changed to a regularly interstratified mica-vermiculite
mixture (curve 4).

Table 12-4 shows the structure factors obtained from the intensities of
10 orders of 00l diffraction peaks in curve 4 and the calculated structure factors for a regularly interstratified mica-vermiculite of the determined chemical composition. The agreement between the calculated and the observed
structure factors supports the contention that regular interstratification of
mica-vermiculite layers is produced by K uptake by vermiculite.

Further K sorption collapsed the vermiculite layers within the interstratified mixture until all vermiculite layers were collapsed and an integral
series of reflections at 10Å were recorded.

B. Mica-Smectite

1. UPTAKE OF K

Although most vermiculites formed regularly interstratified mica-
vermiculite mixtures on K sorption, vermiculite from Llano and several
smectites produced random, not regular, interstratification of expanded and
collapsed layers (Sawhney, 1969b). Lack of regular interstratification in
smectites was attributed to their lower charge density. In a study of layer
collapse on K sorption in a number of vermiculites, Weed and Leonard (1968)
found that development of regular interstratification was not related to gross
surface charge density, but appeared to be positively related to the amount
of Fe in the octahedral positions. They found that K-depleted biotite formed
regular interstratification while K-depleted phlogopite and muscovite showed
a tendency to form random interstratification on K sorption. Ross and Rich
(1973) also observed that K-depleted phlogopite tended to form random
interstratification on K sorption.

Although removal of K from some micas and uptake of K by certain K-depleted micas and vermiculites produced random interstratification, mild treatments for removal of K and uptake from dilute K solution generally produces regular interstratification.

C. Mechanism of Regular Interstratification

Bassett (1959) suggested that when K ions in one interlayer zone of biotite are replaced by Ca or Mg ions, bonding in that zone becomes weaker, because the bond length of the hydrated Ca or Mg ion from the silicate sheet is longer than the bond length of the K ion from the silicate sheet. As a result, bonding in the adjacent interlayer zone becomes stronger. Hence, the next interlayer in which Ca or Mg must enter is not the adjacent, but the one beyond the adjacent interlayer, producing regular interstratification by progressive removal of K from alternate interlayers. Regular interstratification on uptake of K or Cs ions by Mg- or Ca-saturated vermiculite has been essentially explained by the reverse process by a number of workers (Sawhney, 1967, 1969a; Rhoades & Coleman, 1967; and Weed & Leonard, 1968). Incorporation of K ions in one interlayer zone reduces the effective charge on the adjacent layer. Consequently, the bonding energy between the K ion and the reduced charge of the layer may be less than the hydration energy of Ca or Mg ions. Hence, the K ions enter not the adjacent but the interlayer beyond the adjacent interlayer, forming regularly interstratified mica-vermiculite layers.

A more plausible explanation for regular interstratification, based on recent experiments, has emerged and has been discussed by Norrish (1973). This mechanism involves the changes in the angle of the OH bond with the cleavage plane and the effect of the proton of the OH ion on bonding between the interlayer cation and the silicate layer. This viewpoint is based on the following observations:

1) Oxidation of octahedral iron by thermal or chemical methods renders extraction of interlayer K more difficult (Barshad & Kishk, 1968; Farmer & Wilson, 1970).

2) Subsequent K removal results in the formation of regularly interstratified layers rather than expanded vermiculite layers (Wilson & Farmer, 1970; Tomita & Dozono, 1972; Gilkes, 1973).

3) Oxidation of octahedral iron also changes the direction of dipole moment of the OH ion from perpendicular to an inclined position to the cleavage plane (Juo & White, 1969).

4) The angle of inclination in muscovite is 18° (Radoslovich, 1963; Giese, 1971) whereas in biotite the OH bond is perpendicular to the cleavage plane. Thus, when the angle of inclination of the OH bond with the cleavage plane is smaller, as in muscovite and lepidolite relative to biotite, or in oxidized biotite relative to reduced biotite, the greater distance between the proton of the OH and the interlayer K ion causes the K to be held more strongly.

Removal of K from one interlayer zone of mica changes the bond direc-

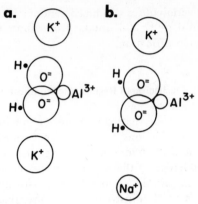

Fig. 12-7. Diagramatic representation of a mechanism describing expansion of alternate
interlayers of mica on K removal: (*a*) Relative positions and orientations of OH and
interlayer K in muscovite, and (*b*) changes in OH orientation on K removal. (Norrish,
1973).

tion to a higher angle, reducing the distance of the proton from OH to the replacing ion, as Na ion in Fig. 12-7. Consequently, the angle of the second
OH bond attached to the same octahedral ion decreases and hence the proton
of the OH is farther away from the K ion in the adjacent interlayer; thus, the
K ion here is more strongly held. This mechanism explains how K is held
more strongly in the interlayer adjacent to the interlayer zone from which K
is replaced. The removal of K by the extracting cation then occurs, not from
the adjacent interlayer, but from the next interlayer zone producing regularly
interstratified mica-vermiculite layer sequences. Conversely, sorption of K in
one interlayer zone of vermiculite must reduce the angle of the OH bond
with this layer and increase the angle of the OH bond with the adjacent
layer, reducing the tendency of the adjacent interlayer to collapse. The K
ion would then enter not the adjacent, but the next interlayer, producing
collapse in alternate interlayers of vermiculite.

V. SUMMARY

     Since layer silicates are structurally similar, they frequently occur as
interstratified minerals in which individual crystals consist of two or more
layer silicates. Interstratification of layers may be regular or random.
     Regularly interstratified layers are readily characterized by the integral
series of basal diffraction peaks on their X-ray diffraction patterns. The 001
basal spacings of the interstratified minerals are equal to the sum of the
spacings of the individual components.
     Randomly interstratified layers give X-ray diffraction patterns which
show nonintegral peaks at positions intermediate between the peaks from the
individual components. The exact positions of the peaks depend upon the
relative proportions of the component layers of different basal spacings and

the order of mixing while their intensities are governed by the structure factors of the component layers and the crystal size when the crystal consists of only few layers.

Brown and MacEwan have constructed curves showing the migration of various combination peaks from random mixtures of common layer silicates containing two components in varying proportions. These curves are based upon the Hendricks and Teller treatment of the diffraction effects from mixing of layers of different spacings. The curves have been extensively used to determine the composition of interstratified mixtures by comparing the positions of the observed peaks with the peaks calculated from the mixtures of assumed composition. The method, however, assumes completely random mixing of layers with the same structure factor, and crystals of infinite size, although limited calculations have been made using different structure factors and crystal size.

Reynolds has recently modified the treatment using a fast digital computer where structure factors, crystal sizes, and the degrees of randomness can be varied rapidly to the point where the calculated and the observed diffraction patterns are similar. The method gives more nearly correct estimates of the composition of interstratified layers. In soil clays, where a number of other minerals in addition to the interstratified layers are frequently present, and the allocation of an exact chemical formula to the interstratified mineral is uncertain, the method can yield only semiquantitative estimates of the composition of interstratified layers.

Both regular and random interstratifications of layer silicates have been produced in the laboratory by the extraction of interlayer K from micas or by the uptake of K by vermiculite or smectite. Although no unequivocal mechanism for the formation of interstratified minerals has emerged, mechanisms have been proposed especially for the formation of regularly interstratified mica-vermiculite layer sequences.

LITERATURE CITED

Alietti, A. 1958. Some interstratified clay minerals of the Taro valley. Clay Miner. Bull. 3:207-211.

Altschuler, Z. S., E. J. Dwornik, and H. Kramer. 1963. Transformation of montmorillonite to kaolinite during weathering. Science 141:148-152.

Bailey, S. W., and S. A. Tyler. 1960. Clay minerals associated with lake Superior iron ores. Econ. Geol. 55:150-175.

Barshad, I. 1948. Vermiculite and its relation to biotite as revealed by base exchange reactions, X-ray analyses, differential thermal curves, and water content. Am. Mineral. 33:655-678.

Barshad, I. 1949. The nature of lattice expansion and its relation to hydration in montmorillonite and vermiculite. Am. Mineral. 34:675-684.

Barshad, I., and F. M. Kishk. 1968. Oxidation of ferrous iron in vermiculite and biotite alters fixation and replaceability of potassium. Science 162:1401-1402.

Bassett, W. A. 1959. The origin of the vermiculite deposit at Libby, Montana. Am. Mineral. 44:282-299.

Baweja, A. S., L. P. Wilding, and E. O. McLean. 1974. Mineralogy and cation exchange properties of Libby vermiculite separates as affected by particle-size reduction. Clays Clay Miner. 22:253-262.

Blatter, C. L., H. E. Roberson, and G. R. Thompson. 1973. Regularly interstratified chlorite-dioctahedral smectite in dike-intruded shales, Montana. Clays Clay Miner. 21:207-212.

Boettcher, A. L. 1966. Vermiculite, hydrobiotite and biotite in the rainy creek igneous complex near Libby, Montana. Clay Miner. 6:283-296.

Bradley, W. F. 1945. Diagnostic criteria for clay minerals. Am. Mineral. 30:704-713.

Bradley, W. F. 1950. The alternating layer sequence of rectorite. Am. Mineral. 35:590-595.

Bradley, W. F., and C. E. Weaver. 1956. A regularly interstratified chlorite-vermiculite clay mineral. Am. Mineral. 41:497-504.

Brindley, G. W. 1956. Allevardite, a swelling double-layer mica mineral. Am. Mineral. 41:91-103.

Brindley, G. W., and Tien-Show Chang. 1974. Development of long basal spacings in chlorites by thermal treatment. Am. Mineral. 59:152-158.

Brown, G., and D. M. C. MacEwan. 1950. The interpretation of x-ray diagrams of soil clays. II. Structures with random interstratification. J. Soil Sci. 1:239-253.

Brown, G., and A. H. Weir. 1963. The identity of rectorite and allevardite. Proc. Int. Clay Conf. (Stockholm) 1:27-35. Pergamon Press, Oxford.

Brydon, J. E., J. S. Clark, and V. Osborne. 1961. Dioctahedral chlorite. Can. Mineral. 6:595-609.

Bystrom, Ann Marie. 1954. Mixed layer minerals from the ordovician bentonite beds at Kinnekulle, Sweden. Nature 173:783-786.

Caillere, S., A. Mathieu-Sicaud, and S. Henin. 1950. Nouvel essai d'identification du mineral de la Table près Allevard, l'allevardite. Bull. Soc. Fr. Mineral. 73:193-201.

Chichester, F. W., C. T. Youngberg, and M. E. Harward. 1969. Clay mineralogy of soils formed on Mazama pumice. Soil Sci. Soc. Am. Proc. 33:115-120.

Cole, W. F. 1966. A study of a long-spacing mica-like mineral. Clay Miner. 6:261-281.

Coleman, N. T., F. H. LeRoux, and J. G. Cady. 1963. Biotite-hydrobiotite-vermiculite in soils. Nature 198:409-410.

Cradwick, P. D., and M. J. Wilson. 1972. Calculated X-ray diffraction profiles for interstratified kaolinite-montmorillonite. Clay Miner. 9:395-405.

DeMumbrum, L. E. 1959. Exchangeable potassium levels in vermiculite and K-depleted micas, and implications relative to potassium levels in soils. Soil Sci. Soc. Am. Proc. 23:192-194.

Earley, J. W., G. W. Brindley, W. J. McVeagh, and R. C. Vanden-Heuvel. 1956. A regularly interstratified montmorillonite-chlorite. Am. Mineral. 41:258-267.

Farmer, V. C., and M. J. Wilson. 1970. Experimental conversion of biotite to hydrobiotite. Nature 226:841-842.

Foscolos, A. E., and H. Kodama. 1974. Diagenesis of clay minerals from lower cretaceous shales of Northeastern British Columbia. Clays Clay Miner. 22:319-335.

Giese, R. F. 1971. Hydroxyl orientation in muscovite as indicated by electrostatic energy calculations. Science 172:263-264.

Gilkes, R. J. 1973. The alteration products of K-depleted oxybiotite. Clays Clay Miner. 21:303-313.

Gilkes, R. J., and F. Hodson. 1971. Two mixed-layer mica-montmorillonite minerals from sedimentary rocks. Clay Miner. 9:125-137.

Grim, R. E., J. B. Droste, and W. F. Bradley. 1960. A mixed-layer clay mineral associated with an evaporite. Clays Clay Miner. 8:228-236.

Gruner, J. W. 1934. The structure of vermiculites and their collapse by dehydration. Am. Mineral. 19:557-578.

Hamilton, J. D. 1967. Partially-ordered mixed-layer mica-montmorillonite from Maitland, New South Wales. Clay Miner. 7:63-78.

Harvey, R. D., and C. W. Beck. 1962. Hydrothermal regularly interstratified chlorite-vermiculite and tobermorite in alteration zone at Goldfield, Nevada. Clays Clay Miner. 9:343-354.

Heller-Kallai, L., and Z. H. Kalman. 1972. Some naturally-occurring illite-smectite interstratifications. Clays Clay Miner. 20:165-168.

Hendricks, S., and E. Teller. 1942. X-ray interference in partially ordered layer lattices. J. Chem. Phys. 10:147-167.

Heystek, H. 1954. An occurrence of regular mixed-layer clay-mineral. Mineral. Mag. 30:400-408.

Heystek, H. 1956. Vermiculite as a member in mixed-layer minerals. Clays Clay Miner. 4:429–434.

Hoda, S. N., and W. C. Hood. 1972. Laboratory alteration of trioctahedral micas. Clays Clay Miner. 20:343–358.

Honeyborne, D. B. 1951. The clay minerals in the Keuper Marl. Clay Miner. Bull. 1: 150–157.

Jackson, M. L. 1963. Interlayering of expansible layer silicates in soils by chemical weathering. Clays Clay Miner. 11:29–46.

Johnson, L. J. 1964. Occurrence of regularly interstratified chlorite-vermiculite as a weathering product of chlorite in a soil. Am. Mineral. 49:556–572.

Jonas, E. C., and T. E. Brown. 1959. Analysis of interlayer mixtures of three clay mineral types by X-ray diffraction. J. Sediment. Petrol. 29:77–86.

Juo, A. S. R., and J. L. White. 1969. Orientation of the dipole moments of hydroxyl groups in oxidized and unoxidized biotite. Science 165:804–805.

Kapoor, B. S. 1972. Weathering of micaceous clays in some Norwegian podzols. Clay Miner. 9:383–394.

Lippmann, F. 1954. Uber einen Keuperton von Zaisersweiher bei Maulbronn. Heidelb. Beitr. Mineral. Petrogr. 4:130–134.

Lucas, J., and G. Ataman. 1968. Mineralogical and geochemical study of clay mineral transformations in the sedimentary Triassic Jura basin (France). Clays Clay Miner. 16:365–372.

MacEwan, D. M. C. 1956. Fourier transform methods for studying X-ray scattering from lamellar systems. I. A direct method for analysing interstratified mixtures. Kolloid Z. 149:96–108.

MacEwan, D. M. C. 1958. Fourier transform methods for studying X-ray scattering from lamellar systems. II. The calculation of X-ray diffraction effects for various types of interstratification. Kolloid Z. 156:61–67.

MacEwan, D. M. C., A. Ruiz-Amil, and G. Brown. 1961. Interstratified clay minerals. p. 393–445. In G. Brown (ed.) X-ray identification and crystal structures of clay minerals. The Mineralogical Soc., London.

Martin-Vivaldi, J. L., and D. M. C. MacEwan. 1957. Triassic chlorites from the Jura and Catalan coastal range. Clay Miner. Bull. 3:177–183.

Mering, J. 1949. L'Interference des rayons X dans les systemes a stratification desordonnee. Acta Cryst. 2:371–377.

Mering, J. 1950. Le reflexions de rayons X par les mineraux argeleux interstratifies. Int. Congr. Soil Sci., Trans. 4th (Amsterdam) 3:21–26.

Milks, J. G., and M. A. Zwarich. 1972. Recognition of interstratified clays. Clays Clay Miner. 20:169–174.

Mortland, M. M. 1958. Kinetics of potassium release from biotite. Soil Sci. Soc. Am. Proc. 22:503–508.

Mumpton, F. A., and R. Roy. 1956. The influence of ionic substitution on the hydrothermal stability of montmorillonoids. Clays Clay Miner. 4:337–339.

Norrish, K. 1973. Factors in the weathering of mica to vermiculite. p. 417–432. In J. M. Serratosa et al. (ed.) Proc. Int. Clay Conf., June 1972. (Madrid). Div. de Ciencas, CSIC, Madrid.

Perry, E., and J. Hower. 1970. Burial diagenesis in Gulf Coast pelitic sediments. Clays Clay Miner. 18:165–177.

Post, J. L., and N. C. Janke. 1974. Properties of "swelling" chlorite in some mesozoic formations of California. Clays Clay Miner. 22:67–77.

Radoslovich, E. W. 1963. The cell dimensions and symmetry of layer lattice silicates. IV. Interatomic forces. Am. Mineral. 48:76–99.

Rausell-Colom, J. A., T. R. Sweatman, C. B. Wells, and K. Norrish. 1965. Studies in the artificial weathering of mica. p. 40–72. In E. G. Hallsworth, and D. V. Crawford (ed.) Experimental pedology. Butterworths, London.

Reynolds, R. C. 1967. Interstratified clay systems: calculation of the total one-dimensional diffraction function. Am. Mineral. 52:661–672.

Reynolds, R. C., and J. Hower. 1970. The nature of interlayering in mixed-layer illite-montmorillonites. Clays Clay Miner. 18:25–36.

Rhoades, J. D., and N. T. Coleman. 1967. Interstratification in vermiculite and biotite produced by potassium sorption. I. Evaluation by simple X-ray diffraction pattern inspection. Soil Sci. Soc. Am. Proc. 31:366–372.

Rich, C. I., and M. G. Cook. 1963. Formation of dioctahedral vermiculite in Virginia soils. Clays Clay Miner. 10:96–106.

Ross, G. J., and C. I. Rich. 1973. Effect of particle size on potassium sorption by potassium depleted phlogopite. Clays Clay Miner. 21:83–87.

Sakharov, B. A., and V. A. Drits. 1973. Mixed-layer kaolinite-montmorillonite: A comparison of observed and calculated diffraction patterns. Clays Clay Miner. 21:15–17.

Sawhney, B. L. 1967. Interstratification in vermiculite. Clays Clay Miner. 15:75–84.

Sawhney, B. L. 1969a. Regularity of interstratification as affected by charge density in layer silicates. Soil Sci. Soc. Am. Proc. 33:42–46.

Sawhney, B. L. 1969b. Cesium uptake by layer silicates: effect on interlayer collapse and cation exchange capacity. Int. Clay Conf. Proc. (Tokyo) 1:605–611.

Sawhney, B. L. 1972. Selective sorption and fixation of cations by clay minerals: A review. Clays Clay Miner. 20:93–100.

Schmehl, W. R., and M. L. Jackson. 1956. Interstratification of layer silicates in two soil clays. Clays Clay Miner. 4:423–428.

Schultz, L. G., A. O. Shepard, P. D. Blackmon, and H. C. Starkey. 1971. Mixed-layer kaolinite-montmorillonite from the Yucatan Peninsula, Mexico. Clays Clay Miner. 19:137–150.

Scott, A. D. 1968. Effect of particle size on interlayer potassium exchange in micas. Int. Congr. Soil Sci., Trans. 9th (Adelaide) 2:649–659.

Shimoyama, A., W. D. Johns, and T. Sudo. 1969. Montmorillonite-kaolin clay in acid clay deposits from Japan. Int. Clay Conf. Proc. (Tokyo) 1:225–231.

Smith, W. W. 1960. Some interstratified clay minerals from basic igneous rocks. Clay Miner. Bull. 4:182–190.

Steiner, A. 1968. Clay minerals in hydrothermally altered rocks at Wairakei, N. Z. Clays Clay Miner. 16:193–213.

Stephen, I. 1952. A study of rock weathering with reference to the soils of the Malvern Hills. II. Weathering of appinite and "Ivyscar rock". J. Soil Sci. 3:219–237.

Stephen, I., and D. M. C. MacEwan. 1951. Some chloritic clay minerals of unusual type. Clay Miner. Bull. 1:157–162.

Sudo, T. 1954. Long spacings at about 30Å confirmed from certain clays from Japan. Clay Miner. Bull. 2:193–203.

Sudo, T., and H. Hayashi. 1956. Types of mixed-layer minerals from Japan. Clays Clay Miner. 4:389–412.

Sudo, T., and H. Kodama. 1957. An aluminum mixed-layer mineral of montmorillonite chlorite. Z. Kristall. 109:379–387.

Sutherland, H. H., and D. M. C. MacEwan. 1962. A swelling chlorite mineral. Clays Clay Miner. 9:451–458.

Tamura, T. 1956. Weathering of mixed-layer clays in soils. Clays Clay Miner. 4:413–422.

Tomita, K., and T. Sudo. 1971. Transformation of sericite into an interstratified mineral. Clays Clay Miner. 19:263–270.

Tomita, K., and M. Dozono. 1972. Formation of an interstratified mineral by extraction of potassium from mica with sodium tetraphenylboron. Clays Clay Miner. 20:225–231.

Tomita, K., and M. Dozono. 1973. An expansible mineral having high rehydration ability. Clays Clay Miner. 21:185–190.

Walker, G. F. 1949. The decomposition of biotite in soils. Mineral. Mag. 28:693–703.

Walker, G. F. 1956. The mechanism of dehydration of Mg-vermiculite. Clays Clay Miner. 4:101–115.

Weaver, C. E. 1956. The distribution and identification of mixed-layer clays in sedimentary rocks. Am. Mineral. 41:202–221.

Weed, S. B., and R. A. Leonard. 1968. Effect of K^+-uptake by K^+-depleted micas on the basal spacing. Soil Sci. Soc. Am. Proc. 32:335–340.

Whittig, L. D., and M. L. Jackson. 1955. Interstratified layer silicates in some soils of Northern Wisconsin. Clays Clay Miner. 3:322–336.

Wiewiora, A. 1971. A mixed layer kaolinite-smectite from Lower Silesia, Poland. Clays Clay Miner. 19:415–416.

Wilson, M. J., and V. C. Farmer. 1970. A study of weathering in a soil derived from a biotite-hornblende rock. II. The weathering of hornblende. Clay Miner. 8:435–444.

Wilson, M. J., and P. D. Cradwick. 1972. Occurrence of interstratified kaolinite-montmorillonite in some Scottish soils. Clay Miner. 9:435–437.

Palygorskite, (Attapulgite), Sepiolite, Talc, Pyrophyllite, and Zeolites[1]

LUCIAN W. ZELAZNY, Virginia Polytechnic Institute and State University,
Blacksburg, Virginia

FRANK G. CALHOUN, University of Florida, Gainesville, Florida

I. INTRODUCTION

The primary objective of this chapter is to review literature on the occurrence of palygorskite, attapulgite, sepiolite, talc, pyrophyllite, and zeolites in the soil environment and to provide data on the physical and chemical properties that they impart to the soil system. The most common denominator throughout this mineral suite is their rare occurrence in sola (A and B horizons). Their occurrences are generally limited to soils with a high pH or where acid weathering has been negligible. The limited occurrence of these minerals has resulted in a scant knowledge of their properties in the soil. Therefore, selected data from geologic deposits are also provided.

Talc and pyrophyllite are layer silicates with zero or negligible interlayer charge and therefore are the "no charge" end-members of the 2:1 type phyllosilicates. Palygorskite, attapulgite, and sepiolite are also designated as true phyllosilicates since they contain continuous two-dimensional tetrahedral sheets in which individual tetrahedra are linked with neighboring tetrahedra by sharing three corners each (Bailey et al., 1971a). "The 1967-CMS Nomenclature Committee . . . recommended that the name 'attapulgite' be relegated to the synonymy, as the name 'palygorskite' is judged to have priority" (Bailey et al., 1971b). Therefore, the name palygorskite will be used in this chapter to refer to literature citations including both names, attapulgite and palygorskite. Sepiolite is structurally distinct from palygorskite. The zeolites include a large number of known minerals although relatively few have been identified in soils. The properties of these minerals will be stressed, especially their unique ionic and molecular sieve characteristics.

[1]Florida Agricultural Experiment Stations Journal Series No. 6167. This work was initiated at the University of Florida and completed at Virginia Polytechnic Institute and State University.

II. PALYGORSKITE AND SEPIOLITE

A. Structural Properties and Mineral Identification

1. STRUCTURE

Palygorskite is an alumino-Mg-silicate with about equal proportions of Al and Mg while sepiolite is a Mg-silicate with only a minor Al component. These minerals are intermediate between a dioctahedral and trioctahedral subgroup, which may be responsible for their fibrous morphology (Millot, 1970). The fibrous morphology results from bands elongated parallel to the c-axis and consist of alternating ribbons with a 2:1 type structure (Fig. 13–1). The ribbons consist of five octahedral positions in palygorskite (Bradley, 1940) and eight positions in sepiolite (Brauner & Preisinger, 1956). These appear as the juxtaposition of two pyroxenic chains in palygorskite and three pyroxenic chains in sepiolite. Although the tetrahedral sheets are continuous, the apices in adjacent bands point in opposite directions. This structure results in a continuous plane of atoms with (i) tetrahedral positions primarily filled with Si atoms, and (ii) octahedral positions primarily filled with Mg or Al atoms alternating to form open channels of fixed dimensions running parallel to the chains. Palygorskite contains channels with a cross section of about 3.8 × 6.3 Å, whereas these channels are about 3.8 × 9.4 Å for sepiolite. The channels contain cations and water molecules. About one-half of the water molecules lie adjacent to the octahedral strip in order to satisfy coordination requirements of the octahedral sheet, while the remainder occupy definite positions in the channels (Bradley, 1940) thus conferring to these minerals zeolitic properties (Henin & Caillere, 1975).

2. ELEMENTAL COMPOSITION

The ideal structural formula for palygorskite is $(OH_2)_4(OH)_2Mg_5Si_8O_{20}4H_2O$ (Bradley, 1940). The ideal structural formula for sepiolite is $(OH_2)_4(OH)_6Mg_9Si_{12}O_{30}6H_2O$ as given by Nagy and Bradley (1955) and $(OH_2)_4(OH)_4Mg_8Si_{12}O_{30}8H_2O$ as given by Brauner and Preisinger (1956). This latter model has been more widely accepted (Gard & Follett, 1968; Nagata et al., 1974; Weaver & Pollard, 1973). Chemical composition of these minerals (Caillere & Henin, 1961a, 1961b; Mumpton & Roy, 1958; Weaver & Pollard, 1973) is rather close to some members of the smectite family.

Palygorskite contains eight tetrahedral positions and five octahedral positions per unit cell of which 4 to 4.25 of the latter positions are filled (Henin & Caillere, 1975). Aluminum occupies 1.13 to 2.34 of these five sites or between 28 to 59% of the occupied sites, while Mg occupies from 29 to 76% of the occupied sites. Tetrahedrally coordinated Al ranges from 0.01 to 0.69 per eight possible positions (Weaver & Pollard, 1973).

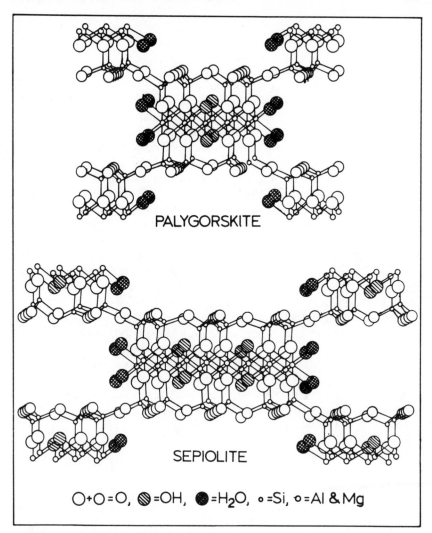

Fig. 13-1. Schematic structure of palygorskite (after Bradley, 1940) and sepiolite (after Brauner & Preisinger, 1956).

Elemental analyses of sepiolite indicate Mg occupies about eight octahedral positions which are all the sites allotted by the Brauner and Preisinger (1956) model and all but one vacant site in the Nagy and Bradley (1955) model, thus about 90 to 100% of the occupied octahedral positions are filled with Mg. Tetrahedrally coordinated Al ranges from 0.04 to 1.05 per 12 possible positions (Weaver & Pollard, 1973).

Substitution in the tetrahedral sheet remains relatively unimportant in both of these minerals (Henin & Caillere, 1975). This deficit of charge created by tetrahedral substitution is generally balanced by an equal replacement of Mg ions by trivalent ions in the octahedral sheet.

3. X-RAY ANALYSIS

The main X-ray diffraction maxima of palygorskite are a strong reflection at 10.5Å, and moderate reflections at 6.44, 5.42, 4.5, 3.68, 3.24, and 2.15Å, while for sepiolite a strong maximum occurs at 12.2Å, with moderate maxima at 4.5, 4.30, 4.0, 3.8, 3.4, and 3.2Å, and diffuse maxima at 7.6 and 5.05Å (Fig. 13-2). X-ray diffraction maxima do not vary with changes in relative humidity or the addition of organic polar molecules.

Heating palygorskite at 210C (483K) for 1 hour decreased the intensity of the reflections at 10.5, 4.5, and 3.23Å, but the 3.68Å reflection increased in intensity (Hayashi et al., 1969). New reflections also appeared at 9.2 and 4.7Å. Further heat treatment to 350C (653K) markedly decreased the 10.5 and 3.23Å reflections, while reflections at 9.2 and 4.7Å further increased. Heating to 600C (873K) completely eliminated the 10.5Å reflection, while the 9.2Å reflection decreased in intensity and the 6.4, 5.4, and 4.5Å reflections increased in intensity. Peak enhancements from rehydrating palygorskite were evident after heating at 350C (623K), but were lost after heating above 540C (813K).

Heating sepiolite at 250C (623K) for 1 hour decreased the intensity of the reflections at 12.2, 4.5, 3.8, and 3.4Å while new reflections appeared at 10.4 and 8.2Å (Hayashi et al., 1969). Further heating to 450C (723K) increased the reflections at 10.4, 8.2, and 4.3Å. These new peaks are persistent up to 600C (873K) for 120 hours (Nagata et al., 1974). The effects of rehydration were evident after heating at 450C (723K) but were unrecognized after heating above 610C (883K).

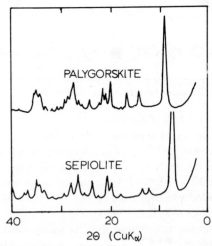

Fig. 13-2. X-ray diffractograms of palygorskite and sepiolite (modified from Hayashi et al., 1969).

4. THERMAL ANALYSIS

Thermogravimetric analysis of palygorskite and sepiolite (Fig. 13-3) show dehydration proceeding in four distinct steps (Hayashi et al., 1969). These steps may be associated with the loss of (i) hygroscopic water, (ii) zeolitic water in structure channels, (iii) structural water bound on edges of octahedral sheets, and (iv) hydroxyl groups associated with octahedral sheets.

Differential thermal analysis curves (Fig. 13-3) for palygorskite have four endotherms at 80-210C (353-483K), 210-325C (483-598K), 350-540C (623-813K), and 690-770C (963-1043K) with an exotherm at about 925C (1198K) (Hayashi et al., 1969). Sepiolite also had four endotherms at 90-200C (363-473K), 325-400C (598-673K), 450-580C (723-853K), and 750-810C (1023-1083K) with an exotherm at about 820C (1093K) (Hayashi et al., 1969). Thermogravimetric analyses indicate an initial rapid loss in weight for both minerals (Fig. 13-3). With increased temperature palygorskite loses more weight at lower temperatures than sepiolite. The suggested steps in the dehydration of sepiolite include loss of hygroscopic water, followed by zeolitic water, and then structural water in two distinct steps. The first dehydration step is rapid and produces a rotation in the original structure, while the second dehydration step is more gentle and results in further distortion of the structure. Further heating results in the loss of structural hydroxyls (Nagata et al., 1974).

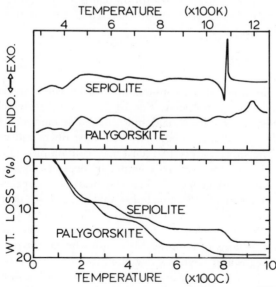

Fig. 13-3. Differential thermal analysis curves and thermal gravimetric curves of palygorskite and sepiolite (modified from Hayashi et al., 1969).

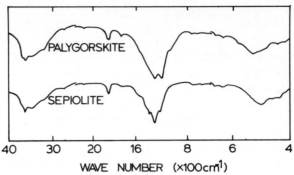

Fig. 13-4. Infrared absorption spectra of palygorskite and sepiolite (modified from van der Marel, 1966).

5. INFRARED ANALYSIS

Palygorskite shows sharp bands in infrared spectra (Fig. 13-4) as follows: 3609 cm^{-1} with shoulders at 3685 cm^{-1} and 3645 cm^{-1}, 3533 cm^{-1} with a shoulder at 3573 cm^{-1}, and 1650 cm^{-1} with a shoulder at 1665 cm^{-1}. Three broad absorption bands also appear with maxima near 3350, 3260, and 3200 cm^{-1} (Hayashi et al., 1969). An absorption band at 1198 cm^{-1} has been assigned to a Si-O vibration (Mendelovici, 1973) and may be characteristic of palygorskite.

Sepiolite displays infrared absorption bands at 3685, 3617 cm^{-1} with a shoulder at 3645; 3571, 3350, 3200, and 1660 cm^{-1} with a shoulder at 1625 cm^{-1} (Fig. 13-4).

6. OPTICAL PROPERTIES

The fibrous nature of these minerals is clearly evident in electron micrographs and is a diagnostic property (Fig. 13-5). The fibers, which result from bundles of rods, are relatively rigid and oriented at random. Two varieties of palygorskite can be identified: one with long fibers and the other with short fibers. The short variety ($\sim$1 μm) may contain more Fe and is more common in soils and the Georgia commercial deposits. Sometimes the individual fibers are bent giving an aggregate mass that appears interwoven. Caldwell and Marshall (1942) concluded that the coarser particles of palygorskite were only bundles of smaller units. Two varieties of sepiolite can also be distinguished by electron microscopy. The alpha variety contains bundles of rods or laths of considerable length whereas the beta variety consists of dense aggregates having a few elongated particles (Marshall, 1964). Robertson (1957) gave the mean dimensions of a sepiolite fiber as 8,000 by 250 by 40Å, and similar dimensions of a palygorskite fiber were described by Haden (1963).

Fig. 13-5. Electron micrographs of (*upper*) palygorskite from Attapulgus, Ga., USA, and (*lower*) Sepiolite from Vallecas, Spain (extracted from Beutelspacher & van der Marel, 1968).

B. Natural Occurrence

1. GEOGRAPHIC EXTENT

Commercial deposits of both palygorskite and sepiolite occur through-out the world associated with nonclastic (mostly chemically precipitated) sediments such as carbonatic rocks, opal, chert, and phosphates. Extensive deposits occur in Miocene-age marine deposits (Hawthorn) of Georgia and Peninsular Florida; the Yucatan Peninsula; the southwestern United States; South Africa (Transvaal); Soviet Union (Ukraine), and Australia (Queensland) as reported by Isphording (1973).

Sedimentary deposits that contain palygorskite, or sepiolite, or both, have been described as forming or formed in normal marine waters, in marine and nonmarine hypersaline waters and in fresh water lakes (Isphording, 1973). The most extensive occurrence of these minerals is in marine sediments (Millot, 1970) with locally important occurrences of lacustrine and hypersaline origin.

2. SOIL ENVIRONMENT

The occurrence of either palygorskite or sepiolite in sola is relatively rare, occurring essentially in alkaline conditions and in the presence of salts and free Si. An aluminous sepiolite was identified in a poorly drained Solonchak of Australia which was developed in marl (Rogers et al., 1956). Although total chemical analysis indicated this species was similar to paly-gorskite, it gave instead the typical 12.2Å first-order reflection of sepiolite. Muir (1951) noted the presence of palygorskite in Syrian Desert soils (Orthids). Vanden Heuvel (1966) reported both sepiolite and palygorskite in the calcareous C horizons of a Paleargid near Las Cruces, New Mexico. Sepiolite and palygorskite were found in caliche (Cca and Ccam) horizons as well as in the sola of many calcareous soils on the Southern High Plains in west Texas and eastern New Mexico (Bingham, 1973[2]; McLean et al., 1972). Palygorskite has been reported in arid soils of Egypt (Elgabaly, 1962), Iraq (Al-Rawi et al., 1969; Barzanji, 1975), Israel (Barshad et al., 1956; Yaalon, 1955) and the Mediterranean (Michaud et al., 1946). Palygorskite occurred in the C horizons of freely drained soils developed in the Hawthorn (Miocene) sediments of peninsular Florida (F. G. Calhoun & L. W. Zelazny, unpublished data). Sepiolite was also identified in the mineral portion of sapric horizons in a Histosol south of Lake Okeechobee in Florida.[3] Millot et al. (1969) noted a close relationship between palygorskite occurrence and calcic horizons in Morocco.

---

[2]J. M. Bingham. 1973. The pedogenic alteration of acicular clay minerals on the semi-arid Texas High Plains. M.S. thesis. Texas Tech University, Lubbock.

[3]Soil Science Society of America. 1972. Tour guide for field study and discussion of selected south Florida soils, 1 Nov. 1972. V. W. Carlisle, tour chairman, Dept. of Soils, Univ. of Florida, Gainesville.

Occurrences of palygorskite and sepiolite in soils appear to be largely inherited from parent material rather than of pedological significance, although Vanden Heuvel (1966) proposed a pedogenic origin. The absence or degraded state of these minerals in the soil environment, even from soils developing from geological materials rich in one or both of these minerals, would suggest a rather high susceptibility to chemical and physical alterations by weathering processes.

Palygorskite and sepiolite are chemically precipitated and crystallized in alkaline sediments with significant quantities of Si and Mg. According to Millot (1970) these minerals would only be stable in this environment, but quite susceptible to decomposition under the leaching environment of most soils. He further contends that even in black tropical clays (tropical Vertisols) palygorskite is extremely unstable outside its sedimentary environment and is transformed quickly into montmorillonite. The persistence of either palygorskite or sepiolite in the sola is dependent upon the maintenance of low H_3O and Al but high Si and Mg concentrations. Further, a relatively closed soil system would be a prerequisite with minimal leaching and little or no influx of Al. This is nearly impossible on terrestrial positions, that would qualify as closed systems, since there would be constant influx of aluminosilicates by water and aerosolic dusts into depressional positions. For these reasons, the identification of sepiolite and occasionally palygorskite in Georgia Paleudults (Fiskell & Perkins, 1970) is both problematic and questionable.

C. Equilibrium Environment and Conditions for Synthesis

In his analysis of geochemical sequences on the Tertiary basins of West Africa, Millot (1970) noted a succession of montmorillonite, palygorskite, and sepiolite with distance from the coastline seaward. He also observed a decreasing Al/Mg ratio with changes from montmorillonite to sepiolite. This ratio, and thus the mineralogical form was in response to the sedimentary environment which varies regularly seaward. Quoting Millot (1970, p. 272):

So long as aluminum is important, it orients the neoformation (direct precipitation) towards montmorillonite. When aluminum decreases, montmorillonite is relayed by attapulgite (palygorskite), and when magnesium occurs alone, sepiolite forms. . . . the most aluminous minerals the most littoral and the most magnesian are the most remote . . . it is alumina, the least soluble element, that becomes exhausted in the direction of the open sea.

Isphording (1973) discussed the origin of the Georgia, Florida, and Yucatan Peninsula palygorskite-sepiolite deposits. He noted that popular hypotheses in past geological literature attributed origin to (i) alteration of volcanic ash, (ii) diagenesis of montmorillonite, and (iii) direct crystallization. Isphording strongly favored the last hypothesis in explaining the origin of the aforementioned deposits. His argument is substantiated by basinward increase in MgO/Al_2O_3 ratios using the rationale offered by Millot (1970).

Marine palygorskite-sepiolite deposits almost invariably occur seaward from terrestrial areas subjected or previously subjected to "tropical" weathering. Intense desilication ("laterization") of the pedosphere adjacent to the Mg-rich oceans would logically produce the Si-Al gradients which in turn results in the mineral sequence of marine sediments observed by Millot. A simplified model (Fig. 13-6) shows terrestrial desilication and concomitant marine silication with the observed principal mineral assemblages associated with the pedosphere and nonexhumed marine sediments.

Isphording (1973) provided an excellent summary of the literature relating to the synthesis of sepiolite. Laboratory synthesis at room temperature was easily obtained by treating saturated solutions of amorphous silica with various concentrations of $MgCl_2$ to which NaOH had been added (Siffert & Wey, 1962). Additions of Al to the system inhibited formation of sepiolite (Mumpton & Roy, 1958). Wollast et al. (1968) indicated that sepiolite–sea water equilibrium was pH-related. With evaporation, the precipitation of sepiolite was as follows:

$$2Mg^{2+} + 3SiO_{2aq} + (n+2)H_2O = Mg_2Si_3O_8(H_2O)_2 + H^+$$

They concluded that (i) sepiolite formation is enhanced by Si-rich, hypersaline environments, and (ii) in the presence of Al, neither sepiolite nor palygorskite is precipitated due to incorporation of Mg into aluminosilicate structures.

In summary, the formation and persistence of palygorskite and sepiolite would be encouraged by high pH (> 8-10), high activities of Si^{4+} and Mg^{2+}, and low to nonexistent Al^{3+} activity.

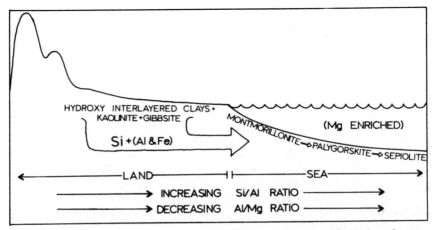

Fig. 13-6. Schematic model of continental desilication and resultant formation of montmorillonite-palygorskite-sepiolite minerals in off-shore ocean sediments.

D. Chemical Properties

Cation exchange capacities (CEC) most characteristic of palygorskite and sepiolite range from 5 to 30 and 20 to 45 meq/100 g, respectively (Weaver & Pollard, 1973). Higher values are sometimes reported but are thought to be due to montmorillonite contaminants. The CEC generally varies little with pH or particle size. Dissociation of K and Ca from palygorskite is reported to be somewhat greater than from any other clay minerals examined (Marshall & Caldwell, 1947).

Sepiolite has been reported to decompose readily in an acid medium whereas palygorskite was more resistant (Marshall, 1964). When palygorskite and sepiolite were treated with excess HCl, reaction rate constants obtained were $Mg > Fe > Al$ (Abdul-Latif & Weaver, 1969). The reaction rate constant of Mg in sepiolite was about 240 times greater than that for palygorskite.

E. Physical Properties

Due to the open packing of fibers, the external surface area of these minerals may be quite large. Grim (1968) reported a surface area of 140 and 392 m^2/g for palygorskite and sepiolite, respectively. Reported surface areas of sepiolite have ranged from 330 m^2/g when measured by hexane (Robertson, 1957) to 60 m^2/g when measured by cetyl pyridinium bromide (Greenland & Quirk, 1964). Using CO_2 adsorption on sepiolite at $-80C$ (193K), Dandy (1969) determined a BET area of 296 m^2/g and an area from the Kaganer equation of 268 m^2/g. Escard (1949) and Kinter and Diamond (1960) observed a surface area of 150 m^2/g for palygorskite by adsorption of N_2 at low temperatures. Adsorption of glycerol provided similar surface area of 165 m^2/g when measured at low temperatures, but increased to 335 m^2/g after heating the mineral to 600C (873K).

The fibrous nature of sepiolite also resulted in the formation of highly porous aggregates in a soil sample from Australia. Rogers et al. (1956) reported an exceptionally high water retention for clay from the Tintinara soil. Water content at 15 bars was 1.32 g H_2O/g of clay and 0.33 g H_2O/g of soil.

F. Quantitative Analysis

Quantitative analysis of palygorskite and sepiolite may be obtained from measurement of peak intensity at 10.5 and 12.2Å, respectively. Palygorskite analysis may be complicated by the presence of halloysite, mica or montmorillonite. Nathan (1970) provided an improved X-ray method for detecting small quantities of palygorskite in clay mineral mixtures. Heating the sample to 150C (423K) produced a marked increase in the peak at 10.5Å; no other clay minerals except smectites and halloysite have reflections in this area that are affected by low temperature heating. The smectite interference could be eliminated by saturating with a quaternary ammonium compound. It is suggested that the magnitude of this change in X-ray reflection intensity with heating can be used to quantify palygorskite.

Sepiolite readily decomposes in acidic conditions whereas palygorskite is more resistant. It is suggested that a differential weight before and after treatment with NaOAc buffered at pH 5.0 be corrected for quantitative sepiolite analysis. Carbonates are usually associated with this mineral and must be analyzed to provide a corrected weight loss equal to that obtained only from sepiolite. However, sepiolite occurring in soil environments appears to be more resistant to acid decomposition and its presence has been detected in soil clays pretreated for carbonate removal.

Caldwell and Marshall (1942) observed that palygorskite was only dispersed by Na-silicate or citrate and not by oxalate or carbonate. Therefore, selected dispersion may provide a means of separation and quantification of these minerals although this technique has not been proven for mixtures of soil minerals. However, it should make one aware of possible problems in the identification of palygorskite and sepiolite in soil clays routinely dispersed with carbonates.

III. TALC AND PYROPHYLLITE

A. Structural Properties and Mineral Identification

1. STRUCTURE

Talc is a trioctahedral, hydrated Mg-silicate composed of two Si tetrahedral sheets with a central octahedral sheet in which all octahedral positions are occupied by Mg (Fig. 13-7). Pyrophyllite is a dioctahedral aluminosilicate with a crystal structure that is similar to talc, but only two of every three octahedra are occupied by Al. These minerals are the simplest members of the 2:1 type phyllosilicates. Each layer is bound to another by relatively weak van der Waals forces rather than through the stronger chemical bonds resulting from electrostatic forces that occur in other phyllosilicates.

2. ELEMENTAL COMPOSITION

Talc has an ideal formula of $Mg_3Si_4O_{10}(OH)_2$ while pyrophyllite has an ideal formula of $Al_2Si_4O_{10}(OH)_2$. Examples of chemical compositions for talc and pyrophyllite are given by Deer et al. (1962); Rayner and Brown (1973), and Haas and Holdaway (1973). It should be observed that these chemical compositions deviate somewhat from the idealized formulas.

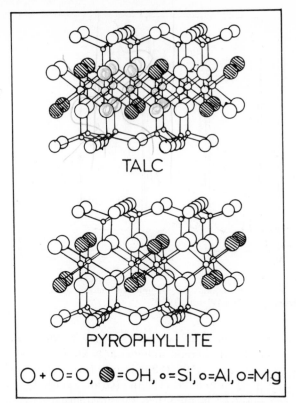

Fig. 13-7. Schematic structure of talc and pyrophyllite.

3. X-RAY ANALYSIS

Talc and pyrophyllite give first-order diffraction spacings of 9.3 and 9.2Å, respectively (Fig. 13-8). These peaks on an X-ray diffractogram are sufficiently unique, in comparison to the more common layer-silicates in soils, to provide positive identification of either of these species. They may also be differentiated on the basis of the 060 reflection which occurs at 1.52Å for talc and 1.49Å for pyrophyllite. Strong diffraction maxima also occur at 3.10Å for talc and 3.07Å for pyrophyllite. X-ray diffractograms of these minerals are unaffected by cation saturation, organic solvation, or heat treatment of the minerals.

4. THERMAL ANALYSIS

Differential thermal analysis as reported by Grim (1968) shows that a talc from Vermont displayed only an endothermic peak between 900–1000C (1173–1273K) (Fig. 13-9) and exhibits very little variability with other samples (Mackenzie & Caillere, 1975). A North Carolina pyrophyllite displayed

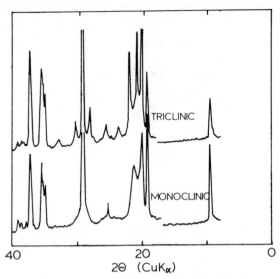

Fig. 13-8.   X-ray diffractograms of triclinic and monoclinic forms of pyrophyllite (modified from Brindley & Wardle, 1970).

three distinct endotherms: (i) an intense dehydration peak at $\sim$ 175C (448K), with a small endothermic shoulder at $\sim$ 200C (473K); (ii) a modest dehydroxylation peak at $\sim$ 550C (823K); and, (iii) a relatively broad, intermediate intensity peak at $\sim$ 850C (1123K) (Grim, 1968).   Considerable variability has been observed in published thermograms for pyrophyllite with

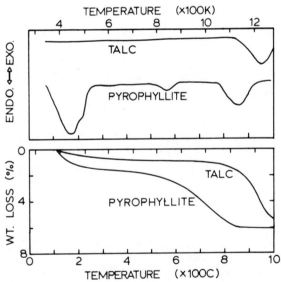

Fig. 13-9.   Differential thermal analysis curves (modified from Grim, 1968), and thermal gravimetric curves (modified from Page, 1943) of talc and pyrophyllite.

some samples showing broad single peaks in the 600–800C (873–1073K) region and other samples broad double peaks occur at 600–800C (873–1073K) and 800–850C (1073–1123K) (Mackenzie & Caillere, 1975).

Thermogravimetric analysis for talc indicates a gradual 1% weight loss to 800C (1073K) and then an additional rapid 4.2% weight loss to 1000C (1273K) (Page, 1943). Pyrophyllite provides a gradual 0.8% weight loss to 450C (723K) and then a more rapid 5.3% weight loss to 800C (1073K) (Fig. 13-9).

5. INFRARED ANALYSIS

Pyrophyllite infrared absorption spectra show a distinct O–H stretching band (Fig. 13-10) at 3675 cm^{-1} with a shoulder at 3410 cm^{-1} and an Al–OH bending band at 950 cm^{-1} (Farmer & Russell, 1964). The three sharp bands at 850, 832, and 810 cm^{-1} present in the spectra reported by Oinuma and Hayashi (1968) are not evident in the infrared pattern given by Farmer and Russell (1964). The Si–O–Si stretching frequency of pyrophyllite occurs at about 1060 cm^{-1} (Farmer & Russell, 1964). Talc displays an O–H stretching band at 3675 cm^{-1}. Absorption bands of very strong intensity at 467 cm^{-1} and 452 cm^{-1} for Mg–O bending were reported for talc by Farmer (1959) and for synthetic talc by Wilkins and Ito (1967). Strong absorption bands also occur at 690 and 668 cm^{-1} (Oinuma & Hayashi, 1968).

6. OPTICAL PROPERTIES

Electron micrographs for talc and pyrophyllite (Fig. 13-11) show platy structures with distinct angular edges. Interference figures are sometimes evident (Beutelspacher & van der Marel, 1968).

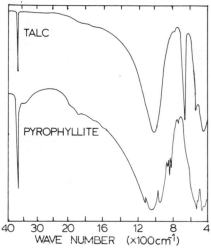

Fig. 13-10. Infrared absorption spectra of talc and pyrophyllite (modified from Oinuma & Hayashi, 1968).

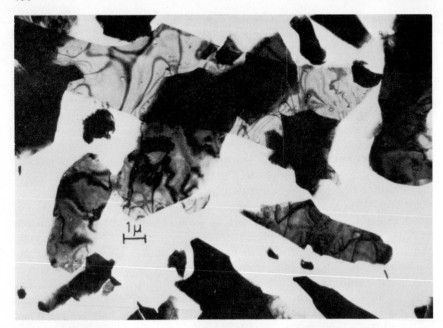

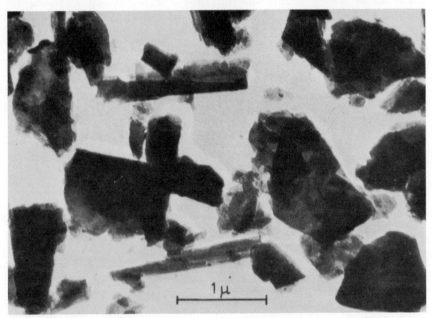

Fig. 13-11. Electron micrographs of (*upper*) pulverized sample of pyrophyllite from
North Carolina, USA, and (*lower*) pulverized sample of talc from Transvaal, South
Africa (extracted from Beutelspacher & van der Marel, 1968).

B. Natural Occurrence

1. GEOGRAPHIC EXTENT

Talc and pyrophyllite are secondary minerals formed through the alteration of Mg- and Al-silicate primary minerals. Talc may occur as pseudomorphs after olivine, pyroxenes, and amphiboles. More commonly talc is
found as a product of hydrothermal alteration of ultrabasic rocks and low
grade metamorphism of siliceous dolomites (Deer et al., 1962). Talc is a
rather common mineral compared to pyrophyllite which is generally crystallized through the hydrothermal alteration of feldspars. Commercial talc deposits in the United States are located in association with the Appalachian
Mountains (Dana, 1959). Significant pyrophyllite deposits are located in
North Carolina, Utah, and Japan.

2. SOIL ENVIRONMENT

A 9.3Å X-ray peak, which is attributed to the presence of pyrophyllite,
has often been identified in several Oxisols of the Llanos Orientales of Colombia (Cortez et al., 1973; Guerrero, 1971[4]; Leon, 1964; Mejia, 1975[5]; and
Weaver, 1974). This peak occurred in both the silt (2–50 μm) and clay ($<$ 2
μm) fractions in fairly uniform amounts throughout the soil profile. Minor
amounts of pyrophyllite have also been identified in two Paleudults and two
Dystropepts in the upper Amazon basin of Colombia (Benavides, 1973[6]) and a
typic Eutrustox from Brazil (Moura Filho & Buol, 1972). On a toposequence
of soils in the Colombian Llanos, preliminary research indicated that pyrophyllite occurrence is independent of landscape position (S. W. Buol, personal
communication). The soil moisture regimes ranged from ustic to udic; however, pyrophyllite occurred in greater amounts in the ustic soils of the Llanos.
Soil temperature regimes in both cases were isohyperthermic. The soils were
developing in weathered alluvial terrace deposits derived from the Andes
mountains. Guerrero proposed either that the formation of pyrophyllite occurred during pedogenesis or it was inherited from the initial material transported from the eastern Cordillera. Pyrophyllite occurred predominantly in
association with kaolinite and hydroxy-interlayered vermiculite in the clay
fractions. In the coarse silt fraction (5–20 μm), it occurred with abundant
quartz and small quantities of kaolinite, hydroxy-interlayered vermiculite,

---

[4]R. Guerrero. 1971. Soils of the Colombian Llanos Orientales—composition and
classification of selected soil profiles. Ph.D. thesis. North Carolina State Univ., Raleigh.

[5]L. Mejia. 1975. Characteristics of a common soil toposequence of the Llanos
Orientales of Colombia. M.S. thesis. North Carolina State Univ., Raleigh.

[6]S. T. Benavides. 1973. Mineralogical and chemical characteristics of some soils of
the Amazonia of Colombia. Ph.D. thesis. North Carolina State Univ., Raleigh.

feldspars, mica, interstratified vermiculite-mica, smectite, and vermiculite. Benavides reported soil SiO_2/Al_2O_3 ratios of 3.2–4.4 and deferrated clay SiO_2/Al_2O_3 ratios of 1.5–3.6 in the pyrophyllite containing pedons of the Colombian Amazon. Goosen (1971) tentatively identified 9.2, 4.7, and 3.11Å reflections by X-ray diffraction as either talc or pyrophyllite in the clay fractions of several Colombian Llanos soils.

Mineralogists have expressed interest in the fact that kaolinite (a 1:1 type clay mineral with little or no ionic substitution) commonly occurs in sediments and soils while pyrophyllite (a 2:1 type clay mineral with little or no ionic substitution) is seldom found in similar types of environments (Millot, 1970). Paragenesis theory would predict the reaction of kaolinite with quartz to form pyrophyllite in sedimentary rocks, but this has not been authenticated. Velde (1968) suggests that reduction of Fe^{3+} to Fe^{2+} during the decomposition of organic material in sediments would enhance the formation of illite and chlorite thus suppressing pyrophyllite formation.

Talc has been identified in several geologic environments. Positional occurrence includes those sediments of lakes and ocean floors interfacing with saline water (Siever & Kastner, 1967) and sites undergoing alteration of serpentines (Chidester, 1968). The occurrence of talc in a pedologic environment is scarce with the only known reports occurring for some Colombian soils (Leon, 1964) and in reduced tideland sediments along the shore of San Pablo Bay, California (Lynn & Whittig, 1966). No explanation was provided for the observed enhancement of the X-ray peak for talc with incubation of these tideland sediments in a reducing environment for 21 weeks.

C. Equilibrium Environment and Conditions for Synthesis

Pyrophyllite is easily synthesized from Al_2O_3–SiO_2–H_2O systems high in Si and is stable between 420–575C (693–848K) under varying water pressures (Deer et al., 1962). Stability limits of pyrophyllite in Al_2O_3–SiO_2–H_2O systems as calculated by Haas and Holdaway (1973), extend from 345 to 380C (618 to 653K) at 2 kbars and from 400 to 420C (673 to 693K) at 7 kbars. Such a limited stability for this species explains to some degree the limited geologic occurrence.

Talc can be synthesized at all temperatures below 800C (1073K) from MgO–SiO_2 mixtures at water vapor pressures between 4 and 22 kg/m^2 (Deer et al., 1962). Stability relations for talc, chrysotile, and brucite in the MgO–SiO_2–H_2O system at 25C (298K) and 1 bar have been given by Bricker et al. (1973).

The conditions for synthesis and stability limits for pyrophyllite seem to be in conflict with its occurrence under the ambient soil climate regime of eastern Colombia. Thermochemical data would indicate considerable instability of pyrophyllite under such conditions. Deer et al. (1962) contend that pyrophyllite is the least stable of the 2:1 type sheet silicates on the basis of its dehydration characteristics upon heating.

D. Chemical Properties

The lack of a layer charge precludes chemical reactivity for talc or pyro-phyllite. The existence of a CEC of < 1 meq/100 g for either talc or pyro-phyllite provides a possible measure of the quantity of edge charge in at least 2:1 type phyllosilicates (Grim, 1968). The difference in CEC between these minerals and kaolinite has been used to predict some substitution in kaolinite or the occurrence of a charged contaminant in kaolinitic deposits.

E. Physical Properties

Both these minerals have prominent basal cleavages as a result of the weak bonds between adjacent layers. Consequently, pyrophyllite and talc have a softness and flexibility caused by ready sliding of successive layers over one another.

F. Quantitative Determination

Estimates for total quantity of either talc or pyrophyllite may be made by measuring the intensity or area of the basal (001) peak at 9.3 and 9.2Å, respectively. Thermal or infrared analysis techniques could be proposed, but have not been substantiated for soil systems.

IV. ZEOLITES

A. Structural Properties and Mineral Identification

1. STRUCTURE

Zeolites include a large group of hydrated aluminosilicates consisting of a tetrahedral framework of O atoms surrounding either a Si or an Al atom ex-tended in a three-dimensional network, which typically provides structural channels. Each O atom is shared between two tetrahedra with no mobile anions present. The resulting structure has a net negative charge due to the presence of Al-centered tetrahedra which are counter-balanced by the pres-ence of alkali and alkaline earth cations within the existing pores. An addi-tional feature of the zeolite group is the presence of water molecules within the structural channel. The water is relatively loosely bound to the frame-work and cations, and like the cations can be replaced or removed without disrupting framework bonds. However, it appears that the cations and water molecules orient themselves in preferred arrangements with respect to the framework. The cation exchange and amount of water present are also subject to steric phenomena resulting from size, number, and disposition of the channels provided by the framework.

The structural channels of the various zeolites are formed by different combinations of linked tetrahedra rings, double rings or large symmetrical polyhedral units. Each ring is composed of either 4, 5, 6, 8, or 12 tetrahedra with specific combinations present within a given zeolite, thereby providing for channels or restrictions of known size. Specific structures of selected zeolites are given in Deer et al. (1963); and Meier and Olson (1971); and those for mordenite and analcite are provided in Fig. 13-12. Greater numbers of tetrahedra per ring result in wider channels. The size of the cation that can be introduced into the structure is dependent on the width of the channel at its narrowest constriction. Zeolites with 8- or 12-membered rings

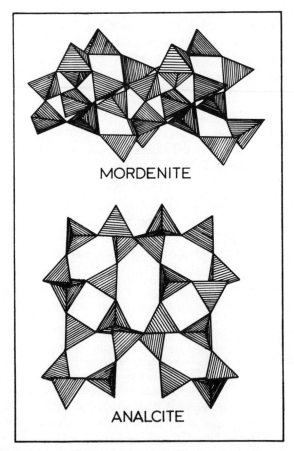

Fig. 13-12. Schematic structure of the characteristic chain of mordenite (after Meier, 1971 and extracted from Kimpe & Fripiat, 1968) and the analcite framework (extracted from Deer et al., 1963).

have channels large enough for admission of organic molecules as well as cations. Thus, zeolites can act as ionic or molecular sieves with characteristic upper limits on the size of the ion or molecule which can permeate these structural channels. However, channel width is not the only criterion for permeability since some cations can block the channels. Both ionic and molecular diffusion is also dependent on water content of the zeolite system.

2. ELEMENTAL COMPOSITION

The zeolite group has the following idealized formula: $(Na_2, K_2, Ca, Ba, Sr, Mg) [(Al, Si)O_2]_n \cdot XH_2O$. They are characterized chemically by an atomic ratio $O/(Al+Si)=2$ and a molecular ratio of $Al_2O_3/(Na_2, K_2, Ca, Ba, Sr, Mg)O=1$. The chemical formula for selected zeolites is provided in Table 13-1 and specific chemical analyses may be found in Deer et al. (1963), and in several zeolite reviews (Gould, 1971a, 1971b; Meier & Uytterhoeven, 1973). The natural range in chemical compositions for the more common zeolites is given by Sheppard (1971).

Table 13-1. Chemical composition of selected zeolites (modified from Deer et al., 1963).

Name	Chemical formula
Analcite	$Na[AlSi_2O_6] \cdot H_2O$
Ashcroftine	$KNaCa[Al_4Si_5O_{18}] \cdot 8H_2O$
Chabazite	$Ca[Al_2Si_4O_{12}] \cdot 6H_2O$
Clinoptilolite	$(Ca,Na_2)[Al_2Si_7O_{18}] \cdot 6H_2O$
Dachiaridite	$(Ca,K_2,Na_2)_3[Al_4Si_{18}O_{45}] \cdot 14H_2O$
Edingtonite	$Ba[Al_2Si_3O_{10}] \cdot 4H_2O$
Epistilbite	$Ca[Al_2Si_6O_{16}] \cdot 5H_2O$
Erionite	$(Na_2,K_2,Ca,Mg)_{4.5}[Al_9Si_{27}O_{72}] \cdot 27H_2O$
Faujasite	$(Na_2Ca)_{1.75}[Al_{3.5}Si_{8.5}O_{24}] \cdot 16H_2O$
Ferrierite	$(Na,K)_4Mg_2[Al_6Si_{30}O_{72}](OH)_2 \cdot 18H_2O$
Gismondine	$Ca[Al_2Si_2O_3] \cdot 4H_2O$
Gmelinite	$(Na_2,Ca)[Al_2Si_4O_{12}] \cdot 6H_2O$
Gonnardite	$Na_2Ca[(Al,Si)_5O_{10}]_2 \cdot 6H_2O$
Harmotome	$Ba[Al_2Si_6O_{16}] \cdot 6H_2O$
Heulandite	$(Ca,Na_2)[Al_2Si_7O_{18}] \cdot 6H_2O$
Laumontite	$Ca[Al_2Si_4O_{12}] \cdot 4H_2O$
Leonhardite	$Ca[Al_2Si_4O_{12}]3.5H_2O$
Levyne	$Ca[Al_2Si_4O_{12}] \cdot 6H_2O$
Mesolite	$Na_2Ca_2[Al_2Si_3O_{10}]_3 \cdot 8H_2O$
Mordenite	$(Na_2,K_2,Ca)[Al_2Si_{10}O_{24}] \cdot 7H_2O$
Natrolite	$Na_2[Al_2Si_3O_{10}] \cdot 2H_2O$
Phillipsite	$(\frac{1}{2}Ca,Na,K)_3[Al_3Si_5O_{16}] \cdot 6H_2O$
Scolecite	$Ca[Al_2Si_3O_{10}] \cdot 3H_2O$
Stilbite	$(Ca,Na_2,K_2)[Al_2Si_7O_{18}] \cdot 7H_2O$
Thomsonite	$NaCa_2[(Al,Si)_5O_{10}]_2 \cdot 6H_2O$

3. X-RAY ANALYSIS

X-ray diffractograms are extremely useful for identifying most zeolites. Specific peak positions and relative intensities for selected zeolites are provided in Deer et al. (1963), Sheppard and Gude (1968), Oinuma and Hayashi (1967), and in Fig. 13–13. Generally these peaks do not coincide with dif-

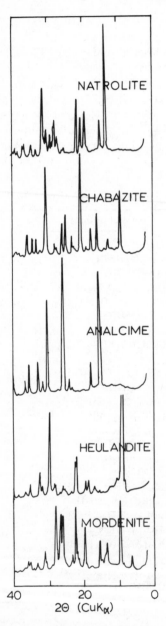

Fig. 13-13. X-ray diffractograms of selected zeolites (modified from Oinuma & Hayashi, 1967).

fraction maxima of the more common soil minerals and can therefore be diagnostic for zeolites. Specific cation saturation, solvation with an organic molecule, or heat treatment usually produce no effect on the specific zeolite framework structure although exceptions do occur. The chabazite framework is sufficiently open and rigid to accommodate Li, Na, Ag, Sr, Ba, K, NH_4, Rb, Ti, and Cs with little or no observable variation in X-ray spacings (Barrer, 1950), however, the analcite structure gave small adjustments in the framework with a similar ionic suite. Heating to 400C (673K) for 3 hours is also used to distinguish clinoptilolite from heulandite on the basis of the absence of an X-ray reflection at 8.3Å for clinoptilolite (Mumpton, 1960).

4. THERMAL ANALYSIS

Zeolites generally give off water somewhat continuously when heated, rather than in separate rates at definite temperatures (Marshall, 1949) and, unlike most soil minerals, then can usually readsorb water to the original content when again exposed to water vapor. The dehydrated zeolite usually can absorb molecules such as alcohol, NH_4, NO_2, H_2S, etc., in the place of the desorbed water (Deer et al., 1963).

Thermal analysis is useful for determining variations in composition and bond strength. Although natrolite, scolecite, and mesolite are isostructural and show markedly similar physical properties, their differential thermal analyses curves and dehydration curves (Fig. 13-14) are different due to variation in the amount of water present and its position in the structure (Peng, 1955).

5. INFRARED ANALYSIS

Infrared spectra of selected zeolites are given in Fig. 13-15 and may be found in Oinuma and Hayashi (1967) and further references given by Flanigen et al. (1971); Moenke (1974); and Ward (1971); The strongest absorption bands occur in the 1200-950 cm^{-1} region and are characterized as antisymmetric Si-O-Si and Si-O-Al stretching vibrations (Moenke, 1974). The second strongest absorption bands occur in the 550-400 cm^{-1} region and can be characterized as O-Si-O bending vibrations. Medium intensity absorption bands generally occur between 850-550 cm^{-1} and are a consequence of the polymerized structure.

Flanigen et al. (1971) proposed that the major structural groups present in zeolites could be detected from their infrared spectra. They proposed the following assignments for (i) internal tetrahedra: 1250-950 cm^{-1} for asymmetrical stretch, 720-650 cm^{-1} for symmetrical stretch, and 500-420 cm^{-1} for tetrahedral cation-O bend; and, (ii) external linkages: 650-500 cm^{-1} for double ring, 420-300 cm^{-1} for pore opening, 820-750 cm^{-1} for symmetrical stretch, and 1150-1050 cm^{-1} shoulder for asymmetrical stretch.

Zeolites generally show infrared absorption bands at about 3750, 3650, 3550 cm^{-1} which are assigned to -OH groups. The 3750 cm^{-1} band is as-

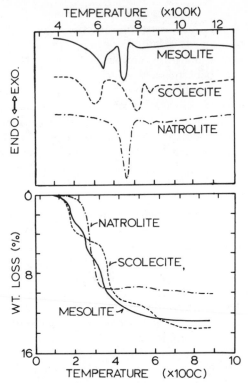

Fig. 13-14. Differential thermal analysis curves and thermal gravimetric curves of selected
zeolites (modified from Peng, 1955).

signed to external crystal –OH groups (Angell & Schaffer, 1965) while the remaining bands are associated with acidic –OH groups within the crystal
(Hughes & White, 1967; Olson & Dempsey, 1969). The 3650 cm^{-1} band has
been assigned to –OH groups in the supercages and the 3550 cm^{-1} band to
–OH groups within the cubooctahedra structure (White et al., 1967).

6. OPTICAL PROPERTIES

The refractive indices for zeolite range between 1.47–1.54 and birefringence between 0–0.015. However, the optical properties of very small zeolite crystals are not readily distinguishable with the polarizing microscope.

Electron optical data for characterization of zeolites are not extensive
in existing literature. The occurrence of zeolites in coarser size fractions
may be responsible for this observation. Zeolites can exist in a platy, blocky,
or fibrous morphology. Mumpton and Ormsby (1975) observed for many
genetic types of deposits that clinoptilolite occurred as aggregates of euhedral
blades and plates displaying characteristic monoclinic symmetry and sometimes as laths. Erionite occurred as individual needles and rods grading into

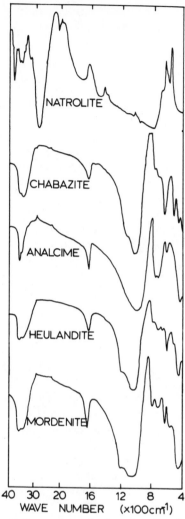

Fig. 13-15. Infrared absorption spectra of selected zeolites (modified from Oinuma & Hayashi, 1967).

flexible fibers, and as stubby bundles or sheaves of closely packed needles. Mordenite also occurred in a filiform habit consisting of reticulated masses of fibers and needles. Chabazite invariably occurred as pseudo-cubes and phillipsite as acicular crystals with many terminated by pyramidal faces. The existence of hammer-shaped crystals has been given as an indication of the presence of both offretite and zeolite L. (Kerr et al., 1970). Electron micrographs have also been used to detect the occurrence of faulting in zeolites (Kerr et al., 1970) and for other morphology transformations (Kimpe & Fripiat, 1968).

B. Natural Occurrence

1. GEOGRAPHIC EXTENT

Known commercial deposits of zeolites occur in saline-lake sediments of Tertiary age in the western United States and in bedded volcanic tuffs of central and northern Japan. Other locations include the Russian platform and Italian volcanic areas. Zeolites were previously known only as well-formed crystals in vugs and cavities of basalts and other traprock formations. However, they are now recognized as major constituents in numerous bedded pyroclastics and are accepted as some of the most widespread and abundant authigenic silicate minerals in sedimentary rocks (Mumpton & Sheppard, 1972). Zeolite occurrences have been recognized in all ages of sedimentary rocks of pyroclastic origin from late Paleozoic to Holocene. The most common zeolites in the United States are clinoptilolite, mordenite, erionite, chabazite, phillipsite, and analcime while the Japanese deposits are characterized by mordenite, clinoptilolite, analcime, heulandite and laumondite (Mumpton & Sheppard, 1972).

2. SOIL ENVIRONMENT

Analcime has been identified in a number of soils in the San Joaquin Valley of California by X-ray diffraction maxima at 5.61, 3.43, and 2.92Å (Baldar & Whittig, 1968; El-Nahal & Whittig, 1973; Kelley, 1957; and Schulz et al., 1964). These strongly alkaline soils have developed from granitic alluvium. The occurrence of analcime was further restricted to soils with a pH above 9 and containing Na_2CO_3. The analcime content correlated well with obvious deposits of Na_2CO_3 on the soil surface and could not be detected below a depth of 122 cm. It was most abundant in the fine silt (5-2 μm) and coarse clay (2-0.2 μm) fractions (Baldar & Whittig, 1968). This mineral could be readily removed below the limits of detection by X-ray diffraction analysis by pretreatment with 0.5N HCl.

The occurrence and distribution of analcime support a probable pedogenic origin. Further evidence is provided by the synthesis of analcime in soil samples extracted with 0.5N HCl and subsequently titrated with NaOH to pH 9 or above (Baldar & Whittig, 1968). Diffraction maxima corresponding to analcime were observed after the samples had been digested at 95C (368K) for only 3 days. No evidence of analcime crystallization occurred at pH 8.5, but maximum formation occurred at higher pH levels. However, the predominant zeolite species which formed in the soil extract was phillipsite. A possible explanation of this phenomenon is the existence of a lower Si/Al ratio in the soil extract (Baldar & Whittig, 1968). This Si/Al ratio ranges between 2.0-2.8 for analcime, but may extend from 1.3-3.4 for phillipsite (Hay, 1966). Travnikova et al. (1973) reported the possible pedogenic synthesis of analcime and phillipsite in soils from Russia containing free Na_2CO_3.

Clinoptilolite was identified in Solonetz and Solonetz-Solonchaks containing appreciable SO_4 and Cl and it was believed to be inherited from the parent rock. No clinoptilolite was identified in Podzolized Chernozems or Gray Forest soils developing on similar parent material, presumably because the acidity of these latter soils rendered this mineral unstable.

C. Equilibrium Environment and Conditions for Synthesis

Zeolites apparently formed from vitric volcanic materials reacting with either meteroic or connate waters of saline, alkaline lakes; however, analcime occurs only as an alteration product of pre-existing zeolites. Zeolite synthesis was related to the Si/Al ratio and the alkalinity of the mixture from which it formed (Mariner & Surdam, 1970). The siliceous zeolites, clinoptilolite, and mordenite, are the most common alteration products of Si-glass in marine and freshwater environments whereas the less siliceous zeolites, phillipsite, and erionite are the dominant alteration products in saline, alkaline lakes. Increasing the OH content in an environment favorable for zeolite formation generally produces a phase with a lower Si/Al ratio which may explain why mordenite formation is altered to analcime formation at higher pH values.

A high pH appears necessary for zeolite formation in order to sustain a high concentration of Al and Si in solution. The nature of the zeolite formed is a function of the Si/Al ratio of the solution and reactive concentrations of alkaline and alkaline earth cations present. Mariner and Surdam (1970) noted a higher rate of Al solubility than that of Si with increasing alkalinity resulting in a Si/Al ratio decreasing with increasing pH. At high concentrations of Si and Al ions in solution, hydroxy gel formation proceeds with polymerization into units of aluminosilicate chains. The charged Al species would probably be accompanied by a hydrated cation in order to maintain electrical neutrality. Mariner and Surdam (1970) considered that the formation of a hydrous cation aluminosilicate gel was essential for zeolite synthesis since it would lead to further dehydration reactions between adjacent chains providing cage-like polyhedral units which are characteristic of zeolites. With subsequent ordering, these polyhedral units are joined by shared O atoms resulting in zeolitic crystals.

D. Chemical Properties

Zeolites have a reported CEC on the order of 100–300 meq/100 g (Grim, 1968). However, the criterion of exchange is dependent on channel width, ionic or molecular diffusion, water content, and hydration. In an example of ion sieving, analcite interchanged Na freely for K, NH_4, Ag, Ti, and Rb, but provided negligible exchange between Rb and Cs (Barrer, 1950). However, chabazite provided free exchange among Li, Na, K, NH_4, Pb, Cs, Ag, Ti, Ca, Sr, and Ba whereas only Mg failed to exchange. Free exchange of very small ions such as Li and Mg apparently was not possible because of the high energy of hydration in solution and changes in degree of hydration during exchange (Barrer, 1950). The selectivity for the heavy metal ions Ag and Ti was common to all zeolites (Helfferich, 1962) and clinoptilolite has been employed for separating and immobilizing radioactive ions from waste solutions. Further examples of cation exchange on zeolites are given by Sherry (1971).

In addition to "sieve action," other ionic-size effects are also observed. Basic cancrinite completely excludes Cs even though its ionic size is considerably smaller than the zeolite channels. The channels in basic cancrinite are not interconnected and are too narrow to allow Cs to pass and exchange with other ions in the channel (Helfferich, 1962). The exchange of Na for the larger K in basic sodalite is completely regular to about 60% conversion at which point it terminates. Likewise, only about 20% of the Na in synthetic faujasite can be replaced by $N(CH_3)_4$. These phenomena result from space requirements since the total available intracrystalline void volume is too small to accommodate stoichiometric exchange (Helfferich, 1962).

Anomalous cation exchange behavior has been reported in soils containing analcime (El-Nahal & Whittig, 1973; Schulz et al., 1964). These soils provide apparent exchangeable Na percentage values of more than 100%. Excellent rice crops have also been obtained on soils having apparent exchangeable Na percentages which should limit plant growth. This dilemma resulted from the unique exchange characteristics of the zeolite analcime. Potassium or NH_4 could readily exchange with Na in analcime, although this type of exchange was restricted for Li, Mg, or Ba (El-Nahal & Whittig, 1973). Apparently only limited exchange of Na from analcime occurred in the soil solution since plant growth was not restricted. However, analcime may serve as a reservoir for Na and certain practices favoring hydrolysis, or exchange with K or NH_4 fertilizer, could release soluble Na to return a soil to a sodic condition after normal reclamation (El-Nahal & Whittig, 1973).

E. Physical Properties

Zeolites generally have a density that ranges between 2.0 and 2.3 g/cm^3 except for the Ba-rich members that range between 2.5 and 2.8 g/cm^3 (Deer et al., 1963).

Generally the zeolite structure is quite open, providing surface areas similar to 2:1 type expandable layer silicates. Ammonium Y zeolites have a surface area range of 880 to 750 m^2/g (Ward, 1970).

Water adsorption by zeolites is high on a unit cell basis. This adsorbed water is dependent on the zeolite species as well as the cation in the channel (Baur, 1964; Helfferich, 1962). Steric effects of channel width and cation size, as well as hydration energy, determine the specific quantity of water adsorbed. The term "zeolitic water" is derived from this physically adsorbed water in these minerals. Generally, zeolitic water is assumed to be removed below 200C (473K) or it can be replaced volume for volume by substances with quite different properties.

F. Quantitative Analysis

Quantitative analysis of zeolites in soil was accomplished by measurement of the intensity of X-ray diffraction maxima (El-Nahal & Whittig, 1973). However, the unique ion exchange and sieve properties of these minerals are also useful criteria. Helfferich (1964) devised an exchange system consisting of an excluded exchanging cation and anion which readily forms a precipitate with a second cation that is on the zeolite exchange position. In this technique, exchange proceeds by hydrolysis with H replacing the exchangeable cation which then precipitates and results in a pH increase No pH change will occur with materials that accept the exchanging cation. This technique has been acceptable for an Ag-analcite exchanged with CsCl solution. It is proposed that layer silicates and zeolites could be quantified by measuring the pH change in a system treated with $AgNO_3$ and subsequently exchanged with tetra-n-butylammonium bromide (Helfferich, 1964). Measurement of exclusion volume of large molecules and selective sorption of small molecules are other diagnostic methods for defining zeolites. Munson (1973) determined gas adsorption isotherms for selected zeolites using methane and chlorotrifluoromethane at elevated pressures. High heats of immersion provided by zeolites can also be used qualitatively to detect the presence of zeolites (Culfaz et al., 1973). Where knowledge of the heat of immersion of the specific zeolitic species is available and zeolite type recognized, quantitative measurements can also be made (Barrer & Cram, 1971; Culfaz et al., 1973).

Characteristically zeolites have a low particle density which can also be utilized for their quantitative analysis. A density separation in a solution of tetrabromoethane and nitrobenzene centrifuged at 500 rpm for 15 min separated cristobalite and zeolites from most of the remaining soil minerals (Henderson et al., 1972). Subsequent determination of the weight difference before and after treatment with acid would then provide a quantitative measurement of the zeolites present.

V. CONCLUSIONS

Palygorskite, sepiolite, pyrophyllite, talc, and zeolites are indeed rare minerals in the soil environment. Their occurrence is generally restricted to soils with a high pH or where the extent of acid weathering has been limited. They are commonly associated with the presence of carbonates. Although their existence in a geological setting is common, the factors of soil formation have quite often resulted in their dissolution and removal. Generally, the formation of these minerals requires a low Fe content. Sepiolite and talc require an environment that is extremely high in Mg content with a concomitant low Al content, but the converse prevails for pyrophyllite. The zeolites clinoptilolite and phillipsite are generally more acid resistant than some of the other common natural zeolites. Likewise, palygorskite is more tolerant to acid conditions than sepiolite.

Soils containing these minerals may have some unique properties. The exceptionally large water holding capacity on a clay basis for the soil reported to contain sepiolite, and the unique ion exchange properties in regard to exchangeable Na percentage values for the soil containing the zeolite, analcime, are noted as examples. Further examples are fragmentary and reportings generally scarce. Although specific occurrences of these minerals may be rare, perhaps another possible explanation for their lack of detection is the almost universal use of pretreatments for mineralogical analysis. Since palygorskite, sepiolite, and zeolites are normally in association with carbonates, the use of acid pretreatments for carbonate removal in mineralogical analysis could destroy these structures if present in soils, and explain the lack of a greater number of detections. Some zeolites also tend to occur in coarser fractions, and could be selectively ignored from analysis unless the silt and sand fractions were analyzed. Likewise, detection of these minerals in an anomalous environment should be questioned. Talc and pyrophyllite are often coated on new laboratory tubing and plastic ware, and these minerals are used as fertilizer fillers.

VI.   PROBLEMS AND EXERCISES

1) What role does pedogenetic weathering play in the distribution of world wide palygorskite-sepiolite deposits?

2) What are the conditions in soil sola which encourage the persistence of palygorskite and/or sepiolite? Would a well-drained soil from New York, developing in acid glacial outwash, be expected to have significant quantities of palygorskite and/or sepiolite? Defend your answer.

3) Why would the occurrence of pyrophyllite in soils developing under a tropical savanna or tropical rainforest be anomalous? What environmental condition would be necessary for this not to be anomalous? Would you expect to find talc under similar circumstances? Zeolites? Palygorskite-Sepiolite?

4) What influence does the presence of analcime in a soil have on: (A) The measured exchangeable Na percentage and the plant response to these saline conditions? (B) Saline reclamation procedures? (C) The exchangeable Na content of the soil after potash fertilization?

5) Develop exchange equations which could separate analcime from montmorillonite in a soil sample. Could quantitative results be obtained?

6) Through structural considerations verify the size of the channels in palygorskite and sepiolite.

7) From thermal analysis data calculate the amount of "zeolitic water" in palygorskite and sepiolite.

8) Calculate the number of terminal OH groups per gram of palygorskite with crystal sizes of 1 μm and 0.1 μm. How could these values be experimentally verified?

VII. SUPPLEMENTARY READING LISTED BY SUBJECT

Palygorskite and Sepiolite

Bonatti, E., and O. Joensuu. 1968. Palygorskite from Atlantic deep sea sediments. Am. Mineral. 53:975-983.

Brindley, G. W. 1959. X-ray and electron diffraction data for sepiolite. Am. Mineral. 44:495-500.

Caillere, S., and S. Henin. 1948. Occurrences of sepiolite in the lizard serpentines. Nature (London) 163:962.

Cannings, F. R. 1968. An infrared study of hydroxyl groups on sepiolite. J. Phys. Chem. 72:1072-1074.

Christ, C. L., P. B. Hostetler, and R. M. Siebert. 1973. Studies in the system MgO-SiO$_2$-CO$_2$-H$_2$O (111); The activity-product constant of sepiolite. Am. J. Sci. 273: 65-83.

Espenshade, G., and C. Spencer. 1963. Geology of phosphate deposits of northern Peninsular Florida. U. S. Geol. Surv. Bull. 1118.

Hathaway, J. C., and P. L. Sachs. 1965. Sepiolite and clinoptilolite from the Mid-Atlantic Ridge. Am. Mineral. 50:852-867.

Heron, S., and H. Johnson. 1966. Clay mineralogy, stratigraphy and structural setting of the Hawthorn Formation, Coosawhatchie District, South Carolina. Southeast. Geol. 7:51-63.

Kerr, P. 1937. Attapulgus clay. Am. Mineral. 22:534-550.

Papke, K. G. 1972. A sepiolite-rich playa deposit in southern Nevada. Clays Clay Miner. 20:211-215.

Parry, W. T., and C. C. Reeves, Jr. 1968a. Clay mineralogy of pluvial lake sediments, Southern High Plains, Texas. J. Sediment. Petrol. 38:516-529.

Parry, W. T., and C. C. Reeves, Jr. 1968b. Sepiolite from Pluvial Mound Lake, Lynn and Terry Counties, Texas. Am. Mineral. 53:984-993.

Rogers, L. E., A. E. Martin, and K. Norrish. 1954. The occurrence of palygorskite, near Ipswich, Queensland. Mineral. Mag. 30:534-540.

Roy, R., and E. F. Osborn. 1954. The system Al$_2$O$_3$-SiO$_2$-H$_2$O. Am. Mineral. 39:853-885.

Tien, Pei-Lin. 1973. Palygorskite from Warren Quarry, Enderby, Leicestershire, England. Clay Miner. 10:27-34.

Talc and Pyrophyllite

Fonarev, V. I. 1967. The thermodynamic constants of pyrophyllite. Geokhimiya 12: 1505-1508.

Henderson, G. V. 1970. The origin of pyrophyllite rectorite in shales of north central Utah. Clays Clay Miner. 18:239-246.

Rayner, J. H., and G. Brown. 1964. Structure of pyrophyllite. Clays Clay Miner. 13: 73-84.

Turobova, Z. V. 1966. Talc rock weathering, Nikolayev Massif, Ukranian SSR. (Translated from K voprosy o vyvetrivanii tal'kovykh porod, Sov. Geol., 1966, No. 5, p. 101-113). Int. Geol. Rev. 9(3):359-370.

Wardle, R., and G. W. Brindley. 1972. The crystal structure of pyrophyllite, 1Tc, and its dehydroxylate. Am. Mineral. 57:732-750.

Zen, E-An. 1972. Gibbs free energy, enthalpy, and entropy of the rock-forming minerals: calculations, discrepancies, implications. Am. Mineral. 57:524-553.

Zeolites

Bennett, J. M., and J. A. Gard. 1967. Non-identity of the zeolites erionite and offretite. Nature (London) 214:1005–1006.

Dempsey, E., G. H. Kühl, and D. H. Olson. 1969. Variation of the lattice parameter with aluminum content in synthetic sodium faujasites. Evidence for ordering of the framework ions. J. Phys. Chem. 73:387–390.

El-Nahal, M. A. 1968. Chemical behavior of a zeolitic sodic soil and its response to laboratory treatments. Ph.D. Thesis Diss. Abstr. 29:4473B.

Kühl, G. H. 1969. Synthetic phillipsite. Am. Mineral. 54:1607–1612.

Liou, Juhn-Guang. 1970. Stability relation of zeolites and related minerals in the system $CaO-Al_2O_3-SiO_2-H_2O$. Ph.D. Thesis Diss. Abstr. 31:1351B.

Moiola, R. J. 1970. Authigenic zeolites and K-feldspar in the Esmeralda formation, Nevada. Am. Mineral. 55:1681–1691.

Mumpton, F. A. 1973. First reported occurrence of zeolites in sedimentary rocks of Mexico. Am. Mineral. 58:287–290.

Sheppard, R. A. 1973. Zeolites in sedimentary rocks. U. S. Geol. Surv., Prof. Pap. no. 820:689–695.

Stamires, Dennis N. 1972. Properties of the zeolite, Faujasite, substitutional series: A review with new data. Clays Clay Miner. 21:379–389.

Tung, S. E. 1970. Zeolite aluminosilicate-3. J. Catal. 17:24–27.

Uytterhoeven, J. B., L. G. Christner, and W. K. Hall. 1965. Studies of the hydrogen held by solids—VIII. The decationed zeolites. J. Phys. Chem. 69:2117–2126.

Wright, A. C., J. P. Rupert, and W. T. Granquist. 1968. High and low silica faujasites: A substitutional series. Am. Mineral. 53:1293–1303.

Wright, A. C., J. P. Rupert, and W. T. Granquist. 1969. High and low silica faujasites: An addendum. Am. Mineral. 54:1484–1490.

LITERATURE CITED

Abdul-Latif, N., and C. E. Weaver. 1969. Kinetics of acid-dissolution of palygorskite (attapulgite) and sepiolite. Clays Clay Miner. 17:169–178.

Al-Rawi, A. H., M. L. Jackson, and F. D. Hole. 1969. Mineralogy of some arid and semi-arid land soils of Iraq. Soil Sci. 107:480–486.

Angell, C. L., and P. C. Schaffer. 1965. Infrared spectroscopic investigations of zeolites and adsorbed molecules. J. Phys. Chem. 69:3463–3470.

Bailey, S. W., G. W. Brindley, W. D. Johns, R. T. Martin, and M. Ross. 1971a. Clay Mineral Society report of nomenclature committee 1969–1970. Clays Clay Miner. 19: 132–133.

Bailey, S. W., G. W. Brindley, W. D. Johns, R. T. Martin, and M. Ross. 1971b. Summary of national and international recommendations on clay mineral nomenclature. Clays Clay Miner. 19:129–132.

Baldar, N. A., and L. D. Whittig. 1968. Occurrence and synthesis of soil zeolites. Soil Sci. Soc. Am. Proc. 32:235–238.

Barrer, R. M. 1950. Ion-exchange and ion-sieve processes in crystalline zeolites. J. Chem. Soc. (London) 3:2342–2350.

Barrer, R. M., and P. J. Cram. 1971. Heats of immersion of outgassed ion-exchanged zeolites. p. 105–131. In R. F. Gould (ed.) Molecular sieve zeolites-II. Adv. Chem. Ser. 102. American Chemical Society, Washington, D. C.

Barshad, I., E. Halevy, H. A. Gold, and J. Hagin. 1956. Clay minerals in some limestone soils from Israel. Soil Sci. 81:423–437.

Barzanji, A. F. 1975. The clay mineralogy of gypsiferous soils of Iraq. Int. Clay Conference Abstr. Mexico City, Mexico. p. 13–14.

Baur, W. H. 1964. On the cation and water position in faujasite. Am. Mineral. 49:697–704.

Beutelspacher, H., and H. W. van der Marel. 1968. Atlas of electron microscopy of clay minerals and their admixtures. Elsevier Publishing Company, New York.

Bradley, W. F. 1940. The structural scheme of attapulgite. Am. Mineral. 25:405-410.

Brauner, K., and A. Preisinger. 1956. Structure of sepiolite. Mineral. Petrogr. Mitt. 6: 120-140.

Bricker, O. P., H. W. Nesbitt, and W. D. Gunter. 1973. The stability of talc. Am. Mineral. 25:405-410.

Brauner, K., and A. Preisinger. 1956. Structure of sepiolite. Mineral. Petrogr. Mitt. 6: 120-140.

Bricker, O. P., H. W. Nesbitt, and W. D. Gunter. 1973. The stability of talc. Am. Mineral. 58:64-72.

Brindley, G. W., and R. Wardle. 1970. Monoclinic and triclinic forms of pyrophyllite and pyrophyllite anhydride. Am. Mineral. 55:1259-1272.

Caillere, S., and S. Henin. 1961a. Palygorskite. p. 343-353. In G. Brown (ed.) The X-ray identification and crystal structures of clay minerals. Mineralogical Society, London.

Caillere, S., and S. Henin. 1961b. Sepiolite. p. 325-342. In G. Brown (ed.) The X-ray identification and crystal structures of clay minerals. Mineralogical Society, London.

Caldwell, O. G., and C. E. Marshall. 1942. A study of some chemical and physical properties of the clay minerals nontronite, attapulgite, and saponite. Univ. Mo. Agric. Exp. Stn. Res. Bull. 354.

Chidester, A. H. 1968. Evolution of the ultramafic complexes of northwestern New England. p. 343-354. In E-an Zen, W. S. White, J. B. Hadley, J. B. Thompson (ed.) Studies of Appalchian geology northern and maritime. Interscience, John Wiley & Sons.

Cortez, R. B., J. Jimenez, and J. A. Rey. 1973. Genesis, classification y aptitud de ex- plotacion de algunos suelos de la Orinoquia y la Amozonia Colombianas. Universi- dad de Bogota. Jorge T. Lozono, Bogota, p. 185.

Culfaz, A., C. A. Keisling, and L. B. Sand. 1973. A field test for molecular sieve zeolites. Am. Mineral. 58:1044-1047.

Dana, J. D. 1959. Dana's manual of mineralogy, 17th ed. (rev. by C. S. Hurlbut, Jr.). John Wiley & Sons, New York.

Dandy, A. J. 1969. The determination of the surface area of sepiolite from carbon di- oxide adsorption isotherms. J. Soil Sci. 20:278-287.

Deer, W. A., R. A. Howie, and J. Zussman. 1962. Rock-forming minerals. John Wiley & Sons, New York.

Deer, W. A., R. A. Howie, and J. Zussman. 1963. Rock forming minerals. Vol. 4, Frame- work silicates. Longmans, Green, & Co., London.

Elgabály, M. M. 1962. The presence of attapulgite in some soils of the western desert of Egypt. Soil Sci. 93:387-390.

El-Nahal, M. A., and L. D. Whittig. 1973. Cation exchange behavior of a zeolitic sodic soil. Soil Sci. Soc. Am. Proc. 37:956-958.

Escard, M. 1949. Méthodes de mesure de la surface des particules. Bull. Groupe Fr. Argiles 1:34-39.

Farmer, V. C. 1959. The infrared spectra of talc, saponite, and hectorite. Mineral. Mag. 31:829-845.

Farmer, V. C., and J. D. Russell. 1964. The infrared spectra of layer silicates. Spectro- chim. Acta. 20:1149-1173.

Fiskell, J. G. A., and H. F. Perkins (eds.). 1970. Selected Coastal Plain soil properties. South. Coop. Ser. Bull. 148.

Flanigen, E. M., H. Khatami, and H. A. Szymanski. 1971. Infrared structural studies of zeolite frameworks. p. 201-228. In R. F. Gould (ed.) Molecular sieve zeolites-I. Advances in chemistry series 101. American Chemical Society, Washington, D. C.

Gard, J. A., and E. A. C. Follett. 1968. A structural scheme for palygorskite. Clay Miner. 7:367-369.

Goosen, D. 1971. Physiography and soils of the Llanos Orientales Colombia. Int. Inst. for Aerial Survey and Earth Sciences (ITC). Enschede, (The Netherlands).

Gould, R. F. (ed.). 1971a. Molecular sieve zeolites-I. Adv. Chem. Ser. 101. American Chemical Society, Washington, D. C.

Gould, R. F. (ed.). 1971b. Molecular sieve zeolites-II. Adv. Chem. Ser. 102. American Chemical Society, Washington, D. C.

Greenland, D. J., and J. P. Quirk. 1964. Determination of the total specific surface areas of soils by adsorption of cetyl pyridinium bromide. J. Soil Sci. 15:178-191.

Grim, R. E. 1968. Clay mineralogy, 2nd ed. McGraw-Hill Book Co., New York.

Haas, H., and M. J. Holdaway. 1973. Equilibria in the system $Al_2O_3-SiO_2-H_2O$ involving the stability limits of pyrophyllite, and the thermodynamic data of pyrophyllite. Am. J. Sci. 273:449-464.

Haden, W. L., Jr. 1963. Attapulgite: properties and uses. Clays Clay Miner. 10:284-290.

Hay, R. L. 1966. Zeolites and zeolitic reactions in sedimentary rocks. Geol. Soc. Am. Spec. Pap. 85.

Hayashi, H., R. Otsuka, and N. Imai. 1969. Infrared study of sepiolite and palygorskite on heating. Am. Mineral. 53:1613-1624.

Helfferich, F. 1962. Ion exchange. McGraw-Hill Book Co., Inc., New York.

Helfferich, F. 1964. A simple identification reaction for zeolites (molecular sieves). Am. Mineral. 49:1752-1754.

Henderson, J. H., R. N. Clayton, M. L. Jackson, J. K. Syers, R. W. Rex, J. L. Brown, and I. B. Sachs. 1972. Cristobalite and quartz isolation from soils and sediments by hydroflousilicic acid treatment and heavy liquid separation. Soil Sci. Soc. Am. Proc. 36:830-835.

Henin, S., and S. Caillere. 1975. Fibrous minerals. p. 335-349. In J. E. Gieseking (ed.) Soil components: Volume 2—Inorganic components. Springer-Verlag, New York.

Hughes, T. R., and H. M. White. 1967. A study of the surface structure of decationized Y zeolite by quantitative infrared spectroscopy. J. Phys. Chem. 71:2192-2201.

Isphording, W. C. 1973. Discussion of the occurrence and origin of sedimentary palygorskite-sepiolite deposits. Clays Clay Miner. 21:391-401.

Kelley, W. P. 1957. Adsorbed Na^+, cation exchange capacity and percentage Na^+ saturation of alkali soils. Soil Sci. 84:473-478.

Kerr, I. S., J. A. Gard, R. M. Barrer, and I. M. Galabova. 1970. Crystallographic aspects of the co-crystallization of zeolite L., offretite and erionite. Am. Mineral. 55:441-454.

Kimpe, C. F. De, and J. J. Fripiat. 1968. Kaolinite crystallization from H-exchanged zeolites. Am. Mineral. 53:216-230.

Kinter, E. B., and S. Diamond. 1960. Pretreatment of soils and clays for measurement of external surface area by glycerol retention. Clays Clay Miner. 7:125-134.

Leon, A. 1964. Estudios quimicos y mineralogicos de diez suelos Colombianos. Agric. Trop. 20:442-451.

Lynn, W. C., and L. D. Whittig. 1966. Alteration and formation of clay minerals during cat clay development. Clays Clay Miner. 14:241-248.

Mackenzie, R. C., and S. Caillère. 1975. The thermal characteristics of soil minerals and the use of these characteristics in the qualitative and quantitative determination of clay minerals in soils. p. 529-571. In J. E. Gieseking (ed.) Soil components: Volume 2—Inorganic components. Springer-Verlag. New York.

McLean, S. A., B. L. Allen, and J. R. Craig. 1972. The occurrence of sepiolite and attapulgite on the southern High Plains. Clays Clay Miner. 20:143-149.

Mariner, R. H., and R. C. Surdam. 1970. Alkalinity and formation of zeolites in saline alkaline lakes. Science 170:977-980.

Marshall, C. E. 1949. The colloid chemistry of the silicate minerals. Academic Press, Inc., Publishers, New York.

Marshall, C. E. 1964. The physical chemistry and mineralogy of soils. I. Soil materials. John Wiley & Sons, Inc., New York.

Marshall, C. E., and O. G. Caldwell. 1947. The colloid chemistry of the clay mineral attapulgite. J. Phys. Colloid Chem. 51:311-320.

Meier, W. M., and D. H. Olson. 1971. Zeolite frameworks. p. 155-169. In R. F. Gould (ed.) Molecular seive zeolites-I. Adv. Chem. Ser. 101. American Chemical Society, Washington, D. C.

Meier, W. M., and J. B. Uytterhoeven (eds.). 1973. Molecular sieves. Adv. Chem. Ser. 121. American Chemical Society, Washington, D. C.

Mendelovici, E. 1973. Infrared study of attapulgite and HCl treated attapulgite. Clays Clay Minerals 21:115-119.

Michaud, R., R. Cerighilli, and G. Dronineau. 1946. Sur les spectres de rayons X des argiles extraites de sols Méditerranéens. C. R. 222:94-95.

Millot, G. 1970. Geology of clays. Springer-Verlag, New York.

Millot, G., H. Paquet, and A. Ruellan. 1969. Néoformation de l'attapulgite dans les sols à carapaces calcaires de la Basse Moulouya (Maroc Oriental). Pédologie. 268:2771-2774.

Moenke, H. H. W. 1974. Silica, the three-dimensional silicates, borosilicates and beryllium silicates. p. 365-382. In V. C. Farmer (ed.) The infrared spectra of minerals. Mineralogical Society Monograph 4. London.

Moura Filho, W., and S. W. Buol. 1972. Studies of a Latosol Roxo (Eutrustox) in Brazil: clay mineralogy. Experientiae 13:218-234.

Muir, A. 1951. Notes on Syrian soils. J. Soil Sci. 2:163-183.

Mumpton, F. A. 1960. Clinoptilolite redefined. Am. Mineral. 45:351-369.

Mumpton, F. A., and W. C. Ormsby. 1975. Morphology of sedimentary zeolites by scaning electron microscopy. 1975. International Clay Conf. Abstr. (Mexico City, Mexico) p. 211-214.

Mumpton, F., and R. Roy. 1958. New data on sepiolite and attapulgite. Clays Clay Miner. 5:136-143.

Mumpton, F. A., and R. A. Sheppard. 1972. Zeolites. Geotimes. 17:16-17.

Munson, R. A. 1973. Properties of natural zeolites. U. S. Bur. Mines Rep. Invest., no. 7744. 13 p.

Nagy, B., and W. F. Bradley. 1955. The structural scheme of sepiolite. Am. Mineral. 40: 885-892.

Nagata, H., S. Shimoda, and T. Sudo. 1974. On dehydration of bound water of sepiolite. Clays Clay Miner. 22:285-293.

Nathan, Y. 1970. Notes: An improved X-ray method for detecting small quantities of palygorskite in clay mineral mixtures. Clays Clay Miner. 18:363-365.

Oinuma, K., and H. Hayashi. 1967. Infrared absorption spectra of some zeolites from Japan. J. Tokyo Univ. (Nat. Sci.) 8:1-12.

Oinuma, K., and H. Hayashi. 1968. Infrared spectra of clay minerals. J. Tokyo Univ. 9: 57-98.

Olson, D. H., and E. Dempsey. 1969. The crystal structure of the zeolite hydrogen faujasite. J. Catal. 13:221-231.

Page, J. B. 1943. Differential thermal analysis of montmorillonite. Soil Sci. 56:273-283.

Peng, C. J. 1955. Thermal analysis of the natrolite group. Am. Mineral. 40:834-856.

Rayner, J. H., and G. Brown. 1973. The crystal structure of talc. Clays Clay Miner. 21: 103-114.

Robertson, R. H. S. 1957. Sepiolite: a versatile raw material. Chem. Ind. (NY) 1492-1495.

Rogers, L. E. R., J. P. Quirk, and K. Norrish. 1956. Occurrence of an aluminum-sepiolite in a soil having unusual water relationships. J. Soil Sci. 7:177-184.

Schulz, R. K., R. Overstreet, and I. Barshad. 1964. Some unusual ionic exchange properties of sodium in certain salt-affected soils. Soil Sci. 99:161-165.

Sheppard, R. A. 1971. Zeolites in sedimentary deposits of the United States—A review. p. 279-310. In R. E. Gould (ed.) Molecular sieve zeolites-I. Adv. Chem. Ser. 101. American Chemical Society, Washington, D. C.

Sheppard, R. A., and A. J. Gude. 1968. Distribution and genesis of authigenic silicate minerals in tuffs of Pleistocene Lake Tecopa, Inyo County, California. Geol. Surv. Prof. Pap. 597. U. S. Government Printing Office. Washington, D. C. p. 1-38.

Sherry, H. S. 1971. Cation exchange on zeolites. p. 350-378. In R. F. Gould (ed.) Molecular sieve zeolites-I. Adv. Chem. Ser. 101. American Chemical Society, Washington, D. C.

Siever, R. A., and M. Kastner. 1967. Mineralogy and petrology of some Mid-Atlantic Ridge sediments. J. Mar. Res. 25:263-278.

Siffert, B., and R. Wey. 1962. Synthese d'une sepiolite à temperature ordinaire. C. R. 254:1460-1462.

Travnikova, L. S., B. P. Gradusov, and N. P. Chizhikova. 1973. Zeolites in some soils. Sov. Soil Sci. 5:251.

Vanden Heuvel, R. C. 1966. The occurrence of sepiolite and attapulgite in the calcareous zone of a soil near Las Cruces, New Mexico. Clays Clay Miner. 13:193-207.

van der Marel, H. W. 1966. Quantitative analysis of clay minerals and their admixtures. Contrib. Mineral. Petrol. 12:96-138.

Velde, B. 1968. The effect of chemical reduction on the stability of pyrophyllite and koalinite in pelitic rocks. J. Sediment. Petrol. 38:13–16.

Ward, J. W. 1970. Thermal decomposition of ammonium Y zeolite. J. Catal. 18:348–351.

Ward, J. W. 1971. Infrared spectroscopic studies of zeolites. p. 380–403. In R. F. Gould Molecular sieve zeolites-I. Adv. Chem. Ser. 101. American Chemical Society, Washington, D. C.

Weaver, R. M. 1974. Chemical and clay mineral properties of highly weathered soil from the Colombian Llanos Orientales. Agronomy Mimeo 72–19. Dep. of Agronomy, Cornell Univ., Ithaca, New York.

Weaver, C. E., and L. D. Pollard. 1973. The chemistry of clay minerals. Elsevier Scientific Publishing Co., New York.

White, J. L., A. N. Jelli, J. M. Andre, and J. J. Fripiat. 1967. Perturbation of OH groups in decationated Y zeolites by physically adsorbed gases. Trans. Faraday Soc. 63:461–475.

Wilkins, R. W. T., and J. Ito. 1967. Infrared spectra of some synthetic talcs. Am. Mineral. 52:1649–1661.

Wollast, R., F. Mackenzie, and O. Bricker. 1968. Experimental precipitation and genesis of sepiolite at earth-surface conditions. Am. Miner. 53:1645–1662.

Yaalon, D. H. 1955. Clays and some non-carbonate minerals in limestones and associated soils of Israel. Bull. Res. Counc. Isr., 5B, 2, Sect. B: Zool.:161–173.

Silica in Soils: Quartz, Cristobalite, Tridymite, and Opal[1]

LARRY P. WILDING, Texas A&M University, College Station, Texas

NEIL E. SMECK, Ohio State University, Columbus, Ohio

LARRY R. DREES, Texas A&M University, College Station, Texas

I. INTRODUCTION

In nature silica occurs as six distinct minerals: quartz, tridymite, cristobalite, coesite, stishovite, and opal. Several varieties of silica glass also occur. Of these minerals, quartz is most abundant; tridymite and cristobalite are widely distributed in volcanic rocks and biogenic sedimentary rocks; coesite and stishovite are rare polymorphic forms produced at high pressures, but found under atmospheric temperature and pressure at sites of meteor impact; and opal is a hydrated "amorphous" silica which comprises the bulk of diatomaceous rocks of biogenic origin and hydrothermal silica of inorganic origin (Frondel, 1962; Deer et al., 1966). Biogenic opal includes silica bodies found in characteristic shapes, sizes, and forms deposited in or secreted by plants and animals. Numerous terms have been given to such bodies such as *Phytolitharien* by Ehrenberg in 1846 (Baker, 1960a), *opal phytoliths* (Baker, 1959a), *plant opal* (Beavers & Stephen, 1958), *grass opal* (Smithson, 1958) and *bioliths* (R. L. Jones & Hay, 1975). The term bioliths is broader in scope and includes microfossils of widely different chemical and mineralogical composition.

This chapter discusses only the first three of the above silica polymorphs (quartz, cristobalite, tridymite) and opal. Collectively, these silica minerals comprise the major weight and volume percentage of most soils. Although quartz is second only to feldspars in its abundance in the earth's crust (Berry & Mason, 1959, p. 461), it is by far the most abundant mineral species in most soils where it often comprises 90 to 95% of all sand and silt fractions (Russell, 1961, p. 76). This reversal in trend can be attributed to the more labile nature of many feldspars under soil weathering environments compared to the stability of quartz (Goldich, 1938).

Silica minerals are classified as *tectosilicates* (Greek root meaning framework). In silica structures, the repeating unit is a SiO_4 tetrahedron in which each O of the tetrahedron is linked to Si atoms of adjacent tetrahedra forming three-dimensional framework structures. The Si/O ratio is 1:2. Limited proxying of Al^{3+} for Si^{4+} which is counter-balanced by alkali and alkaline earth cations has been proposed (Deer et al., 1966; Millot, 1970; Frondel, 1962), but attempts to differentiate such proxying from occluded impurities

[1]Contribution from the Agronomy Department, Ohio Agricultural Research and Development Center and The Ohio State University, Columbus, Ohio.

chapter 14

Table 14-1. Properties of silica polymorphs and opal

Mineral	Specific gravity	Hardness	Optical indicatrix	Optical sign	Refractive indices[†]	Bire-fringence	Crystal[‡] system
Quartz	2.65	7	Uniaxial	(+)	ϵ = 1.553 ω = 1.544	0.009	Trigonal
Cristobalite	2.32–2.38	6.5–7	Uniaxial	(–)	ϵ = 1.482–1.484 ω = 1.486–1.487	0.002–0.005	Tetragonal
Tridymite	2.20–2.28	6.5–7	Biaxial	(+)	α = 1.472–1.479 β = 1.478–1.480 γ = 1.477–1.483	0.002–0.005	Orthorhombic
Opal	1.5–2.3	5.5–6.5	Isotropic	--	1.41–1.47	Nil	None

† White light at room temperature.   ‡ Crystal systems for low (α) form.

are still wanting. Quartz exhibits far less of this phenomenon than the other silica polymorphs.

Although all of the silica polymorphs exhibit a framework tetrahedral motif, the pattern of linkage for each mineral is different as reflected in structural, physical, and chemical properties (Table 14-1). Quartz, cristobalite, tridymite, and opal have progressively more open structures which result in decreasing density, hardness, stability, and refractive index concomitant with increasing porosity, impurities, hydration, and specific surface.

The following sections discuss the properties and utility of silica minerals as soil genesis indicies—stable reference constituents, parent material indicators, and palenological tools. Considering the abundance of silica minerals in soils and their role in soil genesis, attention given them by soil scientists is not commensurate with extent. This likely reflects their assumed physiochemical inactivity though little definitive evidence is available to quantify this aspect.

II. STRUCTURAL PROPERTIES AND MINERAL IDENTIFICATION

A. Crystal Structure

Quartz, cristobalite, and tridymite are crystalline while historically opal has been considered an amorphous mineraloid without crystal structure (Rogers, 1937, p. 321). The latter is really a misnomer; more recent literature suggests several varieties of opal which constitute a continuum varying from no microstructure to short-range ordering of silica microcrystallites, namely disordered cristobalite or tridymite, to larger more crystalline occluded phases including low temperature tridymite and quartz (J. B. Jones et al., 1963, 1964; Frondel, 1962, p. 287–289; Mizutani, 1970; Wilding & Drees, 1974; Sanders, 1975).

The crystalline polymorphs have low-high temperature inversions related to structural linkage. Inversion from lower to higher forms is reversible and is accompanied by increased crystal symmetry. It results in rotation and displacement of tetrahedra without Si-O bond breakage. These structural modifications will be identified as low (α or β), middle (α_1 or β) and high (α, β or β_2) because Greek symbols have been inconsistently applied to a given modification.

1. QUARTZ

The silica tetrahedron is almost symmetrical in quartz and has a Si-O distance of 1.61Å (Frondel, 1962, p. 18). The Si-O bond is about equally ionic and covalent in nature. The structure of quartz can be visualized as a spiral network of silica tetrahedra about the Z-axis (Fig. 14-1). Each tetra-

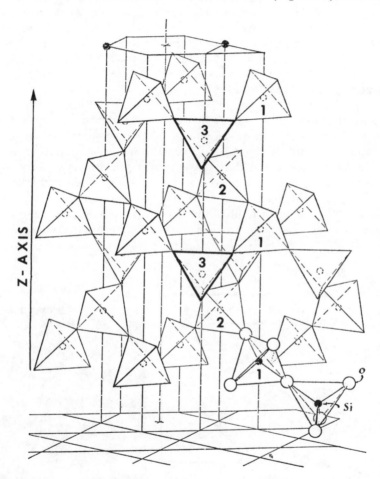

Fig. 14-1. The structure of low quartz, SiO_2, shown as SiO_4 tetrahedra with a small silicon atom at the center of a group of four oxygens in tetrahedron shares a corner (oxygen) with an adjoining tetrahedron. For simplicity most of the tetrahedrons are shown without the circles representing oxygen. The screw axis symmetry (threefold) of the Z-axis is shown by the dot-dash line, repeating each tetrahedron with a rotation of $120°$ and a translation of $c/3$. Two units of structure along Z are shown. Note spiral formed by tetrahedra 1 through 3. Modified from Fig. 2-59, Berry and Mason (1959, p. 112).

Table 14-2. Unit cell dimensions (Å) for natural occurring minerals
(low temperature forms)

Mineral	a	b	c
Quartz	4.90		5.39
Cristobalite	4.97		6.91-6.95
Tridymite	9.90-9.98	17.1-17.26	8.18†

† Polytypes representing 4-sheet (16.3Å), 10-sheet (40.8-40.9Å), and 20-sheet (81.6Å)
structures have been identified (Frondel, 1962, p. 264; Sato, 1964b).

hedron is repeated in the network by a rotation of 120° and a translation of
$c/3$. The tetrahedra are linked in such a way so as to form a hexagonal structure (unit cell dimensions are given in Table 14-2). The crystal class of low
quartz is trigonal trapezohedral which yields a three-fold axis which is perpendicular to three equal horizontal axes intersecting at 120°.

2. CRISTOBALITE AND TRIDYMITE

Most literature that discusses cristobalite and tridymite report the
"idealized" structures (Fig. 14-2A and 14-3A) which are the high temperature forms. The precise crystal structure of low temperature forms is not
well understood; the crystal class of low temperature tridymite is unknown
(Frondel, 1962, p. 264). However, from X-ray diffraction and crystallographic evidence the crystal system (Table 14-1) and unit cell parameters
(Table 14-2) of low temperature forms have been reported. Natural crystals
of tridymite and cristobalite invariably contain impurities that may stabilize

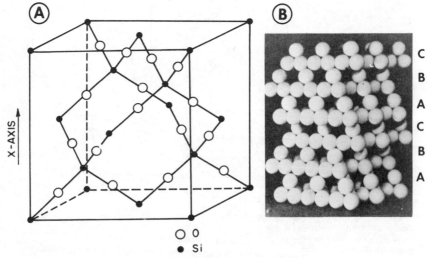

Fig. 14-2. Structure and packing of idealized high cristobalite. *(A)* Structure of high
cristobalite. Modified from Wyckoff (1925). *(B)* Tetrahedral packing in the three-
layer structure of cristobalite. Reprinted from Fig. 10, Eitel (1957) with permission
of American Ceramic Society.

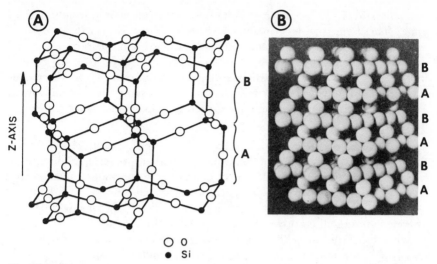

O   O
● Si

Fig. 14-3. Structure and packing of idealized high tridymite. *(A)* Structure of high tridymite. Modified from Gibbs (1927). *(B)* Tetrahedral packing in the two layer structure of tridymite. Reprinted from Fig. 9, Eitel (1957) with permission of American Ceramic Society.

the structure in the high temperature form even under amibent conditions (Eitel, 1954, p. 617). For example, high temperature cristobalite has been observed in geologic opal and bentonites as a crystal phase at ambient temperatures. These cristobalite phenocrysts did not show thermal inversion effects near 220C (493K) confirming their high temperature form (Eitel, 1954, p. 620). While this is apparently widespread, low temperature forms have also been well established.

In idealized structures J. B. Jones and Segnit (1972) consider cristobalite and tridymite to be composed of sheets of silica tetrahedra stacked parallel to the (111) and (0001) planes, respectively. According to them, these sheets consist of "fairly regular six member rings." Vertices of alternate tetrahedra within the ring are inverted so as to form common oxygen bond links perpendicular to adjacent sheets. In cristobalite, the tetrahedra are arranged in a cubic close packing scheme such that every fourth tetrahedral sheet is identical to the first. This yields a three-sheet structure with a repeating ABC-ABC regular stratification (Fig. 14-2B). In contrast tridymite has an alternating AB-AB-AB layer stacking arrangement (Fig. 14-3B) of silica sheets in a hexagonal close-packed arrangement (Eitel, 1957).

3. OPAL

Two different structural concepts of opal are discussed in the following sections. The first, *crystal structure,* deals with the crystal chemistry of opal on an atomic level and the evidence for tetrahedral ordering within the opal structure. The second, *macromolecular structure,* considers the physical ordering of submicron opaline spheres as structural entities which are clearly characteristic of precious opal and responsible for light diffraction from same. These macromolecular units have been formed by growth of colloidal micelles which possess the crystal structural properties of opal; the aggregated domains are at levels of 20- to 1000-fold larger than submicroscopic crystallites discussed under crystal structure. Opaline varieties which are often considered "amorphous" may or may not exhibit crystal structure, but invariably are comprised of submicron aggregates or spheres as macromolecular units.

a. **Crystal Structure**—It is the level of ordering that differentiates opal from other crystalline silica polymorphs rather than presence or absence of crystal structure per se. Because opal consists of a continuum of submicroscopic phases varying from short-range to long-range order, it is difficult to clearly differentiate opal from disordered low-cristobalite or low-tridymite intermediates; the literature is often confused in this regard.

Frondel (1962, p. 288) suggests that opal is a crystalline aggregate composed of a continuum of submicroscopic crystallites of "opal-cristobalite" with a more or less disordered internal structure. This continuum could be envisioned as a sheet-linkage of silica tetrahedra, about 4.1Å thick, which are geometrically stacked so as to constitute randomly interstratified mixed layer assemblages of cristobalite and tridymite. It is postulated on the basis of X-ray diffraction data that cristobalite layers predominate. With increased number and ordering of cristobalite layers disordered cristobalite is recognized instead of opal. Frondel (1962, p. 287) believes that opal is related to cristobalite in much the same relationship as chalcedony is to quartz. Conversely, from recent electron microscopy and diffraction work, Sanders (1975) concludes that the microstructure of precious opal consists of discrete phases of cristobalite and tridymite with no evidence of intergrowths of these two minerals.

Alternatively, opal has been viewed as an X-ray amorphous groundmass with dispersed crystalline phases of low-quartz and low-cristobalite (Teodorovich, 1961). Neither this concept nor that of the continuum considered above are well documented; further work is needed to elucidate the crystal structure of opal and to more adequately define it in relationship to other silica polymorphs.

b. **Macromolecular Structure**—The submicron structure of precious opal is composed of close-packed aggregates of silica spheres arranged hexagonally in layers which are usually stacked randomly (J. B. Jones et al., 1964; Sanders, 1968; J. B. Jones & Segnit, 1969). In some specimens stacking is ordered resulting in face-centered cubic or hexagonal close-packed motifs (Sanders, 1968; Monroe et al., 1969). Greer (1969) using scanning electron microscopy has confirmed the above physical structure.

Spheres (Fig. 14-4) which comprise the submicron aggregate structure of gem opal range from 1,500 to 3,500Å in diameter (J. B. Jones et al., 1964; L. H. P. Jones et al., 1966). These particles in turn are aggregates of spheres of smaller dimensions (Greer, 1969). Aggregation is believed to be in the form

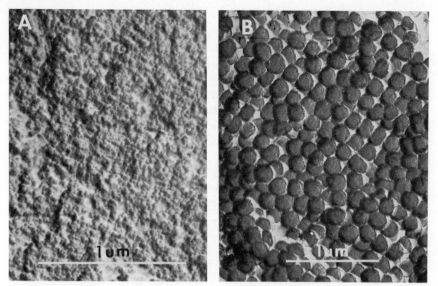

Fig. 14-4. Transmission electron replicas of fractured opal surfaces: *(A)* biogenic opal from bamboo-tabashir; and *(B)* gem opal of inorganic origin. Reprinted from Fig. 1d and 1e, Jones et al. Copyright 1966 by the American Association for the Advancement of Science.

of geometric progression from initial stages of macromolecular spheres a few 10's of Å in diameter through the terminal stages of about 3,500Å (J. B. Jones & Segnit, 1969; Greer, 1971). Pollard and Weaver (1973) report opaline silica spheres from 1,000 to 8,300Å in diameter which occur individually and in small aggregates around cavities of diatom fragments in a silica nodule from Miocene deposits. These spheres are apparently equivalent to those of precious opal, but they attribute differences in sphere size and degree of ordering to a lack of a sedimentation environment during nodular formation. The largest individual spheres reported by Pollard and Weaver (1973) are surprisingly similar to opaline spheres isolated from forest soils (Wilding & Drees, 1973, 1974) and silicified plant remains in acid sulfate soils (Buurman et al., 1973). It is possible these spheres may be of phytolith origin rather than from authigenic inorganic precipitation. The aggregated structure of opal phytoliths generally consists of less regular and smaller diameter microspherules (Fig. 14-4) than those in gem opal (L. H. P. Jones et al., 1966; Wilding & Drees, 1971, 1974).

B. Elemental Composition and Impurities

Ideally, quartz, cristobalite, and tridymite should constitute 100% SiO_2 with opal defined as a hydrated silica ($SiO_2 \cdot nH_2O$). However, these minerals invariably contain impurities which increase in magnitude from quartz to opal due to increased microporosity.

1. QUARTZ

Quartz is one of the purest minerals known; only diamond, graphite, and ice appear to have a smaller compositional variation than colorless quartz (Frondel, 1962, p. 151). The chemical purity of quartz is greater than the other silica polymorphs due to the more closed structure of quartz which inhibits the entry of impurities. Quartz does, however, contain trace quantities of elements other than silicon and oxygen (Fig. 14-5, Table 14-3). Colored quartz varieties and the blackening of colorless quartz due to X-ray bombardment are both attributed to the presence of impurities. Quartz contains impurities of three types: atomic substitution within the crystal structure; interstitial elemental occlusion within the structure; and inclusion of liquid, solid, or gas foreign compounds within crystals.

The elements which are commonly reported as trace occlusions in quartz, either interstitially or as substitutions, are Al, Ti, Fe, Na, Li, K, Mg, Ca, and H (OH) (Frondel, 1962, p. 144-149; Dennen, 1966). Isomorphic substitution proceeds most readily with an ion of similar ionic radii and charge; however, Dennen (1966) and Stavrov (1961) both indicate that the dominant elemental contaminants of natural quartz are generally the most abundant ions in the medium in which quartz crystallizes and not necessarily those of best structural fit. Aluminum tends to be the most frequently encountered elemental contaminant in quartz. The substitution of Al for Si in the crystal structure is the proposed mechanism for Al entry. Titanium, which is common in rose quartz and amethyst, and ferric iron may also substitute for Si in the crystal structure. The substitution of Al or Fe for Si would result in a charge deficiency which is compensated for by the entry

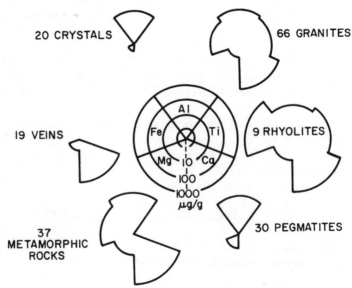

Fig. 14-5. Abundance of chemical impurities in quartz from different environments. Modified from Fig. 2, Dennen (1964).

Table 14-3. Elemental composition of quartz, tridymite, and opal
(values as percentage)

Element	Quartz[†]	Tridymite[‡]	Opal Plants[§]	Opal Soils[¶]	Opal Geological[#]
SiO_2	>99	88.6-99.0	82.8-87.2	76.4-90.5	85.8-96.5
Al_2O_3	0.009-0.06	0.5-2.7	0.02-0.70	0.84-4.7	T-3.22
Fe_2O_3	0.001-0.7	T-1.9	T-0.56	0.18-1.3-	0.08-1.85
TiO_2	T-0.07	0.02-0.86	T	T-0.30	--
CaO	0.0007-0.014	0.2-0.4	T-1.55	<0.1-2.04	0.09-0.96
K_2O	--	0.08-0.75	T-0.90	0.14-0.97	0.75
Na_2O	NF-0.004	0.24-0.80	T-0.50	0.10-3.44	0.18
MgO	0.0008-0.016	T-1.24	T-0.51	T-1.72	T-1.48
C	--	--	5.78	0.86	--
$H_2O^{(+)}$	--	--	3.83-7.61	4.26-12.1	3.26-9.40

[†] Frondel (1962, p. 144-149).
[‡] Grant (1967); Rivalenti and Sighinolfi (1968); Mason (1953), Frondel (1962, p. 267).
[§] Twiss (1967), Arimura and Kanno (1965), Jones and Milne (1963).
[¶] Jones and Beavers (1963), Arimura and Kanno (1965), Buurman et al. (1973).
[#] Range of nine geological opal samples (Frondel, 1962, p. 294).
T = trace; NF = not found; -- = no data reported.

of interstitial ions. Lithium and Na are the most frequently cited interstitial occlusions, but K, Fe, Mn, Mg, and Ca are also reported. However the charge resulting from Al substitution could also be compensated for by the substitution of a hydroxyl ion for an oxygen which would undoubtedly disrupt the structure. The presence of (OH) groups in quartz has been established by infrared absorption by several investigators (Frondel, 1962, p. 150). Since elemental composition is a function of the environment in which crystallization takes place, several investigators (Stavrov, 1961; Dennen, 1967; Suttner & Leininger, 1972) have used elemental compositional as an indicator of provenance.

Inclusions of compounds in quartz, which are almost always present, occur as liquid, solid, or gas phases and commonly are a combination of at least two of these phases. Foreign compounds in quartz consist of phases trapped within the enveloping crystal. Inclusions of this type commonly range in size from microscopic to several milliliters or more. Liquid inclusions consisting of pure water, solutions containing Na, K, Ca, Cl, SO_4, and CO_3 ions, and pure liquid CO_2 have been reported (Frondel, 1962, p. 227). Gaseous inclusions may consist of NH_3, H_2S, and N_2. Of all the solid inclusions, rutile is the best known. Rutile inclusions in quartz result from both mechanical entrapment and exsolution. Tourmaline is also a common solid inclusion. Other solid enclosures sometimes found in quartz crystals are: actinolite, hornblende, byssolite, crocidolite, goethite, chlorite, stibnite, arsenopyrite, pyrrhotite, sphalerite, stephanite, pyrite, hematite, anatase, brookite, pyrochlore, zircon, anhydrite, copper, epidote, hiddenite, ilmenite, magnetite, siderite, calcite, dolomite, topaz, sphene, mica, and gold (Frondel, 1962, p. 236-237).

Although quartz is generally an extremely pure mineral, care should be exercised when basing mineralogic composition on elemental composition even in a soil with a high quartz content. Khangarot et al. (1971) have shown that the concentration of Fe, Ti, and Zr in the light mineral fraction (sp gr < 2.86) cannot be accounted for by mineral grain counts in this isolate; they conclude that these elements represent mineral inclusions or ionic substitution in the predominantly quartz minerals. As much as 75% of the total Ti and 50% of the total Zr in the 20–50 μm soil fraction was found in the light isolate. About 50% of the total Fe in this fraction occurred as free iron oxide coats with the remainder as occluded Fe.

2. CRISTOBALITE AND TRIDYMITE

Cristobalite and tridymite both have a more open framework structure than quartz; thus they can more easily accommodate impurities within the crystal. Data on cristobalite is wanting, but tridymite usually contains < 99% SiO_2. The predominant impurity is Al with lesser amounts of Fe, Ti, K, Na, Ca, and Mg (Table 14-3). Little is known of the specific loci of these impurities (i.e. solid solution, ionic substitution, or occluded compounds).

Beginning in the late 1950's, two schools of thought have prevailed concerning the nature of tridymite and the role of observed impurities. One school holds that tridymite should not be considered a separate silica polymorph, but rather a solid solution component between silica and nonsilica phases. This argument is supported by the ever present impurities and the seemingly inability to synthesize a pure silica phase possessing tridymite structure without "impurities" as a flux in the melt (Eitel, 1954, p. 724, 1957; Rockett & Foster, 1967). The alternate view holds that the impurities are not essential for tridymite formation and stability (Hill & Roy, 1958b; Roy & Roy, 1964), and tridymite is a true silica polymorph. The same arguments can probably be put forth for cristobalite. Berry and Mason (1959, p. 474) indicate that these impurities may in part have a stabilizing influence on cristobalite and tridymite, allowing them to exist in a metastable phase almost indefinitely below their respective phase equilibrium temperatures.

3. OPAL

Opal is a hydrated silica—$SiO_2 \cdot nH_2O$. Water content commonly varies from 4 to 9%, but rarely up to 34% (Frondel, 1962, p. 292; Teodorovich, 1961). Opal is mostly SiO_2 (85 to 95%) with significant amounts of occluded, chemisorbed, or solid solution impurities including Al, Fe, Ti, Mn, P, Cu, N, C, alkalies, and alkaline earths (Table 14-3). The N and C reported in plant opal have been attributed to occlusion of cyptoplasmic material during opal synthesis (R. L. Jones & Beavers, 1963a). It may also be the result of silica impregnation of cellulose and lignin in cellular structures or other organo-silicon compounds (Lewin & Reimann, 1969). Over 50% of the occluded organic carbon is not readily accessible to oxidation (Wilding et al., 1967) and thus phytoliths serve as a potential carbon source for radiocarbon dating purposes (Wilding, 1967). Some of the water reported in opal above 100C (373K) may in fact be due to occluded C which is lost upon ignition, but not considered in partitioning the weight loss (Wilding et al., 1967). Opal impurities markedly influence color, refractive index, stability, and specific gravity to be discussed later.

L.H.P. Jones and Milne (1963) observed higher assays for most accessory elements in opal isolated from plants upon dry-ashing vs. wet acid digestion. They attributed these differences to some deflagration and fusion of silica with external chemical components upon dry-ashing of plant tissues.

C. Physical Identity

1. CRYSTAL FORM AND HABIT

With few exceptions under soil environments, the original crystal form and habit of the silica minerals have been altered either by in situ physical and chemical weathering, or as a consequence of multiple transport, sedimentation, and secondary weathering cycles. Thus, these properties often do not provide definitive criteria for mineral identification.

a. **Quartz**—In soils, quartz generally occurs as anhedral grains; it does not exhibit a short prismatic habit with the prism faces [1010] being terminated by a set of rhombohedral faces [1011 and 0111] characteristic of large euhedral quartz crystals. Krinsley and Smalley (1973) report that quartz grains < 100 μm are almost always shaped like flat plates due to a cleavage mechanism. Quartz grains of sand size frequently are rounded or subrounded, but in other cases may exhibit an angular morphology. Rounding is a result of both physical attrition, particularly during transport, and solution phenomena. Angular grains are generally attributed to mechanical fracturing. Crook (1968) presents an excellent review of work pertaining to the relationship between chemical solution and quartz grain shapes. Many workers have described the shape and surface morphology of quartz grains. Schneider (1970) reviews many of these and cites the work of Cailleux (1942) as the "classical" morphoscopy. The utility of quartz grain surface morphology as an indicator of provenance will be discussed later.

b. **Tridymite and Cristobalite**—Tridymite crystals are usually small (< 1 mm), often platy, and usually characterized by a pseudohexagonal crystal habit which is assumed to be an inversion pseudormorph after high-tridymite. Consequently the true morphology of low-tridymite is concealed (Frondel, 1962, p. 264). Upon inverting from high to low form, tridymite commonly exhibits secondary wedge shaped twins (Fig. 14-6).

The crystal habit of cristobalite has also been identified from small (< 1 mm), often platy grains, found in vesicles of rhyolites, trachytes, and Middle Eocene radiolarian-rich claystones (Klasik, 1975). Crystals are octahedral or pseudooctahedral in habit although cubical forms have been reported (Frondel, 1962, p. 278).

c. **Opal**—Morphology is one of the most diagnostic features of biogenic

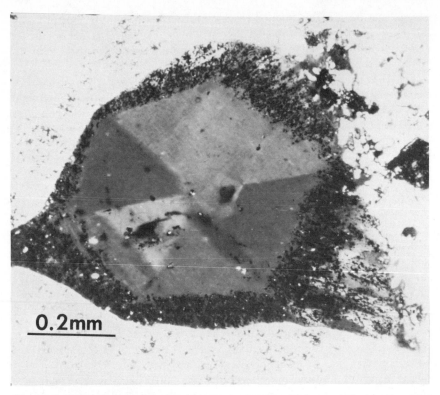

0.2mm

Fig. 14-6. Pseudohexagonal wedge shaped twins in tridymite (crossed nicols). Reprinted from Fig. 2, Friedlaender (1970) with permission of Schweizerischen Mineralogischen und Petrographischen Gesellschaft.

opal, but it may be less useful for opal of inorganic origin. The latter commonly take one of the following macroscopic forms: massive, botryoidal, globular, stalactitic, filamentous or pisolitic (Berry & Mason, 1959, p. 480; Frondel, 1962, p. 288).

Unless subsequently fragmented, biogenic opal closely conforms to the biological cell or structure in which it originates; thus, it can be easily identified by shape as confirmed by other supportive diagnostic criteria including elemental composition, specific gravity, and refractive index. Nomenclature for opal phytoliths has not been standardized. Some investigators prefer generic terms according to loci of opal synthesis (Geis, 1973) while others prefer morphological terms (Rovner, 1971; Twiss et al., 1969; Wilding & Drees, 1973, 1974).

Commonly observed opaline forms derived from grasses (Fig. 14-7 and 14-8) include fan and rectangular tabular forms, hook or shield-shaped bodies, smooth or serrated rods, dumbbells, tabular perforated platelets, conical "hats" or "caps," and discs (Prat, 1936; Metcalfe, 1960; Parry & Smithson, 1964, 1966; Blackman & Parry, 1968; Stewart, 1965; Twiss et al., 1969, Rovner, 1971; Yeck & Gray, 1972).

Commonly observed forms of opal derived from deciduous angiosperm

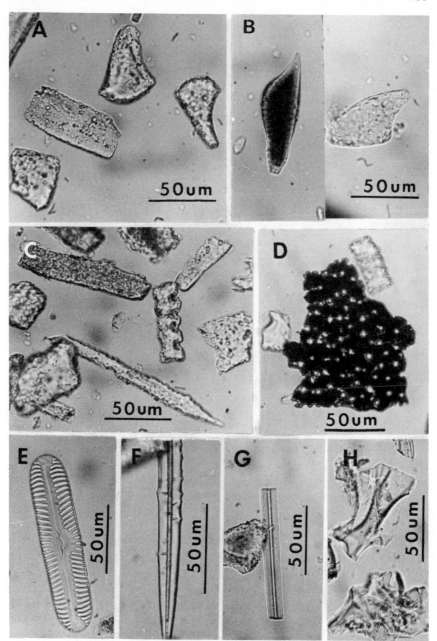

Fig. 14-7.   Micrographs (plane polarized transmitted light) of biogenic opal *(A–G)* and volcanic ash *(H)* isolated from soils: *(A)* fan-shaped bulliform cells; *(B)* shield-shaped trichome cells; *(C)* serrated and smooth rod-shaped fundamental cells, *(D)* perforated root platelet; *(E)* diatom; *(F* and *G)* sponge spicule fragments; and *(H)* Mount Mazoma volcanic ash from Tolo soil, Oregon (courtesy of Dr. J. A. Norgren, Department Soil Science, Oregon State University, Corvallis).

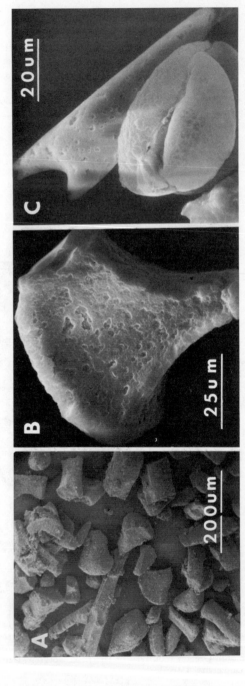

Fig. 14-8. Scanning electron micrographs of grass opal phytoliths isolated from soils: (A) general field; (B) fan-shaped bulliform cell; (C) shield-shaped trichome cell; (D) and (E) serrated and slightly cusped rods from fundamental cells; (F) and (G) dumbbells from silica cells—F from Panicoid and G from Chloridoid; (H) helical xylem vacular element; and (I) perforated root platelet. A and I are reprinted from Fig. 2A (Wilding & Drees, 1971) and Fig. 7A (Wilding & Drees 1973), respectively, with permission of the Soil Science Society of America.

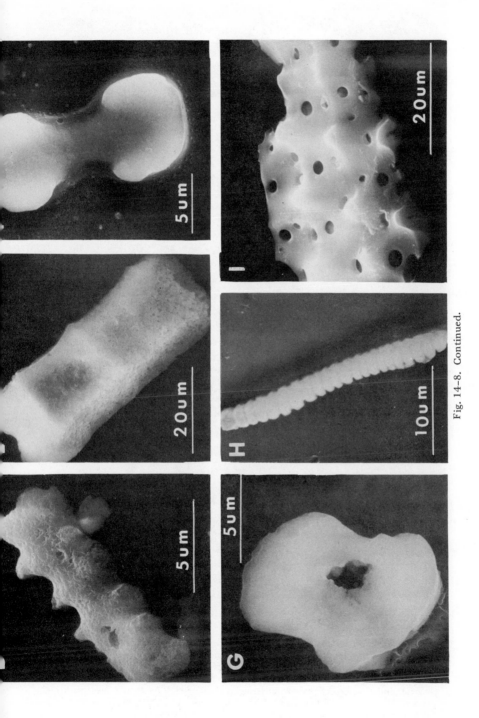

Fig. 14-8. Continued.

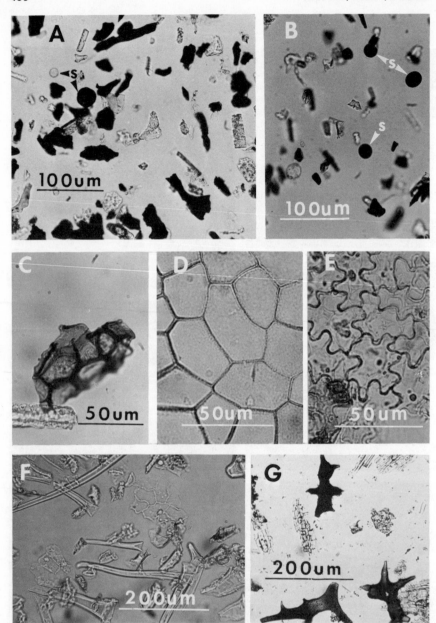

Fig. 14-9. Micrographs (plane polarized transmitted light) of biogenic opal isolated from the surface horizon of forest soils *(A-C)*, deciduous tree leaves *(D-F)*, and conifer needles *(G)*. *A* and *B* general fields with numerous opaque bodies and spheres (s); *C* a cup assemblage; *D* and *E* cup assemblages from white oak *(Quercus alba)* and sugar maple *(Acer saccharum)*, respectively; *F* epidermal hairs from slippery elm *(Ulmus rubra)*, and *G* asterosclereids from Douglas-fir needles *(Pseudotsuga menziesü)*. *D, E,* and *F* reprinted from Fig. 5 (G, E, and F), respectively, Wilding and Drees (1968a) with permission of College of Agric., Univ. of Illinois, Urbana.

tree species (Fig. 14-9 and 14-10) include: spheres, cups or cup assemblages, rods, convoluted sheets, kidney-shaped stomata bodies, tabular mosaic forms, bladed forms, silicified hairs, and hair bases (Geis, 1973; Norgren, 1973; Wilding & Drees, 1973, 1974). Asterosclereids (Fig. 14-9) from Douglas-fir needles are also unique (Brydon et al., 1963; Norgren, 1973).

Because of multiplicity and redundancy in phytolith forms, the precise species contributing biogenic opal is difficult to identify. Present state of knowledge permits differentiation of grass versus forest origins (Wilding & Drees, 1968b, 1971, 1974; Geis, 1973) and perhaps groups of subfamilies within *Gramineae* (Twiss et al., 1969). In general, forest opal consists of incrustations of cellular components with numerous thin sheet structures (Fig. 14-10 and 14-11) while grass opal consists mostly of solid polyhedral structures (Fig. 14-7 and 14-8) resulting from silicification of the entire cell. Loose-packed aggregates of thin, fragile plate-shaped incrustations are also a common form of forest opal (Fig. 14-10 J, K, L, and 14-11 C).

Sponge spicules (Fig. 14-12 B, C, and D), diatoms (Fig. 14-12 A), and Chrysostomatacea shells are often abundant opaline microfossils in soils developed under ponded or poorly drained conditions. Spicules are readily differentiated from rod-shaped grass phytoliths by their acicular shape, axial canal extending the length of the tapering form (Fig. 14-7 I, and 14-12 B, C, and D), relatively nonpitted surface texture, and smooth or only slightly spinose overall conformation. Fragments of spicules are likewise readily identifiable (Fig. 14-7 G). Diatoms, either intact or as fragments, exhibit characteristic sieve pores that are regularly arranged along the valve surface (Patrick & Reimer, 1966). R. L. Jones et al. (1964) reports characteristic shells of Chrysostomatacea that can be identified by their unique form.

Morphologically, plant opal can be mistaken for glass shards (Kanno & Arimura, 1958) or vice versa (Fig. 14-7 H); however, glass shards can usually be positively identified from biogenic opal on the basis of form, host inclusions, and refractive index. [Refractive index of biogenic opal is usually distinctively lower than that for volcanic shards (1.41-1.47 vs. 1.48-1.61).] Shards are usually massive (sometimes vesicular or perlitic), tear-like, fluidal or angular in shape, and contain occluded lacunae (gas bubbles) and spherulites or phenocrysts of feldspars (Rogers & Kerr, 1942, p. 374-375). The shape is often modified by weathering.

Opal of inorganic origin in soils commonly takes the macroscopic form of silica-rich cutans (coatings) separating the structural prisms in the lower part of a duripan (silica-cemented zone). Within the *s*-matrix (soil matrix) opal in the form of diffuse opal flocs a few micrometers in diameter (Fig. 14-13) or durinodes in the form of weakly silica-cemented soil nodules occur during incipient stages of cementation (Flach et al., 1969). In more advanced stages of pan development, continuous sheets and stringers of opal-chalcedony occur within the *s*-matrix and/or interspersed with argillans (Fig. 14-13). Brewer et al. (1972) have also illustrated silans (secondary silica coatings), neosilans (secondary silica impregnated zones within the *s*-matrix), and silica nodules in Red and Brown Hardpan (duripan) soils of western Australia (Fig. 14-14 C and D).

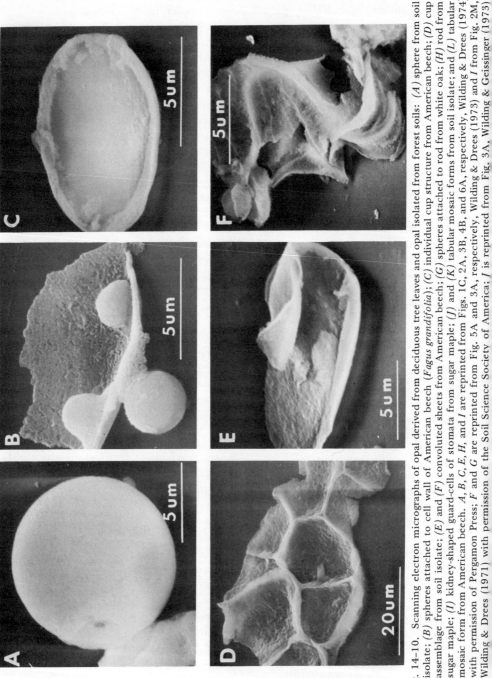

Fig. 14-10. Scanning electron micrographs of opal derived from deciduous tree leaves and opal isolated from forest soils: (A) sphere from soil isolate; (B) spheres attached to cell wall of American beech (Fagus grandifolia); (C) individual cup structure from American beech; (D) cup assemblage from soil isolate; (E) and (F) convoluted sheets from American beech; (G) spheres attached to rod from white oak; (H) rod from sugar maple; (I) kidney-shaped guard-cells of stomata from sugar maple; (J) and (K) tabular mosaic forms from soil isolate; and (L) tabular mosaic form from American beech. A, B, C, E, H, and I are reprinted from Figs. 1C, 2A, 3B, 4B, and 6A, respectively, Wilding & Drees (1974) with permission of Pergamon Press; F and G are reprinted from Fig. 5A and 3A, respectively, Wilding & Drees (1973) and I from Fig. 2M, Wilding & Drees (1971) with permission of the Soil Science Society of America; I is reprinted from Fig. 3A, Wilding & Geissinger (1973)

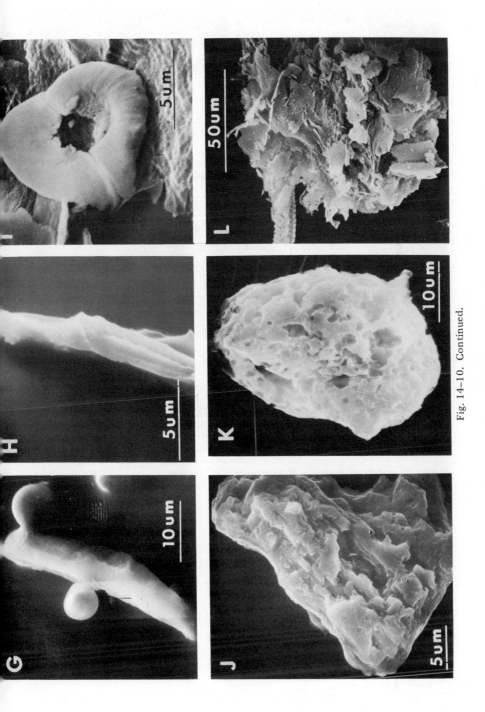

Fig. 14-10. Continued.

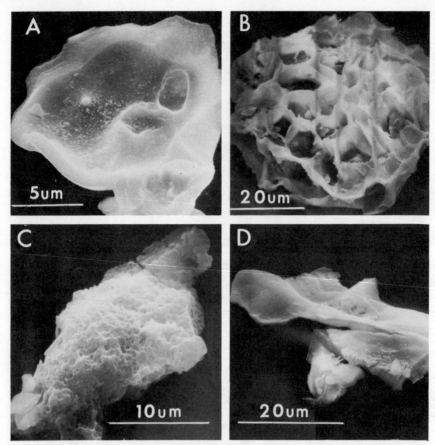

Fig. 14-11. Scanning electron micrographs of opaque opaline forms isolated from a forest soil: *(A)* cup; *(B)* cup assemblage; *(C)* tabular mosaic form; and *(D)* convoluted sheet. *A, B, C,* and *D* are reprinted from Figs. 8C, 8B, 6B, and 5B, respectively. Wilding & Drees (1973) with permission of the Soil Science Society of America.

2. SPECIFIC GRAVITY

There is a systematic decrease in specific gravity in the order quartz, cristobalite, tridymite, and opal (Table 14-1). This decrease is attributed to the increased microporosity of the more open framework structures and to impurities.

a. **Quartz**—The specific gravity of quartz is commonly reported as 2.65 in microcrystalline varieties and somewhat lower (about 2.60) in crypto-crystalline varieties. More explicit values are reported by Frondel (1962, p. 114). The specific gravity of optically pure quartz has been given as 2.647 by Katz et al. (1970).

The specific gravity of quartz is a function of temperature and chemical composition. The specific gravity of quartz increases approximately 0.0001 per 1C of temperature decrease (Frondel, 1962, p. 114). Specific gravity will

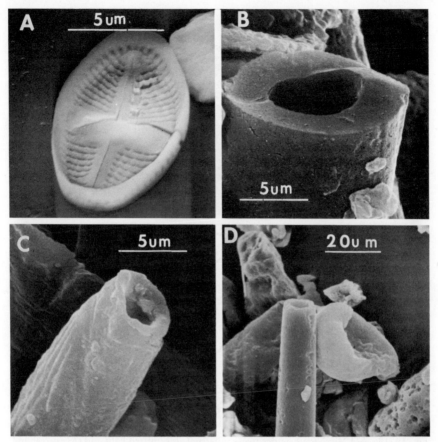

Fig. 14-12. Scanning electron micrographs of a diatom *(A)* and sponge spicules *(B-D)* isolated from a poorly drained soil in Ohio. *B, C,* and *D* reprinted from Fig. 4 (C, B, and A), respectively. Wilding & Drees (1971) with permission of the Soil Science Society of America.

decrease slightly due to the substitution of Al for Si and inclusion of interstitial Li, the increase in cell volume overweighing the increased molecular weight of the cell contents; also various quartz varieties differ slightly due to the type and content of inclusions. Due to the higher porosity and content of inclusions, cryptocrystalline varieties exhibit a wider range in specific gravity than macrocrystalline forms. Occasionally the specific gravity of naturally occurring quartz will be greater than that of pure quartz due to the presence of occluded heavier minerals and coatings of iron oxides.

Several investigators (Ray, 1923; Clelland & Ritchie, 1952; Dempster & Ritchie, 1953) have reported a decrease in the specific gravity of quartz on grinding. This decrease has been attributed to the (i) conversion of the surfaces of quartz to vitreous silica (Ray, 1923; Clelland & Ritchie, 1952), (ii) contamination by agate (Dale, 1924), (iii) adsorption of atmospheric water (Moore & Rose, 1973), and (iv) contamination and adsorbed water (Dempster & Ritchie, 1953). The absorption of water seems to be the most plausi-

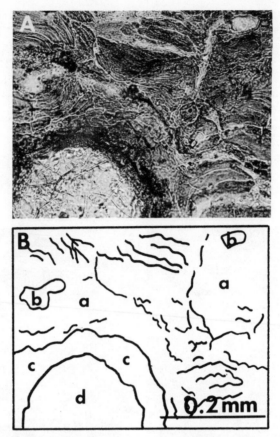

Fig. 14-13. *(A)* Micrograph under plane polarized transmitted light and *(B)* corresponding sketch of pedogenic silica in a duripan (a Durixeralf). *(a)* Clear stringers of silica
cement permeating an argillan; *(b)* sand grains; *(c)* a mass of diffuse silica globules; and
(d) clear isotropic chalcedony-like material. Reprinted from Fig. 1, Flach et al. (1969)
with permission of Williams and Wilkins Co.

ble explanation. Moore and Rose (1973) cite work indicating that the intensity of the OH infrared absorption bonds increase as the grinding period
increases.

The statistical specific gravity of quartz sediments are reported to
change as a function of "maturation" (Katz et al., 1970). The authors state
that either an increase or decrease of the specific gravity from that of optically pure quartz, owing to the occurrence of structural defects, inclusions
or microfissures, entails a greater susceptibility to both mechanical and
chemical destruction. Consequently as the sediment ages, not only does the
content of quartz increase, due to removal of less resistant minerals, but the
statistical specific gravity of quartz approaches that of optically pure quartz.

b. Cristobalite and Tridymite—The specific gravities of cristobalite and
tridymite are less than and considerably more variable than quartz (Table

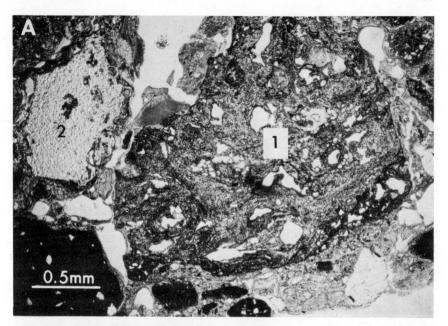

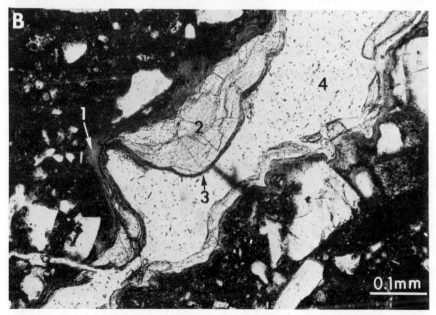

Fig. 14-14. *(A)* Nodules of secondary silica in Red and Brown Hardpan soils of western Australia (ordinary light): *(1)* large nodule of cloudy, somewhat laminated secondary silica with voids, and *(2)* nodule of clear secondary silica; *(B)* argillan-silan-argillan in similar soils: *(1)* argillan overlain by *(3)* argillan, and *(4)* large void. Dark flecks in void are grinding powder. Reprinted from Fig. 7 and 8, respectively, Brewer et al., 1972 with permission of Commonwealth Scientific and Industrial Research Organization, Australia.

Table 14-4. Relationship between specific gravity and occluded organic carbon
in opal isolates

Specific gravity	% Organic carbon	Source
< 2.3	0.9	Jones and Beavers (1963)
< 2.1	1.6	Jones and Beavers (1963)
< 2.3	2.4	Wilding et al. (1967)
2.0-2.3	1.4	Wilding et al. (1967)
1.5-2.0	4.7	Wilding et al. (1967)

14-1). The more common values are 2.32 and 2.26 for cristobalite and tri-dymite, respectively.

The similarity in specific gravities of cristobalite and tridymite makes it difficult to use this property to differentiate the two minerals. There is essentially complete overlap in the specific gravity of opal (1.5 to 2.3) with tri-dymite. However, the specific gravity of cristobalite, tridymite, and opal are sufficiently different from quartz (sp gr 2.65) to quantitatively isolate them.

c. **Opal**—The specific gravity of biogenic opal varies continuously over the range 1.5 to 2.3; median specific gravity varies from 2.10 to 2.15 (R. L. Jones & Beavers, 1963a; Kanno & Arimura, 1958; Baker, 1959a; Wilding et al., 1967). This relatively low specific gravity in contrast to other soil miner-als permits isopycnic density separation of reasonably pure opal isolates. When coupled with morphology and refractive index, specific gravity pro-vides a useful identification criterion.

Specific gravity of opal is a function of the submicron opal structure (porosity due to interstices between stacked spheres), water (Huang & Vogler, 1972), occluded organic matter (R. L. Jones & Beavers, 1963a; Wilding et al., 1967), and microscopic lacunae (R. L. Jones & Beavers, 1963a).

Lighter opal isolates with a specific gravity < 2.0 or < 2.1 contain about 2 to 3 times more organic carbon than the isolate between 2.0 or 2.1 and 2.3 (Table 14-4). However, the light fraction comprises only about 1/3 of the total weight percentage, thus, the amount of organic carbon in the total opal isolate is approximately equal between the light and heavy opal isolates.

3. COLOR

A pure silica mineral should be colorless. Various colors may occur in the silica polymorphs and opal due to chemical impurities within the crystal. These impurities may be in the form of ionic substitution or inclusions (either solid or fluid).

a. **Quartz**—Quartz is usually colorless and transparent or white and has a vitreous luster. However, quartz can occur in almost any color, a fact which may be observed in the coarser sand fractions of many soils if no sur-face coatings are present. Color is the basis for differentiating a number of varieties of quartz, both macrocrystalline and cryptocrystalline. Some of the more common macrocrystalline varieties (rose quartz, smoky quartz, milky quartz, etc.) and cryptocrystalline varieties (agate, flint, chert, jasper, etc.) are discussed by Frondel (1962, p. 170-226) and Berry and Mason 1959, p. 478-479).

b. **Cristobalite and Tridymite**—Tridymite, depending on location and origin, ranges from colorless to white, often transparent (Frondel, 1962, p. 265). Cristobalite however, is usually reported as white to milky-white and ranges from translucent to opaque (Frondel, 1962, p. 277; Howard, 1939).

The variable color in tridymite may be due in part to the incorporation of Al and alkali elements into the crystal structure. Cristobalite often exists as cryptocrystalline blades aggregated into rounded sperulites (Fig. 14-15). This aggregation is most likely the cause of the color and degree of translucency in cristobalite.

c. **Opal**—Color is not diagnostic for opal. It is a function of three mechanisms: pigmentation (occluded chemical impurities), diffraction of light (play of colors), and scattering of light (dispersion). In transmitted light biogenic opal isolated from soils ranges from colorless or light tan to various shades of brown to black (Fig. 14-7 and 14-9). Due to light scattering effects, opaque forms are commonly porcelain colored (pale bluish tint) under reflected light (R. L. Jones & Beavers, 1963a). There is a direct relationship between the percentage of opaque opaline phytoliths and percentage organic carbon while an inverse relationship exists between percentages opaque opal phytoliths and specific gravity (R. L. Jones & Beavers, 1963a; Wilding et al., 1967). Light diffraction in opal is attributed to its submicron structure that serves as a grating surface (Sanders, 1968).

4. CLEAVAGE AND FRACTURE

a. **Quartz**—In most elementary texts, quartz is generally considered to exhibit conchoidal fracture but not cleavage. Berry and Mason (1959, p. 476) indicate that quartz shows an occasional rhombohedral parting. Though cleavage is not ordinarily observed in quartz, Frondel (1962, p. 104-107) and Margolis and Krinsley (1974) both present good reviews of quartz cleavages which have been observed (Fig. 14-16). Cleavage which parallels structural planes is most probable along planes in which the least number of Si-O bonds are broken. Calculations involving the number of bonds broken along various planes indicate that rhombohedral planes are most likely to cleave (Fairbairn, 1939). This is in agreement with observations which reveal that these are easiest to produce and generally are the highest quality cleavages in quartz.

Krinsley and Smalley (1973) indicate that the relative magnitudes of fracture and cleavage in quartz are a function of grain size. They attribute fracture to a statistical distribution of defects in the grain which leads to non-structurally oriented separations. Consequently they indicate that as particle-size is reduced, the number of defects also decreases resulting in a stronger

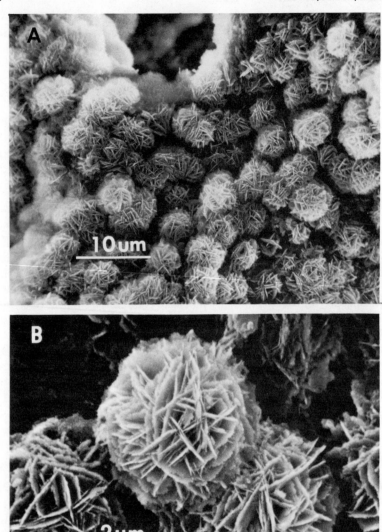

Fig. 14–15. Scanning electron micrographs of thin bladed cristobalite spherules from Cretaceous-Tertiary flint clay *(A)* and magnification showing spherule morphology *(B)*. Reprinted from plate IV, Wise et al. (1972) with permission of Schweiz Geologischen Gesellschraft.

grain which makes fracture more difficult. Apparently a critical particle size exists where cleavage supercedes fracture which Krinsley and Smalley (1973) place at approximately 100 μm. In a later publication, Margolis and Krinsley (1974) indicate that fracture predominates on grains $>$ 500 μm and cleavage on those $<$ 50 μm in diameter.

Fig. 14-16. Cleavage plane, irregular blocks, and conchoidal fracture on small grain, *Modern glacier, Switzerland.* The top surface of the grain represents a cleavage plane. The front edge displays a series of irregular conchoidal fractures and breakage blocks. Reprinted from micrograph 30, Krinsley & Doornkamp (1973) with permission of Cambridge University Press.

b. Opal—Opal is likewise characterized by conchoidal or flat-conchoidal fracture in macroscopic specimens. Although cleavage is not commonly recognized, ultraresolution of fractured or intact opal phytolith surfaces (Fig. 14-10 J, and 14-11 C) suggest a possible plane of weakness parallel to the stacking direction of growth zones or layers (Wilding & Drees, 1971, 1974). Such planes of weakness yield plate-shaped microcrystallites upon weathering and exfoliation of the opal surface (Fig. 14-17 B). It seems reasonable that this phenomenon may reflect the stacked layers of close-packed opaline spheres in the aggregated submicron structure.

D. Light Optical Properties

Refractive indices exhibit a progressive decrease in the order of quartz, cristobalite, tridymite, and opal (Table 14-1). Variability in refractive indices within a given mineral is attributed to mineral impurities; refractive indices generally increase with content of impurities (Frondel, 1962, p. 265). Water

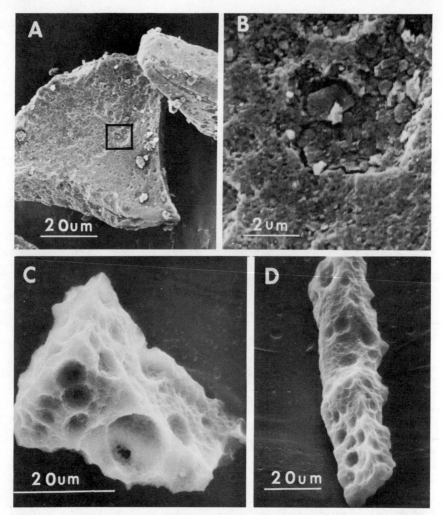

Fig. 14–17. Scanning electron micrographs of: *(A)* opal carbon-dated from a prairie (Mollisol) soil in Ohio (Warsaw, CH-34); *(B)* magnification of square area in A; *(C)* and *(D)* opal isolated from Malaysian paleosol (B-12). Note extreme dissolution effects in opal from paleosol as contrasted to slight effects in opal from Ohio soil. *A* and *B* reprinted from Fig. 2 (B and D) Wilding & Drees (1971) with permission from the Soil Science Society of America.

content and crystallinity also affect the refractive index of opal (Frondel, 1962, p. 292; Huang & Vogler, 1972).

1. QUARTZ

Quartz commonly exhibits an undulatory or wavy extinction ranging from slight to pronounced. The undulatory extinction is attributed to mechanical deformation or strain imposed on the mineral grains. Blatt and Christie (1963) found 15% of plutonic igneous rocks to be nonundulatory and 91% of extrusive igneous rocks to be nonundulatory. Many investigators have utilized the various degrees of undulatory extinction as an indicator of origin. Krynine (1940) and Folk (1968) present two of the most detailed classifications based on extinction. However, from the comprehensive review of this subject by Blatt and Christie (1963), it is apparent that about as many investigators attribute undulatory extinction to grains from metamorphic environments as those who attribute them to igneous environments. Furthermore, Blatt and Christie (1963) indicate that undulatory grains are selectively destroyed by mechanical and chemical agencies during successive sedimentary cycles. Therefore, they conclude that the presence or absence of undulatory extinction is of very limited usefulness in determining the provenance of sediments.

2. CRISTOBALITE AND TRIDYMITE

Cristobalite may show twinning characteristics, often as a result of inversion after the high temperature form. A twin plane has been observed at 45 or 90° to the cubic axis of the original high cristobalite (Sosman, 1965, p. 274). Spinel-type twins on the 111 plane are also common (Henderson et al., 1971).

Tridymite is invariably twinned. In euhedral crystals, a "secondary" wedge-shaped twin commonly occurs as an inversion pseudomorph after high hexagonal tridymite (Fig. 14-6). Each wedge or thin plate of tridymite may in turn give rise to repeated twins; repetition of this twin gives a trilling habit (crystal consisting of three twinned individuals) to tridymite which is so common as to have given the mineral its name (Sosman, 1965, p. 273).

3. OPAL

Characteristically opal is isotropic and lacks birefringence. Exceptions to this generality are specimens which have undergone strain or contain crystalline phases. Examples of the latter have been reported by Beavers and Stephen (1958) in opal phytoliths isolated from a Farmdale loess paleosol in Illinois and by Yarilova (1952) in a Russian Chernozem soil. These workers attributed the birefringent margins, interior zones, or entire bodies to conversion of opal to chalcedony. Wilding and Drees (1974) also noted birefringent zones in opal phytoliths isolated from forest soils but concluded that the crystalline phases were cosynthesized with opal in the tree leaves prior to deposition of the opal in the soil.

Plant opal upon dehydration and crystallization (either naturally or under laboratory conditions) undergoes an increase in refractive index of 0.03 to 0.05 (L. H. P. Jones & Milne, 1963; Yarilova, 1956). Different portions of the same opal phytolith may have different refractive indices, either suggesting partial conversion of opal under ambient soil conditions to crystalline phases (Beavers & Stephen, 1958; Yarilova, 1952) or cosynthesis of crystalline and "amorphous" phases in vegetative tissue prior to deposition in the soil (Wilding & Drees, 1974).

E. X-Ray Diffraction Properties

X-ray diffraction provides a positive means to identify silica poly-morphs. Quartz, cristobalite, and tridymite generally exhibit a unique set of X-ray diffraction maxima (Table 14-5); opal has long been considered X-ray amorphous, but likewise yields unique diffraction effects that will be discussed herein. Disordered cristobalite, tridymite, and opal are often confused in the literature because they are not consistently differentiated.

1. QUARTZ

Quartz is easily identified because it yields a characteristic X-ray pattern with an abundance of intense, well-defined peaks (Table 14-5). For quantitative determinations of quartz, either the 101 (d = 3.34Å) or 100 (d = 4.26Å) peak is generally employed. However, the intensity between these peaks is variable with the intensity ratio of 100 to 101 ranging from 0.23 to 28.5 (Eslinger et al., 1973; Smith, 1962). These differences are attributed to preferred orientation of secondary quartz in the sample. Eslinger et al. (1973) indicate that secondary quartz tends to form well-developed prism faces which results in a greater percentage of grains lying with their Z-axes parallel to the slide than would be expected for a randomly oriented sample. Consequently the authors believe that the ratio I_{100}/I_{101} can be used to distinguish between detrital and secondary quartz in sedimentary rocks.

2. CRISTOBALITE AND TRIDYMITE

Although cristobalite possesses a unique set of diffraction maxima, the expression of these peaks may change depending on location and mode of genesis. Synthetic and volcanic cristobalite usually exhibits a full array of

Table 14-5. X-ray d-spacing (Å) and intensity (I)[†] of major diffraction peaks of quartz, tridymite, cristobalite and orthoclase

Quartz		Tridymite				Cristobalite			Orthoclase	
d	I^1	d	I^2	I^3	I‡	d	I^4	I‡	d	I^5
4.26	35	4.30–4.33	100	100	VS	4.04–4.11	100	VS	4.02	90
3.34	100	4.09–4.11	33	80	VS-S	3.13–3.18	11	W	3.18	100
2.46	12	3.80–3.82	67	80	S-M	2.83–2.85	13	W	2.83	60
2.28	12	3.24–3.25		20	W	2.48–2.52	20	W-M	2.53	70
1.82	17	2.96	17	40	W-M	2.12–2.13	5	VW	2.13	40
1.54	15	2.48	27	60	W-M					

[†] Intensity from Joint Committee on Powder Diffraction Standards: 1—Data card no. 5-490, 2—Data card no. 1-378, 3—Data card no. 3-227, 4—Data card no. 11-695, 5—Data card no. 5-402.
‡ Relative intensities. Tridymite: Grant (1967), Rivalenti & Sighinolfi (1968); Cristobalite: Brindley (1957), Hardjosoesastro (1956). VS = very strong, S = strong, M = medium, W = weak, VW = very weak.

peaks (Table 14–5). However, cristobalite associated with bentonite deposits and soils derived therefrom may not yield a well-defined X-ray pattern. In some bentonites, a broad somewhat asymetrical peak in the range of 4.08 to 4.12Å is the only identifying peak which has been attributed to cristobalite (Brindley, 1957; Reynolds & Anderson, 1967; Henderson et al., 1971). However, Wilson et al. (1974) believe this peak is due to disordered tridymite rather than disordered low cristobalite.

Cristobalite associated with post-Cretaceous cherts generally yields a strong but broad diffuse peak in the range of 4.05 to 4.10Å that is indicative of very fine-grained or poorly-ordered cristobalite (Calvert, 1971). Greenwood (1973) has noted a distinct low angle shoulder at 4.30Å which merges with the main cristobalite peak. This shoulder has been attributed to opal or opal-cristobalite phase(s) within the sample (Peterson & Von der Borch, 1965; Greenwood, 1973). Florke (cited by Wilson et al., 1974) indicates that this diffraction pattern is due to a unidimensionally disordered low-cristobalite structure in which layers of cristobalite and tridymite are irregularly stacked in the (111) direction normal to the layers. Wilson et al. (1974), however, attribute the X-ray diffraction spectra to layers of disordered low-tridymite affected by random transverse displacement normal to the Z-axis (00l) while Jorgenson (1970) postulates they represent intergrowths of cristobalite and tridymite. As seen from the literature, there is some doubt as to the origin of the broad 4.1Å diffraction peak. Unless specific diffraction maxima are present for positive identification, it may be more appropriate to refer to the broad asymetrical peaks in bentonites and cherts as a silica phase rather than a specific mineral species.

Diagenic cristobalite found in a Monterey shale yielded increased X-ray peak sharpness and decreased d (101) spacings (4.11 to 4.04Å) with samples that represented successively greater depths of burial and age (Murata & Nakata, 1974). The authors ascribe these changes to increased ordering of the atomic structure as a consequence of increased temperature and pressure.

In soils, care must be taken not to confuse cristobalite with feldspars. The 4.04, 3.14, 2.84, and 2.49Å peaks of cristobalite superpose or fall very close to respective orthoclase peaks (Table 14–5). If one suspects both cristobalite and orthoclase in a sample, positive differentiation will depend on identifying other characteristic peaks of orthoclase (i.e., the 6.44, 3.90, and 2.93Å peaks). If an intense 4.04Å peak is obtained in combination with a very weak or nondetectable 3.14Å peak, this provides a useful clue that cristobalite rather than orthoclase is present in the sample. The 3.14Å peak of orthoclase is ideally more intense than its 4.04Å peak.

A cursory examination of the limited tridymite X-ray data indicates a wide disparity in the intensity of the major peaks (Table 14–5). Such disparities would diminish the utility of using X-ray procedures to quantify tridymite. Such variations are attributed primarily to sample preparation (inability to attain a truly random sample) with secondary effects due to structural distortions by chemical impurities. A comprehensive survey of tridymite X-ray diffraction data is given by Grant (1967).

3. OPAL

Since earlier work that suggested all opal was X-ray amorphous, (Frondel, 1962, p. 288), it is now generally accepted that opaline silicas fall into two broad categories: (i) those that exhibit diffuse bands of disordered low-cristobalite (Lanning et al., 1958; J. B. Jones et al., 1963, 1964; R. L. Jones & Beavers, 1963a; L. H. P. Jones & Milne, 1963; Arimura & Kanno, 1965; Twiss et al., 1967; Wilding & Drees, 1974); and (ii) specimens which yield characteristic X-ray patterns of crystalline silica polymorphs superposed on the diffuse bands of disordered cristobalite (Lanning et al., 1958; Teodorovich, 1961; J. B. Jones et al., 1963, 1964; Fondel, 1962, p. 288; Wilding & Drees, 1974).

Diffuse bands in the region of 8.8–10Å and/or 4.1–4.3Å have been reported for geologic opal and biogenic opal isolated from soils developed under forest and grass vegetation. These diffraction lines in the 4.1–4.3Å region are essentially the same as observed for < 2 μm selected tree-leaf isolates by Wilding & Drees (1974). Opals with these diffraction effects have been termed "X-ray amorphous" or simply "amorphous." Although diffuse bands may be characteristic of opal, they are not diagnostic nor likely detected in samples when opal is a minor constituent.

Opal isolates from soil, and vegetative materials commonly yield characteristic peaks of low-quartz and low-cristobalite in addition to the diffuse bands of the major opaline phase. These diffraction patterns are very similar to those reported for geological opal. In some cases these crystalline phases are attributed to opal diagenesis upon aging (J. B. Jones et al., 1964) while in others it is believed to represent coprecipitated phases during opal genesis (Wilding & Drees, 1974). It is also well established that heating "amorphous" opaline specimens results in synthesis of crystalline silica phases—low-cristobalite, low-tridymite and low-quartz (J. B. Jones et al., 1964; L. H. P. Jones & Milne, 1963; Arimura & Kanno, 1965; Lanning et al., 1958). Heating opal to 940C (1213K) for 10 hours yields a sharp cristobalite reflection at about 4.04Å (Goldberg, 1958); this treatment is commonly used as an identification criterion. Conversion of plant opal to crystalline polymorphs apparently occurs at much lower temperatures in the presence of alkali and alkaline-earth impurities. These impurities are believed to catalyze the reaction (Arimura & Kanno, 1965). For example, L. H. P. Jones & Milne (1963) indicate that partial conversion of plant opal to cristobalite can occur at temperatures as low as 450C (723K). Thus, isolation of opal from vegetative tissues by ashing may result in crystalline silica artifacts. Low-temperature ashing [60C (333K)] or wet-ashing procedures would be desirable alternatives (Wilding & Drees, 1974).

F. Electron Optical Properties

The transmission electron microscope (TEM) has been widely employed to examine replicas of the surface texture of quartz grains (Biederman, 1962;

Porter, 1962; Krinsley et al., 1962; Krinsley & Takahashi, 1962a, 1962b, 1962c, 1962d, 1964; Kuenen & Perdok, 1962; Shoji & Folk, 1964; Krinsley et al., 1964; Krinsley & Funnell, 1965; Krinsley & Donahue, 1968a, 1968b) and the submicron structure of opal (J. B. Jones et al., 1964); however, it has not found wide application in the identification of silica minerals per se. Likewise, this technique has been employed by Lang (1968) to locate crystal defects in the quartz structure by observation of overlap moiré patterns. When morphology is unique and not readily confounded with other minerals, identification of a component may be made directly by the scanning electron microscope (SEM); this is often the case with characteristic opaline forms. The advantages of the SEM over the TEM are adequately reviewed by Krinsley and Margolis (1969) and Schneider (1970). However, relying solely on morphology fosters misidentification (Wilding & Geissinger, 1973) and thus, the SEM should be complemented whenever possible with more direct means of identification such as energy dispersive X-ray spectroscopy (Klasik, 1975), X-ray diffraction, infrared, TEM, or light microscopy.

1. QUARTZ

The SEM has been used extensively by Krinsley and co-workers to expand and refine the criteria used in the evaluation of surface textures. Krinsley and Doornkamp (1973) recently reported that surface morphology is basically recognition and assessment of the following features: (i) the occurrence of conchoidal breakage patterns; (ii) the occurrence of flat cleavage plates; (iii) the presence of upturned plates on cleavage or crystal faces (cleavage traces); and (iv) solution-precipitation alteration of these features. Examples of these features are shown in Fig. 14-16 and 14-18.

Krinsley and Tovey (1973) have used the cathodoluminescence mode with the scanning electron microscope to study sand grain surface textures. They attribute the presence of noncathodoluminescence areas to the occurrence of a disrupted surface layer due to grinding and believe this mode is a valuable tool for examining the disrupted surface layer on quartz.

2. CRISTOBALITE AND TRIDYMITE

Cristobalite in cherts consists of very fine blades 300 to 500Å thick which agglomerate into spherules 3 to 10 μm in diameter (Fig. 14-15) as identified by Weaver and Wise (1972), Wise et al. (1972), Wise and Kelts (1972), and Wise and Weaver (1973). Oehler (1973) concluded from SEM observation of synthetic microspheres, similar to spherules in cherts, that the basic morphology of the blades comprising the spheres was more characteristic of tridymite than cristobalite. Euhedral crystals with octahedral and tabular hexagonal forms are clearly observed in SEM micrographs of middle Eocene chert deposits; they are interpreted as cristobalite and tridymite, respectively (Klasik, 1975). Although morphology is indicative of high-temperature forms, geothermal evidence suggests chemical precipitation at near ambient temperatures. Electron diffraction of a deep sea "cristobalite

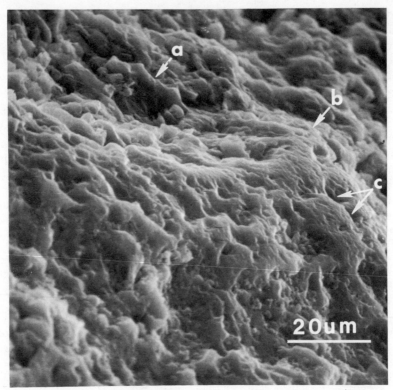

Fig. 14-18. Solution, precipitation and upturned plates of quartz from Roslyn, New York, USA, Dune and Upper Pleistocene. Micrograph shows numerous upturned plates, most of which are relatively unmodified. However, a certain amount of solution and precipitation is evident. Prominent examples of each of these features are marked by arrows: (a) upturned plates; (b) precipitation (light area in vicinity of arrow); and (c) solution pits. Reprinted from micrograph 84, Krinsley & Doornkamp (1973) with permission of Cambridge University Press.

chert" and "cristobalite" from bentonite deposits suggests that the platy particles are hexagonal, indicative of tridymite and not cristobalite (Wilson et al., 1974). Jorgensen (1970) suggests that phases described as "disordered cristobalite" in cherts and "disordered tridymite" as indicated above may actually be intergrowths of the two phases.

Cristobalite in an upper Eocene Helms bentonite consists of crystallites with a porous, spongy surface texture (Henderson et al., 1971) while cristobalite of hydrothermal origin consists of large blocky particles. Cubic forms of cristobalite have been observed in soils and attributed to pedogenic origin (Eswaran & DeConinck, 1971).

3. OPAL

L. H. P. Jones et al. (1966) illustrate differences in submicron structure between an opaline phytolith (tabashir from bamboo cane) and geologic opal.

In gem opal, spheres which comprise the submicron structures are more regular in size, larger, and more systematically arranged than those observed in tabashir (Fig. 14-4).

Using TEM, Shoji and Masui (1971) recognized circular, elliptical, rectangular, and rhombic platy forms of opaline silica in recent volcanic ash soils in Japan. Most of these occurred in the 0.2 to 5 μm size range. Electron micrographs of replicas exhibited discrete areas composed of close-packed aggregates that the authors note are similar to those reported for precious opal. They attribute the origin of these constituents to authigenic precipitation from supersaturated silica solutions in surface horizons. The illustrated opal forms and their increasing concentration toward the soil surface (Shoji & Masui; 1969) both suggest a possible plant origin (Parfitt, 1974).

The TEM provides a potentially useful means to identify opal in the clay fractions of soils and to distinguish between different opaline origins. As yet, its application for this purpose has been limited. Electron diffraction analysis of opal specimens has likewise not been utilized. L. H. P. Jones et al. (1966) have utilized electron diffraction and found a diffuse ring at about 4.0Å for tabashir, opal phytoliths, and silica gel. These silica forms could not be distinguished by this technique. Sanders (1975) concluded from electron diffraction of precious opal that the 4Å ring was resolved into two components: (i) the first consists of a central portion containing many spots corresponding to cristobalite; and (ii) the second comprising more sparsely populated outer spots from a smaller concentration of tridymite. Thus, the ring represents patterns from these two phases superimposed.

Characteristic opaline forms have been identified with the SEM in soil and vegetative opaline isolates of fine silt and clay fractions (Wilding & Drees, 1974). Many of the opaque constituents observed in opaline isolates have been positively identified by SEM (Fig. 14-11) as biogenic opal (Wilding & Drees, 1973); because of opacity, these constituents could not be identified using light microscopy. Figures 14-8, -10, -11, -12, and -17 illustrate characteristic plant opal forms that have been observed with the SEM. Through use of the SEM, differences in morphology of grass vs. forest opal were first recognized (Wilding & Drees, 1971).

Little or no application has been made of the SEM to examine HF-etched biogenic opal surfaces for submicron structure. This may represent a fruitful approach to identification of opal in clay fractions when morphological distinctions are not clearly evident. The SEM has been employed by Greer (1969) to examine the submicron structure of geologic opal.

G. Thermal Properties

At atmospheric pressures silica minerals are stable in the following temperature ranges (Berry & Mason, 1959, p. 472):

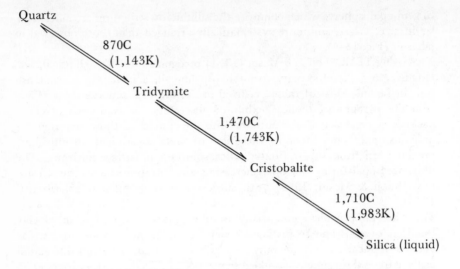

Although tridymite and cristobalite are considered metastable, they may persist indefinitely at the earth's surface; likewise they may be synthesized under ambient temperatures and pressures of soil environments. In addition to pressure, the stability field can also be markedly altered in the presence of alkali and alkaline earth impurities.

From Fenner's (1913) temperature-pressure diagram of the silica system, Mitchell (1975) has presented the following sequence of inversions and conversions:

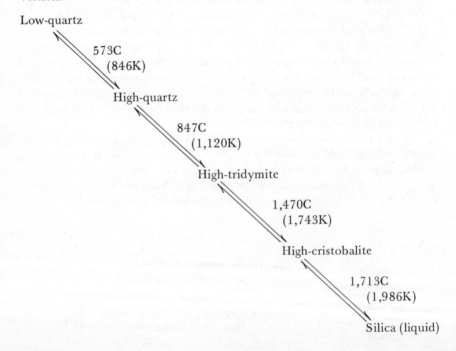

The low-temperature inversions associated with tridymite and cristo-balite are as follows:

Low-tridymite $\underset{117C\ (390K)}{\overset{\longrightarrow}{\rightleftharpoons}}$ Middle-tridymite $\underset{163C\ (436K)}{\overset{\longrightarrow}{\rightleftharpoons}}$ High-tridymite

Low-cristobalite $\underset{180\ to\ 270C\ (453-543K)}{\overset{\longrightarrow}{\rule{0pt}{0pt}}}$ High-cristobalite

Opal does not yield a characteristic inversion phenomenon.

1. QUARTZ

Quartz generally exhibits a sharp endothermic peak near 573C (846K). Faust (1948) suggests that quartz be used for calibration of DTA apparatus due to its constant inversion peak. Keith and Tuttle (1952) conclude from a study of 250 different quartz samples that virtually all quartz undergoes the low-high inversion between 572 and 574C (845 and 847K). The endothermic peak on heating and exothermic peak on cooling for quartz is shown in Fig. 14–19.

The characteristic inversion peaks are commonly not evident for chalce-dony (Frondel, 1962, p. 201; Buurman & Van Der Plas, 1971). At a "high" heating rate of 15C (288K) min^{-1}, St. J. Warne (1970) reported that peaks for agate and chalcedony occurred at somewhat higher temperatures and were smaller and less clearly defined in comparison to quartz (Fig. 14–19). Moore and Rose (1973) report that quartz which was ground for 400 hours lacked an inversion peak when heated at the normal rate of 10C (283K) min^{-1} but showed a pronounced peak when heated at 18C (291K) min^{-1}. They attribute the lack of an inversion peak for finely ground or crypto-crystalline quartz at low heating rates to rapid dissipation of the heat within the sample. A change in sample temperature is not noted because each in-dividual specimen grain undergoes inversion at a slightly different tempera-ture with the heat conducted to the sample as a whole. At higher heating rates, the inversions of all of the quartz grains will occur within a narrower time frame, allowing for less dissipation of the differential temperatures pro-duced and resulting in a greater temperature fluctuation than at lower heat-ing rates. This hypothesis is supported by Smykatz-Kloss (1972) who states that microcrystalline quartz generally shows no sharp inversion point because the inversion takes place over an interval of nearly 50C (323K).

Detection limits for crystalline quartz are on the order of 5% with the highly sensitive commercial DTA units presently available. The detection limits for the microcrystalline varieties are somewhat greater, but by using high heating rates, the characteristic inversion peak can be identified.

2. CRISTOBALITE AND TRIDYMITE

Hill and Roy (1958a) show that poorly ordered cristobalite has a lower inversion temperature than well ordered specimens. Crystal order in cristo-

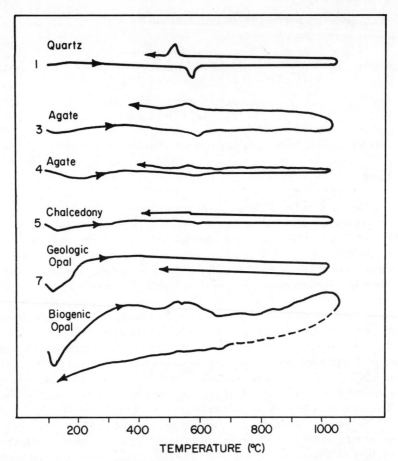

Fig. 14-19. A comparison of DTA curves of quartz, agate, chalcedony, and opal. Biogenic opal of plant origin isolated from 20-50 μm fraction a forest soil. All but biogenic opal modified from Fig. 1 (numbers 1, 3, 4, and 5 and 7 of Warne (1970)).

balite is increased upon heating; this results in an increased inversion temperature and decreased inversion range. Apparently, differences in low-high
temperature inversions are more a consequence of crystal order than structural impurities. The inversion hysteresis effect that occurs between heating
and cooling curves in cristobalite is variable. In general, Sosman (1965, p.
110) states that the higher the inversion temperature, the greater the temperature differential between endothermic and exothermic reactions.

Tridymite exhibits a more complex series of endothermic reactions
than either cristobalite or quartz. Data by Hill and Roy (1958b), considered
in more detail by Sosman (1965), suggest a stable form of tridymite (tri-
dymite-S) that exhibits endothermic reactions at 210 and 475C (483 and
748K) in addition to the well-established endotherms at 117 and 163C

(390 and 436K); the latter two temperatures have been observed to range from 105 to 128C (378-396K) and 145 to 170C (418-443K), respectively (Sato, 1964b; Sosman, 1965, p. 99). Hill and Roy (1958b) believe that differences in thermal reactions of tridymite-S vs. the metastable variety (tridymite-M) are due to differences in polytypism. Alternatively, Florke (cited by Sato, 1964a) believes thermal differences are due to a stacking disorder between cristobalite and tridymite layers; tridymite-M is postulated to contain cristobalite layers in the structure. X-ray and thermal data indicate the existence of two types of tridymite, but it is not known if the difference is due to polytypism or stacking disorder (Sato, 1964a).

3. OPAL

Thermal responses do not provide diagnostic criteria for opal identification. Biogenic opal of plant origin yields thermal effects ranging from none (R. L. Jones & Beavers, 1963a) to one or several endotherms in the region of 95 to 120C (368-393K) (unbound water) and a single broad exotherm in the region of 350 to 500C (623-773K) (L. H. P. Jones & Milne, 1963; Twiss et al., 1967); the latter is likely due to oxidation of occluded organic matter (Fig. 14-19). Buurman et al. (1973) note an endothermic reaction between 130 and 150C (403 and 423K) for recent silicified plant remains in acid sulphate soils which is not associated with an increased weight loss; they believe this represents an inversion in the crystal structure.

J. B. Jones et al. (1963) report DTA analysis for 30 geological opal specimens and broadly classify them into the following categories: (i) those with a very small or no apparent endotherm(s) between 100 and 200C (373 and 473K); (ii) those yielding a prominent rounded endotherm starting at about 90C (363K), with a peak in the range of 125-140C (398-413K); and (iii) those with a strong sharp endotherm starting at 90C (363K) and with a peak at 140C (413K). No correlation was found between water content, composition, and DTA data.

Opal does not yield diagnostic thermogravimetry curves because most of the water is lost continuously over a broad temperature range rather than in characteristic steps (Buurman et al., 1973; Wilding et al., 1967; Twiss et al., 1967). In contrast, L. H. P. Jones and Milne (1963) suggest that water released from plant opal occurs in two distinct temperature ranges [60 to 150C (333-423K) and 570 to 670C (843-943K)], but these two ranges were not evident on accompanying DTA curves. Weight losses cannot be solely attributed to water losses when occluded organic matter is present (Wilding et al., 1967).

J. B. Jones and Segnit (1969) report that physically held water in large capillaries of opal is lost below 150C (432K). Between 150 and 400C (423 and 673K), centered on 200C (473K), the loss is attributed to twin surface OH groups around 200Å submicron spheres; above 400C (673K) the water lost is attributed to single OH groups (silanols) bound to similar surfaces.

H. Infrared Properties

1. QUARTZ

The infrared spectrum of quartz (Fig. 14-20) contains three main elements in the 600 to 5000 cm^{-1} region: a strong broad absorption band at 1085 cm^{-1}; a medium sharp absorption doublet with bandheads at 800 and 780 cm^{-1}; and a weak sharp band at 695 cm^{-1} (Hunt et al., 1950; Chester & Green, 1968). According to Gaskell (1966), another very broad bandhead occurs in the vicinity of 455 cm^{-1}. Gaskell (1966) also indicates that as the temperature is raised, the intensity of all the bandheads decrease and there is a shift of the bandheads to a lower frequency. Chester and Green (1968) selected the doublet at 800 and 780 cm^{-1} for the identification and quantitative estimation of quartz. They conclude that the particle size of the sample must be less than that of the minimum wavelength used for quantitative infrared analysis. This aspect is considered further under Quantitative Determination.

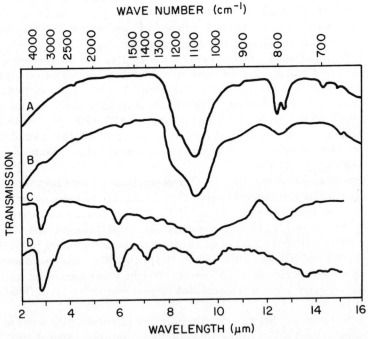

Fig. 14-20. A comparison of infrared spectra of: *(A)* low-quartz, *(B)* geological opal, *(C)* untreated biogenic opal isolated from Warsaw soil, and *(D)* HF residue of biogenic opal from same soil. *A* and *B* modified from Fig. 5, Hunt et al. (1950) and *C* and *D* are modified from Fig. 1, Wilding et al. (1967).

2. OPAL

Infrared absorption spectra for plant opal isolated from soils (Fig. 14–20) and plant tissues (R. L. Jones & Beavers, 1963a; Arimura & Kanno, 1965; Twiss et al., 1967; Wilding et al., 1967) closely resemble those reported for synthetic silica gel (Hunt et al., 1960; Yoshida et al., 1962) and opal of geologic origin (Hunt et al., 1950; R. L. Jones & Beavers, 1963a). Slight differences in exact peak positions are reported among the above spectra, but IR bands are in the following regions: 460 to 470, 785 to 800, 1050 to 1110, 1625 to 1650 and 3425 to 3700 cm^{-1}. Absorption effects at 785 to 800 and 1050 to 1110 cm^{-1} are attributed to the Si–O and Si–O–Si stretching bonds of siloxanes. In this same region (1030 cm^{-1}) a silanol (Si–OH) deformation band occurs. Peaks at about 1650 cm^{-1} have been attributed to the O–H deformation bonds of unbonded pore water (Arimura & Kanno, 1965). This band may also be attributed to C=O stretching vibration and/or C=O aromatic band for those opal specimens which contain occluded carbon (Wilding et al., 1967). Absorption at about 3550 cm^{-1} can be attributed to either bonded O–H stretching vibrations of silanol groups (Arimura & Kanno, 1965; Wilding et al., 1967) or to pore water (R. L. Jones & Beavers, 1963a) depending on method of sample preparation. If samples are not prepared under anhydrous conditions the IR absorption band in this region may represent combined absorbed water and bonded OH groups (Arimura & Kanno, 1965). Detailed work by J. B. Jones and Segnit (1969) on the IR spectra in the region of 3550 to 3700 cm^{-1} indicate that the major absorption peak at 3700 cm^{-1} is due to stretching O–H vibrations of single OH groups and the second weaker one at 3550 cm^{-1} to twin OH groups.

When opal contains appreciable occluded organic matter, additional IR bands similar in many respects to humic acids (Fig. 14–20) may be observed (Wilding et al., 1967).

III. QUANTITATIVE DETERMINATION

Quantification of silica components in soils has entailed optical microscopy, specific gravity fractionation, differential dissolution, X-ray diffraction, thermal analysis, and infrared techniques. Commonly, several of these methods are used in combination. The precise method(s) employed depends upon the quantity and form of silica components in soils, the probable confounding of silica polymorphs, the accuracy desired, and the objectives of the work. The simplest means of quantifying silica components is by microscopic counts of the entire particle size separate. This method is limited by light optical resolution which has a theoretical limit of 0.16 μm, but a practical limit of 5 to 10 μm. Accuracy of this method also suffers when the sample contains only small quantities of the mineral in question. For example, if a separate contains 5% opal, over 7000 counts would be required to achieve a probable error of 10% with a 95% confidence interval. If only 400 grains were counted, the probable error would be 95% (95% confidence interval) or 57% with a 50% confidence interval (Brewer, 1964, p. 47).

A. Quartz

Since quartz is not attacked by HCl, HNO_3, or H_2SO_4, dissolution at room temperature is obtained by the use of HF, warm NH_4F, or strongly alkaline solutions. The latter act slowly at room temperature, but at elevated temperatures and pressures quartz is readily attacked by NaOH, KOH, Na_2CO_3, Na_2SiO_3, and $Na_2B_4O_7$. Quartz as well as other silica minerals is also readily dissolved in fusions of borax, NaOH, Na_2CO_3, and $KHSO_4$, containing added fluorides. One of the most frequently employed dissolution techniques for the isolation and quantitative determination of quartz is that of Trostel and Wynne (1940). By their procedure, the sample is fused with potassium pyrosulfate and extracted with $2.5N$ NaOH. Only quartz and feldspars persist by this treatment. Quartz and feldspars are recovered and determined gravimetrically. Rowse and Jepson (1972) indicate that this technique suffers in several respects. Minerals other than quartz and feldspars are only partially attacked and quartz is lost physically by the passage of fine quartz through the filter paper during the washing stage and chemically through solution in hot $2.5N$ NaOH. Kiely and Jackson (1965) have reported a similar procedure using pyrosulfate which avoids some of these problems. They employed centrifugation rather than filtration during washing of the insoluble residues and used $0.5N$ NaOH which is adequate for dissolution of residual layer silicates but does not dissolve as much quartz and feldspar as the $2.5N$ NaOH used by Trostel and Wynne (1940). In the method of Kiely and Jackson (1965), Na-pyrosulfate is used rather than K-pyrosulfate to avoid introducing K into the plagioclase feldspars at the expense of Na and Ca. They also employ HCl to wash the residues after fusion to remove divalent cations which interfere with the subsequent dissolution of residual layer silicates in hot NaOH (Hashimoto & Jackson, 1960). After the NaOH extraction, the residue is analyzed for K, Na, and Ca with each element being allocated to its respective feldspar. The silica remaining after allocation to the feldspars is considered to be quartz. Whereas Rowse and Jepson (1972) would not advocate the method of Trostel and Wynne (1940) for the determination of small quantities of quartz in clays, Kiely and Jackson (1965) indicate that their method yields an accurate determination of the quartz content in clays and that in samples low in feldspars, quartz can be determined within ± 1%. In a later modification of this method, Chapman et al. (1969) proposed a method to determine quartz directly. After the NaOH treatment they impose a cold hydrofluorsilicic acid treatment which completely removes the feldspars, leaving only quartz. Cristobalite, tridymite and opal would also be measured by above methods, but commonly do not represent serious confounding agents because of their paucity in soil environments.

When present in significant quantities they can be quantitatively isolated as discussed later.

Since chemical techniques are rather time consuming and require considerable analytical ability, other techniques such as X-ray diffraction are frequently employed for the quantitative determination of quartz. Phillippe and White (1950) utilized the ratio of the height of the 3.34Å peak of quartz to the height of the 2.32Å peak of sodium fluoride (an internal standard) as a quantitative determination of the amount of quartz in the silt fraction of soils. They obtained a standard deviation of 2.91%. Using a slight modification of this method, Johnson and Beavers (1959) determined the quartz content of the 2- to 20-μm and 20- to 50-μm fractions of some loessial soils in Illinois; they obtained standard deviations of 2.84% and 4.10%, respectively, for determinations made at different positions on the same slide and 3.56% and 4.72%, respectively, for different slides of the same sample. The use of peaks other than the 3.34Å for quantitative determinations was investigated by Rowse and Jepson (1972). They reported relative standard errors in peak heights of 2.6%, 8.7%, and 12.2% for the peaks reflecting spacings of 3.34Å, 4.26Å, and 1.82Å, respectively.

Differential thermal analysis has also been used for quantitative estimates of quartz content. Grimshaw (1953) indicates that a reproducible rate of temperature change is essential for accurate estimations of quartz by DTA using the low-high inversion peak. Consequently he advocates raising the temperature above the low-high transition temperature [573C (846K)] and monitoring the cooling curve. Using this technique on pure quartz crystals, Grimshaw (1953) obtained (i) excellent reproducibility of peak height, (ii) sharp peaks, (iii) negligible drift, and (iv) a direct proportion between the peak height and quartz content. However, Grimshaw (1953) also revealed that DTA underestimates the quartz content when the sample contains cryptocrystalline varieties of quartz or very finely ground quartz. The reasons for this were considered in the previous discussion of the thermal properties of quartz. Nevertheless, Rowse and Jepson (1972) concluded that DTA is better than either X-ray diffraction or chemical techniques for detecting small quantities of quartz in clay materials. They obtained a relative standard error of approximately 10% with a sample containing 4% quartz.

Quantitative estimates of quartz are also possible by the use of infrared techniques. Hunt and Turner (1953) used spray techniques to introduce the sample into the spectrophotometer and reported a relative accuracy of approximately 10%. Chester and Green (1968) introduced the sample with a KBr disc and employed the infrared doublet at 780 and 800 cm^{-1} for quantitative determination of quartz. They found that the particle-size must be less than the minimum wavelength and that particle-size must be reduced until further reduction does not result in increased absorption of the 800 cm^{-1} bandhead. The sample must be opal free and the technique is not recommended for samples containing < 3% quartz. When these conditions are met, the infrared technique of Chester and Green (1968) compares very favorably with X-ray or chemical methods.

B. Cristobalite and Tridymite

Most soils contain small or negligible quantities of cristobalite and tridymite; thus, quantification of these components necessitates some means of concentration. Commonly particle size fractionation, specific gravity, and/or differential dissolution procedures are used for this purpose.

Specific gravity isolation of cristobalite and tridymite follow procedures similar to those outlined for opal quantification. Cristobalite is separated from heavier soil minerals using a heavy liquid of specific gravity 2.38; tridymite and opal are subsequently separated from the light isolate containing cristobalite by fractionation with a heavy liquid of 2.3. Tridymite and opal cannot be differentiated by specific gravity because of overlapping densities.

Henderson et al. (1971, 1972) described in detail a procedure for cristobalite isolation employing a combination of heavy liquid-differential dissolution treatments. Upon specific gravity separation, the light isolate containing cristobalite is treated with acid ($6N$ HCl), base ($0.5N$ NaOH) and hydrofluosilicic acid to differentially remove more labile mineral contaminants such as feldspars.

Identification of tridymite and cristobalite in soils and sediments has been accomplished principally by X-ray diffraction. Unfortunately, for these minerals wide variations, occur in the intensity and line profile of diffraction peaks; this makes quantification by X-ray diffraction difficult, if not impossible. For example, some disordered low-cristobalites may exhibit only a broad 4.1Å peak with a weaker 2.5Å peak while varieties of volcanic origin may yield their full complement of diffraction maxima (Table 14-5).

C. Opal

Detrital opal in soils can be quantitatively analyzed by gravimetric (R. L. Jones & Beavers, 1964b; Rovner, 1971), differential dissolution (R. L. Jones, 1969), and light optical microscopy techniques (Smithson, 1958; Witty & Knox, 1964; Norgren, 1973). Upon particle-size fractionation of the total soil, opal is determined on one or several of the sand and silt separates. The gravimetric procedure is outlined in detail by Rovner (1971) and more briefly by R. L. Jones and Beavers (1964b); it entails separation of the opal from heavier soil minerals by a sink-float method. The opal is isolated by repeated centrifugation and decantation of the floating minerals in a nitrobenzene-bromoform solution (or other suitable heavy liquid) of specific gravity 2.3. Opal is computed as the light-isolate weight percentage of the total separate after correction is made for opal purity by microscopic examination. This procedure is effective for > 5 μm size fractions. To achieve an opal purity of about 95% requires 5 to 10 centrifugation-decantation cycles with subsequent purification of the light isolate repeating the centrifugation-decantation steps several more times. The specific gravity of the heavy liquid must be adjusted after each 2 or 3 cycles because nitrobenzene differentially volatilizes from the mixture. This method is less effective with the < 5 μm separates because agglomeration of particles occurs. Fine silt and clay-size separates can be dispersed in a surfactant such as 9% polyvinylpyrrolidone in ethanol (Henderson et al., 1972; Francis et al., 1972) to help overcome this limitation. Zonal density gradient centrifugation should also provide a fruitful means of gravimetric determination of opal in fine silt and clay-size fractions of soils (Francis & Tamura, 1972).

R. L. Jones (1969) found a close relationship between silica dissolved in a 20 min digestion of boiling $0.5N$ NaOH and the opal content of the 20 to 50-μm soil fraction. He established that silica dissolution for opal followed first-order kinetics; about 33% of the opal was dissolved in a 20-min reaction period. Although more rapid than gravimetric quantification, this approach does not permit identification of opaline forms nor possible multiple origins of biogenic opal in the sample. The reaction-rate constant will be dependent upon opaline origin (i.e., grass vs. forest) and particle-size. Morphology governs surface area available for dissolution and striking differences occur between forest and grass opal in this regard. A differential dissolution procedure would be a valid means of opal quantification within a limited geographical region where most of the opal has been deposited by similar fauna or flora species. Separate relationships would need to be established for each particle-size analyzed. This procedure is also adapted to quantifying nondetrital opal occurring as a cementing agent in soils.

In summary, an effective quantitative assessment of opal in soils involves a combination of gravimetric and dissolution methods supported by microscopic evaluation to determine opaline form and possible origin.

IV. STABILITY AND SYNTHESIS OF SILICA MINERALS

A. Synthesis of Silica Minerals

In a recent review article, Mitchell (1975) proposes that silica occurs as every intermediate stage in a gradual transition between monomeric silicic acid and solid mineral bodies. In a simplistic model this can be envisioned as:

$$\text{Silicic acid} \rightarrow \text{Hydrosols} \rightarrow \text{Hydrogels} \rightarrow \text{Xerogels}.$$

This concept involves molecular solubility (silicic acid), homogeneously dispersed colloids (hydrosols), nonrigid gels (hydrogels) and rigid gels (xerogels). The system is an irreversible continuum; the boundaries are not sharp; and the entities are only arbitrarily definable.

In going from silicic hydrosols to gels a long time is required for the condensation and polymerization of monomeric to polysilicic acids, a short time for the actual gelation phase, and then a long time (perhaps several million years) for aging of gels from one mineral to another (Mitchell, 1975). There are apparently a number of alternative pathways for silica syn-

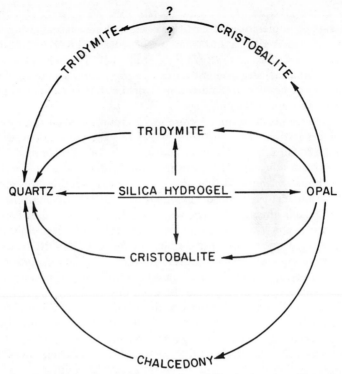

Fig. 14–21. Proposed pathways for conversion of silica hydrogel to silica polymorphs.

thesis, starting with silica hydrogel as a precursor (Fig. 14–21). Refinement and documentation of these routes under given conditions are often difficult if not impossible. A continuum exists from short-range to long-range crystal order. Direct neoformation of quartz, tridymite, cristobalite, and opal are possible as well as secondary transformations through either opal or crystalline intermediates to quartz (Fig. 14–21).

1. QUARTZ

Quartz is formed as both a primary mineral, crystallization from magma, and as a secondary mineral, synthesis at ambient temperatures and pressures. Quartz is one of the last minerals to crystallize from magma; hence it is synthesized under conditions closer to present earth-surface conditions than minerals crystallized earlier. This fact contributes to the high stability of quartz. Goldich (1938) uses quartz as an example of megascopic minerals representing the highest degree of mineral stability. Dense packing of the crystal structure, high activation energy required to alter the Si–O–Si bond (Stöber, 1967), and the low content of impurities contributes to the high stability of quartz relative to the other silica polymorphs.

Evidence for the direct precipitation of quartz from supersaturated solutions at earth-surface conditions is very meager. Morey et al. (1962) concluded that quartz precipitated from a solution which dropped from 37 mg/liter Si to 2.8 mg/liter Si in 30 days, however, he did not document the precipitation of quartz. Recently, Mackenzie and Gees (1971) reported the direct precipitation of quartz from seawater onto existing quartz grains and Harder (1971) reported the precipitation of quartz from very dilute Si solutions in a short time at low temperatures. Mackenzie and Gees (1971) used the scanning electron microscope, with further confirmation by X-ray diffraction, to observe the precipitation of quartz which occurred on quartz grains free of sorbed materials at a steady state silica concentration just slightly above 2.0 mg/liter Si at 20C (293K). They further indicate that the precipitation of quartz may be inhibited in natural environments by quartz grains covered with organic or sesquioxide coatings. Harder (1971) indicates that quartz can be synthesized by the sorption of Si from solutions undersaturated with Si in respect to quartz, onto X-ray amorphous hydroxides of Al, Fe, and Mg.

The formation of quartz at earth-surface conditions is generally attributed to the aging of silica hydrogels (Fig. 14–21). Secondary transformation from opal to chalcedony to quartz is a commonly observed pedogenic transformation in duripans and silica cemented soils (Flach et al., 1969; Brewer et al., 1972). Buurman (1972) suggests the conversion pathway from opal to tridymite to quartz in fossil woods. Other investigations have demonstrated the conversion pathway from amorphous silica to cristobalite to quartz (Ernst & Calvert, 1969; Mizutani, 1970). Ernst and Calvert (1969) suggest the conversion of cristobalite to quartz is a zero-order reaction and at 20C (293K) the transformation would require about 180 million years, but only 4 to 5 million years at 50C (323K). Mizutani (1970) suggests that the transformation is divided into two first-order reaction steps, the first from amorphous silica to cristobalite and the second from cristobalite to quartz.

The conversion of biogenic opal to chalcedony and/or quartz has also been suggested. Beavers and Stephen (1958) and Yarilova (1952) note all stages of opal transformation in soils from unaltered specimens, through opal with marginal chalcedony alteration, to completely altered paramorphs of chalcedony. Yarilova (1952) even reports some prismatic grains reminiscent of "short cells" of grasses that exhibit wavy extinction and a refractive index close to that of quartz. In examining isolates of forest opal, Wilding and Drees (1974) initially believed that these opaline forms were also undergoing crystalline alteration in soils, but after examining opal isolates extracted from forest leaf tissues concluded the birefringent crystalline phases were co-synthesized with opal in the plant prior to deposition in the soil. More recent work in our laboratory (unpublished) confirms the synthesis of quartz and cristobalite in opal isolates from monocotyledons and dicotyledons isolated by either dry-ashing or wet-acid digestion methods (samples courtesy of Dr. J. W. Geis, Department of Forest Botany and Pathology, Syracuse University, N. Y.). Lanning et al. (1958) also reports birefringence and other optical properties characteristic of quartz in silica from lantana (*Lantana camara*

Linn.). Based on the above evidence and the long geologic time-frame commonly cited for conversion of opal to chalcedony or quartz (millions of years, Mizutani, 1970), the concept of paramorph conversion of opal to crystalline phases in soils should be carefully reevaluated.

2. CRISTOBALITE AND TRIDYMITE

Classically, tridymite and cristobalite have been considered a late product of crystallization of magmatic and pyroclastic rocks. Mizota and Aomine (1975) provide recent evidence to support this concept. However, other evidence indicates that these minerals should not necessarily be attributed to high temperature formation. Sosman (1965, p. 54) suggests that cristobalite may form at temperatures as low as 150C (423K). Rogers (1928) considered tridymite a secondary postmagmatic product. Cristobalite in gas cavities of some volcanic rocks is assumed derived by gas transfer after crystallization of the groundmass (Larsen et al., 1936).

In bentonites, shales, deep sea cherts, claystones, and lake clays, cristobalite is often attributed to neoformation from solution (Peterson & Von der Borch, 1965; Henderson et al., 1971; Wise et al., 1972; Weaver & Wise, 1972, 1974; Wise & Weaver, 1973, 1974; Klasik, 1975). Silica solutions for direct cristobalite formation may be derived from the dissolution of biogenic opal, siliceous microfossils, pyroclastic glass, and even quartz (Peterson & Von der Borch, 1965) depending on local environmental conditions and deposits.

As mentioned above, tridymite (Buurman, 1972) or cristobalite (Ernst & Calvert, 1969; Mizutani, 1970) may also form as the immediate precursor to quartz in a conversion sequence from biogenic opal or amorphous inorganic silica to quartz. Evidence for this origin is from hydrothermal studies and the observation that natural cristobalite (and tridymite) are seldom identified in deposits older than Cretaceous.

3. OPAL

Pedogenic opal is synthesized in soils by both organic and inorganic means. Biogenic opal of plant origin is an important component under a wide range of environmental conditions, while inorganic opal as a secondary silica cement is formed only when soil environmental conditions favor supersaturated soluble Si levels. Most of the emphasis herein will be on biogenic opal. Comprehensive reviews regarding silica content, uptake, distribution and loci of opal synthesis in plants are available elsewhere, and except for several general concepts, will not be considered herein (Amos, 1952; Siever & Scott, 1963; L. H. P. Jones & Handreck, 1967; Lewin & Reimann, 1969; Geis & Jones, 1973; Norgren, 1973).

Silicification in cellular tissues and walls of vascular plants is most marked in the *Gramineae* (grasses), *Cyperaceae* (sedges), *Equisetaceae* (horsetails), and *Urticaceae* (nettles). Silicification occurs in the aerial tissues of most plants and in roots and rhizomes of certain grasses (Kanno & Arimura,

1958; L. H. P. Jones et al., 1963; Pease & Anderson, 1969). Root systems of some gramineous species may contribute as much biogenic opal to the soil as the above ground parts (Geis & Jones, 1973).

It is generally accepted that monocotyledons (and *Gramineae* in particular) contain 10 to 20 times the silica content of dicotyledons. Grasses commonly contain 3 to 5% silica on a dry-weight basis, although values of over 20% have been reported (Norgren, 1973). Conifer woods are generally low in silica, but some deciduous woods contain appreciable amounts. Of 400 tree species, Amos (1952, p. 38–53) reports a maximum of 9.4% silica (dry-weight basis) with most ranging from 0.1 to 0.8%. Norgren (1973) reports needles of 10 conifer species that range from 0.2 to 7.9% silica on a dry-weight basis; Wilding and Drees (1971) observed silica contents of seven deciduous tree leaves that range from 0.9 to 8.8%; and Geis (1973) noted the silica content of 22 deciduous angiosperms range from 0.01 to 3.79%.

Uptake of soluble Si by plants is in the form of monosilicic acid, $Si(OH)_4$, through the transpiration stream. In most plants this appears to be a nonselective passive uptake process (L. H. P. Jones & Handreck, 1967), however, in some plants there is evidence for either metabolic exclusion or preferential concentration of Si. Lewin and Reimann (1969) note a number studies that suggest energy from aerobic respiration reactions in the roots is necessary for silicic acid absorption. Silicic acid uptake in diatoms is also apparently linked to aerobic respiration.

B. Solubility of Silica Minerals

Silica solubility decreases as a function of increasing silica tetrahedral packing density and long-range crystal order. Accordingly, decreasing solubilities occur along the following sequence: amorphous silica, opal, cristobalite, and quartz (Fig. 14–22).

The solubility of amorphous silica at room temperature and neutrality (i.e., pH approximately 7) is placed in the range 50 to 65 mg/liter Si by most investigators (White et al., 1956; Krauskopf, 1956, 1959, 1967; Iler, 1973; Frondel, 1962, p. 154; Alexander et al., 1954; Morey et al., 1962). Amorphous silica accumulates as an intermediate in supersaturated solutions because it precipitates more readily than crystalline silica. Even though dissolution and precipitation are faster with amorphous silica than quartz, dissolution and precipitation of amorphous silica are quite slow; equilibrium is not commonly reached for weeks or months (Krauskopf, 1959).

There is great disparity in the literature concerning the solubility of opal (Krauskopf, 1959; Lewin, 1961; Yoshida et al., 1962; Siffert, 1967). Geological varieties of inorganic or fossilized biogenic opal commonly yield solubilities in the range of 10 to 15 mg/liter Si (Fig. 14–23). However, Lewin (1961) reports values as low as 1 mg/liter Si for untreated and 3 to 14 mg/liter Si for acid-cleaned diatomaceous earth ranging in age from upper Cretaceous through Pliocene. Recent diatom silica dissolves readily with

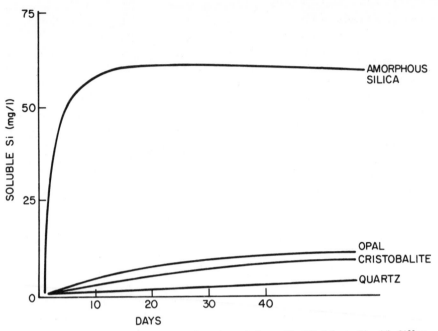

Fig. 14–22. Dissolution of silica as a function of time. Modified from Fig. 15, Siffert (1967).

solubilities approximating that of synthetic silica gel (Lewin, 1961). Kraus-kopf (1956) suggests that geologic opal probably has the same equilibrium solubility as other forms of amorphous silica, but its dissolution is extremely slow. Silica of plant origin is reported to have solubilities similar to that of synthetic silica gel (Yoshida et al., 1962), but little information of this nature is available. Based on observed morphological properties, particle-size, and dissolution evidence one would speculate that between 50 and 75% of silica in plant tissues would yield solubilities in the range of amorphous silica (50–65 mg/liter Si) while the remainder would exhibit markedly lower values. It is quite probable that opal phytoliths isolated from soils would have solu-

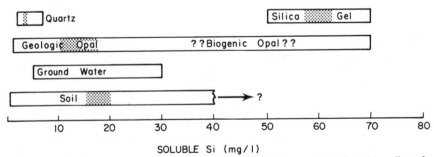

Fig. 14–23. Relationship between the concentration of Si commonly found in soils and ground waters and the solubility of quartz, silica gel, and opal (Stippled areas are commonly reported ranges).

bilities approximating those reported for fossilized opal; the older the opal specimen the lower its predicted solubility because of decreased internal surface area and surface energy (Huang & Vogler, 1972; Lewin, 1961). In addition to above considerations, the solubility of opal would also be a function of crystal structure; depending on degree of ordering, its solubility would be expected to range from values of amorphous silica to those approaching quartz. Further work is needed to quantify opal solubilities under natural soil environments.

The solubility of quartz has been determined by many investigators whose results have been summarized by Henderson et al. (1970). Though the reported range is 0 to 14 mg/liter Si, most of the values are included in a range of 3 to 7 mg/liter Si (Krauskopf, 1959) and it seems that the most probable value is approximately 3 mg/liter Si. Chalcedony is reported by Pelto (1956) to have a higher solubility than more massive quartz, the former is reported to have a solubility of 24 mg/liter Si after 17 hours at 95C (268K). The equilibration of quartz with the surrounding media is extremely sluggish at ordinary temperatures and pressures due to the high activation energy required to alter the Si–O–Si bond (Stöber, 1967). In fact, concentrations of < 3 mg/liter Si can exist in aqueous solution over quartz for months.

The solubilities of the silica minerals suggested previously imply room temperature and neutrality; however, the solubility of silica is a function of temperature, pH, particle size, and the presence of a disrupted surface layer. The solubility of quartz is independent of temperature up to 145C (418K); above which, the solubility increases linearly with temperature (Lidstrom, 1968). The solubility of amorphous silica increases linearly with temperature from 0C (273K) (Lidstrom, 1968; Krauskopf, 1959; Alexander et al., 1954); Okamoto et al., 1957). Krauskopf (1956, 1959) lists the solubility of amorphous silica as 23 to 37 mg/liter Si at 0C (273K), 50 to 65 mg/liter Si at 25C (298K), and 170 to 200 mg/liter Si at 100C (373K).

Many pedologists and mineralogists have mistakenly assumed that the solubility of silica increases as pH increases in the range 3 to 8. This misconception may be attributed to a report by Correns (1941), which indicated that the solubility showed a progressive increase above a pH of 3. However, more recent reports (Okamota et al., 1957; White et al., 1956; Alexander et al., 1954; Krauskopf, 1956, 1959, 1967; Frondel, 1962; Lidstrom, 1968) reveal that the solubility of silica (both crystalline and amorphous) is independent of pH below 9, but increases rapidly above 9 (Fig. 14-24). The rapid rise in solubility above pH 9 is due to ionization of monosilicic acid as illustrated:

$$Si(OH)_4 + (OH)^- \rightleftharpoons Si(OH)_3O^- + H_2O$$

Lidstrom (1968) calculated the solubility of quartz of various grain sizes and showed that quartz solubility increases as particle size decreases (Fig. 14-25). Iler (1955) has reported a similar relationship. Experimentally, quartz grains of sand-size (250–500 μm) showed nil solubility after 200 days where-

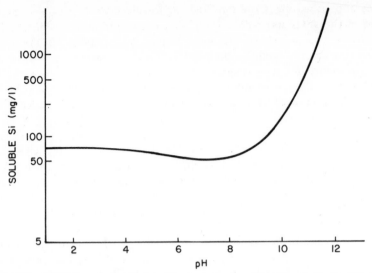

Fig. 14-24. Solubility of amorphous silica in water as a function of pH. Modified from Fig. 2, Alexander et al. (1954).

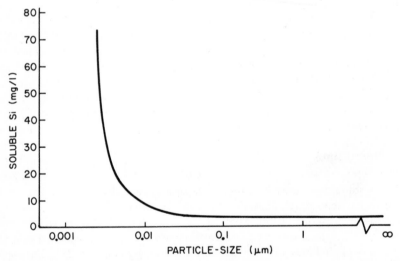

Fig. 14-25. Theoretical solubility of quartz as a function of particle-size; data from Lidstrom (1968, p. 13).

as grains < 5 μm in diameter gave 3 mg/liter Si in solution after 43 days (Siffert, 1967).

Freshly ground quartz or agitated suspensions of quartz commonly show abnormally high solubility. This has been attributed to the formation of a disrupted surface layer which is believed to be amorphous by Siffert (1967), Liberti and Devitofrancesco (1963), Dempster and Ritchie (1952,

1953), and Nagelschmidt et al. (1952) and microcrystalline by Lidstrom (1968) who reports that the microcrystalline layer gradually grades to crystalline quartz on moving into the interior of the grain. Similar surficial features were noted by Ribault (1971) on natural grains of quartz collected from a wide range of soil and geological environments. On the basis of differential dissolution and X-ray diffraction evidence, he concluded the surfaces were amorphous. Since the solubility of the disrupted surface layer, whether it is amorphous or microcrystalline, is considerably greater than that of quartz, the apparently high solubility of quartz in many determinations may, in fact, reflect the solubility of the disrupted surface layer. By continuous agitation of a suspension of quartz for 386 days, Morey et al. (1962) obtained 37 mg/liter Si in solution. Samples of a gravelly loam soil yielded 3 to 5 mg/liter Si after 40 days of standing or perfusing whereas shaking yielded approximately 16 mg/liter Si in same time period (McKeague & Cline, 1963b). After 48 hours of equilibration at pH 7, quartz samples employed by Henderson et al. (1970) yielded 5 to 21 mg/liter Si as contrasted to 1 to 2 mg/liter Si after treatment with NaOH, H_2SiF_6, or HF to remove the disrupted surface layer. The thickness of the disrupted surface layer is quite variable and depends on grinding intensity, or for nautral specimens environmental history; however, approximately 0.03 μm is the most common thickness reported for the disrupted surface layer (Gibb et al., 1953; Gordon & Harris, 1955; Nagelschmidt et al., 1952; Dempster & Ritchie, 1952). Gordon and Harris (1955) indicate that particles < 0.5 μm in diameter are disrupted throughout. Lidstrom (1968) reports that the disrupted thickness ranges from 0.15 to 0.02 μm for dry grinding and is approximately 10Å for wet grinding. Lidstrom (1968) also indicates that quartz which is wet ground will develop a zone of amorphous silica a few Å thick which is probably underlain by microcrystalline quartz. Krinsley and Tovey (1973) have observed the disrupted surface layer on quartz by cathodoluminescence and found that it is irregularly distributed in cross section and on grain surfaces.

C. Dissolution of Silica and Soluble Silica Equilibrium in Soils

In soils, the dissolution of silica minerals is not only affected by solubility, as discussed in preceding section, but also by organic and inorganic soil components, surface coatings, and soil solution chemistry. The presence of organic molecules, particularly alginic acid, ATP and amino acids, have been shown to greatly enhance the dissolution of silica, including quartz (Evans, 1965). After reviewing considerable literature and observing quartz grains from soils, paleosols, and silcretes, Crook (1968) attributes quartz dissolution in soils to leachates rich in organic molecules; quartz passes into solution as a Si-organic molecular complex. Beckwith and Reeve (1964) found that citrate ion promotes the release of native Si in soils and retards Si absorption by soil materials, whereas Iler (1973) found that citrate concentrations ranging from 0 to 2.2 mM had little effect on the rate of dissolution of amorphous silica.

Studies by Cleary and Conolly (1972) reveal that the dissolution of quartz is greatest in the root zone where production of organic complexes affect silica dissolution by complexing monosilicic acid. However, Wilding and Drees (1974) indicate that opal dissolution is retarded by occluded and/or chemisorbed organic carbon which may serve as a protective agent. They found that dark brown to opaque opal bodies persist boiling $0.5N$ NaOH digestion. Opal isolated from Malaysian paleosols supports above observations (samples courtesy of Dr. R. Protz, Dep. of Land Resource Science, Univ. of Guelph, Ontario, Canada). From one paleosol (Payong) the opal was abundant, almost completely opaque and exhibited little evidence of weathering while in the other (B-12) it was sparse, nearly transparent, and exhibited extreme dissolution effects (Fig. 14-17 C and D). It is apparent that additional work is necessary in order to clarify the role of organic compounds in solubilizing silica.

Many metallic ions, such as Al, Fe, Mg, Ca, Ag, Cu, Pb, and Hg, chemisorbed to silica surfaces are reported to inhibit the dissolution of silica due to the formation of relatively insoluble silicate coatings (Krauskopf, 1959; Lewin, 1961; L. H. P. Jones & Handreck, 1963; Beckwith & Reeve, 1963; Lidstrom, 1968). Most of the literature has been concerned specifically with the role of Al^{3+} ion upon silica dissolution; the Al^{3+} ion greatly reduces the dissolution of opal and amorphous silica (Denny et al., 1939; Jephcott & Johnston, 1950; Okamoto et al., 1957; Lewin, 1961; Iler, 1973). Okamato et al. (1957) showed that 20 mg/liter Al^{3+} will reduce the concentration of molecularly dispersed Si to 7 mg/liter whereas 100 mg/liter Al^{3+} will reduce the concentration to approximately 0.5 mg/liter. Evans (1965) attributes the stability of rounded quartz grains to the existence of a molecular layer of Al and Fe on most detrital grains. Iler (1973) indicates that much less than a monolayer of Al on the surface of amorphous silica greatly reduces the rate of dissolution. As little as 5% surface coverage by Al will reduce the concentration of soluble Si from amorphous silica to approximately 25 mg/liter (Iler, 1973). The stability of freshly ground quartz which has been shaken with aluminum sulfate has been shown to be pH dependent (Beckwith & Reeve, 1969). Dissolution of Al-contaminated freshly ground quartz which has the disrupted surface layer intact is greater in acid media than at a pH of 8.0. Beckwith and Reeve (1969) suggest that surface sorbed Al would decrease the dissolution of crushed quartz at neutrality, but might enhance its dissolution in natural acids. However, it seems more probable that the formation of a silicate coating would be retarded in acid media allowing for greater dissolution of the reactive silica surface than at higher pH values, but still less than the dissolution of crushed quartz in the absence of Al. This is consistent with the following reaction:

$$-\overset{|}{\underset{|}{Si}}(OH) + [Al(OH)_2]^+ \rightleftharpoons -\overset{|}{\underset{|}{Si}}OAl(OH)_2 + H^+.$$

$$\begin{pmatrix} \text{reactive} \\ \text{silica} \\ \text{surface} \end{pmatrix} \qquad\qquad \begin{pmatrix} \text{silicate} \\ \text{coating} \end{pmatrix}$$

According to this reaction, formation of the silicate coating will decrease as the pH decreases since the reaction will shift to the left in response to the increasing $[H^+]$ permitting dissolution of disrupted surface layers.

To this point the discussion has concerned systems containing soluble Al and/or Fe and particulate silica. In systems containing particulate sesquioxides, the adsorption of monosilicic acid by the sesquioxides (Fig. 14-26) is greatest at pH's between 8 and 10 (Beckwith & Reeve, 1963, 1964; McKeague & Cline, 1963c; L. H. P. Jones & Handreck, 1963). The pH dependency of the sorption of Si by sesquioxides is evident in the following equations:

$$Si(OH)_4 \rightleftharpoons [SiO(OH)_3]^- + H^+$$

$$[SiO(OH)_3]^- + Fe(OH)_3 \rightleftharpoons Fe(OH)_2OSi(OH)_3 + [OH]^-.$$

In such systems, it is suggested that sesquioxides act as a soluble Si sink and increase the dissolution of silica; in fact, it is suggested that even finely divided quartz is not stable at these pH values in the presence of excess sesquioxides (Beckwith & Reeve, 1964).

In summary, the sorption of Al or Fe onto the surface of amorphous silica, opal or quartz, exhibiting a disrupted surface layer, will decrease silica dissolution whereas the sorption of monosilicic acid by sesquioxides may increase dissolution of amorphous silica or reactive uncoated quartz surfaces. These effects are most pronounced at pH values of 8–10. In either case, the final concentration of soluble Si in Al–Fe–Si systems will be that in equilibrium with an aluminum or iron silicate surface. It is doubtful that sorbed Al or the presence of sesquioxides greatly affects the dissolution of quartz if it does not possess a disrupted surface layer; this is because of the low solubility of crystalline quartz and the sluggishness with which it equilibrates with the surrounding media. As a result of the latter, sorbed iron or aluminum silicate surfaces will control the concentration of soluble Si in Al–Fe–Si systems rather than quartz due to its negligible precipitation rate.

To date little is known concerning the dissolution and stability of biogenic opal in soil environments. Dissolution studies suggest that 35 to 78% of the total opal from tree leaves and grasses (Table 14-6) can be dissolved in a 2.5-min digestion of boiling 0.5N NaOH. In contrast, R. L. Jones (1969) found that only 33% of a soil opaline isolate (mostly of grass origin) could be dissolved in a 20-min digestion period of the same alkali; following first-order dissolution kinetics only 5% of the grass opal would have dissolved after a 2.5-min digestion period. This difference reflects the amount of labile opal as sheet-like incrustations in vegetative isolates (rather than solid polyhedral opaline bodies as found in soils) and the large amount of < 5 μm opal in vegetative isolates. Further, opal isolated from soils will reflect more stable forms that have persisted natural dissolution; they may contain a protective coat of chemisorbed Fe and Al. Disparities in rates of alkali dissolution between forest and grass opal of the same size fraction are attributed to mor-

Table 14-6. Alkali dissolution of opaline constituents isolated from leaves of various deciduous trees and grasses

Species	Size fraction	Weight lost	
		Fraction	Total
	μm	———— % ————	
American beech†	5-20	92	
(*Fagus grandifolia*)	Total		78
White oak†	>50	85	
(*Quercus alba*)	5-20	85	
	Total		60
Sugar maple†	>50	54	
(*Acer saccharum*)	5-20	64	
	Total		63
Idaho fescue‡			
(*Festuca idahoensis*)	Total		63 (39)
Bluebunch wheatgrass‡			
(*Agropyron spicatum*)	Total		69 (44)
Giant wildrye‡			
(*Elymus cinereus*)	Total		58 (35)

† Data from Wilding & Drees (1974); 2.5 min digestion in boiling 0.5N NaOH.
‡ Data from Norgren (1973); 5 min digestion in boiling 0.5N NaOH; values in parenthesis are calculated weight losses after 2.5 min digestion in boiling 0.5N NaOH assuming first-order reaction kinetics.

phological differences among respective opal specimens. Many of the stable < 5 μm forest opal isolates are associated with organic carbon pigmentation and anisotropic crystalline occluded phases (Wilding & Drees, 1973, 1974).

Huang and Vogler (1972) have found that dissolution of opal is inversely related to refractive index and directly with water content. They propose that dehydration of the opal would be accompanied by the growth of relatively large colloidal particles at the expense of smaller ones; thus, as the water content of the opal decreases, particles become larger, surface area and surface energy smaller, dissolution rates decrease and index of refraction increases. If this postulate is valid, one would predict that recently precipitated opal phytoliths should be much more susceptible to dissolution than older ones.

Figure 14-23 illustrates that the concentration of monosilicic acid in ground water generally ranges from 5 to 30 mg Si/liter (Krauskopf, 1967), whereas concentration in soils ranges from 1 to 40 mg Si/liter (McKeague & Cline, 1963b; L. H. P. Jones & Handreck, 1963, 1967; Beckwith & Reeve, 1964; Crook, 1968; Elgawhary & Lindsay, 1972) with 15 to 20 mg Si/liter most commonly present in soils at field capacity (L. H. P. Jones & Handreck, 1967). The low concentration of monosilicic acid in natural waters is attributed to the slowness of the dissolution process as compared to dilution by rainwater (Krauskopf, 1959). The concentration of soluble Si in soil lies approximately halfway between the equilibrium solubility of amorphous silica and quartz which would imply, based only on thermodynamic considerations, that amorphous silica should dissolve and quartz precipitate. This would lead to an equilibrium concentration of approximately 3 mg Si/liter which is rarely encountered. Quartz does not control the equilibrium concentration of monosilicic acid in soils due to its very slow rate of dissolution and negligible precipitation rate. Further, quartz is frequently coated with crystalline iron oxides which prevents the dissolution of silica. Elgawhary and Lindsay (1972) conclude that a solid phase less soluble than amorphous silica and more soluble than quartz controls silica concentrations in soils. Opal may serve this function. Kittrick (1969) presents data which indicate that soluble Si concentrations in equilibria with kaolinite and montmorillonite are of the same magnitude as that in soils. Other investigators suggest that the adsorption and desorption of soluble Si by soil components play a major role in determining the concentration of soluble Si in soils (McKeague & Cline, 1963c; Beckwith & Reeve, 1963, 1964; L. H. P. Jones & Handreck, 1963). All of these investigators indicate that sesquioxides are responsible for much of the capacity of soils to sorb soluble Si, with the maximum occurring between pH's 8 and 10 (Fig. 14-26). Beckwith and Reeve (1963) indicate that clay minerals also have the ability to adsorb monosilicic acids. Besides the dissolution-precipitation equilibria of soil minerals and sorption-desorption of soluble Si by soil components, Kittrick (1969) indicates that

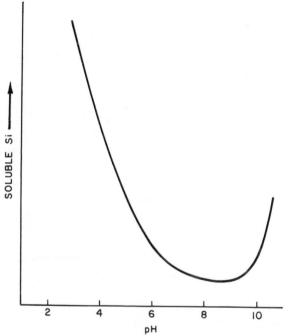

Fig. 14-26. Sorption of soluble Si by soils as a function of pH. Generalized from Fig. 2, Beckwith and Reeve (1964).

leaching of monosilicic acid from the soil and plant uptake are important in determining monosilicic acid concentrations in soils. In reality, soluble Si concentration in soils is undoubtedly dynamic where equilibrium concentrations are the exception rather than the rule. Concentration of Si in solution at any given time is related closer to the relative rates of the above processes than to equilibrium concentrations. Changes in moisture content of the soil will influence the Si concentration in solution faster than the other processes. Sorption-desorption reactions will result in Si equilibration faster than dissolution-precipitation reactions which are considerably slower. In order to initiate the dissolution of quartz lacking a disrupted surface, influxes of moisture due to rainfall or plant uptake must reduce the Si concentration below approximately 3 mg Si/liter. Quartz is lost from soils upon weathering, consequently perturbations of the Si equilibrium must occur which reduces the soluble Si concentrations to low levels.

V. NATURAL OCCURRENCE

A. Quartz

Due to its resistance to weathering and ubiquitous nature, quartz is the most abundant mineral in most soils and occurs in nearly every locality on the earth. Even in areas where the underlying rock is essentially devoid of quartz, the soils commonly contain quartz as a result of pedisediments (mass-wasted deposits on a pediment) from cyclic erosion-deposition or aeolian additions.

1. GEOLOGIC MATERIALS

Quartz constitutes 12 to 20% of an average igneous rock (Clarke, 1924; Pettijohn, 1957), however, it must be emphasized that the silica-rich igneous rocks (granites, rhyolites, and pegmatites) will contain considerably more quartz than the above values whereas basic igneous rocks (basalts and gabbros) and volcanic ash contain very little quartz. Clarke (1924) also reports that quartz constitutes 67% of an average sandstone; however, many sandstones contain over 90% quartz. Generally as the grain size of sedimentary deposits decreases, the quartz content also decreases (Van Straaten, 1954; Ferm, 1962). Van Straaten (1954) attributes this to the greater abrasive resistance of quartz relative to other mineral components comprising most sedimentary deposits. Yaalon (1962) reports that an average shale contains 20% quartz with a range from 11 to 26%, but that it is the second most important mineral.

Recent aquatic sediments also show a decrease in quartz content with particle size. Thomas (1969) found that sediments from Lake Erie and Lake Ontario which contained 27 to 79% quartz showed a high order of correlation between median particle size (median ϕ) and quartz content (correlation coefficient is −0.982). Thomas (1969) attributed the decrease in quartz content with a decrease in particle size to natural sorting during sedimentation. In pelagic sediments of the North Pacific Ocean, Rex and Goldberg (1958) found that quartz constitutes 2 to 24% of the sediments with maximum concentration in the 1 to 20 μm fraction. Rex and Goldberg (1958) attribute much of the pelagic sediments to an eolian origin.

Quartz transported as tropospheric dust is not only a common constituent of pelagic sediment, but also contributes significantly to continental surfaces. Rex et al. (1969) and Jackson et al. (1971) found that up to 34% of the A horizon of some Hawaiian soils was quartz yet the soils occurred over quartz-free mafic rocks. They deduced that the quartz was of tropospheric origin because the quartz content increased with precipitation, the oxygen isotopic abundance of quartz in the soils was similar to those of the north central Pacific Ocean pelagic sediments, and 70% of the quartz particles fell within the 2 to 10 μm fraction (Rex et al., 1969; Jackson et al., 1971). Laterites tend to be relatively low in quartz; however, Sivarajasingham et al. (1962) found that some Indian laterites contain approximately 20% quartz where there were significant aeolian contributions.

In most loessial deposits of the world, quartz is the most abundant mineral. Quartz constitutes > 60% of the silt fraction which comprises at least 70% of all the loess deposits examined by Springer (1948) in Missouri. Johnson and Beavers (1959) report that the 2 to 20 μm and 20 to 50 μm fractions of loess in Illinois contained 45 to 60% and approximately 80% quartz, respectively.

Basal glacial till is reported to have a tri-modal distribution of quartz in its "terminal grade" (final product of glacial comminution, Dreimanis & Vagners, 1971) with maxima occurring in the 0.125 to 0.250, 0.032 to 0.062, and 0.004 to 0.008 mm fractions (Dreimanis & Vagners, 1971). Though the quartz is concentrated in the sand and silt fractions, it is common in the clay-size fraction of Pleistocene tills (Horberg & Potter, 1955; Murray & Leininger, 1956; Tedrow, 1954). However, quartz tends to be restricted to the coarse clay (0.2–2 μm) fraction (Tedrow, 1954). Slatt and Hoskin (1968) could not identify any clay-size quartz in outwash sediments from the Norris Glacier in southeastern Alaska. In Wisconsin, Illinoian, and Kansan tills examined by Tedrow (1954), quartz content in the clay fraction was higher in the Wisconsin till (5 to 10% quartz) than in either of the older tills (approximately 5% quartz). Though the differences are slight, Tedrow (1954) suggests that the difference may be due to weathering of the quartz from the finer fractions in accord with Jackson and Sherman's (1953) weathering sequence for clay minerals.

2. SOILS

In soils, the relationship between quartz content and particle size follows the same trend as previously discussed for geologic materials; that is, the quartz content is generally concentrated in sand and silt fractions with secondary quantities in the clay fraction. Both sand and silt fractions of most soils are dominantly quartz (Springer, 1948; Yaalon, 1955; Hill & Shearin, 1969; Huizing, 1971; St. Arnaud & Whiteside, 1963; Kunze & Oakes, 1957; Borchardt et al., 1968). The parent material of the soil generally dictates which sand or silt subfraction will contain the maximum quartz content. Quartz is highest in the fine silt fraction of soils derived from tropospheric dusts (Rex et al., 1969; Jackson et al., 1971) whereas quartz is evenly distributed throughout the silt fraction (Springer, 1948) or highest in the coarse silt fraction (Johnson & Beavers, 1959) of loess-derived soils. Soils developed in glacial till generally contain the maximum quartz content in coarse silt or one of the sand fractions (St. Arnaud & Whiteside, 1963). The quartz content of the clay fraction generally ranges from 0 to 25% depending on the parent material and degree of weathering (Tedrow, 1954; Borchardt et al., 1966, 1968; Hill & Shearin, 1969; Bleeker, 1972; LeRoux, 1973); but under some conditions may be as high as 75 to 85% (L. W. Zelazny, personal communication). Generally the highly weathered soils contain less quartz in the clay fraction than soils weathered to a lesser extent. Jackson and Sherman (1953) report a 2 μm lower size limit for quartz in strongly weathered soils. Nearly all of the quartz in the total clay fraction (< 2 μm) of soils is concentrated in the coarse-clay fraction (0.2 to 2 μm) (Kunze & Oakes, 1957; Borchardt et al., 1966; Ratliff & Allen, 1970). Quartz has not been detected in the fine clay fraction (< 0.2 μm) in most investigations (St. Arnaud & Whiteside, 1963; Kunze & Oakes, 1957; Borchardt et al., 1966; Stahnke et al., 1969), but has been identified in the A1 horizon of some Aeric Aquods in Florida (L. W. Zelazny, personal communications). However, recent data presented by Jackson and Sayin (1975)[2] indicate that quartz may not be evident in the fine clay fraction (< 0.2 μm) due to masking by clay minerals, the main component. They found that the fine clay fraction (< 0.2 μm) from eight young soils exhibited small quantities of quartz after fusion to remove the clay minerals. Even in the < 0.08 μm fraction, quartz has been identified after fusion to remove phyllosilicates (J. B. Dixon, personal communication).

Two hypotheses can be advanced to explain the apparent lack of quartz in the fine clay fraction of soils: (i) dissolution of < 0.2 μm quartz particles because of their relatively high solubility, and (ii) physical inability to comminute quartz to particles of < 0.2 μm in diameter. The former hypothesis is supported by data of Lucas and Dolan (1939), Iler (1955), Siffert (1967), and Lidstrom (1968) which indicate rapidly increasing solubility of quartz as particle size decreases. The latter hypothesis is supported by the fact that quartz is not found in the clay-size fraction of recent glacial outwash in Alaska (Slatt & Hoskin, 1968) and in the fine clay fraction of pedogenetically unweathered glacial tills (Tedrow, 1954; unpublished data by the authors). Slatt and Hoskin (1968) suggest that physical properties of minerals control the size to which they can be reduced by glacial attrition and that the physical resistance of quartz to abrasion accounts for its absence in the clay frac-

---

[2]M. L. Jackson, and M. Sayin. 1975. The occurrence of quartz in the fine clay fraction of soils. Agron. Abstr. p. 178.

tion. This suggests that the bonding forces in the quartz grain become progressively stronger as its size is reduced; eventually a size may be reached
where the bonding forces limit further subdivision. Bonding forces may become stronger as particle size decreases due to the exclusion of crystal defects. Additional work is needed to determine the size to which quartz can
be physically comminuted.

The quartz depth distribution in soils is a function of parent material
and degree of weathering. The relatively undifferentiated soils, Entisols and
Inceptisols, reflect the quartz content of their parent materials and show
little pedogenic change in quartz content with depth. Inherited forms and
amounts of silica are also conspicuous in Aridisols due to restricted weathering by a lack of moisture (McKeague & Cline, 1963a). Vertisols are dominated
by expanding-layer silicate clays, but may have a significant component of
quartz. Though Mollisols, Alfisols, and Ultisols do not generally contain as
much quartz as Spodosols which develop most readily in highly siliceous
parent materials, quartz is most often the most abundant mineral in the sand
and silt fractions. In moderately weathered soils such as Spodosols, Alfisols,
Mollisols, and Ultisols, the eluvial horizons are enriched in quartz relative to
the parent material due to the weathering and removal of less resistant minerals while the illuvial horizons are lower in quartz than the parent material due
to dilution. Dilution can be effected by the accumulation of silicate clays,
carbonates, or sesquioxides. In the highly weathered Oxisols, quartz along
with other primary minerals in the clay fraction are weathered and leached
from the solum. In contrast, quartz may contribute as much as 85% of silt
and sand fractions in Oxisols. A more detailed discussion of the relationship
between quartz content and distribution in soils is presented by McKeague
and Cline (1963).

B. Cristobalite and Tridymite

1. GEOLOGIC MATERIALS

Unlike quartz and biogenic opal, the occurrence of cristobalite and tridymite is limited to specific geographic regions and rock stratigraphic units.
They are commonly associated with magmatic and volcanic deposits of upper
Cretaceous and Tertiary age, especially andesites, rhyolites, latites, trachytes,
and some basalts (Frondel, 1962, p. 269, 280-281). While these minerals
are usually reported to be of authigenic origin (post-depositional products),
allogenic occurrences in fluvial deposits have been cited (Mejia et al., 1968).
Recent evidence by Mizota and Aomine (1975) suggest that cristobalite and
tridymite associated with some andesitic volcanic ashes are of igneous origin
rather than a post-depositional alteration product.

In the United States, tridymite is abundant in Tertiary rhyolites and
latites of the San Juan Region, Colorado (Larsen et al., 1936). Tridymite is
commonly confined to the groundmass of volcanic rocks (Larsen et al.,

1936; Sato, 1962; Mizota & Aomine, 1975), but it has been reported as a post-depositional cement in some rhyolitic tuffs. This mineral is also a product of wood fossilization (Mitchell, 1967; Mitchell & Tufts, 1973; Buurman, 1972).

Cristobalite has been identified as an integral component of the following deposits; volcanic rocks, bentonites (Gruner, 1940a, 1940b; Brindley, 1957; Reynolds & Anderson, 1967; Papke, 1969; Guven & Grim, 1972; Henderson et al., 1971) recent deep-sea marine cherts (Wise et al., 1972; Wise & Weaver, 1973, 1974); modern lake clays (Peterson & Von der Borch, 1965); Cretaceous shales (Davis, 1970); and Tertiary clay-stones (Reynolds, 1970; Weaver & Wise, 1974). In volcanic rocks it commonly occurs as infillings in gas cavities in the form of euhedral crystals or microcrystalline rounded spherulites (Larsen et al., 1936; Howard, 1939; Kleck, 1970).

2. SOILS

Cristobalite in soils is usually associated with Quaternary volcanism. Cristobalite has been reported primarily in various size fractions (predominantly clay) of several Andepts (Andosols) and soils derived from volcanic ash in South America (Wright, 1964; Mejia et al., 1968; Besoain, 1969; Calhoun et al., 1972; Cortes & Franzmeier, 1972), Java (Hardjosoesastro, 1956), New Guinea (Bleeker & Parfitt, 1974), New Zealand (Swindale & Jackson, 1960), Costa Rica (Tan et al., 1975) and Japan (Mizota & Aomine, 1975). Perhaps the occurrence of cristobalite in soils was first reported by Yaalon (1955) for two soils developed in Israel. Kunze and Oakes (1957) have identified cristobalite in a Lufkin soil from East Texas of deltaic and marine origin. In the above two cases, the deposits are of Eocene to Pliocene age. Wilding and Drees (1974) have also reported cristobalite in opaline constituents of forest soils which are believed to be cosynthesized with biogenic opal.

Cristobalite, if identified at all, usually constitutes a minor component of the total mineral fraction; however, Swindale and Jackson (1960) have reported over 60% cristobalite in the 2–5 μm fraction of a New Zealand soil derived from rhyolite. Initially it was believed that this component was a pedogenic product of podzolization, but Henderson et al. (1972) have recently shown through oxygen isotope analysis that the cristobalite is of volcanic or hydrothermal origin. Mizota and Aomine (1975) have recently reported cristobalite in abundance in every size fraction from coarse sand to clay in soils developed from andesitic volcanic ash. They conclude that the origin of the cristobalite is indigenous to the parent magma just prior to eruption.

C. Opal

Most of the opal found in soils and surficial geological deposits is of biogenic (organic) origin (Siever & Scott, 1963; L. H. P. Jones & Handreck, 1967). However, sometimes substantial quantities of inorganic opal occur as nodules and cementing agents in soils (Flach et al., 1969; Soil Survey Staff, 1975), in modern lake muds (Peterson & Von der Borch, 1965) and in other geologic deposits (Frondel, 1962, p. 287-306). The following will emphasize biogenic opaline forms in geological materials and soils.

1. BIOGENIC OPAL

a. **Biogenic Opal in Geologic Materials**—Persistent opaline microfossils (namely phytoliths, sponge spicules, and radiolaria) have been recognized in calcareous Wisconsin-age loess and till deposits (R. L. Jones et al., 1963; R. L. Jones & Beavers, 1963b; Wilding & Drees, 1968b); in sedimentary rocks (Baker, 1960c; R. L. Jones, 1964; Gill, 1967; Weaver & Wise, 1974), paleosols (Beavers & Stephen, 1958; Dormaar & Lutwick, 1969; Norgren, 1973), and in deep-sea cores (Kolbe, 1957). The sedimentary rocks are of Mesozoic and Cenozoic ages and include Holocene diatomaceous silty clays (Baker, 1960d), opaline Eocene claystones (Weaver & Wise, 1974), Cretaceous shales (R. L. Jones et al., 1963), and Tertiary clays, arenites and other deposits (Baker, 1960d; R. L. Jones, 1964; Gill, 1967). Limestone residuum often contains notable quantities of these microfossils.

Opal phytoliths have also been identified in atmospheric dusts (Baker, 1960d; Folger et al., 1967; Twiss et al., 1969), rain (Baker, 1959b), animal droppings, (L. H. P. Jones & Handreck, 1967, p. 136), and native pottery (Linné, 1965; Evans & Meggers, 1968; Lathrap, 1970, p. 256) that accentuate their universality. Folger et al. (1967) report that opal phytoliths and diatoms comprise about 25% of the silt fraction in dusts collected 500 km off the west coast of Africa. They conclude that phytoliths and fresh water diatoms can be transported as an aerosol $> 1,000$ km in relative proportions similar to those observed in deep sea and pelagic sediments.

b. **Biogenic Opal in Soils**—Biogenic opal is a minor, but ubiquitous constituent of most soils; opal phytoliths represent the major form under non-aquatic environments. Ruprecht in 1866 (Smithson, 1958) first drew attention to the occurrence of opal phytoliths in Russian Chernozem soils. Smithson (1956) introduced western workers to the nature of opal phytoliths and Beavers and Stephen (1958) were among the first in the United States to report the occurrence of opal in Illinois soils.

Although quantities may vary by several orders of magnitude from one geographical area to the next, a number of interrelated factors other than geography likely govern their concentration. These include: plant species, soil (pH, soluble silica, reactive Fe and Al sesquioxides, hydrology, etc.), climate, geomorphology, and opal stability (L. H. P. Jones & Handreck, 1967; Wilding & Drees, 1968a, 1971, 1974). Important opaline components in aquatic environments include sponge spicules, diatoms, and radiolaria (Smithson, 1959; R. L. Jones et al., 1964; Wilding & Drees, 1968b, 1971; Weaver & Wise, 1974).

Amounts of opal phytoliths in soils commonly range from < 0.1 to 3% on a total soil basis. These values are often based on opal extracted from sand and/or coarse-silt fractions recalculated to a total soil basis; they may underestimate actual opal quantities by at least two-fold (Geis & Jones, 1973). As much as 50 to 75% of the total opal contributed to soils by many grass and forest species is in the < 5 μm size fractions (R. L. Jones & Beavers, 1964b; Wilding & Drees, 1971; Geis, 1973). Coarser opaline constituents are more stable and easier to fractionate from soils, but finer ones are often more valuable for taxonomic purposes (Twiss et al., 1969). Over two-thirds of the opal phytoliths in Ustolls (Brunizems in drier regions) of western Oklahoma (Yeck & Gray, 1972) and Udolls (Brunizems) of Minnesota (Verma & Rust, 1969) occur in the 5–20 μm size fraction. In contrast, the 5–20 μm and 20–50 μm fractions of Udolls in eastern Oklahoma (Yeck & Gray, 1972) and in Illinois (R. L. Jones & Beavers, 1964b) contain about equal amounts of opal. In Alfisols (Gray-brown Podzolics) of Ohio and southern Ontario, Canada, the 5–20 μm fraction contains about 5 to 10 times more opal than that in the 20–50 μm fraction (unpublished data by authors).

Comparisons as above should be made on a total soil basis because absolute opal quantities found in any given size-separate simply reflect the particle-size distribution of the total soil. For example, two soils of widely different texture, but developed under the same biotic and weathering conditions would yield strikingly different opal contents in any given-size separate.

Markedly greater concentrations of opal in soils have been reported. For example, Riquier (1960) found a light gray zone initially identified as an A2 horizon in an East African basaltic soil developed under acacia bamboo and fern to be almost completely opal phytoliths. Norgren (1973) has reported opal phytolith concentrations in grassland soils of Oregon to exceed 20%. Kanno and Arimura (1958) noted that opal content of the 20–200 μm fraction of several Japanese soils comprised 30 to 60% of the separate. These unusually high values are attributed to labile, silica-rich parent materials in an environment conducive to high silica uptake by silica-accumulating species. For this reason Andepts (Andosols) would be expected to contain high levels of plant opal.

Quantities of biogenic opal in soils commonly follow a decreasing depth-function. A maximum is usually achieved in the surface or subjacent horizons and a minimum 50 to 100 cm below the surface unless a buried paleosol is encountered (Wilding & Drees, 1968a; R. L. Jones & Beavers, 1964b; Verma & Rust, 1969, Witty & Knox, 1964; Norgren, 1973). Opal contents closely follow organic matter depth distributions if the same plant species or communities have contributed significantly over time (Dormaar & Lutwick, 1969). Conversely, where plant opal and organic matter depth functions display little similarity, different species or plant communities were likely responsible for each. The opal distribution with depth provides useful clue in determining whether the opal is in situ or of allogenic origin (generated elsewhere and transported to the present site). An abrupt increase in opal content in the lower portion of the soil profile is strong evidence for a buried paleosol surface (Verma & Rust, 1969; Norgren, 1973). Alternatively, such an increase may simply reflect differential mixing by flora and fauna but this seems less plausible when opal depth distributions are systematic.

2. INORGANIC OPAL

Inorganic opal (in addition to other forms of silica) serves as a primary cement for indurated soil horizons (duripans) over extensive areas in sub-humid mediterranean and arid climates. These soils have moisture regimes where silica is solubilized and translocated to lower positions of the soil pro-file. Geographically, duripans are restricted largely to areas of volcanism or labile intermediate and basic igneous rocks, materials that provide ready source of soluble Si (Flach et al., 1969; Soil Survey Staff, 1975, p. 37). Duri-pans are well documented in western United States, western Australia, South Africa, and New Zealand. As little as 10% Si as $Si(OH)_4$ is apparently suf-ficient to effectively indurate the horizon (Flach et al., 1969). Duripans oc-cur on old, stable geomorphic surfaces as well as younger, less stable pediment backslope, footslope, and basin positions (Flach et al., 1969; Brewer et al., 1972). Soils on pediment surfaces may not be silica cemented, but still con-tain allogenic detrital silica nodules and other pedogenic silica, features in-herited from cyclic erosion of silicified uplands.

VI. IMPACT ON PHYSICAL AND CHEMICAL SOIL PROPERTIES

Tectosilicates, and silica in particular, exert a secondary influence on most physiochemical properties in soils including surface area, ion exchange, moisture retention, plasticity, cohesion, shrink-swell, and porosity. Silica components are generally considered inert; they do not directly govern plant nutrient mineral-solution equilibrium dynamics, and thus their physiochemi-cal attributes have not been comprehensively examined. Their low physio-chemical activity serves as a dilutent to the much more reactive clay mineral components. Indirectly, silica minerals play a major role in soil behavior be-cause they are closely allied to soil texture; coarse textured soils are generally silica-rich while fine-textured soils commonly contain quantities of silica in-versely proportional to the clay content.

It is the structure behind mineral surfaces that dictate their physio-chemical behavior and surface activity. Silica minerals are framework struc-tures, lack significant isomorphic proxying of lower valency cations for Si, are electrically balanced, and thus exhibit negligible surface charge (Grim, 1968, p. 192). Both anion and cation exchange arise from Si–O broken bonds and Si–OH silanol groups around particle edges that would increase with decrease in particle size. Disrupted surface layers would also tend to increase the number of reactive bonds and the effective surface charge (Grim, 1968, p. 193). In spite of these factors, the surface charge of silica minerals is considered negligible, though little or no effort has been made to quantify this property. Specific surface area for silica minerals (excluding opal) is also small. For example, if quartz particles are assumed to be spherical in shape, the surface area for the 2–0.2 μm clay fraction (mean diameter of 1.1 μm) is calculated to be about 2 m^2/g; if the particles are assumed to be platy (as recent evidence suggests for quartz) with a thickness to width ratio of 1:10, the surface area for the above clay fraction is calculated to be about 3.6 m^2/g (Jackson, 1956, p. 331–334). This is in agreement with krypton adsorption values of 1.04 to 2.85 m^2/g by Jorgensen (1970). In contrast, opal isolated from oat plants (*Avena sativa* L.) has a surface area by ethyl alcohol adsorption of about 14 m^2/g (L. H. P. Jones & Milne, 1963) and the surface area of diatoms is reported to range from 89 to 123 m^2/g by nitrogen adsorption (Lewin, 1961). Acid-digested opal phytoliths have also yielded specific surfaces as high as 122 m^2/g (Peinemann et al., 1970). These relatively high values are attributed to the macromolecular structure of opal which provides internal surface area; such submicron spheres range from 100 to 3,000Å which yield calculated surface areas varying from 8 to 200 m^2/g. However, in gem opal J. B. Jones and Segnit (1969) found that a large part of this surface was not available to nitrogen adsorption; observed surface areas were only about 0.4 m^2/g. Most of the surface area was apparently covered by sorbed single and/or double hydroxy groups.

Because of crystal structure, silica surfaces (opal excluded) are not highly hydrated; only weak adhesive forces bond polar water molecules to the essentially uncharged, nonionic silica surface. Few, if any, adsorbed cations are available for solvation. Except when silica particles are cemented together by organic matter, sesquioxides, silica, carbonates or other cementing agents, the only cohesive forces bonding the skeleton framework together are the relatively weak Van der Waals forces and surface tension of curved menisci at air-water interfaces (Baver et al., 1972; p. 77). When a soil consists primarily of sand and silt grains without sufficient plasma to yield cohesive properties, the soil is said to be cohesionless. Soils that are comprised dominantly of silica minerals are cohesionless and nonplastic (Sowers & Sowers, 1970, p. 27; Baver et al., 1972, p. 93). As the silica content of cohesive soils increases, adhesion, cohesion, shrink-swell, cation exchange capacity, surface area, moisture retention, plasticity limits, capillary porosity, compression and compaction decrease; conversely, noncapillary porosity increases.

In silica-cemented soils (duripans), even small amounts of cement have profound effects on soil physical properties (Flach et al., 1969); such materials are hard to extremely hard when dry, have high unconfined compressive strength (i.e., 778 kg/cm^2 when dry and 289 kg/cm^2 under 10 cm water tension); resist dispersion in sodium hexametaphosphate; reduce shrink-swell potential; and cause low intensity, broad smectite peaks. The presence of silica cement apparently does not affect CEC, surface area, or expansion of smectite clays upon glycerol solvation.

VII. PEDOLOGICAL IMPLICATIONS

A. Quartz as a Stable Reference Mineral

Quartz has been utilized by numerous investigators (Springer, 1948; Cann & Whiteside, 1955; St. Arnaud & Whiteside, 1963; Barshad, 1964; Al-

Janabi & Drew, 1967; Redmond & Whiteside, 1967; Sudom & St. Arnaud, 1971) to access parent material uniformity, as an index for quantitative evaluation of soil formation, and as the stable member of a mineral ratio which expresses the degree of weathering. From a study comparing the suitability of quartz, zirconium, and titanium as references for pedological studies, Sudom and St. Arnaud (1971) concluded that both quartz and zirconium are more reliable indices upon which to base quantitative evaluation of pedogenic changes than titanium. Several characteristics, all of which have been discussed in detail previously, make quartz suitable for such purposes: ubiquitous nature, abundance, resistance to weathering and immobility. The abundance of quartz eliminates statistical errors associated with determinations involving small quantities. The main criterion for any reference mineral is a high stability and the high Si–O bonding energy imparts such a character to quartz. Since quartz occurs mainly in the sand and silt fractions, it is relatively immobile in soils.

However, there are several dangers associated with the use of quartz as a stable reference mineral. St. Arnaud and Whiteside (1963) indicate that quartz does physically disintegrate in soils; consequently, Sudom and St. Arnaud (1971) advocate the use of quartz in the > 2 μm fine earth fraction as a reference. If quartz does breakdown to particle sizes of < 2 μm, errors are incorporated into the estimations. Another reason for use of the > 2 μm fine earth fraction is that preferential distribution of accompanying minerals, such as carbonates, in the various size fractions would result in differential dilution of the quartz and confound reconstruction analysis (Smeck et al., 1968). The occurrence of authigenic quartz in soils (Breese, 1960; Flach et al., 1969; Brewer et al., 1972) also poses problems for the use of quartz as a reference mineral. Quartz of biogenic origin (Wilding & Drees, 1974) would be maximum in surficial horizons and decrease with depth. This contribution would accentuate quartz accumulation in surficial horizons. The magnitude of such an error is unknown at this time, but would probably not exceed 5 to 10%.

B. Silica as an Index of Parent Materials

1. OXYGEN ISOTOPE ABUNDANCE

Oxygen isotope composition of minerals represents a tool for determining the genesis and history of geologic materials; it recently has been applied to soils as an index of parent material origins (Rex et al., 1969; Syers et al., 1969; Henderson et al., 1972; Mokma et al., 1972; Jackson et al., 1971; Jackson et al., 1972). Quartz serves as a very useful mineral for the determination of oxygen isotope composition because ^{18}O concentrates in quartz preferentially to other minerals and because quartz is resistant to weathering. Epstein and Taylor (1967) indicate that ^{18}O is concentrated in minerals with high Si–O bond energies.

The oxygen isotope composition of minerals is dependent upon the tem-

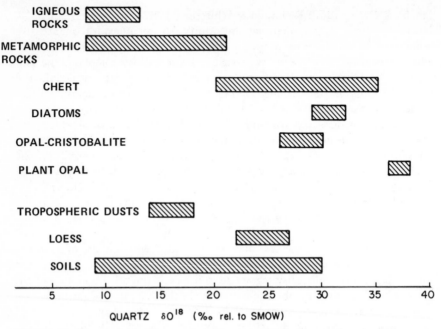

Fig. 14-27. Distributions of oxygen isotopic composition of quartz from various sources.

perature of formation. Delta values[3] (δ ^{18}O) decrease as the temperature of formation increases (Fig. 14-27). Quartz in igneous rocks has delta values on the order of 8 to 13 $^o/oo$ (Clayton et al., 1972) whereas quartz from metamorphic rocks has somewhat higher δ ^{18}O's but shows a wider range due to varying degrees of metamorphism. Minerals formed at ambient temperatures have considerably higher values as illustrated by chert (δ ^{18}O of 20 to 35 $^o/oo$, Fig. 14-27). Henderson et al. (1972) report δ ^{18}O of 29.1 to 32.2 $^o/oo$ for diatoms and 25.8 to 29.5 $^o/oo$ for opal-cristobalite; Jackson et al. (1971) report δ ^{18}O of 37.2 $^o/oo$ for plant opal isolated from a Warsaw soil in Ohio (Fig. 14-27).

At ambient temperatures, quartz is exceedingly resistant to oxygen isotope exchange; consequently, the isotopic composition of quartz in soils, dusts, and detrital sediments can be used as a tracer of its origin (Syers et al., 1969). The delta values of minerals from most sediments, aeolian dusts, and soils are generally higher than that characteristic of igneous and metamorphic origins and lower than that of authigenic minerals. Thus isotopic compositions of quartz in soils, dusts, and sediments are a reflection of the relative contributions of quartz formed under different temperature environments occurring as discrete isotopically homogeneous grains or as isotopically light and heavy zoned grains.

$$^3 \quad \delta\ ^{18}O = \left[\frac{(^{18}O/^{16}O)\ \text{sample}}{^{18}O/^{16}O\ \text{SMOW}} - 1 \right] \times 1000$$

where SMOW = Standard Mean Ocean Water.

By determining the oxygen isotope composition and morphology of quartz from pelagic sediments and tropospheric dusts, as well as the particle size distribution of these materials, the existence of three distinct tropospheric dusts has been established. Quartz from each exhibits a very uniform and distinct oxygen isotope composition reflecting respective source areas: δ ^{18}O of 18 $^O/oo$ for North Pacific dust (Rex et al., 1969); 14 $^O/oo$ for South Pacific dust (Mokma et al., 1972; Clayton et al., 1972); and 16 $^O/oo$ for the Sahara-Bahama-Caribbean-North Carolina dust (Syers et al., 1969). Quartz in the fine silt fraction of loess from North-Central USA exhibits higher delta values (24 ± 3 $^O/oo$ Syers et al., 1969) than any of the tropospheric dusts which indicates a higher proportion of quartz from cherts or carbonate rocks. There is a systematic variation between oxygen isotopic composition and grain size: as the particle-size decreases, δ ^{18}O increases (Syers et al., 1969; Clayton et al., 1972). This implies that quartz from igneous and metamorphic origins can not be comminuted to a grain size as fine as quartz from limestones and dolomites.

Delta values of quartz from soils have been reported ranging from 9 to 30 $^O/oo$ (Fig. 14-27). To date, the main use of oxygen isotope analysis in pedologic studies has been for the identification of aeolian additions to soils (Rex et al., 1969; Syers et al., 1969; Mokma et al., 1972; Jackson et al., 1972). However the full potential of this tool in pedologic studies has not yet been realized.

During oxygen isotope analysis, care should be exercised in extracting biogenic opal from the silica components. Soil opaline forms are likely to persist through alkali dissolution and other pretreatments. Since biogenic opal exhibits a higher delta value than other silica components, its presence in samples could seriously confound the results of oxygen isotope analysis. Opal can be quantitatively isolated from quartz and cristobalite by isopynic density separations. This purification approach is less plausible for separation of opal from tridymite.

2. BIOGENIC OPAL

Opal phytoliths have been used as index to trace the up-slope origin of colluviated sediments that form over-thickened Ah horizons in the foothills of Alberta (Lutwick & Johnston, 1969). Biogenic opal may also serve as a parameter to determine the vector of mixing in soils. Their utility in identifying surface horizons of paleosols has also been well established (Beavers & Stephen, 1958; Dormaar & Lutwick, 1969; Norgren, 1973). Biogenic opal is particularly useful for paleosol recognition when organic matter has been oxidized from the paleosol; under these conditions it not only marks the paleosol, but serves as a datable carbon source (Wilding, 1967).

Depth distributions of sponge spicules have been used as an index of parent material uniformity (Wilding & Drees, 1968b). In landscape positions which preclude an authigenic origin of sponge spicules; their presence in surficial soil horizons high in silt content (50-75%) may imply an aeolian loessial component or admixture with lower solum parent materials (R. L. Jones & Beavers, 1963b; Wilding & Drees, 1968b). Because spicules are not always found in loessial deposits (R. L. Jones & Beavers, 1963b; R. L. Jones et al., 1963) or all aquatic environments, their absence is not confirmative evidence for parent material uniformity. Sponge spicule distributions often are useful for the identification of a lithological discontinuity when field or other laboratory evidence is inconclusive.

Opal of aquatic organisms, particularly a preponderance of diatoms, sponge spicules and siliceous shells of Chrysostomataceae provide direct evidence that the parent material of the soil is of marine or lacustrine origin (Smithson, 1959; R. L. Jones et al., 1964; Wilding & Drees, 1971).

C. Silica as an Index of Environmental History

1. BIOGENIC OPAL

For over 100 years efforts have been made to relate biogenic opal in soils with their vegetative history by comparing amounts, shapes, and sizes of opal phytoliths extracted from soils with those from plant species known or assumed to have been present during pedogenesis. For example, work of this nature has been conducted in Russia (Tyurin, 1937; Usov, 1943; Yarilova, 1956), Great Britain (Smithson, 1958), Australia (Baker, 1959b), New Zealand (Raeside, 1970), Japan (Arimura & Kanno, 1965), Canada (Brydon et al., 1963; Lutwick & Johnston, 1969; Dormaar & Lutwick, 1969), and the United States (Beavers & Stephen, 1958; R. L. Jones & Beavers, 1964a, 1964b; R. L. Jones et al., 1964; Witty & Knox, 1964; Yeck & Gray, 1969, 1973; Verma & Rust, 1969; Wilding & Drees, 1968a, 1971, 1973; Norgren, 1973). From this work it may be concluded that broad differentiation of plant species contributing opal to soils may be made on the basis of opal morphology (i.e. grass vs. forest, or groups of grass subfamilies), but precise taxonomic reconstruction must await further development of opal systematics as reviewed by Geis and Jones (1973). Assignment of opal to a given vegetative type has been accomplished either by unique individual forms (Twiss et al., 1969; Geis, 1973) or relative proportions of group assemblages (Norgren, 1973; Yeck & Gray, 1972).

It is clear that soils developed under long periods of grass vegetation commonly contain 5 to 10 times more biogenic opal than those formed under forest environments (Witty & Knox, 1964; R. L. Jones & Beavers, 1964b; R. L. Jones et al., 1964; Wilding & Drees, 1971; Norgren, 1973). Soils in prairie-forest transition regions and in areas where prairies have recently invaded forests (or vice versa) may yield similar opal quantities (Wilding & Drees, 1971; Verma & Rust, 1969). Attempts to relate opal quantities to vegetative history can be hazardous unless attention is given to those factors which control silica content in plants, soil geomorphic stability, and opal stability. In spite of these variables meaningful interpretations for a limited geographical region can be made from total soil opal accumulations; assumed rates of opal accumulation can be translated into an estimated occupancy period for a given vegetative type accompanying pedogenesis. Opal yields from grass vegetation vary markedly by vegetative species, geography, climate, soil, and particle-size, but total values of the following order of magnitude have been reported: 8 to 10 kg/ha in New Mexico (Pease & Anderson, 1969); 150 kg/ha in Illinois (R. L. Jones & Beavers, 1964a, 1964b); and from 300 kg/ha (Norgren, 1973) to 40 kg/ha (Witty & Knox, 1964) in Oregon. Forest vegetation would likewise be dependent upon above factors, but likely range from < 5 kg/ha to 80 kg/ha.

As a palenological record, opal has an advantage over pollen because of its persistence under oxidizing environments and its resistance to long distance transport (Geis, 1973). The latter decreases the probability of extraneous contamination of phytolith assemblages which are of important consideration of pollen analysis. Though yet untested biogenic opal should prove useful to document the environmental history of archeological sites.

2. SURFACE FEATURES ON QUARTZ GRAINS

With the development of electron microscopy, particularly the scanning electron microscope (SEM), the surface features of quartz grains have been studied by numerous geologists, most extensively by Krinsley and co-workers (Krinsley & Takahashi, 1962a, 1962b, 1962c, 1962d, 1964; Krinsley et al., 1962, 1964; Krinsley & Funnell, 1965; Krinsley & Donahue, 1968a, 1968b; Krinsley & Margolis, 1969; Krinsley & Smalley, 1972; Krinsley & Doornkamp, 1973; Margolis & Krinsley, 1974). The work prior to 1969 was with the TEM and thereafter primarily with the SEM. The objective of these investigations was to elucidate the erosional and depositional history of quartz grains as recorded by surface features. Krinsley and Takahashi (1962a, 1962b, 1962c) concluded that it is possible to distinguish between littoral, aeolian and glacial environments on the basis of surface features. Krinsley and Funnell (1965) suggested criteria for the distinction of glacial, glacial-fluvial, dune, beach, and estuarine environments. They also suggested that it is possible to identify sequential environments under favorable conditions. For example, the surface markings of one quartz grain indicated that it had been abraded by glacial ice, carried by a glacial stream, abraded on a beach, carried by turbidity currents across the continental shelf and finally deposited in the deep sea (Krinsley & Smalley, 1972). To date, information available from the study of quartz grain surfaces by SEM has not been extensively utilized in pedological studies. In order to realize the potential wealth of information concerning environmental history, more attention should be devoted to this tool in pedologic studies.

VIII. SUPPLEMENTAL READING LIST

The following references are suggested to complement this chapter:

Epstein, and Taylor (1967)—a reference text devoted to the principles of $^{18}O/^{16}O$ isotopic dating of rocks and minerals.

Frondel (1962)—a comprehensive mineralogical reference text devoted solely to silica polymorphs and silica glass.

L. H. P. Jones, and Handreck (1967)—a chapter considering disposition of silica in soils, plants and animals.

R. L. Jones, and Hay (1975)—a chapter considering siliceous and non-siliceous microfossils in soils as a paleonological index.

Krinsley, and Doornkamp (1973)—a reference atlas illustrating the SEM surface features of quartz as related to environmental history.

Lidstrom (1968)—a reference monograph regarding theoretical and applied considerations of quartz and silicates.

Lutwick (1969)—a concise chapter summarizing the utility of opal phytoliths as a soil genesis index.

McKeague, and Cline (1963a)—a chapter devoted to forms and amounts of silica components in soils.

Mitchell (1975)—a chapter summarizing forms, synthesis, stability and the determination of oxides and hydrous oxides of silicon in soils.

Siever, and Scott (1963)—a chapter devoted to organic geochemistry of silica with geological emphasis.

Sosman (1965)—a comprehensive text summarizing the physical properties of silica polymorphs—a good review of tridymite thermal properties.

LITERATURE CITED

Alexander, G. B., W. M. Heston, and R. K. Iler. 1954. The solubility of amorphous silica in water. J. Phys. Chem. 58:453–455.

Al Janabi, A., and J. V. Drew. 1967. Characterization and genesis of a Sharpsburg-Wymore soil sequence in Southeastern Nebraska. Soil Sci. Soc. Am. Proc. 31:238–.244.

Amos, G. L. 1952. Silica in timbers. Aust. C. S. I. R. Org. Bull. 267:1–55.

Arimura, S., and I. Kanno. 1965. Some mineralogical and chemical characteristics of plant opal in soils and grasses of Japan. Bull. Kysuhu Agric. Exp. Stn. 11:111–120.

Baker, G. 1959a. Fossil opal-phytoliths and phytolith nomenclature. Aust. J. Sci. 21:305–306.

Baker, G. 1959b. Opal phytoliths in some Victorian soils and "red rain" residues. Aust. J. Bot. 7:64–87.

Baker, G. 1960a. Phytolitharien. Aust. J. of Sci. 22:392–393.

Baker, G. 1960b. Hook-shaped opal phytoliths in the epidermal cells of oats. Aust. J. Bot. 8:69–74.

Baker, G. 1960c. Fossil opal-phytoliths. Micropaleontology 6:79–85.

Baker, G. 1960d. Phytoliths in some Australian dusts. Proc. Roy. Soc. Victoria. 72:21–40.

Barshad, Isaac. 1964. Chemistry of soil development. p. 1–70. In Firman E. Bear (ed.) Chemistry of the soil. Reinhold Publ. Corp., New York.

Baver, L. D., W. H. Gardner, and W. R. Gardner. 1972. Soil physics. 4th edition. John Wiley & Sons, Inc., New York.

Beavers, A. H., and I. Stephen. 1958. Some features of the distribution of plant opal in Illinois soils. Soil Sci. 86:1–5.

Beckwith, R. S., and R. Reeve. 1963. Studies of soluble silica in soils. I. The sorption of silica acid by soils and minerals. Aust. J. Soil Res. 1:157–168.

Beckwith, R. S., and R. Reeve. 1964. Studies on soluble silica in soils. II. The release of monosilicic acid from soils. Aust. J. Soil Res. 2:33–45.

Beckwith, R. S., and R. Reeve. 1969. Dissolution and deposition of monosilicic acid in suspensions of ground quartz. Geochim. Cosmochim. Acta. 33:745–750.

Berry, L. G., and B. Mason. 1959. Mineralogy concepts, descriptions determinations. W. H. Freeman & Co., San Francisco.

Besoain, E. M. 1969. Clay minerals of volcanic ash soils. Para. Bl. 1–1.16. In H. W. Fassbender (Coordinator). Panel on volcanic ash soils in Latin America. Inter-American Inst. of Agric. Sci. of OAS, Turrialba, Costa Rica.

Biederman, E. W., Jr. 1962. Distinction of shoreline environments in New Jersey. J. Sediment. Petrol. 32:181–200.

Blackman, E., and D. W. Parry. 1968. Opaline silica deposition in Rye (Secale Cereale L.). Ann. Bot. 32:199–206.

Blatt, H., and J. M. Christie. 1963. Undulatory extinction in quartz of igneous and metamorphic rocks and its significance in provenance studies of sedimentary rocks. J. Sediment. Petrol. 33:559–579.

Bleeker, P. 1972. The mineralogy of eight latosolic and related soils from Papua New Guinea. Geoderma 8:191–205.

Bleeker, P., and R. L. Parfitt. 1974. Volcanic ash and its clay mineralogy at Cape Haskins, New Britain, Papua, New Guinea. Geoderma 11:123–135.

Borchardt, G. A., M. L. Jackson, and F. D. Hole. 1966. Expansible layer silicate genesis in soils depicted in mica pseudormorphs. In L. Heller and A. Weiss (ed.) (Jerusalem, Israel) Proc. Int. Clay Conf. 1:175–185.

Borchardt, G. A., F. D. Hole, and M. L. Jackson. 1968. Genesis of layer silicates in representative soils in a glacial landscape of southeastern Wisconsin. Soil Sci. Soc. Am. Proc. 32:399–403.

Breese, G. F. 1960. Quartz overgrowths as evidence of silica deposition in soils. Aust. J. Sci. 23:18–20.

Brewer, R. 1964. Fabric and mineral analysis of soils. John Wiley & Sons, Inc. New York.

Brewer, R., E. Bettenay, and H. M. Churchward. 1972. Some aspects of the origin and development of Red and Brown soils of Bulloo Downs, Western Australia. Aust. C.S.I.R.O. Div. Soils Tech. Pap. no. 13:1–13.

Brindley, G. W. 1957. Fullers earth from near Dry Branch, Georgia—a montmorillonite-cristobalite clay. Clay Miner. Bull. 3:167–169.

Brydon, J. E., W. G. Dore, and J. S. Clark. 1963. Silicified plant asterosclereids preserved in soil. Soil Sci. Soc. Am. Proc. 27:476–477.

Buurman, P. 1972. Mineralization of fossil wood. Scripta Geol. 12(1972):1–43.

Buurman, P., N. van Breemen, and S. Henstra. 1973. Recent silicification of plant remains in acid sulfate soils. Neues Jahrb. Mineral. Monatsh. 3:117–124.

Buurman, P., and L. Van der Plas. 1971. The genesis of Belgian and Dutch flints and cherts. Geol. Mijnbouw 50:9–28.

Cailleux, A. 1942. Les actions éoliennes périglaciaires en Europe. Mem. Soc. Geol. France 46:1–176.

Calhoun, F. G., V. W. Carlisle, and C. Luna Z. 1972. Properties and genesis of selected Colombian andosols. Soil Sci. Am. Proc. 36:480–485.

Calvert, S. E. 1971. Composition and origin of North Atlantic deep sea cherts. Contrib. Mineral. Petrol. 33:273–288.

Cann, D. B., and E. P. Whiteside. 1955. A study of the genesis of a Podzol-gray brown Podzolic intergrade soil profile in Michigan. Soil Sci. Soc. Am. Proc. 19:497–501.

Chapman, S. L., J. K. Syers, and M. L. Jackson. 1969. Quantitative determination of quartz in soils, sediments and rocks by pyrosulfate fusion and hydrofluosilicic acid treatment. Soil Sci. 107:348–355.

Chester, R., and R. N. Green. 1968. The infra-red determination of quartz in sediments and sedimentary rocks. Chem. Geol. 3:199–212.

Clarke, F. W. 1924. The data of geochemistry. U. S. Geol. Surv. Bull. 770. 841 p.

Clayton, R. N., R. W. Rex, J. K. Syers, and M. L. Jackson. 1972. Oxygen isotope abundance in quartz from Pacific pelagic sediments. J. Geophys. Res. 77:3907–3915.

Cleary, W. J., and J. R. Conolly. 1972. Embayed quartz grains in soils and their significance. J. Sediment. Petrol. 42:899–904.

Clelland, D. W., and P. D. Ritchie. 1952. Physico-chemical studies on dusts. II. Nature and regeneration of the high-solubility layer on silicious dusts. J. Appl. Chem. 2:42–48.

Correns, C. W. 1941. Uber die Loslichkeit von Kiesselsaure in schwach saurer und alkalischen Losungen. Chem. Erde 13:92–96.

Cortes, A., and D. P. Franzmeier. 1972. Weathering of primary minerals in volcanic ash-derived soils of the central cordellera of Colombia. Geoderma 9:165–176.

Crook, Keith A. W. 1968. Weathering and roundness of quartz sand grains. Sedimentology 11:171–182.

Dale, A. J. 1924. The effect of prolonged grinding on the density of Quartz. Trans. Brit. Ceram. Soc. 23:211–216.

Davis, J. C. 1970. Petrology of cretaceous Mowry shale of Wyoming. Am. Assoc. Petrol. Geol. Bull. 54:487–502.

Deer, W. A., R. A. Howie, and J. Zussman. 1966. An introduction to rock forming minerals. John Wiley & Sons, Inc., New York. 528 p.

Dempster, P. B., and P. D. Ritchie. 1952. The surface of finely ground silica. Nature (London) 169:538–539.

Dempster, P. B., and P. D. Ritchie. 1953. Physicochemical studies on dusts. V. Examination of finely ground quartz by differential thermal analysis and other physical methods. J. Appl. Chem. 3:182–192.

Dennen, W. H. 1964. Impurities in quartz. Geol. Soc. Am. Bull. 75:241–246.

Dennen, W. H. 1966. Stoichiometric substitution in natural quartz. Geochim. Cosmochim. Acta 30:1235–1241.

Dennen, W. H. 1967. Trace elements in quarts as indicators of provenance. Geol. Soc. Am. Bull. 78:125–130.

Denny, J. J., W. D. Robson, and D. A. Irwin. 1939. The prevention of silicosis by metallic aluminum. Can. Med. Assoc. J. 40:213–228.

Dormaar, J. F., and L. E. Lutwick. 1969. Infrared spectra of humic acids and opal phytoliths as indicators of paleosols. Can. J. Soil Sci. 49:29–37.

Dreimanis, A., and U. J. Vagners. 1971. Biomodal distribution of rock and mineral fragments in basal tills. p. 237–250. In Richard P. Goldthwait (ed.) Till—a symposium. Ohio State Univ. Press, Columbus.

Eitel, W. 1954. The physical chemistry of the silicates. The University of Chicago Press, Chicago.

Eitel, W. 1957. Structure anomalies in tridymite and cristobalite. Bull. Am. Ceram. Soc. 36:142–148.

Elgawhary, S. M., and W. L. Lindsay. 1972. Solubility of silica in soils. Soil Sci. Soc. Am. Proc. 36:439–442.

Epstein, S., and H. P. Taylor, Jr. 1967. Variation of $^{18}O/^{16}O$ in minerals and rocks. p. 29–62. In P. H. Abelson (ed.) Researches in geochemistry, Vol. 2. John Wiley & Sons, Inc., New York.

Ernst, W. G., and S. E. Calvert. 1969. An experimental study of the recrystallization of porcelanite and its bearing on the origin of some bedded cherts. Am. J. Sci., Schairer Vol. 267–A:114–133.

Eslinger, E. V., L. M. Mayer, L. Durst, J. Hower, and S. M. Savin. 1973. An X-ray technique for distinguishing between detrital and secondary quartz in the fine-grained fraction of sedimentary rocks. J. Sediment. Petrol. 43:540–543.

Eswaran, H., and F. De Coninck. 1971. Clay mineral formations and transformations in basaltic soils in tropical environments. Pedologie 21:181–210.

Evans, C. B., and B. J. Meggers. 1968. Archeological investigation on the Rio Napo, eastern Ecuador. Washington, D. C. Smithson. Contrib. Anthropol. 6:127.

Evans, W. P. 1965. Facets of organic geochemistry. p. 14–28. In E. G. Hallsworth and D. V. Crawford (ed.) Experimental pedology. Butterworths, London.

Fairbairn, H. W. 1939. Correlation of quartz deformation with its crystal structure. Am. Mineral. 24:351–368.

Faust, G. T. 1948. Differential thermal analysis of quartz and its use in calibration. Am. Mineral. 33:337–345.

Fenner, C. N. 1913. The stability relations of the silica minerals. Am. J. Sci. 36:331–384.

Ferm, J. C. 1962. Petrology of some Pennsylvanian rocks. J. Sediment. Petrol. 32:104–123.

Flach, K. W., W. D. Nettleton, L. H. Gile, and J. C. Cady. 1969. Pedocomentation: Induration by silica, carbonates, and sesquioxides in the Quaternary. Soil Sci. 107:442–453.

Folger, D. W., L. H. Burckle, and B. C. Heezen. 1967. Opal phytoliths in a North Atlantic dust fall. Sci. 155:1243–1244.

Folk, R. L. 1968. Petrology of sedimentary rocks. Hemphill's Book Store, Austin, Texas. 154 p.

Francis, C. W., W. P. Bonner, and T. Tamura. 1972. An evaluation of zonal centrifugation as a research tool in soil science: I. Methodology. Soil Sci. Soc. Am. Proc. 36:366-372.

Francis, C. W., and T. Tamura. 1972. An evaluation of zonal centrifugation as a research tool in soil science: II. Characterization of soil clays. Soil Sci. Soc. Am. Proc. 36: 372-376.

Friedlaender, G. G. I. 1970. Tridymite in the gangue of a Pb–Cu–Zn occurrence. Schweiz. Mineral. Petrogr. Mitt. 50:183-199.

Frondel, C. 1962. Dana's system of mineralogy. Vol. III–Silica minerals. John Wiley & Sons, New York.

Gaskell, P. H. 1966. Thermal properties of silica. I. Effect of temperature on infra-red reflection spectra of quartz, cristobalite and vitreous silica. Trans. Farad. Soc. 62: 1493-1504.

Geis, J. W. 1973. Biogenic silica in selected species of deciduous angiosperms. Soil Sci. 116:113-130.

Geis, J. W., and R. L. Jones. 1973. Ecological significance of biogenic opaline silica. p. 74-85. In D. L. Dindal (ed.) First Soil Microcommunities Conf. USAEC, Soil Microcommunities I.

Gibb, J. G., P. D. Ritchie, and J. W. Sharpe. 1953. Physico-chemical studies on dusts. VI. Electron-optical examination of finely ground silica. J. Appl. Chem. 3:213-218.

Gibbs, R. E. 1927. The polymorphism of silicon dioxide and the structure of tridymite. Proc. Royal Soc. A113:351-368.

Gill, E. D. 1967. Stability of biogenetic opal. Sci. 158:810.

Goldberg, E. D. 1958. Determination of opal in marine sediments. J. Marine Res. 17: 178-182.

Goldich, S. S. 1938. A study in rock weathering. J. Geol. 46:17-58.

Gordon, R. L., and G. W. Harris. 1955. Effect of particle-size on the quantitative determination of quartz by X-ray diffraction. Nature (London) 175:1135.

Grant, R. W. 1967. New data on tridymite. Am. Mineral. 52:536-541.

Greenwood, R. 1973. Cristobalite: its relationship to chert formation in selected samples from the deep sea drilling project. J. Sediment. Petrol. 43:700-708.

Greer, R. T. 1969. Submicron structure of amorphous opal. Nature (London) 224: 1199-1200.

Greer, R. T. 1971. The growth of colloidal silica particles. p. 153-159. In Proc. the Fourth Annu. Scanning Electron Microscopy Symposium, ITT Res. Inst., Chicago, Ill.

Grim, R. E. 1968. Clay mineralogy. 2nd ed. McGraw-Hill Book Co., New York.

Grimshaw, R. W. 1953. Quantitative estimation of silica minerals. Clay Mineral. Bull. 2:2-7.

Gruner, J. W. 1940a. Cristobalite in bentonite. Am. Mineral. 25:587-590.

Gruner, J. W. 1940b. Abundance and significance of cristobalite in bentonites and fuller's earths. Econ. Geol. 35:867-875.

Guven, N., and R. E. Grim. 1972. X-ray diffraction and electron optical studies on smectite and α cristobalite associations. Clays Clay Miner. 20:89-92.

Harder, H. 1971. Quartz and clay mineral formation at surface temperatures. Mineral. Soc. Japan Spec. Pap. 1:106-108.

Hardjosoesastro, R. R. 1956. Preliminary note on cristobalite in clay fractions of volcanic ashes. J. Soil Sci. 7:185-188.

Hashimoto, I., and M. L. Jackson. 1960. Rapid dissolution of allophane and kaolinite-halloysite after dehydration. Clays Clay Miner. 7:102-113.

Henderson, J. H., R. N. Clayton, M. L. Jackson, J. K. Syers, R. W. Rex, J. L. Brown and I. B. Sachs. 1972. Cristobalite and quartz isolation from soils and sediments by hydrofluosilicic acid treatment and heavy liquid separation. Soil Sci. Soc. Am. Proc. 36:830-835.

Henderson, J. H., M. L. Jackson, J. K. Syers, R. N. Clayton, and R. W. Rex. 1971. Cristobalite authigenic origin in relation to montmorillonite and quartz origin in bentonites. Clays Clay Miner. 19:229-238.

Henderson, J. H., J. K. Syers, and M. L. Jackson. 1970. Quartz dissolution as influenced by pH and the presence of a disturbed surface layer. Israel J. Chem. 8:357-372.

Hill, D. E., and A. E. Shearin. 1969. The Charlton soils. Bull. Conn. Agric. Exp. Stn. 706: 49 p.

Hill, V. G., and R. Roy. 1958a. Silica structure studies: V. The variable inversion in cristobalite. J. Ceram. Soc. 41:532–537.

Hill, V. G., and R. Roy. 1958b. Silica structure studies: VI. On tridymites. Trans. Brit. Ceram. Soc. 57:496–510.

Horberg, L., and P. E. Potter. 1955. Stratigraphic and sedimentologic aspects of the Lemont drift of northeastern Illinois. Ill. State Geol. Surv. Rept. Inv. 185: 23 p.

Howard, A. D. 1939. Cristobalite in southwestern Yellowstone Park. Am. Mineral. 24: 485–491.

Huang, W. H., and D. L. Vogler. 1972. Dissolution of opal in water and its water contents. Nature (London) Phys. Sci. 235:157–158.

Huizing, H. G. J. 1971. A reconnaissance study of the mineralogy of sand fractions from East Pakistan sediments and soils. Geoderma 6:109–133.

Hunt, J. M., and D. S. Turner. 1953. Determination of mineral constituents of rocks by infra-red spectroscopy. Anal. Chem. 25:1169–1174.

Hunt, J. M., M. P. Wisherd, and L. C. Bonham. 1950. Infrared absorption spectra of minerals and other inorganic compounds. Anal. Chem. 22:1478–1497.

Iler, R. K. 1955. Colloid chemistry of silica and silicates. Cornell Univ. Press. Ithaca, New York. 324 p.

Iler, R. K. 1973. Effect of adsorbed alumina on the solubility of amorphous silica in water. J. Colloid. Interface Sci. 43:399–408.

Jackson, M. L. 1956. Soil chemical analysis—Advanced course. Published by the author. Dep. of Soils, Univ. of Wis., Madison. 991 p.

Jackson, M. L., F. R. Gibbons, J. K. Syers, and D. L. Mokma. 1972. Eolian influence on soils developed in a chronosequence of basalts of Victoria, Australia. Geoderma 8: 147–163.

Jackson, M. L., T. W. M. Levelt, J. K. Syers, R. W. Rex, R. N. Clayton, G. D. Sherman, and G. Uehara. 1971. Geomorphological relationships of tropospherically derived quartz in the soils of the Hawaiian Islands. Soil Sci. Soc. Am. Proc. 35:515–525.

Jackson, M. L., and G. D. Sherman. 1953. Chemical weathering of minerals in soils. Adv. Agron. 5:219–318.

Jephcott, C. M., and J. H. Johnston. 1950. Solubility of silica and alumina. Arch. Ind. Hyg. Occup. Med. 1:323–340.

Johnson, P. R., and A. A. Beavers. 1959. A mineralogical characterization of some loess-derived soils in Illinois. Soil Sci. Soc. Am. Proc. 23:143–146.

Jones, J. B., and E. R. Segnit. 1969. Water in sphere-type opal. Mineral. Mag. 37:357–361.

Jones, J. B., and E. R. Segnit. 1972. Genesis of cristobalite and tridymite at low temperatures. Geol. Soc. Aust. J. 18:419–422.

Jones, J. B., J. V. Sanders, and E. R. Segnit. 1964. Structure of opal. Nature (London) 204:990–991.

Jones, J. B., E. R. Segnit, and N. M. Nickson. 1963. Differential thermal and X-ray analysis of opal. Nature (London) 198:1191.

Jones, L. H. P., and K. A. Handreck. 1963. Effects of iron and aluminum oxides on silica in solution in soils. Nature (London) 198:852–853.

Jones, L. H. P., and K. A. Handreck. 1967. Silica in soils, plants and animals. Adv. Agron. 19:107–149.

Jones, L. H. P., and A. A. Milne. 1963. Studies of silica in the oat plant. I. Chemical and physical properties of the silica. Plant Soil 18:207–220.

Jones, L. H. P., A. A. Milne, and J. V. Sanders. 1966. Tabashir: an opal of plant origin. Science 151:464–466.

Jones, L. H. P., A. A. Milne, and S. M. Wadham. 1963. Studies of silica in the oat plant. II. Distribution of the silica in the plant. Plant Soil MVIII:358–371.

Jones, R. L. 1964. Note on the occurrence of opal phytoliths in some Cenozoic sedimentary rocks. J. Paleontol. 38:773–775.

Jones, R. L. 1969. Determination of opal in soil by alkali dissolution analysis. Soil Sci. Soc. Am. Proc. 33:976–978.

Jones, R. L., and A. H. Beavers. 1963a. Some mineralogical and chemical properties of plant opal. Soil Sci. 96:375–379.

Jones, R. L., and A. H. Beavers. 1963b. Sponge spicules in Illinois soils. Soil Sci. Soc. Am. Proc. 27:438-440.

Jones, R. L., and A. H. Beavers. 1964a. Variation of opal phytolith content among some great soil groups in Illinois. Soil Sci. Soc. Am. Proc. 28:711-712.

Jones, R. L., and A. H. Beavers. 1964b. Aspects of catenary and depth distribution of opal phytoliths in Illinois soils. Soil Sci. Soc. Am. Proc. 28:413-416.

Jones, R. L., and W. W. Hay. 1975. Bioliths. p. 481-496. In J. E. Gieseking (ed.) Soil components, Vol. 2, Inorganic components. Springer-Verlag, N. Y.

Jones, R. L., W. W. Hay, and A. H. Beavers. 1963. Microfossils in Wisconsin loess and till from western Illinois and eastern Iowa. Science 140:1222-1224.

Jones, R. L., L. J. McKenzie, and A. H. Beavers. 1964. Opaline microfossils in some Michigan soils. Ohio J. Sci. 64:417-423.

Jorgensen, S. T. 1970. The application of alkali dissolution techniques in the study of cretaceous flints. Chem. Geol. 6:153-163.

Kanno, I., and S. Arimura. 1958. Plant opal in Japanese soils. Soil Plant Food 4:62-67.

Katz, M. Y. A., M. M. Katz, and A. A. Rasskazov. 1970. Mineral studies in the gravitation-gradient field. 2. Changes of quartz and density due to natural and experimental "Maturation." Sediment 15:161-177.

Keith, M. L., and O. Tuttle. 1952. High-low inversion of quartz. Am. J. Sci., Bowen 1: 203-252.

Khangarot, A. S., L. P. Wilding, and G. F. Hall. 1971. Composition and weathering of loess mantled Wisconsin- and Illinoian-age terraces in Central Ohio. Soil Sci. Soc. Am. Proc. 35:621-626.

Kiely, P. V., and M. L. Jackson. 1965. Quartz, feldspar and mica determination for soils by sodium pyrosulfate fusion. Soil Sci. Soc. Am. Proc. 29:159-163.

Kittrick, J. A. 1969. Soil minerals in the Al_2O_3-SiO_2-H_2O system and a theory of their formation. Clays Clay Miner. 17:157-167.

Klasik, J. A. 1975. High cristobalite and high tridymite in a middle eocene deep-sea chert. Science 189:631-632.

Kleck, W. D. 1970. Cavity minerals at Summit Rock, Oregon. Am. Mineral. 55:1396-1404.

Kolbe, R. W. 1957. Fresh water diatoms from Atlantic deep-sea sediments. Science 126:1053-1056.

Krauskopf, K. B. 1956. Dissolution and precipitation of silica at low temperatures. Geochim. Cosmochim. Acta 10:1-26.

Krauskopf, K. B. 1959. The geochemistry of silica in sedimentary environments. In H. Andrew Ireland (ed.) Silica in sediments. Soc. Econ. Paleontol. Mineral., Spec. Publ. 7:4-19.

Krauskopf, K. B. 1967. Introduction to geochemistry. McGraw-Hill Book Co., New York.

Krinsley, D. H., and J. Donahue. 1968a. Environmental interpretation of sand grain surface textures by electron microscopy. Geol. Soc. Am. Bull. 79:743-748.

Krinsley, D. H., and J. Donahue. 1968b. Methods to study surface textures of sand grains, a discussion. Sedimentology 10:217-221.

Krinsley, D. H., and J. C. Doornkamp. 1973. Atlas of quartz sand surface textures. Cambridge Univ. Press, Cambridge, England.

Krinsley, D. H., and B. M. Funnell. 1965. Environmental history of quartz sand grains from the lower and middle pleistocene of Norfolk, England. Quart. J. Geol. Soc. London 121:435-461.

Krinsley, D. H., and S. V. Margolis. 1969. A study of quartz sand grain surface with the scanning electron microscope. Trans. New York Acad. of Sci. 31:457-477.

Krinsley, D. H., W. S. Newman, T. Takahashi, and M. L. Silberman. 1962. Electron-microscopic studies of the textures of sand grains in the Atlantic shore of Long Island, N. Y. Geol. Soc. Am. Spec. Papers 73:304.

Krinsley, D. H., and I. J. Smalley. 1972. Sand. Am. Sci. 60:286-291.

Krinsley, D. H., and I. J. Smalley. 1973. Shape and nature of small sedimentary quartz particles. Science 180:1277-1279.

Krinsley, D. H., and T. Takahashi. 1962a. The surface textures of sand grains, an application of electron microscopy. Science 135:923-925.

Krinsley, D. H., and T. Takahashi. 1962b. The surface textures of sand grains, an application of electron microscopy: glaciation. Science 138:1262-1264.

Krinsley, D. H., and T. Takahashi. 1962c. Applications of electron microscopy to geology. Trans. N. Y. Acad. Sci. 25:3–22.

Krinsley, D. H., and T. Takahashi. 1962d. Electron microscopic examination of natural and artificial glacial sand grains. Geol. Soc. Am. Spec. Papers 73:190.

Krinsley, D. H., and T. Takahashi. 1964. A technique for the study of surface textures of sand grains with electron microscopy. J. Sediment. Petrol. 34:423–426.

Krinsley, D. H., T. Takahashi, M. L. Silberman, and W. S. Newman. 1964. Transportation of sand grains along the Atlantic shore of Long Island, New York: an application of electron microscopy. Marine Geol. 2:100–120.

Krinsley, D. H., and N. K. Tovey. 1973. Studies at quartz sand grains by cathodoluminescence. Am. Assoc. Petrol. Geol. Bull. 57:789.

Krynine, P. D. 1940. Petrology and genesis of the third Bradford sand. Penn. State College Mineral Ind. Exp. Stn. Bull. 29, 134 p.

Kuenen, Ph. H., and W. G. Perdok. 1962. Experimental abrasion 5: frosting and defrosting of quartz grains. J. Geol. 70:648–658.

Kunze, G. W., and H. Oakes. 1957. Field and laboratory studies of the Lufkin soil, a planosol. Soil Sci. Soc. Am. Proc. 21:330–335.

Lang, A. R. 1968. X-ray moiré topography of lattice defects in quartz. Nature (London) 220:652–657.

Lanning, F. C., B. W. X. Ponnaiya, and C. F. Crumpton. 1958. The chemical nature of silica in plants. Plant Physiol. 33:339–343.

Larsen, E. S., J. Irving, F. A. Gonyer, and E. S. Larsen, 3rd. 1936. Petrologic results of a study of the minerals from the tertiary volcanic rocks of the San Juan Region, Colorado. Am. Mineral. 21:679–701.

Lathrap, D. W. 1970. The upper Amazon. Praeger, New York.

Le Roux, J. 1973. Quantitative clay mineralogical analysis of Natal Oxisols. Soil Sci. 115:137–144.

Lewin, J. C. 1961. The dissolution of silica from diatom walls. Geochim. Cosmochim. Acta 21:182–198.

Lewin, J., and B. E. F. Reimann. 1969. Silica and plant growth. Annu. Rev. Plant Physiol. 20:289–304.

Liberti, A., and G. Devitofrancesco. 1963. The nature of ground quartz particles. Chem. Ind. 51:1983–1984.

Lidstrom, L. 1968. Surface and bond-forming properties of quartz and silicate minerals and their application in mineral processing techniques. Acta Polytech. Scand. 75:149.

Linné, S. 1965. The ethnologist and the American Indian potter. p. 20–42. In F. R. Matson (ed.) Ceramics and man. Aldine, Chicago.

Lucas, C. C., and M. E. Dolan. 1939. Studies on the solubility of quartz and silicates. Can. Med. Assoc. J. 40:126–134.

Lutwick, L. E. 1969. Identification of phytoliths in soils. p. 77–82. In S. Pawluk (ed.) Pedology and Quaternary research, symposium. National Canadian Research Council and Univ. of Alberta, Canada.

Lutwick, L. E., and A. Johnston. 1969. Cumulic soils of the rough fescue prairie-popular transition region. Can. J. Soil Sci. 49:199–203.

Mackenzie, F. T., and R. Gees. 1971. Quartz: synthesis at earth-surface conditions. Science 173:533–535.

Margolis, S. V., and D. H. Krinsley. 1974. Processes of formation and environmental occurrence of microfeatures on detrital quartz grains. Am. J. Sci. 274:449–464.

Mason, B. 1953. Tridymite and christensenite. Am. Mineral. 38:866–867.

McKeague, J. A., and M. G. Cline. 1963a. Silica in soils. Adv. Agron. 15:339–396.

McKeague, J. A., and M. G. Cline. 1963b. Silica in soil solutions. I. The form and concentration of dissolved silica in aqueous extracts of some soils. Can. J. Soil Sci. 43:70–82.

McKeague, J. A., and M. G. Cline. 1963c. Silica in soil solutions. II. The absorption of monosilicic acid by soil and other substances. Can. J. Soil Sci. 43:83–96.

Mejia, G., H. Kohnke, and J. L. White. 1968. Clay mineralogy of certain soils of Colombia. Soil Sci. Soc. Am. Proc. 32:665–670.

Metcalfe, C. R. 1960. Anatomy of the monocotyledons. I. Gradineae. Clarendon Press, Oxford. p. 731.

Millot, G. 1970. Geology of clays. Springer-Verlag, New York. p. 429.

Mitchell, B. D. 1975. Oxides and hydrous oxides of silicon. p. 395–432. *In* J. E. Gieseking (ed.) Soil Components, Vol. 2, Inorganic Components. Springer-Verlag, New York.

Mitchell, R. S. 1967. Tridymite pseudomorphs after wood in Virginian lower cretaceous sediments. Science 158:905–906.

Mitchell, R. S., and S. Tufts. 1973. Wood opal—A tridymite like mineral. Am. Mineral. 58:717–720.

Mizota, C., and S. Aomine. 1975. Clay mineralogy of some volcanic ash soils in which cristobalite predominates. Soil Sci. Plant Nutr. 21:327–335.

Mizutani, S. 1970. Silica minerals in the early stage of diagenesis. Sedimentology 15: 419–436.

Mokma, D. L., J. K. Syers, M. L. Jackson, R. N. Clayton, and R. W. Rex. 1972. Aeolian additions to soils and sediments in the south Pacific area. J. Soil Sci. 23:147–162.

Monroe, E. A., D. B. Sass, and S. H. Cole. 1969. Stacking faults and polytypism in opal, $SiO_2 \cdot nH_2O$. Acta Crystallagr. 25:578–580.

Moore, G. S. M., and H. E. Rose. 1973. The structure of powdered quartz. Nature (London) 242:187–190.

Morey, G. W., R. O. Fournier, and J. J. Rowe. 1962. The solubility of quartz in water in the temperature interval from 25 to 300°C. Geochim. Cosmochim. Acta 26:1029–1043.

Murata, K. J., and J. K. Nakata. 1974. Cristobalite stage in the diagenesis of diatomaceous shale. Science 184:567–568.

Murray, H. H., and R. K. Leininger. 1956. Effect of weathering on clay minerals. Clays Clay Miner. 456:340–347.

Nagelschmidt, G., R. L. Gordon, and O. G. Griffin. 1952. Surface of finely ground silica. Nature (London) 169:539–540.

Norgren, A. 1973. Opal phytoliths as indicators of soil age and vegetative history. Ph.D. Thesis. Oregon State Univ., Corvallis, Ore. Univ. Microfilms. Ann Arbor, Mich. Diss. Abstr. Int. 33:3421 B.

Oehler, J. H. 1973. Tridymite-like crystals in cristobalite "cherts" Nature. Physiol. Sci. 241:64–65.

Okamoto, G., T. Okura, and K. Goto. 1957. Properties of silica in water. Geochim. Cosmochim. Acta 12:123–132.

Papke, K. G. 1969. Montmorillonite deposits in Nevada. Clays Clay Miner. 17:211–222.

Parfitt, R. L. 1974. Clay minerals in recent volcanic ash soils. R. Soc. N. Z. Bull. 13.

Parry, D. W., and F. Smithson. 1964. Types of opaline silica depositions in the leaves of British grasses. Ann. Bot. 28:169–185.

Parry, D. W., and F. Smithson. 1966. Opaline silica in the inflorescences of some British grasses and cereals. Annu. Bot. 30:525–538.

Patrick, R., and C. Reimer. 1966. The diatoms of the United States. Monographs of the Academy of Natural Sciences of Philadelphis. No. 13, 1:1–688.

Pease, D. S., and J. U. Anderson. 1969. Opal phytoliths in *Bouteloua Eriopoda Torr.* roots and soils. Soil Sci. Soc. Am. Proc. 33:321–322.

Peinemann, N., M. Tschapek, and R. Grassi. 1970. Properties of phytoliths. Z. Pflanzenernaehr. Bodenkd. 127:126–133.

Pelto, C. R. 1956. A study of chalcedony. Am. J. Sci. 254:32–50.

Peterson, M. N. A., and C. C. Von der Borch. 1965. Chert. modern inorganic deposition in a carbonate-precipitating locality. Science 149:1501–1503.

Pettijohn, F. J. 1957. Sedimentary rocks. 2nd ed. Harper, New York.

Phillippe, M. M., and J. L. White. 1950. Quantitative estimation of minerals in the fine sand and silt fractions of soils with the Geiger Counter X-ray Spectrometer. Soil Sci. Soc. Am. Proc. 15:138–142.

Pollard, C. O., Jr., and C. E. Weaver. 1973. Opaline spheres: loosely-packed aggregates from silica nodule in diatomaceous miocene fuller's earth. J. Sediment. Petrol. 43: 1072–1076.

Porter, J. J. 1962. Electron microscopy of sand surface texture. J. Sediment. Petrol. 32: 124–135.

Prat, H. 1936. La systematique des graminees. Ann. Sci. Nature Bot. Biol. Veg. (10) 18: 165–258.

Raeside, J. D. 1970. Some New Zealand plant opals. N. Z. J. Sci. 13:122-132.

Ratliff, L. F., and B. L. Allen. 1970. The mineralogy and genesis of two soils from Trans-Pecos, Texas. Soil Sci. 110:268-277.

Ray, R. C. 1923. The effect of long grinding on quartz (silver sand). Proc. Roy. Soc. A102:640-642.

Redmond, C. E., and E. P. Whiteside. 1967. Some till-derived Chernozem soils in eastern North Dakota: II. Mineralogy, micromorphology, and development. Soil Sci. Soc. Am. Proc. 31:100-107.

Rex, R. W., and E. D. Goldberg. 1958. Quartz contents of pelagic sediments of the Pacific Ocean. Tellus 10, 1:153-159.

Rex, R. W., J. K. Syers, M. L. Jackson, and R. N. Clayton. 1969. Eolian origin of quartz in soils of Hawaiian Islands and in Pacific pelagic sediments. Science 163:277-279.

Reynolds, R. C., Jr., and D. M. Anderson. 1967. Cristobalite and clinoptilolite in ben-tonite beds of the Colville Group, North Alaska. J. Sediment. Petrol. 37:966-969.

Reynolds, W. R. 1970. Mineralogy and stratigraphy of Lower Tertiary clays and clay-stones of Alabama. J. Sediment. Petrol. 40:829-838.

Ribault, L. L. 1971. Presence d'une pellicule de silice amorphe a la surface de cristaux de quartz des formations sableuses. C. R. Acad. Sci. Paris. 272, Serie D:1933-1936.

Riquier, J. 1960. Les phytoliths de certain sols Tropicaux et des podzols. Int. Congr. Soil Sci., Trans. 7th (Madison, Wis.) IV:425-431.

Rivalenti, G., and G. P. Sighinolfi. 1968. Inclusi di tridymite del basalto di Gambellara (Vicenza). Period. Mineral. 37:495-501.

Rockett, T. J., and W. R. Foster. 1967. Thermal stability of purified tridymite. Am. Mineral. 52:1233-1240.

Rogers, A. F. 1928. Natural history of silica minerals. Am. Mineral. 13:73-92.

Rogers, A. F. 1937. Introduction to the study of minerals. McGraw-Hill Book Co. Inc., New York.

Rogers, A. F., and P. F. Kerr. 1942. Optical mineralogy. McGraw-Hill Book Co. Inc., New York. 390 p.

Rovner, I. 1971. Potentail of opal phytoliths for use in paleoecological reconstruction. Quat. Res. 1:343-359.

Rowse, J. B., and W. B. Jepson. 1972. The determination of quartz in clay minerals: A critical comparison of methods. J. Therm. Anal. 4:169-175.

Roy, D. M., and R. Roy. 1964. Tridymite-cristobalite relations and stable solid solutions. Am. Mineral. 49:952-962.

Russell, E. J. 1961. Soil conditions and plant growth. 9th ed. Longmans, Green & Co., New York. p. 688.

Sanders, J. V. 1968. Diffraction of light by opals. Acta. Crystallograph. A24:427-434.

Sanders, J. V. 1975. Microstructure and crystallinity of gem opals. Am. Mineral. 60: 749-757.

Sato, M. 1962. Tridymite crystals in opaline silica from Kusatsu, Gumma Prefecture. Mineral. J. 3:296-305.

Sato, M. 1964a. X-ray study of tridymite (1) on tridymite M and tridymite S. Mineral. J. 4:115-130.

Sato, M. 1964b. X-ray study of low tridymite (2) Structure of low tridymite, type M. Mineral J. 4:131-146.

Schneider, Horst E. 1970. Problems of quartz grain morphoscopy. Sedimentology 14: 325-335.

Shoji, R., and R. L. Folk. 1964. Surface morphology of some limestone types as revealed by electron microscopy. J. Sediment. Petrol. 34:144-155.

Shoji, S., and J. Masui. 1969. Amorphous clay minerals of recent volcanic ash soils in Hokkaido. II. Soil Plant Nutr. 15:191-201.

Shoji, S., and J. Masui. 1971. Opaline silica of recent volcanic ash soils in Japan. J. Soil Sci. 22:101-108.

Siever, R., and R. A. Scott. 1963. Organic geochemistry of silica. p. 579-595. In I. A. Breger (ed.) Organic geochemistry. Pergamon Press, N. Y.

Siffert, B. 1967. Some reactions of silica in solution: Formation of clay. Israel Program for Scientific Translations Ltd., Jerusalem. Translation of Memoires du service de la carte geologique d'Alsace et de Lorraine. Rep. no. 21 (1962).

Sivarajasingham, S., L. T. Alexander, J. G. Cady, and M. G. Cline. 1962. Laterite. Adv. Agron. 14:1-60.

Slatt, R. M., and C. M. Hoskin. 1968. Water and sediment in the Norris Glacier outwash area, upper Taku Inlet, southeastern Alaska. J. Sediment. Petrol. 38:434–456.

Smeck, N. E., L. P. Wilding, and N. Holowaychuk. 1968. Genesis of argillic horizons in Celina and Morley soils of western Ohio. Soil Sci. Soc. Am. Proc. 32:550–556.

Smith, J. V. 1962. Index to the powder data file. American Society for Testing and Materials, Philadelphia, Pa.

Smithson, F. 1956. Silica particles in some British soils. J. Soil Sci. 7:122–129.

Smithson, F. 1958. Grass opal in British soils. J. Soil Sci. 9:148–154.

Smithson, F. 1959. Opal sponge spicules in soils. J. Soil Sci. 10:105–109.

Smykatz-Kloss, W. 1972. The high-low inversion of microcrystalline quartz crystals. Contrib. Mineral. Petrol. 36:1–18.

Soil Survey Staff. 1975. Soil taxonomy—A basic system of soil classification for making and interpreting soil surveys. Agric. Handb. no. 436. U. S. Gov't. Printing Office. Washington, D. C.

Sosman R. B. 1965. The phases of silica. Rutgers Univ. Press. New Brunswick, N. J.

Sowers, G. B., and G. F. Sowers. 1970. Introductory soil mechanics and foundations. 3rd ed. The Macmillan Company. N. Y. 556 p.

Springer, M. Elsworth. 1948. The composition of the silt fraction as related to the development of soils from loess. Soil Sci. Soc. Am. Proc. 13:461–467.

Stahnke, C. R., J. R. Rogers, and B. L. Allen. 1969. A genetic and mineralogical study of a soil developed from granitic gneiss in the Texas central basin. Soil Sci. 108:313–320.

St. Arnaud, R. J., and E. P. Whiteside. 1963. Physical breakdown in relation to soil development. J. Soil Sci. 14:267–281.

St. J. Warne, S. 1970. The detection and identification of the silica minerals quartz, chalcedony, agate and opal, by differential thermal analysis. J. Inst. Fuel 43:240–242.

Stewart, D. R. M. 1965. The epidermal characters of grasses with special reference to east African plains species. Bot. Jb. 84:63–174.

Stöber, W. 1967. Formation of silicic acid in aqueous suspensions of different silica modification. p. 161–182. In R. F. Gould (ed.) Equilibrium concepts in natural water systems. Advanced Chem. Ser. 67. Am. Chem. Soc., Washington, D. C.

Stavrov, O. D. 1961. On the content of rare elements in quartz. Geochemistry 6:542–549.

Sudom, M. D., and R. J. St. Arnaud. 1971. Use of quartz, zirconium and titanium as indices in pedological studies. Can. J. Soil Sci. 51:385–396.

Suttner, L. J., and R. K. Leininger. 1972. Comparison of the trace element content of plutonic, volcanic, and metamorphic quartz from southwestern Montana. Geol. Soc. Am. Bull. 83:1855–1862.

Swindale, L. D., and M. L. Jackson. 1960. A mineralogical study of soil formation in four rhyolite-derived soils from New Zealand. N. Z. J. Geol. Geophys. 3:141–183.

Syers, J. K., M. L. Jackson, V. E. Berkheiser, R. N. Clayton, and R. W. Rex. 1969. Eolian sediment influence on pedogenesis during the Quaternary. Soil Sci. 107:421–427.

Tan, K. H., H. F. Perkins, and R. A. McCreery. 1975. Amorphous and crystalline clays in volcanic ash soils of Indonesia and Costa Rica. Soil Sci. 119:431–440.

Tedrow, J. C. F. 1954. Clay minerals in three podzol profiles. Soil Sci. Soc. Am. Proc. 18:479–481.

Teodorovich, G. I. 1961. Authigenic, principally syngenetic minerals in sedimentary rocks, their characteristics, conditions of formation, and classification. p. 17–21. InAuthigenic minerals in sedimentary rocks. Consultants Bureau, New York.

Thomas, R. L. 1969. A note on the relationship of grain size, clay content, quartz and organic carbon in some Lake Erie and Lake Ontario sediments. J. Sediment. Petrol. 39:803–809.

Trostel, L. J., and D. J. Wynne. 1940. Free silica determination. J. Am. Ceram. Soc. 23:18–22.

Twiss, P. C., R. M. Smith, and R. K. Krauss. 1967. Composition of phytoliths from Andropogon scoparius. Geol. Soc. Am. Abstr. p. 225.

Twiss, P. C., E. Suess, and R. M. Smith. 1969. Morphological classification of grass phytoliths. Soil Sci. Soc. Am. Proc. 33:109–115.

Tyurin, I. V. 1937. On the biological accumulation of silica in soils. Probl. Sov. Soil Sci. 4:3–23.

Usov, N. I. 1943. Biological accumulation of silica in soils. Pochvovedenie no. 9–10: 30–36.

Van Straaten, L. M. J. V. 1954. Composition of recent marine sediments in the Netherlands. Leidse Geol. Meded. 19:1–108.

Verma, S. D., and R. H. Rust. 1969. Observations on opal phytoliths in a soil biosequence in southeastern Minnesota. Soil Sci. Soc. Am. Proc. 33:749–751.

Weaver, F. M., and S. W. Wise. 1972. Ultramorphology of deep sea cristobalitic chert. Nature (London) Phys. Sci. 237:56–57.

Weaver, F. M., and S. W. Wise, Jr. 1974. Opaline sediments of the southeastern coastal plain and horizon A: biogenic origin. Science 184:899–901.

White, D. E., W. W. Brannock, and K. J. Murata. 1956. Silica in hot-spring waters. Geochim. Cosmochim. Acta 10:27–59.

Wilding, L. P. 1967. Radiocarbon dating plant opal in Ohio soil. Science 153:66–67.

Wilding, L. P., R. E. Brown, and N. Holowaychuk. 1967. Accessibility and properties of occluded carbon in biogenetic opal. Soil Sci. 103:56–61.

Wilding, L. P., and L. R. Drees. 1968a. Biogenic opal in soils as an index of vegetative history in the Prairie Peninsula. p. 99–103. In R. E. Bergstrom (ed.) The quaternary of Illinois. University of Ill. Col. of Agric. Spec. Publ. 14.

Wilding, L. P., and L. R. Drees. 1968b. Distribution and implications of sponge spicules in surficial deposits in Ohio. Ohio J. Sci. 68:92–99.

Wilding, L. P., and L. R. Drees. 1971. Biogenic opal in Ohio soils. Soil Sci. Soc. Am. Proc. 35:1004–1010.

Wilding, L. P., and L. R. Drees. 1973. Scanning electron microscopy of opaque opaline forms isolated from forest soils in Ohio. Soil Sci. Soc. Am. Proc. 37:647–650.

Wilding, L. P., and L. R. Drees. 1974. Contributions of forest opal and associated crystalline phases to fine silt and clay fractions of soils. Clays Clay Miner. 22:295–306.

Wilding, L. P., and H. D. Geissinger. 1973. Correlative light optical and scanning electron microscopy of minerals: a methodology study. J. Sediment. Petrol. 43:280–286.

Wilson, M. J., J. D. Russel, and J. M. Tait. 1974. A new interpretation of the structure of disordered α-cristobalite. Contrib. Mineral. Petrol. 47:1–6.

Wise, S. W., Jr., F. B. Buie, and F. M. Weaver. 1972. Chemically precipitated sedimentary cristobalite and the origins of chert. Eclogae Geol. Helv. 65:157–163.

Wise, S. W., Jr., and K. R. Kelts. 1972. Inferred diagenic history of a weakly silicified deep sea chalk. Trans. Gulf Coast Assoc. of Geol., Soc. 22:177–203.

Wise, S. W., Jr., and F. M. Weaver. 1973. Origin of cristobalite-rich tertiary sediments in the Atlantic and gulf coastal plain. Transactions—Gulf coast Assoc. of Geol. Soc. 22:305–323.

Wise, S. W., Jr., and F. M. Weaver. 1974. Chertification of ocean sediments. Int. Assoc. Sedimentologists, Spec. Publ. no. 1, p. 301–326.

Witty, J. E., and E. G. Knox. 1964. Grass opal in some chestnut and forested soils in North Central Oregon. Soil Sci. Soc. Am. Proc. 28:685–688.

Wright, A. C. S. 1964. The "andosols" or "humic allophane" soils of South America. World Soil Resour. Rep.

Wyckoff, R. W. G. 1925. The crystal structure of the high temperature form of cristobalite (SiO_2). Am. J. Sci. 9:448–459.

Yaalon, D. H. 1955. Notes on the clay mineralogy of the major soil types of Israel. Bull. Res. Counc. Isr. Sect. B: Biol. 5-B:162–167.

Yaalon, D. H. 1962. Mineral composition of the average shale. Clay Miner. Bull. 5:31–36.

Yarilova, E. A. 1952. Crystallization of phytolitharia in the soil. Dokl. Akad. Nauk. 83(6):911–912.

Yarilova, E. A. 1956. Mineralogical investigation of a sub-alpine Chernozem on andesite basalt. Kora Vyvetrivaniya. Acad. Sci. USSR 2:45–60.

Yeck, R. D., and F. Gray. 1969. Preliminary studies of opaline phytoliths from selected Oklahoma soils. Proc. Okla. Acad. Sci., Vol. 48:112–116.

Yeck, R. D., and F. Gray. 1972. Phytolith size characteristics between Udolls and Ustolls. Soil Sci. Soc. Am. Proc. 36:639–641.

Yoshida, S., Y. Onishi, and K. Kitagishi. 1962. Histochemistry of silicon in race plant: III. Soil Sci. Plant Nutr. 8:1–5.

Feldspars, Olivines, Pyroxenes, and Amphiboles[1]

P. M. HUANG, University of Saskatchewan, Saskatoon, Canada

I. INTRODUCTION

Feldspars are anhydrous three-dimensional aluminosilicates containing vary-ing amounts of Na, K, and Ca, and occasionally of other large cations such as Ba. They make up an average of 59.5, 30.0, and 11.5% by weight of igneous rock, shale, and sandstone, respectively (Clarke, 1924). Many metamorphic rocks also contain large amounts of feldspars. Therefore, feldspars are found in virtually all sediments and soils and their quantity may vary with the na-ture of parent material and the stage of weathering.

The nature of feldspars may provide useful information on the geologi-cal origin of soil parent material (Somasiri & Huang, 1971) and on the degree and processes of soil formation (van der Plas, 1966). Feldspars are stores of Na, K, Ca, and numerous trace elements such as Cu, Rb, Cs, and Pb (Rankama & Sahama, 1950). Van der Plas (1966) stated that only 50 ppm of a certain element in feldspars which accounts for 10% of a soil is much more sig-nificant in amount than even 1% of the same element in a heavy mineral which occurs in the soil only once in about 10,000 particles. However, at present little is known about the distribution and dynamics of release of these trace elements from feldspars in the soil profiles.

Olivines are olive-green nesosilicates in which Mg and Fe^{2+} are octa-hedrally coordinated by O atoms. Pyroxenes and amphiboles are ferromag-nesian minerals with single and double chain structures, respectively, of linked silica tetrahedra. They account for 16.8% by weight of igneous rocks (Clarke, 1924). Olivines also occur as the principal constituents of some igneous rocks. However, in the upper lithosphere their amount certainly is smaller than the quantity of pyroxenes and amphiboles. These three groups of minerals belong to the accessory minerals of soils and are present in the heavy specific gravity fractions. The variety of isomorphous substitution possible in olivines, pyroxenes, and amphiboles (Bragg & Claringbull, 1965) and their relative ease of weathering (Goldich, 1938; Pettijohn, 1941) make them excellent source minerals for Ca, Mg, and trace elements in soils.

In order to understand pedogenic processes and the status of certain macronutrients and trace elements, it is indispensable to study the mineralogy and chemistry of feldspars, olivines, pyroxenes, and amphiboles of soils with special reference to structural characteristics, natural occurrence and equilib-rium environment, physical and chemical properties, and methods of identi-fication and quantitative determination.

chapter 15

---

[1]Saskatchewan Institute of Pedology Publication No. R192.

II. STRUCTURAL PROPERTIES AND MINERAL IDENTIFICATION

A. Feldspars

1. BASIC STRUCTURAL SCHEME

The structure of feldspars is a three-dimensional framework of linked SiO_4 and AlO_4 tetrahedra, with sufficient opening in the framework to accommodate K, Na, Ca, or Ba to maintain electroneutrality (Taylor, 1933; Laves, 1960; J. V. Smith, 1974). In the framework, chains of four-membered rings of tetrahedra are parallel to the a axis and cross-linked to adjacent parallel chains by shared O atoms. One zigzag chain is formed by superimposing four-membered rings of tetrahedra which share some of their vertices to form new four-membered rings; the remaining vertices form cross-links between chains (Fig. 15-1). In the ideal case, the repeat periodicity along the chain is 8.4Å. The approximate range of the a axis of different feldspars varies from 8.1 to 8.6Å. In the building up of the framework, four-membered rings of tetrahedra are the basic units. These rings are linked together to form

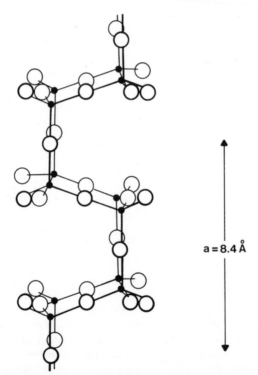

a = 8.4 Å

Fig. 15-1. The essential structural feature of all feldspars projected on the (001) plane (Modified from Taylor, 1965). The small black circles are Si and Al; the large circles O. The O atoms projecting to right and left form the means of linking this chain to its neighbors.

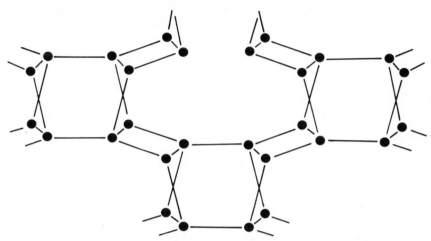

Fig. 15-2. Projection of the Al/Si positions of the $AlSi_3O_8$ framework on the (001) plane (After Laves, 1960).

a honeycomb type arrangement as depicted by the projection of the Si/Al positions on the (001) plane (Fig. 15-2).

Barth (1934), using the structural model proposed by Taylor (1933), reasoned that the distribution of three Si and one Al must be spread over four available sites in such a way so as to account for the different structural symmetries. If Si and Al occur with no preference of site, the resulting symmetry is monoclinic. If the Al takes up a preferred site, then a triclinic structure results. Barth's suggestion is now referred to as the "order-disorder" relation in the structure of feldspars. Detailed structural analysis by X-ray methods confirmed that orthoclase and microcline, two K-feldspars, differ in the degree of ordering of Al/Si in the crystal structure (Bailey & Taylor, 1955; B. E. Brown & Bailey, 1964). The ordered feldspar has a triclinic structure, whereas the disordered feldspar has a monoclinic structure. This relation is illustrated schematically in Fig. 15-3 and 15-4.

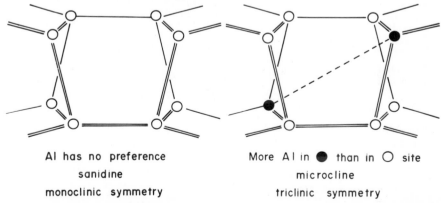

Al has no preference
sanidine
monoclinic symmetry

More Al in ● than in ○ site
microcline
triclinic symmetry

Fig. 15-3. The Al/Si distribution in monoclinic and triclinic K-feldspar structures (After Laves & Goldsmith, 1961).

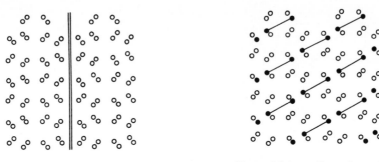

More Al in • than in ○ site

A   B

Fig. 15-4. Schematic representation of the Si/Al distribution in ordered (triclinic) and disordered (monoclinic) feldspars (After Laves & Goldsmith, 1961). *(A)* disordered feldspar: The presence of a symmetry plane brings about a monoclinic structure. *(B)* ordered feldspar: No symmetry plane present, triclinic structure.

As with the K-bearing feldspars, the order-disorder relation also exists in Na-bearing feldspars. The most obvious structural difference between the low temperature and high temperature albites is the high degree of Al/Si order in the former and disorder in the latter (Taylor, 1965).

2. STRUCTURAL CHARACTERISTICS AND CHEMICAL COMPOSITION OF THE PRINCIPAL FELDSPARS

Feldspars represent limited solid solution between the three end members, namely, K-feldspar, albite, and anorthite, with pure K, Na, and Ca, re-

Table 15-1. Unit cell dimensions of selected feldspars (After Taylor, 1965)

Mineral name	Ideal† composition	a (Å)‡	b (Å)	c (Å)	$\alpha°$§	$\beta°$	$\gamma°$
Sanidine		8.564	13.030	7.175	90	115.99	90
Orthoclase		8.562	12.996	7.193	90	116.01	90
Intermediate microcline	$KAlSi_3O_8$	8.578	12.960	7.211	90.30	115.97	89.13
Maximum microcline		8.561	12.966	7.216	90.65	115.83	87.70
Low albite	$NaAlSi_3O_8$	8.138	12.789	7.156	94.33	116.57	87.65
High albite		8.149	12.880	7.106	93.37	116.30	90.28
Anorthite (low-temperature)	$CaAl_2Si_2O_8$	8.177	12.877	14.169	93.17	115.85	91.22
Oligoclase	$Ab_{74}An_{22}Or_4$	8.169	12.836	7.134	93.83	116.45	88.99
Andesine (high-temperature)	$Ab_{52}An_{48}$	8.176	12.879	7.107	93.40	116.17	90.40
Bytownite	$Ab_{20}An_{80}$	8.171	12.869	14.181	93.37	115.97	90.53

† The symbols Ab, An, and Or stand for albite, anorthite, and orthoclase, respectively.
‡ The values of axial length quoted are accurate to within ± 0.005Å.
§ The values of axial angles quoted are accurate to within $\pm 0.03°$.

spectively (J. V. Smith, 1974). The K- and Na-rich members of the group with a small amount of Ca are known as alkali feldspars. Members of this group of minerals rich in Ca and/or Na, but with small amounts of K, are known as plagioclases. The unit cell dimensions for principal feldspars are listed in Table 15-1.

   a. **Alkali Feldspars**—The crystalline phases of alkali feldspars have a range of chemical composition. The end members may be written as $KAlSi_3O_8$ and $NaAlSi_3O_8$. The structures of phases with identical chemical compositions are not necessarily the same. The K-feldspar polymorphs, sanidine, orthoclase, microcline, and adularia, have identical chemical compositions (van der Plas, 1966; Barth, 1969). Sanidine is a monoclinic alkali feldspar with small optic axial angle (2V) and commonly occurs in volcanic rocks. The monoclinic alkali feldspars which have the larger optic axial angle (2V), look homogeneous, and do not show cross-hatched twinning are referred to as orthoclase. Microcline is triclinic, exhibits the typical cross-hatched twinning (Fig. 15-5A) and has the larger optic axial angle (2V). The alkali feldspars which may be either monoclinic or triclinic, but have special crystal habit, and occur in low temperature hydrothermal veins, are referred to as *adularia*. A Na-rich feldspar which is triclinic and shows a very fine cross-hatched twinning is called *anorthoclase*.

   The degree of departure of the α and γ angles of the triclinic structure of microcline from 90° has been referred to as the "triclinicity" (Goldsmith & Laves, 1954a, 1954b) or "obliquity" (Dietrich, 1962; Smithson, 1962; van der Plas, 1966). The relationship between the d-spacing and the unit cell parameters for the triclinic K-feldspar can be expressed by the following equation (Henry et al., 1961; Barth, 1969):

$$d(hkl) = [h^2a^{*2} + k^2b^{*2} + l^2c^{*2} + 2klb^*c^* \cos \alpha^* + 2lhc^*a^* \cos \beta^*$$
$$+ 2hka^*b^* \cos \gamma^*]^{-\frac{1}{2}} \tag{1}$$

where hkl are Miller indices, d is interplanar spacing, and a^*, b^*, c^*, α^*, β^*, and γ^* are reciprocal lattice values. For the monoclinic K-feldspar, Eq. [1] is reduced to Eq. [2]:

$$d(hkl) = [h^2a^{*2} + k^2b^{*2} + l^2c^{*2} + 2lhc^*a^* \cos \beta^*]^{-\frac{1}{2}}. \tag{2}$$

Mackenzie (1954) showed that the 130 and $1\bar{3}0$ X-ray reflections and the 131 and $1\bar{3}1$ reflections do not coincide, i.e., these planes are symmetrically nonequivalent. Monoclinic K-feldspars produce a single peak from each of the above sets of planes. From Eq. [1] it is evident that $d(hkl)$ is different from $d(h\bar{k}l)$. On the other hand, according to Eq. [2], $d(hkl)$ and $d(h\bar{k}l)$ are the same, i.e., they are symmetrically equivalent. Therefore, in the case of triclinic K-feldspar, when the difference between $d(hkl)$ and $d(h\bar{k}l)$ is large enough to be resolved by X-ray, two peaks corresponding to d-spacings of $d(hkl)$ and $d(h\bar{k}l)$ would be obtained. In the diffraction patterns of monoclinic K-feldspar, the hkl reflection coincides with the $h\bar{k}l$ reflection. Mac-

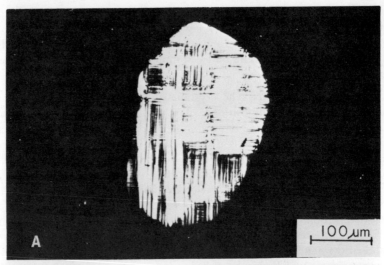

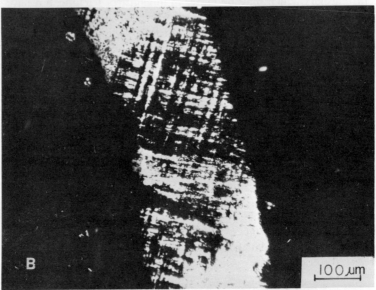

Fig. 15-5. Photomicrographs of K-feldspar crystals from the Ap horizon of an Orthic Brown Haverhill soil in Saskatchewan, Canada. *(A)* A K-feldspar crystal with the cross-hatched pattern of maximum microcline. Fine sand (50–250 μm) fraction. *(B)* A K-feldspar crystal, showing two phases. Coarse to very coarse sand (500–2,000 μm) fraction (After Somasiri & Huang, 1973).

kenzie (1954) suggested that the separation of the 130 and 1$\bar{3}$0 reflections, or the 131 and 1$\bar{3}$1, or both, may be used as a measure of the triclinicity of K-feldspar. Goldsmith and Laves (1954a, 1954b) calculated the triclinicity by the following equation:

$$\Delta = 12.5\,[d(131) - d(1\bar{3}1)]\qquad\qquad [3]$$

where $d(131)$ and $d(1\bar{3}1)$ are the interplanar spacings. The maximum difference between $d(131)$ and $d(1\bar{3}1)$ observed for microcline was 0.08. Therefore, the factor 12.5 was selected to assign the value of 1.0 to the Δ value of maximum microcline. Accordingly, the triclinicity of orthoclase would be 0, whereas microcline can take any value higher than 0 up to a maximum of 1.0. The terms *obliquity* and *triclinicity* have been used synonymously. However, in recent literature, the term *obliquity* seems to be preferable, because it describes a structure more inclined than another without reference to the symmetry aspect of the structure. Accurate structural determinations have been completed for sanidine, orthoclase, intermediate microcline and maximum microcline (Cole et al., 1949; Bailey & Taylor, 1955; Jones and Taylor, 1961; Brown & Bailey, 1964; Finney & Bailey, 1964). However, the structural characteristics of K-feldspars, presently known, are based on the analysis of specimens collected from rocks for petrographical studies. Similar information on K-feldspars of soils and sediments is relatively lacking. Recently, Somasiri and Huang (1971, 1973) studied the nature of K-feldspars of soils in the Canadian Prairies and found that the obliquity values of feldspars in the soils studied ranged from 0.81 to 0.89.

In the K-feldspar structure, one out of every four Si atoms in the framework is replaced by Al. This substitution imparts a negative charge to the framework, which is neutralized by the incorporation of other positively charged ions, e.g., K. Only rarely do K-feldspars of ideal chemical composition ($KAlSi_3O_8$) occur in nature. The K-feldspars contain foreign cations either in fourfold coordination replacing Al and Si or in higher coordination replacing K ions. Lithium and Be have the suitable size relationship to occupy tetrahedral positions (Heier, 1962). Frequently Na substitutes for K in the K-feldspar structure, and a complete solid solution series exists between the end members of composition $KAlSi_3O_8$ and $NaAlSi_3O_8$ (Bowen & Tuttle, 1950). The other ions that may occupy K positions are Rb, Cs, Pb, Tl, Ca, Sr, Cu, Ga, and Ba (Heier, 1962; van der Plas, 1966; Barth, 1969). The presence of trace elements may affect the unit cell parameters of the minerals.

b. **Plagioclase Feldspars**—Plagioclases have a chemical composition between pure albite and pure anorthite. The nomenclature of plagioclase is based on the chemical composition. The plagioclase series is classified in terms of the mole fraction of the albite (Ab) component and the anorthite (An) component as stated below:

albite	Ab_{100} – $Ab_{90}An_{10}$
oligoclase	$Ab_{90}An_{10}$ – $Ab_{70}An_{30}$
andesine	$Ab_{70}An_{30}$ – $Ab_{50}An_{50}$
labradorite	$Ab_{50}An_{50}$ – $Ab_{30}An_{70}$
bytownite	$Ab_{30}An_{70}$ – $Ab_{10}An_{90}$
anorthite	$Ab_{10}An_{90}$ – An_{100}

Albite is a pure Na-feldspar. Tuttle and Bowen (1950) reported that there are two distinct triclinic phases of albite composition, namely, a high and a low temperature form. The intermediate temperature forms of albite composition have also been found in nature. Since the work of Mackenzie (1952), as confirmed by W. L. Brown (1960), a monoclinic form of albite composition is known to exist and is called *barbierite* (Schneider & Laves, 1957).

The mineral anorthite found in nature is never a chemically pure Ca-feldspar, but always contains some Na. The structure of natural anorthites has been studied by many researchers (Gay & Taylor, 1953; Megaw, 1959; Chandrasekhar et al., 1961; Kempster et al., 1962; Megaw et al., 1962; Wainwright & Starkey, 1971). Anorthite is known to have high and low temperature forms, both of triclinic symmetry. The crystal structure of anorthite has been refined using data collected at 410C (683K) and 830C (1103K) (Foit & Peacor, 1973). The results suggest that within this temperature range a process, possibly a variation in domain texture, may be taking place. Ribbe and Colville (1968) studied the structure of transitional anorthite. Gay (1954) reported that the order-disorder relation is significant up to An_{90}. Therefore, this relation does not play a role in anorthite, although it is very important in other feldspars (Gay & Taylor, 1953; Megaw et al., 1962).

The existence of intermediate plagioclases in nature has been proven by X-ray diffraction analyses (Gay, 1954, 1956; J. V. Smith, 1956; J. V. Smith & Gay, 1958) and optical investigations (Karl, 1954; Baskin, 1956). The structure of low temperature plagioclases is not the same throughout the series. The series is also discontinuous in composition. In high temperature regions, plagioclases assume the high albite structure even up to large anorthite contents.

Natural plagioclases vary widely in chemical composition. Besides the most important cations Na and Ca, fair amounts of other elements may be present. Sen (1959) contends that the K content of plagioclases increases along with an increasing temperature of genesis. The trace elements of a number of plagioclases have been reported by Howie (1955), Young (1958), and Heier (1962). In addition to fair amounts of Fe, Mg, Ti, and in some cases Ba, the trace elements like Sr, Li, Rb, Co, Cu, and Pb are nearly always present in plagioclases. The amount of Sr is generally over 500 ppm. The amounts of Li, Rb, Cu, and Pb are in most cases around 20 ppm. Plagioclases present in soils may frequently be much more important reserves of trace elements than heavy minerals.

c. **Perthites, Mesoperthites, Antiperthites, and Peristerites**—The high temperature series of mixed crystals between the end members orthoclase and albite, and albite and anorthite are quite continuous ones. However, under low temperature conditions, these phases undergo structural changes and tend to unmix. The unmixing results in the formation of a lamellar aggregate in which the composition of lamellae is quite similar to that of the end members.

The lamellar aggregate which is composed of a large amount of alkali feldspar and a subordinate amount of albite is referred to as *perthite*. Most naturally occurring alkali feldspars, except authigenic K-feldspar, usually contain varying amounts of Na in the structure and are thus more or less perthitic (Barth, 1969). The separation of the phases into a K-rich phase and a Na-rich phase of feldspars is called *exsolution*. During recent years it has been shown that the formation of perthite is governed by extremely complex combinations of different processes. Therefore a variety of growth patterns and compositional variations in perthite may be expected. The composition and structure of perthites have been studied by Mackenzie and Smith (1962). Morphologically the exsolved phase typically forms small strings of lamellae or blebs of some shape making up a characteristic pattern (Barth, 1969). The classification system of perthites proposed by Tuttle (1952) is considered most adequate for use in soil science (van der Plas, 1966). This system is based on the size of unmixed domains as stated below:

   1) sub-X-ray perthite: size of domains < 15Å;
   2) X-ray perthite: approximately 1 μm;
   3) cryptoperthite: 1–5 μm;
   4) microperthite: 5–100 μm; and
   5) macroperthite: 100–1,000 μm.

Macroperthites are not very important in soil science, because the particle size of most soils is smaller than the width of the perthite lamellae. X-ray diffraction analyses of soil perthitic crystals from the Canadian Prairies (Somasiri & Huang, 1971, 1973) revealed the presence of peaks near 2Θ values (Cu-K_{α}) of 22.1° and 28.0° [$d(\overline{2}01) = 4.02$Å, $d(002) = 3.18$Å]. This indicates the presence of the albite phase. Based on the theory of Marfunin (1962), it is reasoned that the degree of exsolution (phase separation) of the K-feldspars is very high as indicated by the angular separation between the $(\overline{2}01)$ reflection of the K-feldspars and that of the perthitic albite phase (Somasiri & Huang, 1971). Moreover, the high values of obliquity, ranging from 0.81 to 0.89, of the soil K-feldspars studied also indicate a high degree of exsolution. Most of the albite component must be exsolved before the fixation of obliquity values at higher than 0.85 in K-feldspars (Dietrich, 1962). The morpohlogy of soil perthitic crystals which appear to contain a second phase, most likely albite, in a host crystal of K-feldspar is illustrated in Fig. 15-5B. The photographs of the grain mounts, after staining to confirm the above observation, show the presence of two types of crystals (Fig. 15-6). Some crystals show unstained areas, indicating the presence of a second phase in the K-feldspar crystals. More or less completely stained crystals in the soils indicate almost a pure phase of K-feldspar. To extend the data obtained from the petrographic examination and X-ray diffraction analysis, nondestructive elemental determination by electron probe X-ray microanalysis was used. The distribution of Al and K in some feldspar crystals did not match (Fig. 15-7), which further indicates the presence of more than one phase of feldspar. Most commonly, the second phase consists of albite (Barth, 1969). The crystals were cut and polished, thus the examination was mostly on the interior of the crystals. The observed K deficiency zones are essentially in the interior of the crystals rather than in the outer regions (Fig. 15-7A). Therefore the K deficiency patterns are due to

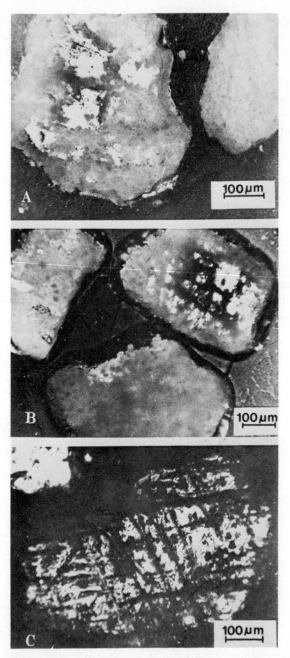

Fig. 15–6. Photomicrographs of K-feldspar crystals after staining with sodium cobaltini-trite (After Somasiri & Huang, 1973). The K-feldspar phase is shown in dark tone and the albite in white. *(A)* K-feldspar crystals from the medium sand (250–500 μm) fraction of the Ap horizon of an Orthic Brown Haverhill soil; *(B)* K-feldspar crystals from the medium sand (250–500 μm) fraction of the Ck horizon of an Orthic Gray Wooded Waitville soil; *(C)* A microcline crystal ($<$250 μm) from Perth, Ontario, Canada.

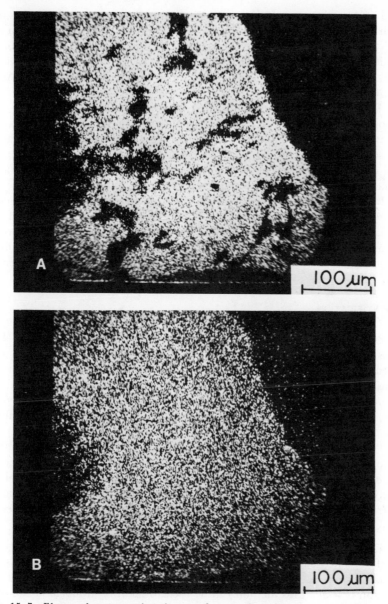

Fig. 15-7. Electron-beam scanning pictures of a part of a K-feldspar crystal of the coarse
to very coarse sand (500–2,000 μm) fraction from the Ap horizon of an Orthic Brown
Haverhill soil in Saskatchewan, Canada, showing the distribution of *(A)* K and *(B)* Al
(After Somasiri & Huang, 1973).

the presence of Na or some trace elements and not due to the effect of
weathering. The use of electron probe X-ray microanalysis for the examination of the distribution of Na and certain trace elements may provide further
information to elucidate the nature of perthitic crystals.

In addition to perthites, a special type of lamellar aggregate exists, in which the amount of the K-rich phase is quite subordinate to the amount of the Na-rich phase. This type is referred to an an *antiperthite*. In antiperthites, the K-rich phase, usually orthoclase, forms thin films, lamellae, strings, or irregular veinlets within the sodic member, usually albite (Am. Geol. Inst., 1957).

A special type of perthite called *mesoperthite* has been defined by Michot (1961). The pattern of mesoperthites is similar to that of microperthites. In mesoperthites it is impossible to see which of the two phases encloses the other, because both alkali feldspar and plagioclase are present in equal quantities. These perthites are assumed to be the result of exsolution of a homogeneous alkali feldspar at temperatures > 600C (873K).

In certain plagioclases, unmixing phenomena may be observed under favorable conditions (Laves, 1954; Ribbe, 1960). Aggregates composed of lamellae with an albite composition occurring next to those with a composition of An_{26} are called *peristerites*. Peristerites are rare in nature and until quite recently, little was known about their presence in soils and sediments (W. L. Brown, 1962; Ribbe, 1962; Ribbe & van Cott, 1962).

3. IDENTIFICATION

Pretreatments of soils to remove carbonates, organic matter, and Fe oxides and separation of various size fractions (Jackson, 1956) should be carried out in order to facilitate efficient mineralogical analysis. Feldspars in the various fractions may be concentrated by specific gravity separation methods, using, for example, mixtures of bromoform and decaline (Doeglas et al., 1965). The specific gravity of a series of feldspars is listed in Table 15-4.

The main approaches used in identification of feldspars are X-ray diffraction, chemical analysis, staining methods, and optical methods. X-ray diffraction and chemical analysis are almost the only two approaches which can effectively provide information on the nature of feldspars in the silt and clay fractions, with the possible exception of the use of phase contrast microscopy.

a.   **X-ray Diffraction Methods**—The most convenient reflections for identifying microcline are 131 and 1$\bar{3}$1 (Goldsmith & Laves, 1954a, 1954b; Marfunin, 1962). In the X-ray diffraction pattern of orthoclase, only the 131 reflection is seen. This method can be used to identify monoclinic and triclinic K-feldspars in soils and sediments (Doeglas et al., 1965; van der Plas, 1966).

There exist various proposals for the measurement of the obliquity of microclines (Mackenzie, 1954; Goldsmith & Laves, 1954a, 1954b). The obliquity values measured by different methods are not exactly the same. However, if authors stipulate what they measure, this will not cause a real problem and large deviations are generally not expected. van der Plas (1966) recommended the use of the 131 and 1$\bar{3}$1 reflections (see Eq. [3]).

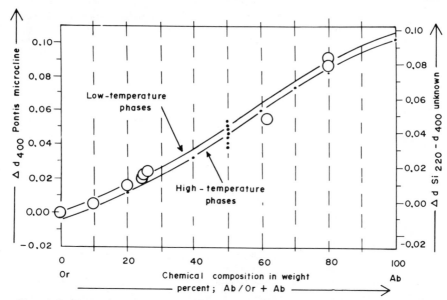

Fig. 15-8. The relation between the 400 reflection and the chemical composition of low- and high-temperature alkali feldspars. The *left* ordinate lists the value of the 400 reflection of Pontis microcline minus the 400 reflection of the phase under investigation. The *right* ordinate lists the value of the 220 reflection of silicon minus the 400 reflection of the phase under investigation (After Goldsmith & Laves, 1961).

The Na content of alkali feldspars may be estimated by using the relationship between $d(400)$ and chemical composition, illustrated in Fig. 15-8 (Goldsmith & Laves, 1961). The other reflections which are related to the Na content are 111 and $\overline{2}01$ (Bowen & Tuttle, 1950; Donnay & Donnay, 1952).

The identification of perthite by X-ray powder diffraction can be carried out after separation of alkali feldspars and perthites from soils and sediments. If the separation is efficient, the presence of the $\overline{2}01$ reflection from albite can be ascribed to the presence of perthites (van der Plas, 1966).

The characteristics of the various structures of plagioclases can be established by X-ray diffraction techniques (J. V. Smith, 1956; J. V. Smith & Gay, 1958). In the study of Smith and Gay (1958), plagioclases are divided into three groups, namely, Na-rich plagioclases, intermediate plagioclases, and Ca-rich plagioclases.

The lattice parameters of high temperature phases of plagioclases with composition between An_0 and An_{20} do not change very extensively with increase in Ca content. Moreover, slight admixtures of K result in similar changes in the pattern as those due to large admixtures of Ca. These high temperature plagioclases are relatively rich in K in some cases (Sen, 1959). Therefore, it is not possible to determine the chemical composition of high temperature plagioclases in this range with X-ray powder methods. However, X-ray powder diffraction analysis can be used to differentiate low and high temperature plagioclases as their diffraction patterns are quite different (J. V.

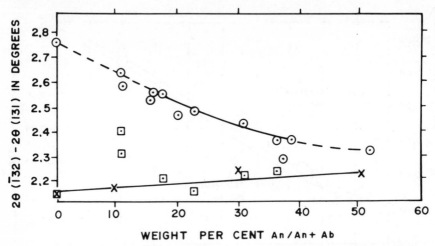

Fig. 15-9. The values of $2\Theta(\bar{1}32) - 2\Theta(131)$ of low- and high-temperature plagioclases plotted against their chemical composition. The plotting is valid for measurements with Cu radiation. ⊙ natural; ▢ heated natural; X synthetic (After J. V. Smith, 1956).

Smith, 1956). Scheidegger (1973) demonstrates that X-ray powder techniques can provide significant information about the order-disorder of calcic plagioclases in the absence of independent compositional information. Since the high temperature phases of plagioclases are less frequently encountered in sediments and soils, emphasis can be given to identification of low temperature phases. These plagioclases show a systematic change in lattice parameters with increasing Ca content. J. V. Smith (1956) proposed that the separation between the $\bar{1}32$ and 131 reflections gives the best estimate of their chemical composition (Fig. 15-9). The $\bar{1}32$ and 131 reflections occur at 2Θ values (Cu-K$_\alpha$) of $33.85 \pm 0.10°$ (2.654-2.638Å) and $31.40 \pm 0.15°$ (2.860-2.833Å). This determination can be checked by measuring the separation between 111 and $1\bar{1}1$. The 111 and $1\bar{1}1$ reflections occur at 2Θ values (Cu-K$_\alpha$) of $23.6 \pm 1.0°$ (3.782-3.751Å) and $23.0 \pm 1.0°$ (3.880-3.847Å). The relation of the separation between 111 and $1\bar{1}1$ and the chemical composition is shown in Fig. 15-10. If the results obtained by these two methods are not in rapport, there is a possibility that phases with an intermediate structure exist.

In the case of plagioclases with a chemical composition between An_{20} and An_{40}, J. V. Smith (1956) proposed using the separation of the 131 and $1\bar{3}1$ reflections to determine their chemical composition (Fig. 15-11). The 131 and $1\bar{3}1$ reflections occur at 2Θ values (Cu-K$_\alpha$) of $31.40 \pm 0.15°$ (2.860-2.833Å) and $29.90 \pm 0.25°$ (3.010-2.942Å). The separation of these spacings may not be used for peristerites (J. V. Smith, 1956). Other methods using the $2\bar{2}0$, $\bar{1}32$, $24\bar{1}$, and $\bar{2}41$ reflections have also been recommended for the estimation of chemical composition. However, J. V. Smith (1956) contends that the use of these spacings has serious limitations for certain plagioclase composition. Nevertheless, under certain circumstances these spacings may be very useful.

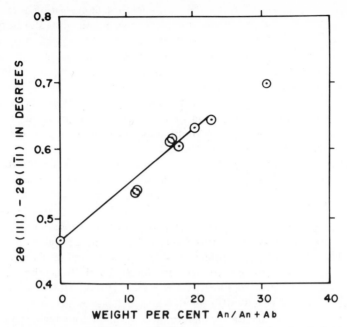

Fig. 15-10. The value of $2\Theta(111) - 2\Theta(1\bar{1}1)$ plotted against the chemical composition for natural low-temperature plagioclases. The plotting is valid for measurements with Cu radiation (After J. V. Smith, 1956).

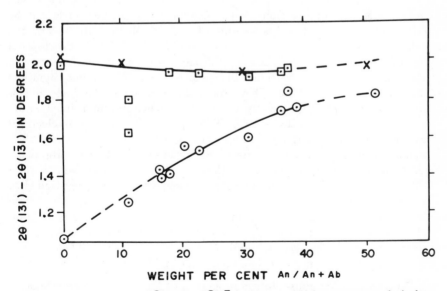

Fig. 15-11. The values of $2\Theta(131) - 2\Theta(1\bar{3}1)$ of low- and high-temperature plagioclases plotted against their chemical composition. The plotting is valid for measurements with Cu radiation. ⊙ natural; ⊡ heated natural; X synthetic (After J. V. Smith, 1956).

For plagioclases with $> 40\%$ An, chemical composition may be estimated by the procedure proposed by J. V. Smith and Gay (1958), based on two variables, Γ or B, where $\Gamma = 2[\Theta(131) + \Theta(220) - 2\Theta(1\bar{3}1)]$ and $B = 2[\Theta(1\bar{1}1) - \Theta(20\bar{1})]$ and the Θ values are for Cu radiation. The chemical composition estimated by these methods varies widely in certain ranges of Γ and B values. Smith and Gay (1958) expressed concern about the practical application of powder methods in this particular case. However, a combination of X-ray powder diffraction data with other information such as the genesis of rocks, the association of minerals in rocks, the index of refraction, and the specific gravity may lead to quite a useful conclusion.

For further X-ray powder data of feldspars, readers are referred to the *Search Manual and Selected Powder Diffraction Data for Minerals* prepared by the Joint Committee on Powder Diffraction Standards (1974).

In summary, the reliability of X-ray diffraction methods in identification of feldspars is in most cases higher than other methods generally used in soil science. For determining tiny fractions of feldspars isolated from soils or sediments, the quadruple Guinier-De Wolff powder camera (De Wolff, 1948, 1950) has been recommended as superior to the Debye-Scherrer powder camera and the diffractometer (van der Plas, 1966). However, the adjustment of the Guinier-De Wolff camera is very delicate and time-consuming. If intensities must be accurately determined, a diffractometer has advantages over cameras, provided preferred orientation can be minimized by using a rotating specimen holder.

b. **Chemical Methods**—The important chemical methods include staining and selective dissolution analysis.

Gabriel and Cox (1929) proposed a method for staining alkali feldspars in rock slabs and thin sections. Alkali feldspars are etched with HF, and the etch-residue is treated with a solution of sodium cobaltinitrite to form the yellow potassium cobaltinitrite. Reeder and McAllister (1957) developed a method for staining the Al ion in feldspars with hemateine after etching. The stained feldspars reveal a lilac blue color. Doeglas et al. (1965) modified the above staining methods. They proposed etching feldspar particles in HF-vapor at 90C (363K) for exactly 1 min. The etched residue is fixed on the surface of the particles by a 400C (673K) heat treatment in an electrical furnace for 5 min. After the etching and heat treatment, the particles are stained in the same way as described. Bailey and Stevens (1960) proposed a method for staining plagioclases. They recommend staining with barium chloride and potassium rhodizonate after etching which gives the plagioclase a brick red color.

Differentiation of feldspars can also be carried out by the determination of K, Na, and Ca in the residue after selective dissolution by sodium pyrosulfate fusion (Kiely & Jackson, 1965). More detailed aspects will be discussed in the section on quantitative determination of feldspars (p. 590).

c. **Optical Methods**—Optical properties of feldspars may provide additional information on the nature of minerals. However, many optical techniques, though quite useful for petrographers, are considered less useful for soil scientists (van der Plas, 1966). The important optical properties include indices of refraction, axial angle, orientation of indicatrix, and twinning. Readers interested in the details of optical methods are referred to the articles by J. R. Smith (1958), Tobi (1961), Marfunin (1962), Slemmons (1962), and van der Plas (1966).

d. **Other Methods**—Electron probe microanalysis can provide a means for examining the chemistry of perthitic crystals and particularly the distribution of trace elements in feldspars (Somasiri & Huang, 1973). Electron-optical techniques, infrared absorption, nuclear magnetic resonance, electron spin resonance and Mössbauer resonance have been used in studying physical properties of feldspars (J. V. Smith, 1974). In addition, the possibility of applying ruby laser technology (Maxwell, 1963) in feldspar identification should be explored.

B. Olivines, Pyroxenes, and Amphiboles

1. STRUCTURE AND CHEMICAL COMPOSITION

a.   **Olivines**—Olivines are nesosilicates forming the forsterite-fayalite series, $(Mg, Fe)_2SiO_4$. The Si–O bond lengths of forsterite are 1.62 to 1.63Å and Mg–O bond lengths are 2.07 to 2.17Å (Bragg & Claringbull, 1965). The structure of forsterite is shown in Fig. 15–12. The O atoms lie in approximately hexagonally close-packed sheets parallel to (100). The SiO_4 tetrahedra point alternately in opposite directions along both the a and b axes.

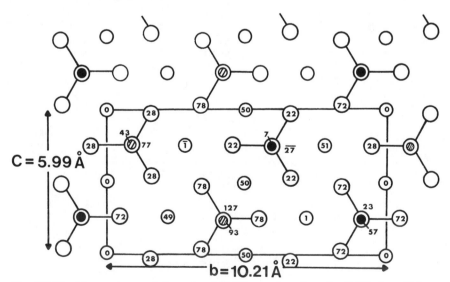

Fig. 15–12. The structure of forsterite, Mg_2SiO_4, projected on the (100) plane (After Bragg & Claringbull, 1965). The large circles are O, the smaller circles Mg, and the black or shaded circles Si; Si–O bonds mark out the SiO_4 tetrahedra. The kinds of positions of atoms are indicated by heights as shown by the numbers inside or beside the circles.

Table 15-2. Nomenclature of forsterite–fayalite series (After Deer & Wager, 1939)

Mineral	Molecular percentage of Fe_2SiO_4
Forsterite (Fo)	0–10
Chrysolite	10–30
Hyalosiderite	30–50
Hortonolite	50–70
Ferrohortonolite	70–90
Fayalite (Fa)	90–100

Table 15-3. The unit cell dimensions of certain members of olivine, pyroxene, and amphibole groups

Mineral name	Specific chemical formula†	Unit cell dimensions†			
		a (Å)	b (Å)	c (Å)	$\beta°$
Olivines					
Forsterite	$(Mg_{0.96}Fe_{0.04})_2SiO_4$	4.758	10.207	5.988	90
Fayalite	Fe_2SiO_4	4.822	10.483	6.095	90
Tephroite	Mn_2SiO_4	4.878	10.560	6.226	90
Monticellite	$CaMgSiO_4$	4.815	11.08	6.370	90
Pyroxenes					
Enstatite	$MgSiO_3$	18.23	8.84	5.19	90
Hypersthene	$(Mg_{0.88}Fe_{0.12})Si_2O_6$	18.325	8.918	5.216	90
Orthoferrosilite	$FeSiO_3$	18.3	9.13	5.20	90
Clinoenstatite	$MgSiO_3$	9.607	8.815	5.169	108.34
Clinoferrosilite	$FeSiO_3$	9.53	9.21	5.15	107.63
Diopside	$CaMgSi_2O_6$	9.761	8.926	5.258	105.79
Pigeonite	$(Ca_{0.04}Mg_{0.45}Fe_{0.48})SiO_3$	9.712	8.959	5.251	108.55
Hedenbergite	$CaFeSi_2O_6$	9.85	9.02	5.26	104.33
Johannsenite	$(Ca_{0.45}Mg_{0.03}Fe_{0.05}Mn_{0.47})SiO_3$	9.90	9.10	5.27	104.08
Acmite-augite	$(Ca,Na)(Fe,Mn,Zn,Mg)(Si,Al)_2O_6$	9.776	8.939	5.270	106.20
Aegirine (Acmite)	$NaFeSi_2O_6$	9.657	8.801	5.291	107.43
Jadeite	$NaAlSi_2O_6$	9.437	8.574	5.225	107.58
Spodumene	$LiAlSi_2O_6$	9.52	8.32	5.25	111.33
Amphiboles					
Anthophyllite	$(Mg,Fe)_7Si_8O_{22}(OH)_2$	18.5	17.9	5.28	90
Cummingtonite	$(Fe_{0.6}Mg_{0.4})_7Si_8O_{22}(OH)_2$	9.534	18.23	5.324	101.97
Tremolite ‡	$Ca_2Mg_5Si_8O_{22}(OH)_2$	9.84	18.02	5.27	104.95
Arfvedsonite	$(Ca,Na,K)_{2.64}Fe_{1.42}(Fe,Mn,Mg,$ $Ti)_{3.54}(Si,Al)_8O_{22}(OH)_{2.15}$	9.94	18.17	5.34	104.4
Glaucophane	$Na_2(Mg,Fe,Al)_5Si_8O_{22}(OH)_2$	9.543	17.726	5.302	103.72
Riebeckite	$(Na,Ca,K)_{2.2}(Fe^{2+},Mg,Mn,Zn,Cu,$ $Li)_{3.0}(Fe^{3+},Al,Ti)_{2.0}(Si,Al)_{8.0}$ $O_{22.1}(OH,F)_{1.9}$	9.769	18.048	5.335	103.59
Hornblende	$(Ca,Na)_{2.26}(Mg,Fe,Al)_{5.15}(Si,Al)_8$ $O_{22}(OH)_2$	9.783	17.935	5.297	104.63

† The specific chemical formulas and the unit cell dimensions are compiled from the Data Book (After Joint Committee on Powder Diffraction Standards, 1974); the α and γ angles of all of the minerals listed are 90°.

‡ The division between tremolite and actinolite is an arbitrary one; tremolite is the low iron end of the series and actinolite comprises the high-iron members.

Every Mg (smaller open circle) is octahedrally coordinated by six O atoms. In the structures described above, Fe^{2+} may replace Mg^{2+} because the two have identical charge and similar ionic radii.

Members of the solid solutions between forsterite and fayalite are given in Table 15-2. Other minerals of the olivine group are tephroite, Mn_2SiO_4; monticellite, $CaMgSiO_4$; glaucochroite, $CaMnSiO_4$; and larsenite, $PbZnSiO_4$. The cell dimensions of selected minerals in the olivine group are given in Table 15-3.

b. **Pyroxenes and Amphiboles**—Pyroxenes and amphiboles are closely related structurally. They are ferromagnesian minerals and classed as inosilicates.

Pyroxenes consist of single chains of linked SiO_4 tetrahedra, each of which shares two O atoms with its neighbor. In amphiboles, double chains of the tetrahedra are linked together by O atoms and the tetrahedra share alternately two and three O atoms (Fig. 15-13). The chains are indefinite in length. When viewed at right angles to the chain and parallel to the base of the tetrahedra, pyroxene and amphibole chains look alike as depicted in Fig. 15-13(c). The pyroxene and amphibole chains are parallel to the c-axis of the crystal and their planes of basal O atoms are perpendicular to (010).

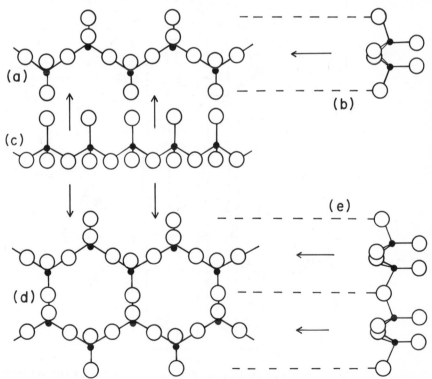

Fig. 15-13. Si–O chains in pyroxenes and amphiboles. *(a) (c)*, *(b)*, the pyroxene chain as seen in plan, in elevation, and end on, respectively. *(d) (c) (e)*, the amphibole chain from the same three aspects (After Bragg & Claringbull, 1965).

The chains of pyroxenes and the double chains of amphiboles are linked together by various cations. The general chemical formula for pyroxenes is $R_2[Si_2O_6]$ and that for amphiboles is $R_{14}[(OH)_4Si_{16}O_{44}]$. In these formulas, R is Mg, Fe^{2+}, or Ca and in many cases, Al, Fe^{3+}, Ti^{3+}, Mn^{3+}, Na, K, or Li. In the augite series, up to one-fourth of the Si ions may be replaced by Al. However, in other pyroxenes, Al rarely substitutes for Si. In amphiboles, up to one-fourth of the Si ions may be replaced by Al, particularly in horn-blende. The OH group in amphiboles may be partly replaced by O or F.

The bond between the O atoms and the cations linking the chains is relatively weaker than that between the O and the Si. Therefore, cleavage takes place diagonally through the crystal and does not rupture the Si-O chains as shown by the heavy black lines (Fig. 15-14). The cleavage planes are thus parallel to (110) for both pyroxenes and amphiboles. There is a difference in cleavage of pyroxenes and amphiboles because of the difference in the width of their Si-O chains. The angle between the cleavage planes is 93° in pyroxene, and 56° in amphibole. These angles between cleavage planes distinguish all pyroxenes from all amphiboles.

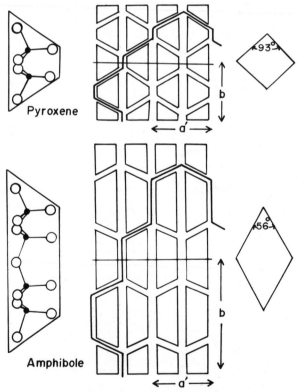

Fig. 15-14. The relation between the structures and cleavages of pyroxene and amphi-bole; the unit cells of diopside and tremolite projected on the (001) plane (After Bragg & Claringbull, 1965).

The principal members of pyroxene and amphibole groups are listed in Table 15-3.

2. IDENTIFICATION

To facilitate the identification in soils of the minerals in the olivine, pyroxene, and amphibole groups, it is common practice to remove carbonates, organic matter, and sesquioxides, to obtain a series of particle size fractions, and to concentrate these heavy minerals (sp gr above 2.8, Table 15-4) by heavy-liquid separation. The pretreatments and particle size fractionations can be carried out by the methods described by Jackson (1956).

Heavy-liquid separations are most effective on clean mineral grains. The presence of organic matter may cause mineral grains to clump together and thus decrease the efficiency of wetting. Iron oxide coatings may increase the specific gravity of heavy grains and coatings of light material may cause them to float. Heavy liquids can be diluted with a number of miscible organic solvents to obtain the preferred density. Mixtures of tetrabromomethane (sp gr 2.98) and nitrobenzene (sp gr 1.20) have proved very convenient (Marshall & Jeffries, 1945). The sizes > 0.10 mm can usually be separated by gravity alone in separating funnels. For smaller size grains, centrifugation is required for separation. Procedures for concentration of olivines, pyroxenes and amphiboles are well treated in standard references (Krumbein & Pettijohn, 1938; Milner, 1962; Muller, 1967). The methods useful for the identification of these mineral groups are briefly outlined below.

Table 15-4. The specific gravity of feldspars, olivines, pyroxenes, and amphiboles

Mineral	Specific gravity
Feldspars (Berry & Mason, 1959; van der Plas, 1966)	
Alkali feldspars	2.54-2.59
Sanidine, orthoclase, microcline	2.56
Plagioclase feldspars	2.59-2.76
Albite	2.61
Anorthite	2.76
Olivines, pyroxenes and amphiboles (Berry & Mason, 1959)	
Olivines	3.22 (Mg_2SiO_4) to 4.39 (Fe_2SiO_4), common olivine about 3.3-3.4
Enstatite, hypersthene	3.2 -3.9†
Diopside, hedenbergite, augite	3.25-3.55†
Aegirine	3.5 -3.6
Jadeite	3.25-3.35
Spodumene	3.1 -3.2
Anthophyllite	2.9 -3.3†
Cummingtonite	3.2 -3.6†
Tremolite-actinolite	2.98-3.35
Hornblende	3.0 -3.4†
Glaucophane-riebeckite	3.0 -3.4†

† Increasing with Fe content.

a. **X-ray Diffraction**—The measurement of $d(130)$ (Yoder & Sahama, 1957) and of $d(062)$ (E. D. Jackson, 1960) can be used to estimate the Mg/Fe ratio of an olivine. The 062 reflection is recommended particularly for olivines in the range Fo_{80}–Fo_{90}. However, the 062 reflection is weaker and there is a greater chance of interference from associated minerals. Further information on the determination of olivine by X-ray diffraction is available from the works by Eliseev (1957), Sahama and Hytönen (1958), Jambor and Smith (1964), and Joint Committee on Powder Diffraction Standards (1974).

X-ray diffraction studies of clinopyroxenes have been carried out by many researchers. The effect of the substitution of Al for Mg and Si on cell parameters was studied by Sakata (1957). The variation of cell parameters over the field diopside–hedenbergite–ferrosilite–clinoenstatite was examined by G. M. Brown (1960) and Viswanathan (1966). Graphs of cell parameters vs. composition have been prepared by Winchell and Tilling (1960) for clinopyroxenes. In the studies of orthopyroxenes by X-ray diffraction analysis, the influence of Mg–Fe replacement and Ca and Al content on cell parameters has been examined (Hess, 1952; Kuno, 1954; Howie, 1962). Pollack and Ruble (1964) studied ordered and disordered enstatite by powder and single crystal X-ray diffraction methods. Some prominent reflections that can be used for deriving the cell parameters of orthopyroxenes are: 12,0,0; 060; 14,5,0; 650; 250; 610; 420; 202; 502; 521; 531; and 11,3,1.

The relationships between the composition of amphiboles and their cell parameters have been examined by powder and single-crystal X-ray diffraction analysis. The important studies include the following: amphiboles in general (Whittaker, 1960), ferrohastingsites (Borley & Frost, 1963), arfvedsonites and ferrohastingsites (Frost, 1963) and cummingtonite-grunerite series (Klein, 1964).

For further information on X-ray powder diffraction data of pyroxenes and amphiboles, readers are referred to the X-ray data compiled by the Joint Committee on Powder Diffraction Standards (1974).

b. **Infrared Absorption**—The infrared spectra of the olivines show absorption maxima in a predictable way with cation substitution. The absorption bands sensitive to cation substitution occur in the spectral range 400–1000 cm^{-1} (Lyon, 1967). Intensive infrared investigation of the olivine group has been carried out by Duke and Stephens (1964).

The infrared spectra of orthopyroxenes clearly show the changing details of a spectrum with substitutions in the Mg–Fe series (Lyon, 1963). The similarities and differences in the infrared spectra for a group of clinopyroxenes, acmite, omphacite, jadeite, diopside, and hedenbergite, have been illustrated (Lyon, 1967).

The amphibole group exhibits a much greater Al/Si substitution than the pyroxene group. Hydroxyl groups are also an essential structural part of the amphiboles, which give rise to additional infrared absorption near 3600 cm^{-1} (Lyon, 1967). The amphiboles which have been studied by infrared absorption spectroscopy are the hornblendes and the alkali amphiboles, glaucophane and riebeckite. The infrared spectra of these minerals are difficult to interpret because of great complexities of possible isomorphous substitution.

c. **Electron Probe Microanalysis**—Electron probe microanalysis provides a tool for showing the distribution of structural cations and for estimating their atomic ratios in olivines, pyroxenes and other ferromagnesian silicates (Keil & Fredriksson, 1964; Long, 1967). Moreover, such measurements serve as a potential method of estimating relative amounts of ions in different oxidation states and different coordination symmetries, since shifts in wavelength of characteristic X-ray emission lines are produced by differences in the electrostatic field surrounding the emitting atoms in different structures.

d. **Optical and Other Methods**—The optical properties of olivines, pyroxenes and amphiboles are useful in their identification. Key references for microscopic identification are Larsen and Berman (1934), Krumbein and Pettijohn (1938), Milner (1962), and Muir (1967). Thermal techniques, emission spectrography, X-ray fluorescence spectrography, and chemical analysis may also provide useful information for identification. Amphiboles have been studied by Wittles (1951, 1952) using DTA. The loss of water which occurs at elevated temperatures distinctly varies with different species. The exothermic peak at about 820C (1093K) is attributed to a crystallographic contraction, which is reversible for tremolite but irreversible for magnesian anthophyllite. In minerals such as amphiboles, structural OH groups may be replaced by F. For differentiating F- from OH-bearing amphiboles, wet chemical methods (P. M. Huang & Jackson, 1967) or physical methods such as emission spectrography (Nicholls, 1967) may be used.

III. NATURAL OCCURRENCE AND EQUILIBRIUM ENVIRONMENT

A. Feldspars

1. NATURAL OCCURRENCE

Feldspars are geochemically the most important group of minerals. The polymorphs of K-feldspars make up 16% of the total earth's crust (Ahrens, 1965). When the alkali feldspars which contain K are taken into account the total K-bearing feldspars make up nearly 31% (Barth, 1969). Microcline is usually formed at lower temperatures than is orthoclase. It is the common K-feldspar of pegmatites and hydrothermal veins and also occurs in metamorphic rocks. Orthoclase is the characteristic K-feldspar of igneous rocks; it occurs both alone and in perthitic intergrowth with albite. Orthoclase also occurs in metamorphic rocks. Sanidine is present in K-rich volcanic rocks such as rhyolite and trachyte. Adularia is formed at relatively low temperatures in veins.

Plagioclase feldspars make up about 29% of the earth's crust (Ahrens, 1965). Anorthite is commonly present in contact with metamorphosed limestones. Relatively pure anorthite is found in volcanic rocks. Bytownite and labradorite are more abundant in the less-siliceous igneous rocks of gabbroic composition and in the anorthosites, andesine in andesites and diorites, and oligoclase in monzonites and granodiorites. Albite occurs in some igneous rocks, but is more common in pegmatites. Albite is also common in metamorphic rocks such as low-grade schists and gneisses. In medium-grade metamorphic rocks, the plagioclase is usually oligoclase or andesine.

A major portion of the feldspars found in sedimentary rocks is of igneous origin. They accumulate in sedimentary environments as a weathering residue of the igneous and metamorphic rocks (Rankama & Sahama, 1950; Mason, 1966). The authigenic feldspars formed at low temperatures and pressures near the earth's surface or at moderate depth are found in sedimentary rocks such as limestones, sandstones, siltstones, and shales (Berg, 1952; Baskin, 1956; Finney & Bailey, 1964). Quantitatively the authigenic feldspars appear to be less significant, constituting about 5% of all feldspars in sediments. This accounts for 0.94% of the total sedimentary mass (Kastner, 1974).

Feldspars are quite stable at elevated temperatures and pressures, but behave differently under the environments that exist on the earth's surface (van der Plas, 1966). They are present in nearly all soils, but their quantities vary with the intensity and capacity factors of weathering reactions. In strongly weathered soils, feldspars are present only in small quantities or completely absent (Hseung & Jackson, 1952), though the parent material contains considerable quantities of those minerals (Stephen, 1953; van der Plas, 1966). However, some feldspars may be found in humid tropical soils which contain relatively fresh rock materials, due to erosional and depositional processes (Jackson & Sherman, 1953). Feldspars are commonly present in the silt and sand fractions of the young to moderately developed soils representing various soil parent materials and soil-forming conditions (de Leenheer, 1950; Hawkins & Graham, 1950; Phillippe & White, 1951; Jeffries et al., 1956; Brydon & Patry, 1961; Pawluk, 1961; Dell, 1963; Somasiri et al., 1971). Alkali feldspars are even present in the clay fraction of soils which are subject to moderate weathering (Hseung & Jackson, 1952; St. Arnaud & Mortland, 1963; P. M. Huang & Lee, 1969).

The crystallization of a K-feldspar as a particular polymorph depends on many factors: temperature of crystallization, pressure, water vapor pressure, composition of the melt, and the rate of cooling (Mason, 1966; Berry & Mason, 1959). The microcline-orthoclase transformation may be caused by metamorphism (Steiger & Hart, 1967), and the extent of the transformation would depend on the grade of metamorphism (Berry & Mason, 1959). Therefore, according to geochemical principles, the proportion of orthoclase to microcline in sediments and soils would be a variable parameter which may provide some useful information on the source or origin of parent material of soils.

According to the Bowen (1922) reaction series and the Goldich (1938) stability series, K-feldspars are more resistant to alteration than albite. P. M. Huang et al. (1968) showed that both orthoclase and microcline have similar K-release characteristics between the pH values of 1 and 7. Under certain conditions of weathering, there may be a difference between the rates of alteration of orthoclase and microcline (Arnold, 1960; van der Plas, 1966). However, this difference would not be greater than the difference between the rates of weathering of albite and quartz. The proportions of orthoclase to microcline in comparable particle size fractions of soils are thus at least as useful an index as the ratio of albite to quartz (Barshad, 1964) in pedological studies.

There have been several studies on the obliquity of K-feldspars in relation to the petrogenesis of their containing rocks (J. R. Smith & Pyke, 1959; Dietrich, 1962; Marfunin, 1962). Information on the obliquity of soil K-feldspars is still quite limited (Somasiri & Huang, 1971, 1973). A question that may interest soil scientists is whether or not the obliquity of soil K-feldspars is a guide to understanding the origin of soil parent material. To provide an answer to such a question, it seems necessary to know the relationship that may exist between the obliquity of K-feldspars and their origin.

2. EQUILIBRIUM ENVIRONMENT AND MINERAL STABILITY

Persistence of feldspars in geological deposits and soils is related to the nature of the minerals, climate, topography, the degree of leaching, chelation, redox potential, and certain solution ionic activities.

According to the Goldich (1938) stability series, the increasing order of stability of feldspars is as follows: anorthite < bytownite < labradorite < andesine < oligoclase < albite < potash feldspars. This order is related to the decrease in Al and Ca contents. Within the K-feldspar group, microcline is more stable than orthoclase. This may be attributed to the smaller volume occupied by the O atoms of microcline due to its triclinic symmetry as compared with the monoclinic symmetry of orthoclase (Barshad, 1964).

Climate is a paramount factor in affecting chemical weathering reactions (Loughnan, 1969). Rainfall and temperature are the two climatic factors; the former controls the supply of water for chemical reactions and for leaching soluble constituents away from the weathering environment, and the latter influences the rate of these chemical reactions. Repeated leaching would hasten the weathering reactions towards completion.

Topography is another important factor affecting the rate of chemical weathering. It modifies the rates of moisture intake by the parent rock, leaching of the soluble constituents, and exposure of fresh mineral surfaces.

The influence of climate, topography, and leaching on the weathering of feldspars to kaolin minerals has been documented in Mesozoic weathered products in Minnesota (Parham, 1969) and in modern weathering products in the Southern Appalachian region (Sand, 1956), in Hawaii (Bates, 1962), in Hong Kong (Parham, 1969), and in Mexico (Keller et al., 1971). However, montmorillonite may form as a product of weathering of feldspars of basaltic flows in semiarid climates or poorly drained basins of even tropical climates (Jackson, 1968).

Some soil organic compounds possess chelating properties (Himes & Barber, 1957). The significance of chelating agents in accelerating chemical weathering reactions has been illustrated (Schatz et al., 1957; Wright & Schnitzer, 1963). Quite recently a series of experimental studies on simulated organo-chemical weathering of feldspars has been carried out (W. H. Huang & Keller, 1970; W. H. Huang & Kiang, 1972). Weathering of feldspars by complexing organic acids may result in a different order of mineral stability than the traditional one of Goldich (1938).

Major elements in feldspars do not exist in more than one valence state, thus the prevailing redox potential of a soil may not be of direct concern to chemical weathering of feldspars. However, as stated before, the weatherability of feldspars can be affected by complexing organic acids which are vulnerable to oxidation. Therefore, stability of feldspars may be indirectly related to the prevailing redox potential of a soil.

Following the establishment of the empirical stability series of Goldrich (1938), relative stabilities of feldspars in aqueous solution at 25C (298K) and 1 bar have been thermodynamically illustrated in stability diagrams. The important thermodynamic parameters in calculating the relative stabilities are the standard free energies of formation of feldspars (Table 15-5), the associated secondary minerals, solution chemical species, and equations of the chemical weathering reactions of feldspars (Garrels & Christ, 1965; Reesman & Keller, 1965; Robie & Waldbaum, 1968; W. H. Huang & Keller, 1972; W. H. Huang & Kiang, 1973). Standard free energies and equilibrium constants of the chemical weathering reactions can be calculated from the above information and the following two equations:

$$\Delta G_r^{\circ} = \Sigma \Delta G_f^{\circ} \text{ (products)} - \Sigma \Delta G_f^{\circ} \text{ (reactants)} \qquad [4]$$

$$\Delta G_r^{\circ} = -RT \ln K \qquad [5]$$

where ΔG_r° is the standard free energy of a reaction, ΔG_f° the standard free energy of formation, K the equilibrium constant, R the molar gas constant, and T the absolute temperature. From equilibrium constants and equations of chemical weathering reactions of feldspars, stability diagrams can be constructed based on activities of the associated solution chemical species. The stability diagrams of K-feldspar and albite shown in Fig. 15-15 and 15-16, respectively, were plotted as a function of the activity of H_4SiO_4 and of the ratio of the activity of each of the alkali ions to that of H ion. From the stability diagrams, it can be predicted that mica and kaolinite are intermediate phases in the weathering of K-feldspar, and that montmorillonite and kaolinite are intermediate phases in the weathering of albite. The ionic compositions of ground waters were plotted on the same diagrams and it is interesting to note that all fall within the stability fields of kaolinite. This indicates that in ground water, kaolinite is the stable mineral phase and mica and montmorillonite are unstable phases and eventually will be transformed to kaolinite. This observation is in harmony with the fact that ground waters tend to kaolinize rock minerals.

The existing stability diagrams of feldspars (Feth et al., 1964; Garrels & Christ, 1965) are constructed by assuming Al_2O_3 as an inert component. In

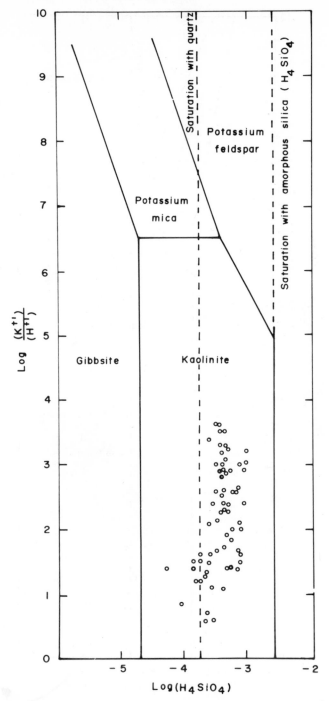

Fig. 15–15. Stability relations of some phases in the systems K_2O–Al_2O_3–SiO_2–H_2O at 25C (298K) and 1 bar pressure as functions of $(K^+)/(H^+)$ and (H_4SiO_4). Circles represent analyses of ground waters (After Feth et al., 1964).

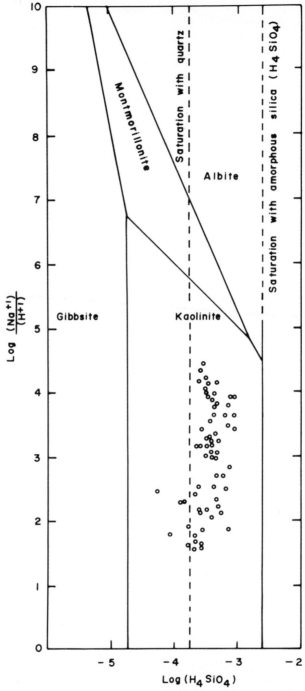

Fig. 15-16. Stability relations of some phases in the systems $Na_2O-Al_2O_3-SiO_2-H_2O$ at 25C (298K) and 1 bar pressure as functions of $(Na^+)/(H^+)$ and (H_4SiO_4). Circles represent analyses of ground waters (After Feth et al., 1964).

chemical weathering transformations, solute species of Al occur; Al therefore should be considered as a mobile, active component rather than a fixed component in the stability diagrams of feldspars (W. H. Huang, 1974). However, even after considering Al as an active component, there are a few more points which need to be considered in applying stability diagrams to soil systems. First of all, standard free energy of formation for feldspars and the associated secondary minerals may vary with particle size, degree of crystallinity and hydration, chemical composition, surface features, etc. Secondly, solution Al may be present as various species of mononuclear and polynuclear Al or as complexes of silicate, phosphate, organic acids, etc. Thirdly, in alkaline soil systems a correction should be made in the activity of H_4SiO_4 for the dissociation of H_4SiO_4. Consequently, the stability boundary between feldspars and the associated secondary minerals would be modified depending on the above factors.

B. Olivines, Pyroxenes, and Amphiboles

1. NATURAL OCCURRENCE

Geochemically the most important olivines are the members of the forsterite–fayalite series. These minerals are less abundant than pyroxenes and amphiboles in the upper lithosphere (Ahrens, 1965). There is complete solid solution between forsterite and fayalite in nature. The Mg-rich olivines are predominant in basic rocks and the Fe-rich varieties in intermediate and acidic igneous rocks. The most abundant olivines in nature probably are hyalosiderites. Tephroite mainly occurs in metasomatic rocks and in Mn deposits. Other minerals of the olivine group are rare in nature.

Three series of pyroxenes, namely, the enstatite–hypersthene, diopside–jadeite and augite series, occur in igneous rocks (Rankama & Sahama, 1950). Pyroxenes found in nature are often mixtures. The composition of the pyroxenes of volcanic rocks usually is more variable than that of the pyroxenes of plutonic rocks. This is attributed to the expanding structures of elevated temperatures allowing a higher degree of diadochic substitution[2] than at lower temperatures.

Among the three series of pyroxenes, the augite series is geochemically more important than are the others. The members of the augite series are among the most important constituents of igneous rocks. Among the pure $Mg-Fe^{2+}$ pyroxenes, the orthorhombic pyroxenes, enstatite, and hypersthene, are geochemically more important than are the monoclinic species, clinoferrosilite and clinoenstatite. The geochemically important members of the enstatite–hypersthene series are usually rich in Mg. Nevertheless, the Fe-rich members of the series exist in common igneous rocks and usually contain notable amounts of Mn. The addition of Mn probably increases the structural stability of Fe hypersthenes. Diopside and hedenbergite belong to

---

[2]Diadochic substitution is the replacement of one atom or ion in a crystal structure by another (Am. Geol. Inst., 1957).

the diopside-jadeite series; they are stable and are common in igneous and contact metamorphic rocks. There are three alkali pyroxenes in the diopside-jadeite series, namely jadeite, aegirine, and spodumene. These alkali pyroxenes are geochemically of minor significance.

Many amphiboles are absent in primary igneous rocks or may occur as minerals of secondary origin. The main primary amphiboles of igneous rocks are the Mg-poor riebeckites and the Si-poor basaltic and common hornblende. Hornblendes are widely distributed in igneous rocks from syenite and granite to gabbro, and in metamorphic rocks such as gneiss, hornblende schist, and amphibolites.

Olivines, pyroxenes, and amphiboles are largely confined to the sand and silt fractions of soils (Marshall & Jeffries, 1945; Huffman, 1954; Mohr & van Baren, 1954). Some of them may occur in the clay fraction of soils which are developed from glacial rock flour and have not been subjected to intensive weathering (Jackson, 1964).

2. EQUILIBRIUM ENVIRONMENT AND MINERAL STABILITY

The weatherability of olivines, pyroxenes, and amphiboles in soils is very high (Jackson, 1964; Loughnan, 1969). The extent of weathering of these minerals is an indication of the intensity of chemical weathering of soils in cool humid regions.

The relative stability of olivines, pyroxenes, and amphiboles appears to be related to the degree of polymerization of the tetrahedra, the ratio of basic cations to Si, and to other factors which induce destruction of the bonds linking the tetrahedra. The stability order of the common members of these mineral groups is olivine $<$ hypersthene $<$ augite $<$ hornblende.

Important factors of the soil environment affecting the stability of these minerals to weathering include redox potential, climate, topography, the degree of leaching, chelation and certain solution ionic activities (Loughnan, 1969; W. H. Huang & Keller, 1970; Goode, 1974).

Olivines may alter to serpentine or trioctahedral smectite (presumably of the saponite type), nontronite and various ferric hydrates and gels where leaching is only moderate (Bates, 1962; Sherman et al., 1962; Craig & Loughnan, 1964). However, in the environment where leaching is intense, such as in the near surface horizons, the degradation products are poorly crystallized smectite, kaolinite, halloysite, and Fe oxides such as goethite, hematite or the noncrystalline precursors of these minerals.

Pyroxenes tend to weather to chlorite or smectite or both through partial dissolution of Mg, Ca, and Fe^{2+}. Calcite may develop in the environment where the rate of dissolution of Ca is higher than that of the complete breakdown of the pyrozene. As the chemical weathering proceeds further, all the Ca and Mg and part of the Si are lost. This results in progressive enrichment of the residue in kaolinite, ferric oxides and hydroxides, and anatase. A generalized sequence of chemical weathering for pyroxenes has been discussed by Loughnan (1969).

Amphiboles appear to have a weathering sequence similar to pyroxenes.

In the Malvern Hills region of Great Britain, hornblende initially alters to chlorite with some sphene, hematite, and epidote. Upon further leaching, chlorite weathers to chlorite-vermiculite mixed layer and then to vermiculite (Stephen, 1952). In the study of the weathering of an amphibolite from the Black Hills area of South Dakota, Goldich (1938) reported that hornblende may alter to beidellite or related clay minerals. On the other hand, the weathering of hornblende in amphibolites from North Carolina is believed to be partially responsible for the genesis of gibbsite in that location (Alexander et al., 1941).

More recently W. H. Huang and Keller (1972) used the dissolution data of olivine and augite to calculate the standard free energies of formation of these minerals (Table 15-5). Additional thermodynamic data of this type

Table 15-5. Gibbs free energies of formation of feldspar, olivine, and pyroxene minerals

Mineral[†]	ΔG_f° (kcal mole^{-1})	
	Feldspar	
Albite		
$Ab_{95}An_1Or_4$	-897.1[‡]	(W. H. Huang & Kiang, 1973)
$Ab_{99}An_1$	-894.7[§]	
Ab_{100}	-884.0[‡]	(Waldbaum, 1966)[¶]
Oligoclase		
$Ab_{75}An_{18}Or_7$	-910.3[‡]	(W. H. Huang & Kiang, 1973)
$Ab_{80}An_{20}$	-906.8[§]	
Labradorite		
$Ab_{50}An_{46}Or_4$	-930.9[‡]	(W. H. Huang & Kiang, 1973)
$Ab_{52}An_{48}$	-928.5[§]	
$Ab_{40}An_{59}Or_1$	-932.4[‡]	(W. H. Huang & Keller, 1972)
$Ab_{40}An_{60}$	-931.1[§]	
Bytownite		
$Ab_{30}An_{67}Or_3$	-948.5[‡]	(W. H. Huang & Kiang, 1973)
$Ab_{31}An_{69}$	-942.7[§]	
Anorthite		
$Ab_{12}An_{86}Or_2$	-959.4[‡]	(W. H. Huang & Kiang, 1973)
$Ab_{12}An_{88}$	-957.7[§]	
An_{100}	-955.6[‡]	(Barany, 1962)
High-K plagioclases		
$Ab_{44}Or_{56}$	-896.4[‡]	(W. H. Huang & Kiang, 1973)
$Ab_{44}Or_{56}$	-896.9[§]	
Microcline	-887.3[‡]	(W. H. Huang & Keller, 1972)
	-892.6[§]	
	-892.8[§]	(Waldbaum, 1968)
	-891.3[§]	(Reesman & Keller, 1965)
	Olivine and augite	
Olivine	-461.6[‡]	(W. H. Huang & Keller, 1972)
	-457.8[§]	
Augite	-679.4[‡]	(W. H. Huang & Keller, 1972)

† The symbols, Ab, An, and Or stand for albite, anorthite, and orthoclase, respectively.
‡ Calculated on the basis of specific mineral formula.
§ Calculated on the basis of ideal structural formula.
¶ D. R. Waldbaum. 1966. Calorimetric investigation of the alkali feldspars. Ph.D. Thesis. Harvard University, Cambridge, Mass.

would be required to construct models or diagrams elucidating the sequence of chemical weathering and stability of olivines, pyroxenes, and amphiboles under varying weathering environments. The methodology outlined in the section on the equilibrium environments and mineral stability of feldspars (p. 577–578) is applicable to the thermodynamic formulations of the weathering stability of these minerals.

IV. PHYSICO-CHEMICAL PROPERTIES

A. Feldspars

Anorthite in soils weathers out rapidly even in temperate regions (Jackson, 1964). The weathering of plagioclases has an important relationship to Ca supply of soils. The presence or absence of plagioclases in the parent rock makes a difference in the productivity in many tropical and subtropical soils. As the weathering intensity and time factors increase, the release of Ca can progressively proceed from the more albitic plagioclases.

The K-feldspars in the clay and finer silt fractions of soils serve as an important source of available K, although they are generally less important than the micas (Jackson, 1964). The fraction of K from feldspar in a series of particle size fractions of moderately weathered soil profiles has been studied (Somasiri et al., 1971) and is illustrated in Table 15–6. The fraction of K from feldspar increases with increasing particle size. The effects of col-luviation processes, coupled with eluviation and illuviation occurring during pedogenesis, on the partition of feldspar K in soil profiles are shown in Fig. 15–17. Moreover, in the prairie soils the proportion of orthoclase as com-pared to microcline increases as the particle size decreases (Table 15–7). The observed variation of the proportion of orthoclase to microcline with particle size is most likely an inherited property of the soil parent material rather than an effect of weathering. The increase of orthoclase in the finer fractions does not seem to be in agreement with the stability and size relationship of minerals. Microcline, the low temperature polymorph, is more stable than orthoclase. Furthermore, more resistant minerals tend to persist in the small size fractions (Jackson & Sherman, 1953). This apparent deviation can be explained. First, orthoclase as the high temperature polymorph would crystallize faster and tend to have smaller grain size, whereas microcline crystallizes in pegmatite and hydrothermal veins at relatively lower tempera-tures and the grain size would be larger (Berry & Mason, 1959). Second, the hardness of K-feldspars varies with the composition and the degree of mixing of Na and K substitution in the K-feldspar structure (Barth, 1969). In addi-tion to the disorder in Al and Si distribution, orthoclase may also be different from microcline in composition and degree of mixing of Na and K. Thus, orthoclase may be less resistant to physical breakdown than microcline. The differences in the genesis of orthoclase and microcline and in their physical properties in response to the grinding action of glacial ice (Somasiri & Huang,

Table 15-6. Distribution of feldspar K in various size fractions of an Orthic Brown Haverhill soil profile in Saskatchewan, Canada (After Somasiri et al., 1971)

Horizon	Particle size	Total K	Feldspar K	Mica K	% fraction of K Feldspar	Mica
	μm	—————————%—————————				
Ap	<0.2	2.11	--	2.11	--	100
	0.2-2	2.61	0.08	2.53	3	97
	2-5	2.01	0.61	1.40	30	70
	5-20	1.73	0.87	0.86	50	50
	20-50	1.54	0.97	0.57	63	37
	50-500	1.27	0.76	0.51	60	40
	500-2,000	1.73	1.73	--	100	--
	<2,000†	1.69	0.68	1.01	40	60
Bm	<0.2	1.83	--	1.83	--	100
	0.2-2	2.99	0.00	2.99	0	100
	2-5	2.46	0.36	2.10	15	85
	5-20	2.04	0.77	1.27	38	62
	20-50	1.47	0.85	0.62	58	42
	50-500	1.42	0.84	0.58	59	41
	500-2,000	1.82	1.82	--	100	--
	<2,000	1.76	0.65	1.11	37	63
Ck	<0.2	1.68	--	1.68	--	100
	0.2-2	2.74	0.21	2.53	8	92
	2-5	2.62	0.87	1.75	33	67
	5-20	2.27	0.92	1.35	41	59
	20-50	1.56	0.83	0.73	53	47
	50-500	1.19	0.92	0.27	77	23
	500-2,000	2.07	2.07	--	100	--
	<2,000	1.77	0.69	1.08	39	61

† Whole mineral fraction.

Table 15-7. Proportion of orthoclase as a percentage of total K-feldspar in selected size fractions of the Ck horizons of Brown, Black to Gray Wooded sequence of soils in Saskatchewan, Canada (After Somasiri & Huang, 1973)

Soil	Parent material	Percent distribution† of orthoclase 20-50 μm	50-250 μm	250-500 μm	500-2,000 μm
Orthic Brown Haverhill	Glacial till	38	40	23	22
Orthic Black Oxbow	Glacial till	33	28	21	18
Orthic Black Melfort	Lacustrine	45	42	--	--
Orthic Gray Wooded Waitville	Glacial till	31	26	17	21

† The standard deviation of the values reported ranged from ±1 to 3 except for the 50-250 μm fraction of the Ck horizon of the Orthic Brown Haverhill, which showed a standard deviation of ±6. The values were computed from eight readings.

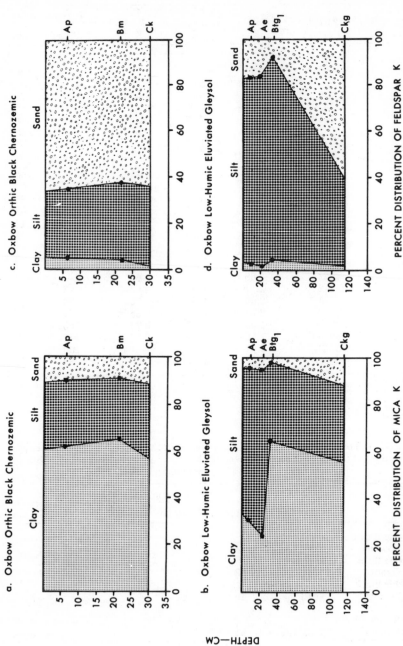

Fig. 15–17. Distribution of feldspar K and mica K among clay, silt and sand fractions of the Orthic Black and Low Humic Eluviated Gleysol profiles in the Oxbow catena in Saskatchewan, Canada (After Somasiri et al., 1971).

1971) apparently account for the observed variation of the proportion of orthoclase to microcline with particle size.

Some studies have been made on dissolution of feldspars in inorganic aqueous solutions (Tamm, 1930; Nash & Marshall, 1956a, b; Correns, 1963; Keller et al., 1963; P. M. Huang, 1966; Burger, 1969). The exposed atoms and ions at the surfaces of feldspars would undergo hydration in aqueous solution through attraction of water dipoles to the charged surfaces (Jenny, 1950). The attractive forces may polarize the water dipoles to dissociate H and OH ions. The dissociated OH ions would bond to exposed cations and the H to O ions and other anions. The metallic ions of the minerals are replaced concurrently by H ions dissociated from water and/or from acidic components (Jenny, 1950; Frederickson, 1951; P. M. Huang et al., 1968). H ions thus play an important role in the breakdown of feldspars. The size of H ions permits easy penetration into crystal structures. The charge/radius ratio of an H ion is higher than for any other ion; this has a marked disrupting effect on the charge balance within the structure. Removal of the bonding cations by H ions leads to instability of the mineral structure and thus mobilizes the silica, presumably through a process of hydrolysis at Al-O-Si bonds (Loughnan, 1962). Upon the removal of Al, the silica at the crystal edges is converted into a silicic acid coating and thence to silicic acid in solution. The increase of the dissolution of Si from feldspars with a decrease of pH from 9 to 2 is considered to be associated with the pH dependence of the removal of the bonding cations (P. M. Huang, 1966). The formation of orthosilicate ions by H ion dissociation from $Si(OH)_4$ and of aluminate ions are regarded as the main reasons for the sharp increase in Si solubility as pH is raised above 9.

Correns and von Engelhardt (1938) reported that there was a formation of a protective layer of hydrous aluminum silicate on feldspar surfaces in aqueous solutions. Garrels and Howard (1959) reacted ground feldspar with water and found that the first result of the reaction was the formation of a surface layer which graded from an outer portion to an inner portion of the feldspar particles. The outer portion was structurally disrupted, whereas the inner portion retained the original silicate structure, but with H substituted for K. The subsequent fate of the outer and inner zones remains to be critically examined.

The chemical dynamics of dissolution of K from K-feldspar by $1N$ HNO_3 solution was studied by P. M. Huang et al. (1968). For the temperature range studied, the rates of K release from muscovite, phlogopite, and biotite are about twice, 9 to 12, and 118 to 190 times, respectively, greater than the rate of K release from microcline (Table 15-8). The release of K from a tektosilicate is more difficult than from the interlayers of the micas. The Arrhenius heat of activation for the release of K from microcline is 22.97 kcal mole^{-1} (96.17 kJ mole^{-1}) which is greater than for the micas. The heat of activation is interpreted as the energy level the structural K must acquire in order to be able to react. As the heat of activation increases, the rate constant decreases, since the mineral with higher heat of activation for the release of structural K would release less K per unit time. The release of K in $1N$

Table 15-8. Apparent rate constant and Arrhenius heat of activation for the release of structural K from K minerals (After P. M. Huang et al., 1968)

Mineral	Rate constant, s^{-1}		Arrhenius heat of activation, kcal mol^{-1}
	28C	38C	
Biotite	4.06×10^{-6}	8.58×10^{-6}	14.00
Phlogopite	2.50×10^{-7}	6.78×10^{-7}	18.57
Muscovite	3.86×10^{-8}	1.15×10^{-7}	20.41
Microcline	2.14×10^{-8}	7.31×10^{-8}	22.97

HNO_3 is not considered to be controlled by film diffusion, since the activation energy for dissolution rates that are controlled by diffusion ranges from 4 to 8 kcal mole^{-1} (17–34 kJ mole^{-1}) (Glasstone et al., 1941). The heat of activation of microcline and micas obtained by P. M. Huang et al. (1968) was in the order of those obtained for a few soils in Indiana by Burns and Barber (1961). The Arrhenius heat of activation and apparent rate constant for K release from the K-bearing minerals (P. M. Huang et al., 1968) are fundamental to understanding the exchangeability and availability of K in feldspars and micas to plants (Reitemeier, 1951; Rich, 1968; S. J. Smith et al., 1968). The reason why K-bearing minerals exhibit different rates of K release can thus be explained by the different crystal structure and atomic bonding of the minerals and the resulting bonding energy with which K is held in the mineral structure.

Although the release of K from K-feldspars is generally more hindered than from micas, the role of K-feldspars in supplying K to plants cannot be ignored. The K-feldspars are the largest natural reserve of K in many soils. The K status of soils was found to be related to the amount of K-feldspars present in certain soils (de Leenheer, 1950; Jeffries et al., 1956). In the usual soil pH range of 5 to 7.5 (Fig. 15–18), as pH decreases, the difference in the amount of released K between the micas decreases, but that between the feldspars and the micas, particularly muscovite, increases (P. M. Huang et al., 1968). This indicates that the difference in K release between micas and feldspars becomes larger as acidity increases. However, as alkalinity develops, the release of K from muscovite and feldspars approaches similar levels. The differential pH dependence of K release from the K-bearing minerals suggests that measurement of the K-supplying power of soils using strong acidic solutions should be interpreted with care. This is particularly significant if the soil has a diverse composition of K-bearing minerals and natural pH conditions.

In the cation exchange reactions of feldspars, there are strong specific effects (Nash & Marshall, 1956a, 1956b, 1957; Marshall, 1964). Na is preferentially liberated by water and acid over K from microcline which contains both Na and K, whereas K is preferentially liberated over Na by NH_4Cl and $SrCl_2$. In the plagioclases, the minor cation tends to be much more easily displaced compared with the major cation. For instance, in albite and oligoclase, Ca is preferentially exchanged over Na; the opposite is true for anorthite. The cation exchange capacities of feldspars are highly variable, de-

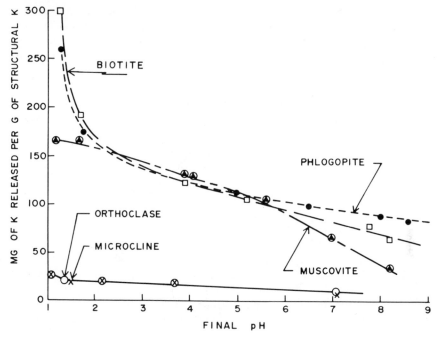

Fig. 15–18. Influence of pH on K release from K minerals common in soils (After P. M. Huang et al., 1968).

pending on the nature of the displacing and released cations. Certain cations have a very strong tendency to be fixed. For example, the NH_4 ion is able to penetrate deeply to displace other cations and is remarkable in its tendency toward fixation in the crystal structure. Experimental studies on cation exchange reactions of feldspar surfaces are relatively few compared to similar studies with phyllosilicates and remain an important subject for investigation.

In addition to H ions dissociated from water and acids in soil solutions, ligands of complexing organic acids may play a significant role in the release of cations from the feldspar structure. The organic acids, in order of increasing effect on dissolution, are acetic, aspartic, salicylic, and citric acids (W. H. Huang & Kiang, 1972). Citric acid is more effective than the other acids in extracting Al and Ca particularly from Ca-rich plagioclases, presumably because of the formation of Al- and Ca-complexes. The reaction of feldspars with complexing organic acids and its relation to nutrient release from soils deserve attention in further studies.

B. Olivines, Pyroxenes, and Amphiboles

Olivines are compact and strong in physical structure. However, the high ratio of divalent cations to Si involves ready chemical attack on outer surfaces. Surface octahedral cations (Mg and Fe^{2+}) of olivines are thus potentially mobile (Marshall, 1964; Loughnan, 1969). The Fe^{2+} readily oxidizes to Fe^{3+} and the Mg has the tendency to be coordinated with OH groups. The loss of these octahedral cations from the mineral surface accelerates the dissolution of silica tetrahedra, thus exposing fresh surfaces for further reactions. Olivines are relatively easily weathered in soils and thus contributed to the fertility levels of Mg and Fe.

Pyroxenes and amphiboles have pronounced cleavages parallel to the silica chains. Access of water through the cleavages promotes dissolution of cations (such as Mg^{2+}, Fe^{2+}, Fe^{3+}, and Al^{3+}), bonding the chains together and thus causes rapid breakdown of the mineral structures (Marshall, 1964; Loughnan, 1969). The variety of isomorphous substitution in these minerals and their easy weatherability make them excellent sources for trace elements and Ca and Mg in soils.

The ionic solubilities of olivines, pyroxenes, and amphiboles in aqueous inorganic solutions and complexing organic acids have been studied (Tunn, 1940; Keller et al., 1963; P. M. Huang, 1966; W. H. Huang & Keller, 1970). The solubilities of Si and the basic cations of these minerals are considered to be highly important in determining the minerals formed upon weathering (P. M. Huang, 1966; Jackson, 1968). However, ionic equilibria and rates and mechanisms of release of structural cations from these minerals as related to soil environments remain as important subjects for systematic studies in the area of physical chemistry and mineralogy of soils. Acquisition of this knowledge is indispensable to understanding the role of these minerals in genetic processes and nutrient transformations of soils.

V. QUANTITATIVE DETERMINATION

A. Feldspars

After the proper pretreatment and preparation of a series of size fractions as described for identification of feldspars (p. 564–569), quantification and interpretation of the data can be facilitated. The methods proposed for quantitative determination of feldspars are X-ray diffraction analysis, selective dissolution, and counting techniques.

1. X-RAY POWDER DIFFRACTION

X-ray powder diffraction analysis has been used for the quantitative determination of feldspars in soils (Jeffries, 1947; Phillippe & White, 1950; Pawluk, 1961). A drawback to this method is the variation of diffraction peak intensity due to factors such as chemical composition, crystal imperfection, the presence of noncrystalline materials, absorption of X-ray radiation, variation in sample packing, particle size, and crystal orientation (Jackson, 1956). Nevertheless, for the estimation of the amount of K feldspar polymorphs and albite in perthites, X-ray diffraction analysis appears to be the most suitable method available (van der Plas, 1966). To facilitate quantifica-

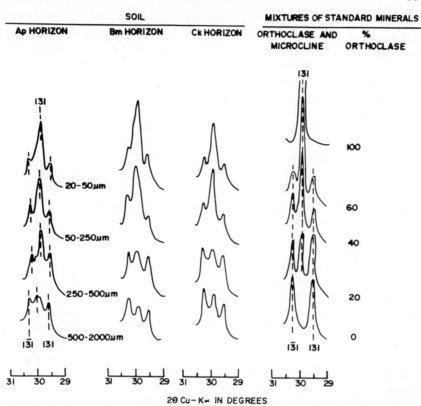

Fig. 15-19. X-ray diffractograms showing the diagnostic peaks of microcline and ortho-clase in K-feldspar concentrates of the selected size fractions of the major genetic hori-zons of the Orthic Brown Haverhill Chernozemic soils (After Somasiri & Huang, 1971).

tion, concentration by specific gravity separation is recommended (see p. 564).

The proportion of orthoclase to microcline in rocks (Steiger & Hart, 1967) and in soils (Somasiri & Huang, 1971, 1973) has been studied by using X-ray diffraction analysis. The proportions of orthoclase to microcline are estimated by comparing the intensity of the 131 reflection with that of the standard mineral mixtures which are prepared for X-ray diffraction in the same manner as the samples. A plot is made of the peak height ratios of the 131 reflection of orthoclase to the 131 reflection of microcline versus the content of orthoclase as a percentage of total K-feldspars. X-ray diffracto-grams showing the relative intensities of the diagnostic peaks of microcline and orthoclase in a series of mixtures of the standard minerals and in K-feldspar concentrates of a Chernozemic soil are given in Fig. 15-19.

If the concentrating technique for the isolation of the alkali feldspar fraction from soils and sediments is efficient, the intensity of the $\overline{2}01$ reflec-tion from albite can be used to estimate the albite content of perthites in the fraction (van der Plas, 1966). If the obliquity of alkali feldspars is close to 1, the minerals are practically of pure K feldspar (Laves & Goldsmith,

1961). Such K feldspars are often perthitic in nature and the Na-rich phase is in most cases close to pure albite. If the alkali feldspars are of intermediate obliquity or even close to monoclinic, the position of the 400 reflection of the alkali feldspar fraction (Goldsmith & Laves, 1961; see Fig. 15-8) indicates the approximate content of albite in the alkali feldspar fraction.

For determining the chemical composition of a series of plagioclases by X-ray powder diffraction, readers are referred to Fig. 15-9, 15-10, and 15-11. The specific gravity of plagioclases (Table 15-4) permits a subdivision, by means of heavy liquids, into a few groups with specific chemical compositions. However, it is important to know that quartz is present in large amounts in soils and sediments and thus contributes a masking effect in the quantitative estimation of feldspars by X-ray diffraction if the intensity is used as the basis for estimation. Nevertheless, the only feldspars present in between the limits of the specific gravity of quartz, namely, 2.63 to 2.67, are the group of oligoclase and part of the sodic andesine.

2. SELECTIVE DISSOLUTION

Pyrosulfate fusion has long been employed for the decomposition of aluminosilicate minerals (J. L. Smith, 1865). More recently, the pyrosulfate fusion using $Na_2S_2O_7$ has been used quite successfully in the quantitative analysis of feldspars (Kiely & Jackson, 1964, 1965). When heated, pyrosulfate decomposes in a quiet fusion liberating acid fumes as follows: $K_2S_2O_7 \rightarrow K_2SO_4 + SO_3$. The fusion thus serves as a high temperature acid treatment, decomposing micas and other phyllosilicates common in soils. The phyllosilicate relics are dissolved by washing with $3N$ HCl and hot $0.5N$ NaOH, leaving feldspars and quartz in the residue. Feldspars are determined by allocation of residual K, Na, and Ca to their equivalent end-members, with suitable correction factors for mineral solubility and for increase of Na from pyrosulfate in K and Ca feldspars. This method is recommended for a range of size fractions between 2-500 μm. Feldspars are extensively dissolved in the 0.2-2 μm fraction, but an estimate of their content may be obtained.

3. COUNTING TECHNIQUES

The quantitative determination of feldspars by counting techniques using the optical microscope is a common method. The concentrating and staining of feldspars facilitates the counting (Doeglas et al., 1965). The grain count method is quite tedious and time-consuming. The optical microscope method cannot be used for a wide range of particle size fractions. The data obtained by the grain counts depend on the counting method used. The results of point counting methods are assumed to represent volume fractions. Other counting procedures give results in undefinable units. Moreover, the determinations made on the finer fractions by X-ray methods are on a weight basis. Thus, compilation of the results obtained by different methods becomes quite difficult. Furthermore, quantitative evaluation of the amount of K-feldspar polymorphs and perthite cannot be made optically. References for the fundamental aspects of counting techniques can be found from the works of Chayes (1956), van der Plas (1959, 1962), and van der Plas and Tobi (1965).

B. Olivines, Pyroxenes, and Amphiboles

The soil pretreatments, particle size fractionation, and heavy-liquid separation (see p. 573) are conventionally carried out prior to quantitative determinations of olivines, pyroxenes, and amphiboles, although critical evaluation of the extent of the effects of the pretreatments on these minerals appears to merit further research. The methods commonly used for quantitative estimation of these groups of minerals are X-ray diffraction analyses and petrographic methods.

1. X-RAY DIFFRACTION ANALYSIS

The amounts of olivines, pyroxenes, and amphiboles in the heavy-mineral concentrates can be estimated by measuring the intensities of the diagnostic reflections of these minerals (see p. 574). Standard mixtures which contain the same minerals are prepared and the variation of the intensity of the diagnostic reflections in the powder pattern is related to percentage composition. A powder pattern of an unknown concentrate may be taken under the same conditions and mineralogical composition determined by interpolation. Further information on the fundamentals of quantitative X-ray diffraction analysis is available in many references (Klug & Alexander, 1954; Jackson, 1956; Brindley, 1961; Whittig, 1965).

2. PETROGRAPHIC METHODS

The percentage composition of olivines, pyroxenes, and amphiboles can be obtained either by counting or estimation methods. It is possible after considerable practice to arrive at reasonably consistent results by estimation methods. However, rarely do estimations by different persons agree. Hence in order to obtain more accurate results, the counting methods are recommended. The drawbacks of counting techniques in quantification of feldspars (see p. 592) are also true in this case. Nevertheless, the quantitative estimation of these minerals by counting techniques is a very common method. For detailed information on petrographic methods refer to Krumbein and Pettijohn (1938) and Milner (1962).

VI. SUMMARY AND CONCLUSIONS

Feldspars are present virtually in all soils and sediments. Many studies have been carried out on structural characteristics, chemical composition, thermodynamic properties, solubilities, kinetics and mechanisms of alteration, and cation exchange reactions of specimen feldspars. The nature and properties of soil feldspars in relation to geological origin of soil parent materials, pedogenic processes, and nutrient status remain as important subjects to be critically explored.

Olivines, pyroxenes, and amphiboles are accessory minerals in soils and sediments. The structure and chemical composition of these minerals have been well documented. Nevertheless, information on the species and thermodynamic and kinetic aspects of these minerals in relation to their weathering stability and nutrient release under varying soil environments is still relatively scarce.

Continued efforts in nondestructive isolation, identification, and quantification of the above mentioned minerals would definitely facilitate the advance of knowledge on their nature, occurrence, distribution, and behavior in the soil environment.

VII. PROBLEMS AND EXERCISES

1) Define the following terms:
   a) order-disorder relation
   b) polymorph
   c) solid solution
   d) obliquity
   e) exsolution
2) Describe the basic structural schemes of feldspars, olivines, pyroxenes, and amphiboles.
3) Describe the nature of perthites in soils.
4) Discuss the significance of pretreatments and heavy-liquid separation in identification of feldspars, olivines, pyroxenes, and amphiboles in soils and sediments.
5) What is the mathematical basis for differentiating microcline from orthoclase by X-ray diffraction analysis?
6) Discuss the relationship between chemical composition of plagioclases and the $d(\bar{1}32)$ and $d(131)$ spacings.
7) Discuss the major reflections for estimating the Mg/Fe ratio of an olivine by X-ray diffraction analysis.
8) Discuss the principle of selective dissolution in quantification of feldspars of soils and sediments.
9) Discuss the rate and heat of activation of K release from K-feldspars and micas in relation to their crystal structure and atomic bonding.
10) In a soil environment where pH is 5, the ionic strength of soil solution is 0.01, and the soil solution concentrations of K^+, Al^{3+} and H_4SiO_4 are 10^{-3}, 10^{-5}, and 10^{-3} M, respectively, is microcline stable? Predict the stability of microcline by assuming that the chemical weathering reaction of microcline can be written as:

$$KAlSi_3O_8 + 4H^+ + 4H_2O = K^+ + Al^{3+} + 3H_4SiO_4$$

Answer: Microcline is stable in the soil environment described.

VIII. SUPPLEMENTARY READING

Carroll, D. 1962. The application of sedimentary petrology to the study of soils and related superficial deposits. p. 457–518. *In* H. B. Milner (ed.) Sedimentary petrography. Vol. 2. The MacMillan Co., New York.

De Vore, G. W. 1959. The surface chemistry of feldspars as an influence on their decomposition products. Clays Clay Miner. 6:24–41.

Farmer, V. C. (ed.). 1974. The infrared spectra of minerals. Mineralogical Society, London.

Goodyear, J., and W. J. Duffin. 1954. The identification and determination of plagioclase feldspars by the X-ray powder method. Mineral. Mag. 30:306–326.

Heinrich, E. Wm. 1965. Microscopic identification of minerals. McGraw-Hill, New York.

Jensen, B. B. 1973. Patterns of trace element partitioning. Geochim. Cosmochim. Acta 37:2227–2242.

Marfunin, A. S. 1962. Some petrological aspects of order-disorder in feldspars. Mineral. Mag. 33:298–314.

Rich, C. I. 1972. Potassium in soil minerals. p. 15–31. *In* Proc. 9th Colloquium, Int. Potash Institute, Landshut, Federal Republic of Germany.

LITERATURE CITED

Ahrens, L. H. 1965. Distribution of the elements in our planet. McGraw-Hill Book Co., New York.

Alexander, L. T., S. B. Hendricks, and G. T. Faust. 1941. Occurrence of gibbsite in some soil forming materials. Soil Sci. Soc. Am. Proc. 6:52–57.

American Geological Institute. 1957. Glossary of geology and related sciences. Am. Geol. Inst., Washington, D. C.

Arnold, P. W. 1960. Nature and mode of weathering of soil potassium reserves. J. Sci. Food Agric. 11:285–292.

Bailey, E. H., and R. E. Stevens. 1960. Selective staining of K-feldspar and plagioclase on rock slabs and thin sections. Am. Mineral. 45:1020–1026.

Bailey, S. W., and W. H. Taylor. 1955. The structure of a triclinic potassium feldspar. Acta Crystallogr. 8:621–632.

Barany, R. 1962. Heats and free energies of formation of some hydrated and anhydrous sodium- and calcium-aluminum silicates. U. S. Bur. Mines Rep. Invest. 5900, 17 p.

Barshad, I. 1964. Chemistry of soil development. p. 1–70. *In* F. E. Bear (ed.) Chemistry of the soil. 2nd ed. Reinhold Publishing Corp., New York.

Barth, T. W. F. 1934. Polymorphic phenomena and crystal structure. Am. J. Sci. 27:273–286.

Barth, T. W. F. 1969. Feldspars. Wiley-Interscience, New York.

Baskin, Y. 1956. Observations on heat treated authigenic microcline and albite crystals. J. Geol. 64:219–224.

Bates, T. F. 1962. Halloysite and gibbsite formation in Hawaii. Clays Clay Miner. 9:307–314.

Berg, R. R. 1952. Feldspathized sandstones. J. Sediment. Petrol. 22:221–223.

Berry, L. G., and B. Mason. 1959. Mineralogy. W. H. Freeman & Co., San Francisco.

Borley, G., and M. T. Frost. 1963. Some observations on igneous ferrohastingsites. Mineral. Mag. 33:646–662.

Bowen, N. L. 1922. The reaction principle in petrogenesis. J. Geol. 30:177–198.

Bowen, N. L., and O. F. Tuttle. 1950. The system NaAlSi$_3$O$_8$-KAlSi$_3$O$_8$-H$_2$O. J. Geol. 58:489-511.

Bragg, W. L., and G. F. Claringbull. 1965. Crystal structures of minerals. G. Bell & Sons, Ltd., London.

Brindley, G. W. 1961. Quantitative analysis of clay mixtures. p. 489-516. In G. Brown (ed.) The X-ray identification and crystal structures of clay minerals. Mineralogical Soc., London.

Brown, B. E., and S. W. Bailey. 1964. The structure of maximum microcline. Acta Crystallogr. 17:1391-1400.

Brown, G. M. 1960. The effect of ion substitution on the unit cell dimensions of the common clinopyroxenes. Am. Mineral. 45:15-38.

Brown, W. L. 1960. Lattice changes in heat-treated plagioclases. The existence of mono-albite at room temperature. Z. Kristallogr. Kristallgeom., Kristallphys., Kristall-chem. 113 (Laue Festschrift):297-330.

Brown, W. L. 1962. Peristerite unmixing in the plagioclases and metamorphic facies series. Norsk. Geol. Tidsskr 42 (Feldspar volume):354-383.

Brydon, J. E., and L. M. Patry. 1961. Mineralogy of Champlain sea sediments and Rideau clay soil profile. Can. J. Soil Sci. 41:169-181.

Burger, D. 1969. Relative weatherability of calcium-containing minerals. Can. J. Soil Sci. 49:21-28.

Burns, A. F., and S. A. Barber. 1961. The effect of temperature and moisture on ex-changeable potassium. Soil Sci. Soc. Am. Proc. 25:349-352.

Chandrasekhar, S., S. G. Fleet, and H. D. Megaw. 1961. The structure of "body-centered" anorthite. Cursillos Conf. Inst. "Lucas Mallada." 8:141.

Chayes, F. 1956. Petrographic model analysis. Chapman and Hall, London.

Clarke, F. W. 1924. The data of geochemistry. U. S. Geol. Survey Bull. 770.

Cole, W. F., H. Sörum, and O. Kennard. 1949. The crystal structures of orthoclase and sanidinized orthoclase. Acta Crystallogr. 2:280-287.

Correns, C. W. 1963. Experiments on the decomposition of silicates and discussion of chemical weathering. Clays Clay Miner. 12:443-460.

Correns, C. W., and W. von Engelhardt. 1938. Neue Untersuchungen über die Verwitter-ung des Kalifeldspates. Chem. Erde 12:1-22.

Craig, D. C., and F. C. Loughnan. 1964. Chemical and mineralogical transformations ac-companying the weathering of basic volcanic rocks from New South Wales. Aust. J. Soil Res. 2:218-234.

Deer, W. A., and L. R. Wager. 1939. Olivines from the Skaergaard intrusion, Kangerd-lugssuak, East Greenland. Am. Mineral. 24:18-25.

de Leenheer, L. 1950. Mineralogical characterization of the sand-fraction in soil profiles. Int. Congr. Soil Sci., Trans. 4th (Amsterdam, The Netherlands) II:84-89.

Dell, C. I. 1963. A study of the mineralogical composition of sand in northern Ontario. Can. J. Soil Sci. 43:189-200.

De Wolff, P. M. 1948. Multiple focussing camera. Acta Crystallogr. 1:207-211.

De Wolff, P. M. 1950. An adjustable curved crystal monochromator for X-ray diffraction analysis. Appl. Sci. Res., Sect. B., 1:119-126.

Dietrich, R. V. 1962. K-feldspar structural states as petrogenetic indicators. Norsk. Geol. Tidsskr. 42 (Feldspar volume):394-415.

Doeglas, D. J., J. Ch. L. Favejee, D. J. G. Nota, and L. van der Plas. 1965. On the identi-fication of feldspars in soils. Meded. Landbouwhogesch. Wageningen, 65(9):1-14.

Donnay, G., and J. D. H. Donnay. 1952. The symmetry change in the high-temperature alkali feldspar series. Am. J. Sci. Bowen Volume:115-133.

Duke, D. A., and J. D. Stephens. 1964. IR investigation of the olivine group minerals. Am. Mineral. 49:1388-1406.

Eliseev, E. N. 1957. X-ray study of the minerals of the isomorphous series forsterite-fayalite. Mem. Soc. Russe Mineral. 86, 657.

Feth, J. H., C. E. Robertson, and W. L. Polzer. 1964. Sources of mineral constituents in water from granitic rocks, Sierra Nevada, California and Nevada. U. S. Geol. Surv. Water Supply Pap. 1535-I.

Finney, J. J., and S. W. Bailey. 1964. Crystal structure of an authigenic microcline. Z. Kristallagr., Kristallgeom., Kristallphys., Kristallchem. 119:437-453.

Foit, F. F., Jr., and D. R. Peacor. 1973. The anorthite crystal structure at 410 and 830°C. Am. Mineral. 58:665–675.

Frederickson, A. F. 1951. Mechanism of weathering. Bull. Geol. Soc. Am. 62:221–232.

Frost, M. T. 1963. Amphiboles from younger granites of Nigeria. Part II. X-ray data. Mineral. Mag. 33:377–384.

Gabriel, A., and P. Cox. 1929. A staining method for the quantitative determination of certain rock minerals. Am. Mineral. 14:290–292.

Garrels, R. M., and C. L. Christ. 1965. Solutions, minerals, and equilibria. Harper & Row, New York.

Garrels, R. M., and P. Howard. 1959. Reactions of feldspar and mica with water at low temperatures and pressures. Clays Clay Miner. 6:68–88.

Gay, P. 1954. The structures of the plagioclase feldspars. V. The heat-treatment of lime-rich plagioclases. Mineral. Mag. 30:428–438.

Gay, P. 1956. The structure of intermediate plagioclase feldspars. VI. Natural intermediate plagioclases. Mineral. Mag. 31:21–40.

Gay, P., and W. H. Taylor. 1953. The structures of the plagioclase feldspars. IV. Variations in the anorthite structure. Acta Crystallogr. 6:647–650.

Glasstone, S., K. J. Laidler, and H. Eyring. 1941. The theory of rate processes. p. 400–401, 522–525. McGraw-Hill, New York.

Goldich, S. S. 1938. A study of rock weathering. J. Geol. 46:17–58.

Goldsmith, J. R., and F. Laves. 1954a. The microcline-sanidine stability relations. Geochim. Cosmochim. Acta 5:1–19.

Goldsmith, J. R., and F. Laves. 1954b. Potassium feldspar structurally intermediate between microcline and sanidine. Geochim. Cosmochim. Acta 6:100–118.

Goldsmith, J. R., and F. Laves. 1961. The sodium content of microclines and the microcline albite series. Cursillos Conf. Inst. "Lucas Mallada" 8:81–91.

Goode, A. D. T. 1974. Oxidation of natural olivines. Nature (London) 248:500–501.

Hawkins, R. H., and E. R. Graham. 1950. Mineral contents of the silt separates of some Missouri soils as these indicate the fertility level and degree of weathering. Soil Sci. Soc. Am. Proc. 15:308–313.

Heier, K. S. 1962. Trace elements in feldspars, a review. Norsk Geol. Tidsskr. 42(Feldspar volume):415–455.

Henry, N. R. M., H. Lipson, and W. A. Wooster. 1961. The interpretation of X-ray diffraction photographs. Macmillan & Co., New York.

Hess, H. H. 1952. Orthopyroxene of the Bushveld type, ion substitutions and changes in unit cell dimensions. Am. J. Sci., Bowen Volume:173–187.

Himes, F. L., and S. A. Barber. 1957. Chelating ability of soil organic matter. Soil Sci. Sco. Am. Proc. 21:368–373.

Howie, R. A. 1955. The geochemistry of the charnockite series of Madras, India. Trans. Roy. Soc. Edinburgh, 62:725–768.

Howie, R. A. 1962. Some orthopyroxenes from Scottish metamorphic rocks. Mineral. Mag. 33:903–911.

Hseung, Y., and M. L. Jackson. 1952. Mineral composition of the clay fraction: III. Of some main soil groups of China. Soil Sci. Soc. Am. Proc. 16:294–297.

Huang, P. M. 1966. Mechanism of reaction of neutral fluoride solution with soil clay minerals and silica solubility scale for silicates common in soils. Ph.D. thesis, Univ. of Wisconsin, Madison (Order No. 66–1281). Univ. Microfilms, Ann Arbor, Mich. (Diss. Abstr. 26:6269–6270).

Huang, P. M., L. S. Crosson, and D. A. Rennie. 1968. Chemical dynamics of potassium release from potassium minerals common in soils. Int. Congr. Soil Sci., Trans. 9th (Adelaide, Australia) II:705–712.

Huang, P. M., and M. L. Jackson. 1967. Fluorine determination in minerals and rocks. Am. Mineral. 52:1503–1507.

Huang, P. M., and S. Y. Lee. 1969. Effect of drainage on weathering transformations of mineral colloids of some Canadian Prairie soils. Proc. Int. Clay Conf. (Tokyo, Japan) I:541–551.

Huang, W. H. 1974. Stabilities of kaolinite and halloysite in relation to weathering of feldspars and nepheline in aqueous solution. Am. Mineral. 59:365–371.

Huang, W. H., and W. D. Keller. 1970. Dissolution of rock forming minerals in organic acids. Am. Mineral. 55:2076–2094.

Huang, W. H., and W. D. Keller. 1972. Standard free energies of formation calculated from dissolution data using specific mineral analyses. Am. Mineral. 57:1152–1162.

Huang, W. H., and W. C. Kiang. 1972. Laboratory dissolution of plagioclase feldspars in water and organic acids at room temperature. Am. Mineral. 57:1849–1859.

Huang, W. H., and W. C. Kiang. 1973. Standard free energies of formation calculated from dissolution data using specific mineral analyses. II. Plagioclase feldspars. Am. Mineral. 58:1016–1022.

Huffman, H. 1954. Mineralogische Untersuchungen au Fünf Bodenprofilen über Basalt, Muschelkalk und Bunstandstein. Heidel. Beit. Mineral. Petrogr. 4:67.

Jackson, E. D. 1960. X-ray determination curve for natural olivines of composition Fo_{80-90}. Prof. Pap. U. S. Geol. Surv. 400B, 432.

Jackson, M. L. 1956. Soil Chemical Analysis—Advanced course. Published by the author, University of Wisconsin, Madison, Wis.

Jackson, M. L. 1964. Chemical composition of soils. p. 71–141. In F. E. Bear (ed.) Chemistry of the soil. Reinhold Publishing Corp., New York.

Jackson, M. L. 1968. Weathering of primary and secondary minerals in soils. Int. Congr. Soil Sci., Trans. 9th (Adelaide, Australia) IV:281–292.

Jackson, M. L., and G. D. Sherman. 1953. Chemical weathering of minerals in soils. Adv. Agron. 5:219–318.

Jambor, J. L., and C. H. Smith. 1964. Olivine composition determination with small-diameter X-ray powder cameras. Mineral. Mag. 33:730–741.

Jeffries, C. D. 1947. The use of X-ray spectrometer in the determination of the essential minerals in soils. Soil Sci. Soc. Am. Proc. 12:135–140.

Jeffries, C. D., E. Grissinger, and L. Johnson. 1956. Distribution of important soil minerals in Pennsylvania soils. Soil Sci. Soc. Am. Proc. 20:400–403.

Jenny, H. 1950. Origin of soils. p. 41–61. In P. D. Trask (ed.) Applied sedimentation. Wiley, New York.

Joint Committee on Powder Diffraction Standards. 1974. Search Manual. Selected powder diffraction data for minerals, Data book, 1st ed., Swarthmore, Pa.

Jones, J. B., and W. H. Taylor. 1961. The structure of orthoclase. Acta Crystallogr. 14:443–445.

Karl, F. 1954. Über Hoch-und Tieftemperaturoptik von Plagioklasen und deren petrographische und geologische Auswertung am Beispiel einiger alpiner Ergussgesteine (Die Existenz von Uebergangslagen). Festband Bruno Sander. Mineral. Petrog. Mitt. 3(4):320–328.

Kastner, M. 1974. The contribution of authigenic feldspars to the geochemical balance of alkali metals. Geochim. Cosmochim. Acta 38:650–653.

Keil, K., and K. Fredricksson. 1964. The iron, magnesium, and calcium distribution in coexisting olivines and rhombic pyroxenes of chondrites. J. Geophys. Res. 69:3487.

Keller, W. D., W. D. Balgord, and A. L. Reesman. 1963. Dissolved products of artificially pulverized silicate minerals and rocks: Part I. J. Sediment Petrol. 33:191–204.

Keller, W. D., E. F. Hanson, W. H. Huang, and A. Cervantes. 1971. Sequential active alteration of rhyolitic volcanic rock to endellite and a precursor phase of it at a spring in Michoacan, Mexico. Clays Clay Miner. 19:121–127.

Kempster, C. J. E., H. D. Megaw, and E. W. Radoslovich. 1962. The structure of anorthite, $CaAl_2Si_2O_8$. I. Structure analysis. Acta Crystallogr. 15:1005–1017.

Kiely, P. V., and M. L. Jackson. 1964. Selective dissolution of micas from potassium feldspars by sodium pyrosulphate fusion of soils and sediments. Am. Mineral. 49:1648–1659.

Kiely, P. V., and M. L. Jackson. 1965. Quartz, feldspar and mica determination from soils by sodium pyrosulphate fusion. Soil Sci. Soc. Am. Proc. 29:159–163.

Klein, C., Jr. 1964. Cummingtonite–grunerite series: A chemical, optical and X-ray study. Am. Mineral. 49:963–982.

Klug, H. P., and L. E. Alexander. 1954. X-ray diffraction procedures for polycrystalline and amorphous materials. John Wiley & Sons, Inc., New York.

Krumbein, W. C., and F. J. Pettijohn. 1938. Manual of sedimentary petrography. Appleton-Century, New York.

Kuno, H. 1954. Study of orthopyroxenes from volcanic rocks. Am. Mineral. 39:30–46.

Larsen, E. S., and H. Berman. 1934. Microscopic determination of non-opaque minerals. Bull. U. S. Geol. Surv. no. 848, 2nd ed.

Laves, F. 1954. The coexistence of two plagioclases in the oligoclase compositional range. J. Geol. 62:409-411.

Laves, F. 1960. Al/Si-Verteilungen, Phasen-Transformationen und Namen der Alkalifeldspäte. Z. Kristallogr., Kristallgeom., Kristallphys., Kristallchem. 113 (Laue Festschrift):265-296.

Laves, F., and J. R. Goldsmith. 1961. Polymorphism, order, disorder, diffusion and confusion in the feldspars. Cursillos Conf. Inst. "Lucas Mallada." 8:71-80.

Long, J. V. P. 1967. Electron probe microanalysis. p. 215-260. In J. Zussman (ed.) Physical methods in determinative mineralogy. Academic Press, New York.

Loughnan, F. C. 1962. Some considerations in the weathering of the silicate minerals. J. Sediment. Petrol. 32:289-290.

Loughnan, F. C. 1969. Chemical weathering of the silicate minerals. American Elsevier Publishing Co., Inc., New York.

Lyon, R. J. P. 1963. Evaluation of infrared spectrophotometry for compositional analysis of lunar and planetary soils. Standford Research Institute, Final Report under contract NASr-49(04). Published by NASA as technical note D-1871.

Lyon, R. J. P. 1967. Infrared absorption spectroscopy. p. 371-403. In J. Zussman (ed.) Physical methods in determinative mineralogy. Academic Press, New York.

Mackenzie, W. S. 1952. The effect of temperature on the symmetry of high temperature soda-rich feldspars. Am. J. Sci. Bowen Volume:319-343.

Mackenzie, W. S. 1954. The orthoclase-microcline inversion. Mineral. Mag. 30:354-366.

Mackenzie, W. S., and J. V. Smith. 1962. Single crystal X-ray studies of crypto-microperthite. Norsk. Geol. Tidsskr. 42(Feldspar Volume):72-104.

Marfunin, A. S. 1962. The feldspars: Phase relations, optical properties, and geological distribution. Akademii Nauk SSSR, Moskva. Translated from Russian. 1966. Israel Program for Scientific Translations, Jerusalem.

Marshall, C. E. 1964. The physical chemistry and mineralogy of soils. Vol. 1: Soil materials. John Wiley & Sons, Inc., New York.

Marshall, C. E., and C. D. Jeffries. 1945. Mineralogical methods in soil research. Soil Sci. Soc. Am. Proc. 10:397-405.

Mason, B. 1966. Principles of geochemistry. 3rd ed. John Wiley & Sons, Inc., New York.

Maxwell, J. A. 1963. The laser as a tool in mineral identification. Can. Mineral. 7:727-737.

Megaw, H. D. 1959. Order and disorder in the feldspars. Mineral. Mag. 32:226-241.

Megaw, H. D., C. J. E. Kempster, and E. W. Radoslovich. 1962. The structure of anorthite, $CaAl_2Si_2O_8$. II. Description and discussion. Acta Crystallogr. 15:1017-1035.

Michot, P. 1961. Struktur der Mesoperthite. Neues Jahrb. Mineral., Abh. 96:213-216.

Milner, H. B. 1962. Sedimentary petrography. Vol. I and II, 4th revised ed. The MacMillan Company, New York.

Mohr, E. C., and F. A. van Baren. 1954. Tropical soils. A critical study of soil genesis as related to climate, rock and vegetation. Interscience, New York.

Muir, I. D. 1967. Microscopy: Transmitted light. p. 31-102. In J. Zussman (ed.) Physical methods in determinative mineralogy. Academic Press, New York.

Muller, L. D. 1967. Laboratory methods of mineral separation. p. 1-30. In J. Zussman (ed.) Physical methods in determinative mineralogy. Academic Press, New York.

Nash, V. E., and C. E. Marshall. 1956a. The surface reactions of silicate minerals. Part I. The reactions of feldspar surfaces with acidic solutions. Univ. Mo. Agric. Exp. Stn. Res. Bull. 613.

Nash, V. E., and C. E. Marshall. 1956b. The surface reactions of silicate minerals. Part II. The reactions of feldspar surfaces with salt solutions. Univ. Mo. Agric. Exp. Stn. Res. Bull. 614.

Nash, V. E., and C. E. Marshall. 1957. Cationic reactions of feldspar surfaces. Soil Sci. Soc. Am. Proc. 21:149-153.

Nicholls, G. D. 1967. Emission spectrography. p. 445-458. In J. Zussman (ed.) Physical methods in determinative mineralogy. Academic Press, New York.

Parham, W. E. 1969. Halloysite-rich tropical weathering product of Hong Kong. Proc. Int. Clay Conf. (Tokyo, Japan) 1:403-416.

Pawluk, S. 1961. Mineralogical composition of some Grey Wooded soils developed from glacial till. Can. J. Soil Sci. 41:228-240.

Pettijohn, F. J. 1941. Persistence of heavy minerals and geologic age. J. Geol. 49:610–625.

Phillippe, M. M., and J. L. White. 1950. Quantitative estimation of minerals in the fine sand and silt fractions of soils with the Geiger counter X-ray spectrometer. Soil Sci. Soc. Am. Proc. 15:138–142.

Phillippe, M. M., and J. L. White. 1951. Acid soluble potassium and microcline content of the silt fractions of 12 Indiana soils. Soil Sci. Soc. Am. Proc. 16:371.

Pollack, S. S., and W. D. Ruble. 1964. X-ray identification of ordered and disordered ortho-enstatite. Am. Mineral. 49:983–992.

Rankama, K., and Th. G. Sahama. 1950. Geochemistry. 5th Printing. The University of Chicago Press, Chicago, Illinois.

Reeder, S. W., and A. L. McAllister. 1957. A staining method for the quantitative determination of feldspars in rocks and sands from soils. Can. J. Soil Sci. 37:57–59.

Reesman, A. L., and W. D. Keller. 1965. Calculation of apparent standard free energies of formation of six-rock forming silicate minerals from solubility data. Am. Mineral. 50:1729–1739.

Reitemeier, R. F. 1951. The chemistry of soil potassium. Adv. Agron. 3:113–164.

Ribbe, P. H. 1960. An X-ray and optical investigation of the peristerite plagioclases. Am. Mineral. 45:626–644.

Ribbe, P. H. 1962. Observations on the nature of unmixing in peristerite plagioclases. Norsk. Geol. Tidsskr. 42(Feldspar Volume):138–152.

Ribbe, P. H., and A. A. Colville. 1968. Orientation of the boundaries of out-of-step domains in anorthite. Mineral. Mag. 36:814–819.

Ribbe, P. H., and H. C. van Cott. 1962. Unmixing in peristerite plagioclases observed by phase contrast and dark-field microscopy. Can. Mineral. 7:278–290.

Rich, C. I. 1968. Mineralogy of soil potassium. p. 79–108. In V. J. Kilmer, S. E. Younts, and N. C. Brady (ed.) The role of potassium in agriculture. Soil Sci. Soc. Am., Madison, Wis.

Robie, R. A., and D. R. Waldbaum. 1968. Thermodynamic properties of minerals and related substances at 298.15°K (25°C) and one atmosphere (1.013 bars) pressure and at higher temperatures. U. S. Geol. Surv. Bull. 1259. 255 p.

Sahama, Th. G., and K. Hytönen. 1958. Calcium-bearing magnesium-iron olivines. Am. Mineral. 43:862–871.

Sakata, Y. 1957. Unit cell dimensions of synthetic aluminian diopsides. Jap. J. Geol. Geogr. 28:161.

Sand, L. B. 1956. On the genesis of residual kaolin. Am. Mineral. 41:28–40.

Schatz, A., V. Schatz, and J. J. Martin. 1957. Chelation as a biochemical weathering agent. Bull. Geol. Soc. Am. 68:1792–1793.

Scheidegger, K. F. 1973. Determination of structural state of calcic plagioclases by an X-ray powder technique. Am. Mineral. 58:134–136.

Schneider, T. R., and F. Laves. 1957. Barbierit oder Monalbit? Z. Kristallogr., Kristallgeom., Kristallphys., Kristallchem. 109:241–244.

Sen, S. K. 1959. Potassium content of natural plagioclases and the origin of antiperthites. J. Geol. 67:479–495.

Sherman, G. D., H. Ikawa, G. Uehara, and E. Okasaki. 1962. Types of occurrence of nontronite and nontronite-like minerals in soils. Pac. Sci. 16:57–62.

Slemmons, D. B. 1962. Observation on order-disorder relations of natural plagioclases. 1. A method of evaluating order-disorder Norsk. Geol. Tidsskr. 42 (Feldspar Volume):533–554.

Smith, J. L. 1865. On the use of the bisulphate of soda as a substitute for bisulphate of potash in the decomposition of minerals, specially the aluminous minerals. Am. J. Sci. Arts 40:248–249.

Smith, J. R. 1958. The optical properties of heated plagioclases. Am. Mineral. 43:1179–1194.

Smith, J. R., and M. W. Pyke. 1959. Microcline from a Precambrian granodiorite. R. Soc. Can. Proc. 3rd Ser., V53, Sec. 4, App. C., p. 22.

Smith, J. V. 1956. The powder patterns and lattice parameters of plagioclase feldspars. 1. The soda-rich plagioclases. Mineral. Mag. 31:47–68.

Smith, J. V. 1974. Feldspar minerals: Vol. I. Crystal structure and physical properties, Vol. II. Chemical and textural properties. Springer-Verlag, New York.

Smith, J. V., and P. Gay. 1958. The powder patterns and lattice parameters of plagioclase feldspars. II. Mineral. Mag. 31:744–762.

Smith, S. J., L. J. Clark, and A. D. Scott. 1968. Exchangeability of potassium in soils. Int. Congr. Soil Sci., Trans. 9th (Adelaide, Australia) II:661–669.

Smithson, S. B. 1962. Symmetry relations in alkali feldspars of some amphibolite facies rocks from the southern Norwegian Precambrian. Norsk. Geol. Tidsskr. 42 (Feldspar Volume):586–600.

Somasiri, S., and P. M. Huang. 1971. The nature of K-feldspars of a Chernozemic soil in the Canadian Prairies. Soil Sci. Soc. Am. Proc. 35:810–815.

Somasiri, S., and P. M. Huang. 1973. The nature of K-feldspars of selected soils in the Canadian Prairies. Soil Sci. Soc. Am. Proc. 37:461–464.

Somasiri, S., S. Y. Lee, and P. M. Huang. 1971. Influence of certain pedogenic factors on potassium reserves of selected Canadian Prairie soils. Soil Sci. Soc. Am. Proc. 35:500–505.

St. Arnaud, R. J., and M. M. Mortland. 1963. Characteristics of the clay fractions in a Chernozemic to Podzolic sequence of soil profiles in Saskatchewan. Can. J. Soil Sci. 43:336–349.

Steiger, R. H., and S. R. Hart. 1967. The microcline-orthoclase transition within a contact aureole. Am. Mineral. 52:87–116.

Stephen, I. 1952. A study of rock weathering with reference to the soils of the Malvern Hills. J. Soil Sci. 3:20–33.

Stephen, I. 1953. A petrographic study of a tropical black earth and grey earth from the Gold Coast. J. Soil Sci. 4:211–219.

Tamm, O. 1930. Experimentelle Studien uber die Vermitterung und tonbildung von Feldspäten. Chem. Erde 4:420–430.

Taylor, W. H. 1933. The structure of sanidine and other feldspars. Z. Kristallogr., Mineral., Petrograph. 85:425–443.

Taylor, W. H. 1965. Framework silicates: The feldspars. p. 293–339. In L. Bragg and G. F. Claringbull. The crystalline state-Vol. IV. Crystal structures of minerals. G. Bell & Sons Ltd., London.

Tobi, A. C. 1961. Pattern of plagioclase twinning as a significant rock property. Proc. K. Ned. Akad. Wet., Ser. B, 64:576–581.

Tunn, W. 1940. Untersuchungen über die verwitterung des tremalit. Chem. Erde 12:275–303.

Tuttle, O. F. 1952. Origin of the contrasting mineralogy of extrusive and plutonic salic rocks. J. Geol. 60:107–124.

Tuttle, O. F., and N. L. Bowen. 1950. High temperature albite and contiguous feldspars. J. Geol. 58:572–583.

van der Plas, L. 1959. Petrology of the northern Adula Region, Switzerland. With an appendix on Wilcoxon's two sample test by A. R. Bloemena. Leidse Geol. Meded. 24:418–602.

van der Plas, L. 1962. Preliminary note on the granulometric analysis of sedimentary rocks. Sedimentology 1:145–157.

van der Plas, L. 1966. The identification of detrital feldspars. Elsevier Publishing Co., New York.

van der Plas, L., and A. C. Tobi. 1965. A graph for judging the reliability of point counting results. Am. J. Sci. 263:87–90.

Viswanathan, K. 1966. Unit cell dimensions and ionic substitutions in common clinopyroxenes. Am. Mineral. 51:429–442.

Wainwright, J. E., and J. Starkey. 1971. A refinement of the structure of anorthite. Z. Kristallogr., Kristallgeom., Kristallphys., Kristallchem. 133:75–84.

Waldbaum, D. R. 1968. High-temperature thermodynamic properties of alkali feldspars. Contrib. Mineral. Petrology 17:71–77.

Whittaker, E. J. W. 1960. The crystal chemistry of amphiboles. Acta Crystallogr. 13:291–298.

Whittig, L. D. 1965. X-ray diffraction techniques for mineral identification and mineralogical composition. In C. A. Black (ed.) Methods of soil analysis, Part 1. Agronomy 9:671–698. Am. Soc. Agron., Madison, Wis.

Winchell, H., and R. Tilling. 1960. Regressions of physical properties on the compositions of clinopyroxenes. Am. J. Sci. 258:529–547.

Wittles, M. 1951. Structural transformations in amphiboles at elevated temperatures. Am. Mineral. 36:851–858.

Wittles, M. 1952. Structural disintegration of some amphiboles. Am. Mineral. 37:28–36.

Wright, J. R., and M. Schnitzer. 1963. Metallo-organic interactions associated with podzolisation. Soil Sci. Soc. Am. Proc. 27:171–176.

Yoder, H. S., and Th. G. Sahama. 1957. Olivine X-ray determinative curve. Am. Mineral. 42:475–491.

Young, R. S. 1958. The geochemistry of cobalt. Geochim. Cosmochim. Acta 13:28–42.

Allophane and Imogolite

KOJI WADA, Kyushu University, Fukuoka, Japan

I. INTRODUCTION

In crystals, atoms are packed together in a regular manner forming a three-dimensional pattern (*long-range order*), whereas in *amorphous* (without form) materials, they are not arranged in a regular manner. There is, however, no sharp dividing line between crystalline and amorphous materials. There are cases where arrangement of atoms is regular in two dimensions or even in only one dimension. For example, each unit layer of halloysite or montmorillonite is a two-dimensional crystallite, but usually they do not form structures with a three-dimensional regularity. Intervention of water molecules between the unit layers makes their arbitrary rotation or shifts inevitable. A similar situation occurs in assemblies of chain units, where one often can state only statistical mean translations which give the probable relative position of adjacent chain units. This type of order is called *paracrystalline* (Vainshtein, 1966). On the other hand, arrangement of atoms is not entirely random even in amorphous materials. Atoms behave as spheres with definite radii and have their own coordination tendency. This results in some order in their arrangement as illustrated by the presence of discrete SiO_4 tetrahedra in vitreous silica. Nevertheless, the order in these materials is local and non-repetitive (*short-range*), as compared with that found in crystalline materials. The word *noncrystalline* rather than amorphous is, therefore, recommended (Bunn, 1961).

Noncrystalline clay materials which are important in soils are oxides and hydroxides of Al, Fe, and Si, and silicates of Al and Fe, all in various combination with water. Brown (1955) proposed a scheme of classification and naming of such clay materials (Table 16-1). The AIPEA Nomenclature Committee has recommended that specific names not be given to poorly defined clay materials or amorphous constituents (Brindley & Pedro, 1972).

The name *allophane* has been used in various ways. Ross and Kerr (1943) used it as a group name of X-ray amorphous clay materials which are often associated with halloysite and consist of essentially a solid solution of silica, alumina, and water. Beutelspacher and van der Marel (1961; quoted by Grim, 1968) concluded that allophane occurs as a coating on crystalline constituents and is very widespread in many soils. Fieldes (1966) proposed the use of the name allophane as any clay size material characterized by structural randomness. Furkert and Fieldes (1968) found allophane in soils derived from either glacial rock flour or basalt and in Spodosols as well as in soils derived from volcanic ash.

chapter 16

Table 16-1. Nomenclature and classification of noncrystalline clay materials
(Brown, 1955)

Group	End member name	Formula
Oxides	Opaline silica	$SiO_2 \cdot nH_2O$
	Limonite	$Fe_2O_3 \cdot nH_2O$
	Kliachite	$Al_2O_3 \cdot nH_2O$
	Wad	$MnO_2 \cdot nH_2O$
Silicates	Allophane	$Al_2O_3 \cdot 2SiO_2 \cdot nH_2O$
	Hisingerite	$Fe_2O_3 \cdot 2SiO_2 \cdot nH_2O$
Phosphates	Evansite	$Al_3PO_4(OH)_6 \cdot nH_2O$
	Azovskite	$Fe_3PO_4(OH)_6 \cdot nH_2O$

Allophane will be used in this chapter as a series name of naturally occurring hydrous aluminosilicate clays. They are characterized by short-range
order and by the predominance of Si-O-Al bonds. Their chemical composition varies, but within the limits necessary to maintain the predominance of
Si-O-Al bonds. This definition is essentially the same as that adopted at the
USA-Japan seminar on amorphous clay materials (van Olphen, 1971).

Brown (1955) used the name *hisingerite* as an Fe analogue of allophane
(Table 16-1). The name hisingerite was also used for an "amorphous" vari-
ant of nontronite (Gruner, 1935), a poorly crystallized iron saponite (Whelan
& Goldich, 1961) and a mica mineral with extensive substitution of Fe for Si
and with interlayer hydronium ions (Lindqvist & Jannson, 1962). Its presence as an Fe analogue of allophane in soils has been suggested, but no solid
information has been available.

The name *imogolite* was not included in Table 16-1. It was first described by Yoshinaga and Aomine (1962b) in a soil derived from glassy volcanic ash known as "imogo." The name imogolite as a new mineral species
has been approved by the AIPEA Nomenclature Committee (Brindley &
Pedro, 1970). Chemically, imogolite is intermediate between allophane and
kliachite in Brown's scheme (Table 16-1), and has an approximate composition $SiO_2 \cdot Al_2O_3 \cdot 2.5H_2O$. It consists of paracrystalline assemblies of a one-
dimensional structure unit. The genesis and properties of imogolite are closely associated with those of allophane rather than crystalline clay minerals.

This chapter deals primarily with allophane and imogolite in soils derived from volcanic ash and in weathered pumices. Intermediates between
them, opaline silica, and noncrystalline hydrous Al and Fe oxides are also included. Very little will be mentioned of noncrystalline clay constituents in
the soils derived from other parent materials. This does not ignore the importance of these other constituents. However, not a great deal is known of
them and their characterization has been hampered by relatively small contents, and possibly by the very nature of these constituents. Emphasis will
be placed on the essential features of allophane and imogolite, and the principles relating to their behavior in soils.

II. STRUCTURAL PROPERTIES AND MINERAL IDENTIFICATION

A. X-ray Diffraction

Because of their short-range order noncrystalline materials do not scatter X-rays directionally so as to produce definite diffraction maxima. Intensity information as a function of 2θ (θ = Bragg angle of reflection), however, permits determination of the magnitude of the interatomic distances as direct consequences of short-range order. The results can be expressed as a radial distribution function, which specifies the density of atoms or electrons as a function of the radial distance from any reference atom in the system (Klug & Alexander, 1954).

Allophane gives an X-ray diffraction pattern characteristic of non-crystalline materials (Fig. 16-1). No enhancement of diffraction intensity due to orientation of the specimen is observed. Okada et al. (1975) obtained differential radial distribution curves of allophane samples which showed maxima at 1.80, 3.00, 4.20, 5.20Å, and so on. By comparing these curves with that of silica gel and by taking the result of X-ray fluorescence spectroscopy for Al coordination, they estimated the Si–O, Al(fourfold coordination)–O, and Al(sixfold coordination)–O distances in allophane to be 1.62, 1.70, and 1.95Å, respectively. Okada et al. (1975) further calculated the differential radial distribution curves of model aluminosilicates consisting of

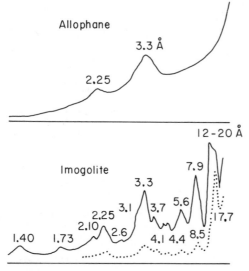

Fig. 16-1. X-ray diffraction patterns of allophane [$<$ 0.2 μm; Andept B horizon (no. 905); Uemura, Kumamoto; H_2O_2, dithionite–citrate, and 2% Na_2CO_3 treated (Yoshinaga & Aomine, 1962a)] and imogolite ["gel films" in weathered pumice; Kitakami, Iwate; H_2O_2, dithionite–citrate, and 2% Na_2CO_3 treated (Wada & Yoshinaga, 1969)]. Dotted line; heated at 300C (573K) and at half the intensity scale of the other curve. Reprinted from Wada and Yoshinaga (1969) Am. Mineral. 54:50–71. Figure 1. Copyright © by the Mineralogical Society of America.

one-, two-, and three-dimensional networks of Si-tetrahedra and Al-octahedra. By referring to these curves, they concluded that the linkages of the tetrahedra and octahedra in allophane most likely produce two-dimensional order.

Both crystalline and noncrystalline materials consisting of extremely small particles with empty spaces between them will show another scattering effect. This arises from electron density variations from particle to particle and is called small-angle scattering. This scattering will be continuous and may or may not show maxima, but it is possible to calculate the particle diameter from the shape of the scattering curve (Klug & Alexander, 1954). Watanabe (1968) applied this procedure to allophane separated from weathered pumice and estimated that allophane consists of particles about 100Å in diameter assuming they are spherical.

The X-ray diffraction pattern of imogolite consists of a number of broad reflections (Yoshinaga & Aomine, 1962b) (Fig. 16-1). Those with maxima at 12-20, 7.8-8.0, and 5.5-5.6Å are enhanced by "parallel" orientation of the specimen, but their spacings do not form an integral series (Wada & Yoshinaga, 1969). A marked intensity change occurs in the broad 12-20Å reflection upon heating at 100-300C (373-573K) resulting in appearance of an 18-19Å peak. This reaction was prevented by the presence of dithionite-citrate extractable constituents (Wada & Tokashiki, 1972), alkylammonium chlorides (Wada & Henmi, 1972), or humified material (Inoue & Wada, 1971), and was attributed to rearrangement of the structure units upon dehydration and partial dehydroxylation. Destruction of the imogolite structure occurs at 350-400C (623-673K) where dehydroxylation takes place (Fig. 16-4). The X-ray pattern of imogolite and its change upon heating show no obvious relations to those of known clay minerals. This serves as a good criterion for distinguishing it from other clay minerals.

B. Electron Microscopy

Important information about the morphology and internal structures of allophane and imogolite is obtained from the use of a high-performance electron microscope. For high resolution electron microscopy, special precautions are taken to minimize the specimen contamination and damage due to irradiation of the electron beam (Jones & Uehara, 1973; Henmi & Wada, 1976). The use of a microplastic grid with many perforations of micrometer sizes (Fukami & Adachi, 1965) is very useful to observe the specimen without interference by a carbon-coated collodion film.

Imogolite appears in the electron microscope as smooth and curved threads varying in diameter from 100 to 300Å and extending several μm in length (Yoshinaga & Aomine, 1962b) (Fig. 16-2a). Figure 16-2b shows that the threads consist of two or more finer filiform units, which are bent, but run parallel with separations of 18-22Å (Yoshinaga et al., 1968). These units never occur singly, but always in pairs. A number of "rings" with inner and outside diameters of about 10 and 20Å appeared on the electron micrograph when imogolite threads were embedded in methacrylate resin and sectioned normal to their axis (Wada et al., 1970). The features indicate that imogolite consists of a tube unit with inner and outside diameters of 10 and 20Å. This tube unit is very useful to identify and detect imogolite even in a trace amount.

Figure 16-2a shows that allophane is composed of fine, rounded particles which form irregular aggregates. Figure 16-2b shows the finer structure of the allophane particles; there are many ring-shaped particles, which are

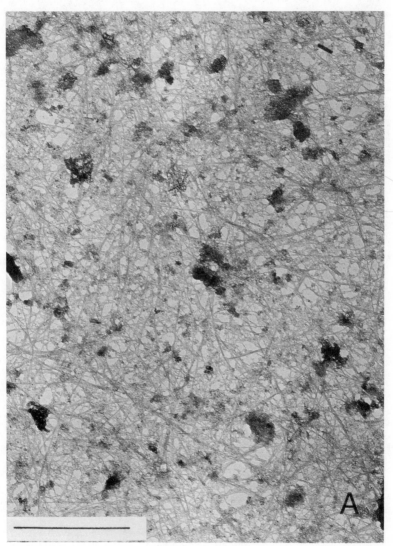

Fig. 16-2. Electron micrographs of allophane and imogolite [<0.2 μm; Andept B horizon (no. 905); Uemura, Kumamoto; H$_2$O$_2$ treated] Scalemaker *(A)* 1 μm and *(B)* 500Å. Reprinted from Henmi and Wada (1976) Am. Mineral. 61:379–390. Figure 1b and 2b. Copyright © by the Mineralogical Society of America (continued on next page).

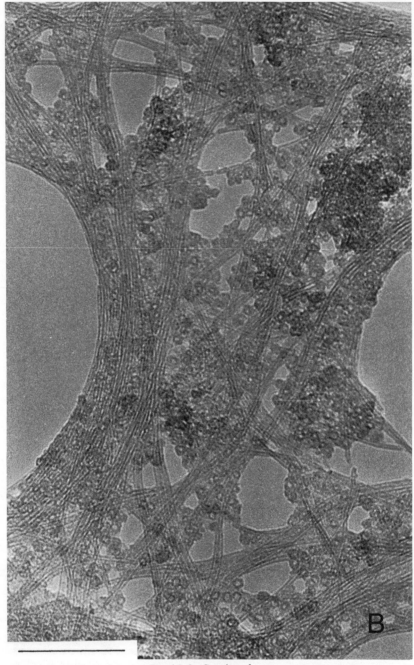

Fig. 16-2. Continued.

often deformed and aggregated, and in three dimensions may be hollow spherules or polyhedrons with diameters of 35-50Å. Similar objects were found in clays separated from weathered pumices and volcanic ashes of different origins and localities (Wada & Yoshinaga, 1969; Kitagawa, 1971; Wada et al., 1972; Henmi & Wada, 1976). Henmi and Wada (1976) noted a tendency for the allophane "spherules" to increase and the imogolite tubes to decrease with increasing SiO_2/Al_2O_3 ratio of the clays (< 0.2 μm) from 0.83 to 1.97, and no particular variation to occur in morphology and size of the allophane "spherules." Allophane was described as material having little or no structural organization and appearing as particles without any definite and regular shape (e.g., Grim, 1968). The high resolution electron micrographs suggest, however, that allophanes must have definite and common structural arrangements and therefore can not be considered amorphous. Variations and indefiniteness of both shape and size of allophane likely arise from aggregation of the allophane "sperules" with themselves and other soil constituents.

C. Electron Diffraction

Allophane gives two broad ring reflections with intensity maxima at 3.3 and 2.25Å on the electron diffraction pattern (Wada & Yoshinaga, 1969). Electron diffraction from imogolite threads without preferred orientation gives a series of ring reflections at 1.4, 2.1, 2.3 (broad), 3.3 (broad), 3.7, 4.1, 5.7 (broad), 7.8 (broad), 11.8 (broad), and 21-23Å (Russell et al., 1969; Wada & Yoshinaga, 1969). A parallel alignment of the threads results in arching and resolving of these ring reflections and makes it possible to assign these reflections kl indices (Fig. 16-3). The 2.1 and 1.4Å reflections indexed as 04 and 06 become sharply defined arcs on a line parallel to the thread

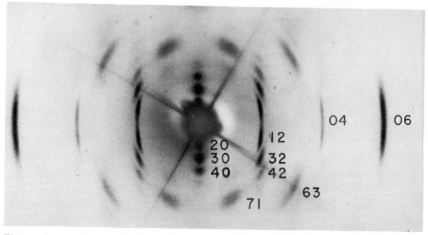

Fig. 16-3. Electron diffraction pattern of imogolite ("gel films" in weathered pumice; Kitakami, Iwate; H_2O_2, dithionite-citrate and 2% Na_2CO_3 treated). Preferred orientation. Figures: kl indices. (modified from Cradwick et al., 1972).

direction. The 11.8, 7.8, and 5.7Å reflections indexed as 20, 30, and 40 are reduced to dots on an axis normal to the thread direction. The reflections 4.1, 3.7, 3.3, 3.1, and 2.3Å indexed as 12, 32, 42, 71, and 63 are resolved into pairs of arcs at various angles to the thread direction. The first group of reflections was interpreted as indicating a repeat unit of 8.4Å along the tube unit axis, and the second group of reflections was related to the lateral arrangement of the tube unit with the interaxial separations of 21 to 23Å (Cradwick et al., 1972). The kl indices in Fig. 16-3 are given on the basis of a unit cell with b = 23Å and c = 8.4Å.

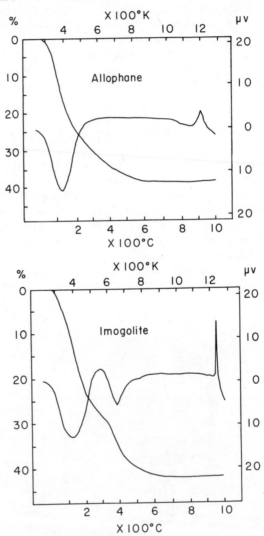

Figure 16-4. DTA and TG curves of *(A)* allophane [$<$ 0.2 μm; weathered volcanic ash (VA); Choyo, Kumamoto; SiO_2/Al_2O_3 ratio = 1.77; H_2O_2 treated] and *(B)* imogolite ($<$ 0.2 μm; "gel films" in weathered pumice; Kurayoshi, Tottori; SiO_2/Al_2O_3 ratio = 1.02; H_2O_2 treated). Heating rate; 20C (293K) per min. Atmosphere; static air. Sample weight; allophane, 15.9 mg; imogolite, 16.6 mg. Analyst: T. Henmi.

D. Thermal Analyses

Figure 16-4 shows differential thermal (DTA) curves and thermogravimetric (TG) curves of allophane and imogolite of established purity. A large endothermic peak between 50–300C (323–573K) on the DTA curves arises from removal of large amounts of adsorbed water. Allophane shows a continuous weight loss due to dehydration and dehydroxylation upon heating, so that there is no endothermic peak at intermediate temperatures on the DTA curve. Imogolite gives a small, but obvious endothermic peak at 390–420C (663–693K) due to dehydroxylation (Yoshinaga & Aomine, 1962b; Wada et al., 1972).

The appearance of an exothermic peak at 900–1,000C (1,173–1,273K) (Fig. 16-4) is attributed to formation or nucleation of mullite and/or gamma alumina in dehydroxylated phases of allophane and imogolite (Tsuzuki & Nagasawa, 1960; Yoshinaga & Aomine, 1962a, 1962b). There are indications that both the appearance of the exothermic peak and the amount of mullite formed at 1,000C (1,273K) from noncrystalline clays and synthetic silico-aluminas correlate with their SiO_2/Al_2O_3 ratios (Wada & Harward, 1974). Thus, the appearance of the high temperature exothermic peak may be taken as an indication of Si-O-Al bond formation in noncrystalline aluminosilicates. On this basis, allophane would be differentiated from noncrystalline oxides and hydroxides of Si, Al, and Fe, and from their mixtures. In order to eliminate undesirable effects of exchangeable "bases" on the reaction, treatment with slightly acid salt solution (e.g., at pH 5) is recommended in preparation of the specimen.

E. Infrared Spectroscopy

Noncrystalline materials give more featureless spectra than do crystalline materials, but absorb as strongly, and spectra of noncrystalline materials reflect atomic composition and atomic bonding as well. Major absorption bands of allophane and imogolite appear in three regions; 2800–3800, 1400–1800, and 650–1200 cm^{-1} (Fig. 16-5). The absorption bands in the first region are due to the OH stretching vibrations of either structural OH groups or adsorbed water. In the second region, an absorption band due to the HOH deformation vibration of adsorbed water appears at 1630–1640 cm^{-1}. There may also appear absorption bands due to the vibrations related with COO (1700–1740 cm^{-1}) or COO^- (1580 and 1425 cm^{-1}) groups. These groups are attributed to (i) humic materials, (ii) their partial decomposition products such as oxalate upon H_2O_2 treatment, or (iii) organic anions such as citrate used for pretreatment, all of which are strongly retained by allophane and imogolite. The absorption bands in the third region are attributed mainly to Si(Al)O stretching vibrations and partly to SiOH and AlOH deformation vibrations.

The principal differences in the spectra of clays in which allophane and imogolite predominate appear in the absorption bands in the 650–1200 cm^{-1} region. Table 16-2 shows that the absorbance at different frequencies in the

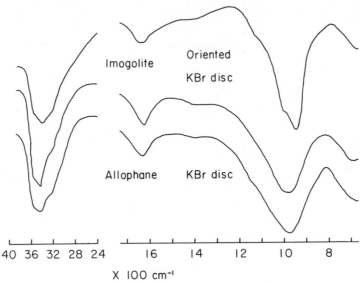

Fig. 16-5. Infrared absorption spectra of imogolite and allophane ("gel films" and nodules in saprolite of basalt, respectively; Maui, Hawaii) (Wada et al., 1972).

Table 16-2. Regression of absorbance at different frequencies in the 650–1200 cm^{-1} region on the contents of SiO_2 and Al_2O_3 in dithionite–citrate, 2% Na_2CO_3, and 0.5N NaOH soluble fractions of clays separated from 10 Andept samples (Tokashiki, 1974)†

| Frequency (cm^{-1}) | Regression equation: Absorbance = a × SiO_2(mg) + b × Al_2O_3(mg) + c | | | Correlation coefficient |
	a	b	c	R
650	0.003	0.034**	0.015	0.660**
800	−0.009*	0.027**	−0.018	0.577**
950	0.056**	0.015	0.110	0.656**
1000	0.090**	0.002	0.052	0.795**
1100	0.069**	−0.009	−0.001	0.886**

* and ** Significant at p = 0.05 and 0.01, respectively.
† The ranges in SiO_2/Al_2O_3 ratio of the dithionite–citrate, 2% Na_2CO_3 and 0.5N NaOH soluble fractions were 0.20–0.81, 0.43–1.40, and 0.85–2.49, respectively.

region reflects the differences in the contents of SiO_2 and Al_2O_3 in allophane, imogolite, and allophanelike constituents. The spectra and composition relationship was also seen in a linear increase in the frequency of the major Si(Al)O absorption maximum in the range of 930–950 cm^{-1} to 1040 cm^{-1} with the increasing SiO_2/Al_2O_3 ratio in the range of 0.2 to 2.6 (Tokashiki, 1974).[1] Similar relationships were found for allophanic soil clays by Kanno et al. (1968), Lai & Swindale (1969), and Henmi and Wada (1976).

---

[1]Y. Tokashiki. 1974. Mineralogical analysis of volcanic-ash soil clays by selective dissolution method. Ph.D. Thesis (Japanese). Kyushu Univ.

The absorption band of imogolite in the 900–1100 cm^{-1} region differs from those of both allophane and layer silicates and shows unique orientation effects. The maximum at 990–1010 cm^{-1} exhibited enhanced intensity when the specimen was "oriented" on a TlBr–TlI plate compared with when pressed with KBr in a disc (Fig. 16-5). Russell et al. (1969) and Cradwick et al. (1972) interpreted an absorption at 930 cm^{-1} as resulting from the presence of isolated orthosilicate groups in imogolite. An enhancement of the 1010 cm^{-1} absorption by random orientation was interpreted in terms of high frequency shift of the 930 cm^{-1} absorption as a consequence of the morphology and dimensions of the imogolite threads.

OH–OD exchange with D_2O at room temperature for allophane and imogolite can be followed by observing the shift of the OH absorption bands to the frequencies of the corresponding OD absorption bands (Wada, 1966; Russell et al., 1969) (Fig. 16-6). The ease of complete OH–OD exchange suggests that all the OH groups and associated water in allophane and imogolite are accessible to the ambient solution. Only partial OH–OD exchange was observed for the OH groups in halloysite and montmorillonite.

The presence of physically adsorbed water in allophane and imogolite in considerable amounts and its almost complete removal by evacuation at room temperature are indicated by the presence and the virtual absence of the 1635 cm^{-1} HOH deformation band on the spectra in Fig. 16-5 and 16-6, respectively. A small absorption band at around 1630 cm^{-1} remains on spectra (a) and (c), which may suggest the presence of difficultly removable water. However, this absorption band appeared equally on spectra (b) and (d) for allophane and imogolite which were almost completely deuterated.

F. Chemical Analyses

A summary of elemental analyses of < 0.2 μm fractions separated from weathered volcanic ash and pumices and pretreated with dithionite–citrate and 2% Na_2CO_3 solutions (Wada & Yoshinaga, 1969) showed that the SiO_2/Al_2O_3 ratio of the clays in which allophane predominated was in a range from 1.3 to 2.0, while that of the clays in which imogolite predominated was in a narrow range from 1.05 to 1.15. The $H_2O(+)$/Al_2O_3 ratio was mostly in a range from 2.5 to 3.0. Yoshinaga (1966) reported that a small but significant amount of Fe (0.3–0.9% as Fe_2O_3) remained in allophane even after 10 dithionite–citrate treatments. Henmi and Wada (1976) found that the SiO_2/Al_2O_3 ratio was in a range from 0.83 to 1.97 for 10 clays which contained allophane as nearly a single constituent and had not been subjected to any dissolution treatment.

Götz and Masson (1971) developed a chemical procedure for differentiating silicate anions possessing low degrees of polymerization. The procedure is based on conversion of the anion to a trimethylsilyl-ether, and subsequent identification and quantitative determination of the volatile ether by gas chromatography. Application of this technique to imogolite gave a high yield of volatile products of which 95% was the orthosilicate ether and 5%

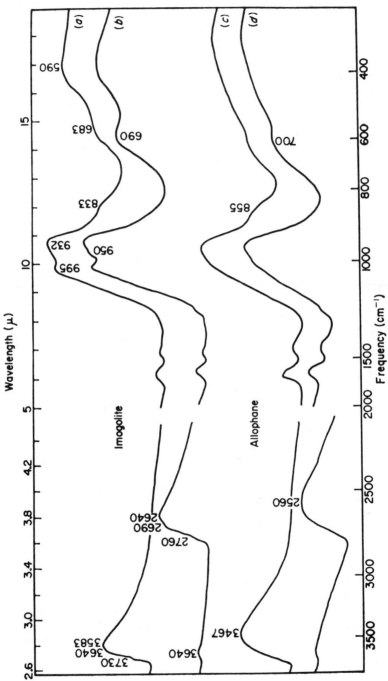

Fig. 16-6. Infrared absorption spectra of normal and deuterated forms of imogolite and allophane [<2 μm; Andept B horizon (no. 905); Uemura, Kumamoto; H_2O_2 and dithionite–citrate treated] following evacuation at 1.33 Pa (10^{-2} mm Hg). (a) OH imogolite, (b) OD imogolite, (c) OH allophane, and (d) OD allophane. (Russell et al., 1969).

the pyrosilicate ether (Cradwick et al., 1972). This furnished evidence in favor of the presence of isolated orthosilicate groups in imogolite.

G. Dissolution Analyses

Noncrystalline clay materials have large specific surface area and high chemical reactivity. These materials are more sensitive to chemical dissolution than crystalline clay minerals, and dissolve depending on their elemental composition. Thus, selective chemical dissolution can be used for determination of noncrystalline clay constituents. Table 16-3 gives a summary of reagents and their dissolution characteristics.

Table 16-3 shows that dissolution with one reagent has some limitation in the specificity of the reaction. Simultaneous characterization of the dissolved material is therefore recommended. For this purpose, an analysis of dissolution kinetics or difference infrared spectroscopy has been used. In the former procedure, the steady state portion of the dissolution—time curve

Table 16-3. Dissolution of Al, Fe, and Si in various clay constituents and organic complexes by treatment with different reagents

Element in: specified component and complex	Treatment with:				
	$0.1M$[†] $Na_4P_2O_7$	Dithionite-[‡] citrate	2%[§] Na_2CO_3	0.15-$0.2M$[¶] Oxalate- oxalic acid (pH 3.0-3.5)	$0.5N$[#] NaOH
Al in:					
Organic complexes	good	good	good	good	good
Hydrous oxides					
Noncrystalline	poor	good	good	good	good
Crystalline	no	poor	poor	no	good
Fe in:					
Organic complexes	good	good	no	good	no
Hydrous oxides					
Noncrystalline	poor	good	no	good	no
Crystalline	no	good	no	no	no
Si in:					
Opaline silica	no	no	poor	no	good
Crystalline silica	no	no	no	no	poor
Al and Si in:					
Allophanelike	poor	good	good	good	good
Allophane	poor	poor	poor	good	good
Imogolite	poor	poor	poor	good-fair	good
Layer silicates	no	no	no	no	poor-fair

[†] McKeague et al. (1971); Wada and Higashi (1976).
[‡] Mehra and Jackson (1960); Wada and Greenland (1970); Tokashiki and Wada (1975).
[§] Jackson (1956); Wada and Greenland (1970); Tokashiki and Wada (1975).
[¶] Schwertmann (1964); Higashi and Ikeda (1973); Wada and Wada (1976).
[#] Hashimoto and Jackson (1960); Wada and Greenland (1970); Tokashiki and Wada (1975).

was taken to indicate a limited attack on crystalline materials (Langston & Jenne, 1964; Follett et al., 1965; Segalen, 1968). In the latter procedure, the spectrum representing the infrared absorption of the material removed by any dissolution treatment was obtained by placing the specimen before and after the treatment at sample and reference side of a spectrophotometer, respectively. This technique demonstrated dissolution of allophanelike constituents, hydrous alumina, and organic material by dithionite–citrate treatment; allophanelike constituents by 2% Na_2CO_3 treatment; and allophane, opaline silica, imogolite, gibbsite, halloysite, and disordered layer silicates by 0.5N NaOH treatment. The allophanelike constituents dissolved by dithionite–citrate and 2% Na_2CO_3 treatments were characterized by the Si(Al)O stretching absorption maximum at 940–980 cm^{-1} and 950–960 cm^{-1}, respectively (Wada & Greenland, 1970; Parfitt, 1972; Tokashiki & Wada, 1972; Wada & Tokashiki, 1972).

H. X-ray Fluorescence Spectroscopy

X-ray fluorescence spectroscopy has been used to obtain data on the coordination number of Al in noncrystalline clay materials. The emission wave length of the K_α line of Al is dependent on its coordination. Egawa (1964), Udagawa et al. (1969), and Okada et al. (1975) obtained data which indicate that Al in allophane is in both four- and sixfold coordinations. Effects of impurities such as volcanic glass were, however, not evaluated. Similar analyses of allophane and imogolite of established purity by Henmi and Wada (1976) have confirmed that Al in fourfold coordination in allophane increases with increasing SiO_2/Al_2O_3 ratio and amounts to 50% of the total Al, and that Al in imogolite is only in sixfold coordination. The presence or absence of Al in fourfold coordination has an important bearing on the differences in the negative charge characteristics and surface acidity between allophane and imogolite.

I. Surface Area and Porosity Measurement

Aomine and Otsuka (1968) measured specific surface areas of allophanic clays (< 2 μm). The obtained values showed a marked variation depending on the nature of adsorbate; polar liquids such as glycerol, ethylene glycol monoethyl ether (EGME), and water gave 435–534 m^2/g, N_2 gas 145–170 m^2/g, and cetylpyridinium bromide and orthophenanthroline zero or very small values. Egashira and Aomine (1973) gave EGME surface areas of 700–900 m^2/g and 900–1,100 m^2/g for allophane and imogolite, respectively. The samples had been freeze dried and then vacuum dried over P_2O_5. They also found that oven drying and exposing the samples to moist air affected EGME retention by allophane and imogolite, but not by montmorillonite.

In all determinations described above, the cross-sectional area of the polar molecule adsorbed on either allophane or imogolite was assumed to be equal to that adsorbed on 2:1 expanding clay minerals. On the other hand, the specific surface area of imogolite was calculated to be 1,400–1,500 m^2/g (Wada & Harward, 1974), using the values of the inner and outside diameters of its tube unit, 7–10 and 17–21Å, and a density of 2.65 g/cm^3.

The shape and size of allophane and imogolite "units" suggest that the system containing them has a high porosity. Wada and Yoshinaga (1969) reported that the density of imogolite determined by displacement with water was 2.6–2.75, whereas that determined by a float-sink test in heavy liquids such as $C_2H_2Br_4$-CH_3OH mixture was 1.70–1.97. This was interpreted in terms of the presence of micropores which can accommodate the smaller water and CH_3OH molecules, but not the larger $C_2H_2Br_4$ molecule. Wada and Henmi (1972) correlated the maximum retention of quaternary ammonium chlorides and water by imogolite with its micropore space. They estimated that the porosity in air-dried imogolite amounts to 55–60% of the total volume, and roughly 50 and 25% of the total porosity belong to the inter- and intra-tube-unit pores, respectively.

J. Structure Models

Knowledge of the structure of clay minerals has great value in predicting and understanding their behavior in soils. Attempts have been made to propound structure models for allophane and imogolite, though many of them have had value only as working hypotheses, as reviewed by Yuan (1973) and by Wada and Harward (1974). The use of high-performance electron microscopy has enabled us to deduce the structure of these minerals on a more solid basis than before.

The most recent version of the imogolite structure was advanced by Cradwick et al. (1972). A part of the postulated cylindrical structure unit is schematically shown in Fig. 16-7. It has a circumference of 10, 11, or 12 gibbsite unit cells, with gibbsite b along the circumference and gibbsite a parallel to the cylinder axis. Each orthosilicate anion displaces H from the three OH groups surrounding a vacant octahedral site in the hydroxide sheet. The fourth Si-O bond points away from the sheet and is neutralized by a proton to form SiOH. This structure requires a considerable shortening of the O–O distance around the vacant octahedral site. This contraction could account for both the shortening of the repeat distance from 8.6Å in gibbsite (a axis) to 8.4Å in imogolite (c-axis) and also for the curling of the hydroxide sheet to form the cylinder.

The resulting structure has a composition $SiO_2 \cdot Al_2O_3 \cdot 2H_2O(+)$, which is compared with the experimental formula $1.1SiO_2 \cdot Al_2O_3 \cdot 2.5H_2O(+)$, and accommodates an orthosilicate anion. The smaller content of OH groups and the absence of water coordinated with Al in the postulated structure should, however, be considered in further refinement. The outside diameter of the cylinder ranges from 18.3 to 20.2Å, consistent with interaxial separations of 21–23Å for an aligned array of the tube units. The electron diffrac-

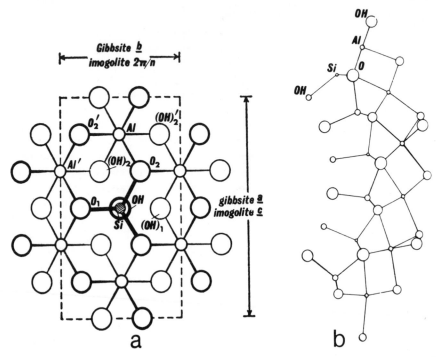

Fig. 16-7. *(a)* Postulated relationship between the structure unit of imogolite and that of gibbsite. SiOH groups which would lie at the cell corners in imogolite are omitted from the diagram. *(b)* Curling of the gibbsite (hydroxide) sheet induced by contraction of one surface to accommodate SiO_3OH tetrahedra; projection along the imogolite *c* axis. (Cradwick et al., 1972).

tion pattern of imogolite, infrared spectra, sixfold coordination of Al, retention of quaternary ammonium chlorides, very weak surface acidity, and unique electrokinetic phenomena for imogolite can also be accounted for by the proposed structure.

High resolution electron microscopy suggested a possibility that allophane consists of "hollow spherules" with diameters of 35–50Å. No particular variation in the size and shape of the "spherules" with variation in the chemical composition was observed. The wall of the "spherules" may be constructed either from a modified imogolite structure or from a kaolinite defect structure. In either structure, Al in fourfold coordination should be included to some extent. That allophane has a sheet structure related to kaolinite has been proposed by Udagawa et al. (1969), Brindley & Fancher (1969), and Okada et al. (1975). A model of allophane, "permutite" core with Al in fourfold coordination plus hydroxy-Al cation coating has also been proposed by de Villiers and Jackson (1967), Cloos et al. (1969), and de Villiers (1971). It seems difficult, however, to reconcile this core and coating differentiation with the observed size and shape of the allophane "spherules."

III. NATURAL OCCURRENCE

A. Geographic Extent

The presence of allophane constitutes the most important feature of soils derived from volcanic ash and called Ando soils (Thorp & Smith, 1949), Andepts (Soil Survey Staff, 1960), or Humic Allophane soils (Kanno, 1962). In these soils, imogolite and allophanelike constituents are also commonly present in association with allophane and impart typical Andept characteristics to the soils. Thus, the established geographic distribution of allophane, imogolite, and allophanelike constituents has been connected with the areas of recent volcanic activity throughout the Pacific ring of volcanism, West Indies, Africa, Italy, and Australia. The presence of noncrystalline clay minerals, specifically allophane in soils derived from parent material other than volcanic ash and pumice has also been suggested by a number of investigators. As reviewed by Wada and Harward (1974), however, it is largely inconclusive and awaits further study.

Andepts occur from the cold subhumid to equatorial tropics except for those in desert and semidesert regions (Wright, 1946; Flach, 1964). Although volcanic ash and pumice are the most common parent materials, allophane and imogolite have been found in soils derived from massive basalt under tropical, humid climate in Africa (Siefferman & Millot, 1969) and in Hawaii (Patterson, 1964; Wada et al., 1972). In Hawaii, Flach (1972) showed that Hydrandepts, the most intensely weathered Andepts, have several properties in common with Spodosols. In such Hydrandepts, noncrystalline Al and Fe hydroxides rather than allophane and imogolite have been found as major clay constituents (Wada & Wada, 1976).

B. Soil Environment

Formation and transformation of noncrystalline clays in weathering of volcanic ash have been studied by many investigators. These studies have been reviewed by Mitchell (1964), Besoain (1969), Wada and Aomine (1973), Yuan (1973), Wada and Harward (1974), and others. Fieldes (1955) and many subsequent investigators pointed out the importance of sequential relationships such as allophane B—allophane AB—allophane A—metahalloysite—koalinite in formation and transformation of noncrystalline clay materials. Wada and Harward (1974) also mentioned the importance of recognizing that some of these processes proceed along with formation and transformation of layer silicates and with accumulation of humus, and are in response to environmental changes during soil development. They presented these processes in soils developed from volcanic ash in a humid, temperate climatic zone in a scheme as shown in Fig. 16-8.

The early stage of weathering of volcanic ash is characterized by the

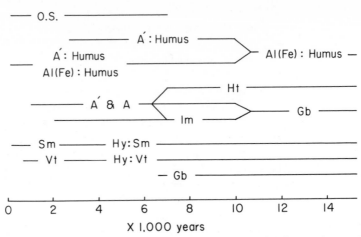

Fig. 16-8. Formation and transformation of clay minerals and their organic complexes in soils developed from volcanic ash in humid, temperate climatic zones. Abbreviations: A, allophane; A', allophanelike constituents; Al(Fe), sesquioxides; Ch, chlorite; Gb, gibbsite; Ht, halloysite; Im, imogolite; Sm, smectite; Hy-Sm, hydroxy interlayered smectite; O.S., opaline silica; Vt, vermiculite; Hy-Vt, hydroxy interlayered vermiculite. Horizontal bars, approximate duration of the respective constituents. (modified from Wada & Harward, 1974).

presence of opaline silica, humus, hydroxy-Al and -Fe ions probably complexed with humus, allophanelike constituents, and montmorillonite, and by absence of allophane and imogolite (Shoji & Masui, 1972; Tokashiki & Wada, 1975; Wada & Higashi, 1976). There seems to be an inverse relationship between opaline silica and allophane in their occurrence in Andepts. An analysis of a series of deposits in two profiles by Tokashiki and Wada (1975) showed that allophane was absent or nearly absent, both in the A1 and in the oldest, buried A1 horizons. All other buried A1 and B horizons contained allophane. Allophanelike constituents were present in all horizons examined except for the oldest, buried A1 horizons. The presence of opaline silica and the accumulation of humus in amounts equivalent to $> 15\%$ C were also noted in the A1 horizons. These observations suggest that Al released by weathering of volcanic ash is retained by organic matter, which is then stabilized against biotic degradation and leaching. Aluminum ion activity in soil solution is suppressed by the formation of Al-humus complexes, thus favoring the formation of opaline silica rather than the formation of allophane and imogolite. This would suggest that a balance between the release rate of Al and the supply rate of organic matter could control the mineral formation in these soils.

Nearly all varieties of volcanic ash (basaltic, andesitic, dacitic, or rhyolitic) produce allophane and allophanelike constituents, though they may be different in nature, stability, and amount (Wada & Harward, 1974). High resolution electron microscopy has also indicated that imogolite, though different in amounts, is always associated with allophane in weatheirng of these volcanic ashes (Henmi & Wada, 1976). Under warm, humid climate, allophane and imogolite have been found most abundantly in the B horizons of soils 5000 to 10,000 years old (Wada & Aomine, 1973). High Al activity in these soils is evidenced by parallel hydroxy-Al interlayering in expandable 2:1 layer silicates (Kawasaki & Aomine, 1966; Masui et al., 1966; Tokashiki & Wada, 1975).

Imogolite may be formed either by transformation from allophane through desilication or by precipitation from weathering solution and subsequent paracrystal formation (Fig. 16-8). Aomine and Mizota (1973) found that imogolite forms in a particular ash or pumice bed only where depositional overburden of volcanic ash serving as a silica source is relatively thin. Occurrence of fairly pure imogolite as "gel films" in several pumice beds in Japan (Miyauchi & Aomine, 1966; and others) and in basalt saprolite in Hawaii (Wada et al., 1972) has been reported. In the pumice bed, allophane was found within the pumice grains, whereas imogolite occurred exclusively as macroscopic "gel films" filling the interstices of pumice grains (Wada & Matsubara, 1968). Electron micrographs indicated that allophane and imogolite formed within and outside of weathered glass fragments, respectively (Henmi & Wada, 1976).

Formation of halloysite and sometimes metahalloysite in old and buried soils derived from volcanic ash has been reported by Fieldes (1955) and a number of subsequent investigators. There are indications that the formation of halloysite is favored by thick depositional overburden as a silica source for resilication of allophane (Mejia et al., 1968; Aomine & Mizota, 1973) and by stagnant moisture regime (Aomine & Wada, 1962; Dudas & Harward, 1975). Halloysite forms from ashes of various compositions, and usually appears as unique spherules with diameters of 0.1-0.5 μm (J. B. Dixon, Chapter 11, this book). Substantial dissolution of these spherules by 0.5N NaOH treatment occurs; dissolution of halloysite, but not any significant amount of allophane was indicated by difference infrared spectroscopy (Wada & Tokashiki, 1972), whereas dissolution of allophane as cores and interlayers was suggested by electron microscopy (Askenasy et al., 1973).

It has been gaining recognition that layer silicates, particularly 2:1 layer silicates and chlorites are present in substantial amounts in some Andepts. Masui et al. (1966) found that the contents of 2:1 layer silicates increased in Andepts in northern Honshu with weathering which was measured by the clay content. They interpreted the result as indicating that "amorphous" clay materials transform to 2:1 layer silicates. Reexamination of these data showed that there were positive correlations between the contents of 2:1 layer silicates, the presence of quartz in the clay, and the albite content in feldspar (Wada & Aomine, 1973). They considered that the presence of large amounts of the layer silicates associated with quartz signifies the effect of dacitic nature of parent ash on weathering rather than the transformation of allophane to these layer silicates. This postulate seems to have been supported by subsequent studies (Mizota & Aomine, 1975; Tokashiki & Wada, 1975). It is likely that the 2:1 layer silicates are at least partly inherited from fine grained mica and hornblende in dacitic ash, but their origin needs further studies.

The relatively short life of allophane, imogolite, allophanelike constituents, and their humus complexes may deserve attention. Allophane and allophanelike constituents would transform to halloysite or gibbsite depending on whether the environment favors resilication or desilication. The formation of halloysite has been described above. The absence of allophane and allophanelike constituents in old Andepts has also been reported; the predominating clay constituents found in a buried A1 horizon about 10,000 years old were hydrous Al and Fe oxides, probably complexed with humus, and hydroxy interlayered layer silicates, while those in a buried B horizon about 30,000 years old were gibbsite and hydroxy interlayered layer silicates (Wada & Greenland, 1970; Tokashiki & Wada, 1975).

C. Sedimentary Occurrence

From the viewpoint of geological history, allophane and imogolite are transition products in transformation from primary to secondary minerals (Fig. 16-8). They may be found in recent aluvium developed in volcanic area, but are not expected in any significant amount in older sediments.

IV. EQUILIBRIUM ENVIRONMENT AND CONDITIONS FOR SYNTHESIS

Clay minerals influence the composition of soil solution and vice versa. This reaction may be more manifest for noncrystalline than crystalline clay constituents. Thus, allophane and imogolite would help to determine the composition of soil solution, and the latter composition would also determine the course of formation and transformation of these minerals in the soil.

The equilibrium environment in which allophane and imogolite are formed may be assessed from analyses of soils in which they are predominating. The pH levels over which allophane and imogolite are usually found range from 5 to 7. The cation-exchange and anion-exchange capacity data indicate that the isoelectric point of allophane and/or imogolite varies from pH 6-8 to below 4 when the SiO_2/Al_2O_3 ratio increases from 1 to 2. There are, however, no particular data indicating whether or not this isoelectric pH has a bearing on the equilibrium pH.

The silica concentration in the equilibrium soil solution may be an important factor which affects the mineral formation. A study of adsorption of monomeric silica by Andepts (Wada & Inoue, 1974) showed that the silica concentration of the ambient solution below which the release of silica from the soil took place was dependent on the major clay mineral species. The values, 11-12, 22, and 26 ppm silica were recorded at pH 6.0-6.6 for allophane and imogolite with the SiO_2/Al_2O_3 ratio of 1.1, allophane with the SiO_2/Al_2O_3 ratio of 2.0, and halloysite, respectively. The effect of activity of Al ions in soil solution on the mineral formation is difficult to evaluate, but there are indications that the high Al ion activity probable at the site of hydrolysis of primary minerals favors formation of allophane and imogolite, whereas its suppression by complex formation with organic matter favors formation of opaline silica.

Towards a better understanding of the nature and properties of allophane, noncrystalline aluminosilicates as an analog have been synthesized and studied by Léonard et al. (1964), Mitchell et al. (1964), and Cloos et al. (1969). Relatively little attention to the process of precipitation of Si and Al and the effects of environmental factors and time on this reaction has been given in these studies. Wada and Kubo (1975) prepared noncrystalline aluminosilicates from solutions containing monomeric silica and Al ions to give a SiO_2/Al_2O_3 ratio ranging from 1 to 8.5. The SiO_2/Al_2O_3 ratio of the precipitate varied in the range from 1 to 3 at pH 6.0-8.0. The pH did not affect the SiO_2/Al_2O_3 ratio of the precipitate, but did its dispersion and flocculation. The amount of proton released in the reaction increased with increasing SiO_2/Al_2O_3 ratios of the precipitates, but the values were generally lower than those predicted from "one-in-four" substitution of Si with Al in fourfold coordination, as suggested by de Villiers (1971). Caution is necessary in using these synthetic products as a model of allophane, particularly in studies of its charge characteristics.

V. CHEMICAL PROPERTIES

A. Ion Exchange

A marked influence of noncrystalline clay materials on ion exchange in soils has been emphasized in the literature. As will be described in this section, charge development on allophane and imogolite depends on environmental conditions such as pH, ion concentration, and temperature. This has important implications on the status of "base" and plant nutrients in Andepts. The values of cation and anion exchange capacities (CEC and AEC) which are quoted here were determined by measuring retention of index ions when the samples had been equilibrated with salt solutions at appropriate concentrations and pH. Excess salt was not removed by washing. Difficulty will be experienced in interpretation of the CEC values obtained by conventional methods in which the excess salt is removed by washing with water and/or alcohol.

The CEC and AEC data for allophane and imogolite of established purity have been scarce. The CEC values of 20-50 meq/100 g and the AEC values of 5-30 meq/100 g clay were obtained by equilibrating allophanic soil clays (< 2 μm) with 0.2N NH_4Cl solution at about pH 7 (Wada & Ataka, 1958; Fieldes & Schofield, 1960; Iimura, 1966). The CEC values of 135 meq/100 g for allophane (SiO_2/Al_2O_3 ratio = 2.0; < 2 μm), and 30 meq/100 g for imogolite were obtained by equilibrating with 0.05N $NaCH_3COO$ at pH 7.0. In this determination the samples were pretreated with dithionite-citrate and 2% Na_2CO_3 (Henmi & Wada, 1974). There are indications that H ions which become dissociable upon treatment with 2% Na_2CO_3 solution

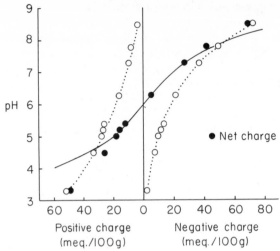

Fig. 16-9. Effect of pH on the negative and positive charges of allophane (< 2 μm; weathered pumice; Ina, Nagano; SiO_2/Al_2O_3 ratio = 1.43). Data; Iimura (1966). Full line; the net charge calculated by using the Nernst and Gouy-Chapman equations (see text).

increase with increasing SiO_2/Al_2O_3 ratios of clays in which allophane and/or imogolite predominate (Wada, 1967).

Figure 16-9 shows the effect of pH on both the CEC and AEC values for an allophanic clay. Similar effects are generally found for other clays containing allophane and imogolite. This development of negative and positive charges in response to pH may be explained by assuming amphoteric dissociation of the surface Si-OH, Al-OH or Fe-OH groups or adsorption of H or OH ions (R. G. Gast, Chapter 2, this book). The net surface charge can be calculated by using the Nernst equation relating the surface electric potential to the pH of the equilibrium solution and the Gouy-Chapman equation relating the net surface charge to the surface potential and electrolyte concentration (Van Raij & Peech, 1972). The value of experimentally determined zero point of charge is used as a reference point in this calculation. The magnitude of agreement between the measured and calculated values of the net surface charge is seen in Fig. 16-9.

Figure 16-10 shows the effect of salt concentration and cation species of the CEC values of samples containing imogolite and allophane at about pH 7. A gradual decrease in CEC with decreasing salt concentration observed for imogolite and allophane with the SiO_2/Al_2O_3 ratio of 1.0 has practical importance, but has not been fully explained. The Gouy-Chapman equation predicts that the net surface charge density is concentration-dependent, being proportional to the square root of the salt concentration (R. G. Gast, Chapter 2, this book). Apparently, this simple relationship does not hold for this system. A similar "hydrolysis" effect was not observed for allophane with the SiO_2/Al_2O_3 ratio of 2.0 (Fig. 16-10) as well as halloysite and montmorillontie (Wada & Harada, 1969).

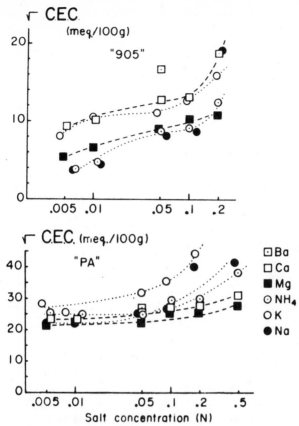

Fig. 16-10. Effects of salt concentration and cation species on the CEC values of a soil sample (no. 905) which contains imogolite and allophane (SiO_2/Al_2O_3 ratio = 1.0) and a weathered pumice sample (PA) which contains allophane (SiO_2/Al_2O_3 ratio = 2.0). (Wada & Harada, 1969).

Different CEC values depending on the cation species occur for soils containing imogolite and/or allophane (Fig. 16-10). No similar effect was found for halloysite and montmorillonite (Wada & Harada, 1969). The proton is involved in the development of the negative charges on imogolite and allophane. The observed effect may therefore be interpreted in terms of differences in relative bonding energies between the proton and the index cation being added. Different values of the AEC depending on the anion species were also found for soils containing imogolite and allophane (K. Wada & S. Tsuji, 1973. Soc. Sci. Soil Manure. (Japan) Abstr. Pap. 18:22). The AEC values increased in the order: $NO_3 < Cl < CH_3COO \ll SO_4$. The much higher value for sulphate ion may be interpreted in terms of specific adsorption.

The temperature coefficient of cation exchange in soils is small (Kelley, 1948). However, the effect of temperature on CEC will vary depending on major cation exchange materials; the range in ratio of the CEC measured at

10-20C (283-293K) to that measured at 50-60C (323-333K) was 0.36-0.59 for allophane, imogolite, and/or humus; 0.62-0.75 for 1:1 caly minerals; and 0.90-0.99 for smectites at pH 7.0 (Wada & Harada, 1971). The trend suggests that the greater the contribution of ionized OH and/or COOH groups to the CEC, the more pronounced the CEC increase at the higher temperature. The increased CEC at the higher temperature was, however, only partly reduced by lowering the temperature again. The exchange reactions attained their equilibria rather slowly, but no particular CEC increase with time was observed at the low compared with the higher temperature. A slight decrease in the AEC values of imogolite and allophane was found when the temperature of the solution was raised.

Previous drying or heating of the sample resulted in an increase in the CEC values of soils containing allophane and/or imogolite (Kubota, 1973; Harada & Wada, 1974). This CEC increase upon drying over P_2O_5 or heating at 105C (378K) amounted to 30-50% of that of the undried sample. The data indicate that the negative charge could develop on dehydration, possibly including changes in the coordination of some surface Al atoms. There are also observations which suggest an increase of the net negative charge upon drying or heating of the sample; Watanabe (1966) observed that the isoelectric point of allophane with a SiO_2/Al_2O_3 ratio of 1.11 was lowered from pH 6.8 to 4.1 by air drying, though Horikawa (1975) did not observe a similar effect. Sadzawka et al. (1972) and Kubota (1973) observed that the pH of Andepts suspended in salt solutions was lowered by about 0.5 pH unit or more when the soils were dried at 110C (383K). On the other hand, the AEC values (Harada & Wada, 1974), the amounts of OH release with NaF (Kubota, 1973), and EGME retention (Egashira & Aomine, 1974) of soils or clays containing allophane and/or imogolite decreased upon heating at 105-110C (378-383K).

The cation selectivity of soils is influenced by the nature of the cation exchange material. Table 16-4 shows retention of Ca and NH_4 as exchange-

Table 16-4. Percentage retention of Ca and NH_4 as exchangeable cation after leaching with a mixture of equinormal Ca and NH_4 salt solution (Yoshida, 1956, 1961; Kato, 1970c)

Material and soil	Major ion-exchange material	Concentration of salt solution			
		1N		0.1N	
		Ca	NH_4	Ca	NH_4
		% retention			
IR-120	Cation-exchange resin	72	28	87	13
Morioka (No. 133)	Humus > allophane	72	28	81	19
Kochi	Allophane and imogolite	--	--	76	24
Miyazaki	Allophane and imogolite	77	23	--	--
Kanuma	Allophane	--	--	51	49
Ina	Allophane	--	--	52	48
Morioka (No. 134)	Montmorillonite	40	60	57	43
Bentonite	Montmorillonite	20	80	55	45

able cations after leaching the sample with a mixture of equinormal Ca and NH_4 salt solution. The soils containing allophane and imogolite, probably with low SiO_2/Al_2O_3 ratios, show higher affinity for Ca than NH_4, while those containing allophane, probably with high SiO_2/Al_2O_3 ratios, show nearly equal affinities for these cations. The Donnan equilibrium theory predicts that the higher the negative charge density the higher the affinity for divalent over monovalent cations (Wiklander, 1964). The observed difference between the different cation exchange materials may partly be accounted for by this difference in the CEC, but may be more closely connected with the difference in the origin of negative charges.

B. Surface Acidity

Imogolite and allophane have weak acid properties. Iimura (1966) found that the pH of electrodialyzed suspensions of allophanic clays is in the range of 4.0 to 6.5 and that these suspensions show little or no exchange acidity. The latter finding was interpreted in terms of balance between exchange acidity and alkalinity. Yoshida (1970; 1971) attempted to characterize the acid sites on clays by treating them with $1N$ $AlCl_3$ solution and determining exchangeable Al and H on these clays. All the exchange sites on allophane and imogolite were occupied by H, whereas > 60% of the exchange sites on layer silicates were occupied by Al. The high affinity of allophane and imogolite for H indicates weak acid properties.

The increasing acidity or proton donating properties of the clay surface with decreasing water content was demonstrated by a number of investigators (Mortland, 1970). Figure 16-11 shows the acid strength of clays treated with H and OH saturated resins and kept at various relative humidities. The acid strength was determined by the color of Hammett indicators with

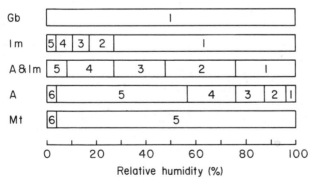

Fig. 16-11. Acid strength of H (Al) saturated clays kept at various relative humidities. Abbreviations: A, allophane; Gb, gibbsite; Im, imogolite; Mt, montmorillonite. Figures, the order of acid strength increasing from 1 to 6. *(1)* pKa 4.0 to 6.8; *(2)* pKa 3.3 to 4.0; *(3)* pKa 2.0 to 3.3; *(4)* pKa 1.5 to 2.0; *(5)* pKa −5.6 to 1.5; *(6)* pKa −8.2 to −5.6. The referred pKa value is that of the indicator which is adsorbed on the clay and shows an acid color. Data from Henmi and Wada (1974).

known pKa values adsorbed on the clays in a nonpolar solvent. The measured weakest and strongest acid strengths correspond to the strengths of 8×10^{-8} to 5×10^{-5} and 71 to 90% sulfuric acid, respectively. The results indicate that there are clear-cut differences in acidity between gibbsite, imogolite, allophane, and montmorillonite, and that imogolite and allophane showed weak acidities in moist environment, but their acid strength was much enhanced in dry environment. The acid sites may develop on allophane and imogolite by changes in the coordination of some surface Al atoms.

C. Anion Sorption

Interaction of anions with noncrystalline aluminosilicates involves three different reactions; nonspecific adsorption, specific adsorption, and decomposition of these aluminosilicates induced specifically by adsorbed anions. The nonspecific adsorption refers to adsorption of anions by simple coulombic interaction with positive charges on $Al-OH_2$ or $Fe-OH_2$ groups, while the specific adsorption refers to incorporation of the anion as a ligand in the coordination shell of an Al or Fe atom (Hingston et al., 1967; Greenland, 1971). These reactions have been extensively studied for hydrous Fe oxides (R. G. Gast, Chapter 2; U. Schwertmann & R. M. Taylor, Chapter 5, this book). The specifically adsorbed anion can not be displaced from the adsorbate simply by exchange with nonspecifically adsorbed anions. The nonspecific adsorption of anions increases with lowering pH, whereas a maximum or inflections of the specific adsorption occurs in the adsorption maximum vs pH curve at pH values corresponding to the pKa values of the acid species formed by the anion. Ortho-, pyro-, and tripoly-phosphates, fluoride, silicate, arsenate, selenite, and molybdate are some of the anions specifically adsorbed.

Nonspecific and specific adsorption of anions occurs extensively on allophane and imogolite. Andepts usually fix large amounts of phosphate. The high value of phosphate sorption obtained by a conventional method ($> 1,500$ mg as P_2O_5 per 100 g of air-dry soil from 200 ml of 2.5% ammonium phosphate solution at pH 7) has often been used in Japan for roughly distinguishing allophanic and nonallophanic soils. This sorption has recently been assigned to the dithionite–citrate soluble "sesquioxidic" constituents in addition to allophane (Kato, 1970a; Miyauchi & Nakano, 1971). Studies on allophanic soils (Miyauchi & Nakano, 1971) and imogolite (Parfitt et al., 1974), both "purified" by dithionite-citrate and 2% Na_2CO_3 or 0.05N HCl treatments, showed that an adsorption maximum of phosphate can be estimated at concentration levels of 10^{-2} to $10^{-3}M$. Adsorption maxima of 280 μmol and 190 μmol phosphate per g of imogolite were obtained at pH 4 and 8, respectively. Tsukada et al. (1967) also found a 1:1 stoichiometric relationship between the amounts of Al extracted at pH 4.8 with an ammonium acetate solution and of phosphate at a "saturation point" in soils including Andepts. A rapid phosphate-induced decomposition of allophane and its transformation of taranakites was demonstrated in weakly acid, ammonium and potassium phosphate solutions (Wada, 1959; Birrell, 1961b).

Large OH release takes place when NaF is added to allophanic soils (Kawaguchi et al., 1954). Fieldes and Perrott (1966) proposed this reaction as a rapid test for allophane. They found that the pH of a suspension containing 1 g of soil in 50 ml of $1N$ NaF was 9.4 after 2 min if its allophane content was appreciable. This test is useful, but it is important to keep in mind that the reagent will react with any available Al associated with hydroxyls and it is not specific for allophane. Fluoride-induced decomposition of allophane and its transformation to cryolite was also reported by Egawa et al. (1960) and Birrell (1961b).

D. Interaction with Organic Compounds

Allophane and imogolite are effective in nonspecific and specific adsorption of organic anions. For example, acetate is adsorbed by coulombic interaction up to the AEC of these clays, while citrate, a strong ligand for Al, is more strongly adsorbed and in a greater amount. In addition, van der Waals interactions contribute to adsorption of high molecular weight organic anions. Much greater adsorption of humified clover extract by allophane and imogolite than by montmorillonite and halloysite was demonstrated by Inoue and Wada (1968).

On the other hand, allophane adsorbed little or no organic cations such as cetylpyridinium (Greenland & Quirk, 1962), orthophenanthroline (Aomine & Otsuka, 1968), and trimethylammonium (Kinter & Diamond, 1960). Birrell (1961a) also reported that the retention of cations by allophane decreased in the order: $NH_4 > (CH_3)_4N > (C_2H_5)_4N$. The positive charges on allophane and the presence of pores which are too small to accommodate these large cations probably prevent their adsorption.

More complex interactions of allophanic clays with organic compounds have been studied. Allophanic clays adsorbed considerable amounts of enzymes such as proteases and amylases and their substrates, and thereby affected the action of these enzymes (e.g., Kobayashi & Aomine, 1967). These authors pointed out the importance of the protective action of allophane against biotic degradation of organic materials which become incorporated in the soils. Allophane also exhibited a catalytic effect on oxidative changes of polyphenols (Kyuma & Kawaguchi, 1964; Kumada & Kato, 1970).

Accumulation of humus constitutes one important feature of Andepts. It is not uncommon for the C content of the well-drained A1 horizon to be 20% in a warm, humid climate with grass vegetation. The important role of allophane and imogolite in the accumulation of humus has been suggested on the basis of their reactivity with organic compounds. In this inference, the preformation of these minerals is implicitly assumed, but there are indications that it is not always true. Accumulation of a considerable amount of humus was noted in young Andept A1 horizons in which little or no allophane was detected by selective dissolution and difference infrared spectroscopy. Kato (1970b) noted the importance of dithionite–citrate soluble sesquioxides, particularly alumina, in the accumulation of humus in Andepts. Wada and Higashi (1976) found a good linear correlation between the amounts of humus and Al, extractable with $0.1M$ $Na_4P_2O_7$, for a number of Andepts, irrespective of presence or absence of allophane. Nevertheless, these observations do not preclude the possibility that allophane, allophane-like constituents, and imogolite contribute to stabilizing organic matter against biotic degradation and leaching.

VI. PHYSICAL AND ENGINEERING PROPERTIES

The physical and engineering properties of soils having a high content of allophane and imogolite are unique. These soils usually have low bulk density, high water holding capacity and high liquid and plasticity limit values, and show slippery, but nonsticky consistence and anomalous compaction behavior. The charge characteristics of allophane and imogolite, the shape and size of their structure "units" and the way in which these structure "units" form secondary aggregates are the major controlling factors. Other factors such as particle size distribution, organic matter content, and geologic history, also affect the physical and engineering properties of the soils.

Allophane and imogolite consist of extremely small "units" with micropores in the order of two to several tens of angstroms (Fig. 16–2b) and have

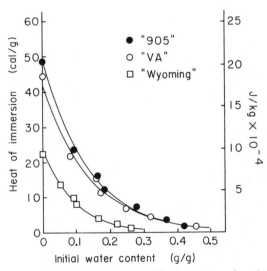

Fig. 16-12. Heat of immersion of Ca saturated clays in water as a function of initial water content. 905, allophane and imogolite. VA, allophane. Wyoming, montmorillonite, API No. 26. (modified from Aomine & Egashira, 1970).

a tremendous amount of potential adsorbing surface. This is illustrated in Fig. 16-12 by heat-of-immersion values measured as a function of the initial water content of the sample for allophane and imogolite, allophane, and montmorillonite (Aomine & Egashira, 1970). The large water sorption capacity of allophane and imogolite, compared even with montmorillonite, is indicated by the high heat-of-immersion values, and by the high initial water content at which the heat of immersion approaches zero.

The "units" of allophane and imogolite can interact with each other and with other soil constituents resulting in formation of stable and porous aggregates. The void ratios of fine-textured soils derived from volcanic ash were in the range of 2 to 5, and mostly 3 to 4. These values were higher than 0.8 to 1.0 for sandy alluvial soils and 1.5 to 2.5 for clayey alluvial soils (Suyama & Oya, 1965). Bulk density values as low as 0.25-0.3 were found for the B horizons of Andepts containing little organic matter (Wada & Aomine, 1973).

Mechanical analysis of moist, air-dried, and oven-dried B horizon samples demonstrated that stable aggregates form upon drying of Andepts, but not of Ultisols, and that allophane contributes to this aggregate formation (Kubota, 1972). A critical pF value was in the range of 3 to 4, and the degree of aggregate formation was dependent on the natural water contents of the soils. The aggregates were stable to H_2O_2 treatment and mechanical shaking, but disrupted by H_2O_2 treatment and sonic oscillation. Aggregates of allophane, imogolite, and humus were not completely disrupted by sonic oscillation alone (Inoue, 1973). A large part, but not all, of humus in air-dried Andepts can be removed by H_2O_2 treatment. Undesirable effects of formation of organic acids such as oxalic acid by the treatment resulting in a partial destruction of the clay constituents were discussed by Mitchell et al. (1964).

The pH of the medium is important to the stability of allophane and imogolite dispersion. An alkaline medium (pH 10) can be used for allophane with a SiO_2/Al_2O_3 ratio of 2 or higher, whereas an acid medium (pH 4) is required for imogolite and allophane with the lower SiO_2/Al_2O_3 ratios. Aomine and Egashira (1968) showed that at pH 4.5-6.5 flocculation of allophane with electrolytes is primarily determined by the valence of anions and that of montmorillonite by the valence of cations. All of these observations indicate that the sign of prevailing charges on clay particles has primary importance in their dispersion. Horikawa (1975) found that imogolite showed zero electrophoretic mobility and flocculation at pH 7 or higher in $0.001M$ NaCl.

A high water content (80 to 180% on oven-dry basis) was common to allophanic soils with low bulk density values (Suyama & Oya, 1965). The subsoils compared with surface soils have high amounts of both free (pF 2.5-4.2) and nonfree water (pF $>$ 4.2). The water content of some subsoils was close to their liquid limit values. They were sensitive to mechanical disturbance, and showed very little strength after being disturbed. Sensitivity is defined as the ratio of the strength of the soil in an undisturbed state to that of the remolded material at the same water content. Values as high as 7 to

Table 16-5. Plasticity limit and liquid limit values of soils (Birrell, 1952, Yamazaki & Takenaka, 1965)

Soil	Major clay minerals	Plasticity limit		Liquid limit	
		Undried sample	Air-dried sample	Undried sample	Air-dried sample
Whennapai	Allophane	131	78	207	85
Mairoa	Allophane and gibbsite	167	--	350	108
New Plymouth	Halloysite and kaolin	73	54	110	70
Utsunomiya	Allophane	76	60	172	76
Tokyo	Allophane	95	81	189	97
Ikuta	Halloysite	50	58	98	90
Hachirogata	Montmorillonite	55	56	186	122

10 were recorded for allophanic soils (Gradwell & Birrell, 1954; Kita et al., 1969).

Table 16-5 shows comparison of plasticity limit and liquid limit values between soils containing allophane and layer silicates. The data indicate that allophane in the natural moisture state has both high plasticity and liquid limit values, and that the reduction of the liquid limit value upon air drying is more pronounced for allophane than layer silicates. This nonreversible drying effect again indicates that allophane contributes to nonreversible bonding which occurs between dispersed soil particles upon drying.

A peculiarity common to Andepts in the standard consolidation test is that preconsolidation load of 18,000–36,000 kg/m^2 far exceeds the present overburden pressure or even the bearing capacity of 9,000–13,500 kg/m^2 (Birrell, 1964; Suyama & Oya, 1965). There was no geologic evidence for this preconsolidation. It was interpreted in terms of aggregation of soil particles with allophane and humus, possibly through drying during weathering of volcanic ash.

The soil normally gives a single compaction curve with a well-defined maximum. The compaction curve shows the dry density values attained at the respective water contents when the soil is subjected to a constant compactive effort. However, an allophanic soil gave a series of curves depending upon the initial water content of the sample, and a clear maximum on the dry density curve appeared only in the moistening cycle but not in the drying cycle (Suyama & Oya, 1965). A high content of nonfree water and its partial transformation to free water upon compaction was considered to be responsible for the observed anomalies.

VII. QUANTITATIVE DETERMINATION

As described in the section on dissolution analyses, noncrystalline and paracyrstalline clay materials can be estimated by the use of selective dissolution treatment, though sometimes there is a question in the specificity of the dissolution treatments. How to estimate the weights of dissolved material is another problem.

Table 16-3 indicates that the amounts of noncrystalline and para-crystalline clay material including oxide and hydroxide minerals can be estimated from the weight loss of the sample either after treatment with 0.15-0.2M oxalate-oxalic acid (pH 3.0-3.5) or after successive treatments with dithionite-citrate and 0.5N NaOH. The former oxalate-oxalic acid procedure will give a better estimate because crystalline constituents, either silicates or hydroxides and oxides, are not dissolved in any substantial amount. Noncrystalline silica minerals are, however, not included in this analysis.

Estimation of a particular noncrystalline mineral or a mineral fraction is much more involved. Combining dissolution treatments in succession, identifying the constituents dissolved by some means, and determining the amount of elements dissolved by the treatments are required. Tokashiki and Wada (1972) compared the weight loss of the clays and the sum of extracted SiO_2, Al_2O_3, and Fe_2O_3 by the successive dithionite-citrate, 2% Na_2CO_3, and 0.5N NaOH treatments. They found that the combined weight loss for the former two treatments and the weight loss for the last treatment can be equated with the amounts of the constituents dissolved by these reagents.

Difficulty has been experienced with physical techniques in estimating the amount of noncrystalline clay constituents. Imogolite in an amount as small as 2% in the specimen can be estimated by DTA utilizing an endothermic dehydroxylation reaction at 420C (693K) (Aomine & Mizota, 1973).

VIII. CONCLUSION

The structure, morphology, and charge characteristics of imogolite and allophane in soils derived from volcanic ash and pumice are fairly well understood. Knowledge about their unique "structure units" and arrangement of these "units" has provided a basis for interpreting their effect on the soil properties which pose a number of problems in practical management and use of soil systems. A better understanding of soil genesis from volcanic ash and pumice has also been obtained by applying advanced methods of mineralogical analysis to these systems which include noncrystalline, para-crystalline, and crystalline clay constituents. One of the problems left to future studies is to characterize noncrystalline constituents which are chemically intermediate between allophane and hydrous oxides of Al, Fe, and Si. There are indications that these noncrystalline constituents as well as allophane play a great role in reactions in Andepts. Another and more important problem is to characterize noncrystalline clay constituents in soils other than Andepts, and to verify their effects on the soil properties.

LITERATURE CITED

Aomine, S., and K. Egashira. 1968. Flocculation of allophanic clays by electrolytes. Soil Sci. Plant Nutr. (Tokyo) 14(3):94-98.

Aomine, S., and K. Egashira. 1970. Heat of immersion of soil allophanic clays. Soil Sci. Plant Nutr. (Tokyo) 16(5):204-211.

Aomine, S., and C. Mizota. 1973. Distribution and genesis of imogolite in volcanic ash soils of northern Kanto, Japan. p. 207-213. *In* J. M. Serratosa (ed.) Proc. Int. Clay Conf. (Madrid, Spain) 1972.

Aomine, S., and H. Otsuka. 1968. Surface of soil allophanic clays. Int. Congr. Soil Sci., Trans. 9th (Adelaide, S. Australia) I:731-737.

Aomine, S., and K. Wada. 1962. Differential weathering of volcanic ash and pumice resulting in formation of hydrated halloysite. Am. Mineral. 47:1024-1048.

Askenasy, P. E., J. B. Dixon, and T. R. McKee. 1973. Spheroidal halloysite in a Guatemalan soil. Soil Sci. Soc. Am. Proc. 37:799-803.

Besoain, E. 1969. Clay mineralogy of volcanic ash soil. Panel on Volcanic Ash soils in Latin America, Training and Research Centre of the IAAIS, Turrialba, Costa Rica B.1.1-1.16.

Beutelspacher, H., and H. W. Van der Marel. 1961. Der elektronenoptische nachweis von amorphen mineral als storendem Faktor bei der quantitativen analyse von tonmineralien in Boden, Second Conf. Clay Mineral. Petrog. (Prague, Czechoslavakia).

Birrell, K. S. 1952. Some physical properties of New Zealand volcanic ash soils. Proc. 1st Aust.-N. Z. Conf. Soil Mech. Found. Eng. 30-34.

Birrell, K. S. 1961a. The adsorption of cations from solutions by allophane in relation to their effective size. J. Soil Sci. 12:307-316.

Birrell, K. S. 1961b. Ion fixation by allophane. N. Z. J. Sci. 4:393-414.

Birrell, K. S. 1964. Some properties of volcanic ash soils. FAO World Soil Resour. Rep. 14:74-81.

Brindley, G. W., and D. Fancher. 1969. Kaolinite defect structures; possible relation to allophanes. p. 29-34. *In* L. Heller (ed.) Proc. Int. Clay Conf. 1969 (Tokyo, Japan) 2:29-34.

Brindley, G. W., and G. Pedro. 1970. Report of the AIPEA nomenclature committee. Assoc. Int. Etud. Argiles (AIPEA) Newsletter 4:3-4.

Brindley, G. W., and G. Pedro. 1972. Report of the AIPEA nomenclature committee. Assoc. Int. Etud. Argiles (AIPEA) Newsletter 7:8-13.

Bunn, C. W. 1961. Chemical crystallography. Clarendon Press, Oxford.

Brown, G. 1955. Report of the clay minerals group sub-committee on nomenclature of clay minerals. Clay Miner. Bull. 2:294-302.

Cloos, P., A. J. Léonard, J. P. Moreau, A. Herbillon, and J. J. Fripiat. 1969. Structural organization in amorphous silicoaluminas. Clays Clay Miner. 17:270-287.

Cradwick, P. D. G., V. C. Farmer, J. D. Russell, C. R. Masson, K. Wada, and N. Yoshinaga. 1972. Imogolite, a hydrated aluminum silicate of tubular structure. Nature Phys. Sci. 240:187-189.

de Villiers, J. M. 1971. The problem of quantitative determination of allophane in soil. Soil Sci. 112:2-7.

de Villiers, J. M., and M. L. Jackson. 1967. Cation-exchange capacity variations with pH in soil clays. Soil Sci. Soc. Am. Proc. 31:473-476.

Dudas, M. J., and M. E. Harward. 1975. Weathering and authigenic halloysite in soil developed in Mazama ash. Soil Sci. Soc. Am. Proc. 39:561-566.

Egashira, K., and S. Aomine. 1974. Effects of drying and heating on the surface area of allophane and imogolite. Clay Sci. 4:231-242.

Egawa, T. 1964. A study on coordination number of aluminum in allophane. Clay Sci. 2:1-7.

Egawa, T., A. Sato, and T. Nishimura. 1960. Release of OH ions from clay minerals treated with various anions, with special reference to the structure and chemistry of allophane. Adv. Clay Sci. (Tokyo) 2:252-262 (Japanese).

Fieldes, M. 1955. Clay mineralogy of New Zealand soils. Part 2. N. Z. J. Sci. Tech. 37B:336-350.

Fieldes, M. 1966. The nature of allophane in soils. Part 1. Significance of structural randomness in pedogenesis. N. Z. J. Sci. 9:599-607.

Fieldes, M., and K. W. Perrott. 1966. The nature of allophane in soils. Part 3: Rapid field and laboratory test for allophane. N. Z. J. Sci. 9:623-629.

Fieldes, M., and R. K. Schofield. 1960. Mechanisms of ion adsorption by inorganic soil colloids. N. Z. J. Sci. 3:563-579.

Flach, K. 1964. Genesis and morphology of ash-derived soils in the United States of America. FAO World Soil Res. Rep. 14:61-70.

Flach, K. W. 1972. The differentiation of the cambic horizon of andepts from the spodic horizon. II. Panel sobre Suelos Volcánicos de America, Instituto Interamericano de Ciencias Agricolas de la O.E.A. y Universidad de Narino. p. 127–138.

Follett, E. A. C., W. J. McHardy, B. D. Mitchell, and B. F. L. Smith. 1965. Chemical dissolution techniques in the study of soil clays: Part 1. Clay Miner. 6:23–34.

Fukami, A., and K. Adachi. 1965. A new method of preparation of self-perforated microplastic grid and its application (1). J. Electron Microsc. 14:112–118.

Furkert, R. J., and M. Fieldes. 1968. Allophane in New Zealand soils. Int. Congr. Soil Sci., Trans. 9th (Adelaide, S. Aust.) 3:133–141.

Götz, J., and C. R. Masson. 1971. Trimethylsilyl derivatives for the study of silicate structures. Part 2: Orthosilicate, pyrosilicate, and ring structures. J. Chem. Soc. Abstr. 686–688.

Gradwell, M., and K. S. Birrell. 1954. Physical properties of certain volcanic clays. N. Z. J. Sci. Tech. 36B:108–122.

Greenland, D. J. 1971. Interactions between humic and fulvic acids and clays. Soil Sci. 111:34–41.

Greenland, D. J., and J. P. Quirk. 1962. Surface areas of soil colloids. p. 79–87. In G. J. Neale (ed.) Trans. Joint Meeting Comm. IV and V, Int. Soc. Soil Sci. (Palmerston, New Zeal.).

Grim, R. E. 1968. Clay mineralogy. McGraw-Hill Book Co., New York.

Gruner, J. W. 1935. The structural relationship of nontronite and montmorillonite. Am. Mineral. 20:475–483.

Harada, Y., and K. Wada. 1974. Effect of previous drying on the measured cation- and anion-exchange capacities of ando soils. Int. Congr. Soil Sci., Trans. 10th (Moscow, USSR) 1974, II:248–256.

Hashimoto, I., and M. L. Jackson. 1960. Rapid dissolution of allophane and kaolinite-halloysite after dehydration. Clays Clay Miner. 7:102–113.

Henmi, T., and K. Wada. 1974. Surface acidity of imogolite and allophane. Clay Miner. 10:231–245.

Henmi. T., and K. Wada. 1976. Morphology and composition of allophane. Am. Mineral. 61:379–390.

Higashi, T., and H. Ikeda. 1974. Dissolution of allophane by acid oxalate solution. Clay Sci. 4:205–212.

Hingston, F. J., R. J. Atkinson, A. M. Posner, and J. P. Quirk. 1967. Specific adsorption of anion. Nature (London) 215:1459–1461.

Horikawa, Y. 1975. Electrokinetic phenomena of aqueous suspensions of allophane and imogolite. Clay Sci. 4:255–263.

Iimura, K. 1966. Acidic properties and cation exchange of allophane and volcanic ash soils. Bull. Nat. Inst. Agric. Sci. (Japan) B17:101–157 (Japanese).

Inoue, T. 1973. A preliminary study of organo-mineral complexes in a glassy volcanic ash soil containing imogolite and allophane as principal clay minerals. Pedologist 17:26–36.

Inoue, T., and K. Wada. 1968. Adsorption of humified clover extracts by various clays. Int. Congr. Soil Sci., Trans. 9th (Adelaide, S. Aust.) 1968, 3:289–298.

Inoue, T., and K. Wada. 1971. Reactions between humified clover extract and imogolite as a model of humus-clay interaction. Part 2. Clay Sci. 4:71–80.

Jackson, M. L. 1956. Dispersion of soil minerals. p. 31–95. In M. L. Jackson (ed.) Soil chemical analysis—advanced course. Published by the author. Madison, Wis.

Jones, R. C., and G. Uehara. 1973. Amorphous coatings on mineral surfaces. Soil Sci. Soc. Am. Proc. 37:792–798.

Kanno, I. 1962. Genesis and classification of Humic Allophane soil in Japan. p. 422–427. In G. J. Neale (ed.) Trans. Joint Meeting Comm. IV and V, Int. Soc. Soil Sci. (Palmerston, New Zeal.).

Kanno, I., Y. Onikura, and T. Higashi. 1968. Weathering and clay mineralogical characteristics of volcanic ashes and pumices in Japan. Int. Congr. Soil Sci., Trans. 9th (Adelaide, S. Aust.) III:111–122.

Kato, Y. 1970a. Changes in phosphorus absorptive coefficient of "Kuroboku" soils through successive H_2O_2-deferration-Tamm's treatments. J. Sci. Soil Manure (Japan) 41:218–224 (Japanese).

Kato, Y. 1970b. A model for amorphous matter of humic soils in Japan—a preliminary report. Pedologist 14:16–21 (Japanese).

Kato, Y. 1970c. A preliminary report on relation between properties of cation exchange materials and amorphous matters in "Kuroboku" soils of Tokai district. J. Sci. Soil Manure (Japan) 41:257–261 (Japanese).

Kawaguchi, K., H. Fukutani, H. Murakami, and T. Hattori. 1954. Ascension of pH values of neutral NaF extracts of allitic soils and semiquantitative determination of active alumina by the titration method. Bull. Res. Inst. Food Sci. Kyoto Univ. 14:82–91 (Japanese).

Kawasaki, H., and S. Aomine. 1966. So-called 14Å clay minerals in some Ando soils. Soil Sci. Plant Nutr. (Tokyo) 12:144–150.

Kelley, W. P. 1948. Cation exchange in soils. Reinhold Publishing Corp., New York.

Kinter, E. B., and S. Diamond. 1960. Pretreatment of soils and clays for measurement of external surface area by glycerol retention. Clays Clay Miner. 7:125–134.

Kita, D., R. Nakata, and M. Harada. 1969. Geochemical study on ashy soil in Kyushu District and its lime stabilization. Nendo Kagaku 9:28–40 (Japanese).

Kitagawa, Y. 1971. The "unit particle" of allophane. Am. Mineral. 56:465–475.

Klug, H. P., and L. E. Alexander. 1954. X-ray diffraction procedures. John Wiley & Sons, Inc., New York.

Kobayashi, Y., and S. Aomine. 1967. Mechanism of inhibitory effect of allophane and montmorillonite on some enzymes. Soil Sci. Plant Nutr. (Tokyo) 13:189–194.

Kubota, T. 1972. Aggregate formation of allophanic soils: Effect of drying on the dispersion of the soils. Soil Sci. Plant Nutr. (Tokyo) 18:79–87.

Kubota, T. 1973. Studies on surface properties of soil particles and formation of soil structure. p. 37–45 (Japanese). In Soc. Sci. Soil Manure, Japan (ed.) Dojo Hiryo no Kenkyu, Vol. 4. Yokendo, Tokyo.

Kumada, K., and H. Kato. 1970. Browning of pyrogallol as affected by clay minerals (I). Soil Sci. Plant Nutr. (Tokyo) 16:195–200.

Kyuma, K., and K. Kawaguchi. 1964. Oxidative changes of polyphenols as influenced by allophane. Soil Sci. Soc. Am. Proc. 28:371–374.

Lai, S., and L. D. Swindale. 1969. Chemical properties of allophane from Hawaiian and Japanese soils. Soil Sci. Soc. Am. Proc. 33:804–808.

Langston, R. B., and E. A. Jenne. 1964. NaOH dissolution of some oxide impurities from kaolins. Clays Clay Miner. 12:633–647.

Léonard, A., S. Suzuki, J. J. Fripiat, and C. De Kimpe. 1964. Structure and properties of amorphous silicoaluminas. I. Structure from X-ray fluorescence spectroscopy and infrared spectroscopy. J. Phys. Chem. 68:2608–2617.

Lindqvist, B., and S. Jannson. 1962. On the crystal chemistry of hisingerite. Am. Mineral. 47:1356–1362.

McKeague, J. A., J. E. Brydon, and N. M. Miles. 1971. Differentiation of forms of extractable iron and aluminum in soils. Soil Sci. Soc. Am. Proc. 35:33–38.

Masui, J., S. Shoji, and N. Uchiyama. 1966. Clay mineral properties of volcanic ash soils in the northeastern part of Japan. Tohoku J. Agric. Res. 17:17–36.

Mehra, O. P., and M. L. Jackson. 1960. Iron oxide removal from soils and clays by a dithionite-citrate system with sodium bicarbonate buffer. Clays Clay Miner. 7:317–327.

Mejia, G., H. Kohnke, and J. L. White. 1968. Clay mineralogy of certain soils of Colombia. Soil Sci. Soc. Am. Proc. 32:665–670.

Mitchell, B. D., V. C. Farmer, and W. J. McHardy. 1964. Amorphous inorganic materials in soils. Adv. Agron. 16:327–383.

Miyauchi, N., and S. Aomine. 1966. Mineralogy of gel-like substance in the pumice bed in Kanuma and Kitakami districts. Soil Sci. Plant Nutr. (Tokyo) 12:187–190.

Miyauchi, N., and A. Nakano. 1971. P-adsorption of the volcanic ash soil—Effects of pH and of the concentration of PO_4 in the added P solution on the amount of P adsorbed. Bull. Fac. Agric. Kagoshima Univ. 21:143–152 (Japanese).

Mizota, C., and S. Aomine. 1975. Relationships between the petrological nature and the clay mineral composition of volcanic ash soils distributed in the suburbs of Fukuoka-City, Kyushu. Soil Sci. Plant Nutr. (Tokyo) 21:93–105.

Mortland, M. M. 1970. Clay-organic complexes and interactions. Adv. Agron. 22:75–117.

Okada, K., S. Morikawa, S. Iwai, Y. Ohira, and J. Ossaka. 1975. A structure model of allophane. Clay Sci. 4:291–303.

Parfitt, R. L. 1972. Amorphous material in some Papua New Guinea soils. Soil Sci. Soc. Am. Proc. 36:683–686.

Parfitt, R. L., A. D. Thomas, R. J. Atkinson, and R. St. C. Smart. 1974. Adsorption of phosphate on imogolite. Clays Clay Miner. 22:455-456.

Patterson, S. H. 1964. Halloysitic underclay and amorphous inorganic matter in Hawaii. Clays Clay Miner. 12:153-172.

Ross, C. S., and P. F. Kerr.' 1934. Halloysite and allophane. U. S. Geol. Surv. Prof. Pap. 185 G:135-148.

Russell, J. D., W. J. McHardy, and A. R. Fraser. 1969. Imogolite: a unique alumino-silicate. Clay Miner. 8:87-99.

Sadzawka, R. M. A., A. E. Melendez, and S. Aomine. 1972. The pH of Chilean volcanic ash soils "Trumaos." Soil Sci. Plant Nutr. (Tokyo) 18(5):191-197.

Schwertmann, U. 1964. The differentiation of iron oxides in soil by extraction with ammonium oxalate solution. Z. Pflanzenernaehr. Dueng. Bodenkd. 105:194-202.

Segalen, P. 1968. Note sur une méthode de determination des produits minéraux amorphes dans certains sols á hydroxydes tropicaux Cah. ORSTOM, Sér. Pédol. 6: 105-126.

Shoji, S., and J. Masui. 1972. Amorphous clay mineral of recent volcanic ash soils. Part 3: Mineral composition of fine clay fractions. J. Sci. Soc. Manure (Japan) 43:187-193 (Japanese).

Sieffermann, G., and G. Millot. 1969. Equatorial and tropical weathering of recent basalts from Cameroon; allophane, halloysite, metahalloysite, kaolinite and gibbsite. *In* L. Heller (ed.) Proc. Int. Clay Conf. (Tokyo, Japan) 1969. 1:417-430.

Soil Survey Staff. 1960. Soil classification, a comprehensive system. 7th Approximation. SCS-USDA, U. S. Gov't. Printing Office, Washington, D. C. p. 139-141.

Suyama, K., and A. Oya. 1965. Applied geology of Kanto loam. p. 335-357 (Japanese). *In* Kanto Loam Study Group (ed.) Kanto loam. Tsukiji-Shokan, Tokyo.

Thorp, J., and G. D. Smith. 1949. Higher categories of soil classification: order, sub-order, and great soil groups. Soil Sci. 67:117-126.

Tokashiki, Y., and K. Wada. 1972. Determination of silicon, aluminum and iron dissolved by successive and selective dissolution treatments of volcanic ash soil clays. Clay Sci. 4:105-114.

Tokashiki, Y., and K. Wada. 1975. Weathering implications of the mineralogy of clay fractions of two Ando soils, Kyushu. Geoderma 14:47-62.

Tsukada, T., T. Nakano, and M. Deguchi. 1967. Two reaction stages of adsorption and precipitation of phosphorus in the soil. J. Sci. Soil Manure (Japan) 38:232-238 (Japanese).

Tsuzuki, Y., and K. Nagasawa. 1960. A study of the exothermic reaction of allophane. Adv. Clay Sci. (Tokyo) 2:377-384 (Japanese).

Udagawa, S., T. Nakada, and M. Nakahira. 1969. Molecular structure of allophane as revealed by its thermal transformation. *In* K. Heller (ed.) Proc. Int. Clay Conf. (Tokyo, Japan) 1969. 1:151-159.

Vainshtein, B. K. 1966. Diffraction of X-rays by chain molecules. Elesevier Publishing Co., Amsterdam.

van Olphen, H. 1971. Amorphous clay materials. Science 171:90-91.

van Raij, B., and M. Peech. 1972. Electrochemical properties of some Oxisols and Alfisols of the tropics. Soil Sci. Soc. Am. Proc. 36:587-593.

Wada, K. 1959. Reaction of phosphate with allophane and halloysite. Soil Sci. 87: 325-330.

Wada, K. 1966. Deuterium exchange of hydroxyl groups in allophane. Soil Sci. Plant Nutr. (Tokyo) 12(5):176-182.

Wada, K. 1967. A structure scheme of soil allophane. Am. Mineral. 52:690-708.

Wada, K., and S. Aomine. 1973. Soil development on volcanic materials during the Quaternary. Soil Sci. 116:170-177.

Wada, K., and H. Ataka. 1958. The ion-uptake mechanism of allophane. Soil Plant Food (Tokyo) 4:12-18.

Wada, K., and D. J. Greenland. 1970. Selective dissolution and differential infrared spectroscopy for characterization of "amorphous" constituents in soil clays. Clay Miner. 8:241-254.

Wada, K., and Y. Harada. 1969. Effects of salt concentration and cation species on the measured cation-exchange capacity of soils and clays. *In* L. Heller (ed.) Proc. Int. Clay Conf. (Tokyo, Japan) 1969 1:561-571.

Wada, K., and Y. Harada. 1971. Effects of temperature on the measured cation-exchange capacities of Ando soils. J. Soil Sci. 22:109–117.

Wada, K., and M. E. Harward. 1974. Amorphous clay constituents of soils. Adv. Agron. 26:211–260.

Wada, K., and T. Henmi. 1972. Characterization of micropores of imogolite by measuring retention of quaternary ammonium chlorides and water. Clay Sci. 4:127–136.

Wada, K., T. Henmi, N. Yoshinaga, and S. H. Patterson. 1972. Imogolite and allophane formed in saprolite of basalt on Maui, Hawaii. Clays Clay Miner. 20:375–380.

Wada, K., and T. Higashi. 1976. The categories of aluminum- and iron-humus complexes in Ando soils determined by selective dissolution. J. Soil Sci. 27:357–368.

Wada, K., and A. Inoue. 1974. Adsorption of monomeric silica by volcanic ash soils. Soil Sci. Plant Nutr. (Tokyo) 20(1):5–15.

Wada, K., and H. Kubo. 1975. Precipitation of amorphous aluminosilicates from solutions containing monomeric silica and aluminum ions. J. Soil Sci. 26:100–111.

Wada, K., and I. Matsubara. 1968. Differential formation of allophane, "imogolite" and gibbsite in the Kitakami pumice bed. Int. Congr. Soil Sci., Trans. 9th (Adelaide, S. Aust.) 3:123–131.

Wada, K., and Y. Tokashiki. 1972. Selective dissolution and difference infrared spectroscopy in quantitative mineralogical analysis of volcanic-ash soil clays. Geoderma 7: 199–213.

Wada, K., and S. Wada. 1976. Clay mineralogy of the B horizon of two Hydrandepts, a Torrox and a Humitropept in Hawaii. Geoderma 16:139–157.

Wada, K., and N. Yoshinaga. 1969. The structure of imogolite. Am. Mineral. 54:50–71.

Wada, K., N. Yoshinaga, H. Yotsumoto, K. Ibe, and S. Aida. 1970. High resolution electron micrographs of imogolite. Clay Miner. 8:487–489.

Watanabe, T. 1968. The study of clay minerals by small-angle X-ray scattering. Am. Mineral. 53:1015–1024.

Watanabe, Y. 1966. A study on the electrokinetic phenomena of clay particles in soil. Bull. Nat. Inst. Agric. Sci. Japan B16:91–148 (Japanese).

Whelan, J. A., and S. S. Goldich. 1961. New data for hisingerite and neotocite. Am. Mineral. 46:1412–1423.

Wiklander, L. 1964. Cation and anion exchange phenomena. p. 163–205. In F. E. Bear (ed.) Chemistry of the soil. Reinhold Publishing Corporation, New York.

Wright, A. C. S. 1964. The "Andosols" or "Humic Allophane" soils of South America. FAO World Soil Resour. Rep. 14:9–22.

Yamazaki, F., and H. Takenaka. 1965. On the influence of air-drying on Atterberg's limit. Trans. Jap. Soc. Irrig. Drain. Reclam. Eng. 14:(Nogyo Doboku Gakkai Ronbunshu)46–48 (Japanese).

Yoshida, M. 1956. Studies on adsorptive ability of the soil. Part 2: Adsorptive intensity of soil for calcium and ammonium ions. J. Sci. Soil Manure (Japan) 27:241–244 (Japanese).

Yoshida, M. 1961. Base adsorptive ability of the soil. Kagaku (Tokyo) 31:310–313 (Japanese).

Yoshida, M. 1970. Acidic properties of montmorillonite and halloysite. J. Sci. Soil Manure (Japan) 41:483–486 (Japanese).

Yoshida, M. 1971. Acidic properties of kaolinite, allophane and imogolite. J. Sci. Soil Manure (Japan) 42:329–332 (Japanese).

Yoshinaga, N. 1966. Chemical composition and some thermal data of eighteen allophanes from Ando soils and weathered pumices. Soil Sci. Plant Nutr. (Tokyo) 12(2):47–54.

Yoshinaga, N., and S. Aomine. 1962a. Allophane in some Ando soils. Soil Sci. Plant Nutr. (Tokyo) 8(2):6–13.

Yoshinaga, N., and S. Aomine. 1962b. Imogolite in some Ando soils. Soil Sci. Plant Nutr. (Tokyo) 8(3):22–29.

Yoshinaga, N., H. Yotsumoto, and K. Ibe. 1968. An electron microscopic study of soil allophane with an ordered structure. Am. Mineral. 53:319–323.

Yuan, T. L. 1973. Chemistry and mineralogy of Andepts. Soil Crop Sci. Soc. Fla. Proc. 33:101–108.

Phosphate Minerals

WILLARD L. LINDSAY, Colorado State University, Fort Collins, Colorado

PAUL L. G. VLEK, International Fertilizer Development Center, Muscle Shoals, Alabama

I. INTRODUCTION

Phosphorus deficiency in plants is second only to N as the major soil fertility problem throughout the world. For this reason, much work has been undertaken to characterize and understand the behavior of P in soils. In spite of this effort, the mineralogy of phosphates in soils is still poorly understood.

The direct determination of phosphate minerals in soils is very difficult. First of all, P is present as a minor constituent in soils, usually ranging from 0.02 to 0.5% with an average content of 0.05%. Thus, phosphate minerals comprise only a small part of the total inorganic matrix of soils. A considerable fraction of this P is present as organic forms or as adsorbed P, which diminishes that present as minerals. Microbiological transformations continually add to or subtract from the P in soil solution. Attempts to identify phosphate minerals through solubility relationships are often met with problems caused by sluggish equilibrium relationships involving numerous and complex mineral transformations.

In spite of these problems, progress is being made to understand which phosphate minerals can form in soils and how they influence the solubility of P.

The phosphate minerals that are important to soils are briefly described in this chapter and the solubility relationships of several of these minerals in soils are demonstrated.

II. CLASSIFICATION AND OCCURRENCE OF PHOSPHATE MINERALS

Phosphate minerals form under a wide variety of environmental conditions ranging from silicate melts, to natural soils, to ocean floors. In nearly all naturally occurring phosphates, P is pentavalent, even though tri, quadri, and hexavalent P compounds are readily synthesized.

The crystal structures of most orthophosphates have been worked out by X-ray diffraction. In all cases the central P atom is surrounded by four O atoms forming an approximate tetrahedral arrangement (space group Td_4). This configuration is possible because of the formation of four σ-bonds after sp_3 hybridization and additional π-bonding using d-orbitals. The structural formulas of these compounds are represented as having a double bond in order to satisfy classical valency requirements, but some sharing of the multiple-bond character occurs among the four O atoms (Fig. 17-1).

In nature, the formation of stable atomic structures containing PO_4-

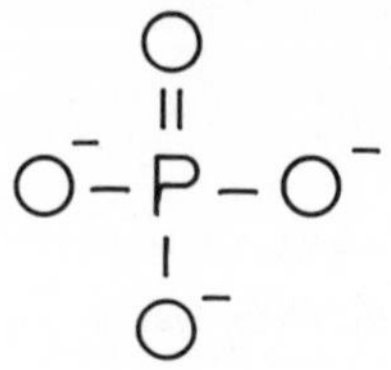

chapter 17

Fig. 17-1. Orthophosphate ion.

tetrahedra is accomplished through the high affinity of PO_4 for cations, particularly those exhibiting eightfold coordination. This results in mineral species that can be classified by a system similar to that used for silicate minerals—that is, framework, chain, and layer phosphates (Povarennykh, 1972). Each of these groups can be subdivided according to hydration and other complexities in chemical composition. Complex minerals contain more than one kind of cation that may or may not be isomorphically substitutional, whereas simple minerals contain only one such cationic species. This system of classification was adopted for use in this text. It is less complex than the outdated approach of Dana's System of Mineralogy (Palache et al., 1951), or the more recent classification of Strunz (1970), neither of which uses the mineral structure (framework, chain, or layer) as a classifying criterion.

The elements with which phosphate reacts to form minerals and the number of minerals in which each element occurs are shown in the periodic table of Fig. 17-2. At the bottom of the table are shown the numbers of minerals found in each of the structural categories.

Under natural conditions, the PO_4-tetrahedra do not polymerize. Many modern fertilizers, however, contain condensed phosphates in which the PO_4-tetrahedra are linked by a common O atom (Fig. 17-3). A phosphate tetrahedron can share up to three of its O atoms yielding one-, two-, or three-dimensional networks and still retain the fundamental Td_4 arrangement. Linear polyphosphates such as pyrophosphates (P_2O_7) and triphosphate (P_3O_{10}) are most common in polyphosphate fertilizers. The polyphosphates are metastable in soils and eventually hydrolyze to orthophosphates as the P-O-P linkages are ruptured.

Normally, substances classified as minerals are compounds that form and persist in nature on a geological time scale. The polyphosphates are considered man made and are, therefore, not included in mineralogical systems. Other reactive transitory compounds forming in nature, such as soil-fertilizer reaction products, tend to elude detection as a distinctive mineral substance. These unclassified "mineral" substances have been studied extensively in recent years, and many of their properties have been reported by Lehr et al. (1967).

Povarennykh's structural classification of phosphate minerals is presented in Table 17-1. Included in this table are minerals from the Dana's System of Mineralogy classification as well as some unclassified phosphate "minerals" that are considered to be important in soils. The most important soil minerals are italicized and discussed separately in the text that follows.

PHOSPHATES

	Ia	IIa	IIIa	IVa	Va	VIa	VIIa	VIIIa	Ib	IIb	IIIb	IVb	Vb	VIb	VIIb	VIIIb
1															H 154	
2	Li 7	Be 10										C 1	N 5	O 172	F 11	
3	Na 19	Mg 14									Al 49	Si 1	P 172	S 5	Cl 2	
4	K 5	Ca 49	Sc 1				Mn 32	Fe 51, Ni 1	Cu 12	Zn 10						
5		Sr 6	Y 2													
6		Ba 5										Pb 10	Bi 1			
7				Th 4	U 25	Ce 4										

	COORDINATION		FRAMEWORK		RING		INSULAR		CHAIN		LAYER	
	SIMPLE	COM-PLEX	SIMPLE	COM-PLEX	SIMPLE	COM-PLEX	SIMPLE	COM-PLEX	SIMPLE	COM-PLEX	SIMPLE	COM-PLEX
			8	12			33	67	6	4	29	13

Fig. 17-2. Elements of the periodic table found in naturally occurring phosphate minerals. The numbers indicate the number of minerals containing each element (from Povarennykh, 1972, p. 526).

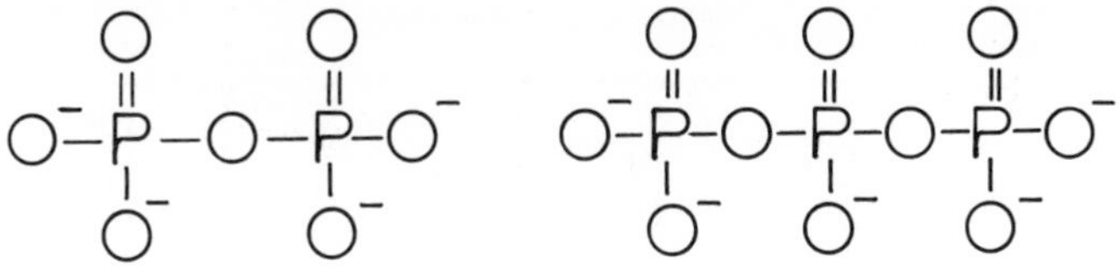

[1]   [2]

Fig. 17-3. Pyrophosphate [1] and triphosphate [2] ions.

Table 17-1. Structural classification of phosphate minerals (modified after Povarennykh, 1972, p. 526)

Anhydrous	Hydrated
Framework minerals	
Berlinite group, $Al(PO_4)$	*Variscite* group, $(Al,Fe)(PO_4)\cdot2H_2O$
Beryllonite group, $NaBe(PO_4)$	Kehoeite group, $CaZn_3(H_3AlP)_8O_{48}\cdot16H_2O$
Hurlbutite group, $CaBe_2(PO_4)_2$	*Hopeite* group, $ZnZn_2(PO_4)_2\cdot4H_2O$
Amblygonite group, $LiAl(PO_4)(OH,F)$	Anapaite group, $Ca_2Fe(PO_4)_2\cdot4H_2O$
	Eosphorite group, $(Mn,Fe)Al(PO_4)(OH)_2\cdot H_2O$
	Turquoise group, $Cu(Al,Fe)_6(PO_4)_4(OH)_8\cdot 4H_2O$
Insular minerals	
(Simple)	(Simple)
Heterosite group, $(Mn,Fe)(PO_4)$	Rhabdophane group, $Ce(PO_4)\cdot H_2O$
Xenotime group, $Y(PO_4)$	Phosphoferrite group, $(Mn,Fe)_3(PO_4)_2\cdot3H_2O$
Monazite group, $Ce(PO_4)$	Hureaulite group, $H_2Mn_5(PO_4)_4\cdot4H_2O$
Whitlockite group, $Ca_3(PO_4)_2$	Beraunite group, $Fe_3(PO_4)_2(OH)_3\cdot2H_2O$
Lithiophosphate group, $Li_3(PO_4)$	Wavellite group, $Al_3(PO_4)_2(OH)_3\cdot5H_2O$
Augelite group, $Al_2(PO_4)(OH)_3$	Veszelyite group, $(Zn,Cu)_3(PO_4)(OH)_3\cdot2H_2O$
Libethenite group, $CuCu(PO_4)(OH)$	*Octocal* group, $Ca_8H_2(PO_4)_6\cdot5H_2O$
Pseudomalachite group, $Cu_5(PO_4)_2(OH)_4$	
Cornetite group, $Cu_3(PO_4)(OH)_3$	
Triplite group, $(Mn,Fe)_2(PO_4)F$	
Apatite group, $Ca_2Ca_3(PO_4)_3(OH,F)$	
(Complex)	(Complex)
Graftonite group, $CaMn_2Fe_3(PO_4)_4$	*Struvite* group, $NH_4Mg(PO_4)\cdot6H_2O$
Arrojadite group, $Na_2(Mn,Fe)_5(PO_4)_4$	Overite group, $Ca_3Al_8(PO_4)_8(OH)_6\cdot15H_2O$
Hagendorfite group, $Na_2(Mn,Fe)_2Fe(PO_4)_3$	Morinite group, $Na_2Ca_4Al_4(PO_4)_4O_2F_6\cdot 5H_2O$
Triphylite group, $Li(Mn,Fe)(PO_4)$	Wardite group, $NaAl_3(PO_4)_2(OH)_4\cdot2H_2O$
Crandallite group, $CaAl_3(PO_4)_2(OH)_5(H_2O)$	*Minyulite* group, $KAl_2(PO_4)_2(OH)\cdot4H_2O$
Dufrenite group, $Fe_3Fe_6(PO_4)_4(OH)_{12}$	Roscherite group, $CaMnFeBe_3(PO_4)_3(OH)_3\cdot 2H_2O$
Rockbridgeite group, $(Mn,Fe)Fe_4(PO_4)_3(OH)_5$	
Lazulite group, $(Mg,Fe)Al_2(PO_4)_2(OH)_2$	
Palermoite group, $SrAl_2(PO_4)_2(OH)_2$	
Brazilianite group, $NaAl_3(PO_4)_2(OH)_4$	
Griphite group, $Na_3CaMn_4Al_2(PO_4)_5(OH)_4$	
Belovite group, $NaSr_3Ce(PO_4)(OH)$	
Isokite group, $CaMg(PO_4)F$	

(continued on next page)

Table 17-1. Continued.

Anhydrous	Hydrated

Chain minerals

Väyrynenite group, $MnBe(PO_4)(OH)$

Monetite group, $CaHPO_4$

Moraesite group, $Be_2(PO_4)(OH) \cdot 4H_2O$

Ludlamite group, $Fe_2Fe(PO_4)_2 \cdot 4H_2O$

Fairfieldite group, $Ca_2(Mn,Fe)(PO_4)_2 \cdot 2H_2O$

Layer minerals

Parsonsite group, $Pb(UO_2)(PO_4)_2$

Herderite group, $CaBe(PO_4)(OH,F)$

Uranium mica group, $R(UO_2)_2(PO_4)_2 \cdot nH_2O$

Churchite group, $Y(PO_4) \cdot 2H_2O$

Vivianite group, $Fe_{3-x} Fe_x (PO_4)_2(OH)_x$ $(H_2O)_{8-x}$

Taranakite group, $K_3H_6Al_5(PO_4)_8 \cdot 18H_2O$

Phosphuranylite group, $Ca(UO_2)_4(PO_4)_2$ $(OH)_4 \cdot 8H_2O$

Dumontite group, $Pb_2(UO_2)_3(PO_4)_2(OH)_4 \cdot$ $3H_2O$

Metavauxite-laueite group, $FeAl_2(PO_4)_2(OH)_2 \cdot 8H_2O$ $MnFe(PO_4)_2(OH)_2 \cdot 8H_2O$

Polyphosphates

Unclassified

1. BERLINITE GROUP—Framework Mineral, Trigonal

Berlinite (alphite) $AlPO_4$

This mineral is homostructural with quartz. The PO_4 and AlO_4 tetrahedra alternate and are linked along vertices.

2. VARISCITE GROUP—Framework Structure, Orthorhombic and Monoclinic

Barrandite $(Al,Fe)[PO_4](H_2O)_2$

Metavariscite $Al[PO_4](H_2O)_2$

Metastrengite $Fe[PO_4](H_2O)_2$

Koninckite $Fe[PO_4](H_2O)_3$

In the variscite group, Al and Fe show a coordination number of six (four O atoms and two water molecules). The four O of the (Al, Fe) octahedron are shared with four different P-tetrahedra. Barrandite has perfect Al–Fe isomorphism, and is the most widely distributed form of the variscite-strengite series. Variscite and strengite are the end members of this isomorphous series. Metavariscite and metastrengite have analogous chemical compositions, but the packing arrangement is of lower (monoclinic) symmetry. Variscites are widely distributed in nature and are found as reaction products of phosphate fertilizers in soils.

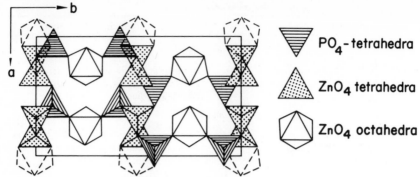

Fig. 17-4. Structure of hopeite: framework of Zn polyhedra and P tetrahedra (from Povarennykh, 1972b, p. 533).

3. HOPEITE GROUP—Framework Structure, Orthorhombic, Monoclinic, or Triclinic

Hopeite	$ZnZn_2[PO_4](H_2O)_4$
Phosphophyllite	$FeZn_2[PO_4](H_2O)_4$
Parahopeite	$ZnZn_2[PO_4](H_2O)_4$

The hopeite structure is related to that of goethite and is built from PO_4 tetrahedra, ZnO_4 tetrahedra, and octahedral $ZnO_4(H_2O)_2$ groups, as can be seen from Fig. 17-4. Hopeite is widely distributed as a secondary mineral in Zn deposits.

4. WHITLOCKITE GROUP—Insular Structure, Trigonal

Whitlockite	$Ca_3[PO_4]_2$
Beta trical	$\beta\text{-}Ca_3[PO_4]_2$
Alpha trical	$\alpha\text{-}Ca_3[PO_4]_2$
Alphaprime trical	$\alpha'\text{-}Ca_3[PO_4]_2$

Whitlockite has been found to form in nature and is an important constituent of dental calculus. The formation and stability of whitlockite are enhanced by the presence of cations smaller than Ca^{2+}, such as Fe^{2+} and Mg^{2+}. The structurally slightly different $\beta\text{-}Ca_3(PO_4)_2$ is formed at higher temperatures used in some fertilizer manufacturing processes (Gregory et al., 1974). Alpha trical is formed at temperatures exceeding 1180C (1453K) whereas formation of α' trical requires temperatures of over 1430C (1703K). Complete structural formulas for these minerals have not been published.

5. APATITE GROUP—Insular Structure, Hexagonal

Apatite	$Ca_2Ca_3[PO_4]_3(OH,F)$
Dahllite	$Ca_4Ca_6[PO_4]_6CO_3$
Francolite	$Ca_4Ca_6[PO_4]_{6-x}(CO_3)_x(F,OH)_{2+x}$
Chlorapatite	$Ca_2Ca_3[PO_4]_3Cl$

Voelckerite	$Ca_4Ca_6[PO_4]_6O$
Stronapatite	$Sr_3Ca_2[PO_4]_3F$
Plumapatite	$Pb_2Pb_3[PO_4]_3OH$

The apatite structure consists of columns of Ca and O atoms parallel to the unique axis forming trigonal prisms. Calcium(I) lies on a threefold axis and is surrounded by six plus three O forming somewhat twisted trigonal prisms, one standing on the other. The six nearest atoms in the Ca(II) polyhedron lie at vertices of a trigonal prism as triplets, linked via edges into columns along the c-axis. These columns of two types of prisms are connected via PO_4 tetrahedra (Fig. 17–5). The F, OH, Cl, and O lie at the axis of the prisms. A detailed description of the structure of apatite including a unit-cell diagram is given by Kay et al. (1964).

From the synthesis of apatites, as well as structural and compositional studies of natural materials, it was learned that many substitutions are possible in the general formula $A_{10}(XO_4)_6Z_2$ of apatite:

A = Ca, Sr, Mn, Pb, Mg, Ba, Zn, Cd, and possibly Fe and Cu among divalent cations,

A = Na, K, Rb, and possibly Ag among monovalent cations,

A = Sc, Y, Bi, and possibly Ti among trivalent cations, and

A = U, and probably Th and Zr among quadrivalent cations.

XO_4 = PO_4, SiO_4, SO_4, AsO_4, VO_4, CrO_4, BeO_4, probably CO_3F and CO_3OH, BO_4, and possibly GeO_4, AlO_4, and FeO_4.

Z = F, OOH, Cl, Br, I, and possibly O in Voelckerite, which replace one O in the Ca(II) trigonal prism.

These substitutions are not equally permissible. Many substitutions are coupled. In the anionic substitution in the Ca(II) trigonal prisms, only OH and F show perfect isomorphism resulting in the end-members of the isomorphous series, hydroxyapatite, and fluorapatite. An extensive listing of apatite subgroups is given by Kreidler (1967).

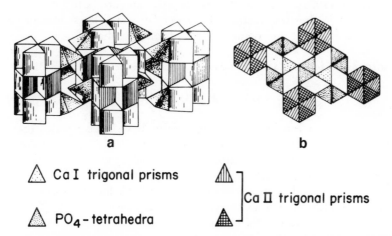

$\triangle$ Ca I trigonal prisms

$\triangle$ PO$_4$ - tetrahedra

Ca II trigonal prisms

Fig. 17–5. Structure of apatite in polyhedra: *(a)* general view, and *(b)* projection on (0001) (from Povarennykh, 1972, p. 542).

Only a few varieties of apatite are found in the widespread primary phosphates in the lithosphere. Fluorapatite is the accessory mineral of igneous rocks and the primary enriched phase of carbonatites. Francolite is the primary marine phosphate in sedimentary rock. Dahllite makes up the mineral matter of fossil bone, while hydroxyapatite is found in guano-altered limestone (Altschuler, 1973).

Weathering of primary phosphates gives rise to the formation of an interesting sequence of transitory phosphate minerals (Altschuler, 1973), some of which are incidentally identified in soils. An example of such a para-genetic sequence is the Quarternary weathering of argillaceous phosphates which essentially consists of the progressive replacement of apatite and clay Ca–Al phosphates.

Apatite inclusions were identified in several primary minerals of soil parent material (Cescas & Tyner, 1970). Secondary hydroxyapatite and fluor-apatite may form as fertilizer reaction products (Lindsay et al., 1962).

6. CRANDALLITE GROUP—Insular Structure, Trigonal

Plumbogummite Subgroup:

Plumbogummite	$PbAl_3[PO_4]_2(OH)_5H_2O$
Crandallite	$CaAl_3[PO_4]_2(OH)_5H_2O$
Goyazite	$SrAl_3[PO_4]_2(OH)_5H_2O$
Gorceixite	$BaAl_3[PO_4]_2(OH)_5H_2O$
Lusungite	$SrFe_3[PO_4]_2(OH)_5H_2O$

Minerals of the plumbogummite subgroup are part of an isostructural series in which there are three distinct structural positions for cations. Phosphorus occupies a tetrahedral position; an octahedral position is filled with Fe or Al; and the third position is filled by any of a large number of cations, e.g., Na, K, NH_4, Ag, H_3O, Ca, Sr, Ba, Pb, etc.

Norrish (1957), and later others, showed that gorceixite and crandallite were present in the clay fraction of a variety of soils. These minerals must be regarded as one of the common forms of soil phosphate.

7. OCTOCALCIUM PHOSPHATE (OCP)—Insular Structure, Triclinic

$$OCP \quad Ca_8H_2[PO_4]_6(H_2O)_5$$

The structure is related to hydroxyapatite (HA), but is not isostructural. OCP contains an alternating "nonapatite" layer with water molecules. Despite the relationship with hydroxyapatite, solid solutions of octocalcium phosphate and hydroxyapatite are unlikely. The relationship between the structure and chemical properties of OCP and HA has been discussed by Brown et al. (1962). The existence of the water-rich layer in OCP (see Fig. 17-6) explains the ease with which this compound loses water and the difficulty of establishing a definite degree of hydration.

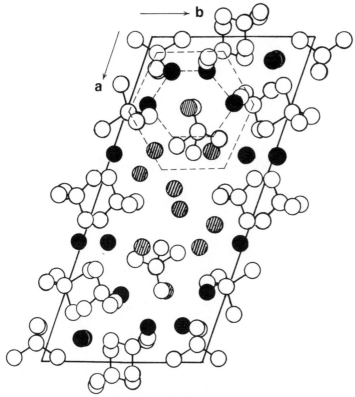

Fig. 17-6. Structure of octacalcium phosphate $Ca_8H_2(PO_4)_6 \cdot 5H_2O$. Triclinic (c) axis projection. Full circles denote Ca and shaded circles, water molecules. Open circles denote P or O in a tetrahedral arrangement. Approximate part-hexagons of P atoms and Ca ions are denoted by broken lines (after Corbridge, 1966).

Octocalcium phosphate is found as a fertilizer reaction product and as a product in the ammoniation of HNO_3-extracts of phosphate rock (Brown et al., 1962): It is also found as a constituent in bones and teeth.

8. STRUVITE GROUP—Insular Structure, Orthorhombic, Monoclinic, Triclinic

Newberyite	$HMg[PO_4](H_2O)_3$
Phosphorrosslerite	$HMg[PO_4](H_2O)_7$
Struvite	$NH_4Mg[PO_4](H_2O)_6$
	$KMg[PO_4](H_2O)_6$
Schertelite	$(NH_4)_2H_2Mg[PO_4]_2(H_2O)_4$
Stercorite	$NH_4NaH[PO_4](H_2O)_8$
Hannayite	$(NH_4)_2H_4Mg_3[PO_4]_4(H_2O)_8$

Structural relationships have been established only for newberyite, which has $Mg(H_2O)_6$-octahedra linked via three O vertices of PO_4-tetrahedra. The three H_2O molecules at the other vertices are linked by H bonds to O atoms in PO_4 radicals. Minerals of the struvite group occur in guano deposits. Struvite, schertelite, hannayite, and $MgKPO_4$ form as soil-fertilizer reaction products. A dehydration product of struvite, $NH_4MgPO_4 \cdot H_2O$, is a principal constituent in some slowly soluble commercial fertilizers (Lehr et al., 1967).

9. MINYULITE GROUP—Insular Structure, Orthorhombic or Monoclinic

Minyulite $KAl_2[PO_4]_2OH(H_2O)_4$
Leucophosphite $KFe_2[PO_4]_2OH(H_2O)_2$

The structures of these minerals have not yet been published. Leucophosphite has been reported from Liberia, and minyulite from Western Australia. Both are possible soil-fertilizer reaction products. Leucophosphite is present in some mixed fertilizers.

10. MONETITE GROUP—Chain Structure, Triclinic

Monetite $CaH[PO_4]$

The most characteristic feature of monetite or anhydrous dicalcium phosphate (DCPA) is the double chain type structure; the PO_4 tetrahedra are linked via H in the bent parts of the discontinuous chains, which are connected by Ca with coordination of 9.

Monetite is an important phosphate in ammoniated superphosphates, nitric phosphates, and feed-grade phosphates. It is reported as a fertilizer reaction product in soils.

11. CHURCHITE GROUP—Layer Structure, Monoclinic

Brushite $CaH[PO_4](H_2O)_2$
Ardealite $Ca_2H[PO_4][SO_4](H_2O)_4$
Monocalcium phosphate monohydrate $CaH_4[PO_4]_2H_2O$

Brushite (dicalcium phosphate dihydrate) and ardealite are homostructural with gypsum. Brushite has corrugated $Ca(PO_4)$ layers perpendicular to the b-axis, consisting of chains that lie at different levels and are firmly connected via Ca–O bonds. The layers have H_2O molecules that provide the hydroxyl-hydrogen bonds that connect them, as can be seen in Fig. 17–7. The mineral is found in guano deposits and some other locations. Synthetic brushite occurs in large amounts in ammoniated superphosphates, nitric phosphates, and feed-grade phosphates along with anhydrous dicalcium phosphate (monetite). It has been found as a major initial reaction product of phosphate fertilizers in soils.

Monocalcium phosphate monohydrate (MCP) is built from sheets similar to brushite. In MCP the sheets are held together by PO_4 ions that serve the same function as one of the water molecules in the brushite structure (see Fig. 17–8). Monocalcium phosphate monohydrate is an important

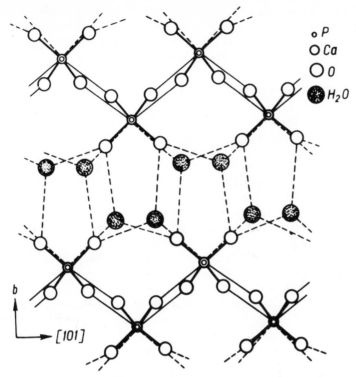

Fig. 17-7. Structure of brushite; the hydroxyl–hydrogen bonds to H_2O molecules are shown by broken lines (Povarennykh, 1972, p. 542).

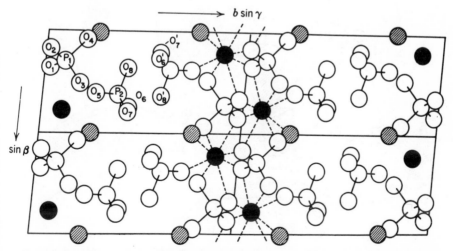

Fig. 17-8. The structure of monocalcium phosphate monohydrate $Ca(H_2PO_4)_2 \cdot H_2O$. Contents of four unit cells projected down the (a) axis. Coordination of Ca ions shown by broken lines and water molecules denoted by shaded circles (from Corbridge, 1966).

constituent of superphosphate produced by acidulation of phosphate rock with H_2SO_4, H_3PO_4, or HNO_3. It is too soluble to persist in soils longer than a few hours, but does give rise to an intriguing sequence of chemical reactions of phosphates in soils (Lindsay & Stephenson, 1959a, b).

12. VIVIANITE GROUP—Layer Structure, Monoclinic

Vivianite $(Fe^{2+}_{3-x}, Fe^{3+}_x)[PO_4]_2(OH)_x(H\,O)_{8-x}$

Bobierrite $Mg_3[PO_4]_2(H_2O)_8$

The structure of vivianite is constructed from PO_4 tetrahedra, $FeO_2(OH_2)_4$ and $Fe_2O_6(OH)_4$ octahedra, the latter being a double group in which two octahedra share a common edge. The single and paired octahedra are linked via PO_4 tetrahedra into uneven layers parallel to (010) which are connected by weak H_2O–H_2O hydroxyl bonds (c.f. Fig. 17-9). Bobierrite has a similar structure.

Oxidation causes a continuous composition change in vivianite, Fe^{2+} being isomorphously replaced by Fe^{3+}, valency being compensated by replacement of H_2O by OH.

Bobierite is a well-known secondary mineral found mainly in guano and rarely in sedimentary apatitic phosphate deposits. Vivianite is a product of weathering associated with Fe-bearing minerals, waterlogged clays, sedimentary deposits, and alluvium containing organic matter (Palache et al., 1951).

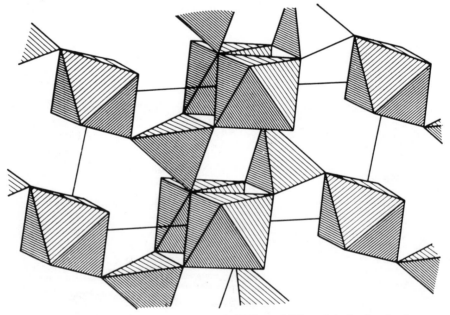

Fig. 17-9. Structure of vivianite, with $FeO_2(OH)_4$ and PO_4 polyhedra forming layers parallel to (010) (from Povarennykh, 1972, p. 558).

13. TARANAKITE GROUP—Layer Structure, Trigonal

NH$_4$-taranakite $(NH_4)_3H_6Al_5[PO_4]_8(H_2O)_{18}$
Taranakite $K_3H_6Al_5[PO_4]_8(H_2O)_{18}$
Englishite $K_2Ca_4Al_8[PO_4]_8(OH)_{10}(H_2O)_9$

The structure of the taranakites is yet unknown, but layers presumably consist of Al and PO$_4$ tetrahedra, perhaps with K atoms, between which water molecules are situated (Smith & Brown, 1959).

Ammonium and potassium taranakite have been found as reaction products of ammonium and potassium phosphate fertilizers in soils (Lindsay et al., 1962). Potassium taranakite has been reported in some of the arid soils of Australia (Bannister & Hutchinson, 1947).

14. POLYPHOSPHATES

Polyphosphates are formed by dehydrating orthophosphoric acid at elevated temperatures, and the anionic species so formed may assume linear, cyclic, or combination configurations (also referred to as ultraphosphates). Linear polyphosphates can be represented by $(P_nO_{3n+1})^{n+2(-)}$, with the simplest and most important example pyrophosphate, $(P_2O_7)^{4-}$. A variety of Na, K, NH$_4$, Ca, Mg, and Al salts have thus been formed. Interest in these salts arises mainly from the successful development of commercial condensed phosphates as fertilizers and from the introduction of superphosphoric acid that contains polyphosphates. Extensive research has been carried out by scientists of the Tennessee Valley Authority (TVA) and many of the crystalline polyphosphates have been characterized. A compilation of polyphosphate crystallographic properties was published by Lehr et al. (1967).

The pyrophosphate ion consists of two PO$_4$ tetrahedra sharing a common O atom, or "bridge O." This symmetrical anion is subject to a variety of distortions in solid-state pyrophosphate compositions, depending upon packing arrangement and coordination requirements of the cation species. The P–O–P bridge bond may be bent ($< 180°$), the terminal tetrahedra may rotate to either a *cis*- or *trans*-configuration, and some or all of the P–O bond lengths may be altered from their normal value.

This variety of possible stereo-configurations gives rise to the frequently observed polymorphism of pyrophosphate compounds, where the differing degrees of asymmetry due to P–O distortions are readily seen on the infrared P–O absorption spectra.

The bent configuration is probably present in most crystalline pyrophosphate salts. Infrared spectra from solutions and Raman spectra from molten salts also indicate a nonlinear pyrophosphate configuration. It has been inferred from IR measurements that the high temperature form of the divalent salt Mg$_2$P$_2$O$_7$ contains anions which are centrosymmetrical and linear (or at least statistically so), whereas the low temperature form contains bent anions with P–O–P angles of around $160°$.

Many ammonium pyro- and polyphosphates are of interest as possible products of the ammoniation of superphosphoric acid. Several of these salts have been prepared by Frazier et al. (1965). Important members of this group are $(NH_4)_4P_2O_7$, $(NH_4)_4P_2O_7 \cdot H_2O$, $(NH_4)_3HP_2O_7$, $(NH_4)_3HP_2O_7 \cdot H_2O$, and $(NH_4)_2H_2P_2O_7$. Addition of these highly soluble pyrophosphates to soils may result in the formation of various products, e.g., $Al(NH_4)_2P_2O_7(OH) \cdot 2H_2O$, $Ca(NH_4)_2P_2O_7 \cdot H_2O$, $Ca(NH_4)HP_2O_7$, $Ca_2(NH_4)H_3(P_2O_7)_2 \cdot 3H_2O$, $Mg(NH_4)_2P_2O_7 \cdot 4H_2O$, and $Mg(NH_4)_2P_2O_7 \cdot 2H_2O$, depending on the solubility of the associated cations in the fertilized soil.

Brown et al. (1963) prepared 25 calcium pyrophosphates, many of which may be found in fertilizers prepared from condensed phosphates or their hydrolysis products. Eight of these salts were $CaNH_4$ pyrophosphates and 10 were CaK pyrophosphates.

As with most other divalent pyrophosphates, $Ca_2P_2O_7$ can occur in various forms depending on temperature and other factors.

$$CaHPO_4 \xrightarrow{} \gamma\text{-}Ca_2P_2O_7 \xrightarrow{} \beta\text{-}Ca_2P_2O_7 \xrightarrow{} \alpha\text{-}Ca_2P_2O_7$$

$$\begin{array}{ccc} 200\text{-}240C & 700\text{-}750C & 1140C \\ (473\text{-}513K) & (973\text{-}1023K & (1413K) \end{array}$$

Other calcium pyrophosphates of importance as fertilizer or fertilizer reaction products are the monoclinic and triclinic dimorphs of $Ca_2P_2O_7 \cdot 2H_2O$, orthorhombic $Ca_2P_2O_7 \cdot 4H_2O$, $CaH_2P_2O_7$, and $Ca_3H_2(P_2O_7)_2 \cdot H_2O$.

III. EQUILIBRIA OF PHOSPHATE MINERALS IN SOILS

Aslying (1954) used phosphate and lime potentials to relate P solubility to that of calcium phosphates. Similar approaches were used for aluminum phosphates (Clark & Peech, 1955; Lindsay et al., 1959), and later Lindsay and Moreno (1960) developed a unified solubility diagram of soils that included calcium, aluminum, and iron phosphates on the same diagram. The use of phosphate solubility diagrams has become well established.

In this chapter, solubility diagrams are used to depict the stability of phosphate minerals in soil environments. New and revised stability constants have been selected from the literature to provide up-to-date information. Examples are given to show the effects of redox potential and hydrolysis reactions on mineral transformations of phosphates in soils.

The equilibrium reactions used to develop the diagrams are summarized in Table 17–2. As future investigations provide more accurate constants, the revised values should be substituted for those used herein.

Table 17-2. Equilibrium constants for various phosphate reactions at 25C (298K)

Reaction no.	Equilibrium reaction	Log K°	Reference
	Orthophosphoric acid		
1	$H_3PO_4^0 \rightleftharpoons H^+ + H_2PO_4^-$	−2.15	NBS†
2	$H_2PO_4^- \rightleftharpoons H^+ + HPO_4^{2-}$	−7.21	NBS
3	$HPO_4^{2-} \rightleftharpoons H^+ + PO_4^{3-}$	−12.35	NBS
4	$H_2PO_4^- \rightleftharpoons 2H^+ + PO_4^{3-}$	−19.56	NBS
	Pyrophosphoric acid		
5	$H_4P_2O_7^0 \rightleftharpoons H^+ + H_3P_2O_7^-$	−1.54	NBS
6	$H_3P_2O_7^- \rightleftharpoons H^+ + H_2P_2O_7^{2-}$	−2.27	NBS
7	$H_2P_2O_7^{2-} \rightleftharpoons H^+ + HP_2O_7^{3-}$	−6.67	NBS
8	$HP_2O_7^{3-} \rightleftharpoons H^+ + P_2O_7^{4-}$	−9.31	NBS
	Calcium phosphates		
9	$Ca(H_2PO_4)_2 \cdot H_2O \rightleftharpoons Ca^{2+} + 2H_2PO_4^- + H_2O$	−1.21	NBS
10	$CaHPO_4 \cdot 2H_2O \rightleftharpoons Ca^{2+} + HPO_4^{2-} + 2H_2O$ (brushite)	−6.60	Patel et al. (1974)
11	$CaHPO_4 \rightleftharpoons Ca^{2+} + HPO_4^{2-}$ (monetite)	−6.90	McDowell et al. (1971)
12	$Ca_8H_2(PO_4)_6 \cdot 5H_2O \rightleftharpoons 8Ca^{2+} + 2H^+ + 6PO_4^{3-}$ (octocalcium phosphate) $+ 5H_2O$	−93.96	NBS
13	$Ca_3(PO_4)_2 \rightleftharpoons 3Ca^{2+} + 2PO_4^{3-}$ (β-tricalcium phosphate)	−28.92	Gregory et al. (1974)
14	$Ca_5OH(PO_4)_3 \rightleftharpoons 5Ca^{2+} + OH^- + 3PO_4^{3-}$ (hydroxyapatite)	−58.20	Avnimelech et al. (1973)
15	$Ca_5F(PO_4)_3 \rightleftharpoons 5Ca^{2+} + F^- + 3PO_4^{3-}$ (fluoroapatite)	−60.43	Farr & Elmore (1962)
16	$Ca_5Cl(PO_4)_3 \rightleftharpoons 5Ca^{2+} + Cl^- + 3PO_4^{3-}$ (chloroapatite)	−43.86	Duff (1972b), NBS
17	$Ca_5Br(PO_4)_3 \rightleftharpoons 5Ca^{2+} + Br^- + 3PO_4^{3-}$ (bromoapatite)	−42.78	Duff (1972b), NBS
18	$Ca_{10}O(PO_4)_6 + 2H^+ \rightleftharpoons 10Ca^{2+} + H_2O + 6PO_4^{3-}$ (oxyapatite)	−67.75	Duff (1972a), NBS
19	$Ca_4O(PO_4)_2 + 2H^+ \rightleftharpoons 4Ca^{2+} + H_2O + 2PO_4^{3-}$ (hilgenstockite)	−14.67	Duff (1971b), NBS
20	$Ca_2FPO_4 \rightleftharpoons 2Ca^{2+} + F^- + PO_4^{3-}$ (spodiosite)	−18.74	Duff (1971c), NBS
21	$Ca_2ClPO_4 \rightleftharpoons 2Ca^{2+} + Cl^- + PO_4^{3-}$	−18.64	Duff (1972b), NBS
22	$Ca_2BrPO_4 \rightleftharpoons 2Ca^{2+} + Br^- + PO_4^{3-}$	−19.39	Duff (1972d), NBS
	Aluminum phosphates		
23	$AlPO_4 \cdot 2H_2O \rightleftharpoons Al^{3+} + PO_4^{3-} + 2H_2O$ (variscite)	−22.52	Taylor & Gurney (1964)

(continued on next page)

Table 17-2. Continued.

Reaction no.	Equilibrium reaction	Log K°	Reference
24	$H_6K_3Al_5(PO_4)_8 \cdot 18H_2O \rightleftharpoons 6H^+ + 3K^+ + 5Al^{3+}$ (K-Taranakite) $\quad 8PO_4^{3-} + 18H_2O$	−178.7	Taylor & Gurney (1964)
25	$H_6(NH_4)_3Al_5(PO_4)_8 \cdot 18H_2O \rightleftharpoons 6H^+ + 3NH_4^+$ (NH$_4$-Taranakite) $\quad + 5Al^{3+} + 8PO_4^{3-} + 18H_2O$	−175.5	Taylor & Gurney

Iron phosphates

26	$FePO_4 \cdot 2H_2O \rightleftharpoons Fe^{3+} + PO_4^{3-} + 2H_2O$ (strengite)	−26.43	Nriagu (1972a)
27	$Fe_3(PO_4)_2 8H_2O \rightleftharpoons 3Fe^{2+} + 2PO_4^{3-} + 8H_2O$ (vivianite)	−36.0	Nriagu (1972b)

Magnesium phosphates

28	$MgHPO_4 \cdot 3H_2O \rightleftharpoons Mg^{2+} + HPO_4^{2-} + 3H_2O$ (newberyite)	−5.82	Taylor & Gurney
29	$Mg_3(PO_4)_2 \cdot 8H_2O \rightleftharpoons 3Mg^{2+} + 2PO_4^{3-} + 8H_2O$ (bobierrite)	−25.20	Taylor & Gurney
30	$Mg_3(PO_4)_2 \cdot 22H_2O \rightleftharpoons 3Mg^{2+} + 2PO_4^{3-} + 22H_2O$	−23.10	Taylor & Gurney (1963a)
31	$Mg_3OH(PO_4)_3 \rightleftharpoons 5Mg^{2+} + OH^- + 3PO_4^{3-}$ (Mg-hydroxyapatite)	−47.37	Duff (1971d)
32	$Mg_5F(PO_4)_3 \rightleftharpoons 5Mg^{2+} + F^- + 3PO_4^{3-}$ (Mg-fluoropatite)	−42.93	Duff (1971d)
33	$Mg_4O(PO_4)_2 + 2H^+ \rightleftharpoons 4Mg^{2+} + H_2O + 2PO_4^{3-}$ (Mg-oxyapatite)	−13.79	Duff (1971d)
34	$Mg_2F(PO_4) \rightleftharpoons 2Mg^{2+} + F^- + PO_4^{3-}$ (wagnerite)	−17.20	Duff (1971d)
35	$MgNH_4PO_4 \cdot 6H_2O \rightleftharpoons Mg^{2+} + NH_4^+ + 6H_2O$ (struvite)	−13.15	Taylor et al. (1963b)
36	$MgKPO_4 \cdot 6H_2O \rightleftharpoons Mg^{2+} + K^+ + PO_4^{3-} + 6H_2O$	−10.62	Taylor et al. (1963b)

Polyphosphates

37	$Ca_2P_2O_7 \cdot 2H_2O \rightleftharpoons 2Ca^{2+} + P_2O_7^{4-} + 2H_2O$	−14.7	Sillen & Martell (1964)
38	$Ca^{2+} + OH^- + P_2O_7^{4-} \rightleftharpoons CaOHP_2O_7^{3-}$	8.9	Sillen & Martell (1964)
39	$Ca^{2+} + HP_2O_7^{3-} \rightleftharpoons CaHP_2O_7^-$	3.6	Sillen & Martell (1964)
40	$H_2P_2O_7^{2-} + H_2O \rightleftharpoons 2H_2PO_4^-$	2.31	NBS

Other phosphates

41	$Zn_3(PO_4)_2 \cdot 4H_2O \rightleftharpoons 3Zn^{2+} + 2PO_4^{3-} + 4H_2O$	−35.3	Nriagu (1973a)

(continued on next page)

Table 17-2, continued

Reaction no.	Equilibrium reaction	Log K°	Reference
42	$Cd_3(PO_4)_2 \rightleftharpoons 3Cd^{2+} + 2PO_4^{3-}$	−32.61	NBS
43	$PbHPO_4 \rightleftharpoons Pb^{2+} + HPO_4^{3-}$	−11.43	Nriagu (1972c)
44	$Pb_3(PO_4)_2 \rightleftharpoons 3Pb^{2+} + 2PO_4^{3-}$	−44.4	Nriagu (1972c)
45	$Pb_5OH(PO_4)_3 \rightleftharpoons 5Pb^{2+} + OH^- + 3PO_4^{3-}$	−76.8	Nriagu (1972c)
46	$Pb_5Cl(PO_4)_3 \rightleftharpoons 5Pb^{2+} + Cl^- + 3PO_4^{3-}$	−84.4	Nriagu (1973b)
47	$Pb_5F(PO_4)_3 \rightleftharpoons 5Pb^{2+} + F^- + 3PO_4^{3-}$	−71.6	Nriagu (1973c)
48	$Pb_5Br(PO_4)_3 \rightleftharpoons 5Pb^{2+} + Br^- + 3PO_4^{3-}$	−78.1	Nriagu (1973c)

Phosphate ion pairs

Reaction no.	Equilibrium reaction	Log K°	Reference
49	$Ca^{2+} + H_2PO_4^{2-} \rightleftharpoons CaH_2PO_4^+$	0.71	Gregory et al. (1970)
50	$Ca^{2+} + HPO_4^{2-} \rightleftharpoons CaHPO_4^0$	2.41	Gregory et al. (1970)
51	$Fe^{2+} + H_2PO_4^+ \rightleftharpoons FeH_2PO_4^+$	2.7	Nriagu (1972b)
52	$Fe^{2+} + HPO_4^{2-} \rightleftharpoons FeHPO_4^0$	3.6	Nriagu (1972b)
53	$Fe^{3+} + H_2PO_4^- \rightleftharpoons FeH_2PO_4^{2+}$	5.43	Nriagu (1972a)
54	$Mg^{2+} + HPO_4^{2-} \rightleftharpoons MgHPO_4^0$	2.91	Taylor et al. (1963a)
55	$Zn^{2+} + H_2PO_4^- \rightleftharpoons ZnH_2PO_4^+$	1.6	Nriagu (1973a)
56	$Zn^{2+} + HPO_4^{2-} \rightleftharpoons ZnHPO_4^0$	3.3	Nriagu (1973a)
57	$Pb^{2+} + H_2PO_4^- \rightleftharpoons PbH_2PO_4^+$	1.5	Nriagu (1972c)
58	$Pb^{2+} + HPO_4^{2-} \rightleftharpoons PbHPO_4^0$	3.1	Nriagu (1972c)
59	$Na^+ + HPO_4^{2-} \rightleftharpoons NaHPO_4^-$	0.85	Patel et al. (1974)

Other reactions

Reaction no.	Equilibrium reaction	Log K°	Reference
60	$H_2O \rightleftharpoons H^+ + OH^-$	−14.00	NBS
61	$Soil\text{-}Ca \rightleftharpoons Soil + Ca^{2+}$	−3.00	Selected (see text)
62	$CaF_2(fluorite) \rightleftharpoons Ca^{2+} + 2F^-$	−9.84	NBS
63	$Al(OH)_3(gibbsite) \rightleftharpoons Al^3 + 3OH^-$	−33.96	Singh (1974)
64	$Fe(OH)_3(soil) \rightleftharpoons Fe^{3+} + 3OH^-$	−39.5	Norvell (1971)
65	$Fe_3O_4(magnetite) + 8H^+ + 2e^- \rightleftharpoons 3Fe^{2+} + 4H_2O$	29.76	NBS
66	$FeSiO_3 + 2H^+ \rightleftharpoons SiO_2(soil) + Fe^{2+} + H_2O$	8.18	Sillen & Martell (1964) Elgawhary & Lindsay (1972)
67	$Fe_2SiO_4(fayalite) + 4H^+ \rightleftharpoons SiO_2(soil) + 2Fe^{2+} + 2H_2O$	18.33	NBS Elgawhary & Lindsay (1972)
68	$SiO_2(quartz) \rightleftharpoons SiO_2(soil)$	−0.90	Elgawhary & Lindsay (1972)
69	$Fe^{3+} + e^- \rightleftharpoons Fe^{2+}$	13.01	NBS
70	$CaCO_3(calcite) \rightleftharpoons Ca^{2+} + CO_2(g) + H_2O$	9.85	NBS

† National Bureau of Standards, Wagman et al. (1968, 1969, 1971), and Parker et al. (1971).

A. Phosphate Species in Soil Solution

Numerous phosphate ions, ion pairs, and complexes are present in a soil solution. The orthophosphates are the most stable forms of phosphate in soils. The pH-dependency of the orthophosphoric acid species is shown by Reactions 1-4 of Table 17-2 and in Fig. 17-10. In the normal pH range of soils, $H_2PO_4^-$ and HPO_4^{2-} are the dominant species. At pH 7.21, the two ions have equal activities and their ratio changes by a factor of 10 for each unit change in pH. Knowledge of the activity of any one of the orthophosphoric acid species and pH permits calculation of the activity of other species (Reactions 1-4 of Table 17-2).

Polyphosphates also form several ionic species in solution. The pH-dependency of the pyrophosphate ions can be seen from Reactions 5-8 in Table 17-2. When pyrophosphates are present, the $H_2P_2O_7^{2-}$ and the $HP_2O_7^{3-}$ ions are the predominant species in solution in the pH range of most soils. At pH 6.6, these two ions have equal activities and their ratio changes by a factor of 10 for each unit change in pH. Polyphosphate species such as triphosphate, tetraphosphates, etc. also have pH-dependent ionic species. Although many modern phosphate fertilizers contain polyphosphates, these are metastable in soils and eventually hydrolyze to orthophosphate, as will be shown later in this chapter.

Phosphate ions combine with many of the divalent and trivalent cations to form ion pairs and complexes. Some of the more important complexes

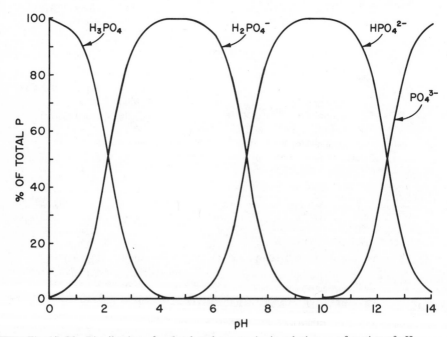

Fig. 17-10. Distribution of orthophosphate species in solution as a function of pH.

whose formation constants are known are listed in Reactions 49–59 of Table 17-2. Whenever ionic activities are calculated from total P in solution, these ion pairs and complexes must be considered. Phosphorus is also a component in soil organic matter and to the extent that soluble complexes are present, they, too, contribute to the total P in soil solution.

B. A Unified Phosphate Solubility Diagram

A unified phosphate solubility diagram for soils can be developed if appropriate estimates are made for the activities of other constituent ions of the phosphate minerals. The estimated activities of such ions in soils are shown in Fig. 17-11. The activities of Ca^{2+}, Mg^{2+}, K^+, Na^+, NH_4^+, and Cl^- are taken at $10^{-3}M$. Leaching losses in soils generally prevent the activities of these soluble ions from going much higher. In calcareous soils, Ca^{2+} activity is depressed by the solubility of $CaCO_3$. Increasing CO_2 further depresses Ca^{2+} activity as demonstrated in Fig. 17-11. The reference level of Fe^{3+} in soils is considered to be that imposed by $Fe(OH)_3(soil)$ corresponding to a $pK_{sp} = 39.5$ (Norvell, 1971). The reference level for Al^{3+} is taken as that in equilibrium with $Al(OH)_3(gibbsite)$. The activity of F^- is limited in soils by the presence of $CaF_2(fluorite)$ when Ca^{2+} is determined as indicated above.

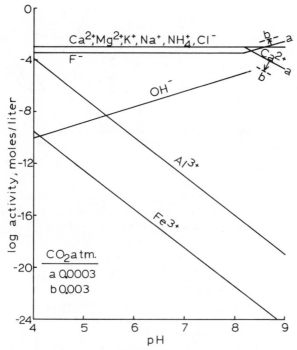

Fig. 17-11. The activities of various ions in soils used for comparing the solubilities of phosphate minerals on a unified phosphate solubility diagram.

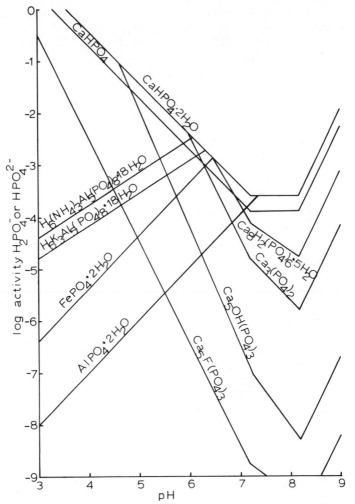

Fig. 17-12. A unified solubility diagram for various Ca, Fe, and Al phosphate minerals in soils.

The reference activity of NH_4^+ is shown as $10^{-3}M$, but in well-oxidized soils it may drop considerably below this level due to the oxidation of NH_4^+ to NO_3^-.

The solubilities of various iron, aluminum, and calcium phosphate minerals in soils are plotted on the unified solubility diagrams in Fig. 17-12 under the conditions that the other constituent ions are fixed as indicated in Fig. 17-11. This diagram expresses phosphate solubility in terms of the predominant ionic species, that is, $H_2PO_4^-$ below pH 7.21 and HPO_4^{2-} above this pH. In acid soils, $AlPO_4 \cdot 2H_2O$(variscite) is shown as the stable mineral followed by $FePO_4 \cdot 2H_2O$(strengite) and then $H_6K_3Al_5(PO_4)_8 \cdot 18H_2O$(potassium taranakite) and $H_6(NH_4)_3Al_5(PO_4)_8 \cdot 18H_2O$(ammonium taranakite), respectively. In alkaline soils calcium phosphates are the most stable minerals.

They decrease in solubility in the order $CaHPO_4 \cdot 2H_2O$(brushite) $> CaHPO_4$ (monetite) $> Ca_8H_2(PO_4) \cdot 5H_2O$(octocalcium phosphate $> \beta$-$Ca_3(PO_4)_2(\beta$-tricalcium phosphate) $> Ca_5OH(PO_4)_3$(hydroxyapatite) $> Ca_5F(PO_4)_3$(fluorapatite).

Points falling above a given line in Fig. 17-12 indicate soil solutions that are supersaturated with respect to the mineral represented by the line, and points falling below that line indicate undersaturation. Supersaturation permits precipitation of the mineral while undersaturation permits its dissolution. Often the rates of mineral transformation are very slow and difficult to predict except through experimentation. The presence of $CaCO_3$ depresses the activity of Ca^{2+} and allows HPO_4^{2-} activity in equilibrium with each of the calcium phosphates to increase by 100-fold for each unit rise in pH. These relationships are clearly illustrated in Fig. 17-12. Increasing CO_2 depresses the activity of Ca^{2+} and allows the activity of HPO_4^{2-} to increase in calcareous soils.

If the activity of F^- in soils is less than that imposed by CaF_2(fluorite), then the $Ca_5F(PO_4)_3$ solubility line would move upward and may rise above that of $Ca_5OH(PO_4)_3$. Under such conditions, fluorapatite would be less stable than hydroxyapatite.

Initial precipitates of iron and aluminum hydroxides are often more soluble than $Fe(OH)_3$(soils) and $Al(OH)_3$(gibbsite) shown in Fig. 17-11. This causes the $FePO_4 \cdot 2H_2O$ and the $AlPO_4 \cdot 2H_2O$ isotherms in Fig. 17-12 to shift downward temporarily.

Many soils do not contain gibbsite, and their Al^{3+} activity is governed by the formation of secondary clay minerals such as montmorillonite or kaolinite (Rai & Lindsay, 1975). Under these conditions, the $AlPO_4 \cdot 2H_2O$ isotherm in Fig. 17-12 would rise accordingly.

Liming acid soils generally increases the solubility of P, but overliming can be expected to depress P solubility due to the formation of more insoluble calcium phosphates. The pH of maximum phosphate solubility in soils is generally in the range of 6.0 to 6.5. The rates of precipitation of strengite, variscite, hydroxyapatite, and fluorapatite are generally slow and often include imperfect crystals with isomorphous substitutions. The resulting precipitates show a range of solubilities somewhat above that of the pure crystalline minerals. Slowly with time the solubilities of the amorphous products may be expected to approach the lines indicated here. Reversing conditions such as changes in pH, redox, or other factors may temporarily reverse the direction of the reactions and tend to keep phosphates from reaching the very low energy states represented by the more insoluble minerals indicated in Fig. 17-12.

Whenever other phosphate minerals are identified and reliable solubility constants are available, they, too, can be included in Fig. 17-12. Recently, Duff (1971a-c; 1972a-d) reported solubility values for several additional phosphates including $Ca_5Cl(PO_4)_3$(chloroapatite), $Ca_5Br(PO_4)_3$(bromoapatite), $Ca_{10}O(PO_4)_6$(oxyapatite), $Ca_4O(PO_4)_2$(hilgenstockite), Ca_2FPO_4(spodiosite), Ca_2ClPO_4, and Ca_2BrPO_4. Since his values for the log K_{sp} for hydroxyapatite (−48.32) and for fluorapatite (−47.99) differed greatly from those included in Table 17-2 (Reactions 14 and 15), it was considered premature to plot them until this anomaly is clarified. Duff's reported values for the above compounds are included in Table 17-2 as Reactions 16-22.

The solubility of $H_6(NH_4)_3Al_5(PO_4)_8 \cdot 18H_2O$ (ammonium taranakite) is slightly greater than that of the K analog (Fig. 17-12). Under the conditions that the activity of NH_4^- drops somewhat below $10^{-3}M$ as may result during nitrification, the ammonium taranakite would be even less stable. In the vicinity of fertilizer granules containing $NH_4H_2PO_4$ or $(NH_4)_2HPO_4$, the activity of NH_4^+ may temporarily be several molar, which would depress the ammonium taranakite isotherm of Fig. 17-12 and cause ammonium taranakite to precipitate as a major initial reaction product (Lindsay et al., 1962).

C. Solubility of Magnesium Phosphates

Magnesium forms a number of orthophosphate minerals similar to those of Ca. The solubilities of several of the magnesium phosphates are given in Reactions 28-36 of Table 17-2 and they are plotted in Fig. 17-13. In this plot, only the HPO_4^{2-} ion was included since it is the predominant phosphate

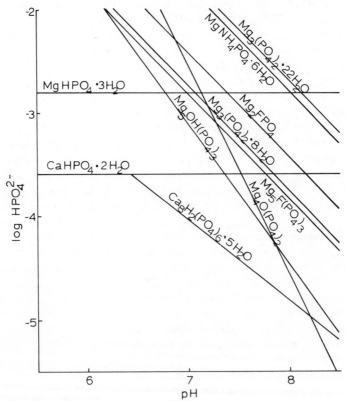

Fig. 17-13. The solubility of magnesium phosphates in soils compared to the solubility of $CaHPO_4 \cdot 2H_2O$ and $Ca_8H_2(PO_4)_6 \cdot 5H_2O$.

species in the pH range where magnesium phosphates are most stable. The solubilities of $CaHPO_4 \cdot 2H_2O$ and $Ca_8H_2(PO_4)_6 \cdot 5H_2O$ are included for comparison. It is obvious that the magnesium minerals which include $MgHPO_4 \cdot 3H_2O$(new beryite), $Mg_3(PO_4)_2 \cdot 22H_2O$, $Mg_3(PO_4)_2 \cdot 8H_2O$(bobierrite), Mg_5OH $(PO_4)_3$, $Mg_5F(PO_4)_3$, $Mg_4O(PO_4)_2$, $Mg_2F(PO_4)$(wagnerite), $MgNH_4PO_4 \cdot 6H_2O$ (struvite), and $MgKPO_4 \cdot 6H_2O$ are considerably less stable (more soluble) than the calcium phosphates. The mineral $MgKPO_4 \cdot 6H_2O$ is too soluble to be included in Fig. 17-13.

Many of these magnesium phosphates form as initial reaction products of phosphate in soils (Lindsay et al., 1962; Taylor et al., 1963a,b), but later disappear as more stable minerals lower the P below their equilibrium levels. The importance of magnesium phosphates as permanent fixation products of P in soils can be discounted. On the other hand, magnesium phosphate fertilizer can be expected to supply readily available P to plants.

D. Effect of Redox on Phosphate Stability

When the supply of O_2 to soils is cut off, the soils become reduced and the stabilities of many minerals are affected. Such changes can affect the solubility and stability of many phosphate minerals. An example of these changes is illustrated for $FePO_4 \cdot 2H_2O$(strengite) and for $Fe_3(PO_4)_2 \cdot 8H_2O$ (vivianite) in Fig. 17-14. Iron in strengite is present as Fe^{3+}, whereas in vivianite it is present as Fe^{2+}.

The parameter, pe + pH, is useful for indicating the redox conditions in soils. The term "pe" is the –log of electron activity (e^-) similar to the use of pH for the –log of proton activity (H^+). A pe + pH of zero represents a highly reduced environment corresponding to 1 atm of H_2. The pe + pH corresponding to 1 atm of O_2, on the other hand, is 20.78, and that in equilibrium with air with 20% O_2 is approximately 20.61. Thus, natural environments are limited to the pe + pH range of 0 to 20.61.

By combining appropriate reactions from Table 17-2, it is possible to show that $Fe(OH)_3$(soil) is stable when pe + pH $>$ 16.77 and that Fe_3O_4 (magnetite) is stable below this point. The presence of $Fe(OH)_3$(soil) fixes the solubility of ferric phosphates such as strengite, but the presence of magnetite permits Fe^{3+} to drop, allowing the activity of $H_2PO_4^-$ in equilibrium with strengite to increase as shown for pe + pH of 16.77, 15, and 12 in Fig. 17-14. With reduction, the activity of Fe^{2+} increases and thereby depresses $H_2PO_4^-$ in equilibrium with vivianite as shown for pe + pH values ranging from 6 to 2 in Fig. 17-14. Thus, moderate reducing conditions cause strengite to dissolve, but very strong reductions are necessary for vivianite to precipitate. Caution must be used in interpreting the conclusions indicated here too literally. They are based on present knowledge of values given for the solubility of the various solid phases involved. Future research will undoubtedly require that many of these relationships be recalculated. but the principles involved are valid. Other chemical transformations in addition to those mentioned here also occur when soils are reduced and some of these re-

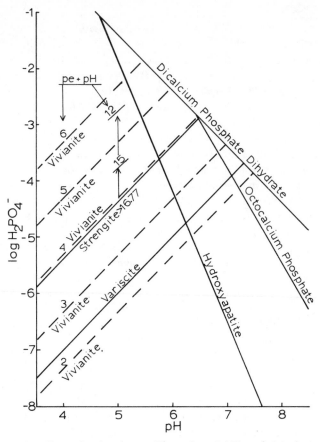

Fig. 17-14. The effect of redox (pe + pH) on the solubility of vivianite and strengite compared to that of Ca phosphates.

actions aré expected to affect either directly or indirectly the stabilities of phosphate minerals. Further research is needed to elucidate this area of work.

E. Polyphosphate Equilibria in Soil

Polyphosphates are metastable in soils and slowly hydrolyze to ortho-phosphates. The reaction for the hydrolysis of pyrophosphates can be shown as

$$H_2P_2O_7^{2-} + H_2O \rightleftharpoons 2H_2PO_4^{-} \tag{1}$$

This reaction can be combined with the solubility expression for $Ca_2P_2O_7 \cdot 2H_2O$ to show the instability of this pyrophosphate mineral in soils:

$$\log K°$$

$$Ca_2P_2O_7 \cdot 2H_2O + H_2O \rightleftharpoons 2Ca^{2+} + P_2O_7^{4-} + 2H_2O \qquad -14.70$$
$$2Ca^{2+} + 2\ Soil \rightleftharpoons 2\ Soil\text{-}Ca \qquad 2(3.00)$$
$$P_2O_7^{4-} + 2H^+ \rightleftharpoons H_2P_2O_7^{2-} \qquad 15.98$$
$$H_2P_2O_7^{2-} + H_2O \rightleftharpoons 2H_2PO_4^- \qquad 2.31$$

$$Ca_2P_2O_7 \cdot 2H_2O + Soil + 2H^+ \rightleftharpoons 2\ Soil\text{-}Ca + 2H_2PO_4^- + H_2O \qquad 9.59 \quad [2]$$

A plot of Eq. [2] is shown as the uppermost line in Fig. 17–15. It is obvious that the activity of $H_2PO_4^-$ needed for equilibrium with $Ca_2P_2O_7 \cdot 2H_2O$ exceeds the solubility of even $CaHPO_4 \cdot 2H_2O$. Consequently, the reaction depicted by Eq. [2] continues to the right until $Ca_2P_2O_7 \cdot 2H_2O$ is completely dissolved. Similar relationships can be developed for other polyphosphates insofar as reliable solubility and hydrolysis data are available.

The solubility of pyrophosphates in soils can also be seen from Fig. 17–15 where the pH-dependency of the $P_2O_7^{4-}$, $HP_2O_7^{3-}$, $H_2P_2O_7^{2-}$, $H_3P_2O_7^{-}$, and $H_4P_2O_7^0$ are included.

The mobility of pyrophosphate per se is greatest in acid soils where the

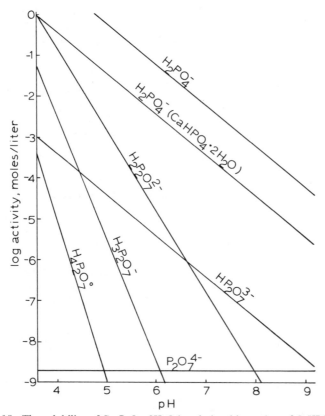

Fig. 17–15. The solubility of $Ca_2P_2O_7 \cdot 2H_2O$ in relationship to that of $CaHPO_4 \cdot 2H_2O$.

protonated pyrophosphate species are most abundant. The hydrolysis of polyphosphates is also greatest under acid soils where the reactant species such as $H_2P_2O_7^{2-}$ as depicted in Eq. [1] are most abundant. Equivalent hydrolysis reactions could be given for the other pyrophosphate species. It must be recognized that the relationships depicted here are thermodynamic relationships. Kinetic relationships having to do with the rates of hydrolysis are known to be affected by microorganisms.

F. Equilibria of Trace Element Phosphates

Many trace elements in soils form insoluble phosphates. Some of the trace element phosphates of Zn, Cd, and Pb whose solubilities have been recently reported are given in Reactions 41-48 of Table 17-2. In order to depict the solubilities and mineral transformations of these phosphates in soils, it is first necessary to know how the activities of the trace elements contained in these minerals are governed in soils. The details of these reactions are too numerous to be included here.

Trace elements are generally present in soils at levels of a few parts per million, whereas the average P content exceeds 500 ppm. This means that even though trace element phosphate minerals form in soils, the trace elements would most likely not govern phosphate solubility, but it is possible that such phosphates may govern the activity of certain trace elements. For example, $Pb_5Cl(PO_4)_3$ is sufficiently insoluble as to depress the activity of Pb^{2+} to approximately $10^{-10}M$ in the pH range of most soils. This is sufficiently low as to account for the low solubility and immobility of Pb in soils. Further investigations are needed to explore these many interesting possibilities with the trace element phosphates as reliable solubility information becomes available.

IV. IDENTIFICATION OF PHOSPHATE MINERALS IN SOILS

The content of P minerals in soils is generally too low to permit direct identification by petrographic methods, X-ray diffraction, or IR-spectroscopy. These traditional methods have been used successfully to identify P minerals in rock by comparison to the crystallographic properties of reference minerals. Little progress has been made in devising means of identifying phosphate minerals in soils.

The detailed crystallographic properties required to identify phosphate minerals are not included in this text. Instead, the reader is referred to compilations such as the *Powder Diffraction Card File* (J. V. Smith, 1969), *The Inorganic Index to Powder Diffraction File* (J. V. Smith, 1968), *KWIC* (key word in context) *Guide to Powder Diffraction File* (G. G. Johnson & Vand, 1968), and *Crystallographic Properties of Fertilizer Compounds* (Lehr et al., 1967). The latter reference includes many of the polyphosphate minerals.

Physical techniques for obtaining high-grade mineral samples are discussed by Muller (1967). The use of these techniques for separating P minerals in soils is often disappointing since most of the inorganic P in soils is in the clay fraction. A particle-size fractionation is often used to concentrate phosphate minerals with the clay fraction. Further concentration of clay-size phosphates is tedious and time consuming (Norrish, 1968). Only a few phosphate minerals have been successfully hand-picked from natural soils. Once P compounds have been separated from the soil, direct petrographic examination is useful for their identification (Lehr & Brown, 1958).

Lack of success in using direct methods for identifying phosphate minerals in soils has led to the use of indirect methods based on solubility relationships, synthesis of phosphate minerals, and the differential solubility of minerals in selected extracting solutions (Chang & Jackson, 1957). With the recent introduction of the microprobe and electron microscope, new attempts have been made for direct identification of phosphate minerals in soils.

Lindsay and Stephenson (1959) developed a method to identify the reaction products of phosphate fertilizers in soils without the difficult task of having to separate reaction products from the soil material. Fertilizer solutions were reacted with soil to simulate the zone of reactions occurring near fertilizer bands. The solution phase removed during these reactions was often supersaturated to several phosphate minerals and these minerals continued to precipitate following filtration. Reaction products obtained in this way were easily amenable to chemical, X-ray, and petrographic examination. Lindsay et al. (1962) thus identified approximately 30 crystalline phosphate compounds as well as colloidal precipitates of variable composition as reaction products following the addition of various fertilizer solutions to soils and soil constituents.

Bell and Black (1970) compared various methods of mineral identification of crystalline phosphates produced by interaction of orthophosphate fertilizer with soils. Optical examination of the soil-fertilizer reaction zone was found to be most successful followed by the optical examination of glass fiber or filter paper inclusions in the soil. X-ray diffraction examination of either soil or paper inclusion was less effective in identifying reaction products.

A semiquantitative method for direct measurement of soil apatite was developed by Skipp and Matelski (1960). After concentrating the apatites in the heavy mineral fraction by heavy liquid separation columns, this fraction was treated with a few drops of concentrated sulfuric acid. Needle-like $CaSO_4$ crystals developed on surfaces of the apatite particles, which could be readily observed and counted.

Norrish (1968) devised a method to concentrate plumbogummite minerals in kaolinite rich soils by concentrating the P mineral in the fine-clay fraction of the soil and then dissolving the kaolinite in hydrofluoric acid. By allowing HF attack on the clay for only 2 min, dissolution of the phosphates was minimized. The effectiveness of this method of concentration is demon-

Table 17-3. Separation of P in soils by particle size separation and HF attack

Soil parameter	Soil number			
	1	13	31	2
% P in soil	0.25	0.018	0.11	0.34
% P in 0.5-5 μm	1.10	0.090		
Yield of 0.5-5 μm, % of soil	17.5	9.55		
% P of soil in 0.5-5 μm	77.0	48.0		
% P in HF residue, 0.5-5 μm	5.4	4.2	3.5	2.5
Residue as % of soil	2.74	0.15	0.75	1.66
% P of soil in residue	59.0	35.0	24.0	12.0

strated in Table 17-3. Identification of plumbogummite minerals was thus possible by X-ray diffraction patterns of the clay residue (Norrish, 1968). The chemical complexity of the soil phosphates prevented identification beyond the mineral group. However, in a few cases the Ca member, vandallite, was identified by chemical analyses of the residue and by X-ray fluorescence. Although plumbogummite minerals have been identified in soils of low P content, most of the soils containing these minerals have higher than average amounts of P.

Sarma and Krishna Murti (1969) devised a method of concentrating phosphate minerals of the plumbogummite group in soils that contain a considerable amount of micaceous minerals and quartz. Hydrofluoric acid was allowed to react for 24 hours with a small amount of sample previously freed of iron oxides. The supernatant was centrifuged off, and the residue was washed with 6N HCl, methanol, and benzene, respectively. X-ray patterns revealed the presence of plumbogummite minerals in samples that showed no sign of these minerals prior to this treatment.

A powerful new tool for the identification of phosphate minerals is the electron microprobe analyzer. In this method, a sample is bombarded with energetic electrons that generate X-rays characteristic of the element to be examined in the sample. The back-scattering of incident electrons produces an image of the sample that is suitable for qualitative analyses. The chemical composition of regions of samples as small as 1 μm^3 can be obtained without destruction of the sample. This method is particularly useful for examining fertilizer precipitates around granules and small particles. Supplemental petrographic examination of the sites of interest should be conducted to provide additional information regarding crystallization sequence relationships.

Norrish (1968) used the microprobe to study the association of P with Ba and various rare earth elements. A soil aggregate was impregnated with plastic and sectioned. The analytical line scans of this sample are given in Fig. 17-16, which show that P is associated with Ba and Ca.

Qureshi et al. (1969) conducted microprobe analyses on an Fe–Mn concretion and on an orange-brown cutan of oriented clay. In both pedologic features it was found that P showed closest association with Fe. In some soils Ca was found associated with P as in plumbogummite type minerals, detrital apatite, or monazite.

Sawhney (1973) studied the P mineralogy of some soils and lake bot-

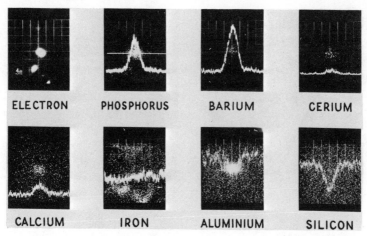

Fig. 17-16. Electron microprobe analyses of a natural soil aggregate, showing the association of P with Ca, Ba, and rare earths in coarse clay grains (from Norrish, 1968).

tom sediments with the electron microprobe. After particle-size fractionation of samples, he imbedded a few sand grains in resin, then polished and carbon-roasted them. The morphology of the grain was pictured on a cathode-ray oscilloscope, and multi-element analyses of the total grain were carried out. Most phosphatic grains were complex phosphate precipitates comparable to the plumbogummite minerals found by Norrish (1968) and characterized by the pattern shown in Fig. 17-17. Some grains consisted of Ca and P with a pattern similar to that of apatite (Fig. 17-18).

Approximately 30 phosphate minerals have been identified in soils. However, the chemistry of soil phosphates and their transformation are very complex and identification of these minerals by either direct or indirect methods needs further investigation.

V. CONCLUSIONS

Much progress has been made in identifying and characterizing various phosphate minerals. Most of this work has been done in regard to mineralogical studies of rocks and minerals, and very little progress has been made with direct identification of phosphate minerals in soils. The complexity of phosphate minerals in soils and the fact that P comprises $< 1\%$ of the total soil mass add to the problem of identification.

There is considerable need to search for additional phosphate minerals and to measure their thermodynamic properties. This information is needed to provide solubility values with which to relate P solubility in soils. Further detailed study of P chemistry in soils, ion-pair formation, and organic complexes of accompanying cations and anions is needed. This information will permit a rigorous examination of phosphate mineral solubility and stability in soils.

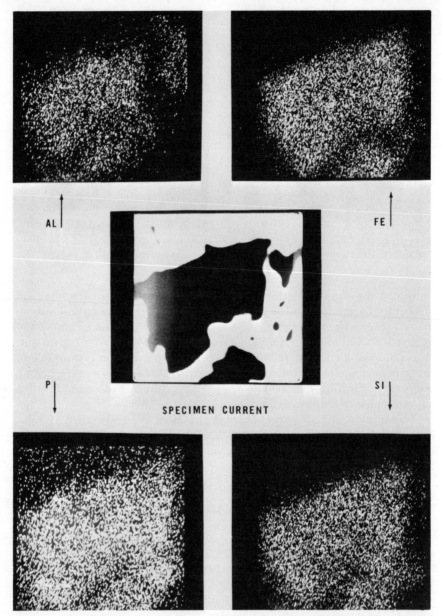

Fig. 17-17. The distribution of different elements in a phosphate grain, approximately 20 μm by 15 μm. Center-specimen current picture showing the morphology of the grain. The sample was scanned with the electron beam to give approximately 10,000 counts for each element (from Sawhney, 1973).

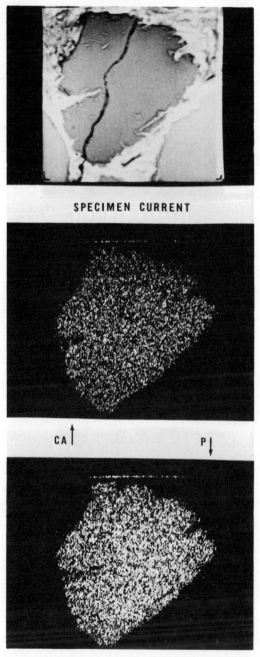

Fig. 17-18. Specimen current picture *(top)* and X-ray images showing the distribution of Ca *(middle)* and P *(bottom)* in calcium phosphate crystal. The sample was scanned with the electron beam to give about 10,000 counts for each element (from Sawhney, 1973).

LITERATURE CITED

Altschuler, Z. S. 1973. The weathering of phosphate deposits—Geochemical and environmental aspects. p. 33–96. *In* E. J. Griffith, A. Beeton, J. M. Spencer, and D. T. Mitchell (ed.) Environmental phosphorus handbook. John Wiley & Sons, New York.

Aslyng, H. C. 1954. The lime and phosphate potentials of soils; the solubility and availability of phosphates. Roy. Vet. Agric. Coll., Copenhagen, Denmark Yearbook reprint p. 1–50.

Avnimelech, Y., E. C. Moreno, and W. E. Brown. 1973. Solubility and surface properties of finely divided hydroxyapatite. J. Res. Nat. Bur. Stand. 77A:149–155.

Bannister, F. A., and G. E. Hutchinson. 1947. Minervite and palmerite = taranakite. Mineral. Mag. 28:31–35.

Bell, L. C., and C. A. Black. 1970. Comparison of methods for identifying crystalline phosphates produced by interaction of orthophosphate fertilizers with soils. Soil Sci. Soc. Am. Proc. 34:579–582.

Brown, W. E., J. P. Smith, J. R. Lehr, and A. W. Frazier. 1962. Octacalcium phosphate and hydroxyapatite. Nature 196:859–1055.

Brown, E. H., J. R. Lehr, J. P. Smith, and A. W. Frazier. 1963. Preparation and characterization of some calcium phosphates. J. Agric. Food Chem. 11:214–222.

Cescas, M. P., and E. H. Tyner. 1970. Distribution of apatite and other mineral inclusions in rhyolitic pumice ash and beach sands from New Zealand: An electron-microprobe study. J. Soil Sci. 21:79–84.

Chang, S. C., and M. L. Jackson. 1957. Fractionation of soil phosphorus. Soil Sci. 84:133–144.

Clark, J. S., and M. Peech. 1955. Solubility criteria for the existence of calcium and aluminum phosphates in soils. Soil Sci. Soc. Am. Proc. 19:171–174.

Corbridge, D. E. C. 1966. The structural chemistry of phosphorus compounds. p. 57–394. *In* Topics in phosphorus chemistry, Vol. 3. Interscience Publishers, New York-London-Sidney.

Duff, E. J. 1971a. Orthophosphates. Part II. The transformations brushite→fluoroapatite and monetite→fluoroapatite and aqueous potassium fluorite solution. J. Chem. Soc. (A):33–38.

Duff, E. J. 1971b. Orthophosphates. Part III. The hydrolysis of secondary calcium orthophosphates. J. Chem. Soc. (A):917–921.

Duff, E. J. 1971c. Orthophosphates. Part IV. Stability relationships of orthophosphates within the systems $CaO-P_2O_5-H_2O$ and $CaF_2-CaO-P_2O_5-H_2O$ under aqueous conditions. J. Chem. Soc. (A):921–926.

Duff, E. J. 1971d. Orthophosphates. Part VIII. The transformation of newberryite into bobierrite in aqueous alkaline solutions. J. Chem. Soc. (A):2736–2740.

Duff, E. J. 1972a. Orthophosphates VII. Thermodynamic considerations concerning the stability of oxyapatite, $Ca_{10}O(PO_4)_6$, in aqueous media. J. Inorg. Nucl. Chem. 34:859–871.

Duff, E. J. 1972b. Orthophosphates IX. Chloroapatite: Phase relationships under aqueous conditions along the $Ca_5F(PO_4)_3-$ and $Ca_5Cl(PO_4)_3$ and $Ca_5OH(PO_4)_3-$ $Ca_5Cl(PO_4)_3-$ joins of the system $CaO-CaCl_2-CaF_2-P_2O_5-H_2O$. J. Inorg. Nucl. Chem. 34:859–871.

Duff, E. J. 1972c. Orthophosphates X. The stability of the magnesium analogues of fluoro- and hydroxy-apatite under aqueous conditions. J. Inorg. Nucl. Chem. 34:95–100.

Duff, E. J. 1972d. Orthophosphates XI. Bromoapatite: Stability of solid solutions of bromoapatite with other calcium apatites under aqueous conditions. J. Inorg. Nucl. Chem. 34:101–108.

Elgawhary, S. M., and W. L. Lindsay. 1972. Silica solubility in soils. Soil Sci. Soc. Am. Proc. 36:439–442.

Farr, T. D., and K. L. Elmore. 1962. System $CaO-P_2O_5-H_2O$: thermodynamic properties. J. Phys. Chem. 66:315–318.

Frazier, A. W., J. P. Smith, and J. R. Lehr. 1965. Characterization of some ammonium polyphosphates. J. Agric. Food. Chem. 13(4):316–322.

Gregory, T. M., E. C. Moreno, and W. E. Brown. 1970. Solubility of $CaHPO_4 \cdot 2H_2O$ in the system $Ca(OH)_2-H_3PO_4-H_2O$ at 5, 15, 25, and 37.5°C. J. Res. Nat. Bur. Stand. 74A:461-475.

Gregory, T. M., E. C. Moreno, J. M. Patel, and W. E. Brown. 1974. Solubility of β-$Ca_3(PO_4)_2$ in the system $Ca(OH)_2-H_3PO_3-H_2O$ at 5, 15, 25, and 37°C. J. Res. Nat. Bur. Stand. 78A:667-674.

Johnson, G. G., Jr., and V. Vand. 1968. KWIC guide to the powder diffraction file. Am. Soc. for Testing and Materials, Philadelphia.

Kay, M. I., R. A. Young, and A. M. Posner. 1964. Crystal structure of hydroxyapatite. Nature 204:1050-1052.

Kreidler, E. R. 1967. Stoichiometry and crystal chemistry of apatite. The Penn. State Univ., Univ. Microfilm no. 68-3549.

Lehr, J. R., and W. E. Brown. 1958. Calcium phosphate fertilizers: II. A petrographic study of their alteration in soils. Soil Sci. Soc. Am. Proc. 22:29-32.

Lehr, J. R., E. H. Brown, A. W. Frazier, J. P. Smith, and R. D. Thrasher. 1967. Crystallographic properties of fertilizer compounds. Chem. Eng. Bull. 6, Nat. Fert. Div. Ctr. TVA, Muscle Shoals, Alabama. p. 166.

Lindsay, W. L., A. W. Frazier, and H. F. Stephenson. 1962. Identification of reaction products from phosphate fertilizers in soils. Soil Sci. Soc. Am. Proc. 26:446-452.

Lindsay, W. L., and E. C. Moreno. 1960. Phosphate phase equilibria in soils. Soil Sci. Soc. Am. Proc. 24:177-182.

Lindsay, W. L., M. Peech, and J. S. Clark. 1959. Solubility criteria for the existence of variscite in soils. Soil Sci. Soc. Am. Proc. 23:357-360.

Lindsay, W. L., and H. F. Stephenson. 1959a. Nature of the reactions of monocalcium phosphate monohydrate in soils: I. The solution that reacts with the soil. Soil Sci. Soc. Am. Proc. 23:12-18.

Lindsay, W. L., and H. F. Stephenson. 1959b. Nature of the reactions of monocalcium phosphate monohydrate in soils: II. Dissolution and precipitation of reactions involving Fe, Al, Mn, and Ca. Soil Sci. Soc. Am. Proc. 23:18-22.

McDowell, H., W. E. Brown, and J. R. Sutter. 1971. Solubility study of calcium hydrogen phosphate. Ion-pair formation. Inorg. Chem. 10:1638-1643.

Muller, L. D. 1967. Mineral separation. p. 1-31. In J. Zussman (ed.) Physical methods in determinative mineralogy. Academic Press, London-New York.

Norrish, K. 1968. Some phosphate minerals of soils. Int. Congr. Soil Sci., Trans. 9th (Adelaide, Aust.) II:713-723.

Norvell, W. A. 1971. Solubility of Fe^{3+} in soils. Ph.D. Thesis. Colo. State Univ. Univ. Microfilms. Ann Arbor, Mich. (Diss. Abstr. 31:5111-B).

Nriagu, J. O. 1972a. Solubility equilibrium constant of strengite. Am. J. Sci. 272:476-484.

Nriagu, J. O. 1972b. Stability of vivianite and ion-pair formation in the system $Fe_3(PO_4)_2-H_3PO_4-H_2O$. Geochim. Cosmochim. Acta 36:459-470.

Nriagu, J. O. 1972c. Lead orthophosphate I. Solubility and hydrolysis of secondary lead orthophosphate. Inorg. Chem. 11:2499-2503.

Nriagu, J. O. 1973a. Solubility equilibrium constant of α-hopeite. Geochim. Cosmochim. Acta 37:2357-2361.

Nriagu, J. O. 1973b. Lead orthophosphate II. Stability of chloropyromorphite at 25°C. Geochim. Cosmochim Acta 37:367-377.

Nriagu, J. O. 1973c. Lead orthophosphate III. Stabilities of fluoropyromorphite and bromopyromorphite at 25°C. Geochim Cosmochim. Acta 37:1735-1743.

Palache, C., H. Berman, and C. Frondel. 1951. The systems of mineralogy. Vol. II. John Wiley & Sons; and, Chapman & Hall, LTD, London.

Parker, V. B., D. D. Wagman, and W. H. Evans. 1971. Selected values of chemical thermodynamic properties. Nat. Bur. Stand. Tech. Note 270-6.

Patel, P. R., T. M. Gregory, and W. E. Brown. 1974. Solubility of $CaHPO_4 \cdot 2H_2O$ in the quaternary system $Ca(OH)_2-H_3PO_4-NaCl-H_2O$ at 25°C. J. Res. Nat. Bur. Stand. 78A:675-681.

Povarennykh, A. S. 1972. Crystal chemical classification of minerals. Vol. 1 and 2, Plenum Press, New York-London.

Quershi, R. H., D. A. Jenkins, R. I. Davies, and J. A. Rees. 1969. Application of microprobe analyses to the study of phosphorus in soils. Nature 221:1142-1143.

Rai, Dhanpat, and W. L. Lindsay. 1975. A thermodynamic model for predicting the formation, stability, and weathering of common soil minerals. Soil Sci. Soc. Am. Proc. 39:991-996.

Sarma, V. A. H., and G. S. R. Krishna Murti. 1969. Plumbogummite minerals in Indian soils. Geoderma 3:321-327.

Sawhney, B. L. 1973. Electron microprobe analyses of phosphates in soils and sediments. Soil Sci. Soc. Am. Proc. 37:658-660.

Sillen, L. G., and A. E. Martell. 1964. Stability constants of metal-ion complexes. Spec. Pub. no. 17. The Chemical Society, London. 754 p.

Sillen, L. G., and A. E. Martell. 1971. Stability Constants Supplement no. 1. Publication no. 25, The Chemical Society, London.

Singh, S. S. 1974. The solubility product of gibbsite at $15°$, $25°$, and $35°C$. Soil Sci. Soc. Am. Proc. 38:415-417.

Skipp, R. F., and R. P. Matelski. 1960. A microscopic determination of apatite and a study of phosphorus in some Nebraska soil profiles. Soil Sci. Soc. Am. Proc. 24: 450-452.

Smith, J. P., and W. E. Brown. 1959. X-ray studies of aluminum and iron phosphates containing potassium or ammonium. Am. Mineral. 44:138-142.

Smith, J. V. (ed.). 1968. Index (inorganic) to the powder diffraction file. Am. Soc. for Testing and Materials, Philadelphia.

Strunz, H. 1970. Mineralogische tabellen. Akademische Verlaggesellschaft, Geest & Portig K.-G Leipzig.

Taylor, A. W., A. W. Frazier, E. L. Gurney, and J. P. Smith. 1963a. Solubility products of di- and trimagnesium phosphates and the dissociation of magnesium phosphate solutions. Trans. Faraday Soc. 59:1585-1589.

Taylor, A. W., A. W. Frazier, and E. L. Gurney. 1963b. Solubility products of magnesium ammonium and mangesium potassium phosphates. Trans. Faraday Soc. 59: 1580-1584.

Taylor, A. W., and E. L. Gurney. 1964. Solubility of variscite. Soil Sci. 98:9-13.

Wagman, D. D., W. H. Evans, V. B. Parker, I. Halow, S. M. Bailey, and R. H. Schumm. 1968. Selected values of chemical thermodynamic properties. Nat. Bur. Stand. Tech. Note 270-3.

Wagman, D. D., W. H. Evans, V. B. Parker, I. Halow, S. M. Bailey, and R. H. Schumm. 1969. Selected values of chemical thermodynamic properties. Nat. Bur. Stand. Tech. Note 270-4.

Wagman, D. D., W. H. Evans, V. B. Parker, I. Halow, S. M. Bailey, R. H. Schumm, and K. L. Churney. 1971. Selected values of chemical thermodynamic properties. Nat. Bur. Stand. Tech. Note 270-5.

Titanium and Zirconium Minerals

JOHN T. HUTTON, CSIRO (Australia), Glen Osmond, South Australia

I. INTRODUCTION

The elements titanium (Ti), zirconium (Zr), and hafnium (Hf), which fall in group IVA of the *Periodic Table,* occur in minerals that are very resistant to weathering in normal soil environments. Thus, these minerals play little or no part in the soil-plant nutrition system, but because of their persistence they are most valuable for soil development studies. Generally there are more of these resistant minerals present in soils than in rocks.

The element Hf is so close to Zr in its chemistry and ionic size (actually slightly smaller) that it always accompanies Zr in nature and most Zr minerals have a Hf/Zr ratio of about 1:50. In view of this and the fact that no independent Hf mineral has been identified, no further reference will be made to this element.

The elements Zr and Ti are both normally quadrivalent and undergo similar chemical reactions, particularly the marked hydrolysis of aqueous solutions which gives rise to titanyl $[TiO]^{2+}$ or zirconyl $[ZrO]^{2+}$ radicals, but differ markedly in ionic radius, Ti being 0.68Å and Zr 0.79Å (Ahrens, 1952). Thus, these two elements are found in different types of primary minerals. The amount of Ti in the earth's crust, about 0.44%, is considerably more than that of Zr, about 0.02% (Day, 1963; Goldschmidt, 1954; Mason, 1966). Data from 70 Australian soil profiles (Stace et al., 1968), excluding some abnormal profiles, show an average Ti content in surface and subsurface soils of 0.6% Ti. For Zr, unpublished data of CSIRO Division of Soils show an average Zr content of about 0.035% Zr. Some tropical soils have much higher Ti content, e.g., Queensland, 3.4% Ti (Stace et al., 1968), Hawaii, 15% Ti (Sherman, 1952), and Norfolk Island 15% Ti (Hutton & Stephens, 1956), and in these soils the Ti minerals form a significant part of the mineral fraction.

II. PROPERTIES OF TITANIUM AND ZIRCONIUM MINERALS

The principal Ti minerals which occur in soils are:

Rutile	TiO_2	High temperature polymorph with extreme birefringence and fairly high density of 4.2 increasing to 5.5 g/cm^3 for Ta-rich varieties.
Anatase	TiO_2	Low temperature polymorph, often fine-grained, low birefringence, and density of 3.8 to 4.0 g/cm^3.

Brookite	TiO_2	Considered to be of secondary origin and found in metamorphic rocks. It is rarer than either rutile or anatase and intermediate between these minerals in birefringence and density (4.1 to 4.2 g/cm^3).
Leucoxene	TiO_2	A nonspecific name for fine-grained Ti minerals (rutile, anatase, brookite, or sphene). The use of the name is not recommended.
Ilmenite	$FeTiO_3$	Very common, and weakly magnetic opaque mineral of density 4.7 to 4.8 g/cm^3. Ilmenite and hematite are isostructural, and so called titaniferous-hematite can range in composition from pure ilmenite to pure hematite due to crystal intergrowth, but up to 10% TiO_2 may be accommodated in the hematite crystal.
Pseudorutile	$Fe_2Ti_3O_9$	A natural alteration product of ilmenite (Teufer & Temple, 1966; Grey & Reid, 1975), with a definite X-ray diffraction pattern and composition. The Fe^{2+} in ilmenite is replaced by Fe^{3+} in this mineral and there are other Fe-Ti oxide minerals.
Sphene (Titanite)	$CaTiSiO_5$	The only Ti-Si mineral found to any extent in rocks and occasionally reported in soils. Density 3.5 g/cm^3.

The principal Zr minerals that occur in nature are:

Zircon	$ZrSiO_4$	Very common mineral in rocks, sediments, and soils. Density 4.5–4.7 g/cm^3.
Baddeleyite	ZrO_2	Very rare but has been recorded in African soils (Baldock, 1968). Density 5.4–6.0 g/cm^3.

Further descriptions of these minerals are given in Deer et al., (1962), Deer et al., (1966), and Palache et al., (1944).

The Ti and Zr minerals are present in both the coarse and fine fractions of soils and actual identification of the minerals and quantitative determination in either fraction are difficult. Haseman and Marshall (1945) in their investigation of the use of heavy minerals in soil studies give full details for separation and examination of the minerals in the fine sand and silt fractions, but draw attention to the problem of coatings on the grains which prevent their satisfactory identification, particularly the Ti minerals. For a quantitative measure of the zircon present, these authors resorted to chemical analysis of the heavy fraction for Zr because of the statistical error in quantitative work based on grain counting.

For identification of the Ti minerals in the clay size fraction ($< 2\mu m$), X-ray diffraction techniques (Zussman, 1967) are used. As the Ti content of the clay fraction of most soils is of the order of 1% Ti; the intensity of the diffraction pattern of any Ti minerals is usually too weak to identify and so it is necessary to remove most of the silicate minerals by hydrofluoric acid attack as described later. The concentration of the Zr minerals is even lower than that of the Ti minerals and the recognition of zircon and/or baddeleyite in the residue after hydrofluoric acid attack is rarely possible, because X-ray absorption and fluorescence effects increase background and reduce line intensity. Again, for quantitative estimation of Ti and Zr minerals in the clay fraction of soils, chemical analysis for the elements is recommended.

Analyses by classical wet chemical methods are slow whereas modern physical methods particularly X-ray fluorescence spectrography are rapid and enable soils to be analyzed for most elements. Fusion techniques such as that of Norrish and Hutton (1969) which eliminate particle size problems enable very accurate analyses for elements numbered 12 through 26 (Mg to Fe) in the range 0.05 to 100% to be made by X-ray spectrography. Elements such as Zr present in range 5 to 1,000 ppm suffer from dilution if fused and so analyses are made directly on pressed discs of the finely powdered sample (Norrish & Chappell, 1967).

The identification of the actual mineral is not essential in most soil studies using persistent minerals and most workers now use the total Ti and/or Zr content of their samples.

III. TITANIUM AND ZIRCONIUM MINERALS IN ROCKS

Rutile, which has the smallest molecular volume of the mineral forms of TiO_2 is found widely in most igneous rocks as a fine-grained accessory mineral and often occurs in quartz grains as fine needles. It is also found extensively in metamorphic rocks, e.g., gneiss and schist.

Ilmenite, like rutile, is found well dispersed as fine grains in most igneous and metamorphic rocks. As both ilmenite and rutile are dense minerals, (density > 4 g/cm³), they readily separate by gravity from other minerals (density < 3 g/cm³). Thus, rocks, particularly those of sedimentary origin, that appear uniform in the hand specimen may not be so uniform in distribution of rutile and ilmenite and, in some rocks, banding is obvious. Hence, if Ti and Zr minerals are to be used in pedogenic studies, adequate sampling must be made to cover variations in the quantity of these heavy minerals.

Anatase is generally considered to be of secondary origin in many sediments, having been derived from sphene or other minerals containing Ti. It is present in some igneous and metamorphic rocks, but much less abundant than rutile.

Sphene, like rutile, is present in igneous and some metamorphic rocks. It is particularly abundant in granitic rocks, and can occur as quite large crystals. It is neither as dense nor as stable as rutile and, hence, is less common in sedimentary rocks.

Perovskite, essentially $CaTiO_3$, is found in some basic igneous rocks rich in Ca and low in Si. Usually, there is considerable substitution of Nb and Ta for Ti, and Ce and Na for Ca. Calzirtite ($CaTiZr_3O_9$) is also known and with perovskite been recorded in a residual soil (Baldock, 1968).

In addition to these specific Ti minerals, many silicates in rocks contain considerable Ti. Deer et al. (1962) give data for the trioctahedral micas, phlogopite and biotite, showing they may contain 3–6% Ti. These authors also give data for titanaugite, a pyroxene, and kaersutite, an amphibole, with similar amounts of Ti.

The greatest concentration of Ti minerals occurs in "heavy" beach sands found in many places, particularly Australia, New Zealand, and Florida. These are the main sources of Ti for industrial purposes.

Zircon is also concentrated in "heavy" beach sands, but its primary occurrence is in granitic rocks as small crystals. Gem varieties of zircon have been known for a long time and they have a high light dispersion, comparable with that of diamond. Zircon also occurs in metamorphic and some sedimentary rocks and, again, adequate sampling of rocks must be made to ascertain the variations in zircon and/or total Zr content.

Although generally considered to be very insoluble under normal weathering conditions, zircon crystals are often found in a *metamict* state, "a state of lattice disorder and imperfection in externally well crystallised natural minerals, which may be annealed out by moderate heating" (Graham & Thornber, 1974). The radius of the Zr^{4+} ion is 0.8Å and this suggests that the most stable structures of compounds of Zr^{4+} with O^{2-} would have sixfold coordination. However, in zircon we have eightfold coordination with Zr–O distances of 2.15Å for four bonds and 2.29Å for the other four (Krstanovic, 1958). The fact that these distances for the Zr–O bond lie between the usual limits for 6- and 8-fold coordination has been suggested as a source of potential instability and the presence of radioactive elements, e.g., U or Th in zircon leads to a breakdown in structure due to self-irradiation by α-particles. The defective crystal, metamict zircon, yields no X-ray diffraction patterns attributable to normal crystalline zircon, has a lower density of 4.0 g/cm^3 as against 4.6 g/cm^3, and a lower refractive index of 1.79 as against 1.92. Vance and Anderson (1972) have found tetragonal zirconia in metamict zircons, but no crystalline silica, and calculate that a mixture of amorphous silica and tetragonal zirconia would have a density of 3.9 g/cm^3. Very little is known about the behaviour of metamict zircon in soil development.

IV. MOBILITY OF GROUP IVA ELEMENTS

A. Mobility of Titanium

The basis of the main reason for the study of Ti minerals in soil environments is that they remain when most other primary minerals have disappeared as weathering proceeds. Jackson in a number of publications, e.g. Jackson et al. (1948); Jackson and Sherman (1953), has identified a mineral weathering sequence, for clay size minerals, and his most stable stage is number 13—"anatase (also zircon, rutile, ilmenite, leucoxene, corundum, etc)." No mineral, however, is completely insoluble and Livingstone (1963), discussing the very limited data on the amount of Ti measured in river waters, concludes there is about 0.02 ppm Ti in solution. This is about 0.002 of the concentration of Si in rivers. As part of a study of the uptake of Si by wheat plants (*Triticum aestivum* L.) (Hutton & Norrish, 1974), Ti was also measured and about 50 ppm Ti was found in plants with 5% Si. The Si content of the soil solution is about 30 ppm Si and these data suggest the solubility of Ti in soil solution to be 0.03 ppm, the same order of magnitude as found in rivers. Goldschmidt (1954) commented on the insolubility of Ti in nature by pointing out that no insoluble complexes of titanium dioxide and phosphoric acid are observed in nature. If Ti was available in solution such complexes could be expected. In the laboratory when Ti and P are brought into solution as in the determination of P by acid solution of soils high in Ti very insoluble titanium phosphates are formed. K. Norrish (private communication), using an electron probe microanalyzer, has found an association of P and Ti to be rare in the large range of soils examined.

Hutton and Stephens (1956), discussing the paleopedology of Norfolk Island, interpret the high Ti, 16% Ti (Norrish & Hutton, 1969, Table 9), of the surface horizon of Middlegate clay as due to the loss of Si, Fe, and Al in relation to Ti as the result of the great maturity of this soil. Pedro (1964) studied the weathering of rocks, including basalt, by means of repeated extractions with hot water. The analysis of "croute ocre" (ferruginous coating) obtained in one of his experiments, when expressed in terms of the ratio of element in the coating/element in original basalt, was identical with that reported by Hutton and Stephens (1956) for Rooty Hill clay, another Norfolk Island soil. Rooty Hill clay is not as highly weathered as Middlegate clay, but at a depth of 19-30 cm Ti had concentrated in the soil relative to parent rock by a factor of 2.5, whereas for Al the factor was 1.7, for Fe 2.8, and for Si 0.33. Pedro's factors were 1.9, 1.7, 2.5, and 0.5, respectively (Pedro, 1964, p. 185).

Silcrete, a hard indurated crust composed of quartz clasts cemented by microcrystalline silica, is found extensively in Australia and South Africa. Many specimens have more than 2% Ti, about 0.1% Zr, 0.5% Fe, < 0.2% Al, and very angular and embayed quartz grains, and are thus good examples of stage 13 of Jackson's weathering sequence. Hutton et al. (1972) describe examples of silcrete containing up to 15% Ti and as can be seen in the X-ray scanning photographs in Fig. 18-1, the Ti is concentrated between the quartz clasts virtually to the exclusion of Si. The figure of about 0.02 ppm Ti as the natural solubility of Ti is discussed in relation to the Ti in the cementing matrix and Hutton et al. (1972) state "although this is a very low concentration, too low to be effective for the transport of Ti, nevertheless repeated successive periods of wetting and drying could allow sufficient solution and deposition to form a hard network of fine crystals of anatase over a long period of time." Subsequently Milnes and Hutton (1974) obtained transmission electron micrographs of the anatase separated by successive hydrofluoric acid attacks of crushed silcrete and these photographs showed a network of crystals intergrown in typical twin relationship (Fig. 18-2A). These crystals are < 0.5 μm in length and so are comparable to the size of the anatase found

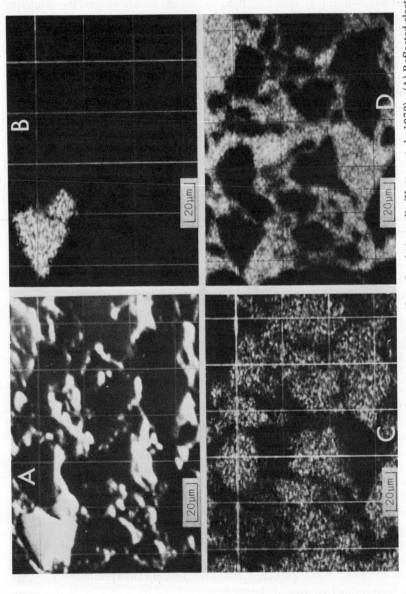

Fig. 18–1. Electron probe microphotographs of a section of silcrete from South Australia (Hutton et al., 1972): (A) Reflected electrons; (B) Zr distribution; (C) Si distribution; and (D) Ti distribution.

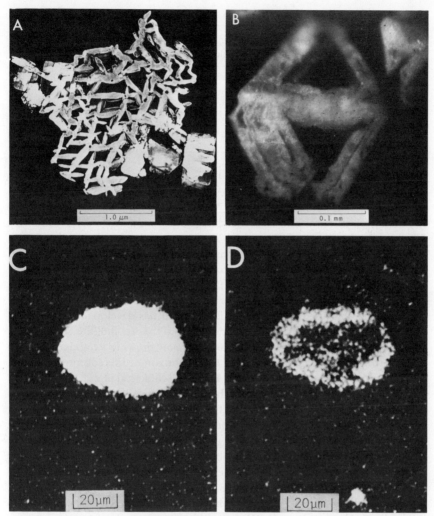

Fig. 18-2. (A) Electron micrograph of anatase crystals separated from silcrete (Milnes & Hutton, 1974); (B) Microphotograph of anatase from dyke clay by courtesy of F. C. Loughnan and H. G. Golding (1957), University of New South Wales and reproduced with permission of Royal Society of New South Wales; (C) Electron probe microphotograph of Zr in a zircon grain in a soil; and (C) Electron probe microphotograph of P distribution in same section of soil as C.

by Campbell (1973) in a New Zealand soil, but much smaller than those shown in Fig. 18-2B which were obtained by Loughnan and Golding (1957) from deeply weathered and leached dyke clays that occur in sandstone near Sydney, New South Wales.

Further evidence for limited solution of Ti in soils is available from consideration of the results of analysis of a duplex soil (soil with considerable texture contrast between A and B horizon), namely Urrbrae loam, a South

Table 18-1. Ratio Ti/Zr in a deep soil.

Horizon	Depth, cm	No. of samples	Ti/Zr whole soil mean value	SD of mean value
A	0-30	6	20.0	0.3
B	30-120	9	22.9	0.5
C	120-180	6	26.2 ⎤	0.2 ⎤
	180-220	4	24.1 ⎬ 25.3	0.3 ⎬ 0.2
	220-310	9	25.3 ⎪	0.3 ⎪
	310-380	7	25.4 ⎦	0.4 ⎦

Australian Red-brown Earth, (Calcic Luvisol).[1] The profile was sampled to a depth of 3.80 m in 10-cm steps except at the surface and at the junction of the A and B horizons. Oertel (1974) has discussed some aspects of the chemistry of this soil and also the change from an average clay content of 30% in the A horizon to 70% in the B horizon, but the detailed analysis available enables other aspects to be considered. In particular an indication of the mobility of Ti can be obtained from a consideration of the Ti/Zr ratio of the whole soil samples (Table 18-1). It would appear that relative to Zr, some Ti has been lost from the A horizon and some from the B horizon, but the colluvial shale sediments on which the soil has developed have a fairly uniform Ti/Zr ratio despite a variation of 0.033% to 0.026% in total Zr. There is no significant evidence for any horizon showing a build up of Ti in respect to Zr. In the clay fraction of the A horizon of this soil there is, however, an increase in the Ti content to 1.5% Ti (whole soil about 0.8%) and an increase of Ti with respect to Zr. The Ti/Zr ratio is 50:1 while in the other horizons it is 35:1. This suggests that clay size anatase crystals are being formed in the soil with a corresponding lowering of the Ti/Zr ratio in the coarse fraction. In the A horizon the loss of clay size layer silicates by weathering produces an increase in absolute amount of Ti and Zr in the clay fraction. Further examination of the detailed chemical analysis of this profile indicates that the Al/Ti ratio is 6.4 (standard deviation [SD] 0.8) in the A horizon and 17.2 (SD 4.4) in the B horizon with an average of 10.4 (SD 0.6) for samples between 1 and 4 m. Aluminum would thus appear to be more mobile than Ti in the development of this soil profile and this is reflected in the Jackson (Jackson et al., 1948) weathering sequence of kaolinite, gibbsite, hematite, anatase in order of increasing stability. The basis of the argument of Craig and Loughnan (1964) for mobility of Ti is contrary to this as they choose to consider Al as immobile under normal weathering conditions. Also the evidence given by Sherman (1952) for the mobility of Ti as "the hydrated colloidal oxide form" is insufficient, for no detailed profile

---

[1] In order to achieve some uniformity in nomenclature of soils, a term from the legend (FAO-Unesco, 1974) for the *Soil Map of the World* has been added in parenthesis after each local name.

descriptions are given to enable the possibility of through drainage to be assessed nor any analysis of the parent material. If results for another element, e.g., Zr were available, a better assessment of the relative mobility of Ti, Fe, and Al could be made.

Another aspect of the mobility of Ti in the soil environment is the extent that Ti can enter into the structure of the clay sized layer silicates found in soils. As discussed earlier it appears that 3% Ti may be part of the structure of trioctahedral micas such as biotite and phlogopite and it is common for $>1\%$ Ti to be found in the clay fraction of many soils, e.g., Urrbrae loam. It is, however, difficult to decide whether the Ti is present as free titania or is held in the octahedral sheet of the silicate structures. Dolcater et al. (1970) found that dihydrogen hexafluorotitanate (H_2TiF_6) selectively dissolved silicate minerals and did not attack rutile or anatase. The reagent nevertheless did dissolve a low Ti ilmenite, possibly a titaniferous hematite. It also tended to precipitate $MgTiF_6$ and, hence, results obtained on some soil materials are difficult to interpret. Mineral kaolinites with high Ti content were found to have no substitution of Ti for Al and this could be due to the large difference in atomic radii of these elements. In the case of the $20-5\mu m$ size fraction of a biotite, all the Ti was dissolved and in this mica there is considerable Mg and Fe in the octahedral layer and the atomic radii of these cations are nearer that of Ti. Evidence available to date suggests Ti is not entering into clay-sized layer silicates in the soil environment but forms clay-sized anatase (Bain, 1976), and that although it is slightly soluble, it is not mobile over a distance of more than a few centimeters.

B. Mobility of Zirconium

Little evidence appears to be available on the mobility of Zr in soils. Some microscopic studies of zircon grains from soils have suggested that zircon is weathering, but evidence of any Zr minerals, other than zircon, being present in soils appears to be lacking. The fate of metamict zircon in soils has not been studied, but Vance and Anderson (1972) in their studies of metamict zircon from Sri Lanka (Ceylon) claim that the zirconia and silica, that may result from the break down of zircon, remain with the zircon grain. Kimura and Swindale (1967) claim that Zr is leached, probably by cheluviation, from the upper horizons of the more developed Hawaiian soils they studied. Unfortunately no work was done to determine the mineral form of the Zr although some soils on andesite contained 0.25% Zr and some rock samples 0.15% Zr. Evidence from the study of the Zr in the clay fractions of Australian soils, unfortunately none on volcanic rocks, shows no increase in Zr above that of the remainder of the soil in any horizon. If Zr did become mobile during the weathering of zircon or any other mineral that may contain Zr, it would probably appear as very fine grained material in the profile as hydrolysis of solutions of Zr compounds occurs so readily.

V. SOIL GENESIS STUDIES BASED ON TITANIUM, ZIRCONIUM, AND THEIR MINERALS

Soil genesis studies always suffer from the fundamental difficulty that the parent material of any part of a soil profile is no longer in existence—it has been changed into the part of the soil being studied. A nearby rock outcrop may satisfactorily represent the soil parent material, but it is possible that the outcrop is standing above the soil surface because it is different from the rock that weathered to produce the soil. Fortunately, in many situations, there is ample evidence of uniformity in the parent material, so that confidence can be placed in a sample taken beyond the depth of weathering. Green (1973) gives results of analysis of 18 samples of basalt collected at different places on Norfolk Island and the mean value of the Ti content is 1.17% Ti with a standard deviation of 0.07 for individual values. Non-uniformity in igneous material is illustrated in a recent weathering study (Hutton et al., 1977) of an exposed mass of norite in South Australia. Twenty samples of hard rock close to joints were very uniform with an Al content of 8.7% (SD 0.1), an Fe content of 7.7% (SD 0.1) and a Ti content of 0.75% (SD 0.05) whereas six other samples taken within 10 cm had 11.3% (SD 0.3) of Al, 4.4% (SD 0.3) of Fe and 0.44% (SD 0.06) of Ti. Another example of variability in Ti and Zr content of rock is given in the data of Stephenson (1973). Ten samples of fine grained granitic Precambrian country rock from Albany, Western Australia showed uniform Al (6.6% Al, SD 0.4) and K (4.0%, SD 0.5) but variable Ti (0.6% Ti, SD 0.3) and Zr (150 ppm, SD 75). Further, attention has already been drawn to the concentration of Ti and Zr minerals in heavy mineral bands in sands and so a study of soil genesis, particularly from sedimentary rocks, must be planned very carefully and adequate sampling and analyses made to cover rock variability. The variability in the data above for Ti and Zr in granitic rock is reduced considerably if Ti/Zr ratios are considered, the relative standard deviation becoming 0.2 rather than 0.5.

Despite these possible difficulties much valuable work on the genesis of soils has been published. Some of this work will be discussed below, but for a detailed consideration of studies of soil development, Barshad (1964) should be consulted. Haseman and Marshall (1945) using zircon as a persistant mineral were able to show that the loessial parent material of the Grundy silt loam is uniform in origin both geologically and depositionally and that there has been considerable loss of material from the A horizon and gain in the B horizon down to a depth of 1 m. Brewer (1955) extended this persistent mineral technique of studying the genesis of a soil by including thin section studies of the various horizons. He worked on a duplex Yellow Podzolic soil (Planosol) developed on granodiorite in New South Wales and one conclusion was that at the maximum depth sampled (4.5 m) considerable weathering (about 40%) of the hornblende, feldspar, and biotite had occurred. In the surface A horizon about 10% of the amount of these minerals in the parent granodiorite remained. Brewer also shows that in the top of the B horizon, with 43% material $< 5\mu m$, there has been considerable decomposition of primary minerals and subsequent removal from the profile of the elements which could have formed clay minerals because the zircon count was still above that of the deeper samples and the rock.

Beavers (1960) drew attention to the potential of chemical analysis by X-ray fluorescence spectrography for the study of soil genesis and gave an example using the CaO/ZrO_2 ratio to show the degree of weathering of Hosmer silt loam, a Fragochrult (Acrisol) developed on loess in southern Illinois. Ahmad et al. (1968) in their studies of two West Indian soils, a Brown Earth (Kastanozem) and a Latosol (Ferralsol), separated both heavy and light mineral fractions from the 50–250 μm fraction and examined them under the microscope, but calculated the degree of weathering on analytical results for K_2O, CaO, TiO_2, and ZrO_2. X-ray spectrography was also used by Kimura and Swindale (1967) to determine Zr in Hawaiian soils. This figure together with total Ni was used to obtain a highly significant discriminant function to separate soils developed on andesite from those developed on olivine basalt. Oertel and Blackburn (1970) studied the pedogenesis of a Solodized Solonetz from Penola, South Australia and showed that the Zr content of both the whole soil and the sand fraction of the A horizon was about two-thirds of that of the B horizon. Additional, unpublished data showed a greater difference in Ti content and so they concluded that the very sandy A horizon (4% clay) and the B horizon (35% clay) resulted from the weathering of distinct sedimentary layers.

Chemical analysis by X-ray fluorescence spectrography has also been used by Hutton (1974) to study the changes that have occurred in volcanic ash ejected about 5,000 years ago from Mt. Gambier, South Australia. Ten samples containing varying amounts (10–60%) of ash were analyzed for 13 elements. Results showed that about 50% of the minerals present in the ash had changed to clay minerals—1:1, mica and interstratified material, with loss of 70% of the Na, Mg, and Ca and a relative gain of 20% in Ti and Al. These fairly marked changes had taken place in this Chernozem (Chernozem) in 5,000 years (data from buried charcoal) under a rainfall of 700 mm and daily maximum temperatures ranging from 285K to 298K.

Table 18-2. Titanium and other elements in arid soil containing carbonates and sulphates.

Depth, cm	Percent				
	Ignited soil				Carbonate and sulphate free soil
	Ti	Ca	S	Mg	Ti
3–17	0.49	0.49	0.08	1.15	0.51
26–40	0.41	2.19	0.19	3.79	0.46
55–65	0.33	16.44	2.19	1.41	0.48
95–115	0.20	18.54	13.90	0.52	0.53
145–155	0.41	6.15	4.60	0.68	0.52
165–175	0.40	6.65	4.98	0.50	0.52

Total Ti can also be used to check the continuity of soil profiles, i.e., are the profiles layered or are they autochthonous profiles (profiles developed in situ) into which material has been introduced. In arid regions, soils often have marked accumulations of carbonates and salts. A Desert Loam (Solonetz) (FAO-Unesco, 1974) from Whyalla, South Australia, for example, has 30% calcite and 10% gypsum at 55-65 cm and 75% gypsum at 95-105 cm. Analytical data for this soil are given in Stace et al. (1968) (p. 66) and Table 18-2 gives results for Ti, Ca, S, and Al together with the recalculated value for Ti content of the soil free from carbonates and salts. The calculated range from 0.46 to 0.53% Ti in 1.75 m of soil suggests a uniform soil material with carbonate and salts introduced.

VI. OUTLINE OF SOME METHODS APPLICABLE TO THE QUANTITATIVE STUDY OF TITANIUM AND ZIRCONIUM MINERALS IN SOILS

A. Separation of Coarse Particles for Grain Counts

The details given by Haseman and Marshall (1945) for dispersion and separation of various size fractions are basic and should be satisfactory for a study of the minerals in soils. Concentration by density separation (bromoform density 2.89; tetrabromoethane density 2.96) is necessary in order to examine the sand separates for Ti and Zr minerals and further separation by an electromagnetic separator helps in identification and counting (Muller, 1967). The precision of mineral analysis by counting is determined by the number of grains of the mineral counted and the standard deviation (SD) of the count is equal to the square root of the number observed, i.e., a count of 100 has an SD of 10. The accuracy is dependent on the positive identification of all grains which, because of oxide coatings, may be difficult. To assist in this regard the separated grains may be analyzed for total Ti and Zr by the methods discussed later.

B. Separation of Titanium and Zirconium Minerals from Other Minerals in Fine Fractions and their Identification

Density and/or magnetic separations are not possible with mineral fractions < 50 μm and concentration by some form of preferential dissolution is recommended. Raman and Jackson (1965) discuss the removal of siliceous minerals by treatment with hydrofluoric acid (HF). In their technique they treat 0.5 g of the lightly ground material with 30 ml of 48% HF at room temperature for 24 hours. For a more controlled attack on silicate minerals and less attack on Ti minerals Dolcater et al. (1970) used a solution of hydrofluotitanic acid (H_2TiF_6) at 318K for 2 days. Campbell (1973) showed that 20% HF reduced the intensity of the X-ray diffraction pattern of clay size anatase in relation to that of rutile if the attack was prolonged for more than

Table 18-3. X-ray diffraction data on Ti and Zr minerals found in soils

	Rutile	Anatase	Brookite	Ilmenite	Pseudorutile	Sphene	Zircon
Crystal d-spacings, Å	3.25	3.52	3.51	2.74	1.69	3.23	3.30
	1.69	1.89	2.90	1.72	2.49	2.99	4.43
(Five strongest	2.49	2.38	3.47	2.54	2.19	2.60	2.52
reflections in	2.19	1.70	1.89	1.86	1.44	2.06	1.71
order of intensity)	1.62	1.67	1.66	1.50	2.76	1.64	2.07
JCPDS† card no.	21-1276	21-1272	16-617	3-781	19-635‡	11-142	2-266

Possible interfering minerals						
	Hematite	Feldspar	Quartz	Goethite	Gibbsite	Kaolinite	Mica
Crystal d-spacing of strongest line, Å	2.70	3.20	3.34	4.18	4.85	7.16 $(d/2 = 3.58)$	10.1 $(d/3 = 3.35)$

† Joint Committee on Powder Diffraction Standards, Pennsylvania.
‡ Data from Grey and Reid (1975) considered as well.

2 hours and concluded that all HF digestions should be as short as possible. For some years J. G. Pickering of the CSIRO Division of Soils has used 24% HF and stops the reaction after 10 minutes by adding excess water (personal communication).

The result of these preferential dissolution techniques is to concentrate the resistant minerals, which will include the Ti and Zr minerals, so that their X-ray diffraction pattern will be much clearer and interference from kaolinite 002 and mica 003 reflections eliminated. The d-spacings corresponding to the five strongest reflections of the Ti minerals found in soils and of zircon are given in Table 18-3 together with the strongest reflection of some other soil minerals.

C. Elemental Analysis of Soils and Soil Fractions for Titanium and Zirconium

Neither classical chemical nor modern atomic absorption flame spectrometric methods are very satisfactory for the determination of either Ti or Zr, whereas arc emission spectrography (Nicholls, 1967) is quite suitable. The latter method has quite good sensitivity and uses only small amounts of finely ground sample.

In recent years X-ray spectrography has developed extensively in the field of soil and rock analysis and it is an excellent method for the determination of Ti and Zr. There are fundamental physical problems, particularly the absorption of X-rays by all elements present in the sample, that necessitate special pretreatment of the sample and calculation of results (Norrish & Chappell, 1967). Very accurate and precise results for Ti can be obtained by fusion of the sample in lithium borate and correcting for absorption due to other elements by calculation (Norrish & Hutton, 1969) or by a "double dilution" technique (Tertian & Géninasca, 1972). For many soil genesis studies, satisfactory results (within 10–20% relative) for Ti might be obtained by grinding the sample very finely in a ring and puck mill[2], pressing the sample into a suitable size pellet, and calibrating against similar samples analyzed by a fusion technique. For Zr, using $K\alpha$ radiation, "particle size" effects are not so critical and a pellet of the finely ground sample is used. It is necessary to correct the peak counts of $ZrK\alpha$ for background and contribution by $SrK\beta$, and relate the net counts to the net counts on a suitable standard, knowing the mass absorption coefficient of both sample and standard.

Electron probe microanalysis (Long, 1967) may also be used for analysis of soils and rocks for Ti and Zr and in particular the distribution of these elements on a microscale. Figure 18-1 shows the distribution of Si, Ti, and Zr in a section from a silcrete from South Australia. In the top left hand corner of the section there is a grain of zircon shown clearly by the photograph taken from $ZrL\alpha$ radiation. The identification of zircon is supported by the low intensity of $SiK\alpha$ radiation and the absence of $TiK\alpha$ radiation in that position. The uniform intensity of $TiK\alpha$ radiation in the areas between the Si-rich areas indicates fine particles of Ti-rich minerals and the microcrystals of anatase (Fig. 18-2A) were obtained from such an area.

VII. FUTURE OUTLOOK

The speed and accuracy with which soils, rocks, etc. can be analyzed for Ti and Zr by X-ray spectrography should enable further soil genesis studies to be undertaken. It is quite easy to extend the X-ray analysis to include other elements, e.g., Y which may be associated with other resistant minerals. The so-called "rare earth" elements are not so rare, but their oxides have limited solubility and a study of their behaviour in the soil environment may also be profitable.

Attention has already been drawn to the comment by Goldschmidt (1954) that no titanium phosphorus complexes are found in nature and yet they are well known in the analytical laboratory. The electron probe microanalyzer can examine very fine particles, about $1\mu m$, and a constant association of Ti and P would indicate the presence of a titanium phosphate mineral if such did occur, but so far no such result has been reported. The association of Zr and P is shown in Fig. 18-2C and D, but this is probably an association that is present in the original grain of zircon for Deer et al. (1962) report the analysis of zircons containing 1.5% P.

An electron beam, similar to that used in the electron probe microanalyzer, can be used to obtain photographs of the surface of mineral grains with a resolution of about $0.01\mu m$ and a satisfactory depth of focus in the scanning electron microscope. A detailed study of the surface of grains of zircon and other resistant minerals may indicate the degree of weathering of such minerals. With many scanning electron microscopes it is also possible

---

[2]Scheibenschwingmühle, Siebtechnik GMBH, Mülheim, Germany or Shatterbox, Spex Industries Inc., Metuchen, USA.

to analyze the specimen for the major elements present and some interesting results should be forthcoming from such studies of the resistant minerals, particularly those of Ti and Zr in the soil environment.

LITERATURE CITED

Ahmad, N., R. L. Jones, and A. H. Beavers. 1968. Genesis, mineralogy and related properties of West Indian soils. J. Soil Sci. 19:1-19.

Ahrens, L. H. 1952. The use of ionization potentials. Part I. Ionic radii of the elements. Geochim. Cosmochim. Acta 2:155-169.

Bain, D. C. 1976. A titanium-rich soil clay. J. Soil Sci. 27:68-70.

Baldock, J. W. 1968. Calzirtite and the mineralogy of residual soils from the Bukusu carbonatite complex, south-eastern Uganda. Mineral. Mag. 36:770-774.

Barshad, I. 1964. Chemistry of soil development. p. 1-70. *In* F. E. Bear (ed.) Chemistry of the soil. Reinhold Publishing Corp., New York.

Beavers, A. H. 1960. Use of X-ray spectrographic analysis for the study of soil genesis. Int. Congr. Soil Sci. Trans. 7th (Madison, Wis.) II:1-9.

Brewer, R. 1965. Mineralogical examination of a Yellow Podzolic soil formed on granodiorite. CSIRO Aust. Soil Publ. 5.

Campbell, A. S. 1973. Anatase and rutile determination in soil clays. Clay Miner. 10: 57-58.

Craig, D. C., and F. C. Loughnan. 1964. Chemical and mineralogical transformation accompanying the weathering of basic volcanic rocks from New South Wales. Aust. J. Soil Res. 2:218-234.

Day, F. H. 1963. The chemical elements in nature. George G. Harrap & Co. Ltd., London.

Deer, W. A., R. A. Howie, and J. Zussman. 1962. Rock-forming minerals. Vol. 1, 2, 3, 4, and 5. Longmans, Green and Co. Ltd., London.

Deer, W. A., R. A. Howie, and J. Zussman. 1966. An introduction to the rock-forming minerals. Longmans, Green and Co. Ltd., London.

Dolcater, D. L., J. K. Syers, and M. L. Jackson. 1970. Titanium as free oxide and substituted forms in kaolinites and other soil minerals. Clays Clay Miner. 18:71-79.

FAO—Unesco. 1974. Soil map of the world. Vol. 1. Legend. Unesco, Paris.

Goldschmidt, V. M. 1954. Geochemistry. Oxford University Press, London.

Graham, J., and M. R. Thornber. 1974. The metamict state. Am. Mineral. 59:1047-1050.

Green, T. H. 1973. Petrology and geochemistry of basalts from Norfolk Island. J. Geol. Soc. Aust. 20:259-272.

Grey, I. E., and A. F. Reid. 1975. The structure of pseudorutile and its role in the natural alteration of ilmenite. Am. Mineral. 60:898-906.

Haseman, J. F., and C. E. Marshall. 1945. The use of heavy minerals in studies of the origin and development of soils. Univ. of Missouri, Agric. Exp. Sta. Res. Bull. 387. Columbia, Mo.

Hutton, J. T. 1974. Chemical characterization and weathering changes in holocene volcanic ash in soils near Mount Gambier, South Australia. R. Soc. S. Aust. Trans. 98:179-184.

Hutton, J. T., and K. Norrish. 1974. Silicon content of wheat husks in relation to water transpired. Aust. J. Agric. Res. 25:203-212.

Hutton, J. T., and C. G. Stephens. 1956. The paleopedology of Norfolk Island. J. Soil Sci. 7:255-267.

Hutton, J. T., D. S. Lindsay, and C. R. Twidale. 1977. The weathering of norite at Black Hill, South Australia. J. Geol. Soc. Aust. 24:37-50.

Hutton, J. T., C. R. Twidale, A. R. Milnes, and H. Rosser. 1972. Composition and genesis of silcretes and silcrete skins from the Beda valley, southern Arcoona Plateau, South Australia. J. Geol. Soc. Aust. 19:31-39.

Jackson, M. L., and G. D. Sherman. 1953. Chemical weathering of minerals in soils. Adv. Agron. 5:221-318.

Jackson, M. L., S. A. Tyler, A. L. Willis, G. A. Bourbeau, and R. P. Pennington. 1948. Weathering sequence of clay-size minerals in soils and sediments. J. Phys. Colloid Chem. 52:1237–1260.

Kimura, H. S., and L. D. Swindale. 1967. A discriminant function using zirconium and nickel for parent rocks of strongly weathered Hawaiian soils. Soil Sci. 104:69–76.

Krstanovic, I. R. 1958. Redetermination of the oxygen parameters in zircon. Acta Crystallogr. 11:896–897.

Livingston, D. A. 1963. Chemical composition of rivers and lakes. p. 49. In M. Fleischer (ed.) Data of Geochemistry. 6th edit. U. S. Geol. Surv. Prof. Paper 440. Washington, D. C.

Long, J. V. P. 1967. Electron probe microanalysis. p. 215–260. In J. Zussman (ed.) Physical methods in determinative mineralogy. Academic Press, London. (2nd ed., in press.)

Loughnan, F. C., and H. G. Golding. 1957. The mineralogy of the commercial dyke clays in the Sydney district, New South Wales. J. Proc. R. Soc. N. S. W. 91:85–91.

Mason, B. 1966. Principles of geochemistry. 3rd ed. John Wiley and Sons, Inc., New York.

Milnes, A. R., and J. T. Hutton. 1974. The nature of microcryptocrystalline titania in silcrete skins from the Beda Hill area of South Australia. Search 5:153–154.

Muller, L. D. 1967. Laboratory methods of mineral separation. p. 1–30. In J. Zussman (ed.) Physical methods in determinative mineralogy. Academic Press, London, (2nd ed., in press.)

Nicholls, G. D. 1967. Emission spectrography. p. 445–458. In J. Zussman (ed.) Physical methods in determinative mineralogy. Academic Press, London. (2nd ed., in press.)

Norrish, K., and B. W. Chappell. 1967. X-ray fluorescence spectrography. p. 161–214. In J. Zussman (ed.) Physical methods in determinative mineralogy. Academic Press, London. (2nd ed., in press.)

Norrish, K., and J. T. Hutton. 1969. An accurate X-ray spectrographic method for the analysis of a wide range of geological samples. Geochim. Cosmochim. Acta 33:431–453.

Oertel, A. C. 1974. The development of a typical Red-brown Earth. Aust. J. Soil Res. 12:97–105.

Oertel, A. C., and G. Blackburn. 1970. Pedogenesis of a Solodized Solonetz, based on duplicate soil profiles. Aust. J. Soil Res. 8:59–70.

Palache, C., H. Berman, and C. Fondel. 1944. The system of mineralogy. Vol. 1. John Wiley and Sons, Inc., New York.

Pedro, G. 1964. Contribution a l'étude expérimentale de l'altération géochimique des roches cristallines. Ann. Agron. 15:85–191, 243–333, 343–456.

Raman, K. V., and M. L. Jackson. 1965. Rutile and anatase determination in soils and sediments. Am. Mineral. 50:1086–1092.

Sherman, G. D. 1952. The titanium content of Hawaiian soils and its significance. Soil Sci. Soc. Am. Proc. 16:15–18.

Stace, H. C. T., G. D. Hubble, R. Brewer, K. H. Northcote, J. R. Sleeman, M. J. Mulcahy, and E. G. Hallsworth. 1968. A handbook of Australian soils. Rellim Technical Publications, Adelaide, South Australia.

Stephenson, N. C. N. 1973. The petrology of the Mt. Manypeaks adamellite and associated high grade metamorphic rocks near Albany, Western Australia. J. Geol. Soc. Aust. 19:413–439.

Tertian, R. and R. Géninasca. 1972. Analyse des roches par un méthode de fluorescence X comportant une correction précise pour l'effet interélément. X-Ray Spectrom. 1:83–92.

Teufer, G., and A. K. Temple. 1966. Pseudorutile—a new mineral intermediate between ilmenite and rutile in the natural alteration of ilmenite. Nature (London) 211:179–181.

Vance, E. R., and B. W. Anderson. 1972. Study of metamict Ceylon zircons. Mineral. Mag. 38:605–613.

Zussman, J. 1967. X-ray diffraction. p. 261–334. In J. Zussman (ed.) Physical methods in determinative mineralogy. Academic Press, London. (2nd ed., in press.)

Shrinking and Swelling of Clay, Clay Strength, and Other Properties of Clay Soils and Clays

KIRK W. BROWN, Texas A&M University, College Station, Texas

I. INTRODUCTION

The influence of clay on the physical properties of soil is of interest to both the agriculturalist and the soils engineer. The agriculturalist is concerned mainly with the properties of the soil within a depth of 1 or 2 m. The amount and species of clay found in a soil used for agricultural production influence the infiltration rate, the rate of internal drainage, the water holding capacity, the structural stability; the depth of water available for roots, the need and frequency of irrigation, the water contents at which soil can best be tilled, the type of tillage equipment required, and the amount of power required to perform the tillage operation.

The soils engineer often needs data on the physical properties of soil close to the surface, but may in some cases be interested in the physical properties of soil at tens or even hundreds of meters below the surface. The soils engineer is interested in the ability of the soil to support the roads, runways, houses, factories, piers, derricks, storage tanks, pipelines, and other structures that he designs. Engineers have developed a series of tests, some of which yield empirical parameters, to provide data on which to base design criteria. The parameters utilized include the bearing capacity, the coefficient of linear extensibility (COLE), the shear strength, the potential volume change (PVC), the permeability, the thermal conductivity, and the Atterberg constants including the liquid limit, the plastic limit, and the plasticity index.

Attempts by agriculturalists to alter the physical properties of soils in the field by manipulating the properties or proportions of the clay constituents have been infrequent because of the expense of treating extensive areas of land. Since the areas requiring treatment for engineering purposes are usually small and since the cost of treatment may represent only a small fraction of the cost of a project which might otherwise be jeopardized, treatments and amendments are commonly used to stabilize clay soils for constructive purposes.

Many approaches have been utilized to study the influence of the amount and species of clay on the physical properties of the soil. Investigations have been made using pure clay systems, clays extracted from soils, pure clays mixed with different fractions of silt or sand to simulate soils, and ground, sieved, and packed soil samples. Undisturbed cores have been utilized by some researchers and others have made measurements on soil in the field. Each of these approaches has its advantages and disadvantages. Detailed

689

mineralogical analyses of the behavior of pure systems can yield information on the mechanisms which may influence the bulk physical properties of the soil. While the same mechanisms may be active, direct extrapolation of laboratory data collected on pure systems to the behavior of field soils has met with only limited success. The heterogeneity of texture and mineralogy plus the complexities introduced by the organic fractions limit the usefulness of direct extrapolations. Soil particles usually clump together and are grouped into associations called aggregates or peds, in which the particles are oriented in more or less random fashion. Furthermore, the clay fraction in most soils represents more than one clay species in addition to other minerals which have undergone various degrees of weathering and which may occur in different stages of crystallization.

Soil physical properties depend not only on the type and amount of clay present, but on the concentration and species of electrolytes in the soil solution. In addition, the development of certain soil properties, especially after amendments have been incorporated, may be time dependent. For some processes, soil temperature plays an important role, while for others, the physical properties are nearly temperature independent. The physical and chemical properties of the clay minerals commonly found in soil differ greatly. Thus, the behavior of a soil containing mixtures of clays may be dominated by only a small fraction of the total clay.

While it is possible to make some predictions of the physical behavior of certain soils from mineralogical analysis, the results are not entirely satisfactory. Thus, for engineering purposes, the use of empirical information has persisted. Perhaps in the future mineralogical analyses of samples will be determined more frequently so that a better understanding of the mechanisms by which clays influence the properties of soil can be achieved.

Emphasis in this chapter is on bulk physical properties. While reference is made to possible mechanisms, detailed consideration of the physical-chemical processes on the clay surfaces which ultimately regulate the bulk physical properties are given in Chapter 2 of this book. A complete literature review is not intended, but efforts have been made to include citations to document the concepts presented. These should serve as a guide to those interested in greater detail.

II. SHRINKING AND SWELLING

The change in volume of soils as the water content changes is influenced greatly by the amount and kind of clay present. When soils containing certain clay minerals are dried, shrinkage and cracking occur. When the soils are rewetted, they swell as water is absorbed by the soil. In soils and most natural deposits of clays, volume changes occur equally in all three dimensions. Certain clay deposits in which the particles have a preferred orientation exhibit anisotropic swelling with maximum change in dimension perpendicular to the major axis of tabular particles (Warkentin & Bozozuk, 1961).

Particle orientation may occur as a result of overburden pressure or in some cases where the deposit resulted from sedimentation of dispersed particles in fresh water.

A soil with a large fraction of smectitic clay exhibits large cracks in the horizontal dimensions on drying. Measurements against a deep benchmark reveal that shrinkage also occurs in the vertical dimension. Elevation changes as great as 8 cm have been reported in the field (Atchison & Holmes, 1953). Soils are rarely uniform in the vertical or horizontal dimension nor is the distribution of water uniform within them. The differential shrinking and swelling associated with these heterogeneties is responsible for much of the structural damage to roads, foundations, and masonary walks built on such soil. The shrinking and swelling of agricultural soils is important in the formation of aggregates which are essential to the continued productivity of the soil.

The swelling of soils have been correlated with other selected properties. Russell (1961) noted that swelling is proportional to the cation exchange capacity (CEC) of clay soils. A correlation between soil shrinkage and both CEC and specific surface was reported by Gill and Reaves (1957). Other factors including the kind of exchangeable cations, the content of organic matter and the content of Fe are important (Davidson & Page, 1956). Overburden pressure and the degree of compaction may also influence the change in volume with change in water content.

Evaporative losses from bare, wet field soils occur at the surface. Once a dry layer forms on the surface, further drying is reduced considerably by the low hydraulic conductivity of the layer. When vegetation is present, the roots permeate the soil, first removing water from near the surface and then at increasing depths. A cropped soil will dry much more rapidly than a bare soil due to water extraction by plant roots, hence shrinkage cracks will increase in number much more rapidly than in bare soil. The shrinkage cracks that form will be widest near the surface and may extend to depths of several meters. Woodruff (1936) demonstrated that as a clay soil dries, the change in vertical dimension decreases with depth. Except for depths very near the surface, field soils do not reach soil water potentials much less than –15 bars. Ritchie et al. (1972) demonstrated that a sorghum crop (*Sorghum bicolor* Moench.) extracted moisture to the –15 bar level to a depth of 1 m in the soil. Clay soils dried to this potential still retain sufficient water to cover all soil particle surfaces with several molecular layers of water. Further drying by artificial means (such as oven drying) results in additional decreases in volume which are never experienced in the field.

Change in volume with decreasing water content for undisturbed peds of two smectitic soils is shown in Fig. 19-1. The first increments of water removed from a saturated soil come from the large pores between the aggregates. Very little volume change results from such removal. As the moisture decreases from 38 to 32% in the Austin soil shown in Fig. 19-1, there is little volume change. The Houston Black clay shown in Fig. 19-1 may not have started at complete saturation or may have lacked an appreciable noncapillary pore volume. Reduction in volume for both soils is then nearly linear over a

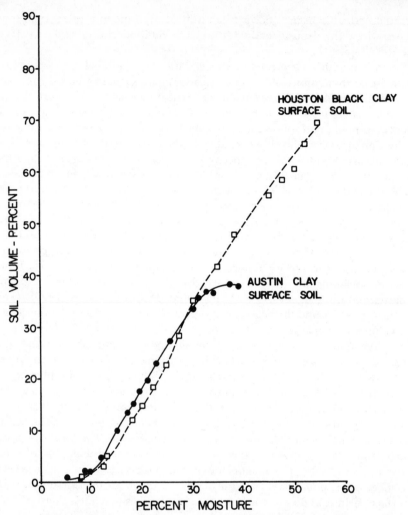

Fig. 19-1. The percent volume change of two soils with reference to the oven dry volume as a function of the percent water by weight expressed as a percent of the oven dry weight. From Johnston and Hill (1944).

wide range of moisture contents. In the intermediate range, water is re-moved from between the particles and the volume change is typically propor-tional to and nearly equal to the volume of water lost. At low water con-tents, below 12% for the soils shown in Fig. 19-1, the loss of volume is again much less than the loss of water. At these water contents, the repulsive forces between particles limit further collapse. Soils containing clays with less surface area than those shown in Fig. 19-1 will exhibit less volume change upon drying and would fall below the curves shown. Soils containing very little clay will exhibit little or no change in volume as water is removed.

Laboratory tests have been utilized to classify the swelling characteristics of soils. Grossman et al. (1968) proposed a technique of calculating the coefficient of linear extensibility, COLE $= (L_m - L_d)/L_m$, where L_m is the length of the sample when moist and L_d is the length of the oven dry sample. Franzmeier and Ross (1968) reported significant relationships between COLE and the amount and kind of clay in several types of soils. While such tests may be useful for classifying soils, the values obtained cannot be used to predict volume changes of field soils since oven drying far exceeds the conditions to which such soils are exposed.

The majority of data available on volume changes of soils as a function of water content has been collected on samples as they shrink. By maintaining a slow rate of water loss, the water content of a sample can be controlled so it is nearly uniform throughout. The unsaturated hydraulic conductivity of fine textured soils is small and it is, thus, difficult to establish uniform moisture contents as a sample is rewetted. By taking special precautions, however, Haines (1923) and Chang and Warkentin (1968) demonstrated that the dependence of volume on water content is hysteretic. Rewetted samples occupied less volume at a particular water content during the rewetting than they did during drying. Chang and Warkentin (1968) also reported that both the volume occupied by a given weight of soils and the hysteretic effect were less for compacted samples.

Several forces are involved in determining volume changes as soils dry. As water is lost from the soil, pressure differences develop across the air-water interfaces which occur where water bridges the particles. As water is removed from the system, water will be drawn by potential pressure gradients from other locations to help minimize pressure differences. As the water is withdrawn, the air-water interface becomes more concave and the particles are drawn closer together. Water is held between the particles by matric forces resulting from adhesion of water to the particle surfaces and cohesion between adjacent water molecules and by osmotic forces resulting from attraction of ions for water. As water is withdrawn from between the particles to satisfy the atmospheric demand, the particles will move closer together until repulsion prevents further collapse.

The species of ions present on the exchange sites influence the amount of shrinking and swelling that a soil will exhibit. Sodium saturated montmorillonite, illite, and kaolinite swelled more than when the same clays were saturated with Ca (Mielenz & King, 1955). The greater swelling associated with monovalent ions is a result of greater expansion of the diffuse double layer between adjacent particles. The presence of other materials including Fe hydroxides stabilizes the clay and prevents nearly all volume change.

Several models have been advanced to predict the swelling volume and pressure of clays as they change water content. Osmotic forces resulting from ions in the diffuse double layer have been calculated for dilute electrolytes assuming a Boltzman distribution. The relation between the water film thickness and the concentration of ions can then be calculated from the concentration of electrolytes in the equilibrium solution. For platy particles, the swelling pressures can be equated to the van't Hoff pressure which is calculated as a function of concentration of electroyltes. Reasonable agreement between theoretical values of swelling pressure and actual measurements has been found (Bolt & Miller, 1955; Warkentin et al., 1957). More recently McNeal (1970), working with mixed ions, developed a demixed-ion model which he reported predicts interlayer swelling better than the double layer model. The use of such models to predict the swelling characteristics of soils is hampered by the complexities of the soil systems.

III. SWELLING PRESSURE

The swelling of soils as water content increases can be prevented by the application of confining pressures. The pressure required to prevent swelling depends on the same factors that influence the volume change of unconfined samples. Only small increases in soil volume are necessary to decrease the pressure developed within the soil. Since many devices available to measure pressure changes depend on small volume increases for accuracy, great care must be taken to achieve meaningful results.

The development of maximum swelling pressures is dependent on the time required for water to completely penetrate and hydrate the sample. Palit (1953) reported that over 60 days were required to develop maximum pressures. The swelling pressures developed by many clay soils will far exceed overburden pressures and pressures resulting from manmade structures. The 3.3 kg/cm^2 swelling pressure developed by the soil investigated by Palit (1953) is approximately 30 times greater than the overburden pressure which would be applied to the soil by a typical domestic structure.

The swelling pressures for soils containing nonexpansive clays are much smaller and, in the extreme case, the addition of water to some soils may cause a small decrease in volume.

IV. SOIL STABILIZATION

Lime and, in some cases, gypsum have been mixed with soils which have high shrink swell potentials to minimize the change in volume as water contents change. Several theories have been advanced to explain the stabilization effect of these soil amendments. The quantity of lime needed to achieve the desired changes in physical properties of clay soils depend on the mineralogy. Typically 4 to 5% $CaCO_3 \cdot 6H_2O$ is needed to stabilize soils containing large fractions of Na smectite. The Ca added to the soil in these amounts of lime is more than adequate to replace the Na on the exchange sites of the treated soils. The flocculation which would be expected from the replacement of Na was one of the earliest mechanisms of stabilization proposed (Goldberg & Kline, 1952). Diamond and Kinter (1965), however, pointed out that since some soils in need of stabilization are already saturated with Ca or Mg and already flocculated, this mechanism is not completely satisfactory. The hypothesis that stabilization of soil may be a result of $CaCO_3$ bonding between particles (Mateos and Davidson, 1963) was also refuted by Diamond and Kinter (1965). They pointed out that stabilization takes place even in the absence of the CO_2 needed for the formation of $CaCO_3$. Eades et al. (1962) proposed that stabilization is a result of the slow formation of calcium silicate hydroxide bonds between particles. The rapid formation of small amounts of tetra-calcium aluminate hydrate by reactions of $Al(OH)_4$ groups exposed at the edge of clay surfaces was postulated by Diamond and Kinter (1965) as the ameliorating mechanism. They further suggested that slower pozzolanic reactions are responsible for the increased stability of lime-treated soils with time.

The effectiveness of lime treatments on soils may be demonstrated by adjustments in a variety of parameters. X-ray diffraction studies have been used by Pettry and Rich (1971) to demonstrate the reduction in interlayer spacing of smectitic soil clays. They also reported that the swelling pressure, liquid limit, and the plastic index decreased after addition of 5% lime to soils containing different clay minerals. The largest reduction in swelling pressure occurred in the soils containing large amounts of smectite. The swelling pressure of treated Houston Black clay soil containing 52% clay, predominantly tabular halloysite and kaolinite, was decreased by only 1/10 as a result of the lime treatment. These differences are attributed to mineralogy.

While the influence of stabilization is nearly instantaneous the adjustment of certain parameters continues for a long period of time. Clare and Crunchley (1957) found that after 12 months, the liquid and plastic limits were still increasing. McDowell (1953) reported that soil strength continued to increase over a period of 4 years.

V. STRENGTH

The *shear strength* of a soil is the force that must be applied to cause relative movement between particles. Depending on the composition of a soil and, in some cases, the water content, a soil may either fracture or exhibit plastic flow upon application of a force. A variety of tests have been developed to determine the shear strength of soils. Laboratory tests and their implications have been described for undisturbed and remolded soils in several general soil mechanics books including Wu (1966) and Terzaghi and Peck (1968). Generally a force is applied to a sample which may or may not be subjected to a pressure normal to the shear plane. Devices have also been developed for in situ measurement of shear strength. The *field vane shear method* described by the American Society for Testing and Materials (1964) is the most widely used. It consists of measuring the force required to twist a vane which has been pushed into the soil.

The shear strength of soils may be attributed to the cohesive forces between adjacent soil particles. The strength of the bond between particles will depend on the shape and size of the particles involved, as well as the sur-

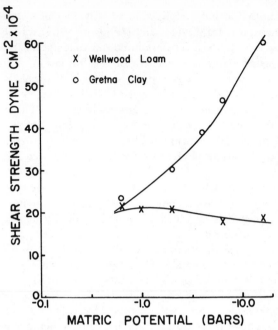

Fig. 19-2. Shear strength of two soils as a function of matric potential. After Williams and Shaykewich (1970).

face charge density of the particles. Thus, the amount and type of clay are important in determining the shear strength.

Grim (1962) reported that interparticle forces contributed to 80% of the shear strength of bentonite samples, from 40 to 50% of the shear strength of illite samples, and < 20% for the kaolinite samples. The influence of the water potential on the shear strength of soils was investigated by Williams and Shaykewich (1970). Their results shown in Fig. 19-2, reveal that the shear strength of Gretna clay, containing 55% clay, more than doubled as the soil dried from –1 to –10 bars while that of the Wellwood loam which contained 27% clay was nearly insensitive to changes in water potential. Trask and Close (1958) and Langston et al. (1958) showed that for a given water content less than saturation, illite had a greater shear strength than kaolinite, and montmorillonite had a greater shear strength than illite. When the clays were very wet, however, the shear strength had an inverse relation to surface area (Trask, 1959). Montmorillonite saturated with Ca exhibits a lower shear strength than when it is saturated with Na (Fig. 19-3). Warkentin and Yong (1962) claim that the lower strength of the Ca-saturated clay results from a lower surface area of interaction and lower repulsion found with the divalent ion.

The complexity of factors influencing the measurement of shear strength and the variety of measuring techniques in use have made it difficult to compare results between researchers. Bell et al. (1969), however, reported simu-

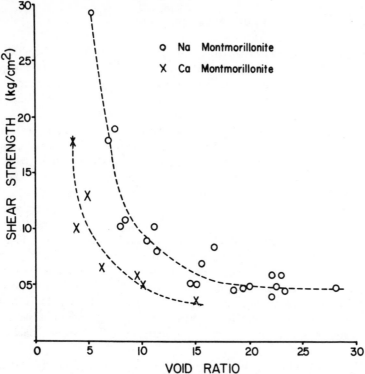

Fig. 19-3. Shear strength of Na and Ca montmorillonite as a function of void ratio, where void ratio is the volume of voids divided by the volume of solids on a percent basis. From Warkentin and Yong (1962).

larity between field and laboratory results when proper precautions were taken.

Several other parameters used to characterize soil strength have been correlated with clay content. Rogowski et al. (1968) showed that the rupture stress, rupture strain, and rupture modulus of soil aggregates were significantly correlated with the clay content. The tensile strength of montmorillonite films saturated with different ions was investigated by Dowdy and Larson (1971). They reported that Fe-saturated montmorillonite had the greatest tensile strength, Na montmorillonite had intermediate strength, and Ca montmorillonite had the least tensile strength. The lower tensile strength of the Ca-saturated clay is in agreement with the lower shear strength for divalent ions reported by Warkentin and Yong (1962).

The presence of clay in the soil significantly influences the stability of trench walls and slopes. In general, as the fraction of clay increases, the fraction of sand decreases, and the stability increases. However, the presence of cracks or slickensides in soils containing shrinking swelling clays makes them very unstable and necessitates the use of props to support the walls even in shallow trenches.

VI. ATTERBERG LIMITS

Early investigators of the physical properties of soils realized that texture alone did not provide an adequate characterization. Clay mineralogy was still in its infancy; thus, alternative means of correlating soil behavior with laboratory analyses were sought. Atterberg (1911) defined four characteristics of soil which vary as a function of their water content.

Originally, Atterberg's (1911) concepts were used on agricultural soils. Since then engineers have widely adopted and perfected the technique to serve their purposes. Casagrande (1932) spent considerable time developing laboratory techniques which would yield reproducible results.   Terzaghi (1955), another prominent soils engineer, has described in detail the equipment necessary to achieve repeatability. The classification system was later adopted and refined by the U. S. Army Corps of Engineers (1960) and renamed the "Unified Soil Classification System." In its present form, the system is widely used as an index of the behavior of soil as an engineering construction material.   Thus, while limited use of Atterberg's (1911) concepts are still being made to classify agricultural soils, the major use in the United States is for engineering purposes. Despite the simplicity of these empirical tests, the results have given satisfactory correlations with more basic engineering parameters including shear strength, bearing capacity, and unconfined compressibility.

The *liquid limit* is defined (Allen, 1942) as the water content, expressed as percent by weight of the oven dry soil, at which the soil just begins to flow when slightly jarred. Subsamples of the soil are wet up to different water contents and the samples, one at a time, are placed in a brass cup. A standard groove is made through the sample and the cup is raised and dropped a standard distance onto a hard surface. This process is repeated with each subsample until the liquid limit is found. The liquid limit is taken as the water content at which the soil sample must be dropped 25 times to close the groove. The *plastic limit* is defined (Allen, 1942) as the lowest water content at which the soil can be rolled into threads 0.32 cm (1/8 inch) in diameter without breaking into pieces. Soils containing significant amounts of sand cannot, of course, be rolled into threads at any water content and are classified as nonplastic. In practice, subsamples of the soil at different water contents are rolled by hand on a piece of etched glass. The *plasticity index* is defined as the difference in water content between the liquid limit and the plastic limit. At water contents greater than the liquid limit, the soil behaves as a liquid. At contents between the liquid and plastic limits, the soil behaves like a plastic and will deform under load without breaking. At lower water contents, the soil will not deform but instead shears or breaks when a load is applied.

While standard equipment and techniques are used to conduct the laboratory tests, the results obtained by different laboratories may differ by a factor of 2 and the coefficient of variability may be as great as 13% (Burmister, 1967). Some of the variability may be attributed to different methods of preparing samples before analyses. Drying, kneading, or soaking of soils may alter results of subsequent tests.

No particular plastic or liquid limit values have been found for any given clay mineral (White, 1949). The variability is a result of not only the different ions present, but also the variability of particle size and the composition of the clay minerals themselves. White's (1949) results reveal that both the liquid and plastic limits are greatest for smectites, medium for illites, and lowest for kaolinites. For Na-saturated montmorillonites the plastic limits are much greater than those of Ca-saturated montmorillonites. The liquid limits of Na-saturated montmorillonites are greater than those of Ca-saturated montmorillonites. As a result, the plastic index of Na-saturated montmorillonites is very large. The plasticity index generally is greatest for montmorillonite and lowest for kaolinites while attapulgite and illite fall between these two extremes.

The influence of electrolyte concentrations on the liquid limit of clay minerals was demonstrated by Warkentin (1961). His data shows that the liquid limit of Na montmorillonite decreased as the concentration of NaCl increased. The liquid limit of Ca montmorillonite was insensitive to the concentration of $CaCl_2$. The author suggests that at high salt concentrations the repulsion between particles decreases, face-to-face orientation increases, and as a result, the particles are free to move at lower water contents. The liquid limit of kaolinitic samples tested were much lower and less sensitive to the electrolyte concentrations. Systematic differences were evident, however, and may be associated with the degree of flocculation.

While the same general relationships exist for soils containing a clay mineral which dominates the soil behavior, the relationships are complicated by the presence of other clay minerals, particles of larger size, the organic fraction, variability of cations present, and the influence of soil structure. The influence of soil structure can be minimized by conducting tests only on remolded samples (Wu, 1966). The influences of the other factors are difficult to eliminate or even characterize and often result in considerable scatter in relations between clay content and liquid and plastic limits for natural soils.

The data of Dumbleton and West (1966) for a group of smectitic soils and artificial mixtures are shown in Fig. 19-4. As the clay content increases, the liquid limit, the plastic limit, and the plasticity index increase both for the soils and the artificial mixtures. Similar results with slightly more scatter were obtained for a group of kaolinitic soils and artificial mixtures. Dumbleton and West (1966) report that the liquid limit of smectitic mixtures is twice the value obtained for kaolinitic mixtures of the same clay content while differences between soils of different mineralogies are not as great.

Odell et al. (1960) investigated the relation between Atterberg limits and other soil properties. They reported multiple correlation coefficients of 0.959, 0.887, and 0.938 between liquid limit, plastic limit, and plasticity index using percent organic C, percent clay (< 0.002 mm), and percent smec-

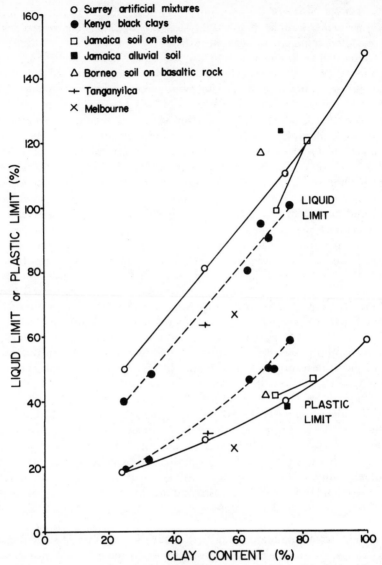

Fig. 19-4. Liquid and plastic limit of artificial mixtures of smectitic clays with coarser fractions and a series of mixtures and soils containing montmorillonite as a function of the clay content. From Dumbleton and West (1966).

tite in the clay as independent variables. Of the three variables, the clay content had the greatest influence on the liquid limit and plasticity index while the organic C had the greatest influence on the plastic limit. Lower, but significant, correlations were reported between the Atterberg parameter and the CEC. Significant correlations between CEC and Atterberg limits were also reported by Gill and Reaves (1957) for another group of soils.

VII. SOIL STRUCTURE

Soil structure is a physical property of great importance to agricultural endeavors. The importance of structure in promoting infiltration and permeability and in providing a good seedbed and rooting medium cannot be over emphasized. Several researchers have investigated the influence of clay and, in some cases, different types of clay on the stability of soil aggregates. Mazurak (1950) reported that the clays having greater surface area resulted in the formation of more stable aggregates than did equal quantities of low surface area clays. The influence of clay content on the aggregate stability of soil samples taken throughout the western United States was reported by Kemper (1966). He defined *aggregate stability* as the dry weight percent of aggregates remaining in a 0.25-mm sieve after it is shaken up and down in distilled water a given distance at a given rate for a period of time. To minimize biasing the results, he subtracted the dry weight of primary particles > 0.25 mm in diameter from both the weight of the aggregates and that of the total soil used in the calculations. His results, summarized in Fig. 19-5, show increasing aggregate stability with increasing clay content. The curve is steepest at low clay contents indicating the greater value of an increment of clay when the clay fraction in the soil is small.

The bulk density (oven-dry weight of soil per unit volume) of surface soils generally decreases as the clay content increases. The increased stability of soil aggregates at higher clay contents results in greater soil porosity. At greater depths in a soil profile where aggregate forming factors are less influential, the effect of clay on the bulk density is not as evident.

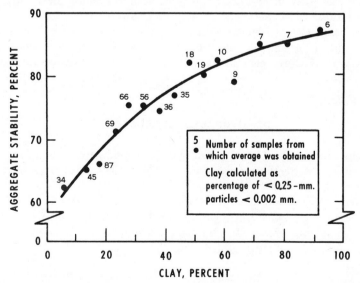

Fig. 19-5.   Soil aggregate stability as a function of the clay content.   From Kemper (1966).

VIII. WATER HOLDING CAPACITY

The type and amount of clay in the soils has a great influence on its water holding capacity. In the field the presence of a subsurface horizon or layer having a higher clay content than the surface will slow the movement of water thus allowing it to be retained for a longer time within the profile. In addition the water retention of each horizon in a profile will depend to a large degree on the species and amount of clay. The influence is minimal when the soil is near saturation since the water is held in larger pores. As soils drain, however, the influence of clay becomes more evident. At a potential of $-1/3$ bar, the water retention of soils is directly related to the clay content. Unger (1975) reported a correlation coefficient of 0.93 between clay content and water retention at $-1/3$ bar. As the soil dries to lower potentials, the influence of clay species becomes more evident. At a potential of -15 bars the water retention of soils with similar portions of different clay species will be approximately proportional to the surface area of the clay. Soils with large fractions of clays with large surface areas may retain more than 25% moisture by weight at -15 bars. Since this water is unavailable to plants, such soils may have smaller available water holding capacity (taken as the difference between $-1/3$ and -15 bars) than soils of less clay content (Longwell et al., 1963) or soils with clays of lower surface area.

El-Swaify and Henderson (1967) and Saffaf (1969)[1] demonstrated that the water retention of pure clay is dependent on the concentration and species of ions in the equilibrium dialyzate. The water retention for soils containing a variety of clays and salts increased at low electrolyte concentrations. The increase was greatest for smectite, less for vermiculite, and nearly negligible for kaolinite. The experimental results of El-Swaify and Henderson (1967) were in reasonable agreement with water retention calculated from the diffuse double layer model. At concentrations of 1 meq/liter the samples treated with NaCl retained about twice as much water as those treated with $CaCl_2$.

IX. HYDRAULIC CONDUCTIVITY

The saturated and unsaturated hydraulic conductivity (K) of soils is influenced by the fraction of clay present. The smaller pores in a soil containing a large fraction of clay will slow the movement of water when the soil is saturated. As the soil dries, however, and the larger pores drain, the small pores which predominate in a clay soil will continue to conduct water and will do so at a rate greater than that of a coarser textured soil, which has few interconnected, small, water-filled pores. The unsaturated hydraulic conduc-

---

[1]A. Y. Saffaf. 1969. The effect of solution composition on unsaturated hydraulic conductivity of some Texas soils. Ph.D. Dissertation submitted to the Texas A & M University Graduate College.

tivities of clay soils at −1 bar are typically 10 times greater than those for a sand.

Adequate internal drainage of soils must be maintained to facilitate leaching of salts so the soil can continue to be productive. Mismanagement of clay soils with water containing inappropriate concentrations of electrolytes may result in marked decreases in both saturated and unsaturated hydraulic conductivity.

The influence of ion concentrations and species on the saturated hydraulic conductivity of clay soils was reported by Quirk and Schofield (1955). They found decreases in saturated K starting immediately after the application of water containing low concentrations of electrolytes. McNeal et al. (1966) demonstrated that the decrease in conductivity of soils was related to the expansion of clay. This was shown by extracting clay from the soil and exposing it to the same concentrations of electrolytes. The influence of electrolyte concentration and composition on the unsaturated K of soils containing predominantly smectitic and kaolinitic clays was studied by Saffaf (1969)[1]. Smectitic soils exhibited a decrease in unsaturated K with a decrease in electrolyte concentration and with an increase in the Na adsorption ratio of the leaching water. The latter response is shown in Fig. 19–6. The kaolinitic soils were much less sensitive to changing electrolyte composition. The influence of electrolytes on both the saturated and unsaturated conductivity of soils having swelling clays may be explained by the decrease in pore volume which occurs during clay expansion and the blockage of small pores by the clay particles which may disperse as the thickness of the double layer increases.

Cracks which form in soils high in shrinking-swelling clays influence the other water relations of the soil. The cracking found under forage or broadcast crops is generally randomly orientated. As the soil shrinks, the cracks grow wider and deeper isolating adjacent peds and disrupting root systems. As drying progresses, cracks form at the weakest point in the matrix. This may be either a previous cleavage plane or at a place where the soil remains wetter. A common occurrence in vertisols is the formation of a predominant crack midway between rows where the soil water is greatest and the cohesive forces least (Johnson, 1962). As drying progresses, smaller randomly orientated cracks also develop. The formation of shrinkage cracks makes it nearly impossible to furrow irrigate such soils once they have become dry.

The infiltration rate of soils containing 2:1 clays is influenced by the size and depth of the cracks and the rate at which they close in relation to the precipitation rate (Adams et al., 1959; Hartman et al., 1960). When the soil is dry, much of an intense rain may enter the cracks before they close. The same amount of rain spread over a long period may result in considerably more runoff since the cracks are no longer evident. Ritchie et al. (1972) have shown that wet field soils may have saturated hydraulic conductivities 25 times greater than the same soil which has been repacked in the laboratory. Apparently considerable water movement occurs in old crack channels and slickensides which are no longer present in repacked soil.

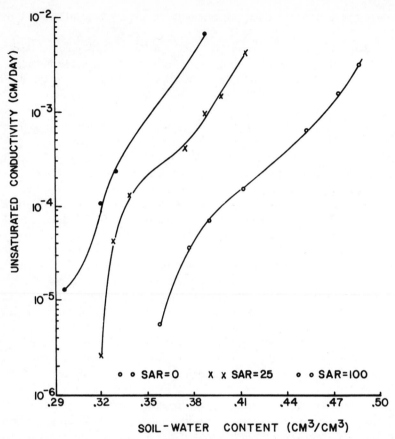

Fig. 19-6. The unsaturated hydraulic conductivity of Houston Black clay soil leached with water containing 100 meq/liter of salts with different Na adsorption ratios as a function of soil water content. After Saffaf (1969).[1]

The evaporative loss of water from the soil is influenced by the presence of shrinkage cracks. Ritchie and Adams (1974) demonstrated that the majority of evaporation from cracked soils occurred from the walls of the cracks even though the rate of evaporative water loss under bare soil was low.

X. SUMMARY

The influence of the quantity of clay and the mineralogy of the clay fractions is evidenced in the bulk physical properties of all soils. High surface area clays and especially those saturated with monovalent ions have the greatest influence on both the engineering and agricultural characteristics of a soil. The physical-chemical mechanisms by which clays influence most of the physical properties of soils are fairly well understood, although predictions of the exact response of a given soil are not possible because of the complexity of the system.

Since the same underlying mechanisms are apparently responsible for the influence of clays on the physical properties of soil, it is not surprising that good correlations are generally found between many of the parameters which have been used to quantify the physical properties of clay soils.

LITERATURE CITED

Adams, J. E., R. C. Henderson, and R. M. Smith. 1959. Interpretations of runoff and erosion from field scale plots on Texas Blackland soil. Soil Sci. 87:232-238.

Aitchison, G. D., and J. W. Holmes. 1953. Aspects of swelling in the soil profile. Aust. J. Appl. Sci. 4:244-259.

Allen, H. 1942. Classification of soils and control procedures used in construction of embankments. Public Roads 22:263-282.

American Society for Testing and Materials. 1964. Laboratory shear testing of soils. Am. Soc. Test. Mater. Spec. Tech. Publ. No. 361.

Atterberg, A. 1911. On the investigation of the physical properties of soils and on the plasticity of clays. Int. Mitt. Bodenkd. 1. p. 10–43.

Bell, J. R., J. D. Clarke, and E. L. Johnson. 1969. Lessons from an embankment failure analysis utilizing vane shear strength data. Seventh Ann. Eng. Geol. Soils Eng. Symp., Moscow, Idaho. p. 199–210.

Bolt, G. H., and R. D. Miller. 1955. Compression studies of illite suspensions. Soil Sci. Soc. Am. Proc. 19:285-288.

Burmister, D. M. 1967. Study and evaluation of the soil test results from the 1964 standard soil sample program of the American Council of Independent Laboratories. Mimeogr. Rep. to ASTM Comm. D-18. Am. Soc. Test. Mater. Philadelphia, Pa. p. 60.

Casagrande, A. 1932. Research on the Atterberg limits. Public Roads 13:121-130.

Chang, R. K., and B. P. Warkentin. 1968. Volume change of compacted clay soil aggregates. Soil Sci. 102:106-111.

Clare, K. E., and A. E. Cruchley. 1957. Laboratory experiments in the stabilization of clays with hydrated lime. Geotechnique 7:97–111.

Davidson, S. E., and J. B. Page. 1956. Factors influencing swelling and shrinking in soils. Soil Sci. Soc. Am. Proc. 20:320-324.

Diamond, S., and E. B. Kinter. 1965. Mechanisms of soil-time stabilization. Highway Res. Rec. 92:83-95.

Dowdy, R. H., and W. E. Larson. 1971. Tensile strength of montmorillonite as a function of saturating cation and water content. Soil Sci. Am. Proc. 35:1010-1014.

Dumbleton, M. J., and G. West. 1966. Some factors affecting the relation between the clay minerals in soils and their plasticity. Clay Miner. 6:179-193.

Eades, J. L., F. P. Nichols, Jr., and R. E. Grim. 1962. Formation of new minerals with lime stabilization as proven by field experiments in Virginia. Highw. Res. Board Bull. 335. p. 31–39.

El-Swaify, S. A., and W. D. Henderson. 1967. Water retention by osmotic swelling of certain colloidal clays with varying ionic composition. J. Soil Sci. 18:223-232.

Franzmeier, D. P., and S. J. Ross, Jr. 1968. Soil swelling: laboratory measurement and relation to other soil properties. Soil Sci. Soc. Am. Proc. 32:573-577.

Gill, W. R., and C. A. Reaves. 1957. Relationships of Atterberg limits and cation-exchange capacity to some physical properties of soil. Soil Sci. Soc. Am. Proc. 21:491-494.

Goldberg, I., and A. Klein. 1952. Some effects of treating expansive clays with calcium hydroxide. Am. Soc. Test. Mater. Symp. 142:53-74.

Grim, R. E. 1962. Applied clay mineralogy. McGraw-Hill Book Co., Inc., New York. p. 204-277.

Grossman, R. B., B. R. Brasher, and D. P. Franzmeier. 1968. Linear extensibility as calculated from natural-clod bulk density measurements. Soil Sci. Soc. Am. Proc. 32:570-573.

Haines, W. B. 1923. The volume changes associated with variations of water content in soils. J. Agric. Sci. 13:296-310.

Hartman, M., R. W. Baird, J. B. Pope, and W. G. Knisel. 1960. Determining rainfall-runoff-retention relationships. Texas Agric. Exp. Sta. Misc. publ. 404.

Johnson, W. C. 1962. Controlled soil cracking as a possible means of moisture conservation on wheat lands in the southwestern Great Plains. Agron. J. 54:323-325.

Johnston, J. R., and H. O. Hill. 1944. A study of the shrinking and swelling properties of Rendzina soils. Soil Sci. Soc. Am. Proc. 9:24-29.

Kemper, W. D. 1966. Aggregate stability of soils from Western United States and Canada. Tech. Bull. 1355. USDA-ARS, Washington, D. C. 52 p.

Langston, R. B., P. D. Trask, and J. A. Pask. 1958. Effect of mineral composition on strength of central California sediments. Calif. J. Mines Geol. 54:215-235.

Longwell, T. J., W. L. Parks, and M. E. Springer. 1963. Moisture characteristics of Tennessee soils. Univ. of Tenn. Agric. Exp. Sta. Bull. 367.

McDowell, C. 1953. Roads and laboratory experiments with soil/lime stabilization. p. 103-128. In Recent developments on lime stabilization in Texas. Proc. 51st Annu. Conv. Nat. Lime Assoc. (10 June 1953; Hot Springs, Va.).

McNeal, B. L. 1970. Prediction of interlayer swelling clays in mixed-salt solutions. Soil Sci. Soc. Am. Proc. 34:201-206.

McNeal, B. L., W. A. Norvell, and N. T. Coleman. 1966. Effect of solution composition on the swelling of extracted soil clays. Soil Sci. Soc. Am. Proc. 30:313-317.

Mateos, M., and D. T. Davidson. 1963. Compaction characteristics of soil-lime-fly ash mixtures. Highw. Res. Rec. 29:27-40.

Mazurak, A. P. 1950. Aggregation of clay separates from bentonite, kaolinite and a hydrous-mica soil. Soil Sci. Soc. Am. Proc. 15:18-24.

Mielenz, R. C., and M. E. King. 1955. Physical-chemical properties and engineering performance of clays. Calif. Div. Mines Bull. 169. p. 196-254.

Odell, R. T., T. H. Thornburn, and L. J. McKenzie. 1960. Relationships of Atterberg limits to some other properties of Illinois soils. Soil Sci. Soc. Am. Proc. 24:297-300.

Palit, R. M. 1953. Determination of swelling pressure of black cotton soil—A method. Proc. Int. Conf. Soil Mech. Found. Eng. 1:170-172.

Pettry, D. E., and C. I. Rich. 1971. Modification of certain soils by calcium hydroxide stabilization. Soil Sci. Soc. Am. Proc. 35:834-838.

Quirk, J. P., and R. K. Schofield. 1955. The effect of electrolyte concentration on soil permeability. J. Soil Sci. 6:163-177.

Ritchie, J. T., and J. E. Adams. 1974. Field measurement of evaporation from soil shrinkage cracks. Soil Sci. Soc. Am. Proc. 38:131-134.

Ritchie, J. T., E. Burnett, and R. C. Henderson. 1972. Dryland evaporative flux in a subhumid climate: III. Soil water influence. Agron. J. 64:168-173.

Ritchie, J. T., D. E. Kissel, and E. Burnett. 1972. Water movement in undisturbed swelling clay soil. Soil Sci. Soc. Am. Proc. 36:874-879.

Rogowski, A. S., W. C. Moldenhaver, and D. Kirkham. 1968. Rupture parameters of soil aggregates. Soil Sci. Soc. Am. Proc. 32:720-724.

Russell, E. J. 1961. Soil conditions and plant growth. 9th ed. p. 125-135. John Wiley & Sons, Inc., New York, N. Y.

Terzaghi, K. 1955. Influence of geological factors on the engineering properties of sediments. Econ. Geol. 50:557-618. (50th Anniv. Vol.)

Terzaghi, K., and R. B. Peck. 1968. Soil mechanics in engineering practice. John Wiley & Sons, Inc., New York. 729 p.

Trask, P. D. 1959. Effect of grain size on strength of mixtures of clay, sand, and water. Bull. Geol. Soc. Am. 70:569-579.

Trask, P. D., and J. E. H. Close. 1958. Effect of clay content on strength of soils. Conf. on Coastal Eng., Proc. 6th (Gainesville, Fla.) p. 827-843.

Unger, P. W. 1975. Relationships between water retention, texture, density and organic matter content of west and south central Texas soils. Tex. Agric. Exp. Sta., Misc. Publ. MP-1192C. p. 1-20.

U. S. Army Corps of Engineers. 1960. The unified soil classification system. Tech. Mem. No. 3-357. Vol. 1. U. S. Army Eng. Waterways Exp. Sta., Vicksburg, Miss.

Warkentin, B. P. 1961. Interpretation of the upper plastic limit of clays. Nature 190: 287-288.

Warkentin, B. P., G. H. Bolt, and R. D. Miller. 1957. Swelling pressure of montmorillonite. Soil Sci. Soc. Am. Proc. 21:495-497.

Warkentin, B. P., and M. Bozozuk. 1961. Shrinking and swelling properties of two Canadian clays. Proc. 5th Int. Conf. Soil Mech. Found. Eng. (Paris, France) 3A/49: 851-855.

Warkentin, B. P., and R. N. Yong. 1962. Shear strength of montmorillonite and kaolinite related to interparticle forces. Clays Clay Miner. 9:210-218.

White, W. A. 1949. Atterberg plastic limits of clay minerals. Am. Mineral. 34:508-512.

Williams, J., and C. F. Shaykewich. 1970. The influence of soil water matric potential on the strength properties of unsaturated soil. Soil Sci. Soc. Am. Proc. 34:835-840.

Woodruff, C. M. 1936. Linear changes in the Shelby loam profile as a function of soil moisture. Soil Sci. Soc. Am. Proc. 1:65-70.

Wu, T. H. 1966. Soil mechanics. Allyn & Bacon, Inc., Boston, Mass.

Reactions of Minerals with Organic Compounds in the Soil

ROBERT D. HARTER, University of New Hampshire, Durham, New Hampshire

I. INTRODUCTION

In any treatise discussing the role of minerals in the soil environment, mineral-organic reactions cannot be overlooked, since complexes resulting from these reactions are present in most soils of the earth's crust. These reactions play an important part in the genesis of soils, particularly in the weathering of primary minerals. The complexes formed affect the type of plant and animal organisms that may exist in any given situation, and even influence quality of the human environment. In fact, a role of primordial mineral-organic complexes in the development of life itself has been postulated (Degens & Matheja, 1967).[1]

Soil genesis and morphology are substantially affected by the organic compounds present and their reactions with the mineral constituents. The rate of parent material weathering is influenced by organic molecules through buffering of the soil solution and complexing of weathering products. Organic acids are also known to effectively attack primary minerals, particularly micas. Finally, mineral-humus complexes are integrally involved in aggregation of the soil, consequently affecting aeration, permeability, water-holding capacity, etc. These complexes will be discussed in detail in Chapter 21.

Many biologic processes within the soil are also affected by the reactions of minerals with organic compounds. Both laboratory and field studies have shown that biodegradability of many organic compounds is reduced upon reaction with and adsorption to a mineral surface. Likewise, the activity of enzymes and the phytotoxicity of pesticides have been shown to be adversely affected upon adsorption to mineral surfaces.

Conversely, the fact that many minerals will react with and adsorb organic compounds is of benefit to mankind. For example, many organic pollutants that might otherwise contaminate foodstuffs or water supplies can be bonded by minerals and slowly degraded. The loss of pesticide phytotoxicity can be a benefit to mankind by doing the job for which it was designed, then being quickly detoxified by mineral adsorption.

Mineral-organic complexes have been studied by numerous researchers under a multitude of conditions. To fully elucidate the current knowledge

---

[1]Egon T. Degens, and Johann Matheja. 1967. Molecular mechanisms on interactions between oxygen coordinated metal polyhedra and biochemical compounds. Reference No. 67-57. Woods Hole Oceanographic Institution, Woods Hole, Mass. Unpublished manuscript.

chapter 20

of soil mineral–organic reactions would require a full treatise, rather than a single chapter. Therefore, rather than discussing specific types of organic compounds and how they react with minerals, the emphasis of this chapter will be on the conditions leading to adsorption and some of the properties of mineral–organic complexes. Specific examples will be used only for illustrative purposes. Since soils are aqueous systems, reactions in non-aqueous systems will not be discussed, and reactions in anhydrous systems only briefly mentioned. It is hoped that this discussion will stimulate the reader to further study in the area, and serve as a basis for understanding the many and varied reaction processes that have been reported.

II. METHODS OF STUDYING MINERAL–ORGANIC REACTIONS AND COMPLEXES

While, theoretically, any technique used by organic chemists and mineralogists might be usable in the study of mineral–organic reactions and complexes, the properties of minerals severely limit the application of most tools useful for the study of organic molecules. The most commonly used techniques, therefore, are those applicable to mineral systems, with the majority of present information being derived from three techniques.

A. Adsorption Isotherms

Adsorption isotherms are normally obtained by adding solutions of progressively higher adsorbate concentrations to a standard amount of mineral, and measuring loss of the adsorbate from solution. In certain instances, this may lead to physical entrapment or excessive adsorption of organic material not really bound by the clay surface, and easily removed by water washing (Harter & Stotzky, 1971). It is possible, however, to modify the procedure by incrementally adsorbing small amounts of organic material, thus effectively altering the volume, rather than concentration, while keeping the mineral-solution ratio constant (Harter, 1975). The latter technique is also of value when the solubility of the organic compound is limited. In any case, either method is usable in constructing adsorption isotherms.

The results of adsorption studies have been graphically presented in several ways, depending upon the point a particular author wishes to make, but a plot of amount of organic adsorbed vs. equilibrium concentration of organic in solution has been the most common presentation. Giles et al. (1960) have classified the plots into four major classes (S, L, H, and C) and five subgroups (Fig. 20-1). The S (cooperative adsorption) curve indicates an increased affinity for the adsorbate after a few molecules have been adsorbed, and is most probably caused by strong intermolecular bonds. The L (Langmuir) curve is the expression of the opposite condition. As sites are filled, it becomes increasingly more difficult for adsorbate molecules to find

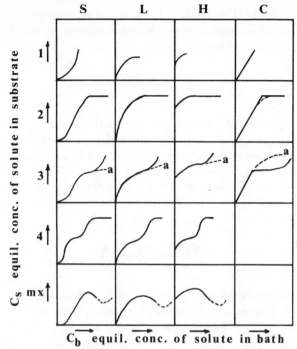

Fig. 20–1. System for isotherm classification (after Giles et al., 1960).

a vacant site. The H, or high affinity curve is a special case of the L curve in which adsorbate in dilute solutions is completely adsorbed, so that the initial part of the curve is vertical. Finally, the C curve occurs as a result of constant partition of adsorbate between solution and adsorbent to the maximum possible adsorption. Each curve class indicates specific characteristics of the adsorbate and adsorbent. Additional information that can be gained from the isotherms includes a variation in the initial slope, which depends on the rate of change of site availability with increased adsorption and a variation in plateau length (subgroup 3 & 4), which is indicative of the energy barrier to be overcome before the surface will adsorb more adsorbate.

An adsorption equation developed by Langmuir (1916, 1918) to theoretically quantify adsorption of gas monolayers by solid surfaces has been the basis for many attempts to quantitize the adsorption phenomena. Langmuir's equation takes the form

$$n = (kbP)/(1 + bP) \qquad [1]$$

where n is the amount of gas adsorbed per unit area, k is the adsorption capacity per unit area, P is the gas pressure, and b is a term which includes the probability of the gas being adsorbed or desorbed from the surface under a given set of circumstances.

In using the Langmuir theory, one must remember that it was originally developed for gas-solid systems, rather than for solids in an aqueous solvent. A modification of this equation was, however, adapted by Boyd et al. (1947) for ion exchange in solution

$$\left(\frac{x}{m}\right)_1 = \frac{kb_1C_1}{1 + b_1C_1 + b_2C_2} \tag{2}$$

where subscripts 1 and 2 refer, respectively, to the exchanging ion in solution and the exchanged ion previously on the surface, x/m is the amount of electrolyte adsorbed per unit area of exchanger, and C_1 and C_2 are the equilibrium concentrations (activities) of the two ions in solution. When solution concentrations are high enough that the quantity unity in the denominator is small compared to the other two terms, the equation can be transformed into the linear form

$$\frac{C_1/C_2}{(x/m)_1} = \frac{b_2}{b_1k} + \frac{1}{k}\frac{C_1}{C_2} \tag{3}$$

Thus, if adsorption obeys the Langmuir theory, a plot of $(C_1/C_2)/((x/m)_1)$ vs. C_1/C_2 will give a straight line, the slope of which is the reciprocal of the adsorption capacity, and the intercept of which may be related to the relative bonding energies of the two species. A simplifying assumption that $(C_1)/(C_2) \simeq C_1$ has been used by a number of researchers. However, this assumption is valid only under limited circumstances, and may be an important reason for the frequently reported lack of conformance to the Langmuir theory.

When adsorption data do not fit the Langmuir theory, they will frequently fit the empirical Freundlich equation

$$x/m = kC^n \tag{4}$$

where k and n are constants which may be related to bonding energy, but not to adsorption capacity. According to this equation, the amount adsorbed increases indefinitely. Thus, it frequently does not satisfactorily explain adsorption when most adsorption sites are saturated. Since the researcher usually wants to know the adsorption capacity of an adsorbent, the absence of a term for maximum adsorption is a major drawback in usage of the Freundlich equation.

A third equation, the BET (Brunauer, Emmett, and Teller) equation bears mentioning. This equation is of the form,

$$\frac{P}{V(P_o-P)} = \frac{1}{V_mC} + \frac{(C-1)P}{V_mCP_o} \tag{5}$$

where P is the equilibrium vapor pressure, P_o the saturation vapor pressure, V is the volume of gas adsorbed, V_m is the volume adsorbed for a monomolecular layer, and C is a constant related to bonding energy. Like the Langmuir equation, the BET equation is theoretically derived for adsorption of gases on solid surfaces, and it does contain a term comparable to Langmuir's adsorption maximum. The BET equation differs from the Langmuir equation in allowing multiple-layer adsorption. This equation has not been adapted for adsorption from aqueous solutions, but it has been useful to mineralogists making surface area measurements by gaseous adsorption of organic molecules (Mortland & Kemper, 1965).

B. X-ray Diffractometry

X-ray diffractometry is one of the major tools in the study of mineral–organic complexes. Its' usefulness in these studies is due to the relative ease with which the researcher can ascertain whether adsorption only occurs on external mineral surfaces (measurable adsorption, but no expansion of mineral), or if the organic molecule is capable of intercalating the mineral. Furthermore, if interlayer adsorption is regular, the orientation of organic molecules between layers can be predicted from known dimensions of the organic molecule and the c-dimension expansion resulting from adsorption. For example, Weiss (1963, 1969) has demonstrated that basal expansion can be related to the orientation of organic molecules parallel, normal, or at an angle to the mineral plane.

Used alone, X-ray does, however, have certain shortcomings in the study of organo–mineral complexes. For example, at intermediate levels of basal expansion distinction between low angle adsorption or parallel adsorption of multiple layers may not be possible. Furthermore, basal expansion resulting from interlayer adsorption of macromolecules, such as protein, is nearly impossible to relate to molecular orientation, since the molecule may be convoluted due to an unknown number of secondary and tertiary intramolecular bonds (Harter & Stotzky, 1973). Additionally, it is not always possible to separate solvent effects on expansion from those of the solute. To circumvent such problems, multi-instrumental approaches to the study of organo–mineral complexes are commonly used.

In most cases, a researcher uses X-ray diffractometry for information it will provide, and recognizes its' limitation. However, if the organic–mineral complex has sufficient regularity to provide several orders of the (001) spacings, electron density maps can be constructed (Fourier transform), and additional orientation and bonding information thereby obtained. For example, an air-dry 6-amino hexanoic acid-vermiculite complex has been found to have a c-dimensional spacing of 17.45 ± 0.04Å (Kanamaru & Vand, 1970). This spacing would allow several orientations of the organic molecules in the interlayer space. A one-dimensional electron density map (Fig. 20-2), however, provides clear evidence that the organic molecules are lying parallel to the surface with both the hydroxyl group of the carboxyl and the amine group

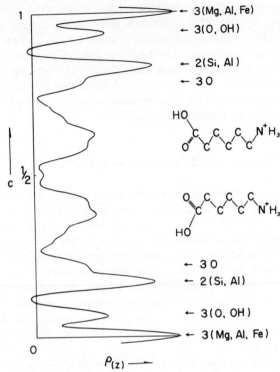

Fig. 20-2. One-dimensional electron density map of air-dry 6-amino hexanoic acid-
vermiculite complex (after Kanamaru & Vand, 1970).

being inclined toward the mineral surface. Furthermore, reflections arising
from other planes can be used to construct electron density projections on
the (010) or (100) planes, with resultant ability to determine interatomic
distances (Fig. 20-3). In the case of the 6-amino hexanoic acid–vermiculite
complex, an O–H–O bond length of 2.82Å and a N–H–O bond length of
2.86Å indicates a probable H-bonding of both the carboxyl O and amine N
to the mineral surface (Kanamaru & Vand, 1970).

C. Infrared Spectroscopy

Although it is a comparatively recent addition to the tools with which a
mineralogist works, infrared spectroscopy has become a major technique by
which clay–organic bonding mechanisms are studied. By the use of infrared
spectroscopy, several postulated bonding mechanisms have been proved,
while others have been disproved, and a few unsuspected mechanisms found
to be important (see Farmer, 1971, and Mortland, 1970). The power of this
technique arises from the fact that the vibration frequency of many intra-

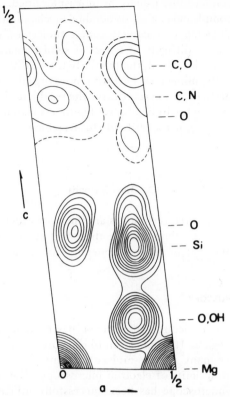

Fig. 20-3. Electron density projection of air-dry 6-amino hexanoic acid-vermiculite complex on the (010) plane: solid contours are drawn at unit electron intervals and dashed contours at zero electron density (after Kanamaru & Vand, 1970).

molecular ion pairs lies in the infrared region. Thus, observed absorption of infrared radiation provides information about intramolecular bond strengths. More important to the study of mineral–organic reactions, however, is that when an intermolecular bond is formed (such as between an organic and a mineral), several intramolecular bonds will be altered. This perturbation of bonds can be observed as shift of their respective infrared absorption bands. The direction of shift is indicative of whether a particular bond is strengthened or weakened, and the amount of shift indicates the change in bond energy. By careful observation of these parameters, along with possible extinction or appearance of absorption bands, a researcher can ascertain the atom(s) of the organic molecule which is (are) reacting with the mineral surface and make an estimation of relative bond strength.

Despite its value, infrared spectroscopy does have several major limitations. For example, water is a strong absorber of infrared radiation in the regions $3650-2930$ cm^{-1}, $1750-1580$ cm^{-1}, and $930-650$ cm^{-1}. Unfortunately, many ion pairs of interest in mineral–organic bonding also vibrate in these regions, so the samples must usually be at least air dried. However, drying a mineral–organic complex may alter its bonding mechanism(s) and/or strength. Second, use is restricted to relatively simple molecules, since alteration of bond characteristics is difficult to detect if similar unaltered ion pairs are present on the molecule. Third, the relatively low sensitivity of IR frequently requires loading the mineral surface with more organic than would normally be found in the field. In certain cases the resultant bonding mechanisms could be different than might be observed with less loading. These problems have however, proved not too difficult to overcome in interpreting natural systems.

D. Other Methods

While the above three techniques are the most widely used, others have been used from time to time. While these techniques are quite useful in specific studies, their adaptability is generally not sufficient for widespread usage.

1. ELECTRON MICROSCOPY

The transmission electron microscope has become a standard tool of many clay mineralogists. Its application to clay–organic studies has, however, been limited, since many organic molecules of interest have a molecular size below the resolution limits of electron microscopy. Nevertheless, the transmission electron microscope has been successfully utilized in certain clay–organic studies such as adsorption of proteins by smectite (Albert & Harter, 1973; Harter & Stotzky, 1973) and interstratification of layer structures by laurylamine–HCl (Yoshida, 1973).

The scanning electron microscope is of use to clay mineralogists wishing to examine three dimensional structures. This tool has been used to study the structure of flocules resulting from the reaction of polyethylene glycol with clays (O'Brien, 1971).

2. DIFFERENTIAL THERMAL ANALYSIS

Although DTA should be more generally applicable to mineral–organic studies than is electron microscopy, it appears not to be a widely used technique. The endothermic peaks for organic compounds have been shown to shift upon adsorption of the compounds by minerals (Suito et al., 1966). This observation leads to the suspicion that DTA might be used to determine relative stabilities of different complexes. In studying a series of organo-ammonium compounds adsorbed by montmorillonite and hectorite, Chou and McAtee (1969) were also able to relate DTA data to geometry of adsorption, thus indicating that this technique might be a valuable supplement to X-ray diffractometry.

3. GAS CHROMOTOGRAPHY

One of the problems in developing a complete understanding of mineral–organic reactions is the difficulty in obtaining thermodynamic properties of the reactions and complexes. By passing neutral, polar organic molecules through gas chromotographic columns packed with clay minerals, it has been possible to compute the heat of adsorption (Bissada & Johns, 1969). Since adsorption is from the vapor phase, the applicability of the technique is generally limited to gaseous reactions in the soil.

4. ULTRAVIOLET SPECTROSCOPY

Ultraviolet spectroscopy has certain advantages over infrared spectroscopy. In particular, water is not a strong absorber of ultraviolet radiation, the spectra are generally simpler, and the surface need not be heavily loaded with adsorbate to obtain well-resolved spectra (Bailey & Karickhoff, 1973a, b). However, absorption bands in this region (due to molecular electron excitation) are somewhat more difficult to interpret than are the molecular vibration absorption bands of infrared spectroscopy. In addition, colloidal particles tend to scatter the shorter wavelength ultraviolet light to a greater extent, creating analytical difficulties. Some of these difficulties are being worked out and this tool is proving useful for certain applications (Bailey & Karickhoff, 1973b; Bailey et al., 1973).

5. ELECTRON SPIN RESONANCE

When a molecule or an ion contains an unpaired electron, a resonance can be observed which results from the unpaired electron's spin mode. In certain circumstances, this resonance can be useful in studying clay–organic interactions (Pinnavaia & Mortland, 1971; Fenn et al., 1973; Rupert, 1973). Limitations of this tool are that oriented samples are ordinarily required, and the samples must be at least air dry. For example, Clementz et al. (1973) have observed that the signal for Cu adsorbed by layer silicates only differs from that normally observed for Cu when water content of the clay does not exceed a few layers.

6. MICROCALORIMETRY

Microcalorimetry provides a technique whereby the heat of reaction can be obtained in aqueous phase. However, the requirement of equilibrating, then mixing the mineral and organic inside the calorimeter cell has been a major limitation, particularly since many organic molecules cause the clay to flocculate. The recent development (Harter & Kilcullen, 1976) of a cell which will combine and mix the two phases with low mechanical energy input is promising for the future.

III. PROPERTIES OF MINERALS WHICH AFFECT ORGANIC ADSORPTION

For adsorption of an organic molecule by a mineral to occur, the mineral must have physical and chemical properties which are conducive to reaction. In the study of various mineral–organic reactions, certain characteristics of the minerals stand out as being particularly important.

A. Cation Exchange

As will be explained in the discussion of organic properties, many organic compounds are cationic in the environment of mineral surfaces, so coulombic interaction is one important mechanism by which organic molecules are bonded to the mineral. When the mineral is saturated with a basic cation, adsorption of small organic molecules is usually via a cation exchange reaction and is limited by the CEC of the mineral. Furthermore, selectivity coefficients for such exchange have been measured (Theng et al., 1967). The cation exchange reaction and the variable selectivity have been of use to mineralogists and to soil chemists. For example, stoichiometric exchange of K by dodecylammonium has been used to measure the CEC of micas (Mackintosh et al., 1971), and stoichiometric exchange of nicotine by several acid clays has been used in titrations to obtain sharper end-points than those normally observed with NaOH titrations (Khan & Singhal, 1967). As an example of selectivity, sodium tetraphenylboron is capable of extracting a portion of the K from K-bearing layer silicates, the amount extracted being correlated with plant uptake of the element (Wentworth & Rossi, 1972).

The adsorption of large organic cations is usually limited by size rather than by negatively charged adsorption sites on minerals. In fact, Kown and Ewing (1969b) found that the relatively small methylene blue molecule (which has frequently been used for CEC determinations) was too large to occupy all charge sites on Mississippi montmorillonite.

B. Surface Charge Density

For many years, mineralogists debated whether the structural charge was uniformily distributed over the mineral surface or occurred as discrete charges. The latter view is now generally accepted. The correctness of the discrete charge model has been confirmed by certain organic adsorption studies, which, in converse, illustrate that surface charge density does affect the amount of organic material which can be adsorbed by minerals. Adsorption of diquat, 1,1'-ethylene-2,2'-dipyridylium, and paraquat, 1,1'-dimethyl-4,4'-dipyridinium, (two similar organic divalent cations which differ in that the charges are separated by 3-4Å and 7-8Å, respectively) to a number of minerals has indicated that the cation whose charge centers most nearly approached the surface charge sites was preferentially adsorbed (Weed & Weber, 1968; Philen et al., 1970 and 1971). Thus, paraquat was preferred by the lower-charged smectites, since the greater distance of charge separation more nearly matched the distance between surface charge sites. Diquat, with less charge separation, was preferred by the more highly charged micas. The lack of match between charge sites on the organic and mineral surface is also the primary reason for the lack of stoichiometric exchange previously cited.

Surface charge density affects the orientation of adsorbed organics as well as the amount of adsorption. For example, Serratosa (1966) has shown that pyridinium (monovalent organic cation) is essentially parallel to the surface when adsorbed by smectite $(d(001) = 12.5Å)$, whereas it is perpendicular to the surface when adsorbed by vermiculite $(d(001) = 13.8Å)$. This difference was attributed to the fact that the more highly charged vermiculite adsorbed more pyridinium, and the resultant interlayer "crowding" did not permit the individual molecules to lie flat on the surface. Apparently, this configuration was thermodynamically more favorable than was adsorption in multiple layers.

C. Cation on the Exchange Complex

Adsorption has also been found to be dependent upon the cation saturating the exchange complex. The degree of dependence and effect of cation on adsorption will depend on the type of bond formed, and, to a lesser extent, on the size of the organic molecule. If adsorption is by a cation exchange reaction, the number of organic molecules that can be adsorbed will depend upon the ability of the organic molecules to compete with inorganic cations for surface charge sites. Competition for the surface charge sites is usually most efficient when the mineral is saturated with low charge, low ionic radius cations. For example, in studying ion effect on lysozyme adsorption by smectite, Harter (1975) found that 68% of the variation in lysozyme adsorbed could be explained by valence and ionic radius alone (Fig. 20-4). Furthermore, 93% of the variation could be explained by inclusion of Pauling electronegativity and ionization potential in the regression equation. All variables would affect the strength with which inorganic cations are bonded to the mineral surface.

Where other types of bonding are involved, however, the relationship to saturating cation will be of a different nature. For example, bonding may be directly to the saturating cation, in which case the relationship is reversed, and minerals saturated with the higher valence ions, particularly the transition metals, will form stronger bonds (Tahoun & Mortland, 1966b; Lailach et al., 1968). Bowman (1973) has found that the product of valence times atomic weight is a fairly accurate predictor of adsorption where dasanit, *O,O-*

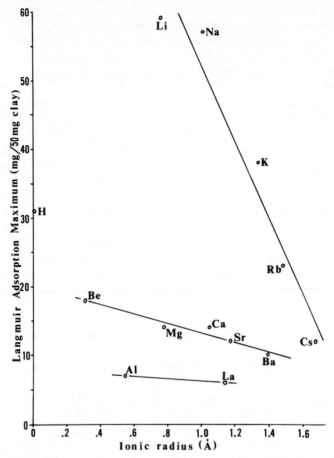

Fig. 20-4. Influence of ionic radius and valence of ions saturating the exchange complex on the maximum amount of lysozyme that can be adsorbed by smectites (after Harter, 1975).

diethyl-O-[p-(methyl sulfinyl) phenyl], was adsorbed by minerals saturated with transition metals, alkaline earths, and alkali metals.

D. Origin of Charge

Mineral–organic reactions and complex formation are also dependent upon whether the mineral charge originates in the octahedral or tetrahedral sheet. At a given surface charge density, electrostatic force at the clay surface will be greater in minerals whose primary charge source is the tetrahedral sheet, thereby affecting the types of bonds that can be formed. Swoboda and Kunze (1968) found, for example, that clays with tetrahedral charges reacted with much weaker organic bases than did clays with predominantly octahedral charges. They attributed this to a greater surface acid strength of the minerals having tetrahedral charges. Martin-Rubi et al. (1974) have reported that, when the charge originated in the tetrahedral sheet, the $-NH_3^+$ group of butylammonium was capable of keying into interlayer ditrigonal cavities of the mineral. The resultant butylammonium–mineral complex should be very stable. The greater localization of charge arising in the tetrahedral sheet (Mortland, 1970) may also contribute to the preferential adsorption of certain organic compounds such as that found by Weed and Weber (1968). Variations in surface electrostatic force due to source of the charge would also affect bonding strength between the mineral and inorganic cations, which in turn could affect the ability of organic molecules to compete for adsorption sites.

E. Expandibility

The amount of organic materials that can be adsorbed at mineral surfaces depends upon the surface area exposed to solution. Thus, expanding minerals will usually adsorb more organic molecules than will nonexpanding minerals. Perhaps less obvious, however, is the fact that expanding minerals differ in their susceptability to swelling pressures. This variation is a function of hydration energy of the interlayer cation. According to Norrish (1954), smectite saturated with Li, H, and Na is capable of swelling to 130Å and, perhaps, greater distances. For all practical purposes, this represents complete suspension. However, as hydration energy of the interlayer ion decreases, less expansion occurs, with little likelihood that smectite saturated with monovalent ions heavier than K will expand to beyond 20Å. In addition, expansion variations with valence of saturating ion can be explained by the so-called *Schultze-Hardy* rule, which says that, as valence of the saturating ion increases, the clay will be more strongly flocculated, i.e. resistent to dispersion. Thus, whether a particular organic molecule can migrate to interlayer adsorption sites will be a function of molecular size and expandibility of the mineral. It should be noted, however, that nonexpandible minerals, such as kaolinite, illite, and micas, can adsorb organic molecules in interlayer positions (Wada, 1959, 1961; Ledoux & White, 1965; Weiss, 1963; Boyle et al., 1967; Olejnik et al., 1971a, b). While intercalation of these minerals has generally required rather harsh laboratory treatment by nonaqueous solvents, the fact that it has been observed does have implications in postulating natural weathering mechanisms.

IV. PROPERTIES OF ORGANIC COMPOUNDS THAT AFFECT ADSORPTION

Not only must the mineral have surface properties conducive to adsorption, but the organic molecule must have characteristics that are compatible with bond formation. The more important characteristics are discussed here.

A. Molecular Charge

If an organic molecule carries an intrinsic charge, it is most commonly anionic in nature. However, certain organic compounds can be positively charged. These compounds usually contain one or more N atoms. The N atom can, under certain circumstances form four bonds with concomitant electron loss to give a positively charged ion (Cotton & Wilkinson, 1966). Examples of positively charged compounds would include the substituted ammoniums, R_4N^+, and similar, but less common oxomiums, R_3O^+. The diazoniums, $Ar-N \equiv N:^+$ also represent an important group of positively charged organic molecules. Many metal–organic compounds are also positively charged.

This is not to say, however, that only positively charged organic molecules can be adsorbed by clay surfaces. If this were the case, there would be no reason to include this chapter in a soil mineralogy text, since explanation of mineral–organic reactions could be explained as simply a coulombic interaction. A significant portion of the chapter is, in fact, devoted to an explanation of why and how commonly anionic organic molecules are adsorbed by negatively charged mineral surfaces.

B. Lone Pair Electrons

Lone pair electrons on organic molecules appear to be of particular importance in the reactions of the molecules with minerals. Although four atoms common in organic molecules (N, O, S, and P) have lone pair electrons, S and P do not appear to be substantially involved in organic–mineral reactions. The lone pair electrons of O are important, particularly in polarity of the organic molecules and in formation of coordination bonds, but the majority of reactions involve N-containing organic molecules. This is not surprising, since the nitrogen lone pair electrons are a particularly active Bronsted base site. In aqueous solution, these lone pair electrons are quite easily protonated to form a positively charged molecule.

C. pK_a

The pK_a of an organic molecule may be considered to be a mathematical description of the organic molecule's pH-dependence. For any given functional group, the pK_a can be related to pH according to the relationship.

$$pK_a = pH - \log(a_{\text{base}}/a_{\text{acid}}) \qquad [6]$$

where a_{base} and a_{acid} indicate the activity of base and acid in solution. As a first approximation, concentration can be used for activity. Thus, if the pK_a of a compound is known, the relative proportions of its dissociated (base)

Table 20-1. The effect of the relationship between pH and pK_a on the proportions of molecules in the dissociated (base) and undissociated (acid) forms

$pH - pK_a$	Base form	Acid form	Ratio base/acid
	— % —		
2	99	1	100:1
1	91	9	10:1
0	50	50	1:1
−1	9	91	1:10
−2	1	99	1:100

and undissociated (acid) forms can be predicted (Table 20-1). Thus, for a N-containing organic molecule, when pH = pK_a − 1, 91% will be protonated; and at pH = pK_a − 2, 99% will be protonated. The amount of adsorption can be predicted accordingly. If, on the other hand, the functional group is an organic acid, base represents the dissociated carboxylate form, having a negative charge.

Thus, the probability of adsorption would decrease when the pH of the compound's environment is above its pK_a. The acid form of the same compound would be an undissociated carboxylic acid and H bonds could

form between the organic and the clay surface.

D. Isoelectric Point

For compounds having both acidic and basic functional groups, the isoelectric point (IP) concept is of more value in predicting adsorption. The IP is the pH at which each molecule of the compound has a net zero charge. At pH values above this point, the molecules have a net negative charge, and at pH values below this point, the molecules have a net positive charge. Although the IP might be considered a summation of all individual pK effects, no precise equation for its determination can be given, and it remains an empirical value only. The lack of a precise definition means it is also less predictive of adsorption, since, for example, large charged molecules such as proteins might retain individual positive charges at pH values well above the IP of the compound. In such a situation, minerals could theoretically adsorb the compound, even though net charge characteristics are unfavorable for adsorption.

E. Polarity

Uncharged molecules can frequently be adsorbed by mineral surfaces as a result of differences in electron density within the individual organic molecules. The amount of charge transfer (dipole moment) within the molecule can be quantitatively measured. Organic compounds may have permanent dipole moments (frequently the result of lone pair electrons) while, in others, a polarity can be induced by the presence of a charge field such as that occurring around clay particles. This charge transfer can play an important role in adsorption of organics by minerals.

F. Molecular Weight

Steric hindrance is always a consideration when organic molecules react with minerals. The ability of a mineral to adsorb a maximum of any given organic compound will depend upon whether individual molecules can fit each adsorption site and whether they can move into interlayer positions. The former indicates an interaction between molecular size and adsorption site density, and the latter an interaction between molecular size and maximum expansion of a mineral.

V. ENVIRONMENTAL FACTORS THAT AFFECT ADSORPTION

A. pH Buffering

As has been indicated in the discussions of pK_a and isoelectric point of the organic molecules, the pH of the adsorbate solution is of considerable importance in determining the form of the organic molecules, thus, whether adsorption will, in fact, take place. Although the electric field around a charged mineral surface is known to alter solution characteristics, including pH, the inclusion of buffering salts will depress the double-layer thickness to the point that pH characteristics near the mineral surface will be little different from those in bulk solution (see Marshall, 1964, Chap. 9). Thus, adsorption can be predicted from knowledge of the bulk solution characteristics.

In systems buffered against pH change, maximum adsorption usually occurs within one pH unit of the organic compound's pK_a or the IP. As the pH increases above this point, the molecules become progressively less positively charged when protonation of nitrogen lone pair electrons is involved, or more negatively charged when dissociation of carboxyl protons is involved. Thus, the probability of an interaction is progressively decreased with increase in pH. A decrease in adsorption as the suspension pH is decreased below the pK_a or the IP has also been reported (Armstrong & Chesters, 1964; Thompson & Brindley, 1969; and Albert & Harter, 1973). When single functional groups are involved, this decrease has been attributed to a competition for negative adsorption sites by protons or Al ions liberated from the mineral structure (Thompson & Brindley, 1969). If both carboxyl and amine functional groups are involved, thus requiring use of the IP concept, the organic molecule becomes more positively charged as the pH decreases, and fewer molecules are required to satisfy the mineral charge sites (Armstrong & Chesters, 1969).

In unbuffered solutions, the pH–pK_a (or IP) relationship is substantially altered. The H ion activity near charged surfaces is almost always greater than that of bulk solution (Mortland et al., 1963; Mortland, 1966; Mortland & Raman, 1968). Even a mineral saturated with basic cations, such as Na, may have an increased H ion activity near the surface due to hydrolysis. Since the OH ion will subsequently migrate into the bulk solution, the result of this hydrolysis will be an increased suspension pH (perhaps even mildly alkaline), and a H ion activity near the mineral surface which is substantially greater than expected from suspension pH measurements (Kamil & Shainberg, 1968). Thus, in unbuffered suspensions, only approximations of expected adsorption can be made from bulk solution pH characteristics, and adsorption can occur at several pH units above that which might be expected from solution pH-isoelectric point (or pK_a) considerations alone.

B. Solution Ionic Strength

Any increase in solution ionic strength will alter the electric field near a charged surface by decreasing the thickness of the electric double layer. The effect of this alteration on adsorption of organic molecules by the surface will depend on the nature of the organic molecule. An effect of ionic strength is seldom reported unless the salt concentration is at least 0.5 to 1.0N. Above that level, however, increasing the solution ionic strength may be expected to decrease adsorption of those organic compounds that are otherwise adsorbed and increase adsorption of compounds not otherwise adsorbed.

For example, although most substituted ureas are readily adsorbed by acid mineral surfaces, much less adsorption may be expected when the mineral is saturated with basic cations. If the salt concentration is increased, however, increased adsorption may be observed, with the amount of organic adsorbed becoming dependent on the solution ionic strength. This effect has been attributed to increased activity coefficients at salt levels above 1.0N (Van Bladel & Moreale, 1974). On the other hand, proteins having high IP

are readily adsorbed by charged mineral surfaces, apparently via a cation exchange reaction. In this case, when solution ionic strength is increased above
0.5 to 1.0N, the amount of protein adsorbed decreases. Since adsorption involves a coulombic interaction, the decrease in thickness of the electric
double layer apparently makes adsorption less probable. Effectively, the cations in solution are competing with the organic molecules for adsorption
sites (Harter, 1975).

C. Equilibration Temperature

Strictly thermodynamic considerations lead to the conclusion that increased temperature should result in decreased adsorption. However, as with
ionic strength, this has proved to be an oversimplification, and the effect of
temperature can only be predicted by considering the nature of the adsorbate
and adsorbent as well as the type of bonds formed. Generally, when adsorption is from the vapor phase, the predicted relationship is obeyed. For example, temperature is of critical importance when glycolating minerals for
surface area measurements (McNeal, 1964). However, when adsorption is
from solution phase, the response is considerably less predictable. In studies
with pesticides, smectite has been shown to adsorb more monuron, atrazine,
CIPC, DNBP, and simizine at 0C (273K) than at 50C (323K), as predicted.
However, the same studies showed diquat, 2,4-D, and amiben to be adsorbed
in similar quantities at both temperatures (Harris & Warren, 1964). Since exchange reactions tend to be temperature independent, the lack of temperature effect was assumed to be evidence that adsorption of the latter three
pesticides was via an exchange reaction. In other instances, the temperature
effect is exactly opposite that which is expected. For example, increased adsorption of polyvinylpyrrolidone by several clays with increasing temperature
has been observed (Francis, 1973). For large polymers such as this, the increased adsorption could be the result of increased solubility at higher temperatures.

D. Concentration of Organic in Solution

As illustrated in the discussion of adsorption isotherms, there is always
a partition of the adsorbate between solution and adsorbing surface, and increased concentration of adsorbate in solution will usually lead to increased
adsorption by the mineral. As long as the adsorbate concentration is maintained at a level below that required for saturation (or near-saturation) of the
surface, this response is a useful tool in studying mineral–organic interactions.
However, if the concentration is allowed to increase above that required to
saturate the surface, additional, easily removed organic molecules may be
loosely adsorbed by the complex (Harter & Stotzky, 1971). On a theoretical
basis, this simply leads to a subgroup 3 or 4 curve in the Giles et al. (1960)

classification system. However, problems may be encountered where attempts are being made to relate the adsorption information to the "real world," since very high concentrations of organics in aqueous solution are not often found in nature, and conclusions drawn in the laboratory may be invalid (though not always) when applied to natural systems.

VI. MINERAL-ORGANIC BONDING MECHANISMS

A. Ion Exchange

The simplest and most easily understood bonding mechanism is an exchange of a charged organic molecule for a like-charged ion on the mineral surface. Although a certain amount of anion exchange may occur on the edges of some minerals, most exchange adsorption will be cationic, with positively charged organic molecules displacing inorganic cations occupying negative adsorption sites on the clay surface. Although, as indicated, some organic molecules have an intrinsic positive charge, adsorption via cation exchange is usually preceded by protonation of one or more lone pair electrons to form a positively charged molecule. If the protonation occurs in solution, the adsorption mechanism may be considered an ion exchange reaction, but it more frequently occurs at the clay surface, as discussed in section VI-B-1.

B. Hydrogen Bonding

Prior to the use of infrared spectroscopy, many organic molecules were thought to be adsorbed to mineral surfaces via hydrogen bonding to the surface oxygens (Farmer, 1971). However, Farmer (1971) has cited high stretching frequencies of ammonium and substituted ammonium N-H bonds as evidence that surface oxygens can form only weak hydrogen bonds (with considerable dependence upon whether the charge arises in the tetrahedral or octahedral sheet of the mineral). Although hydrogen bonding does, in fact, play an important role in mineral-organic reaction, direct hydrogen bonding to surface oxygen is of minor importance.

1. PROTONATION AT THE MINERAL SURFACE

Protonation at the mineral surface may be the most common method by which minerals in nature adsorb organic molecules. This type of bond is not actually a H bond, but is discussed here because a redissociated organic molecule will leave the proton behind. Protonation occurs when an uncharged molecule approaches an acid mineral surface. Bonding is the result of lone pair electron protonation and concomitant attachment to the mineral surface by coulombic forces. This type of bonding has frequently been reported where amine groups are involved. Tahoun and Mortland (1966a) have indicated that both the carbonyl groups and amine groups of proteins and peptides may be protonated in acid clay systems.

2. WATER BRIDGING

A second important type of H bonding can occur whereby organic molecules are linked to cations on the exchange complex through a water molecule in the primary hydration shell of the cation. This type of bonding is particularly favorable in low water content systems. It differs from the preceding in that it can occur in other than acid systems, and in that the lone pair electrons are not really protonated. Thus, bond formation is between the H of the water molecule and the lone pair electrons, rather than between H on the organic molecule and the mineral surface. Mortland (1968) found that surface acidity increased as the mineral system is dehydrated, and is related to the electronegativity of cations saturating the mineral surface (Fig. 20-5). In any primary hydration shell, the water molecules are slightly polarized; the degree to which they are polarized will depend on the electro-negativity of the hydrated cation. When water molecules are removed from the primary hydration shell during drying, the hydration energy of the cation must be shared with fewer water molecules, resulting in increased polarization of the remaining molecules, thus accounting for the increased acidity observed as the system was dehydrated. It was also noted that de-

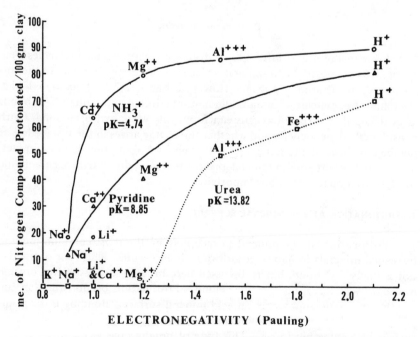

Fig. 20-5. Extent of proton donation by residual water in montmorillonite to three different bases (ammonia, pyridine, urea) as related to the electronegativity (Pauling) of the exchangeable cation (after Mortland, 1968).

hydration and saturation with cations of higher electronegativity will lead to adsorption of weaker bases than might otherwise be expected. Van Bladel and Moreale (1974) have shown a direct relationship between adsorption and the free energy of hydration (or polarizing ability) of interlayer cations. Both adsorption and free energy of hydration increased in the order Na < Ca < Mg < Al. The greater polarity of hydration shell water has been shown to be particularly important in the bonding of substituted ureas by minerals. Mortland and coworkers have extensively studied both bonding mechanisms of amines and the acid nature of water near mineral surfaces by observing the infrared spectra of substituted ureas adsorbed by the minerals.

3. BONDING TO MINERAL HYDROXYL GROUPS

All silicate minerals contain exposed OH groups which are potential sites for H bonding of organic molecules. In most 2:1 minerals, these sites are probably of minor importance, since OH groups are exposed only at broken crystal edges. They may, however, provide one of the few sites for organic bonding to uncharged 2:1 minerals such as pyrophyllite or talc. On the other hand, OH groups cover approximately 50% of 1:1 mineral surfaces. Organic reactions with these minerals have not been extensively studied, since they tend to be nonexpanding and, therefore, are relatively poor adsorbers of organic molecules. Wada (1969, 1961) showed, however, that, under certain circumstances, molecules such as acetates were capable of expanding halloysite and kaolinite. Using infrared spectrographic techniques, Ledoux and White (1965) demonstrated that interlayer bonding of acetate to kaolinite was the result of acetate oxygen forming H bonds with inner-surface OH. Thus, in certain cases, bonding of organic molecules to mineral OH groups can be an important adsorption mechanism.

4. C-H . . . O-Si BONDS

Although Farmer (1971) has indicated that energy relationships are not favorable for H bonding to surface oxygens, some evidence for their existence has been presented. For example, Francis (1973) observed a shift in the IR deformation bond for C-H when polyvinylpyrrolidone was adsorbed by minerals, indicating that weak H bonds had been formed. Clapp and Emerson (1972) also felt that H bonding to surface O played a role in adsorption of uncharged polymers by minerals. In all probability, however, H bonding of organic molecules to the surface O plane is of relatively minor importance as a mineral–organic bonding mechanism.

C. Van der Waals Forces

Van der Waals forces are weak electrical forces resulting from oscillating charges which produce synchronized dipoles that attract each other. The energy of interaction is small when compared to that of covalent or ionic bonds and is inversely proportional to the sixth power of the distance between interacting atoms. Although the forces between any two atoms are small, they are additive. The total interaction between two semi-infinite plates is inversely proportional to the second power of the distance between the plates (Stumm & Morgan, 1970). Thus, the contribution of van der Waals forces to bonding of organic molecules by minerals increases as the molecular size increases, and may be of considerable importance in bonding of large polymeric molecules. Farmer (1971) has indicated that van der Waals forces between the C-H groups of nitrobenzene and the surface oxygen atoms of minerals are energetically comparable with weak H bonds, so it is also possible that these forces play a role in bonding certain smaller organic molecules.

D. Coordination

Various authors have attempted to distinguish between coordination bonds and ion–dipole bonds, but reading more than one author's interpretations usually leads to utter confusion. One can only conclude that differences between the two types of bonding are so subtle that they are best left to the most advanced inorganic chemistry courses. Therefore, the definition used here will be that of Stumm and Morgan (1970) who do not attempt to differentiate between the two bond types. According to them, coördination (or complex formation) can be either electrostatic or covalent, or mixtures of the two, and complexes are formed by combinations of metal cations with molecules or anions (ligands) having lone pair electrons. This definition encompasses both coordination and ion–dipole bonds.

The most common and pervasive type of coordination complex is the hydrate, in which the central cation is surrounded by water molecules. Thus, whether a given organic molecule is capable of forming a complex with any given ion depends upon the molecules ability to compete with water for cation sites. Many molecules can form coordination complexes in the absence of water, but they may be displaced upon rehydration of the system. For example, Yariv et al. (1966) found that, in anhydrous systems, smectite saturated with Na, K, Li, NH_4, Ca, and Al was capable of adsorbing benzoic acid via complex formation. The benzoate ion was, however, completely removed from all complexes by subsequent water washing, indicating an inability of benzoate to compete with water for ligand sites.

Coordination does provide one of the few mechanisms by which negatively charged organic molecules can be bonded to mineral surfaces. Clapp and Emerson (1972) found, for example, that very strong bonds could be formed between negatively charged carboxylate groups and Ca ions on the mineral exchange complex. Some of these bonds could be broken only by pyrophosphate treatment of the complex. They further reported that the strength of bonding was inversely proportional to the distance between the organic carbonyl groups, and they felt that two neighboring carbonyls must be attached to a single Ca ion for a strong bond to be formed. Dowdy and

Mortland (1968) found that glycolation of clay surfaces occurred due to co-ordination, rather than by O-H $\cdots$ O-Si bonding, with the resultant complex being less energetic in the Ca system than in the Cu or Al system. (They do not, however, discount the possibility of H bonding to the surface when ions of low solvation energy, such as Na or NH_4 are on the mineral. This may explain the formation of two monomolecular layers when Ca-minerals are glycolated, whereas Na-minerals form one monomolecular layer.)

Jang and Condrate (1972a, b, c, d) have indicated that amino acids may bond to the clay surface via chelation of adsorbed cations. This appeared to occur at pH values above levels where protonation of the amino acids would be expected. They indicated that adjacent amine and carboxylate groups were involved, forming a five-member ring such as

$$
\begin{array}{ccc}
H_2O & & NH_2-CH-R \\
\diagdown & \diagup & | \\
& M & | \\
\diagup & \diagdown & | \\
H_2O & O-C & \\
& & \diagdown \\
& & O
\end{array}
$$

Adsorption by chelate formation appeared to be particularly favorable when the mineral was saturated with transition metal ions, which have a high affinity for electrons.

E. π-Bonds

These are of relatively little importance in nature, since they probably occur only in completely anhydrous systems of the type seldom found, even in deserts. Their existance is pointed out here only for the recognition that they do exist. π-bond formation also requires the presence of transition metals on the mineral exchange complex, since the unique properties of their d-orbitals allow bonding through donation of π-electrons of the organic compound (Mortland, 1970).

F. Si-O-C Bonds

The direct bonding of C to surface O atoms may be of even less importance, in nature, than are π-bonds, although they could occur in geologic materials. Devel et al. (1950) have postulated the reactions of certain organics, such as alcohols and oxiranes, with mineral surfaces to form Si-O-C bonds. However, the creation of these bonds has been possible only by use of special laboratory techniques, and generally require extremely low water vapor partial pressures (Weiss, 1969).

VII. CHEMICAL AND PHYSICAL PROPERTIES OF MINERAL-
ORGANIC COMPLEXES

In many cases, the bonding of organic molecules to minerals is not a truly reversible process, so the mineral–organic complexes might be considered to be a separate species in nature. Therefore, a discussion of their properties is in order, since they do differ in many respects from the reacted minerals.

A. Electro-chemical Properties

Since the mineral surface is at least partially covered by the organic molecules, it is not surprising that electric field characteristics are altered. The effect on the electric field has been demonstrated in several ways.

1. ELECTROPHORETIC MOBILITY

A general understanding of surface electric properties can be obtained through observation of movement in an electric field (cataphoresis). If the velocity, v, of a particle under a potential gradient, H, is measured, the zeta potential, ζ, can be calculated by

$$\zeta = (4\pi\eta v)/(HD) \qquad\qquad [7]$$

where D is the dielectric constant and η the viscosity of the liquid. In practice, however, both viscosity and the dielectric constant of liquid are altered by inclusion of the colloid. In addition, clays and clay–organic complexes differ in their ability to internally conduct charge. As a result, some authors have preferred to report only the cataphoretic velocity or, more commonly, the electrophoretic mobility.

Interestingly, the adsorption of organic molecules by minerals may either increase or decrease the electrophoretic mobility of the mineral. For example, adsorption of proteins has been shown to decrease the velocity of smectite saturated with H, Na, and Th, whereas the velocity of both Al- and La-saturated smectites increased (Fig. 20–6) (Harter & Stotzky, 1973).

The divergent responses of the different clays apparently reflect differences in the charge density or electro-kinetic potential at the plane of shear which affect the ability of proteins to compete for exchange sites and to break tactoids (packets of clay layers resistant to dispersion). The decreased velocity of the H- and Na-saturated clays was attributed to a shielding of the charged surfaces by adsorbed organics, whereas the increased velocity of Al- and La-saturated clays was attributed to a breaking of tactoids without concomitant adsorption of organic molecules on the surfaces. The organic

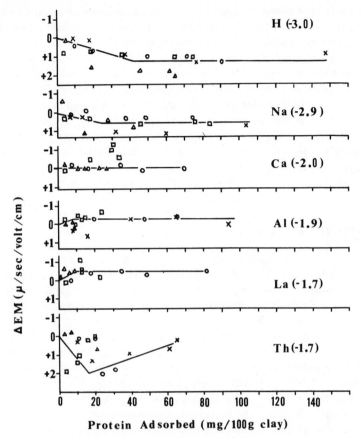

Fig. 20-6. Changes in electrophoretic mobility of homoionic smectite after adsorption of proteins: catalase (X), casein (O), chymotrypsin (□), and ovomucoid (△). Numbers in parentheses refer to the electrophoretic mobility of clays prior to protein adsorption (modified from Harter & Stotzky, 1973).

molecules were apparently not able to break tactoids of Th-saturated clays, and shielded the particles of this clay from the electric field by adsorption of an insulating "blanket" on external surfaces. Since many reactions are affected by the mineral electrokinetic potential, the manner in which the surface is shielded by adsorbed organics may be of value in explaining other properties of the complexes.

2. CATION EXCHANGE CAPACITY

Although it may be possible for a mineral–organic complex to have a higher CEC than the unreacted mineral, the change is usually in the opposite direction due to organic reaction with negative charge sites on the mineral or to physical coverage of sites by large organic molecules. Ensminger and

Gieseking (1941) noted that the adsorption of proteins by montmorillonite caused a reduction in the mineral CEC, with gelatin reducing it from 88 to 25 meq/100 g.   Other researchers have noted the same trend for many mineral–organic combinations.   The amount of CEC reduction depends upon the mechanism by which the organic molecules are adsorbed and the extent to which they cover the surface.   For example Kown and Ewing (1969b) reported that methylene blue reduced the CEC of Mississippi montmorillonite both by an exchange adsorption and physical blocking of exchange sites. Conversely, the CEC of complex organic molecules can also be reduced by adsorption to minerals.   DeSilva and Toth (1964) found that adsorption of several natural organic materials (with CEC of 150 and 720 meq/100 g at pH 7.0) to smectite, kaolinite, and illite caused up to a 94% reduction of the combined clay + organic colloid CEC.

3. ION SELECTIVITY

The energy involved in the exchange of one metal ion for another on negative exchange sites is independent of the proportions of the two ions on the exchange sites. However, if one of the ion pairs is an organic ion, the picture changes. The relative affinity, or selectivity, of the adsorbent for the organic ion decreases as the ion occupies more sites. This change in selectivity may be attributed to the contribution of forces other than electrostatic forces to the bonds being formed (Kown & Ewing, 1969a).   Theng et al. (1967) have indicated that these forces are probably van der Waals forces. They found that the affinity of alkylammonium ions for montmorillonite surfaces increased in the order $R_1NH_3^+ < R_2NH_2^+ < R_3NH^+$. This was interpreted as indicating that the increased contribution of van der Waals forces in adsorption of the di- and tri-substituted ammoniums more than compensated for the loss of H bonds between the cation and the mineral surface.

B. Physical Properties

Physical, as well as chemical, properties of mineral colloids are frequently altered upon reaction with organic molecules. This is particularly true of water flow parameters, since adsorbed organic molecules frequently form physical barriers in the interlayer spaces, and may even change mineral surfaces from hydrophilic to organiphilic (Kown & Ewing, 1969a). Kutilek and Slangerova (1966) reported that both the hydraulic conductance and the initial hydraulic conductance were greater and the non-Darcian behavior of water was restricted when organic molecules were adsorbed by clay surfaces. This response was attributed to destruction of "quasi-crystalline water" near the clay surface, as well as to alteration of sample tortuosity and porosity. At the same time, rheological properties, such as viscosity, sedimentation rate, sediment volume, etc., may be affected by adsorption of organic molecules by mineral surfaces (e.g., DeSilva & Toth, 1964; Chang & Anderson, 1968; Conley & Lloyd, 1974).

Many polyelectrolytes are adsorbed on external, rather than interlayer surfaces. Where this occurs, the large molecules will usually form bonds with several mineral particles, increasing aggregate stability. Using tensile strength as a measure of aggregate stability, Dowdy (1972) found a direct relationship between molecular weight and stability. (Attempts to use this property for improving the physical condition of soils has generally proved not to be economically feasible, however, since the externally adsorbed organics are too susceptable to biological degradation.)

Most physical properties of minerals will, in fact, probably be affected by adsorption of organic molecules. Some of these changes have been experimentally quantified, and some are merely observed in the laboratory while conducting adsorption studies. Expansion of the basal spacing has received considerable attention by mineralogists, and color changes have been frequently reported (Theng, 1971, has published an excellent review on colored clay–organic complexes.) However, changes in such characteristics as viscosity, thixotropy, sedimentation volume, etc. are often observed merely in the context of how easily a sample may be centrifuged from suspension, resuspended, or otherwise handled.

VIII. SUMMARY

The development, during the last half-century, of experimental tools such as X-ray diffractometry and infrared spectroscopy, as well as the refinement of methods for studying adsorption processes has provided the means for studying the formation and properties of mineral–organic complexes. This ability to study these complexes has been abetted by a growing interest in knowing the fate of organic pollutants modern society introduces into the environment. The study of mineral–organic complexes is scarcely out of its infancy stage, and much remains to be learned, particularly in the area of physico-chemical properties of stable complexes. However, considerable strides have been made in understanding the conditions which lead to adsorption and the mechanisms by which organic molecules are bonded to mineral surfaces. From knowledge of mineral properties such as surface charge density, cation on the exchange complex, and expandibility, organic molecule properties such as molecular charge, pK_a and polarity, and environmental characteristics such as pH buffering, temperature, and solution ionic strength, the modern scientist is able to reasonably predict whether adsorption will occur, as well as the type of bonding that might be expected. Therefore, the scientist has a capability for predicting the fate of a given organic compound added to the environment. Further refinement of this capability will, of course, always be necessary to understand systems that don't obey the predictions.

IX. SUPPLEMENTARY READING LIST

Bailey, George W., and Joe L. White. 1964. Review of adsorption and desorption of organic pesticides by soil colloids, with implications concerning pesticide bioactivity. J. Agric. Food Chem. 12:324–332.

Brindley, G. W. 1966. Ethylene glycol and glycerol complexes of smectites and vermiculites. Clay Miner. 6:237–259.

Brindley, G. W. 1970. Organic complexes of silicates. Mechanisms of formation. Reun. Hisp.-Belga Miner. Arcilla, An. 1970:55–66.

Farmer, V. D. 1971. The characterization of adsorption bonds in clays by infrared spectroscopy. Soil Sci. 112:62–68.

Greenland, D. J. 1965a. Interaction between clays and organic compounds in soils. Part I. Mechanisms of interaction between clays and defined organic compounds. Soils Fert. 28:415–425.

Greenland, D. J. 1965b. Interaction between clays and organic compounds in soils. Part II. Adsorption of soil organic compounds and its effect on soil properties. Soils Fert. 28:521–532.

Helling, Charles S., Philip C. Kearney, and Martin Alexander. 1971. Behavior of pesticides in soils. Adv. Agron. 23:147–240.

Mortland, M. M. 1968. Protonation of compounds at clay mineral surfaces. Int. Congr. Soil Sci., Trans. 9th (Adelaide, Aust.) I:691–699.

Mortland, M. M. 1970. Clay-organic complexes and interactions. Adv. Agron. 22:75–117.

Solomon, D. H. 1968. Clay minerals as electron acceptors and/or electron donors in organic reaction. Clays Clay Miner. 16:31–39.

Stotzky, G. 1967. Clay minerals and microbial ecology. Trans. N. Y. Acad. Sci. 30:11–21.

Theng, B. K. G. 1970. Interaction of clay minerals with organic polmers. Some practical applications. Clays Clay Miner. 18:357–362.

Theng, B. K. G. 1971. Mechanisms of formation of colored clay-organic complexes. A review. Clays Clay Miner. 19:383–390.

Theng, B. K. G. 1974. The chemistry of clay-organic reactions. John Wiley & Sons, Inc., New York. p. 343.

Weber, Jerome B. 1970. Mechanisms of adsorption of s-triazines by clay colloids and factors affecting plant availability. Residue Rev. 32:93–130.

Weber, Jerome B. 1972. Interaction of organic pesticides with particulate matter in aquatic and soil systems. Adv. Chem. 111:55–120.

Weiss, Armin. 1969. Organic derivatives of clay minerals, zeolites, and related minerals. p. 737–781. In G. Eglinton, and M. T. J. Murphy (eds.) Organic geochemistry. Springer-Verlag, New York.

LITERATURE CITED

Albert, Jenoe T., and Robert D. Harter. 1973. Adsorption of lysozyme and ovalbumin by clay: Effect of clay suspension pH and clay mineral type. Soil Sci. 115:130–136.

Armstrong, David E., and G. Chesters. 1964. Properties of protein-bentonite complexes as influenced by equilibrium conditions. Soil Sci. 98:38–52.

Bailey, G. W., and S. W. Karickhoff. 1973a. UV-Via spectroscopy in the characterization of clay mineral surfaces. Anal. Letter. 6:43–49.

Bailey, G. W., and S. W. Karickhoff. 1973b. An ultraviolet spectroscopic method for monitoring surface acidity of clay minerals under varying water content. Clays Clay Miner. 21:471–477.

Bailey, G. W., D. S. Brown, and S. W. Karickhoff. 1973. Competitive hydration of Quinazaline at the montmorillonite-water interface, Science 182:819–821.

Bissada, K. K., and W. D. Johns. 1969. Montmorillonite-organic complexes—Gas chromatographic determination of energies of interaction. Clays Clay Miner. 17: 197–204.

Bowman, Bruce. 1973. The effect of saturating cations on the adsorption of Dasnit, O,O-diethyl O-[p-(methyl sulfinyl) phenyl] phosphorothioate by montmorillonite suspensions. Soil Sci. Soc. Am. Proc. 37:200–207.

Boyd, G. E., J. Schubert, and A. W. Adamson. 1947. The exchange adsorption of ions from aqueous solutions by organic zeolites. I. Ion-exchange equilibrium. J. Am. Chem. Soc. 69:2818–2829,

Boyle, J. R., G. K. Voigt, and B. L. Sawhney. 1967. Biotite flakes: Alteration by chemical and biological treatment. Science 155:193–195.

Chang, C. W., and J. U. Anderson. 1968. Flocculation of clays and soils by organic compounds. Soil Sci. Soc. Am. Proc. 32:23–27.

Chou, Chung Chi, and James L. McAtee, Jr. 1969. Thermal decomposition of organoammonium compounds exchanged onto montmorillonite and hectorite. Clays Clay Miner. 17:339–346.

Clapp, C. E., and W. W. Emerson. 1972. Reactions between Ca-montmorillonite and polysaccharides. Soil Sci. 114:210–216.

Clementz, D. M., Thomas J. Pinnavaia, and M. M. Mortland. 1973. Sterochemistry of hydrated copper (II) ions on interlamellar surfaces of layer silicates. An electron spin resonance study. J. Phys. Chem. 77:196–200.

Conley, Robert F., and Mary K. Lloyd. 1971. Adsorption studies on kaolinites-II. Adsorption of amines. Clays Clay Miner. 19:273–282.

Cotton, F. Albert, and G. Wilkinson. 1966. Advanced inorganic chemistry, a comprehensive text. 2nd ed. John Wiley & Sons, Inc., New York.

DeSilva, J. A., and S. J. Toth. 1964. Cation exchange reactions, electrokinetic, and viscometric behavior of clay-organic complexes. Soil Sci. 97:63–73.

Devel, H., G. Huber, and R. Iberg. 1950. Organische derivative von tonmineralin. Helv. Chim. Acta. 33:1229–1232.

Dowdy, R. H. 1972. Effects of hydroxy-containing organics on the strength-energy characteristics of montmorillonite. Soil Sci. Soc. Am. Proc. 36:162–166.

Dowdy, R. H., and M. M. Mortland. 1968. Alcohol-water interactions on montmorillonite surfaces: Ethylene glycol. Soil Sci. 105:36–43.

Ensminger, L. E., and J. E. Gieseking. 1941. The adsorption of proteins by montmorillonite clays and its effect on base exchange capacity. Soil Sci. 51:125–132.

Farmer, V. C. 1971. The characterization of adsorption bonds in clays by infrared spectroscopy. Soil Sci. 112:62–68.

Fenn, D. B., M. M. Mortland, and Thomas J. Pinnavaia. 1973. The chemisorption of anisole on Cu(II) Hectorite. Clays Clay Miner. 21:315–322.

Francis, C. W. 1973. Adsorption of polyvinylpyrrolidone on reference clay minerals. Soil Sci. 115:40–54.

Giles, C. H., T. H. MacEwan, S. N. Nakhwa, and D. Smith. 1960. Studies in adsorption. Part XI. A system of classification of solution adsorption isotherms, and its use in diagnosis of adsorption mechanisms and in measurement of specific surface areas of solids. J. Chem. Soc. 786:3973–3993.

Harris, C. I., and G. F. Warren. 1964. Adsorption and desorption of herbicides by soil. Weeds 12:120–126.

Harter, Robert D. 1975. Effect of exchange cations and solution ionic strength on formation and stability of smectite-protein complexes. Soil Sci. 120:174–181.

Harter, Robert D., and Brian M. Kilcullen. 1976. Microcalorimeter adaption for measuring heats of adsorption at solid-solution interfaces. Soil Sci. Soc. Am. J. 40:612–614.

Harter, Robert D., and G. Stotzky. 1971. Formation of clay-protein complexes. Soil Sci. Soc. Am. Proc. 35:383–389.

Harter, Robert D., and G. Stotzky. 1973. X-ray diffraction, electron microscopy, electrophoretic mobility, and pH of some stable smectite-protein complexes. Soil Sci. Soc. Am. Proc. 37:116–123.

Jang, Sung Do, and Robert A. Condrate, Sr. 1972a. The infrared spectra of valine adsorbed on Ca-Montmorillonite. Am. Mineral. 57:494–498.

Jang, Sung Do, and Robert A. Condrate, Sr. 1972b. The I. R. spectra of lysine adsorbed on several cation-substituted montmorillonites. Clays Clay Miner. 20:79–82.

Jang, Sung Do, and Robert A. Condrate, Sr. 1972c. Infrared spectra of 2-alanine adsorbed on Cu-montmorillonite. Appl. Spect. 26:102–104.

Jang, Sung Do, and Robert A. Condrate, Sr. 1972d. The infrared spectra of glycine adsorbed on various cation substituted montmorillonites. J. Inorg. Nucl. Chem. 34: 1504–1509.

Kamil, J., and I. Shainberg. 1968. Hydrolysis of sodium montmorillonite in sodium chloride solutions. Soil Sci. 106:193–199.

Kanamaru, F., and V. Vand. 1970. The crystal structure of a clay-organic complex of 6-amino hexanoic acid and vermiculite. Am. Mineral. 55:1550–1561.

Khan, Samiullah, and J. P. Singhal. 1967. Titrations of hydrogen clays with nicotine. Soil Sci. 104:427–432.

Kown, B. T., and B. B. Ewing. 1969a. Effects of the counter-ion pairs on clay ion-exchange reactions. Soil Sci. 108:231–240.

Kown, B. T., and B. B. Ewing. 1969b. Effects of the organic adsorption on clay ion-exchange property. Soil Sci. 108:321–325.

Kutilek, Miroslav, and Jana Slangerova. 1966. Flow of water in clay minerals as influenced by adsorbed quinolinium and pyridinium. Soil Sci. 101:385–389.

Lailach, G. E., T. D. Thompson, and G. W. Brindley. 1968. Adsorption of pyrimidines, purines, and nucleosides by Co-, Ni-, Cu-, and Fe(III)-montmorillonite (clay-organic studies XIII). Clays Clay Miner. 16:295–301.

Langmuir, Irving. 1916. The constituents and fundamental properties of solids and liquids. Part 1. Solids. J. Am. Chem. Soc. 38:2221–2295.

Langmuir, Irving. 1918. The adsorption of gases on plane surfaces of glass, mica, and platinum. J. Am. Chem. Soc. 40:1361–1382.

Ledoux, Robert L., and Joe L. White. 1965. Infrared studies of the hydroxyl groups in intercalated kaolinite complexes. Clays Clay Miner. 13:289–315.

Mackintosh, E. E., D. G. Lewis, and D. J. Greenland. 1971. Dodecylammonium-mica complexes—I. Factors affecting the exchange reaction. Clays Clay Miner. 19: 209–218.

McNeal, Brian L. 1964. Effect of exchangeable cations on glycol retention by clay minerals. Soil Sci. 97:96–102.

Marshall, C. Edmund. 1964. The physical chemistry and mineralogy of soils. Volume I: Soil materials. John Wiley & Sons, Inc., New York.

Martin-Rubi, J. A., J. A. Rausel-Colom, and J. M. Serratosa. 1974. Infared adsorption and X-ray diffraction study of butylammonium complexes of phyllosilicates. Clays Clay Miner. 22:87–90.

Mortland, M. M. 1966. Urea complex with montmorillonite: an infrared adsorption study. Clay Miner. 6:143–156.

Mortland, M. M. 1968. Protonation of compounds at clay mineral surfaces. Int. Congr. Soil Sci., Trans. 9th. (Adelaide, Aust.) I:691–699.

Mortland, M. M. 1970. Clay-organic complexes and interactions. Adv. Agron. 22:75–117.

Mortland, M. M., J. J. Fripiat, J. Chaussidon, and J. Uytterhoeven. 1963. Interactions between ammonia and the expanding lattices of montmorillonite and vermiculite. J. Phys. Chem. 67:248–258.

Mortland, M. M., and W. D. Kemper. 1965. Specific surface. In C. A. Black (ed.) Methods in soil analysis, Part 1. Agronomy 9:532–544. Am. Soc. of Agron., Madison, Wis.

Mortland, M. M., and K. V. Raman. 1968. Surface acidity of smectite in relation to hydration, exchangeable cation, and structure. Clays Clay Miner. 16:393–398.

Norrish, K. 1954. The swelling of montmorillonite. Discuss. Faraday Soc. 18:120–134.

O'Brian, Neil R. 1971. Fabric of kaolinite and illite floccules. Clays Clay Miner. 19: 353–359.

Olejnik, S., A. M. Posner, and J. P. Quirk. 1971a. The I. R. spectra of interlamellar kaolinite-amide complexes—I. The complexes of formamide, N-methyl formamide and dimethylformamide. Clays Clay Miner. 19:83–94.

Olejnik, S., A. M. Posner, and J. P. Quirk. 1971b. The I. R. spectra of interlamellar kao-linite-amide complexes—II. Acetamide, N-methylacetamide and dimethylacetamide. J. Colloid Interface Sci. 37:356–547.

Philen, O. D., Jr., S. B. Weed, and J. B. Weber. 1970. Estimation of surface charge density of mica and vermiculite by competitive adsorption of diquat^{2+} vs. paraquat^{2+}. Soil Sci. Soc. Am. Proc. 34:527–531.

Philen, O. D., Jr., S. B. Weed, and J. B. Weber. 1971. Surface charge characterization of layer silicates by competitive adsorption of two organic divalent cations. Clays Clay Miner. 19:295–302.

Pinnavaia, Thomas J., and M. M. Mortland. 1971. Interlamellar metal complexes on layer silicates I copper(II)-arene complexes on montmorillonite. J. Phys. Chem. 75: 3957–3962.

Rupert, J. Paul. 1973. Electron spin resonance spectra of interlamellar copper(II)-arene complexes on montmorillonite. J. Phys. Chem. 77:784–790.

Serratosa, J. M. 1966. Infrared analysis of the orientation of pyridine molecules in clay complexes. Clays Clay Miner. 19:385–391.

Stumm, Werner, and James J. Morgan. 1970. Aquatic chemistry; an introduction emphasizing chemical equilibria in natural waters. John Wiley & Sons, Inc., New York.

Suito, Eiji, Masafumi Arakawa, and Tunoru Yoshida. 1966. Adsorbed state of organic compounds in organo-bentonite II. Differential thermal analysis. Bull. Inst. Chem. Res., Kyoto Univ. 44:325–334.

Swoboda, R., and G. W. Kunze. 1968. Reactivity of montmorillonite surfaces with weak organic bases. Soil Sci. Soc. Am. Proc. 32:806–811.

Tahoun, Salah A., and M. M. Mortland. 1966a. Complexes of montmorillonite with primary, secondary, and tertiary amides: I. Protonation of amides on the surface of montmorillonite. Soil Sci. 102:248–254.

Tahoun, Salah A., and M. M. Mortland. 1966b. Complexes of montmorillonite with primary, secondary, and tertiary amides: II. Coordination of amides on the surface of montmorillonite. Soil Sci. 102:248–254.

Theng, B. K. G. 1971. Mechanisms of formation of colored clay-organic complexes. A review. Clays Clay Miner. 19:383–390.

Theng, B. K. G., D. J. Greenland, and J. P. Quirk. 1967. Adsorption of alkylammonium cations by montmorillonite. Clay Mineral. 7:1–17.

Thompson, T. D., and G. W. Brindley. 1969. Absorption of pyrimidines purines, and nucleosides by Na-, Mg-, and Cu(II)-illite. (Clay-Organic studies XVI) Am. Mineral. 54:858–868.

Van Bladel, R., and A. Moreale. 1974. Adsorption of Fenuron and Monuron (substituted ureas) by two montmorillonite clays. Soil Sci. Soc. Am. Proc. 38:244–249.

Wada, K. 1959. Oriented penetration of ionic compounds between silicate layers of halloysite. Am. Mineral. 44:153–165.

Wada, K. 1961. Lattice expansion of kaolin minerals by treatment with potassium acetate. Am. Mineral. 46:78–91.

Weed, S. B., and J. B. Weber. 1968. The effect of adsorbent charge on the competitive adsorption of divalent organic cations by layer-silicate minerals. Am. Mineral. 53: 478–490.

Weiss, A. 1963. Organic derivatives of mica-type layer-silicates. Angew. Chem. 2:134–144.

Weiss, A. 1969. Organic derivatives of clay minerals, zeolites, and related minerals. p. 737–781. *In* G. Eglinton and M. T. J. Murphy (eds.) Organic geochemistry. Springer-Verlag, New York.

Wentworth, Sally A., and N. Rossi. 1972. Release of potassium from layer silicates by plant growth and by NaTPB extraction. Soil Sci. 113:410–416.

Yariv, S., J. D. Russell, and V. C. Farmer. 1966. Infrared study of the adsorption of benzoic acid and nitrobenzine in montmorillonite. Isr. J. Chem. 4:201–213.

Yoshida, Tsunoru. 1973. Elementary layers in the interstratified clay minerals as revealed by electron microscopy. Clays Clay Miner. 21:413–420.

Reactions of Minerals with Soil Humic Substances[1]

M. SCHNITZER and H. KODAMA, Soil Research Institute, Agriculture Canada, Ottawa, Ontario, Canada

I. INTRODUCTION

Humic substances are the principal organic components of soils and waters in which they interact with metal ions, metal oxides, metal hydroxides, and clay minerals to form associations of widely differing chemical and biological stabilities. These interaction products affect the moisture and aeration regime, exchange capacity, nutrient availability, chemical and biological degradation as well as many other reactions that occur in these systems. One potentially important reaction that soil and water scientists have so far tended to underestimate is the known ability of inorganic surfaces to catalyze organic reactions. Thus, in soils and waters where relatively large amounts of humic materials are adsorbed on hydrous oxide and clay surfaces, catalytic reactions may play more important roles in the synthesis, alteration, and degradation of humic substances than one would conclude from the literature where most of these reactions are considered to be of biological nature. The importance of metal-humic and clay-humic interactions in predominantly inorganic soils, with the exception of extremely sandy ones, is illustrated by data presented by Greenland (1965) which show that between 52 and 98% of the organic carbon in a wide range of soils is associated with the clay fraction. It is likely that most of the remaining organic carbon is linked to metal oxides and hydroxides. Thus, a more adequate knowledge of the synthesis, stability, and characteristics of inorganic-organic complexes would aid pedologists, geochemists, and water scientists in better understanding the physical, chemical and biological properties of soils and waters.

II. HUMIC SUBSTANCES

A. Definitions

The organic matter of soils and waters consists of a mixture of plant and animal products in various stages of decomposition, of substances synthesized chemically and biologically from the breakdown products, and of micro-

chapter 21

---

[1]Contribution no. 527 of the Soil Research Institute, Agriculture Canada, Ottawa, Ontario.

organisms and small animals and their decomposing remains. This complex system is usually simplified by dividing it into nonhumic and humic substances.

Nonhumic substances include those with still recognizable physical and chemical characteristics such as carbohydrates, proteins, peptides, amino acids, fats, waxes, alkanes and low-molecular weight organic acids. Most of these compounds are attacked relatively readily by microorganisms in the soil and have a short survival rate.

The major portion of the organic matter in most soils and waters, however, consists of humic substances. These are amorphous, dark colored, hydrophilic, acidic, partly aromatic, chemically complex organic substances that range in molecular weights from a few hundred to several thousand.

Based on their solubility in alkali and acid, humic substances are partitioned into three main fractions: (i) humic acid (HA), which is soluble in dilute alkali but is precipitated by acidification of the alkaline extract; (ii) fulvic acid (FA), which is that humic fraction which remains in solution when the alkaline extract is acidified, that is, it is soluble in both dilute alkali and dilute acid, and (iii) humin which is that humic fraction that cannot be extracted from the soil or sediment by dilute base and acid. From analytical data published in the literature (Schnitzer & Khan, 1972) it becomes apparent that structurally the three humic fractions are similar, but differ in molecular weight, ultimate analysis, and functional group content, with FA having a lower molecular weight but higher content of O-containing functional groups (CO_2H, OH, C=O) per unit weight than the other two humic fractions. Important characteristics exhibited by all humic fractions are resistance to microbial degradation and ability to form stable water-soluble and water-insoluble complexes with metal ions and hydrous oxides and to interact with clay minerals.

B. Analytical Characteristics

Table 21-1 shows analytical data for three relatively well developed, naturally occurring humic materials: HA and humin originating from the A1 horizon of a Haploboroll and FA extracted from the Bh horizon of a Podzol (Spodosols).

As shown in Table 21-1, elementary and functional group analyses of HA and humin are similar but different from those for FA in the following respects: (i) HA and humin contain more C, H, N, and S but less O than does FA; (ii) the total acidity and CO_2H-content of FA are approximately twice as great as those of HA and humin; (iii) the ratio of CO_2H to phenolic OH groups is about 3 for FA but only approximately 2 for both HA and humin; and, (iv) E_4/E_6 ratios and ESR data also indicate similarities between HA and humin.

Table 21-1. Analytical characteristics of HA, FA and humin†

	HA	FA	Humin
Elementary composition, % (dry, ash-free)			
Carbon	56.4	50.9	55.4
Hydrogen	5.5	3.3	5.5
Nitrogen	4.1	0.7	4.6
Sulfur	1.1	0.3	0.7
Oxygen	32.9	44.8	33.8
Oxygen-containing functional groups (meq/g dry, ash-free)			
Total acidity	6.6	12.4	5.9
Carboxyl	4.5	9.1	3.9
Total hydroxyl	4.9	6.9	--
Phenolic hydroxyl	2.1	3.3	2.0
Alcoholic hydroxyl	2.8	3.6	--
Total carbonyl	4.4	3.1	4.8
Quinone	2.5	0.6	--
Ketonic carbonyl	1.9	2.5	--
Methoxyl	0.3	0.1	0.1
Other characteristics			
E_4/E_6 ratio‡	4.3	7.1	5.2
Free radicals (spins/g $\times 10^{-18}$)	0.8	0.2	0.8
Line Width (G)	3.5	5.0	5.5
g-value	2.0029	2.0031	2.0029

† HA and humin from the A1 horizon of a Haploboroll and FA from the Bh horizon
   of a Spodosol.
‡ Ratio of optical densities at 465 and 665 nm.

C. Special Role of Fulvic Acid

Of the three principal humic fractions only FA is completely soluble in water when ash-free at pH values prevailing in most soils. FA extracted from Podzol (Spodosol) Bh horizons and then purified is also practically free of carbohydrates and low in N, so that it essentially is a "pure" humic material. On the basis of these characteristics and the preponderance of CO_2H and OH groups, this FA is a most suitable humic material for investigating metal-organic interactions in soils and waters. FA's extracted from soils belonging to other great soil groups, however, may contain between 10 and 20% (by weight) of carbohydrates + protein-like materials (Griffith & Schnitzer, 1975). In soils, FA accounts for between 25 and 75% of the total organic matter; in swamp water it has been reported to constitute up to 85% of the organic matter content (Schnitzer, 1971). For more detailed information on the chemical structure and reactions of humic substances the reader is directed to a recent monograph (Schnitzer & Khan, 1972).

III. REACTIONS OF HUMIC SUBSTANCES WITH METAL OXIDES, METAL HYDROXIDES, AND SILICATE MINERALS

A. Reactions with Metal Oxides and Metal Hydroxides

Several workers have reported on the strong solvent activity of humic substances toward minerals which has significant effects on the weathering cycle and on soil genesis. Baker (1973) leached hematite (Fe_2O_3) and pyrolusite (MnO_2) with a 0.1% (wt/vol) aqueous HA solution (pH 3.4) for 24 h in a perfusion apparatus. The HA was extracted from an Australian podzolic soil. The HA removed from one unit weight of hematite 340 μg of Fe as compared with 20 μg by distilled water equilibrated with atmospheric CO_2. From pyrolusite HA removed 2,100 μg of Mn as compared to 30 μg by distilled water.

In a similar experiment, Schnitzer and Skinner (1963) investigated the uptake by FA of Fe and Al from goethite, gibbsite, and a soil sample rich in "free" Fe and Al. On continuous wetting and leaching for 1 week, 1.0 mole of FA (670 g) dissolved 1.0 mole of Fe from goethite. Metal uptake on standing and shaking was considerably lower. At pH 3, more Al was extracted from gibbsite than Fe from goethite; at pH 5 and 7 the reverse was observed. The remarkable extracting and complexing power of FA was demonstrated by the finding that shaking 1 g of soil with 25 ml of 0.04% (wt/vol) aqueous FA solution for 1 week at room temperature removed 1.13 mg of Fe (9% of total Fe) and 2.64 mg of Al (6% of total Al). On this basis, 1 g of FA would extract all of the Fe and 63% of the Al in 1 g of soil.

A mixture of HA and FA was found to be more effective in complexing Mn from Mn^{3+}- and Mn^{4+}-oxides and occluded Mn^{2+}-hydroxides than was $10^{-3}M$ EDDHA and EDTA at pH 9 (Rosell & Babcock, 1968). In soils of high pH, where Mn normally forms the hydroxide, it is humic substances that are mainly responsible for maintaining sufficiently high concentrations of water-soluble Mn for satisfactory plant growth.

Freshly precipitated Fe and Al hydroxides adsorb HA and FA from aqueous solutions, with Al hydroxides having a higher adsorption capacity than Fe hydroxides (Levashkevich, 1966).

Iron and Al hydroxides are weak bases. The dissociation of $R(OH)_3$, where R = Fe or Al, produces $Fe(OH)_2^+$, $Fe(OH)^{2+}$, $Al(OH)_2^+$ and $Al(OH)^{2+}$ ions, all of which can react with humic substances. Alexandrova (1960), and more recently, Levashkevich (1966) describe interactions of HA and FA with Fe and Al hydroxides with the aid of the following equations:

$$(a) \quad R \underset{\diagdown (OH)_m}{\overset{\diagup (COOH)_n}{}} + [Fe(OH)_3] \quad \underset{Fe(OH)^{2+} \quad m-1(HO)}{\overset{Fe(OH)_2^+ \quad n-2(HOOC)}{\longrightarrow}} \quad R - COO \underset{\diagup}{\overset{\diagdown COO - Fe(OH)_2}{}} Fe(OH)$$

(b) $R \underset{(OH)_m}{\overset{(COOH)_n}{\Big\langle}} + [Al(OH)_3] \begin{array}{c} Al(OH)_2^+ \\ \\ Al(OH)^{2+} \end{array} \longrightarrow \quad n-2(HOOC) \underset{m-1(HO)}{\overset{COO-Al(OH)_2}{\diagdown}} R-COO \underset{O}{\overset{}{\diagdown}} Al(OH)$

R stands for the HA or FA "nucleus" and OH is a phenolic hydroxyl group.

It has also been suggested (Levashkevich, 1966) that aluminum hydroxide can interact with HA or FA to form the following hydrated complex:

$$n-2(HOOC) \diagdown \quad \diagup COO$$
$$R-COO-[Al(H_2O)_6]$$
$$m-1(HO) \diagup \quad \diagdown O$$

Schwertmann (1966) reports that soil organic compounds soluble in dilute alkali (humic substances) inhibit the crystallization of amorphous ferric hydroxide, most likely by adsorption of negatively-charged HA or FA on positively-charged hydroxide surfaces. In the presence of humic substances, natural and synthetic geothite as well as amorphous ferric hydroxide are not directly dehydrated to hematite on heating but first to maghematite (Schwertmann, 1966). The involvement of humic materials may be responsible for the failure to detect sharp exothermic peaks in DTA curves of amorphous hydroxides from soils. The hardening of laterites may also arise from interactions of humic substances with soil hydroxides. Evans and Russell (1959) have emphasized the remarkable capacity of sesquioxides to adsorb HA and FA. Their data are listed in Table 21-2. Although more of both fractions is adsorbed from the more concentrated HA and FA suspensions, increases in adsorption are not proportional to increases in concentrations. Both lepidocrocite and goethite remove, per unit weight, more HA and FA

Table 21-2. Amounts of humic- and fulvic-acid carbon adsorbed from suspension by sesquioxide minerals as g C/100 g mineral (Evans & Russell, 1959)

	pH	FA		HA	
		0.0159% C	0.0318% C	0.0159% C	0.0318% C
Goethite	3.2	4.26	4.51	4.36	4.59
Lepidocrocite	3.2	4.43	4.89	4.56	4.92
Lepidocrocite	5.0	3.23	--	3.31	--
Lepidocrocite	7.0	2.97	--	3.08	--
β Fe$_2$O$_3$·H$_2$O	3.2	2.53	3.15	2.61	3.23
Gibbsite	3.2	1.41	1.63	1.90	2.12
Gibbsite	5.0	3.42	--	3.48	--
Gibbsite	7.0	2.28	--	2.31	--
Boehmite	3.2	0.0	0.0	0.0	0.0

from solutions at the lower concentrations than do montmorillonite and kaolinite. The reduction in adsorption of HA and FA by lepidocrocite with rising pH is smaller than expected, while with gibbsite adsorption increases with rise in pH. These data suggest that sesquioxides play an important role in the removal of HA and FA from aqueous systems in soils and waters. The data in Table 21-2 also indicate that, even at neutral pH, hydrous Fe and Al oxides can still adsorb substantial amounts of HA and FA.

Recently, attention has been drawn to a possibly important role for FA in the synthesis of kaolinite at room temperature and atmospheric pressure (Linares & Huertas, 1971). The synthesis of kaolinite under mild conditions, requires that Al be present in sixfold coordination in a metal-organic complex or in the form of pregibbsitic hydroxide. Since FA is known to readily form stable, water-soluble complexes with Al^{3+} (Schnitzer & Hansen, 1970), Linares and Huertas (1971) prepared an Al-FA complex by adding 1 meq of Al^{3+} to 10 ml of FA solution with a C content of 1 g/liter. Solutions of monomeric silica were then prepared from sodium silicate at pH 2.5, and varying volumes of Al-FA solution were added to provide for SiO_2 to Al_2O_3 ratios ranging from 0.1 to 10. Portions of each solution were adjusted to pH 4, 5, 6, 7, 8, and 9. Precipitated products were allowed to age for 1 month at room temperature, separated, washed, dried and examined by X-ray, IR, DTA, and chemical analyses which revealed the presence of gibbsite, bayerite, boehmite, kaolinite, and amorphous silicoaluminum gels. These results emphasize the importance of FA in reactions with Al in soils. For example, FA promotes the hydrolysis of silicates by the formation of soluble complexes with Fe and Al in Spodosols until ultimately only a pure silica residue remains in the eluvial horizon. The Al-FA complex migrates downward and under acidic conditions kaolinite and gibbsite are formed. Hem and Lind (1974) investigated the synthesis of kaolinite at 25C (298K) by adding quercetin to aqueous solutions of silica and Al and adjusting the pH to values ranging between 6.5 to 8.5. After 6 to 16 months of aging, 1:1 aluminosilicate precipitates were isolated which contained up to 5% of well-formed kaolinite plates. Similar solutions containing no organic ligand produced amorphous precipitates with very little crystalline material even after 2 years of aging. Hem and Lind (1974) believe that the presence of organic constituents such as FA or quercetin slows the polymerization of polynuclear Al hydroxide species in the pH range 6.5 to 8.5, where polymerization is normally very rapid. This permits more reactants to follow the slow reaction path leading to kaolinite that is better crystallized. In addition to its effects on kinetic factors, the Al-organic complex may facilitate the formation of Al-O-Si bonds because it contains Al-O bonds and can provide the Al as Al-O. Regardless of which mechanism is the more correct one, sound experimental evidence for the participation of FA in the synthesis of kaolinite and possibly other silicates is now at hand and further interesting developments along these lines can be expected.

B. Reactions with Nonexpanding Layer Silicate Minerals

1. ADSORPTION BY KAOLINITE

According to Evans and Russell (1959), H- and Ca-saturated kaolinite adsorbs greater amounts of HA than does the K-saturated form. FA is adsorbed to a greater extent by H-kaolinite than is HA, but the latter is more adsorbed on the Ca-system. The relatively high adsorption of HA by Ca-saturated kaolinite may be due to exchangeable Ca at the structural edges acting as bridge linkages between clay and HA.

Arshad and Lowe (1966) fractionated and characterized a naturally occurring organo-clay complex separated from a Solonetz (Udic Natroboroll) Bnt horizon. They found that the largest portion of the humic materials was associated with the coarse clay fraction which contained relatively large amounts of kaolinite; only a small percentage of the humic materials occurred in the fine clay fraction where kaolinite occurred only in trace amounts.

The adsorption of FA by kaolinite was investigated by Kodama and Schnitzer (1974a). Adsorption decreased with an increase in pH and was directly proportional to the surface areas of the kaolinite fractions. Maximum amounts of FA adsorbed by 1 g of kaolinite at low pH were 73.0 mg for the < 1 μm fraction and 37.0 mg for the 5–2 μm fraction. Since X-ray analysis failed to show any changes in the original structures of the kaolinite samples, all of the FA must have been adsorbed on external surfaces and edges.

2. ADSORPTION BY MUSCOVITE

Kodama and Schnitzer (1974a) found that the amount of FA adsorbed by muscovite also decreased with an increase in pH. The maximum amount of FA adsorbed by 1 g of muscovite at pH 2.5 was 49.0 mg; at pH 6.5, only 27.0 mg was adsorbed under the same conditions.

3. DISSOLUTION OF CHLORITE MINERALS BY FULVIC ACID

The dissolution of two chlorite minerals by FA has been studied by Kodama and Schnitzer (1973). The two minerals selected were leuchtenbergite, rich in Mg,

$$(Mg_{4.94}Al_{0.71}Fe^{2+}_{0.54})(Si_{2.88}Al_{1.12})O_{10}(OH)_8$$

and thuringite, rich in Fe,

$$(Fe^{2+}_{2.51}Al_{0.84}Mg_{0.70}Mn_{0.52}Fe^{3+}_{0.15})(Si_{2.54}Al_{1.46})O_{10}(OH)_8.$$

Table 21-3. Dissolution of cations from chlorites by 0.2% aqueous FA solution and by acidified H_2O, both initially at pH 2.5 (expressed in mg per 1 g of sample (Kodama & Schnitzer, 1973)

Time, hours	Leuchtenbergite FA			Acidified H_2O			Thuringite FA			Acidified H_2O		
	Al	Fe	Mg	Al	Fe	Mg	Al	Fe	Mg	Al	Fe	Mg
3							3.50	8.63	0.84	0.90	1.20	0.37
40							11.30	32.00	3.08	3.21	6.86	0.95
48	2.00	0.40	3.50	1.10	0.40	1.70						
72							13.80	36.20	3.89			
96							16.80	47.20	4.54	4.63	9.48	1.26
120	3.20	0.90	5.70	1.20	0.30	2.90						
144							19.30	53.10	5.30	5.00	10.10	1.37
192	3.20	1.00	6.90	1.30	0.70	2.70	21.90	61.10	6.03			
312	4.20	1.00	8.50	1.90	0.50	4.40	23.70	65.30	6.73	6.00	12.24	1.72
360	4.80	1.20	9.00									

The experimental procedure consisted of shaking at room temperature 25 mg of chlorite with 25 ml of aqueous 0.2% (wt/vol) FA solution (pH 2.5) for various lengths of time. Extracted Fe, Al, and Mg were then determined. For comparative purposes, similar experiments were run with distilled water adjusted with dilute HCl to the same pH as that of the 0.2% FA solution (pH 2.5). The results obtained are shown in Table 21-3. As 1 g of initial leuchtenbergite contained 84.15 mg of Al, 51.61 mg of Fe, and 204.81 mg of Mg, the FA solution, after 312 hours of contact, decomposed approximately 4% of the crystal structure of the initial leuchtenbergite, whereas acidified distilled water decomposed only 2% (Table 21-3). On the other hand, 1 g of thuringite contained 90.23 mg of Al, 273.45 mg of Fe and 24.73 mg of Mg, so that the FA solution dissolved about 26% of the crystal structure of the initial thuringite, but acidified distilled water dissolved only 6%. Although both extractants decomposed the Fe-rich species more readily than the Mg-rich one, FA was 6.5 times more effective in decomposing thuringite than leuchtenbergite. FA was about twice as efficient in decomposing the latter as was acidified distilled water. Because previous investigations by Schnitzer and Hansen (1970) had shown that FA becomes insoluble when it complexes large amounts of Fe and Al, Kodama and Schnitzer (1973) examined the residues after the reactants had been shaken for 312 hours. X-ray diffraction analysis showed no changes in structures from those of the initial samples. Infrared, however, indicated remarkable changes. In the IR spectrum of the untreated FA, the ratio of absorbances at 1725 and 1620 cm^{-1} was 2.0, it approached 1.1 in the leuchtenbergite-FA residue and reached 0.7 in the thuringite-FA residue. The decrease in absorbance at 1720 cm^{-1} and the increase in absorbance at 1620 cm^{-1} indicate conversion of COOH to COO^- groups, which then interact with the metal ions (M^+) to form electrovalent COO^-M bonds. Thus, FA forms water-soluble as well as water-insoluble complexes with Fe and Al. Kodama and Schnitzer (1973) concluded that FA can decompose chlorite minerals, especially those rich in Fe, in a relatively short time, thus bringing substantial amounts of Al, Fe, and Mg into aqueous solution and so enhancing the mobility of these metal ions in soils and waters.

C. Reactions with Other Silicate Minerals

While silicates are generally more resistant to attack by HA and FA than nonsilicate minerals referred to in the previous paragraphs, amounts of metals removed by humic substances from silicates are nonetheless substantial as shown in Table 21-4 (Baker, 1973). Relatively high concentrations of Na, K, Mg, Ca, Fe and Al (but little Si) are extracted by HA. Baker's (1973) results are in agreement with those of Ponomareva and Ragim-Zade (1969) who also found that HA and FA removed more K, Na, Fe, and Mg from feldspar and biotite than Si and Al. After observing the effectiveness of HA in degrading the silicates listed in Table 21-4, Baker (1973) leached Portland cement with a dilute aqueous HA solution. After 24 hours of perfusion, large amounts of Ca were extracted from the cement. These findings led Baker (1973) to suggest that reinforced concrete piping would have a comparatively short life-span if buried in an environment rich in humic materials. The activity of humic materials with respect to metals should be of interest to corrosion specialists who often ascribe the destruction of concrete structures, piping,

Table 21-4. Reaction of water and HA with silicates (Baker, 1973)

Mineral	Element determined	μg metal extracted in 5 days	
		H_2O/atmos CO_2	0.1% HA
Feldspar	Na	340	1,260
(perthite)	K	190	640
$(Na,K)AlSi_3O_8$	Al	<10	220
	Si	< 5	<5
Biotite	K	60	100
$K(Mg,Fe)$	Mg	20	90
$(AlSi_3O_{10})(OH)_2$	Fe	< 5	240
	Al	<10	300
	Si	< 5	<5
Enstatite	Mg	80	620
$(Mg,Fe^{2+})SiO_3$	Fe	< 5	220
	Si	< 5	<5
Actinolite	Mg	25	80
$Ca(Mg,Fe^{2+})_5$	Fe	< 5	960
$(Si_8O_{22})(OH,F)_2$	Ca	750	2,900
	Si	< 5	40
Epidote	Ca	1,800	10,700
$Ca Fe^{2+} Al_2O.OH$	Fe	10	1,300
$(Si_2O_7)(SiO_4)$	Al	40	320
	Si	< 5	<5

etc. buried in soils to bacterial corresion rather than to dissolution and degradation by humic substances.

The absorption of HA by gypsum and chalk has been investigated by Gorbunov et al. (1971). They conclude that during the interaction ionic bonds are formed between Ca and negatively-charged functional groups on the HA leading to the formation of insoluble Ca-humates. They also note from studies on interactions of HA and FA with a number of clay minerals that the adsorption of humic substances by clays is related to the chemical composition of the surface layers, especially to the amounts of Fe and Al contained in them, the microrelief of the surface, the presence of nonsilicate sesquioxide films, and of free and adsorbed Ca and Mg, pH, and moisture content. Poor dispersion, the presence of much quartz, amorphous silica, adsorbed and free Na and K, an alkaline pH and a high moisture content tend to reduce adsorption of humic substances by clays. Fe and Al play a dual role: under some conditions they fix humic substances, under other conditions they form mobile chelates (Gorbunov et al., 1971). The order of adsorption of HA by nonexpanding soil minerals was found to be as follows: gypsum > Al(OH)$_3$ > chlorite > muscovite = desmine > calcite > kaolinite. Quartz and amorphous silica do not adsorb HA (Gorbunov et al., 1971).

D. Reactions with Expanding Layer Silicate Minerals

1. EFFECT OF pH ON THE ADSORPTION OF HUMIC ACID AND FULVIC ACID BY MONTMORILLONITE

Decreased adsorption on external surfaces of HA and FA by montmorillonite with increase in pH has been observed by several workers (Demolon & Barbier, 1929; Evans & Russell, 1959). Schnitzer and Kodama (1966) were the first to report the interlayer adsorption of FA by montmorillonite. At a later date, Martin and Rodriguez (1969) showed that HA (apparently a low molecular weight sample) could also penetrate the interlayer spaces of montmorillonite.

Effects of pH on the $d(001)$ spacing and adsorption of FA (mg C × 2) by Na-montmorillonite are shown in Fig. 21-1. The $d(001)$ of the original Na-montmorillonite in dry air is 9.87Å; after interaction with FA at pH 2.5 it increases to 17.60Å (Schnitzer & Kodama, 1966). The interlayer spacing is pH-dependent and decreases with increase in pH, with the steepest decrease occurring between pH 4 and 5 (see Fig. 21-1). Since the apparent overall pK of the FA is 4.5, it appears that the magnitude of $d(001)$ is related to the degree of the ionization of the functional groups, particularly CO$_2$H groups, in the FA. At pH < 4, relatively few of these groups are ionized, so that the FA behaves like an uncharged molecule that can penetrate interlayer spaces and displace water from between silicate layers of the montmorillonite. As the pH rises, more and more functional groups ionize, which results in an increased negative charge on the FA, so that at pH > 5, $d(001)$ is < 11Å, in-

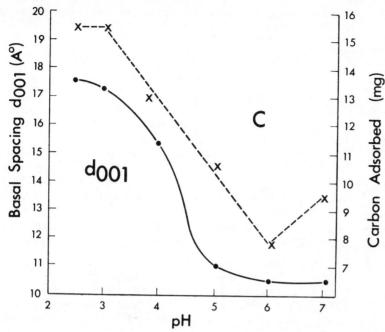

Fig. 21-1. Effect of pH on $d(001)$ and adsorption of FA (% C $\times$ 2) (Schnitzer & Kodama, 1966). Reproduced with the permission of the American Association for the Advancement of Science. Copyright © 1966 by the American Association for the Advancement of Science.

dicating repulsion of negatively charged FA by negatively charged montmorillonite. The curve depicting the adsorption of C (mg C $\times$ 2 = mg FA) in Fig. 21-1 follows that showing $d(001)$ values. At pH 3, 40 mg of Na-montmorillonite adsorbed 31 mg of FA. Above pH 5, amounts of FA adsorbed by 40 mg of Na-montmorillonite ranged between 19 and 16 mg, but $d(001)$ spacings remained more or less constant (Schnitzer & Kodama, 1966). The Debye-Scherrer powder pattern of untreated FA exhibited a halo with a broad maximum near 4Å (Kodama & Schnitzer, 1967). Thus, the increase (7.73Å) in interlayer spacing in the complex prepared at the lowest pH over that in the untreated clay corresponds approximately to two layer thicknesses of FA.

Martin and Rodriguez (1969) report that the interlayer adsorption of Black Earth HA by Na-montmorillonite increases the $d(001)$ spacing up to 30Å.

2. EFFECTS OF HUMIC ACID AND FULVIC ACID CONCENTRATIONS AND REACTION TIME ON ADSORPTION

Shaking Na-montmorillonite with increasing FA concentrations increases the extent of adsorption but there is no direct proportionality between the two (Demolon & Barbier, 1929; Evans & Russell, 1959; Schnitzer & Kodama,

Table 21-5. Effect of time of shaking on the d001 spacing and on the amount of FA adsorbed at pH 2.5 by 40 mg of montmorillonite (Schnitzer & Kodama, 1967)

Time of shaking	d(001) spacing	FA adsorbed
	Å	mg
1 min	15.5	27.2
1 hour	16.5	29.2
3 hours	17.0	--
5 hours	17.6	31.2
14 hours	17.6	--
18 hours	17.6	33.2

1967; Martin & Rodriguez, 1969). Amounts of humic materials adsorbed from more dilute solution appear to be retained by the clay more energetically than those removed from more concentrated solutions. Adsorption by Na-montmorillonite occurs rapidly as illustrated by the data in Table 21-5. After 1 min of shaking 40 mg of clay with 20 ml of 0.5% aqueous FA solution, the d(001) spacing increases from 9.9 to 15.5Å and 27.2 mg of FA is adsorbed. After 5 hours of shaking, the spacing increases to 17.6Å and remains constant up to 18 hours of shaking (Schnitzer & Kodama, 1967). The findings of the latter workers are in agreement with those of Evans and Russell (1959) who report that the adsorption of HA by montmorillonite is rapid and at least 90% completed within the first hour of contact, and virtually 100% completed in 24 hours.

3. EFFECT OF TEMPERATURE ON ADSORPTION

Varying the temperature between 25 and 60C (298 and 333K) does not significantly affect the adsorption of HA and FA on external montmorillonite surfaces (Evans & Russell, 1959; Schnitzer & Kodama, 1969), nor does it affect the interlayer adsorption of FA (Schnitzer & Kodama, 1969).

4. DESORPTION OF FULVIC ACID FROM A FULVIC ACID-Na-MONTMORILLONITE COMPLEX

The desorption of FA from a laboratory-prepared FA-Na-montmorillonite complex (containing 33.2 mg of FA and 40.0 mg of clay) by distilled

Table 21-6. Desorption of FA from FA-clay complex by H_2O adjusted to pH 2.5 and by 0.1N NaOH solution (Schnitzer & Kodama, 1967)

	Desorbent			
	H_2O		0.1N NaOH	
Time of desorption	FA desorbed	d(001)	FA desorbed	d(001)
	mg	Å	mg	Å
10 min	5.4	15.8	28.2	10.0
6 hours	6.6	15.5	30.8	10.0
18 hours	7.4	15.2	30.8	10.0

water (acidified to pH 2.5) and 0.1N NaOH solution is illustrated in Table 21-6. Amounts of FA desorbed by acidified water even after 18 hours of shaking are small. By contrast, desorption by 0.1N NaOH is rapid. After 10 min of contact 86% of the FA is desorbed, after 18 hours it is 93%; the $d(001)$ spacing decreases to that of the initial clay after only 10 min of shaking (Schnitzer & Kodama, 1967). This indicates that 0.1N NaOH rapidly desorbs FA from both external and internal clay surfaces.

5. EFFECTS OF INTERLAYER CATIONS ON ADSORPTION OF FULVIC ACID BY MONTMORILLONITE

Effects of saturating montmorillonite with different cations on interlayer adsorption have been investigated by Kodama and Schnitzer (1968). As shown in Fig. 21-2, $d(001)$ spacings of cation-saturated clays increase after interaction with FA to between 15.1 and 19.2Å. The magnitude of the spacing is proportional to the amounts of the FA adsorbed. As shown in Fig. 21-2, the $d(001)$ spacings decrease in the following order:

$$Pb^{2+} > Cu^{2+} > Na^+ > Zn^{2+} > Co^{2+} > Mn^{2+} > Mg^{2+} > Ca^{2+} > Fe^{3+} > Ni^{2+}.$$

This order differs from that for metal-FA stability constants which is:

$$Fe^{3+} > Cu^{2+} > Ni^{2+} > Co^{2+} > Pb^{2+} \approx Ca^{2+} > Zn^{2+} > Mn^{2+} > Mg^{2+}$$

(Schnitzer and Hansen, 1970). Thus, variations in $d(001)$ spacings are not a function of the capacity of the different cations to form stable complexes with FA nor are they related to the ionization potentials of the metal ions or to metal-FA complexes having different molecular sizes. The high interlayer adsorption of FA by clays saturated with Pb^{2+}, Cu^{2+}, and Na^+ may be related to the relative ease with which FA can displace water molecules from these cations. In this respect, FA behaves like an uncharged molecule that can penetrate interlayer spaces and displace water molecules associated with counter ions from between silicate layers of montmorillonite. In this manner, FA and water compete for ligand positions around the exchangeable cations (Kodama & Schnitzer, 1968).

IV. SOME ASPECTS OF INTERLAYER FA-CLAY INTERACTIONS AS REVEALED BY THE STUDY OF A Cu^{2+}-MONTMORILLONITE-FA COMPLEX

To obtain more detailed information on the interlayer adsorption of FA by montmorillonite, Schnitzer and Kodama (1972) and Kodama and Schnitzer (1974b) investigated reactions between FA and Cu^{2+}-montmorillonite by chemical, X-ray, IR, ESR, thermal, and electron microscopic methods.

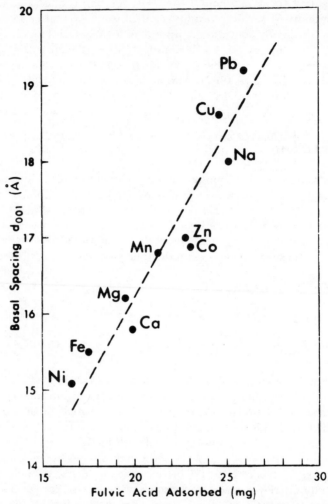

Fig. 21-2. Relationship between d(001)-spacings and amounts of fulvic acid adsorbed by 40 mg of montmorillonite saturated with different cations (Kodama & Schnitzer, 1968). Reproduced with the permission of the Williams and Wilkins Co., Baltimore, Md. Copyright © 1968 The Williams and Wilkins Co., Baltimore, Md.

Since FA is known to form relatively stable, water-soluble complexes with Cu^{2+} (Schnitzer & Hansen, 1970; Gamble et al., 1970) amounts of Cu^{2+} extracted by FA from Cu^{2+}-montmorillonite were determined with the aid of a specific Cu^{2+}-ion electrode, which permits measurements of "free" or "uncomplexed" Cu^{2+} aside from "total" Cu^{2+}. Thus, the difference between "total" and "free" Cu^{2+} is due to "complexed" Cu^{2+}. As shown in Table 21-7, amounts of "free" Cu^{2+} (Cu_f) in the supernatant solutions decrease significantly as the pH increases, although the total Cu-content (Cu_t) increases. Above pH 3.5, close to 100% of the Cu^{2+} brought into solution is

Table 21-7. Amounts of Cu^{2+} solubilized by shaking 20 mg of Cu^{2+}-montmorillonite for 18 hours with 25 ml (50.0 mg) of FA solution (Kodama & Schnitzer, 1974b)

pH	μg Cu^{2+}-solubilized		$Cu_f/Cu_t \times 100$ (= %)
	Cu_t	Cu_f	
2.5	125.8	7.8	6.8
3.5	193.0	4.8	2.5
4.5	233.8	1.3	0.6
5.5	249.8	0.2	0.07
6.5	257.8	0.001	0.0009

complexed, attesting to the high complexing power of FA for Cu^{2+}. Increases in Cu^{2+}-complexing with rise in pH are most probably due to increased ionization of O-containing functional groups (CO_2H and phenolic OH) in the FA. The data in Table 21-7 show that the mechanism by which FA removes Cu^{2+} from the clay is almost exclusively complexation, most likely chelation. Since 20 mg of Cu^{2+}-montmorillonite contain 540 μg of Cu^{2+}, the proportions of Cu_t removed by the FA range from 23.3% (at pH 2.5) to 48.7% (at pH 6.5) of those in the initial clay samples. Thus, one treatment with dilute aqueous FA solution (containing 50 mg of FA) at pH 6.5 suffices to remove one-half of the Cu_t in and on 20 mg of Cu^{2+}-saturated montmorillonite. These data illustrate the remarkable solvent action of aqueous FA solutions on naturally occurring minerals, especially at pH above 3.5.

In addition to water-soluble Cu^{2+}-complexes, FA also forms water-insoluble complexes with Cu^{2+}-montmorillonite. This point is illustrated in the IR spectra shown in Fig. 21-3. The IR spectrum of untreated FA (curve a) shows the following major bands in the 1900 to 1400 cm^{-1} region, the most informative one for assessing metal-FA interactions: 1730 (C=O of CO_2H, C=O stetch of ketonic C=O), 1625 (aromatic C=C, H-bonded C=O of carbonyl, double bond conjugated with C=O, COO^-) and 1415 cm^{-1} (COO^-). The Cu^{2+}-montmorillonite spectrum (curve b) shows a strong band near 1635 cm^{-1}, due to adsorbed water. The IR spectrum of the Cu^{2+}-montmorillonite–FA complex prepared at pH 2.5 (curve c) shows bands at 1720, 1620, and 1425 cm^{-1}, whereas spectra of clay–FA complexes prepared at pH > 2.5 exhibit additional bands at 1850 and near 1520 cm^{-1} (curves d to g). The latter two bands tend to increase with increase in pH and are absent in the IR spectra of the untreated FA and of the clay. According to Nakamoto (1963) Cu^{2+}-, Pd^{2+}- and Al^{3+}-acetylacetonates absorb between 1525 and 1535 cm^{-1} and the absorption arises from the C$\cdots$O stretch of the two ketone groups. A β-diketone structure, of which the acetylacetone anion ($R_1=R_3=CH_3$; $R_2=H_2$) is the simplest representative, is shown in Fig. 21-4. The chelate ring is planar and symmetrical and the two C$\cdots$O bonds are equivalent as are the C$\cdots$C bonds in the ring (Nakamoto, 1963), which provides strong possibilities for resonance. It is noteworthy that de Serra and Schnitzer (1972) have recently provided evidence for the occurrence of β-diketone structures in FA

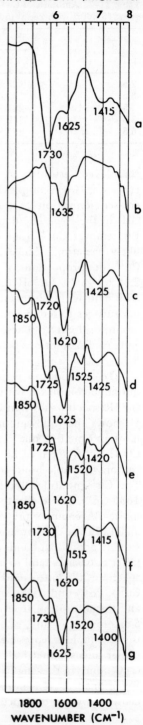

Fig. 21-3. Infrared spectra of: *(a)* untreated FA; *(b)* Cu^{2+}-montmorillonite; *(c)* FA-Cu^{2+}-montmorillonite complex prepared at pH 2.5; *(d)* FA-Cu^{2+}-montmorillonite complex prepared at pH 3.5; *(e)* FA-Cu^{2+}-montmorillonite complex prepared at pH 4.5; *(f)* FA-Cu^{2+}-montmorillonite complex prepared at pH 5.5; *(g)* FA-Cu^{2+}-montmorillonite complex prepared at pH 6.5 (Schnitzer & Kodama, 1972). Reproduced with the permission of Pergamon Press Ltd.

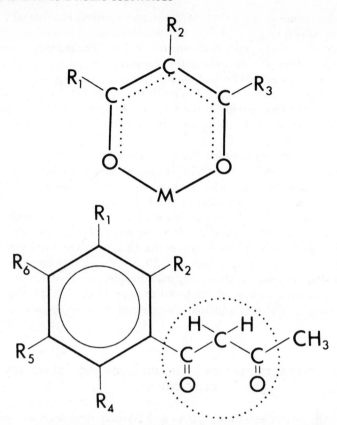

Fig. 21-4. *(Top)* β-diketone metal chelate; *(bottom)* Compounds produced by the permanganate oxidation of methylated FA. Compound I: $R_1 = CH_2CO_2CH_3$, $R_2 = OCO_2CH_3$, $R_4 = OCO_2CH_3$, $R_5 + R_6 = CO.O.CO$. Compound II: $R_1 = OCO_2CH_3$, $R_2 = COCO_2CH_3$, $R_4 = OCO_2CH_3$, $R_5 + R_6 = OCO_2CH_3$. (de Serra & Schnitzer, 1972).

and for their likely participation in reactions with metals by the isolation of two complex anhydrides with β-diketone groups on the aromatic ring (Fig. 21-4). To obtain further evidence on the participation of C=O groups in reactions with Cu^{2+} on and in the clay, C=O groups in FA were blocked by the formation of oximes and semicarbazones and the modified FA's interacted with Cu^{2+}-montmorillonite. Infrared spectra of the interaction products showed disappearance of 1850 and 1525 cm^{-1} bands, indicating that blocking of the C=O groups had prevented these groups from reacting with Cu^{2+}.

These data show that on clay surfaces and in interlayers the reaction between FA and Cu^{2+} (and possibly also with most other metals ions) differs from that in aqueous solution in that it also involves β-diketone groups in the FA to form chelates which are similar to acetylacetonates. In aqueous systems, in the absence of clays, FA is thought to interact with di- and tri-valent metals ions by two types of reactions to form stable complexes: (i) a major one, involving simultaneously both CO_2H and phenolic OH groups, and (ii) a minor one, in which adjacent CO_2H groups participate (Schnitzer & Skinner, 1965; Gamble et al., 1970). Water-soluble complexes are usually formed if molar metal/FA ratios are 1:1 or less, but if the ratios exceed 1:1, the complexes become water-insoluble (Schnitzer, 1971). The main conclusion that can be drawn from the data discussed above is that montmorillonite affects the conformation of the FA polymer in a manner that favors reactions between C=O groups and metal ions. The IR spectra in Fig. 21-3 also show the involvement of COO⁻ groups in the FA. The reaction of COO⁻ groups with metal ions is most likely to be most prominent on the outer clay surfaces (Schnitzer & Kodama, 1972).

The intensity with which FA is retained by Cu^{2+}-montmorillonite is illustrated by the finding that $> 75\%$ of the weight of the FA complexed by the clay resisted decomposition even when heated up to 1000C (1273K). Surface area measurements showed the following external areas: 2.5 m²/g for FA, 8.0 m²/g for Cu^{2+}-montmorillonite but 16.1 m²/g for the Cu^{2+}-montmorillonite–FA complex (Kodama & Schnitzer, 1974b). Assuming that no interaction had taken place between the first two, the total surface area should have been 7.2 m²/g, which is $< 50\%$ of the surface area observed experimentally. It appears, therefore, that the adsorption of FA on clay surfaces is associated with conformational changes in the FA structure, as has been pointed out above, and that these result in an expansion of the FA surface.

V. CHARACTERIZATION OF HUMIC ACID- AND FULVIC ACID- CLAY COMPLEXES

For the characterization of HA- and FA–clay complexes a large variety of methods have been used. These include chemical, X-ray, spectrophotometric, spectrometric, thermal, and electron microscopic procedures. Some of these procedures will be discussed in the following paragraphs.

1. CHEMICAL METHODS

Most chemical analyses on HA- or FA–clay complexes are concerned with the analysis of C which is often firmly retained by the mineral part of the complex even when heated up to 1000C (1273K) (Schnitzer & Kodama, 1972). To overcome these difficulties, C in FA–clay complexes can be accurately determined by dry-combustion provided that the sample is covered by a catalyst such as WO_3 (Schnitzer & Kodama, 1966) or V_2O_5 (Morris & Schnitzer, 1967).

2. X-RAY ANALYSIS

Measurements of $d(001)$ spacings are usually made on oriented aggregates on glass slides with the aid of a diffractometer. In order to obtain reproducible d-spacings, measurements should be made under relative humidities of $< 1\%$ (Schnitzer & Kodama, 1967). This can be accomplished by passing a stream of air or N_2 first through columns of Drierite before allowing it to flow over the sample.

3. SPECTROPHOTOMETRIC METHODS

The most widely used spectrophotometric method is IR analysis. FA–clay complexes can be mixed with KBr and then pressed into pellets (Schnitzer & Kodama, 1967). The KBr-pellet method was also employed by Inoue and Wada (1971) for obtaining IR spectra of humified clover extract-imogolite complexes. FA–clay complexes can also be prepared for IR analysis by allowing aqueous suspensions to evaporate at room temperature and atmospheric pressure on Al-foil. The resulting, thin, self-supporting films are then peeled away from the Al-foil surface and mounted in a holder in the IR spectrophotometer (Schnitzer and Kodama, 1972). This method avoids KBr and possible interactions between KBr and the FA–clay complex.

Narkis et al. (1970) measured absorbances between 300 and 800 nm to study interactions between HA and montmorillonite and kaolinite. Changes in absorbance as a result of HA–clay interactions were ascribed to changes in the size of clay tactoids. Na-humate was found to cause the dissociation of montmorillonite tactoids in suspension, but had no effect on kaolinite particles.

4. SPECTROMETRIC METHODS

Electron Spin Resonance (ESR) spectometry has been used to examine FA–Cu^{2+}-montmorillonite complexes (Schnitzer & Kodama, 1972). The ESR spectra were single symmetrical lines without fine splitting. Line widths were about 125 G and g-values 2.03. The spectra indicated that no Cu^+ or organic cation radicals were formed during the FA–clay interaction. Spectra were recorded on finely ground powders.

5. THERMAL ANALYSIS

The major thermal methods that have so far been employed for the analysis of HA- and FA-clay complexes are thermogravimetry (TG), differential thermogravimetry (DTG), differential thermal analysis, (DTA), and isothermal heating (Schnitzer & Kodama, 1967, 1969, 1972; Kodama & Schnitzer, 1969, 1970, 1971; Inoue & Wada, 1971; Rashid et al., 1972; Lutwick, 1972).

Figure 21-5 illustrates the use of TG and DTG for the analysis of a FA–Na-montmorillonite complex (Kodama & Schnitzer, 1969). Curve A shows that the thermal decomposition of FA is completed by 530C (803K). Four main reactions can be distinguished, with maxima at 100, 200, 300, and 465C (373, 473, 573, and 738K) (see DTG curve). These are due to dehydration, dehydrogenation, a combination of decarboxylation and dehydroxyla-

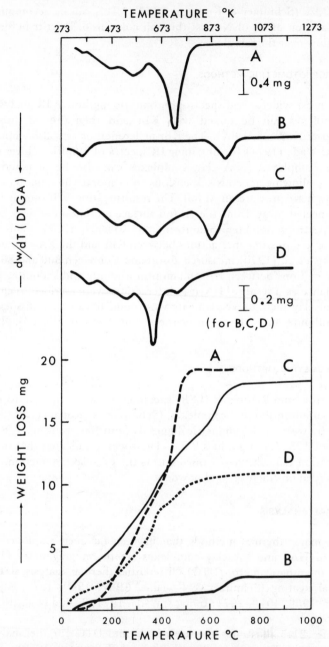

Fig. 21-5. Thermogravimetric (TG) and differential thermogravimetric (DTG) curves for untreated FA *(A)*, montmorillonite *(B)*, FA-unheated montmorillonite complex *(C)*, FA-heated montmorillonite complex *(D)* (Kodama & Schnitzer, 1969).

tion and to oxidation of the FA "nucleus", respectively. The corresponding weight losses are 2.5, 24.5, 15.5, and 57.0%.

The DTG curve of the FA–Na-montmorillonite complex (curve C) shows maxima at 90, 230, 370, and 610C (363, 503, 643, and 883K). For the FA-heated ("collapsed")-Na-montmorillonite complex (curve D), maxima appear at 120, 265, 370, and 470C (393, 538, 643, and 743K). The main difference between curves C and D is the disappearance of the strong DTG peak at 610C (883K) in the curve of the "collapsed" clay-FA complex, which suggests that this peak is due to the decomposition of FA adsorbed in the interlayers and that maxima at 275 and 370C (548 and 643K) arise from FA adsorbed on external clay surfaces. From weight loss curves C and D one can calculate that about one-half of the FA is adsorbed on external clay surfaces and the remaining one-half in interlayers.

The use of DTA is illustrated in Fig. 21-6. Peaks in DTA curve A (untreated FA) can be assigned in the following manner (Kodama & Schnitzer, 1969): the shallow and broad endotherm at 100C (373K) is due to dehydration; the shoulder-like exotherm at about 330C (603K) and the prominent exotherm at 450C (723K) arise from decarboxylation and oxidation of the FA "nucleus," respectively.

The endotherm near 80C (353K) in the DTA curve of Na-montmorillonite (curve B) is due to dehydration, the endotherm near 675C (948K) arises from dehydroxylation of the clay, and the exotherm at 925C (1198K) is associated with a change in crystal structure of the clay.

The DTA pattern for the physical mixture of FA and Na-montmorillonite (curve C) is essentially a composite of the two constituents, except that the peak at the temperature of the main exotherm that occurs at 450C (723K) in the untreated FA is lowered to 400C (673K) in the FA–clay mixture, possibly due to catalytic action of the clay surface.

The DTA curve for the FA–Na-montmorillonite (unheated) complex (curve D) differs from that of the physical FA–clay mixture in a number of ways: the exotherm in the 400 to 460C (673 to 733K) region is much broader, indicating that decomposition takes place over a wider temperature range. Near 670C (943K) an exotherm appears instead of an endotherm, and this is indicative of the presence of an interlayer complex, and is similar to DTA data for clay–protein and similar clay–organic complexes. In addition, curve D shows a small, but well-defined exotherm near 930C (1203K), which suggests that the reorganization of the Na-montmorillonite is accompanied by combustion of some carbonaceous residue derived from FA that has persisted up to this temperature.

The curve for FA-("collapsed") Na-montmorillonite complex (curve E) exhibits a shoulder-like exotherm at 330C (603K) and a symmetrical exotherm of medium size near 425C (698K) followed by a minor hump in the 500 to 550C (773 to 823K) region, probably due to loss of water from rehydroxylated clay.

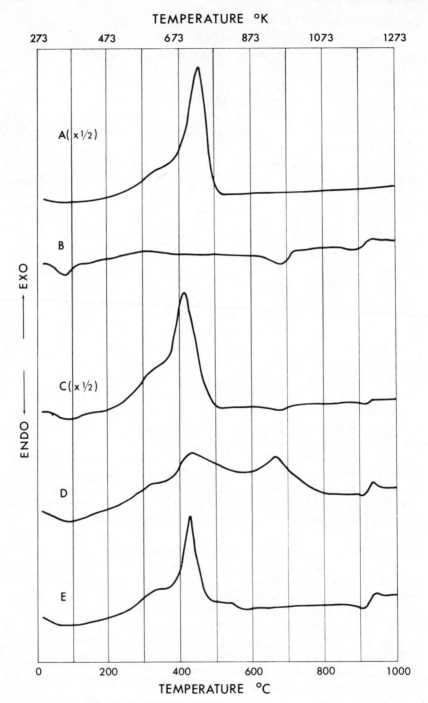

Fig. 21-6. DAT curves for untreated FA (A), montmorillonite (B), physical mixture of untreated FA and montmorillonite (C), FA-unheated montmorillonite complex (D), and FA-heated montmorillonite complex (E) (Schnitzer & Kodama, 1969).

A comparison of curves D to E shows that in curve E the first endotherm, due to dehydration, is less pronounced, that no exotherm is detectable near 670C (943K), but that a similar S-shaped exotherm is discernible in the 900 to 950C (1173 to 1223K) region. Since only curve D shows an exotherm near 670C (943K), this exotherm, which extends from 550 to 789C (823 to 1072K) is assigned to interlayer FA. The reaction extending from 350 to 550C (623 to 823K) is most likely associated with the combustion of FA adsorbed by Na-montmorillonite on its external surfaces. The DTA results are in agreement with TG and DTG data discussed above. Thus, the two principal adsorption reactions (that is, on external surfaces and in interlayers) are well separated in curve D.

Isothermal heating at 350, 370, 390, and 410C (623, 643, 663, and 683K) has been used by Kodama and Schnitzer (1969) to study the kinetics of thermal decomposition of a FA–Na-montmorillonite complex. It was found that the decomposition reaction of FA adsorbed on Na-montmorillonite differed significantly from that of unadsorbed FA. The clay retarded the decomposition reaction, which may explain the observed stability of HA–clay and FA–clay complexes in nature.

6. ELECTRON MICROSCOPY

A number of workers (Beutelspacher, 1955; Brydon & Sowden, 1959; Arshad & Lowe, 1966; Dudas & Pawluk, 1970; Kodama & Schnitzer, 1974b) have used the electron microscope to look at HA– and FA–clay complexes. As illustrated in Fig. 21-7a (Kodama & Schnitzer, 1974b) the electron micrograph of FA shows two types of particles: relatively uniform disks and a transparent, shapeless material. Figure 21-7b is the electron micrograph of Cu^{2+}-montmorillonite. It shows the presence of thin, flaky, tissue-paper-like material, typical of montmorillonite. The electron micrograph of the FA–Cu^{2+}-montmorillonite complex (Fig. 21-7c) shows the preservation of the major montmorillonite features. Outlines in Fig. 21-7c show more contrast than those in Fig. 21-7b due, most likely to interlayer adsorption of FA. Figure 21-7c also shows adsorption on outer surfaces. Figure 21-7d is a lattice image taken from the edges of a FA–Cu^{2+}-montmorillonite complex. The observed periodicity of the lattice image ranges from 13 to $\approx$ 18Å, which is well within the range of the 15Å $d(001)$ spacing observed by X-ray analysis (Schnitzer & Kodama, 1972). Thus, information provided by the electron microscope indicates that at low pH, FA is adsorbed in interlayer spaces as well as on external surfaces of Cu^{2+}-montmorillonite.

VI. EVIDENCE FOR THE FORMATION OF INTERLAYER HUMIC ACID–CLAY AND FULVIC ACID–CLAY COMPLEXES IN NATURE

Evidence for the presence of humic substances in interlayers of expanding naturally occurring clays has been reported by Satoh and Yamane (1971), Lowe and Parasher (1971), Parasher and Lowe (1970), and Kodama and

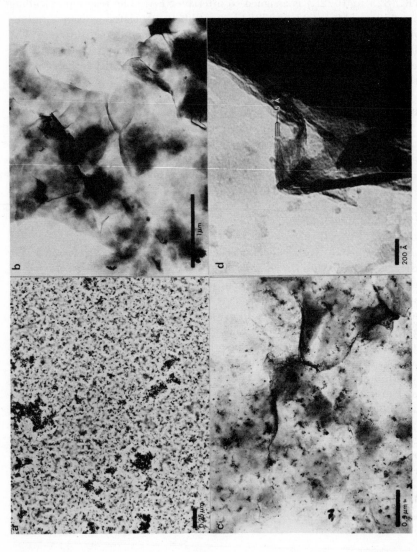

Fig. 21-7. Electron micrographs of: (a) FA; (b) Cu²⁺-montmorillonite; (c) FA-Cu²⁺-montmorillonite complex prepared at pH 2.5; (d) enlargement taken from edges of FA-clay complex shown in (c) (Kodama & Schnitzer, 1974b). Reproduced with the permission of Pergamon Press Ltd.

Schnitzer (1971). Satoh and Yamane (1971) isolated clay–organic complexes from an acidic volcanic ash soil. The dominant mineral in the complex was smectite which complexed humic materials in interlayers. Kodama and Schnitzer (1971) emphasize that unambiguous evidence for the interlayer adsorption of humic substances, especially FA, by soil clays can be obtained by using a variety of diagnostic procedures, which include DTA, X-ray, IR and chemical methods, followed by comparisons of experimental data with those obtained on laboratory-prepared FA–clay complexes. Kodama and Schnitzer (1971) showed that the fine clay fraction (< 0.2 μm) isolated from the Ae horizon of an Orthic Humo-Ferric Podzol contained large amounts of FA, some of which was adsorbed in interlayers of an expandable layer silicate which was an interstratified clay mineral consisting of three-component layers of mica-vermiculite-smectite with a ratio of 0.53:0.27:0.20. It is likely that in the near future more evidence of this nature will appear in the literature.

VII. MECHANISMS OF PRINCIPAL TYPES OF REACTIONS BETWEEN HUMIC ACID AND FULVIC ACID AND SOIL MINERALS

From the data presented in this chapter a number of tentative conclusions on the principal reaction mechanisms governing HA- and FA-soil mineral interactions can be drawn.

A. Dissolution

In view of their ability to complex di- and trivalent metal ions and their appreciable solubility in water at pH levels prevailing in soils, FA's and, to some extent, low-molecular weight HA's can attack and degrade soil minerals to form water-soluble and water-insoluble metal complexes, depending on how much metal per unit weight of HA or FA is dissolved. If the metal/humic ratio is low, the complex is water-soluble. Thus, low-molecular weight humic substances do not only dissolve metals but can also transport them within and between soil horizons and so affect practically all reactions that occur in these systems.

B. Adsorption on External Surfaces

These reactions will vary with the physical and chemical characteristics of the mineral surface, with the pH of the system, and with the water content. Again, the ability of humic substances to form metal complexes via CO_2H and phenolic OH groups will play a significant part in these interactions. Not all complexes are necessary chelates. In some instances only one HA- or FA-CO_2H group may react with one metal ion and the resulting association is relatively weak. Thus, one can visualize the formation of a wide range of humic–mineral associations, involving chemical bonding, with widely differing stabilities. Another mechanism of considerable relevance as far as surface adsorption is concerned is H-bonding, the occurrence of which is clearly indicated in IR spectra of FA–mineral and FA–clay complexes. These reactions are likely to involve H or O of CO_2H and OH groups in HA or FA and O or H at mineral surfaces and edges. It is possible, as proposed by Yariv et al. (1966) that cations with high solvation energies react via water bridges with HA- and FA-functional groups, with the link between the O and the water (in the hydration shell of the cation) being a H-bond. It is probable that van der Waal's forces also contribute to the adsorption of humic substances on mineral surfaces, although we tend to agree with Mortland (1970) that the principal van der Waal's interactions occur between adsorbed molecules rather than between adsorbed molecules and the surface.

The importance of surface geometry and surface chemistry has been pointed out by Inoue and Wada (1971). They studied reactions between humified clover and imogolite and assigned the adsorption of a humified clover extract on imogolite surfaces to the following two mechanisms: (i) incorporation of CO_2H groups into the coordination shell of Al atoms, and (ii) relatively weak interaction between the humified materials with imogolite units by H-bonding and van der Waal's forces. The two adsorption mechanisms are thought to assist each other and the number of bonds formed depend on the orientation of the humic molecules on the clay. Kodama and Schnitzer (1974a) report high adsorption of FA on sepiolite surfaces. Sepiolite has a channel structure formed by joining of edges of long and slender talc-like layers. In untreated-sepiolite the channels are occupied by bound and/or zeolitic water which apparently can be displaced by undissociated FA. Infrared spectra of FA–sepiolite interaction products suggest the formation in the channels of COO^- groups which are linked to Mg^{2+} at edges that have been exposed by displacement of water by FA (Kodama & Schnitzer, 1974a).

According the Edwards and Bremner (1967), soil organic matter and clay particles are linked to each other via polyvalent metal cations to form microaggregates. The latter can be dispersed by shaking with Na-resin which exchanges Na^+ for polyvalent ions such as Ca^{2+}, Mg^{2+}, Fe^{3+}, and Al^{3+}. Mortland (1970) believes that the C-P-OM complexes of Edwards and Bremner (where C=clay, P=polyvalent cation and OM=organic matter) are constituted more likely of $C-P-H_2O-OM$, where H_2O stands for a water bridge. Mechanistically the formation of microaggregates is governed by the same types of chemical interactions as those described above.

C. Adsorption in Clay Interlayers

The evidence presently available shows that the interlayer adsorption by expanding clay minerals of FA's and of low-molecular weight HA's is pH-dependent, being greatest at low pH and no longer occurring at pH > 5.0.

FA cannot be displaced from clay interlayers by leaching with $1N$ NaCl solution and an inflection occurs in the adsorption-pH curve near the pH corresponding to the pK of the acid species of the FA. On the basis of these criteria, the adsorption reaction could be classified as a "ligand-exchange reaction" (Greenland, 1971). In this type of adsorption the anion is thought to penetrate the coordination shell of the dominant cation in the clay and displace water coordinated to the dominant cation in the clay interlayer. The ease with which water can be displaced will depend on the affinity for water of the dominant cation with which the clay is saturated and also on the degree of dissociation of the FA. Since the latter is very low at low pH, interlayer adsorption of FA is greatest at low pH levels. Concurrently the FA can dissolve a proportion of the dominant cation in the clay by forming a soluble complex and replacing the removed cation by H^+. If this process continues over long periods of time, the FA will eventually degrade the clay structure. However, in short-term experiments (up to 24 hours) this kind of deterioration is not significant.

VIII. CONCLUSIONS

Humic substances, the major organic components of soils and waters, which contain, per unit weight, relatively large numbers of O-containing functional groups (CO_2H, phenolic OH and C=O) are naturally-occurring polyelectrolytes. Humic compounds are capable of attacking and degrading soil minerals by complexing and dissolving metals and transporting them within soils and waters. Especially active in this regard are FA and low-molecular weight HA, which are water-soluble. At low metal-to-FA or HA ratios, the metal-organic complexes are water-soluble; at high ratios, they are no longer soluble in water.

Humic substances are readily adsorbed on soil mineral surfaces. The extent of adsorption depends on the geometry and chemistry of the surface, the pH of the system, and the water content. Surface adsorption may, in addition to metal complexing, also involve the formation of H-bonding between CO_2H, OH, and C=O groups in HA and FA and OH and O on minerals.

FA and low-molecular weight HA are adsorbed on external surfaces and in interlayers of expanding clay minerals. The magnitude of interlayer adsorption depends on the pH, being greatest at low pH, and no longer occurring at pH > 5.0. The main reaction governing the interlayer adsorption of low-molecular weight humic materials by clay minerals appears to be the ability of relatively undissociated humic materials to displace water from interlayers. The ease with which water can be displaced depends on the affinity for water of the dominant cation on the clay mineral and the degree of dissociation of the FA.

Chemical, X-ray, spectrophotometeric, spectrometric, thermal, and electron microscopic methods can be used for investigating interactions between soil minerals and humic substances. As our knowledge of the chemical structure and reactions of humic substances increases, interactions between humic substances and soil minerals will be better understood. In view of the considerable importance of these interactions in soils and waters, it is hoped that in the future soil and water scientists will devote more time and effort to investigating them than they have done in the past.

ACKNOWLEDGMENT

We are grateful to B. Johnston for typing this manuscript.

LITERATURE CITED

Alexandrova, L. N. 1960. On the composition of humus substances and the nature of organic-mineral colloids in soil. Int. Congr., Soil Sci. Trans. 7th (Madison, Wis.) II:74-81.

Arshad, M. A., and L. E. Lowe. 1966. Fractionation and characterization of naturally occurring organo-clay complexes. Soil Sci. Soc. Am. Proc. 30:731-735.

Baker, W. E. 1973. Role of humic acids from Tasmanian podzolic soils in mineral de-degradation and metal mobilization. Geochim. Cosmochim. Acta 37:269-281.

Beutelspacher, H. 1955. Interaction between the inorganic and organic colloid in soil. Z. Pflanzenernaehr Dueng. 69:108-115.

Brydon, J. E., and F. J. Sowden. 1959. A study of the clay-humus complexes of a chernozemic and podzol soil. Can. J. Soil Sci. 39:136-143.

Demolon, A., and G. Barbier. 1929. Conditions de formation et constitution du complexe argilo-humique des sols. C. R. Acad. Sci. 188:654-656.

Dudas, M. J., and S. Pawluk. 1970. Naturally occurring organo-clay complexes of orthic black chernozems. Geoderma 3:5-17.

Edwards, A. P., and J. M. Bremner. 1967. Microaggregation in soils. J. Soil Sci. 18:64-73.

Evans, L. T., and E. W. Russell. 1959. The adsorption of humic and fulvic acids by clays. J. Soil Sci. 10:119-132.

Gamble, D., M. Schnitzer, and I. Hoffman. 1970. Cu^{2+}-fulvic acid chelation equilibration in $0.1M$ KCl at $25.0°C$. Can. J. Chem. 48:3197-3204.

Gorbunov, N. I., G. L. Erokhina, and G. N. Shchurina. 1971. Relationship between soil minerals and humic substances. Pochvovedenie no. 7, p. 117-128.

Greenland, D. J. 1965. Interaction between clays and organic compounds in soils. II. Adsorption of soil organic compounds and its effects on soil properties. Soils Fert., Commonw. Bur. Soil Sci. 28:521-532.

Greenland, D. J. 1971. Interactions between humic and fulvic acids and clays. Soil Sci. 111:34-41.

Griffith, S. M., and M. Schnitzer. 1975. Analytical characteristics of humic and fulvic acids extracted from tropical volcanic soils. Soil Sci. Soc. Am. Proc. 39:861-867.

Hem, J. D., and C. J. Lind. 1974. Kaolinitic synthesis at $25°C$. Science 184:1171-1173.

Inoue, T., and K. Wada. 1971. Reactions between humified clover extract and imogolite as a model of humus-clay interaction. Part II. Clay Sci. 4:61-70.

Kodama, H., and M. Schnitzer. 1967. X-ray studies of fulvic acid, a soil humic compound. Fuel 47:87-94.

Kodama, H., and M. Schnitzer. 1968. Effects of interlayer cations on the adsorption of a soil humic compound by montmorillonite. Soil Sci. 106:73-74.

Kodama, H., and M. Schnitzer. 1969. Thermal analysis of a fulvic acid-montmorillonite complex. Proc. Int. Clay Conf. (Tokyo, Japan) 1:765-774.

Kodama, H., and M. Schnitzer. 1970. Kinetics and mechanism of the thermal decomposition of fulvic acid. Soil Sci. 109:265-271.

Kodama, H., and M. Schnitzer. 1971. Evidence for interlamellar adsorption of organic matter by clay in a Podzol soil. Can. J. Soil Sci. 51:509-512.

Kodama, H., and M. Schnitzer. 1973. Dissolution of chlorite minerals by fulvic acid. Can. J. Soil Sci. 53:240–243.

Kodama, H., and M. Schnitzer. 1974a. Adsorption of fulvic acid by nonexpanding clay minerals. Int. Congr. Soil Sci. Trans. 10th (Moscow) II:51–56.

Kodama, H., and M. Schnitzer. 1974b. Further investigations on fulvic acid-Cu^{2+}-montmorillonite interactions. Clays Clay Miner. 22:107–110.

Linares, J., and F. Huertas. 1971. Koalinite: synthesis at room temperature. Science 171:896–897.

Levashkevich, G. A. 1966. The interaction of humic acids with hydroxides of iron and aluminum. Pochvovedenie no. 4, p. 58–65.

Lowe, L. E., and C. D. Parasher. 1971. Observations on clay-size organo-mineral complexes isolated from soil by ultrasonic dispersion. Can. J. Soil Sci. 51:136–137.

Lutwick, L. E. 1972. Thermal decomposition reactions of clay-organic matter complexes and organic matter separated from a Black Chernozemic soil. Can. J. Soil Sci. 52: 417–425.

Martin, F. Martinez, and Perez Rodriguez. 1969. Interlamellar adsorption of a black-earth humic acid on Na-montmorillonite. Z. Pflanzenernaehr. Dueng. Bodenk. 124: 52–57.

Morris, G. F., and M. Schnitzer. 1967. Rapid determination of carbon in soil organic matter by dry-combustion. Can. J. Soil Sci. 47:143–144.

Mortland, M. M. 1970. Clay-organic complexes and interactions. Adv. Agron. 22:75–117.

Nakamoto, K. 1963. Infrared spectra of inorganic and coordination compounds. John Wiley & Sons, New York.

Narkis, N., M. Rebhun, N. Lahav, and A. Banin. 1970. An optical-transmission study of the interaction between montmorillonite and humic acids. Israel J. Chem. 8:383–389.

Parasher, C. D., and L. E. Lowe. 1970. Isolation of clay-size organo-mineral complexes from soils of the Lower Fraser Valley. Can. J. Soil Sci. 50:403–407.

Ponomareva, V. V., and A. L. Ragim-Zade. 1969. Comparative study of fulvic and acids as agents of silicate mineral decomposition. Sov. Soil Sci. no. 2. p. 157–166.

Rashid, M. A., D. E. Buckley, and K. R. Robertson. 1972. Interactions of a marine humic acid with clay minerals and a natural sediment. Geoderma 8:11–27.

Rosell, R. A., and K. L. Babcock. 1968. Precipitated manganese isotopically exchanged with ^{54}Mn and chelated by soil organic matter. p. 453–469. In Isotopes and radiation in soil organic matter studies. Int. Atomic Energy Agency, Vienna, Austria.

Satoh, T., and I. Yamane. 1971. On the interlamellar complex between montmorillonite and organic substance in certain soil. Soil Sci. Plant Nutr. 17:181–185.

Schnitzer, M. 1971. Metal-organic interactions in soils and waters. p. 297–315. In S. J. Faust and J. V. Hunter (ed.) Organic compounds in aquatic environments. Marcel Dekker, New York.

Schnitzer, M., and E. H. Hansen. 1970. Organo-metallic interactions in soils: 8. An evaluation of methods for the determination of stability constants for metal-fulvic acid complexes. Soil Sci. 109:333–340.

Schnitzer, M., and S. U. Khan. 1972. Humic substances in the environment. Marcel Dekker, New York.

Schnitzer, M., and H. Kodama. 1966. Montmorillonite. Effect of pH on its adsorption of a soil humic compound. Science 153:70–71.

Schnitzer, M., and H. Kodama. 1967. Reactions between a Podzol fulvic acid and Na-montmorillonite. Soil Sci. Soc. Am. Proc. 31:632–636.

Schnitzer, M., and H. Kodama. 1969. Reactions between fulvic acid, a soil humic compound, and montmorillonite. Israel J. Chem. 7:141–147.

Schnitzer, M., and H. Kodama. 1972. Reactions between fulvic acid and Cu^{2+}-montmorillonite. Clays Clay Miner. 20:359–367.

Schnitzer, M., and S. I. M. Skinner. 1965. Organo-metallic interactions in soils: 4. Carboxyl and hydroxyl groups in organic matter and metal retention. Soil Sci. 99: 278–284.

Serra, de, M. I., and M. Schnitzer. 1972. The chemistry of humic and fulvic acids extracted from Argentine soils. II. Permangante oxidation of methylated humic and fulvic acids. Soil Biol. Biochem. 5:287–296.

Schwertmann, U. 1966. Inhibitory effect of soil organic matter on the crystallization of amorphous ferric hydroxide. Nature (London) 212:645–646.

Yariv, S., J. D. Russell, and V. C. Farmer. 1966. Infrared study of the adsorption of benzoic acid and nitrobenzene in montmorillonite. Israel J. Chem. 4:201–213.

Mineralogy and Soil Taxonomy

B. L. ALLEN, Texas Tech University, Lubbock, Texas

I. INTRODUCTION

Mineralogy has not been extensively used as a differentiating criterion in soil classification systems until the advent of the taxonomic classification currently used in the United States. However, in several systems, classes have been separated by morphological features strongly influenced by mineralogy, e.g., cracking and swelling. Soil-forming factors, dominant pedogenic processes, and gross morphology have been, and still are, emphasized in most schemes. Although composition per se has not been commonly used as a criterion, many physical and chemical analyses have been performed to elucidate processes and to determine causes of morphological differences among soils.

Spectacular progress has been made in developing methods for determination of soil mineralogical composition within the last five decades. Composition can now be determined with sufficient accuracy so that it can be used at specified levels in soil taxonomy. Nevertheless, problems remain in the quantification of mineralogical composition and also in the correlation of soil physical and chemical properties with mineralogy.

The taxonomic soil classification used in the United States (Soil Survey Staff; 1960, 1975) is based upon soil properties that can be observed or measured. Soil genesis enters into the system only as guides to relevance and the weighting of soil properties. The primary objective of the soil taxonomy is to have hierarchies of classes that will permit an understanding within the limits of existing knowledge, the relationships between soils, and also between soils and the factors responsible for their properties (Soil Survey Staff, 1975). A decision was made early in the development of the classification to incorporate existing knowledge about soil mineralogy. Consequently, soil mineralogy is used as a differentiating criterion at several levels in the system.

Taxa are separated in the highest categories of the soil taxonomy by the presence of diagnostic horizons and features. Surface horizons ("epipedons"), subsurface horizons, and other features are diagnostic at various categorical levels. Different sets of diagnostic properties and horizons are used for mineral and organic soils.

II. DIAGNOSTIC HORIZON–MINERALOGY RELATIONSHIPS

Mineralogy enters directly into diagnostic horizon criteria in only a few cases, but many of the properties on which the definitions are based are strongly influenced by the mineralogical composition. Consequently, many relationships can be deduced.

chapter 22

A. Surface Horizons (Epipedons)

Diagnostic surface horizons, termed *epipedons* (Soil Survey Staff, 1960) may, in addition to the part of the profile considered to be A horizon(s), include the upper part of the subsoil under specified conditions.

1. MOLLIC EPIPEDONS

Mollic epipedons are the dark upper profiles so commonly observed in the grasslands of humid, subhumid, and semiarid climates. They must meet specified color, organic matter, and thickness requirements (Soil Survey Staff, 1975). Base saturation percentage must be $> 50\%$. Phyllosilicates of the 2:1 type, often with considerable smectite, tend to predominate in the clays of such horizons.

2. UMBRIC EPIPEDONS

Umbric epipedons meet all the requirements for mollic epipedons except that base saturation percentage is $< 50\%$. They have variable mineralogies because of the range of conditions in which they occur. Those in cool moist climates at high altitudes usually have a clay mineral suite composed mostly of nonexpanding 2:1 phyllosilicates, except where amorphous materials, as in soils formed in pyroclastic ejecta, predominate. Some in coastal areas are dominantly smectitic, whereas others in tropical areas have large quantities of 1:1 minerals and hydrous oxides.

3. ANTHROPIC EPIPEDONS

Anthropic epipedons, which have been strongly altered because of the activities of man, meet all the requirements of mollic epipedons, but in addition have > 250 ppm of citrate-soluble P_2O_5.

4. PLAGGEN EPIPEDONS

Plaggen epipedons, unknown in the United States, are much thickened horizons produced by additions of sod, manure, etc. during long periods of cultivation. The limited data on the composition of such horizons do not suggest marked differences in mineralogy from associated, less disturbed soils.

5. OCHRIC EPIPEDONS

Ochric epipedons are the exceedingly common surface horizons that do not meet the color, organic matter, or thickness requirements of other epipedons. Many A1 horizons, especially those overlying A2 horizons, are sufficiently dark and have enough organic matter to meet the criteria of mollic or umbric horizons, but are too thin.

Since ochric epipedons occur under climates ranging from desert to perhumid in many different parent materials, and exhibit a variety of drainage conditions, clay mineralogy is exceedingly variable. In cool humid climates, 2:1 phyllosilicates of limited expandability and hydroxy interlayered types seem to predominate. The latter are common in warm temperate humid climates, especially in soils associated with older landscapes, such as the southeastern United States. Kaolinite and hydrous oxides are common in the humid tropics. Poorly crystallized 2:1 types with considerable expandability, often together with calcite and sometimes gypsum, predominate in arid regions.

6. HISTIC EPIPEDONS

Histic epipedons are layers, usually at the surface, in which the properties are primarily determined by organic materials. The high organic C content requirement varies depending on clay content. They are either saturated for a specified length of time during the year or have been artificially drained. Although they are little influenced by the inorganic fraction because of the high organic matter, their mineralogy could become of considerable importance under long continued use as the organic matter decomposes.

B. Subsurface Horizons

Diagnostic "master" subsurface horizons commonly occur in the position of what is generally regarded as "subsoil."

1. ARGILLIC HORIZONS

Argillic horizons are those in which there has been significant illuviation of silicate clays. Specific requirements for the amount of clay in the horizon relative to that in the eluviated horizon depends on the clay percentage and the thickness of the eluviated horizons and the solum as well as the texture of the illuviated horizon. Most argillic horizons have evidence of silicate clay illuviation as coatings on ped surfaces, channel walls, or sand grains. These coatings have been variously referred to, among others, as "clay skins" (Buol & Hole, 1959) and "cutans," or more specifically as "illuviation argillans" (Brewer, 1964). Although there are no specific mineralogical requirements for argillic horizons, their very nature precludes dominance of certain minerals. The clay mineralogy of argillic horizons is most often mixed, but smectite, vermiculite, illite, glauconite, or kaolinite may predominate (Soil Survey Staff, 1975, unpublished material). Some contain large quantities of hydrous oxides (Fiskell & Perkins, 1970; Cook, 1973; Bryant & Dixon, 1963).

Buol and Hole (1959) found only small differences in the mineralogical composition of argillans relative to that of the clay in ped interiors in Wisconsin soils. However, more smectite should be in the films than in the matrix clay of the ped interior in soils that contain moderate amounts of smectite. Smectite tends to be concentrated in the fine clay (< 0.2 μm), the subfraction that tends to move preferentially during pedogenesis. Khalifa and Buol (1968) reported the clay in the cutans of a kaolinitic soil (Cecil) to have poorer crystallinity than the clay from the whole peds in the argillic horizon.

Illuviation argillans are not always present, although the horizons may indeed have significant quantities of illuviated clay. Clay content and mineralogical composition are primary determinants of shrink-swell potential and, indirectly, whether cutans are present. Whenever moderate amounts of smectite are present and the clay content is high, illuviation argillans are often lacking because of destruction during shrinking and swelling. Instead, stress argillans (due to differential movement) may be present on ped surfaces.

Nettleton et al. (1969) found that evidence for illuviation argillans may be meager at best in many soils in arid and Mediterranean climates although all other criteria suggest an argillic horizon. It seems that the argillans are more prone to destruction in such climates with pronounced dry seasons than in higher rainfall areas. Gile and Grossman (1968) also noted the lack of argillans on ped surfaces in the southwestern United States desert soils and attributed their destruction to authigenic calcite crystal growth in addition to high energy moisture changes. They found the argillans to be more stable as sand grain and pebble coatings than on ped surfaces. Allen and Goss (1974) also found pronounced argillan disruption by calcite crystal growth in buried argillic horizons in West Texas.

Soils with a mineralogy dominated by amorphous aluminosilicates ordinarily do not have argillic horizons because of the resistance to dispersion of such materials. Neither do argillic horizons commonly develop in strongly calcareous materials; however, they are exceedingly common in the southwestern United States that now have calcite coatings on ped surfaces and pores, probably as a result of surface additions of calcareous dust.

Tubular halloysite can be translocated to form argillans on peds or in pores; however, the particles cannot orient to produce an optical effect similar to that of layer silicates when viewed in thin section (Soil Survey Staff, 1975).

Differences in clay mineralogy between argillic horizons and overlying eluviated horizons are often pronounced, but may be only slight in other soils. Differences in mineralogy may be due to: (i) differential translocation, (ii) in situ transformations, or (iii) depositional differences, including stratification in parent materials or additions, especially eolian, during pedogenesis.

2. NATRIC HORIZONS

Natric horizons have properties similar to argillic horizons, but also must meet certain chemical criteria. Exchangeable Na is $\geq 15\%$, or the sum of exchangeable Na and Mg is more than the sum of exchangeable Ca and H, or the SAR (sodium adsorption ratio) is $\geqslant 13$. Columns are often present, the interiors of which are often very dark because of dispersed humus. Natric horizons have mineralogies similar in most respects to those of argillic horizons. Considerably more smectite has been reported in horizons that would qualify as natric as now defined than in the overlying eluviated horizons (Whittig, 1959; Goss & Allen, 1968).

3. SPODIC HORIZONS

Spodic horizons are illuvial horizons in which the translocated materials consist primarily of humus and free sesquioxides (used in a general sense for Fe and Al compounds). Amorphous Al seems to always be present, but Fe oxides may or may not be present. Humus and sesquioxides apparently move as complexes. Pelletized aggregates of humus and sesquioxides are often a characteristic micromorphological feature. Pronounced segregation may occur within the subsoil to produce subhorizons dominated by humus (Bh horizons) and by Fe compounds (Bir horizons). Alumina maxima may coincide with either horizon, but may occur deeper in the profile.

Despite the apparent secondary role of crystalline phyllosilicates in spodic horizons, their nature and pedogenic transformation has interested a number of investigators. Vermiculite, chlorite, and hydroxy interlayered vermiculite have been reported as the dominant phyllosilicate mineral in B horizons, which would mostly now qualify as spodic horizons in the classification of many northern United States and Canadian Podzols (Brown & Jackson, 1958; Ross & Mortland, 1966; and Brydon et al., 1968). Coen and Arnold (1972) found vermiculite and hydroxy interlayered vermiculite as the principal clay in New York spodic horizons.

Franzmeier et al. (1965) determined that the CEC of field-identified spodic horizons was markedly decreased upon heating to 250C, whereas the decrease in other horizons, except those known to be high in allophane or vermiculite, was much less pronounced. Specific alternative criteria involving CEC/clay ratios, thickness, and particle size class have been designated by the Soil Survey Staff (1975).

4. PLACIC HORIZONS

Placic horizons are thin wavy layers in materials ranging from clays to sands that are cemented by Fe, Fe and Mn, or by Fe-humus complexes.

5. SOMBRIC HORIZONS

Sombric horizons are illuvial horizons in which the translocated materials consist mostly of humus, but which do not have the high CEC/clay ratios characteristic of spodic horizons.

6. AGRIC HORIZONS

Agric horizons, which are unknown in the United States, are illuvial horizons formed under the influence of long-continued cultivation. Illuviated humus, clay, and silt form thick coatings on channels or form thick lamellae underneath an Ap horizon. Little data are available on the mineralogy of placic, sombric, or agric horizons.

7. CAMBIC HORIZONS

Cambic horizons are those in which considerable pedogenesis has occurred, but which do not show evidence of significant illuviation of silicate clays or sesquioxides. Soil structure has developed sufficiently so that rock structure and bedding planes of stratified sediments have been destroyed. They are extremely variable in mineralogy because of their pedogenic youthfulness and occurrence under widely differing conditions. However, some mineralogical limitations are implied because of the requirement that they have sufficient 2:1 phyllosilicate or amorphous clays to give a CEC of > 16 meq/100 g clay or significant amounts of weatherable primary minerals. The clay mineralogy is rarely kaolinitic or oxidic except in tropical areas where such soils have often developed in reworked sediments apparently derived from old land surfaces. Some cambic horizons that meet the CEC requirement have siliceous sand and silt mineralogies. Cambic horizons in most soils strongly reflect the lithology of the parent material.

Cambic horizons in arid regions may be dominated by carbonates or gypsum. Their phyllosilicate mineralogy usually is mixed. Often poorly ordered 2:1 minerals, various illite-vermiculite-smectite intergrades, are common components. Cambic horizons with carbonatic mineralogy also occur in more humid regions, primarily in soils derived from high-lime parent materials. The chemical and physical properties of cambic horizons developed in volcanic pyroclastics are usually influenced strongly by amorphous aluminosilicates, in which case they are of sufficient significance to be considered at high categorical levels in the taxonomy.

Soils developed in pyroclastics have been extensively studied, especially in Japan, New Zealand, Chile, Hawaii, and the northwestern United States because of their widespread occurrence and unusual physicochemical properties. Among reports of large amounts of allophane (or X-ray amorphous material) in parts of the soil profile that would qualify for cambic horizons, those of Birrel and Fieldes (1952) and Fieldes (1955) in New Zealand, Aomine and Yoshinaga (1955) and Aomine and Wada (1962) in Japan, and Besoain (1964, 1969) in Chile may be cited. Tamura et al. (1953) were among the first to report significant amounts of amorphous material in Hawaiian soils. Whittig et al. (1957) detected considerable amorphous colloids in the Cascade and Powell soils of Oregon. More recently, Mejia et al. (1968); Calhoun et al. (1972); and Cortes and Franzmeier (1972) have reported dominantly amorphous material in some Colombian soils. Besides the dominant amorphous material reported in the latter studies; many other minerals, including halloysite (often listed as a common constituent), vermiculite, smectite, kaolinite, imogolite, hydroxy interlayered vermiculite, illite, cristobalite, and gibbsite were reported.

Few detailed mineralogical studies on cambic horizons from humid regions, other than those developed in volcanic materials, have been reported. McCracken et al. (1962) and Losche et al. (1970) reported mixed mineralogies with a major hydroxy interlayered vermiculite component, resulting from alteration of parent rock chlorite, in soils from the Great Smoky Mountains and Southern Appalachians.

8. OXIC HORIZONS

Oxic horizons are of fine or medium texture and are dominated by 1:1 phyllosilicates, or sesquioxides, or both. They are strongly weathered and cannot have more than traces of weatherable primary aluminosilicates or glass. Oxic horizons occur primarily in tropical climates and are unknown in the United States except in Hawaii.

The mineralogy of oxic horizons is indirectly limited by defining criteria. They must retain $\leqslant 10$ meq of NH_4^+ from an unbuffered $1N$ NH_4Cl solution per 100 g clay or have < 10 meq NH_4OAc-extractable bases plus Al extractable with $1N$ KCl per 100 g clay. They nearly always have a CEC $\leqslant 16$ meq/100 g clay. These criteria limit the phyllosilicate mineralogy to 1:1 clays, or to a mixture of 1:1 types and hydroxy interlayered 2:1 types. Nearly all contain appreciable quantities of hydrous Al and Fe oxides and many are dominated by such minerals. Significant amounts of hydrated manganese oxides sometimes occur. Minerals highly resistant to weathering such as quartz, zircon, and anatase may be present. Oxic horizons are essentially devoid of weatherable minerals, e.g., feldspars, mica, ferromagnesian species, glass, etc., in all textural fractions.

Development of illuviation argillans is minimal in oxic horizons since clay translocation is not an active pedogenic process because of strong aggregation. Eswaran (1972) described degraded, Fe-enriched cutans in oxic horizons in Nicaraguan soils as "relicts" from a former argillic horizon stage. He also presented evidence for authigenic quartz in the horizons.

Many subsoils of former "Latosols" do not meet "oxic" criteria. Many are in fact argillic horizons; others are cambic horizons. As a result, definitive data for oxic horizons are rare. Tamura et al. (1953) characterized the subsoils, still apparently considered as oxic horizons, of the Molokai and Wahiawa "Low Humic Latosols" in Hawaii as dominantly kaolinitic and oxidic. Most oxic horizons have a mineral suite characteristic of weathering stages 10 to 13 described by Jackson et al. (1948).

In the characterization of Puerto Rican soils the Soil Survey Staff (1967) reported both kaolinitic (or halloysitic, or both) and oxidic (gibbsite and goethite) dominated horizons. Traces of a 14Å mineral, mica, and considerable quartz were present in some pedons. Le Roux (1973) found appreciable hydroxy interlayered 2:1 type clay, along with much kaolinite and alkali-extractable amorphous constituents, in some Natal oxic horizons. Lepsch and Buol (1974) reported that two oxic horizons, which showed minimal illuviation argillan development, in a toposequence from Sao Paulo State, Brazil, were predominantly kaolinitic. Considerable gibbsite and amorphous material (by dissolution analysis), quartz, and smaller amounts of hydroxy interlayered vermiculite were also present.

C. Other Horizons

In addition to the diagnostic subsurface horizons, other horizons, mostly in the substratum, and other pedologic features have been used in the development of the United States soil taxonomy.

1. FRAGIPANS

Fragipans, although sometimes in the substratum, usually are in the solum. They are brittle, have high bulk densities in comparison to adjacent horizons, form in loamy textures, and are slowly permeable. The mineralogy of fragipans has been extensively studied because of the possibility of thereby explaining their unusual physical properties. Despite the general lack of success in providing such answers, considerable mineralogical data on fragipans have been collected.

Numerous investigators in the eastern United States (Knox, 1957; Anderson & White, 1958; Grossman et al., 1959; Hutcheson et al., 1959; Glenn, 1960; Yassaglou & Whiteside, 1960; Jha & Cline, 1963; Nettleton et al., 1968; Petersen et al., 1970; Miller et al., 1971; Ritchie et al., 1974) have reported a mixed clay mineralogy for fragipans. Lozet and Herbillon (1971) reported similar findings on Belgian fragipans. Illite (or mica), or some intergradient combination of illite, vermiculite, or chlorite has been most commonly reported as the dominant clay. Hutcheson et al. (1959) and Glenn (1960) reported more smectite than any other one clay mineral. Glenn (1960) and Nettleton et al. (1968) reported evidence for appreciable amorphous material. Small amounts of kaolinite were often detected.

Fragipans often contain illite that exhibits little weathering in contrast to more weathered vermiculite or hydroxy interlayered vermiculite in overlying horizons. Some fragipans contain considerable silt- and sand-size feldspar, but others are almost devoid of any weatherable minerals. Nettleton et al. (1968) postulated that feldspar was the source of amorphous alumina in the fragipans studied on the North Carolina Coastal Plain.

2. CALCIC AND PETROCALCIC HORIZONS

Calcic horizons are horizons of significant pedogenic carbonate accumulation relative to adjacent horizons. They often are in the substratum, but may be within the solum. Soils containing calcic horizons within the solum are usually on ancient land surfaces where leaching probably once obtained to depths greater than the present climate permits. Recalcification of parts of argillic horizons has subsequently taken place.

Petrocalcic horizons are strongly indurated horizons that are cemented primarily with large amounts of $CaCO_3$, sometimes approaching 90% $CaCO_3$ equivalent. Obviously, the mineralogy of petrocalcic horizons and many calcic horizons is dominated by carbonate minerals.

Numerous unpublished analyses indicate that the noncarbonate mineralogy of both calcic and petrocalcic horizons is usually very similar to that of the overlying horizons in the absence of lithologic discontinuities. X-ray diffractograms of phyllosilicate clays extracted from calcic and petrocalcic horizons sometimes show better defined peaks than overlying noncalcic horizons; however, other measurements such as CEC of the clay usually do not indicate pronounced mineralogical differences.

Calcic horizons may contain measurable gypsum or more soluble salts and accumulations of secondary silica minerals sometimes occur in petrocalcic horizons.

3. GYPSIC AND PETROGYPSIC HORIZONS

Gypsic horizons are those that have significant accumulations of gypsum relative to adjacent horizons. Petrogypsic horizons are those that are distinctly indurated because of accumulation of gypsum in large amounts. When developed in highly gypsiferous parent sediments, pedogenically derived gypsic horizons may be difficult to distinguish from the underlying gypsite.

Gypsic and petrogypsic horizons are analagous to calcic and petrocalcic horizons in that their very nature is somewhat dependent upon mineralogy, i.e. gypsum content. Some gypsic horizons may have a relatively small gypsum content if their thickness is sufficient, but many have large amounts. Petrogypsic horizons, although little studied in the United States, usually have at least 60% gypsum and may have as much as 85%. Either of the horizons may have significant quantities of salts more soluble than gypsum, or carbonates, or both. However, the maximum accumulation of either often does not coincide with the most gypsiferous horizon in the profile. Limited analyses of the clay mineralogy of gypsic and petrogypsic horizons suggest little mineralogical difference between them and nongypsiferous horizons.

4. DURIPANS

Duripans are indurated horizons that are primarily cemented with secondary silica, mostly chalcedony and opal. Carbonates and Fe oxides are often present as accessory minerals (Soil Survey Staff, 1975). Despite a considerable number of analyses of overlying horizons, practically no data have been published on the phyllosilicate mineralogy of the pans per se. Duripans commonly occur in soils that now contain, or have contained, considerable volcanic glass.

5. SALIC HORIZONS

Salic horizons have accumulations of salts more soluble than gypsum. Chlorides and sulfates of Na, K, and Mg are the principal salts. They occur under a wide variety of conditions; many are in deserts of continental interiors, but some are in soils near seacoasts. Most studies of soil salts have been directed toward determining the ionic composition of solution extracts and almost no data are available concerning the mineral species present. Perhaps many of the same minerals, halite (NaCl), mirabilite ($Na_2SO_4 \cdot 10\ H_2O$), thenardite (Na_2SO_4), bloedite [$Na_2Mg(SO_4)_3 \cdot 4\ H_2O$], epsomite ($MgSO_4 \cdot 7\ H_2O$), and hexahydrite ($MgSO_4 \cdot 6\ H_2O$), identified by Driessen and Schoorl (1973) from surface soil efflorescences in Turkey by X-ray diffractometry, also occur in salic horizons.

Salic horizons often contain considerable quantities of gypsum and carbonates. A silicate clay mineralogy similar to that of nonsalic horizons in the profile and of associated nonsaline soils is indicated by the few analyses performed on salic horizons.

6. SULFURIC HORIZONS

Sulfuric horizons are extremely acidic and the presence of jarosite [$KFe_3(OH)_6SO_4)_2$] imparts distinctive yellowish mottles.

7. ALBIC HORIZONS

Albic horizons are eluvial horizons in which the color (hue and chorma) is primarily dependent on the color of uncoated, often highly siliceous, sand and silt particles. Color value depends upon the amount (and physicochemical state) of the humus relative to the mineral constituents. Albic horizons are typically underlain by argillic or spodic horizons.

Few studies have been directed toward the phyllosilicate mineralogy of albic horizons per se. Most studies have been primarily of comparisons of the mineralogy with that of the associated illuvial horizon.

Brown and Jackson (1958); Franzmeier et al. (1963); and Coen and Arnold (1972) reported well-crystallized smectite to be dominant in the eluvial horizons associated with spodic horizons. Ross and Mortland (1966) and Brydon et al. (1968) found evidence of some interstratification in the smectite in similar horizons. Conversely, albic horizons sometime exhibit poorly crystalline smectite relative to that in associated argillic horizons.

8. PLINTHITE

Plinthite is defined as an Fe-rich, humus-poor mixture of clay with quartz and other diluents that usually forms platy, polygonal, or reticulate patterns and which tends to harden irreversibly upon exposure (Soil Survey Staff, 1975). It occurs in soils with wide-ranging mineralogies in such areas as the southeastern United States. Consequently, such minerals as illite, chlorite, and hydroxy interlayered vermiculite, in addition to the common kaolinite, gibbsite, and Fe minerals, can be expected. Few analyses are available on materials fitting the definition of plinthite despite a considerable number reported for other Fe-rich related materials, such as laterites, ferric duricrusts, nodules, concentrations, etc. The lack of data can partially be explained because the definition excludes material already irreversibly hardened. Ahmad and Jones (1969) found kaolinite to be the principal clay-size mineral in plinthite from Trinidad. Smaller amounts of mica and quartz were present. The large amounts of Fe oxide present were X-ray amorphous although goethite and lepidiocrocite were identified in the horizon immediately overlying the plinthic layer. Less kaolinite but more goethite and hematite has been described in the plinthic mottles relative to the matrix in Georgia soils (B. W. Wood & H. F. Perkins. 1974. Some chemical and physical changes during plinthite formation. p. 164. Agron. Abstr.).

III. CATEGORIES OF THE TAXONOMY

Ten orders, the highest category, in the classification are distinguished on the basis of diagnostic horizons or features. When all known soils are classed into only 10 orders there necessarily must be considerable heterogeneity in each except for the properties that have been used to separate them. Five categorical levels of decreasing rank below the order are recognized in the system. They are suborder, great group, subgroup, family, and series. The series concept has been essentially retained from earlier classifications although many new taxa have been established and others modified extensively.

A. Orders

Inorganic (mineral) materials predominate in Alfisols, Aridisols, Entisols, Inceptisols, Mollisols, Oxisols, Spodosols, Ultisols, and Vertisols. Organic materials are dominant in Histosols. Mineralogy is not definitive for the orders; nevertheless, some orders, e.g., Oxisols and Vertisols, are defined so that only soils with selected mineralogies occur (Table 22–1). The intent in describing the orders here is to convey to the reader a "central concept" of the taxon. Obviously, a definitive description would be too detailed for presentation. The reader interested in more precise descriptions is referred to *Soil Taxonomy* (Soil Survey Staff, 1975).

1. ENTISOLS

Entisols are weakly formed soils and an ochric epipedon is generally the only diagnostic horizon present. An albic horizon may be present in some sandy materials. Certain diagnostic horizons, including argillic and spodic

Table 22-1. Distribution of mineralogy families among soil orders in the United States, Puerto Rico, and the Virgin Islands†

| Soil order | Applied to any particle size class | | | | | | Applied to sandy, sandy skeletal, loamy, or loamy skeletal particle size classes | | | Applied to clayey particle size classes; determined on <0.002-mm fraction | | | | | | |
| | Whole soil (<2 mm or <20 mm), whichever has higher total of carbonates plus gypsum | | Whole soil (<2 mm) | | | | 0.02- to 20-mm fraction | 0.02-to 2-mm fraction | | | | | | | | |
	Carbonatic	Gypsic	Ferritic	Oxidic	Serpentinitic	Glauconitic	Micaceous	Siliceous	Mixed	Halloysitic	Kaolinitic	Illitic	Smectitic	Vermiculitic	Chloritic	Mixed
Alfisols	5			2	4			79	330		13	21	101	4		152
Aridisols	44	5							335		1	2	90			38
Entisols	37	7		1	3		1	25	501		1	2	72			41
Inceptisols	21			6	5		6	35	491	1	9	10	37			75
Mollisols	67						1	9	810	1	12	7	292	3	1	182
Oxisols			3	8							6					
Spodosols				1				28	133							2
Ultisols	1		1	21			3	103	56		26	3				52
Vertisols						3					2		57			9

† Does not include families of Histosols or of mineral soil orders where mineralogy is used as differentiae among taxa above the family. (Source: Soil Survey Staff, 1975; unpublished material).

horizons, may occur at such depths that they are not considered significant for most uses or else they are considered to be buried horizons.

Entisols occur on many deposits of light-colored alluvium and are extensive on eolian deposits where development has either been inhibited because of the lack of weatherable minerals, or of insufficient time, or both. Many soils developing on consolidated rock have epipedons that are too thin or too light in color, or have sola that are too weakly developed to be placed in any of the other orders. Entisols are widespread, but are more extensive in the mountainous west, the western Great Plains, and poorly drained areas of the southeastern United States.

The mineralogy of Entisols is very variable and primarily reflects that of the parent material. However, significant quantities of authigenic minerals may be present in the thin sola of Entisols developing in bedrock residuum.

2. INCEPTISOLS

Inceptisols are more strongly developed than Entisols, as evidenced by the presence of a cambic horizon, but they are less developed than soils of most of the other orders. Essentially, they have a subsoil, primarily distinguished by color or structure, but they do not have one with appreciable illuviation. The epipedon is usually ochric or umbric. The definition of Inceptisols is, of necessity, somewhat complex because of the great variety of soils included.

The lack of development may be due to a number of factors including a seasonally high water table, strongly sloping topography, young land surfaces, or episodic deposition of materials, e.g., volcanic ash.

Inceptisols are extensive, but are more common in the Pacific Northwest, Alaska, Hawaii, the Appalachian region, and the broad alluvial plains of the southern states of the USA. They are restricted from aridic climates.

The varied mineralogies found in Inceptisols (Table 22–1) are mostly inherited from parent materials except where they have developed in volcanic ejecta. However, alteration of some minerals, e.g., chlorite and illite, has been pronounced in some soils.

3. SPODOSOLS

Spodosols have a spodic horizon or, in a few soils, a cemented placic horizon overlying a fragipan. Many have pronounced horizonation. Most unplowed Spodosols have an albic horizon and are strongly acidic. In many cases the albic (A2) horizon directly underlies an organic (O) horizon. Ochric, histic, or umbric epipedons are generally present. A fragipan or a lower sequum with an argillic horizon may be present below the spodic horizon. Spodosols have mostly formed in coarse-textured quartzose sediments on late Pleistocene or Holocene surfaces.

Spodosols occur mostly in cool humid climates, but they may be present in warm humid or even tropical climates. Some have a seasonally high ground water table. Although they form under a variety of vegetation types, they seem to be most extensive under coniferous forests and heaths. In the United States, Spodosols are most extensive in the northern Lake states, New England, New York, the Atlantic Coastal Plain, the Pacific Northwest, and Alaska.

4. ALFISOLS

Alfisols have argillic horizons and hence exhibit appreciable horizon differentiation. An essential part of the Alfisol concept is absence of strong leaching as reflected by a moderate to high base saturation status ($>$ 35% at specified depths) depending on the morphology of the profile. With very few exceptions, they have ochric epipedons.

The high base status of Alfisols may be due to one or more of several factors. In subhumid and semiarid regions they often occur on old land surfaces, but still have a high base status because precipitation is not sufficient to leach bases from the profile. Well-drained soils in humid regions may still retain moderate to high amounts of bases because the parent materials were strongly basic, or the land surfaces on which the soils occur are relatively young, or both. Other soils in humid regions have a high base status because of restricted drainage. Most Alfisols, except those that are poorly drained, in humid climates, and many in subhumid regions, have albic horizons. Such horizons in semiarid regions are uncommon.

Alfisols are most extensive in the lower Midwest, the upper South, the subhumid and semiarid Southwest, and California. By definition, Alfisols are restricted from aridic moisture regimes. Some Alfisols are calcareous throughout the argillic horizon and a few even have carbonatic mineralogy (Table 22–1). In such soils an argillic horizon likely developed in a more moist climate than that prevailing today. Carbonates were subsequently deposited by eolian processes on the surface and have since been translocated into the argillic horizon.

5. ULTISOLS

Ultisols are similar to Alfisols in morphology; however, they are more strongly leached as evidenced by lower base status ($<$ 35%) at specified depths within the profile. They have a diagnostic argillic horizon or a fragipan that also meets the requirements of an argillic horizon. They mostly have ochric or umbric epipedons and many have albic horizons.

Ultisols are mostly on relatively old, stable surfaces in a variety of moisture regimes except aridic. They are most extensive in the Southeast, but significant areas are present in California, Oregon, and Washington.

6. MOLLISOLS

Mollisols, an extensive order in the United States, have as their primary diagnostic horizon a mollic epipedon. Although mollic epipedons may occur under restricted conditions in some other orders, they are the unifying feature of Mollisols. They may have argillic, natric, or cambic subsurface horizons, or they may be devoid of any diagnostic subhorizon, especially when overlying bedrock at shallow depths. They often have calcic horizons, and some are underlain by petrocalcic horizons or duripans.

Mollisols are the ubiquitous dark, high base status soils of the grasslands, yet they may have forest, savannah, or even mixed grass-desert shrub vegetation. They may occur in any moisture regime, but in aridic regimes they usually are in positions that receive runoff water. Most in humid forested areas are poorly drained.

They are most extensive in the central Midwest and the Great Plains from Canada to West Texas; however, they occur to some extent in almost all of the United States.

7. ARIDISOLS

Aridisols are soils with aridic moisture regimes and have more mature profiles than Entisols. (Moisture regimes more moist than aridic are permitted if there is a saline layer as determined by conductivity of the saturation extract.) They have ochric epipedons and most have an argillic, natric, or cambic horizon. Calcic, petrocalcic, gypsic, petrogypsic, salic, or duripan horizons may be present.

Most Aridisols with argillic horizons, especially those underlain by petrocalcic horizons, are on old land surfaces. Others, however, are on such recent surfaces that only weakly developed subhorizons, e.g., cambic or calcic, are present.

Vegetation is mostly desert shrub or, in more moist situations, short grass. Aridisols are most extensive in the intermountain West and the drier parts of the Plains states.

8. OXISOLS

Oxisols have an oxic horizon or plinthite that forms a continuous phase within prescribed depth limits. Most have ochric epipedons, but some have umbric epipedons. Oxisols occur on old land surfaces in tropical or subtropical regions that have moisture regimes ranging from aridic to perhumid. Those in drier climates perhaps formed under previously higher rainfall conditions. Because of great thickness in a deeply weathered regolith and the difficulty in determining a lower boundary, a 2-m arbitrary depth is used for classification purposes.

Oxisols are only known to occur in Hawaii in the United States and the classification has been developed primarily with soils from there and Puerto Rico in mind. They are extensive in South America, Africa, and southeast Asia. As more information becomes available from other regions, it is likely that different groupings will be needed and that new subclasses will be added.

9. VERTISOLS

Vertisols, often referred to as "cracking clays," differ from the order previously described because placement in the order does not depend upon diagnostic horizons. Instead, other criteria such as the extent of cracking, gilgai microrelief, slickensides, and wedgeshaped peds within prescribed depths are used.

The central concept of Vertisols is one of moderately deep to deep soils with sufficient clay that has enough expandability to meet the designated criteria. Since by definition they must have > 30% clay, only clayey families are recognized.

Although not diagnostic, most Vertisols have mollic epipedons. Some, mostly in drier regions, have ochric epipedons. Horizonation is weakly developed because of the mixing process. Some have sufficient carbonate accumulation in the substratum to qualify as calcic horizons, and a few may have gypsic horizons.

Vertisols may occur under varied moisture regimes, but the greatest concentration is in the drier zones of humid regions and in subhumid areas. A dry season seems to be essential to their development. They also are present in semiarid and arid regions, but primarily in topographic positions that receive additional water.

Predictably, most Vertisols are dominantly smectitic (Table 22-1). Nevertheless, there are some that have a mixed clay mineralogy and still meet the necessary morphological criteria.

The greatest extent of Vertisols is in Texas. Smaller areas are present in Mississippi, Alabama, Oklahoma, Arizona, California, and Hawaii.

10. HISTOSOLS

Histosols are those in which organic materials play a dominant role. Because horizons in the sense used for mineral soils are not ordinarily present, differentiae other than diagnostic horizons are used. Specific criteria for organic C content depend on clay percentage; soils with more clay must have more organic C to meet the criteria. There are also specific depth and thickness requirements.

Most Histosols, especially in the United States, are saturated with ground water most of the year, unless they have been drained. They are commonly called *peats* or *mucks*. Some, formed in very humid, cool climates, may blanket the landscape and are referred to as *high moors, blanket bogs,* or *raised peats* (Soil Survey Staff, 1975). They usually overlie consolidated or fragmented rock. Histosols generally have very low bulk densities and more than half by volume is organic material. Since most Histosols have been influenced by a high water table, they occur in a variety of climates.

The classification of Histosols is still tentative and some of the differentiae are subject to change. A somewhat different approach has been taken in the development of mineralogy families from that followed in the other orders. "Materials" rather than "minerals" are used in some cases in designating families as discussed later. Mineralogy families seem to be appropriate for only selected subgroups.

B. Suborders, Great Groups, and Subgroups

Forty-seven suborders, second highest category in the taxonomy, have been defined. Forty-four have been recognized in the United States (Soil Survey Staff, 1975; unpublished material). Differentiae used to distinguish suborders, primarily those considered to be of special importance to plant growth and those that reflect a certain degree of genetic homogeneity, are not applied uniformly in each of the orders.

Soil climate, especially moisture, has been used extensively as a differentiating criterion for the suborders. For example, four Alfisol suborders (Aqualfs, Udalfs, Ustalfs, and Xeralfs) are distinguished on the basis of soil moisture characteristics. The fifth suborder (Boralfs) is defined by soil temperatures. All Vertisols are distinguished at the suborder level on the basis of the season when cracking occurs and the amount of time the cracks remain open, properties that primarily reflect moisture status. In other orders, e.g., Spodosols, Entisols, and Inceptisols, moisture status has been used less extensively. Characteristics of the spodic horizon are primarily used as distinguishing criteria in Spodosols. Entisols, it should be remembered, have minimal development and may be present in any climate. Primarily, differentiae that reflect reasons for lack of development are used. A variety of criteria, including the dominance of amorphous materials in one suborder (Andepts), is used as differentiae in Inceptisols. Moisture status is not used as a criterion in Aridisols, since by definition all (except some highly saline ones) have limited soil moisture most of the year. Hence, other features, i.e., an argillic (or natric) horizon vs. a cambic horizon, are used to distinguish the two suborders (Argids and Orthids).

Mineralogy is only used directly as a criterion for one suborder, the aforementioned Andepts. Mineralogical limitations are inferred in definitions of additional suborders, e.g., Rendolls (high $CaCO_3$ equivalent in or immediately below a mollic epipedon) and Ferrods (high free Fe/C ratios).

Great groups compose the third category of the taxonomic system. Approximately 225 great groups have been defined and 185 recognized in the United States (Soil Survey Staff, 1975). The assemblage of horizons, as reflected by the whole soil profile, forms the basis for distinguishing great groups. More horizons, along with additional morphological features and chemical properties, are considered than the few used as differentiae in the two higher categories. One example will suffice to illustrate. Ustalfs, the Alfisols of subhumid and semiarid areas, are divided into six great groups on the basis of morphological properties: (i) Durustalfs, those with a duripan; (ii) Plinthustalfs, those containing plinthite; (iii) Natrustalfs, those having a natric horizon; (iv) Paleustalfs, those with a morphology suggesting advanced maturity; (v) Rhodustalfs, those redder than a designated limit; and (vi) Haplustalfs, those without any of the other differentiae. (Durustalfs and Plinthustalfs are not known to occur in the United States.)

Where moisture characteristics are not used to define the suborder, they are often used as differentiae at the great group level. For example, the Orthent suborder of Entisols is classified into Udorthents, Ustorthents, Xerorthents, and Torriorthents on moisture criteria. Troporthents have both moisture and temperature limits and the Cryorthents are distinguished by soil temperature only.

Mineralogy is used directly as a distinguishing criterion for two great groups, Andaquepts (amorphous) and Quartzipsamments (quartz and other highly resistant minerals). Mineralogy enters into the definition, indirectly and by inference, in a number of other great groups, e.g., Calciaquolls (very high $CaCO_3$ equivalent), Gypsiorthids (high gypsum content in selected parts of the profile), Gibbsiorthox (high gibbsite), Sideraquods (high Fe accumulation relative to humus in a spodic horizon), and Acrorthox (extremely low CEC, indicating very low amounts of phyllosilicates).

Somewhat $> 1,000$ subgroups have been defined (Soil Survey Staff, 1975) and approximately 975 have been identified in the United States and Puerto Rico (Soil Survey Staff, 1975; unpublished material). They are defined on the basis of a typic member, i.e., the taxon that fits the central concept of the class. Other soils in the subgroups have, in addition to features diagnostic of the class, characteristics indicative of processes that are dominant in another great group. That is, they have properties that are transitional to other orders, suborders, great groups, or even to "not-soil" materials. Paleustalfs may be used as an example to illustrate the concept. Besides the typic subgroup, there are 14 other subgroups of Paleustalfs. There are, among others, such diverse soils as Aquic Paleustalfs (having wetness mottles, suggesting a transition to Aqualfs), Calciorthidic Paleustalfs (transitional to the Calciorthid great group of Aridisols), Petrocalcic Paleustalfs (a variation from the typic subgroup because of a petrocalcic horizon), and Udertic Paleustalfs (transitional to Vertisols typical of humid climates).

Mineralogy enters directly into definitions of andic, andeptic, and andequeptic (all strongly influenced by amorphous materials) subgroups. Mineralogical influences are indicated in the naming of many subgroups, e.g., calcic (containing considerable $CaCO_3$), petrocalcic, duric (having a duripan), sulfic (sulfidic materials), oxic (clayey soils with low CEC), petroferric (having an ironstone layer), and quartzipsammentic (sandy with high amounts of quartz).

C. Families

Approximately 5,000 families (Soil Survey Staff, 1975; unpublished material) comprise the fifth category of the taxonomy. Some subgroups contain many families, but others contain only one. The primary intent in developing the family category has been to group soils with similar responses to management, and to some extent, for engineering and related uses. The groupings are empirical; little consideration has been given to processes responsible for the properties used in making the groupings. It is at the family level that mineralogy enters into the classification of almost all mineral soils unless it has been used as a criterion in a higher category, e.g., Andepts, Quartzipsamments, etc. Particle size and soil temperature are the other differentiae used for family placement of most soils. Additional criteria, e.g., depth and reaction, are used for some taxa.

The properties used for family placement are those of a selected part of the profile called the "control section." The part of the soil profile considered in mineral soils for determination of both particle size and mineralogy classes depends on profile morphology and the nature of the underlying material. In shallow soils, those underlain by either hard or soft rock, a fragipan, duripan, petrocalcic horizon, or permafrost at a depth of < 25 cm, the control section is the entire soil profile from the surface to the contact with the restricting layer. In deep soils that do not have an argillic (or natric) horizon the control section extends from 25 cm (or base of an Ap horizon, if present) to a 1-m depth unless one of the restricting layers, identified above, is present, in which case it extends only from 25 cm to the top of the layer. In soils that have an argillic (or natric) horizon, the control section is the upper 50 cm of the horizon if > 50 cm thick; if < 50 cm, the control section is the entire horizon. Many variations from these criteria occur and the reader is referred to Soil Survey Staff material (1975) for more specific information.

Because considerable variation in particle size distribution may occur within the control section, the class named is a weighted average except in strongly contrasting particle size classes as defined by the Soil Survey Staff (1975). In such cases, both classes are named.

Eleven particle size classes are used to determine family placement of most mineral soils, but only seven are used in selected subgroups. Fragmental, sandy-skeletal, loamy-skeletal, clayey-skeletal, sandy, loamy, and clayey comprise the seven classes. The loamy class is often subdivided into coarse-loamy, fine-loamy, coarse-silty, and fine-silty families, and the clayey class may be subdivided into fine and very fine families. Precise definitions and limits for each class may be obtained from Soil Survey Staff material (1975).

Approximate mineralogical composition of selected size fractions of the same control section considered for particle size modifiers is used for mineralogy family determination. In soils with layers of strongly contrasting particle size classes the mineralogy family is defined by the composition of the upper layer only. A key for determination of mineralogy families in mineral soils is given in Table 22-2. Since it is a key, it must be used as such, i.e. a soil is placed in the first class for which the criteria are met. It is evident that mineralogy family placement can only be done after the particle size class is determined.

Substitute terms that connote both texture and mineralogy as defined by the Soil Survey Staff (1975), are used for some taxa, e.g., the Andept suborder, as well as the Andaquept and Quartzipsamment great groups. The terms are also used for selected taxa among andic, andaqueptic, andeptic, and cryic subgroups. "Cindery" and "ashy" serve as substitute terms at the family level. Mineralogy classes are not named in the Calciaquoll great group because carbonates overshadow other mineralogical variations.

The approximate distribution of mineralogy families by soil order is shown in Table 22-1 (Soil Survey Staff, 1975; unpublished material). It is emphasized that numbers in the table apply to families, not series. Although some families contain only one series, many contain a number of series. The reader is cautioned that the numbers in Table 22-1 are only approximate and are subject to change as more data become available and as new families are added. Insufficient mineralogical data have precluded precise placement of many soils. Most soils have been placed by extrapolation of data from soils with similar physical and chemical properties of which the mineralogy is

Table 22-2. Dichotomous key to mineralogy families†

No.	Description	Family
1a.	Whole soil, <2mm or <20 mm, whichever has the higher total of $CaCO_3$ plus gypsum; $>40\%$ $CaCO_3$ plus gypsum (2)	
1b.	Whole soil, <2 mm or <20 mm, whichever has the higher total of $CaCO_3$ plus gypsum; $<40\%$ $CaCO_3$ plus gypsum (3)	
2a.	$CaCO_3$ is $>65\%$ of sum of $CaCO_3$ and gypsum	Carbonatic
2b.	$CaCO_3$ is $<65\%$ of sum of $CaCO_3$ and gypsum	Gypsic
3a.	Whole soil (<2 mm); more than 40% Fe oxide expressed as Fe_2O_3 (or 28% Fe) extractable by citrate-dithionate	Ferritic
3b.	Less than 40% Fe_2O_3 (4)	
4a.	Whole soil (<2 mm); $>40\%$ hydrated Al oxides (gibbsite and boehmite)	Gibbsitic
4b.	Less than 40% hydrated Al oxides (5)	
5a.	Less than 90% quartz; $<40\%$ any other mineral listed below in the table; and the ratio: ‡ $$\frac{\text{extractable } Fe_2O_3\ (\%) + \text{gibbsite } (\%)}{\text{clay } (\%)\S} \geqslant 0.2$$	Oxidic
5b.	Not as above (6)	
6a.	Whole soil (<2 mm); $>40\%$ serpentine minerals (antigorite, chrysotile, etc.) and talc	Serpentinitic

(continued on next page)

Table 22-2. Continued

No.	Description	Family
6b.	Less than 40% serpentine minerals and talc (7)	
7a.	Whole soil ($<$2 mm); $>$40% glauconite	Glauconitic
7b.	Less than 40% glauconite (8)	
8a.	Less than 35% clay (sandy, sandy-skeletal, loamy, and loamy-skeletal particle size classes) (9)	
8b.	More than 35% clay (clayey particle-size classes)¶ (11)	
9a.	0.02- to 20-mm fraction; $>$40% mica (estimated from grain counts)	Micaceous
9b.	0.02- to 20-mm fraction; $<$40% mica (estimated from grain counts) (10)	
10a.	0.02- to 2-mm fraction, $>$90% silica minerals (quartz, chalcedony, opal, etc.), and other minerals extremely resistant to weathering (estimated from grain counts)	Siliceous
10b.	0.02- to 2-mm fraction; other soils	Mixed
11a.	Clay fraction ($<$0.002 mm); $>$50% halloysite (tubular forms) and with smaller amounts of allophane and/or kaolinite	Halloysitic
11b.	Not as above (12)	
12a.	Clay fraction; $>$50% kaolinite, nacrite, dickite, and tabular halloysite; and with smaller amounts of other 1:1 or non-expanding 2:1 minerals, or gibbsite; and $<$10% smectite	Kaolinitic
12b.	Not as above (13)	
13a.	Clay fraction; $>$50% smectite or more smectite than any other clay mineral	Smectitic
13b.	Not as above (14)	
14a.	Clay fraction; $>$50% illite (hydrous mica); commonly $>$4% K_2O	Illitic
14b.	Not as above (15)	
15a.	Clay fraction; $>$50% vermiculite or more vermiculite than any other clay mineral	Vermiculitic
15b.	Not as above (16)	
16a.	Clay fraction; $>$50% chlorite or more chlorite than any other clay mineral	Chloritic
16b.	Not as above	Mixed

† Adapted from Soil Survey Staff (1975). The parenthesis and numbers in the key indicate the section to which the reader must progress. For example, after testing section 1, the reader progresses to either section 2 or 3, depending on whether the soil has more than or less than 40% $CaCO_3$ plus gypsum. The size fraction on which the mineralogy is determined is given at the outset of the class description. All percentage figures are by weight.

‡ For the class the 0.02- to 2-mm fraction for quartz and other minerals is used; for the ratio of Fe oxide and gibbsite to clay, the whole soil ($<$2 mm) is used.

§ Percentage of clay or percentage of 15-bar water times 2.5, whichever is greater, provided the ratio of 15-bar water to clay is 0.6 or more in at least one-half of the control section.

¶ A clay content of only 30% is needed to qualify as "clayey" in Vertisols.

known. As indicated in Table 22-1, mineralogy family placement depends upon the particle size class of the soil control section.

A number of families are rare, e.g., ferritic and glauconitic (applied on a whole soil basis) as well as halloysitic and chloritic (determined on the < 0.002-mm fraction). On the other hand, mixed mineralogies are exceedingly common in both coarser (sandy, sandy-skeletal, loamy, and loamy-skeletal) particle size classes (the mineralogy of the 0.02- to 2-mm fraction of the control section used for family placement) and the clayey particle size class (the < 0.002-mm fraction of the control section used for family placement). Although a gibbsitic class is provided in the key (Table 22-2), no gibbsitic families have been identified in the United States and, consequently, are omitted from Table 22-1.

Certain distribution trends among mineralogy families are evident (Table 22-1). Carbonatic mineralogy occurs only in orders that are in drier regions and gypsic mineralogy has only been described from two orders, Aridisols and Entisols. In the latter, gypsic families have only been identified in aridic moisture regimes. The number of families having oxidic mineralogy in Ultisols is surprisingly large. Some pedologists believe the criteria for the class should be altered because the oxidic mineralogy does not properly indicate the behavior of many soils that are now so classified.

Siliceous mineralogy occurs widely, but the greatest number is in the intensively weathered Ultisols. The class is excluded from Vertisols because only clayey particle size classes are therein recognized, whereas siliceous families are restricted to the coarser size classes (Table 22-1). Siliceous mineralogy has not been described in Aridisols, apparently because of low weathering intensity. All established Oxisol series in the United States have been described as having clayey control sections, thus siliceous mineralogy has been excluded.

Kaolinitic mineralogy families, as would be expected, are more common in Ultisols. Although kaolinite is the most common silicate clay in Oxisols, the organization of the key (Table 22-2) permits many families to fall into the oxidic class. Probably most of the kaolinitic mineralogies in Alfisols, Inceptisols, and Mollisols are inherited; however, several of the families are in intergradient ultic and oxic subgroups. These subgroups may have considerable authigenic kaolinite. Kaolinitic mineralogies in two Vertisol families, both in Hawaii, would seem to be anomalous. Illitic and smectite families are confined to orders where high weathering intensity does not obtain except for three illitic families, each having only one series, in Ultisols.

The mixed mineralogy class for clayey (> 35% clay) soils is perhaps defined too broadly. For example, a soil with roughly equal amounts of illite, smectite, and vermiculite will have considerably different properties from a kaolinitic-illitic soil containing a minor smectite component. Smith and Wilding (1972) proposed a division into "mixed, expandable" and "mixed, nonexpandable" classes because of the wide range of the mixed class as it is now defined. Clayey mixed families occur in all orders (Table 22-2) except in Oxisols. However, only two families of Spodosols, each containing only one series, and nine families of Vertisols, only one containing more than one series, are so listed. At first it may seem anomalous for Vertisols to have a mixed mineralogy, yet it is possible for pronounced movement to take place, especially in very clayey soils, but still not have enough smectite to meet the criteria for that family.

Perhaps other mineralogy classes will in time be added. Although sepiolitic families are not now listed in the key, they should be named if found (Soil Survey Staff, 1975). Soils in which the clay fraction is dominated by sepiolite, or palygorskite, or both, almost surely exist in the semiarid and arid Southwest (McLean et al., 1972) and possibly in the Southeast (Fiskell & Perkins, 1970). However, their extent and identification by series are presently unknown.

Somewhat different criteria from those used for mineral soils are employed in determination of family classes in Histosols. The control section of Histosols has been arbitrarily defined as being either 130 cm or 160 cm thick depending on the type of material (percentage and kind of fibers) if there is no restricting layer, e.g., water or consolidated rock, within that depth (Soil Survey Staff, 1975). When a restricting layer is present, the control section extends only to the top of that layer.

Four mineralogy families; ferrihumic, coprogenous, diatomaceous, and marly, are recognized. Perhaps some other term besides "mineralogy" would be better since a variety of materials are included. *Ferrihumic* families contain at least 2% (by weight) of "bog iron" (hydrated Fe oxides with variable amounts of organic material). *Coprogenous, diatomaceous,* and *marly* classes are used when limnic materials (mostly inorganic precipitates or deposits produced by the activity of aquatic organisms) are at least 5 cm thick in the control section. Coprogenous sediments, relatively high in organic matter, contain many fecal pellets. Diatomaceous materials have intermediate color values and are composed mostly of diatoms, often coated with organic matter. Marls are light-colored sediments that have a high carbonate content.

Family modifiers are not used for all great groups and subgroups of Histosols. Also, the same mineralogy families used for mineral soil orders are used for some Histosol subgroups (Soil Survey Staff, 1975).

Little attention has previously been given in the United States to the mineralogy of the inorganic fraction in materials classed as Histosols except where some undesirable property, e.g., pronounced acidity, could be traced to a specific mineral. However, the mineral fraction of Histosols will surely receive increasing future attention.

The construction of family names, as well as the relationship among different categorical levels of the mineral soil orders is illustrated by the classification of selected soils (Table 22-3). As explained earlier, mineralogy classes would not be listed for Quartzipsamments (Entisols) and Andepts (Inceptisols). The term "coated" is used in Quartzipsamment families if there are sufficient fine particles to give a moisture equivalent to $\geq 2\%$. The term "medial" in the Andept family name is one of the substitute terms referred to earlier.

Table 22-3. Classification of selected soils

Soil series	Order	Suborder	Great group	Subgroup	Family
Amarillo	Alfisol	Ustalf	Paleustalf	Aridic Paleustalf	Fine loamy, mixed, thermic
Au Gres	Spodosol	Aquod	Haplaquod	Entic Haplaquod	Sandy, mixed, frigid
Decatur	Ultisol	Udult	Paleudult	Rhodic Paleudult	Clayey, kaolinitic, thermic
Clarion	Mollisol	Udoll	Hapludoll	Typic Hapludoll	Fine loamy, mixed, mesic
Houston Black	Vertisol	Ustert	Pellustert	Udic Pellustert	Fine, smectitic, thermic
Lozier	Aridisol	Orthid	Calciorthid	Lithic Calciorthid	Loamy-skeletal, carbonatic, thermic
Lakeland	Entisol	Psamment	Quartzipsamment	Typic Quartzipsamment	Thermic, coated
Matanzas	Oxisol	Orthox	Eutrorthox	Tropeptic Euthrothox	Clayey, oxidic, isohyperthermic
Waimea	Inceptisol	Andept	Eutrandept	Typic Eutrandept	Medial, isothermic

LITERATURE CITED

Ahmad, N., and R. L. Jones. 1969. A Plinthaquult of the Aripo savannas, North Trinidad: II. Mineralogy and genesis. Soil Sci. Soc. Am. Proc. 33:765–768.

Allen, B. L., and D. W. Goss. 1974. Micromorphology of paleosols from the semiarid Southern High Plains of Texas. p. 511–525. In G. K. Rutherford (ed.) Fourth international working meeting on soil micromorphology. The Limestone Press, Kingston, Ont., Canada.

Anderson, J. U., and J. L. White. 1958. A study of fragipans in some southern Indiana soils. Soil Sci. Soc. Am. Proc. 22:450–454.

Aomine, S., and N. Yoshinaga. 1955. Clay minerals of some well drained volcanic ash soils in Japan. Soil Sci. 79:349–358.

Aomine, S., and K. Wada. 1962. Differential weathering of volcanic ash and pumice resulting in formation of halloysite. Am. Mineral. 47:1024–1048.

Besoain, M. E. 1964. Clay formation in some Chilean soils derived from volcanic materials. New Z. J. Sci. 7:79–86.

Besoain, E. M. 1969. Clay mineralogy of volcanic ash soils. p. B.1.1–B.1.12. In H. E. Fassbender (Coordinator). Panel on volcanic ash soils in Latin America.

Birrel, K. S., and M. Fieldes. 1952. Allophane in volcanic ash soils. J. Soil Sci. 3:156–171.

Brewer, R. 1964. Fabric and mineral analysis of soils. John Wiley & Sons, New York. 470 p.

Brown, B. E., and M. L. Jackson. 1958. Clay mineral distribution in the Hiawatha sandy soils of northern Wisconsin. Clays Clay Miner. 7:213–266.

Bryant, J. P., and J. B. Dixon. 1963. Clay mineralogy and weathering of a Red-Yellow Podzolic soil from quartz mica schist in the Alabama Piedmont. Clays Clay Miner. 12:509–521.

Brydon, J. E., H. Kodama, and C. J. Ross. 1968. Mineralogy and weathering of the clays in Orthic Podzols and other podzolic soils in Canada. Int. Cong. Soil Sci. Trans. 9th (Adelaide, Aust.) III:41–51.

Buol, S., and F. D. Hole. 1959. Some characteristics of clay skins on peds in the B horizons of a Gray-Brown Podzolic Soil. Soil Sci. Soc. Am. Proc. 23:239–241.

Calhoun, F. G., V. W. Carlisle, and C. Luna. 1972. Properties and genesis of selected Colombian Andosols. Soil Sci. Soc. Am. Proc. 36:480–485.

Coen, G. M., and R. W. Arnold. 1972. Clay mineral genesis of some New York Spodosols. Soil Sci. Soc. Am. Proc. 36:342–350.

Cook, M. G. 1973. Compositional variations in three Typic Hapludults containing mica. Soil Sci. 115:159–169.

Cortes, A., and D. P. Franzmeier. 1972. Climosequence of ash-derived soils in the Central Cordillera of Colombia. Soil Sci. Soc. Am. Proc. 36:653–659.

Driessen, P. M., and R. Schoorl. 1973. Mineralogy and morphology of salt efflorescences on saline soils in the Great Konya Basin, Turkey. J. Soil Sci. 24:436–442.

Eswaran, H. 1972. Micromorphological indicators of pedogenesis in some tropical soils derived from basalts from Nicaragua. Geoderma 7:15–31.

Fieldes, M. 1955. Clay mineralogy of New Zealand soils. II. Allophane and related mineral colloids. New Z. J. Sci. Tech. 37:336–350.

Fiskell, J. G. A., and H. F. Perkins (eds.). 1970. Selected Coastal Plain soil properties. Southern Coop. Series Bull. 148.

Franzmeier, D. P., E. P. Whiteside, and M. M. Mortland. 1963. A chronosequence of Pod- zols in northern Michigan. III. Mineralogy, micromorphology, and net changes oc- curring during soil formation. Mich. Agric. Exp. Stn. Quart. Bull. 46:37–57.

Franzmeier, D. P., B. F. Hajek, and C. H. Simonson. 1965. Use of amorphous material to identify spodic horizons. Soil Sci. Soc. Am. Proc. 29:737–743.

Gile, L. H., and R. B. Grossman. 1968. The morphology of the argillic horizon in desert soils of southern New Mexico. Soil Sci. 106:6–12.

Glenn, R. C. 1960. Chemical weathering of layer silicate minerals in loess-derived Loring silt loam of Mississippi. Int. Congr. Soil Sci. Trans. 7th (Madison, Wis.) IV:523–531.

Goss, D. W., and B. L. Allen. 1968. A genetic study of two soils developed on granite in Llano County, Texas. Soil Sci. Soc. Am. Proc. 32:409–413.

Grossman, R. B., I. Stephen, J. B. Fehrenbacher, A. H. Beavers, and J. M. Parker. 1959. Fragipan soils of Illinois: II. Mineralogy in reference to parent material uniformity of Hosmer silt loam. Soil Sci. Am. Proc. 23:70–73.

Hutcheson, T. B., Jr., R. J. Lewis, and W. A. Seay. 1959. Chemical and clay mineralogical properties of certain Memphis catena soils of western Kentucky. Soil Sci. Soc. Am. Proc. 23:474–475.

Jackson, M. L., S. A. Tyler, A. L. Willis, A. L. Bourbeau, and R. P. Pennington. 1948. Weathering sequence of clay-size minerals in soils and sediments. I. Fundamental generalizations. J. Phys. Colloid. Chem. 52:1237–1260.

Jha, P. P., and M. G. Cline. 1963. Morphology and genesis of a Sol Brun Acide with fragi- pan in uniform silty material. Soil Sci. Soc. Am. Proc. 27:339–344.

Khalifa, E. M., and S. W. Buol. 1968. Studies of clay skins in a Cecil (Typic Hapludult) soil: I. Composition and genesis. Soil Sci. Soc. Am. Proc. 32:857–861.

Knox, E. G. 1957. Fragipan horizons in New York soils: II. The basis of rigidity. Soil Sci. Soc. Am. Proc. 21:326–330.

Lepsch, I. F., and S. W. Buol. 1974. Investigations in an Oxisol-Ultisol toposequence in Sao Paulo State, Brazil. Soil Sci. Soc. Am. Proc. 38:491–496.

Le Roux, J. 1973. Quantitative clay mineralogical analysis of Natal Oxisols. Soil Sci. 115:137–144.

Losche, C. K., R. J. McCracken, and C. B. Davey. 1970. Soils of steeply sloping land- scapes in the southern Appalachian Mountains. Soil Sci. Soc. Am. Proc. 34:473– 478.

Lozet, J. M., and A. J. Herbillon. 1971. Fragipan soils of Condroz (Belgium): miner- alogical, chemical, and physical aspects in relation with their genesis. Geoderma 5: 325–343.

McCracken, R. J., R. E. Shanks, and E. C. Clebsch. 1962. Soil morhpology and genesis at higher elevations of the Great Smoky Mountains. Soil Sci. Soc. Am. Proc. 26:384– 388.

McLean, S. A., B. L. Allen, and J. R. Craig. 1972. The occurrence of sepiolite and at- tapulgite on the southern High Plains. Clays Clay Miner. 20:143–149.

Mejia, G., H. Kohnke, and J. L. White. 1968. Clay mineralogy of certain soils of Colum- bia. Soil Sci. Soc. Am. Proc. 32:665–670.

Miller, F. P., L. P. Wilding, and N. Holowaychuk. 1971. Canfield silt loam, a Fragiudalf: II. Micromorphology, physical, and chemical properties. Soil Sci. Soc. Am. Proc. 35:324–331.

Nettleton, W. D., R. J. McCracken, and R. B. Daniels. 1968. Two North Carolina Coastal Plain catenas: II. Micromorphology, composition, and fragipan genesis. Soil Sci. Soc. Am. Proc. 32:582–587.

Nettleton, W. D., K. W. Flach, and B. R. Brasher. 1969. Argillic horizons without clay skins. Soil Sci. Soc. Am. Proc. 33:121–125.

Petersen, G. W., R. W. Ramney, R. L. Cunningham, and R. P. Matelski. 1970. Fragipans in Pennsylvania soils. Soil Sci. Soc. Am. Proc. 34:719–722.

Ritchie, A., L. P. Wilding, G. F. Hall, and C. R. Stahnke. 1974. Genetic implications of B horizons in Aqualfs in northeastern Ohio. Soil Sci. Soc. Am. Proc. 38:351-358.

Ross, G. J., and M. M. Mortland. 1966. A soil beidellite. Soil Sci. Soc. Am. Proc. 30: 337-343.

Smith, H., and L. P. Wilding. 1972. Genesis of argillic horizons in Ochraqualfs derived from fine textured till deposits of northwestern Ohio and southeastern Michigan. Soil Sci. Soc. Am. Proc. 36:808-815.

Soil Survey Staff. 1960. Soil classification, a comprehensive system; 7th approximation. Soil Conservation Service, USDA.

Soil Survey Staff. 1967. Soil survey laboratory data and descriptions for some soils of Puerto Rico and the Virgin Islands. Soil Survey Investigations Report No. 12. Soil Conservation Service, USDA.

Soil Survey Staff. 1975. Soil taxonomy: A basic system of soil classification for making and interpreting soil surveys. Agric. Handb. no. 436. U. S. Government Printing Office, Washington, D. C.

Tamura, R., M. L. Jackson, and G. D. Sherman. 1953. Mineral content of low humic, humic, and hydrol humic Latosols of Hawaii. Soil Sci. Soc. Am. Proc. 17:343-346.

Whittig, L. D., V. J. Kilmer, R. C. Roberts, and J. G. Cady. 1957. Characteristics and genesis of Cascade and Powell soils of northwestern Oregon. Soil Sci. Soc. Am. Proc. 21:226-232.

Whittig, L. D. 1959. Characteristics and genesis of a Solodized-Solonetz of California. Soil Sci. Soc. Am. Proc. 23:469-473.

Yassoglou, N. J., and E. P. Whitside. 1960. Morphology and genesis of some soils containing fragipan in northern Michigan. Soil Sci. Soc. Am. Proc. 24:396-407.

Preparation of Clay Samples for X-ray Diffraction Analysis

CHARLES I. RICH,[1] Virginia Polytechnic Institute and State University, Blacksburg, Virginia

RICHARD I. BARNHISEL, University of Kentucky, Lexington, Kentucky

I. INTRODUCTION

A large number of techniques have been used to prepare samples for X-ray diffraction analysis. The best method is largely a function of the type of sample and what use is to be made of the X-ray diffraction data. The techniques may be classified into two subgroups (i) randomly oriented specimens and (ii) parallel or "preferred" oriented samples. In the latter case the experimenter prefers that the clay minerals be oriented so that the a-b axis is parallel to the plane of the glass slide or ceramic tile. The degree to which this may occur is somewhat dependent on the sample, its pretreatment, and the technique of applying the sample to the mount or supporting media.

In most cases, the experimenter desires proportional or equal representation of the minerals present in the sample to be exhibited in the X-ray diffraction pattern. It is not easy to obtain such diffraction patterns from randomly oriented specimens due to the plate-like nature of phyllosilicates. Achievement of ideal "preferred" orientation of clay particles is not always obtained either for a variety of reasons that will be discussed later. One should always include some method of periodically testing the degree of orientation when preparing specimens for X-ray diffraction analysis by using either an internal standard or standard clay samples. This practice will also lend itself to checking the operation of the X-ray diffractometer over a period of time.

II. SAMPLE PREPARATION

A. Pretreatment

The rationale for pretreatment of whole samples prior to preparing specimens for X-ray diffraction may be attributed to two major problems: (i) some samples may contain a wide distribution of minerals of differing particle sizes, and (ii) some samples may contain cementing agents or compounds that prevent and otherwise restrict the clay particles from achieving a parallel orientation.

---

[1]Dr. Rich contributed to the earlier preparation of the manuscript. He died 18 Sept. 1975 after a long illness.

Organic matter, free Fe oxides, and carbonates are examples of compounds that may be found in samples that, as a result of the questions asked, are often removed prior to doing the X-ray diffraction analysis. Removal of these compounds may be necessary on one hand to achieve solutions to the two major problems listed above, but in doing so, the removal treatments may alter the properties of some of the phyllosilicates in the sample, e.g., Harward et al., 1962; Douglas and Fiessinger, 1971.

Fractionation or separation of a sample into various ranges of particle sizes is commonly useful as a pretreatment technique. Such a range of particle distributions may be restricted to sand > 50 μm, silt 50–2 μm, and clay < 2 μm, whereas in some laboratories these major fractions are further subdivided. This process often concentrates like minerals and allows a person to more easily recognize trace amounts. In order to achieve this separation, removal of cementing agents is often required, and chemical as well as mechanical dispersion is commonly practiced. Potentially any treatment may alter the properties of the minerals, but without such pretreatments, mineral dispersion during particle fractionation may be impossible. Therefore, quantitative identification may also be impossible.

Removal of the cementing agents and X-ray amorphous compounds often aids in identification of minerals present in a specimen by not only increasing better orientation, but also reducing the back scatter of X-rays from the specimen. However, some X-ray diffractometers may be equipped to reduce this effect electronically.

B. Preparation of Clay Specimens

The most diagnostic X-ray diffraction peaks for clay minerals are those from the basal plane, (001). Other reflections, such as the 060 peak and certain prismatic reflections are useful but not as frequently as the 00l reflections. Thus, emphasis in this discussion is given to techniques which enhance the 00l reflections.

The X-ray diffraction patterns of clay minerals, particularly those of expansible layer silicates, are affected greatly by the cation saturating the exchange sites, by the kind and amount of solvating liquid, and by temperature. Diagnosis of the kind of mineral(s) present is based on the extent of swelling along the c* axis (perpendicular to the (001) plane) under conditions favorable for swelling and the extent of contraction under conditions that favor shrinkage along the c* axis.

Several methods have been proposed to orient the clay so that the (00l) planes are parallel to the sample holder. These include smearing the clay on a glass slide with a spatula, Theisen and Harward (1962). A slurry of clay can be sedimented and dried on a glass slide, Jackson (1956). The slurry can be applied to a ceramic tile and the excess water removed by suction or centrifugation, Kinter and Diamond (1956) and Rich (1969).

These methods and others were studied by Gibbs (1965). Methods which relied on sedimentation, either by gravity or by centrifugation, were found to emphasize the reflections of those minerals which settled more slowly than the others. For example, montmorillonite particles because of their small size and thin sheets, settle last and therefore X-ray patterns may have abnormally strong montmorillonite peaks because of the position of the particles in the prepared specimen. Thus X-ray diffraction patterns from samples prepared in such a way are difficult to use for quantitative estimation of mineral species.

The method which is recommended here for mounting specimens is the rapid suction method employing ceramic tiles held in a simple apparatus described by Rich (1969). This technique will be described more fully in a following section. Other methods may be equally good for many other situations as the one to be described, but the use of the ceramic tile-suction technique has other distinct advantages: Ease of ionic sautration, solvation, heating, uniform and known clay film thickness. This latter point will be elaborated upon more fully in a following section.

C. Cation Saturation

Clays are commonly Mg- and K-saturated for diagnostic purposes. Saturation may be accomplished for bulk samples by a centrifuge procedure or done directly on the porous ceramic tiles. Removal of organic matter and carbonates is recommended by methods described by Kunze (1965) if the samples were not so treated prior to particle fractionation. Chloride (N) salts of Mg and K are commonly used to produce respective saturated samples. The salt is mixed with the sample, the clay suspension is centrifuged, and the clear supernatant liquid is decanted. This procedure is repeated four more times and then water is mixed with the sample and centrifuged and the clear supernatant liquid is decanted. This procedure is repeated until the decanted liquid no longer shows the presence of Cl⁻ ions as indicated by adding a few drops of a $AgNO_3$ solution. A high-speed angle centrifuge may have to be employed in the later washes to shorten the time necessary to centrifuge out the clay.

The reason for using the centrifuge method rather than Mg-saturating the clay on the tile is that the Mg-saturated specimen can also be used for other methods such as cation-exchange capacity (CEC), infrared analysis (IR), and differential thermal analysis (DTA). However, one should keep the clay moist until tiles (or slides) are prepared, thus allowing good dispersion and distribution of oriented particles on the tile.

Potassium-saturated specimens (and Mg-saturated if desired) may be prepared directly on the tiles by simply passing adequate amounts (> 40 ml) of the respective $1N$ chloride salt. The excess salts may be removed with an equal amount of ion-free water.

III. SUCTION APPARATUS FOR MOUNTING CLAY SPECIMENS
ON CERAMIC TILE

The suction and centrifuge methods were proposed by Kinter and Diamond (1956) for preparation of clay specimens on unglazed ceramic tile. The centrifuge method requires special holders, and, according to Gibbs (1965), centrifugation or gravity sedimentation of a clay suspension may result in segregation of mineral components, particularly if their size range differs. According to the study of Gibbs, there is little or no segregation with the suction method. With the suction apparatus proposed by Kinter and Diamond, individual specimens may require 15 or more minutes for preparation. The difficulty was traced to air leakage through the sides of the tile. The apparatus illustrated in Fig. 23-1 overcomes the air leakage problem and permits the filtration of 5 ml of a 4% clay suspension in about 1 min.

The tile holder is prepared by cutting a rectangular hole in a size 15 rubber stopper at least as deep as the thickness of the ceramic tile to be used and about 5 mm larger on each edge of the tile. A ceramic tile is then coated with a thin film of stopcock grease and inserted in the recess of the stopper, its flat side up. The cavities along the sides of tile are then filled with liquid silicon rubber, the type commonly used for sealing tile-bathtub joints. A second tile, greased at the edges, is placed, flat side down, on top of the first

Fig. 23-1. Suction assembly showing the recess in the rubber stopper molded in silicon rubber (*white*). A tile is in place.

tile. Additional silicon rubber is added so as to build a sloping wall around the second tile. This allows for sufficient suspension capacity once the preparation of the assembly is completed. The assembly is allowed to set for a few hours and then placed in an oven at 100C (383K) overnight. After the assembly is cooled, the tile are easily removed. A hole for 1 cm diameter is made through the center of the recess and through the bottom of the stopper. A second smaller hole may be made at the side of the recess and fitted with a glass tube for obtaining suction. The stopper is placed in a suitable heavy-walled bottle as illustrated. An alternative method is to place the rubber stopper apparatus in a Büchner funnel of suitable size, mounted on a suction flask.

A. Preparation of Clay Specimens

An unglazed ceramic tile (not greased) is soaked in water and inserted in the suction apparatus and suction applied. A suspension containing at least 225 mg of Mg-saturated clay is pipetted on to the tile. The free water is extracted in about 1 min and other solutions may be passed through the system, e.g., 5 ml of 20% glycerol solution for glycerol solvation. The tile may be removed from the apparatus by a thin spatula being placed between the end of the tile and the silicon rubber after the suction has been released.

If clay specimens are to be saturated after being deposited on the tiles, at least 40 ml of $1N$ $MgCl_2$ (or KCl, etc.) insures complete saturation with the respective cation. The excess salts may be removed by passing at least 20 ml of ion-free water through the system.

The amount of clay deposited on the ceramic tile should be sufficient to prevent diffraction from the crystalline material in the tile, as well as to insure that the relative intensities of low and high angle peaks, between $0°$ and $30°$ 2θ, are not a function of specimen thickness. This thickness factor is equally important if clay specimens are prepared by other methods and/or on glass slides. According to Bradley (1964), "the solid path length should be in the order of 200 microns." Assuming the density of clay to be 2.6, and since the tile used has surface dimensions of 28.6 by 58.7 mm (16.8 cm^2), the amount of clay per tile for various scanning angles should be:

Degrees 2θ	mg
5	38
10	76
15	114
30	225
40	296
60	436

The problem of the amount of clay needed is covered in a later section of this chapter. Since most diffraction patterns obtained with CuKα radiation of soil clays are terminated at $30°$ 2θ, the 225 mg would be sufficient.

B. Optimum Amount of Clay Needed for X-ray Diffraction[2]

The proper amount of clay in a specimen mounted for X-ray diffraction analysis should be adequate not only to prevent diffraction peaks or effects originating from the supporting medium, e.g., the ceramic tile, but also to produce diffraction peaks from minerals in the specimen in their proper relative intensities.

From the basic equations reported by Cullity (1967, p. 271), the following equations can be derived:

$$mg/cm^2 = \frac{-500[\ln(1-Gx)]\sin(2\theta/2)}{\mu^*}$$

where mg/cm^2 is the concentration of clay on a tile or glass slide surface, Gx

---

[2]Condensed from a paper by Rich (1975b).

Table 23-1. Concentrations of clay on a slide surface from which 95%, 99%, and 99.9% of the observed diffraction is produced. These concentrations are calculated for various mass absorption coefficients and for a range of 2θ values (Rich, 1975b)

Degrees 2θ	Mass absorption coefficient						
	30	40	50	75	100	200	250
	—mg/cm^2—						
95% of the observed diffraction							
10	4.5	3.4	2.7	1.8	1.4	0.7	0.5
20	8.7	6.5	5.2	3.5	2.6	1.3	1.0
30	12.9	9.7	7.8	5.2	3.9	1.9	1.6
40	17.1	12.8	10.2	6.8	5.1	2.6	2.0
50	21.1	15.8	12.7	8.4	6.3	3.2	2.5
60	25.0	18.7	15.0	10.0	7.5	3.7	3.0
80	32.1	24.1	19.2	12.8	9.6	4.8	3.8
99% of the observed diffraction							
10	6.7	5.0	4.0	2.7	2.0	1.0	0.8
20	13.3	10.0	8.0	5.3	4.0	2.0	1.6
30	19.9	14.4	12.9	7.9	6.0	3.0	2.4
40	26.2	19.7	15.8	10.5	7.9	3.9	3.2
50	32.4	24.3	19.5	13.0	9.7	4.9	3.9
60	38.4	28.8	23.0	15.4	11.5	5.8	4.6
80	49.3	37.0	29.6	19.7	14.8	7.4	5.9
99.9% of the observed diffraction							
10	10.0	7.5	6.0	4.0	3.0	1.5	1.2
20	20.0	15.0	12.0	8.0	6.0	3.0	2.4
30	29.8	22.3	17.9	11.9	8.9	4.5	3.6
40	39.4	29.5	23.6	15.7	11.8	5.9	4.7
50	48.7	36.5	29.2	19.5	14.6	7.3	5.8
60	57.6	43.2	34.5	23.0	17.3	8.6	6.9
80	74.0	55.5	44.4	29.6	22.2	11.1	8.9

is the fraction of the thickness of the sample from which the observed diffraction is produced, 2θ is the angle of diffraction indicated by the spectrogoniometer, and μ^* is the mass absorption coefficient (Brindley, 1961).

The concentration of clay, mg/cm^2, for Gx values of 0.95, 0.99, and 0.999 is presented by Rich (1975b) for a wide range of mass absorption coefficients and for 2θ values ranging from 10 through 80° 2θ. For convenience, this table[2] is reproduced in Table 23-1. Less material is needed at low angles because the path of the X-rays through the sample is longer. Less sample is also needed when the mass absorption coefficient is high.

If one uses a ceramic tile of 16.8 cm^2 area and 225 mg kaolinite, somewhat more than 95% of the radiation is diffracted compared to a sample infinitely thick (mass absorption coefficient of kaolinite for CuKα equals 30). To obtain 99% with the same size tile, 334 mg of kaolinite would be needed at 30° 2θ, according to the above equation. For convenience the mass absorption coefficients calculated by Carroll (1970) are reproduced in Table 23-2.

Rich (1975b) compared the intensities of 001 and 002 reflections of kaolinite applied to ceramic tiles in varying amounts. With a 250-mg sample, the quartz in the tile was no longer detected. The intensities of the two kaolinite peaks increased with higher amounts of clay, but the ratio of 001/002 intensities had leveled off.

Table 23-2. Mass absorption coefficients for CuKα and FeKα radiation of common minerals (Carroll, 1970)

Mineral	Mass absorption coefficient	
	CuKα	FeKα
Quartz	35	67
Orthoclase	48	101
Albite	34	78
Anorthite ($An_{10}Ab_0$)	52	98
Calcite	71	140
Dolomite	50	93
Siderite	162	47
Pyrite	200	117
Kaolinite	30	60
Metakaolin	32	64
Muscovite	44	85
Illite	55	73
Glauconite	117	75
K-bentonite	52	74
Mixed-layer	59	72
Biotite	94	87
Chlorite (Mg)	54	57
Chlorite (intermediate)	104	61
Chlorite	133	62
Montmorillonite	42	66
Gibbsite	24	48
Mullite	32	64
Iron oxide, Fe_2O_3	231	58
Magnetite	238	60
Ilmenite	189	135

When the analyst has only a limited amount of sample for a one-dimensional Fourier analysis, several holders can be made so that the area covered by the specimen decreases as 2θ increases and the sample remains nearly infinitely thick. This would require removal of the sample and re-mounting on a smaller ceramic tile as large 2θ angles are needed. The mass absorption coefficients given in Table 23-2, which are based on 99.9% of the diffraction intensity derived from the sample, would be used to calculate the size of this holder for the quantity of sample available using the equation given earlier.

Different quantities of clay/cm^2 have been suggested in the literature for specimens for X-ray diffraction analysis. A commonly used amount is 25 mg deposited on a petrographic slide 2.6 by 4.6 cm. If one assumes that a sample was composed of montmorillonite and kaolinite, Mg-saturated, glycerol-solvated, the $00l$ reflections would appear at about 4.7 and 12.4° 2θ, respectively, giving corresponding d-spacing of 18.8Å and 7.15Å. For this case, the relative thickness of material at the 2θ angle of 4.7° would approximate 98% of an "infinitely thick" sample, whereas at 12.4° 2θ, the position at which the kaolinite reflection is expressed, the sample would only approximate 69% that of an "infinitely thick" sample.

In making semiquantitative analyses of clay and in making one-dimensional Fourier analyses, an "infinitely thick" sample should be employed. The use of Tables 23-1 and 23-2 should permit the analyst to select the proper amount of clay for such analyses.

C. Inhibition of Curling of Clays Mounted on Slides for X-ray Diffraction Analysis[3]

Curling, peeling, and/or breaking up of clay films mounted on glass slides or porous ceramic tiles may occur. These imperfections not only lead to lowered intensities of the $00l$ reflections, but also incorrect d-spacings since the sample surface is no longer in the plane of the sample holder on the spectrogoniometer.

The breaking of the clay films may be caused by a variety of reasons, e.g., (i) drying of the clay more rapidly at the surface whereas the tile remains moist; (ii) large changes in volume of clay film upon drying; (iii) differential expansion between the clay film and the slide or tile upon heating; and (iv) thick clay films necessary to satisfy requirements for semiquantitative or one-dimensional Fourier analysis.

Placing a second slide and weight on the slide during the drying or heating processes is a partial solution to this problem. Differential drying of necessarily thick specimens remains as a cause of curling and peeling of the film with some clays, particularly those high in smectite.

The incorporation of short rods of glass wool in the slurry of clay sig-

---

[3]Condensed from article by Rich (1975a).

Fig. 23-2. The effect of glass wool rods on the curling of <2 μm Ca-saturated Oklahoma montmorillonite on ceramic tile. (*Left*)—No glass wool, 225 mg of montmorillonite applied as a suspension in water. (*Middle*)—25 mg of glass wool was first applied in suspension to the tile and then 225 mg of montmorillonite in suspension was added to the slide. (*Right*)—Same as left except 25 mg of glass wool mixed in the suspension.

nificantly inhibits curling and loosening of the clay film. The action is visualized as comparable to the addition of reinforcing rods in concrete.

The effect of glass wool in Camargo, Oklahoma, montmorillonite is shown in Fig. 23-2. The breaking of the clay after drying at room temperature is seen. Application of the glass wool before adding the clay or adding it with the clay made no apparent difference. The X-ray diffraction patterns for these specimens before drying are shown in Fig. 23-3. The patterns were made before the clay had dried at room temperature, otherwise the clay without glass wool would have fallen off the tile. Heating the clay plus glasswool (105-550C; 378-823K) did not disturb the specimen uniformity.

D. Procedure

About 25 g of glass wool in 500 ml of water was ground in an Omnimixer for 30 min at 14,000 RPM to break the long fibers into short rods. This suspension was diluted so that a suspension of 225 mg clay containing 25 mg of ground glass wool was pipetted on the unglazed tile. The glass settles first, but with washing, the clay tends to migrate through the glass fibers. The specimen was dried in a dust-free compartment at room temperature.

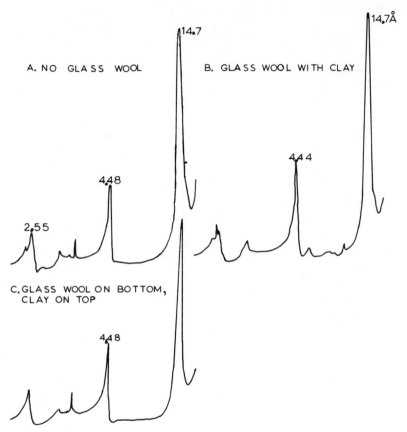

Fig. 23-3. X-ray diffraction patterns (CuKα) of slides prepared with and without glass wool and partially dried at room temperature (Rich, 1975a).

IV. PREPARATION OF ORIENTED CLAY FILMS FOR 060 REFLECTIONS

The position of the 060 reflection is useful in distinguishing between dioctahedral and trioctahedral clay minerals. A simple and rapid method for measuring this reflection with a counter type X-ray diffraction instrument has been proposed by Rich (1957). The basis of the method is the orientation of clay particles so as to emphasize scattering along the (oko) reciprocal lattice line. The method orients the clay particles in only one dimension and there is random orientation in the other two dimensions, but the removal of one degree of randomness increases the intensity of the 060 reflection. Orientation is accomplished by sedimenting the clay (7.5 to 10 mg/cm^2) on a thin Al foil. The film is then mounted in the sample holder of the X-ray instrument in a plane perpendicular to the X-ray beam when the X-ray detector is at zero degrees 2θ. Thus, diffracted X-rays are transmitted through the sample and Al foil. The sample is scanned between 55 and 65° 2θ if CuKα radiation is used.

Heat treatment may be employed to distinguish further between some of the clay minerals. For instance, the 060 reflection of kaolinite is at 1.49Å and of dioctahedral illite at 1.50Å. In mixtures there may be some doubt as to the mineral causing a reflection in this region. The 060 reflection of kaolinite is removed by heating the clay to 550C (823K) while that of illite remains. The assembly described may be heated to 550C (823K) and although the cellophane tape (Scotch) is burned, the Al foil remains in place. There may be some loosening of the clay flake, but this may be secured at the edges by a suitable cement. Thinner flakes are less likely to loosen, but there is some decrease of intensity of the reflections by reducing the sample thickness.

The samples mounted on tiles or glass slides, heated at 550C (823K) can be a source of a clay specimen for the determination of the 060 reflection. Scotch tape can be placed over the clay on the tile and rubbed with a finger until there is good adhesion. The tape is then peeled, taking with it a part of the oriented sample. This can be cut to fit the "window" prepared for the regular mounts. Additional strips of tape secure the sample in the window.

V. PREPARATION OF RANDOM SAMPLES

Sometimes it is advantageous to obtain the *hkl* reflections of clays. This is particularly useful in the determination of the polytypes. Of the methods that have been published, that of Jonas and Kuykendall (1966) appears to give the most randomly oriented samples. The clay suspension is sprayed into a heated chamber where the water is evaporated and the clay in the spray droplets remain as tiny balls. This dust is packed lightly into holders for diffraction cameras or the holders for instruments equipped with a spectrogoniometer and an electronic recording system. Pressure tends to orient the clay parallel to the (00*l*) planes and thus should be avoided as much as possible.

The chapter by Brindley (1961) on "Experimental Methods" in the monograph by Brown should be consulted for techniques for randomly oriented samples.

VI. SUPPLEMENTAL READING LIST

Jackson, M. L. 1956. Soil chemical analysis. Advanced course. Published by the author, Dep. of Soils, Univ. of Wis., Madison, WI.

Jackson, M. L. 1964. Soil mineralogical analysis. p. 245–253. *In* C. I. Rich and G. W. Kunze (ed.) Soil clay mineralogy. Univ. of North Carolina Press, Chapel Hill.

Kittrick, J. A., and E. W. Hope. 1963. A procedure for the particle-size separation of soil for X-ray diffraction analysis. Soil Sci. 96:319–325.

Whittig, L. D. 1965. X-ray diffraction techniques for mineral identification and mineral composition. *In* C. A. Black (ed.) Methods of soil analysis. Part I. Physical and mineralogical properties including statistics of measurement and sampling. Agronomy 9:674–687. Am. Soc. Agronomy, Madison, WI.

LITERATURE CITED

Bradley, W. F. 1964. X-ray diffraction analysis of soil clays and structure of clay minerals. p. 113–124. *In* C. I. Rich and G. W. Krunze (ed.) Soil clay mineralogy. Univ. of North Carolina Press, Chapel Hill.

Brindley, G. W. 1961. Quantitative analysis of clay mixtures. p. 489–516. *In* G. Brown (ed.) The X-ray identification and crystal structures of clay minerals. Mineralogical Society, London.

Carroll, Dorothy. 1970. Clay minerals: a guide to their X-ray identification. Special paper 126, The Geological Society of America. Boulder, Colorado. p. 80.

Cullity, B. D. 1967. Elements of X-ray diffraction. Addison-Wesley Publ. Co. Inc., Reading, Mass. p. 514.

Douglas, L. A., and F. Fiessinger. 1971. Degradation of clay minerals by H_2O_2 treatments to oxidize organic matter. Clays Clay Miner. 19:67–68.

Gibbs, R. J. 1965. Error due to segregation in quantitative clay mineral X-ray diffraction mounting techniques. Am. Mineral. 50:741–751.

Harward, M. E., A. A. Theisen, and D. D. Evans. 1962. Effect of iron removal and dispersion methods on clay mineral identification by X-ray diffraction. Soil Sci. Soc. Am. Proc. 26:535–541.

Jackson, M. L. 1956. Soil chemical analysis—Advanced course. Published by the author, Dep. of Soils, Univ. of Wisconsin, Madison.

Jonas, E. C., and J. R. KuyKendall. 1966. Preparation of montmorillonites for random powder diffraction. Clay Miner. 6:232–236.

Kinter, E. B., and S. Diamond. 1956. A new method for preparation and treatment of oriented-aggregate samples of soil clays for X-ray diffraction analysis. Soil Sci. 81: 111–120.

Kunze, G. W. 1965. Pretreatment for mineralogical analysis. *In* C. A. Black (ed.) Methods of soil analysis. Part I. Physical and mineralogical properties including statistics of measurement and sampling. Agronomy 9:568–577. Am. Soc. of Agronomy, Madison, WI.

Rich, C. I. 1957. Determination of (060) reflections of clay minerals by means of a counter type X-ray diffraction instrument. Am. Mineral. 42:569–570.

Rich, C. I. 1969. Suction apparatus for mounting clay specimens on ceramic tile for X-ray diffraction. Soil Sci. Soc. Am. Proc. 33:815–816.

Rich, C. I. 1975a. Inhibition of curling of clays mounted on slides for X-ray diffraction analysis. Soil Sci. Soc. Am. Proc. 39:155–156.

Rich, C. I. 1975b. Determination of the amount of clay needed for X-ray diffraction analysis. Soil Sci. Soc. Am. Proc. 39:161–162.

Theisen, A. A., and M. E. Harward. 1962. A paste method for preparation of slides for clay mineral identification by X-ray diffraction. Soil Sci. Soc. Am. Proc. 26:90–91.

Preparation of Specimens for Electron Microscopic Examination

T. R. MC KEE, Arizona State University, Tempe, Arizona

J. L. BROWN, Georgia Institute of Technology, Atlanta, Georgia

I. INTRODUCTION

Many techniques have been developed to prepare various types of samples for examination in an electron microscope. The optimum method to use in each case depends upon the nature of the sample and the type of information desired. The same criteria also determine the type of electron microscope to be used. There are two basic types of electron microscopes and microscopy, transmission (TEM) and scanning (SEM). Generally dispersions of individual clay size particles would be studied by TEM and larger particles or aggregates would be studied by SEM. In some cases, both TEM and SEM are used jointly. Interpretation of SEM images is straightforward; however, interpretation of TEM images requires experience and an understanding of TEM image formation. Electron microscopy is usually most fruitful when combined with other techniques such as X-ray diffraction and thermal and infrared analyses.

II. TRANSMISSION ELECTRON MICROSCOPY

A. Microscope Characteristics

A diagram of the optics of a TEM is shown in Fig. 24-1. The basic geometry of the TEM is similar to that of the optical microscope. The source of the electron illumination is a small heated tungsten filament at the top of an evacuated column (10^{-5} Torr or $1.33 \times 10^{-3} \, N/m^2$). The electrons are emitted from the hot filament and are propelled down the column at constant speed by a high electric potential between the filament and the anode. The electron beam is focused by an electromagnetic condenser lens onto the specimen. Here diffraction and scattering processes occur, and the resultant electron beams are focused to an initial image by the objective lens. Most microscopes employ a removable objective aperture beneath the objective lens. It enhances the contrast of the final image by reducing the background from scattered electrons. The intermediate and projector lenses provide additional magnification until a final image is projected on a fluorescent viewing screen or a photographic plate (Siegel, 1964, p. 1-78). Some disadvantages of an electron microscope are the fact that the specimen is exposed to a high intensity electron beam and must be in a high

chapter 24

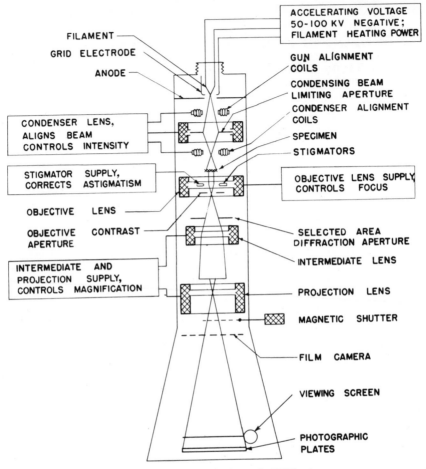

Fig. 24-1. Schematic diagram of a TEM column.

vacuum which will desiccate the sample (Chute & Armitage, 1968). These conditions call for special preparation of the sample before insertion into the electron microscope or use of special environmental sample chambers (Fullam, 1972; Parsons, 1974). Several different sample preparations for TEM and SEM examination of clay minerals are shown for comparisons in Fig. 24-2. The penetrating power of the electron beam is quite low and TEM samples must consist of either very thin films of material or very small particles supported upon thin film substrates. The examination of opaque surfaces requires a SEM or the preparation of a TEM thin-film replica which duplicates the topography of the original surface. To resolve fine surface detail of minerals, a replica is usually made by evaporating a thin film of metal and carbon directly upon the surface to be replicated and then dissolving the sample while retaining the metal-carbon replica to be viewed by TEM.

Advantages of the TEM are the high resolution and great depth of field due to the small angular aperture of the objective (about 10^{-3} rad). An important analytical capability is the formation of an electron diffraction pattern from a crystalline sample. These patterns are similar to those obtained by X-ray diffraction, and interpretation may be made by reference to standard tables of X-ray crystallographic data. The method can be used to identify particles as small as 1 μm or thin crystalline films below the detection limit of X-ray methods.

B. Preparation of Supporting Substrates

The thin supporting membrane or the thin surface replica is so delicate that it is in turn supported by a fine-mesh, metal grid about 3 mm in diameter. A modern grid is an electrolytically made one-piece disk available in a wide variety of patterns and mesh sizes (commonly stated as a certain mesh, meaning mesh per inch). The usual grid is made of copper because of its high thermal and electrical conductivity, but grids of nickel, titanium, gold, silver, nylon, and other materials are available for special purposes.

The supporting membranes should satisfy a number of requirements. The following are among the most important: (i) strength and stability in the electron beam; (ii) transparency to electrons; and (iii) a minimum of interfering visible structure.

Membranes may be made from both organic and inorganic materials. In general, the organic materials show the most structure and surface detail, particularly when the contrast is enhanced by shadowing (Fig. 24-2). (Shadowing techniques will be discussed later in the chapter). Inorganic films are more stable and have the advantage of insolubility in the various organic solvents which are sometimes used in specimen preparation. Organic films have the advantage of ease of preparation without special equipment, but will degrade during extended periods of storage and exposure to light.

1. ORGANIC SUBSTRATES

Thin membranes may be made from a number of organic materials. The two most frequently used are cellulose nitrate (Parlodion, a Mallinckrodt trade name for a solid form) and polyvinyl formal (Formvar, trade name, Shawinigan Resins Corp.). Formvar is somewhat more stable in the electron beam than Parlodion. In present-day electron microscopy, the question is somewhat academic since organic substrates usually are coated with carbon to increase their strength. Many ways of preparing films are given in the literature (Bradley, 1965, Chap. 3, Comer, 1971, p. 81–87) but one of the best methods involves dissolving the solid cellulose nitrate in a solution of amyl acetate. This can be done as follows: Parlodion is dissolved in amyl acetate to form a 0.5% by weight solution. Dissolution is slow and the mixture may be left standing overnight. Three or four drops of the solution are placed on a clean 72 by 55 mm glass microscope slide held at one end be-

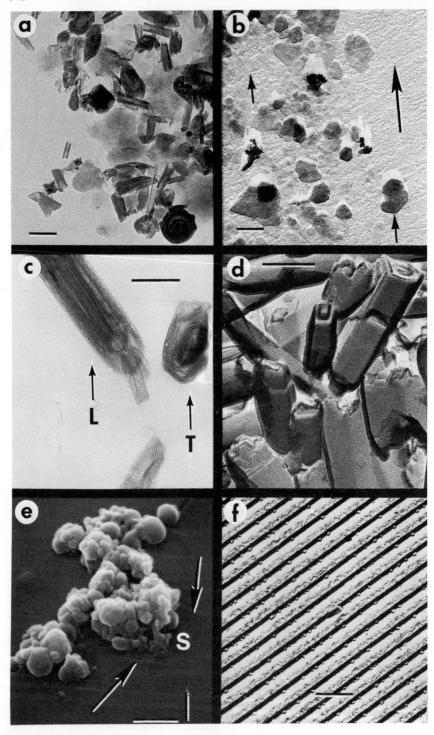

(Fig. 24-2)

tween the fingers, and the slide is rocked and tilted until the solution covers about two-thirds of the surface. The solution should not contact the fingers or it will be contaminated. The slide is then held vertically and excess solution drained onto a filter paper. Drying should take place in a location free of draft, dust, and moisture. After drying, the edge of the glass slide is scraped lightly with a scalpel to sever the film, the film surface is breathed on to form a light coating of moisture, and the slide is slowly immersed at a slight angle into distilled water. The Parlodion film will float onto the water surface. Concentration of the Parlodion-amyl acetate solution may be changed to suit individual techniques and produce films of the proper thickness. If the film is the proper thickness, it will appear to be silvery gray in color when viewed by reflection from the surface of the water. If only a few grids are to be prepared at one time, it is simple to scribe the surface of the Parlodion film on the glass slide with a sharp tool to produce a pattern of 3-mm squares. As the film is floated off the slide the individual squares will float on the surface of the water. The individual squares of film may be picked up one at a time by clamping a grid at the rim in a pair of sharp tweezers and bringing the grid up from beneath the water surface under the film surface. Excess water should be carefully blotted with filter paper at the grid rim.

If a number of supporting substrates are to be produced at one time, it

← ———

Fig. 24-2. Micrographs showing different sample preparation techniques for TEM (2a, 2b, 2c, 2d, 2f) and SEM (2e) as follows: (2a) drop-mounted suspension on carbon film; (2b) drop-mounted suspension on Formvar film, shadowed with 80% Pt/20% Pd (arctan = 0.3) with the shadowing direction shown by arrows; (2c) ultrathin section of material embedded in Spurr's epoxy resin and cut with a diamond knife showing both longitudinal and transverse cross sections; (2d) single stage 80% Pt/20% Pd shadowed carbon replica shadowed from two directions; (2e) drop-mounted suspension on Formvar film, shadowed with 60% Au/40% Pd and viewed by SEM (at a 45° tilt angle), after sputter coating with another layer of Au/Pd about 100Å thick, the arrows indicate the edge of the shadow marked (S); and (2f) commercially available shadowed replica of an optical diffraction grating with 28,800 lines/inch (1134 lines/mm) commonly used for magnification calibration for both TEM and SEM.

The shadowing in (2b) enhances both the thin leading edges of small particles (small arrows) and "worse than normal" surface detail of the Formvar film visible in the upper right (large arrow). The accelerating voltages and magnification scales, respectively, are: (2a) 100kV, 0.1μm; (2b) 50kV, 0.2μm; (2c) 100kV, 0.1μm; (2d) 100kV, 0.2μm; (2e) 25kV, 1.0μm (note: taken at 45° tilt with the top tilted away from the viewer, hence the vertical scale marker is foreshortened); and (2f) 50kV, 2.0μm. The sample designations are: (2a) 0.2-0.08μm clay from halloysite-rich river sediment, Rio Jamapa, Mexico (T. R. McKee, unpublished); (2b) <0.2μm clay from Falba soil no. 2, B horizon, Catahoula Formation, Walker Co., Texas, (J. B. Dixon unpublished, micrograph T. R. McKee); (2c) <2μm tubular halloysite, Wagon Wheel Gap, Colorado (McKee, 1972); (2d) 2-0.2μm tubular halloysite from Wagon Wheel Gap, Colorado (Dixon & McKee, 1974); and (2e) 2-0.2μm spheroidal halloysite from Washoe County, Nevada (McBride et al., 1976;[1] micrograph T. R. McKee).

[1]K. C. McBride, J. B. Dixon, and T. R. McKee. 1976. Spheroidal and tubular halloysite in a volcanic deposit of Washoe County, Nevada. 25th Annual Clay Miner. Conf. Abstr., Corvallis, Oregon.

is easy to place grids beneath the surface of water in a large petri dish resting the grids upon a fine-mesh wire gauze about the size of a microscope slide. The Parlodion film may then be floated from the glass slide onto the surface of the water in the petri dish. By utilizing a pipette, the water is slowly removed from the petri dish while the plastic film is maneuvered in such a way that it will come down on top of the wire gauze and trap all of the grids. The wire gauze containing the grids is blotted for excess water and carefully lifted out of the petri dish and allowed to dry. Individual grids may then be removed from the wire gauze by scribing around the edge of the grid with a sharp needle to cut the film.

Most electron microscope laboratories have access to a high vacuum evaporator. The plastic substrates previously prepared would then usually be overcoated with carbon to strengthen the plastic and form a stable substrate. In order to insure that support films are free from contaminating particles, several films from each batch should be saved and examined in the TEM as a control. The procedure discussed by Davies (1967) is an extreme example of the precautions taken for support films to be used in study of particulates. The subject of vacuum evaporation, carbon coating, and shadowing will be covered in succeeding sections.

2. INORGANIC SUBSTRATES

The most commonly used substrates today are prepared from carbon (Bradley, 1965). At magnifications above 200,000X the carbon film may show granularity. Films prepared from silicon monoxide or dioxide (Comer, 1971) are used by some investigators who object to the granularity, although the carbon films are easier to prepare. Carbon films are commonly prepared by evaporation onto the surface of a sheet of freshly cleaved mica. The carbon film is then transferred to grids in a manner similar to the organic films. A large number of carbon films may be prepared at one time and stored for future use since they will not degrade with time.

C. Preparation of Particle Dispersions

If a crude soil sample is to be examined in the TEM it usually is first fractionated by standard techniques. The < 2-μm equivalent spherical diameter fraction represents the upper limit for most useful TEM viewing. Although the specific type of ion saturation is not extremely important for TEM, use of Na or K should reduce the possibility of problems related to hydration and salt precipitation. If possible an aliquot of the same specimen prepared for other analyses (Chapter 23) should be used for TEM also. The aliquot should be prepared by Na- or K-saturation and chloride removal by ultracentrifugation as outlined in Chapter 23.

An aliquot of the saturated clay sample (wet or dry) which contains 5 to 10 mg of clay should be dispersed in about 3 to 5 ml of distilled, deionized water in a labeled test tube. The final concentration of clay to be used must usually be determined empirically since the optimum suspension concentration depends upon the particle size range, type of clay, and the desired examination of the prepared sample. A common problem encountered with stored clay suspensions is contamination by bacteria and fungi. This problem is averted by repeating a hydrogen peroxide treatment as employed for soil organic matter removal before TEM sample preparation.

Of all the sample materials examined in the TEM, clay minerals are among the easiest to disperse. Dispersing agents and wetting agents should be used with caution since they may alter the morphology of clay particles (McAtee & Henslee, 1969) or cause background precipitates. The diluted clay aliquot may usually be dispersed by vigorous manual agitation of the test tube. If an ultrasonic generator is available, the clay may be dispersed by agitation in the ultrasonic field. The generator frequency is not critical but a power level of 500 W can break quartz grains and very high frequencies of 10^6 Hz or higher may produce particle degeneration. Porter (1962) reported that sonication at 21 kHz and 60 W for 80 minutes increased the quartz content of a combined silt and clay fraction from 20 to 80%. The frequency range usually encountered in ultrasonic cleaners (20 to 100 kHz) will produce sufficient mechanical agitation. The time necessary to produce dispersion will vary with the individual conditions and power output, but times of 30 seconds to a few minutes are usually adequate. With longer agitation times, heating may be a problem and should be checked. Mechanical blenders may also be used (Brown, 1964).

After dispersion, a trial amount of suspension may be removed from mid-depth in the suspension with a disposable pipette and droplets of different sizes placed on several filmed grids and placed in a dust-free desiccator to dry. The different size droplets will give a choice of population density on the substrates. If the support films to be used are not dry, the droplets will run off of the grid. As a convenient means of holding the filmed grids to receive the droplets, two strips of double-face tape are placed on a microscope slide and spaced just under 3 mm apart. The grids are placed between the strips and are held at their edges.

When the sample suspension droplets are allowed to dry on the support films, aggregation or flocculation may occur. The flocculation may be minimized by spray mounting. The suspension is propelled through an atomizer or nebulizer by pressurized gas and the small droplets impinge onto specimen supports set at a certain distance in the path of the droplets (Backus & Williams, 1950). Rapid drying of the small droplets minimizes flocculation and suspension-film wetting problems. Green et al. (1974) discuss the application of an adsorption mounting technique. A positively charged film on a carbon-coated support film is used and produces excellent results for problematic fine fractions. The aggregation will be most severe for carbon support films since they are hydrophobic. Organic films are hydrophilic, with Formvar being the least hydrophilic type. Carbon films may be rendered hydrophilic by ion bombardment from a glow discharge source.

D. Metal Vapor Shadowing

Where higher contrast is needed in thin specimens, shadowing techniques are valuable (Fig. 24-2b). Although the higher contrast will obscure some types of detail it is particularly valuable to interpretation of surface structure (Jones & Uehara, 1973). On the original negative or a reversal print the sample appears to have been illuminated from the side by a strong light. Surface configurations may be interpreted, and with a known shadow angle, the thickness of particles may be computed (Preuss, 1965).

The shadowing process consists of depositing a thin film of metal on the sample surface from a distant source placed at an acute angle to the sample surface. Varying inclinations of the sample surface to incoming metal atoms result in a film of differing thickness. A particle projecting above the substrate will intercept the beam of metal atoms and produce a "shadow" free of deposited metal.

When the sample is examined in the electron microscope with the electron beam passing perpendicularly through the substrate, the varying shadow film thickness causes differential scattering of electrons and reveals the gradations of the original surface.

The geometry of the shadowing process is shown in Fig. 24-3. The source is an electrically heated filament holding the shadowing metal. From simple geometry the following equation may be derived for the thickness of the metal deposited on the substrate:

$$t = K M \sin \theta / 4 \, \pi D^2 d.$$

In this equation t equals the shadowing metal thickness, M is the mass of the metal evaporated, θ is the angle of inclination of the filament, D is the distance to the filament, d is the density of the metal used, and K is a con-

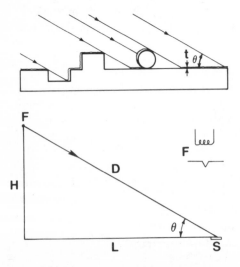

Fig. 24-3. Geometry of the shadowing process. Either the angle θ may be measured or the perpendicular distances H and L, since H/L = tan θ. The enlarged view of the sample area (S) shows the variation in metal thickness due to local changes in the angle of deposition in response to sample topography. The thickest metal layer will be deposited on surfaces most nearly perpendicular to the impinging metal atoms (note thin crystal edges in Fig. 24-2b). An aperture may be placed between the sample and filament (F) to produce a better effective point source and to reduce radiant heating of the sample.

stant (0.75) to allow for metal loss upon evaporation (Preuss, 1965). This equation may be used to estimate the necessary amount of metal to evaporate for a given film thickness. It serves as a starting point for empirical determinations. Inaccuracies arise from the assumption of spherical propagation from the evaporation source. No filament is an ideal point source and the radiation pattern is a function of the filament geometry (Preuss, 1965). Furthermore, the densities of thin metal films may not be the same as their bulk density.

1. SHADOWING VACUUM REQUIREMENTS

Shadowing must be carried out under high vacuum conditions. The mean free path of the residual gases in the vacuum should be greater than the distance from the filament to specimen. This requires a pressure of 10^{-4} Torr (1.33×10^{-2} N/m^2) or better for usual shadowing geometry. Frequently the pressure may rise during the shadowing process, due to outgassing caused by the temperature rise in the vacuum chamber. When shadowing is done under poor vacuum conditions, the resulting shadows are indistinct due to scattering of the metal atoms.

Most modern vacuum systems have a cold trap which provides liquid nitrogen cooled baffles for trapping the carbonaceous oils from the diffusion pump. This prevents their deposition on the sample surface during the vacuum pumping cycle and will help give a shadow coating with less visible detail.

2. SHADOWING METALS AND FILAMENTS

Metals used for shadowing should have a high atomic number in order to yield contrasting shadows from the thinnest possible films. Other requirements to be met by a shadowing metal are lack of granularity, stability in the electron beam, and ease of evaporation. In general, the easiest metals to evaporate, because of lower melting points, do not produce high quality thin films. Gold is dense, easily evaporated and available as wire, but crystallites in the film aggregate under an intense electron beam and give a grainy appearance to the final image. An alloy of 60% gold–40% palladium is commonly used since it is easily evaporated and has less granularity than the gold films.

Platinum is probably the best metal for shadowing, considering all requirements. Its chief drawback is difficulty in evaporation. It alloys with tungsten at high temperatures and may destroy the filament before evaporation is completed. This can be avoided by placing the required amount of platinum in foil or wire form on a 2 turn helix of 0.75 mm diameter tungsten wire, dividing the load equally on each coil. Under normal shadowing conditions, the amount of platinum on each coil is insufficient to cause breaking of the filament.

An alloy wire of 80% platinum–20% palladium is available and com-

Table 24–1. Approximate thickness of various shadowing materials necessary to produce
visible shadows or films. The recommended filament types are listed as V for "v"-
shaped filaments and H for helical-shaped filaments (after Bradley, 1965).

Material	Specific gravity	Thickness, Å	Filament type
Pt	21.5	5–10	V, H
80% Pt/20% Pd	19.4	5–10	V
60% Au/40% Pd	16.1	5–10	V
Pt/C	13.7	20–40	arc
C	2	100–200	arc

monly used for shadowing. The palladium coats the tungsten wire and prevents the platinum-tungsten problem. The granularity of the film is not as
good as platinum but better than 60% gold–40% palladium. The gold or alloys can be evaporated from the 0.5-mm diameter tungsten baskets which are
commercially available. Larger diameter tungsten wire may be shaped into a
V-form which provides a smaller effective source and a more resistant filament.

Goodhew (1972) points out that the granularity of the shadowing metal
may be substantially reduced by placing an aperture between the filament
and the sample during evaporation. This also allows better control of the
shadowing angle. Approximate thicknesses required for perceptible shadows
from various metals are given in Table 24–1. It should be noted that the contrast of the samples will vary with the accelerating potential of the electron
beam. The thickness of metal required for an optimum shadow will therefore change slightly with changes in accelerating potential.

At the present time the shadowing technique producing the least granularity involves the use of platinum and carbon. Bradley (1969) discusses
evaporating platinum and carbon simultaneously from the same source forming a composite film. Cook (1952) describes a simple technique using platinum wire which is evaporated from a groove in a carbon rod in an assembly
similar to that shown in Fig. 24–4. Apparently the carbon is deposited
simultaneously, inhibiting the growth of the platinum crystallites during deposition. Electron diffraction comparisons of the film structure indicates a

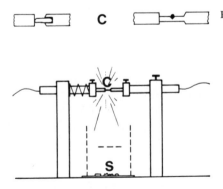

Fig. 24–4. Arrangement used for carbon arc
coating and/or Pt-C shadowing of samples
(S). The carbon rods (C) may be removed
from the holder to be sharpened or modified for shadowing as shown in the detailed insets. For shadowing, the Pt-C
pellet may be inserted or Pt metal laid
into a groove in the narrowed rod as
shown. Again an aperture may be desired
since heating from the carbon arc is a
problem for some samples. In order to
produce carbon films with a minimum of
structure at high magnification, nondirect
deposition is obtained by inserting an
open-topped cylinder and a small shield to
prevent direct deposition (dashed lines).

very small crystallite size in the combination film. This shadowing technique has been used in the carbon replication process to reveal surface structures on the order of 10Å units (Kohn et al., 1971).

Commercial sources of microscopy supplies furnish ready-made platinum-carbon pellets which may be held between the butting ends of the carbon rods for evaporation as shown in Fig. 24-4. To hold the Pt-C pellet the ends of the carbon rods are hollowed slightly with a small drill point or scalpel. A small object is placed on a piece of white glass or paper near the microscope sample. Evaporation of the pellet is continued until a perceptible shadow is formed on the white surface. More advanced systems have piezoelectric sensors to monitor the thickness of the film.

3. SHADOWING FOR HEIGHT DETERMINATION

The geometry for determining specimen height from shadow length is shown in Fig. 24-3. When shadowing a dispersion of particles on a substrate for the purpose of measuring particle height or thickness, care must be taken that the shadowing angle is known at each point on the substrate. One way of doing this is to disperse the particles on a Parlodion-coated glass slide where they may be shadowed in situ and the film floated off and mounted on grids for examination. The measured angle between selected areas of the slide and the filament may be used as the shadowing angle. If the particles are deposited on a film-coated grid, the shadow angle may vary from point to point due to wrinkles in the substrate. If particles are near a tear in the support film, any measurements should be held suspect or discarded entirely.

Two methods are available for determining shadow angle. If a small amount of polystyrene latex suspension is added to the original particle suspension before deposition on the substrate, some small latex spheres will be visible in each field of view. Measurement of sphere diameter versus shadow length will give the shadow angle for that particular field of view. The thickness of the metal coating on the shadowed particles should not necessarily be neglected since it may be significant. If the polystyrene latex is not available, the sample may be shadowed from two diametrically opposed sources, and the height computed from the average shadow length. Care must be exercised when using this technique with nonsymmetric particles.

E. Replication of Surfaces

A replica is a thin film of material transparent to electrons and duplicating the surface contours of a specimen. The films may be made of either organic or inorganic materials, but the latter show less visible detail. The replica method most useful to the soil clay mineralogist is the pre-shadowed platinum-carbon replica. The platinum-carbon replica is capable of high resolution and is most suitable to clays, minerals, and most ceramics (Comer & Turley, 1955; Kohn et al., 1971).

The first step in preparing a replica is to shadow the surface with platinum. Bulk materials such as lump clays or minerals are broken to yield fresh fracture faces. Some samples may require drying in a vacuum desiccator to avoid a long pump-down time in the shadowing unit. No other treatment is applied except removal of any loose material from the fracture surface. Hulbert and Bennett (1975) describe an electrostatic technique which is effective in removing loose particles. Fine particulate materials may be deposited from suspension onto a glass slide or mica surface using dispersion techniques described previously. The dry dispersion on the slide serves as the sample surface to be replicated.

When the sample has a rough and uneven surface, it is necessary to shadow with platinum from two mutually perpendicular angles (Fig. 24-2d), or if the equipment is available, the sample may be rotated during the shadowing process using a single evaporation source. In some areas, if the two-shadow method is used, both shadows may appear while in other areas only one of the evaporated films may have deposited on the specimen.

The shadowed specimen is next coated at normal incidence with an evaporated film of carbon using the assembly shown in Fig. 24-4. Using spectrographic carbon rods, with the tip necked-down to 1 mm in diameter, a 2-mm tip length is a satisfactory average for most specimens when the source-to-specimen distance is 12 to 13 cm. Small particles on a glass substrate will require only a 1-mm tip, rough porous surfaces may require as much as 4 mm. The final film on a porous surface is relatively uniform. Apparently there is some molecular redistribution of the film and possibly reflection from adjacent surfaces during deposition. Care must be exercised, however, to avoid unnecessarily thick films since contrast and resolving power decrease as the thickness is increased.

The sides of the replica are first scraped to remove undesired carbon. The replica and embedded sample are then floated off the substrate by lowering it at an angle into a container of water. Hydrofluoric acid is added to dissolve the clay from the floating replica. After about 1 hour the replica is transferred to fresh acid solution for an equivalent length of time. It may then be transferred to distilled water, removed, and allowed to dry. There is usually considerable breakage when the replica is transferred into water. It should be removed from the water on 500 mesh per inch (20 mesh per mm) grids. It may be necessary to use grids with substrates to avoid further breakage of the film upon drying. Addition of a wetting agent such as Kodak Photoflo (one drop to 250 ml of water) or Teepol to the water will aid in removal of the films. Depending upon the composition of the sample, precipitates may form on the replica, in which case aqua regia should be added before the hydrofluoric acid. For bulk samples and samples with irregular surfaces, it may be necessary to support the replica film with a plastic base to avoid replica breakage (Comer, 1971).

In many cases, use of the SEM may provide the same information much quicker and easier than replication. However, in other cases, replication is still preferable (see Fig. 24-2) as small detail may be more visible and in particular show more contrast in the replica.

F. Ultramicrotomy Applications

1. ULTRAMICROTOMY

Ultramicrotomy techniques have been used in the field of biological electron microscopy almost since the beginning of microscopy. The basic principle of all ultramicrotomes is to move the sample past a fixed, sharp knife edge. In biological work, glass knives are commonly used (Pease, 1964, p. 149–159), but these are not hard enough for applications in mineral studies except for pretrimming purposes. Precisely sharpened diamond knives are available for cutting very hard materials such as minerals that may be encountered in soil clay studies.

In the basic operation of the ultramicrotome a water trough behind the diamond knife edge is filled with water or water with 10% acetone to reduce surface tension. The level is adjusted with a pipette or syringe until the meniscus touches and is below the diamond knife edge. As the sample advances past the knife edge, sections are cut and will float onto the surface of the liquid. The sections may be observed through the stereo microscope attached to the microtome and fished out on specimen grids for observation in the electron microscope. Sections which have silver to gray color by reflection, range from 800 to 500Å thick. Gold sections are 900 to 1200Å thick and are at the limit of penetration for 100 kV electrons. Extensive details of the technique are given by Pease (1964). Ultramicrotomes utilize mechanical advancement or preferably thermal expansion advancement to advance the specimen during cycles past the knife edge.

In order to observe the interior detail of a sample, the first step is to permeate the sample with a suitable resin. Complete kits of embedding materials suitable for this type of specimen preparation are available from microscopy supply companies. Once the sample is embedded in the resin, the desired portion may be sawed out with a jeweler's saw and epoxied to a short piece of lucite rod for mounting in the ultramicrotome. Depending upon the sample hardness, it is usually not possible to cut a section larger than about 0.5 mm in width. For difficult samples, it may be necessary to cut a sample only 200 to 300 μm wide.

2. FABRIC STUDIES

In order to study the interior detail or fabric of a soil aggregate the sample must first be permeated with a suitable resin while preserving the original fabric. Foster and De (1971) carried out a detailed investigation of impregnating techniques using a series of diffusing liquids to replace the original water with the final resin which was then hardened. They showed minimal artifacts but the technique was very time consuming. Freeze-drying and impregnation have been used by numerous investigators (Bowles, 1968; O'Brien, 1971) but Gillott (1969) indicated that ice crystallization during the freeze-drying may be a problem and that air-drying is accompanied by a large vol-

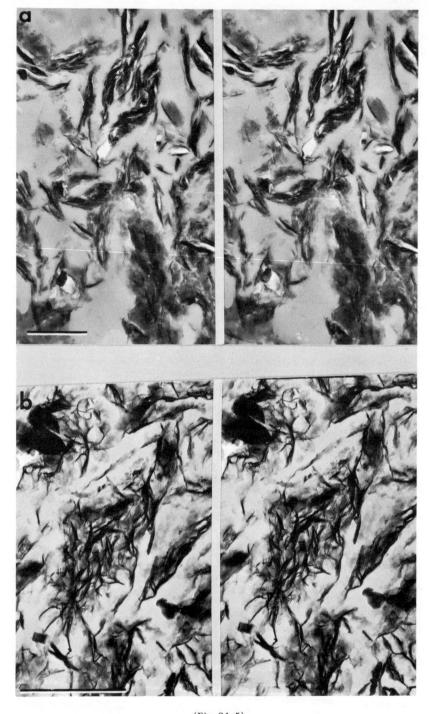

(Fig. 24-5)

Fig. 24-5. Ultrathin sectioned high smectite marine clays are shown in (5a) and (5b). The samples were processed to preserve the relative interparticle association or fabric. The three-dimension character of the well oriented, high void ratio central Pacific sample is shown in stereo (5a). A contrasting nonoriented, high void ratio fabric from the Mississippi River Delta is shown in (5b) indicating that the fabric observed is not an artifact of preparation. The two prints in (5b) are duplicates from a single negative and illustrate a false stereo effect. The micrographs were taken at 80kV with 12° relative tilt for the stereo pair. Sample locations are: (5a) Deep Sea Drilling Project Site 163A, 143-m sediment depth; and (5b) Mississippi Delta borehole B-2, 1.4-m sediment depth (both from Bennett et al., 1977). Scale markers are 1.0μm.

ume decrease. Naymik (1974) has clearly shown that the critical point-drying technique is far superior to either air-drying or freeze-drying of clay-rich specimens. Figure 24-5 shows the interparticle association preserved by critical point drying and embedding (Bennett et al., 1977).

3. INTRAPARTICLE STUDIES

Ultramicrotomy is also very useful for studying particle interiors or particles having morphologies that prevent viewing them in certain orientations when they are prepared by conventional dispersion methods. Figures 24-6 and 11-11 show the application of this technique to a halloysite clay sample (Dixon & McKee, 1973, 1974). The internal structure of the individual particles is shown to good advantage. Figure 24-6 also shows some of the problems encountered with this technique. Mechanical artifacts and particle distortion are evident since the nearly spherical particles are elongated in a common direction. The embedding medium and the clay particles do not have exactly the same physical properties, hence they cut differently and the embedding medium often tears away from the particles. In extreme cases, entire particles fall out of the thin section leaving holes. If more massive particles such as quartz or feldspar are encountered during the thin sectioning, they may chip, break, or fracture.

Figure 24-7 shows the technique applied at a much higher magnification and resolution revealing the layer structure of clay particles. In Fig. 24-7a, it was desired to obtain an electron microscope view along the (00l) planes which would not be possible by normal sedimentation preparation. To prepare the sample, droplets of the chlorite suspension in water were dried on a cured epoxy block to obtain parallel orientation of the clay particles. The position of the clay-coated area was circled with india ink to facilitate locating the areas for ultramicrotoming. Additional epoxy solution was then added and heat cured at 38C (311K) overnight to form an epoxy sandwich. A small block was sawed from the epoxy, suitably shaped by filing to center the ink marks, then sectioned by means of a Reichert ultramicrotome using a diamond knife. The section was cut perpendicular to the sediment plane to cut across the flat faces of the platy particles. The particle platelets and (00l) crystal planes in the section thus were approximately normal to the section surface. Further details of the microscopy involved

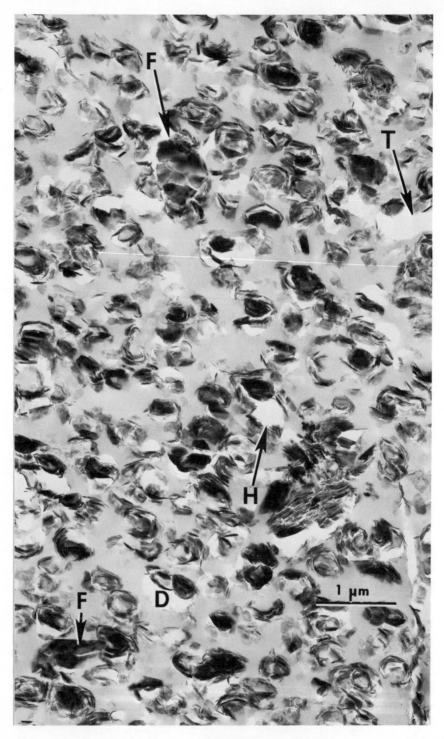

(Fig. 24–6)

Fig. 24-6. Ultrathin section of spheroidal halloysite illustrating some of the problems inherent with thin sectioning. The shearing force of the cutting process (*from lower left to upper right*) obviously compresses the nearly spherical particles as illustrated by the two distorted particles at (*D*). Fractured particles (probably quartz) are indicated at (*F*). Holes appear as white areas in the gray embedding medium. The holes most commonly occur on the leading or trailing edges of particles and indicate tearing of the embedding medium during sectioning (*T* and *D*) or voids created by missing particle interiors (*H*). Sample of 2–0.2μm spheroidal halloysite from Redwood County, Minnesota (Dixon & McKee, 1974). Micrograph taken at 100kV.

◄——— (Fig. 24-6 is on page 824)

(Fig. 24-7 is on page 826) ———►

Fig. 24-7. Illustration of high resolution TEM (HRTEM) applied to the study of sheet silicate structure. The fringes in (*7a*) commonly called lattice fringes, represent the 13.9Å fringes formed by aligning the (001) crystallographic planes of chlorite parallel to the electron beam (ultrathin section of $<2\mu$m clay, mafic chlorite from Benton, Arkansas; Brown & Jackson, 1973). The degrading effect of amorphous phase contrast interference upon the lattice image of 4.4Å fringes in ultrathin sectioned rectorite is illustrated in (*7b*) where the clear lattice image on the left is slowly obscured toward the right due to the increasing thickness of embedding epoxy (Rectorite, Fort Sandeman, Pakistan; unpublished, T. R. McKee). The image of the epoxy alone (*E*) is shown at the right.

The HRTEM images can be interpreted as structure images or projections of the atomic arrangements only under very specific and well-defined conditions regarding crystal orientation to the beam (axes aligned within 2–3 milliradians of the incident beam), crystal thickness (50–100Å for most crystals), and critical underfocus. The thin edge of a wedge-shaped biotite crystal is shown in (*7c*). A rim (*R*) of beam damaged and contaminated material has formed at the thinnest edge of the crystal; however, fringes are visible in the thicker portions of the crystal. At the proper crystal thickness the image from the boxed area of (*7c*), shown enlarged in (*7d*), represents the crystal structure with the white spots representing the "tunnels" between the interlayer potassium sites. The series of black dots and the white line drafted onto the photograph designate the positions of "equivalent" interlayer positions and are the basis for the 1M, 2M, and 3T polytype designations shown near the top of (*7d*) (biotite, Mitchell Co., N Carolina; Iijima & Buseck, 1977).

are given in Brown and Jackson (1973). Figure 24-7c shows a similar application of ultramicrotomy to biotite.

G. Filter Mounts

In some cases, it is desirable to observe particles from a liquid or air suspension. Appropriate quantities of the suspension may be filtered onto a membrane filter and transferred to a specimen grid while dissolving the filter (Harris et al., 1972). If desired, a portion of the same filter may also be observed by SEM. Care must be exercised to prevent overloading the filter with material if individual particles are to be studied. If interparticle association or fabric is to be studied, attention must be given to the drying technique. Whitehouse and McCarter (1958) and Jernigan and McAtee (1975) have observed that critical point drying is the only reliable method to retain morphology of individual dispersed particles, which applies equally well to associated particles.

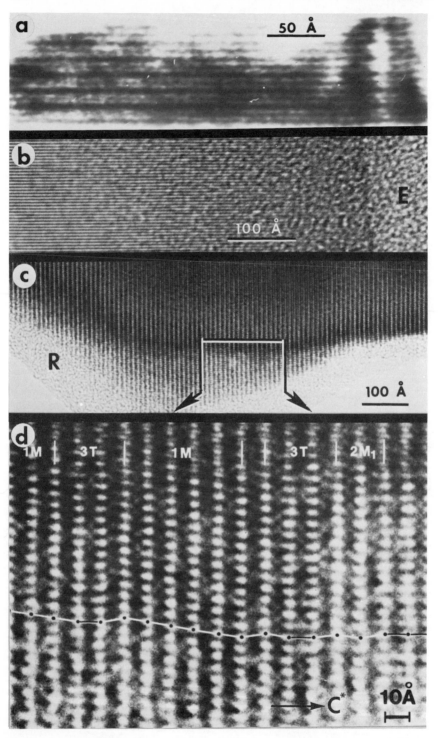

(Fig. 24–7)

H. Ion Thinning

Some materials are too brittle to be successfully sectioned with a diamond knife. As previously pointed out, if the various constituents of a sample are not of the same relative hardness, ultramicrotomy will result in particle loss or extraction and the shear stress may induce defects in the crystals. It has recently been possible to prepare sections of these materials thin enough for electron transmission by using ion bombardment thinning of a ground and polished slice. The sample is placed between two ion guns in a vacuum. A direct current, argon ion beam thins the sample in one region such that a small hole is produced and the edges of the hole are thin enough to transmit the electron beam (Paulus et al., 1975; Phakey et al., 1972).

I. Transmission Electron Microscopy Sample Examination

1. DISPERSED PARTICLES

In examining a dispersion of particles on a thin film substrate, dense or thick particles will show only a profile view. No information may be obtained from the surface. Thinner particles will transmit the electron beam sufficiently to show some surface or interior features. Common microscopes usually have a choice of accelerating potentials up to 100 kV. The selection of a higher potential will permit a greater penetration into the sample. High voltage electron microscopes can attain acceleration potentials of 1,000 to 3,000 kV. Sample penetration at 1,000 kV may increase as much as 7 to 10 times the penetration at 100 kV. Figure 24-8 shows the variation in penetration of halloysite particles at three acceleration potentials. In addition to penetration increases, specimen damage is reduced by employing higher acceleration potentials (Cosslett, 1970).

When focusing the microscope on the edge of a particle, a dark or a light fringe may be seen surrounding the edge. The fringe contrast will change with the position of focus and exact focus is obtained when the fringe almost disappears. These are known as *Fresnel fringes* and are due to diffraction effects around the edge of objects in the image formation of the microscope (Thomas, 1962, p. 88-92).

There are several other diffraction effects which may be observed in thin crystals. If these are not understood they may be misinterpreted as being real features of the object. Mineral particles thin enough to transmit the electron beam are frequently very flexible. In examining these thin particles, dark bands may be seen to move across the surface. These are due to a warping or bending of the plate from thermal or electrostatic effects and are caused by diffraction contrast through the crystal structure. The dark bands are due to electrons which are selectively scattered outside the objective aperture of the microscope and do not contribute to the final image (Heidenreich, 1964).

Thin crystal plates which overlap each other may frequently display a set of parallel fringes with spacings much greater than the actual structural plane spacing along the viewing direction in the crystal. These are called *moiré patterns* and are caused by double diffraction through the crystal structures of the overlapping plates (Hirsch et al., 1967).

There are several types of moiré patterns. The parallel type occurs where the two crystals have slightly different structural spacings resulting in an alternating fit and misfit. In the rotational type, two crystals of the same structural spacing may overlap at a slight angle (Gard, 1971, p. 31).

Figure 24-9 is a view of a mica particle which shows both bending contours and moiré patterns on the same particle. The bending contours are the circular configurations with an approximate central cross. These are caused by blisters within the layer structure and can be made to expand and contract by changing the intensity of the illuminating beam. The bottom edge of the particle shows a slight angular displacement between two of the layers which results in the multiplicity of moiré patterns shown about the surface of the crystal.

With continued examination of a sample in the microscope, a rim or coating may sometimes appear on a particle. This is due to contamination occurring in the microscope itself (Hart et al., 1970). All microscopes using oil diffusion pumps contain some degree of pump oil vapor circulating in the vacuum system. The electron beam tends to transform the vapor forming a polymerized carbonaceous layer on the sample surface. This is particularly true when using the double condenser illumination system due to the low energy input to the sample. The low sample temperature greatly increases the rate of contamination. This contamination would look very much like the amorphous coatings on mineral surfaces shown by Jones and Uehara (1973) but the contamination would be on top of any shadowing metal rather than under it. To avoid this contamination most modern microscopes have some type of cold trap located in the specimen chamber region. This cold trap is cooled externally by liquid nitrogen and offers a cold surface upon which contamination may deposit since there is a thermal gradient from the sample toward the cold surface.

(Fig. 24-8 is on page 829) ——————>

Fig. 24-8. Increased penetration of spheroidal halloysite as a function of increased acceleration potential (*8a*) 50kV, (*8b*) 200kV, and (*8c*) 800kV. With increased penetration it becomes apparent that the spheroidal particles have void or lower density cores, with the remainder of the particles being composed of roughly concentric bands of crystalline material. The poor resolution of the 800kV images (*8c*) represents a temporary operational problem. The use of additional procedures such as ultrathin sectioning (Fig. 24-6) show that the apparent concentric bands were in fact a series of overlapping flakes and not continuous bands. Sample of $<2\mu m$ spheroidal halloysite from Redwood County, Minnesota (Dixon & McKee, 1974). The 200kV micrograph was provided by D. F. Harling, JEOL (USA) Inc., Medford, Mass. and the 800kV micrograph by U. S. Steel Corp. Research Center, Monroeville, Pa. Scale markers are 0.2 μm.

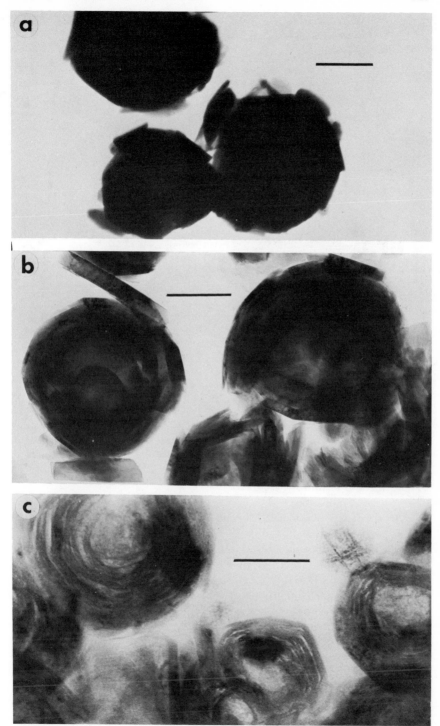

(Fig. 24-8)

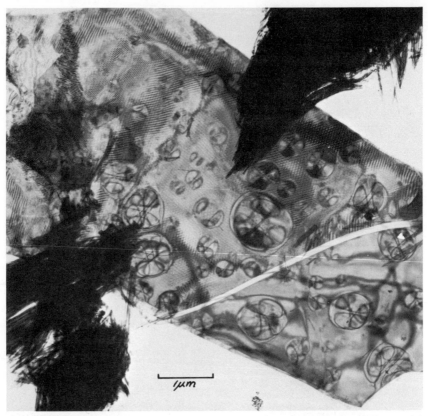

Fig. 24-9. A bright field TEM image of an attapulgite sample showing a mica particle
exhibiting *moiré patterns* and circular bend contours (sample from Attapulgus, GA;
John L. Brown).

2. LATTICE AND STRUCTURE IMAGES

If the TEM (Poppa, 1975) and sample preparation are of sufficient
quality, it is sometimes possible to see electron optical fringes, which vary
with the periodicity of the crystal lattice, superimposed on the crystal image.
The instinctive interpretation of these fringes (commonly called lattice
fringes) is attractively simple. Unfortunately, in most cases, little is known
concerning the exact conditions of formation of the individual images and
they are easily susceptible to misinterpretation (Allpress & Sanders, 1973).
Alignment of the electron beam with respect to the lattice planes in the
crystal is very critical to the production of such images. Conditions for resolving the crystal lattice require that at least two beams, the main (undif-
fracted) or zero-order beam and the first-order diffracted beam, pass through
the objective aperture and be recombined at a critical focus setting. An example of this has been shown in Fig. 24-7. Numerous examples of lattice
images of clay minerals have appeared in the recent literature (Brown & Jack-

son, 1973; Brown & Rich, 1968; Lee et al., 1975a, 1975b; McKee et al., 1973; Uyeda et al., 1973; Yada, 1967, 1971; and Yoshida, 1973).

Electron structure images (Fig. 24-7d) show the structural arrangement or layer sequence rather than just simple periodicities (Buseck & Iijima, 1974; Cowley & Iijima, 1975). The structure images are even more difficult to obtain than lattice images due to more severe experimental conditions and physical constraints.

Specimen preparation for lattice and structural imaging is more demanding than for normal TEM. The samples are normally deposited in some manner on a holey carbon film which should have open holes covering a large portion of its area. The individual particles which are suspended over the open holes are then selected for imaging in order to reduce the phase interference of the carbon support film (Cowley & Iijima, 1975; Melvin, 1970).

3. REPLICA EXAMINATION

Examining replicas in the electron microscope is much simpler than examining fine mineral particles. The surface topography of the sample is shown as it originally existed within the limits of the technique. Some artifacts to be aware of involve the appearance of a thick rim around some features. The rim can be due to an extra heavy carbon deposit in the replication process or to the same type of carbonaceous contamination discussed previously. Also, an extension of the replica surface along the line of sight can give this effect since the beam is passing through a greater thickness of replica film. Sometimes if the shadowing metal is excessively thick on the sample, a grainy effect may be produced which can be misinterpreted as fine structure. This effect will usually vary with surface topography and may become apparent only after exposure to the electron beam for a period of time.

4. ELECTRON DIFFRACTION

Electron diffraction in the electron microscope is very much analogous to X-ray diffraction. As the electron wave passes through the crystalline particles, the alignment of various planes in the crystals can cause diffraction in accordance with the Bragg equation, $n\lambda = 2\,d\,\sin\theta$. Due to the short electron wavelengths, this equation may be approximated by $d = \lambda L/r$ where λL is the camera constant, r is the ring radius, and d the interplanar spacing. The TEM provides a removable aperture called a *selected area diffraction* or *field limiting aperture* (Fig. 24-1) to limit the field of view to a single particle or group of particles. The lenses are then adjusted to obtain a diffraction pattern in the focal plane rather than an image of the field of view. This technique is usually referred to as selected area diffraction. Some microscopes may also provide a special chamber lower in the electron optical column where the sample may be placed for electron diffraction examination in order to produce a better resolved pattern. Here the electron beam passes through many more grid openings and gives a better statistical sampling of

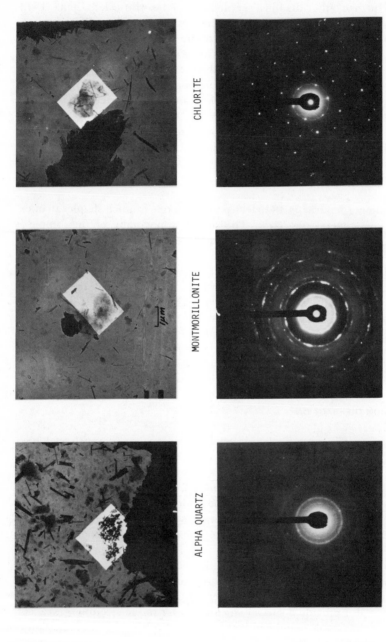

CHLORITE

MONTMORILLONITE

ALPHA QUARTZ

Fig. 24–10. Selected area electron diffraction identification of fine grained minerals in a palygorskite rich sample. The bright rectangle in the center of the TEM images is formed by the field limiting aperture and indicates the portion of the sample contributing to the diffraction patterns. When the selected area is larger than the crystals a movable beam stop is often used to block the intense unscattered electron beam. (micrographs, John L. Brown).

the crystalline particles on the grid. This chamber also has facilities for mounting the sample so that the beam may graze the surface to produce a reflection diffraction pattern. This technique is useful in the examination of thin films of material or coatings on particles.

Figure 24-10 shows some minerals which have been identified in a sample consisting primarily of palygorskite. The camera constant was obtained by making a diffraction pattern of a standard sample of thallium chloride at the same lens settings used for the unknowns. Using the value of this constant and measuring the diameter of the rings in the pattern, it is possible to determine a d value for each ring corresponding to the interplanar spacing. In the case of the chlorite pattern, consisting primarily of spots, the radii on which the spots occur were used instead of rings. Identification is often aided by observing the morphology of the sample. Particle shape and orientation may cause missing spots or rings as compared to X-ray diffraction results for the same material.

A detailed discussion of electron diffraction applied to clay minerals is given in Gard (1971, Chapter 2) and Zvyagin (1967). The precision of diffraction identification may be improved by evaporating a standard such as aluminum or thallium chloride directly onto the substrate along with the sample. Identification may be made by referring to the standard tables of d values for X-ray diffraction. The intensity values listed for X-ray diffraction will be of little value in electron diffraction. A straightforward set of worked examples of solved diffraction patterns is available in the appendix of Andrews et al. (1971).

III. SCANNING ELECTRON MICROSCOPY

The SEM has had a great impact on the field of microscopy. The basic principles of the SEM have been known for some 30 years, but it has only been recently that certain technical problems have been overcome to make the instrument commercially feasible. These microscopes have been commercially available since 1965 and have been rapidly applied to many problems in the field of soil mineralogy.

A schematic diagram of the SEM is shown in Fig. 24-11. The SEM forms an image by scanning the surface of the sample with a finely focused electron beam about 10^{-2} μm in diameter. The high energy beam stimulates the emission of secondary electrons, backscattered electrons, X-rays, and sometimes light photons from the sample surface. The emission amplitude of the radiation varies with surface topography and the atomic number of the elements present. In the normal mode of scanning microscopy, the electrical signal derived from the collected secondary or back-scattered electrons is used to form a television-type image of the surface being examined. The X-rays emitted are characteristic of the elements present in the sample and they can be collected and analyzed. Therefore, the chemical identity and spacial distribution of the elements comprising the area under the electron beam can

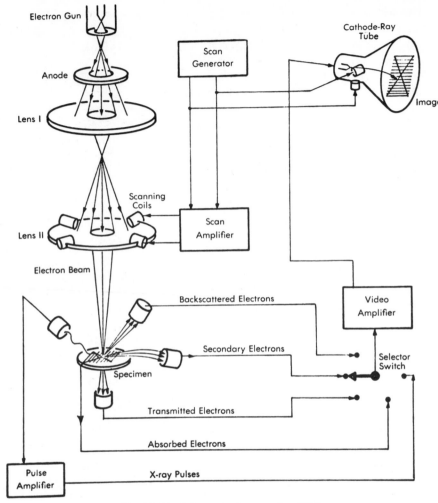

Fig. 24–11. Schematic diagram of an SEM.

be determined. In the SEM, the X-rays are usually detected with a lithium-drifted silicon detector, and the signals from this device are sorted and counted with a multi-channel analyzer. This instrument is known as an energy dispersive X-ray analyzer. The X-ray spectrum is normally displayed on a cathode ray tube (CRT). A teletype print-out of the X-ray counts in each channel or the area of each peak can be obtained if desired. This microscopic application of X-ray spectroscopy is called *X-ray microanalysis* (Goldstein & Yakowitz, 1975).

One advantage of the SEM is its great depth of field. This sometimes makes it more useful than the light microscope, even at low magnifications of 25X to 100X. The range of magnification available depends upon the instrument being used but is usually from about 20X to 50,000X or greater.

The wide range of magnification allows correlation of optically visible features with features well beyond the resolution of optical systems. Another advantage is ease of sample preparation.

A. Sample Preparation

Versatility of the SEM allows the direct use of a wide range of sample types from large crystals or soil aggregates to fine particulate materials. Gross soil specimens may be used directly after proper conditioning for the vacuum. If the soil fabric or particle arrangement is important, then a dehydration technique must be used which will preserve that information. As mentioned previously, critical point drying or rapid freeze-drying is considered much better than air or oven drying (Erol et al., 1976). If sand- and silt-sized materials are to be studied in the SEM the nature of the individual study will dictate the preparation technique. If particle surface textures are to be studied, the material must be specially cleaned (Krinsley & Doornkamp, 1973, p. 7) and ultrasonic vibration may damage the surface (Porter, 1962). Preparation of specimens for SEM is simpler than for TEM. With rare exception, samples must be coated in some manner to make them electrically conductive.

1. SAMPLE MOUNTING

The sample must first be attached to a metal specimen mount or stub (usually brass or aluminum). The nature of the mounting material used depends on the size and chemistry of the sample (Johari & DeNee, 1972). Clay or fine silt sized particles may be mounted directly onto a smooth stub surface by spraying or dropping the suspension onto the clean surface. Conducting silver paint may be used as an adhesive for mounting larger particles onto the specimen stub. The surface tension of the paint may cause it to "creep" over the surface of smaller particles. Double-face tape is often used in these cases. A strip of conducting silver paint must then be applied from the stub onto the tape surface. For larger particles, crystals or aggregates, the conducting paint may be used alone as an adhesive (Brown & Teetsov, 1976). Figure 24–2e is an SEM micrograph of a clay sample which was mounted on a Formvar film for TEM viewing. The specimen grid was then attached to a brass SEM stub with conducting paint, coated with a "sputtered" layer of gold-palladium, and viewed in the SEM.

2. SAMPLE COATING

There are two possible ways to reduce specimen charging of nonconducting specimens in the SEM: (i) apply a conductive coating, or (ii) work at very low accelerating potentials and beam currents. Working at low potentials does not allow high resolution surface studies making the conductive coating a necessity (Echlin & Hyde, 1972). For samples observed at low magnification, the type and nature of the conductive coating is not critical. For high resolution work, the thickness and nature of the coating are very important.

Two coating techniques are commonly available: sputtering and evaporation. Sputtered coatings are highly uniform and reproducible and give the best coverage of rough samples. Sputtering is accomplished at a very poor vacuum, scattering the gold or gold-palladium metal into non-line-of-sight locations. Sputtering, however, does produce more heat than evaporation and may damage sensitive samples. Evaporation is achieved in a manner similar to shadowing, except that the sample is being rapidly turned and tilted. For rough surfaces, a thin layer of carbon must first be applied in order to cover the entire surface, followed by a layer of metal such as gold or gold-palladium (200 to 400Å thick). Samples should not be coated until just before use as the coatings may degrade upon storage. All dried and coated samples must be stored in a desiccator to preserve the integrity of the coating (Ingram et al., 1976). If very loosely compacted aggregates of soil are being examined, it may be necessary to paint the sides of the sample with conducting silver paint, excluding only the area desired for SEM examination.

If X-ray microanalysis is to be done on the sample along with micrographs, the sample may be mounted on a special stub such as graphite or beryllium and coated with carbon or a noninterfering metal. For quantitative energy dispersive analysis, the type of coating and its thickness are very important and must match those of the standards used. Micrographs obtained from a sample coated with carbon usually lack the contrast and definition of those coated with gold-palladium or other conductive metals.

B. Scanning Electron Microscopy Sample Examination

One of the many advantages of the SEM is the simplicity of image interpretation. The perspective of viewing is from the direction of the scanning beam and the "illumination" of the object is from the direction of the electron collector; therefore, the image may be grossly interpreted as in optical macroscopic examination.

Contamination occurs in the SEM in the same manner as previously described for TEM. Elimination of the effect by cold traps is not as simple as for the TEM. In the SEM, contamination is not a serious problem unless high resolution SEM or quantitative energy dispersive data are desired. It is best to work rapidly to minimize exposure of the sample to the scanning beam.

A common artifact called "charging" occurs when all portions of the sample do not remain at the same electrical potential. This is a frequent effect in the observation of nonconducting samples such as minerals even when the sample has been coated. The sample surface may be so rough that the coating is not continuous. Regions of charge appear excessively bright and may cause lines or streaks on the micrograph in the direction of scan. In extreme cases, the entire image may shift (Hearle et al., 1973, p. 205-208). Charging can be reduced by using lower scanning beam potentials and rapid scan rates.

Thin edges or small projections on a sample surface may also appear excessively bright, obscuring surface detail. In very small features the mean free path for the escape of secondary electrons is less, resulting in increased emission. The electric field between the collector and sample is more intense at small features and this adds to the effect. A reduced scanning beam voltage will also minimize this effect.

Figure 24-12a shows a gross soil sample obtained in a study of the underlying foundation soil of a proposed nuclear power reactor site. The sample is glued to a specimen mounting stub used in a Stereoscan SEM. The SEM micrograph in Fig. 24-12b was obtained with secondary electron emission imaging and shows the depth profile of the sample. An X-ray map of the same identical area (Fig. 24-12c) indicates the muscovite distribution by $K\alpha$ X-ray imaging. Figure 24-13 shows an SEM secondary electron image of a fractured soil sample showing the various morphologies present. Spot X-ray microanalysis spectra are shown as displayed on a CRT indicating the presence of barium sulfate and silica in separate phases with distinct morphologies.

IV. IDENTIFICATION OF MINERAL SPECIES

The morphology of particles as seen in the microscope is an aid to their identification but is by no means a final identification. For example, halloysite has been found in spherical, tubular, and platy forms (Askenasy et al., 1973; Bates et al., 1950; Souza Santos et al., 1966). The association of X-ray microanalysis with both TEM and SEM is a big step forward since morphology and composition may be associated on an individual particle basis. Bassin (1975) has used X-ray microanalysis to identify different types of clays on membrane filters. In addition to morphology and composition, structure and crystal orientation may be determined in TEM. In an "analytical" TEM relatively complete characterization may be accomplished for individual particles if the time and expense are justified. There are many references in the literature showing micrographs of various types of clays and minerals. Several of the most helpful for TEM are Bates (1971), Beautelspacher & van Der Marel (1968), Davis et al. (1950), and Gard (1971); and for SEM are Borst and Keller (1969), Krinsley and Doornkamp (1973), and Mumpton and Ormsby (1976). Probably the single most useful reference is McCrone and Delly (1973) which includes TEM, SEM, optical microscopy, X-ray microanalysis, and many other techniques, all of which are directed toward identification of unknown particles.

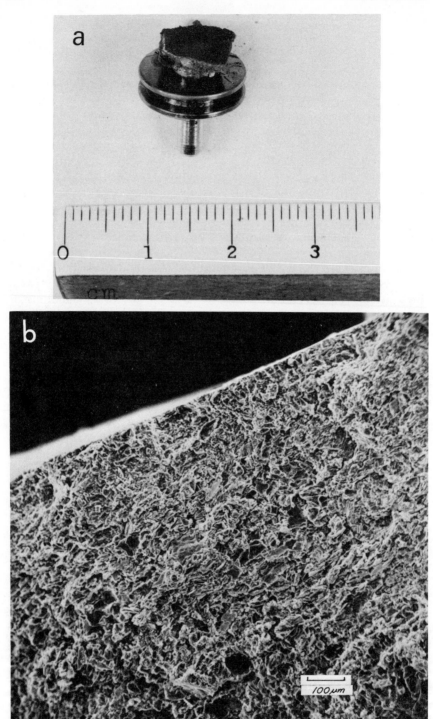

Fig. 24-12

V. STEREOSCOPIC MICROGRAPHS

Stereoscopic views may be made of particle dispersions, or ultra thin sections (Fig. 24-5), but they are much more effective in the study of samples with high relief. They are a great aid to the interpretation of particle shape and surface topography in both TEM and SEM. Stereoscopic pictures are made by tilting the sample and making an exposure at two different tilt settings. Most TEM's provide tilts of 6° to 12° while SEM's usually tilt at least 45°. To make stereoscopic micrographs it is best to utilize only that part of the sample which lies along the axis of tilt. Otherwise, as the sample is tilted, the tilted portion may move along the optical axis to such an extent that a focal change is required which will cause a rotation of the image in TEM and make it very difficult to align the stereoscopic pairs. A good review is available in Nemanic (1974) and Clarke (1975).

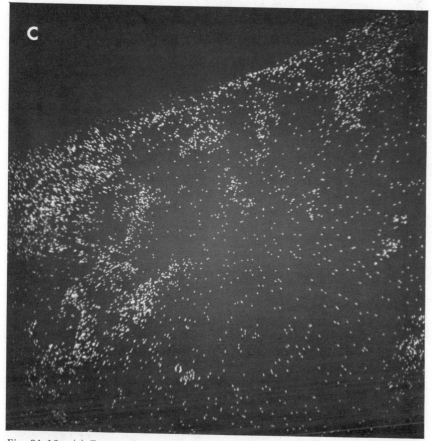

Fig. 24-12. (a) Fractured soil specimen attached onto an SEM sample mount; (b) an SEM micrograph of the soil specimen; and (c) an X-ray element map of potassium from the same area shown in (b). (micrographs, John L. Brown).

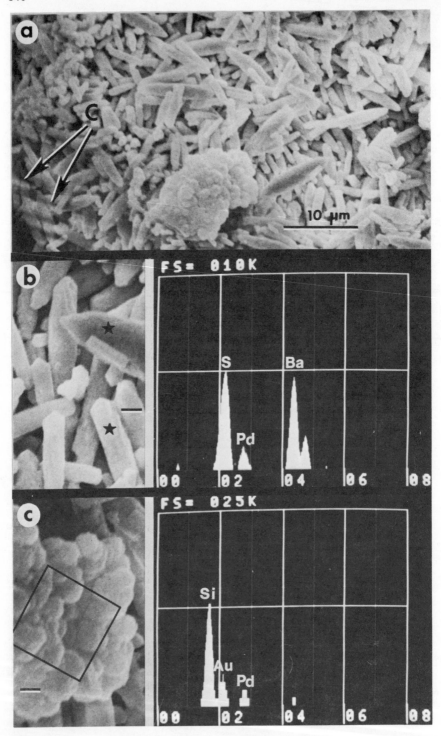

(Fig. 24-13)

Fig. 24–13. Secondary electron SEM images of an internal surface of a fractured barite nodule showing morphologies of the various crystals present. The streaks in (13a) marked (C) are due to charging of the sample. The use of an energy dispersive X-ray analyzer allowed identification of the various morphologies. Spot mode analyses at the points marked by stars in (13b) indicated the presence of Ba and S whereas analysis of the marked area in (13c) indicated the presence of Si. The displays to the right of the micrographs in (13b) and (13c) represent the CRT images of the total counts per channel versus channel energy of the X-rays analyzed. The small Au and Pd peaks are from the thin conductive coating applied to the sample. Sample from Falba no. 3 subsoil, Catahoula Formation, Walker Co., Texas (unpublished, T. R. McKee). Scale markers for (13b) and (13c) are 1.0 μm.

◀——— (Fig. 24–13 is on page 840)

VI. RECENT ADVANCES

The field of electron microscopy and SEM in particular is advancing at a very rapid rate as technology moves forward. In the limited space of this chapter we have attempted to introduce the most important and most common aspects of electron microscopy relevant to soil studies. As we have mentioned, there are many different techniques to be utilized and each sample will have a slightly different optimum sequence of necessary techniques in order to produce the maximum information.

The serious student of electron microscopy will have difficulty staying abreast of new developments. A list of references at the end of this chapter is only a beginning of the areas to be covered. One would do well to follow the proceedings of at least the three following annual meetings:

1) Scanning Electron Microscopy/(year). This meeting began in 1968 and is sponsored by IIT Research Institute, 10 West 35th Street, Chicago, IL 60616.

2) *Electron Microscopy Society of America (EMSA) Proceedings;* society established in 1942; published as a bound volume since 1967 with abstracts previously published in the *Journal of Applied Physics* (Claitor's Publishing Division, 3165 Acadian at I-10, P. O. Box 239, Baton Rouge, LA 70821).

3) Microbeam Analysis Society (MAS) formerly Electron Probe Analysis Society of America (EPASA), meeting since 1966 (Dr. J. I. Goldstein, Metallurgy & Material Science Department, Lehigh University, Bethlehem, PA 18015).

There are many technological advancements which are currently being developed that should greatly advance our research capabilities. Scanning transmission electron microscopy (STEM) is one important advancement which has significantly increased the usable specimen thickness without going to expensive high voltage EM. A normal TEM is adapted with a scanning beam which is much smaller than the normal TEM beam. SEM type image processing is used past the objective lens system. The individual portions of the sample are then exposed to the beam for very short periods of time reducing the beam damage. The new high-brightness electron sources (lanthanum hexaboride and field emission emitters) will significantly extend secondary electron detection capabilities and coupled with other advances in technology have made SEM resolution of 25Å a reality (Broers, 1973). Johari (1976) reports on numerous developments of the STEM for physical applications.

ACKNOWLEDGMENT

TRM acknowledges support from the following during manuscript preparation: Office of Naval Research, contract N00014-68-A-0308-002, and Earth Science Division of National Science Foundation grant EAR77-00128 (to P. R. Buseck).

VII. SUPPLEMENTAL READING LIST

Bahr, G. R., and E. H. Zeitler (ed.). 1965. Quantitative electron microscopy. Williams and Wilkins, Baltimore, Md.

Gard, J. A. 1971. Electron-optical investigation of clays. Mineral. Soc., London.

Heidenreich, R. D. 1964. Fundamentals of transmission electron microscopy. Interscience, New York.

Kay, D. H., ed. 1965. Techniques for electron microscopy. 2nd edition. Blackwell Scientific Pub. Ltd., F. A. Davis Co., Philadelphia, Pa.

Thomas, G. 1962. Transmission electron microscopy of metals. John Wiley & Sons, New York.

Wells, O. C. 1974. Scanning electron microscopy. McGraw-Hill, New York.

VIII. SELECTED SOURCES OF ELECTRON MICROSCOPY SUPPLIES

Ernest F. Fullam, Inc., P. O. Box 444, Schenectady, NY 12310

Ladd Research Industries, Inc., P. O. Box 901, Burlington, VT 05401

Walter C. McCrone Associates, Inc., 2820 South Michigan Ave., Chicago, IL 60616

Ted Pella Co., P. O. Box 510, Tustin, CA 92680

Polysciences, Inc., Paul Valley Industrial Park, Warrington, PA 18976

SPI Supplies, Division of Structure Probe, Inc., P. O. Box 342, West Chester, PA 19380

Tousimis Research Corp., P. O. Box 2189, Rockville, MD 20852

LITERATURE CITED

Allpress, J. G., and J. V. Sanders. 1973. The direct observation of the structure of real crystals by lattice imaging. J. Appl. Crystallogr. 6:165–190.

Andrews, K. W., D. J. Dyson, and S. R. Keown. 1971. Interpretation of electron diffraction patterns. 2nd edition. Hilger and Watts, London.

Askenasy, P. E., J. B. Dixon, and T. R. McKee. 1973. Spheroidal halloysite in a Guatemalan soil. Soil Sci. Soc. Am. Proc. 37:799–803.

Backus, R. C., and R. C. Williams. 1950. The use of spraying methods and of volatile suspending media in the preparation of specimens for electron microscope. J. Appl. Phys. 21:11–15.

Bassin, N. J. 1975. Suspended clay mineral identification by SEM and energy dispersive x-ray analysis. Limnol. Oceanogr. 20:133–137.

Bates, T. R. 1971. The kaolin minerals. p. 109–157. In J. A. Gard (ed.) Electron-optical investigations of clays. Mineral. Soc., London.

Bates, T. F., F. A. Hildebrand, and Ada Swineford. 1950. Morphology and structure of endellite and halloysite. Am. Mineral. 35:463–483.

Bennett, R. H., W. R. Bryant, and G. H. Keller. 1977. Clay fabric and geotechnical properties of selected submarine sediment cores from the Mississippi Delta. NOAA Prof. Paper no. 9, U. S. Dep. of Commerce. NOAA-Environmental Research Laboratory Publishers, Miami, Florida. 96 p.

Beautelspacher, H., and H. W. van der Marel. 1968. Atlas of electron microscopy of clay minerals and their admixtures. Elsevier, New York.

Borst, R. L., and W. D. Keller. 1969. Scanning electron micrographs of API reference minerals and other selected samples. p. 871–901. In Proc. Int. Clay Conf., Tokyo, Vol. I.

Bowles, F. A. 1968. Microstructure of sediments: Investigation with ultra-thin section. Science 159:1236–1237.

Bradley, D. E. 1965. The preparation of specimen support films. p. 58–74. In D. H. Kay (ed.) Techniques for electron microscopy. Blackwell Sci. Pub., F. A. Davis Co., Philadelphia, Pa.

Bradley, D. E. 1969. High-resolution shadow-casting technique for the electron microscope using the simultaneous evaporation of platinum and carbon. Br. J. Appl. Phys. 10:198–203.

Broers, A. N. 1973. High-resolution scanning electron microscopy of surfaces. p. 83–121. In C. A. Anderson (ed.) Microprobe analysis. Wiley-Interscience, New York.

Brown, J. A., and A. Teetsov. 1976. Some techniques for handling particles in SEM studies. p. 386–392. In O. Johari (ed.) Scanning electron microscopy/1976, IIT Research Institute, Chicago, Ill.

Brown, J. L. 1964. Laboratory techniques in the electron microscopy of clay minerals. p. 148–169. In C. I. Rich and G. W. Kunze (ed.) Soil clay mineralogy. Univ. of North Carolina Press, Chapel Hill, N. C.

Brown, J. L., and M. L. Jackson. 1973. Chlorite examination by ultramicrotomy and high resolution electron microscopy. Clays Clay Miner. 21:1–7.

Brown, J. L., and C. I. Rich. 1968. High resolution electron microscopy of muscovite. Science 161:1135–1137.

Buseck, P. R., and S. Iijima. 1974. High resolution electron microscopy of silicates. Am. Mineral. 59:1–21.

Chute, J. H., and T. M. Armitage. 1968. Alteration of clay minerals by electron irradiation. Clay Miner. 7:455–457.

Clarke, I. C. 1975. Stereographic techniques. p. 154–194. In M. A. Hayat (ed.) Principles and techniques of scanning electron microscopy, Vol. 3. Van Nostrand Reinhold, New York.

Comer, J. J. 1971. Specimen preparation. p. 81–87. In J. A. Gard (ed.) Electron-optical investigation of clays. Mineral. Soc., London.

Comer, J. J., and J. W. Turley. 1955. Replica studies of bulk clays. J. Appl. Phys. 26: 346–350.

Cook, C. F., Jr. 1962. New method for producing platinum preshadowed carbon substrates. Rev. Sci. Instr. 33:359–361.

Cosslett, V. E. 1970. Recent progress in high voltage electron microscopy. p. 341–376. In S. Amelinckx, R. Gevers, G. Remaut, and J. Van Landuyt (ed.) Modern diffraction and imaging techniques in material science. North-Holland, Amsterdam.

Cowley, J. M., and S. Iijima. 1975. The direct imaging of crystal structures. p. 123–136. In H. R. Wenk (ed.) Electron microscopy in mineralogy. Springer-Verlag, New York.

Davies, D. M. 1967. Clean electron microscope substrate films used for micrometeorite collection and study. p. 366–367. In D. J. Arceneaux (ed.) Proc. EMSA 25th annual meeting. Claitors, Baton Rouge, La.

Davis, D. W., T. G. Rochow, F. G. Rowe, M. L. Fuller, P. F. Kear, and P. K. Hamilton. 1950. Electron micrographs of reference clay minerals. Am. Pet. Inst. Proj. 49, Preliminary Report No. 6, Columbia Univ., New York.

Dixon, J. B., and T. R. McKee. 1973. Spherical halloysite formation in a volcanic soil of Mexico. Int. Congr. Soil Sci. Trans. 10th (Moscow, U.S.S.R.) VII:115-126.

Dixon, J. B., and T. R. McKee. 1974. Internal and external morphology of tubular and spheroidal halloysite particles. Clays Clay Miner. 22:127-137.

Echlin, P., and P. J. W. Hyde. 1972. The rationale and mode of application of thin films to nonconducting minerals. p. 137-146. In O. Johari (ed.) Scanning electron microscopy/1972, IIT Research Institute, Chicago, Ill.

Erol, D., R. A. Lohnes, and T. Demirel. 1976. Preparation of clay-type, moisture-containing samples for SEM. p. 769-776. In O. Johari (ed.) Scanning electron microscopy/1976, IIT Research Institute, Chicago, Ill.

Foster, R. H., and P. K. De. 1971. Optical and electron microscopic investigation of shear induced structures in lightly consolidated (soft) and heavily consolidated (hard) kaolinite. Clays Clay Miner. 19:31-47.

Fullam, E. F. 1972. A closed wet cell for the electron microscope. Rev. Sci. Instr. 43: 245-247.

Gard, J. A. 1971. Interpretation of electron micrographs. p. 27-78. In J. A. Gard (ed.) Electron-optical investigation of clays. Mineral. Soc., London.

Gillott, J. E. 1969. Study of the fabric of fine-grained sediments with the scanning electron microscope. J. Sediment. Petrol. 39:90-105.

Goldstein, J. I., and H. Yakowitz, ed. 1975. Practical scanning electron microscopy. Plenum Press, New York.

Goodhew, P. J. 1972. Specimen preparation in materials science. p. 1-180. In A. M. Glauert (ed.) Practical methods in electron microscopy, Vol. I. North-Holland/ American Elsevier, New York.

Greene, R. S. B., P. J. Murphy, A. M. Posner, and J. P. Quirk. 1974. A preparative technique for electron microscopic examination of colloid particles. Clays Clay Miner. 22:185-188.

Harris, J. E., T. R. McKee, R. C. Wilson, and U. G. Whitehouse. 1972. Preparation of membrane filter samples for direct examination with an electron microscope. Limnol. Oceanogr. 17:784-787.

Hart, R. K., T. R. Kassner, and J. K. Maurin. 1970. The contamination of surfaces during high-energy electron irradiation. Philos. Mag. 21:453-467.

Hearle, J. W. S., J. T. Sparrow, and P. M. Cross. 1972 (reprinted 1973). The use of the scanning electron microscope. Pergamon Press, Oxford.

Heidenreich, R. D. 1964. Fundamentals of transmission electron microscopy. Interscience, New York.

Hirsch, P. B., A. Howie, R. B. Nicholson, D. W. Pashley, and M. J. Whelan. 1965 (revised 1967). Electron microscopy of thin crystals. Butterworths, London.

Hulbert, M. H., and R. H. Bennett. 1975. Electrostatic cleaning technique for fabric SEM samples. Clays Clay Miner. 23:331.

Iijima, S., and P. R. Buseck. 1977. Experimental study of disordered mica structure by high resolution electron microscopy. Acta. Crystallogr., Sec. A, Vol. 33 (in press).

Ingram, P., N. Morosoff, L. Pope, F. Allen, and C. Tisher. 1976. Some comparisons of the techniques of sputter (coating) and evaporative coating for SEM. p. 75-82. In O. Johari (ed.) Scanning electron microscopy/1976, IIT Research Institute, Chicago, Ill.

Jernigan, D. L., and J. L. McAtee. 1975. Critical point drying of electron microscope samples of clay minerals. Clays Clay Miner. 23:161-162.

Johari, O., ed. 1976. Scanning electron microscopy/1976, IIT Resaerch Institute, Chicago, Ill.

Johari, O., and P. B. DeNee. 1972. Handling, mounting and examination of particles for scanning electron microscopy. p. 249-256. In O. Johari (ed.) Scanning electron microscopy/1972 (Part 1), IIT Research Institute, Chicago, Ill.

Jones, R. C., and G. Uehara. 1973. Amorphous coatings on minerals surfaces. Soil Sci. Soc. Am. Proc. 37:792-798.

Krinsley, D. H., and J. C. Doornkamp. 1973. Atlas of quartz sand surface textures. Cambridge Univ. Press, London.

Kohn, J. A., D. W. Echart, and C. F. Cook. 1971. Crystallography of the hexagonal fer-rites. Science 171:519-525.

Lee, S. Y., M. L. Jackson, and J. L. Brown. 1975a. Micaceous occlusions in kaolinite ob-served by ultramicrotomy and high resolution electron microscopy. Clays Clay Miner. 23:125-129.

Lee, S. Y., M. L. Jackson, and J. L. Brown. 1975b. Micaceous vermiculite, glauconite, and mixed-layered kaolinite-montmorillonite examination by ultramicrotomy and high resolution electron microscopy. Soil Sci. Soc. Am. Proc. 39:793-800.

McAtee, J. L., Jr., and W. Henslee. 1969. Electron-microscopy of montmorillonite dis-persed at various pH. Am. Mineral. 54:869-874.

McCrone, W. C., and J. G. Delly. 1973. The particle atlas, 2nd edition. Ann Arbor Sci-ence Pub., Ann Arbor, Mich.

McKee, T. R. 1972. Electron microscopical study of halloysite spherules. p. 526-527. In C. J. Arceneaux (ed.) Proc. EMSA 30th annual meeting. Claitors, Baton Rouge, La.

McKee, T. R., J. B. Dixon, U. G. Whitehouse, and D. F. Harling. 1973. Study of TePuke halloysite by a high resolution electron microscope. p. 200-201. In C. J. Arceneaux (ed.) Proc. EMSA 31st annual meeting. Claitors, Baton Rouge, La.

Melvin, C. 1970. Elimination of substrate scattering in electron diffraction. p. 554-555. In C. J. Arceneaux (ed.) Proc. EMSA 28th annual meeting. Claitors, Baton Rouge, La.

Mumpton, F. A., and W. C. Ormby. 1976. Morphology of zeolites in sedimentary rocks by scanning electron microscopy. Clays Clay Miner. 24:1-23.

Naymik, T. G. 1974. The effects of drying techniques on clay-rich soil texture. p. 466-467. In C. J. Arceneaux (ed.) Proc. EMSA 32nd annual meeting. Claitors, Baton Rouge, La.

Nemanic, M. 1974. Preparation of stereo slides from electron micrograph stereopairs. p. 135-148. In M. A. Hayat (ed.) Principles and techniques of scanning electron microscopy, Vol. I. Van Nostrand Reinhold, New York.

O'Brien, N. R. 1971. Fabric of kaolinite and illite floccules. Clays Clay Miner. 19:353-359.

Parsons, D. F. 1974. Structure of wet specimens in electron microscopy. Science 186:407-414.

Paulus, M., A. Dubon, and J. Etienne. 1975. Application of ion-thinning to the study of the structure of argillaceous rocks by transmission electron microscopy. Clay Miner. 10:417-426.

Pease, D. C. 1964. Histological techniques for electron microscopy. Academic Press, New York.

Phakey, P. P., C. D. Curtis, and G. Oertel. 1972. Transmission electron microscopy of fine-grained phyllosilicates in ultra-thin rock sections. Clays Clay Miner. 20:193-197.

Poppa, H. 1975. High-resolution transmission electron microscopy. p. 215-279. In J. W. Matthews (ed.) Epitaxial growth, Part A. Academic Press, New York.

Porter, J. J. 1962. Electron microscopy of sand surface texture. J. Sediment. Petrol. 32:124-135.

Preuss, L. E. 1965. Shadow casting and contrast. p. 181-194. In G. F. Bahr and E. H. Zeitler (ed.) Quantitative electron microscopy. Williams and Wilkins, Baltimore, Md.

Siegel, B. M. 1964. The physics of the electron microscope. p. 1-78. In B. M. Siegel (ed.) Modern developments in electron microscopy. Academic Press, New York.

Souza Santos, P. de, H. de Souza Santos, and G. W. Brindley. 1966. Mineralogical studies of kaolinite-halloysite clays: Part IV. A platy mineral with structural swelling and shrinking characteristics. Am. Mineral. 51:1640-1648.

Thomas, G. 1962. Transmission electron microscopy of metals. John Wiley & Sons, New York.

Uyeda, N., P. T. Hang, and G. W. Brindley. 1973. The nature of garnierites—II. Electron-optical study. Clays Clay Miner. 21:41-50.

Whitehouse, U. G., and R. McCarter. 1958. Diagenetic modification of clay mineral types in artificial sea water. Clays Clay Miner. 5:81-119.

Yada, K. 1967. Study of chrysotile asbestos by high resolution electron microscope. Acta. Crystallogr. 23:704-707.

Yada, K. 1971. Study of microstructure of chrysotile asbestos by high resolution electron microscopy. Acta. Crystallogr. A27:659–664.

Yoshida, T. 1973. Elementary layers in the interstratified clay minerals as revealed by electron microscopy. Clays Clay Miner. 21:413–420.

Zvyagin, B. B. 1967. Electron-diffraction analysis of clay mineral structures. Plenum, New York.

Preparation of Specimens for Infrared Analysis

JOE L. WHITE, Purdue University, West Lafayette, Indiana

I. INTRODUCTION

Some of the principles and potential uses of infrared techniques in soil studies have been pointed out previously (Fripiat, 1960; Lyon, 1964; Mortensen et al., 1965; Chaussidon, 1972a, 1972b; Fieldes et al., 1972; Mitchell et al., 1964). Farmer (1968), Farmer and Russell (1967), Farmer et al. (1968), and White (1971) have reviewed some of the earlier advances in this area. Tuddenham and Stephens (1971) have discussed in a thorough manner the application of infrared techniques to the solution of geochemical and mineralogical problems. The most comprehensive references on infrared studies of minerals include the monograph, *The Infrared Spectra of Minerals*, edited by Farmer (1974), a chapter on "Characterization of Soil Minerals by Infrared Spectroscopy" by Farmer and Palmieri (1975), and an *Atlas of Infrared Spectroscopy of Clay Minerals and Their Admixtures*, by van der Marel and Beutelspacher (1976). Those planning to undertake infrared studies of soils minerals will find these authoritative treatments invaluable aids.

It must be emphasized that infrared spectroscopy is most powerful as a research tool in soil mineralogy when it is used in conjunction with X-ray diffraction and other techniques. It corroborates some of the information provided by other analytical techniques as well as furnishing unique information about composition, orientation of bonds, etc., in specific cases.

A. Origin of Spectra

No attempt will be made to consider the details of the theoretical principles of infrared spectroscopy. The reader should consult texts such as those by Conley (1972) and Cross and Jones (1969). The monograph edited by Farmer (1974) also provides an up-to-date treatment of the theory of the interaction of infrared radiation with crystals and the application of this in infrared studies of minerals.

This section will briefly describe the origin of spectra and the terms used to describe a position in the infrared range of the electromagnetic spectrum. This is followed by a short statement and references relative to factors which influence the interaction of soil minerals with infrared radiation.

The region of the electromagnetic spectrum of interest in the application of infrared spectroscopy to the study of minerals is that region consist-

chapter 25

847

ing of radiant energies of slightly greater wavelengths than those associated with visible light. The terms used to specify a position in the infrared range are the wavelength (λ), in units of micrometers (μm) per wavelength, and a so-called frequency or wavenumber (ν), in units of waves per centimeter (cm^{-1}). The relationship of wavenumber to wavelength may be expressed as follows:

$$\nu \ (cm^{-1}) = 1/\lambda \ (cm) = 10^4/\lambda \ (\mu m).$$

True frequency ($\bar{\nu}$) is related to so-called frequency or wavenumber (ν) in the following manner: $\bar{\nu} \ (sec^{-1}) = c\nu$ where c is the velocity of light (3×10^{10} cm/sec). The so-called frequency unit ν (cm^{-1} or waves per cm) is used rather than the more fundamental unit of true frequency $\bar{\nu}$ (cycles or waves per sec) as a matter of convenience. *Wavenumber* is the preferred term for the frequency designation. The region of primary interest in the infrared range for mineral studies is between wavelengths of 2.5 and 50 μm (wavenumbers of 4000 and 200 cm^{-1}).

The frequency of oscillation of atoms and molecules about their equilibrium positions is 10^{13} to 10^{14} cycles per sec. Infrared radiation has frequencies in this range and promotes transitions in a molecule between rotational and vibrational energy levels of the ground electronic energy state.

There are two types of bond vibration modes in simple molecules, stretching and bending (deformation); the former refers to the periodic stretching of the bond A–B along the bond axis. Bending modes for the A–B bond involve displacements which occur at right angles to the bond axis. These vibrations produce periodic displacements of atoms with respect to one another, causing a change in interatomic distance. When vibrations are accompanied by a change in dipole moment they give rise to absorption of radiation in the infrared region.

B. Spectra-Composition-Structure Relationships

The interaction of soil minerals with infrared radiation is dependent upon chemical composition, crystal structure, crystallinity, and other related factors. For discussions of assignments of infrared absorption bands to structural elements and groups the reader is referred to the monograph on *The Infrared Spectra of Minerals* edited by Farmer (1974). Published studies of spectra-composition relationships for specific groups of minerals include those for micas and smectites (Stubican & Roy, 1961a, 1961b; Vedder, 1964; Farmer & Russell, 1964; Wilson, 1966, 1970; Farmer et al., 1967; Wilkins, 1967; Ishii et al., 1969; Russell et al., 1970; Akhmanova & Alekhina, 1971; Russell & Fraser, 1971; Farmer et al., 1971), chlorites (Hayashi & Oinuma, 1965, 1967), zeolites (Oinuma & Hayashi, 1967; Flanigen et al., 1971), and carbonates, sulfates, nitrates, and phosphates (Moenke, 1962, 1966; Lehr et al., 1967).

C. Instrumentation

Most infrared spectrometers produced currently use grating or prism-grating systems and have much improved resolution compared to earlier prism instruments. The wavelength range has also been extended to limits ranging from 25 to 50 μm (wavenumbers 400 to 200 cm^{-1}). These improvements are important since spectral features in the 400 to 600 cm^{-1} region are very useful in the identification and characterization of minerals.

Prism instruments are relatively inexpensive, but are being replaced by grating instruments of various degrees of sophistication and cost. The use of a "slave recorder" accessory on some of the inexpensive prism instruments to expand the wavenumber scale can give resolution that is satisfactory for purposes of identification.

For details of spectrometer operations, optimization of recording parameters, calibration of the spectrometer, etc., the reader should consult operation manuals provided by the instrument manufacturer as well as practical guides to infrared instrumentation and techniques (Potts, 1963; Miller & Stace, 1972).

II. EXPERIMENTAL TECHNIQUES

A. Sample Pretreatment

Soils are usually subjected to various pretreatments designed to remove organic matter, iron oxides, etc. and enhance dispersion. It should be emphasized that these treatments may produce artifacts (Farmer & Mitchell, 1963) as well as change the properties of the minerals so that the infrared spectra are not representative of the mineral components in their original condition. As an example, oxidation treatments used to decompose organic matter will also oxidize Fe^{2+} in biotite.

In the case of soils whose general composition is unknown it may be worthwhile to make a preliminary examination of the infrared spectra of the clay fraction from some representative horizons of the soil profile before application of routine treatments for dispersion, organic matter removal, deferration, etc. A sample of the whole soil, after appropriate grinding treatment, may also be used in the preliminary examination (Fieldes et al., 1972). Identification of the predominant minerals can provide a basis for application of the most appropriate treatments for more detailed studies.

One of the major problems encountered in the application of infrared spectroscopy to the study of soil minerals is the distortion of absorption bands due to light scattering by the mineral particles. Prerequisites for obtaining satisfactory infrared spectra include having mineral samples whose particle size is less than the wavelength of the infrared radiation to be used. Since the usual wavelength range is from 2.5 to 50 μm (wavenumbers 4000 to 200 cm^{-1}), the clay fraction ($<$ 2 μm) is of appropriate size and requires no further reduction in particle size. Silt and sand fractions need to be reduced in size by careful grinding.

This light scattering due to particle size is referred to as the Christiansen effect; the effect diminishes as particle size is reduced (Duychaerts, 1959). Polycomponent mineral systems whose components differ in particle size cannot be quantitatively analyzed by infrared techniques because of the Christiansen effect. This difficulty cannot be overcome by grinding the sample to reduce the particle size of the components having the larger particle size since grinding differentially affects the other components because of variations in hardness of the minerals (Jonas et al., 1973).

Particle size reduction may be accomplished by filing, by grinding, and by sonification.

Filing is most appropriate for preparing samples from layer silicate minerals of hand specimen dimensions, such as chlorite and mica. Filing may be accomplished by using a small glass file and using a filing motion that is perpendicular to the planar surfaces of the mineral. Sedimentation may be used to separate the various particle size fractions. Filing tends to minimize structural modifications in layer silicates whereas the pressure and heat generated in grinding, especially dry grinding, can result in drastic changes in structure and properties.

There are two techniques for grinding: dry grinding and wet grinding. Dry grinding, whether by mortar and pestle or by ball-milling, causes destruction of the mineral structure and modification of the chemical and physical properties and should not be used for particle size reduction of samples to be examined by infrared spectroscopy. Wet grinding, in which a liquid such as acetone or ethanol is used to reduce the friction and pressure of the grinding operation, minimizes structural changes in minerals. Wet grinding may be carried out by placing 10 to 15 mg of the sample in a mullite or agate mortar, then adding 10 to 15 drops of ethanol to the mortar with an eye dropper. The sample is then ground with a vigorous rotary motion, confining the sample to 1/3 to 1/2 of the mortar surface until the ethanol evaporates completely. Do not continue grinding after the sample becomes dry. The wet grinding step may need to be repeated several times in order to reduce the particle size to $<$ 2 μm. Levitt and Condrate (1970) have described a wet grinding procedure in which the mineral particles are impacted with themselves in a test tube containing acetone in a Wig-L-Bug vibrator.

Particle-size reduction may also be achieved by sonification—treatment of a suspension of the mineral particles with ultrasonic frequencies. The frequency and time period of treatment need to be adjusted through experimentation in order to minimize any structural modifications.

Farmer (1974, p. 354) suggests that clays should be saturated with K to prevent the retention of cation hydration water to high temperatures which may occur with smaller exchangeable cations such as Ca and Mg. It is recommended that clay fractions to be used in preparation of KBr disks be freeze dried or dried from benzene after washing in ethanol.

B. Techniques of Sample Presentation

1. ABSORPTION

The most commonly employed technique for the infrared study of minerals involves absorption of infrared radiation by a thin layer of the sample placed directly in the infrared beam, the orientation of the sample layer or film being essentially perpendicular to the direction of the infrared beam. Several methods for preparation and presentation of mineral specimens for infrared absorption measurements are described below.

a. Sedimented Films on Infrared Transparent Windows—Material which has been reduced in size to < 2 μm is dispersed in water and an amount of suspension sufficient to give a concentration of 1 to 2 mg/cm^2 is placed on infrared transparent windows, such as AgCl or Irtran, and allowed to dry. Glass microscope slides and cover glasses are useful window materials for the region in which most OH-stretching frequencies occur (wavenumbers 4000 to 3000 cm^{-1}). Use of a duplicate glass slide or cover glass will compensate the OH absorption bands of the glass. Volatile organic liquids, e.g. isopropyl alcohol, may also be used as a suspending medium to hasten drying.

b. Self-supporting Films—Self-supporting films of many smectites and vermiculites may be formed by evaporating aqueous suspensions on foil or plastic films and separating the mineral film by pulling the foil or plastic sheet over a sharp edge (Farmer, 1968). The mineral film is more rigid than the plastic or foil and will separate, much in the manner of certain packaged self-adhesive labels. The mineral films should have a density of 1 to 2 mg/cm^2.

Smooth polyethylene film such as used in storage of frozen foods, etc. may be used. A piece of film 15 to 20 cm in width and of a convenient length is placed on a level sheet of plate glass on which has been placed several milliliters of distilled water. A straight-edge ruler covered with a piece of the plastic is used as a squeegee to smooth the plastic film so that the capillary film of water serves to hold the plastic film flat until the mineral suspension dries. A convenient volume (10 to 25 ml) of a 2% suspension of the mineral colloid is placed on the plastic film using a pipette held vertically at a fixed position. This will give a film having a circular shape with a uniform thickness everywhere except at the edge of the circle.

Mylar film may also be used for preparation of self-supporting films. Mylar film is much stronger and smoother than polyethylene and may be fastened to the sheet of plate glass by use of cellophane tape, pulling it taut to make it as smooth as possible. The suspension is added in the manner described in the preceeding paragraph.

After the mineral film is air dry the plastic with the attached film is carefully separated from the sheet of plate glass. The mineral film is then detached from the plastic film by slowly pulling the plastic film over a sharp edge at an angle of 90° or greater.

The clay film may then be cut into flakes of the desired size and shape with a sharp razor blade or cork borer. Because of the tendency of the clay films to curl, they should be stored between sheets of stiff paper until they are ready to be used.

The clay films may be conveniently mounted in the infrared spectrophotometer by placing between two pieces of blotter paper (5.0 cm by 7.6 cm) in which openings of the desired size and in the correct position to intercept the infrared beam have been cut. When the clay film has been positioned in the proper place, a staple may then be placed in the blotter paper just outside each edge of the film. The clay film will be held firmly in position and can be removed for further treatment and studies if desired.

If the clay film is to be used in studies involving dehydration or heating in a vacuum infrared cell, it may be mounted in a holder fashioned of Al foil. The clay film in the holder can be easily moved from one part of the cell to another by tilting the cell or gently tapping the cell until the holder is in the proper position for recording the spectra.

Self-supporting films of amorphous silica-alumina gels and synthetic zeolites can be prepared by pressing the powders between polished steel faces at a pressure of about 10,000 psi (700 kg/cm^2; 68.9MPa) (McDonald, 1958). Such films are very fragile and must be handled with great care.

c. **Alkali Halide Disk Technique**—Potassium bromide powder can be pressed at about 10,000 psi (700 kg/cm^2; 68.9 MPa) into clear disks having high transmission throughout the 4000 to 250 cm^{-1} range. Mineral samples may be mixed with the KBr powder prior to pressing at a concentration level of 0.1 to 2% and their spectra obtained in the KBr matrix. As in the mulling technique, the sample must be < 2 μm in order to reduce scattering losses and absorption band distortions.

A word of caution concerning the KBr technique must be given at this point. Modification in the spectra of some minerals may occur due to ion exchange with the KBr matrix, pressure effects, or transformations to other crystalline forms or polytypes. These possibilities must be kept in mind when comparing unknown spectra obtained as KBr disks with known or reference spectra obtained by other techniques.

The amount of sample and KBr to be used in making the disk depends upon the size of the disk desired. For a disk 1 mm in thickness and 13 mm in diameter, approximately 0.5 to 3 mg of sample in 300 mg of KBr will suffice. Intimate mixing of the sample and the KBr may be accomplished by placing the sample-halide mixture in a steel capsule with steel balls then vibrating in a Wig-L-Bug vibrator for 2 min.

Satisfactory mixing may also be carried out by hand, using a mortar and pestle. The appropriate amount of the mineral sample (0.5 to 3 mg) is weighed on a microbalance and placed in a clean 50- or 65-mm (OD) agate or mullite mortar. Next, 300 mg ± 5 mg of dry KBr powder (infrared quality such as Harshaw or Isomet) which has been dried at 105C (378K) for 12 hours and stored in a desiccator over P_2O_5, is weighed and 5 to 10 mg of the

KBr is added to the mortar containing the sample. The pestle is used to mix the KBr and the sample with a gentle rubbing motion. The purpose of this manipulation is only to mix the sample and the KBr; *do not grind the KBr* during this mixing step as it will lead to adsorption of water on the KBr. Fifteen milligrams of KBr is added and mixed as before. This is followed by the addition of another amount of KBr approximately equal to the total quantity in the mortar (i.e., add about 30 mg) and further mixing. Further addition and mixing of KBr in the above manner are continued until all the KBr has been added to the mortar. This procedure will produce a homogeneous mixture of the sample and the KBr and minimize the adsorption of water by the KBr. The mixture may be dried in an oven or vacuum oven for 1 hour at 378 to 383K before pressing the disk. A small camel's hair brush is used to quantiatively transfer the mixture to the die for pressing the disk.

The disk is pressed as follows: the lower plunger of the die is placed in position at the bottom of the die body. Sufficient mixed sample-halide powder to give a disk 1-mm thick is poured in and tamped lightly with the upper plunger. The small upper plunger is then placed on top of the compacted powder, the upper plunger inserted, and the die placed in a hydraulic press. Very slight pressure is placed on the die so as to produce a good air seal. A vacuum pump is attached and the air is exhausted from the die for 5 min; following this a pressure of 9,000 to 11,000 kg (20,000 to 24,000 lb) total force is applied to the press and maintained for about 5 min. The vacuum is first released, then the pressure on the hydraulic press is released and the disk removed from the die by forcing the plunger to extrude the disk from the bottom of the die. The disk should be transparent; care should be taken not to touch the disks with the hands.

Potassium bromide disks are not suitable for studies involving absorbed water or surface cations. The possibility of cation exchange reactions must be considered in cases where this may alter the initial state of the mineral. The hygroscopic nature of KBr can be minimized by drying the disk in a vacuum oven at 50C (323K); the disk may be dried at higher temperatures, but will need to be repressed in the die to restore the transparency of the disk. Nortia and Kontas (1973) have described an improved KBr pellet method for the measurement of infrared spectra of hydroscopic compounds.

Farmer (1968) recommends recording spectra at two concentrations in KBr disks. Depending on the specific mineral, concentrations of from 0.5 to 5 mg of sample in 300 mg of KBr for a 13-mm diameter disk should be sufficient. The use of wedge-shaped KBr pellets mounted on a vernier sample holder permits one to produce spectra for various concentrations with a single sample.

A rapid procedure that does not require an expensive die is as follows: the KBr-sample mixture is placed in a 13-mm circular cavity in 100-lb blotting paper with a layer of Al foil above and below the sample in the cavity and this sandwich is pressed between two stainless steel die (20-mm diameter) in a hydraulic press at a total pressure of 9,000 to 11,000 kg (20,000 to 24,000 lb).

d. **Oil Mulls**—One of the most useful techniques for preparing a soil mineral sample for infrared analysis is mulling. The reduction of particle size to an average diameter < 2 μm is most important when preparing a mull. This usually is not necessary for the clay fraction of soils (< 2 μm), but is required for silts, sands, and for samples of whole soil.

The most common mulling agent is Nujol (paraffin oil, obtainable from a drug store) which for many spectra can be used over the full wavelength range of the instrument. To eliminate the possibility of overlapping of bands of the Nujol and bands for certain minerals, i.e. carbonates, in the 1400 cm^{-1} region, it may be desirable to use Fluorolube LG 160 (Hooker Chemical), a perfluorohydrocarbon, over the range from 4000 cm^{-1} to 1350 cm^{-1}, and Nujol from 1380 cm^{-1} to 650 cm^{-1}. Nujol absorbs strongly in the 3000 to 2800 cm^{-1} region as well as at 1460 and 1375 cm^{-1}, but is transparent elsewhere. Fluorolube has essentially no absorption from 4000 to 1400 cm^{-1}.

Place about 10 to 15 mg of the mineral powder in a 50- or 65-mm diameter mullite or agate mortar. Add a small drop of the mulling agent to the mortar by using the end of a partially unfolded paper clip. Grind the sample with a vigorous rotary motion until all the material is suspended in the mulling agent. It may be necessary to add another small drop of the mulling agent as this grinding is continued. The consistency of the final mixture is about that of vaseline.

The mull may be removed from the mortar and transferred to an infrared-transparent window (Irtran, AgCl, NaCl, etc.) by use of a rubber policeman that has been carefully cleaned to remove all traces of talc remaining from its manufacture. The mull may also be collected by means of a microspatula having a curved blade; the mull is scraped off the surface of the mortar by holding the spatula with the convex side up and collecting the mull at the leading edge on the convex side. After the mull is transferred to one window, a second window is placed on top of the first, and the sample evenly distributed between the plates. The concentration of the mull can be adjusted by gently squeezing out the excess mull, or by separating the the windows to give approximately one-half the initial concentration on each window. The film between the windows should appear slightly translucent and should be free of air bubbles when examined in front of a light.

It should be noted that specimens prepared as films and mulls will have some degree of preferred orientation, whereas specimens prepared in KBr disks will tend to show random orientation of particles. This difference is of importance in samples which show dichroic behavior because of the orientation of OH dipoles or other bonds.

2. ATTENUATED TOTAL REFLECTION (ATR)

Attenuated total reflection, also referred to as frustrated multiple internal reflection (FMIR), represents a different approach to sampling and was designed primarily for materials that are opaque to infrared radiation.

In this method, a sample held in close optical contact with a prism surface acts to absorb a portion of the radiation striking the prism surface. The reflected spectrum is characteristic of the sample. Harrick (1967) has described this technique in his book *Internal Reflection Spectroscopy*. ATR spectra of several rocks and minerals have been obtained on powder samples (Harrick & Riderman, 1965).

3. REFLECTION

Reflection is concerned with specular or diffuse reflection from polished or powdered samples. Hidalgo and Serratosa (1961) showed that clay minerals have distinctive reflection spectra in the 200 to 650 cm^{-1} region. Lyon (1962) used specualr reflection spectra to broadly classify rocks and minerals. Lindberg and Snyder (1972) have recently published diffuse reflectance spectra of several clay minerals.

III. IDENTIFICATION

Most soil mineral fractions are polycomponent systems and unless one mineral is predominant, the complexities of the overlapping infrared vibrations may preclude a straightforward interpretation.

The identification of soil minerals by infrared spectroscopy is facilitated by reference to spectra of well-characterized specimens. Valuable collections of infrared spectra of minerals include those published by Moenke (1962, 1966), Afremow and Vandeberg (1966), Afremow et al. (1970), Farmer (1974), and van der Marel and Beutelspacher (1976). The monograph edited by Farmer (1974) also includes lists of references of spectra for oxides (p. 196-197), orthosilicates (p. 298-299), chain and ring silicates (p. 328-329), layer silicates and related minerals (p. 358-359), phosphates (p. 420-422), and sulfates (p. 439-440). Farmer (1968) published an excellent list of references to infrared spectra of minerals that occur in soils. Lehr et al. (1967) have published infrared spectra for 207 compounds that are constituents in fertilizers or which may form in reactions with the soil; the majority of these are phosphate compounds.

Kodama and Oinuma (1963) have applied infrared spectroscopy in the identification of minerals in polycomponent mineral systems. They showed that the high frequency OH-stretching band at about 3698 cm^{-1} was effective in distinguishing kaolinite from other clay minerals in sediments in which chlorite and kaolinite were dominant. Information on the type of kaolinite, degree of crystallinity, and particle shape may be obtained through techniques developed by Parker (1969).

Spectral features characteristic of chlorites of a wide range of composition have been described by Hayashi and Oinuma (1965, 1967) and Oinuma and Hayashi (1966).

Farmer and Russell (1964) published infrared spectra for both dioctahedral and trioctahedral micas which are very useful in the identification of micaceous minerals.

The characterization of mixed-layer minerals (Oinuma & Hayashi, 1965; Veniale & van der Marel, 1969) and hydroxy interlayers in expanding minerals (Ahlrichs, 1968; Brydon & Kodama, 1966; Weismiller et al., 1967) has been facilitated by infrared techniques.

The infrared spectral features of aluminum hydroxide minerals have been well illustrated by Caillère and Pobeguin (1965); this reference includes infrared spectra of many well-characterized specimens of bauxites in which gibbsite, boehmite, diaspore, kaolinite, goethite, and hematite were found to occur.

Since the characteristic infrared vibrations of a particular group, e.g. hydroxyls, begin to occur as soon as nucleation of a specific structural environment commences, infrared spectroscopy is very useful in detecting short-range order in poorly crystalline materials. With increasing attention being given to the amorphous components in soils, infrared techniques are being used more frequently to detect and characterize these materials (Kanno et al., 1968; Léonard et al., 1967; Léonard et al., 1964; Mejia et al., 1968; Mitchell et al., 1964; Wada & Greenland, 1970; Wada & Tokashiki, 1972).

Characteristic infrared bands for selected crystalline soil minerals are tabulated in Table 25-1. Chemical and structural variations will determine the values for a specific mineral. The wavenumber values given in Table 25-1 are not to be considered as invariant.

In addition to band positions, the sharpness or diffuseness of bands may be helpful in identification of mineral components. For example, the presence of a very broad OH-stretching band in the region 3000–3600 cm^{-1}, the absence of well-defined OH-stretching vibrations characteristic of the kaolinite and smectite groups, together with a very broad deformation band for water in the 1640 cm^{-1} region, make possible the detection of high surface area amorphous components. This is illustrated in Fig. 25–1 in the spectrum of a Columbian soil formed from volcanic ash (Chinchina profile no. 3 (Mejia et al., 1968)). The above spectral features plus the diffuse band in the 1050 cm^{-1} region suggest the material is probably allophane.

The ease of identification of well-crystallized kaolinite and gibbsite by infrared absorption is apparent in Fig. 25-2 which shows the spectrum of a Columbian soil formed on unconsolidated sediments (Buenaventura profile no. 13 (Mejia et al., 1968)). The broad and moderately intense band at 1640 cm^{-1} suggests the presence of some high surface area amorphous materials along with the crystalline phases.

The characteristic spectral features of halloysite are shown in Fig. 25-3 which is the infrared spectrum of the clay fraction of a Columbian soil developed on volcanic ash (Tesorito profile no. 2 (Mejia et al., 1968)). These features include the two broad OH-stretching vibrations at 3690 and 3620 cm^{-1} and a very broad OH-stretching vibration for adsorbed water at about 3400 cm^{-1}.

Table 25-1. Infrared absorption bands for selected soil minerals (White, 1971)

Mineral	Wavenumber (cm^{-1})																													
Kaolinite	3695	3670	3650	3620									1108		1038	1012		940	915				700			540	472		
Halloysite	3695			3620			3400†		1640‡				1100		1040	1020			918				695			545	474		
Montmorillonite				3620			3400†		1640‡				1100		1040	1020			915							520	470		
Nontronite					3560		3400†		1640‡				1130		1050	1020											490	430	
Muscovite				3628									1120							827		750				535	480		
Biotite				3658	3550											1020		928		828		760	690				465	445	
Vermiculite					3550		3380†		1640‡							1000	985							670			480		
Chlorite Al-rich				3620		3520	3340									1004					812		692			528	475		
Gibbsite				3610		3525	3445	3395										940			825	745		670		560 540			
Quartz											1172	1102	1084	1030		975					800	780	697	670		540			
Microcline													1110	1030		1000					800	769	727		647	512	462		
Calcite										1435										877					712				

† OH-stretching frequency for water molecules.
‡ OH-bending frequency for water molecules.

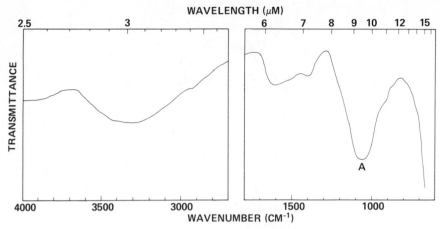

Fig. 25-1. Infrared spectrum of clay fraction of a Colombian soil formed from volcanic ash (Chinchina profile no. 3, 0–35 cm) (Mejia et al., 1968) (A = allophane).

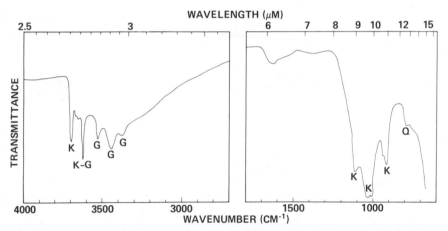

Fig. 25-2. Infrared spectrum of clay fraction of a Colombian soil formed on unconsolidated sediments (Buenaventura profile no. 13, 50–60 cm) (Mejia et al., 1968) (K = kaolinite; G = gibbsite; Q = quartz).

The above samples were prepared as films deposited on Irtran-2 windows.

Confirmatory tests may be necessary to further establish the identity and properties of the components in a sample after cursory identification. These tests may include recording infrared spectra of the sample following dehydration and dehydroxylation treatments (Russell & Farmer, 1964; Tarasevich, 1969). In addition, surface reactions such as ease and extent of deuteration of water and structural hydroxyls, may provide supporting information (Ledoux & White, 1965). The adsorption of water and other molecules by minerals and the nature of the interactions as revealed by infrared absorption techniques have been reviewed by Farmer (1971). Little

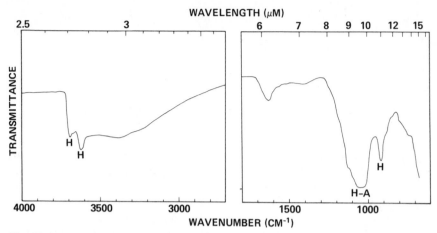

Fig. 25-3. Infrared spectrum of clay fraction of a Colombian soil formed from volcanic ash (Tesorito profile no. 2, 85–125 cm) (Mejia et al., 1968) (H = halloysite; A = allophane).

(1966) has made a very useful survey of surface studies of minerals by infrared techniques. Dichroic studies of oriented specimens may aid in determining the chemical composition of the octahedral layer and the orientation of the dipole moments of hydroxyl groups in minerals (Juo & White, 1969). Some of the above confirmatory tests will necessitate the use of special cells similar to those described by Russell (1974).

IV. QUANTITATIVE ANALYSIS

Application of infrared techniques in quantitative estimation of the composition of mineral mixtures has been discussed by Lyon et al. (1959). Quantitative analysis of liquids by infrared spectroscopy is quite feasible as can be shown by both theoretical and experimental considerations. Minerals, however, do not lend themselves so readily to quantitative analysis by infrared techniques because they are particulate in nature and light scattering phenomena (the Christiansen effect) impose inherent limitations on the relationship between the concentration of a given mineral component and the infrared radiation (Duychaerts, 1959). Problems associated with particle-size reduction in mineral mixtures (Tuddenham & Lyon, 1960) as well as the lack of "standard" reference minerals having the same structure, particle size distribution, composition and spectral features as the component in the sample (Jonas et al., 1973) further limit the cases in which quantitative determinations can be successfully made. A recent example of quantitative analysis of minerals has been given for the favorable case of α-quartz by Mangia (1975). Use of a membrane support technique gave better results than did the use of KBr disks. The KBr pressed disk technique is most commonly used in quantitative studies; adsorbed water may be removed by heating at 100C (373K)

(Farmer et al., 1968). Lyon (1964) should be consulted for details of this procedure.

V. FUTURE DEVELOPMENTS

In a manner somewhat parallel to the development of chemical and heat treatments for the identification and characterization of minerals by X-ray diffraction techniques, we may expect the development of routine chemical and thermal treatments for the identification of specific minerals by infrared techniques. Gas or vacuum cells with arrangements for convenient heating and treating of several samples simultaneously will expedite the development and use of such techniques. Infrared techniques such as those used for characterization of zeolite surfaces (Ward, 1971) may be directly applicable to soil mineral systems. Some of the adsorption reactions discussed by Farmer (1971) are also useful for this purpose.

The compilation of a catalog of infrared spectra of well-characterized minerals by a cooperative effort within or among the several mineralogical organizations would provide considerable impetus for more extensive use of infrared spectroscopy in mineralogical studies. The Clay Minerals Society has initiated such a program for a limited number of reference clay minerals.

LITERATURE CITED

Afremow, L. C., K. E. Isakson, D. A. Netzel, D. J. Tessari, and J. T. Vandeberg. 1970. IR spectroscopy. Its use in the coatings industry. Fed. Soc. Paint Tech., Philadelphia, Pa.

Afremow, L. C., and J. T. Vandeberg. 1966. High resolution spectra of inorganic pigments and extenders in the mid-infrared region from 1500 cm^{-1} to 200 cm^{-1}. J. Paint Tech. 38:169-202.

Ahlrichs, J. L. 1968. Hydroxyl stretching frequencies of synthetic Ni-, Al-, and Mg-hydroxy interlayers in expanding clays. Clays Clay Miner. 16:61-71.

Akhmanova, M. V., and L. G. Alekhina. 1971. Infrared spectroscopic study of isomorphism in minerals. p. 243-267. In A. P. Vinogradov (ed.) Probl. izomorinykh zameschchenii at. krist. (Russ) "Nauka", Moscow, USSR.

Brydon, J. E., and H. Kodama. 1966. The nature of aluminum hydroxide-montmorillonite complexes. Am. Mineral. 51:875-889.

Caillère, S., and Th. Pobeguin. 1965. Considérations générales sur la composition minéralogique et la genèse des bauxites du midi de la France. Mem. Mus. Nat. Hist. Nat. Paris Ser. C 12(4):125-212.

Chaussidon, J. 1972a. Application of infrared spectroscopy in the study of clay minerals. Ceramurgia 3:210-213 (Ital.).

Chaussidon, J. 1972b. Application of infrared spectroscopy to the study of mineral weathering. p. 53-56. Potassium soil, Proc. Colloq. Int. Potash Inst., Bern, Switzerland.

Conley, R. T. 1972. Infrared spectroscopy. 2nd ed. Allyn & Bacon, Inc., Boston, Mass.

Cross, A. D., and R. A. Jones. 1969. Introduction to practical infrared spectroscopy. Plenum, New York.

Duyckaerts, G. 1959. The infrared analysis of solid substances. A review. Analyst 84:201-214.

Farmer, V. C. 1968. Infrared spectroscopy in clay mineral studies. Clay Miner. 7:373-387.

Farmer, V. C. 1971. The characterization of adsorption bonds in clays by infrared spectroscopy. Soil Sci. 112:62–68.

Farmer, V. C. (ed.). 1974. The infrared spectra of minerals. Mineral. Soc., London.

Farmer, V. C., and B. D. Mitchell. 1963. Occurrence of oxalates in soil clays following hydrogen peroxide treatment. Soil Sci. 96:221–229.

Farmer, V. C., and F. Palmieri. 1975. The characterization of soil minerals by infrared spectroscopy. p. 573–670. In J. E. Gieseking (ed.) Soil components. Vol. 2, Inorganic components. Springer-Verlag, Berlin.

Farmer, V. C., and J. D. Russell. 1964. The infrared spectra of layer silicates. Spectrochim. Acta 20:1149–1173.

Farmer, V. C., and J. D. Russell. 1967. Infrared absorption spectrometry in clay studies. Clays Clay Miner. 15:121–142.

Farmer, V. C., J. D. Russell, J. L. Ahlrichs, and B. Velde. 1967. Vibration du groupe hydroxyle dans les silicates en couches. Bull. Groupe Fr. Argiles 19:5–10.

Farmer, V. C., J. D. Russell, and J. L. Ahlrichs. 1968. Characterization of clay minerals by infrared spectroscopy. Int. Congr. Soil Sci. Trans. 9th (Adelaide, Aust.) 3:101–110.

Farmer, V. C., J. D. Russell, W. J. McHardy, A. C. D. Newman, J. L. Ahlrichs, and J. Y. H. Rimsaite. 1971. Evidence of loss of protons and octahedral iron from oxidized biotites and vermiculites. Mineral. Mag. 38:121–137.

Fieldes, M., R. J. Furkert, and N. Wells. 1972. Rapid determination of constituents of whole soils using infrared absorption. New Z. J. Sci. 15:615–627.

Flanigen, E. M., H. Khatami, and H. Szymanski. 1971. Infrared structural studies of zeolite frameworks. Adv. Chem. Ser. 101:201–229.

Fripiat, J. J. 1960. Application de la spectroscopie infrarouge à l'étude des mineraux argileux. Bull. Groupe Fr. Argiles, 12(7):25–41.

Harrick, N. J. 1967. Internal reflection spectroscopy. Interscience Publishers, Inc., New York.

Harrick, N. J., and N. H. Riederman. 1965. Infrared spectra of powders by means of internal reflection spectrometry. Spectrochim. Acta 21:2135–2139.

Hayashi, H., and K. Oinuma. 1965. Relationship between infrared absorption spectra in the region of 450–900 cm^{-1} and chemical composition of chlorite. Am. Mineral. 50:476–483.

Hayashi, H., and K. Oinuma. 1967. Si–O absorption band near 1000 cm^{-1} and OH absorption bands of chlorite. Am. Mineral 52:1206–1210.

Hidalgo, A., and J. M. Serratosa. 1961. Infrared reflection spectra of clay minerals. An. R. Soc. Esp. Fis. Quim. 57A:225–230.

Ishii, M., M. Nakahira, and H. Takeda. 1969. Far infrared absorption spectra of micas. Proc. Int. Clay Conf., Tokyo. 1:247–259.

Jonas, K., K. Solymar, and M. Orban. 1973. Phase analysis and characterization of bauxites and red muds by infrared spectrophotometry. Proc. 3rd Int. Congr. Int. Com. for the Study of Bauxites and Aluminum Oxides-Hydroxides, Nice. p. 325–330.

Juo, A. S. R., and J. L. White. 1969. Orientation of the dipole moments of hydroxyl groups in oxidized and unoxidized biotite. Science 165:804–805.

Kanno, I., Y. Onikura, and T. Higashi. 1968. Weathering and clay mineralogical characteristics of volcanic ashes and pumices in Japan. Int. Congr. Soil Sci. Trans. 9th (Adelaide, Aust.) 3:111–131.

Kodama, H., and K. Oinuma. 1963. Identification of kaolin minerals in the presence of chlorite by X-ray diffraction and infrared absorption spectra. Clays Clay Miner. 11:236–249.

Ledoux, R. L., and J. L. White. 1965. Infrared studies of the hydroxyl groups in intercalated kaolinite complexes. Clays Clay Miner. 13:289–315.

Lehr, J. R., E. H. Brown, and A. W. Frazier. 1967. Crystallographic properties of fertilizer compounds. TVA, Chem. Eng. Bull. no. 6, Muscle Shoals, Ala.

Léonard, A. J., F. Van Cauwelaert, and J. J. Fripiat. 1967. Structure and properties of amorphous silicoaluminas: III. Hydrated aluminas and transition aluminas. J. Phys. Chem. 71:695–708.

Léonard, A. J., S. Suzuki, J. J. Fripiat, and C. De Kimpe. 1964. Structure and properties of amorphous silicoaluminas. I. Structure from X-ray fluorescence spectroscopy and infrared spectroscopy. J. Phys. Chem. 68:2608–2617.

Levitt, S. R., and R. A. Condrate, Sr. 1970. The preparation of fine mineral powders for infrared spectroscopy. Am. Mineral. 55:522–524.

Lindberg, James D., and David G. Snyder. 1972. Diffuse reflectance spectra of several clay minerals. Am. Mineral. 57:485–493.

Little, L. H. 1966. Infrared spectra of adsorbed species. Academic Press, New York.

Lyon, R. J. P. 1962. Evaluation of infrared spectrophotometry for compositional analysis of lunar and planetary soils. Stanford Res. Inst., Palo Alto, Calif.

Lyon, R. J. P. 1964. Infrared analysis of soil minerals. p. 170–199. In C. I. Rich and G. W. Kunze (eds.) Soil clay mineralogy. Univ. North Carolina Press, Chapel Hill.

Lyon, R. J. P., W. M. Tuddenham, and C. S. Thompson. 1959. Quantitative mineralogy in 30 minutes. Econ. Geol. 54:1047–1055.

McDonald, R. S. 1958. Surface functionality of amorphous silica by infrared spectroscopy. J. Phys. Chem. 62:1168–1178.

Mangia, A. 1975. Determination of α-quartz in atmospheric dust: A comparison between infrared spectrometry and X-ray diffraction techniques. Anal. Chem. 47: 927–929.

Mejia, G., H. Kohnke, and J. L. White. 1968. Clay mineralogy of certain soils of Colombia. Soil Sci. Soc. Am. Proc. 32:665–670.

Miller, R. G. T., and B. C. Stace. 1972. Laboratory methods in infrared spectroscopy. 2nd ed. Heyden & Son, Ltd., London.

Mitchell, B. D., V. C. Farmer, and W. J. McHardy. 1964. Amorphous inorganic materials in soils. Adv. Agron. 16:327–383, Academic Press, New York.

Moenke, H. 1962. Mineralspektren, Akademie Verlag, Berlin.

Moenke, H. 1966. Mineralspektren II. Akademie Verlag, Berlin.

Mortensen, J. L., D. M. Anderson, and J. L. White. 1965. Infrared spectrometry. In C. A. Black (ed.) Methods of soil analysis. Part I. Agronomy 9:743–770. Am. Soc. Agron., Madison, Wis.

Nortia, T., and E. Kontas. 1973. An improved potassium bromide pellet method for the measurement of i.r. spectra of hygroscopic compounds. Spectrochim. Acta 29A: 1493–1495.

Oinuma, K., and H. Hayashi. 1965. Infrared study of mixed-layer clay minerals. Am. Mineral. 50:1213–1227.

Oinuma, K., and H. Hayashi. 1966. Infrared study of clay minerals from Japan. J. Tokyo Univ. General Educ. (Nat. Sci.), 6:1–15. Tokyo, Japan.

Oinuma, K., and H. Hayashi. 1967. Infrared absorption spectra of some zeolites from Japan. J. Tokyo Univ. General Educ. (Nat. Sci.) 8:1–12, Tokyo, Japan.

Parker, T. W. 1969. A classification of kaolinites by infrared spectroscopy. Clay Miner. 8:135–141.

Potts, W. J. 1963. Chemical infrared spectroscopy. Vol. 1, Techniques. John Wiley, New York.

Russell, J. D. 1974. Instrumentation and techniques. p. 11–25. In V. C. Farmer (ed.) The infrared spectra of minerals. Mineral. Soc., London.

Russell, J. D., and V. C. Farmer. 1964. Infrared spectroscopic study of the dehydration of montmorillonite and saponite. Clay Miner. Bull. 5:443–464.

Russell, J. D., V. C. Farmer, and B. Velde. 1970. Replacement of OH by OD in layer silicates, and identification of the vibrations of these groups in infrared spectra. Mineral. Mag. 37:869–879.

Russell, J. D., and A. R. Fraser. 1971. Infrared spectroscopic evidence for interaction between hydronium ions and lattice OH groups in montmorillonite. Clays Clay Miner. 19:55–59.

Stubican, V., and R. Roy. 1961a. Isomorphous substitution and infrared spectra of the layer lattice silicates. Am. Mineral. 46:32–51.

Stubican, V., and R. Roy. 1961b. Infrared spectra of layer-structure silicates. J. Am. Ceram. Soc. 44:625–627.

Tarasevich, Yu. I. 1969. Investigation of infrared spectra of heavy water sorbed by montmorillonite and vermiculite saturated with different cations. Proc. Int. Clay Conf., Tokyo, 1:261–269 (Russ. English summary).

Tuddenham, W. M., and R. J. P. Lyon. 1960. Infrared techniques in the identification and measurement of minerals. Anal. Chem. 32:1630–1634.

Tuddenham, W. M., and J. D. Stephens. 1971. Infrared spectrophotometry. p. 127–168. *In* R. E. Wainerdi (ed.) Modern methods of geochemical analysis. Plenum, New York.

van der Marel, H. W., and H. Beutelspacher. 1976. Atlas of infrared spectroscopy of clay minerals and their admixtures. Elsevier, Amsterdam.

Vedder, W. 1964. Correlations between infrared spectrum and chemical composition of mica. Am. Mineral. 49:736–768.

Veniale, F., and H. W. van der Marel. 1969. Identification of some 1:1 regular interstratified trioctahedral clay minerals. Proc. Int. Clay Conf., Tokyo, 1:233–244.

Wada, K., and D. J. Greenland. 1970. Selective dissolution and differential infrared spectroscopy for characterization of amorphous constituents in soil clays. Clay Miner. 8:241–254.

Wada, K., and Y. Tokashiki. 1972. Selective dissolution and differential infrared spectroscopy in quantitative mineralogical analysis of volcanic-ash soil clays. Geoderma 7:199–213.

Ward, J. W. 1971. Infrared spectroscopic studies of zeolites. Adv. Chem. Ser. 101:380–404.

Weismiller, R. A., J. L. Ahlrichs, and J. L. White. 1967. Infrared studies of hydroxy-aluminum interlayer material. Soil Sci. Soc. Am. Proc. 31:459–463.

White, Joe L. 1971. Interpretation of infrared spectra of soil minerals. Soil Sci. 112:22–31.

Wilkins, R. W. T. 1967. The hydroxyl-stretching region of the biotite mica spectrum. Mineral. Mag. 36:325–333.

Wilson, M. J. 1966. The weathering of biotite in some Aberdeenshire soils. Mineral. Mag. 35:1080–1093.

Wilson, M. J. 1970. A study of weathering in a soil derived from a biotite-hornblende rock: I. Clay Miner. 8:291–303.

Thermal Analysis of Soils

K. H. TAN, University of Georgia, Athens, Georgia

BEN F. HAJEK, Auburn University, Auburn, Alabama

I. THERMAL APPARATUS

Thermal analysis is a term covering a group of analyses that determine some physical parameter such as energy, weight, dimension, and evolved volatiles as a function of temperature. A large number of techniques, related to each other, are available (Fig. 26-1). For detailed information reference is made to Wendlandt (1974), Smothers and Chiang (1966), Schultze (1969), and Mackenzie (1970). Among the many methods listed in Fig. 26-1, three of the major types that find frequent application in soil research will be discussed: differential thermal analysis, differential scanning calorimetry, and thermogravimetry.

A. Differential Thermal Analysis

This method, usually called DTA is probably the most important and widely used technique (Mackenzie, 1970). It has been employed in geology, in ceramic, glass, polymer, cement, and plaster industries, and in research in chemistry and catalysis. Recently it has been applied to studies of organic matter, explosives and radioisotopes. In soil research, DTA was first employed by Matejka (1922) for determination of kaolinite. Russell and Haddock (1940) and Hendricks and Alexander (1940) were perhaps among the first American scientists to recognize DTA as an important tool for analysis of clays and clay minerals in soils. A review of developments in DTA during 1940-1960 was published by Murphy (1962).

Differential thermal analysis determines the differences in temperature between a sample and reference material as the two are heated at a controlled rate. When the sample undergoes a transformation the heat effect causes a difference in temperature between the sample and reference material. This difference in temperature is usually plotted against the temperature, at which this difference occurs (Fig. 26-2). If the temperature of the sample falls below that of the reference (ΔT is negative) an endothermic peak develops. When the temperature of the sample rises above the reference (ΔT is positive) an exothermic peak develops. The portion of the curve, for which ΔT is approximately zero, is considered to be the base line (Mackenzie, 1969; 1972).

The DTA instrument can be simple or very complex. For detailed descriptions of commercially available DTA instruments, see Mackenzie (1970), Wendlandt (1974), and Smothers and Chiang (1966). All of the instruments have in common the following basic components: (i) a furnace or heating

chapter 26

THERMAL ANALYSIS

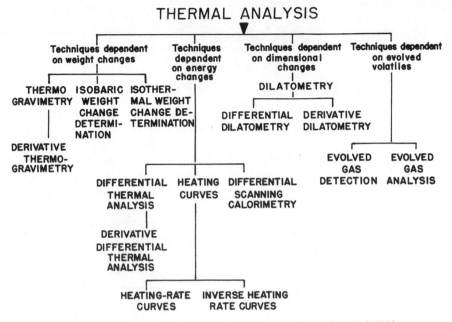

Fig. 26-1. Major types of thermal analysis according to Mackenzie (1970).

unit, (ii) a specimen holder, and (iii) a temperature regulating and measuring system. A differential thermocouple is used to measure the temperature difference between sample and reference material, whereas the temperature, at which the difference occurs, is determined by an additional thermocouple located in the test assembly.

B. Differential Scanning Calorimetry

This instrument is based on the same principles of DTA. However, differential scanning calorimetry (DSC) maintains the sample at the same temperature as the reference material or furnace block. The amount of energy required to establish zero temperature difference between sample and reference material is then recorded as a function of temperature or time. For more details about different DSC instrumentation see Wendlandt (1974) and Smothers and Chiang (1966).

C. Thermogravimetric Analysis

The new term proposed by the International Committee for Standardization of Thermal Analysis (ICTA) is *thermogravimetry* (Redfern, 1970; McAdie, 1969; Mackenzie, 1969). This is a technique whereby a sample is continuously weighed as it is being heated or cooled at a controlled rate.

IDEALIZED DTA CURVE

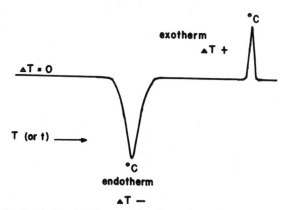

Fig. 26-2. Idealized DTA curve showing endo- and exothermic peaks.

Two major types of techniques are recognized in this category of analysis: (i) thermogravimetry (TG), which in the past was called integral thermal analysis (Jackson, 1956), or dyanamic thermogravimetric analysis, and (ii) derivative thermogravimetry, in the past also known as differential thermogravimetry or differential thermogravimetric analysis.

In thermogravimetry, weight changes are recorded as a function of temperature (T) or time (t), i.e.,

$$\text{weight} = f \, (T \text{ or } t).$$

The recording is called a thermogravigram or a thermogravimetric curve (Mackenzie et al., 1972). In derivative thermogravimetry, the difference of the weight change with respect to time is recorded as a function of temperature (T) or time (t):

$$d_w/d_t = f(T \text{ or } t).$$

The curve obtained is in fact the first derivative of the weight-change curve and as pointed out by Mackenzie et al. (1972) this type of thermal analysis should be called *derivative thermogravimetry,* instead of *differential thermogravimetry.* Use of the latter term can result in confusion, since the term "differential" has a different meaning in thermogravimetry and in DTA.

The principles in thermogravimetry require that the sample be continuously weighed during the heating process. For this purpose the following basic components are necessary: (i) a furnace, (ii) a temperature regulator and measuring system, and (iii) a thermobalance. The latter is probably the most important part of the apparatus. It is a balance that is capable of weighing a sample continuously, while the latter is being heated or cooled. Essentially it is a good quality analytical balance with rugged construction and a high degree of mechanical and electronic stability. For more details

on construction and models of thermogravimetric instruments and thermo-
balances, see Wendlandt (1974).

II. SAMPLE PREPARATION

A. Soil Fractions

Generally, DTA can be performed on any liquid or solid sample. Soil
is complex because it is a mixture of mineral and organic matter, varying
widely in particle size, and in physical, chemical, and biological properties.
One question of concern is which fraction of soil should be analyzed. Should
the "whole" soil be analyzed, since this is the medium to which plants and
management practices react; should the organic matter be destroyed by H_2O_2;
is it sufficient to analyze only individual mineral and organic fractions?
Particle size separates can be obtained by fractionation procedures for other
types of mineralogical investigation such as X-ray analysis (Jackson, 1956).
However if size separates are to be analyzed by DTA, one must remember
that particle size and crystallinity have a pronounced influence on size, shape,
and temperature of endothermic peaks (Smothers & Chiang, 1966).

1. WHOLE SOILS

No pertinent data are available in the literature concerning DTA of
whole soil samples perhaps because soil organic matter may obscure characteristic endothermic peaks by strong exothermic reactions. DTA curves of
whole soils (untreated and H_2O_2 pretreated), as shown in Fig. 26-3, show
that organic matter caused a strong exothermic reaction culminating into a
peak at approximately 300C (573K) which obscured any endothermic peak
that may have developed between 250-400C (523-673K). In soils with low
organic matter content, this is not a problem (Fig. 26-3, No. 2). Removal of
organic matter with H_2O_2 or analysis in an inert N_2 atmosphere usually increases the intensity of endothermic peaks at 270C (543K), 490C (764K),
and at 573C (846K), for gibbsite, kaolinite, and quartz, respectively (Fig.
26-3, No. 3). In addition to organic matter effects, peaks may also be obscured because of dilution by quartz.

When whole soils are analyzed, the < 2 mm fraction should be treated
with 30% H_2O_2, washed with distilled water, dried and ground again to pass
a 2-mm or smaller sieve. In general the analysis of whole soils gives peaks of
low intensity. These same peaks are very large and intense if only the clay
fractions are analyzed (Fig. 26-3, No. 4), however the quartz α-β inversion
peak at 573C (846K) is often absent.

2. SAND FRACTION

For use in DTA, the sand fraction, 2.0 mm to 0.05 mm, can be used
directly or this fraction can be separated further into very coarse sand (2.0-

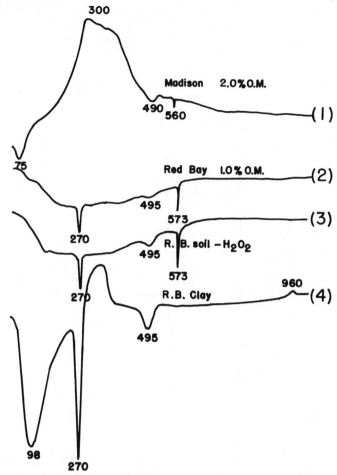

Fig. 26-3. DTA curves of whole soil and soil clay: *(1)* nontreated Madison surface soil (2.0% organic matter); *(2)* nontreated Red Bay surface soil (1.0% organic matter); *(3)* Red Bay surface soil pretreated with 30% H_2O_2; and *(4)* Clay (<2 μm) of Red Bay surface soil. Tracor-Stone instrument, 3 mg of sample in Pt cups. Temperature in °C.

100 mm), coarse sand (1.0-0.50 mm), medium sand (0.50-0.25 mm), fine sand (0.25-0.10 mm), and very fine sand (0.10-0.05 mm). If small sample sizes are required the finer sand fractions are more suitable. Where a relatively large amount of sample can be used, as is the case with DTA instruments equipped with well-type sample holders, it is immaterial whether coarse or fine sand is used, since particle size has little influence on the DTA curve of sand. As can be noticed from Fig. 26-4 (No. 1 and 2) the DTA curve of the composite sand fraction (2.0-0.05 mm) was identical to that of the fine sand. In both cases, only a strong endothermic peak of quartz at 573C (846K), was the main characteristic of the curve. Analysis of the sand fraction by DTA is only of importance in investigation of primary minerals and/ or Fe-Mn concretions. In soils of the southeastern United States and in other

CECIL SOIL

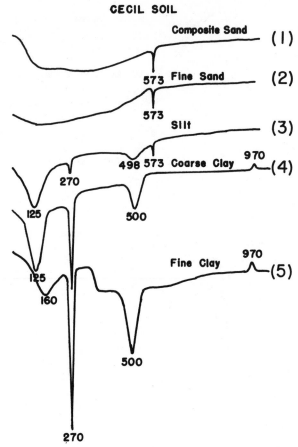

Fig. 26-4. DTA curves of sand fractions separated from Cecil soil A3 horizon: *(1)* composite and fraction, 2.0–0.05 mm; *(2)* fine sand, 0.25–0.10 mm; *(3)* silt, 0.05–0.002 mm; *(4)* coarse clay, 2.0–0.2 μm; and *(5)* fine clay, $<$ 0.2 μm. Tracor-Stone instrument, 3 mg of sample in Pt cups. Temperature in $^{\circ}$C.

parts of the continental United States, quartz is the dominant component of the sand fraction. Therefore, DTA of the sand fraction of these soils results in featureless DTA curves, except the quartz peak.

3. SILT

The silt fraction of soils (0.05–0.002 mm) can be analyzed directly and usually yields curves showing more complexity than that of sand (Fig. 26-4, no. 3). Often the amount of sesquioxides and kaolinite in the silt size fraction is large enough to yield detectable endothermic peaks at 270 and 498C (543 and 771K), respectively, in addition to the quartz peak at 573C (846K). Generally the DTA curve of silt resembles that of whole soil. However, the peak intensities are less than in whole soil.

4. CLAY

The clay fraction of soils ($<$ 2 μm) can be used directly, or can be separated into coarse clay (2.0-0.20 μm) and fine clay ($<$ 0.2 μm) fractions. The choice of size fraction to be used depends on many factors, e.g., purpose, objective, precision, and type and quantity of minerals present. A number of investigators preferred the use of clay separated into various fractions (Jackson, 1956); however, the present authors found that, for general purposes the clay fraction ($<$ 2 μm) gives results that are satisfactory for qualitative and quantitative interpretations. However, it was noted that the finer clay fraction exhibited DTA curves with more intense peaks and this sample showed evidence of interlayering of expandable 2:1 minerals that was not detected in analysis of the total clay.

B. Exchange Saturation and Hydration

1. CATION SATURATION

Saturation of the sample with a known cation prior to DTA has been proposed for clay or material with cation exchange properties. Clay in soils is expected to be saturated with a variety of cations (Na, K, Ca, Mg, Al). Hydration properties of these cations which vary considerably, markedly affect the results of DTA. Mackenzie (1970) recommended that for comparative work each sample should have received identical pretreatments. The authors have found that Ca saturation of the samples is usually satisfactory. To determine, in detail, the effect of different cation saturation on DTA curves, kaolinite and montmorillonite were saturated with Na, Ca, Al, and H by shaking for 1 hour with $0.1N$ NaCl, $CaCl_2$, $AlCl_3$, and HCl solutions and allowing them to stand overnight. After washing with distilled water the samples were dried at 45C (318K), ground and stored in a desiccator over $CaCl_2$. DTA curves of these clay samples are shown in Fig. 26–5 and 26–6. The curves in Fig. 26–5 show that different cations affected both the size and shape of endo- and exothermic peaks for kaolinite. However, no change in peak temperature occurred. Ca-kaolinite had the lowest peak intensity for both the 530C (803K) and 1,000C (1,273K) peaks. Peak intensity increased from Na-kaolinite, H-kaolinite to Al-kaolinite (Fig. 26–5, no. 2-4). Saturation with Al made the main endothermic peak at 530C (803K) very sharp and slender and resulted in an additional strong endothermic peak at 125C (398K). The latter suggests that Al-kaolinite retains considerable amounts of water even after overnight storage in a desiccator over $CaCl_2$.

Saturation of montmorillonite with H ions gave curves with a broad peak at 675C (948K) and a S-shape curve at 950C (1,223K) (Fig. 26–6, no. 1). Calcium–montmorillonite resulted in a curve with its main endothermic peak shifted to 700C (973K) (Fig. 26–6, no. 2). The presence of two peaks

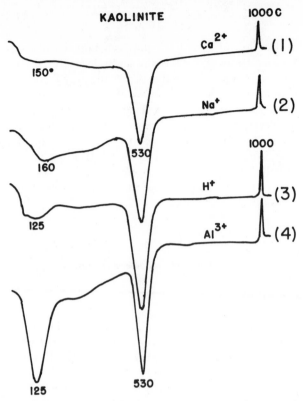

Fig. 26-5. DTA curves of kaolinite (no. 2, Birch Pit, Macon, Ga.) saturated with different cations: *(1)* Ca-kaolinite; *(2)* Na-kaolinite; *(3)* H-kaolinite; and *(4)* Al-kaolinite. Tracor-Stone instrument, 2 mg of sample in Pt cups. Temperature in °C.

at 75 and 140C (348 and 413K) for Ca-montmorillonite is in agreement with that reported in the literature for montmorillonite treated with Ca (Mackenzie, 1970; Barshad, 1965). Curve no. 3 of Fig. 26-6 shows DTA features of a soil smectite for comparison. This sample was obtained from a Houston Black soil (Vertisol). The curve resembles that exhibited by Ca-montmorillonite with the difference that the main endothermic peak at 695C (968K) was very pronounced.

2. HYDRATION AND SOLVATION

As discussed in the preceeding section, hydration of samples prior to DTA may result in changes of the low temperature endothermic peaks at 0–200C (273–473K). Soil colloids are very reactive due to large surface areas and negative charge and may adsorb considerable amounts of water. Moreover, the various types of soil colloids are known to have different capacities for adsorption of moisture; differences in these factors may result in different DTA curve features. Samples should be equilibrated at constant relative

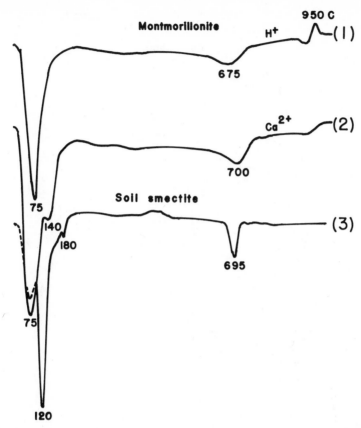

Fig. 26-6. DTA curves of montmorillonite (Oklahoma Geological Survey) saturated with different cations: *(1)* H-montmorillonite; *(2)* Ca-montmorillonite; and *(3)* Ca-clay from ($<$2 μm) a Houston Black soil. Tracor-Stone instrument, 3 mg of sample in Pt cups. Temperature in $^{\circ}$C.

humidity to insure that amorphous material and expandible 2:1 type minerals exhibit low temperature endothermic curves that can be used for meaningful comparisons (Jackson, 1956). Equilibration is usually carried out in a desiccator over a compound developing a stable relative humidity. Compounds such as $Mg(NO_3)_2 \cdot 6H_2O$, $Mg(NO_3)_2$, and H_2SO_4 have been used (Barshad, 1965; Jackson, 1956). The authors have found that keeping samples overnight over $CaCl_2$ produces satisfactory results and will not lead to disappearance of characteristic low temperature endothermic peaks (Fig. 26-7, no. 3). Keeping samples for prolonged periods over $CaCl_2$ may, however, prove to be a disadvantage. Figure 26-7 also shows DTA curves of Ca-kaolinite kept overnight over $CaCl_2$, and kept open in contact with air in the laboratory (air-dry). The curve of $CaCl_2$ equilibrated Ca-kaolinite shows practically no low temperature reaction, which is, of course, normal. However, air-dry Ca-kaolinite gave a DTA curve with a strong endothermic peak at 125C (398K) (Fig. 26-7, no. 1 and 2).

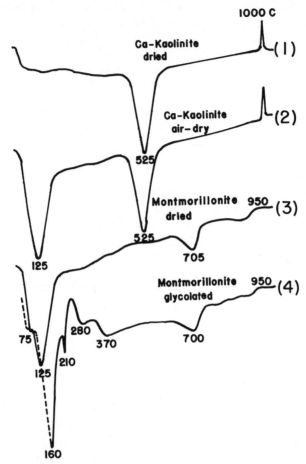

Fig. 26-7. DTA curves of: *(1)* Ca-kaolinite, dried overnight over CaCl₂; *(2)* Air dry Ca-kaolinite; *(3)* nontreated montmorillonite stored overnight over CaCl₂; and *(4)* montmorillonite glycolated with ethylene-glycol. Tracor-Stone instrument, 3 mg of sample in Pt cups. Temperature in °C.

III. THE THERMAL ANALYSIS

A. Differential Thermal Analysis

1. REFERENCE

The reference material, sometimes called standard material, is a known substance that is thermally inert over the temperature range under investigation (Mackenzie et al., 1972). It is a "neutral" body or comparison standard against which the temperature of the sample is measured. For a detailed listing of preferred properties that a reference material should have, see the progress report of the ICTA by McAdie (1969). A number of compounds, both organic and inorganic, such as octanol, benzol, tributyrin, Al₂O₃, clay, quartz,

NaCl, KCl, glass beads, MgO, Al foil, etc. have been used as reference material, (Schultze, 1969; Smothers & Chiang, 1966). Among the compounds listed, the most commonly used reference material is α- or γ-alumina, or alumina calcined to 1,200C (1,473K). For the latter, Al-oxide powder is ignited in a Pt crucible, cooled, and stored in a desiccator. Clay especially kaolinite, preheated this way is also used as a reference in DTA. However, caution is required in the use of α- or γ-alumina or kaolinite. It has been observed (Arens, 1951) that calcined alumina may become hydroscopic after use in DTA, and may need to be replaced after two or three runs. Calcined clay creates a similar problem and may contain components with reversible thermal reactions. According to Arens (1951) the large difference in thermal conductivity between kaolinite (0.72×10^{-3} cal cm^{-1} sec^{-1} °C^{-1} or 0.30 watts m^{-1} °K^{-1}) and calcined kaolinite (1.6×10^{-3} cal cm^{-1} sec^{-1} °C^{-1} or 0.67 watts m^{-1} °K^{-1}) is a decisive factor for rejection of calcined kaolinite as reference material.

2. SAMPLE SIZE

The size of sample to be analyzed is in many cases dictated by the type of the DTA instrument available. Most instruments, especially those equipped with well-type sample holders, require sizeable amounts of sample ($>$ 100 mg) to fill the hole. In this case dilution, by mixing the sample with an inactive material, becomes beneficial. Dilution is also suggested to reduce base-

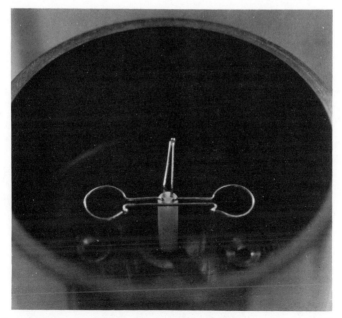

Fig. 26-8. Nickel alloy sample holder of Tracor-Stone DTA instrument with ring type Pt-Pt/Rh thermocouples (5 mm in diameter).

line drift and/or increase accuracy. For the merits and objections of dilution, reference is made to Mackenzie (1970).

Instruments, such as the Tracor-Stone apparatus equipped with ring-type thermocouples (Fig. 26-8) upon which a small Pt pan can be seated, or the Dupont 990 Thermal Analyzer (plate-type, Al-pans) require very small samples. Although the amount required for an analysis will vary according to type and thermal characteristics of the material, the size is usually in the order of 1-10 mg; as little as 0.5 mg of sample can be used. These small samples approach the ideal postulated by Mackenzie (1970), i.e., the ideal sample should be an infinitely small sphere surrounding the thermocouple junction.

In qualitative analysis, it is not necessary to weight the sample for DTA, although comparison of curves should be made with curves obtained from identical amounts of samples. However, in quantitative analysis, the amount used for DTA must be weighed accurately. As can be seen in Fig. 26-9, the main endothermic peak height (or area) of kaolinite increases proportionally with sample size.

3. PACKING

Packing the sample is required when instruments with well-type sample holders are used. A number of methods have been proposed including hand

KAOLINITE ENDOTHERMIC PEAK

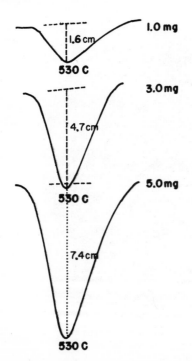

Fig. 26-9. Main endothermic peaks of kaolinite (Mesa Alta, New Mexico) obtained by analysis using different sample sizes: (i) 1 mg, with heating rate of 15C (288K)/min; (ii) 3 mg, with heating rate of 15C (288K)/min; and (iii) 5 mg, with heating rate of 15C (288K)/min. Tracor-Stone instrument; Pt cups.

tapping the sample or the block and layer packing (Arens, 1951; Schultze, 1969; Smothers & Chiang, 1966; Mackenzie, 1970). In the latter, the sample is packed around the thermocouple junction as a sandwich between two layers of reference material. The sandwich method may yield peaks of approximately one-half the intensity of those obtained by filling the whole cavity with the sample.

Reproducibility in packing is of importance, since differences in packing may create differences in sample density, leading in turn to differences in heat conductivity and thermal diffusivity of the sample. The latter is a problem in the low temperature ranges, where heat transfer is mainly controlled by conduction. In the high temperature region, the effects of packing are less apparent, because heat transfer is mostly through radiation. Both reference and sample materials should be of similar aggregate size and should be packed to a similar bulk density in order to avoid base-line drift. In plate or ring-type instruments, where a small amount of sample is placed in Al or Pt cups, packing effects are negligible. Minimum zero drift is usually attained by balancing carefully the relative amounts of reference and/or test sample in the cups.

4. FURNACE ATMOSPHERE

Some control of furnace atmosphere is desirable and often necessary. The furnace atmosphere affects DTA through one or a combination of the following reactions (Schultze, 1969): (i) A change in furnace atmosphere creates a change in partial pressure inducing a shift of peak temperature, (ii) interactions occurring between gas used and gaseous products of the sample may change furnace atmosphere and obscure the appearance of characteristic DTA peaks, or (iii) peaks are enhanced because oxidation reactions can be eliminated. A number of methods such as use of vacuum, static air, static gas, dynamic gas, gas-flow over the sample, gas-flow through the sample, and self-generated gas have been tested. For the advantages and disadvantages of one method over the other, reference is made to Mackenzie (1970).

For most analyses, it is sufficient to run DTA in the presence of a self-generated gas atmosphere. However, it is necessary to place a close-, but loose-fitting lid on top of the sample holder to maintain uniform pressure and thermal conditions around sample and reference material. By using a lid, direct radiation of both the specimens by the furnace will also be avoided.

5. HEATING RATE

The heating rate is defined as the rate of temperature increase, expressed in degrees centrigrade per minute. Correspondingly, the cooling rate is the rate of temperature decrease. The heating or cooling rate is constant when the temperature/time curve is linear (Mackenzie et al., 1972). The heating must be controlled at a uniform and steady rate through the analysis. Heat-

KAOLINITE ENDOTHERMIC PEAK

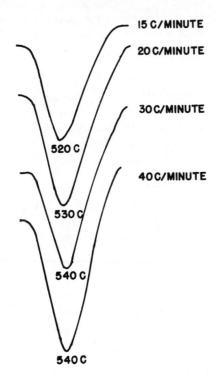

15 C/MINUTE

20 C/MINUTE

30 C/MINUTE

520 C

40 C/MINUTE

530 C

540 C

540 C

Fig. 26-10. Main endothermic peaks of 3 mg kaolinite (Mesa Alta, New Mexico) obtained by analysis using different heating rates: (i) 15C (288K)/min; (ii) 20C (293K)/min; (iii) 30C (303K)/min; and (iv) 40C (313K)/min. Tracor-Stone instrument, Pt cups.

ing rates used vary from 0.1 to 200C/min (273.1-473K). For a review of the effects of slow and fast heating rates on DTA curves, see Mackenzie (1970). For most purposes rates of 5 to 20C/min (278-293K) are used.

Although many investigators have stressed the fact that heating rates affect peak height, peak width, peak temperature, and peak area, the authors have found that heating rate is seldom a serious problem.

The differences in DTA curves obtained by different heating rates, as shown in Fig. 26-10, indicated that peak height increases gradually and consistently, when heating rates are increased from 15 to 20, 30, and to 40C/min (288 to 293, 303, and 313K). The most significant differences is obtained between heating at 15 and 40C/min (288 and 313K). Kaolinite heated at a rate of 15C (288K)/min yielded a broad and relatively shallower endothermic peak at 520C (793K). When heated at a rate of 40C (313K)/min a more intense, but slender endothermic peak was obtained, with the peak temperature shifted to 540C (813K). The peak temperatures, at 520C (793K) or 540C (813K) are well within the range for kaolinite. Since the shape and size of endothermic peaks obtained by heating at 15C or 40C/min (288 or 313K) are well within detection limits, the choice of a heating rate between 15-40C/min (288-313K) makes little difference.

B. Differential Scanning Calorimetry

1. REFERENCE, SAMPLE SIZE, AND HEATING RATE

There are two types of instruments which are named "differential scanning calorimeter." They are of completely different design. Wendlandt (1974) describes these as: (i) differential scanning calorimeters which are heat-flow-recording instruments such as the Perkin-Elmer, and (ii) differential scanning calorimeters which are actually differential-temperature-recording or DTA instruments such as the DuPont and Stone. DSC curves (Fig. 26–11) are very similar to DTA. The area under the DSC curve is proportional to the enthalpy and the ordinate value at any given temperature (or time) is directly proportional to the differential heat flow between a sample and reference material. Thus the ordinate is in terms of Δq (cal or mcal/inches) recorded in relation to time since heating started.

Reference material, sample size, and heating rate effects are similar to those discussed for DTA of soils and clays. The most commonly used references for DSC analysis are either Al_2O_3 or an empty sample pan. Empty Al-pan references have given satisfactory DSC curves on the DuPont 990 instrument. All that is required is a baseline adjustment due to the mass-heat capacity difference between the sample and reference. For most quantitative determinations, 1- to 5-mg samples give DSC curves with peak areas that can be measured accurately. Larger samples result in some sensitivity loss because of heat conductivity effects of the sample. Heating rates of 10 to 20C/ (283–293K) give satisfactory curves for soil minerals that can be quantitatively determined by DSC.

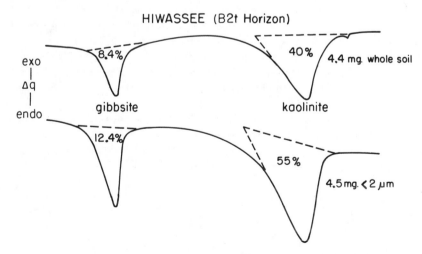

Fig. 26–11. DSC curves for whole soil and the clay fraction of soil from the B2t horizon of a Hiwassee loam. Gibbsite and kaolinite contents on a whole soil basis for the <2 μm fraction are 8.5 and 38%, respectively. DuPont 990 Thermal Analyzer, Al pans.

2. CALORIMETRIC AND QUANTITATIVE DETERMINATIONS

The determination of ΔH (heat of reaction) and the mass of the reacting part of the sample is a commonly used procedure for DSC. The exact equation used for ΔH depends on the instrument being used. The simple form is,

$$\Delta Hm = KA$$

in which m is the reactive mass, K is a constant, and A is the peak area (Wendlandt, 1974; Mackenzie, 1970; Barshad, 1965). In the past the equation was further simplified for m by including ΔH with K, yielding,

$$m = KA.$$

A standard curve of m vs. A was then developed for comparison with unknowns.

Differential scanning calorimetry can be used in the same manner, however, ΔH or m are usually obtained by using the peak area and substitution into an equation of the form:

$$\Delta H = (A/m) (BE \Delta qs)$$

in which ΔH, A, and m are as defined previously, B is the chart speed in inches/min, E is the cell calibration coefficient, and Δqs is the instrument sensitivity setting in cal sec^{-1} $inch^{-1}$. The cell calibration coefficient E is somewhat dependent on temperature for the DuPont DSC, consequently E must be determined as a function of temperature by analyzing samples of known ΔH and solving for E. Only a single temperature calibration is required for instruments such as the Perkin-Elmer (Wendlandt, 1974). To find A (peak area) a line is drawn from the point where the thermogram departs from the baseline to the point where it returns. The area is measured with a planimeter for best results.

As pointed out by Mackenzie (1970) for DTA, quantitative DSC is most useful for soils that have undergone intensive weathering. Usually kaolinite, gibbsite, and hydrous Fe oxides dominate the clay fraction, and quartz is predominant in the sand. Figure 26-11 shows whole soil and clay fraction DSC curves for the B2t horizon of a Hiwassee soil. Soils in this series are clayey and representative of highly weathered soils in the southeastern USA. The points of departure and return to baseline were used for gibbsite determination. However because of the presence of highly hydroxy-interlayered vermiculite the area for kaolinite was drawn to exclude the endotherm effect of interlayer hydroxyl loss. When the quantities are compared on a whole soil basis the difference between clay and whole soil are well within the limits of error. For routine analysis, if minerals can be quantitated by DSC and the soils contain $>$ 35% clay, fractionation is seldom necessary. At times Fe oxide removal may be required (Mackenzie, 1970). The quartz α-β inversion peak is easily excluded from the peak area of kaolinite. In many soils smectite will give endotherms in the kaolinite region thus preventing quantitative determination of kaolinite. In addition some highly hydroxy-interlayered vermiculite clays can interfere.

Figure 26-12 shows DSC and the derivative curve (ΔT sample/ΔT) of untreated air-dry whole soils. The low temperature peaks allow positive identification of smectite. Kaolinite can be identified, but cannot be quantitated.

C. Thermogravimetric Analysis

Thermogravimetry (TG) has been used to study many clay minerals, soils and compounds. A very comprehensive and current coverage of theory and application of TG is given in Wendlandt (1974).

Since DTA or DSC alone are often not adequate on their own, TG and derivative thermogravimetry are valuable complementary techniques (Mackenzie, 1970). In soil studies TG is essential for the determination of adsorbed water and structural water. The exact temperature at which soil

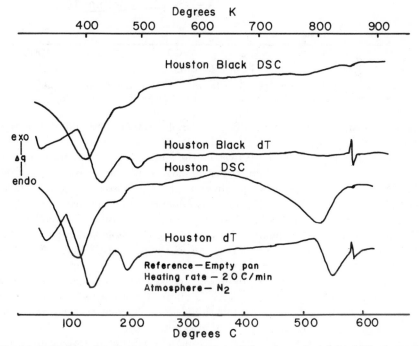

Fig. 26-12. DSC and derivative curves for air dry whole soil samples of the AC horizon of a Houston clay (Alabama) and A1 horizon of a Houston Black clay (Texas). DuPont 990 Thermal Analyzer, Al pans.

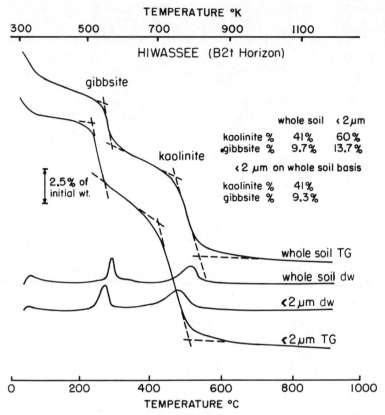

Fig. 26–13. TG and derivative TG curves for whole soil and the clay fraction of soil from the B2t horizon of a Hiwassee loam. DuPont 990 Thermal Analyzer.

minerals lose water varies. In addition, sample pretreatment and heating rate will affect the weight loss region. The latter variables may or may not be serious depending on the objectives of the study, type of minerals present, and the capability of the thermal instrument used. The temperatures of de-hydroxylation and desorption and theoretical weight losses are given by Jackson (1956) and Barshad (1965) for most minerals found in soils.

As with DTA and DSC analysis untreated whole soils can often be used for qualitative and semiquantitative analysis. The whole soil curves for the highly weathered Hiwassee soil (Fig. 26–13) and the two Vertisols in Fig. 26–14 (Houston from Alabama and Houston Black from Texas) were obtained by grinding the air-dry whole soil to pass a 140-mesh sieve and using a 10-mg sample for analysis. The clay fraction of Hiwassee was obtained by fractionation methods given by Jackson (1956). All analyses were at a heating rate of 10C (283K)/min in a N_2 atmosphere.

The TG curves for Hiwassee confirm the quantitative DSC analysis for gibbsite and kaolinite for both the whole soil and clay fraction. No quantitative attempt was made for the Vertisols. Smectite, kaolinite, and calcite can

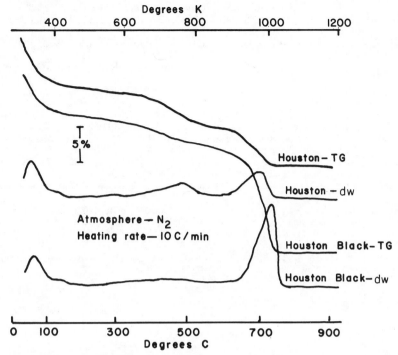

Fig. 26-14. TG and derivative TG curves for whole soil samples of the AC horizon of a Houston clay (Alabama) and A1 horizon of a Houston Black clay (Texas). DuPont 990 Thermal Analyzer.

be identified in the Houston soil. Smectite and calcite can be identified in the Houston Black, however, additional analysis of the clay fraction would be required to confirm the presence of kaolinite.

The current availability of TG instruments with sensitive solid-state automatic recording capability and low mass rapid cooling furnaces, have made TG a rapid, accurate, and relatively simple analytical technique.

LITERATURE CITED

Arens, P. L. 1951. A study on the differential thermal analysis of clays and clay minerals. Proefschrift Landbouw Hogeschool, Wageningen. Excelsiors drukkery, s'Gravenhage. 131 p.

Barshad, I. 1965. Thermal analysis techniques for mineral identification and mineralogical composition. In C. A. Black (ed.) Methods of soil analysis. Part 1. Am. Soc. of Agron. Agronomy 9:699-742.

Hendricks, S. B., and L. T. Alexander. 1940. Semiquantitative estimation of montmorillonite and clays. Soil Sci. Soc. Am. Proc. 5:95-99.

Jackson, M. L. 1956. Soil chemical analysis—advanced course. Published by the author. Dept. of Soils, Univ. of Wisconsin, Madison. 991 p.

McAdie, H. G. 1969. Progress towards thermal analysis standards. A report from the committee on standardization international confederation for thermal analysis. In R. F. Schwenker and Garn (eds.) Thermal analysis. Academic Press, London. Vol. I:693-706.

Mackenzie, R. C. 1969. Nomenclature in thermal analysis. Talanta 16:1227-1230.

Mackenzie, R. C. (ed.). 1970. Differential thermal analysis. Academic Press, London and New York. 775 p.

Mackenzie, R. C. 1972. How is an acceptable nomenclature system achieved. J. Therm. Anal. 4:215-221.

Mackenzie, R. C. et al. 1972. Nomenclature in thermal analysis—II. Talanta 19:1079-1081.

Matejka. 1922. Thermal analysis as a means of detecting kaolinite in soils. Chem. Listy. 16:8-14.

Murphy, C. B. 1962. Differential thermal analysis. A review of fundamental developments in analysis. Anal. Chem., Review issue 34:298R-301R.

Redfern, J. P. 1970. Complementary methods. p. 123-158. In R. C. Mackenzie (ed.) Differential thermal analysis. Academic Press.

Russell, M. B., and J. I. Haddock. 1940. The identification of clay minerals in five Iowa soils by the thermal method. Soil Sci. Soc. Am. Proc. 5:90-94.

Schultze, D. 1959. Differential thermo-analyses. Verlag Chemie. GmbH. Weinheim/Bergstr. 335 p.

Smothers, W. J., and Yao Chiang. 1966. Handbook of differential thermal analysis. Chemical Publishing Co., Inc., New York. 633 p.

Soil Survey Staff. 1951. Soil survey manual. USDA handb. No. 18. U. S. Gov't Printing Office, Washington, D. C. 503 p.

Wendlandt, W. W. 1974. Thermal methods of analysis. 2nd ed. Interscience, John Wiley & Sons Div., New York. Part I. 424 p.

AUTHOR INDEX

The numbers in italic refer to the pages on which the complete references are listed.

SUBJECT INDEX